# Shock Compression of
# Condensed Matter — 2005

# Proceedings in the Series of
## Conferences of the American Physical Society Topical Group on Shock Compression of Condensed Matter

| Year | Held in | Publisher | ISBN |
|------|---------|-----------|------|
| 2005 | Baltimore, Maryland, USA | AIP Conference Proceedings 845 | 0-7354-0341-1 |
| 2003 | Portland, Oregon, USA | AIP Conference Proceedings 706 | 0-7354-0181-0 |
| 2001 | Atlanta, Georgia, USA | AIP Conference Proceedings 620 | 0-7354-0068-7 |
| 1999 | Snowbird, Utah, USA | AIP Conference Proceedings 505 | 1-56396-923-8 |
| 1997 | Amherst, Massachusetts, USA | AIP Conference Proceedings 429 | 1-56396-738-3 |
| 1995 | Seattle, Washington, USA | AIP Conference Proceedings 370 | 1-56396-566-6 |
| 1993 | Colorado Springs, Colorado, USA | AIP Conference Proceedings 309 | 1-56396-219-5 |
| 1991 | Williamsburg, Virginia, USA | North-Holland | 0-444-89732-1 |
| 1989 | Albuquerque, New Mexico, USA | North-Holland | 0-444-88271-5 |
| 1987 | Monterey, California, USA | North-Holland | 0-444-87097-0 |
| 1985 | Spokane, Washington, USA | Plenum Press | 0-306-42276-X |
| 1983 | Santa Fe, New Mexico, USA | North-Holland | 0-444-86904-2 |
| 1981 | Menlo Park, California, USA | AIP Conference Proceedings 78 | 0-88318-177-0 |

To learn more about these titles, or the AIP Conference Proceedings Series, please visit the webpage
**http://proceedings.aip.org/proceedings**

# Shock Compression of Condensed Matter — 2005

Proceedings of the Conference of the American Physical Society
Topical Group on Shock Compression of Condensed Matter
held in Baltimore, Maryland, July 31 – August 5, 2005

## PART TWO

*Edited by:*

MICHAEL D. FURNISH
*Sandia National Laboratories*
*Albuquerque, New Mexico, USA*

MARK ELERT
*U.S. Naval Academy*
*Annapolis, Maryland, USA*

THOMAS P. RUSSELL
*Naval Surface Warfare Center*
*Indian Head, Maryland, USA*

CARTER T. WHITE
*Naval Research Laboratory*
*Washington, DC, USA*

*SPONSORING ORGANIZATION*
American Physical Society

◎ CD-ROM INCLUDED

Melville, New York, 2006
AIP CONFERENCE PROCEEDINGS ■ 845

*75 Years of Service*

EDITORS

Michael D. Furnish
Sandia National Laboratories
MS 1168, P.O. Box 5800
Albuquerque, NM 87185-1168
USA
E-mail: mdfurni@sandia.gov

Mark Elert
Chemistry Department
U.S. Naval Academy
572 Holloway Rd.
Annapolis, MD 21402
USA
E-mail: elert@usna.edu

Thomas P. Russell
NSWC-IH, Code 90
101 Strauss
Indian Head, MD 20640
USA
E-mail: thomas.p.russell@navy.mil

Carter T. White
Naval Research Laboratory
4555 Overlook Ave., S.W.
Washington, DC 20375
USA
E-mail: carter.white@nrl.navy.mil

L.C. Catalog Card No. 2006928108
ISBN 0-7354-0341-4
ISSN 0094-243X

Printed in the United States of America

# CONTENTS

## PART ONE

### CHAPTER I

### PLENARY

### CHAPTER II

### EQUATION OF STATE: NONENERGETIC MATERIALS

## CHAPTER III

### EQUATION OF STATE: ENERGETIC MATERIALS

## CHAPTER IV

## PHASE TRANSITIONS

# CHAPTER V

## MODELING, THEORY, AND SIMULATION: NONENERGETIC MATERIALS

# CHAPTER VI

## MOLECULAR DYNAMICS MODELING: NONREACTIVE MATERIALS

# CHAPTER VII

## MODELING, THEORY AND SIMULATION: ENERGETIC MATERIALS

## CHAPTER VIII

### MOLECULAR DYNAMICS MODELING: ENERGETIC MATERIALS

## CHAPTER IX

### SPALL, FRACTURE, AND FRAGMENTATION

CHAPTER X

CONSTITUTIVE AND MICROSTRUCTURAL PROPERTIES OF METALS

## PART TWO

### CHAPTER XI

#### MECHANICAL PROPERTIES OF POLYMERS AND COMPOSITES

# CHAPTER XII

## MECHANICAL PROPERTIES OF CERAMICS, GLASSES, IONIC SOLIDS, AND LIQUIDS

## CHAPTER XIII

## MECHANICAL PROPERTIES OF REACTIVE MATERIALS

## CHAPTER XIV

### DETONATION AND BURN PHENOMENA

## CHAPTER XV

## EXPLOSIVE AND INITIATION STUDIES

# CHAPTER XVI

## SHOCK-INDUCED MODIFICATIONS AND MATERIALS SYNTHESIS

# CHAPTER XVII

## INSTRUMENTATION

CHAPTER XVIII

EXPERIMENTAL TECHNIQUES

## CHAPTER XIX

### ISENTROPIC COMPRESSION EXPERIMENTS

## CHAPTER XX

### OPTICAL AND ELECTRICAL MEASUREMENTS

# CHAPTER XXI

## IMPACT PHENOMENA, BALLISTICS, HYPERVELOCITY STUDIES, AND EXOTIC SHOCK CONFIGURATIONS

## CHAPTER XXII

## LASER-DRIVEN SHOCKS AND INTERACTIONS OF LIGHT WITH MATERIALS

## CHAPTER XXIII

## GEOPHYSICS AND PLANETARY PHYSICS

# CHAPTER XXIV

## DYNAMIC FRICTION AND EXOTIC CONFIGURATIONS AND MATERIALS

CHAPTER XI

# MECHANICAL PROPERTIES OF POLYMERS AND COMPOSITES

CP845, *Shock Compression of Condensed Matter - 2005,*
edited by M. D. Furnish, M. Elert, T. P. Russell, and C. T. White
© 2006 American Institute of Physics 0-7354-0341-4/06/$23.00

# COMPOSITIONAL EFFECTS ON THE SHOCK COMPRESSION AND RELEASE PROPERTIES OF ALUMINA-FILLED EPOXY

**M. U. Anderson, D. E. Cox, S. T. Montgomery, and R. E. Setchell**

*Sandia National Laboratories, Albuquerque, NM, 87185*

**Abstract.** Alumina-filled epoxy is used for encapsulation in explosively driven pulsed power devices. Its shock compression and release properties have a strong influence on device performance. Previous studies using a material containing 43% by volume alumina showed a complex behavior characterized by extended wave profiles and high release-wave velocities. In recent studies, these properties have been examined while changing the total alumina volume fraction, the alumina particle size and morphology, and the epoxy constituents. Reducing the alumina volume fraction in steps from 43% to 0% had anticipated effects on Hugoniot states, compressive wave profiles and velocities, and release-wave velocities, although release velocities changed more rapidly. Only minor effects were observed when the alumina volume fraction was held constant while varying alumina particle characteristics or the host epoxy. Thin-pulse experiments showed combined compression and release effects resulting from decreasing the alumina volume fraction.

**Keywords:** composites, alumina-filled epoxy, encapsulants
**PACS:** 62.50+p, 46.40.Cd, 46.35.+z

## INTRODUCTION

Alumina-filled epoxy (ALOX) is used as an encapsulant in explosively driven pulsed power supplies. In such a device, the stress histories experienced by the active elements are strongly influenced by the shock compression and release properties of the encapsulant. An early study by Munson et al. (1) examined the shock and release behavior of an epoxy using Epon 828 resin (2) and Z hardener (3) with different volume fractions of added alumina powder. They found that compressive waves had extended rise times and dispersive rounding near peak values, and that release waves displayed unusually high velocities. The highest volume percent of alumina in their study was 43%, and this particular composition was examined in a recent study (4) that extended the characterization to higher shock pressures and

examined the viscous wave structure in more detail. In the current study, the effects of compositional changes on ALOX shock compression and release properties are examined. Of interest are changes in the total alumina volume fraction, the alumina particle size and morphology, and the epoxy constituents.

## SHOCK AND RELEASE EXPERIMENTS

To obtain useful insights into compositional effects without requiring an excessive number of gas gun experiments, two types of planar-impact configurations were chosen. In the first type, an ALOX impactor backed by carbon foam is accelerated into a thicker target of the same ALOX composition having a fused silica buffer/window backing. A 0.025-mm-thick piezoelectric PVDF gauge is attached to the impact surface to provide accurate impact timing. VISAR instrumentation is used to record the profiles of the transmitted compressive wave and the subsequent release wave. Details on this configuration have been given previously (4). An experiment of this type was conducted with each ALOX composition of interest while keeping sample dimensions and impact velocity constant. This produced shock-compressed states at a common peak particle velocity, corresponding to nearly equal strains. In addition to the compressive-wave and release-wave profiles, each experiment provided an average compressive-wave velocity, a final Hugoniot state, and a release-wave velocity.

The test matrix used for this portion of the study is shown in Table I. The first material containing 43% T64 tabular alumina (5) is the same

**Table 1.** Summary of ALOX compositions studied.

| Epoxy | Alumina | Volume % | Density - g/cm³ |
|---|---|---|---|
|  |  |  |  |
| 828/Z | T64 | 43 | 2.377 |
| 828/Z | AA18 | 43 | 2.389 |
| 828/Z | AA5 | 43 | 2.391 |
| 826/custom | T64 | 43 | 2.339 |
| 828/Z | T64 | 38 | 2.233 |
| 828/Z | T64 | 34 | 2.121 |
| 828/Z | T64 | 20 | 1.750 |
| 828/Z | T64 | 0 | 1.200 |

composition studied previously (1,4), and will be denoted as the baseline material. Alumina particle size varies from 5-50 μm in this powder. The next two materials used different alumina powders having a more spherical particle shape and nominal diameters of 18 and 5 micron, respectively (6), at the same 43% volume fraction. The next material used a 43% volume fraction of T64 alumina in a different epoxy consisting of Epon 826 resin (2) and a non-commercial hardener. The remaining materials had volume fractions of T64 alumina from 38% to 0% in 828/Z epoxy. The impactor and target ALOX samples had 50 mm diameters and were 4.5 mm and 9.5 mm thick, respectively. Impact velocity was 0.741±0.005 km/s.

The results from varying the alumina particle size and morphology are shown in Fig.1. The curves are

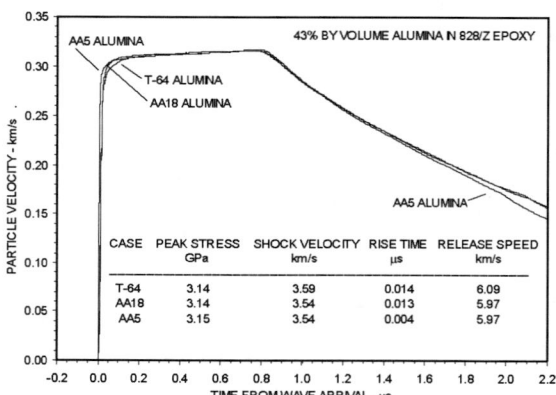

**FIGURE 1.** Transmitted wave profiles for ALOX with varying alumina particle size and morphology.

VISAR-measured transmitted wave profiles at the ALOX/window interface. Peak stress, shock and release speeds, and wave rise time (to half the peak particle velocity) are also listed in the figure. Only minor differences generally were observed as the alumina particle characteristics were changed from the tabular T64 to the powders with smaller, spherical particles. An exception was the short rise time seen in the case with 5 μm alumina particles.

The results from varying the host epoxy are shown in Fig. 2. A slightly smaller peak stress

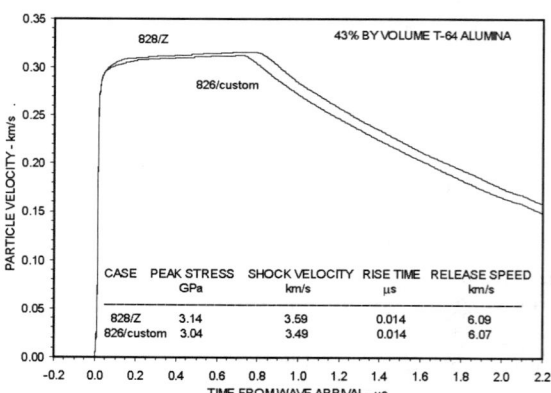

**FIGURE 2.** Transmitted wave profiles for materials with different host epoxies.

and shock velocity were observed for the material with the second epoxy. This material has a reduced density which results from the second epoxy having a smaller density (1.14 g/cm³) than 828/Z (1.20

790

g/cm³). No differences beyond expected density effects are apparent.

Figure 3 shows the results from varying the volume fraction of T64 alumina. As the fraction of alumina decreases, the peak axial stress, the average

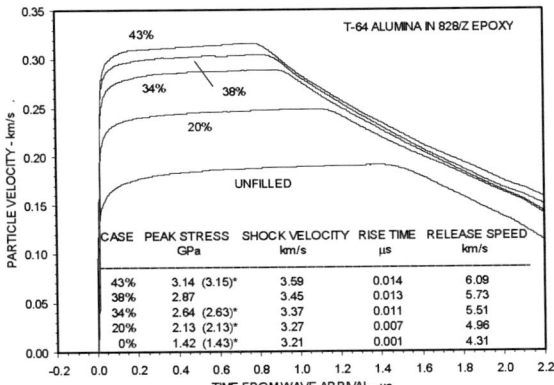

**FIGURE 3.** Transmitted wave profiles for materials with different alumina volume fractions. Peak stress values in parentheses are predictions based on previous studies.

wave velocity, the rise time, and the release-wave velocity all decrease. These trends are expected, and peak stress values are quantitatively consistent with predictions (parenthetical values in Fig. 3) based on previous studies (1,4,7).

Some significant aspects of the effects of alumina loading fraction can be identified by examining these results in more detail. Figure 4 shows the compressive wave profiles from Fig. 3 plotted on an expanded time scale, with each

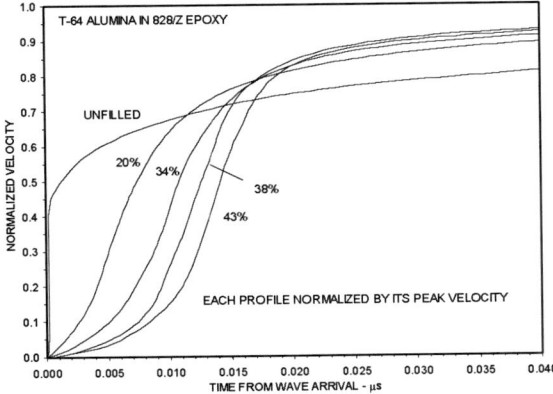

**FIGURE 4.** Wave profiles from Fig. 3 shown on an expanded time scale.

profile normalized by its peak value. The addition of alumina filler to the epoxy results in a change from viscoelastic behavior with an initial elastic shock to a viscoplastic behavior that shows no evidence of an elastic precursor. As previously found for the baseline material (4), the viscoplastic wave profile for each material containing alumina would likely evolve to a viscoelastic profile at a sufficiently high value of peak stress.

Figure 5 shows the release wave and shock wave velocities measured in these experiments as a

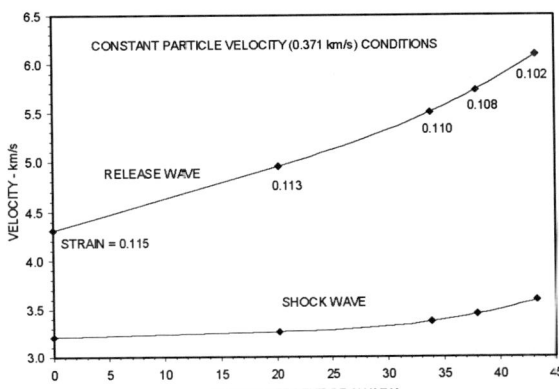

**FIGURE 5.** Measured shock and release velocities for materials with different alumina volume fractions.

function of alumina volume fraction. The release velocity decreases faster than the shock velocity with decreasing alumina, so the ratio of these velocities decreases as well. This ratio is a measure of how rapidly downstream unloading events can overtake and attenuate a propagating shock wave, which can be important in an encapsulation application. For a given alumina volume fraction, this ratio will also decrease somewhat with decreasing shock pressure.

## "THIN-PULSE" EXPERIMENTS

The second type of experiment used in this study is a "thin-pulse" configuration in which a 2-mm thick, unfilled 828/Z impactor backed by carbon foam is accelerated into a 20-mm thick ALOX sample having a fused silica buffer/window backing. These dimensions allow the release wave to overtake and attenuate the shock wave prior to reaching the window interface. All conditions except the target ALOX material were fixed, with the impact velocity $1.054 \pm 0.005$ km/s. The alumina volume fraction of the target material was again varied from 43% to 0%. The instrumentation is identical to the first experiment type, with an impact surface PVDF gauge for timing and VISAR for transmitted wave profiles at the ALOX/window interface.

The results of these experiments are shown in Fig. 6. In addition to the transmitted profiles, the

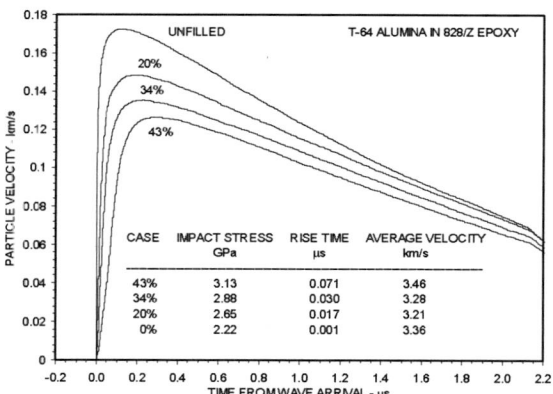

**FIGURE 6.** Transmitted wave profiles in "thin-pulse" experiments using materials with different alumina volume fractions.

figure lists the predicted impact stress and the measured average wave velocity and wave rise time for each case. Although the impact stress decreased significantly as the alumina fraction was decreased, the peak particle velocity of the transmitted wave progressively increased. This trend is a direct consequence of the decreasing ratio of release velocity to shock velocity as alumina fraction is decreased (Fig. 5). Even though the impact stress is lower when the alumina fraction is reduced, the corresponding reduction in the release-wave velocity is such that the transmitted wave experiences less attenuation. The wave interaction with the fused silica window then produces a stronger wave in the window. This shows an important and possibly surprising effect of alumina volume fraction in an encapsulation material required to transmit a shock wave to second material following a fixed, short-duration input stimulus.

## SUMMARY

Compositional effects on the shock compression and release behavior of alumina-filled epoxy have been examined in a limited set of gas gun experiments. In shock and release experiments, only minor effects were observed when the alumina volume fraction was held constant while varying the alumina particle characteristics or the host epoxy. Changing the alumina volume fraction had anticipated effects on peak stresses, compressive-wave velocities and rise times, and release-wave velocities. The ratio of release to shock velocities decreased significantly as alumina fraction was reduced. "Thin-pulse" experiments examined the combined effects of compression and release properties in materials with different alumina volume fractions. The results showed that the rapid reduction in release velocity with decreasing alumina fraction could produce stronger transmitted waves following a fixed input condition. This could be important for optimizing the properties of an ALOX encapsulant.

## ACKNOWLEDGMENTS

The authors would like to thank Miriam Hilborn at Sandia for the ALOX sample processing. Sandia is a multiprogram laboratory operated by Sandia Corporation, a Lockheed Martin Company, for the United States Department of Energy's National Nuclear Security Administration under Contract DE-AC04-94AL85000.

## REFERENCES

1. Munson, D. E., Boade, R. R. and Schuler, K. W., *J. Appl Phys.* **49**, 4797-4807 (1978).
2. A product of Resolution Performance Products.
3. Previously available from Shell Chemical Company.
4. Setchell, R. E., and Anderson, M. U., *J. Appl Phys.* **97**, 083518 (2005).
5. A product of Alcoa World Chemicals.
6. Products of Sumitomo Chemical Company.
7. Munson, D. E., and May, R. P., *J. Appl Phys.* **43**, 962-971 (1972).

CP845, *Shock Compression of Condensed Matter - 2005*,
edited by M. D. Furnish, M. Elert, T. P. Russell, and C. T. White
© 2006 American Institute of Physics 0-7354-0341-4/06/$23.00

# COLLAPSE OF HOLLOW CYLINDERS OF PTFE AND ALUMINUM PARTICLES MIXTURES USING HOPKINSON BAR

**Jing Cai[1] and Vitali F. Nesterenko[1, 2]**

[1]*Materials Science and Engineering Program*
[2]*Department of Mechanical and Aerospace Engineering*
*University of California, San Diego, CA 92093-0411*

**Abstract.** Hopkinson bar based thick walled cylinder (TWC) method was developed to collapse hollow cylinders with small mass about 0.5 gram made from the mixtures of PTFE and Al particles of different sizes (2 and 95 μm). Different media (water, suspension of alumina particles in water, and glycerol) in different geometrical configurations were investigated to ensure the collapse of hollow cylinders with a single pressure pulse under pressure/time conditions achievable in Hopkinson bar tests. Raman spectroscopy of the samples of PTFE and 2 μm aluminum particles mixtures demonstrated the evidence of the decomposition or reaction of PTFE inside the shear localization area or cracks.

**Keywords:** Hollow cylinder, collapse, shear localization, PTFE, Al, reaction, Hopkinson bar
**PACS:** 06.60.Ei, 06.60.Jn, 33.20.Fb, 62.20.Fe, 82.30.Lp

## INTRODUCTION

PTFE (polytetrafluoroethylene) has an excellent combination of electric and mechanical properties that make it suitable for many applications [1]. It is also one of components of energetic mixtures [2-6]. A few studies were conducted to investigate the dynamic behavior of this polymer [7-12].

The athermal mechanism of strain softening (e.g., crazing), instead of thermal softening typical for metals, causes the initiation and subsequent patterning of shear bands observed in the explosively driven collapse of PTFE cylinders [13] and in Hopkinson bar based tests with samples of smaller size [14]. The dynamic shear localizations in polycarbonate and polymethyl methacrylate also exhibited similar behaviors [15].

The creation of a hot spot by the oxidation of Al particles in Teflon[AF] led to shock-induced decomposition into monomers [16] and resulted in the condensation of carbon [17]. Dienes [18] analyzed four hot spot mechanisms: void collapse

with a closing shock, void collapse with uniform, nonlocalized plastic flow, shear banding with plastic flow, and shear cracking with frictional heating as mechanisms of detonation observed when propellant cylinders were fired at low speeds against a steel plate. He proposed that interfacial heating due to friction inside closed cracks is the most likely mechanism of XDT detonation observed below critical velocity characteristic for Shock to Detonation Transition,.

This paper presents results of experimental tests which allow characterization of the critical conditions for the start of shear induced instability in non-traditional (like PTFE/Al or their simulants) heterogeneous materials. The method generates an array of self-organized shear bands under controlled boundary conditions and the identification of the major mesomechanical mechanisms of softening of heterogeneous energetic materials which may destabilize uniform plastic flow. We used a different diameter of Al particles in the mixture to determine the influence of metal particle size on the initiation of shear instability, shear band patterning and chemical reaction.

## EXPERIMENTAL PROCEDURE

Details and geometry of the experimental set-up for Hopkinson bar based thick walled cylinder tests are described elsewhere [14].

The samples were prepared by mixing of PTFE powder (DuPont, PTFE 9002-84-0, type MP 1500J) with fine Al powder (2-3 $\mu$m, Valimet, H-2) with purity 99.7 wt%) or by mixing of PTFE powder with coarse Al particles (H-95, -100 mesh with 95 wt. % min. and with purity 99.7 wt.%). The cold isostatic pressing at 345 MPa and normal temperature was used to form 2.23 g/cm$^3$ high-density, high accuracy composite samples of PTFE and Al (76wt% and 24wt%, respectively). Inner and outer diameters of the hollow cylindrical samples were 4.8 mm and 10.4 mm respectively. The weight varied from 1 to 0.5 g.

Different liquids were used as driving media in the sample's chamber to ensure a controlled pressure in the chamber with small displacement of the incident bar: (a) water, (b) suspension of alumina particles, 20% by volume, in water with overall density 1.6 g/cm$^3$, particle size less than 10$\mu$m (ALDRICH) (c) Glycerol (EMD, GR ACS, density 1.26 g/cm$^3$). The typical pressure pulses in the chamber filled with different liquids (with no sample inside) detected by gauges on the outside surface are shown in Fig. 1. They were generated by an impact with a 254 mm-long, 19 mm-diameter striker bar with a velocity of 14 m/s. Ideal pressure transmitted liquid should have a small compressibility which will ensure a larger pressure in the chamber with a small displacement of the incident bar. The compressibility of water, alumina suspension in the water, and glycerol are equal to $45.8 \cdot 10^{-11}$ Pa$^{-1}$, $36.7 \cdot 10^{-11}$ Pa$^{-1}$, and $21 \cdot 10^{-11}$ Pa$^{-1}$ [19]. At the similar striker velocity, the largest pressure was achieved with glycerol (Fig. 1(c)) due to the smallest compressibility of this liquid.

Time of the pulse reverberation inside the chamber, when the length of the cavity filled by liquid equal 22 mm, is about 30 microseconds for water (sound speed $C_0$=1500 m/s), 34 $\mu$s for the suspension ($C_0$=1300 m/s) and 23 $\mu$s for glycerol ($C_0$=1904 m/s). This reverberation time is causing

a step-like increase of the stress in the transmitted bar. It is less noticeable in the measured stresses on the outside wall of thick walled chamber.

Comparison of averaged pressure in the chamber and its duration based on stresses on its surface and stresses in the transmitted bar demonstrated a reasonable agreement.

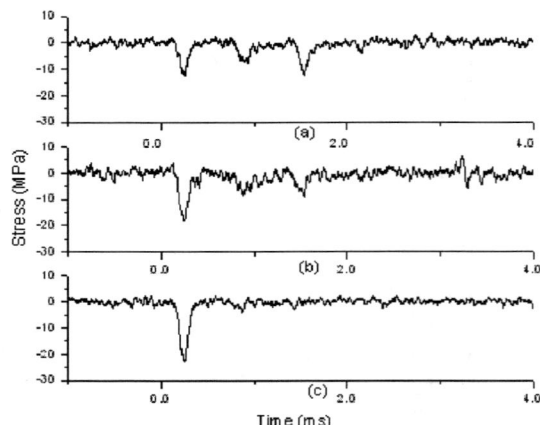

**FIGURE 1.** The stresses detected by the strain gages on the chamber wall with different liquid medium inside: (a) Water, (b) Alumina suspension in the water, (c) Glycerol.

The higher speed of sound in glycerol facilitates smoother increase of pressure that creates a higher stress amplitude. The total duration of the pulses is about 200-250 $\mu$s, which is significantly longer than reverberation times and do not depend significantly on the nature of the liquids inside.

The Raman spectra of initial and collapsed samples were collected on a Renishaw Raman Spectometer with a Melles Griot laser. The 514.5 nm excitation line was chosen with an incident power varying from 10 mW to 25 mW.

## RESULTS AND DISCUSSION

The Hopkinson bar based test allowed monitoring of the pressure history in the chamber using gages on the transmitted bar and on the outside wall of the chamber. The detected pulses are shown in Fig. 2 when sample with a height of 2.97 mm was placed inside the chamber. This sample was totally collapsed and shattered. Other samples in a similar condition of loading were also almost completely collapsed, but were held together by the polymer holder. They usually had a few radial and spiral cracks (shear bands) (Fig. 3).

In totally collapsed samples, black, flaky residues were observed (Fig. 4) in areas adjacent to the cracks/shear bands. The Raman spectra of the original samples before testing, black solid chunks and black flaky residues after testing (Fig. 5) show the decomposition of PTFE in the latter. The peaks in the samples before testing and in the black solid chunks after testing are the same, indicating no reaction. New sp2 and sp3 peaks from the black flaky residues verify the decomposition of PTFE into graphitic carbon.

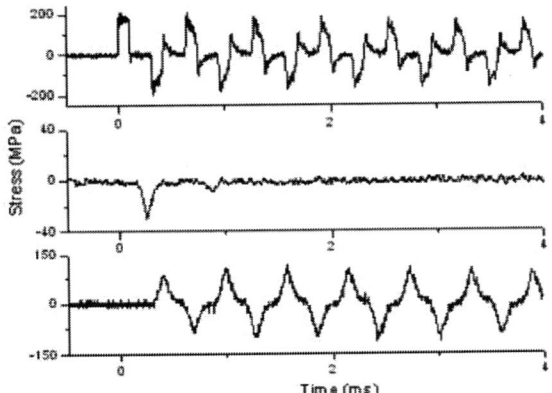

FIGURE 2. (a) Stresses in the incident bar; (b) one main stress peak corresponding to pressure 100 MPa inside the chamber is detected in the chamber wall; (c) stresses in the transmitted bar.

No lines in the Raman spectra of the black flaky residue corresponded to $AlF_3$ (or other similar compounds).

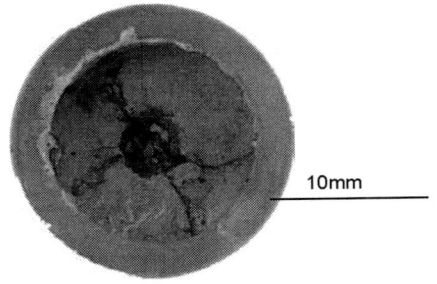

FIGURE 3. The nearly collapsed sample from the mixture of PTFE and 2-μm Al in polymer jacket. Note the characteristic four cracks/shear bands.

The black residue was also observed in laser initiated reactions in Teflon[AF] (Dupont Teflon, Amorphous Fluoropolymer which is composed of the chemical monomers tetrafluoroethylene and a dioxole monomer) and its mixture with nanoparticles (500 nm diameter) of Al [17]. Raman spectra of the Teflon[AF] and Teflon[AF]/Al samples indicate that the recovered opaque material is graphitic carbon. The proof of reaction was based on the change of the diameter of Al particles measured using time resolved absorption spectroscopy.

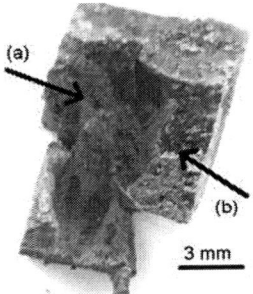

FIGURE 4. Part of collapsed sample in experiment with PTFE/Al particles with 2-μm diameter. Surfaces on the left and right correspond to the initial side of the sample. (a) Black flaky residue identified as graphitic carbon; (b) solid black chunk with the same composition as the sample before testing.

Cylindrically symmetrical collapse of samples in the Hopkinson bar based TWC test was observed with negligible axial strain (0.007). Estimated radial strain rate about $10^3 \sim 10^4$ s$^{-1}$.

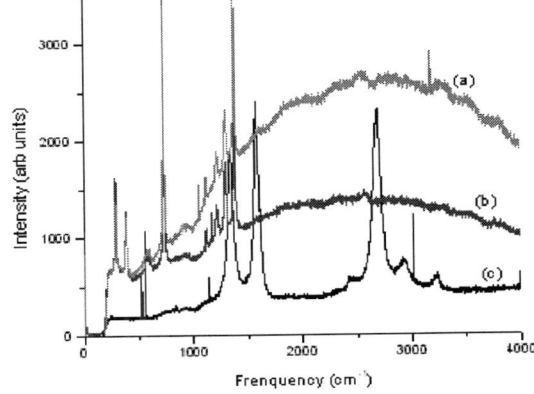

FIGURE 5. The Raman spectra of (a) original samples before testing, (b) solid black chunks after testing, and (c) black flaky residues after testing.

The solid PTFE sample was symmetrically collapsed with the effective strain of 0.22 on the inner surface without developing shear instability [14]. The start of shear instability at larger strain resulted in a loss of cylindrical symmetry. This is in agreement with data for an explosively collapsed PTFE sample where critical strain for localized shear propagation was 0.4 [13, 14].

The black flaky residue was not observed in pure PTFE, in the mixture of PTFE and 95 μm Al particles or in the mixture with Tin particles under the same conditions of deformation. This behavior strongly indicates that the observed phenomenon is due to the reaction of PTFE and 2 micron size Al particles inside the shear zone or on the sliding sides of cracks during collapse of hollow cylindrical samples. The reaction did not propagate into the bulk of the sample, being restricted to the area adjacent to a shear band or crack.

## CONCLUSIONS

A few liquids were investigated as pressure transmitted media and glycerol was demonstrated as the best choice. CIP (Cold Isostatic Press) based method was used to prepare high-accuracy small cylinders (mass about 0.5 gram) from PTFE and Al particles. The dynamic collapse of solid PTFE/Al samples with different particle sizes was accomplished with the shear localization bands and cracks. The mechanism of the shear localization phenomenon is athermal. Only samples with 2 μm Al particles demonstrated a black flaky residue which was identified using Raman Spectroscopy as carbon. This phenomenon was not detected under collapse of cylinders made from pure Teflon or from mixtures of PTFE and large aluminium particles or in mixtures of PTFE and Sn particles, indicating that the reaction between PTFE and 2 μm Al particles was initiated inside the shear band (crack). It did not propagate into the rest of the sample.

## ACKNOWLEDGMENTS

The support for this project provided by ONR (Program Officer Dr. Judah M. Goldwasser, N00014-02-1-0491) is highly appreciated. Authors are grateful to Professor Marc A. Meyers for help with experiments on Hopkinson bar and to Eric Herbold for Guassian smoothing of the wrought experimental data and paper preparation, and to Ralf Brunner for help with Raman spectroscopy analysis.

## REFERENCES

1. Brydson, J. A. *Plastics Materials,* Butterworths, London-Boston, 1989.
2. Davis, J. J., Lindfors, A. J., Miller, P.J., Steve Finnegan, S., and Woody, D. L. In: *11th International Detonation Symposium,* Office of Naval Research, ONR 33300-5, 1007, 1998.
3. Holt, W. H., Mock, W. Jr., and Santiago, F. *J. Appl. Physics,* **88**, 5485 (2000).
4. Parker, L. J., Ladouceur, H. D., and Russell, T. P. In: *Shock Compression of Condensed Matter-1999,* 941, 2000.
5. Davis, J. J. and Lindfors, A. J. In: *Shock Compression of Condensed Matter-1997,* 663, 1998.
6. Woody, D. L., Davis J. J. and Deiter, J. S., In: *Shock Compression of Condensed Matter-1997,* 667, 1998.
7. Zerilli, F. J. and Armstrong, R. W., In: *Shock Compression of Condensed Matter,* 2001, 657, 2002.
8. Khan, A. and Zhang, H., *Int. Journal of Plasticity,* **17**, 1167, 2001.
9. Fried, L. E. and Howard, W. M., In: *Shock Compression of Condensed Matter-1999,* 57, 2000.
10. Jones, H. D., Zerilli, F. J., Holt, W.H., Mock, W., Jr., Miller, P. J. and Lindfors, A. J. In: *Shock Compression of Condensed Matter-1999,* 137, 2000.
11. Gray, G. T. III, Cady, C. M. and Blumenthal, W. R., In: *Proc. Plasticity '99, Seventh Int. Symp. On Plasticity and Its Current Applications,* 955, 1998.
12. Kerley, G. I., In: *Shock Compression of Condensed Matter-1997,* 671, 1998.
13. Nesterenko, V. F., *Dynamics of Heterogeneous Materials,* Springer-Verlag, New York, 2001.
14. Gu, Y., Nesterenko, V. F. and Cai, J., In: *Shock Compression of Condensed Matter-2003,* 775, 2004.
15. Fleck, N. A., Strong, W.J. and Liu, J.H., *Proc. R. Soc. Lond, A* **429**, 459, 1990.
16. Nakamura, K. G., Wakabayashi, K. and Kondo, K. *AIP Conf. Proc.* **20**, 1259, 2001.
17. Parker, L. J., Ladouceur, H. D. and Russell, T. P., *AIP Conf. Proc.* **505**, 941, 1999.
18. Dienes, J. K., *Mat. Res. Soc. Symp. Proc.* **24**, 373, 1984.
19. Young, H. D., Freedman, R. A., Sandin, T. R. and Ford, A. L., *Sears and Zemansky's University Physics,* 10th Ed., Section 11-6, Addison-Wesley, 2000.

CP845, *Shock Compression of Condensed Matter - 2005,*
edited by M. D. Furnish, M. Elert, T. P. Russell, and C. T. White
© 2006 American Institute of Physics 0-7354-0341-4/06/$23.00

# HIGH STRAIN RATE RESPONSE OF AN EPOXY AND A VINYL ESTER

**Rodney J. Clifton***, **Petch Jearanaisilawong**[†] **and Tong Jiao***

*Division of Engineering, Brown University, Providence, RI 02912*
[†]*Department of Mechanical Engineering, MIT, Cambridge, MA 02139*

**Abstract.** Pressure-shear plate impact experiments are used to study the nonlinear dynamic response of an epoxy and a vinyl ester at shearing rates of $10^5$ - $10^6$ $s^{-1}$. Samples with thicknesses in the range $10\mu m$ - $100\mu m$ are formed between two hard steel plates. Because of its higher wave speed, the longitudinal wave generated at impact reaches the sample first and, after a few reverberations through the thickness of the sample, subjects the sample to a state of uniaxial strain compression. Once the sample is fully compressed the shear wave arrives and imposes a simple shearing deformation. From the transverse velocity, measured interferometrically at the rear surface of the sandwich target, the shear stress and the transverse velocity at the rear surface of the sample are determined. These measurements provide an indication of the shearing resistance of the material under pressure. Because the sample bonds to the bounding plates, the shearing of the sample continues even after longitudinal unloading waves arrive from the rear surface of the target and reduce the nominal pressure in the sample to zero. Thus, from a single experiment, one obtains the response of the sample in simple shear – both under pressure and without pressure. From such experiments a pressure-sensitivity of inelastic shearing resistance is found for both the epoxy and two vinyl esters.
**Keywords:** Epoxy, Vinyl Ester, Pressure-Shear Plate Impact
**PACS:** 43.35.Ei, 78.60.Mq

## PRESSURE-SHEAR PLATE IMPACT

The pressure-shear plate impact configuration used for the experiments reported here is shown in Figure 1 of [1]. A sample is cast between two hard, AL6XN stainless steel plates. The target assembly is impacted by an AL6XN stainless steel flyer plate. The time-distance diagram for the principal wavefronts is similar to that shown in Figure 2 of [1], except the flyer plate is much thicker than the rear plate. Therefore, one can ignore the effect of the reflected wave from the free surface of the flyer plate. At impact, both longitudinal waves and shear waves are generated. These waves propagate both forward into the front plate of the target assembly and backward into the flyer. No slip at the impact plane is assured by keeping the skew angle $\theta$ sufficiently small (22° is used) and roughening slightly the impact faces.

Then, from symmetry, the initial velocity at the impact face is one-half the in-coming velocity of the flyer plate. Details of the techniques involved in imposing the pressure-shear loading and in recording the free surface normal displacement interferometer (NDI) and transverse displacement interferometer (TDI) are given in [2]. Throughout the loading and reverse loading of the sample the traction on the rear face of the sample can be inferred from the velocity-time profiles monitored at the rear surface of the target. From the relations that hold along characteristics the compressive and shear components of this traction are related to the normal and transverse components of the free surface velocity by

$$\sigma(t) = -\rho c_1 \frac{u_{fs}(t+h_3/c_1) - u_{fs}(t-h_3/c_1)}{2} \quad (1)$$

**TABLE 1.** Shot Summary

| Shot No. | Sample (h) | | Thickness (mm) | | | Diameter (mm) | Impact Velocity (m/s) |
|---|---|---|---|---|---|---|---|
| | | | Flyer $(h_1)$ | Front $(h_2)$ | Rear $(h_3)$ | | |
| PJ0101 | Hysol EA9394 | 0.05 | 11 | 4 | 4 | 50 | 36.3 |
| PJ0102 | | 0.05 | 11 | 3.5 | 3.5 | 50 | 32.1 |
| PJ0201 | | 0.01 | 11 | 3.8 | 3.8 | 50 | 34.6 |
| AF0401 | Vinyl Ester | 8084 | 0.076 | 10.813 | 4.039 | 3.734 | 50 | 110 |
| AW0501 | | | 0.0762 | 10.516 | 3.835 | 3.886 | 50 | 111.7 |
| AW0502 | | 411-350 | 0.075 | 10.414 | 4.013 | 3.962 | 50 | 111.1 |

where $u_{fs}$ is the normal component of the free surface velocity and

$$\tau(t) = -\rho c_2 \frac{v_{fs}(t + h_3/c_2) - v_{fs}(t - h_3/c_2)}{2} \quad (2)$$

where $v_{fs}$ is the transverse component of the free surface velocity. In these equations, $\rho$ is the mass density, and $h_3$ is the thickness of the rear plate of the target assembly. The longitudinal and shear wave speeds in this steel plate are denoted by $c_1$ and $c_2$, respectively. The second term in the numerators of these equations is zero until the corresponding elastic wave has reflected from the free surface of the target and has returned again to its rear surface. The density and wave speeds for AL6XN are taken to be: $\rho = 7900 kg/mm^3$, $c_1 = 5.839 mm/\mu s$, and $c_2 = 3.121 mm/\mu s$ [3]. The shearing resistance can be obtained from the transverse velocity-time profiles for which the shear stress at the back of the sample is given by Eqn. (2). Once sufficient reverberations of shear waves have occurred to make the shear stress in the sample nominally uniform through the thickness, the shear strain rate in the sample is given by

$$\dot{\gamma}(t) = \frac{v_{fs}(t + h_3/c_2) - v_0}{h} \quad (3)$$

where $v_0$ is the transverse component of the velocity of the flyer plate and $h$ is the thickness of the sample. Equation (4) can be integrated to obtain the shear strain corresponding to the shear stress of Eqn. (2). The result gives a shear stress vs. shear-strain curve at the nominal strain rate given by Eqn. (3) and the pressure given by Eqn. (1). Table 1 gives the shot summary.

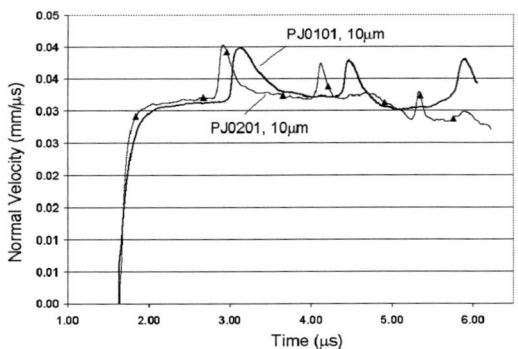

**FIGURE 1.** Normal velocity history: Hysol EA9394.

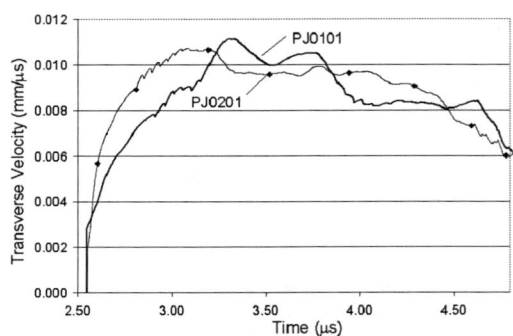

**FIGURE 2.** Transverse velocity history: Hysol EA9394.

## EXPERIMENTAL RESULTS

### Hysol EA9394

Velocity-time profiles for the normal velocity for two Hysol EA9394 samples having thicknesses of $50 \mu m$ and $10 \mu m$, respectively, are shown in Fig.1. In view of the relatively small thicknesses of the samples, the normal velocity rings up quickly – especially for the $10 \mu m$ thick sample. In either case the pre-compression of the sample reaches a uniform state before the shear wave arrives at approximately $2.55 \mu s$. The normal stress in the sample is obtained from the plateau value of the free surface normal velocity by means of Eqn. (1). The second pulses arriving at approximately $3 \mu s$ are due to the return of the wave reflected from the rear surface of the target and subsequently from the low impedance sample. Again, the duration of the rise and fall of these pulses

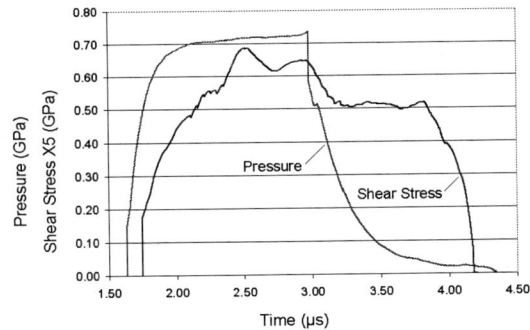

**FIGURE 3.** Shear stress decrease with decreasing pressure: Hysol EA9394.

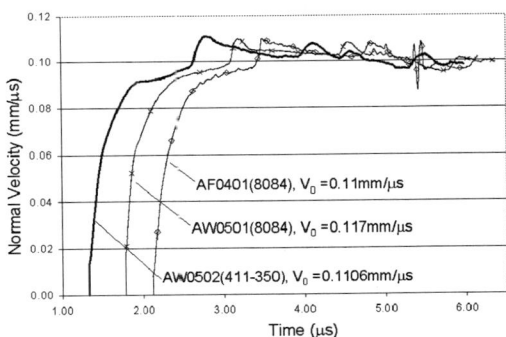

**FIGURE 4.** Normal velocity history: vinyl esters.

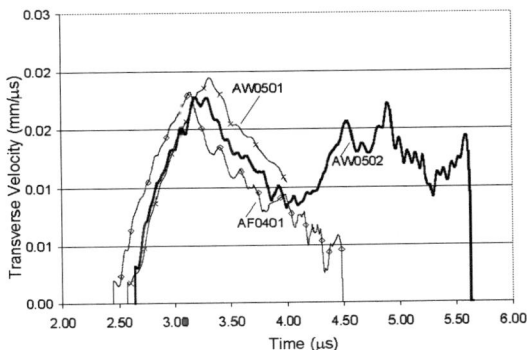

**FIGURE 5.** Transverse velocity history: vinyl esters.

is related to the thicknesses of the samples. The subsequent pulses are due to reflected waves that have made an additional roundtrip through the thickness of the rear plate of the target assembly.

The corresponding velocity-time profiles for the transverse component of the rear surface motion are shown in Fig. 2. Again the risetimes are related to the thicknesses of the samples. In both cases the transverse velocity has essentially reached a plateau velocity before the normal stress begins to decrease due to the arrival of the reflected unloading wave from the rear surface of the target. This plateau velocity is substantially less than the transverse component of the projectile velocity, indicating that shearing of the sample is continuing after the state of stress in the sample has become nominally uniform through the thickness of the sample. The shear stress in the sample is given by Eqn. (2). Much of the rise in the early part of the curves is associated with the ring up of the low-impedance sample sandwiched between two high impedance plate. Nevertheless, at least the last half of the curves should provide a good indication of the shearing resistance of the sample. In view of the similarity of the plateau levels of the transverse particle velocity the corresponding shear stress does not vary much over the strain rate range from $233,000 \ s^{-1}$ for the 10 $\mu m$ thick sample to $61,000 \ s^{-1}$ for the 50 $\mu m$ thick sample. The shearing resistance decreases as the pressure decreases as shown in Fig. 3 where the time scales are shifted to provide simultaneous records of pressure (actually normal stress) and shear stress at the sample. For Fig. 3, the records used are those for the 50 $\mu m$ samples shown in Figs. 1 and 2. Begin-

ning at approximately $t = 3 \mu s$ the shear stress falls by approximately 25 $MPa$ as the pressure falls by approximately 700 $MPa$, indicating a pressure sensitivity of the shearing resistance of approximately $25/700 \approx 0.036$.

## Vinyl Ester

Experimental results for two different vinyl esters are reported here. One is Derakane 8084, the other is Derakane Momentum 411-350. Both are from Dow Chemical Company. Velocity-time profiles for the normal velocity for three vinyl ester samples having thicknesses of approximately $75 \mu m$ are shown in Fig.4. Because the thicknesses of these samples are greater than those for the Hysol EA9394 samples, the risetimes are significantly larger. Also, based on elastic moduli reported for these vinyl esters as well

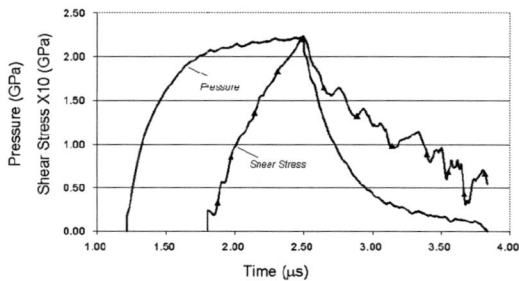

**FIGURE 6.** Shear stress decrease with decreasing pressure: vinyl ester.

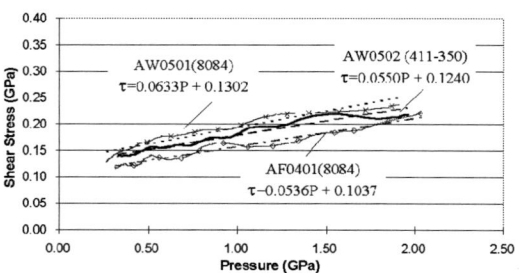

**FIGURE 7.** Dependence of shearing resistance on pressure: vinyl esters.

as those estimated from the roundtrip transit times shown in the ring up of the longitudinal wave fronts in Fig. 4, the longer risetimes are due in part to slower wave speeds in the vinyl esters. Nevertheless, the pre-compression of each of the samples reaches a uniform state before the shear wave arrives at approximately $2.55\mu s$. The normal stress in the sample is again obtained from the plateau value of the free surface normal velocity by means of Eqn. (1). The reflected second pulses arriving at approximately $2.5 - 2.7\mu s$ are similar to those obtained for Hysol EA9394 except that the fall time is substantially longer due to the greater thickness of the sample. The third pulses again represent waves that arrive after an additional roundtrip through the thickness of the rear plate of the target assembly.

The corresponding velocity-time profiles for the transverse component of the rear surface motion are shown in Fig. 5. The greater risetimes are related to the increased thicknesses of the samples and the reduced wave speeds. In all cases the transverse velocity has not reached a plateau velocity before the normal stress begins to decrease due to the arrival of the reflected unloading wave from the rear surface of the target. Thus, while the sample is under pressure, the shear stress does not become uniform through the thickness of the sample. Consequently, a dynamic stress-strain curve, at pressure, cannot be obtained from these velocity-time profiles. However, the velocity-time profiles do, through Eqn.(2), provide the shear stress at the rear surface of the sample as the ring up of the shear stress continues. The maximum shear stress at the rear surface provides a lower bound on the shearing resistance of the sample that would be established if there were sufficient time for the ring up of the shear stress to be complete. When unloading of the normal stress occurs, the ring down of the shear stress is nearly complete as shown, for example, in Fig. 6. This figure shows a very strong decrease in the shearing resistance as the normal stress is reduced. Although uniform shear stress is not established through the thickness of the samples while the samples are under pressure, the change in shear stress at the rear surface is indicative of the pressure-sensitivity of the shearing resistance. A pressure-sensitivity obtained in this way is shown in Fig. 7. From the linear fits to the curves shown here it appears that these vinyl esters have a pressure-sensitivity of the shearing resistance that is at least 0.055.

## ACKNOWLEDGMENTS

The authors gratefully acknowledge the financial support of ONR. They also appreciate the valuable assistance of Steve Grunschel in executing the plate impact experiments.

## REFERENCES

1. Clifton, R. J., and Jiao, T., "High Strain Rate Response of an Elastomer," in *APS-SCCM2005*, Baltimore, 2005.
2. Klopp, R., and Clifton, R., "Pressure-Shear Plate Impact Testing," in *Metals Handbook, 9th Edition*, American Society for Metals, Metals Park, OH, 1985, vol. 8, pp. 230–239.
3. Jearanaisilawong, P., *Pressure and Shearing Response of Hysol Epoxy at High Strain Rate*, Sc. B. thesis, Brown University, Providence, RI, 2002.

CP845, *Shock Compression of Condensed Matter - 2005*,
edited by M. D. Furnish, M. Elert, T. P. Russell, and C. T. White
© 2006 American Institute of Physics 0-7354-0341-4/06/$23.00

# THE BEHAVIOUR OF A GLASS-FIBRE EPOXY COMPOSITE DURING PLATE IMPACT

**M. Eatwell, J.C.F. Millett, N.K. Bourne\* and Y. Meziere**

*Defence Academy of the United Kingdom, Cranfield University, Shrivenham, Swindon,
SN6 8LA. United Kingdom.*

*\*School of Mechanical, Aerospace and Civil Engineering, University of Manchester, Sackville Street,
Manchester, M60 1QD. United Kingdom.*

**Abstract.** The response to shock loading of a glass-fibre epoxy composite has been investigated in terms of stress, particle velocity, shock velocity, release velocity and reload signal due to spall. The shock velocity has a linear relationship with particle velocity, and shock stress lies a little above the corresponding hydrodynamic pressure. Results show that the likelihood of spallation increases with pulse duration, suggesting that damage in the material accumulates behind the shock front. This would seem confirmed by release wave speed measurements, that show the zero particle velocity intercept is lower than either $c_0$ (for the shock velocity – particle velocity relationship) or the longitudinal sound speed ($c_L$).

**Keywords**: shock, composite, Hugoniot, release, spall
**PACS**: 62.50

## INTRODUCTION

The response of fibre based composite materials to shock loading is becoming of increasing interest, both from their potential as light-weight armour systems, and from the aerospace community as low density alternatives to traditional aluminium alloys. In the latter, both low velocity impact behaviour, for example a bird strike, or higher impact velocity behaviour from weapons are sought. These materials are formed from woven planes of fibres, infiltrated with a polymer matrix. Some work has investigated these materials under shock loading conditions [1, 2], and this paper aims to add to that data in terms of equation of state and constitutive information. This material will be anisotropic in its mechanical behaviour, so we investigate its response normal to the plane of the fibres.

## EXPERIMENTAL

The material tested in this programme was an E glass (similar to soda-lime glass; SL) epoxy (LY564 Ceiba-Geigy) with the fibres in an *x-y* lay up. In all cases, the impact axis was normal to the plane of the fibres. All shots were performed on a 5 m long, 50 mm bore single-stage gas gun. Two types of experiment were performed. The first was designed to determine the equation of state. Manganin stress gauges (MicroMeasurements type LM-SS-125CH-048) were placed between 7 mm tiles of composite, using a low viscosity epoxy adhesive. A second gauge (the 0 mm position) was supported on the front of the target assembly with a 1 mm plate of either dural (aluminium alloy 6061-T6) or copper, matched to the material of the flyer plate. Thus the equation of state could be determined in terms of stress amplitude ($\sigma_x$), particle velocity ($u_p$) and shock velocity through the gauge spacing ($\Delta w$) and temporal separation ($\Delta t_{shock}$), $U_s = \Delta w / \Delta t_{shock}$. Impactors were either 5 mm flyer plates of either dural or copper fired in the velocity range 200 to 837 m s$^{-1}$. In the second series of experiments, a single stress gauge was supported on the back of either a 3.8 or 9.8 mm tile of composite with a block of 12 mm thick polymethylmethacrylate (PMMA), and impacted with 3 and 10 mm dural flyers respectively at *ca.* 500 m s$^{-1}$. In this way, the shape of the pulse could be monitored as a function of thickness as it moved through the target. Gauge calibrations were according to Rosenberg *et al.* [3] Acoustic wave speeds were measured using quartz transducers at 5 MHz with a Panametrics 5077PR pulse receiver. These are presented in Table 1, along with the corresponding properties of epoxy, soda-lime glass and PMMA.

**Table 1**. Acoustic Properties of GFRP and constituents.

|       | $c_L$ (mm/µs) | $\rho_0$ (g/cm$^3$) | $Z = c_L \rho_0$ |
|-------|---------------|--------------------|------------------|
| GFRP  | 4.62±0.03     | 2.05±0.03          | 9.47±0.06        |
| SL    | 5.84±0.03     | 2.49±0.03          | 14.54±0.06       |
| LY564 | 2.54±0.03     | 1.14±0.03          | 2.71±0.06        |
| PMMA  | 2.72±0.03     | 1.19±0.03          | 3.24±0.06        |

A schematic diagram of the target assembly is presented in Fig. 1.

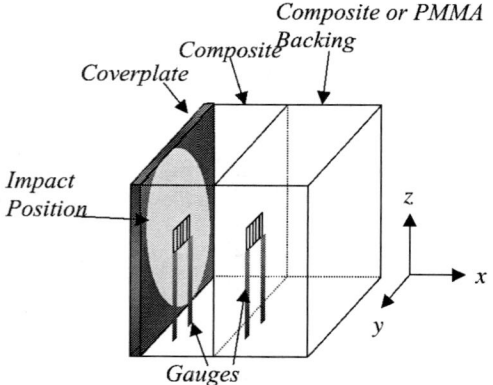

**Figure 1.** Gauge placement and specimen configuration.

## RESULTS AND DISCUSSION

Typical gauge records from the embedded configuration are presented in Fig. 2. It can be seen

that there is some noise superimposed on the signals. We believe this to be due to the impedance differences between the glass fibre layers and the epoxy. However, this noise is not sufficiently great to prevent the measurement of an impact stress, and this demonstrates that manganin stress gauges are capable of making valid stress measurements in these materials. Also observe that the amplitude of the traces at both positions within the target assembly show no evidence of attenuation.

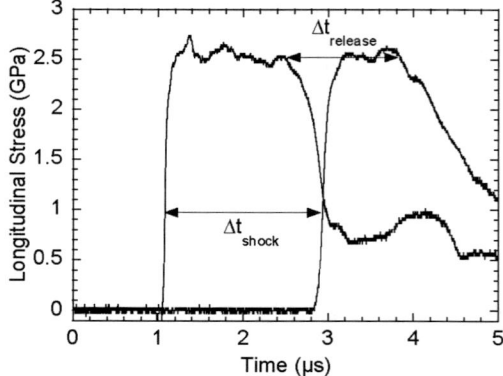

**Figure 2.** Gauge traces from the embedded configuration. The impact conditions are 5 mm dural flyer at 500 m s$^{-1}$.

In Fig. 3, we present the results of back surface gauge measurements.

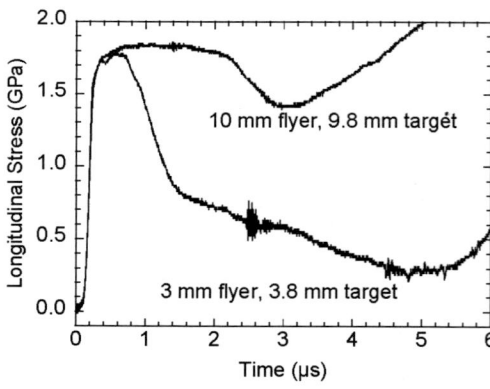

**Figure 3.** Gauge traces from the back surface configuration. The impact conditions are dural flyer at 500 m s$^{-1}$.

There are a number of features that are worthy of discussion. Firstly, note that the initial part of the pulse (the rise) and the amplitude are identical, thus agreeing with the observations made in the previous figure. However, significant differences can be observed on the release phase. In the 3.8 mm sample, the stress releases quickly until a time of ca. 1.4 µs, before continuing to release at a slower rate. In contrast, the 9.8 mm sample releases until a time of ca. 3 µs, before increasing in amplitude. These experiments were designed such that the partial release from the composite / PMMA interface would meet the release from the rear of the flyer plate in the middle of the composite plate itself. The results therefore indicate that in the thicker sample, the observed reload is due to spallation in the material, whilst no such events occur in the thinner target. This further suggests that the spall strength in this particular material is pulse duration dependent, decreasing as pulse duration increases. Therefore, this would be a first indication that the material damages behind the shock front, and that this accumulates as the pulse duration increases. This issue is explored further in the following figure.

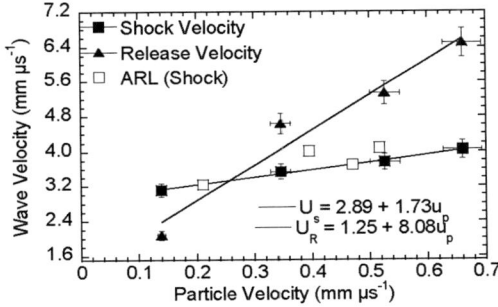

**Figure 4.** Shock and release velocities versus particle velocity.

The relationship between shock velocity and particle velocity reveals itself to be linear over the experimental range of this investigation, in common with many other materials [4]. We have also included the results of Dandekar et al. [1] from a similar material, showing close agreement between the two data sets. However, the relationship between the release velocity ($U_R$) and particle velocity is more revealing. This was calculated using the temporal spacing ($\Delta t_{release}$) between the pulses where release immediately began, as shown in Fig. 2, thus we are assuming that at this point, the release velocity is elastic in nature, and in fact represents the one-dimensional elastic wave speed as a function of pressure. In calculating this, we have also had to account for the fact that the release wave

is moving into previously shock-compressed material, thus,

$$U_R = \frac{\Delta w}{\Delta t_{release}}\left(1 - \frac{u_p}{U_S}\right). \qquad 1.$$

A simple linear curve fit has been used to determine the relationship with particle velocity of the form,

$$U_R = A + Bu_p, \qquad 2.$$

with $A$=1.25 mm µs$^{-1}$ and $B$=8.08. Note that the value of $A$ (the release speed at zero particle velocity and pressure) is considerably lower than $c_0$ (the shock velocity at zero particle velocity) and the ambient longitudinal sound speed ($c_L$). One might expect that $A$ should be near equal to the value of $c_L$, but a possible explanation presents itself. From Fig. 3, we observed that the spall strength of this material appears to have a negative dependence on pulse duration, implying that damage increases behind the shock front. Thus the behaviour of the release velocity would also suggest a degree of damage has occurred. Therefore, if the release wave is moving into previously damaged material, it would have a lower velocity than might otherwise be expected. This raises the interesting possibility that the release velocity is itself pulse duration dependent, and further work is under way to investigate this. We also point out that Dandekar et al. [1] have noted density decreases during shock loading, implying a damage process, although in that work, this was attributed to the release phase of the loading cycle.

Finally, in Fig. 5 we present the Hugoniot of the composite in stress-particle velocity space. A curve fit has been made according to,

$$P_{HD} = \rho_0 U_S u_p, \qquad 3.$$

using the $U_s - u_p$ relationship determined from Fig. 4, and where $P_{HD}$ is the hydrodynamic pressure. As with the previous figure, we have included the data of Dandekar et al. [1] It is clear that the calculated hydrodynamic pressure is lower than the measured stress, showing qualitatively at least that this material maintains a degree of shear strength behind the shock front. Also observed that the stress measurements of Dandekar *et al.* [1] lie within errors of our own.

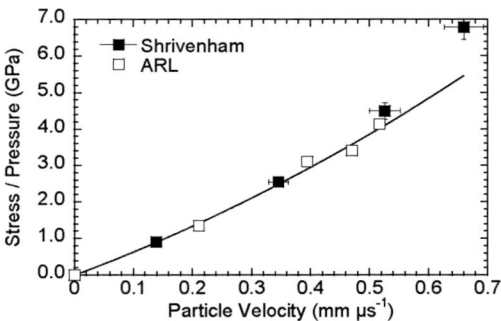

**Figure 5.** Hugoniot of glass-fibre – epoxy composite in terms of stress versus particle velocity.

Such behaviour has been observed in a monolithic epoxy resin [5], and confirmed by independent measurements of the shear strength [6]. However, other results in this work suggest that the material undergoes damage behind the shock front and thus these two sets of results must be reconciled. Previous work on the shock response of soda-lime glass, similar to the E-glass used in this material, has shown that it remains elastic up to a stress of *ca.* 6 GPa [7]. Thus it is likely that the majority of the load applied during shock loading is bourn by the glass fibres themselves, thus enabling the material to carry significant stresses. This is possible, even if some degree of material failure, presumably delamination between the fibres and the epoxy matrix, occurs.

## CONCLUSIONS

The response of an *x-y* glass fibre – epoxy composite has been investigated in terms of the shock stress, shock velocity, spall behaviour and release velocity. The relationship between shock velocity and particle velocity has been shown to be linear over the experimental range of this investigation. We have used the shock velocity to calculate the hydrodynamic pressure and compared it to the measured shock stress. We observe good agreement between the two at lower particle velocities, but there is a divergence between them as particle velocity increases, with the stress being the higher. This suggests that the shear strength increases with shock stress, a response that has been observed previously in a 'pure' epoxy, indicating that in part the shock response of this material is controlled by the epoxy matrix. The spall response has been shown to be pulse duration dependent, with damage becoming more prevalent as pulse duration increases. This indicates that the material damages behind the shock and this increases in severity with increasing time at pressure. We also observe that the initial release wave velocity also has a linear relationship with particle velocity. More significantly, we also note that the release speed at zero particle velocity ($A$) is lower than the value of $c_0$ from the shock velocity measurements. We believe that this is also an indication of damage generation behind the shock front, and also suggests, in combination with the spall response, that release wave velocity itself is pulse duration dependent.

## ACKNOWLEDGEMENTS

The authors acknowledge Gary Cooper and Ivan Knapp for valuable technical support.

## REFERENCES

1. Dandekar, D. P., Hall, C. A., Chhabildas, L. C. and Reinhart, W. D., *Composite Structures*, **61**, 51 (2003)
2. Zaretsky, E., Botton, G. D. and Perl, M., *Int. J. Solids Struct.*, **41**, 569 (2004)
3. Rosenberg, Z., Yaziv, D. and Partom, Y., *J. Appl. Phys.*, **51**, 3702 (1980)
4. Marsh, S. P., LASL Shock Hugoniot data, University of California Press (Los Angeles), 1980
5. Barnes, N., Bourne, N. K. and Millett, J. C. F., in Shock Compression of Condensed Matter - 2001, edited by Furnish, M. D., Thadhani, N. and Horie, Y., American Institute of Physics, Melville, NY, 2002, pp 135
6. Millett, J. C. F., Bourne, N. K. and Barnes, N. R., *J. Appl. Phys.*, **92**, 6590 (2002)
7. Bourne, N. K. and Rosenberg, Z., in Shock Compression of Condensed Matter-1995, edited by Schmidt, S. C. and Tao, W. C., American Institute of Physics, Seattle WA, 1996, pp 567

CP845, *Shock Compression of Condensed Matter - 2005*,
edited by M. D. Furnish, M. Elert, T. P. Russell, and C. T. White
© 2006 American Institute of Physics 0-7354-0341-4/06/$23.00

# DYNAMIC MECHANICAL BEHAVIOR CHARACTERIZATION OF EPOXY-CAST Al+Fe$_2$O$_3$ MIXTURES

## Louis Ferranti, Jr.,* Naresh N. Thadhani,* and Joel W. House**

*School of Materials Science and Engineering, Georgia Institute of Technology, Atlanta, GA 30332-0245
**US Air Force Research Laboratory, Munitions Directorate-AFRL/MNMW, Eglin AFB, FL 32542

**Abstract.** Dynamic mechanical property measurement experiments were conducted on epoxy-cast Al+Fe$_2$O$_3$ powder mixture specimens using the classic Taylor anvil test at impact velocities up to 200 m/s. Reverse Taylor anvil impact experiments were also conducted at lower velocities using a single stage gas gun by impacting a rigid anvil onto a stationary specimen. Dynamic deformation, fracture, and viscoelastic response of the cast specimens (containing 20-50 and 100% epoxy) were captured in real time utilizing high-speed photography. Detailed image analysis of transient deformation reveals a significant elastic strain contribution to the total strain, which complicates the calculation of a constant dynamic yield strength value for the composite material.

**Keywords:** Taylor impact, aluminum-iron oxide, epoxy.
**PACS:** 61.41.+e,62.20.Fe, 81.05.Qk, 81.70.Bt

## INTRODUCTION

Multifunctional structural energetic materials that can simultaneously release energy while providing high structural strength are of great interest [1,2]. Thermite mixtures (Al+Fe$_2$O$_3$) undergoing oxidation-reduction reactions provide an opportunity of multi-functionality if these can also be processed as structural materials. Design of such materials requires characterization of energetic and mechanical properties under dynamic loading conditions. In this paper, the viscoelastic and plastic response of epoxy-cast Al+Fe$_2$O$_3$ composites, characterized using the Taylor anvil [3] impact experiments is described.

In the one-dimensional rigid-plastic analysis performed by Taylor [3] and modified by Wilkins and Guinan [4], the dynamic yield strength is related to the fractional length change of a recovered deformed rod. In the present work, direct and reverse Taylor impact tests were performed on epoxy-cast Al+Fe$_2$O$_3$ composites to extend the analysis on such viscoelastic, brittle composites.

## EXPERIMENTAL PROCEDURE

Epoxy-cast Al+Fe$_2$O$_3$ composite materials were prepared using a stoichiometric mixture of aluminum and iron-oxide powders combined with 20 to 50 wt.% epoxy. The mixture consists of spherical aluminum powder (Valimet H-2 grade) with an average particle size of 2 μm, and platelet shaped iron-oxide (Fisher Scientific anhydrous grade) powder of sub-micron particle size. The epoxy was prepared from Epon 826 resin combined with diethanolamine (DEA) curing agent with a 12:1 mixing ratio. Additionally, 100% epoxy specimens were also prepared.

Specimens were first prepared by pouring the particle filled mixture into aluminum molds, degassing to remove trapped air bubbles, and curing in an oven at 70°C for 24 hours. Samples for the Taylor tests were machined into rods (7.6 mm diameter by 50.8 mm length for direct and 19 mm diameter by 75.1 mm or 63.7 mm in length for reverse Taylor tests). Ends of samples were lapped to within 0.001". In the "direct" Taylor anvil test, specimens were launched using 30 caliber powder gun (at Eglin AFRL) at a 76.2 mm stand-off distance from the anvil (made from high-strength 4340 steel heat-treated to $HR_C$ 50 hardness). A Cordin 330A high-speed film framing camera was used to image transient deformation (Fig. 1a) profiles, typically with a 3 μs framing interval. Backlighting was used to enhance the contrast and clearly trace specimen edges. The reverse Taylor anvil test was performed using the 80 mm diameter gas gun at Georgia Tech. A 9.5 mm thick high-strength AISI 1018 steel ($HR_C$ 55) anvil plate, mounted on an aluminum sabot, was launched to impact a stationary specimen held in a plastic ring. An Imacon 200 digital framing camera was used to obtain backlit images (Fig. 1b) of the specimen's deformation history.

Detailed analysis was conducted on each transient deformation image to obtain incremental lengths and diameters. Each image was calibrated using a fudicial marker attached to the anvil face as shown in Fig. 1(a) for direct Taylor experiments. Reverse Taylor experiments used a static image taken before each experiment to calibrate from known specimen dimensions. Captured images from direct and reverse Taylor experiments, had a 14 and 12 pixel/mm resolution, respectively. The sample configuration, impact velocity, and length changes for each direct and reverse Taylor impact experiment is shown in Table 1.

## RESULTS AND DISCUSSION

True stress-strain curves of epoxy-cast Al+$Fe_2O_3$ composites obtained from quasistatic compression tests performed according to ASTM

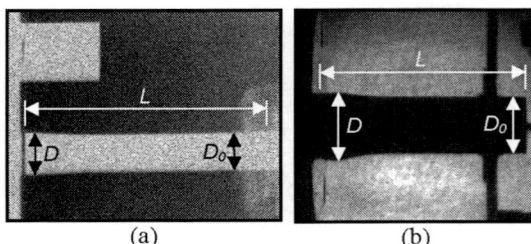

**Figure 0** High-speed camera images capturing the transient deformation for (a) direct impact (RM-22) $L$ = 49.02 mm, $D$ = 9.42 mm, $D_0$ = 7.48 mm and (b) reverse impact (0330) $L$ = 72.82 mm, $D$ = 24.50 mm, $D_0$ = 19.21 mm. (a) shows specimen rebounding off anvil face.

standard D695 for rigid plastics, shown in Fig. 2, illustrate the effect of epoxy concentration on the mechanical response. The slope of the elastic region increases as epoxy concentration decreases. The yield point behavior is some what scattered, with 30 wt.% epoxy sample showing highest strength, and 20 wt.% sample having lowest strength, due to relatively high porosity (~ 4.5%).

The direct and reverse Taylor anvil impact experiments conducted at 80 to 200 m/s (Table 1) showed typical "mushroom" shaped deformation region at the impact end of the specimen (Fig. 1). Application of Wilkins criteria for dynamic yield strength (based on length change) was only possible for three experiments; RM-23 (105±15 m/s, 20 wt.%), RM-22 (110±11 m/s, 30 wt.%), and RM-21 (111±14 m/s, 40 wt.%). The corresponding dynamic yield strengths were respectively, 587±172, 423±83, and 573±142 MPa, which appear to be almost similar and within the range of the error bar. The similarities reveal a response different from that observed in static

**Table 1** Data summary for Taylor impact experiments performed on epoxy-cast Al+$Fe_2O_3$ specimens with epoxy concentrations ranging from 20 to 50 wt.% and 100 wt%.

| Shot Number | Epoxy wt.% | Specimen Density [g/cm³] | TMD [%] | Initial Length [mm] | Final Length [mm] | Impact Velocity [m/s] | Test Method |
|---|---|---|---|---|---|---|---|
| RM-23 | 20% | 2.6866 | 95.65 | 50.77 | 49.50 | 105 ± 15 | direct |
| RM-26 | 20% | 2.6888 | 95.72 | 50.74 | fracture | 169 | direct |
| RM-24 | 20% | 2.6872 | 95.67 | 50.71 | fracture | 210 | direct |
| RM-22 | 30% | 2.3761 | 98.84 | 50.76 | 49.07 | 110 ± 11 | direct |
| RM-21 | 40% | 2.0517 | 97.64 | 50.80 | 49.70 | 111 ± 14 | direct |
| RM-14 | 50% | 1.8388 | 98.27 | 50.75 | fracture | 87 | direct |
| 0330 | 50% | 1.8640 | 99.89 | 75.09 | fracture | 79 ± 8 | reverse |
| 0328 | 100% | 1.1974 | 100 | 63.67 | fracture | 94 ± 9 | reverse |

measurements, which clearly show the 30 wt.% sample having the highest strength.

A comparison of direct Taylor experiments performed on specimens with 20 wt.% epoxy, shows brittle fracture occurring at 169 and 210 m/s, while a fully intact specimen was recovered at 105 m/s impact velocity. The 30 and 40 wt.% epoxy cast samples were also recovered intact at similar impact velocity (110 m/s). Fracture was however observed for specimens with 50 wt.% (RM-14 and 0330) and 100 wt.% (0328) epoxy concentrations at relatively low (90 m/s) impact velocities in both the direct and reverse experiments. For the 50 wt.% epoxy direct Taylor test, the specimen appeared to fracture at the moment of impact. Numerous large fragments were recovered with fracture surfaces showing brittle failure. In contrast, the 100 wt.% reverse Taylor experiment showed the specimen initially plastically deforming and then fracturing, as deformation exceeded a certain limit.

Axial and areal transient strains were measured from high-speed images taken over a 100 μs time interval. Axial strain was measured for discrete instants of time using $\varepsilon = ln(L/L_0)$, where $L$ and $L_0$ are the incremental and initial length respectively.

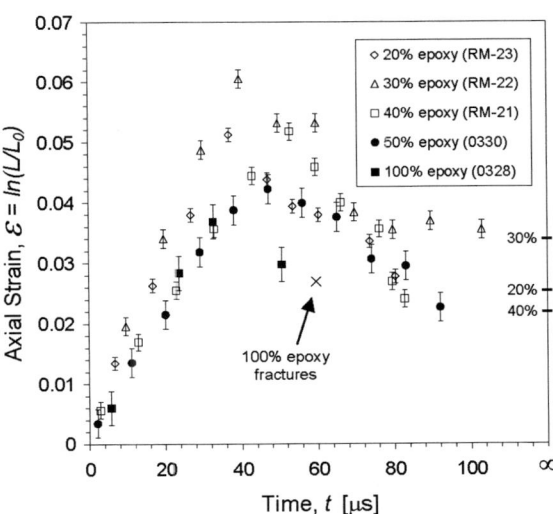

**Figure 2** Axial strain measured from high-speed camera images capturing transient deformation during direct (open points) and reverse (filled points) Taylor impact experiments. Evidence of significant elastic recovery is observed 40 to 50 μs after impact. Axial strain measured from recovered specimens are shown at $t = \infty$.

Similarly, areal strain was measured using $\varepsilon = 1 - (A_0/A)$, where $A$ and $A_0$ are incremental and initial areas respectively. Results of axial and areal strains showing the deformation response of each composition, plotted as a function of time, are shown in Fig. 3 and 4, respectively, for tests performed at low impact velocities (79 to 111 m/s). The measurements reveal that the specimens undergo significant elastic and plastic deformation during both loading and unloading stages. As shown in Fig. 3, each composition reveals a similar strain response reaching a maximum strain where the elastic and plastic strains in the specimen are equivalent, around 40 to 50 μs after impact. Beyond this stage, the samples undergo significant elastic recovery with the final permanent strain measured from recovered sample geometry, shown in Fig. 3 at $t = \infty$. The strain rate history is also evident from this figure, revealing an almost linear (constant) strain rate up to the maximum strain for each composition. The 20 and 30 wt.% epoxy samples show a strain rate response distinctly different from the 50 and 100 wt.% samples.

Areal strain measurements in Fig. 4, show that strain increases continuously over the time interval

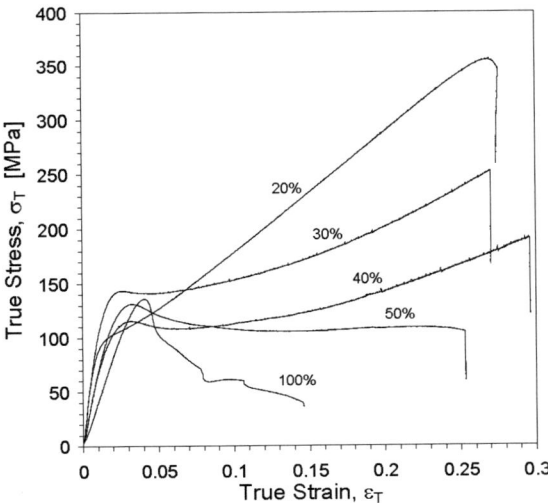

**Figure 1** Static compression curves for Al+Fe$_2$O$_3$ with 20 to 50 wt.% and 100 wt% epoxy. Results indicate elastic modulus and strain hardening scale with decreasing epoxy concentration. However, yield points for each composition have a less obvious behavior.

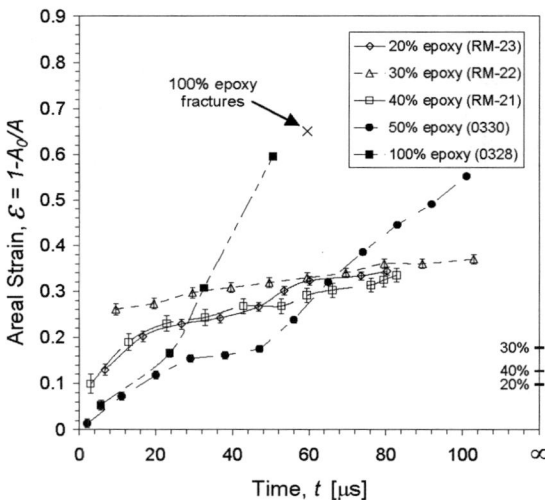

**Figure 3** Areal strain measured from high-speed camera images capturing transient specimen deformation during direct (open points) and reverse (filled points) Taylor impact experiments. Radial elastic recovery occurs after final high-speed camera image is captured. Areal strain measured from recovered specimens are shown at $t = \infty$.

investigated, with evidence of elastic recovery observed only from recovered specimens. The 20, 30, and 40 wt.% epoxy samples experience combined peak elastic and plastic strains above 35% and recover to final plastic areal strains ranging from 10 to 18%, as shown in Fig. 3. Their overall mechanical response is quite similar (open points in figure), and they also experience comparable axial strain-rates. The reverse Taylor test experiments on 50 and 100 wt.% epoxy samples show a distinctly different response (Fig. 4). The 50 and 100 wt%. epoxy samples achieve peak strains of $\sim$ 60%. The 100 wt.% epoxy sample fractures at $\sim$ 60 $\mu$s after impact, while the 50 wt.% epoxy specimen remains unfractured up to 100 $\mu$s, but ultimately fractures during recovery.

## SUMMARY

Taylor impact experiments using direct and reverse configurations conducted on epoxy cast Al+Fe$_2$O$_3$ composites, show effects of varying epoxy content on their viscoelastic and fracture response during impact loading. Measurements of dynamic strength based on Wilkins criteria is only possible at low velocities, however errors in

velocity and length change measurements lead to large errors in measured values. High-speed digital imaging shows that the materials undergo mushrooming behavior at low impact velocities, and the samples fracture at higher velocities. An important effect observed from the measured transient deformation histories over a 100 $\mu$s time interval, is the significant elastic recovery in the axial direction, $\sim$ 40 to 50 $\mu$s after impact. Elastic recovery is also observed in the radial direction and was not directly observed from transient images.

Application of the Taylor impact tests on epoxy-cast composites to understand their dynamic deformation response is thus, complicated by the overlapping viscoelastic behavior, which is difficult to separate from the total strain, for evaluation of the dynamic strength. Our future work involves use of additional diagnostics, including sample back surface velocity measurements with VISAR, and correlating these along with digitally-captured transient deformation profiles to simulate velocity and deformation states calculated based on constitutive equations.

## ACKNOWLEDGEMENTS

Funding was provided in part by AFOSR-MURI (Grant No. F49620-02-1-0382); and by a graduate research internship (for L. Ferranti) through EGLIN-AFRL/MNME (Contract No. F08630-03-C-001. The authors thank the staff at the HERD and AWEF facilities at Eglin AFRL, for their assistance with direct Taylor impact tests.

## REFERENCES

1. Jordan, J.L., Dick, R.D., Ferranti, L., and Thadhani, N.N., Austin, R.A., McDowell, D.L., and Benson, D.J., *Shock Compression of Condensed Matter (this volume)*.
2. Jordan, J.L., Dick, R.D., Ferranti, L., and Thadhani, N.N., *Intl. J. Impact Eng.* (*accepted for publication*).
3. Taylor, G.I., *Proc. R. Soc. Lond. A* **194**, 289-299 (1948).
4. Wilkins, M.L. and Guinan, M.W., *J. Appl. Phys.* **44** [3], 1200-1206 (1973).

CP845, *Shock Compression of Condensed Matter - 2005*,
edited by M. D. Furnish, M. Elert, T. P. Russell, and C. T. White
© 2006 American Institute of Physics 0-7354-0341-4/06/$23.00

# HIGH STRAIN RATE RESPONSE OF AN ELASTOMER

## Tong Jiao *, Rodney J. Clifton* and Stephen E. Grunschel*

*Division of Engineering, Brown University, Providence, RI 02912*

**Abstract.** Pressure-shear plate impact experiments are used to study the nonlinear dynamic response of an elastomer at shearing rates of $10^5$ - $10^6$ $s^{-1}$. Samples with thicknesses in the range 100 $\mu m$ - 400 $\mu m$ are cast between two hard steel plates. Because of the comparatively low impedance of the elastomer, longitudinal waves reverberating through the thickness of the sample – and recorded with a laser interferometer – are used to determine the isentrope of the material under uniaxial strain compression. Once the sample is fully compressed a shear wave arrives and imposes a simple shearing deformation. From the transverse velocity, measured interferometrically at the rear surface of the sandwich target, the shear stress and the transverse velocity at the rear surface of the sample are determined. These measurements provide an indication of the shearing resistance of the material under pressure. When the longitudinal unloading wave arrives from the rear surface of the target, these same measurements provide an indication of the shearing resistance of the material at zero pressure. Because the sample adheres to the bounding plates the reflection of unloading waves from both the rear surface of the flyer and the rear surface of the target allows the sample to be strained in uniaxial extension. Thus, from a single experiment, one obtains the response of the elastomer in uniaxial strain compression, simple shear and uniaxial strain extension.

**Keywords:** Elastomer, Isentrope, Shearing Resistance, Pressure-Shear Plate Impact
**PACS:** 43.35.Ei, 78.60.Mq

## INTRODUCTION

Improved understanding of the dynamic response of elastomers is important to understanding their effect in mitigating damage in impact loading applications. These applications often involve large deformations at high strain rates. While some understanding of the dynamic response of elastomers at strain rates of $10^3$ $s^{-1}$ has been obtained using split-Hopkinson (or Kolsky) bar techniques [1] [2] [3] there appears to have been no systematic study of their response at the higher rates that may occur in some applications and that are accessible by means of pressure-shear plate impact experiments. Herein such a study is reported for one elastomer: Versathane P-1000, a polyurea.

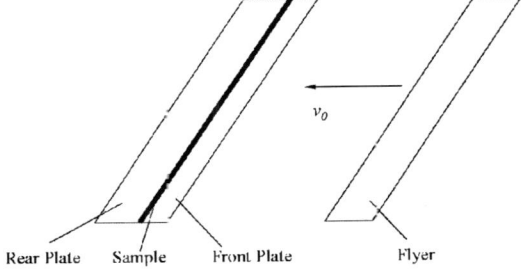

**FIGURE 1.** Schematic of sandwich impact.

## PRESSURE-SHEAR PLATE IMPACT

The pressure-shear plate impact configuration used for the experiments reported here is shown in Figure 1. A polyurea sample is cast between two hard, tool-steel plates. The target assembly is impacted by

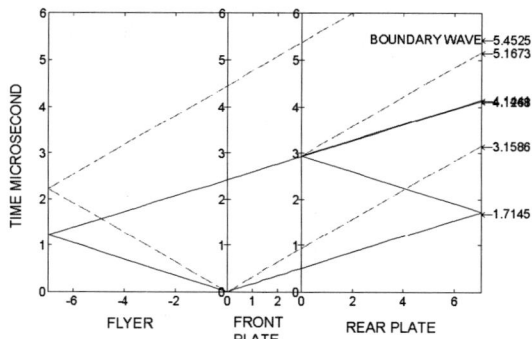

**FIGURE 2.** Time-distance diagram.

**TABLE 1.** Shot Summary.

| Shot No. | Thickness (mm) | | | | Diameter (mm) | Impact Velocity (m/s) | Tilt Angle (mrad) |
|---|---|---|---|---|---|---|---|
| | Sample | Flyer ($h_1$) | Front ($h_2$) | Rear ($h_3$) | | | |
| 401 | 0.2 | 6.881 | 2.794 | 6.858 | 57.15 | 116 | 0.18 |
| 402 | 0.4 | 6.923 | 2.845 | 6.860 | 57.15 | 114 | — |
| 403 | 0.3 | 6.844 | 2.807 | 6.859 | 57.15 | 114 | 0.2 |
| 404 | 0.11 | 6.991 | 2.896 | 7.041 | 60 | 112.6 | 0.3 |
| 405 | 0.43 | 6.832 | 2.808 | 6.840 | 57.15 | 115.7 | 0.28 |
| 501 | 0.154 | 7.078 | 2.823 | 7.077 | 57.15 | 120.9 | — |
| 502 | 0.147 | 6.998 | 2.871 | 6.927 | 60 | 139.8 | 0.8 |

a tool-steel flyer plate. The time-distance diagram for the principal wavefronts is shown in Figure 2. At impact, both longitudinal waves (shown by solid lines)and shear waves (shown by dashed lines) are generated. These waves propagate both forward into the front plate of the target assembly and backward into the flyer. No-slip at the impact plane is assured by keeping the skew angle $\theta$ sufficiently small (18° is used) and roughening slightly the impact faces. Then, from symmetry, the initial velocity at the impact face is one-half the in-coming velocity of the flyer plate. Details of the techniques involved in imposing the pressure-shear loading, and in recording the free surface velocities by means of a combined normal displacement interferometer (NDI) and transverse displacement interferometer (TDI), are given in [4]. Table 1 gives the shot summary.

When the incident longitudinal wave in the target arrives at the sample, longitudinal waves reverberate back and forth through the thickness of the sample until the stress becomes essentially uniform through the sample thickness. Then, the shear wave arrives

and the sample begins to ring up toward a uniform state of stress. For the thinnest samples this nominally uniform state of shear stress is established; these shots are the ones for which dynamic stress-strain curves in shear are presented. After the longitudinal wave passes through the sample it passes through the rear plate of the target assembly, reflects from the rear surface, and returns to the sample to unload the normal stress. Almost simultaneously, the longitudinal wave reflected from the rear surface of the flyer arrives at the sample. The combined effect of these two unloading waves impinging on the sample is to tend to load the sample in tension. Throughout the loading and reverse loading of the sample the traction on the rear face of the sample can be inferred from the velocity-time profiles monitored at the rear surface of the target. From the relations that hold along characteristics, the compressive and shear components of this traction are related to the normal and transverse components of the free surface velocity by

$$\sigma(t) = -\rho c_1 \frac{u_{fs}(t + h_3/c_1) - u_{fs}(t - h_3/c_1)}{2} \quad (1)$$

where $u_{fs}$ is the normal component of the free surface velocity and

$$\tau(t) = -\rho c_2 \frac{v_{fs}(t + h_3/c_2) - v_{fs}(t - h_3/c_2)}{2} \quad (2)$$

where $v_{fs}$ is the transverse component of the free surface velocity. In these equations, $\rho$ is the mass density, and $h_3$ is the thickness of the rear plate of the target assembly. The longitudinal and shear wave speeds in this steel plate are denoted by $c_1$ and $c_2$, respectively. The second term in the numerators of these equations is zero until the corresponding elastic wave has reflected from the free surface of the target and has returned again to its rear surface. The density and wave speeds for tool steel are taken to be: $\rho = 7880 kg/mm^3$, $c_1 = 5.767 mm/\mu s$, and $c_2 = 3.146 mm/\mu s$, based on elastic moduli for tool steel from http://www.matweb.com.

## EXPERIMENTAL RESULTS

Velocity-time profiles for the normal velocity for the seven experiments are shown in Fig.3. Profiles for the three thinnest samples are those on the left that

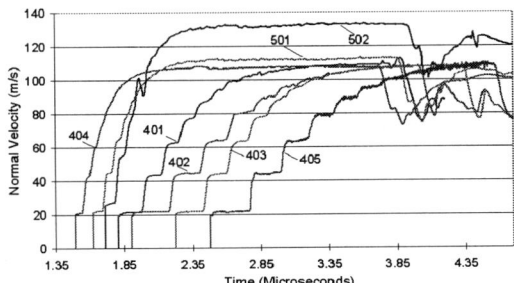

**FIGURE 3.** Normal velocity profiles.

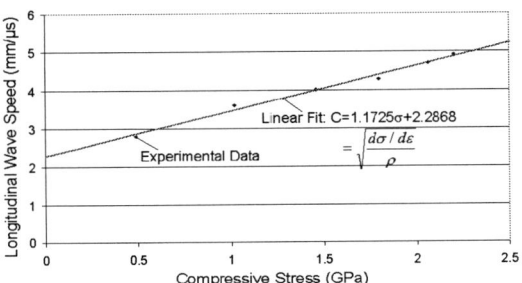

**FIGURE 4.** Lagrangian wave speeds.

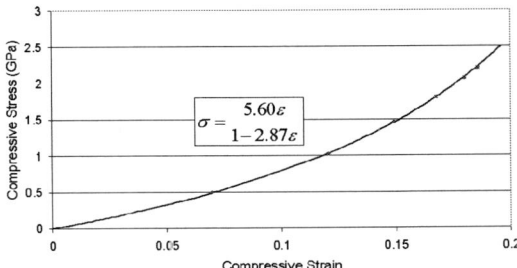

**FIGURE 5.** Isentropic uniaxial strain compression.

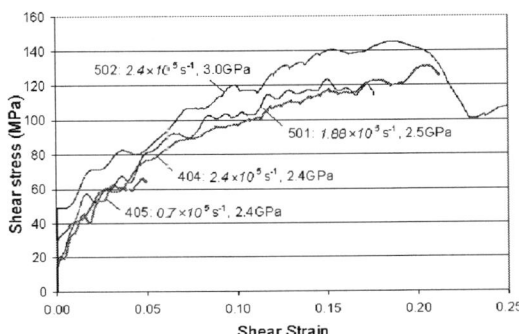

**FIGURE 6.** Shearing resistance at high shear rates.

ring up the quickest. Conversely, those on the right that ring up the slowest are those for the thicker samples. The latter profiles are used for determining the stress dependence of the longitudinal sound speed. The sound speed for the stress at each of the velocity plateaus is computed from $2h_3/\Delta t$ where $\Delta t$ is the round trip transit time indicated by the duration of the plateau. The stress dependence of the Lagrangian longitudinal wave velocity obtained in this way is shown in Fig. 4, where the experimental points are fit reasonably well by a straight line. Then, the stress-strain curve for isentropic, uniaxial-strain compression – shown in Fig. 5 – can be obtained by integrating the relationship between the slope of the stress-strain curve and the wave speed, i.e.

$$d\varepsilon = \frac{d\sigma}{\rho c^2(\sigma)}. \qquad (3)$$

The shearing resistance can be obtained from the transverse velocity-time profiles for which the shear stress at the back of the sample is given by Eqn. (2). Once sufficient reverberations of shear waves have occurred to make the shear stress in the sample

nominally uniform through the thickness, the shear strain rate in the sample is given by

$$\dot{\gamma}(t) = \frac{v_{fs}(t + h_3/c_2) - v_0}{h} \qquad (4)$$

where $v_0$ is the transverse component of the velocity of the flyer plate and $h$ is the thickness of the sample. Equation (4) can be integrated to obtain the shear strain corresponding to the shear stress of Eqn. (2). The result gives a shear-stress vs. shear-strain curve at the nominal strain rate given by Eqn. (4) and the pressure given by Eqn. (1). The resulting shear stress vs. shear-strain curves are shown in Fig. 6. Only the shearing deformation record before the arrival of the reflected longitudinal wave is included in Fig. 6 in order to preclude the effects of transverse motion associated with the arrival of the reflected longitudinal wave. Much of the rise in the early part of the curves is associated with the ring up of the low-impedance sample sandwiched between two high impedance plates. Nevertheless, at least the last half of the curves should provide a good indication

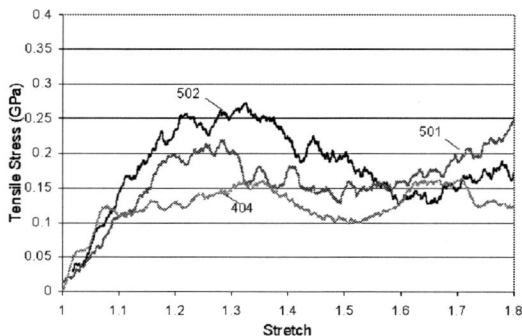

**FIGURE 7.** Tensile response in uniaxial strain.

of the shearing resistance of the sample. This shearing resistance is relatively insensitive to the shearing rate at shearing rates of approximately $2 \times 10^5 s^{-1}$, but appears to increase significantly with increasing pressure. The latter trend is confirmed by the reduction in shear stress that occurs when the unloading waves arrive and the normal stress is reduced to approximately zero.

When the reflected unloading waves arrive at the sample, as shown in the latter part of the normal velocity-time profiles in Fig. 3, the particle velocity is first reduced by the unloading wave from the back surface of the flyer and then increased by the unloading wave from the rear surface of the target assembly. Steps in the unloading velocity-time profiles correspond to reverberations through the thickness of the sample both during the initial loading and during the unloading. From these steps it is evident that the unloading is elastic with very little dispersion. Once the velocity rises again to essentially a plateau, the level of the plateau is lower than that of the plateau preceding the arrival of the unloading waves. Then, from Eqn. (1), the stress in the sample is tensile. The stretching rate is given by the velocity difference across the sample divided by the initial sample thickness. Once the stress is uniform across the thickness of the sample, this stretching rate is simply the rear surface velocity of the second plateau divided by the initial sample thickness. Integration of this stretching rate to give the stretch gives the plots of stress vs. stretch shown in Fig. 7. From these plots it is evident that the sample undergoes large stretch, and consequently large volume expansion, at tensile stresses of approximately $150 MPa$.

## DISCUSSION

The relationship between sound speed and compressive stress presented in Fig. 4 is based on an approximation that the shock wave speed averaged over the round trip is equal to the sound speed at the stress level of the plateau. A more nearly exact analysis [5] that accounts for different wave speeds for the two shock waves that are invovled in the round trip tends to increase the inferred sound wave speeds at the lower stresses. For this reason, and because data was not obtained for stresses less than 0.5GPa, the sound wave speeds shown by linear extrapolation to low stresses in Fig. 4 are less well established than those at higher stresses. Consequently, accurate data on sound speeds at lower stresses are required to improve the accuracy of the initial slope of the stress-strain curve in Fig. 5.

## ACKNOWLEDGMENTS

The authors gratefully acknowledge the financial support of ONR as well as the contributions of Jeffrey Fedderly and David Owen at the Naval Research Laboratory, Carderock Division, who cast the samples used in these experiments.

## REFERENCES

1. Nemat-Nasser, S., McGee, J., Kang, W., Amirkhizi, A., Amini, M., Isaacs, J., Lischer, D., Guo, W., and Delaurison, N., "Experimental Characterization of Polyurea with Consititutive Modeling and Simulations," in *ERC ACTD Workshop*, Cambridge, Nov. 2004.
2. Yi, J., and Boyce, M., "Stress-Strain Behavior of Polyurea and Polyurethane: Preliminary Experiments and Molding Results," in *ERC ACTD Workshop*, Cambridge, Nov. 2004.
3. Song, B., and Chen, W., *Journal of Engineering Materials and Technology*, **126**, 213–217 (2004).
4. Klopp, R., and Clifton, R., "Pressure-Shear Plate Impact testing," in *Metals Handbook, 9th Edition*, American Society for Metals, Metals Park, OH, 1985, vol. 8, pp. 230–239.
5. Clifton, R. J., and Jiao, T., "High Strain Rate Response of a Polyurea," in *ONR Airlie Workshop*, Airlie, 2005.

CP845, *Shock Compression of Condensed Matter - 2005*,
edited by M. D. Furnish, M. Elert, T. P. Russell, and C. T. White
© 2006 American Institute of Physics 0-7354-0341-4/06/$23.00

# LATERAL STRESS MEASUREMENTS AND SHEAR STRENGTH OF AN ALUMINA-FILLED EPOXY

## K. Kos, J.C.F. Millett*, N.K. Bourne[‡] and D. Deas

*AWE, Aldermaston, Reading, RG7 4PR. United Kingdom.*

*\*Defence Academy of the United Kingdom, Cranfield University, Shrivenham, Swindon, SN6 8LA. United Kingdom.*

[‡]*University of Manchester, Sackville Street, Manchester, M60 1QD. United Kingdom.*

**Abstract.** The variation of shear strength with impact stress in an alumina-filled epoxy has been measured with lateral stress gauges. At lower stresses, a degree of hardening behind the shock front has been observed, which diminishes as shock stress increases. It is believed that this is due to a transition from a viscous response dominated by the epoxy matrix, to a more viscoplastic response. The measured shear strength has also been observed to reach a near constant level, as was suggested in a previous paper. We have also used these results to make an estimation of the HEL at *ca.* 2.0 GPa.

**Keywords**: Shock, composite, shear strength, lateral stress
**PACS**: 62.50

## INTRODUCTION

The shock response of ceramic particulate – polymer matrix composites, and in particular alumina-epoxy based systems is of interest as they are used as potting compounds for electronic components. Munson *et al.* [1] examined a number of such materials with varying alumina contents. They showed that there was no clear indication of a Hugoniot Elastic Limit (HEL) on the rising part of rear surface interferometry traces. This was explained in terms of the vastly different natures of the alumina particles and epoxy matrix resulting in a "smeared" shock front as it travelled through the target. Release speeds were seen to increase with both increasing particle velocity and alumina content, as the alumina particles were brought closer together, either by shock compression of the

microstructure or simply increasing their concentration. Our own measurements [2] on similar materials showed a transition from a typical polymeric response in the epoxy binder, to a behaviour more commonly seen in metals or ceramics as alumina content increased. Shock velocity – particle velocity ($U_s$-$u_p$) relationships were observed to be linear. These were used to calculate the hydrodynamic pressure ($P_{HD}$), in combination with the ambient density ($\rho_0$), thus,

$$P_{HD} = \rho_0 U_s u_p. \qquad 1.$$

Comparison with measured shock stresses ($\sigma_x$) showed close agreement between the two in the composites, suggesting that the shear strength ($\tau$) of these filled epoxies would be near constant with increasing shock stress, above the HEL.

## EXPERIMENTAL

All shots were performed on a 50 mm bore, 5 m long single stage gas gun [3]. Specimens were supplied in the form of matched pairs of tiles, 28 mm x 57 mm x 10 mm, lapped flat and parallel to ± 5 μm. Target assemblies were made by introducing a manganin stress gauge (MicroMeasurements type J2M-SS-580SF-025) between two of the provided tiles, 4 mm from the impact face, using a low viscosity epoxy adhesive. The impact faces were then relapped to ensure flatness. Impact stresses in the range 1.14 to 6.40 GPa were generated with 5 mm flyer plates of dural (aluminium alloy 6082-T6) or copper in the velocity range 221-811 m s$^{-1}$. Voltage – time data were converted to lateral stress-time using the methods of Rosenberg and Partom [4], with a modified analysis that does not require prior knowledge of the impact conditions [5]. A schematic of the target assembly is shown in Fig. 1.

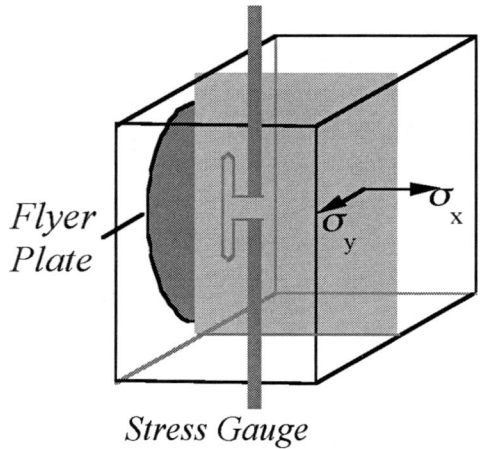

**Figure 1.** Gauge placement and specimen configuration.

## MATERIALS

The material in this investigation was provided by Sandia National Laboratories. It consisted of 43% by volume of alumina in epoxy, as studied previously by Setchell and Anderson [6], referred hence forth as ALOX. The Hugoniot ($\sigma_x$-$u_p$) is presented in Fig. 2. It was observed that it could be fitted with a simple quadratic relation,

$$\sigma_x = 6.83u_p + 4.73u_p^2. \qquad 2.$$

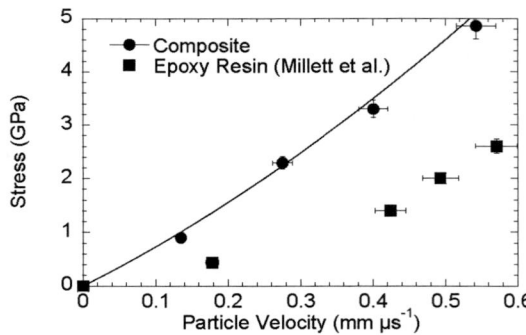

**Figure 2.** Hugoniot of ALOX [6] (solid curve). The data points are from a previous investigation [2] of a similar material, along with those of the epoxy binder [7].

Material properties are presented in Table 1.

**Table 1.** Material properties of ALOX and epoxy resin.

|  | $c_L$ mm/μs | $c_s$ mm/μs | $\rho_0$ g/cm$^3$ | $\nu$ |
|---|---|---|---|---|
| ALOX | 3.31±.03 | 1.76±.03 | 2.36±.06 | .303 |
| Epoxy | 2.38±.03 | 1.20±.03 | 1.14±.06 | .329 |

## RESULTS AND DISCUSSION

Lateral stress traces are shown in Fig. 3.

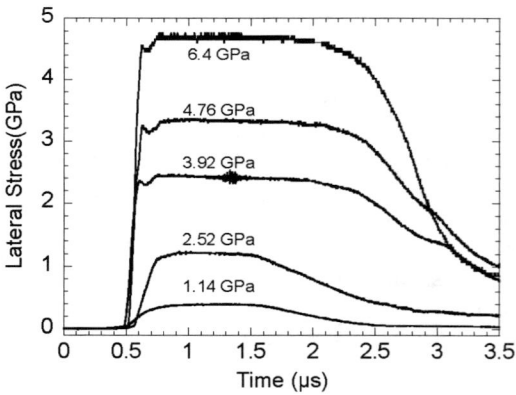

**Figure 3.** Lateral stress gauge traces from ALOX. Each trace is labelled with the imposed longitudinal shock stress.

Each trace displays a rapid rise to its final amplitude, a near constant stress level, of a duration defined by the flyer plate material and width, and then a drop in stress as releases from the rear of the flyer plate enter the gauge location. Careful examination of these traces shows that those shocked to 3.92 and 4.76 GPa display a slight decrease in lateral stress. From the well known relation,

$$2\tau = \sigma_x - \sigma_y, \qquad 3.$$

this suggests that under these conditions, ALOX has a slight increase in strength behind the shock front. However, at higher stress levels (6.40 GPa), the trace is flat topped behind the shock. In contrast, the response of a pure epoxy [7] is somewhat different, as shown in figure 4.

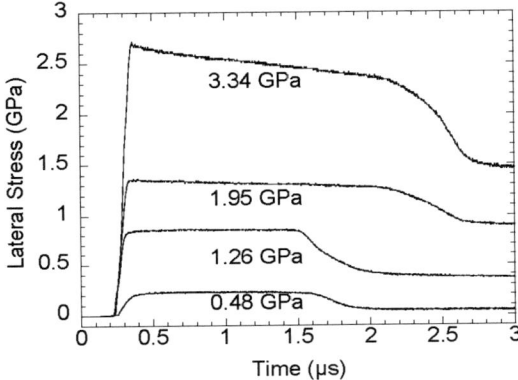

**Figure 4.** Lateral stress gauge traces from an unfilled epoxy resin from a previous investigation [7].

In the pure epoxy, the drop in lateral stress is much more pronounced. The most likely reason for the differences between these two materials has been identified by Setchell and Anderson [6]. They demonstrated that as the impact stress increased, the rise time in the longitudinal response in ALOX decreased, as deformation transformed from the viscous behaviour controlled by the epoxy matrix at lower stresses, to a more viscoplastic behaviour at higher stresses. Thus, if the epoxy is controlling the deformation at lower stresses, it would be expected that the lateral stress behaviour would be more akin to the pure epoxy response, and hence a drop in lateral stress behind the shock front should be noted, as is the case discussed above. However, it should also be bourn in mind that as the volume fraction of alumina is so large, it will have a "diluting" effect upon this "hardening" type behaviour, hence making it less prevalent.

Finally, we present data showing how the shear strength of ALOX varies with shock stress. We also include the equivalent data from the pure epoxy studied earlier [7].

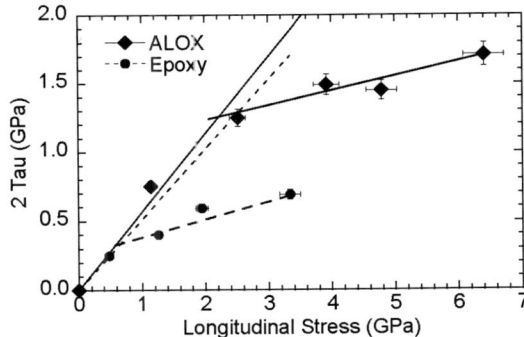

**Figure 5.** Shear strength versus longitudinal stress for ALOX and epoxy. The steeper lines are the calculated elastic response according to equation 4.

Two sets of straight lines have been fitted to the data; the solid lines represent the ALOX and the dotted represent the epoxy resin. The steeper lines are the calculated elastic response, according to,

$$2\tau = \frac{1 - 2\nu}{1 - \nu}\sigma_x, \qquad 4.$$

using the appropriate values of Poisson's ratio ($\nu$) from Table 1. The two shallower lines have been fitted through the higher impact stress data, where it has been assumed that the material response is inelastic. It can be seen that at lower impact stresses, the measured shear strengths of both materials agree well with the calculated response. As impact stress increases, the shear strength deviates below the elastic calculation as other deformation mechanisms come into operation. As would be expected, the presence of alumina within the epoxy gives a microstructure with a significantly higher strength.

By making the assumption that a straight line can be fitted to the inelastic data, an estimation of the HEL can be made where the elastic and inelastic responses intersect. In polymethylmethacrylate (PMMA) [8] we used this method to identify the HEL, showing close agreement with the work of others [9], who determined it at *ca.* 0.75 GPa, using different techniques. We have also used this method to determine the HEL in pure epoxy, at *ca.* 0.6 GPa. Therefore by putting a simple linear fit through the higher stress points, and extrapolating back so that it intercepts with the elastic response will give a reasonable estimate of the HEL. In ALOX, this gives us a value of *ca.* 2.0 GPa. This value may seem somewhat high, but it should be remembered that ALOX has an alumina content of 43% by volume (a weight ratio of *ca.* 3:1 alumina to epoxy). AD 995, a nominally pure alumina itself has an HEL of *ca.* 6.7 GPa [10], hence the relatively high estimation of the HEL in ALOX is not unreasonable.

## CONCLUSIONS

Manganin stress gauges have been placed in samples of an alumina – epoxy composite in such orientation that renders them sensitive to the lateral component of stress. Slight decreases in lateral stress behind the shock front have been noted, but have been shown to diminish as shock stress increases. It is believed that this correlates with the observation (in previous work) that shock front itself becomes less dispersive with increasing shock stress, as the material moves from a mechanism dominated by the viscous behaviour of the epoxy, to a more viscoplastic response as stress increases. In a previous paper, from the close agreement between the measured longitudinal stress and the calculated hydrodynamic pressure, it was suggested that the shear strength would be near constant with increasing shock stress. The results presented in this paper would appear to confirm that statement. The high alumina loading in ALOX has been shown to have a significant strengthening effect upon alumina, as would be expected. Finally, we have used these results to make a first estimation of the HEL of ALOX, at *ca.* 2.0 GPa.

## ACKNOWLEDGMENTS.

We would like to thank Ivan Knapp of Cranfield University for performing the shock loading experiments discussed in this paper. We are also grateful to Sandia National Laboratories for provision of experimental material and helpful discussions throughout the progress of this programme.

## REFERENCES.

1. Munson, D. E., Boade, R. R. and Schuler, K. W., *J. Appl. Phys.*, **49**, 4797 (1978)
2. Millett, J. C. F., Bourne, N. K. and D.Deas, *J. Phys. D. Applied Physics*, **38**, 930 (2005)
3. Bourne, N. K., *Meas. Sci. Technol.*, **14**, 273 (2003)
4. Rosenberg, Z. and Partom., Y., *J. Appl. Phys.*, **58**, 3072 (1985)
5. Millett, J. C. F., Bourne, N. K. and Rosenberg, Z., *J. Phys. D. Applied Physics*, **29**, 2466 (1996)
6. Setchell, R. E. and Anderson, M. U., *J. Appl. Phys.*, **97**, 083518 (2005)
7. Millett, J. C. F., Bourne, N. K. and Barnes, N. R., *J. Appl. Phys.*, **92**, 6590 (2002)
8. Millett, J. C. F. and Bourne, N. K., *J. Appl. Phys.*, **88**, 7037 (2000)
9. Barker, L. M. and Hollenbach, R. E., *J. Appl. Phys.*, **41**, 4208 (1970)
10. Dandekar, D. P. and Bartkowski, P., in High Pressure Science and Technology 1993, edited by Schmidt, S. C., Shaner, J. W., Samara, G. W. and Ross, M., American Institute of Physics, New York, 1994, pp 733

CP845, *Shock Compression of Condensed Matter - 2005*,
edited by M. D. Furnish, M. Elert, T. P. Russell, and C. T. White
© 2006 American Institute of Physics 0-7354-0341-4/06/$23.00

# DYNAMIC MECHANICAL BEHAVIOR OF NICKEL-ALUMINUM REINFORCED EPOXY COMPOSITES

## M. Martin[1], S. Hanagud[2], N. N. Thadhani[1]

[1] *Materials Science and Engineering, Georgia Institute of Technology, 771 Ferst Dr., Atlanta, GA 30332*
[2] *Aerospace Engineering, Georgia Institute of Technology, Atlanta, GA 30332*

**Abstract.** Epoxy-based composites reinforced with micron-sized Ni and micron or nano-sized Al powders were fabricated by casting/curing. The mechanical behavior of the composites was evaluated using elastic and plastic property measurements performed on rod-shaped samples. Dynamic reverse Taylor anvil-on-rod impact tests gave qualitative and quantitative information about the transient deformation and failure response of the composites. The composite containing 20wt% epoxy and nano-sized Al powder showed the most superior mechanical properties in terms of elastic modulus, static compressive strength, and dynamic incremental areal and axial strains, as compared to the other cast materials. The results illustrate that nano-sized Al particles alter the deformation response of the composite and provide significant enhancement to the strength by dispersing in the epoxy and generating a nano Al-containing epoxy matrix with embedded Ni particles.

**Keywords:** mechanical properties, Taylor test, powder reinforced composites, Ni-Al, epoxy, nano powder
**PACS:** 62.20.-x, 62.20.Fe, 62.25.+g, 62.50.+p

## INTRODUCTION

Traditional classes of materials typically have conflicting properties; for example, energetic materials have no structural strength, whereas structural materials have no energy releasing ability. Multifunctional Energetic Structural Materials (MESMs) are a class of materials with both strength and energy release capabilities. Such materials have great technological potential, but the conflicting trends of reactivity and strength make development a challenge.

Alloys based on the intermetallic compounds of Ni and Al possess an ideal combination of mechanical and physical properties (e.g., high melting point, high strength-to-weight ratio, low density, and high thermal conductivity [1, 2]) that make them attractive candidates for many high-temperature applications. The Ni-Al system also has the advantage of existence of large differences in the heats of reaction of its various intermetallic compounds (e.g., $NiAl_3$, $Ni_2Al_3$, $NiAl$, and $Ni_3Al$) [3]. Thus, the Ni-Al system was studied for potential use in MESMs. Additionally, since strength and reactivity of materials have been shown to change with particle size [2, 4], both nano- and micron-sized Al powders were used in fabrication of bulk materials.

Whereas aggregated powder mixtures do not have the mechanical properties to perform as a structural material, particle reinforced polymer composites have many favorable properties including desirable mechanical properties and the ability to form into bulk shapes. Therefore, this study involved fabrication of Ni+Al+epoxy composites in which the Al particle size and amount of epoxy binder was varied. The static and dynamic structural/mechanical properties of the fabricated Ni+Al-reinforced epoxy composites will be reported here.

## EXPERIMENTAL PROCEDURE

Equivolumetric (76.6wt% Ni, 23.4wt% Al) dry powder mixtures of Ni (-325 mesh, Cerac) and Al (-325 mesh, Cerac) or nano Al (avg. 56.3 nm, Technanogy) powders were prepared by mixing in a V-blender. Cylindrical samples of Ni+Al-reinforced epoxy were fabricated by casting/curing with epoxy (92.3wt% Epon Resin 826, Miller-Stephenson; 7.7wt% Diethanolamine hardener, Sigma-Aldrich). Fabricated samples were 20 and 30wt% epoxy, with the balance equivolumetric Ni+Al/nano Al powder mixture.

Static and dynamic mechanical properties were evaluated using static compression tests and dynamic reverse anvil-on-rod Taylor [5] impact tests. Static compression tests were performed on each cast material (pure epoxy, Ni+Al+20%epoxy, Ni+nano Al+20% epoxy, Ni+nano Al+30% epoxy), which allowed for evaluation of the effects of Al particle size and epoxy content.

Reverse Taylor anvil-on-rod impact tests were performed using a gas gun at impact velocities ranging from 61-152 m/s. A schematic of the reverse Taylor impact test setup can be seen in Fig. 1. The projectiles consisted of an Al sabot with a maraging steel rigid anvil secured to the front surface. Cylindrical rods of each cast material were mounted to a PMMA target ring in the experiment chamber, where they served as the target. An Imacon-200 High Speed Digital Camera captured images (16 frames) of the transient deformation of the specimen following impact.

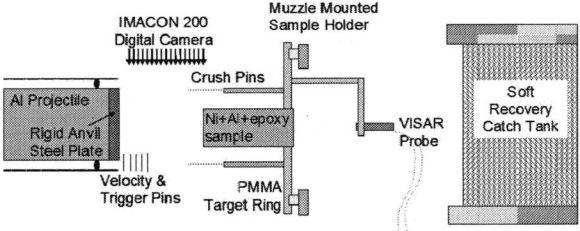

**Figure 1.** Schematic of the reverse Taylor impact test.

## RESULTS AND DISCUSSION

Static compression tests yielded the engineering stress-strain curves shown in Fig. 2. A graphical comparison of theoretical (based on the Rule of Mixtures) and experimental yield strengths

and elastic moduli is shown in Fig. 3. The yield strength of epoxy was found to be only slightly improved by the addition of Ni+micron Al powders, but greatly improved by Ni+nano Al powders. Ni+nano Al+20% epoxy exhibited the highest yield strength (~157% of the yield strength of epoxy), followed by Ni+nano Al+30% epoxy (~130% of the yield strength of epoxy), and then Ni+Al+20% epoxy (~100% of the yield strength of epoxy). Ni+nano Al+20% epoxy also exhibited the highest elastic modulus, followed by the other two reinforced epoxy samples. The stress-strain curves in Fig. 2 also reveal a different deformation (work hardening) response for the nano Al-containing composites in comparison to the pure epoxy and micron Al-containing composites.

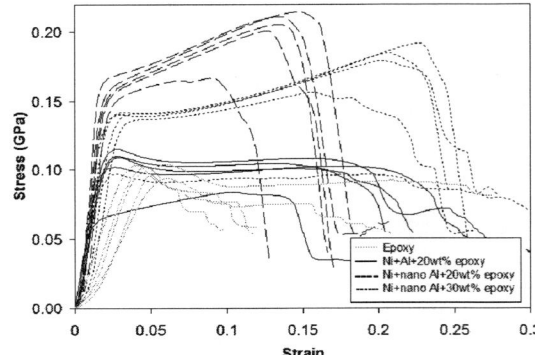

**Figure 2.** Engineering stress-strain curves generated from static compression tests at a loading rate of 0.585 lb/min.

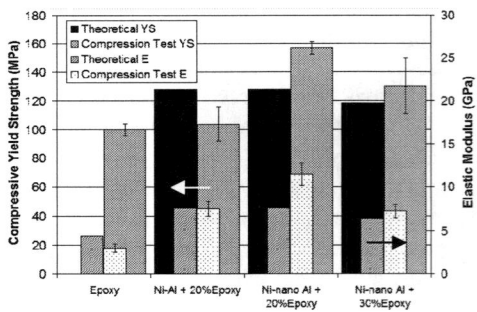

**Figure 3.** Comparison of experimental (compression test) and theoretical (Rule of Mixtures) static compressive yield strengths and elastic moduli.

Fig. 4 shows an example of the transient profiles isolated from images captured by the camera throughout the deformation event in the reverse Taylor test of (a) a pure epoxy sample and (b) a Ni+nano Al+20wt% epoxy sample. These images were used to analyze the dynamic behavior of the materials by calculating incremental strain at the time of each image capture. Both areal and axial incremental strains were calculated based on the analyses originally described by Taylor [5] and Wilkins and Guinan [6], respectively. The expression used for areal strain calculations is given in Equation 1, and the expression for axial strain is given in Equation 2.

$$\varepsilon = 1 - \frac{A_0}{A} \qquad (1)$$

$$\varepsilon = \ln\left(\frac{L_f}{L_0}\right) \qquad (2)$$

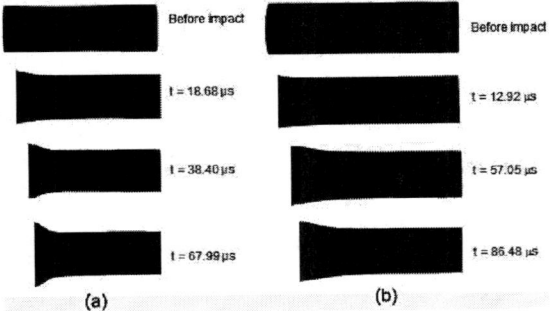

**Figure 4.** 4 of 16 deformation profiles captured using during impact of: (a) pure epoxy at 152 m/s, and (b) Ni+nano Al+20wt% epoxy at 100 m/s.

A plot of incremental areal strains is shown in Fig. 5, and a plot of incremental axial strains is shown in Fig. 6. To eliminate any variability caused by effects of specimen density ($\rho$) or impact velocity (U), the incremental strains were normalized by $\rho U^2$, since dynamic yield strength has been shown to be a function of these parameters, according to the definitions by both Taylor [5] and Wilkins [6]. Both plots demonstrate increasing strain with time after impact, with the Ni+nano Al+epoxy composites exhibiting lesser strain than the pure epoxy and the composite containing micron-sized Al. The Ni+nano Al composite containing 20wt% epoxy exhibits the

least areal and axial strain throughout the captured duration. If the strains, normalized by $\rho U^2$, are used as an inference of strength, then it can be seen that the dynamic strength of the nano Al-containing composites is superior to those of pure epoxy and Ni+micron-sized Al composites, with the lower epoxy content (20wt%) composite showing lesser strain, and thus slightly higher strength than the 30wt% composite.

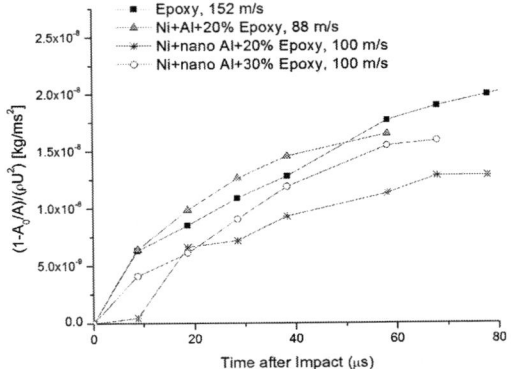

**Figure 5.** Areal strain, normalized by $\rho U^2$, as a function of time after impact. The Ni+nano Al+20% epoxy sample exhibits the least amount of areal strain, leading to the conclusion that it exhibits the highest dynamic strength since it is most resistant to radial expansion.

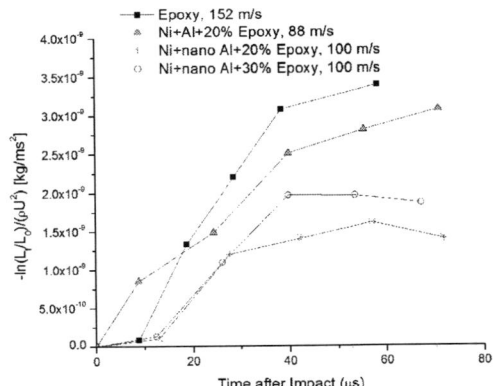

**Figure 6.** Axial strain, normalized by $\rho U^2$, as a function of time after. The Ni+nano Al+20% epoxy sample exhibits the least amount of axial strain, leading to the conclusion that it exhibits the highest dynamic strength since it is most resistant to decrease in length.

Several explanations were explored for the observed difference in behavior. Only samples with nano Al showed strain hardening (Fig. 2), which suggests effects of not only composition, but also the nano-scale size of the Al particles. The dispersion of nano Al particles in the epoxy seems to be altering the structure of the epoxy, causing a change in its deformation response that leads to strain hardening. Van Melick et al. [7] found that strain hardening in polymers is proportional to network density, regardless of whether the density increase is caused by chemical cross-links or physical entanglements. The dispersion of the nano Al particles throughout the epoxy matrix can provide physical entanglements since nanoparticle/matrix interactions lead to a loss in mobility of the chain segments [8]. Nano particles have a very large surface area, and due to the increased contacts between the epoxy and nano particles, the cohesive strength of the epoxy increases and leads to a higher mechanical strength [9]. This is partially due to the mechanical interlocking resulting from the extensive contact between the epoxy and the filler particles [9, 10]. Nano-sized particles are known to have different effects on mechanical properties than micron-sized particles, and the increased strength and strain hardening seen in this study are likely caused by the entanglements and interlocking caused by the dispersion of nano Al particles in the epoxy matrix.

## CONCLUSIONS

The structural behavior of Ni+Al+epoxy composites was evaluated using static compression tests and dynamic reverse Taylor anvil-on-rod impact tests performed using a gas gun equipped with high-speed digital imaging. The material with the lowest epoxy content and nano-sized Al powder showed the most superior mechanical properties, with an elastic modulus of 11.4 GPa and a static compressive yield strength of 156 MPa. Samples containing nano-sized Al powder exhibited strain hardening in addition to increased strength. The superior mechanical properties exhibited by Ni+nano Al+20wt% epoxy can be explained by increased polymer network density caused by physical entanglements from the dispersion of nano particles in the epoxy matrix.

## ACKNOWLEDGEMENTS

Funding for this research was provided by AFOSR/MURI Grant No. F49620-02-1-0382.

## REFERENCES

1. Makino, A. and C.K. Law, *SHS combustion characteristics of several ceramics and intermetallic compounds*. Journal of the American Ceramic Society, 1994. **77**(3): pp. 778-786.
2. Hunt, E.M., K.B. Plantier, and M.L. Plantoya, *Nano-scale reactants in the self-propagating high-temperature synthesis of nickel aluminide*. Acta Materiala, 2004. **52**(11): pp. 3183-3191.
3. Thadhani, N.N., *Shock-induced chemical reactions and synthesis of materials*. Progress in Materials Science, 1993. **37**: pp. 117-126.
4. Wronski, *The size dependence of the melting point of small particles of tin*. British Journal of Applied Physics, 1967: pp. 1731-1737.
5. Taylor, G., *The use of flat-ended projectiles for determining dynamic yield stress. I: Theoretical considerations*. Proceedings of the Royal Society of London A, 1948. **194**: pp. 289-299.
6. Wilkins, M.L. and M.W. Guinan, *Impact of cylinders on a rigid boundary*. Journal of Applied Physics, 1973. **44**(3): pp. 1200-1206.
7. van Melick, H.G.H., L.E. Govaert, and H.E.H. Meijer, *On the origin of strain hardening in glassy polymers*. Polymer, 2003. **44**: pp. 2493-2502.
8. Wetzel, B., F. Haupert, and M.Q. Zhang, *Epoxy nanocomposites with high mechanical and tribological performance*. Composites Science and Technology, 2003. **63**: pp. 2055-2067.
9. Meguid, S.A. and Y. Sun, *On the tensile and shear strength of nano-reinforced composite interfaces*. Materials and Design, 2004. **25**: pp. 289-296.
10. Hussain, M., A. Nakahira, and K. Niihara, *Mechanical property improvement of carbon fiber reinforced epoxy composites by $Al_2O_3$ filler dispersion*. Materials Letters, 1996. **26**: pp. 185-191.

CP845, *Shock Compression of Condensed Matter - 2005*,
edited by M. D. Furnish, M. Elert, T. P. Russell, and C. T. White
© 2006 American Institute of Physics 0-7354-0341-4/06/$23.00

# THE SHOCK RESPONSE, SIMULATION AND MICROSTRUCTURAL DETERMINATION OF AN INERT SIMULANT

## S.A. McDonald[1], N.K. Bourne[1], P.J. Withers[1], J.C.F. Millett[2], K. Bennett[3] and A.M. Milne[3]

[1]*University of Manchester, PO Box 88, Manchester, M60 1QD, UK.*
[2]*Defence Academy of the UK, Cranfield University, Shrivenham, Swindon, SN6 8LA, UK.*
[3]*Fluid Gravity Engineering, 83 Market Street, St. Andrews, Fife, KY16 9NX, UK.*

**Abstract.** Assessing microstructural details in a polymer matrix composite is important in addressing safety issues in energetic materials. The generation of three-dimensional microstructure, using a non-invasive method of high resolution will advance knowledge in a range of fields. An inert composite analogous to plastic bonded explosives (PBXs) has been studied, and both X-ray microtomography for microstructural investigation in 3-D and a parallel series of shock experiments (with associated modelling) have been conducted. The experimental aims of this study lay in several areas. Firstly, to adequately define the bulk morphology, secondly, to determine the geometry of defects that might lead to sites for accidental ignition within the material and finally, to demonstrate a direct linkage into the finite element prediction of mechanical response. This work is the first step in providing a coordinated capability to understand accidental ignition within insensitive high explosives (IHEs).

**Keywords:** shock, composite, X-ray tomography, Hugoniot
**PACS number:** 62.50

## INTRODUCTION

Understanding the response of energetic materials, such as plastic bonded explosives (PBXs), to high strain rate loading is of concern given that this can result in reactions that can ultimately lead to detonation. The response of such composite materials, consisting of the particulate explosive (or energetic) crystals bound in an inert or energetic polymer binder phase, to purely mechanical loading (shock loading) is therefore of interest from a safety point of view. The dynamic behaviour of energetic materials is related to their bulk morphology and the behaviour of each of the individual phases contained within. Thus it is important to investigate and define their microstructure, information necessary both to describe the morphology of the material and also to define the form of defects. In the former case this

will refine understanding of the fines present within the binder phase, and in the latter to resolve voids or cracks that might lead to hot spots by collapse or shear. It is the detection of these defects that requires increased resolution.

X-ray microtomography is a non-destructive evaluation technique that can reveal detailed information about the internal structure of an object. This is carried out by reconstructing the three dimensional spatial distribution of the local X-ray attenuation coefficients of the materials contained within the object from a series of 2-D radiographs taken at different orientations. Linking the three dimensional data obtained directly to simulation can lead to a computation design capability.

The composite that has been studied is a soda-lime glass, hydroxy-terminated polybutadiene (HTPB) system, for which the shock response of both materials has been determined [1, 2]. The approach linking testing, modelling and microstructural investigation has been taken to understand the response of real systems.

**EXPERIMENTAL AND NUMERICAL**

A composite containing equal proportions of coarse and fine soda-lime glass spheres was made, with average particle size distributions of 300 and 30 μm respectively. The HTPB binder system, has been studied previously [1]. These were mixed with 60 wt % of the glass beads and cured at 60°C for one week.

X-ray microtomography

Measurements were carried out on an X-Tek HMXST 225 kV 8 μm source X-ray radiography and tomography set [3], using an X-ray energy of 50 kV and a copper anode target. This, in conjunction with the use of a beryllium windowed detector, ensured attenuation of the X-rays through such low density materials. The 3-D tomographic volumes were reconstructed using a cone-beam extension of the filtered back projection method [4] from 470 radiographs acquired at 0.4° sample rotation intervals. The characteristic voxel size of the reconstructions was 7 μm.

Shock testing

Plate impact experiments were performed to determine the shock response of the composite material using a 50 mm bore, 5 m single-stage gas gun [5]. This involved the measurement of the Hugoniot in terms of the shock (impact) stress, shock velocity and the particle velocity. The target was cast in the manner used previously [6]. The two gauges mounted in the sample allowed the shock velocities ($U_s$) to be extracted and the stresses to be measured. Impact velocities were measured from the electrical shorting of sequentially mounted pairs of pins to an accuracy of 0.1%. Voltage data from the gauges were converted to stress according to the methods of Rosenberg *et al.* [7]. Particle velocities ($u_p$) were determined from the measured Hugoniot stresses ($\sigma_x$) and the impact velocities of the flyer plates using impedance matching techniques. The method has been successfully used to measure the shock properties of soda-lime glass-HTPB composites

and cyclotrimethylene trinitramine (RDX)-based plastic-bonded explosive and sugar-based simulants [6, 8]. The utility of the simultaneous measurement of the various state variables is that strength effects have been noted by varying particle size.

In the regimes chosen for investigation the binder phase yields inelastically but the glass filler phase is within its elastic regime. An analogous material has been recovered and examined and the following observations were noted [6]. The glass spheres show no signs of damage. The binder phase was seen to have delaminated from the spheres. There was some evidence of apparent flow of the binder through the glass spheres.

Simulations

The tomographic images were read directly into the mesh generators for three-dimensional simulation. The code employed was an Eulerian, multimaterial hydrocode Eden [9]. The material descriptions employed a Murnaghan equation of state and an elastic-perfectly plastic constitutive description. The simulation was a reverse ballistic geometry onto a rigid boundary across the base plane.

**RESULTS AND DISCUSSION**

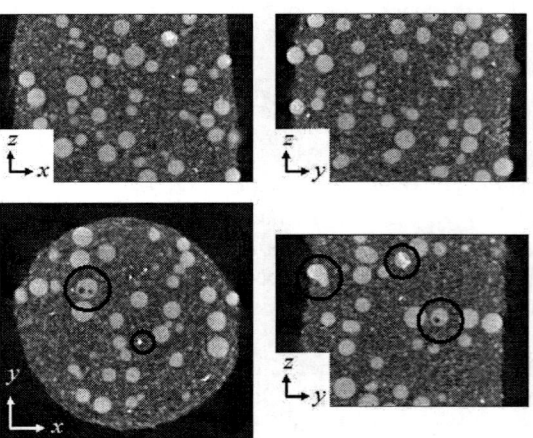

**Figure 1.** Representative tomographic slices through the three orthogonal directions of the composite sample.

Fig. 1 shows virtual greyscale slices through the three orthogonal directions of the reconstructed volume of the glass composite material. A number of observations can be made from the tomographic images regarding the microstructure of the material, as highlighted on the images. The glass spheres are clearly resolved and are uniformly distributed throughout the material – both the larger spheres and smaller ones embedded in the binder. A few brighter particles are observed, thought to be inclusions introduced into the material. The presence of voids within the glass is also highlighted, although this was only in a few of the spheres.

Fig. 2 shows representative traces from plate impact experiments on the composite material. The impact conditions are ca. 437 m s$^{-1}$, with a 10 mm Dural flyer plate. The initial overload pulse results from equilibration at the Dural coverplate. The later amplitude of the first trace (ca. 1.6 GPa) has been used in combination with the particle velocity (deduced from impedance matching) to determine the Hugoniot in stress-particle velocity space.

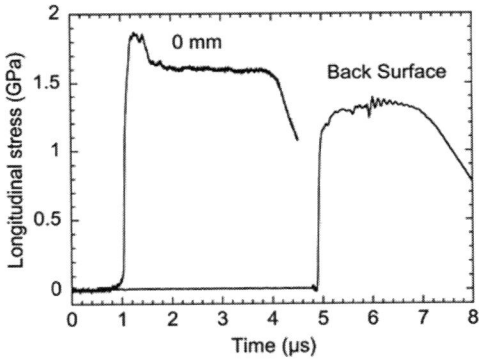

**Figure 2.** Representative stress gauge traces from the composite. Impact conditions are 10 mm Dural flyer at 437 m s$^{-1}$.

The back surface trace is lower than the 0 mm trace as it was measuring the stress in the polymethylmethacrylate (PMMA) backing. This has a lower shock impedance than the composite, resulting in a lower stress amplitude.

The temporal spacing between the histories, along with the known target thickness was used to calculate the shock velocity. These results are presented in Fig. 3. This composite appears to show a linear relation between $U_s$-$u_p$ which is somewhat surprising for this material. Previous studies [6] showed a non linear relationship. These differences maybe due to the limited number of experimental data points in this work. The fitted values for $c_0$ and

$S$ are shown in the figure caption. These are not similar to values previously obtained for an analogous material [6]. The shock parameters appear to be dependent on the distribution of phases through the microstructure.

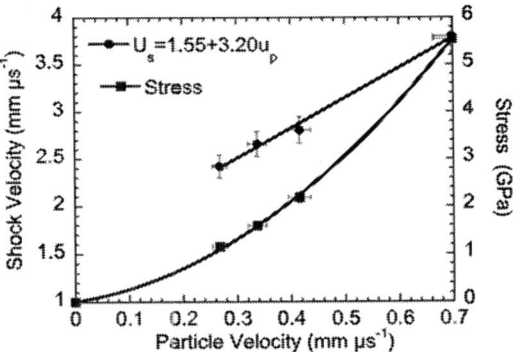

**Figure 3.** Hugoniots for the composite in $\sigma_x$-$u_p$ and $U_s$-$u_p$ space.

## NUMERICAL SIMULATIONS

Numerical simulations have been carried out using the tomography data as input to the hydrocode EDEN.

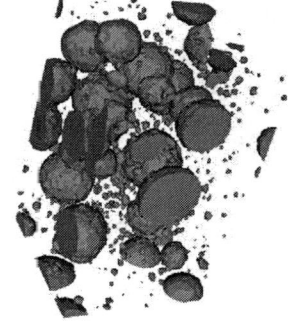

**Figure 4.** Reconstruction of glass beads in a sample of the target.

The voxel data is on a linear scale proportional to the mass of the material contained in the voxel. From the known ingredient densities and the measured mixture density [6] one can produce volume fraction distributions of each of the components. The spatial resolution of the data presented here is sufficient to clearly identify the large spheres. The smaller spheres are not so well resolved and are observed to form clusters. This means care must be taken to ensure accurate reconstruction of an interface. We invoke threshold contours to identify the edge between a sphere and the binder and vary these thresholds (within narrow bands) to ensure that the calculated mixture density matches the measured value. Fig. 4 shows the reconstructed glass sphere interfaces in a square cross section portion of the full target.

Given this initial data we can use the code to carry out numerical impact experiments. In Fig. 5(a) a section of the sample at an early stage after it has been impacted into a rigid wall at the base of the calculation is shown. In this figure we have extracted a square cross section sample of the data and applied rigid boundary conditions at the edges and inflow boundary conditions at the top. The whole sample was subjected to an initial applied velocity of 1000 m s$^{-1}$ into the rigid wall. The reconstruction of the glass spheres in the binder in the unshocked region as well as the early stages of the shock wave moving back from the impact face is illustrated. Fig. 5(b) shows the density contours at a later stage. The shock has progressed through the composite and significant structure is apparent.

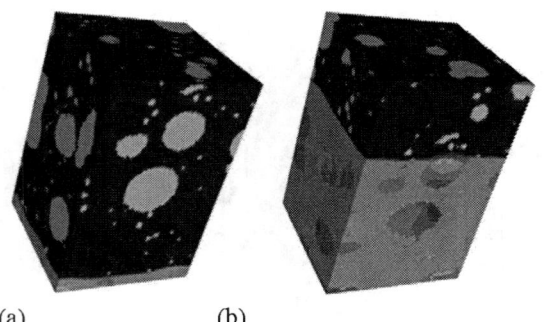

(a)                    (b)

**Figure 5.** Density plots of the same section of the sample at an (a) early and (b) later stage of a 1000 m s$^{-1}$ impact.

This analysis complements alternative means of diagnosing the shock properties in a composite [10]. Details of the shock interaction on the meso-scale are important in understanding ignition processes.

## CONCLUSIONS

This work has provided a direct link between microstructural analysis, experimental investigation and numerical simulation. Effort was put into both resolving features at appropriate length scales, and into reading these across onto the numerical platform. It was noted that there was processing debris in the target which will be removed in further work. Shock data has been gathered for the composite and differs from that collected for a similar material previously [6]. We believe that the shock response of this class of materials is sensitive to the distribution of phases within the microstructure. Our aim now is to complete studies on this inert stimulant and then move towards applying the same techniques to real explosive samples.

## ACKNOWLEDGEMENTS

We gratefully acknowledge the support of Dr. Peter Barnes of the UK Defence Ordnance Safety Group (DOSG) who funded part of this work.

## REFERENCES

1. Millett, J. C. F., Bourne, N. K. and Akhavan, J., J. Appl. Phys., **95**, 4722 (2004)
2. Bourne, N., Millett, J., Rosenberg, Z. and Murray, N., J. Mech. Phys. Solids, **46**, 1887 (1998)
3. McDonald, S. A., Thesis, University of Manchester, (2004)
4. Feldkamp, L. A., Davis, L. C. and Kress, J. W., J. Opt. Sci., **1**, 612 (1984)
5. Bourne, N. K., Meas. Sci. Technol., **14**, 273 (2003)
6. Millett, J. C. F., Bourne, N. K., Akhavan, J. and Milne, A. M., J. Appl. Phys., **97**, 043524 (2005)
7. Rosenberg, Z., Yaziv, D. and Partom, Y., J. Appl. Phys., **51**, 3702 (1980)
8. Millett, J. C. F. and Bourne, N. K., J. Phys. D. Applied Physics, **37**, 2613 (2004)
9. Milne, A. M., EDEN user manual, Fluid Gravity Engineering Ltd, St. Andrews, 2004
10. Milne, A. M., Bourne, N. K., Millett, J. C. F, these proceedings.

CP845, *Shock Compression of Condensed Matter - 2005,*
edited by M. D. Furnish, M. Elert, T. P. Russell, and C. T. White
© 2006 American Institute of Physics 0-7354-0341-4/06/$23.00

# THE BEHAVIOUR OF A CARBON-FIBRE EPOXY COMPOSITE UNDER SHOCK LOADING.

## R. Vignjevic, J.C.F. Millett*, N.K. Bourne[+], Y. Meziere* and A. Lukyanov.

*Cranfield University,Cranfield, Bedfordshire, MK43 0AL. United Kingdom.*

*\*Defence Academy of the United Kingdom, Cranfield University, Shrivenham, Swindon, SN6 8LA. United Kingdom.*

[+] *University of Manchester, Sackville Street, Manchester, M60 1QD. United Kingdom.*

**Abstract.** The behaviour of a carbon fibre – epoxy composite has been investigated as a function of specimen thickness. Results show that both the shock velocity and the shock stress are unaffected by transit distance. The relationship between $U_s$ and $u_p$ over the range of measurements made in this investigation shows a higher value of $c_0$ and lower value of $S$ than that of a similar material within the literature. We believe that this is due to the slightly higher proportion of carbon fibres within the microstructure reducing the compressibility. Differences between the Hugoniot stress and the calculated hydrodynamic pressure suggests that the shear strength of this material increases with shock stress.
**Keywords**: shock, composite, CFRP, Hugoniot.
**PACS**: 62.50

## INTRODUCTION

Reductions in density of aerospace materials lead to greater efficiency and cost effectiveness in air transportation. However, a major requirement is that they display a high resistance to impact damage, such as bird strike, hail stones *etc*. One group of materials that has attracted attention is that of fibre-based composites. For example, the structural elements of the Airbus A380 are composed of 25 % of composite materials (carbon fibre, glass fibre epoxy composites, Glare © *etc*). Previously, such materials have been used as secondary structural elements (non load bearing) but are now considered for primary areas such as the leading edge of airfoil sections. Lower impact velocity events such as tool drop or foreign object

damage have been studied [1-3], whilst higher impact velocities (such as protection from fragment attack on military vehicles) has been studied by Justo *et al.* [4] Plate impact studies have been performed on a number of fibre-based composites including glass based by Dandekar *et al.* [5] and ourselves in a parallel paper [6], and Reidel *et al.* [7] on both carbon and Kevlar fibre composites. In this work, we report on the response of a carbon-fibre epoxy composite to one-dimensional shock loading.

## EXPERIMENTAL

Samples of a carbon-fibre reinforced epoxy composite (CFRP) of thicknesses 2.3, 3.8 and 5.7 mm thick were lapped flat and parallel to ± 5 µm. A manganin stress gauge (MicroMeasurements LM-

SS-025CH-048) was supported on the back of the specimen plate with a 12 mm block of polymethylmethacrylate (PMMA). In addition, the gauge was also backed into the PMMA by *ca.* 1.5 mm to act as extra protection for the gauge. Gauge calibration was according to Rosenberg *et al.* [8] A second gauge (the 0 mm position) was supported on the front of the target assembly with a 1 mm plate of either dural (aluminium alloy 6082-T6) or copper, insulated with 25 μm of mylar on either side of the gauge. Shock stresses were induced with dural or copper flyer plates impacted in the velocity range 225 to 1125 m s$^{-1}$, using a single stage gas gun. The impact axis was normal to the plane of the fibres. Impact velocities were measured from the shorting of sequentially mounted pairs of pins to an accuracy of *ca.* 0.5%. A schematic of the target assembly and gauge placement is shown in Fig. 1.

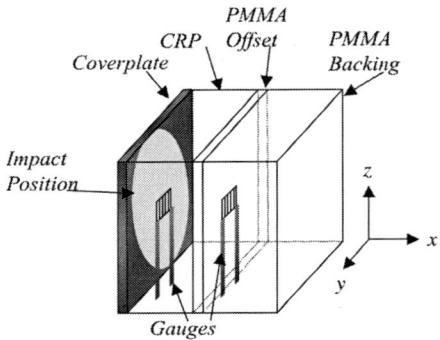

**Figure 1.** Target assembly.

## MATERIALS

Experimental material was provided by Short Brothers PLC of Belfast. The fibres were Hexcel 5HS in a woven lay up of orientation 0/90, ±45, 0/90, ±45... The arial weight was 370 g m$^{-2}$. The resin was an epoxy, Hexcel RTM6. The composite was cured at 180°C under a pressure of 100 psi (6.7 atm) for 1 hour 40 minutes. The microstructure is shown in Fig. 2. The longitudinal sound speed ($c_L$) was 3.02 mm μs$^{-1}$, and the ambient density ($\rho_0$) was 1.50 g cm$^{-3}$.

## RESULTS AND DISCUSSION

In Fig. 3, we present representative gauge traces

from the composite. There is some noise superimposed on both gauge traces, most likely due to complex wave interactions between the carbon fibres and the epoxy matrix.

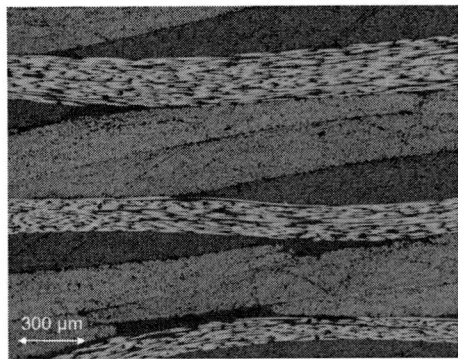

**Figure 2.** Microstructure of composite.

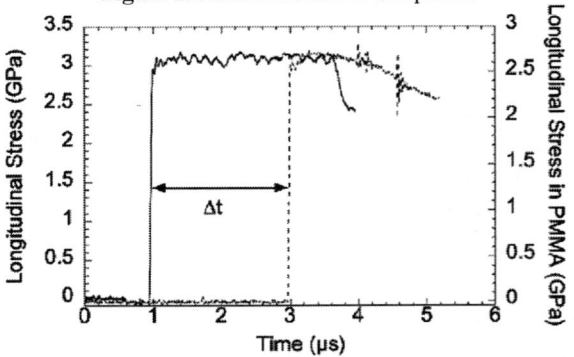

**Figure 3.** Representative gauge traces for carbon-fibre composite. The solid curve is the 0 mm position, the dotted the back surface trace. The impact conditions are – 5.7 mm thick specimen, 7.4 mm copper flyer at 627 m s$^{-1}$.

We have used the amplitude of the 0 mm trace to determine the Hugoniot in terms of stress ($\sigma_x$) and particle velocity ($u_p$) using standard impedance matching techniques. Even though the second gauge is backed into PMMA, there is a square shape to the pulse, similar to previous experience of other fibre based composites [6]. We have used the temporal spacing of the gauge traces ($\Delta t$), taking into account that the back surface gauge is backed *ca.* 1.5 mm PMMA [9] and the thickness of the sample ($\Delta w$) to determine the shock velocity of the composite ($U_s = \Delta w/\Delta t$). The results are presented in the following figure.

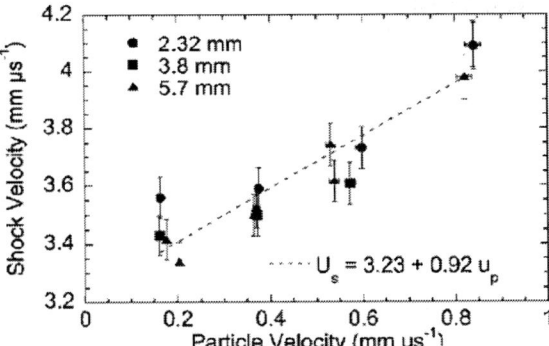

**Figure 4.** $U_s$-$u_p$ for the composite, showing the variation with specimen thickness.

It can be seen that there is a degree of scatter in this data. However, we can see no obvious variation between shock velocity and specimen thickness. Therefore, we have fitted a straight line through the data of the form,

$$U_s = c_0 + Su_p,\qquad\qquad 1.$$

based on the assumption that the relationship between $U_s$ and $u_p$ is linear, as it is for many materials [10]. This yields values of $c_0$ and $S$ (the shock parameters) of 3.23 mm $\mu s^{-1}$ and 0.92 respectively. The value of $c_0$, in metallic materials at least, has been shown to agree with the measured ambient bulk sound speed, $c_B$. In the case of this composite, $c_0$ is larger than the measured longitudinal sound speed of 3.02 mm $\mu s^{-1}$. This behaviour has been observed in some monolithic polymers [9], including an epoxy resin [11]. We have used this relation to calculate the hydrodynamic pressure ($P_{HD}$), according to,

$$P_{HD} = \rho_0 U_S u_p,\qquad\qquad 2.$$

and have compared it to the measured Hugoniot stress, as shown in Fig. 5. As with the previous figure, we can see no obvious variation with specimen thickness, although there is some degree of scatter in the data. It can also be seen that there is an increasing divergence between the measured Hugoniot stress and the calculated pressure, with the stress being the greater. As the Hugoniot stress has hydrostatic ($P$) and shear strength ($\tau$) components, thus,

$$\sigma_x = P + \frac{4}{3}\tau,\qquad\qquad 3.$$

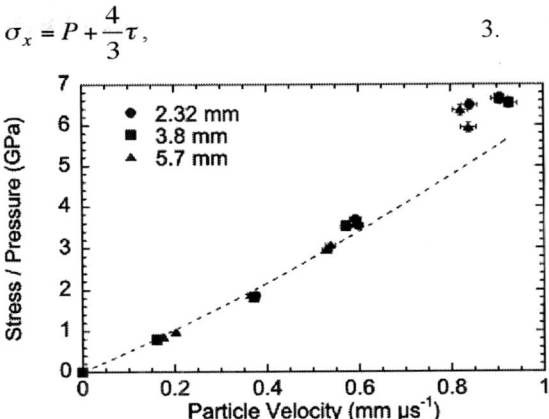

**Figure 5.** Hugoniot of the carbon-fibre epoxy composite. The dotted curve is calculated from equation 2.

suggesting that the shear strength of this material increases with increasing stress. We have observed this behaviour, both in other composites [6] and an epoxy resin [11] where in the latter material, this was confirmed by independent shear strength measurements. As it is likely that under the loading regimes studied in both composites that the fibres remain elastic, it seems likely that such a response is due to the epoxy matrix itself.

Finally, in Fig. 6 we present a comparison of our own data with that of other workers.

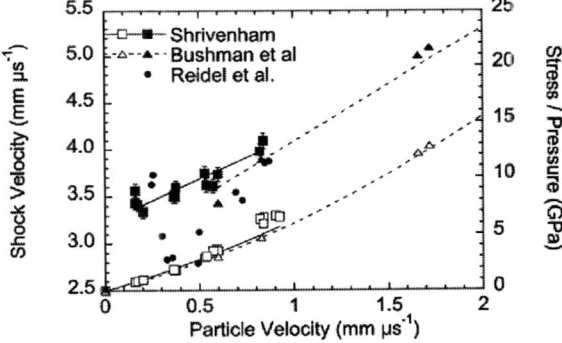

**Figure 6.** Comparison of the shock response of carbon-fibre composites with Reidel *et al.* [7] and Bushman *et al.* [12]. The filled symbols are the shock velocity and the open the shock stress.

In the case of Reidel *et al.* [7] we can say very little as materials properties data were not included in the original paper. They quote values of $c_0$ and $S$ of 1.56 mm $\mu s^{-1}$ and 2.10 respectively. Without knowing more about the material, little further comparison can be made. The case with the data of Bushman *et al.* [12] is more interesting. At a density of 1.46 g cm$^{-3}$, their material is similar to ours, although the slight difference suggests that the fraction of carbon fibres may be a little lower. Values of $c_0 = 2.86$ mm $\mu$s$^{-1}$ and $S = 1.22$ have been extracted. These are somewhat different than the equivalent values for our own material. The value of $S$ has been suggested to be dependent upon the first pressure derivative of bulk modulus [13], thus a higher value indicates a higher compressibility with pressure. Whilst it would be difficult to make such a comparison between vastly different materials (for example between a polymer and a metal), a valid comparison can be made between related materials. In a previous work, we showed that a commercial explosive binder (hydroxyterminated polybutadiene – HTPB) with added plasticsizers had a higher $S$ than a "pure" laboratory composition, which correlated with the commercial material having a lower density. We can also observe differences between our measured stresses and those of Bushman *et al* [12], with our data being slightly higher. We believe that this is due to the fact the Russian data was determined from rear surface velocity measurements, and converted to pressure using equation 2, whilst our own takes into account the materials strength properties. We have used the $U_s$-$u_p$ relations from both materials to calculate the hydrodynamic response (equation 2) to perform a more accurate comparison. It can be seen that in this case, both materials have similar behaviour, although our data appears be slightly higher in magnitude. This is most likely due to the slight differences in compressibility discussed previously. However, it also suggests that if the balance of fibres to epoxy is similar, and that the material is shocked normal to the fibre plane, then the Hugoniot of the material is largely unaffected by either fibre diameter or lay up morphology.

## CONCLUSIONS

Plate impact experiments have been performed on a carbon-fibre epoxy composite. Results show that both the shock velocity and the shock stress are unaffected by the thickness of the sample. The relationship between the shock velocity and particle velocity is linear over the experimental range of this investigation. Differences between the measured Hugoniot stresses and calculated hydrodynamic pressure indicate that the shear strength of this material increases with increasing shock stress. Finally, we have compared our results with those of a similar material within the literature. The density of that material was slightly lower, suggesting that volume fraction of carbon fibres was lower as well. The lower value of $c_0$ and higher $S$ in that work (when compared to our own) also indicated a higher shock compressibility, which would be consistent with a higher proportion of epoxy within the microstructure.

## ACKNOWLEDGEMENTS

The authors acknowledge Gary Cooper and Ivan Knapp for valuable technical support. Funding was through EPSRC Grant No. GR/S33994/01.

## REFERENCES

1. Ray, R., Sarkar, D. K. and Bose, N. R., *Bull. Mater. Sci.*, **24**, 137 (2001)
2. Symons, D. D. and Davis, G., *Composites Sci. Technol.*, **60**, 379 (2000)
3. Gustin, J., Joneson, A., Mahinfalah, M. and Stone, J., *Composite Structures*, **69**, 396 (2005)
4. Justo, J. and Marques, A. T., *J. Phys IV*, **110**, 651 (2003)
5. Dandekar, D. P., Hall, C. A., Chhabildas, L. C. and Reinhart, W. D., *Composite Structures*, **61**, 51 (2003)
6. Eatwell, M., Millett, J. C. F., Bourne, N. K. and Meziere, Y., these proceedings,
7. Riedel, W., Nahme, H. and Thoma, K., in Shock compression of condensed matter, edited by Furnish, M. D., Gupta, Y. M. and Forbes, J. W., America Institute of physics, Portland, 2004, pp 701
8. Rosenberg, Z., Yaziv, D. and Partom, Y., *J. Appl. Phys.*, **51**, 3702 (1980)
9. Carter, W. J. and Marsh, S. P., Hugoniot equation of state of polymers, Los Alamos National Laboratory, LA-12006-MS, 1995
10. Marsh, S. P., LASL Shock Hugoniot data, University of California Press (Los Angeles), 1980
11. Millett, J. C. F., Bourne, N. K. and Barnes, N. R., *J. Appl. Phys.*, **92**, 6590 (2002)
12. Bushman, A. V., Efremov, V. P., Lomonosov, I. V., Fortov, V. E. and Utkin, A. V., *Teplofiz. Vys. Temp.*, **28**, 1232 (1990)
13. Davison, L. and Graham, R. A., *Physics Reports*, **55**, 255 (1979)

CP845, *Shock Compression of Condensed Matter - 2005*,
edited by M. D. Furnish, M. Elert, T. P. Russell, and C. T. White
© 2006 American Institute of Physics 0-7354-0341-4/06/$23.00

# FRACTURE STUDIES OF PBX SIMULANT MATERIALS

## D. M. Williamson, S. J. P. Palmer, W.G. Proud

*University of Cambridge, Cavendish Laboratory, Physics & Chemistry of Solids Group,
Madingley Road, Cambridge, CB3 0HE, UK*

**Abstract.** Fracture studies have been performed on three inert PBX simulants; PBS 9501 which consists of sugar bound in Estane and is a PBX 9501 simulant. EDC1037 and EDC1032 which consist of barium sulphate and melamine bound in NC/K10 and Viton-A respectively, and are simulants for EDC37 and EDC32. The effect of microstructure, geometry and testing rate are investigated, and through the application of elastic-plastic fracture mechanics, energy release rates have been calculated. Such data are required for the development and validation of accurate failure models.

**Keywords:** PBX, fracture, J-resistance.
**PACS:** 62.20.Fe, 62.20.Mk.

## INTRODUCTION

Polymer Bonded eXplosives (PBXs) are particulate composite materials formed from crystalline explosive grains bound in comparatively soft polymeric binder systems. Knowledge of the fracture properties of PBXs is essential for the correct understanding of how these materials respond to mechanical stimulus.

Despite the obvious importance fracture plays in PBX failure, only a relative few studies appear in the literature, with most examining the microstructural interaction behaviour between filler particle and binder at quasi-static rates of deformation [1,5]. The conclusion drawn is that failure is initiated at the sites of the largest crystals, by adhesive failure between the crystals and the binder material. Such sites then link up and the crack front moves forward. It should be noted that this description implies non-linear processes.

Within the literature concerning more global observations of PBXs in specific fracture experiment geometries, two approaches are utilized: either Linear Elastic Fracture Mechanics (LEFM) or else Elastic-Plastic Fracture Mechanics (EPFM). LEFM can be applied to some of the more brittle energetic composite compositions, and the results show that the measured stress intensity factors at room temperature and quasi-static loading rates are low, typically > 1 MPa·m$^{1/2}$ [6,7].

In general, however, PBXs are widely accepted to be visco-elastic. Observations made of the fracture of Polymer Bonded Sugar (PBS) and PBX 9501 indicate that a relatively large process zone is in operation at the crack tip [8, 9, 10]. These features suggest that EPFM is the appropriate analysis framework; this approach is adopted in this study.

In view of the potentially hazardous nature of the PBX materials, tests have been performed on three inert simulant materials with a view to transferring to PBXs at a later stage. Within each simulant, the binder material is the same as that of the parent PBX, but the energetic component has been replaced with an inert one. Table 1 gives simulant compositions. Note that PBS 9501 is a mechanical simulant, whereas EDC1037 and EDC1032 are density simulants, however sufficient mechanical data exist on these materials to make the results immediately useful.

**Table 1.** Composition of inert simulant PBXs

| Product | Filler | | Binder | | Plasticizer | |
|---------|--------|--|--------|--|-------------|--|
| | Material | Content % w/w | Material | Content % w/w | Material | Content % w/w |
| PBS 9501 | Sugar | 95 | Estane 5703 | 2.5 | BDNPA-F | 2.5 |
| EDC1032 | Melamine $BaSO_4$ | 61 24 | Viton A | 15 | - | - |
| EDC1037 | Melamine $BaSO_4$ | 62 29 | Nitrocellulose | 1 | K10 | 8 |

## EXPERIMENTAL PROCEDURE

Compact Tension (CT) samples and testing procedures were taken from the European Structural Integrity Society (ESIS) protocol for conducting J-crack growth resistance curve tests on plastics. Crack tips were generated using a modified 3/16"×6" cutting wheel on which the cutting teeth had been precision-ground to cut a slot terminating in a V-shape of 60° included angle. Each notch was then further razor sharpened to produce the sharpest possible notch condition.

In addition to the standard measurements of load and displacement, two other diagnostics were used. On one side of the sample, crack tip position, and therefore velocity, measurements were made by the propagating crack break conducting elements consisting of ten thin, > 20 μm thickness, deposited silver strips, of 4 mm pitch. On the other side of the sample, photographs were taken of the area surrounding propagating crack, to allow digital image speckle correlation measurements to be made. The results of this second diagnostic are not stated here but are to be published elsewhere, under the title of 'Digital image correlation method applied to crack tip strain fields in PBX materials'. The presence of these additional diagnostics is not expected to modify the fracture data.

Samples were evaluated in a screw-driven Instron testing machine, in displacement control mode. Experimental variables were loading rate and sample dimensions. The latter is of interest because it is generally accepted that fracture experiments are size dependent, and the aim is that the final PBX CT samples will have a mass restriction. Sensitivity to sample dimensions over a wide range can therefore only be evaluated on

simulant materials. Loading rate was investigated as viscoelastic materials such as PBXs have strong temperature/ time dependencies. Loading rates varied between 0.135 and 403 mm/min.

## RESULTS AND DISCUSSION

Figure 1 is a typical experimental result. $\Delta a$ is the crack growth measured using the strip breaking technique. The first $\Delta a$ point is based on the intuitive assumption that crack propagation begins when the applied load reaches a maximum. The fit to the $\Delta a$ data shown is intended only to aid the eye and is based on a quadratic best fit. Using a linear least squares fit to the first and last three $\Delta a$ points, the measured crack speed is initially $0.32 \pm 0.07$ mm/s, falling to $0.17 \pm 0.04$ mm/s by the end of the test. The crack propagation is stable, and the running crack could have been arrested at any point by stopping the loading process.

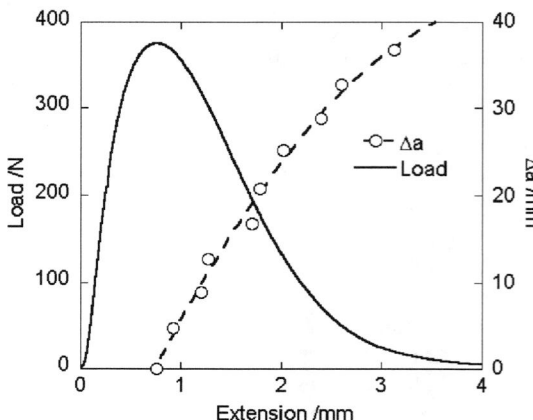

**Figure 1.** Load, extension and crack position data taken from an EDC1032 sample tested at 1 mm/min.

This is a consequence of the low stresses required for crack propagation – no substantial amount of elastic energy is stored in the bulk to be released in the form of new surface area. Stable crack growth was observed in all the samples studied.

### Inter-material comparison

Figure 2 shows results from all three material types taken from samples of identical dimensions at a loading rate of 1 mm/min. There are clear differences in strength between the three materials. This ranking agrees with that of the tensile moduli of their binder systems which have been measured as 1.7, 0.3 [1] and 0.01 MPa respectively.

Another effect is samples with finer particle size distributions tend to be stronger, all else being equal, since the stresses required to debond the particles are necessarily greater [11, 12]; this is likely to be the dominant failure mode. Within the three materials, EDC1032 has a melamine grain size of 20 μm, PBS 9501 has large sugar grains circa. 250 μm and EDC1037 has a more complex microstructure composed of mm sized melamine agglomerations, which on closer inspection by SEM are revealed to be concentrations of melamine grains, circa. 20 μm. The loading density ought to play a secondary role; those samples with the least binder should be stronger, assuming a given set of conditions and

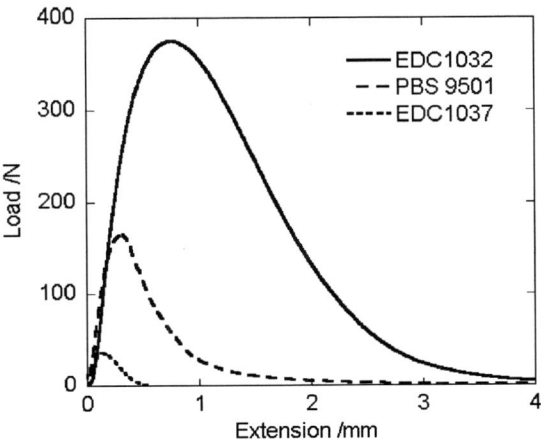

**Figure 2.** Load extension data for all material types with identical geometries tested at 1 mm/min.

a relatively homogeneous mixture of filler and binder. This is because thinner binder coatings layers have greater adhesive strength [13].

The ranking of strength in the materials are in agreement with these observations.

### Loading rate dependence

In general, the mechanical properties of polymers are temperature and time dependent, and within limits, the two variables are interchangeable; this is the basis for the well-known WLF formula. The fracture properties of the simulants are no exception: Fig. 3 shows peak load as a function of loading rate for EDC1037 and EDC1032 obtained from samples of identical geometry. The materials are clearly rate dependant, and so they will also be temperature dependant.

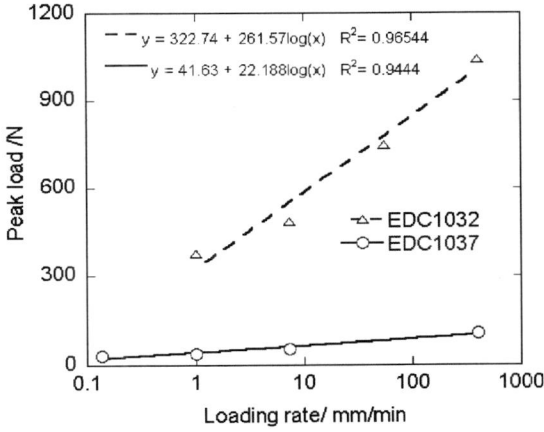

**Figure 3.** Loading rate dependency of EDC1037 and EDC1032 tested with identical geometries.

### J-resistance

The load, extension and crack position data are important to the modeller, as they represent an easily quantified and direct measure of material response. It is also of interest to apply EPFM to obtain the non-linear elastic energy release rate, J, as a function of crack extension; known as the J-resistance curve, Fig. 4 shows this. Note the extensive $\Delta a$ range shown in Fig.4 is not strictly within the bounds of the test protocol. Compared to polymers in general, the measured J-initiation values are low; they are not 'tough' materials, cf. polycarbonate circa 3 kJm$^{-2}$. Doubling sample thickness in EDC1037, all other parameters constant, resulted in a lower J-initiation value due to the stress state approaching the plain-strain condition [6], cf. $4.3 \pm 0.3$ Jm$^{-2}$ for a 7 mm thick sample to $3.5 \pm 0.2$ Jm$^{-2}$ for 14 mm.

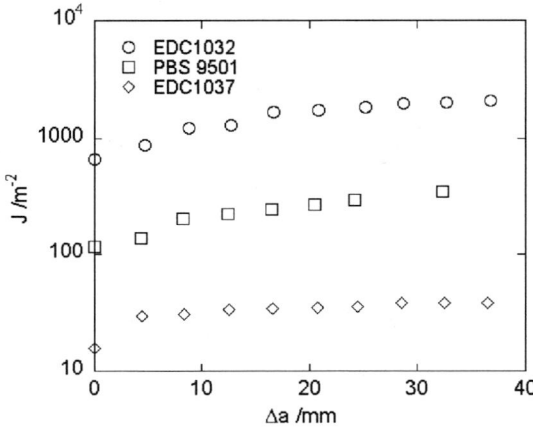

**Figure 4.** J-resistance curves for all material types with identical geometries tested at 1 mm/min.

## CONCLUSIONS

Experiments performed on three PBX simulants have shown their measured fracture properties are dependent on geometry and loading rate (and therefore temperature). Fracture behavior appears to be dominated by binder properties. EPFM has been successfully applied to measure J initiation values. Crack positional data give velocity and facilitate calculation of J-resistance curves.

## ACKNOWLEDGEMENTS

Funding was supplied by EPSRC and AWE, who additionally supplied sample materials. R. Flaxman and D. Johnson of the Cavendish Workshop provided technical support. Prof. J.E. Field gave invaluable advice and discussions.

## REFERENCES

1. Palmer, S.J.P., Field, J.E., and Huntley, J.M., "Deformation, strength and strains to failure of polymer bonded explosives", Proc. R. Soc. Lond. A 440, pp. 399-419, 1993.
2. Rae, P.J., Goldrein, H.T., Palmer, S.J.P., Field, J.E., and Lewis, A.L., "Studies of the failure mechanisms of polymer bonded explosives by high resolution moiré interferometry and environmental scanning elenctron microscopy", in 11$^{th}$ Detonation Symposium, Snowmass, C0, 31 August – 4 September, pp. 66-75, 1998.
3. Rae, P.J., Palmer, S.J.P., Goldrein, H.T., Field, J.E., and Lewis, A.L., "Quasi-static studies of the deformation and failure of β-HMX based polymer bonded explosives", Proc. R. Soc. Lond. A 458, pp. 743-762, 2002.
4. Rae, P.J., Palmer, S.J.P., Goldrein, H.T., Field, J.E., and Lewis, A.L., "Quasi-static studies of the deformation and failure of PBX 9501", Proc. R. Soc. Lond. A 458, pp. 2227-2242, 2002.
5. Rae, P.J., Goldrein, T.H., Palmer, S.J.P., and Field, J.E., "The use of digital image cross-correlation (DICC) to study the mechanical properties of a polymer bonded explosive (PBX)", in 12$^{th}$ International Detonation Symposium, San Diego, CA, 11 – 16 August, to be published.
6. Kinlock, A.J., and Gledhill, R.A., "Propellant failure: a fracture-mechanics approach", J. Spacecraft 18-4, pp. 333-337, 1981.
7. Li, M., Zhang, J., Xiong, C.Y., and Fang, J., "Fracture analysis of a plastic bonded explosive by digital image correlation technique", Proc. SPIE 4537, pp. 107-110, 2002.
8. Liu, C., Stout, M.G., and Asay, B.W., "Stress bridging in a heterogeneous material", Eng. Frac. Mech. 67, pp. 1 -20, 2000.
9. Liu, C., "Fracture of the PBX 9501 high explosive", In 13th APS topical meeting on shock compression of condensed matter, Portland, Oregon, 20th -25th July (eds. M.D Furnish, Y.M. Gupta, and J.W. Forbes).
10. Liu, C., and Browning, R., "Fracture in PBX 9501 at low rates", in 12th International Detonation Symposium, San Diego, CA, 11 – 16 August, to be published.
11. Nicholson, D.W., "On the detachment of a rigid inclusion from an elastic matrix", J. Adhesion 10, pp. 255-260, 1980.
12. Gent, A.N., "Detachment of an elastic matrix from a rigid spherical inclusion", J. Mater. Sci 15, pp. 2884 – 2888, 1980.
13. Bickerman, J.J., The science of adhesive joints, 1$^{st}$ edn. Academic Press: New York, 1968.

CHAPTER XII

# MECHANICAL PROPERTIES OF CERAMICS, GLASSES, IONIC SOLIDS, AND LIQUIDS

CP845, *Shock Compression of Condensed Matter - 2005*,
edited by M. D. Furnish, M. Elert, T. P. Russell, and C. T. White
© 2006 American Institute of Physics 0-7354-0341-4/06/$23.00

# EVOLUTION OF SHOCK WAVES IN SILICON CARBIDE RODS

## I. A. Balagansky[1], A. I. Balagansky[1], S. V. Razorenov[2], A. V. Utkin[2]

[1]*Novosibirsk State Technical University, K. Marx Ave, 20, 630092, Novosibirsk, Russia*
[2]*Institute of Problems of Chemical Physics, 142432, Chernogolovka, Moscow Region, Russia*

**Abstract.** Evolution of shock waves in self-bonded silicon carbide bars in the shape of 20 mm x 20 mm square prisms of varying lengths (20 mm, 40 mm, and 77.5 mm) is investigated. The density and porosity of the test specimens were 3.08 $g/cm^3$ and 2%, respectively. Shock waves were generated by detonating a cylindrical shaped (d=40 mm and l=40 mm) stabilized RDX high explosive charge of density 1.60 $g/cm^3$. Embedded manganin gauges at various distances from the impact face were used to monitor the amplitude of shock pressure profiles. Propagation velocity of the stress pulse was observed to be equal to the elastic bar wave velocity of 11 km/s and was independent of the amplitude of the impact pulse. Strong fuzziness of the stress wave front is observed. This observation conforms to the theory on the instability of the shock formation in a finite size elastic body. This phenomenon of wave front fuzziness may be useful for desensitization of heterogeneous high explosives.

**Keywords:** Shock waves, silicon carbide, fuzziness of the wave front.
**PACS:** 46.40.Cd, 47.40.-x.

## INTRODUCTION

The desensitization phenomenon of heterogeneous high explosives under loading with a shock wave having a precursor of smaller amplitude is now well-known [1]. Thereby behavior of high explosive as an inert substance can be achieved under considerably high pressures in the main shock wave. If the initial impulse has a precursor with amplitude insufficient for detonation initiation, pores in heterogeneous HE

collapse, and charges do not detonate even if pressure in the main shock wave considerably exceeds the level of reliable initiation.

Because of that, usage of high modulus ceramic materials for desensitization of HE charges looks very promising. Table 1 shows characteristics of several ceramics, interesting from this point of view.

In ceramics, just like in any other materials, elastic waves can propagate with amplitudes not exceeding the Hugoniot elastic limit, which for

**TABLE 1.** Characteristics of several ceramics

| Material | Density $\rho_0$, $g/cm^3$ | Bulk speed of sound $C_b$, km/s | Thin rod speed of sound $C_R$, km/s | Hugoniot elastic limit HEL, GPa |
|---|---|---|---|---|
| BN | 2.30 | 13.5 | 18.8 | - |
| $B_4C$ | 2.52 | 9.9 | 13.7 | 15 – 20 |
| $Al_2O_3$ | 3.96 | 7.9 | 10.1 | 11.2 |
| BeO | 3.01 | 9.1 | 11.4 | 8.2 |
| SiC | 3.22 | 8.5 | 11.2 | 8 –16 |
| $AlB_{12}$ | 2.54 | 7.4 | 11.7 | 8.7 |

silicate glass lies within the range of 6-10 GPa [2], and reaches 16 GPa for silicon carbide [3].

Such amplitudes are typical for shock waves in solid bodies. Presence of free boundaries influences the character of elastic shock wave evolution. It is known that propagation of elastic waves in finite size bodies exhibits peculiarities related to influence of cross-section inertia [4-6].

If the boundary condition at the end of semi-infinite elastic rod with radius $r$ is defined as a step function of time, then this step is smoothed due to propagation along the rod and we will not have a discontinuous solution in $x$, $t$ plane.

The size $d$ of front fuzziness area can be estimated using the formula $d \approx \sqrt[3]{r^2 x}$ [7]. Here $r$ is the radius of the rod; $x$ is the run of the front. Here the propagation velocity of maximum amplitude area is equal to the thin rod speed of sound in the material.

The possibility of initiation of a stabilized RDX charge using cross section silicon carbide rods of various lengths located between active and passive HE charges was investigated in work [8]. Active and passive HE charges were pressed of stabilized RDX. The experimental assembly is shown in Fig. 1.

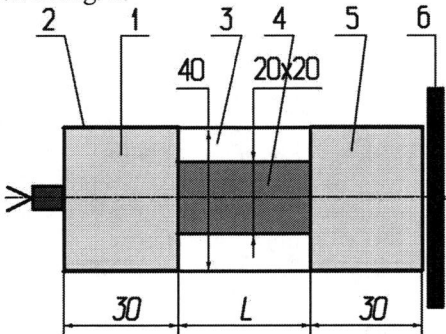

**Figure 1.** Schematic of experimental assembly: 1- active charge (stabilized RDX of density 1.64 g/cm³); 2 – aluminium shell; 3 -air; 4 –ceramic rod; 5 - passive charge (stabilized RDX of density 1.64 g/cm³); 6 – PMMA

Rotating-mirror camera was used to record detonation in the passive charge. Results of these experiments testify that the elastic wave in a ceramic rod has very low initiating ability. Detonation is absent even if length of a rod $L$ equals to 10 mm. Air shock wave and detonation products do not cause detonation as well.

Generally, the distance between charges necessary for reliable propagation of detonation in air is approximately 75 mm for stabilized RDX with density 1.60 g/cm³. Similar distances for water, steel, and aluminum are within 15-20 mm [9].

The goal of presented work is the experimental research of shock waves evolution in these ceramic rods.

## EXPERIMENTAL PROCEDURE

Evolution of shock waves in square rods made of self-bonded silicon carbide and similar ones used in paper [8] has been investigated. Several experiments were conducted to measure the shock wave profiles in ceramic rods. The scheme of experimental assembly for shock wave parameters measurement in silicon carbide rod is shown in Fig. 2.

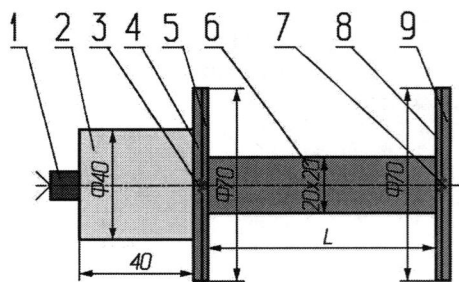

**Figure 2.** Assembly for measurement of shock wave parameters in silicon carbide rod: 1-detonator; 2-RDX charge, $\rho = 1.60$ g/cm³; 3,7- manganin gauges; 4,5,8,9- copper screens; 6-ceramic rod

Square rods made of self-bonded silicon carbide were used; rod thickness was 20 mm x 20 mm, rod length – 20.0, 40.0, 77.5 mm. Density of the samples was 3.08 g/cm³, porosity – less than 2%, free silicon content – 9.5%, free carbon content – 3%. Ends of all rods were polished. The value of Hugoniot elastic limit was estimated at 10 GPa. Shock waves in the rods were initiated by detonating a stabilized RDX high explosive charge 40 mm in height, 40 mm in diameter, having density of 1.60 g/cm³. Thin copper screens were placed between the charge and the rod, and a manganin gauge was mounted between those screens. On the other end of the rod, there were also copper screens with the second gauge mounted between them. The gauges were 30 μm thick, and

**TABLE 2.** Main results of measurements with manganin gauges

| Experiment number | Length of rod $L$, mm | Thickness of copper screens | | | | Maximum pressure in the first gauge $P_1$, GPa | Maximum pressure in the second gauge $P_2$, GPa | Time between the first and the second gauges $\Delta t$, µs |
|---|---|---|---|---|---|---|---|---|
| | | $h_4$, mm | $h_5$, mm | $h_8$, mm | $h_9$, mm | | | |
| 1 | 20.0 | 1.4 | 1.4 | 1.4 | 5.0 | 36.2 | 8.2 | 2.44 |
| 2 | 40.0 | 1.4 | 1.4 | 1.4 | 5.0 | 35.6 | 5.7 | 4.32 |
| 3 | 77.5 | 3.0 | 2.0 | 2.0 | 3.0 | 32.0 | 2.5 | 7.92 |

3. Results for ceramics and high explosive

| Length of rod $L$, mm | Wave height in the rod $P$, GPa | Wave height in the passive HE charge $P_{HE}$, GPa | Experimental value of front fuzziness area $d$, mm | Wave velocity in the rod $D$, km/s | Calculated value of front fuzziness area $d_c$, mm |
|---|---|---|---|---|---|
| 10.0 | ~20.0 | ~8.0 | - | - | 10.8 |
| 20.0 | 7.5 | 2.2 | 3.3 | 11.0 | 13.7 |
| 40.0 | 5.7 | 1.6 | 5.3 | 10.9 | 17.2 |
| 77.5 | 2.5 | 0.7 | 11.0 | 11.1 | 21.5 |

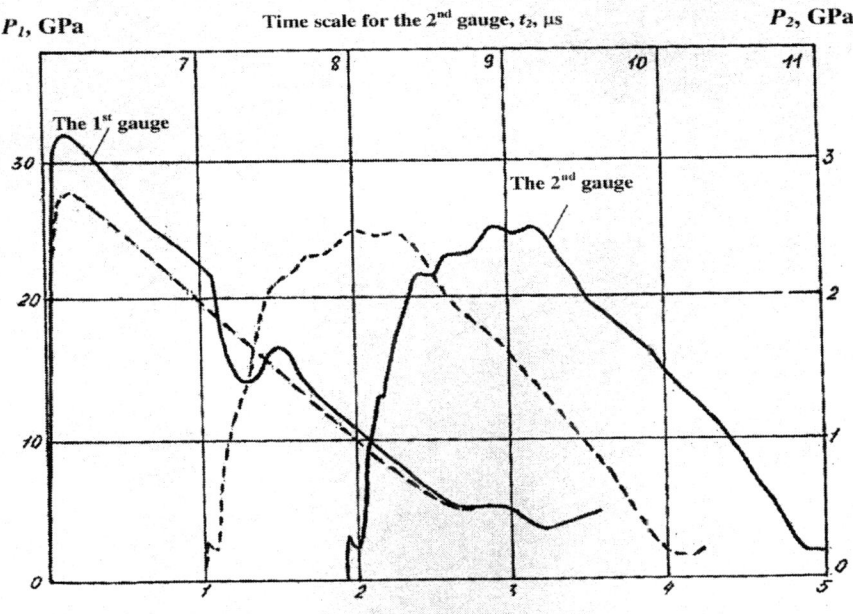

**Figure 3.** Wave profiles on the ends of 77.5 mm ceramic rod:
$P_1$ – pressure for the first gauge; $P_2$ - pressure for the second gauge

were isolated from the screens with 80 µm fluoroplastic coating.

Time measurement error was less than 0.01 µs, pressure measurement error less than 5%, time resolution was about 0.1 µs. Manganin gauges

[10, 11] were used to measure the pressure profiles during propagation of a shock wave along the ceramic rods of different lengths.

Main results of measurements that were recorded by manganin gauges are given in the table 2. Results that were calculated from those experiments for ceramics and high explosive are given in the table 3. Calculations for high explosive were provided to estimate parameters of initiating impulses in passive HE charge for experiments described in [8] (see introduction).

Fig. 3 shows wave profiles measured with manganin gauges on the ends of 77.5 mm ceramic rod. Pressure profiles recorded by the first and the second gauge in copper screens are given as solid contours; profiles calculated for ceramics from experiment are given as dashed contours.

It should be noted that in most experiments of work [8] amplitudes of initial shock waves in passive HE charges (see table 3) exceed the critical pressure required for detonation initiation. The value of critical pressure in stabilized RDX of density 1.67 $g/cm^3$ with shock wave without precursor equals about 1.5 GPa.

## CONCLUSIONS

Analysis of the results allows drawing the following conclusions.

It has been observed that the propagation velocity of compressed impulse is constant along the length of ceramic rod. The velocity does not depend on the pressure amplitude and amounts to 11.0 km/s, which equals to elastic wave velocity in the rod.

Strong fuzziness of the wave front is observed, which conforms to theoretical conception of impossibility of shock wave existence in a finite size elastic body.

Numerical estimates of the size $d$ of front fuzziness area as a function of traversed path conform by order of magnitude to experimentally measured values. For example, for a 77.5 mm rod experimentally measured value $d$=11.0 mm, calculated value $d$=21.5 mm.

It should be noted that significant attenuation of wave amplitude in the initial 20 mm section (more than four times) should be revised by taking into account an important observation on stress wave propagation along bars. The shock wave generated by the explosive propagates in the bar initially as a 1-d strain wave, which transforms to 1-d stress wave along the length bar at about 2-3 times the lateral dimensions of the bar. As this transition of the stress from 1-d strain to 1-d stress happens the bar undergoes lateral tension and begins to fracture. Axial stress amplitude begins to attenuation as reported in the Brar and Bless paper [6]. Further attenuation of the stress wave is relatively small.

The phenomenon of fuzziness of the wave front may be used for desensitization of heterogeneous high explosives.

## REFERENCES

1.  Campbell J. R., Davis W. C., Ramsay J. B., Travis J. R., Phys. of Fluids, vol. 4, no 4, p. 511, 1961.
2.  Kanel G. I., Molodetz A. M., Journal of Technical Physics, vol. 56, no 2, p. 398, 1976.
3.  Grady D. E., "Shock-wave strength properties of boron carbide and silicon carbide", Journal de Physique IV, vol. 4, (C8), pp. 385-391, 1994.
4.  Rayleigh J. W. S., The Theory of Sound, Moscow, Gostechizdat, 1955.
5.  Love A. E. H., A Treatise on the Mathematical Theory of Elasticity, New York, Dover Publications, 1944.
6.  Brar N.S., Bless S.J., "Dynamic fracture and failure mechanisms of ceramic bars", Shock Wave and High-Strain-Rate Phenomena in Materials, pp. 1041-49, Eds. M. A. Meyers et al., Marcel Dekker, Inc., 1992.
7.  Rabotnov Yu. N. The Mechanics of Deformation of The Solid Bodies. Moscow, Nauka, 1979.
8.  Balagansky I. A., Gryaznov E. F. "Desensitization of RDX-Charges after Preshocking by Compression Wave in SiC-Ceramic Rod", Proceedings of International Conference on Combustion, Moscow, vol. 2, pp. 476-478, 1994.
9.  Baum F. A., Stanyukovich K. P., Shechter B. I. The Physics of Explosion, Moscow, Fizmatgiz, 1959.
10. Lyle J. W., Scrivener R. L., and McMillan A. R., "Dynamic Piezoresistive Coefficient of Manganin to 392 kbar", J. Appl. Phys., vol. 40, no. 11, pp. 4663-4664, 1969.
11. Kanel G. I., Utkin A. V., and Fortov V. E., "The Equations of State and Macrokinetics of Decomposition of Solid Explosives in Shock and Detonation Waves", Thermal Physics Reviews, vol. 3, part 3, pp. 1-86.

CP845, *Shock Compression of Condensed Matter - 2005,*
edited by M. D. Furnish, M. Elert, T. P. Russell, and C. T. White
© 2006 American Institute of Physics 0-7354-0341-4/06/$23.00

# NEW PHENOMENA OBSERVED IN PLATE IMPACTS ONTO ALUMINA BARS

## T. Beno[1], S. Bless[1], and S. Nichols[2]

[1]*Institute for Advanced Technology, The University of Texas at Austin, Austin, TX 78759*
[2]*Department of Mechanical Engineering, The University of Texas at Austin, Austin, TX 78712*

**Abstract.** Steel flyer plates were used to impact alumina bars at 275 m/s, nominally. Manganin gauges were used to monitor stress waves in the bars. Geometry of the impact was varied in an attempt to extend gauge records. Gauge life was best improved by careful alignment. The longest gauge records indicated that alumina retains a strength level of about 2 GPa after initial failure. Stress levels of over 5 GPa were obtained with impact-zone confinement.

**Keywords:** Alumina, bar impact, spall.
**PACS:** 47.40.Nm, 62.20.Fe

## EXPERIMENT SETUPS

The experiments performed investigated two different types of alumina, AD995 and AD998. These are trade names of CoorsTek and are 99.5% and 99.8% $Al_2O_3$, respectively.

Three setups were used to test the material. Setup 1, testing both AD995 and AD998, involved steel plate impact onto an unconfined bar. Setup 2 involved a padded steel plate impacting an unconfined bar. AD998 was the material tested. Setup 3 involved a steel plate impact onto a bar that was confined on the impact end. The confinement was provided by a tapered steel cone that slid over the end of the bar. Bar geometry was 12.7 mm in diameter and 101 mm in length. Figure 1 shows a schematic of Setup 1. All impactors were steel plates 50.8 mm diameter by 9.5 mm thick.

Instrumentation for each shot consisted of a manganin gauge to measure in situ stress waves and a high-speed camera to capture fracture phenomena.

## EXPERIMENT RESULTS

All test results are summarized in Table 1. The purpose of Setup 1 was to establish baseline results for alumina. Agreement with previous accepted results [1, 2] gives credibility to later results obtained with new experimental techniques. Gauge and photo results are given in Figure 2.

The purpose of Setup 2 was to try to reduce impact strain rate by padding the flyer plate. It was hoped that these "pillow" impactors would suppress failure damage, thereby reducing attenuation, and promote full stress-wave propagation. The impactor in each experiment was a layered assembly consisting of 0.4 mm PMMA, 0.4 mm aluminum, 0.4 mm copper and 5 mm tungsten heavy alloy. Gauge and photo results are given in Figure 3.

The purpose of Setup 3 was to achieve high stress-wave propagation by reducing impact zone damage. Unlike Setup 2, this setup used mechanical impedance in the form of a confinement cone to suppress failure. The confinement cone was made of steel and covered the front 32 mm of the bar. Gauge and photo results are given in Figure 4.

**TABLE 1.** Shot data summary

| Set-Up | Shot | Target | Peak Gauge Stress (GPa) | Peak Gauge Stress (GPa) | Rise time (μs) |
|--------|------|--------|-------------------------|-------------------------|----------------|
| 1 | 1 | AD995 | 270 ± 5 | 3.6 | 2 |
| 1 | 2 | AD998 | 281 ± 7 | 3.6 | 1.8 |
| 1 | 14 | AD998 | 276 ± 9 | 3.5 | 2.3 |
| 2 | 5 | AD998 | 278 ± 4 | 3.2 | 2.4 |
| 2 | 11 | AD998 | 273 ± 8 | 3.9 | 2.2 |
| 3 | 8 | AD998 | 271± 8 | 3.9 | 2.3 |

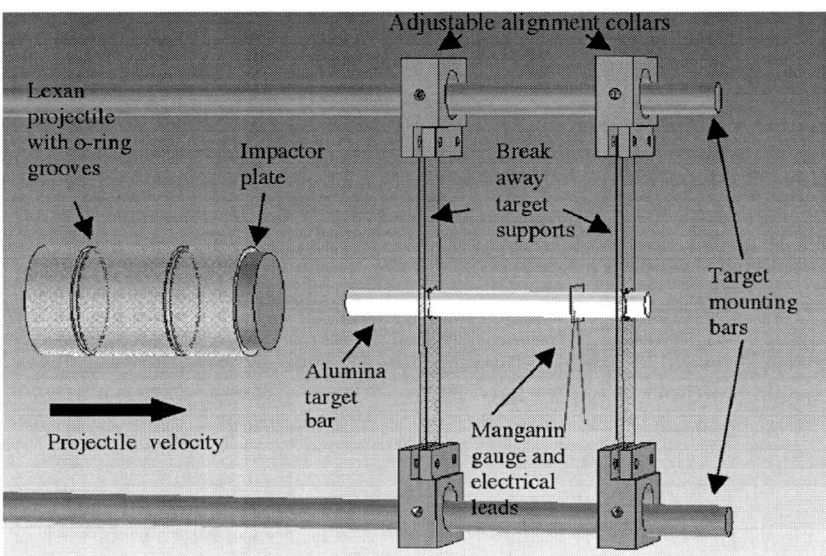

**Figure 1.** Bar impact test setup.

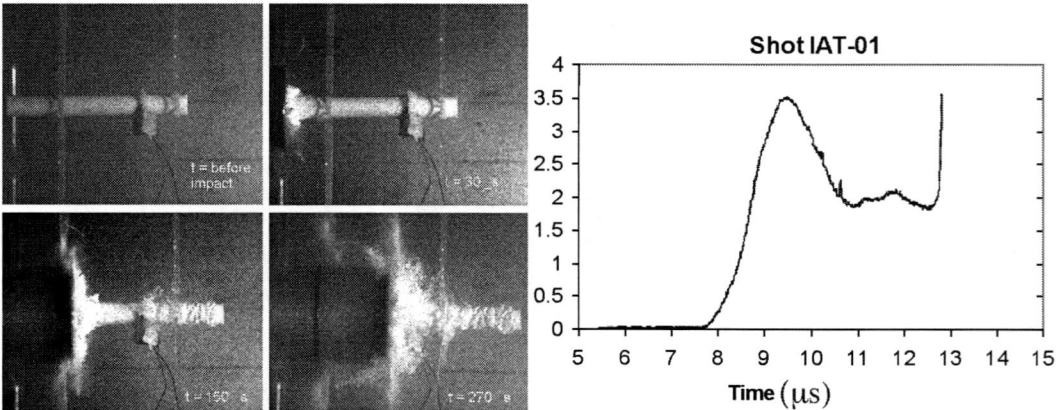

**Figure 2.** Photographs and gauge record from Shot IAT-01.

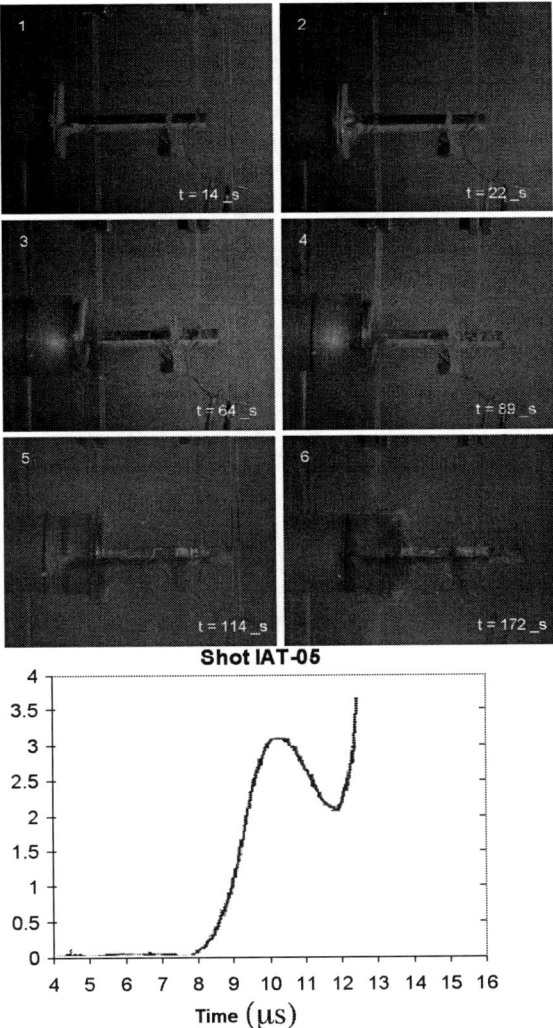

**Shot IAT-05**

Figure 3. Photographs and gauge trace from Shot IAT-05.

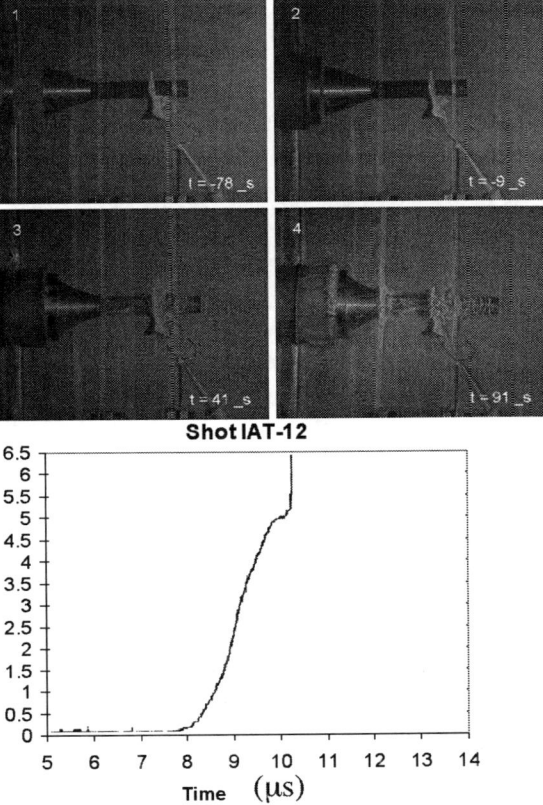

**Shot IAT-12**

Figure 4. Photographs and gauge trace from Shot IAT-13.

## DISCUSSION

Early photographs for Setups 1 and 2 show very similar phenomena. Impact between target and plate drives a radial expansion of the target visible as early as 14 μs in Shot IAT-05. By 30 μs in Shot IAT-01, the outward bulging of the front end of the bar is significant. As the expansion continues, axial cracks in the direction of projectile motion help further expand the target bar. As material near the projectile pulverizes, the particulate alumina is ejected outward. Soon after this initial compressive failure, remarkable failure patterns begin to form in the back of the bar. First seen as singular transverse cracks, failure zones quickly materialize. In Setups 1 and 2, this failure zone is a series of distinct spall layers. As the impact event continues, these layers separate and spread out. In Setup 3, however, this failure zone is not as organized. Individual transverse cracks form, but instead of simply giving rise to spall cracks, longitudinal cracks also develop. This curious "checker pattern" failure was reproducible in each of the confinement cone experiments.

Another point of interest in this rear failure zone is the presence of an "end cap" in each of the photograph series. As the spall region develops, the piece at the very end of the bar stays mostly intact. In Setup 1, this cap is seen flying downrange in late-time photographs. This end-cap phenomena has also been seen in previous bar impact experiments using the brittle plastic Homalite as the target material [3]. End cap velocities were measured: 283 m/s in Setup 1, 173–154 m/s in Setup 2, and 170–256 m/s in Setup 3.

The gauge record for Setup 1 (Shot IAT-01) turned out to be the most complete of this entire investigation. Peak stress measured 3.6 GPa, which agreed with earlier results [1, 2]. After peak stress, the bar is unloaded to roughly 2 GPa, where it levels off. This "relaxed" level may in fact be the strength level of the comminuted alumina material. After residing at this level for a time, the trace goes off record as the gauge is destroyed.

The other gauge records of greatest interest are from Setup 3. Peak stress was clearly over 5 GPa in some shots. This is higher than the 1-D stress yield strength predicted from the HEL using either von Mises or Griffith theory.

If spall occurs, than the spall stress is related to the cap velocity $u$ and gauge stress by $\sigma_s = 2(\sigma_{gauge} - \sigma_{fs})$, where $\sigma_{fs} = \frac{1}{2}Cu$, and $C$ is the sound speed. This formula gives wide scatter. It may be more reasonable to interpret $\sigma_{fs}$ as the amplitude of the stress wave reaching the back of the bar. In that case, these results, combined with data from [4] and two experiments on longer AD998 bars, give the results in Figure 5. This can be interpreted as indicating stress decays with unsupported bar length in these tests, out to a distance of about 150 mm.

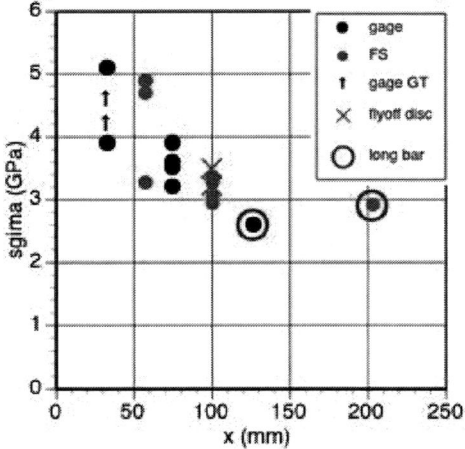

**Figure 5.** Stress vs. unsupported bar length, both gauge data and stress computed from spall caps and fly-off disks [4].

**ACKNOWLEDGMENT**

The research reported in this document was performed in connection with Contract number DAAD17-01-D-0001 with the US Army Research Laboratory. The views and conclusions contained in this document are those of the authors and should not be interpreted as presenting the official policies or positions, either expressed or implied, of the US Army Research Laboratory or the US Government unless so designated by other authorized documents. Citation of manufacturer's or trade names does not constitute an official endorsement or approval of the use thereof. The US Government is authorized to reproduce and distribute reprints for Government purposes notwithstanding any copyright notation hereon.

**REFERENCES**

1. Simha, C. H. M., Bless, S. J., and Bedford, A., "What is the Peak Stress in the Ceramic Bar Impact Experiment?" Shock Compression of Condensed Matter–1999, ed. M. D. Furnish, L. C. Chhabildas, R. S. Hixson, Am Institute of Physics, 615–618, 2000.
2. Cazamias, J. U., Reinhart, W. D., Konrad, C. H., Chhabildas, L. C., and Bless, S. J. "Bar Impact Tests on Alumina (AD995)." *Shock Compression of Condensed Matter 2001*, ed. M. D. Furnish, N. N. Thadhani, Y. Horie, AIP Conference Proceedings 620, pp. 787–790, Melville, New York, 2002.
3. Russell, R., Bless, S., and Beno, T., "Impact Induced Failure Phenomenology in Homalite Bars," Shock Compression of Condensed Matter–2001, American Institute of Physics, 2002.
4. Bless, S. J., Bourne, N. K. "The Effect of Shock Rise Time on Strength in Alumina in 1-D Stress and 1-D Strain." Topical Conference on Shock Compression of Condensed Matter, Portland, OR, July 20–25, 2003.

CP845, *Shock Compression of Condensed Matter - 2005*,
edited by M. D. Furnish, M. Elert, T. P. Russell, and C. T. White

# CONFINED ALUMINA BAR-ON-BAR IMPACT EXPERIMENTS

## N. S. Brar[1] and A. M. Rajendran[2]

[1]*Mechanical and Aerospace Engineering and Research Institute, University of Dayton, OH 45469-0182*
[2]*US Army Research Office, Research Triangle Park, NC 27709*

**Abstract.** In an earlier study on unconfined alumina bar-on-bar impact measured velocity history (using VISAR) data at an impact velocity of 100 m/s showed that the material response is elastic[1]. At higher impact velocities of 220 m/s and 300 m/s, the data suggested the material behavior is inelastic. This study is extended to confined alumina bars. Alumina bars (12.7-mm diameter) were shrunk fit into 3.17 mm thick steel sleeves to provide confinement stress. Axial velocity histories at the far end of the confined AD998 target bar are measured at nominal impact speeds of 200 m/s, 300 m/s, and 500 m/s. Lateral expansion of the confinement sleeve around the impactor and target bars during impact is photographed using a high-speed (Imacon) camera. Peak axial velocities increase from 0.135 mm/µs for unconfined bars to 0.170 mm/µs for confined bars at a nominal impact velocity of 200 m/s. At an impact velocity of ~300 m/s peak axial velocity of confined bar increase to 0.200 mm/µs from 0.170 mm/µs for unconfined bar. At ~500 m/s the confinement shatters on impact and peak axial velocity is measured to be almost same as that for ~200 m/s. These results show that the confinement provided by a 3.17-mm thick steel sleeve to alumina bar enhances its impact response for impact velocities to ~300 m/s and confined alumina behaves as inelastic at the lowest impact velocity of 200 m/s.

**Keywords:** Alumina bar, high speed photography, material strength, VISAR studies, dynamic fracture
**PACS:** 07.35.+k, 07.68.+m, 52.50.+p, 62.30.+d

## INTRODUCTION

Bar impact has been widely used to study failure in brittle materials. Bar impact experiments are conducted by impacting a specimen bar, about 8-10 diameters long, either with a flyer plate or a bar impactor (bar-on-bar impact) of the same material or of a material of known Hugoniot. Strain rates produced in the bar target ($\sim 10^{3\text{-}4}$/s) bridge the gap between the strain rates achieved in split Hopkinson bar ($10^2$-$10^3$/s) and flyer plate impact experiments ($10^5$-$10^7$/s).

Wise and Grady[2] performed impact tests on unconfined and confined (in close-fitting tantalum sleeve) Coors 99.5% alumina bars with aluminum flyer plates launched at velocities ranging from 1035 to 2182 m/s. The magnitude of confinement stress was not quantified since the bars were not shrunk fit into metal sleeves.

They measured a maximum in-situ axial stress of 3.15 GPa in unconfined bars, irrespective of impactor velocity, as observed by Cosculluela et al.[3] on similar alumina bars. Furthermore, the maximum in-situ stress value of 3.15 GPa was lower than the dynamic yield strength of 4.3 GPa of the material, suggesting a (confining) pressure dependence of the yield strength of alumina. Maximum in-situ axial stress in confined alumina bars is about twice (6.1 GPa) the value for the unconfined bars. This increase in measured axial stress was interpreted by the authors in terms of confinement, consistent with an expected upper dynamic limit equal to the HEL of alumina (6.2 GPa). Chhabildas et al.[4] extended Wise and Grady's study to determine the effect of confinement (in shrunk-fit steel cylinder) on in-situ axial stress for 99.5% alumina bars impacted by steel plates at 300-366 m/s. Measured in-situ axial stress in confined alumina bars was in the range of 4.6-5.1 GPa, significantly higher than that in the case of unconfined bars, in agreement with earlier studies. Tensile waves generated as a result of lateral release in a sleeved bar target are partially eliminated due to confinement during the initial loading. Simha [5] and Simha et al. [6-7] also reported an increase in measured in-situ axial stress (4.2 GPa) in 12.7-mm diameter confined (in a steel sleeve) Coors 99.5% alumina (6 diameters long) bars compared to that (3.7 GPa) in unconfined bars.

The results presented in this paper are an extension of the recently completed series of unconfined AD998 alumina bar-on-bar impact experiments at impact velocities in the range 100 -300 m/s[1]. High speed (Imacon) camera was employed to examine the fracture and failure of both impactor and target bars, in conjunction with, in-situ axial velocity history measurements of the target bar using VISAR. Velocity history data at an impact velocity of 100 m/s suggested the response of alumina is inelastic. In the present experiments alumina imactor and target bars are confined in 3.17 mm thick steel sleeves. Measured axial velocity histories in the confined bar-on-bar impact experiments would provide comprehensive data for generating a constitutive material model for the bar material. High-speed photographs will provide valuable information on the role of confinement stress on axial splitting of impacted ceramic bars.

## EXPERIMENTAL PROCEDURE

Alumina AD998 bars of 12.7-mm diameter were purchased from Coors Ceramic Company, Golden, CO. Both the impactor (50-mm long) and the target (125-mm long) alumina bars were shrink fitted in 3.17-mm thick 4340 steel sleeves at the University of Dayton Research Institute. The inside diameter of the steel sleeves was 0.025-mm smaller than the diameter of the alumina bars. A schematic of the alumina bar-on-bar impact experiment is shown in Figure 1.

Four experiments were performed at nominal impact speeds of 200 m/s, 300 m/s, and 500 m/s at the gas gun facility of the University of Cambridge, UK. Axial velocity histories at the far end of the confined target bar were measured using VISAR. Lateral expansion of the confinement sleeve around the impactor and target alumina bars resulting from tensile fracture of the alumina bar during impact was monitored using a high speed (Imacon) camera. The shot matrix is summarized in Table 1.

Table 1. Summary of the impact shots.

| Shot No. | Impactor Velocity (m/s) | Peak Axial Velocity (mm/µs) |
|---|---|---|
| 040519A | 203 | 0.170 |
| 040519B | 293 | 0.200 |
| 040520B | 300 | 0.200 |
| 040520A | 511 | 0.180 |

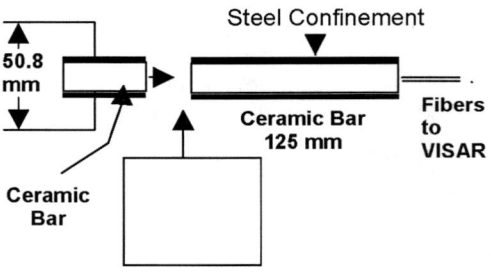

**FIGURE 1.** Schematic of the confined alumina bar-on-bar impact experiment.

## RESULTS AND DISCUSSION

In this paper we present a brief version of results and discussion; the details on experimental configuration, photographs of the impactor and target bars before and after impact, and a sequence of high speed photographs of the impactor and target bars during impact are given in Reference 8. A representative sequence of high speed photographic images at $10^5$ frames/s from shot 040520B (~300 m/s) are shown in Figure 2. This photograph and the high speed photographs taken during other shots are intended to be used by the material modelers to validate the material model for alumina through numerical simulations of the impact experiments. Recovered confined impactor and target bars at the impact velocity of 203 m/s show bulging of both the bars at the impact faces. In the two shots at 293 m/s the confining steel sleeve was cracked and mushroomed near the impact face. At the highest impact velocity of 511 m/s the steel confinement sleeve was severely split from its impact face down to the middle of the bar.

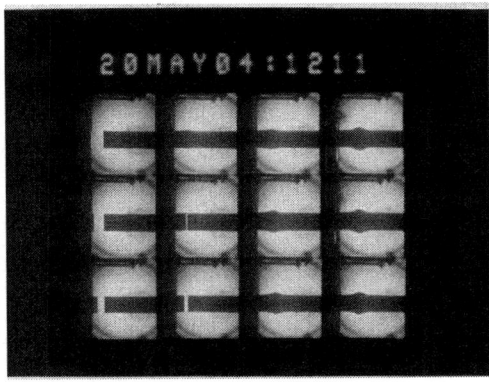

**FIGURE 2.** High speed photographic images of the impactor bar shot at 300 m/s and target alumina bar taken at $10^5$ f/s. Frames begin from the top left of the photograph.

Measured axial (particle) velocity-time data for the four shots are shown in Figure 3. The peak axial velocity of a confined bar at the lowest impact velocity of 203 m/s is about 0.170 mm/μs compared to 0.145 mm/us for an unconfined alumina bar at a similar impact velocity of 220 m/s (Fig.4). Peak axial velocities for the two shots (040519B and 040520B) at impact velocity of 293 m/s were about 0.2mm/μs and the values for the two shots agree within 2%. These data are compared to those for unconfined bar (Fig.5). For the shot at the highest impact velocity of 511 m/s the peak axial velocity was measured to be about 0.18 mm/μs (Fig. 3). In the previous study on unconfined alumina bar-on-bar impact the response of an alumina bar was elastic only at the lowest nominal impact velocity of 100 m/s. At the higher nominal impact velocities of 200 m/s and at 300 m/s the

response was inelastic. In the present study on confined alumina bar-on-bar impact we find that the response of confined alumina is inelastic at impact velocities of 203 m/s, 293 m/s, and 511 m/s.

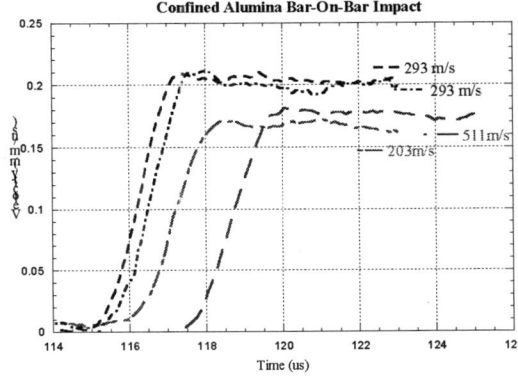

**FIGURE 3.** Axial velocity histories at impact velocities of 203 m/s, 293 m/s, and 511 m/s.

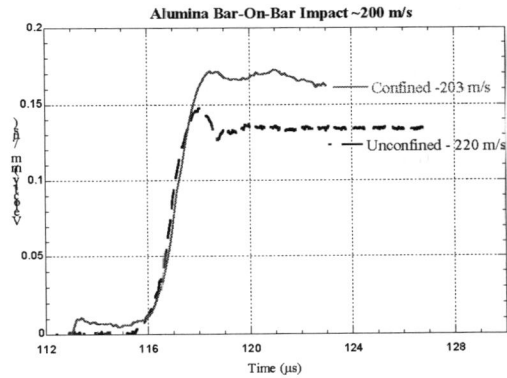

**FIGURE 4.** Comparison of axial velocity histories of confined (203 m/s) and unconfined (220 m/s) alumina bar-on-bar impact.

These results provide unambiguous evidence that the impact response of alumina bars is significantly enhanced by the confinement provided by 3.17-mm thick steel sleeve for impact velocities to ~300 m/s. At higher impact velocity (~500 m/s) the confinement shatters due to axial splitting of alumina bar on impact and the measured axial velocity of the confined bar is about the same as for impact velocity of ~200 m/s.

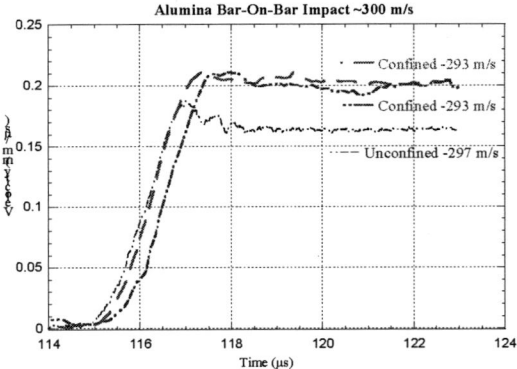

**FIGURE 5.** Axial velocity histories of confined and unconfined aluimna bars at ~300 m/s.

## CONCLUSIONS

High speed photographs of the confined alumina bar-on-bar impact experiments are presented. Axial velocity history data on confined alumina bars obtained using a VISAR at impact velocities of ~200 m/s and ~300 m/s are compared to those obtained for the unconfined bars[1]. Axial peak velocity data for the two confined bar-on-bar impact shots at an impact velocity of 293 m/s (Figure 3) agree within 2 %. Average peak particle velocities at the three impact velocities exhibit an inelastic response for confined AD-998 alumina bars. Axial velocity results suggest that confinement provided by 3.17 thick steel sleeve around alumina bar improves its impact response for impact velocities to ~300 m/s. Axial velocity data and high speed photographs in impact experiments on confined alumina bars combined with similar data on unconfined alumina bars will be useful to validate the material model for AD-998 alumina.

## ACKNOWLEDGEMENTS

Research support by US Army Research Office, under Grant DAAD19-03-1-0291 is sincerely acknowledged.

## REFERENCES

1. Brar, N. S., Ceramic bar impact experiments for improved material model," Shock Compression of Condensed Matter-2003, Eds. M. D. Furnish et al., AIP Conference Proceedings 706, pp. 727-30, 2004.
2. Wise, J. L. and Grady, D. E., Dynamic multi-axial impact response of ceramic rods, High Pressure Science and Technology-1993, Eds. S. C. Schmidt et al., AIP Conference Proceedings 309, New York, pp. 777-80, 1992.
3. Cosculluela, A., Cagnoux, J., and Collombet, F., Two types of experiments for studying uniaxial dynamic compression of alumina, Shock Compression of Condensed Matter-1991, Eds. S. C. Schmidt et al., Elsevier Science Publishers, pp. 951-54, 1992.
4. Chhabildas, L. C., Furnish, M. D., Reinhart, W. D., and Grady, D. E., Impact of AD995 alumina rods, Shock Compression of Condensed Matter-1997, Eds. S. C. Schmidt et al., AIP Conference Proceedings 429, New York, pp. 505-08, 1998.
5. Simha, C. M., High rate loading of a high purity ceramic-1-D stress experiments and constitutive modeling, Ph. D. Thesis, University of Texas, Austin, 1998.
6. Simha, C. M., Bless, S. J., and Bedford, A., What is the peak stress in ceramic bar impact experiments?, Shock Compression of Condensed Matter-1999, Eds. M. D. Furnish et al., AIP Conference Proceedings 505, pp. 615-18, 2000.
7. Simha, C. M., Bless, S. J., and Bedford, A., Computational modeling of the penetration response of a high-purity ceramic, Int. J. of Impact Engineering, **27**, pp. 65-86, 2002.
8. Brar, N. S., Confined Alumina bar-on-bar impact experiments, University of Dayton Research Institute technical Report, UDR-TR-2004-00157, October, 2004.

CP845, *Shock Compression of Condensed Matter - 2005*,
edited by M. D. Furnish, M. Elert, T. P. Russell, and C. T. White
© 2006 American Institute of Physics 0-7354-0341-4/06/$23.00

# DELAYED FAILURE IN A SHOCK LOADED ALUMINA.

## G.A. Cooper, J.C.F. Millett, N.K. Bourne* and D.P. Dandekar[‡].

*Defence Academy of the UK, Cranfield University, Shrivenham, Swindon, SN6 8LA. United Kingdom.*

*\*University of Manchester, Sackville Street, Manchester, M60 1QD. United Kingdom.*

[‡] *Army Research Laboratory, Weapons Materials Directorate, Aberdeen Proving Ground, Maryland 21005-5069. U.S.A.*

**Abstract.** Manganin stress gauges have been used to measure the lateral stress in a shock-loaded alumina. In combination with known longitudinal stresses, these have been used to determine the shear strength of this material, behind the shock front. The two-step nature of the lateral stress traces shows a slow moving front behind the main shock, behind which shear strength undergoes a significant decrease. Results also show that this front decreases markedly in velocity as the HEL is crossed, suggesting that limited plasticity occurs during inelastic deformation. Finally, comparison of measured shear strengths with other aluminas shows a high degree of agreement.

**Keywords**: Shock, alumina, failure, shear strength
**PACS**: 62.50

## INTRODUCTION

The shock response of alumina-based ceramics has attracted attention for the past few decades due to their use as armour materials. One of the more important features concerning the shock response of glasses is that of the failure wave. This was first observed by Razorenov *et al.* [1] who noticed small reload signals superimposed on the main velocity trace (VISAR) in K19 glass. This they interpreted as the result of release from the rear surface interacting with a moving boundary behind the main shock. As this was recorded as a reload, it was suggested that this was due to a reduction of impedance across the boundary, caused by failure of the material. Subsequent measurements of spall strength and shear strength in soda-lime glass [2] showing that both reduced significantly behind this front confirmed its existence. With regard to the latter,

shear strengths were determined by measuring the lateral component of stress ($\sigma_y$), and in combination with the known Hugoniot stress ($\sigma_x$), the shear strength ($\tau$) be determined thus,

$$2\tau = \sigma_x - \sigma_y, \qquad 1.$$

where the failure wave manifests itself as an increase in lateral stress trace. Recently we have observed such signals in polycrystalline ceramics such as SiC [3] and alumina [4]. Previous work by Rosenberg and Yeshurun [5] also observed reload signals superimposed on rear surface measurements in a low purity alumina, although at the time these were not interpreted as failure waves.

In this work, we investigate the lateral stress response in AD 995, a high purity alumina, and the possibility that failure waves may be a result of the shock loading of this material.

## EXPERIMENTAL

All shots were performed using a 50 mm bore, 5 m long single stage gas gun [6]. 60 mm x 60 mm x 8 mm plates of AD995 alumina were sectioned in half and Manganin stress gauges (MicroMeasurements J2M-SS-580SF-025), introduced such that they were sensitive to the lateral component of stress. The targets were reassembled using a low viscosity epoxy adhesive and held in a special jig for a minimum of 12 hours. The gauges were 2 mm from the impact face. Voltage data from the gauges were converted to stress using the methods of Rosenberg and Partom [7] with a modified analysis that does not require prior knowledge of the impact conditions [8]. Afterwards, the impact surfaces were lapped to a flatness of 5 optical fringes over 50 mm. Impact stresses were generated using 5 and 10 mm dural (aluminium alloy 6082-T6) and copper flyers, flat and parallel to ±5 μm, in the range 4.6 to 18.4 GPa, with velocities from 402 to 962 m s$^{-1}$. A schematic diagram of the target assembly is presented in Fig. 1. Acoustic properties were measured using quartz transducers operating at 5 MHz with a Panametrics 5077 PR pulse receiver. These are summarised in Table 1.

**Table 1**. Acoustic Properties of AD 995 Alumina

| $c_L$ (mm μs$^{-1}$) | $c_S$ (mm μs$^{-1}$) | $\rho_0$ (g cm$^{-3}$) | $v$ |
|---|---|---|---|
| 10.66±0.03 | 6.28±0.03 | 3.89±0.01 | 0.234 |

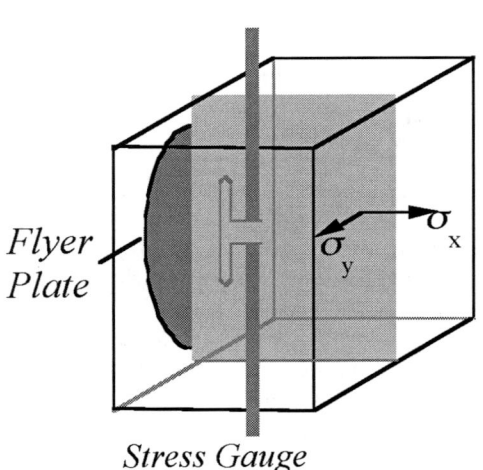

*Flyer Plate*

*Stress Gauge*

**Figure 1.** Gauge placement and specimen configuration.

In Fig. 2 we present the Hugoniot of AD 995 alumina [9] in stress ($\sigma_x$) particle velocity ($u_p$) space.

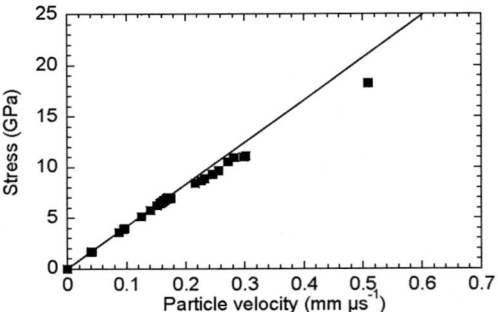

**Figure 2.** Stress versus particle velocity for AD 995.

The straight line fit is according to the elastic response,

$$\sigma_x = \rho_0 c_L u_p, \qquad\qquad 2.$$

where $c_L$ is the longitudinal sound speed and $\rho_0$ is the ambient density.

## RESULTS AND DISCUSSION

In Fig. 3, we show lateral gauge traces taken over the impact stress range 4.6 to 10.4 GPa.

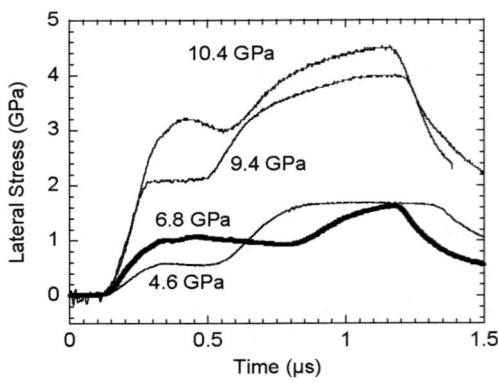

**Figure 3.** Lateral gauge traces for AD 995. Gauges are 2 mm from impact.

In all cases, lateral gauge traces show the two-step nature that is indicative of the presence of a front across which the shear strength ($\tau$) reduces, *i.e.* the failure wave, as shown through equation 1. The most interesting feature of this figure concerns the differences between the gauge traces taken at 4.6 and 6.8 GPa. One might expect that the failure wave velocity would increase as the impact stress increases, and this indeed has been observed in other aluminas [4]. However, it can be seen that in contrast, failure wave velocity appears to decrease as stress increases. It is interesting to note that this occurs as the quoted Hugoniot Elastic Limit (HEL) of *ca*. 6.7 GPa [9] is crossed, and suggests a possible explanation. Grady [10] has proposed that ceramics may undergo two failure regimes, in a brittle manner at lower strain-rates, with some degree of plasticity at higher strain-rates. Thus, if a degree of plasticity does occur at higher strain-rates (such as those above the HEL), it would be possible for that plasticity to reduce the failure velocity, thus increasing the time spent at the first step in lateral stress, as shown in Fig. 3. Limited dislocation generation in alumina [11] and other ceramics [12] have been noted previously, hence this hypothesis is possible. Bourne [13], in collating the spall response of a number of different aluminas, has also observed that in purer aluminas (99.5% pure and above) in some cases a finite spall strength has been maintained at up to three times the HEL. This would further suggest that deformation in such materials is at least in part accommodated by a non-brittle fracture based mechanism. Finally, it is interesting to note that exactly the same phenomenon has also been observed in a silicon carbide [3] although it was not commented upon at the time.

In Fig. 4 we show a single lateral gauge trace for an impact at a stress of 18.4 GPa.

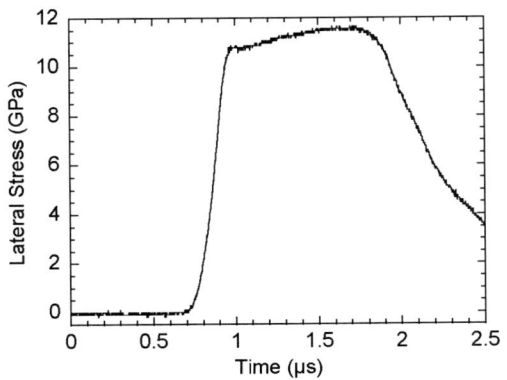

**Figure 4.** Lateral gauge traces for AD 995. Impact stress is 18.4 GPa.

We have chosen to show this particular trace separately to aid clarity. Unlike the previous traces, this shows no evidence of the two-wave structure that suggests the presence of a failure wave. In contrast, it would seem likely that failure occurs in the shock front itself, given the higher impact stress. Study of other brittle materials such a filled lead glasses [14] and other alumina compositions [4] indicate that this response occurs somewhere around twice the HEL. Given the high stress at which this measurement was made, it would seem a possibility.

Finally, in Fig. 5, we have taken the lateral stress measurements and used them to determine the shear strength of AD 995 ahead of and behind the failure wave. We have included similar measurements from two other aluminas as a comparison.

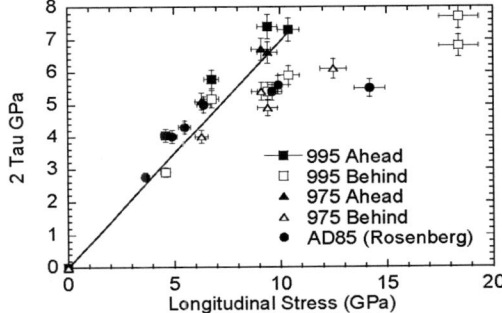

**Figure 5.** Shear strength versus impact stress for three aluminas.

The elastic response has been included as a straight line, calculated via the Poisson's ratio, thus,

$$2\tau = \frac{1-2\upsilon}{1-\upsilon}\sigma_x \qquad 3.$$

It can be seen that the shear strengths ahead of the failure wave in both AD 995 and the lower purity 975 are in good agreement with the calculated elastic response. The agreement between these two materials is also good behind the failure front as well. We note the agreement with the measurements of Rosenberg [15] in AD 85 although in this latter work, failure waves were not observed, although this may have been due to the precise placement of the gauges within the target assembly. However, Rosenberg and Yeshurun [5] did observe small reload signals on back surface gauge traces in this material, another indication of failure waves, although these were not interpreted as such at the time. We would also point out that these results are near identical to those of Munson and Lawrence [16] who determined shear strength in a 99.9% pure alumina via rear surface interferometry. The similarity between compositions was commented upon in a previous paper [4], where it was suggested that this be due to shear strength being controlled by the alumina grains, with other microstructural features such as porosity and glassy phases having a secondary effect.

## CONCLUSIONS

The shock response of a high purity alumina, AD 995 has been monitored using manganin stress gauges, mounted in such orientation as to be sensitive to the lateral component of stress. Between impact stresses of 4.6 and 10.4 GPa, a two-step gauge trace is noted, indicating that a second front travels behind the main shock, behind which shear strength drops, the failure wave. We note that as the HEL is crossed, the failure wave velocity appears to decrease when compared to a similar trace just below the HEL. We have suggested that this be due to a limited amount of plasticity (dislocation motion and generation and/or twinning) above the HEL that decreases the amount of cracking and thus reduces the failure wave velocity. At higher stresses (18.4 GPa), the lateral stress measured shows only a single step. In this case failure occurs within the shock front itself. Finally we compare the shear strengths ahead of and behind the failure wave in AD 995 with those taken from other aluminas of differing purities. Results indicate that the degree of agreement in both cases is excellent. We have suggested that the shear strength of shock-loaded alumina is controlled by the alumina grains themselves, with other factors such as glassy phases or porosity having a secondary effect.

## ACKNOWLEDGMENTS.

The authors would like to thank Richard Hall of Cranfield University for valuable technical assistance.

## REFERENCES.

1. S. V. Razorenov, G. I. Kanel, V. E. Fortov and M. M. Abasemov, High Press. Res., 1991; 6: 225-32.
2. N. S. Brar, Z. Rosenberg and S. J. Bless, Journal de Physique IV, 1991; Colloque C3: 639-44.
3. N. Bourne, J. Millett, N. Murray and Z. Rosenberg, J. Mech. Phys. Solids, 1998; 46: 1887-908.
4. J. C. F. Millett and N. K. Bourne, J. Mater. Sci., 2001; 36: 3409-14.
5. Z. Rosenberg and Y. Yeshurun, J. Appl. Phys., 1986; 60: 1844-6.
6. N. K. Bourne, Meas. Sci. Technol., 2003; 14: 273-8.
7. Z. Rosenberg and Y. Partom., J. Appl. Phys., 1985; 58: 3072-6.
8. J. C. F. Millett, N. K. Bourne and Z. Rosenberg, J. Phys. D. Applied Physics, 1996; 29: 2466-72.
9. D. P. Dandekar and P. Bartkowski, In: S. C. Schmidt, J. W. Shaner, G. A. Samara and M. Ross, S. C. Schmidt, J. W. Shaner, G. A. Samara and M. Rosss. High Pressure Science and Technology 1993. New York: American Institute of Physics; 1994. p. 733-6
10. D. E. Grady, In: S. C. Schmidt and W. C. Tao, S. C. Schmidt and W. C. Taos. Shock Compression of Condensed Matter 1995. Woodbury, New York: American Institute of Physics; 1996. p. 9-20
11. J. Cagnoux and F. Longy, In: S. C. Schmidt and N. C. Holmes, S. C. Schmidt and N. C. Holmess. Shock Waves in Condensed Matter 1987. New York: North-Holland; 1988. p. 293-6
12. W. D. Winkler and A. J. Stilp, In: S. C. Schmidt, R. D. Dick, J. W. Forbes and D. G. Tasker, S. C. Schmidt, R. D. Dick, J. W. Forbes and D. G. Taskers. Shock Compression of Condensed Matter 1991. Amsterdam: North-Holland; 1992. p. 475-46
13. N. K. Bourne, Proc. R. Soc. Lond. A, 2001; 457: 2189-205.
14. N. K. Bourne, J. C. F. Millett and Z. Rosenberg, J. Appl. Phys., 1996; 80: 4328-31.
15. Z. Rosenberg, D. Yaziv, Y. Yeshurun and S. J. Bless, J. Appl. Phys., 1987; 62: 1120-2.
16. D. E. Munson and R. J. Lawrence, J. Appl. Phys., 1979; 50: 6272-82.

CP845, *Shock Compression of Condensed Matter - 2005*,
edited by M. D. Furnish, M. Elert, T. P. Russell, and C. T. White
© 2006 American Institute of Physics 0-7354-0341-4/06/$23.00

# THE SHOCK INDUCED EQUATION OF STATE OF TWO FERROELECTRIC CERAMICS.

## D. Deas, J.C.F. Millett*, N.K. Bourne‡.

*AWE, Aldermaston, Reading, RG7 4PR. United Kingdom.*

**Defence Academy of the United Kingdom, Cranfield University, Shrivenham, Swindon,
SN6 8LA. United Kingdom.*

‡*University of Manchester, Sackville Street, Manchester, M60 1QD. United Kingdom.*

**Abstract.** Manganin stress gauges have been used to determine the Hugoniots of two ferroelectric ceramics, lead zirconium titanate (PZT) and a similar material modified with tin (PSZT). Comparison with previously published data shows close agreement between our results for PZT and earlier work. The Hugoniot Elastic Limit has been determined, and also agrees with previous data. In the case of PSZT, the Hugoniot in terms of stress and particle velocity is similar to PZT. In terms of elastic wave velocity – particle velocity, results show an overall increase, in contrast to PZT, where wave speed was observed to decrease with increasing particle velocity.

**Keywords**: Shock, PZT, PSZT, Hugoniot, ferroelectric
**PACS**: 62.50

## INTRODUCTION

The behaviour of ferroelectric materials under shock loading conditions is of interest as this can result in the release of high electric charges in suitably poled specimens. One group of materials that has attracted attention is based on solid solutions of lead zirconate and lead titanate, with small additions of niobium or tin, in particular, with a composition ratio of Zr:Ti of 95:5, with approximately 2% niobium. This material has been studied under quasi-static conditions [1], showing that it has a rhombohedral structure with a ferroelectric (FE) nature under ambient conditions, but transforming to an orthorhombic, antiferroelectric (AFE) structure at around 0.3 GPa [2]. Shock loading experiments have also been performed. Dick and Vorthman [3] investigated a relatively porous material at 1.6 GPa, showing that

around 85% of the bound charge was released, which they interpreted as an incomplete transformation to the AFE phase. Setchell [4] examined the mechanical shock response as a function of porosity. His results showed that even below the Hugoniot Elastic Limit (HEL), the calculated shock response using measured elastic properties over estimated the actual results, as these were collected above the FE-AFE transformation. He also demonstrated that the HELs of these materials were sensitive to porosity, increasing from 1.5 GPa to 3.5 GPa over the density range 6.94 to 7.66 g cm$^{-3}$. In this work, we measure the Hugoniots of two such materials, with and without additions of tin, and compare the results to existing data.

## EXPERIMENTAL

Specimens of PZT and PSZT were supplied in

the form of lapped plates 25 mm x 50 mm x 10 mm, hence two plates were glued together to form a single plate. A manganin stress gauge (MicroMeasurements type LM-SS-125CH-048) was supported on the back of the ceramic plate with 12 mm of polymethylmethacrylate (PMMA). The gauge was further backed off by *ca*. 1.8 mm of PMMA to provide extra protection. A second gauge was supported on the front of the target assembly with a 1 mm plate of either dural (aluminium alloy 6082-T6) or copper. Gauge calibrations were according to Rosenberg *et al.* [5] Shock stresses were generated by the impact of 5 mm flyer plates of dural and copper, in the velocity range 144 to 487 m s$^{-1}$. Impact velocities were measured via the shorting of sequentially mounted pairs of pins to an accuracy of *ca*. 0.5%. A schematic of the target assembly is shown in Fig. 1.

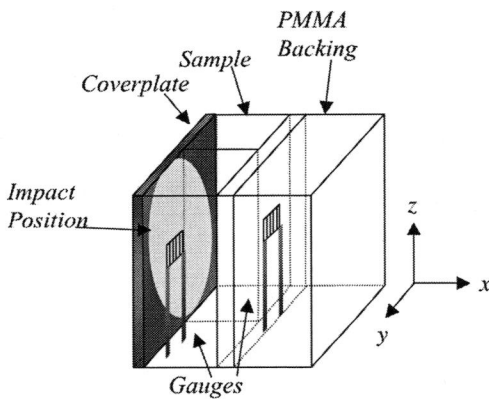

**Figure 1.** Target assembly and gauge placement.

## MATERIALS

PZT was a lead zirconium titanate with small additions of niobium. Its composition in atomic ratios was $Pb_{0.991}(Zr_{0.940}Ti_{0.042})Nb_{0.018}O_3$. PSZT was modified with quantities of tin to give a composition of $Pb_{0.993}(Zr_{0.802}Ti_{0.049})Nb_{0.016}Sn_{0.133}O_3$. Elastic properties are shown in Table 1.

**Table 1**. Acoustic Properties of ferroelectric ceramics.

|      | $c_L$ (mm $\mu s^{-1}$) | $c_S$ (mm $\mu s^{-1}$) | $\rho_0$ (g cm$^{-3}$) |
|------|-------------------------|-------------------------|------------------------|
| PZT  | 4.35±0.03               | 2.61±0.03               | 7.71±0.01              |
| PSZT | 4.32±0.03               | 2.60±0.03               | 7.84±0.01              |

## RESULTS AND DISCUSSION

In Fig. 2, we present representative gauge traces from plate impact experiments on PZT and PSZT.

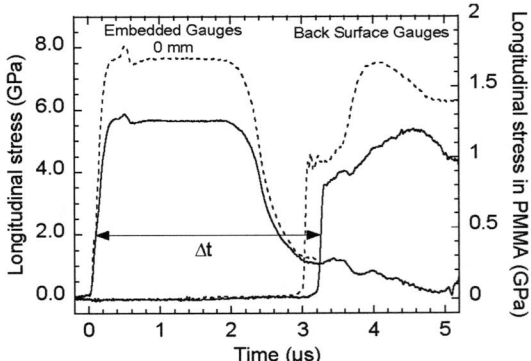

**Figure 2.** Gauge traces from PZT and PSZT. The solid traces are PZT (5 mm copper flyer at 361 m s$^{-1}$) and dotted, PSZT (5 mm copper flyer at 484 m s$^{-1}$).

Both sets of traces have features in common. At the 0 mm position, only the amplitude of the signal can be used since the shock has only passed through *ca*. 1 mm of coverplate material. However, the height, in combination with the particle velocity ($u_p$) determined from impedance matching considerations have been used to measure the Hugoniot in terms of shock stress ($\sigma_x$) and $u_p$, discussed later in the text. The back surface traces, however show a break in slope at *ca*. 0.8 GPa for PZT and 0.9 GPa for PSZT. We believe that these represent the transition from elastic to inelastic behaviour, *i.e.* the HEL. These values have of course been measured in PMMA ($\sigma_{PMMA}$), and thus have to be converted to in-material values using the known impedances of the sample ($Z_P$) and PMMA ($Z_{PMMA}$) from the relation,

$$\sigma_{HEL} = \frac{Z_P + Z_{PMMA}}{2Z_{PMMA}} \sigma_{PMMA}. \qquad 1.$$

This is further complicated by the fact that all experiments were carried out above the phase transformation stress of 0.3 GPa [2], and thus ambient impedances could not be used. Instead, the dynamic property was used, based on the known positions of the gauges through the specimen thickness, and taking into account the dynamic response of PMMA [6] as the back surface gauge was backed into the PMMA backing by 1.8 mm. Thus wave speed (*c*) was determined as a function of particle velocity, and the results presented in Fig. 3.

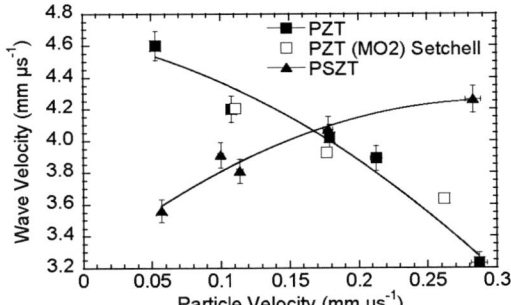

**Figure 3.** Wave speed versus particle velocity for PZT and PSZT.

Note that we have included the equivalent data from Setchell [4], selecting his mixed oxide-2 (MO2) material as being closest in density (7.66 g cm$^{-3}$) to our own PZT. In the case of both PZT materials, it can be seen that wave velocity drops with increasing particle velocity, and that there is close agreement between the two sets of data. We believe that this is a manifestation of the FE/AFE transformation. At 1.6 GPa, Dick and Vorthman [3] demonstrated that the transformation was incomplete. Therefore, as the particle velocity (*i.e.* stress) increases, there will result in an increasing amount of the AFE phase within the microstructure. If the wave speed in the AFE phase is significantly lower than that of the FE phase, an increase in AFE volume fraction will result in a decrease in the overall wave speed of the sample. In contrast, with PSZT, it can be seen that wave velocity *increases* with particle velocity. It is likely that the presence of tin modifies the properties of the AFE phase with regard to its wave speed, hence in this case if the tin modified AFE phase has a higher wave speed than its FE counterpart, then an increase in wave speed with increasing stress (and particle velocity) would be expected. However, further comments cannot be made as the influence of tin in PZT is as yet unknown.

In Fig. 4, we present the Hugoniots of PZT and PSZT in stress-particle velocity space.

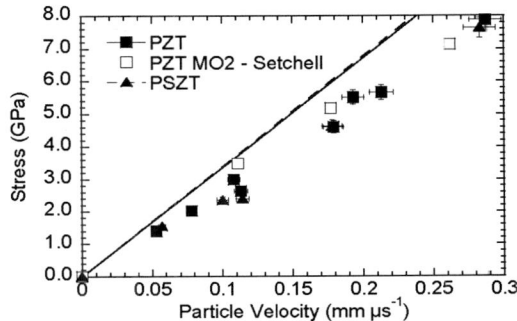

**Figure 4.** Shock Hugoniots of PZT and PSZT. The straight lines are the calculated elastic response using the ambient pressure properties.

From this figure, it can be seen that the Hugoniots of both materials, and that of Setchell's MO2 material [4] are in close agreement. Also observe that in all three materials, the measured stresses lie below the calculated elastic response, according to,

$$\sigma_x = \rho_0 c_L u_p, \qquad\qquad 2.$$

where $\rho_0$ and $c_L$ are the ambient densities and longitudinal sound speeds respectively. This is in agreement with the observations of Setchell [4] who pointed out that as the FE/AFE phase transformation occurs at *ca.* 0.3 GPa, measurements made above this stress cannot be expected to agree with elastic calculations made using properties made at ambient pressures, as the shock measurements will have been made in material that has already begun to transform.

Finally, in Fig. 5, we address the issue of the HEL in these materials. Earlier in the text we pointed out that as these tests were performed above the phase transformation pressure, the pressure rather than the ambient wave speed would have to be used. From the PZT gauge traces in Fig. 2, the HEL occurs at a stress, measured in PMMA of 0.86 GPa. The wave speed measured for this experiment was 3.89 mm μs$^{-1}$. The density for this particular experiment was not measured, hence the ambient density (7.71 g cm$^{-3}$) was used instead to give an impedance of 29.99. However, as the volume change at the phase transformation has been measured (under hydrostatic conditions) at *ca.* 0.9% [2] we do not believe that this will introduce a significant error.

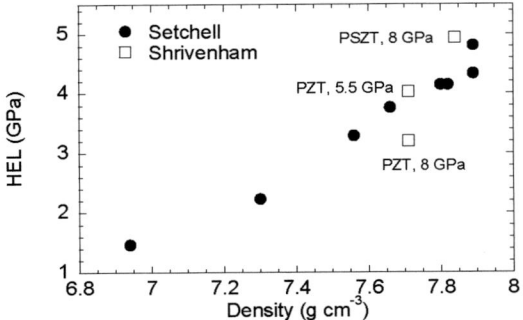

**Figure 5.** Variation of the HEL for PZT and PSZT with respect to density.

In combination with the corresponding dynamic impedance of PMMA [7] (3.59), and equation 1, this gives an in-material value of the HEL of 4.02 GPa. Using the same considerations for PSZT (Fig. 2), the measured HEL, from the back surface gauge, is 0.95 GPa, which converts to an in-material value of 4.94 GPa. Therefore, at first analysis, it appears that the presence of tin has a strengthening effect in PZT based ceramics. However, Setchell [4], in collating the effect of density on the HEL showed that there was an increasing trend in HEL with density, with other factors such as composition having a much smaller (if any) effect. Both results from this paper, in combination with others from this investigation have been plotted against the ambient density of the material. The HEL from our PZT lies between 3.20 and 4.02 GPa. Whilst the spread of these two values is quite large, as a ceramic, PZT is a statistical material, thus more experiments are necessary to establish an average. However, the upper limit of HEL at least agrees with the trends established by Setchell [4]. In the case of PSZT, although the HEL is greater than PZT, it also has a higher density, and again agrees well with previous data trends. Therefore, our results agree with the observations of Setchell [4], showing that the major influence on the HEL of PZT and its derivatives is the density, with factors such as composition having a much lesser effect.

## CONCLUSIONS

The shock response of the ferroelectric ceramics PZT and PSZT have been measured in terms of shock stress, particle velocity and wave velocity. In the case of PZT, close agreement in terms of wave speed – particle velocity and stress – particle velocity between our own results and previously published data has been noted. We have also been able to determine an HEL for this material, again in agreement with that work. In contrast, whilst the Hugoniot of PSZT in terms of the shock stress is similar to PZT, the wave velocity variation with particle velocity shows an increasing trend, as opposed to PZT where the opposite is true. We believe that this may be due to the effect of tin modifying the characteristics of the high-pressure AFE phase. In both PZT and PSZT, it was noted that the calculated elastic response, determined from ambient elastic properties consistently over estimated the Hugoniot, even below the HEL. It is proposed that as all measurements were made above 0.3 GPa (the FE/AFE transition), ambient considerations would not apply. It was also possible to determine the HEL of PSZT at 4.94 GPa. Comparison with existing data has lead us to the conclusion that this increase in strength compared to PZT is due to the higher density of PSZT rather than a strengthening effect of the tin itself.

## ACKNOWLEDGMENTS.

The authors would like to thank Ivan Knapp of Cranfield University for performing the shots in this paper.

## REFERENCES.

1. Fritz, I. J. and Keck, J. D., *J. Phys. Chem. Solids*, **39**, 1163 (1978)
2. Fritz, I. J., *J. Appl. Phys.*, **49**, 4922 (1978)
3. Dick, J. J. and Vorthman, J. E., *J. Appl. Phys.*, **49**, 2494 (1978)
4. Setchell, R. E., *J. Appl. Phys.*, **94**, 573 (2003)
5. Rosenberg, Z., Yaziv, D. and Partom, Y., *J. Appl. Phys.*, **51**, 3702 (1980)
6. Barker, L. M. and Hollenbach, R. E., *J. Appl. Phys.*, **41**, 4208 (1970)
7. Marsh, S. P., LASL Shock Hugoniot data, University of California Press (Los Angeles), 1980

CP845, *Shock Compression of Condensed Matter - 2005*,
edited by M. D. Furnish, M. Elert, T. P. Russell, and C. T. White
© 2006 American Institute of Physics 0-7354-0341-4/06/$23.00

# BOND MODULUS AND STABILITY OF COVALENT SOLIDS

## John J. Gilman

*6532 Boelter Hall, MSE Department*
*University of California, Los Angeles, CA 90095*

**Abstract.** The chemical stabilities of molecules are determined by their LUMO-HOMO energy gaps. For solids the analogs of these are their energy band gaps. However, solids are poly-molecules (i.e., polymers). But, the stabilization energy of a monomer cannot be used to describe the stability of a polymer. An intensive parameter is needed. Such a parameter is the gap energy per molecular volume. The author has coined the name "bond modulus" for this parameter because it tends to be proportional to elastic moduli and it has the same dimensions. It applies primarily to covalent solids with localized bonding (i.e., Group IV elements, III-V compounds, and II-VI compound. A related parameter is electronegativity difference density. It correlates dislocation mobilities, indentation hardnesses, and critical compressions for structure transformations. It is proportional to chemical hardnesses, and bulk moduli, as well as octahedral shear moduli, and inverse polarizabilities.

**Keywords:** modulus, stability, covalent, hardness.
**PACS:** 64.10.+h, 64.30.+t

## INTRODUCTION

For individual covalent molecules, stability is determined by the size of the gap between the HOMO and LUMO molecular orbitals; otherwise known as the *chemical hardness* [1]. However, this measure cannot be used for aggregates of molecules (dimers, trimers, polymers, etc.) because it is independent of the number of atoms, or groups of atoms. Instead, the chemical hardness per molecular volume, or *bond modulus* must be used. This is sometimes called the *physical hardness*. The gap in the bonding energy spectrum in this latter case is the minimum band gap, $E_g$, and the bond modulus is: BM = $E_g/V_m$ where $V_m$ is the molecular volume (for numerical consistency, for homopolar crystals, it is the volume of a dimer). For non-covalent crystals (salts and metals), the heat of formation per molecule may be more appropriately used, instead of $E_g$.

The name "bond modulus" has been chosen for this parameter because it has the same dimensions as an elastic modulus, and is related to elastic stability.

What is meant by stability in this case? Stability is the resistance of a covalent solid to inelastic shear. Solids are quite resistant to volumetric compression which changes their volumes reversibly. They are not so resistant to inelastic shear which induces irreversible deformation or phase transformations, or chemical reactions.

Irreversible deformation is commonly measured by means of indentation hardness. Two versions of this are mostly used for physical property studies: the Vickers method, and the nano-hardness methods. Both measure the ratio of the force applied to an indenter, and the area of the permanent indentation that is produced. This yields a "hardness number" whose units are energy/volume.

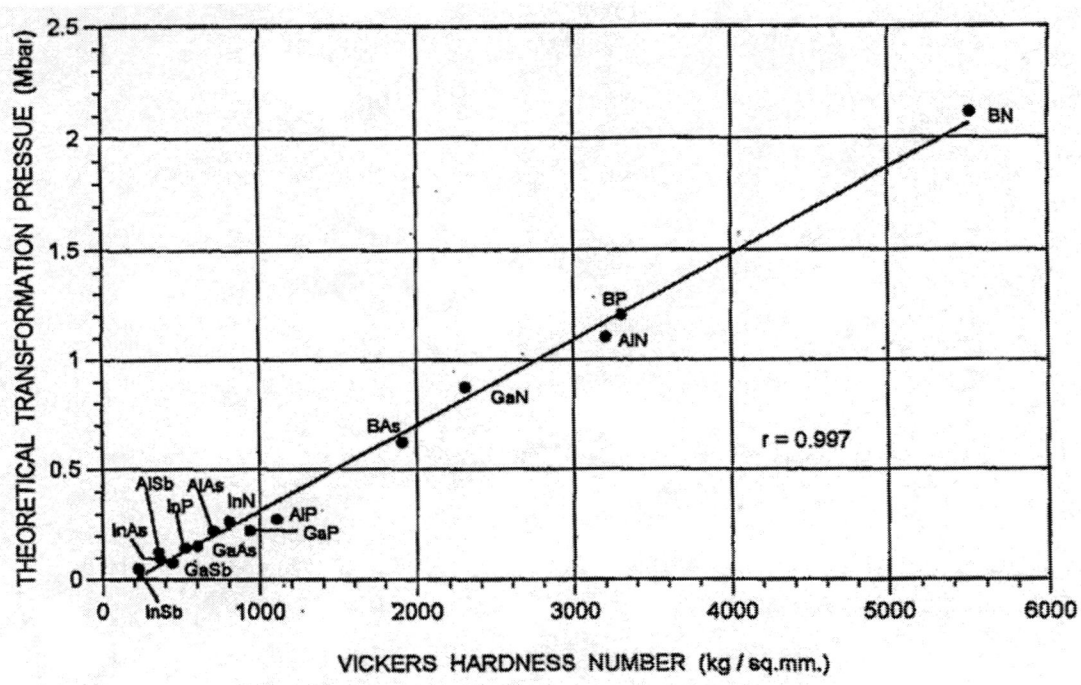

**FIGURE 1.** Theoretical transformation compressions (nominal hydrostatic pressures) vs. indentation hardness numbers for tetrahedral semiconductors (Group IV and III-V crystals). "r" = correlation coefficient.

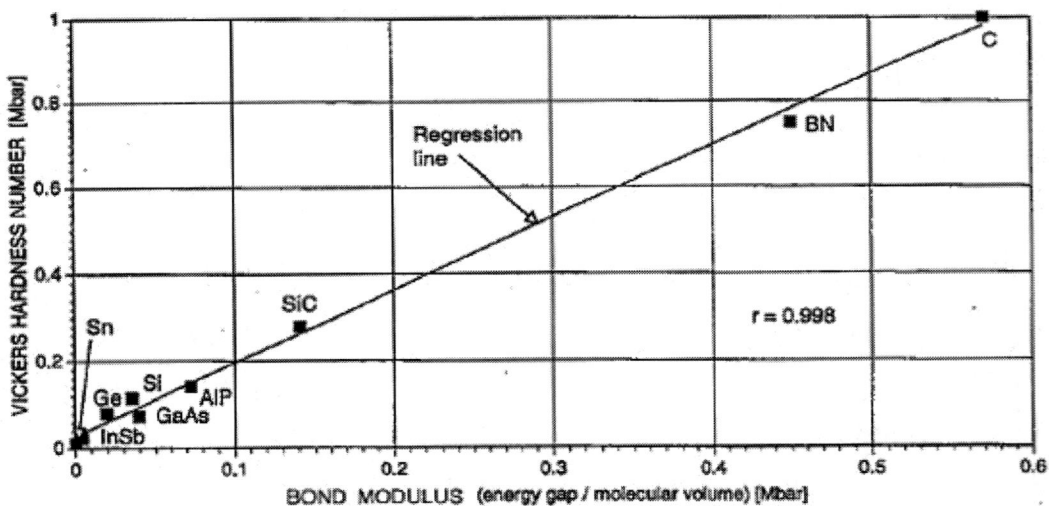

**FIGURE 2.** Indentation hardness numbers vs. bond moduli for some tetrahedral semiconductors (Group IV and isoelectronic III-V crystals).

Mechanically induced phase transformations are commonly produced by applying "pressure" to a specimen of solid in a pressure cell. The state of the specimen is monitored by means X-rays that pass through the cell, or by recording the applied pressure together with the nominal specimen volume, or a property such as electrical resistance. Deviations from smooth variation are taken to indicate phase changes. However, solid-solid phase changes require changes of shape (few, if any involve volume changes alone) [2]. By definition, therefore, shears are required. Thus, applying hydrostatic pressure is an inept way of inducing solid-solid phase changes. Indentation is a better way, as will be demonstrated here.

Chemical reactions induced by mechanical potentials, are discussed elsewhere [3], so they will not be discussed in detail here. They are associated with elastic shear strains which reduce the energy gaps in covalent crystals thereby making the anti-bonding states accessible to the bonding electrons which are then able to rearrange (react).

## INDUCTION OF PHASE CHANGES

The values of the critical compressions ("pressures") at which phase transformations occur in pressure cells scatter badly. The reason is that dilatation and shear do not couple, and the symmetry change required for a phase transformation creates internal stresses in specimens so local stresses are no longer hydrostatic [2]. This suggests that the compression associated with hardness indentations might yield more consistent critical parameters. Phase changes at indentations have been observed for many years [4]. The numerical values of hardness numbers, and critical "pressures" coincide.

To test this idea, calculated values of the critical pressures are used. The theoretical values for the transformation "pressures" calculated by just one author (Van Vechten [5]) were used to improve consistency. As the data in Fig. 1 show, a good linear correlation results with considerably less scatter than if values reported from experiments had been used. This tends to confirm that hardness numbers are better monitors of phase transformations than pressure cell measurements.

The shear deformations at indentations, and the reduced confinement under an indenter (compared with a pressure cells) may account for the good consistency of the data of Fig. 1.

## MEASURES OF STABILITY

Indentation hardness is a measure of dislocation mobility at low temperatures (below the Debye temperature). It also may measure the stress for twinning, and the stress for phase transformations. In other words it measures the resistance to inelastic shear deformation. Thus it is related polarizability, and to the strengths of chemical bonds (in both extension and bending).

A measure of resistance to bond-bending is the energy gap density. This is proportional to the curvature of the bond energy vs. bond length curve at the equilibrium bond length. It has the same dimensions as an elastic modulus. Therefore, it is called the "bond modulus".

Fig. 2 shows that that hardness number (Vickers) is proportional to the bond modulus for several tetrahedral semiconductors. Figs. 1 and 2 are both linear showing that the critical compressions for phase changes are proportional to the bond moduli; also called the chemical hardness density. Thus, these measure structural stability.

The proportionality of the hardness to the bond moduli indicate that dislocation mobility at low temperatures (below the Debye temperature) is also related to the chemical hardness density (sometimes called the "physical hardness").

## REFERENCES

1. Pearson, R. G., Chemical Hardness, Wiley-VCH, New York, 1997.
2. Gilman, J. J., "Phase Transformation Induced by Mechanical Compression", in High Pressure Surface Science and Engineering, 2004, (Y. Gogotsi and V. Domnich, eds.), Chap. 1, Inst. Phys. Publ., Philadelphia.
3. Gilman J. J., "Shear Induced Chemical Reactivity", in Metal-Insulator Transitions Revisited, 1995, (P. P. Edwards and C. N. R. Rao, eds.), Taylor & Francis, London, UK., 269.
4. Trefilov, I.V., and Mil'man, Y. V., Soviet Phys. Dokl., 1964, 8, 1240.
5. Van Vechten, J. A., Quantum Dielectric Theory of Electronegativity in Covalent Systems, III., phys. Rev. B, 1973, 7 (4) 1479.

CP845, *Shock Compression of Condensed Matter - 2005,*
edited by M. D. Furnish, M. Elert, T. P. Russell, and C. T. White
2006 American Institute of Physics 0-7354-0341-4/06/$23.00

# FAILURE FRONTS IN BRITTLE MATERIALS AND THEIR MORPHOLOGICAL INSTABILITIES

## M. A. Grinfeld, S. E. Schoenfeld, T.W. Wright

*U. S. Army Research Laboratory, Aberdeen Proving Ground, MD, 21005-5069*

**Abstract.** There are various observations and experiments showing that in addition to standard shock-wave fronts, which propagate with trans-sonic velocities, other much slower wave-fronts can propagate within glass or ceramic substances undergoing intensive damage. These moving fronts propagate into intact substance leaving intensively damaged substance behind them. They have been called failure fronts or waves. Failure fronts can be modeled either as sharp interfaces separating two states - the intact and comminuted states – or, alternatively, as continuous traveling waves with large spatial gradients of a damage parameter. Our approach is motivated by the analogy between failure fronts and fronts of slow combustion. In this paper we present two main theoretical results that require experimental verification. One of them concerns the speed of a failure wave generated by oblique impact of a brittle target. The other establishes a criterion for morphological instability of failure fronts.

**Keywords**: brittle fracture, ceramics, failure fronts, morphological instability.
**PACS**: 62.50.+p, 68.35.-p

## INTRODUCTION

Extensive experiments with glasses and brittle ceramic materials made by different groups in Britain, Russia, and the USA (see, for instance, [1] – [8]) brought the experimenters to the conclusion of existence of, so called, failure waves. This conclusion is backed by other observations in geomechanics and engineering [9] – [12].

The problem of failure waves demands significant progress of the relevant experiment, theory and numerical modeling, and these three pillars grow simultaneously and in close interaction with each other. The experimenters find an analogy of such waves with solid-solid phase transformation fronts [13] or with fronts of slow combustion [14]. These two analogies are not antagonistic: modeling of both phenomena includes changes in the thermodynamic potentials of the solid states involved. Although

damage is definitely not a solid/solid phase transformation both problems have the same roots: it is the minimization of accumulated energy by means of big changes in the microstructure. For us the analogy with combustion looks somewhat more appealing.

Our two-states model is based on the analogy with the *simplified* theory of slow combustion. In the theory of slow waves with phase transformations such an approach was quite successfully applied about two decades ago in geophysics and celestial physics [15] and in low temperature physics [16]. According to the simplified theory, an "internal structure" of the front is ignored and the front is treated as a mathematical surface (relevant discussions can be found, for instance, in the classical monographs [17, 18]). Certain difficulties of the simplified theory of phase transformation waves have been discussed in [19, 20]. The discussions given there are equally relevant for modeling of failure waves.

## FORMAL STATEMENT OF THE PROBLEM FOR TWO-STATE MEDIA

We limit our study with 2D propagation and consider an initially resting uniform half-plane $x \geq 0$ experiencing an impact at $x = 0$ by the oblique force $F$.

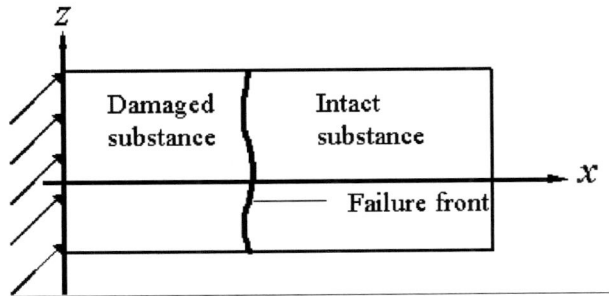

**Figure 1**. Oblique impact of a brittle substance.

Within each of the domains the energy densities $e_{int}$ and $e_{dam}$ per unit mass of the intact and damaged states, respectively, are given by the formulas $\rho e_{in}(u_{m,n}) = c_{int}^{ijkl} u_{i,j} u_{k,l} / 2$ and $\rho e_{dam}(u_{m,n}) = c_{dam}^{ijkl} u_{i,j} u_{k,l} / 2 + q_b$; here $u_{m,n}$ is the displacement gradient, and $c_{in}^{ijkl}$ and $c_{dam}^{ijkl}$ are the elasticity tensors of the two states, $\rho$ is the original mass density ($a_{,i}$ is the symbol of differentiation with respect to the spatial coordinates $x^i$). Limiting ourselves with the approximation of linear elasticity we ignore all effects of mass density change. The positive constant $q_b$ takes into account the energy required to produce various defects (interfaces, vacancies, shear bands, holes, etc.) distributed within unit volume of the bulk of damaged substance. This term is the analogy of the energy release/consumption term used in the theory of slow combustion. See the paper [21] for a detailed discussion of the model.

In addition to appropriate initial and (external) boundary conditions: the master system includes a) the bulk momentum equation within each of the bulk domains, b) the displacement continuity equation across the failure front, c) the traction continuity across the failure front:

$$\text{a)} \quad \rho \frac{\partial^2 u^i}{\partial t^2} = \frac{\partial p^{ji}}{\partial x^j} \tag{1}$$

$$\text{b)} \quad \left[ u^i \right]_{-}^{+} n_j = 0, \quad \text{c)} \quad \left[ p^{ji} \right]_{-}^{+} n_j = 0, \tag{2}$$

where $p^{ji} = \rho \partial e(u_{m,n}) / \partial u_{i,j}$ - the stress tensor, $n_i$ is the unit normal to the failure front.

The last equation across the failure front describes the kinetics of damage at the interface:

$$c = -K \left[ \mu_{.k}^{j} \right]_{-}^{+} n_j n^k, \tag{3}$$

where $c$ is the velocity of the failure-front, $K$ is a positive (kinetic) constant or function with dimension $[velocity]^{-1}$, and the tensor $\mu_{.k}^{j}$ is defined as $\mu_{.k}^{j} = e \delta_k^j - \rho^{-1} p^{ji} \left( \delta_{ik} + u_{i,k} \right)$. The last tensor plays the same role as the scalar Gibbs chemical potential of a liquid substance. We assume that the displacements and traction remain continuous across the interface. This is not the only option. Another reasonable option, especially when dealing with pulverized states, would be the model of a friction-free interface with discontinuous displacements. In this case the last - kinetic - constitutive equation (2) should be modified as well (a similar system was analyzed in papers [16], [19] in the context of phase transformations).

## THE VELOCITY OF A FAILURE FRONT

The system (1), (2) allows piece-wise linear solutions: $u_+^i(x, z, t) = d_+^i t + a_+^i z$ at $x \geq ct$ and $u_-^i(x, z, t) = d_-^i t + a_-^i z$ at $x \leq ct$, where $d_\pm^i$, $a_\pm^i$, and $c$ are certain constants. The signs $+(-)$ mark the quantities related to the intact (damaged) state. The constant $c$ is a speed of the failure front. In this case, the system (1) - (3) leads to the following formula of the velocity of the failure front [22]:

$$\frac{\rho}{K}c = -\frac{\Pi^2}{2}\left[(\lambda + 2\mu)^{-1}\right]_-^+ - \frac{T^2}{2}\left[\mu^{-1}\right]_-^+ - q_b \quad (4)$$

where $\Pi$ and $T$ are the normal and tangential tractions at the failure front. The boundary condition (2) and the formula (4) assume that the velocity of failure front is considerably smaller than the velocities of the bulk waves.

The simple formula (4) allows various generalizations for additional effects which will be accounted for in future validation.

## MORPHOLOGICAL INSTABILITY OF FAILURE FRONTS

In order to explore morphological stability of the above mentioned piece-wise linear solution we decompose the full elastic displacement $u_\pm^i(x, z, t)$ and velocity $c(z, t)$ into the unperturbed fields $u_\pm^{\circ i}(x, t)$, $c^\circ$ and their small disturbances $\tilde{u}_\pm^i(x, z, t)$, $\tilde{c}(z, t)$:

$u_\pm^i = u_\pm^{\circ i} + \tilde{u}_\pm^i$ and $c = c^\circ + \tilde{c}$. This decomposition should be substituted in the bulk equations and boundary conditions (1) - (3). This master system, in turn, implies linear equations for small disturbances. Then, we look for solutions of the linear system in the form ($k$ is the in-plane wave-number and $\eta$ is the rate of growth): $\tilde{u}_\pm^i(x, z, t) = W_\pm^i(x - c^\circ t)e^{ikz+\eta t}$ and $\tilde{c}(z, t) = Se^{ikz+\eta t}$, where the functions $W_\pm^i(X)$ exponentially decay at $X \to \pm\infty$. The general dispersion equation for $\eta$ is too lengthy to be presented here. It becomes more compact and transparent when $V_- = 1/2$ and $V_+ = 1/3$; here $V_\pm = \lambda_\pm / 2(\lambda_\pm + \mu_\pm)$ are the Poisson ratios. Then, the dispersion relation reads:

$$\frac{\rho}{8Kk}\eta = \frac{\mu_+\left(\mu_+^2 - \mu_-^2\right)}{5\mu_-^2 + 8\mu_-\mu_+ + 3\mu_+^2} -$$

$$4\frac{T^2}{\Pi^2}\frac{\mu_+\left(\mu_+ - \mu_-\right)^2}{\mu_-}\frac{4\mu_- + 3\mu_+}{5\mu_-^2 + 8\mu_-\mu_+ + 3\mu_+^2} \quad (5)$$

When the rate $\eta$ has positive real part the failure front is morphologically unstable. The last formula shows that shear stresses $T$ play a stabilizing role like in the case of morphological instabilities of solid-solid phase interfaces [19]). According to (5), at $T = 0$, the failure front is morphologically unstable when the shear modulus $\mu_-$ of the damaged state is less than the shear modulus $\mu_+$ of the intact state:. Intuitively, the assumption $\mu_- < \mu_+$ is quite natural. Speaking loosely, the instability has a simple physical meaning. It means that penetration of fingers of damaged material into intact material is the fastest way of releasing accumulated elastic energy from the system.

The morphological instability of failure-fronts has the potential of explaining appearance of roughness at the disintegration fronts within the Prince Rupert drops. Such roughness was photographed by Chandrasekar and Chaudhri [4]. This interpretation, however, requires further (numerical) studies of deeply nonlinear stage of the instability. Another appealing possibility of the morphological instability concerns the roughness of failure fronts detected in numerical modeling of [23].

## A DYNAMIC MASTER SYSTEM FOR CONTIMUUM MODELS OF DAMAGE

Any damaged brittle substance is actually a solid substance with numerous cracks, holes, voids, vacancies, etc. Of course, all these defects have different individual morphologies and sizes. But, because our goal is to provide a simple description with the most robust features accounted for, it makes sense to assume that all the defects have identical morphologies and sizes and can be characterized by a minimum number of parameters. The volume element is assumed considerably bigger than individual defects but smaller than the wavelengths of interest.

Following the idea of order parameter, the level of damage can be characterized by a single scalar – the damage parameter $\kappa$. Damaged substance is described by energy density of the form

$$\rho e(u_{i|j}, \kappa) = e^{\circ}(\kappa) + c^{ijkl}(\kappa)u_{i|j}u_{k|l}/2 .$$

When describing damaged substances, we can no longer ignore the bulk energy ingredient $e^{\circ}(\kappa)$ associated with the energy of broken chemical bonds appearing in cracking. The damage has an effect on both elements: on $e^{\circ}(\kappa)$ and on the "homogenized" elasticity tensor $c^{ijkl}(\kappa)$. It is intuitively clear that with growth of damage the substance becomes weaker. It is somewhat less obvious that with growth of damage the term $e^{\circ}(\kappa)$ becomes greater: this might even look counter-intuitive.

Different specific models of the above mentioned sort have been discussed in [21, 25, 26]. One of them is described by the following energy density function

$$e(u_{i|j}, \kappa) = \frac{\xi}{2}\left(\kappa - \kappa^{\circ}\right)^2 +$$
$$\mu\left(1 - c_{min}\frac{\kappa}{\kappa^*}\right)\left(\frac{v}{1-2v}u_{,i}^i u_{,j}^j + u_{(i,j)}u_{,\cdot}^{ij}\right) \quad ; \quad (6)$$

here $\xi, c_{min}, \kappa^{\circ}$, and $\kappa^*$ are positive constants.

## REFERENCES

1. Kanel, G. I., Molodets, A. M., and Dremin, A. N., "Investigation of singularities of glass strain under intense compression waves", Combust. Explos. Shock Waves 13, 772 , 1977.
2. Brar, N. S., Rozenberg, Z., and Bless, S. J., "Spall strength and failure waves in glass", *J. Phys.* IV *France (Colloq. C3, DYMAT 91)* 1, 639, 1991.
3. Brar, N. S., and Bless, S.J., "Failure waves in glass under dynamic compression", High Pressure Res. 10, 773, 1992.
4. Bless, S.J., Brar, N.S., Kanel', G., and Rozenberg, Z., "Failure waves in glass", J. Amer. Ceramic Soc., 1992, 75 (4), 1002-1004.
5. Dandekar, D. P., and Beaulieu, P. A., "Failure wave under shock wave compression in soda-lime glass " in Metallurgical and Materials Applications of Shock-Wave and High-Strain-Rate Phenomenon, edited by Murr, L. E., Staudhammer, K. P., and

Meyers, M. A. (Elsevier Science, New York, 1995), pp. 211- 218.
6. Bourne, N., Rosenberg, Z., and Field, J.E., "High-speed photography of compressive failure waves in glasses", J. of Applied Physics 78, 3736, 1995.
7. Bourne, N., Millett, J., and Rosenberg, Z., "On the origin of failure waves in glass", J. of Applied Physics 81, 6670, 1997.
8. Chaudhri, M. M., "Crack bifurcation in disintegrating Prince Rupert's drops", Phil. Mag. Lett., 78(2), 153, 1998.
9. Grigoryan, S. S., "Some problems of the mathematical theory of deformation and fracture of hard rocks", Prikl. Mat. Mekh., 31, 643, 1967.
10. Slepyan, L.I , "Models in the theory of brittle fracture waves", MTT, 1, 181, 1977.
11. Chandrasekar, S., and Chaudhri, M. M., "The explosive disintegration of Prince Rupert drops" Phil. Mag. 70, 1195, 1994.
12. Grady, D. E., "Shock-wave compression of brittle solids", "Mech Mat. 29, 181, 1998.
13. Clifton, R. J., "Analysis of failure waves in glasses", Appl. Mech. Rev. 46, 540, 1993.
14. Kanel, G. I., Bogatch, A. A., and Razorenov, S. V., and Zhen Chen, "Transformation of Shock Compression Pulses in Glass due to the Failure Wave Phenomena", J. Appl. Phys. 92, 5045, 2002.
15. Grinfeld, M. A., "Ramsey-like planets", Dokl. AN SSSR 262, 1339, 1982.
16. Parshin, A. Ya., "Crystallization waves in He" in Low Temperature Physics, Mir, Moscow, 1985.
17. Landau, L. D., and Lifshitz, E. M., Fluid Mechanics (Pergamon Press, New York, 1987).
18. Courant, R., and Friedrichs, K. O., Supersonic Flow and Shock Waves, Springer, New York, 1948.
19. Grinfeld, M. A. , Thermodynamic Methods in Theory of Heterogeneous Systems, Longman, 1991.
20. Glimm, J., "The continuous structure of discontinuities" in PDEs and Continuum Models of Phase Transitions (M. Rascle, D. Serre, and M. Slemrod, eds.), Lect. Notes in Phys., 344, 177 1989.
21. Grinfeld, M. A., and Wright, T. W., "Morphology of Fractured Domains in Brittle Fracture", Metall. Mater. Trans. A, 35(9), 2651, 2004.
22. Grinfeld, M. A., Schoenfeld, S. E., and Wright, T. W., Failure wave propagation in brittle substances, Proc.29th International Conference on Advanced Ceramics and Composites, Cocoa Beach, 2005 (in press).
23. Resnyansky, A. D., Romensky, E. I., Bourne, N. K., "Constitutive modeling of fracture waves", J. of Applied Physics 93, 1537, 2003.

CP845, *Shock Compression of Condensed Matter - 2005,*
edited by M. D. Furnish, M. Elert, T. P. Russell, and C. T. White
2006 American Institute of Physics 0-7354-0341-4/06/$23.00

# APPLICATION OF A BRITTLE DAMAGE MODEL TO NORMAL PLATE-ON-PLATE IMPACT

## Martin N. Raftenberg

*U.S. Army Research Laboratory, Attn: AMSRD-ARL-WM-TD, Aberdeen Proving Ground, MD 21005-5069*

**Abstract.** The brittle damage model of Grinfeld, Schoenfeld, and Wright [1] was implemented in the LS-DYNA finite element code and applied to the simulation of normal plate-on-plate impact. The damage model introduces a state variable measure of damage that evolves in proportion to the elastic strain energy. The model degrades the elastic shear modulus in proportion to the state variable's current level. In a simulation of normal plate-on-plate impact, the model produces a gradient in elastic properties within the initially homogeneous target, and this gradient leads to a partial reflection of the unloading wave. For a range of values for the material constants introduced by the damage model, the target's free-surface velocity showed a gradual increase over time following the arrival of the initial compressive shock.

**Keywords:** Brittle failure, shock loading, damage evolution.
**PACS:** 62.20.Mk, 62.50.+p.

## INTRODUCTION

Kanel et al. [2] shock-loaded K-19 glass in a normal plate-on-plate impact test, and the VISAR measurement of normal velocity at the free surface contained a second plateau that they interpreted as evidence of a failure wave (see their Fig. 1c). Impact by the flyer plate introduces a compressive shock into the target plate. This compressive shock traverses the target plate and reaches the free surface, there producing an unloading wave that travels back towards the impacted surface. The hypothesis of Kanel et al. was that, in K-19 glass, before reaching the impacted surface the unloading wave encounters a slower-moving failure front. The abrupt change in shock impedance at the failure front causes a partial reflection of the unloading wave. The reflected wave, upon reaching the free surface, produces a second plateau in the velocity signal.

To produce this hypothesized phenomenon in a simulation, a damage model should contain two features. First, the damage evolution equation should introduce one or more time scales, thereby allowing the damage front to lag the initial compressive shock. Second, the shock impedance should be substantially altered from its pre-damaged level.

In the immediately preceding paper, Grinfeld, Schoenfeld, and Wright presented a damage model that contains these features [1]. The model applies to a material that in its pre-damaged state obeys isotropic linear elasticity. The pre-damaged strain energy density $W_0$ is

$$W_0(\mathbf{e}) = \mu \cdot \left( \frac{\nu}{1-2\nu} e_{ii} e_{jj} + e_{ij} e_{ij} \right) \quad (1)$$

where $\mu$ is the elastic shear modulus, $\nu$ is the Poisson ratio, $\mathbf{e}$ is the infinitesimal strain tensor, and the summation convention applies. Strain is related to the displacement vector $\mathbf{u}$ by

$$e_{ij} = \frac{1}{2}\left(\frac{\partial u_i}{\partial x_j} + \frac{\partial u_j}{\partial x_i}\right) \qquad (2)$$

The Grinfeld-Schoenfeld-Wright damage model introduces the state variable $\kappa$, a local measure of the degree of damage. $\kappa$ is initialized to zero and evolves according to the equation

$$\frac{\partial \kappa}{\partial t} = -C\frac{\partial E}{\partial \kappa} \qquad (3)$$

Here, $C$ is a material constant and $E$ is the internal energy density. The latter is related to the strain energy by

$$E(\mathbf{e}, \kappa) = \phi(\kappa) \cdot W_0(\mathbf{e}) \qquad (4)$$

$\phi(\kappa)$ is the degradation function, related to $\kappa$ by

$$\phi(\kappa) = 1 - (1 - \phi_{min}) \cdot \frac{\kappa}{\kappa_{max}} \qquad (5)$$

and is sketched in Fig. 1. The $\phi(\kappa)$ function introduces two additional material constants, $\phi_{min}$ and $\kappa_{max}$. At time $t = 0$, $\kappa = 0$, $\phi = 1$, and the internal energy reduces to the elastic strain energy. Thereafter, as $\kappa$ evolves, $\phi$ and $E$ are determined by eqs. 5 and 4, respectively. The Cauchy stress tensor $\boldsymbol{\sigma}$ is evaluated from

$$\sigma_{ij} = \frac{\partial E}{\partial e_{ij}} = \phi(\kappa) \cdot \frac{\partial W_0}{\partial e_{ij}} \qquad (6)$$

so that

$$\sigma_{ij} = 2\,\phi(\kappa)\,\mu \cdot \left(\frac{v}{1-2v} \cdot e_{kk}\delta_{ij} + e_{ij}\right) \qquad (7)$$

This completes the description of the model. This model was implemented in the LS-DYNA finite element code by means of the user material interface [3]. The remainder of this paper deals with the application of this model to a uniaxial strain problem relevant to plate-on-plate impact.

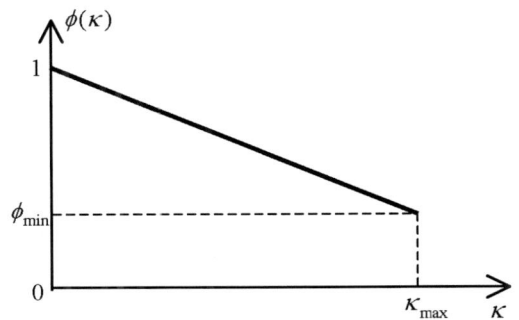

**Figure 1.** The degradation function introduced by the damage model.

## SIMULATION OF NORMAL PLATE-ON-PLATE IMPACT

Consider the case of "symmetric impact", in which one face of a stationary target plate of thickness $L$ is impacted by a flyer plate composed of the same material. Let the target's face opposite to its impacted face be stress-free. The target plate is in a state of uniaxial strain with the boundary condtions shown in Fig. 2, in which $v_0$ is one-half the impact speed of the flyer plate.

Deformation within the target plate is governed by equations 1–5 and 7, and in addition the momentum equation

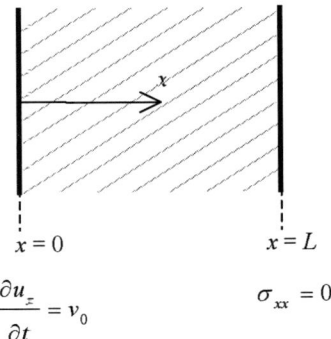

$$\frac{\partial u_z}{\partial t} = v_0 \qquad\qquad \sigma_{xx} = 0$$

**Figure 2.** The boundary conditions imposed on the target in the uniaxial strain model of normal plate-on-plate impact.

$$\rho \frac{\partial^2 u_x}{\partial t^2} = \frac{\partial \sigma_{xx}}{\partial x} \qquad (8)$$

$\rho$ is the mass density.

We define the following scaled quantities.

$$\hat{x} = \frac{x}{L}, \qquad \hat{t} = \frac{c_0 t}{L}, \qquad \hat{u} = \frac{c_0 u_x}{v_0 L} \qquad (9a)$$

$$\hat{v} = \frac{v_x}{v_0}, \qquad \hat{\kappa} = \frac{\kappa}{\kappa_{max}}, \qquad \hat{\sigma} = \frac{\sigma_{xx}}{\rho c_0 v_0} \qquad (9b)$$

where

$$c_0 = \sqrt{\frac{2(1-\nu)\mu}{(1-2\nu)\rho}} \qquad (10)$$

The scaled initial-value, boundary-value problem is

DE

$$\frac{\partial^2 \hat{u}}{\partial \hat{t}^2} = \frac{\partial^2 \hat{u}}{\partial \hat{x}^2} - (1-\phi_{min}) \cdot \left( \hat{\kappa} \cdot \frac{\partial^2 \hat{u}}{\partial \hat{x}^2} + \frac{\partial \hat{\kappa}}{\partial \hat{x}} \cdot \frac{\partial \hat{u}}{\partial \hat{x}} \right) \qquad (11)$$

$$\frac{\partial \hat{\kappa}}{\partial \hat{t}} = \Pi \cdot \left( \frac{\partial \hat{u}}{\partial \hat{x}} \right)^2 \qquad (12)$$

where

$$\Pi = \frac{(1-\phi_{min})C}{\kappa_{max}^2} \cdot \frac{\rho}{2c_0} \cdot L \cdot v_0^2 \qquad (13)$$

BC

$$\frac{\partial \hat{u}}{\partial \hat{t}}(0,\hat{t}) = 1, \qquad \frac{\partial \hat{u}}{\partial \hat{x}}(1,\hat{t}) = 0 \qquad (14)$$

IC

$$\hat{u}(\hat{x},0) = 0, \qquad \frac{\partial \hat{u}}{\partial \hat{t}}(x,0) = 0, \qquad \hat{\kappa}(\hat{x},0) = 0 \qquad (15)$$

$$\hat{x} \in [0,1], \qquad \hat{t} \geq 0$$

The scaled system contains two dimensionless parameters, $\phi_{min}$ and $\Pi$. The latter measures the material's damage sensitivity to strain energy. In the following, $\phi_{min}$ is fixed at 0.1 and $\Pi$ is varied.

## RESULTS AND DISCUSSION

The solution for the case of $\Pi = 0$ (no damage) is obtained using Laplace transforms:

$$\hat{u}_0 = (\hat{t}-\hat{x})H(\hat{t}-\hat{x}) + (\hat{t}+\hat{x}-2)H(\hat{t}+\hat{x}-2)$$
$$+ (\hat{t}-\hat{x}-2)H(\hat{t}-\hat{x}-2)$$
$$\hat{t} \in [0,3) \qquad (20)$$

H is the Heaviside unit step function.

The solutions for $\Pi = 0.1, 0.2, 0.3, 0.4$ were obtained using the LS-DYNA implementation. The mesh consisted of 1600 8-node brick elements spanning the domain $\hat{x} \in [0,1]$. The solutions are evaluated at $\hat{t} = 1.5$ (about when the unloading wave should encounter the failure front) in Figs. 3 and 4. In Fig. 3 $\hat{\kappa}$ and $\phi$ vary gradually with $\hat{x}$ across the target. This gradient in elastic properties causes distributed reflection of the unloading wave.

Fig. 4 presents $\hat{\sigma}$, $\hat{\kappa}$, and $\phi$ as functions of $\hat{x}$ for $\hat{t} = 1.5$ and $\Pi = 0.4$. The evolved damage has had two effects. First, damage has decreased the speed of the unloading front. I.e., the $\hat{\sigma}$ front is located at $\hat{x} = 0.53$ for $\Pi = 0.4$, whereas it would be located at $\hat{x} = 0.50$ for $\Pi = 0$. Second, damage has decreased (in absolute value) the amplitude of $\hat{\sigma}$ at the front from a value of $-1$ for $\Pi = 0$ to $-0.65$ for $\Pi = 0.4$.

Fig. 5 presents, as a function of time, $\hat{v}$ at $\hat{x} = 1$, the scaled normal velocity at the target's rear surface. For $\Pi = 0$, $\hat{v}$ jumps from zero to 2 at $\hat{t} = 1$ and remains at this plateau until $\hat{t} = 3$. As $\Pi$ increases to 0.1, 0.2, 0.3 and 0.4, the magnitude of the $\hat{v}$ plateau at $\hat{t} = 1$ decreases (Table 1).

Consider Fig. 5 for $\hat{t} \in (1,3)$. For $\Pi = 0.1$ and 0.2, $\hat{v}$ increases with time slightly from its value at $\hat{t} = 1$. For $\Pi = 0.3$ and 0.4, $\hat{v}$ decreases from its value at $\hat{t} = 1$ (Table 1). Evidently, for $\Pi = 0.1$ and 0.2, the contribution to $\hat{v}$ at $\hat{x} = 1$ from recompression due to unloading-wave reflection dominates over the diminution of the signal due to damage. For $\Pi = 0.3$ and 0.4, the reverse is the case.

864

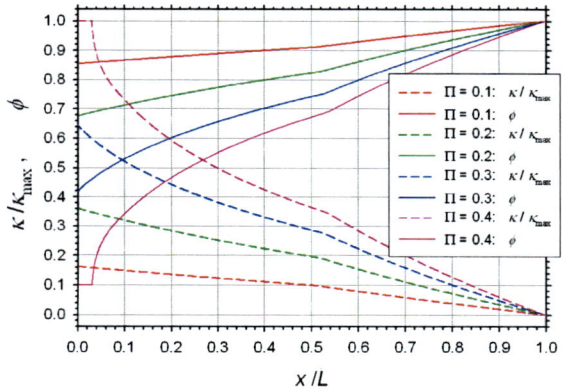

**Figure 3.** Results for $\hat{\kappa}$ and $\phi$ as functions of $\hat{x}$ at $\hat{t} = 1.5$ and for various $\Pi$. ($\phi_{min} = 0.1$).

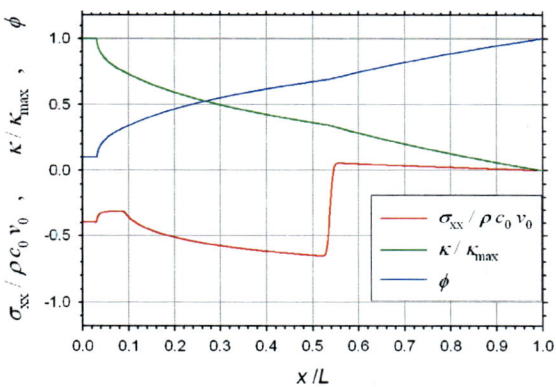

**Figure 4.** Results for $\hat{\sigma}_{xx}$, $\hat{\kappa}$, and $\phi$ as functions of $\hat{x}$ at $\hat{t} = 1.5$ and for $\Pi = 0.4$. ($\phi_{min} = 0.1$).

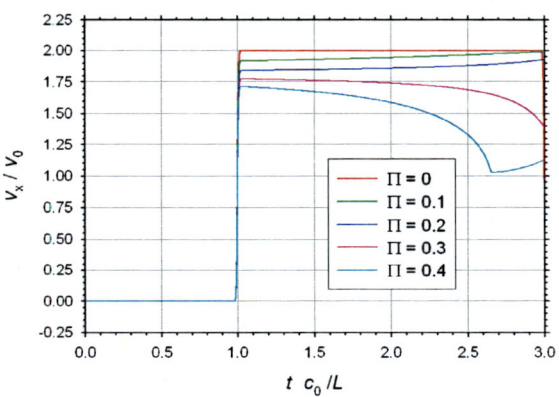

**Figure 5.** Results for $\hat{v}$ at $\hat{x} = 1$ as a function of $\hat{t}$ and for various $\Pi$. ($\phi_{min} = 0.1$).

**TABLE 1.** $\hat{v}$ at $\hat{x} = 1$. ($\phi_{min} = 0.1$)

| $\Pi$ | $\hat{t} = 1.02$ | $\hat{t} = 2.00$ |
|---|---|---|
| 0 | 2.00 | 2.00 |
| 0.1 | 1.91 | 1.94 |
| 0.2 | 1.84 | 1.86 |
| 0.3 | 1.77 | 1.74 |
| 0.4 | 1.71 | 1.59 |

## CONCLUSIONS

We noted in Fig. 5 that, for certain values of $\Pi$, the damage model had the effect of producing an increase in rear surface velocity following the initial rise corresponding to the arrival of the compressive shock. However, the additional rise subsequent to the shock's arrival was gradual (distributed over time) and thus did not closely resemble the abrupt rise to a second plateau that was observed by Kanel et al. [2]. Perhaps a more abrupt rise can be achieved by replacing the linear $\phi(\kappa)$ of eq. 5 and Fig. 1 with a different functional form that changes more abruptly with $\kappa$.

## ACKNOWLEDGEMENTS

The author thanks Michael A. Grinfeld of the Army Research Laboratory for his helpful comments on the interpretation of the damage model.

## REFERENCES

1. Grinfeld, M. A., S. E. Schoenfeld, and T. W. Wright, "Failure fronts in brittle materials and their morphological instabilities", in Shock Compression of Condensed Matter 2005.
2. Kanel, G. I., S. V. Rasorenov, and V. E. Fortov, "The failure waves and spallations in homogeneous brittle materials", in Shock Compression of Condensed Matter 1991 (S.C. Schmidt, R.D. Dick, J.W. Forbes, D.G. Tasker, eds.), 451, 1992.
3. Livermore Software Technology Corporation, LS-DYNA Keyword User's Manual, Version 970, Livermore, CA, 2003.

CP845, *Shock Compression of Condensed Matter - 2005,*
edited by M. D. Furnish, M. Elert, T. P. Russell, and C. T. White
2006 American Institute of Physics 0-7354-0341-4/06/$23.00

# COMPUTATIONAL DESIGN STUDY FOR RECOVERY OF SHOCK DAMAGED SILICON CARBIDE

## K. Iyer and D. Dandekar

*U. S. Army Research Laboratory*
*ATTN: AMSRD-ARL-WM-TD*
*Aberdeen Proving Ground, MD 21005-5069*

**Abstract.** The paper presents a computational study for design of experimental configurations that may permit the recovery of weak-shock loaded high-strength brittle ceramics such as silicon carbide with controlled amounts of damage. A set of 8 configurations involving momentum traps, and subjected to a nominal shock pressure of 4 GPa, is analyzed using finite element analysis with linear elastic and damage material models. The analyses identify influences of: (i) introducing a hole in the specimen center, (ii) specimen size, and (iii) impedance graded trapping.

**Keywords:** recovery, AD995, silicon carbide, finite element analysis, damage.
**PACS:** 62.50.+p, 47.40.Nm.

## INTRODUCTION

Post-test recovery of a high-strength brittle material subjected to any type of dynamic impact loading is non-trivial owing to complete fragmentation. But it is necessary for identifying constitutive damage mechanisms and relating them to the load-unload history. The only known successful recovery of a macrocracked yet contiguous high-strength ceramic specimen is by Bourne et al [1], in which alumina AD-995 discs were subjected to shock compressive stresses up to ~8 GPa. A cross-sectional view of the successful cylindrical configuration is shown in Fig. 1(a) -- the target consisted of an alumina disc (20 mm dia. x 6 mm thk.) that was encased by a bronze cup and cover plate, and 5 copper discs (3mm thk.) for momentum trapping. The entire configuration was impedance matched. A cross-sectional view of the macrocracks generated in the alumina specimen is shown in Fig. 1(b).

With the aim of developing general principles for successful recovery of plate impacted ceramics with controlled amounts of damage, the present work analyzes the Bourne et al. configuration and examines the influence of 3 design modifications – specimen size, introducing a hole in the specimen and impedance graded trapping – on computed stress and strain histories and damage under a nominal shock stress of 4 GPa.

## COMPUTATIONAL ANALYSES

The configuration shown in Fig. 1(a) was analyzed using linear elastic and inelastic material models – the latter was a pressure-dependent deviatoric strength model with damage for alumina AD-995 [2]. The findings from these analyses were used to develop a matrix of linear elastic design calculations for silicon carbide, listed in Table 1. These examine the influence of introducing a hole in the ceramic specimen, specimen size and impedance graded trapping, in which the copper flyer and last 4 copper momentum traps were replaced by tungsten carbide, and tin, aluminum alloy 6061-T6, magnesium and PMMA, respectively.

**TABLE 1.** The 5 silicon carbide specimen configurations/dimensions considered. In all cases, the bronze cup wall (including back) is 3 mm thick. When present (models *ARL4* and *ARL4-IG*), the hole diameter is 10 mm. $D/t \sim 6.7$ in all models.

| Model | Diameter $D$ (mm) | Thickness $t$ (mm) | Flyer Velocity (m/s) |
|-------|-------------------|--------------------|----------------------|
| *ARL2* | 38 | 5.6 | 200 |
| *ARL3* | 76.2 | 11.4 | 200 |
| *ARL4* | 76.2 | 9.9 | 200 |
| *ARL3_IG* | 76.2 | 11.4 | 135 |
| *ARL4_IG* | 76.2 | 9.9 | 135 |

-*ARL2-4* were impedance matched designs
-*IG* denotes impedance graded design

In all cases, the flyer velocity was chosen so as to generate a nominal shock pressure of 4 GPa. Table 2 lists the linear elastic material properties used in the analyses. All calculations were performed with axisymmetric finite element models[*].

**TABLE 2..** Elastic modulus, $E$, Poisson's ratio, $v$, and density, $\rho$, values used in the analyses.

| | $E$ (GPa) | $v$ | $\rho$ (kg/m³) | $Z$ (MPa.s/m) |
|---|------|------|--------|------|
| alumina | 374 | 0.23 | 3880 | 41 |
| silicon carbide | 448 | 0.16 | 3220 | 39 |
| copper | 117 | 0.34 | 8800 | 40 |
| bronze | 130 | 0.33 | 8800 | 41 |
| tungsten carbide | 691 | 0.20 | 15545 | 109 |
| magnesium | 44 | 0.28 | 1740 | 10 |
| tin | 60 | 0.32 | 7287 | 25 |
| Al 6061-T6 | 69 | 0.33 | 2704 | 16 |
| PMMA | 6 | 0.33 | 1185 | 3 |

Linear elastic analyses were performed with the ABAQUS code and damage analyses were performed with the EPIC code. A Coulomb friction

---

[*] with the 4 levels of mesh refinement considered (50 μm x 50 μm, 50 μm x 62.5 μm, 62.5 μm x 75 μm, or 158.75 μm x 127 μm elements), differences between theoretical and numerical values of particle velocity was found to be negligible in all cases. The difference between the values of shock pressure was found to be 2-8%.

model with a friction coefficient, $\mu = 0.1$, was assumed for all interfaces.

## RESULTS AND DISCUSSION

Computations indicate advancing damage fronts from the radial boundaries of the specimen as well as the impact surface near the symmetry axis 2 μs after impact in the AD995 specimen (Fig. 2(a)). The damage contours at 20 μs (Fig. 2(b)), however, do not fully reflect the discrete macrocrack pattern observed experimentally (Fig. 1(b)). The longitudinal stress history approximately midway through the depth of the specimen at 3 radial distances is shown in Fig. 3(a). A separate linear elastic calculation of the same configuration revealed the presence of dynamic tensile stress concentrations that traverse the symmetry axis of the configuration. This phenomenon is visible in the recurring peaks in the longitudinal stress and lateral (radial) strain histories at the specimen's geometric center, shown in Fig. 3(b). It is also responsible for the damage along the symmetry axis seen in Fig. 2(a).

Detailed analyses, not presented here, showed that the severity of the axial reverberating tensile stress and strain concentrations were related to dynamic changes in the interfacial contacts (chatter), which in turn was related to the deformation modes and dimensions of the individual components of the configuration. These findings stimulated the pursuit of design modifications that could minimize contact chatter by increasing the rapidity with which the momentum traps and flyer plate would begin to separate from each other. Minimizing the magnitude and duration of the dynamic stress and strain concentrations in the ceramic specimen was, in fact, the motivation behind the 3 particular design modifications considered in this work.

Figure 4 shows the influence of size, introducing a hole in the ceramic specimen and impedance grading on the computed longitudinal stress and lateral strain concentration histories at the geometric center of silicon carbide specimens. Impedance grading (Fig. 4(d)) is found to be most effective in minimizing the magnitudes of the tensile concentrations.

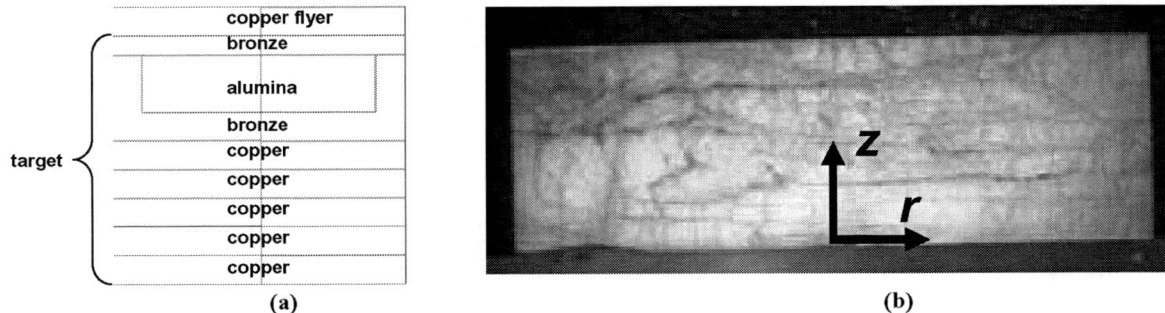

**Figure 1.** (a) Configuration used by Bourne et al. [1]. (b) Fracture pattern observed in a recovered, sectioned alumina specimen that was subjected to a 4 GPa shock (micrograph provided by J. McCauley) [1].

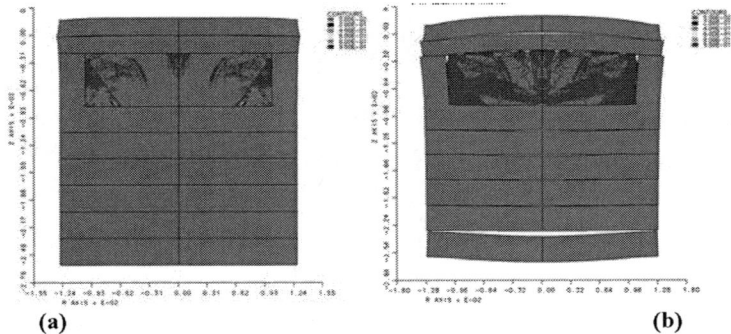

**Figure 2.** Computed damage contours using the JH model [2] in the alumina specimen for an impact velocity of 200 m/s at: (a) 2 μs, and (b) 20 μs.

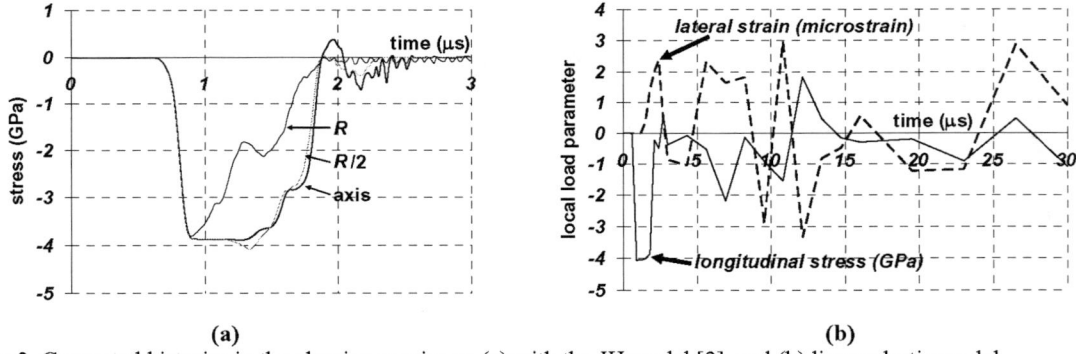

**Figure 3.** Computed histories in the alumina specimen: (a) with the JH model [2], and (b) linear elastic model.

Consistent with the results from the linear elastic calculations, damage calculations in silicon carbide shown in Figure 5 reflect less damage with an impedance graded configuration. The damage is also localized around the symmetry axis, where the elastic stress and strain concentrations are present.

The impact strength and failure of high-strength ceramics is influenced by microstructural defects, which is not accounted for here. This work, however, does treat those aspects of failure that originate from the dynamic continuum mechanics of finite plate impact, which are neglected in the traditional uniaxial-strain-based design paradigm. Based on the present findings, a matrix of experiments to produce controlled amounts of damage in silicon carbide is underway.

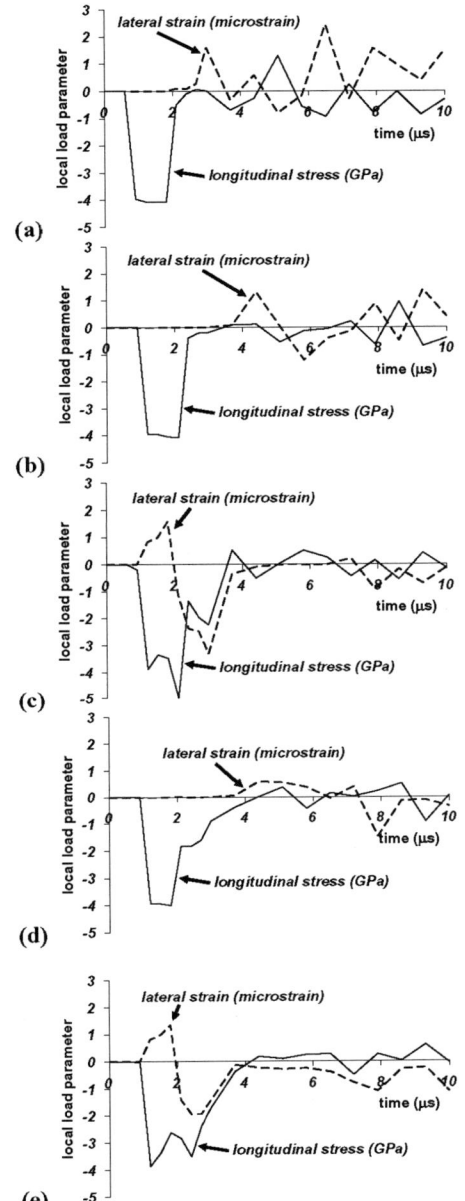

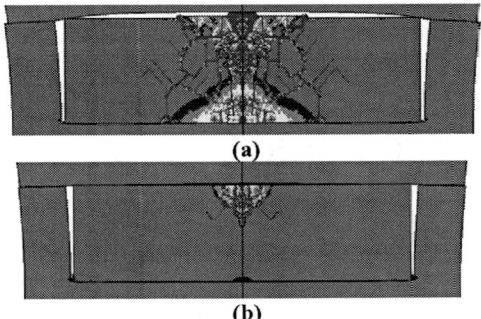

**Figure 5.** Computed damage contours at 20 μs, based on the JHB model [3], in the silicon carbide specimen shocked to 4 GPa in the Bourne et al. [1] configuration with: (a) impedance matching, and (b) impedance grading.

## CONCLUSIONS

Continuum computations for shocked (4 GPa) silicon carbide indicate a correlation between linear elastic tensile stress and strain parameters, and damage calculated using a state-of-the-art pressure-dependent strength material model. Impedance graded trapping and specimen size are promising control parameters for modulating damage in plate impact tests with high-strength ceramics.

## REFERENCES

1. Bourne, N. K., Green, W. and Dandekar, D. (2005), unpublished results.
2. Anderson, C. E., Johnson, G. R. and Holmquist, T. J. (1995), "Ballistic Experiments and Computations of Confined 99.5% Al2O3 Ceramic Tiles," Proc. 15th Intnl. Symp. on Ballistics, Jerusalem, Israel.
3. Holmquist, T. J. and Johnson, G. R. (2005), "Characterization and evaluation of silicon carbide for high velocity impact," J. Appl. Phys., Vol. 97 (093502).

**Figure 4.** Computed histories at the geometric center of a linear elastic silicon carbide specimen in configuration: (a) *ARL2*, (b) *ARL3*, (c) *ARL4*, (d) *ARL3_IG*, and (e) *ARL4_IG*.

CP845, *Shock Compression of Condensed Matter - 2005*,
edited by M. D. Furnish, M. Elert, T. P. Russell, and C. T. White
© 2006 American Institute of Physics 0-7354-0341-4/06/$23.00

# FAILURE WAVES IN SHOCK-COMPRESSED GLASSES

## G.I. Kanel

*Institute for High Energy Densities of Russian Academy of Sciences
IVTAN, Izhorskaya 13/19, Moscow, 125412 Russia*

**Abstract.** The failure wave is a network of cracks that are nucleated on the surface and propagate into the elastically stressed body. It is a mode of catastrophic fracture in an elastically stressed media whose relevance is not limited to impact events. In the paper, main properties of the failure waves are summarized and discussed. It has been shown that the failure wave is really a wave process which is characterized by small increase of the longitudinal stress and corresponding increments of the particle velocity and the density. The propagation velocity of the failure wave is less than the sound speed; it is not directly related to the compressibility but is determined by the crack growth speed. The failure wave is steady if the stress state ahead of it is supported unchanging. In some sense the process is similar to a subsonic combustion wave. Computer simulations based on the phenomenological combustion-like model reproduces well all kinematical aspects of the phenomenon.

**Keywords:** Failure wave, glass, compressive fracture, shock compression, spall strength.
**PACS:** 61.43.Fs, 62.20.Mk, 62.50.+p, 83.60.Uv.

## INTRODUCTION

The impact loading of a glass and, probably, other brittle materials can result in the appearance of a failure wave. The failure waves present a mode of catastrophic fracture in elastically compressed media that is not limited to impact events. It is regular self-propagating process. One may hope that the investigations of failure waves provide information about the mechanisms and general rules of nucleation, growth, and interaction of the multiple cracks under compression.

The term "failure wave" has been introduced in sixtieths [1,2] when a detonation-like model of fracture of stressed brittle materials was developed. The model supposes an ability of fragmentation occurring within relatively thin layer which propagates through undamaged material with the sound speed. The first theoretical models did not provide a base for correct estimations of kinematical parameters of the failure waves.

A similar fracture mode under compression was revealed in shock-wave experiments. The history and some preliminary results of observations of the failure wave phenomena were recently reviewed [3-5]. In this paper the discussion is concentrated on main properties of the failure waves. Before this discussing, it looks reasonable to remind features of compressive behavior of glasses.

## BEHAVIOR OF GLASSES UNDER SHOCK COMPRESSION

Silicate glasses exhibit high yield strength and low fracture toughness. The fracture of glasses under compression occurs by axial splitting. At high pressures, brittle glasses become ductile. Glasses show gradual structural changes resulting in increased density [6]. It is supposed that the irreversible densification of the silicate structure is responsible for the plastic flow properties of glasses under high pressure. Irreversible densification of occurs also under shock compression above the HEL [7].

Figure 1 presents the waveforms for K8 crown glass which were measured at two different impact velocities [8]. The rise time of compression wave gradually increases with the increase of the propagation distance that is a result of anomalous compressibility of the glass below its elastic limit and the stress relaxation above it. The waveforms do not exhibit a distinct transition from the elastic to plastic response. Spallation was not observed in these shots, which means that the spall strength of the glass exceeds 6.8 GPa below the HEL and remains very high above it. For comparison, the static tensile strength of glasses is around 0.1 GPa. The reason for such a large discrepancy is that the fracture nucleation sites in homogeneous glass are concentrated on the surface. These incipient microcracks determine the strength magnitude in the static measurements, whereas spall strength is an intrinsic property of matter.

In Fig. 2 the free surface velocity histories of several different glass samples are shown. Most of silicate glasses have anomalous longitudinal compressibility within the region of elastic compression where the longitudinal sound speed decreases as the compressive stress increases that, in turn, causes broadening of the elastic compression wave with its propagation. As a result of an anomalous behavior of longitudinal sound speed, a rarefaction shock wave should be formed in glass at unloading from shock-compressed state

if the compression is completely reversible. Since the reversibility of stress–strain processes is a main attribute of elastic deformations, observation of the rarefaction shock (demonstrated by the waveform 2 in Fig. 3) may be considered as evidence of an elastic regime of deformation. Above the HEL the unloading wave speed becomes greater than the compression wave speed that is demonstrated by the waveform 1 in Fig. 3.

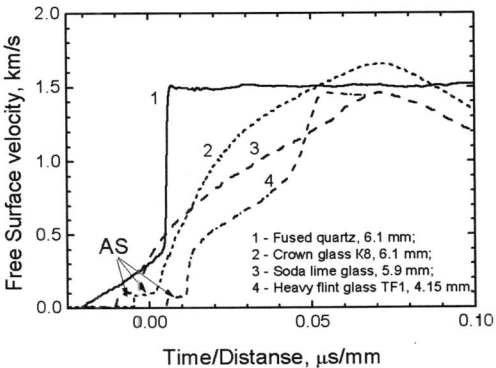

**Figure 2.** Structure of the compression waves in fused quartz [9], K8 crown glass [8], soda lime glass [10] and TF1 heavy flint glass [9]. The weak velocity steps AS is the result of an air shock ahead of the flyer plate.

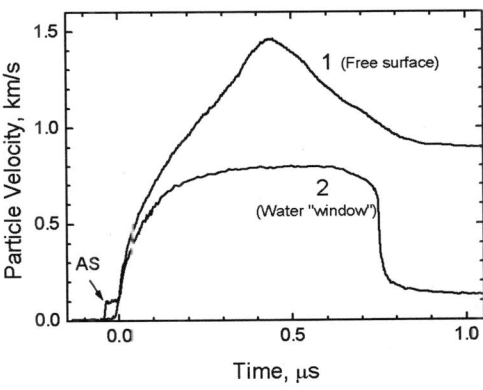

**Figure 3.** Particle velocity histories of soda lime glass plates of thickness 5.9 mm [10]. The wave profile 1 corresponds to impact by aluminum flyer plate 2 mm thick backed by paraffin, with the impact velocity being 1.90±0.05 km/s. The wave profile 2 corresponds to impact by aluminum flyer plate of 2.1 mm thick at the impact velocity 0.97±0.03 km/s, measured through a water window.

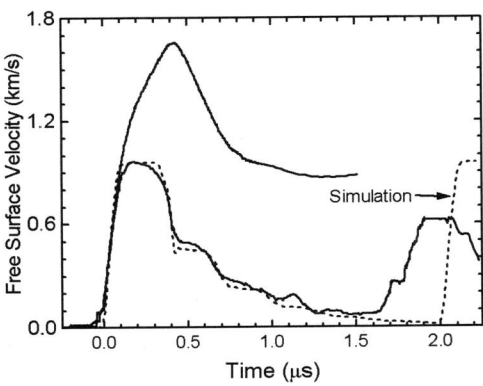

**Figure 1.** The free surface velocity histories of the K8 crown glass samples at different peak stresses [8]. The dashed line shows results of computer simulations.

## OBSERVATIONS OF THE FAILURE WAVES

### Cracking of glass near the impact surface

In the Fig. 1 the results of the measurements are compared with the computer simulation for the shot of K8 glass target impacted by a low-velocity steel plate. Simulation has been done supposing purely elastic behavior for the glass, and without fracture under both compression and tension. Whereas the computed first velocity pulse is in a reasonable agreement with the measured one, the second velocity pulse arrives at the rear surface later as compared to the measurements. This difference means that the observed second velocity pulse is actually a reflection of the rarefaction wave from a near-surface layer which is not able to sustain tension. In other words, the layer of glass near the impact surface has been failed to the moment when the reflected tensile pulse reached it. Expansion of the cracked layer from the impact surface has been treated as propagation of the failure wave. No any evidences of cracking were observed at peak stress above the HEL.

As a result of cracking, a glass looses its optical uniformity that gives a possibility of optical recording these processes. Using this circumstance, Bourne et al. [11] and Senf et al. [12] have photographed failure waves in transmitted light.

### Initiating conditions of the failure waves

Raiser et al. [13] have found that a surface roughness of aluminosilicate glass does not appear to play a significant role in the formation of a failure wave. Independently of the surface roughness, they observed high spall strength of the glass when the compressive stress was around 3.5 GPa whereas, at peak stresses of 7.5–8.4 GPa, the spall strength was high ahead of the failure front and was low behind it. Bourne et al. [14] have demonstrated that roughening the surface speeds the fracture of a glass. Although the failure waves are formed in glass independently on the roughness of its surface, the shock-wave behavior of lapped glass plates is much more reproducible than that of as-received plates with mirror-like surfaces. [10].

### The failure wave speed

Whether the failure wave is steady or it decays and stops at some distance is an important issue for understanding the mechanism and nature of the phenomenon. Figure 4 presents the measured free surface velocity histories of soda lime glass plates at various stress levels where the time is normalized by the sample plate thickness. The wave profiles contain small recompression pulses which are due to the wave reflection from a failed region inside the sample [15, 16, 17]. It follows from consideration of the time–distance diagram shown in Fig. 5 that the failure wave speed $c_f$ may be determined by means of measurement of the time interval $t_r$ between the arrivals of the initial compression wave and the recompression pulse front at the plate free surface. For constant speed of the failure wave the ratio $t_r/\delta$, where $\delta$ is the glass plate thickness, should not depend on the plate thickness.

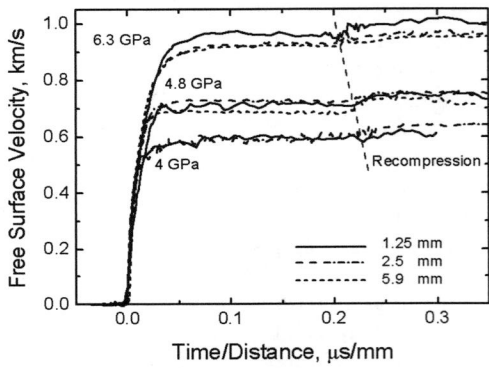

**Figure 4.** Free surface velocity histories of the soda lime glass plates of different thicknesses at three different stress levels of shock compression [9].

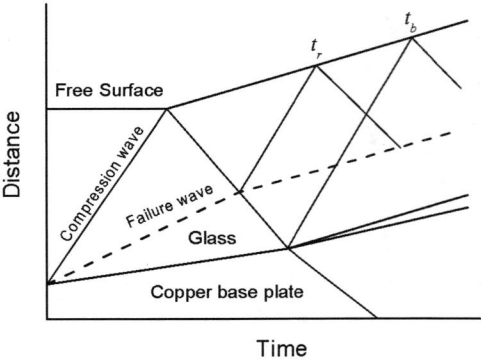

**Figure 5.** Distance-time diagram of experiments shown in Fig. 4.

872

As it is seen in Fig. 4, the failure waves indeed propagate at a constant speed which decreases from 1.58±0.06 km/s at 6.3 GPa of compressive stress to 1.35±0.06 km/s at 4 GPa. The observed constant speed of the failure wave is in agreement with the data by Dandekar and Beaulieu [17]. The stress dependence of the failure wave speed explains its apparent deceleration that was found in the first observation by Razorenov et al. [15] where the glass samples were loaded by decaying stress pulses. The failure wave process becomes unstable, and stops at the stress level near the failure threshold.

Comparison of the time $t_r$ of arrival of re-reflected pulses in Figures 1 and 4 shows that the reflected signal arrives later in the case of short loading pulse. The latter observation indicates that the unloading decreases the failure wave velocity or even arrests the failure wave propagation.

### State of a glass behind the failure wave

Brar et al. [16,18] and Bourne et al. [19] have shown by direct measurements on different glasses that behind the failure wave the tensile strength drops to zero and the transverse stress increases, indicating a decrease in shear strength. The failure waves were recorded in the longitudinal stress range 4–10 GPa. At peak stresses exceeding 10 GPa, the densification processes start in glass. This produces shear stress relaxation without cracking.

### Kinematics of the failure waves

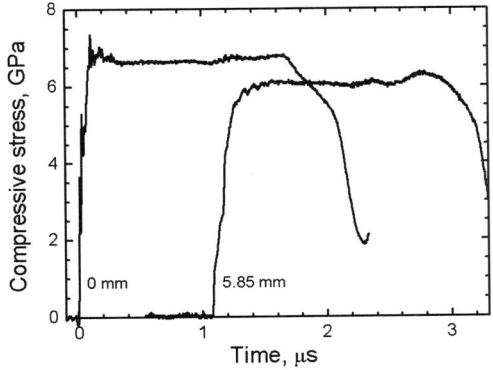

**Figure 6.** Stress histories on the input surfaces of glass sample and at the distance of 5.85 mm from it [10].

Since the first experimental observations of failure waves in shock-compressed glass it was believed that the failure wave is accompanied by increasing lateral stresses and is not accompanied by any change in longitudinal stresses. However, experiments of Dandekar [20] and of Millet et al. [21] revealed a disagreement between the longitudinal stresses measured on different distances from the impact surface. Although the recorded wave profiles did not contain second compression wave, the measured stresses at some distance were less than at the impact surface when the incident shock amplitude exceeded some threshold. These observations may be treated as an evidence of an unrecorded second compression wave.

Figure 6 presents stress histories measured by Kanel et al. [10] on input and output surfaces of glass samples. These data confirm the difference between peak stresses measured on the impact surface of a shock-loaded glass plate and the stresses measured at some distance from the impact surface, as observed in [20,21].

Experiments with layered glass samples [10] have shown that the network of growing microcracks in shock-compressed glass may indeed be considered as a wave with a small stress increment which obeys the Rankine-Hugoniot conservation laws. The estimated final state behind the failure wave agrees with direct measurements of the principal stress difference.

Kinematics of the failure waves differs from those of elastic–plastic waves. The shock compression wave in an elastic–plastic body becomes unstable at sudden decrease of longitudinal compressibility that occurs when yielding begins. As a result, the wave splits into an elastic precursor wave and a plastic shock wave. The peak stress behind the elastic precursor front is determined by the yield stress. The propagation velocities of the elastic precursor wave front and the second compression wave are determined by the longitudinal and bulk compressibility, respectively. In contrast to that, the propagation velocity of the failure wave is determined by the crack growth speed, which is not directly related to the compressibility.

On the other hand, the final longitudinal stress in the comminuted glass behind the failure wave is determined by the boundary conditions of impact loading, whereas the deviator stress component is controlled by the post-failure properties of the glass. Thus, since the propagation velocity of a failure wave and the final stress are fixed, the stress ahead of the failure wave should be governed by these values and should not necessarily be equal to the failure threshold.

## Shock response of glass piles

Since the failure wave nucleates on the glass surface, the magnitude of the leading elastic wave in the shocked specimen consisting of layered glass plates should decrease as a result of its decomposition into two waves at each interface. The decrease of elastic wave amplitude repeats at each interface until the failure threshold is reached. Hence, for a sufficiently large number of layered glass plates, an elastic precursor wave with its amplitude close to the failure threshold could be formed. Figure 7 presents results of two shots where free surface velocity histories were recorded for one thick glass plate and layered assemble of 8 thin glass plates, subjected to the same impact loading. The shot with a pile shows the waveform that is typical for elastic–plastic solids. The final magnitude of the free surface velocity is practically equal to that of a single glass plate. The response of a layered assembly of thin brittle plates as compared to that of one thick plate is a simple way to diagnose nucleation of the failure process on the plate surfaces and determine the failure threshold.

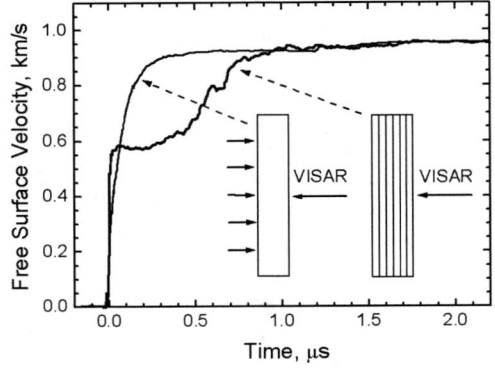

**Figure 9**. The free surface velocity histories recorded in two shots with layered assemblies of 8 soda lime glass plates of 1.2 mm average thickness in comparison with the data for single glass plate 5.9 mm thick.

## Response of comminuted glass

Response of shock-compressed glass behind the failure wave was probed in the plane impact experiments with copper witness plate placed behind the glass sample plate. Results of these experiments, which are presented in Fig. 10 in comparison with the data for free glass plates, confirm lowered dynamic impedance of the cracked glass behind the failure front.

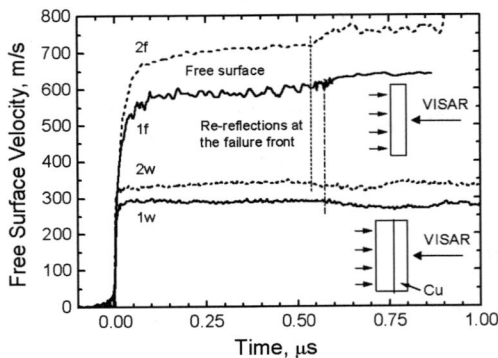

**Figure 10**. Free surface velocity histories of soda lime glass plate 2.5 mm in thickness (waveforms 1f and 2f) and copper witness plate placed behind the glass sample (1w and 2w). The shock stress was 4 GPa in shots 1f and 1w, and 4.8 GPa in shots 2f and 2w.

## MODELS

The glass surface plays an important role in the failure-wave process because the surface is a source of cracks. In this sense the process is similar to diffusion [22]. When the stressed state is maintained, the subsonic failure wave may evidently propagate in a self-supported mode like a combustion wave.

Figure 11 presents examples of computer simulations of the failure wave phenomena based on the combustion-like model. Expansion of a cracked layer at compressive stresses above the failure threshold $\sigma_f = \sigma_{f0} + k_1\sigma_y$ was described by the wave equation [23]

$$\left(\frac{\partial D}{\partial t}\right)_h = c_f \left|\frac{\partial D}{\partial h}\right|_t$$

where $c_f$ is the failure wave speed and $h$ is the Lagrangian space coordinate. Parameter $D$, $0 < D < 1$, characterizes the degree of material fracture. It was assumed that the plate surfaces are

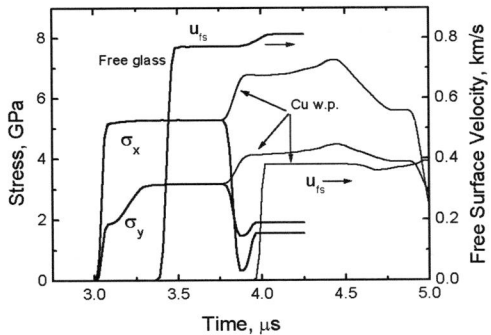

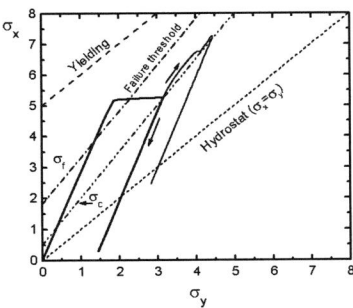

**Figure 11**. Results of simulations of shock waves in free glass plate and a glass plate backed by copper witness plate. Upper diagram presents the stress histories inside the glass plate and the free surface velocity histories of the glass plate and the copper witness plate. Lower diagram shows calculated trajectories of the stress state

potential sources of the failure waves. When the material becomes failed the stress difference relaxes to $\sigma_D = (1-D)\sigma_f + D\sigma_c$, where $\sigma_c$ is the post-failure inter-particle friction in comminuted matter: $\sigma_c = \sigma_{c0} + k_2\sigma_y$. At the stress difference $|\sigma_x - \sigma_y| \le \sigma_D$ the material response is elastic.

The model is obviously simplified but nevertheless it helps to systematize the conclusions made from experimental observations and it provides quite reasonable agreement of simulations with experimental data.

### REFERENCES

1. Galin L.A. and Cherepanov G.I. *Dokl. AN SSSR*, **167**(3), 543 (1966). (In Russian).
2. Grigoryan S.S. *Mekhanika Tverdogo Tela*, 1977, no 1, 173; Slepyan L.I., ibid, 181 (In Russian).
3. Brar, N.S. In: Shock Compression of Condensed Matter – 1999, eds: M.D. Furnish et al., AIP CP **505**, 601 (2000).
4. Kanel G.I., Razorenov S.V., Fortov V.E. Shock-Wave Phenomena and the Properties of Condensed Matter. New York, Springer, 320 pp. (2004).
5. S. J. Bless and N. S. Brar. Failure Waves and their Effects on Penetration Mechanics in Glass and Ceramics. In: High-Pressure Shock Compression of Solids. New York, Springer, in print (2005)
6. Arndt J. and Stoffer D. *Phys. and Chem. of Glasses*, **10**(3), 117 (1969).
7. Gibbons R.V. and Ahrens T.J. *J. Geophys. Res.*, **76**(23), 5489 (1971).
8. Kanel G.I., Razorenov S.V., Utkin A.V., Hongliang He, Fuqian Jing, and Xiaogang Jin. *High Pressure Research*, **16**, 27 (1998).
9. Kanel G.I., Bogach A.A., Razorenov S.V., Savinykh A.S., Chen Z., and Rajendran A. A. In: Shock Compression of Condensed Matter – 2003, Eds. M. D. Furnish et al., AIP CP **706**, pp. 739 (2004).
10. Kanel G.I., Bogatch A.A., Razorenov S.V., Zhen Chen. *J. Appl. Phys.*, **92**(9), 5045 (2002).
11. Bourne N.K., Rosenberg Z., Field J.E. *J. Appl. Phys.* **78**, 3736 (1995).
12. Senf H., Strausburger E., and Rothenhausler H. In: *Metallurgical and Material Applications of Shock-Wave and High-Strain-Rate Phenomena* (eds. L.E. Murr et al.) Elsevier, Amsterdam, 163 (1995).
13. Raiser G., Wise J.L., Clifton R.J., Grady D.E., and Cox D.E. *J. Appl. Phys.* **75**(8), 3862 (1994).
14. Bourne N. K., Millet J.C.F., and Rosenberg Z. *J. Appl. Phys.* **81**(10), 6670 (1997).
15. Rasorenov S.V., Kanel G.I., Fortov V.E., Abasehov M.M. *High Pressure Research*, **6**, 225-232 (1991).
16. Brar N.S. and Bless S.J. *High Pressure Research*, **10**, 773 (1992).
17. Dandekar D.P. and Beaulieu P.A. In: *Metallurgical and Material Applications of Shock-Wave and High-Strain-Rate Phenomena*. Edited by L. E. Murr et al. Elsevier, Amsterdam, Netherlands, 211 (1995).
18. Brar N.S., Bless S.J., and Rozenberg Z. *Appl. Phys. Lett.* **59**(26), 3396 (1991).
19. Bourne N. K, Rosenberg Z., and Millet J.C.F. In: *Structures under Shock and Impact IV*, N. Jones et al., eds. Computational Mechanics Publications, Southampton, 553 (1996).
20. Dandekar D.P. *J. Appl. Phys.* **84**(12), 6614 (1998).
21. Millet J., Bourne N., and Rozenberg Z. *J. Appl. Phys.* **84**(2), 739 (1998).
22. Z. Chen, R. Feng, X. Xin, L. Shen. *Internat. J. Numerical Methods in Engineering*, **56**(14), 1979 (2003).
23. Y. Partom. *Int. J. Impact Engng.* **21**(9), 791 (1998)

CP845, *Shock Compression of Condensed Matter - 2005*,
edited by M. D. Furnish, M. Elert, T. P. Russell, and C. T. White
© 2006 American Institute of Physics 0-7354-0341-4/06/$23.00

# A STUDY OF THE FAILURE WAVE PHENOMENON IN GLASSES AT PEAK STRESSES EXCEEDING THE HEL

## G.I. Kanel[1], S.V. Razorenov[2], A.S. Savinykh[2], A. Rajendran[3], and Zhen Chen[4]

[1] *Institute for High Energy Densities, IVTAN, Izhorskaya 13/19, Moscow, 125412 Russia*
[2] *Institute of Problems of Chemical Physics, Chernogolovka, 142432 Russia*
[3] *U.S. Army Research Office, RTP, NC 27709-2211, USA*
[4] *Department of Civil and Environmental Engineering, University of Missouri-Columbia, Columbia, Missouri, 65211-2200, U.S.A.*

**Abstract.** Shock-wave experiments with two glasses of different hardness have been carried out at shock stress levels above the Hugoniot elastic limit. A comparison between the measured free surface velocity histories from two plate impact experiments performed at approximately the same shock stress level (one with a single thick target plate, and the other with several adjacent target plates of total thickness equal to that of the thick target plate) revealed: 1) at shock loading the failure wave is not formed at stress levels above the HEL, indicating suppression of the fracture process by plasticity, 2) at gradual compression the failure wave process occurs as the stress increases above the failure threshold up to the stress at which plastic deformation begins.

**Keywords:** Glass, failure wave, compressive fracture, shock compression
**PACS:** 61.43.Fs, 62.20.Mk, 62.50.+p, 83.60.Uv.

## INTRODUCTION

The failure wave is a network of cracks that are nucleated on the glass surface and propagate with a subsonic speed into the stressed specimen (see [1-4] and references herein). The existing experimental data are not sufficient to identify what the threshold conditions are for the failure wave phenomena, and what the relationship is between the ductile yielding and compressive fracture of the glasses. In this study two different glasses were tested at the peak stresses both below and above the corresponding Hugoniot elastic limit (HEL). Plane wave experiments are performed with single thick glass plates and layered assembles of thin plates. Measuring the response of layered thin plates, as compared with that of a single thick plate, was performed to diagnose the formation of a failure wave at different loading conditions. It has been shown [5] that in glass laminates the magnitude of the leading elastic wave decreases as a result of its decomposition into two waves at each interface.

## MATERIALS

The materials tested were soda lime glass of 2.45 $g/cm^3$ density having longitudinal sound speed $c_l = 5.72\pm0.08$ km/s and TF1 heavy flint glass of 3.86 $g/cm^3$ density and $4.04\pm0.06$ km/s sound speed. The glass plates were 1 mm to 8 mm in thickness. The lateral dimensions were large enough to provide the condition of uniaxial shock compression. Since it was found earlier [5] that the shock-wave behavior of lapped glass is much more reproducible than that of as-received plates, the glass plates were lapped with SiC powder of 40 μm grain size. When thin plates were assembled into layered specimens, the difference between the measured thickness of the layered specimens and the sum of the plate thicknesses was due to the gaps (~3±1 μm) between layered plates.

## EXPERIMENTAL PROCEDURE

Plane shock waves were created by impacting the specimen with a flyer plate. The aluminum impactor plates 2 mm to 7 mm in thickness were launched, using explosive devices, with the velocities from 0.7 km/s to 1.9 km/s. In addition, the peak stress in the glass specimen was varied using intermediate base plates of different dynamic impedances. The free-surface velocity profiles or the velocity histories of the interface between the specimen and a water window were recorded with the VISAR. The laser beam of the VISAR was reflected by an aluminum foil of 8 μm in thickness which was glued upon the sample surface.

## RESULTS OF MEASUREMENTS

Figure 1 presents results [6] of two shots with soda lime glass at peak stresses below and above the HEL. The transition through HEL is indicated by sharp increase of the velocity of unloading front. Spallation was not observed in these shots, which means that the spall strength of glass remains high after shock compression both below the HEL and above it. On the other hand, the wave reverberation time $t_r = 1.95$ μs in the shot at lower peak stress is distinctly less than expected time $\Delta t = 2h/c_l = 2.33$ μs, where $h$ is the plate thickness. The decrease of the reverberation time is a result of reflection of the tensile pulse in glass from cracked layer near the impact surface. At peak stress above the HEL, there are not any evidences of cracking.

Figure 2 presents the experimental results with assemblies of thin soda lime glass plates impacted at peak stresses of 6.3 GPa [5] and 8.5 GPa, respectively. In both cases, the initiation of a failure wave at each plate surface transformed the elastic compression wave in glass into typical two-wave configuration. The time interval between the leading elastic compression wave and following "plastic" compression wave in Fig. 2 is not proportional to the thickness of glass plate assemblies. In other words, the propagation speed of the second compressive wave in the assembly grows with increasing the peak stress.

In the shot at 8.5 GPa the peak stress was in the region of plasticity of the soda lime glass. This is confirmed by a small interval of time between the compression and rarefaction fronts in this shot: the unloading at glass/water interface begins at 0.66 μs after the compression front whereas calculated pulse duration at the impact surface is a little bit more than 1 μs. In other words, the propagation speed of the unloading front essentially exceeds that of the compression front that means irreversible behavior of the sound speed.

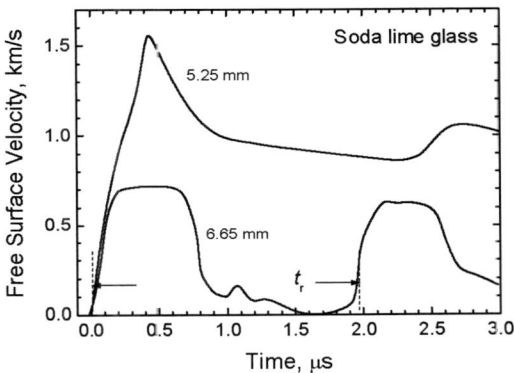

**Figure 1.** Free surface velocity histories of soda lime glass specimens impacted by aluminum flyer plates with the velocities 0.7 km/s and 1.9 km/s [6]. In the later case the impactor was backed by thick paraffin layer.

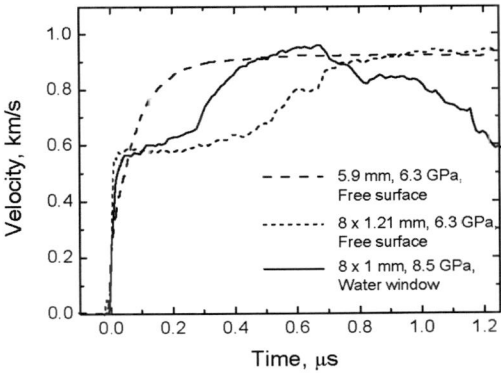

**Figure 2.** Experimental results with assemblies of soda lime glass plates of different thicknesses at different peak stresses in comparison with the free surface velocity history of a single glass plate of 5.9 mm in thickness.

Figure 3 shows the free surface velocity history with the assembly of 5 thin soda lime glass plates in comparison with the results for single glass plates at the same and lower peak stresses. The shock pulse obviously decayed to approximately 9.5 GPa near the sample rear surface because of the relatively small thickness of the flyer plate. With the same total sample thickness and the same peak stress, the total rise time is less for the glass plate assembly than that for a single glass plate. A reason of this discrepancy is partially due to the thin gaps between thin glass plates. The total time of propagation of the wave front through the assembly is the sum of the individual time of wave propagation through each plate and the time of closing these gaps. Since the wave speed is much higher than the speed of closing the gaps, even a very thin gap markedly reduces the average propagation velocity of the wave front. The gaps become closed ahead of the upper part of compressive wave so the velocity of the latter in the assembly is the same as that in a single plate.

The waveform for the glass plate assembly demonstrates a steeper plastic part than that for the single plate. The velocity "pullback" in the unloading part of the stress pulse is less for the assembly than that for the single plate because the assembly can not sustain tension at the interfaces between plates. The velocity oscillations in the residual part of the waveform are the result of wave reverberations inside the last plate of the assembly. However, the period of these oscillations is about 0.17 μs whereas its expected value for undamaged plate is $\Delta t_{exp} = 2\delta_{lp}/c_l = 0.42$ μs (where $\delta_{lp}$ is the thickness of last plate). We may conclude that a part of the last glass plate in the assembly has been damaged by the failure wave. In this case, a steeper plastic part of the waveform implies that the comminuted glass has less viscosity than undamaged material. On the other hand, the reverberation time of 0.17 μs in the shot at the peak stress of 9.5 GPa is less than 0.25 μs of the time when the recompression signal appears in the free surface velocity history at the 6.3 GPa peak stress. We may assume a higher velocity and longer propagation time of the failure waves in the shot at high stress.

Figure 4 summarizes the experimental results with thick single plates and assemblies of thin plates of the TF1 heavy flint glass. Possible HEL value of this glass corresponds to the free surface velocity between 0.45 km/s and 1.0 km/s or to the stress level between 3.5 GPa and 7 GPa. Response of this glass to the stress above 7 GPa is certainly plastic. Thus, all shots have been done at the peak stresses in the vicinity of the HEL or above it.

No signatures of the failure wave process were

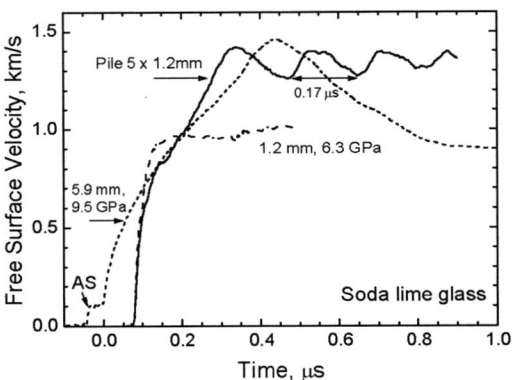

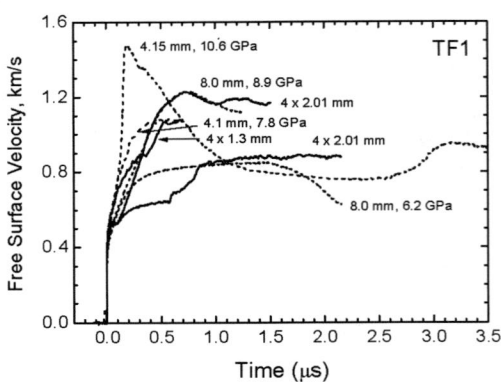

**Figure 3**. The free surface velocity history with the assembly of 5 soda lime glass plates of 1.2 mm in thickness impacted by an aluminum flyer plate 2 mm thick at 1.9±0.05 km/s impact velocity, in comparison with the data for single plates at the same and lower peak stresses. A weak velocity step AS is due to an air shock in front of the flyer plate.

**Figure 4**. The free surface velocity histories of single plates and assemblies of the TF1 heavy flint glass shocked at different peak stresses. The flyer plates for the peak stresses of 10.6 GPa and 8.9 GPa were 2 mm and 4 mm in thickness and their impact velocities were 1.9±0.05 km/s and 1.5±0.05 km/s, respectively. At lower stresses Al impactors 7 mm thick were used.

observed in the shot at 10.6 GPa peak stress. The shots at the peak stress 8.9 GPa were done with a 4-mm-thick aluminum flyer plate backed by paraffin. The small period of the velocity oscillations in the shot with the assembly is similar to that observed for the soda lime glass, and may be considered as a possible indicator of the failure wave process. Further lowering the peak stress resulted in increasing the difference between the waveforms for single plates and assemblies. The distinct two-wave configuration is formed in the assembly at the peak stresses of 8 GPa and less. It appears that the failure wave processes occurred, at least partly, in all three shots below 9 GPa. The failure threshold is in the vicinity of 4 GPa for this glass, which is very close to the HEL value.

## DISCUSSION AND CONCLUSION

The new experimental results with single glass plates impacted at the peak stress much above the HEL do not reveal any evidence of fracture in this stress range. Under gradual compression with a smoothed wave, however, the cracking may occur within a short time interval between the moment when the failure threshold is reached and the time when the plastic yielding begins.

Figure 5 illustrates the propagation of a compressive wave and the failure waves through a glass plate assembly. As a result of anomalous compressibility, the shock discontinuity is transformed to a ramped compressive wave. The

failure waves are nucleated at each plate surface in the assembly when the failure threshold is reached, and are stopped when they meet the part of compressive wave where plastic yielding occurs. The time and distance of propagation of the failure waves are the smallest for the first plate in the assembly but they increase as the compressive wave propagates through the assembly. In other words, contribution of the failure wave phenomenon is negligible near the impact surface, and increases with the propagation of the compressive wave through the assembly.

The failure waves moving in the impact direction meet the yielding threshold after a longer time of propagation than that for the failure waves moving in the opposite direction. This produces an asymmetry in the distribution of microcracking with respect to the interfaces between plates that agrees with the asymmetry in localized inelastic deformation following the shock compression of a glass above its HEL observed earlier [1,7].

## ACKNOWLEDGEMENTS

The work was supported in part by the US Army Research Office under contracts numbers N62558-02-M-6020 and N62558-03-M-0038, and by the Basic Research Program of Russian Academy of Sciences "Physics and mechanics of strongly compressed matter and the problems of interior structure of earth and planets."

## REFERENCES

1. G.I. Kanel, S.V. Razorenov, V.E. Fortov. "Shock-Wave Phenomena and the Properties of Condensed Matter." Springer, New York, 2004, 320 pp.
2. Brar, N.S. In: Shock Compression of Condensed Matter – 1999, eds: M.D. Furnish et al., AIP CP **505**, 601-606 (2000).
3. Bourne, N. K., Rosenberg, Z., Field, J. E. *J. Appl. Phys.* **78**, 3736-3739 (1995).
4. Bourne, N. K. and Millett, J.C.F. *Proc. R. Soc. Lond.* **A 456**, 2673-2688 (2000)
5. Kanel, G.I., Bogatch, A.A., Razorenov, S.V., Zhen Chen. *J. Appl. Phys.*, **92**(9), 5045-5052 (2002).
6. Rasorenov, S.V., Kanel, G.I., Fortov V.E., Abasehov, M.M.. *High Pressure Research*, **6**, 225-232 (1991).
7. Kanel, G.I., and Molodets, A.M. *Sov. Phys.–Tech. Phys.* **21**(2), 226–232 (1976).

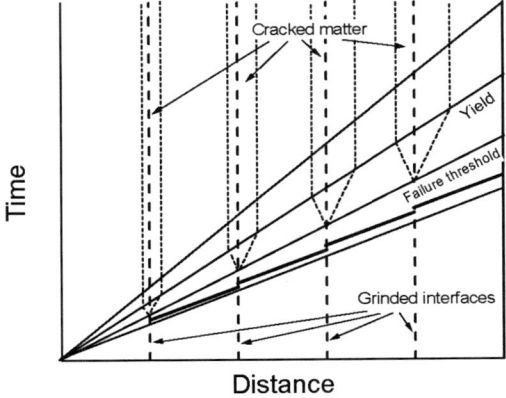

**Figure 5.** Assumed time–distance diagram of the failure wave phenomena in a glass plate assembly impacted above the HEL.

CP845, *Shock Compression of Condensed Matter - 2005*,
edited by M. D. Furnish, M. Elert, T. P. Russell, and C. T. White
© 2006 American Institute of Physics 0-7354-0341-4/06/$23.00

# STUDY OF COMPRESSIVE FAILURE OF ALUMINA IN IMPACT EXPERIMENTS WITH DIVERGENT FLOW

## V.E. Paris and E.B. Zaretsky

*Ben-Gurion University of the Negev, P.O.B. 653, Beer-Sheva 84105, Israel*

**Abstract.** Axisymmetric divergent flow characterized by increasing (with propagation distance) difference between longitudinal and radial stress was produced in the plane-parallel alumina samples by impact of spherical (R=200-600 mm) copper impactors having velocities 210 to 260 m/s. The velocity of the interface between the impacted 5-mm alumina samples and 6-mm sapphire windows was continuously monitored by VISAR. Preliminary AUTODYN simulations show that such impact is capable of producing in the sample the stress states which cannot be produced by planar impact loading and which may result in the brittle failure of alumina. Actually, the waveforms recorded in these experiments contain distinct signatures of the alumina failure. AUTODYN numerical simulations of the experiments allow one finding the alumina failure threshold, the path of the increasingly damaged alumina, the kinetics of this damaging and the locus of the states of comminuted material in the principal stress space. Possible applications of the developed experimental/numerical technique and its limitations are discussed.

**Keywords:** Brittle compressive failure threshold, divergent flow, AUTODYN simulations, alumina.
**PACS:** 83.60. Uv, 62.20.Mk.

## INTRODUCTION

Predicting the response of a brittle material on the three-dimensional impact loading requires the knowledge of its compressive failure threshold, failure kinetics, and post-failure behavior as a function of the acting stresses. Planar impact experiments are widely used for obtaining quantitative information concerning the dynamic inelastic response of materials. This information, however, does not allow unambiguous attributing of the material transition at Hugoniot Elastic Limit (HEL) to either ductile or brittle mode of inelastic behavior. Moreover, due to the loading conditions of these experiments (1-D strain, parallel material flow) the relation between transversal $\sigma_2$ and longitudinal $\sigma_1$ principal stresses stays, below HEL, constant and equal to $\sigma_2/\sigma_1 = \nu/(1-\nu)$, while

in the real impact it may vary within wide limits even though the Poisson's ratio $\nu$ is independent of stress.

The objective of this work is to extend the shock-wave testing capabilities towards generating in the studied material the divergent, non-parallel, flow providing a possibility to vary the $\sigma_2/\sigma_1$ ratio and, thus, to obtain experimental information about the mode (brittle or ductile) and conditions of the material compressive failure. Such divergent loading of alumina samples was realized using spherical copper impactors with relatively large (200 mm to 600 mm) radius of curvature. The stress states corresponding to the material failure and post-failure behavior were determined via numerical simulations (AUTODYN™-2D) of the VISAR [1] records of the waveforms generated by the divergent loading in the alumina samples.

## DESIGN OF EXPERIMENT WITH DIVERGENT FLOW

Hot pressed alumina ceramic was chosen for impact testing under divergent loading conditions. It has been shown [2] that under planar impact loading the inelastic alumina response, starting above $\sigma_1 = \sigma_{HEL} \approx 6$ GPa and $\sigma_2 = \sigma_3 \approx 1.6$ Gpa, is ductile. (Here and in follows the stress is positive in compression.) On the other hand, Heard and Cline found that under static compression accompanied by lateral confinement up to 1.25 GPa ($\sigma_1 \approx 6$ GPa, $\sigma_2 = 1.25$ GPa) the inelastic behavior of the alumina stays brittle [3]. These data, together with those obtained in rod impact experiments [4, 5] ($\sigma_2 = 0$, $\sigma_1 = 2.8 \div 3.6$ GPa) allow us to locate, at the principal stress space, possible compressive failure threshold of the alumina, Fig. 1. Consequently, the impact experiments with the divergent material flow should provide crossing the threshold line by the material trajectories; under planar impact loading, Fig. 1, the crossing is impossible.

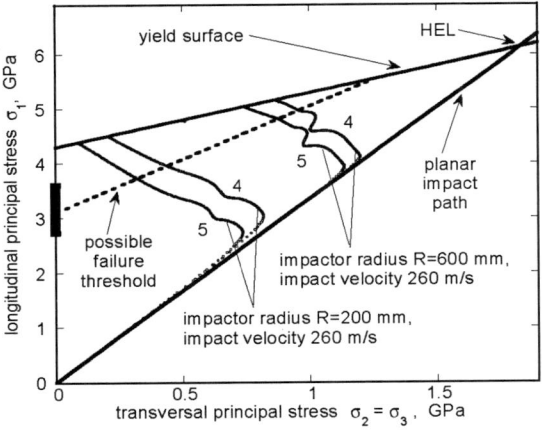

**Figure 1.** Planar impact path $\sigma_1 = \sigma_2(1-\nu)/\nu$ and yield surface $\sigma_1 = Y + \sigma_2$ of alumina in the principal stress space. Solid rectangle corresponds to the literature data on the alumina strength measured by rod impact technique [4, 5]. The trajectories $\sigma_1(\sigma_2)$ correspond to the divergent material flow at the points located on the sample axis 4 and 5 mm apart the impacted surface.

Such possibility was verified in preliminary AUTODYN-based numerical simulations of the stress states produced by the curved copper impactor in the flat alumina target. No compressive failure in alumina was permitted in these runs. AUTODYN library constitutive description was used for copper, while the constitutive parameters of alumina were corrected (see the next Section) to fit the real alumina samples.

The preliminary simulations, Fig. 1, showed that at different points of the sample axis the stress states capable of triggering the compressive failure of alumina may be achieved by varying either the velocity of the impact or the radius of curvature of impactor.

The simulations also showed a surplus outcome of the use of curved impactors: the conservation of momentum requires simultaneous generation of the shear wave together with the longitudinal compressive wave. Stress difference $\Delta\sigma = \sigma_1 - \sigma_2$, corresponding to the divergent flow behind the longitudinal wave and responsible for the failure initiation, is growing with the propagation distance while the arrival of the shear wave halts the increase of $\Delta\sigma$. Thus, the compressive failure is always initiated ahead of the shear wave front, in immediate proximity to it; behind the shear wave no compressive failure can be triggered. On the other hand, the damaged region should evolve, with the failure initiation, an unloading signal propagating through the sample with the longitudinal speed of sound. Knowing the speeds of sound (longitudinal and shear) in the material and the time of the arrival of the unloading signal at the sample surface yields the location of the point where the failure was initiated.

## MATERIALS AND EXPERIMENTAL

The studied samples were 25-mm diameter and 5-mm thickness alumina disks precisely cut (2-$\mu$ tolerance, 0.1 mrad parallelism) from the rod of hot-pressed alumina (99% $Al_2O_3$, Microceramica Ltd., Carmiel, Israel) having density equal to $\rho_0 = 3.902 \pm 0.007$ g/cm$^3$. The longitudinal $C_l$ and the shear $C_s$ speeds of sound, and the Poisson's ratio $\nu$ were measured and found equal to $C_l = 10580 \pm 60$, $C_s = 6230 \pm 40$ m/sec, and $\nu = 0.235 \pm 0.01$.

The experiments were performed with 58-mm gas gun [2]. Planar impacts (2 shots for accurate determination of the yield strength of the alumina, equal to $Y = 4.3 \pm 0.25$ GPa) were performed using impactors made of 2-mm OFNC copper disks glued to 15-mm PMMA backing. For impacts with convex impactors the copper disks, the PMMA backing and the hollow aluminum projectile were glued together prior to the precise CNC-machining of the projectile head for needed radius of curvature. Six such experiments, with impact velocities ranged from 208 to 268 m/s with impactors having radii of curvature $200^{+30}$, $300^{+60}$ and $600^{+90}$ mm, were performed.

In order to prevent premature fracture of the alumina free surface by the tensile stress appearing with the arrival of the spherical wave, the alumina samples were backed by 6-mm sapphire window. The velocity of the alumina-sapphire interface was monitored by VISAR having interferometer constant 216 m/sec per fringe. The VISAR data were corrected using the derivative $k_0 = 0.7864$ [6] of sapphire refractive index on compression. The difference between literature data on sapphire refractive index $n_0 = 1.771763$ [6] ($\lambda = 532$ nm) and that for actual VISAR light $(\lambda = 514$ nm) was neglected as well as the influence of the wave curvature on the index correction. The latter is justified by negligibly small difference $\mu_0 - k_0 = n_0 - 1 - k_0$, to which the curvature correction is proportional [7]. In all the experiments the non-concentricity of sample and impactor axes and the impactor-sample misalignment did not exceed 0.5 mm and 2 mrad, respectively.

## RESULTS OF THE EXPERIMENTS

Typical interface velocity profiles obtained after impact loading by convex impactors are shown in Fig. 2. The profiles contain distinct failure signatures. The fast initial ramp of the interface velocity corresponding, in the principal stress space, to the material path almost coinciding with that of planar impact, Fig. 1, is followed by a slow velocity increase accompanied by the increase of the stress difference $\sigma_1 - \sigma_2$ (owing to the decrease of the radial stress $\sigma_2$ in the divergent flow) terminated by the failure. At the absence of the failure the interface velocity should continue to grow (solid line in Fig. 2) up to the instant of the arrival of the shear wave at the interface.

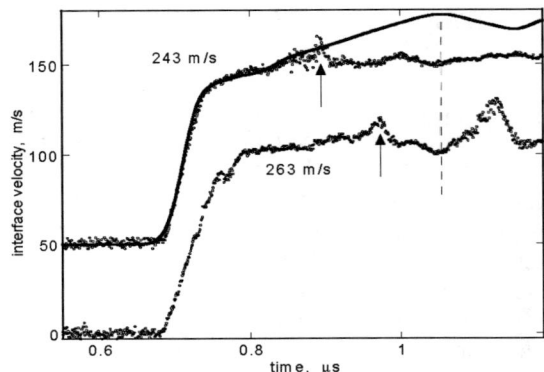

**Figure 2.** VISAR records of the velocity of alumina-sapphire interface after impact by copper impactors having 600-mm radius of curvature. The arrows correspond to the arrival of the failure initiation signal at the interface. Solid line corresponds to the simulation of the 243-m/s experiment with no failure. Dashed line shows the time of the arrival of the shear wave.

## FAILURE MODELING. ALUMINA FAILURE THRESHOLD

The failure signatures of the waveforms obtained after impact loading of the alumina samples by curved copper impactors were reproduced in AUTODYN-2D simulations using linear form of compressive failure threshold suggested by Ashby and Sammis [8]:

$$\sigma_{1br} = \sigma_{0br} + f \cdot \sigma_2. \tag{1}$$

Eq. (1) corresponds to the possible compressive failure threshold of Fig.1. The description of behavior of the comminuted alumina was based on the Mohr-Coulomb post-failure friction criterion:

$$\sigma_{1fr} = \sigma_{0fr} + c \cdot \sigma_2. \tag{2}$$

Parameters $\sigma_{0br}$, $\sigma_{0fr}$, $c$, $f$ are considered as the material constants [8]. When the failure criterion is satisfied, the compressive cracking is initiated. It leads to the increase of the material damage parameter $D(\varepsilon_{ie})$, which varied between zero (no failure) and one (complete failure) and proportional to the increment of inelastic strain $\varepsilon_{ie}$:

$$\delta D = k_D \cdot \delta \varepsilon_{ie}. \tag{3}$$

The coefficient $k_D$ characterizes the damage kinetics. The inelastic strain increment $\delta\varepsilon_{ie}$ is produced by the excess of the principal stress $\sigma_1$ with respect to the compressive failure threshold $\sigma_{1br}$:

$$\delta\varepsilon_{ie} = |\sigma_1 - \sigma_{1br}|/2G \qquad (4)$$

Here G is a shear modulus. The local failure threshold $\sigma_D$ decreases with fracture from $\sigma_{1br}$ and moves towards $\sigma_{1fr}$:

$$\sigma_D = D\cdot\sigma_{1fr} + (1-D)\cdot\sigma_{1br}. \qquad (5)$$

The description of alumina failure, Eqs. (1) - (5), was added to the AUTODYN problem in a form of FORTRAN subroutine. The best-fit curves were obtained for each shot after several iterations started from the draft threshold expression having form of Eq. (1), $\sigma_{1br} = 3.39$ GPa $+ 2.12\cdot\sigma_{1br}$ (GPa). The latter was obtained by locating the failure site from the failure signature time of arrival and, in turn, by finding, from AUTODYN runs without failure, the stress state corresponding to the failure initiation at the site. An example of such best-fit curve is shown in Fig. 3.

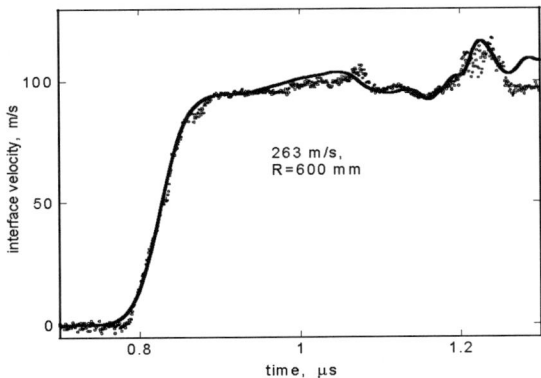

263 m/s,
R=600 mm

**Figure 3.** The velocity of the alumina-sapphire interface recorded after impact of copper impactor having 600-mm radius of curvature and velocity 263 m/s (circles), and corresponding best-fit curve.

Finally, averaging the best-fit parameters found for each shot yield, respectively, the alumina compressive failure threshold and the trajectory of the comminuted alumina particles

$$\sigma_{1br} = \sigma_{0br} + f\cdot\sigma_2 = 3.50\pm0.20 +1.85\cdot\sigma_2 \text{ [GPa]}, \qquad (6)$$
$$\sigma_{1fr} = \sigma_{0fr} + c\cdot\sigma_2 = 1.33\pm0.30 +1.85\cdot\sigma_2 \text{ [GPa]}. \qquad (7)$$

The failure kinetics coefficient is found equal to $k_D = 260\pm40$ and corresponds to relatively fast kinetics of alumina damaging. The failure threshold (6) intersects the ductile yield (von Mises) surface $\sigma_1 - \sigma_2 = 4.3\pm0.25$ GPa at the point having coordinates $\sigma_1 = 5.2\pm0.4$ GPa and $\sigma_2 = 1.0\pm0.25$ GPa. This means that the response of the alumina should become ductile under radial compression of about 1.0 GPa. This figure agrees with the static data of Heard and Cline [3] who found that up to radial compression of about 1.25 GPa the response of alumina stays brittle.

## CONCLUSIONS

The obtained results demonstrate the possibility of accurate determination of compressive failure threshold of brittle materials by such combined experimental/numerical technique. The method, although with lesser accuracy, allows one both to locate, in the principal stress space, the surface at which the comminuted brittle material finally arrives and to obtain important information concerning the brittle failure kinetics. This technique, with some improvement by tightening the control of the impactor shape and by detail description of the brittle failure process, may be used for study of constitutive behavior of brittle solids.

## REFERENCES

1. Barker L. M. and Hollenbach R.E., Journ. Appl. Phys., 43, 4669, 1972.
2. Zaretsky E. B. and Kanel G.I., Appl. Phys. Lett., 81, 7, 2002.
3. Heard H.C. and Cline C.F., J. Mat. Sci., 15, 1889-1897, 1980.
4. Cazamias J.U., Reinhart W.D., Konrad C.H., Chhabildas L.C., and Bless S.J., AIP 2002, 787-790.
5. Simha C.H.M., Ph.D. Thesis, U.T. Austin, 1998.
6. Setchell R.E., J. Appl. Phys., 91, 2833-41, 2002.
7. Wackerle J., Stacy H.L., and Dallman J.C., SPIE Vol. 832, High Speed Photography, Videography, and Photonics V-1987, 72-82.
8. Ashby M.F. and Sammis C.G., PAGEOPH, 133 (3), 490-521, 1990.

CP845, *Shock Compression of Condensed Matter - 2005,*
edited by M. D. Furnish, M. Elert, T. P. Russell, and C. T. White
© 2006 American Institute of Physics 0-7354-0341-4/06/$23.00

# THE DYNAMIC BEHAVIOR OF A FILLED GLASS

## D.D. Radford [1], K. Tsembelis [2] and W.G. Proud [2]

*[1] Dept. of Engineering, University of Cambridge, Trumpington St., Cambridge, CB2-1PZ, UK*
*[2] PCS, Cavendish Laboratory, Madingley Road, Cambridge, CB3 0HE, UK.*

**Abstract.** The in-material longitudinal and lateral stress histories in a filled, silica-based glass (Schott SF-57) have been measured during plate impact experiments using embedded manganin stress gauges. Longitudinal stress histories are presented and used to provide the principal Hugoniot for the material. Lateral stress measurements show a two-wave structure indicative of shock-induced failure. High-speed photographic sequences taken in both configurations provide insight into the pressure-dependent nature of failure, consistent with recent studies on other dense glasses.

**Keywords:** Filled glass, shock induced failure, in-material response.
**PACS:** 62.20-x, 62.20-Mk, 62.50+p.

## INTRODUCTION

The response of silica-based glasses to shock loading has received considerable attention over the past 20 years [1], and one of the ongoing research topics involves shock induced failure in plates and rods [2,3]. Recent work on dense, or filled, glasses has provided new insight into the failure process in brittle materials[4-6] with applications to ballistic protection and brittle fracture.

In the current study, plate impact experiments on a glass denoted Schott SF-57 (SF-57) are presented, and include measured longitudinal ($\sigma_x$) and lateral ($\sigma_y$) stress histories. Also, high-speed photography is used with corresponding gauge records to examine the development of failure.

## MATERIAL DESCRIPTION

SF-57 is a silica-based glass with a filled microstructure resulting in a considerable increase in density ($\rho_0$) compared to open-structured borosilicate, as detailed in Table I. The elastic

wave speeds in the SF-57 were measured with quartz transducers in both the longitudinal and transverse orientations, using a Panametrics 5052PR pulse receiver operating at 5 MHz. The measured longitudinal ($c_L$) and transverse, or shear, wave ($c_S$) speeds, and the density were used to calculate the elastic properties, which are also detailed in Table I. The elastic properties of borosilicate and two dense glasses [4,5] are included and show that the wave speeds in the dense materials are lower, as compared to the borosilicate glass. In addition, the elastic impedance, calculated by $\rho_0 c_L$ is larger in the dense glasses, thereby allowing higher in-material stresses to be attained for a given impact velocity.

## EXPERIMENTAL METHODS

Experiments were performed using the plate impact facility at the Cavendish Laboratory, University of Cambridge [7], which consists of a single-stage 50-mm-bore, 5-m-long light gas gun. The specimen configurations used in this study are standard [5], with longitudinal stress measurements

**TABLE 1.** Elastic Properties of Glasses.

| Glass Type | Density (g cm$^{-3}$) | $c_L$ (mm µs$^{-1}$) | $c_S$ (mm µs$^{-1}$) | $c_0$ (mm µs$^{-1}$) | Poisson's ratio | Impedance (10$^3$ kg m$^{-2}$s$^{-1}$) |
|---|---|---|---|---|---|---|
| **Borosilicate** | 2.23±0.05 | 6.05±0.01 | 3.69±0.01 | 4.30±0.01 | 0.20±0.005 | 13.49 |
| **DEDF** | 5.18±0.05 | 3.49±0.01 | 2.02±0.01 | 2.60±0.01 | 0.25±0.005 | 18.08 |
| **SF-57** | 5.53±0.02 | 3.42±0.01 | 1.98±0.01 | 2.53±0.02 | 0.25±0.005 | 18.91 |
| **Pilkington-RW62** | 5.94±0.06 | 3.27±0.03 | 1.86±0.01 | 2.47±0.02 | 0.26±0.005 | 19.42 |

obtained with embedded piezo-resistive manganin gauges (Micro-measurements type LM-SS-210FD-050) between tiles of the target materials (10-mm-thick), and lateral measurements made with embedded piezo-resistive manganin gauges (Micromeasurements type J2M-SS-580SF-025) in the lateral orientation at ~ 3 mm and 6 mm from the impact face. Gauges were calibrated for each experiment according to the work of Rosenberg et al. [8,9], and the target plates (100-mm-diameter) were assembled using a low viscosity epoxy adhesive. Specimen alignment was fixed to be less than 1 m rad by means of an adjustable specimen mounting.

All flyer plates used were 10-mm-thick (type 101-Cu), and impact velocities were measured by the shorting of sequential pairs of pins to an accuracy of ± 1 %. High-speed photographic sequences were obtained using an Imacon UltraNAC-501 programmable image-converter camera capable of taking $2 \times 10^7$ frames per second. The outer surfaces of the specimens were polished to allow visualization of the shock propagation and failure process within the material, and an exposure time of 100 ns was used for all photographs.

## LONGITUDINAL RESPONSE

A total of seven experiments were performed on the SF-57 material in the longitudinal configuration with impact velocities $v_0$ ranging from 191 to 831 m s$^{-1}$. Measured longitudinal stress histories for experiments conducted at $v_0 = 191$ m s$^{-1}$, 500 m s$^{-1}$, and 831 m s$^{-1}$ are plotted in Fig. 1, which resulted in equilibrium stresses of 2.58 GPa, 5.46 GPa, and 10.54 GPa, respectively.

From Fig. 1 it is seen that there is a change in the rate of increase at ~ 4.5 GPa for the experiments performed at $v_0 = 500$ m s$^{-1}$ and 831 m s$^{-1}$, which was taken to represent the HEL for the material. This value agrees closely with the measured HEL for the two other filled glasses [4,5,10] listed in Table I. The trace for the experiment performed at 831 m s$^{-1}$ shows a two-wave structure prior to reaching the equilibrium stress, which is similar to that observed in the DEDF glass [5,10] above the HEL. The dense glass (Pilkington-RW62), however, showed a unique densification behavior under shock compression for impact pressures in the range 4 GPa to 10 GPa [4].

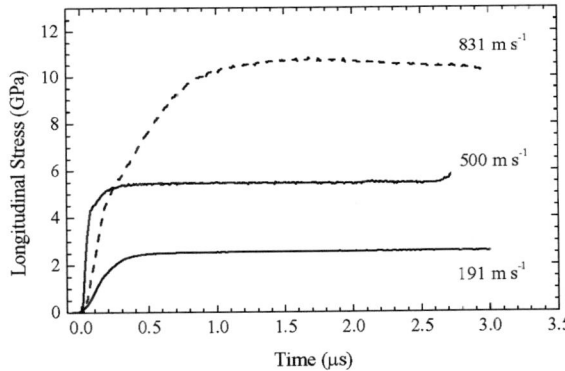

**Figure 1.** Longitudinal stress histories measured in SF-57 at impact velocities of 191 m s$^{-1}$, 500 m s$^{-1}$, and 831 m s$^{-1}$.

The principal Hugoniot for the SF-57 glass is plotted in stress-particle velocity ($u_p$) space in Fig. 2. Below the estimated HEL of 4.5 GPa the data follow the elastic line ($z = \rho_0 c_L u_p$). Above the HEL, the data lie below the elastic line until ~ 10 GPa where it crossed the principal Hugoniot. For the Pilkington-RW62 [5,10] glass listed in Table I, it was observed that Hugoniot exceeded the elastic response at ~ 15GPa. Included in Fig. 2 are data from a similar dense glass to the Schott SF-57 tested in this study, which is denote Schott Optical glass in reference [11] having a lower density of 5.085 g cm$^{-3}$. It is seen that the two dense glasses have a similar Hugoniot, over the range of pressures considered.

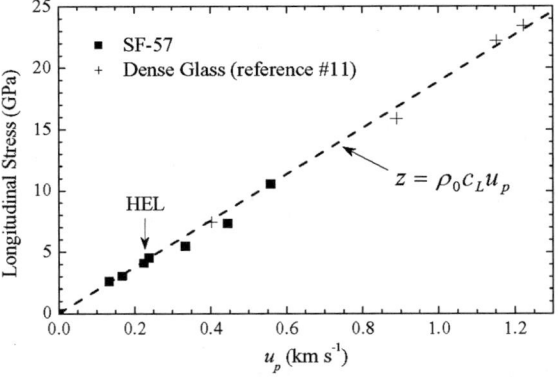

**Figure 2.** Comparison of the measured Principal Hugoniots for SF-57 and another dense glass (Scott Optical Glass[11]).

A high-speed sequence taken during an experiment performed at 831 m s$^{-1}$ in the longitudinal configuration is shown in Fig. 3, where frame 1 (F1) shows the target plate $\sim$ 100 ns

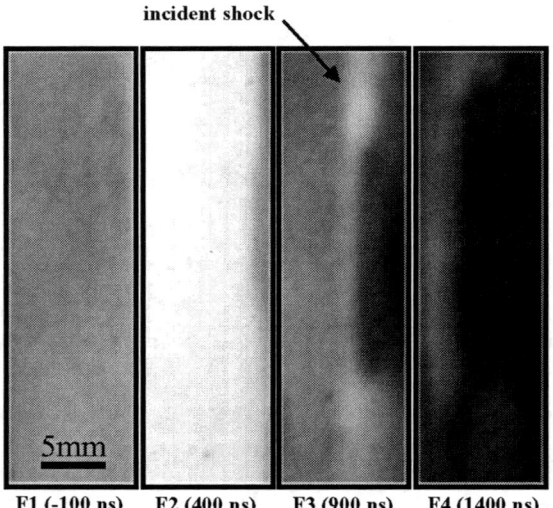

F1 (-100 ns)    F2 (400 ns)    F3 (900 ns)    F4 (1400 ns)

**Figure 3.** High-speed photographic sequence of longitudinal experiment ($v_0 = 831$ m s$^{-1}$) showing development of failure behind incident shock. Times relative to impact indicated.

prior to impact. In F2, it is seen that the entire target plate is illuminated, due fracto-emission at the impact face similar to observations in DEDF [5]. In F3, the incident shock is visible as a light zone with material behind illuminated due to the formation of cracks. However, as the cracks in the comminuted material open the failed material appears opaque (F3 and F4), eventually filling the region behind the incident shock and propagating at $c_L$. In this specific experiment ($v_0 = 831$ m s$^{-1}$) the dark zone propagated with the incident shock as the gauge was traversed, resulting in the equilibrium stress being recorded in failed material, consistent with observations in reference [5].

## LATERAL RESPONSE

The measured lateral stress histories at $\sim$ 3 mm and 6 mm from the impact face for an experiment performed at $v_0 = 503$ m s$^{-1}$ are plotted in Fig. 4. At 3 mm (G-1) it is seen that $\sigma_y$ initially rises quickly to $\sim$ 2 GPa and then gradually continues to increase, finally reaching $\sim$ 5.3 GPa after $\sim$ 2 $\mu$s. At 6 mm (G-2), a distinct two-wave structure is evident with an initial plateau at $\sim$ 1.8 GPa followed by an increase to a plateau at $\sim$ 5.2 GPa. This behavior is consistent with observations in other filled glasses [6], and is associated with the propagation of a failure front behind the shock.

Corresponding high-speed photographs are presented in Fig. 5, with the times of the photos indicated by F1 to F4 in Fig. 4. Frame 1 shows the location of the gauges indicated as G-1 and G-2, with the incident shock approaching from the right side (impact face). The opaque zone behind the shock is comminuted, or failed, material. Frame 2 shows that as the shock traverses G-1, the trailing failure zone is consistently dark. As the shock passes G-2 (F3 and F4), however, it is seen that the material behind has a lighter region followed by the dark zone. Referring to the traces in Fig. 4, it is seen that the arrival of the shock at G-2 causes an increase in lateral stress to the first plateau, and the arrival of the darkest zone (failure front) is the onset of the rise to the second plateau. These results are consistent with observations in DEDF [5], and represent the transient nature of the failure process in filled glasses.

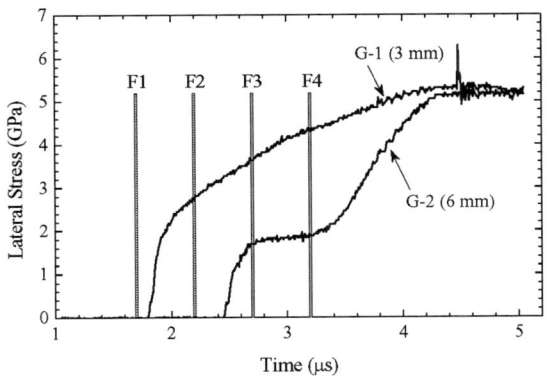

**Figure 4.** Lateral stress histories in SF-57 at ~ 3 mm (G-1) and 6 mm (G-2) from the impact face ( $v_0$ = 503 m s$^{-1}$). Corresponding photographs (Fig. 5) taken at times F1 to F4.

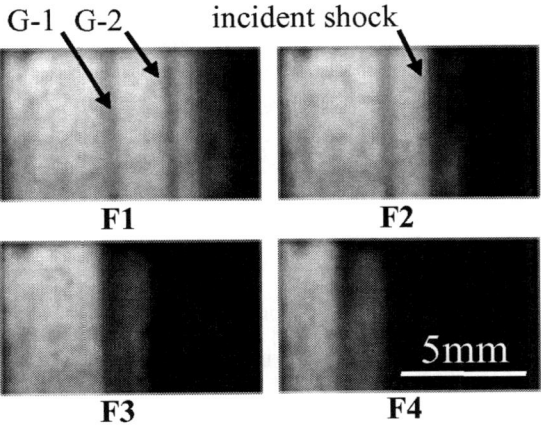

**Figure 5.** High-speed photographs taken during lateral experiment ( $v_0$ = 503 m s$^{-1}$) showing shock and failure front as it propagates past gauge locations G-1, G-2 (Fig. 4).

## CONCLUDING REMARKS

In-material longitudinal and lateral stress histories were measured and show the pressure-dependent nature of failure, consistent with previous studies on dense glasses. The use of high-speed photography and simultaneous gauge records have further shown that the equilibrium stress used to construct the principal Hugoniot above the HEL is measured in failed material. Similar to observations in DEDF glass, the development of a failure zone, or front, occurs as a result of shock induced crack initiation and coalescence.

## ACKNOWLEDGEMENTS

Financial support by QinetiQ, and the continued encouragement of T. Andrews, P.D. Church, and I.G. Cullis is acknowledged. We also thank Prof. J.E. Field for useful comments, and acknowledge D.L.A. Cross and R.P. Flaxman for technical support. High-speed photography at the Cavendish laboratory is supported by the EPSRC.

## REFERENCES

1. N S. Brar and H. D. Espinosa, Chem. Phys. Rep. **17**, 317-342 (1998).
2. N. S. Brar, in *Shock Compression of Condensed Matter - 1999*, edited by R. S. Hixson (American Institute of Physics, Melville, New York, 2000), p. 601-606.
3. G. R. Willmott and D. D. Radford, Journal of Applied Physics **97**, (2005), 093522-1.
4. D. D. Radford, K. Tsembelis, and W. G. Proud, Proceedings of the Royal Society (Trans A.) **(submitted)** (2005).
5. D. D. Radford, Journal of Applied Physics **(accepted for publication)** (2005).
6. D. D. Radford, W. G. Proud, and J. E. Field, in *Shock Compression of Condensed Matter - 2001*, edited by M. D. Furnish, N. N. Thadhani, and Y. Horie (American Institute of Physics, Melville, NY, 2002), p. 807-810.
7. N. K. Bourne, Z. Rosenberg, D. J. Johnson, J. E. Field, A. E. Timbs, and R. P. Flaxman, Meas. Sci. Technol. **6**, 1462-1470 (1995).
8. Z. Rosenberg, Y. Partom, and B. Keren, J. Appl. Phys. **54**, 2824-2826 (1983).
9. Z. Rosenberg, D. Yaziv, and Y. Partom, J. Appl. Phys. **51**, 3702-3705 (1980).
10. N. K. Bourne, J. C. F. Millett, and Z. Rosenberg, Proc. R. Soc. Lond. A **452**, 1945-1951 (1996).
11. S. P. Marsh, *LASL Shock Hugoniot Data* (University of California Press, Berkeley, California, 1980).

CP845, *Shock Compression of Condensed Matter - 2005*,
edited by M. D. Furnish, M. Elert, T. P. Russell, and C. T. White
© 2006 American Institute of Physics 0-7354-0341-4/06/$23.00

# COMPRESSIVE FRACTURE OF BRITTLE MATERIALS UNDER DIVERGENT IMPACT LOADING

## A. S. Savinykh[1], G. I. Kanel[2], S. V. Razorenov[1], A. Rajendran[3]

*[1]Institute of Problems of Chemical Physics of Russian Academy of Sciences,
Chernogolovka, Moscow region, 142432 Russia*
*[2]Institute for High Energy Densities of Russian Academy of Sciences, Moscow, 125412 Russia*
*[3]U.S. Army Research Laboratory, ARO, RTP, NC 27709-2211*

**Abstract.** The main objective of this work was to extend the techniques of shock-wave testing of brittle materials upon divergent loading conditions in order to vary the relationship between longitudinal and transversal stresses and to obtain experimental information about the conditions of the compressive fracture thresholds. Experiments with plane and divergent shock loading of alumina and boron carbide ceramic plates have been carried out. The results of measurements outlined the range of stressed states which are below the failure criterion.

**Keywords:** Ceramics, Hugoniot elastic limit, compressive fracture, alumina, boron carbide
**PACS:** 07.35.+k, 62.50.+p, 81.70.Bt.

## INTRODUCTION

When impact or explosive attack happens, inelastic deformation, fracture, and fragmentation occur under conditions of divergent flows. Planar impact experiment is a widely used way to obtain quantitative information about the resistance of materials to inelastic deformation at high-rate compression. However, the uniaxial strain condition of planar impact tests does not allow the variation of relationship between longitudinal and transversal stresses whereas in divergent flows transversal stresses are decreasing with strains faster than longitudinal ones. The main objective of this work was to extend the techniques of shock-wave testing of brittle materials upon divergent loading conditions in order to vary the relationship between longitudinal and transversal stresses and to obtain experimental information about the conditions of compressive fracture. This work is not a first study of divergent shock loading of brittle materials. Grady [1] and Fowles [2]

analyzed propagation of spherical waves. Measurements under conditions of divergent shock loading were done by Tranchet and Collombet [3] for spherical samples and by Kanel et al. [4] for tube-like cylindrical samples using explosive facilities.

## EXPERIMENTAL PROCEDURE

The divergent impact loading was realized using convex flyer plates. The convex shape of impactors was formed during its launching by explosive facilities shown in Fig. 1. Collision of spherical impactor with plane target creates divergent shock wave in the latter as it is shown in Fig. 2. During initial phase of collision the shock waves in the target and in the impactor are emanated from the intersection line between the impactor and the target. For spherical impactor the intersection line moves in accordance with the relationship

$$r^2 + (R_i - u_i t)^2 = R_i^2$$

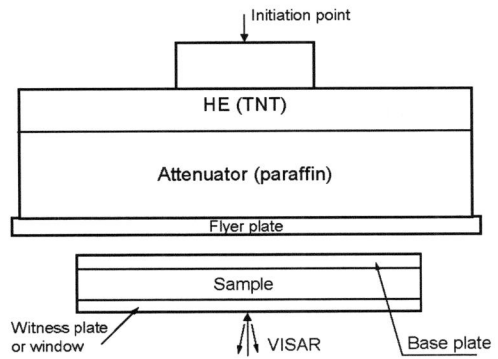

**Figure. 1.** Scheme of experiment with divergent impact loading of ceramic samples.

where $r(t)$ is the radius of intersection line, $u_i$ is the impact velocity, $R_i$ is the impactor radius. If to assume spherical shape of the elastic shock wave in a target, it can be shown that initial radius of curvature at the impact axis is

$$R_S = R_i u_i / c_l$$

where $c_l$ is the propagation speed of longitudinal elastic wave in the target.

At stresses below the elastic limit a collision of convex flyer plate with plane elastic target generates not only divergent longitudinal wave but also shear wave in the target. The flow geometry is illustrated in Fig. 3. The shear wave separates a region 1 of highly divergent flow behind the spherical longitudinal shock wave and a region 2 of low-divergent flow [5]. Roughly, the divergence of flow in the region 1 is characterized by the radius of curvature of the longitudinal wave, which is of order of a few centimeters in the experiments discussed below. Divergence of flow in the region 2 is characterized by the curvature of impact surface which is of order of a few tens of centimeters. The leading longitudinal shock wave decays in reverse proportion to its radius. In the region 2 the stresses decrease much slower.

The facility shown in Fig. 1 launched convex-shape copper impactors 5.5 mm in thickness with the curvature radius 36.5 cm and the impact velocity 555 m/s, and aluminum impactors 7 mm in thickness with the curvature radius 19.5 cm and the impact velocity 1475 m/s. In order to decrease the peak stress of shock compression, the ceramic samples were placed on intermediate aluminum,

Teflon, or PMMA base plates which were 4 mm to 6 mm in thickness. In the experiments, the free surface velocity histories of copper witness plates 2.6 mm in thickness were recorded. The use of copper witness plates placed behind the samples allowed to support compressive stresses in the sample and to improve the quality and reliability of the registrations.

## MATERIALS

The materials tested were alumina and boron carbide ceramics. B6 cold pressed and sintered alumina ceramic of 1800 HV hardness contains 96% of $AlB_2O_3$, has 0% porosity, the density $\rho = 3.85$ g/cm$^3$, longitudinal sound speed $c_l = 10.15$ km/s. The boron carbide ceramic was hot-pressed, of 1 μm to 10 μm grain size, of 2.52 g/cm$^3$ density and of 13.62±0.1 km/s longitudinal sound speed. Tested samples were grinded plates with lateral

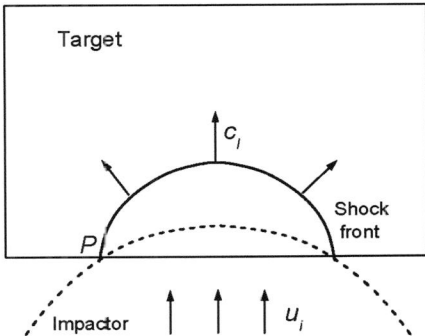

**Figure 2.** Generation of divergent shock wave in a plane target by spherical impactor.

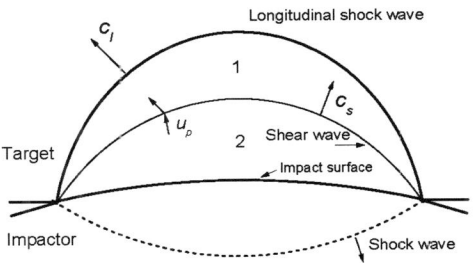

**Figure 3.** A wave configurations at impact of convex flyer plate upon elastic target plate.

dimensions of 80×80 mm² and 4 mm to 10 mm in thickness.

## RESULTS AND DISCUSSIONS

Figure 4 presents results of experiments with B6 alumina samples subjected to plane and divergent impacts. The free surface velocity history of B6 ceramic plate 5.85 mm in thickness impacted by a plane aluminum flyer plate 2 mm thick at 1.9±0.05 km/s of the impact velocity is similar to that recorder for alumina ceramics earlier. Compressive part of the waveform demonstrates the elastic precursor wave and plastic shock wave with ramped transition between them. The free surface velocity behind the elastic precursor front is 333 m/s that corresponds to the HEL = 6.5 GPa. The velocity pullback in the unloading part of the

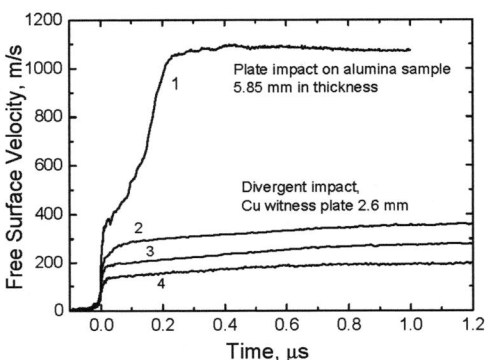

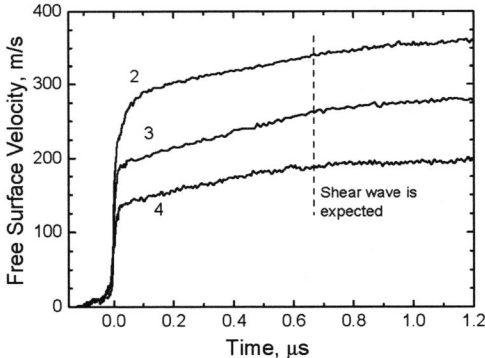

**Figure 4.** Results of experiments with B6 alumina samples 10 mm in thickness subjected to plane (curve 1) and divergent (curves 2, 3, and 4) impacts. In the latter case the free surface velocity histories of copper witness plates were recorded.

free surface velocity history does not exceed 15 m/s that means that the spall strength of material does not exceed 0.3 GPa.

In the shot with divergent impact marked as 2 in Fig. 4 the radial stresses near the input surface of ceramic sample exceeded the HEL. The radial stresses near the input surface in shots 3 and 4 did not exceed the HEL. All waveforms at divergent impact loading demonstrate gradual increase of the velocity behind the longitudinal compression wave and sudden decrease of their slops near the time moment when output of the shear wave is expected. The waveforms do not indicate unambiguous signatures of compressive fracture.

Although the fracture events were not recorded, results of measurements may be used to outline the range of stressed states which are below the failure criterion. Since the measured waveforms demonstrate growing radial stresses behind the compression wave, it is natural to assume for rough estimations that the spherical extension occurs at constant mean stress. In this case it may be shown

$$\dot{S}_\theta = -2G\frac{u_p}{r}, \quad \dot{S}_r = -2\dot{S}_\theta$$

where $S_r$ and $S_\theta$ are the radial and hoop deviatoric stresses, correspondingly. Figure 5 presents thus estimated trajectories of stressed states of outer layers of the alumina samples. Initial states on these trajectories correspond to measured free surface velocity right behind the elastic compression wave. The time interval $\Delta t = 0.65$ μsec of estimated spherical extension is

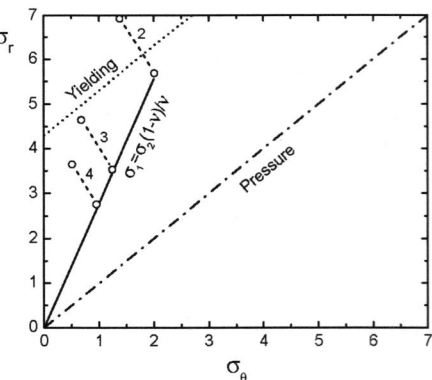

**Figure 5.** Estimated field of stressed states reached in the experiments with divergent impact loading of alumina ceramic samples. The indexes correspond to Fig. 4.

the delay time between longitudinal and shear waves at the sample surface. The flow radius $r = 31$–$35$ mm was estimated accounting for thicknesses of the base plates and the samples.

Estimated stressed state in the shot 2 is beyond the Von Mises criterion of ductility that means that plastic deformation should occur during divergent extension. Probably this explains why the free surface velocity history in this shot has less relative slope behind the compression wave than that in shots 3 and 4. On the other hand, the stresses are obviously overestimated in this case because of the assumption of purely elastic response. Data of shots 3 and 4, in general, do not contradict to the data by Heard and Cline [6].

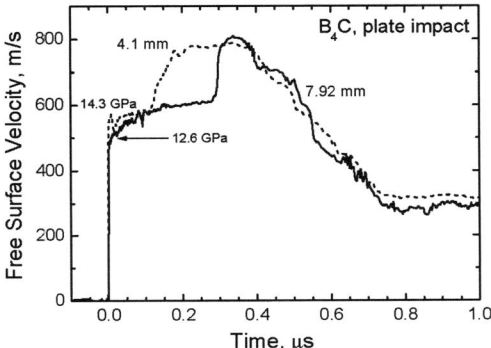

**Figure 6.** Wave profiles in the boron carbide samples impacted by plane aluminum flyer plates 2 mm in thickness at 1.9±0.05 km/s of the impact velocity. The VISAR measurements with LiF window.

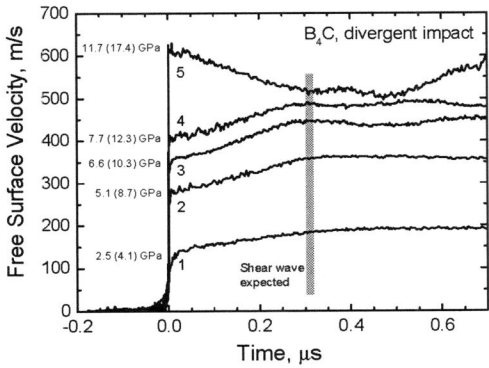

**Figure 7.** The free surface velocity histories of $B_4C$ plates 8 mm in thickness. The numbers at shock front indicate the shock stresses corresponding to measured free surface velocities and to estimated input stress.

Figure 6 presents results of plane impact experiments with the boron carbide ceramic. The HEL of our $B_4C$ samples is around 14 GPa. Figure 7 summarizes results of the divergent impact experiments with $B_4C$ ceramic plates over wide range of the peak stresses. The waveform measured at maximum peak stress radically differs from that at low stress: instead of increase the measurements show decrease of stress behind the compression wave front. At lower peak stress the stress decrease behind shock wave is replaced by stress increase. However, whereas in the case of alumina the stress growth behind the shock wave was practically linear, the $B_4C$ waveforms demonstrate growth with acceleration. We associate these features with slow fracture process in shock compressed boron carbide.

Performed preliminary study has demonstrated that divergent impact loading, in general, is realistic and promising way of varying the stressed state and determining the failure conditions for brittle materials. In following, 2-D computer simulations should be attracted for more accurate estimations of stressed states and for optimizing the experimental conditions.

## ACKNOWLEDGEMENT

The work was supported by the US Army Research Office under contract number N62558-03-M-0038, and by the Program of Basic Research of Russian Academy of Sciences "Damage accumulation, fracture, wear-out and structural transformations of materials under intense mechanical, thermal and radiative attacks".

## REFERENCES

1. Grady, D.E *J. Geophys. Res.*, **78**(8), 1299 (1973).
2. Fowles, R. *J. Appl. Phys.*, **41**, 2740 (1970)
3. Tranchet, J-Y., Collombet, F. In: *Metallurgical Applications of Shock-Wave and High-Strain-Rate Phenomena*, edited by L.E. Murr et al., Els. Science B.V., pp. 535-542 (1995).
4. Kanel, G.I., S.V. Razorenov, A.V. Utkin, S.N. Dudin, V.B. Mintsev, S. Bless, and C.H.M. Simha. in: *Shock Compression of Condensed Matter - 1997* (edited by S.C. Schmidt et al.), AIP CP **429**, pp. 489-492 (1998).
5. Paris, V., Zaretsky, E. Study of compressive failure of alumina in impact experiments with divergent flow. *This issue*.
6. Heard, H.C. and Cline, C.F. *J. Mat. Sci.*, **15**, 1889 (1980).

CP845, *Shock Compression of Condensed Matter - 2005*,
edited by M. D. Furnish, M. Elert, T. P. Russell, and C. T. White

# HIGH-SPEED PHOTOGRAPHIC STUDY OF WAVE PROPAGATION AND IMPACT DAMAGE IN FUSED SILICA AND AlON USING THE EDGE-ON IMPACT (EOI) METHOD

## E. Strassburger[1], P. Patel[2], J. W. McCauley[2] and D. W. Templeton[3]

*[1]Fraunhofer-Institut für Kurzzeitdynamik, Ernst-Mach-Institut (EMI),*
*Am Klingelberg 1, 79588 Efringen-Kirchen, Germany*
*[2]US Army Research Laboratory, Aberdeen Proving Ground, MD 21005, USA*
*[3]US Army TARDEC, Warren, MI, USA*

**Abstract..** An Edge-on Impact (EOI) technique, developed at the Ernst-Mach-Institute (EMI), coupled with a Cranz-Schardin high-speed camera, has been successfully utilized to visualize dynamic fracture in many brittle materials. In a typical test, the projectile strikes one edge of a specimen and damage formation and fracture propagation is recorded during the first 20 µs after impact. In the present study, stress waves and damage propagation in fused silica and AlON were examined by means of two modified Edge-on Impact arrangements. In one arrangement, fracture propagation was observed simultaneously in side and top views of the specimens by means of two Cranz-Schardin cameras. In another arrangement, the photographic technique was modified by placing the specimen between crossed polarizers and using the photo-elastic effect to visualize the stress waves. Pairs of impact tests at approximately equivalent velocities were carried out in transmitted plane (shadowgraphs) and crossed polarized light.

**Keywords:** Transparent ceramic, AlON, Fused Silica, fracture, visualization, damage propagation
**PACS:** 62.50.+p

## INTRODUCTION

When a high-speed projectile hits a brittle material like glass or ceramic severe fragmentation can be observed, preceding the penetration of the projectile. Several types of glass [1,2] ceramic [3] and a glass-ceramic [4] have already been studied at EMI by means of the Edge-on Impact test. Fused silica and AlON are materials being considered for a variety of transparent armor, sensor window and radome applications. AlON is a polycrystalline ceramic that fulfills the requirements of transparency and requisite mechanical properties for transparent armor against armor piercing ammunition. [5]. AlON has a cubic crystal structure (Fd3m) that can be processed to transparency in a polycrystalline microstructure. It

differs from glasses which do not have any periodic crystalline order, but is akin to polycrystalline opaque ceramics such as aluminum oxide. In the current study, two different optical configurations were employed. A regular transmitted light shadowgraph set-up was used to observe wave and damage propagation and a modified configuration, where the specimens were placed between crossed polarizers and the photo-elastic effect was utilized to visualize the stress waves. Pairs of impact tests at approximately equivalent velocities were carried out in transmitted plain (shadowgraphs) and crossed polarized light. AlON and fused silica specimens were impacted using solid cylinder steel projectiles with velocities ranging from 270 to 925 m/s. The nucleation of crack centers was observed ahead of the apparent fracture front, growing from the impacted edge of the specimens. A comparison of the shadowgraphs to photographs recorded in a reflected light configuration with a coated AlON specimen at the same impact conditions, indicated fracture nucleation in the interior of the ceramic.

## EXPERIMENTAL SET-UP

The specimens were tested in the so-called Edge-On Impact configuration, where the projectile strikes one edge of a specimen and damage formation is recorded during the first 10-30 µs after impact. Twenty photographs were recorded in each test. Fused silica specimens measuring 100 x 100 x 13 mm and AlON specimens of the dimensions 100 x 100 x 10 mm were impacted using solid cylinder steel projectiles of 30 mm diameter and 23 mm. With fused silica the impact velocities were ranging from 125 m/s to 350 m/s, whereas the impact velocities ranged from 270 m/s to 925 m/s with AlON. Figure 1 shows a schematic of the experimental configuration.

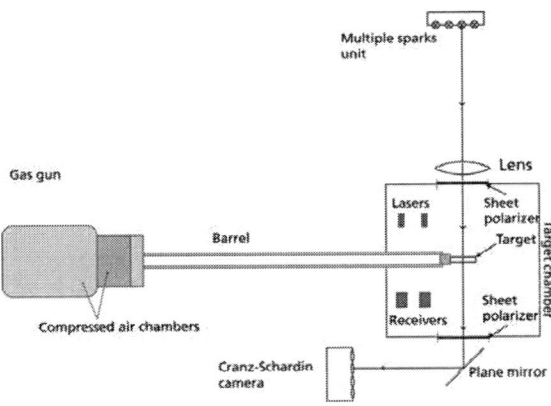

**Figure 1.** Schematic of the experimental set-up

In the velocity range from 100 m/s to 400 m/s a gas gun was used for the acceleration of the projectiles. The transparent targets were placed at a distance of 1 cm in front of the muzzle of the gas gun in order to achieve reproducible impact conditions and a high precision. In this set-up the rear of the projectile is still guided by the barrel of the gun when the front hits the target. In order to achieve impact velocities between 400 m/s and 950 m/s a 30 mm powder gun had to be used for

the acceleration of the projectiles. Due to the resulting muzzle flash and fumes, the specimens had to be placed 170 cm from the muzzle.

The sheet polarizers were fixed to the glass windows of the target chamber when needed. The polarizers were of the type HN32, manufactured by ITOS Ltd. (Mainz, Germany) and had a thickness of 0.7 mm.

## TEST RESULTS

With both materials the test series was designed as pairs of tests at different impact velocities. The first test was conducted in a regular shadowgraph arrangement and the second test was conducted at the same conditions (impact velocity, impactor) with crossed polarizers, respectively. A detailed description of the test series and results with fused silica is given in [6] and for AlON in [7].

### Fused Silica

Figure 2 shows two shadowgraphs and the corresponding crossed polarizers photographs of two tests conducted at 350 m/s. Note that damage appears dark on the shadowgraphs and the zones with stress birefringence are exhibited as bright zones in the crossed polarizers photographs. The shadowgraphs and crossed polarizers photographs are aligned one below the other, allowing for a direct comparison. The time of each pair of photographs is denoted in the crossed polarizers photographs. The moment of impact (t = 0 µs) was determined by means of a short circuit between two trigger foils at the impact edge of the specimens, generated by the projectile.

On the shadowgraphs it can be seen that damage starts first where the edge of the projectile impacts the specimen. Triangularly shaped damage zones spread towards the upper and lower edge of the specimen. The photographs also show the rapid growth of separated, damage zones ahead of the projectile, seemingly due to crack nucleation and growth apparently created by the stress wave interaction with pre-existing processing defects or structural inhomogenieties in the fused silica. The stress waves itself exhibits a relatively plain front in the center and a curved shape outwards.

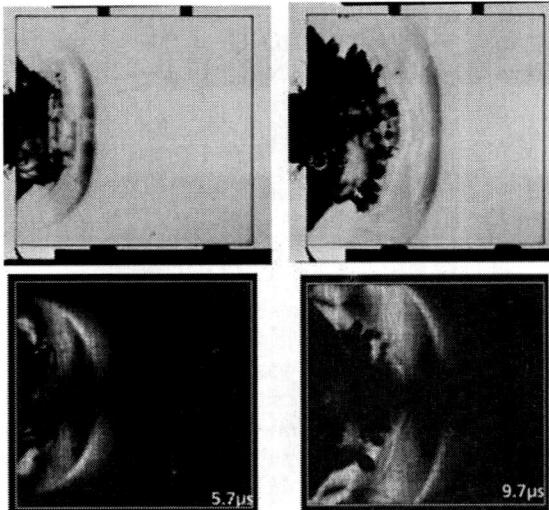

**Figure 2.** Selection of two shadowgraphs and corresponding crossed polarizers pictures from impact on fused silica at 350 m/s

Unlike the damage front exhibited in the shadowgraphs, the crossed polarizers photographs exhibit an approximately semicircular stress wave front, which is further advanced compared to the front visible in the shadowgraphs at the same time. However, the stress wave is not as clearly defined (especially in the center) as the actual damage front in the shadowgraphs.

The photographs taken with the two different recording techniques reveal different processes. In the crossed polarizers arrangement those zones of the specimen are visible, where the stresses are high enough to cause birefringence, so that enough light passes through in order to expose the film. Thus, in the crossed polarizers configuration basically the stress birefringence field is visualized. In the regular shadowgraph arrangement those zones of the specimen appear dark, where the material is either damaged or fractured and, therefore, blocking light transmission or where the light is absorbed more strongly due to a pressure induced change in refractive index.

Linear regression of the position-time data of the wave front in the shadowgraphs yielded an average wave speed of 5823 m/s. The average velocity of the stress wave front in the crossed polarizers view was 5491 m/s. The expansion of a few crack centers at the front of the damage zone was also analyzed and the slope of the straight line through the nucleation sites, which is denoted damage velocity $v_D$, was 5121 m/s.

## AlON

Figure 3 shows a selection of two regular shadowgraphs along with the corresponding crossed polarizers photographs for tests with AlON at 380 m/s. The high-speed photographs show rapidly growing darkened to opaque regions, which reflect changes in the optical transmission due to pressure induced refractive index changes, damage and fractured zones within the specimen. In addition, the nucleation of crack centers ahead of the crack front is clearly visible 8.7 $\mu$s after impact. In contrast to the shadowgraphs, where a wave front is not discernible, the crossed polarizers configuration reveals an approximately semicircular wave front which is a little further advanced compared to the damage front visible in the shadowgraphs at the same time.

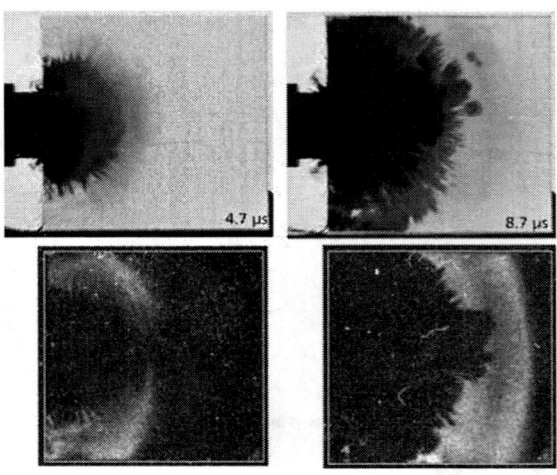

**Figure 3.** Selection of two shadowgraphs and corresponding crossed polarizers pictures from impact on AlON at 380 m/s

From the crossed polarizers photographs a wave front velocity of 9367 m/s was determined. The coherent damage/fracture front initiated at the impacted edge of the specimen propagated at an average velocity of 8381 m/s.

The observed crack centers were generated in the interior of the specimens. This was validated with a test on an aluminum coated AlON specimen to mimic the observations from previous work on opaque ceramic. Figure 4 shows a direct comparison of damage patterns and wave positions in the different optical configurations.

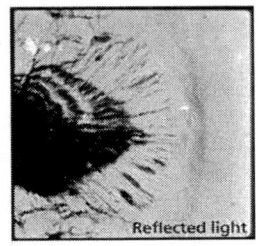
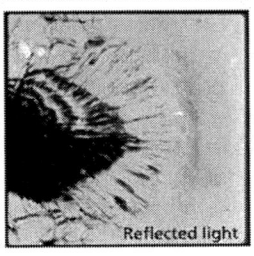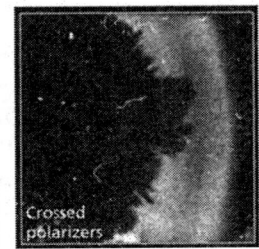

**Figure 4.** Comparison of shadowgraph, reflection mode photograph and crossed polarizers view for AlON impacted with blunt cylinder at 380 m/s

The damage front visible in the reflected light mode is approximately parallel with the most advanced crack centers which appear in the shadowgraph configuration; isolated crack centers were not observed in the reflected light mode. Therefore, it may be concluded that the crack centers were generated in the interior of the specimen, possibly by the passage of the stress wave, over stress concentrating regions of processing defects (pores, inclusions etc.) similar to what has been observed in previous work on glass [2].

## SUMMARY

Fused silica and the transparent polycrystalline ceramic, AlON, were tested in an Edge-On Impact configuration. Stress wave propagation could be visualized when the specimens were placed between crossed polarizers. Damage propagation was observed in a regular shadowgraph set-up.

The nucleation of crack centers was observed ahead of the coherent fracture front, growing from the impacted edge of the specimens in both materials. A comparison of the shadowgraphs to photographs recorded in a reflected light configuration with a coated AlON specimen at the same impact conditions, indicated damage/fracture and isolated crack nucleation in the interior of the ceramic. This suggests that the investigation into AlON offers insight into the damage evolution not only in AlON, but also in other opaque ceramics where only surface damage can be observed with optical methods.

## ACKNOWLEDGEMENTS

This work was performed under a contract from the European Research Office supported by the U. S. Army Tank Automotive Research, Development and Engineering Center and the Army Research Laboratory.

## REFERENCES

1. Senf, H., Strassburger, E., Rothenhäusler, H., "Stress wave induced damage and fracture in impacted glasses", J. PHYS. IV, C8, Vol. 4, pp. 741-746, 1994.
2. Senf, H., Strassburger, E., Rothenhäusler, "Visualization of fracture nucleation during impact in glass", Metallurgical and Materials Applications of Shock-Wave and High-Strain-Rate Phenomena, pp. 163-170, 1995.
3. Strassburger, E., "Visualization of Impact Damage in Ceramics Using the Edge-On Impact Technique", Int. J. of Appl. Cer. Technology, Vol. 1, Number 3, pp. 235-242, 2004.
4. Senf, H., Strassburger, E., Rothenhäusler, "A Study of Damage during Impact in Zerodur", J. PHYS IV, C3, Vol. 7, pp. 1015-1020, 1997.
5. Patel, P. J., Gilde, G. A., "Transparent Armor Materials: Needs and Requirements"; Ceramic Transactions Vol. 134, pp. 573-586, 2001.
6. Strassburger, E., Patel, P., McCauley, J.W., Templeton, D.W., "High-Speed Photographic Study of Wave and Fracture Propagation in Fused Silica", Proc. 22nd Int. Symp. on Ballistics, 14-18 Nov., Vancouver, BC, Canada, 2005
7. Strassburger, E., Patel, P., McCauley, J.W., Templeton, D.W., "Visualization of Wave Propagation and Impact Damage in a Polycrystalline Transparent Ceramic – AlON", Proc. 22nd Int. Symp. on Ballistics, 14-18 Nov., Vancouver, BC, Canada, 2005

CP845, *Shock Compression of Condensed Matter - 2005,*
edited by M. D. Furnish, M. Elert, T. P. Russell, and C. T. White

# TENSION OF ETHYL ALCOHOL AND HEXADECANE BY SHOCK WAVES

## A. V. Utkin, V. A. Sosikov, V. E. Fortov

*Institute of Problems of Chemical Physics Russian Academy of Science.*

**Abstract**. The influences of strain rate and shock wave amplitude on the negative pressure in ethyl alcohol, and hexadecane have been investigated. The method of spall strength measurements was applied and wave profiles were registered by laser interferometer VISAR. Unlike other liquids the process of destruction in methyl alcohol and hexadecane are double staged. At the first stage formation of cavities starts and there is a kinked at free velocity profile was observed. At the second stage the cavity grow rate increases and the spall pulse occurs. The dependence of negative pressure from the strain rate was instigated. The value of the negative pressure correspondent to the kinked at free velocity profile was practically constant and equal to 14MPa for methyl alcohol, and the maximal strength value may be much higher and equal to about 50MPa. Theory of homogeneous bubble nucleation was used to explain the experimental results.

**Keywords**: ethanol, hexadecane, cavitation, spalling, shock waves, impulsive tension
**PACS**: 62.10.+s, 62.50.+p, 64.60.My, 64.60.Qb

## INTRODUCTION

According to theoretical concepts, liquids can endure high tensile stresses of up to 1 GPa [1]. It is assumed that discontinuity of the material results from pore formation by a homogeneous nucleation mechanism. At the same time, considerably smaller values are observed [2] in practice under static test conditions, which is explained by the presence of heterogeneous centers in real liquids, at which pore growth is initiated. Conditions of liquid fracture during homogeneous nucleation can be obtained using dynamic tension. In the present study, this is done by analyzing the spalling phenomena involved in the reflection of compression pulses from the free surface of examined materials [3]. Impulsive tension under shock-wave loading has been used previously to study the cavitation of liquids [4-6]. It has been shown, in particular, that the kinetics of pore formation and, as a consequence, the nature of the dependence of

strength on strain rate are largely determined by the physicochemical properties of the liquids. To elucidate the general features of cavitation, it is of interest to compare the fracture patterns of liquids of various structures. The present paper considers the results of experiments on determining the spall strength of hexadecane and ethanol and discusses the possibility of using the homogeneous nucleation model to interpret the results obtained.

## EXPERIMENTAL PROCEDURE

A diagram of the experiments on impulsive tension of liquids is presented in Fig. 1. Shock waves were produced by collision of an aluminum impactor (1) accelerated by explosion products to a velocity of 600 m/sec with a PMMA dish bottom (shield 2) 2 mm thick. The loading conditions were varied by changing the thicknesses of the impactor $h_{imp}$ and the liquid layer $h_{liq}$ (3) and are listed in Table 1. The velocity was recorded by a VISAR laser interferometer.

## Table 1. Loading conditions.

| NN | $W_0$, m/s | $P_0$, MPa | $h_{imp}$, mm | $h_{liq}$, MM | $\Delta W$, m/s | $P_{s0}$, MPa | $P_{sm}$, MPa | $P_s$, MPa | $\varepsilon*10^{-4}$, sec$^{-1}$ |
|---|---|---|---|---|---|---|---|---|---|
| | | | *Conditions of experiments with ethanol* | | | | | | |
| 1 | 200 | 246 | 0,4 | 8 | - | - | - | - | 8,6 |
| 2 | 400 | 246 | 0,4 | 8 | 28 | 12,8 | 46 | 28,8 | 8,6 |
| 3 | 720 | 533 | Cu | 8 | 30 | 13,3 | 64 | 27,6 | 6,2 |
|  | 722 | 535 | Cu | 8 | 27 | 12,4 | 60 | 26,6 | 4,5 |
| 4 | 269 | 360 | 0,4 | 4 | - | - | - | - | 17,2 |
| 5 | 544 | 365 | 0,4 | 4 | 32 | 14,7 | 50 | 32,5 | 17,2 |
| 6 | 766 | 581 | 0,4 | 1 | 33 | 15,1 | 15,1 | 15,1 | 38,6 |
|  | 777 | 592 | 0,4 | 1 | 35 | 16 | 16 | 16,0 | 43,0 |
| 7 | 323 | 460 | 0,4 | 2 | - | - | - | - | 21,1 |
| 8 | 642 | 456 | 0,4 | 2 | 30 | 13,7 | 37 | 20,1 | 21,1 |
|  | 650 | 464 | 0,4 | 2 | 29 | 13,3 | 36 | 19,5 | 20,5 |
| 9 | 899 | 730 | 2,0 | 8 | 20 | 9,2 | - | 30,2 | 17,4 |
|  | 901 | 730 | 2,0 | 8 | 19 | 8,7 | - | 29,0 | 17,4 |
| | | | *Conditions of experiments with hexadecane* | | | | | | |
| 1 | 578 | 426 | 0,4 | 4 | 20 | 10 | - | 19 | 19,0 |
| 2 | 485 | 340 | 0,2 | 1,5 | 32 | 16 | - | 24 | 18,0 |
| 3 | 400 | 267 | 0,4 | 8 | 39 | 20 | - | 28 | 7,5 |
| 4 | 284 | 177 | 0,38( polyethylene impactor ) | 2 | 67 | 34 | - | 34 | 11,8 |

The laser beam was reflected from an aluminum foil 7 μm thick (4), which separated the liquid from air. The geometrical dimensions of the assembly provided one-dimensional loading conditions. By the moment of arrival at the free surface, the compression pulse had the shape of a triangle, as was determined in separate experiments similar to those shown schematically in Fig. 1 but with the foil placed inside the liquid. In the experiments, ethanol of density $\rho_0 = 0.786$ g/cm$^3$ at an initial temperature of 19$^0$C was used and the sound velocity was $c_0 = 1.165$ km/sec. The particle velocity profiles plotted from the results of the

experiments are given in Figs. 2–3. Particle velocities in the incident compression pulses were measured in experiment Nos. 1, 4 and their corresponding free-surface velocities in experiment Nos. 2, 5. A comparison of these profiles shows that the velocity-doubling rule is satisfied with good accuracy. The dashed curves in Figs. 2–3 show how the free-surface velocity would vary in

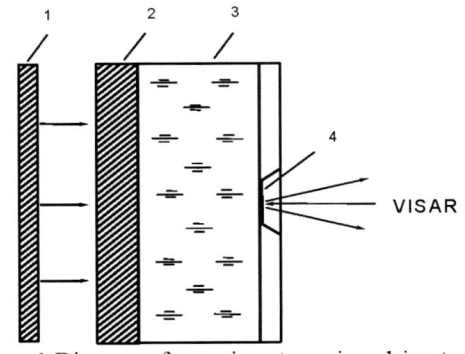

**Figure 1** Diagram of experiments on impulsive tension of liquids: 1) impactor; 2) shield; 3) liquid; 4) aluminum foil.

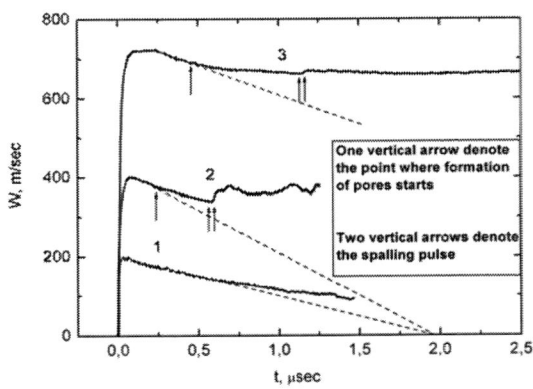

**Figure 2**. Free-surface velocity profiles (2 and 3) and mass-velocity profile (1) in experiment No. 2 (the dashed curves are extrapolations of the free-surface velocity in the absence of cavitation).

the absence of fracture. Measurements of the incident-pulse parameters allow one to uniquely determine the effect of the fracture kinetics on rarefaction. For example, in the phase of decreasing free-surface velocity (experiment No. 3), an inflection (denoted in Fig. 2 by a vertical arrow) is recorded in approximately 0.25 μsec after the arrival of the shock wave. A comparison of profiles 1 and 2 shows that the incident compression pulse does not have an inflection in the neighborhood of 0.25 μsec; therefore, its occurrence on the free-surface velocity profile is due to the pore-growth kinetics. Moreover, a similar inflection was observed in all experiments (except in experiment No. 6), and its corresponding negative pressure $P_{s0}$ should be treated as the fracture initiation threshold, which is determined from the free-surface velocity $W_{s0}$ at the inflection point [3]:

$$P_{s0} = 0.5\rho_0 c_0 \Delta W, \qquad (1)$$

where $\Delta W = W_0 - W_{s0}$. The fracture initiation threshold calculated in such a manner is given in Table 1. It is evident that $P_{s0}$ is almost constant in the entire range of strain rates and is equal to (14.5±1.5) MPa.

After the beginning of cavitation, the negative pressures continue to increase inside the liquid as the rarefaction wave reflected from the free surface propagates into the depth of the sample. In this case, volume fracture occurs rather slowly and shows up on the velocity profile as a decrease in the absolute value of the velocity gradient behind the inflection point [7] rather than as a spalling pulse. This proceeds until the pore growth rate exceeds a certain critical value [7], which leads to the formation of a spalling pulse. In Figs. 2–3, the time of arrival of the spalling pulse on the free surface is denoted by two arrows. As the shock-wave amplitude becomes larger, the deviation of the velocity profile from the dashed curve increases and, simultaneously, the spalling pulse approaches the inflection point up to their coincidence in experiment No. 6. A further increase in the amplitude leads to degeneration of the spalling pulse into a horizontal line, followed by an almost monotonic decrease in the velocity.

Besides, experiments with hexadecane of density $\rho_0 = 0.773$ g/cm$^3$ at an initial temperature of 19$^0$C were made (Fig. 4). The sound velocity was $c_0$=1.330 km/sec. The experimental setup was similar to that in experiments with ethanol, except for the experiment N3 in which polyethylene impactor was used. Typical experimental results are listed in table 1 and shown in fig. 4. As in the experiments with ethanol, a double staged character of destruction is observed in hexadecane. It should be mentioned that double staged character of destruction become apparent when the amplitude of shock-wave exceeds 177MPa. When shock-wave amplitude is less then the 177MPa then usual single staged character of destruction was observed (experiment N4) with abrupt spalling-pulse front.

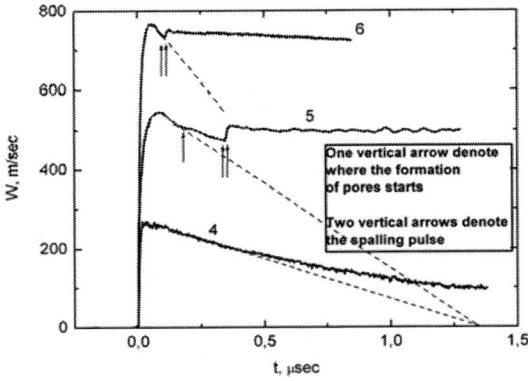

**Figure 3**. Free-surface velocity profiles (5 and 6) and mass-velocity profile (4) in experiment No. 5 (the dashed curves are extrapolations of the free-surface velocity in the absence of cavitation)

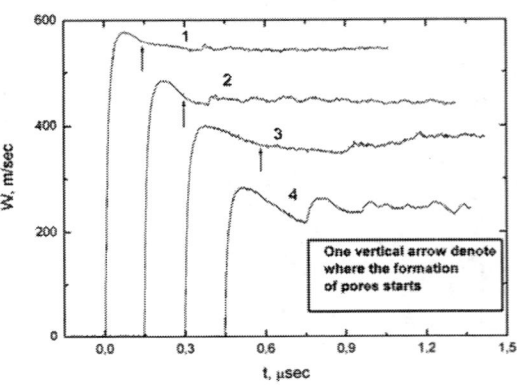

**Figure 4**. Free-surface velocity profiles in experiments with hexadecane.

## DISCUSSION OF EXPERIMENTAL RESULTS.

Let us consider the effect of homogeneous nucleation on the porosity increase in the case of spalling in ethanol. According to thermodynamic fluctuation theory [1], the number of pores of the critical radius $R_C$ formed per unit volume in unit time J under negative pressure P is described by the kinetic equation

$$J = N_0 \frac{\sigma}{\eta} \sqrt{\frac{\sigma}{kT}} \exp(-\frac{16\pi\sigma^3}{3P^2 kT}), \qquad (2)$$

where $N_0$ is the number of molecules per unit volume of the liquid, $\sigma$ is the surface tension, $\eta$ is the viscosity, T [K] is the temperature, and k is Boltzmann's constant. From Figs. 2–4, it is evident that after the fracture initiation, the absolute value of the velocity gradient in the rarefaction part of the pulse decreases severalfold. Investigation of the effect of the pore-growth kinetics on the dynamics of wave interactions during spallation shows [7] that this is possible if the porosity growth rate, which is proportional to $R_c^3$ J [6], exceeds a certain critical value dependent on the strain rate in the rarefaction part of the incident pulse: $R_c^3$ J$= \gamma\dot\varepsilon$, where $\gamma \sim 1$.

Assuming that the tough growth of pores, i.e., the nucleation process described by relation (2), is the determining process for the porosity growth, it is possible to find the dependence of the fracture initiation threshold on the strain rate [7]:

$$P_{s0} \approx A / \sqrt{\ln(B / \dot\varepsilon)} \qquad (3)$$

Here A and B are constants that depend on temperature both explicitly and via viscosity and surface tension. This relationship gives very slow dependence of $P_{s0}$ from $\dot\varepsilon$ as it was found in experiment.

To estimate the maximum negative pressures in ethanol we assume that the initial fracture rate is equal to zero. Then, as shown in [7], the maximum negative pressures $P_{sm}$ occur behind the rarefaction-wave front reflected from the free surface and has the same value as in the intact sample. The value of $P_{sm}$ is calculated by formula (1) but instead of $W_{s0}$ we use the free-surface velocity value $W_{sm}$ that would be reached in front of the spalling pulse if there was no inflection on the free-surface velocity profile. The thus obtained values of $P_{sm}$ are listed in Table 1, and, as can be seen, they are several time higher than $P_{s0}$ and vary in the interval 40–60 MPa. The exception is experiment No. 6, in which the inflection point coincides with the spalling pulse ($P_{sm} = P_{s0}$). Thus, in contrast to the previously studied liquids, the fracture of ethanol and hexadecane is a two-stage process. In case of ethanol the first stage at negative pressures of about 14MPa, pore formation begins, which proceeds at a slow velocity and is manifested as an inflection on the free-surface velocity profile. In the second stage, the porosity growth rate becomes higher, leading to formation of a spalling pulse. The homogeneous nucleation model explains the experimental weak dependence of the fracture initiation threshold on the strain rate for of ethanol.

The work was supported by Russian Science Support Foundation.

## REFERENCES

1. Zel'dovich, Ya. B. "On the theory of new phase formation. Cavitation," Zh. . Eksp. Teor. Fiz., 12, Nos. 11/12, 525–538 (1942).
2. Skripov, V. P. Metastable Liquid [in Russian], Nauka, Moscow (1972).
3. Kanel', G. I., Razorenov, S. V., Utkin, A. V. and Fortov, V. E., Shock-Wave Phenomena in Condensed Media [in Russian], Yanus, Moscow (1996).
4. Utkin, A. V., Sosikov, V. A. and Bogach, A. A., "Impulsive tension of hexane and glycerol under shock-wave loading," J. Appl. Mech. Tech. Phys., 44, No. 2, 174–180 (2003).
5. Dremin, A. N., Kanel', G. I. and Koldunov, S. A. "Spalling in water, ethanol, and Plexiglas, in: Combustion and Explosion, Proc. III All-Union Symp. on Combustion and Explosion (Leningrad, July 5–10, 1972), Nauka, Moscow (1972), pp. 569–574.
6. Bogach, A. A. and Utkin, A. V. "Strength of water under pulsed loading," J. Appl. Mech. Tech. Phys., 41, No. 4, 752–759 (2000).
7. Utkin, A. V. "Determination of the constants of spall-fracture kinetics of materials using experimental data," J. Appl. Mech. Tech. Phys., 38, No. 6, 952–961 (1997).488

CP845, *Shock Compression of Condensed Matter - 2005,*
edited by M. D. Furnish, M. Elert, T. P. Russell, and C. T. White
© 2006 American Institute of Physics 0-7354-0341-4/06/$23.00

# COMMINUTED PARTICLES ORIGINATING FROM CATASTROPHIC FAILURE OF A GLASS BAR

**P. Zeinert, S. J. Bless, and T. Beno**

*Institute for Advanced Technology, The University of Texas at Austin, 3925 W. Braker Lane, Austin, TX 78759*

**Abstract.** Particles were recovered from bar impact tests on glass. Examination reveals a dense network of fast-moving cracks resulting in multi-faceted particles.

**Keywords:** Glass, failure waves
**PACS**: 81.40.Np, 62.20.Mk

## INTRODUCTION

The goal of this effort was to characterize the particles produced by dynamic failure of glass bars. The impact geometry was the same as [1], except that there was no embedded gauge, and the target was borosilicate glass 152 cm long and 12.7 mm in diameter. Two tests are discussed in this study. The first was a bar that was photographed with a Cooke high-speed camera. The other was a capture test, in which all but the leading 50 mm of the sample bar was confined in 50 mm of wood and then a steel tube. Impact velocity was about 280 m/s.

## EXPERIMENT

Figure 1 shows one of the frames from the high-speed camera record. Qualitatively, the damage is very similar to what was seen earlier in homalite [2]. There are three distinct regions of damage in the bar. The forward region is swept by a failure wave. The speed of this wave was > 520 m/s; it is bounded by the photographs because in the first frame it has not yet begun to propagate, and in the third it has stopped. The middle portion of the bar fails very late. The rear

portion of the bar has experienced an "explosive" fracture that is presumably tensile. The free end of the bar is launched at a speed of 151 m/s. Assuming a wave speed of 6 mm/μs, this corresponds to a stress of 1.0 GPa. If there was no decrease due to spall, this is the amplitude of the elastic wave produced by the impact, which presumably corresponds to the elastic strength of this glass. See [1] for the case of finite spall stress.

Figure 2 illustrates the appearance of the bar when the wood confinement was split open. Also indicated are the regions of the bar from whence samples were obtained.

Recovered particles were examined using the conventional framework of fractography of brittle materials. Observations were with an optical microscope, using either transmitted or reflected light.

In general, every particle exhibited Wallner lines and hackle marks on every face. Mirror and mist regions, which are usually associated with the slow start-up phase of crack growth, were only minimally present. Many particles retained sections of the original bar surface as one of their faces.

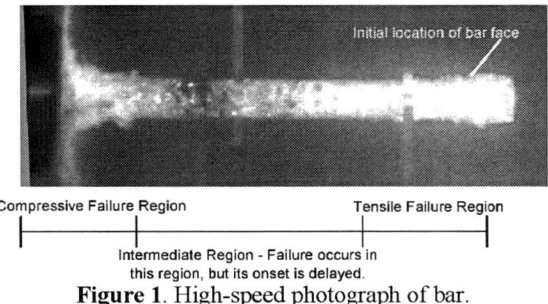

**Figure 1**. High-speed photograph of bar.

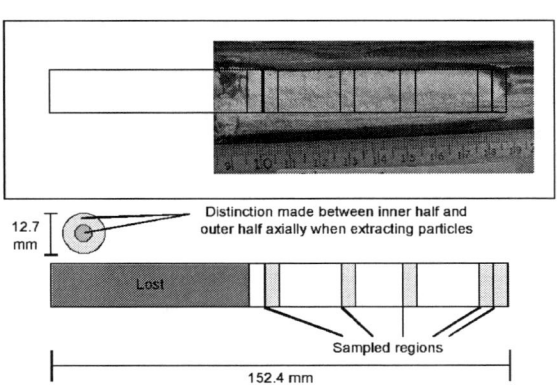

**Figure 2**. Initial appearance of recovered bar. Sampled regions are divided into zones 1–6, depending on their order from the impact face.

Figure 3 illustrates breakout of a crack onto a section of the original bar surface. The Wallner lines show that the crack was moving left to right. A piece of the surface is left hanging over the crack; the depth of focus shows that the intersection angle was very shallow.

Figure 4 is an example of mirror/hackle sequences; it also illustrates the chaotic nature of the fracture. The hackle pattern shows that the crack rapidly rotated out of plane, and there is even new hackle from the crack associated with twisting of the original crack. There are also chipping regions, which are caused by repeated crack branching as the primary crack speeds up and tries to change direction.

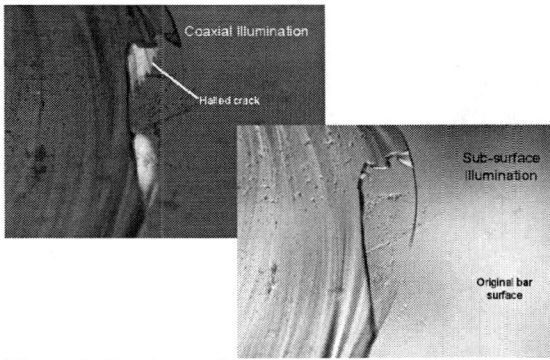

**Figure 3**. Breakout of a crack onto the lateral surface of the bar.

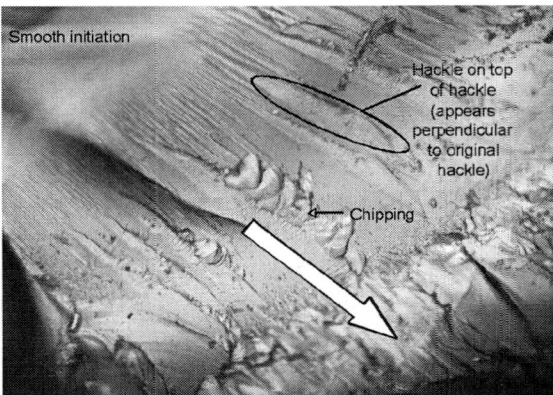

**Figure 4**. Evidence for crack acceleration and rapid change of direction.

Figure 5 is a example from zone 2 that shows how particles are formed from intersecting cracks. Every face of this particle is related to a different fracture event. The upper level and lower level cracks ran generally bottom to top. Figure 6 is a clear example of a fracture that intersected a preceding fracture surface.

Particles recovered from the tensile end of the bar, zone 6, differed from those near the impact end in that they were totally microcracked within. They had a frosted appearance. Interior structure was not visible. Figure 7 is an example of such a particle, taken from the axial region, which measures about a millimeter in length.

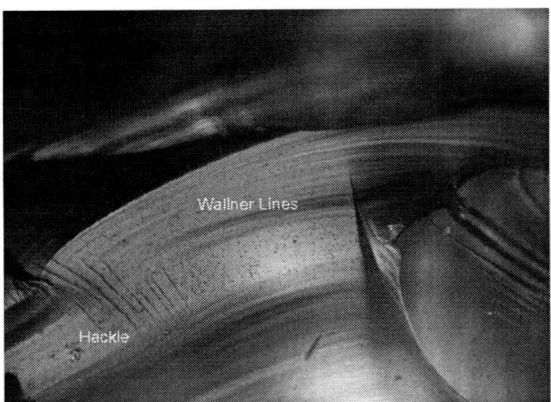

**Figure 5.** Intersecting crack planes.

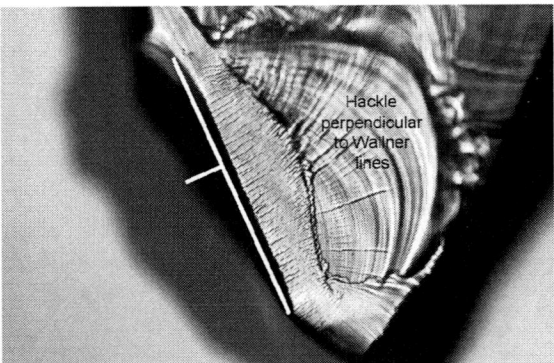

**Figure 6.** A fracture that broke through an existing fracture plane.

**Figure 7.** Particle from tensile region of the target bar.

The size distribution of the particles from all the zones was measured with sieves, of which the smallest was 1.1 mm. The results of those measurements, weighted by volume, are shown in Figure 8. Most of the glass consisted of sub-millimeter particles.

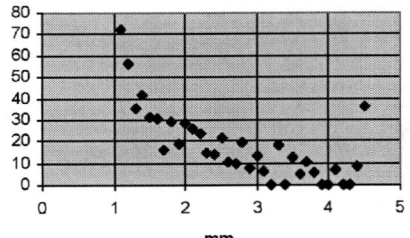

**mm**

**Figure 8.** Mass distribution of particles.

In summary, all of the classic features associated with glass fracture were observed in these particles. The interpretation is that cracks were initiated throughout the volume. There are relatively few lenticular particles. Faces on particles intersect at much higher angles then are associated with crack branching. Most faces are dominated by hackle, indicated rapid crack growth. The tensile region seems to have a much denser crack network, resulting in a frosted appearance of the particles. Most particles are considerably smaller than 1 mm. Material from the compressive failure wave region of the bar was not recovered.

## ACKNOWLEDGMENT

The research reported in this document was performed in connection with Contract number DAAD17-01-D-0001 with the US Army Research Laboratory. The views and conclusions contained in this document are those of the authors and should not be interpreted as presenting the official policies or position, either expressed or implied, of the US Army Research Laboratory or the US Government unless so designated by other authorized documents. Citation of manufacturers or trade names does not constitute an official endorsement or approval of the use thereof. The US Government is authorized to reproduce and distribute reprints for government purposes notwithstanding any copyright notation hereon.

## REFERENCES

1. Beno, T., Bless, S., and Nichols, S. "New Phenomena Observed in Plate Impacts," onto Alumina Bars, this volume.
2. Russell, R., Bless, S., and Beno, T., "Impact Induced Failure Phenomenology in Homalite Bars," Shock Compression of Condensed Matter-2001, American Institute of Physics, 2002.

CHAPTER XIII

# MECHANICAL PROPERTIES OF REACTIVE MATERIALS

CP845, *Shock Compression of Condensed Matter - 2005*,
edited by M. D. Furnish, M. Elert, T. P. Russell, and C. T. White
© 2006 American Institute of Physics 0-7354-0341-4/06/$23.00

# HIGH STRAIN RATE CHARACTERISATION OF A POLYMER BONDED SUGAR

## P.R. Laity[1], C.R. Siviour[2], P.D. Church[3] and W.G. Proud[2]

[1]*Department of Materials Science and Metallurgy, University of Cambridge, Cambridge, CB2 3QZ, UK*
[2] *PCS Group, Cavendish Laboratory, Cambridge, CB3 0HE, UK*
[3] *QinetiQ, Fort Halsted, Sevenoaks, Kent, TN14 7BP*

**Abstract.** The mechanical properties of a polymer bonded sugar consisting of 78% sugar crystals, of modal particle size 310 μm, dispersed in an HTPB binder have been characterized in a split Hopkinson pressure bar system at a strain rate of $10^3$ s$^{-1}$ and temperatures from −100 to +20 °C. These high rate experiments were supplemented by further experiments in an Instron at $10^{-3}$ s$^{-1}$. The material behavior is compared to that of other polymer bonded explosives and simulants. In order to further understand the structural deformation mechanisms specimens of both pristine material and that after Instron testing were examined using X-ray microtomography.

**Keywords:** Polymer bonded explosive (PBX), Hopkinson Bar, X-ray microtomography.
**PACS:** 62.20.Fe

## INTRODUCTION

Polymer bonded explosives (PBXs) consist of explosive crystals held together by a polymer binder. They have a number of advantages over pure explosive crystals: castability, machinability and low impact sensitivity; which allow their use in a wide range of applications.

Understanding the deformation mechanisms in PBX systems is of great importance for the assessment of explosive properties. Despite containing up to 95% by mass of explosive crystals, the mechanical properties of the composite are strongly affected by those of the binder. The modulus of the binder may be of the order 1000 times less than that of the crystals. This means that the binder takes up virtually all of the imposed deformation, which has implications for localization of strain and strain rate in the composite Moreover, as the binder has a strong dependence on strain rate, temperature and age, so does the composite [1-3].

In addition to the properties of the pure binder, the loading density and size distribution of the crystals play important roles in both the quantitative and qualitative behavior of the material [4]. Previous research has shown that for a given loading density and binder type smaller crystals lead to a stronger composite; the flow stress of a polymer-bonded explosive varies as the inverse square root of the modal particle size [5].

Surprisingly, there is little data in the literature on the properties of PBXs at high strain rates. Furthermore, most of the data that is available is on particular service compositions. The research presented in this paper is part of a wider study to understand the behavior of a polymer bonded sugar (PBS) material. The binder and composite have been characterized using Hopkinson bar [6,7] and Instron systems as well as DMTA and X-ray tomography; the latter technique allowing non-destructive measurement of internal structure and deformation mechanisms within the material. The data presented here are from high strain rate measurements of stress-strain behavior at temperatures from −100 to +25 °C, and Instron experiments at room temperature. Images from X-ray tomography of pristine material and specimens recovered from the Instron experiments are also shown.

## MATERIAL STUDIED

The PBS used consisted of 75% by mass of Caster sugar, modal particle size 310 μm, dispersed in an HTPB binder. Specimens were machined to a number of sizes: 10 mm by 5 mm, 8 by 4 and 6 by 3. Pristine specimens were translucent and slightly creamy in color – after deformation they were white. In all compression experiments specimens were lubricated using paraffin wax [8].

## RESULTS AND DISCUSSION

### Hopkinson bar experiments

Stress strain curves at a range of temperatures are shown in Fig.1. These curves show the evolution of behavior as the binder passes through its glass transition temperature. At room temperature, the material yields and begins to flow at a strain of approximately 0.05. Reducing the temperature to 0 °C increases the strength slightly, but does not affect the shape of the curve; however, at −20 °C the strength has increased dramatically; it increases further at −40 °C. At −60 °C the first indication of strain softening after yield is seen; at −80 °C this becomes more dramatic; and at −100 °C the material behaves in a brittle manner.

### Instron experiments

Whilst it is usually the case for polymer materials that reducing the strain rate is equivalent to increasing the temperature, the qualitative behavior of PBXs at low rates is often similar to that in low temperature Hopkinson bar experiments. This is the case in experiments on this material, Fig.2.

It was noted that the material changed color during deformation; this color change is associated with debonding of the crystals from the binder. Independent visual observations recorded the

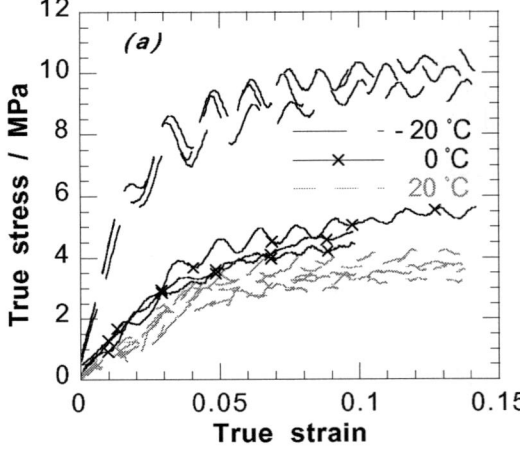

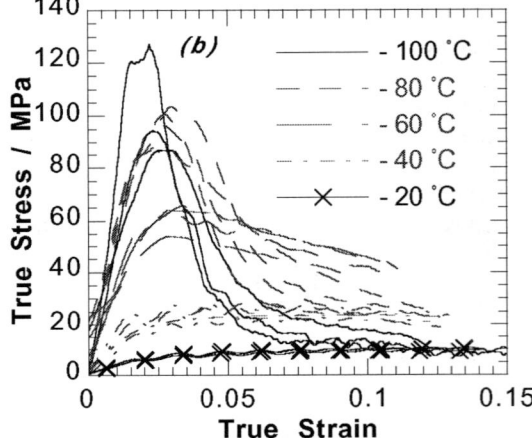

**Figure 1** *(a) and (b).* Stress-strain curves for the PBS material at a range of temperatures. The strain rate was $2000 \pm 200$ s$^{-1}$.

change as starting and ending at strains of approximately 0.15, and 0.3 respectively. This is consistent with the falling region of the stress-strain curve. After the experiments specimens recovered such that the residual strain was 3% or less; however, there was a volume increase of 16.5 $\pm$ 2 % relative to the pristine specimens.

### X-ray microtomography

X-ray tomography was performed on pristine material and a recovered specimens from the Instron experiments. Comparison of Figs 3 and 4, tomographs of an untested and a tested Instron specimen respectively, indicate that debonding of

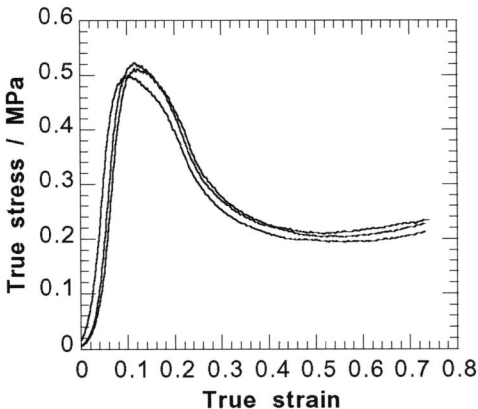

**Figure 2.** Stress-strain curves for the PBS at a strain rate of 0.0021 s⁻¹.

the crystals from the binder is the main failure mechanism in these experiments. This is shown by the opening up of gaps along the edges of crystals, and of paths running along crystal faces. As expected, there is no visible evidence of crystal fracture; the tensile strength of sugar crystals is not well established but is ~50 MPa, much greater than the stress achieved in these experiments.

In order to examine the crystals more closely, a tomographic scan was also performed on a smaller piece of pristine material, Fig.5. There is some damage and debonding around the edges of the piece from its preparation. An interesting feature of this image is the pores seen in some crystals; by scanning through the reconstructed images it is possible to confirm firstly that these are *internal* defects, and secondly that they are not 'noise' in the picture. Whilst these are sugar crystals, porosity is thought to significantly increase sensitivity of explosives to shock and mechanical stimulation. It is therefore important to be aware that this technique can be used to visualize such features.

### Discussion

The results from Hopkinson bar experiments show an evolution of behavior with temperature that is typical of polymer bonded explosives. This behavior is strongly modified by the glass transition in the binder, which occurs between −60 and 0 °C. The rapid strain softening which occurs at very low temperatures may be due to the onset

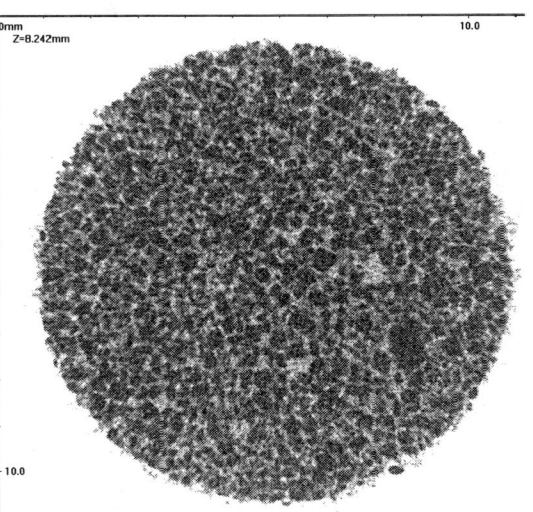

**Figure 3.** X-ray tomography: Slice through a specimen of pristine PBS material.

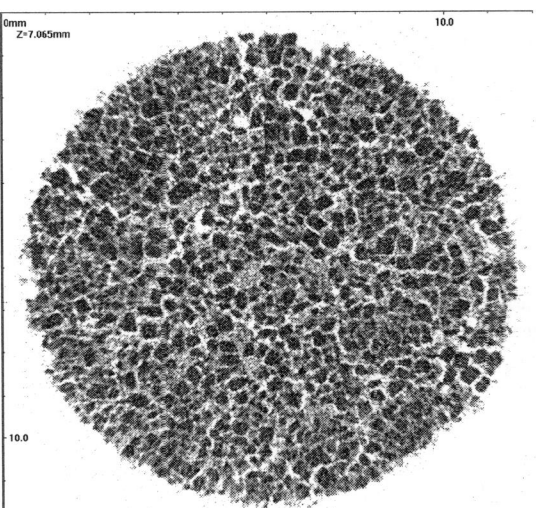

**Figure 4.** Tomographic slice through a specimen of material recovered from an Instron experiment.

of particle–particle interactions, which are no longer moderated by a soft binder material. It is noted that in the more highly loaded materials examined by Gray [1], the behavioral changes occurred at higher temperatures; however, the binders in the two materials are not the same.

Measurements in the Instron and with X-ray tomography indicate that the main deformation mechanism in low strain rate loading is debonding of the crystals from the binder. The importance of

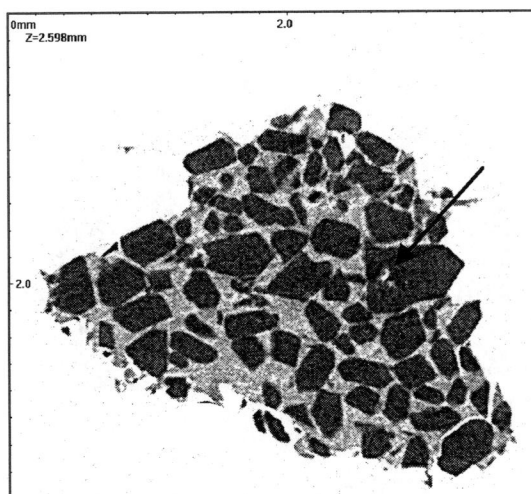

**Figure 5.** Tomographic slice through a smaller specimen of pristine polymer bonded sugar. A flaw in one of the sugar crystals is visible, indicated by an arrow on the figure

debonding over crystal fracture has been noted by other authors [9,10]. In this case, there was also an increase in specimen volume at the end of the experiment, although no measurement of volume as a function of strain has been performed. Further investigation of this volume change is required, including tomographic scans of the evolution of damage with strain in the specimen.

The damage mechanism in Hopkinson bar loading is not yet known; interrupted experiments are required to investigate this more fully. The dramatic strain softening at low temperatures probably has a different cause to that in Instron experiments. Gray *et al.* reported brittle crystal fracture and glassy binder failure in similarly behaved specimens of PBX 9501. This would agree with the completely brittle behavior of the current material at −80 ˚C.

## CONCLUSIONS

Measurements have been made of the stress-strain behavior of a polymer bonded sugar (PBS) at high and low strain rates. Overall, the evolution of behavior with temperature agrees qualitatively with the work of previous authors. X-ray tomography measurements of damaged and undamaged material indicate that debonding is the main damage mechanism in Instron loading; however,

further experiments are required to extend these observations to Hopkinson bar strain rates.

## ACKNOWLEDGEMENTS

This work has been supported by MoD and QinetiQ, who support WG Proud as a QinetiQ Senior Research Fellow. CR Siviour would like to thank EPSRC, [dstl] and The Worshipful Company of Leathersellers for supporting his research. DM Williamson assisted with the Instron experiments, and we would like to thank JE Field for his advice and encouragement.

## REFERENCES

1. Gray III, G.T., et al., "High- and low-strain rate compression properties of several energetic material composites as a function of strain rate and temperature", in Proc. 11th Int. Detonation Symposium, 2000 (J.M. Short and J.E. Kennedy, eds), pp. 76-84.
2. Idar, D.J., et al., "Influence of polymer molecular weight, temperature, and strain rate on the mechanical properties of PBX 9501", in Shock Compression of Condensed Matter, 2001 (M.D. Furnish, N.N. Thadhani, and Y. Horie, eds), pp. 821-824.
3. Goldrein, H.T., et al., "Ageing effects on the mechanical properties of a polymer bonded explosive", in Ageing Studies and Lifetime Extension of Materials, (L.G. Mallinson, ed.) 2001, New York: Plenum. pp. 129-136.
4. Siviour, C.R., et al., "Particle size effects on the mechanical properties of a polymer bonded explosive", J. Mater. Sci. 39, pp. 1255-1258, 2004
5. Balzer, J.E., et al., "Behaviour of ammonium perchlorate-based propellants and a polymer-bonded explosive under impact loading". Proc. R. Soc. Lond. A, 460, pp. 781-806, 2004.
6. Gray III, G.T. and W.R. Blumenthal, "Split-Hopkinson pressure bar testing of soft materials", in ASM Handbook. Vol. 8: Mechanical Testing and Evaluation, 2000 (H. Kuhn and D. Medlin, eds). pp. 488-496.
8. Trautmann, A., et al., "Lubrication of polycarbonate at cryogenic temperatures in the split Hopkinson pressure bar", Int. J. Impact Engng, 31, pp. 523-544, 2005
9. Palmer, S.J.P., J.E. Field, and J.M. Huntley, "Deformation, strengths and strains to failure of polymer bonded explosives", Proc. R. Soc. Lond. A, 440, pp. 399-419, 1993

CP845, *Shock Compression of Condensed Matter - 2005*,
edited by M. D. Furnish, M. Elert, T. P. Russell, and C. T. White
2006 American Institute of Physics 0-7354-0341-4/06/$23.00

# MECHANICAL BEHAVIOR OF TNAZ/Hytemp EXPLOSIVES DURING HIGH ACCELERATION

## Y. Lanzerotti[1], J. Sharma[2], and C. Capellos[1]

*[1]U. S. Army ARDEC, Picatinny Arsenal, NJ 07806-5000*
*[2]Naval Surface Warfare Center, Carderock Division, West Bethesda, MD 20817-5700*

**Abstract**. The mechanical behavior of TNAZ/Hytemp (1,3,3-trinitroazetidine/polyacrylic elastomer) explosives subjected to high acceleration has been studied in an ultracentrifuge. Pressed plastic-bonded TNAZ/Hytemp was studied as a function of the percentage of Hytemp at -10°C and 25°C. The percentage of Hytemp in the samples varied from 1% (weight percent) to 2% (weight percent). Failure occurs when the shear or tensile strength of the explosive is exceeded. The fracture acceleration of pressed plastic-bonded TNAZ/Hytemp decreases with increasing percentage of Hytemp in the explosive at -10°C and 25°C. The fracture acceleration of pressed plastic-bonded 98%/2% (weight percent) TNAZ/Hytemp at 25°C is about 1/3 that at -10°C.

**Keywords:** fracture, high acceleration, TNAZ/Hytemp explosive
**PACS:** 62.20.-x, 62.20.Mk

## INTRODUCTION

We have been studying the mechanical behavior of energetic materials during high acceleration by using an ultracentrifuge [1-5]. Energetic materials are of significant interest for scientific and practical reasons in the extraction (mining) industry, structure demolition, space propulsion, and ordnance. In these applications the materials can be subjected to high, fluctuating and/or sustained acceleration. The nature of the fracture process of such materials under high acceleration is of particular interest, especially in ordnance and propulsion applications. For example, explosives in projectiles are subjected to setback forces as high as 50,000 $g$ ($g$ = 980.6 cm/s$^2$) during the gun launch. These high setback forces can cause fracture [6,7] and premature ignition of explosives.

Fundamental understanding of the behavior of energetic materials subjected to high acceleration is a key to better practical ordnance designs that solve the problems of abnormal propellant burning and premature ignition of explosives during gun launch. The pressure gradient that is experienced by the explosive during acceleration in the gun and under $g$-loading in the ultracentrifuge is unique and will produce different kinds of behavior and failure than under other material test conditions. [8-14] The present work is particularly relevant to the future development of insensitive energetic materials to be used in devices with higher acceleration.

The experimental work discussed herein is most applicable to understanding the structure and fracture of energetic materials that are subject to high or sustained acceleration conditions. A centrifuge is used to provide controlled acceleration conditions that can be applied to different material samples. The experiments discussed here have loading rates (order 500 $g$/sec) that are not comparable to the loading rates for a gun launch [5] (order 500 $g/$ $10^{-5}$ sec). Nevertheless, the high accelerations achieved in the centrifuge can provide conditions whereby the energetic material samples can be systematically studied. Further, the centrifuge loading conditions can be tailored to be comparable to other conditions in which loading rates are lower, such as in some space propulsion and demolition.

Previously, [1-4] we have used an ultracentrifuge to study the fracture behavior of explosives at 25°C and -10°C. In the present work we studied the fracture behavior at high acceleration in a ultracentrifuge at -10°C and 25°C of pressed plastic-bonded 99%/1% TNAZ/Hytemp and pressed-plastic bonded 98%/2% TNAZ/Hytemp (1,3,3-trinitroacetidine/polyacrylic elastomer) [15-16]. All percentages stated in this paper are weight percentages.

## TECHNIQUE

A Beckman ultracentrifuge model L8-80 with swinging-bucket rotor model SW 60Ti was used to rotate the samples up to 60,000 rpm (about 500,000 $g$). The distance of the specimen from the axis of rotation ranged from 6 to 12 cm.

The samples were machined into the shape of the frustrum of a cone. The large diameter was typically 11 mm and the small diameter 9 mm. The angle between the base and the side was 80°. Each sample was fitted into a 5-mm long, 11-mm o.d. aluminum cylinder. At one end the i.d. was 11 mm and the other end 9 mm. The angle between the inner and outer sides of the sleeve was 10°.

The 9-mm diameter top of the sample faced away from the axis of rotation.

For an experiment the acceleration was increased up to a maximum and then held there until the total elapsed time reached 5 min. The sample was then decelerated smoothly. In a series of experiments on a material, the initial maximum acceleration was then increased systematically in each successive 5-min run. The sample fractured when the shear or tensile strength of the material was exceeded, causing particles to break loose from the exposed surface and transfer to the closed-end of the tube. A hemispherical fracture surface resulted.

## RESULTS

The fracture acceleration of pressed plastic-bonded 99%/1% TNAZ/Hytemp and pressed plastic-bonded 98%/2% TNAZ/Hytemp at 25°C are compared in Table 1. Table 1 shows that pressed plastic-bonded 98%/2% TNAZ/Hytemp fractured at an average value of ~ 37 kg at 25°C. This pressed plastic-bonded 98%/2% TNAZ/Hytemp was an average value of 97.5% of its theoretical maximum density (TMD) of 1.821 g/cc.

Three samples of 99%/1% TNAZ/Hytemp were investigated, with three different values of its %TMD of 1.830 g/cc. The results shown in Table 1, indicate that the fracture acceleration was a function of the %TMD of the material. With a %TMD of 97.3, no fracture was found for an acceleration of 133 kg. The lower the value of the %TMD, the lower was the acceleration level at which fracture occurred.

**TABLE 1.** Fracture Acceleration of Pressed Plastic-Bonded TNAZ/Hytemp in an Ultracentrifuge at 25°C.

| Explosive | Fracture Acceleration, k$g$ | %TMD |
|---|---|---|
| 98%/2% TNAZ/Hytemp | 33 | 97.3 |
| 98%/2% TNAZ/Hytemp | 41 | 97.7 |
| 99%/1% TNAZ/Hytemp | no fracture at 133 | 97.3 |
| 99%/1% TNAZ/Hytemp | 126 | 96.3 |
| 99%/1% TNAZ/Hytemp | 93 | 95.2 |

The fracture acceleration of pressed plastic-bonded 99%/1% TNAZ Hytemp and pressed plastic-bonded 98%/2% TNAZ/Hytemp at -10°C are compared in Table 2. Pressed plastic-bonded 99%/1% TNAZ/Hytemp did not fracture at the largest acceleration employed in these experiments, 133 kg. The average %TMD of the two 99%/1% TNAZ/Hytemp samples was 97.3%. Table 2 also shows that pressed plastic-bonded 98%/2% TNAZ/Hytemp fractured at an average value of 103 kg at -10°C. The average %TMD of the two 98%/2% TNAZ/Hytemp samples was 97.7%.

**TABLE 2**. Fracture Acceleration of Pressed Plastic-Bonded TNAZ/Hytemp in an Ultracentrifuge at -10°C

| Explosive | Fracture Acceleration, kg | %TMD |
|---|---|---|
| 98%/2% TNAZ/Hytemp | 120 | 97.4 |
| 98%/2% TNAZ/Hytemp | 87 | 98.0 |
| 99%/1% TNAZ/Hytemp | no fracture at 133 | 97.2 |
| 99%/1% TNAZ/Hytemp | no fracture at 133 | 97.4 |

## SUMMARY

The fracture acceleration of pressed plastic-bonded TNAZ/Hytemp tends to decrease with increasing percentage of Hytemp at both temperatures. This is illustrated in Fig. 1 where the average fracture accelerations are shown for both temperatures as a function of the percentage of Hytemp. For those samples that did not fracture, the acceleration was taken as 133 kg, the maximum in the experiments.

The fracture acceleration of pressed plastic-bonded 98%/2% TNAZ/Hytemp at 25°C is about 1/3 that at -10°C, as is summarized in Fig. 2. The relationship of fracture acceleration to sensitivity needs to be investigated.

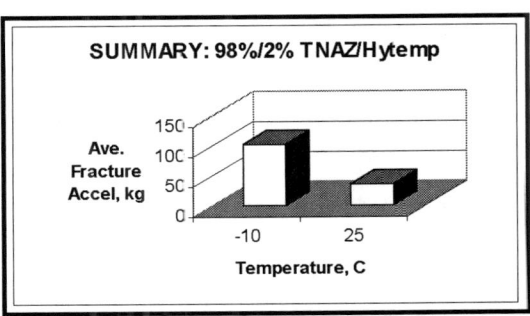

**Figure 2.** Average fracture acceleration of 98%/2% TNAZ/Hytemp as a function of temperature.

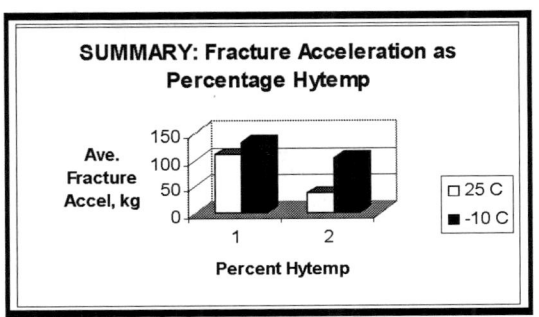

**Figure 1.** Average fracture acceleration as a percentage of Hytemp.

## ACKNOWLEDGMENT

We thank Dr. D. Wiegand, U. S. Army ARDEC for helpful comments.

# REFERENCES

1. Lanzerotti, Y. D., and Sharma, J, "Brittle Behavior of Explosives During High Acceleration", Appl. Phys. Letters **39**, 455-457 (1981).

2. Lanzerotti, Y. D., and Sharma, J., "Mechanical Behavior of Energetic Materials During High Acceleration in an Ultracentrifuge", in Grain Size and Mechanical Properties – Fundamentals and Applications (M.A. Otooni, R. W. Armstrong, N. J. Grant, and K. Ishizaki, eds.) Mat. Res. Soc. Proc. **362**, Pittsburgh, Pennsylvania, 1995, pp. 131-135.

3. Lanzerotti, Y. D., and Sharma, J., "Mechanical Behavior of Energetic Materials During High Acceleration", in Twelfth International Detonation Symposium, pp. 303-305, 2002.

4. Lanzerotti, Y. D., Sharma, J., Capellos, C., and Travers, B., "Mechanical Behavior of TNAZ/CAB Explosives During High Acceleration", in Shock Compression of Condensed Matter, 2003 (M. D. Furnish, Y. M. Gupta, and J. W. Forbes, eds.), part II, pp. 783-785.

5. Lanzerotti, Y. D., "Fracture Phenomena of Energetic Materials During High Acceleration", in Thermomechanical Properties of Energetic Materials and Their Effects on Munitions Survivability, Naval Weapons Center, China Lake, CA, March, 1985, Vol. II, pp. 1-43.

6. Williamson, D., Palmer, S. J. P., Grantham, S. G., Proud, W., and Field, J. E., "Mechanical Properties of PBX9501", in Shock Compression of Condensed Matter, 2003 (M. D. Furnish, Y. M. Gupta, and J. W. Forbes, eds.), part II, pp. 816-819.

7. Liu, C., "Fracture of the PBX 9501 High Explosive", in Shock Compression of Condensed Matter, 2003 (M. D. Furnish, Y. M. Gupta, and J. W. Forbes, eds.), part II, pp. 786-791.

8. Wiegand, D. and Reddingius, B., "Mechanical Properties of Plastic Bonded Composites As A Function of Hydrostatic Pressure", in Shock Compression of Condensed Matter, 2003 (M. D. Furnish, Y. M. Gupta, and J. W. Forbes, eds.), part II, pp. 812-815.

9. Wiegand, D., "The Influence of Confinement On The Mechanical Properties of Energetic Materials", in Shock Compression of Condensed Matter, 2003 (M. D. Furnish, Y. M. Gupta, and J. W. Forbes, eds.), part I, pp. 675-678.

10. Wiegand, D. "Mechanical Failure of Composite Plastic Bonded Explosives and Other Energetic Materials", in Eleventh International Detonation Symposium, pp. 744-750, 1998.

11. Wiegand, D., "Mechanical Failure Properties of Composite Plastic Bonded Explosives", in Shock Compression of Condensed Matter, 1997 (S. C. Schmidt, C. P. Dandekar, and J. W. Forbes, eds.), pp. 599-602, and "Mechanical Failure of Composite Plastic Bonded Explosives and Other Explosives", in 20th Army Science Conference, pp. 63-67, 1996.

12. Field, J. E., Whalley, S. M., Proud, W. G., Balzer, J. E., Gifford, M. J., Grantham, S. G., Greenaway, W. M., and Siviour, C. R. in "Synthesis, Characterization and Properties of Energetic/Reactive Nanomaterials" (R. Armstrong, N. Thadhani, W. Wilson, J. Gilman, and R. Simpson, eds.), Mat. Res. Soc. Proc. **800**, Warrendale, Pennsylvania, 2004, pp. 179-190.

13. Rae, P. J., Goldrein, H. T., Palmer, S. J. P., Field, J. E., and Lewis, A. L., "Studies Of The Failure Mechanisms of Polymer-Bonded Explosives By High Resolution Moire Interferometry And Environmental Scanning Electron Microscopy", in Eleventh International Detonation Symposium, pp. 66-75, 1998.

14. Christopher, F. R., Foster, J. C., Wilson, L. L., and Gilland, H. L., "The Use of Impact Techniques To Characterize The High Rate Mechanical Properties Of Plastic Bonded Explosives", in Eleventh International Detonation Symposium, pp. 286-292, 1998.

15. Archibald, T. G., Gilardi, R., Baum, K. and George, C., "Synthesis and X-ray Crystal Structure of 1,3,3-Trinitroazetidine", J. Org. Chem. **55**, 2920-2924 (1990).

16. Iyer, S., Eng, S., Joyce, M., Perez, R., Alster, J., and Stec, D., "Scaled-up Preparation of 1,3,3 – Trinitroazetidine (TNAZ)", in Joint International Symposium on Compatibility of Plastics and Other Materials with Explosives, Propellants, Pyrotechnics and Processing of Explosives, Propellants, and Ingredients, pp. 80-84, 1991.

912

CP845, *Shock Compression of Condensed Matter - 2005*,
edited by M. D. Furnish, M. Elert, T. P. Russell, and C. T. White
© 2006 American Institute of Physics 0-7354-0341-4/06/$23.00

# A RATE-DEPENDENT VISCOELASTIC DAMAGE MODEL FOR SIMULATION OF SOLID PROPELLANT IMPACTS

## E. R. Matheson[1] and D. Q. Nguyen[1]

*[1]Flight Sciences Directorate, Lockheed Martin Missiles & Space, Sunnyvale, CA 94089*

**Abstract.** A viscoelastic deformation and damage model (VED) for solid rocket propellants has been developed based on an extensive set of mechanical properties experiments. Monotonic tensile tests performed at several strain rates showed rate and dilatation effects. During cyclic tensile tests, hysteresis and a rate-dependent shear modulus were observed. A tensile relaxation experiment showed significant stress decay in the sample. Taylor impact tests exhibited large dilatations without significant crack growth. Extensive modifications to a viscoelastic-viscoplastic model (VEP) necessary to capture these experimental results have led to development of the VED model. In particular, plasticity has been eliminated in the model, and the multiple Maxwell viscoelastic formulation has been replaced with a time-dependent shear modulus. Furthermore, the loading and unloading behaviors of the material are modeled independently. To characterize the damage and dilatation behavior, the Tensile Damage and Distention (TDD) model is run in conjunction with VED. The VED model is connected to a single-cell driver as well as to the CTH shock physics code. Simulations of tests show good comparisons with tensile tests and some aspects of the Taylor tests.

**Keywords:** damage, computations, viscoelasticity, solid propellant, impact
**PACS:** 46.35.+z, 81.70.Bt

## INTRODUCTION

There is a need to address hazardous environments and events associated with solid rocket propellants. Mechanical behavior governing internal damage evolution due to a mechanical insult is investigated for this class of materials. The ultimate goal is to predict hazardous events with chemical kinetics that are driven by damage processes.

In References 1 and 2, a viscous internal damage model (VID) was developed to study unknown-to-detonation transition (XDT). The VID model is a continuum model composed of two fully coupled modules: a viscous-elastic-plastic module (VEP) and a tensile damage module. In References 3 and 4, an internal damage model for viscoelastic-viscoplastic energetic materials was based on the mixture theory of Baer and Nunziato[5,6]. The theory was extended to include viscoelastic-viscoplastic behavior of the solid phase. The VEP model of Olsen, et al., in Reference 1 was adopted for this purpose. These models were inserted into the CTH hydrocode to study the behavior of granulated energetic materials during compaction and release such as might occur in impact tests and accidents.

In this paper, a viscoelastic deformation and damage model (VED) for solid propellants based on an extensive set of experiments to determine mechanical properties has been developed. Extensive modifications to VEP necessary to capture these experimental results have led to development of the VED model. In particular, plasticity has been eliminated, and the Maxwell formulation has been replaced with a time-dependent shear modulus. Also, material loading and unloading behaviors are modeled separately. To characterize damage and dilatation behaviors, the Tensile Damage and Distention (TDD) model is run in conjunction with VED. The VED model is implemented in the CTH shock physics code and comparisons to test data are made.

## MODEL DESCRIPTION

Per our review of VEP, a number of problems were identified that lead to substantial modifications. The modifications became so extensive that we decided to create the VED model.

In the VED model, the Maxwell formulation has been replaced with a time-dependent shear modulus. It can be shown that the time-dependent modulus produces the mathematical equivalent of viscosity. The shear stress tensor is defined as follows:

$$\boldsymbol{\tau} = 2G(t)\mathbf{e} \qquad (1)$$

where, $\mathbf{e}$ is the deviatoric strain tensor, and $G(t)$ is the time-dependent shear modulus. Differentiating Equation 1, we obtain:

$$\dot{\boldsymbol{\tau}} = 2G(t)\dot{\mathbf{e}} + \left(\frac{\dot{G}}{G}\right)\boldsymbol{\tau} \qquad (2)$$

The first term in Equation 2 is the response of the shear stress due to the applied strain rate, and the second term allows for stress relaxation. The time-dependent modulus in the first term provides the equivalent behavior of the Maxwell terms in VEP, and the second term mimics the behavior of the viscous terms in the Maxwell model. The time-dependent modulus is based on a Prony series:

$$G(t) = \sum_i G_i' e^{-\beta_i'(t - t_{ref})} \qquad (3)$$

where, $G_i'$ and $\beta_i'$ are the effective shear modulus and relaxation constant for the *ith* term in the series. In Equation 3, $t_{ref}$ is the time at which either loading or unloading begins for a given loading cycle. For this formulation, distinct Prony series parameters are used for loading and unloading. Moreover, using $t_{ref}$, the time is reset to zero when the strain rate changes sign.

## TEST RESULTS AND VED PREDICTIONS

A number of simulations have been performed to evaluate the VED model. Fig. 1 shows the results of monotonic tensile tests at two different rates. Note that there is a sudden change in the slope of the stress strain curve in the monotonic tensile tests. With no plasticity in the VED model, another mechanism was required to predict this sudden change in slope. The data also shows a strain-hardening effect. To account for these effects, the initial shear moduli in the Prony series are multiplied by the following factor:

$$f_G = \max\left[0, 1 - \left(\frac{D_{eff}}{D_G^o}\right)^{n_G^D}\right] + H_G \max\left(0, \varepsilon_t^{prin}\right)^{n_G^H} \quad (4)$$

where, $D_{eff}$ is the effective damage from TDD, $e_t^{prin}$ is effective principal total strain computed in VED, and $D_G^o$, $n_G^D$, $H_G$, and $n_G^H$ are model parameters. The first term in Equation 4 causes the initial

change in slope of the stress strain curve when particle/binder decohesion occurs. The second term causes strain hardening to occur as the polymer chains in the binder are stretched. It was found that this modification was not sufficient to explain all the trends observed in the monotonic tests especially since the highest rate test exhibited a large stress drop after decohesion occurs (see Fig. 2). The stress drop is a relaxation effect, and therefore, it became apparent that the viscosities in the VED terms needed to be modified due to damage effects by the following factor:

$$f_\mu = \left[1 + H_\mu (1 - \phi_s)^{n_\mu}\right]^{-1} \qquad (5)$$

where, $\phi_s$ is the solid volume fraction from TDD, and $H_\mu$ and $n_\mu$ are model parameters. The effect of Equation 5 is to reduce the viscosity as porosity evolves allowing the stress to relax more rapidly. The viscosity for each Prony series term is calculated by first equating the Prony series terms to the VEP Maxwell terms:

$$\mu_i = \frac{2G_i}{\beta_i} \qquad (6)$$

The effective shear modulus and viscosity for each term in the Prony series follow:

$$G_i' = f_G G_i \qquad (7)$$

$$\mu_i' = f_\mu \mu_i \qquad (8)$$

Finally, the effective relaxation constant for each Prony series term is:

$$\beta_i' = \frac{2G_i'}{\mu_i'} \qquad (9)$$

With the addition of these damage effects, very good fits to the monotonic tensile bar tests were obtained (see Fig. 1).

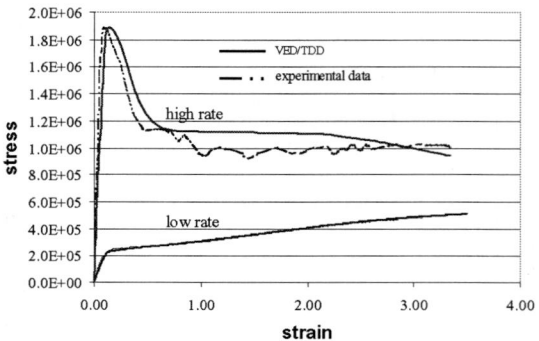

**FIGURE 1.** VED/TDD Simulation of a Low and High Rate Monotonic Tensile Bar Test.

Cyclic tensile bar data is also used to evaluate the VED model. Fig. 2 shows comparison of a simulation against experimental data. In Fig. 3, a simulation of a static relaxation test is compared with data. The material in the relaxation test is similar but not identical to the material in the monotonic and cyclic tests, and therefore, we can match the trend of the experimental data but not the stress levels. Our result is an underestimate of the experimental data.

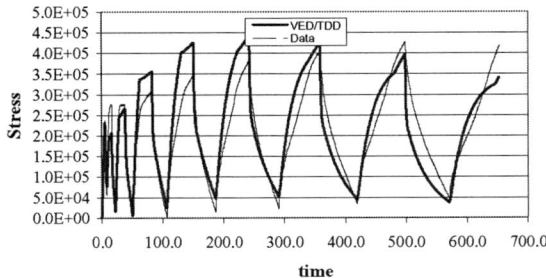

**FIGURE 2.** VED/TDD Stress-Time Curve for a Cyclic Test.

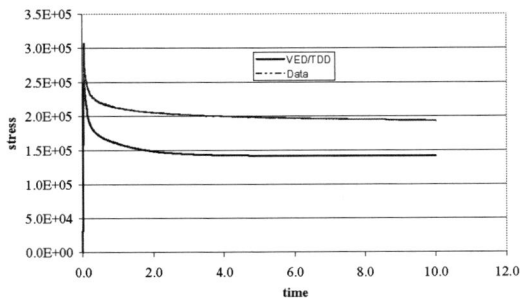

**FIGURE 3.** Comparison of VED/TDD to Experimental Relaxation Data.

The VED and TDD models are also implemented in the CTH code in order to simulate more complex tests. The first test simulated with VED/TDD in CTH is the wave profile test. In this test, three particle velocity gages are embedded in a cylindrical propellant sample. A thin disk of PMMA is fired at the sample generating a 1-D shock wave followed by a release wave. No damage occurs in this test, and the model is exercised at much higher (and compressive rather than tensile) strain rates. In Fig. 4, our VED simulation compares well to the velocity gage records in the wave profile test.

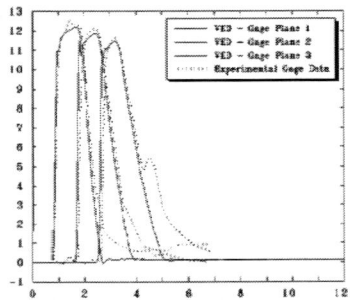

**FIGURE 4.** Correlation of VED to Velocity Gage Records in a Wave Profile Test.

The second test modeled in CTH is the interrupted spall test. In this test, a thick PMMA flyer is impacted onto a disk of propellant at successively higher velocities until full spallation occurs. Fig. 5 shows a 1-D damage profile from TDD for the critical flyer impact velocity that yields full spallation. TDD models damage using separate processes for particle-binder decohesion and binder scission. The first process cannot cause full damage, as the binder is intact and can support much larger stresses and strains than the particle-binder interface. The total stresses used to drive the damage processes are computed from pressure provided by the CTH equation of state package[7] and the stress deviator tensor computed using VED. In Fig. 5, decohesion damage was limited to 80% in TDD. Notice a region where damage exceeds the 80% damage for decohesion. In this region, scission damage has occurred and has gone to 100%.

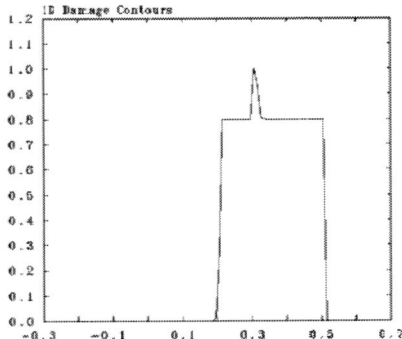

**FIGURE 5.** VED/TDD 1-D Damage Profile for a Spall Test Simulation.

Fig. 6 shows shadowgraphs of a deformed propellant rod at discrete times during a Taylor rod impact test. For comparison, deformed shapes from a CTH simulation of the Taylor test are shown in Fig. 7. The foreshortening of the rod compares well, but spreading at the impact face is significantly under-predicted, and consequently, dilatation is greatly under-predicted. The cause of the large dilatation in the test is not understood and will be the focus of future studies. However, these three CTH calculations demonstrate the ability of VED/TDD to simulate high strain rate impacts.

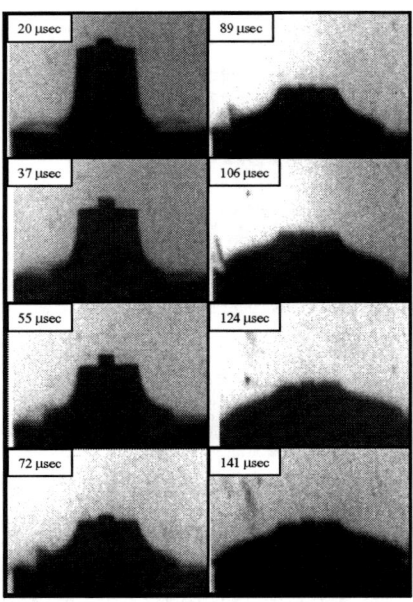

**FIGURE 6.** Taylor Test Shadowgraph Sequence.

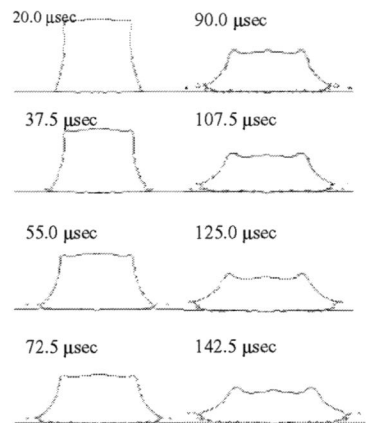

**FIGURE 7.** Simulation of the Taylor Rod Impact Experiment using VED/TDD.

## CONCLUSIONS

In this study, we have shown that plasticity is not necessary to explain the observed stress drop in the monotonic tensile bar tests. Instead, the damage and porosity evolution models in TDD give the correct behavior for not only the stress-strain response but also for the dilatation. Some permanent strain is observed in tensile bars, but this could be due to plasticity associated with pore opening/closure rather than plastic strain in the matrix. We haven't studied this effect, but TDD does distinguish between elastic and plastic pore opening and could be used to study this. We have used these models with relatively good success to study low-rate solid propellant behavior. However, it is not clear that these behaviors will be manifested at the high rates characteristic of the Taylor rod and XDT impact tests. The impact tests will be studied in greater detail in the near future.

## REFERENCES

1. Olsen, E. M., Rosenberg, J. T., Kawamoto, J. D., Lin, C. F., Seaman, L., "XDT Investigation by Computational Simulation of Mechanical Response Using a New Viscous Internal Damage Model," Eleventh Symposium (International) on Detonation, Snowmass, CO, 31 August-4 September, 1998, pp. 170-178.
2. Seaman, L., Simons, J. W., Erlich, D. C., Olsen, E. M., Rosenberg, J. T., Matheson, E. R., "Development of a Viscous Internal Damage Model for Energetic Materials Based on the BFRACT Microfracture Model," Eleventh Symposium (International) on Detonation, Snowmass, CO, 31 August-4 September, 1998, pp. 632-639.
3. Matheson, E. R., Drumheller, D. S., and Baer, M. R., "An Internal Damage Model for Viscoelastic-Viscoplastic Energetic Materials," in *Shock Compression of Condensed Matter - 1999*, edited by M. D. Furnish, L. C. Chhabildas, and R. S. Hixson, AIP Conference Proceedings 505, New York, 1999, pp. 691-694.
4. Matheson, E. R., Drumheller, D. S., and Baer, M. R., "A Viscoelastic-Viscoplastic Distention Model for Granulated Energetic Materials," *Proceedings of the JANNAF 18th Propulsion Systems Hazards Subcommittee (PSHS) Meeting*, Cocoa Beach, FL, 18-21 October, 1999.
5. Baer, M. R. and Nunziato, J. W., "A Two-Phase Mixture Theory for the Deflagration-to-Detonation Transition (DDT) in Reactive Granular Materials," *International Journal of Multiphase Flow* **12**, 861-889 (1986).
6. Baer, M. R., Hertel, E. S., and Bell, R. L., "Multidimensional DDT Modeling of Energetic Materials," in *Shock Compression of Condensed Matter - 1995*, edited by S. C. Schmidt and W. C. Tao, AIP Conference Proceedings 370, New York, 1996, pp. 433-436.
7. Kerley, G. I., "CTH Reference Manual: The Equation of State Package," SAND91-0344, 1991.

CP845, *Shock Compression of Condensed Matter - 2005,*
edited by M. D. Furnish, M. Elert, T. P. Russell, and C. T. White
© 2006 American Institute of Physics 0-7354-0341-4/06/$23.00

# ADHESION STUDIES BETWEEN HMX AND EDC37 BINDER SYSTEM

## S. J. P. Palmer, D. M. Williamson, W.G. Proud

*University of Cambridge, Cavendish Laboratory, Physics & Chemistry of Solids Group, Madingley Road, Cambridge, CB3 0HE, UK*

**Abstract.** EDC37 is a PBX which is composed of 91 % HMX and 9 % NC/K10 by weight. Previous studies have shown that damage under quasi-static conditions occurs preferentially via the adhesive failure of the HMX/ binder interface. Single crystals of HMX have been grown for use in an idealised experiment in which HMX/ binder joints are broken in a simple tension geometry instrumented with a loadcell of millinewton sensitivity. This to quantitatively determine the parameters involved in this important failure mode. Such data are required for the development and validation of accurate microstructural models of PBXs. This paper outlines the current state of research and details the important observations to date.

**Keywords:** EDC37, HMX, adhesion.
**PACS:** 68.35.Np, 81.70.Bt, 81.10.Dn

## INTRODUCTION

EDC37 is a polymer bonded explosive (PBX) which is composed of 91 % HMX and 9 % NC/K10 by weight. It has been well established in the literature that failure in tension of PBX materials, including EDC37, begins with adhesive failure at multiple sites between the largest crystals of HMX and the polymeric binder system [1,5]. These sites then link up resulting in global failure of the sample under consideration.

It is, therefore, of obvious interest to measure the forces and energies which are required to cause adhesive failure between HMX and the binder. Previous quantitative studies have investigated the thermodynamic work of adhesion between various energetic materials and binder systems using contact angle measurements [6, 7]. The drawback of these kinds of measurements is that they greatly underestimate the mechanical work of adhesion, which due to viscoelastic loss mechanisms can be orders of magnitude larger.

An experimental procedure is being developed in which purpose grown HMX crystals can be adhered to controlled thicknesses of the NC/K10 binder system, and then pulled off in tension whilst measuring the load required to do so with millinewton sensitivity. Similar experiments were previously reported on HTPB/AP systems [11].

This is an idealized geometry and it is the intention that the data obtained should feed directly into computational models of microstructural failure. Despite the simplicity of the experimental configuration, of which there are two forms, there are many experimental parameters including: binder layer thickness and area, strain rate and temperature, surface roughness and joint preparation. The last parameter is listed as there exists the possibility that the adhesive properties of the binder are pressure sensitive, in which case the preload and its application time would be important in determining adhesive strength.

## EXPERIMENTAL PROCEDURE

Essentially this study can be split into four stages: crystal growth, binder layer deposition, system calibration and adhesive experiment.

### Crystal growth

β-HMX crystals were grown by evaporation from supersaturated solutions of HMX in acetone at room temperature. Based on LANL data [8] the HMX mass evolution rate, $\partial M/\partial t|_T$, in an evaporator at fixed temperature $T$ is related to the evaporation rate $\mathrm{d}V/\mathrm{d}t$ by the following expression:

$$\left.\frac{\partial M}{\partial t}\right|_T = (0.0112 + 0.0004T)\frac{\mathrm{d}V}{\mathrm{d}t}. \quad (1)$$

Seed crystals were spontaneously grown at evolution rates of 6.3 μg s$^{-1}$ and then singled out for growth at 34.5 μg s$^{-1}$. Slow growth is associated with high quality crystals; imperfections in the seed are magnified in the final crystal [9]. This approach was found to be an acceptable compromise between quality and growth time.

Crystals were mounted on dural stubs with low-shrink Araldite 2014 for grinding and polishing; necessary as most crystals exhibited growth steps on their faces. The final polishing stage utilised a 0.04 μm colloidal grit. The same dural stub was used to mount the crystals during the experiments.

### Binder layer deposition

Binder thicknesses between crystals in EDC37 are of the order of a few microns thick. A representative experiment must be able to reproduce this. The approach used here was to spin-coat the binder. By varying the dilution of binder solution, from undiluted to 1:1 by mass, and the rotational speed from (1-4)× 10$^3$ rpm for 30 s operations, thicknesses are spun down to 500 nm.

### System calibration

Load measurements were made using an amplified Novatech F301 3N loadcell. This is a temperature compensated bending beam type load-cell with millinewton sensitivity. Vertical displacements are applied and measured by mounting the load-cell in a screw driven Instron machine. By fixing the load-cell head and flexing the beam a known amount, the compliance of the system is measured and the loadcell calibrated.

### Experiment

Two configurations are under investigation. In the first, binder material is spin-coated onto a microscope slide and the HMX crystal introduced into the binder. In this configuration, an optical path exists to view the adhesive joint along the axis of loading. Point of failure can be identified; this system can also be used to detect binder cavitation. In the second configuration a large HMX crystal is the substrate, it is spin-coated with binder, and a smaller HMX crystal is introduced and pulled off. No optical path exists in this case. Results reported in this paper are based on the second configuration.

## Results and Discussion

### Crystal growth

Figure 1 is an image of a crystal prior to polishing. The crystal faces are identifiable, in this case a {010} type face is uppermost, the dimensions are typical.

**Figure 1.** Single crystal of HMX prior to polishing with {010} type face uppermost. Scale-bar minor graduations are in millimeters.

## Binder layer deposition

Figure 2 shows the thicknesses of binder produced by spin coating with dilution. As one might expect, for a given spin duration, the resulting thickness of binder is a function of dilution and spin speed.

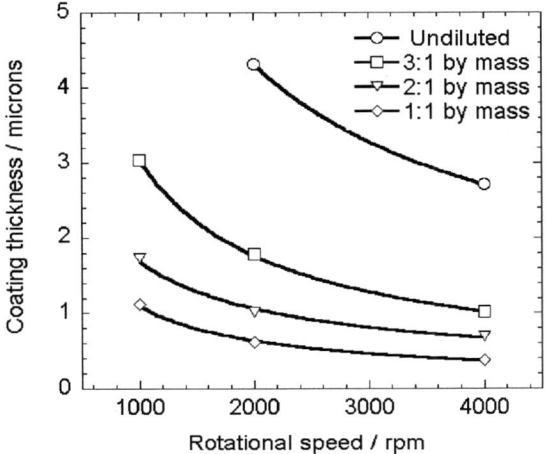

**Figure 2.** EDC37 binder thicknesses after 30 s of spin-coating as a function of spin speed and binder dilution.

## System Calibration

Figure 3 shows system calibration, eight load cycles are shown, no hysteresis is observed.

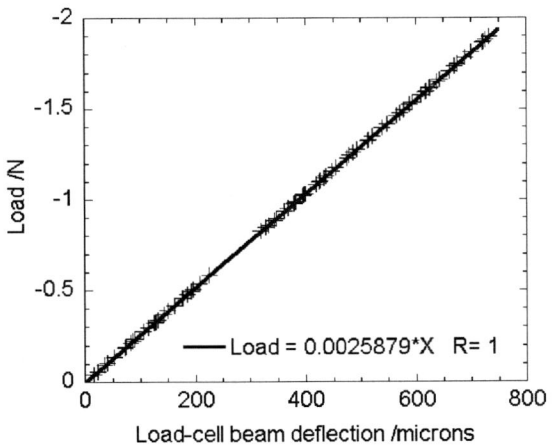

**Figure 3.** Loadcell force-deflection calibration.

Within the limits of signal resolution, the loadcell output is a linear function of deflection. Figure 3 can be used to generate load displacement data of millinewton, micrometer scale.

## Experiment

Figure 4 shows load-extension curves due to two experiments with the following approximate parameters: binder thickness 20 μm, adhesive area 12.8 mm$^2$, preload 40 kPa for 16 hours, surfaces prepared with 0.04 μm grit, loading rates 0.1 and 1.0 mm/min. For clarity not all data points are indicated, a Stineman curve of best fit is shown. Moving from a state of joint precompression to tensile failure occurs in ~10 μm of displacement.

In each case, failure was abrupt and brittle-like over most of the contact area, but some binder was left spanning the crystal faces in the form of a film. On continued loading, holes appeared in the film and grew, reducing it to fibrils. Fibril formation during fracture is often reported [1-5, 10]. The observed fibrils did not have detectable load bearing capacity. Figure 5 shows a sequence of images taken from the 0.1 mm/min loading experiment. It is thought that brittle failure will be more complete at higher rates of loading, and visa-versa, due to the binder's viscoelastic nature.

The stresses, strains and initial elastic moduli (approx. 0.2 MPa) are all comparable with measured data from other binder systems [1, 11].

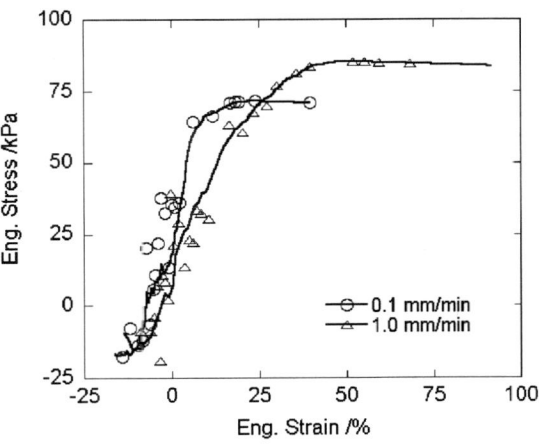

**Figure 4.** Adhesive joint engineering stress strain curves for two different loading rates.

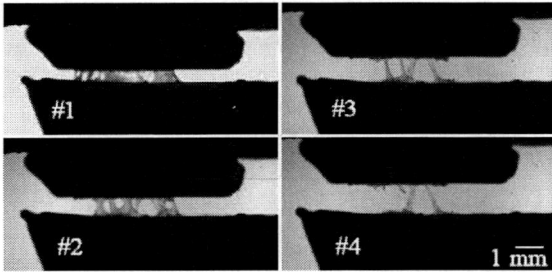

**Figure 5.** Sequence of images taken after initial failure. An element of binder film spanning the two HMX crystals subsequently decomposes into fibrils, # 1 - 4 in chronological order, approximately 20 s interframe time.

## CONCLUSIONS

This program of study is only in the early stages and the technical challenges are very demanding. Even so, the initial results appear promising. β-HMX crystals can be routinely grown with acceptable quality with linear dimension of the order of 10 mm. NC/K10 binder can be spun on to the crystals post polishing to thicknesses of the order of microns. Stress and energy measurements are readily obtained in this idealized configuration via a calibrated low-load load-cell. Preliminary results are comparable to published data on other energetic and binder systems. Future efforts will seek to refine the experiment further and exploit the system to obtain as much quantitative failure data in terms of stresses and energies as possible. Experiments of this nature will feed data directly into models of EDC37 microstructural failure, and may potentially be used to evaluate future PBX compositions.

## ACKNOWLEDGEMENTS

Funding was supplied by EPSRC and AWE. AWE also supplied sample materials. R. Flaxman and D. Johnson of the Cavendish Workshop provided technical support. Prof. J.E. Field gave invaluable advice and discussions.

## REFERENCES

1.  Palmer, S.J.P., Field, J.E., and Huntley, J.M., "Deformation, strength and strains to failure of polymer bonded explosives", Proc. R. Soc. Lond. A 440, pp. 399-419, 1993.
2.  Rae, P.J., Goldrein, H.T., Palmer, S.J.P., Field, J.E., and Lewis, A.L., "Studies of the failure mechanisms of polymer bonded explosives by high resolution moiré interferometry and environmental scanning elenctron microscopy", in 11th Detonation Symposium, Snowmass, C0, 31 August – 4 September, pp. 66-75, 1998.
3.  Rae, P.J., Palmer, S.J.P., Goldrein, H.T., Field, J.E., and Lewis, A.L., "Quasi-static studies of the deformation and failure of β-HMX based polymer bonded explosives", Proc. R. Soc. Lond. A 458, pp. 743-762, 2002.
4.  Rae, P.J., Palmer, S.J.P., Goldrein, H.T., Field, J.E., and Lewis, A.L., "Quasi-static studies of the deformation and failure of PBX 9501", Proc. R. Soc. Lond. A 458, pp. 2227-2242, 2002.
5.  Rae, P.J., Goldrein, T.H., Palmer, S.J.P., and Field, J.E., "The use of digital image cross-correlation (DICC) to study the mechanical properties of a polymer bonded explosive (PBX)", in 12th International Detonation Symposium, San Diego, CA, 11 – 16 August, to be published.
6.  Bower, J.K., Kolb, J.R., and O. Pruneda, C., "Polymeric coatings effect on the surface activity and mechanical behavior of high explosives", Ind. Eng. Chem. Prod. Res. Dev. 19, pp. 326-329, 1980.
7.  Rivera, T., and Matuszak, M.L., "Surface properties of potential plastic-bonded explosives", J. Colloid Interface Sci., 93, pp. 105-108, 1983.
8.  Gibbs, T.R., and Popolato, A., LASL explosive property data, University of California Press, 1980.
9.  Teipel, U., Energetic Materials Particle Processing & Characterization, Wiley-VCH: Weinheim, 2005.
10. Liu, C., and Browning, R., "Fracture in PBX 9501 at low rates", in 12th International Detonation Symposium, San Diego, CA, 11 – 16 August, to be published.
11. Hori., K., and Iwama. A., "On the adhesion between hydroxyl-terminated polybutadiene fuel-binder and ammonium perchlorate. Performance of bonding agents", Propellants, Explosives, Pyrotechnics, 10, pp. 176-180, 1985.

CP845, *Shock Compression of Condensed Matter - 2005,*
edited by M. D. Furnish, M. Elert, T. P. Russell, and C. T. White
© 2006 American Institute of Physics 0-7354-0341-4/06/$23.00

# SHOCK-INDUCED CHEMICAL REACTION IN ORGANIC AND SILICON BASED LIQUIDS

## S. A. Sheffield*, D. M. Dattelbaum*, R. R. Alcon*, D. L. Robbins*, D. B. Stahl* and R. L. Gustavsen*

*DX-2, MS-P952, LANL, Los Alamos, NM 87545*

**Abstract.** Shock-induced chemical reactions remain an area in shock physics that needs further investigation, particularly for determining the influence of pressure, temperature, and chemical structure on reactivity. Several studies have been done in the past that indicate dimerization, polymerization, and decomposition take place in different shock-produced pressure and temperature regimes depending on chemical functionality. We present results obtained from single-shock experiments in which liquids were studied using embedded multiple magnetic gauges to make *in-situ* measurements of the particle velocity profiles at up to ten Lagrangian positions in the liquid. One of the liquids was organic ( *tert*-butylacetylene) and the other was a closely related silicon-based material (ethynyltrimethylsilane). Multiple wave structures were measured in each liquid when the input pressure was above a certain threshold. Here, the reactivity of these materials are compared.
**Keywords:** shock-induced chemistry, ethynyltrimethylsilane, *tert*-butylacetylene, liquid, magnetic gauges
**PACS:** 82.40.Fp,62.50.+p

## INTRODUCTION

A number of liquid organic materials have been identified as having shock-induced reactions. A basic understanding of why the materials react, what the reactions are or even the shock conditions under which the liquids undergo reaction, has yet to be developed. The purpose of this study is to compare an organic liquid to a closely related silicon-based liquid with the hope of gleaning information about how the reactivity of carbon-carbon bonds compare to carbon-silicon bonds in the shock environment. Because of earlier work done on phenylacetylene ($C_6H_5C{\equiv}CH$ – an acetylene group hooked to a benzene ring) in which a two-wave structure was observed[1], we anticipated that simpler materials with an acetylene group would also react in a shock. We chose *tert*-butylacetylene (($CH_3)_3CC{\equiv}CH$)(TBA) as the carbon based material and ethynyltrimethylsilane (($CH_3)_3SiC{\equiv}CH$)(ETMS) as the silicon based material as shown in Fig. 1.

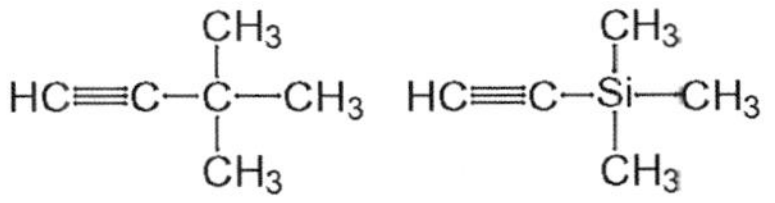

**FIGURE 1.** Structure of ethynyltrimethylsilane (ETMS) on the right and *tert*-butylacetylene (TBA) on the left.

We completed six magnetic gauge experiments on these two materials and observed multiple wave structures indicative of shock-induced chemistry in both materials. The experiments will be described in some detail. Comparisons are made between the wave structures obtained, the input pressure threshold for the onset of reaction, and the shock compression energy which leads to reaction. Preliminary indications show that the shock pressure above a certain level disrupts the acetylene bond.

**TABLE 1.** Properties of *tert*-butylacetylene (TBA) and ethynyltrimethylsilane (ETMS).

| Material | Property | Value |
|----------|----------|-------|
| TBA | Molecular Weight | 82.15 g/mol |
| | Boiling Point | 37-38 °C |
| | Density | 0.667 g/cm$^{-3}$ |
| | Sound Speed$^{22\pm2°C}$ | 0.99 mm/$\mu$s |
| ETMS | Molecular Weight | 98.22 g/mol |
| | Boiling Point | 52 °C |
| | Density | 0.709 g/cm$^{-3}$ |
| | Sound Speed$^{22\pm2°C}$ | 0.94 mm/$\mu$s |

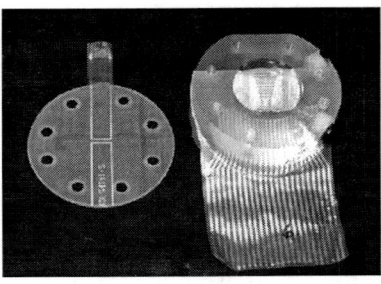

**FIGURE 3.** Cell before the top is glued and screwed on.

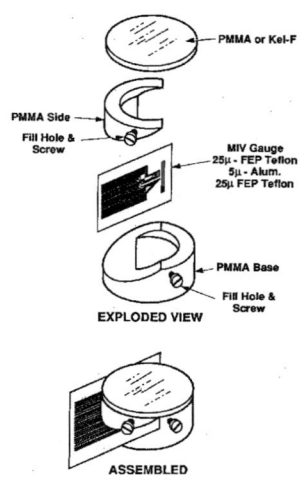

**FIGURE 2.** Cell used to contain the liquids in this study – both an exploded view and the assembled unit.

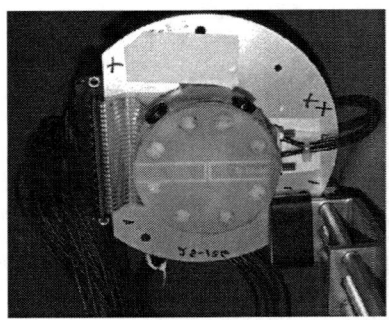

**FIGURE 4.** Finished cell mounted on the target plate. Liquid is loaded through the two screw holes on the top of the cell. Then, screws are inserted and covered with epoxy.

## EXPERIMENTAL SETUP

The liquids in this study were used as received from the suppliers. Ethynyltrimethylsilane (ETMS) was obtained from Gelest Inc. (Gelest No. SIE4904.0). *Tert*-butylacetylene (TBA) was obtained from Aldrich (Aldrich No. 24,439-2, 98%). The properties of these materials are given in Table 1.

Sound speeds were measured using an ultrasonic transducer system, Panasonic Model 5058 pulser/receiver. The thickness of the liquid (the sound wave was reverberating in) was measured with a digital height gauge with a readout accuracy of 0.001mm. The sound speed measurements are accurate to about ±1%.

The liquid was contained in a plastic cell shown in Fig. 2. The top and bottom angled pieces of the cell are made of polymethylmethacralate (PMMA) and are coated with epoxy to prevent the liquid from interacting with the PMMA. The gauge membrane (~60 $\mu$m thick) is a sandwich of FEP Teflon and 5 $\mu$m thick aluminum. It is epoxied on the 30° angle of the bottom; then the top angled piece is epoxied to it. A single element magnetic gauge, termed a stirrup gauge, is epoxied to the Kel-F 81 top (carefully made flat and parallel) and then it is glued and screwed to the cell using nylon screws. A picture of the cell before the top is attached is shown in Fig. 3. The finished cell, mounted on the target plate with all gauge connections hooked up, is shown in Fig. 4.

All experiments were conducted on a gas-driven two-stage gun with a launch tube bore of 50 mm and a maximum velocity capability of about 3.5 km/s. The gun has an electromagnet mounted in the target chamber that produces a uniform field of 1200 gauss in gauge region. The multiple magnetic gauge method used in this study is discussed in Ref. 2.

922

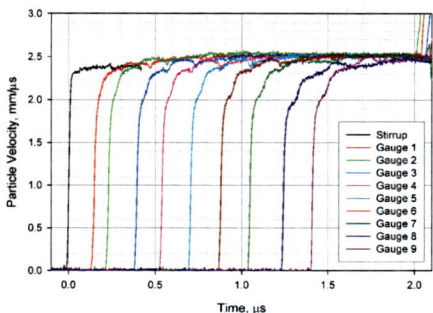

**FIGURE 5.** Particle velocity waveforms obtained from Shot 2s-173 on TBA. An evolving two-wave structure is clearly evident in the successive gauge waveforms.

The gauge membrane had nine particle velocity gauges and three shock tracker gauges. The stirrup gauge, in contact with the liquid, provided an additional particle velocity measurement at the input interface. Using these gauges it was possible to measure the shock velocity and particle velocity independently. Projectile velocity was measured to an accuracy of ~0.1% in all the experiments. All the gun projectiles were made from Lexan with a Kel-F 81 impactor on the front.

## RESULTS AND DISCUSSION

Three experiments were done on each material, one below and two above the reaction condition. Representative particle velocity waveforms are shown in Fig. 5 obtained from TBA Shot 2s-173 (see Table 2).

Since particle velocity, shock velocity, and projectile velocity are all measured, the Hugoniot state is overdetermined and the internal consistency of the data from each experiment can be scrutinized. Shock velocity is determined using the shock tracker gauges on each side of the particle velocity gauges. The data are plotted in a distance vs. time plot and, because there are a large number of points that can be fitted, the shock velocity value is determined to about 1 %.

The membrane particle velocity gauges produce waveforms that are low in voltage by more than 10% because of slippage that occurs at the gauge plane as the shock moves through it. They are low because the shock impedance of the liquid is lower than that of the FEP Teflon gauge. This effect has been in-

vestigated and reported in a previous APS Meeting.[3] These experiments, because of the redundant data, have allowed us to determine the *actual* particle velocity and the waveforms have been modified by multiplying the ETMS waveforms by 1.112 and the TBA waveforms by 1.125. Impedance matching was used to determine these correction factors.

The stirrup gauge data does not have this problem because it is parallel to the shock front. However, there is some question about the length of the active end of the stirrup gauge. These experiments allowed us to determine that we needed to multiply the stirrup gauge length by 1.023 to make the data consistent. This is in agreement with a study done by Koller in 1978 who determined that the length should be $1.027\pm0.03$ times the average of the inside and outside lengths.[4]

Experimental data from the six experiments performed in this study are shown in Table 2. Data from the ETMS shots indicate that the first wave state has a pressure of 6.6 GPa, i.e., this is the incipient reaction condition for ETMS. This compares to the TBA which has an incipient reaction condition of 6.1 GPa. We have, in the past, considered the compression energy $(1/2\ P\Delta V)$ at the incipient reaction condition to be a measure of when reaction should occur in certain related materials. The compression energy for TBA is 1.99 GPa cm$^3$/g and 2.02 GPa cm$^3$/g for ETMS. We interpret this to mean the acetylene bond is breaking first and whether it is hooked to a carbon in one case and a silicon in the other case makes very little difference.

In one experiment (see Fig. 5) the first wave was apparently overdriven, i.e., the shock velocity was greater than that of the first wave in the lower input pressure experiment (4.9 mm/$\mu$s compared to 4.59 mm/$\mu$s; last column in Table 2). When the gauge profiles are plotted on top of each other, the second wave is not falling behind the first wave after the initial evolutionary process. This means we have measured a steady wave with a "notch" which is related to the reaction process occurring, i.e., the shape does not change as the wave propagates. This condition was discussed by Dremin et al., in 1965[5] as applied to phase transitions. It was later applied to chemical reactions in 1978.[6] To our knowledge, this is the first time a steady wave with a notch has been measured in a shock-induced chemical reaction experiment where the products are more dense than the reactants.

**TABLE 2.** Shot data for all experiments in this study of ETMS and TBA

| Mat'l | Shot No. | Proj. Vel. km/s | Impact $u_p$ mm/$\mu$s | Impact P, GPa | First Wave State P, GPa | $u_p$,mm/$\mu$s | $U_s$,mm/$\mu$s |
|---|---|---|---|---|---|---|---|
| ETMS | 2s-156 | 2.695 | 1.88 | 5.8 | 5.8 | 1.88 | 4.36 |
| ETMS | 2s-157 | 3.258 | 2.24 | 8.0 | 6.6 | 2.0 | 4.61 * |
| ETMS | 2s-158 | 3.405 | 2.3 | 8.9 | 6.6 | 2.0 | 4.63* |
| TBA | 2s-172 | 3.114 | 2.16 | 7.4 | 6.1 | 2.08 | 4.59* |
| TBA | 2s-173 | 3.448 | 2.4 | ~8.4 | ~8.4 | 2.6 | 4.9* † |
| TBA | 2s-176 | 2.006 | 1.42 | 3.6 | 3.6 | 1.42 | 3.82 |

* Shock-induced reaction in these experiments
† The initial reactive wave is apparently overdriven

Another interesting phenomena was observed in these experiments relating to the release of energy during the reaction. Each experiment has a heavy cylindrical aluminum shroud around the target to protect the electromagnet from shrapnel and debris during the experiment. In the case of a nonreactive experiment, the shroud is essentially undamaged – this was the case in these experiments when no reaction was observed. When there was reaction, considerable damage to the shroud was observed, indicating that an exothermic reaction occurred sometime during the reaction process. We don't know what reaction caused this damage. The measured waveforms indicate the reaction products are more dense than the reactants. We believe the energetic reaction must be occurring at late times – after our measurements are over.

## CONCLUSIONS

Both liquids in this study were observed to react in a shock environment where the input pressure was above 6.1 GPa for TBA and 6.6 GPa for ETMS. The compression energy at these conditions was about the same for both materials; we believe this means the acetylene bond was where the reaction started. The fact that the acetylene group was attached to a carbon in one case and a silicon in the other case was of no consequence. In one experiment, we observed a steady wave with a reaction-related notch – the first time this has been measured. Additionally, the reactive process released enough energy to significantly damage a heavy aluminum shroud.

## ACKNOWLEDGMENTS

We thank Pete Chavez and Joe Lloyd for shooting the two-stage gun. Valuable discussions with Ray Engelke on shock chemistry are also appreciated.

## REFERENCES

1. Sheffield S. A., and Alcon, R. R. *Shock-Induced Reaction In Several Liquids*, in Shock Waves in Condensed Matter – 1989, North-Holland, New York, 1990, p. 683.

2. Sheffield, S. A., Gustavsen, R. L., and Alcon, R. R. *In-Situ Magnetic Gauging Technique Used at LANL – Method and Shock Information Obtained*, in Shock Waves in Condensed Matter – 1999, AIP Conference Proceedings 505, New York, 2000, p. 1043.

3. Gustavsen, R. L., Sheffield, S. A., and Alcon, R. R. *Response of Inclined Electromagnetic Particle Velocity Gauges in Shocked Liquids*, in High Pressure Science and Technology – 1993, AIP Conference Proceedings 309, New York, 1994, p. 1703.

4. Koller, L. R. *Generation and Measurement of Simultaneous Compression-Shear Waves in Arkansas Novaculite*, Ph.D. Thesis at Washington State University, Pullman, WA (1978), p. 48.

5. Dremin, A. N., Persin, S. V., and Pogorelov, V. F. Comb. Expl. and Shock Waves 1(4)(1975) p. 1.

6. Sheffield, S. A. *Shock-Induced Reaction in Carbon Disulfide*, Ph.D. Thesis at Washington State University, Pullman, WA (1978), p. 41.

CHAPTER XIV

# DETONATION AND BURN PHENOMENA

CP845, *Shock Compression of Condensed Matter - 2005*,
edited by M. D. Furnish, M. Elert, T. P. Russell, and C. T. White
© 2006 American Institute of Physics 0-7354-0341-4/06/$23.00

# ELECTRICAL CONDUCTIVITY MEASUREMENTS IN REACTING METASTABLE INTERMOLECULAR COMPOSITES

**B. W Asay, D. G. Tasker, J. C. King, V. E. Sanders, and S. F. Son**

*Los Alamos National Laboratory, Box 1663, Los Alamos, NM 87545*

**Abstract.** Metastable Intermolecular Composite (MIC) materials are comprised of a mixture of oxidizer and fuel with particle sizes in the nanometer range. To better understand the reaction mechanisms of burning MIC materials, dynamic electrical conductivity measurements have been performed on a MIC material for the first time. Simultaneous optical measurements of the wave front position have shown that the reaction and conduction fronts are coincident within 160 µm. Unlike detonating high explosives (HE) where the conductivity profile is represented by an initial peak followed by an exponential decay of conductivity, the MIC conductivity profile is a gradual, irregular ramp which increases from zero over many microseconds. This suggests that the reaction zone thickness is different in MICs compared to detonating HE. Static measurements of conductivity of pressed MIC pellets suggest that the conduction is associated with chemical reaction in the MIC.

**Keywords:** Metastable Intermolecular composites, conductivity, reaction zone, nanoenergetics
**PACS:** 82.33.Vx, 65.80.+n

## INTRODUCTION

Metastable Intermolecular Composites (MICs), also known as super-thermites, are comprised of a mixture of oxidizer and fuel with particle sizes in the nanometer range. These small particle sizes promote relatively fast kinetics with reaction front velocities of the order of 1 km/s [1]. In typical applications these materials are packed to densities of the order of 10% to 30% of the theoretical maximum density (TMD). To better understand the mechanism(s) of reaction propagation in these loose compacts a systematic study of the mode of propagation was reported [1]. Radiation, convection, conduction, and acoustic/compaction modes were considered; of these four, convection was deemed the most likely mechanism of propagation.

In this work, dynamic electrical conductivity measurements have been performed on a reacting MIC mixture of aluminum and molybdenum trioxide ($MoO_3$), adapting a technique first developed for measuring the electrical conductivity of detonating explosives [2]. MIC mixtures were prepared from 38% 80 nm Nanotechnologies aluminum plus 62% $MoO_3$ by weight and the pressing densities, for the experiments reported here, were 0.44 g/cc (12% TMD) [4].

The original technique is due to Ershov [3] who measured the conductive region associated with the detonation wave of a solid explosive as it swept past a narrow slit in one of a pair of metal electrodes. In this work a ~3-mm thick layer of MIC was sandwiched between a pair of plane, parallel electrodes and initiated with an electric match at one end. One electrode had one or more slits set in it; in this way conductivity structure of the reaction front as it passed each slit could be measured. Fiber optic probes were used to determine the position of the reaction front relative to the conductivity front.

## EXPERIMENTAL PROCEDURE

For a complete description of the equations and theory of this technique, see [4]. We will only present the experimental configuration here. A 3.175 mm (1/8-inch) square cross-section column of the MIC under investigation was loaded between parallel, plane electrodes having first determined that the unreacted MIC will not breakdown in the applied electric field. A voltage was applied across the electrodes thus introducing an electric field E into the unreacted MIC.

The MIC was ignited with an electric match, causing a reaction wave to sweep from right to left under the electrodes as shown in Fig. 1. As the reaction entered the electrodes the circuit was completed by conduction through the conduction zone. Current, I, then began to flow through the reaction zone of the MIC and along the electrodes. The bottom electrode had a 25 μm-wide slit in it, which was shunted with a low inductance loop (~2.5 nH). The rate of change of current, dI/dt, was measured in that loop with a Rogowski coil. The data we show later suggest that the conduction zone is more than 10 mm wide, so the slit is negligibly thin by comparison.

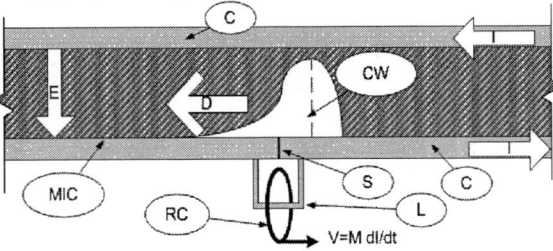

**Figure 1**. Schematic of conductivity measurement technique. The brass conductors C apply an electric field E across the MIC and carry a current I through the conduction wave, CW. The wave moves from right to left with a wave speed D. S is a slit; L is a low inductance loop bridging the slit, and RC is a Rogowski coil.

If the Rogowski coil is connected as shown in Fig. 1 it produces a voltage MdI/dt, where M is the mutual inductance of the coil with the shunt loop. As E is independent of x we can express the measured Rogowski voltage Vr(t) in terms of the conductivity as a function of time. The reaction front wave speed, D, was found to be constant from the optical measurements.

The field E(t) is obtained from the applied potential difference between the electrodes, P(t), divided by h, the separation of the plates, (E = P/h), and h, M and D are constant. Hence we can measure the conductivity $\sigma(t) = \sigma (x/D)$.

In the various experiments one to four slits were placed in the bottom electrode to observe the conduction wave as it swept from left to right in the test cell. To correlate the time of emission of light with the arrival of the conduction front, optical fibers were inserted through cylindrical cavities in a poly(methyl methacrylate) (PMMA) insulator adjacent to each slit, fast photodiodes were attached to the other ends of the fibers to detect the optical signals. To prevent electrical breakdown across the end a 3.175-mm thick PMMA or polyethylene wafer was inserted at the match end.

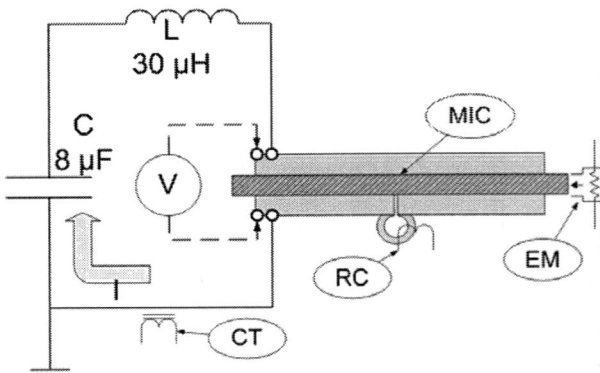

**Figure 2.** Electrical circuit: V is the voltage measurement circuit; I is the current in the circuit; CT the current transformer that measures I; EM the electric match; RC the Rogowski coil adjacent to the slit; L and C are the inductance and capacitance. Just one of the four slits and Rogowski coils are shown

A resonant series LC circuit was designed to have a half period comparable to the expected transit time of a combustion wave in the test cell; in fact we found the period to be too short as reported later. Fig. 2 shows the 8-μF capacitor wired in series with a 30-μH inductor and the cell. Before igniting the MIC with an electric match the capacitor was charged to 2500 V and this voltage was thus applied to the electrodes of the test cell.

Once the match was fired the reaction swept across the electrodes and current flowed in the circuit. The voltage across the cell was measured using an isolated resistor and current monitor circuit; the power supply and voltage probe were attached separately in a four-point configuration to avoid contact resistance errors. The total current in the circuit was measured with a calibrated commercial current transformer. Great care was taken to isolate all peripheral equipment in the experiment from the main test cell circuit to avoid ground loops. In particular, all diagnostics were electrically isolated from the main circuit by using magnetically coupled or optical sensors. Several different grounding arrangements were tried before the data reported here were obtained.

## RESULTS

The results of the data analysis are shown in Fig. 3 for two consecutive experiments in nominally the same material. We were only able to collect conductivity data for the first two of the four slits. This was because the cell resistance was high enough to dissipate all the electrical energy in the circuit after the wave had traveled 40 $\mu$s or 50 $\mu$s past the first slit, i.e., before the combustion wave arrived at the last two slits. By that time the electric field $E(t)$ between the top and bottom electrodes had decayed to zero.

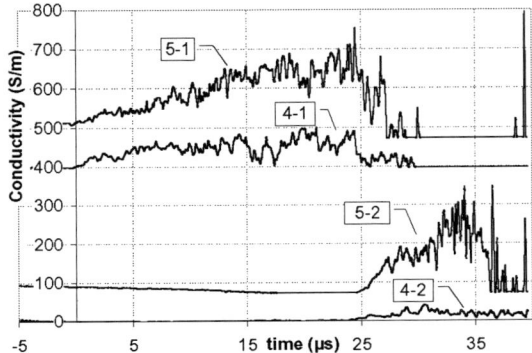

**Figure 3.** Comparison of conductivity data in S/m versus time for shots 42104-4 & 42104-5. Shot 42104-4 data for the first and second slits are labeled 4-1 and 4-2; similarly for shot 42104-5 they are 5-1 and 5-2. The data have been offset for clarity.

The measurements obtained in the first two slits were of reactive waves that had probably not reached steady-state. The conductivity rose steadily behind the reaction front over a period of 15 to 25 $\mu$s and then fell more rapidly, see Discussion.

Figure 3 shows the conductivity data on a common plot for direct comparison of the profiles. The times between arrivals of the waves at the two slits are the same, 25 $\mu$s, and the shapes are similar, but the magnitudes are significantly different. We surmise that the magnitudes differ because of small scale differences in the packing of the material adjacent to the slits, but as yet we have no proof of this.

It originally appeared that the conductivity fronts were detected before the arrival optical signals, but that was incorrect. The optical data were long slowly-rising signals, which made it difficult to detect the onset of light emission. However, by differentiating the optical data, it appears that the signals for both shots arrived approximately simultaneously, at least within the resolution of the optical data.

The time resolution of the Ershov technique is determined by the electrical response of the Rogowski coil (< 1 ns) and the time it takes the reaction wave to sweep across the width of the slit, dx. It turns out that sweep time is the dominant factor. The measured wave speeds D for the mixtures reported here were ~650 m/s. For dx = 25 $\mu$m that translates to a time resolution ~40 ns.

The time resolution of the optical fibers is determined by the width of the cone of light that the fibers admitted, and light scatter. Light scatters ahead of the luminous reaction front because the material is not optically dense at the loading densities used in these experiments. Also, the reaction front is fairly diffuse (as evidenced by the large reaction zone length, approximately 25 $\mu$s or 15 mm). For the shots reported here the resolution was ~250 ns at a wave speed of 650 m/s; this corresponds to a spatial resolution of ~ 160 $\mu$m.

The wave velocities of the reactive fronts, as determined from the optical data, were 605 m/s and 704 m/s for the two shots. However, there were only three points recorded for Shot 42104-4 and four points for Shot 42104-5 and the standard errors for the two velocities were 20 and 49 m/s respectively, Given the relatively large errors and the small samples the two velocities were not significantly different.

## DISCUSSION

From the data we see that the reaction and conduction fronts are coincident within the resolution of the experiment, i.e., ~160 μm. Unlike detonating explosives where the conductivity profile is represented by a sharply rising initial peak, followed by an exponential decay of conductivity [3], the MIC conductivity profile is a gradual, irregular ramp which increases from zero over many microseconds. The results are consistent with those of [1], specifically that the reactive wave appears very different from classical detonation. In detonating HE it is thought that the conduction is due to the coagulation of unburned carbon in the product gases [5]. It is not clear what the conduction mechanism is in MICs, it could be due to either chemical or physical processes, or both. But if MIC reaction occurs more slowly, then the products would be released more slowly, leading to a slower rise in conductivity. We note that the conductivity measurements obtained in the first two slits were of reactive waves that had probably not reached steady-state.

Spectral analysis of a conductivity plot shows a noisy spectrum up to ~50 MHz, the limit of the slit resolution, with no dominant resonances. SEM images show that the $MoO_3$ has sheet dimensions and spacing of tens of microns. So this 50 MHz spectrum is also consistent with the time it would take for the combustion wave to sweep across the $MoO_3$ particles. However, the interaction with aluminum particles (circa 80 nm across) would be too fast to resolve because each particle would be swept by the wave in ~40 ps and the Rogowski coil and slit combination has a resolution of 20 to 40 ns. So the "noise" may be structure due to the stochastic reaction associated with individual $MoO_3$ flakes. Some structure of the conductivity profiles is seen in the optical data, which suggests that contact irregularities do not account for the noise.

For the two experiments reported here, the times between arrivals of the waves at the two slits are the same, 25 μs, and the shapes are similar, but the magnitudes are significantly different. We surmise that the magnitudes differ because of small scale localized differences in the packing of the material adjacent to the slits, but as yet we have no proof of this. Some variations can be explained because the reactive waves had probably not reached steady-state in the first two slits.

Static conductivity measurements of the MIC pellets showed that they do not conduct when pressed to 20 MPa, i.e., at pressures comparable to those observed in dynamic combustion experiments. The initial evidence, although not conclusive, suggests that the conduction observed in the dynamic experiments is due to reaction in the MIC, not compaction. It will be interesting to continue the dynamic conductivity studies on MICs with different particle sizes and compositions in an effort to gain further insight into the conduction and reaction processes.

## ACKNOWLEDGEMENTS

The authors acknowledge the excellent technical assistance rendered by Alan Novak. This study was supported in part by the US Department of Defense, Joint Munitions Program.

## REFERENCES

1. Asay B.W., Son S.F., Busse J.R., Oschwald D.M., "Ignition Characteristics of Metastable Intermolecular Composites," Propellants, Explosives, Pyrotechnics, 29, 4 , 216 - 219
2. Tasker D.G. and Lee R.J., "The Measurement of Electrical Conductivity in Detonating Condensed Explosives", Procs. Of Ninth Symposium (International) on Detonation, Portland, OR, 1989, p.396.
3. Ershov A.P., Zubkov P.I., and Luk'yanchikov L.A., "Measurements of the Electrical Profile in the Detonation Front of Solid Explosives," Combustion, Explosion and Shock Waves, 10 (6), 1974, p.776.
4. Tasker D. G., Asay B. W., King J. C., Sanders V. E., and Son S. F., " Dynamic measurements of electrical conductivity in metastable intermolecular composites," submitted to Journal of Applied Physics, May 2005.
5. Hayes B., "On Electrical Conductivity in Detonation Products," Procs. Of Fourth Symposium (International) on Detonation, White Oak, MD, Oct 1965, p. 595.

CP845, *Shock Compression of Condensed Matter - 2005,*
edited by M. D. Furnish, M. Elert, T. P. Russell, and C. T. White
© 2006 American Institute of Physics 0-7354-0341-4/06/$23.00

# DIRECT NUMERICAL SIMULATION OF DETONATION[1]

## Tariq D. Aslam

*Los Alamos National Laboratory, Los Alamos, New Mexico 87545 USA*

**Abstract.** The last decade has been witness to a thousand fold gain in computational power [1], in addition to comparable gains from improved computational algorithms such as adaptive mesh algorithms (AMR) [2] [3]. But, even with these gains, there are many detonation phenomena which are beyond the current and foreseeable capabilities of direct simulation of resolved reaction zones. There are many other issues that arise in the simulation of detonation which lead to ill-posed computations. Many of these issues, as well as an overview of accomplishments in the field, current state of the art, and future work on detonation simulation will be discussed.
**Keywords:** detonation, numerical simulation
**PACS:** 47.40.-x, 47.40.Nm, 47.40.Rs, 47.27.ek

## INTRODUCTION

The focus of this paper is to examine some of the mathematical and computational issues that arise in the simulation of detonation reaction zones. For this work, it is assumed that a continuum model of the inviscid Euler equations with reaction are appropriate. These are given by the conservation of mass, momentum, and energy and a reaction rate law as follows in two dimensional Cartesian coordinates:

$$(\rho)_t + (\rho u)_x + (\rho v)_y = 0, \quad (1)$$
$$(\rho u)_t + (\rho u^2 + p)_x + (\rho uv)_y = 0, \quad (2)$$
$$(\rho v)_t + (\rho uv)_x + (\rho v^2 + p)_y = 0, \quad (3)$$
$$(E)_t + (uE + up)_x + (vE + vp)_y = 0, \quad (4)$$
$$(\rho \lambda)_t + (\rho u\lambda)_x + (\rho v\lambda)_y = \rho r, \quad (5)$$

where subscripts denote partial derivatives with respect to time, $t$ and space $x$ and $y$. The values of density, $\rho$, particle velocities, $u$ and $v$, pressure, $p$, reaction progress, $\lambda$ and total energy

$$E = \rho e + \frac{\rho}{2}(u^2 + v^2)$$

are taken. Furthermore, $r$ is the reaction rate and $e$ is the specific internal energy. In addition to initial and boundary conditions, the equation of state (EOS) and reaction rate law are required to close the above system of equations.

## RESOLUTION REQUIREMENTS

As an illustrative example, let's examine the resolution requirements to compute a simple model sandwich test [4] [5]. Here, the high explosive (HE) is given by $\rho_0 = 1$ and $p_0 = 10^{-6}$, and an EOS given by the polytropic gas:

$$e = \frac{p}{\rho(1 - \gamma)} - q\lambda$$

with $\gamma = 3$ and $q = 0.0625$. These parameters yield a Chapman-Jouguet detonation velocity, $D_{CJ}$, of unity. The reaction rate law is given by:

$$r = 0.31434(1 - \lambda)^{1/2} H(p - 0.05) \quad (6)$$

---

[1] Work supported by the U.S. Department of Energy

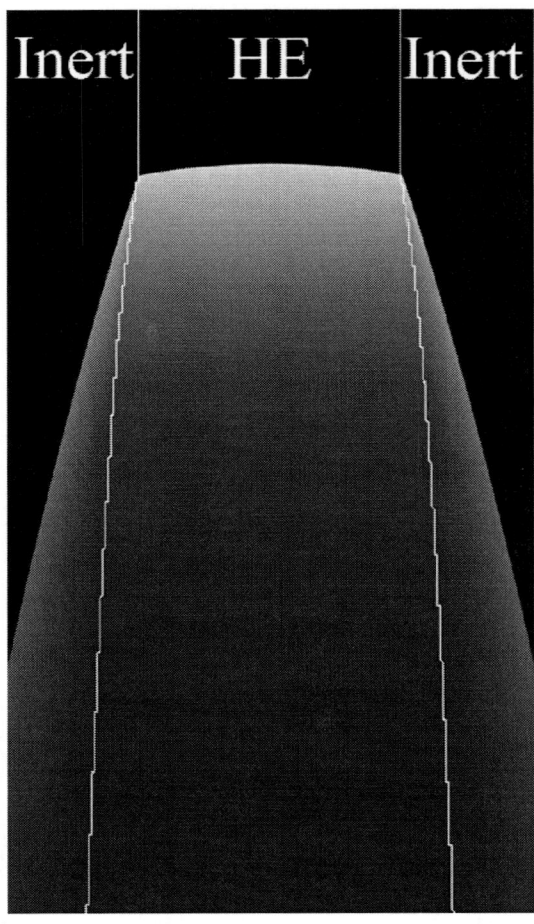

**FIGURE 1.** Pressure gray-scale plot for sandwich test; White lines indicate the material interface between HE and inert.

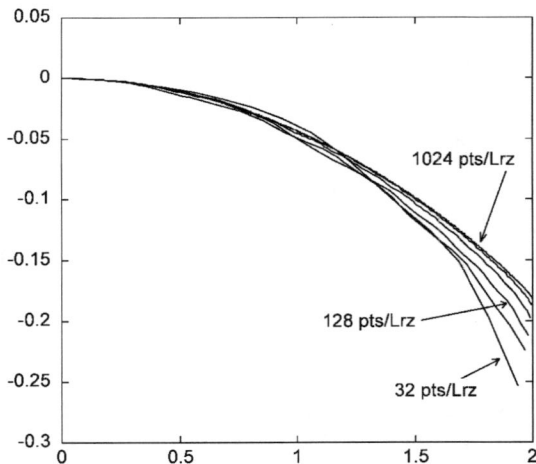

**FIGURE 2.** Shock shapes of sandwich test for various resolutions from 32 to 1024 (by factors of 2) computational cells per $L_{RZ}$.

where $H$ is the Heaviside function. This sets the complete ZND reaction zone length, $L_{RZ}$, to unity. The thickness of the HE, in the sandwich, is set to unity, with a surrounding inert polytropic gas with $\gamma = 3$ and $\rho_0 = 10$. A plot of the resulting pressure and material interface is given in Fig. 1.

A series of these calculations, varying the grid spacing, $\Delta x$, was performed, and the resulting shock shapes are given in Fig. 2. The lowest resolution case, $\Delta x = 1/32$, corresponds to 32 computational zones in the steady 1D ZND reaction zone structure. The highest resolution contains 1024 computational zones per $L_{RZ}$. From Fig. 2, it is evident that the solution requires roughly 64 computational zones in $L_{RZ}$

before the curves are quantitatively agreeing. Note that the shocks at the centerline of the sandwich have all been shifted to $z = 0$, so as to make the comparison easier. This argues for a resolution minimum of roughly 64 computational zones in $L_{RZ}$.

If one takes this resolution of 64 cells per $L_{RZ}$ and attempts to compute an engineering system of say 200mm across in each dimension, we can estimate the computational time given an AMR grid over say a 1mm $L_{RZ}$. First the two dimensional case will be examined, followed by the three dimensional case.

If a planar wave of 1mm in depth resolved at $\Delta x = 1/64$ in each of the two dimensions, and if it is assumed to take roughly $13\mu sec$ of computer time to update a cell, then for the $(200 \times 64)^2 \times 64 = 8.2 \times 10^{10}$ cells to update with $200 \times 64 \times 3 \times 2 = 10^5$ timesteps would take roughly 6 days to compute on a single processor machine.

For a planar wave of 1mm in depth resolved at $\Delta x = 1/64$ in each of the three dimensions, and assumed to take roughly $20\mu sec$ of computer time to update a cell, then for the $200 \times 64 \times 64 = 8.2 \times 10^5$ cells to update with $200 \times 64 \times 2 \times 2 = 76800$ timesteps would take roughly 507 years to compute on a single processor machine. If one could achieve perfect parallelization over 1000 processors, the computational time would be brought down to roughly 1/2 a year. For now it is apparent that direct numerical simulation of the reaction zone on a

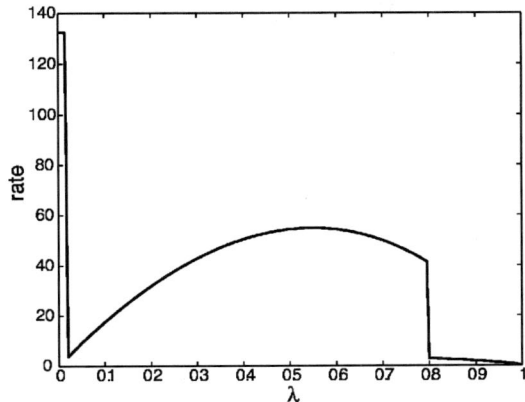

**FIGURE 3.** Reaction rate through ZND structure versus reaction progress, $\lambda$, for the Ignigion and Growth model.

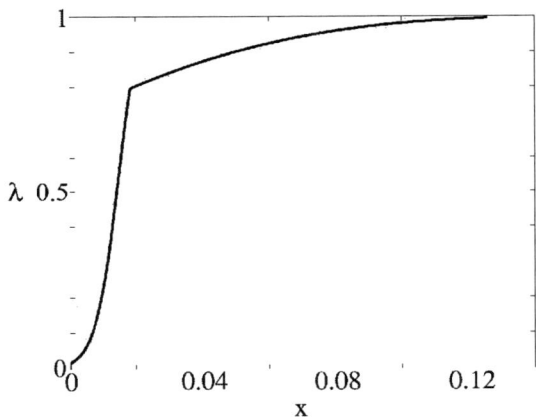

**FIGURE 4.** Reaction progress, $\lambda$, versus $x$ for the Ignigion and Growth model.

routine basis for three dimensional detonation is not realistic.

Furthermore, for realistic EOS and reaction rate models, such as Ignition and Growth [6], the reaction zone structure is much more complicated than in Eqn. 6. This further strains the computational requirements. Fig. 3 and Fig. 4 show a typical ZND structure for the Ignition and Growth model for the insensitive HE LX-17. Note that there are discreet jumps in the rate, and thus kinks in $\lambda$ and the other field variables.

## PRESSURE-TEMPERATURE EQUILIBRATION

Another complication that often arises in computation of non-ideal EOS, is the assumption of pressure and temperature equilibration between the reactants and products in the reaction zone. For the Ignition and Growth model, both the unreacted EOS and products EOS are given by a JWL form [7]. Unfortunately, this leads to a nonlinear algebraic problem when computing $p(e, \rho, \lambda)$, which is required during the computation of the momentum equation. Once one assumes the temperatures of each phase are equal, then one can look for the zero root of $\Delta p$ between each phase (reactant and product) as a function of scaled volumes, $\beta$. One such case is shown in Fig. 5 for LX-17 at the midpoint of reaction, $\lambda = 0.5$ in the ZND structure. Of note, there are three roots to this equation, near $\beta = 0.04, 0.48, 0.98$. Only the root near $\beta = 0.48$ makes any sense physically; the other two correspond to negative temperature and imaginary sound speed. So, these other two roots are unimportant, except that the typical Newton method for finding the zero in Fig. 5 can go astray and either not converge or converge to a physically meaningless result. It has been pointed out in [8] that for $p - T$ equilibrium mixtures, that maximizing the entropy subject to constraint on energy and density (that is partitioning energy and volume among components) yields pressure and temperature equilibrium. Thermodynamic stability requires entropy to be concave function of $1/\rho$ and $e$. If the entropies of the component EOS are concave then the sum or entropy of the system for fixed mass fraction is also concave. And since a concave function has a unique maximum, this is usually not a problem, and the $\beta = 0.04, 0.98$ roots in Fig. 5 are thus non physical in the sense that either the reactants or products EOS has gone outside the range of thermodynamic stability.

## SHOCK-STATE SENSITIVE RATE LAWS

Many reactive flow models for detonation and deflagration to detonation transition contain a term in the rate law that is related to the shock state that particle experienced [9] [10] [11] [12]. There is no real mathematical modeling issue associated with this type of rate law, assuming one does not consider acoustic

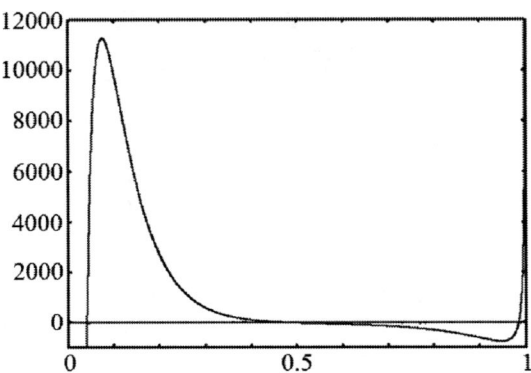

**FIGURE 5.** $\Delta p$ versus scaled volume fraction, $\beta$.

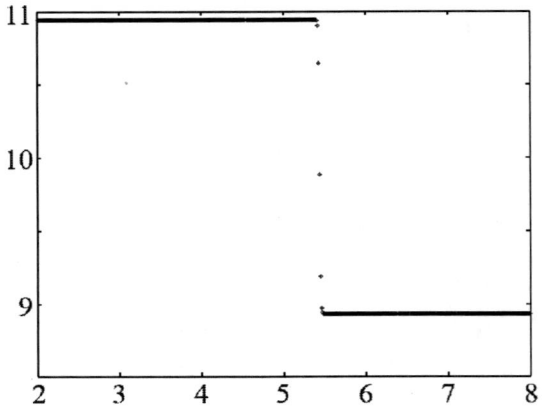

**FIGURE 6.** Computed density, $\rho$ versus $x$.

waves as shock waves. But, there is an issue with computing the shock state itself in a shock capturing scheme. This applies to Lagrangian, Arbitrary Lagrangian Eulerian (ALE) and pure Eulerian schemes equally. Very often the numerical details of how one is computing the shock state are not given. Furthermore the shock state density, pressure, etc. are all simply related though the Hugoniot jump conditions to the shock speed, which is not typically part of the solution. This typically causes an $O(1)$ error in these values, since the underlying shock capturing scheme has $O(1)$ errors in the $O(\Delta x)$ vicinity of the shock. Only away from the shock by an $O(1)$ distance is the solution $O(\Delta x)$. This presents the conundrum. Only a scheme which takes this into account, such as [13], will have a chance at obtaining convergence to such problems.

For example, if one computes a shock in copper with a post-shock particle velocity of $1mm/\mu s$. One can see upon examination of Fig. 6 and Fig. 7 that there are large errors in the vicinity of the shock, and this results in an error in setting the shock state for each particle downstream of the shock, which feeds into errors in the rate.

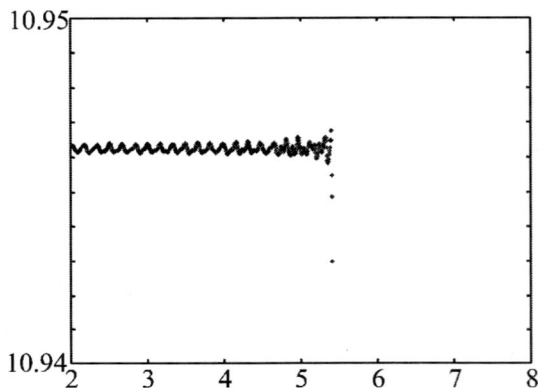

**FIGURE 7.** Computed density, $\rho$, versus $x$, zoomed into post shock state.

as $p - T$ equilibration and having shock state sensitive rate laws. Given all this, care must be taken in the computation of detonations within engineering systems.

## CONCLUSIONS

It has been demonstrated that there are several issues in computing the reaction zone in detonation. Firstly, there are stiff requirements to obtain a qualitatively correct numerical solution, due to resolution requirements of detonation. Also, there are several numerical issues with many detonation models, such

## REFERENCES

1. *"Top 500 List for June 2005,"* http://www.top500.org/ (2005)
2. Berger, M.J. and Colella, P *"Local adaptive mesh refinement for shock hydrodynamics,"* Journal of Computational Physics, Vol. 82, (1989)
3. Quirk, J.J., *"A parallel adaptive grid algorithm for computational shock hydrodynamics,"* Applied Numerical Mathematics, Vol. 20, No. 4, (1996)

4.  Aslam, T.D.,Hill, L.G. and Bdzil, J.B. *"Analysis of the LANL detonation-confinement test,"* Shock Compression of Condenssed Matter (2003)

5.  Hill, L.G. and Aslam, T.D., *"The LANL detonation-confinement test: prototype development and sample results,"* Shock Compression of Condenssed Matter (2003)

6.  Lee, E.L. and Tarver, C.M., *"Phenomenological model of shock initiation in heterogeneous explosives,"* Physics of Fluids, Vol. 23. (1980)

7.  Fickett, W., and Davis, W. C., *Detonation*, University of California Press (Berkeley) (1979)

8.  Menikoff, R.T., *"Empirical Equations of State for Solids,"* Shock Waves Encyclopedia: Shock Waves in Solids, Y. Horie, editor, Springer Verlag (2006)

9.  Xu, S.J., Aslam, T., and Stewart, D.S., "High Resolution Numerical Simulation of Ideal and Non-Ideal Compressible Reacting Flows with Embedded Internal Boundaries," *Combustion Theory and Modelling*, Vol. 1, No. 1, 1997, pp. 113-142.

10. Johnson, J.N.; Forest, C.A.; Tang, P.K., *"Shock-wave initiation of heterogeneous reactive solids,"* Journal of Applied Physics; 1 May 1985; vol.57, no.9, p.4323-4334

11. Aminov, YA; Es'kov, NS; Nikitenko, YR; Rykovanov, GN, *"Calculation of the reaction-zone structure for heterogeneous explosives,"* COMBUSTION EXPLOSION AND SHOCK WAVES; MAR-APR 1998; v.34, no.2, p.230-233

12. Wescott, B.L., Stewart, D.S. and Davis, W.C., *"A wide-ranging equation of state and reaction rate for condensed-phase explosives calibrated to PBX-9502,"* Los Alamos Report LA-UR-04-6054 (2004)

13. Aslam T.D. and Bdzil J.B., *"Numerical and theoretical investigations on detonation-inert confinement interactions,"* 12th Det. Symp. (2002)

14. Marsh, S.P. ed.; LASL shock Hugoniot data (1980)

CP845, *Shock Compression of Condensed Matter - 2005,*
edited by M. D. Furnish, M. Elert, T. P. Russell, and C. T. White
© 2006 American Institute of Physics 0-7354-0341-4/06/$23.00

# MODELING A MATERIAL'S INSTANTANEOUS VELOCITY DURING ACCELERATION DRIVEN BY A DETONATION'S GAS-PUSH

## Joseph E. Backofen

*BRIGS Co., 4192 Hales Ford Road, Moneta, VA 24121*

**Abstract.** Analysis of cylinder test data reveals that an explosive's gas-dynamic-driven acceleration of a boundary material from its initial free-surface velocity up to the final steady-state velocity can be described by an equation derived by fitting data for twenty explosives. The time derivative of this velocity equation can represent pressure as well as acceleration with results comparing favorably to Jones-Wilkins-Lee equation-of-state model calculations performed using published parameters.

**Keywords:** Detonation, gas dynamics, equation of state, cylinder test
**PACS:** 47.40.-x, 06.30.Gv

## INTRODUCTION

This paper presents both scientific findings and a model developed in order to quantify a boundary material's instantaneous velocity versus position, time or the expansion ratio of an explosive's gaseous products while being accelerated. An equation is derived to represent the second stage of the BRIGS Two-Step Detonation Propulsion Model wherein: 1) initial motion is imparted by a brisant shock-dominated process that depends upon intimate contact of an explosive with the propelled material, and 2) subsequent acceleration is by a gas-push (gas-dynamic) process [1,2]. Initial motion is envisioned as being caused by the higher-pressure region of a detonation front (i.e. envision the von Neumann spike or reaction zone region as being a finite thickness of solid material squeezed at high pressure). The gas-push process is envisioned similar to that assumed by Gurney modeling – a gaseous product expansion from a homogeneous "all burnt" condition while pushing the confining boundaries to steady-state velocities [3].

## ANALYSIS AND DISCUSSION

Published experimental data for detonation-driven explosive devices show that there is a limited set of data that can be and have been measured beyond the initial geometry and material properties. These data include:

- The detonation velocity (D),
- The initial free-surface velocity (Vi), and
- Velocities occurring at either specific distances or times (Vx or Vt),

where these velocities are measured in km/s.

Detonation velocities are well represented in the literature for many explosives and their mixtures. Furthermore, the final "steady-state" velocity (Vf) achieved either at a late time or after a long travel distance by an inert material bounding device geometry represents one of the more easily measured and most consistent experimental data found throughout published literature. Vf can be used to calculate an explosive's Gurney Energy (Eg) using the standard Gurney formula appropriate for the device's geometry.

Previously, the brisant 1st propulsion stage's initial velocity during a cylinder test was shown to be a function of six quantities:

$$Vi = F ( \rho_{ex}, \rho_{cyl}, R_{ex}, t_{cyl}, D, \Gamma ) \quad (1)$$

Where $\rho_{cyl}$ and $\rho_{ex}$ are the cylinder and explosive densities in g/cm$^3$ respectively, $t_{cyl}$ and $R_{ex}$ represent the cylinder wall thickness and explosive radius in mm, and $\Gamma$ is the non-dimensional adiabatic coefficient for modeling expanding explosive gaseous products [1,2].

According to Gurney modeling, the final velocity is also a function of these same quantities:

$$Vf = F ( \rho_{ex}, \rho_{cyl}, R_{ex}, t_{cyl}, Eg ) \quad (2)$$

Where D and $\Gamma$ can replace Eg using [3]:

$$(2Eg)^{1/2} \cong 0.605 D / [\Gamma - 1] \quad (3)$$

Previous publications revealed that Vi was not only a major portion of Vf but also that they and their ratio (Vi/Vf ) were affected by the properties and thickness of the boundary materials [2,4,5].

The findings for the ratio Vi/Vf suggested that propulsion during the gas-push stage should also be examined by normalizing the instantaneous velocity (Vx, Vt, or Ver) respectively coordinated to distance, time or the explosive's gaseous product expansion ratio using Vf. Figure 1 presents data selected from eight of the over twenty analyzed explosives so that the plotted data points might be spread out enough that they would be discernible when printed [6-9]. The data demonstrate the commonality by which copper cylinders are propelled when Ver is normalized by Vf as an explosive's gaseous products expand from 1.0 < Expansion Ratio (ExR) < 13. However, the commonality might actually be much better than shown in Fig.1 since its data are normalized using published Vf values – values mostly measured at arbitrary "final" expansion ratios (ExRf), such as 6.5 or 7. Such published Vf do not necessarily represent the maximum velocity that could have been reached during experiments.

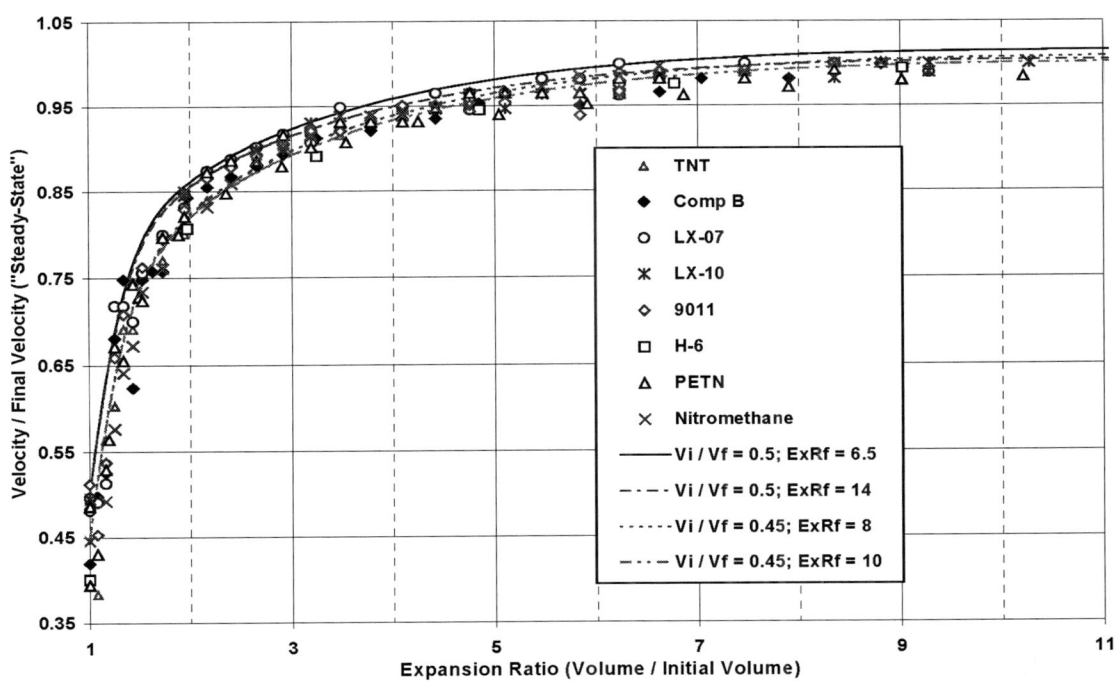

**Figure 1.** Normalized velocity data for eight explosives driving copper cylinders

937

The data shown in Fig.1, as well as those for twelve other explosives, demonstrate a rapid velocity rise from the initial free-surface velocity during $1 < \text{ExR} < $ about 1.5. After this, the velocity rises at a less steep slope until eventually reaching Vf at ExRf. Thus, two terms were chosen to represent all the data from experiments involving twenty liquid, cast, and pressed polymer-bonded explosives:

$$\text{Ver} / \text{Vf} = \text{V1/Vf} + \text{V2/Vf} \qquad (4)$$

$$\text{V1/Vf} = (\text{Vi} / \text{Vf}) \{ (e^{--\,\text{ExR}} - e^{-\,\text{ExR}^{\wedge}3}) + ([\text{ExR} / \text{ExRf}]^{-0.5} / [1 / \text{ExRf}]^{-0.5}) \}$$

$$\text{V2/Vf} = [1 - \{ (\text{Vi} / \text{Vf}) / (1 / \text{ExRf})^{-0.5} \}] (\text{ExR} / \text{ExRf})^{-0.333} [\log (\text{ExR}) / \log (\text{ExRf})]$$

The first term (V1/Vf) appears to represent a lingering effect from the 1st propulsion stage; and the second term (V2/Vf) appears to represent acceleration by detonation product pressure during their expansion. Some experimental data exhumed from published literature appear to support that typical "ideal" explosives continue pushing out to expansions beyond 10 and "non-ideal" explosives towards 14 as in the case of some aluminized explosives. This means that most data in textbooks and technical papers probably do not provide a sound basis for accurately modeling an explosive's gas-dynamic propulsion since the data were almost always only measured up to or normalized to ExRf = 6.5. (In other words, cylinder wall movement was normalized to travel of 19 mm during a "standard" 1-inch copper cylinder test.) Nevertheless, Eqn.4 offers a means to estimate velocity versus time, expansion ratio or boundary movement by means of direct measurement of Vi and Vf at ExRf. The former velocity can be measured using a Fabry-Perot interferometer and the latter velocity using a streak camera. However, Vi and Vf also can be estimated after measuring the geometry and materials by using the explosive's detonation rate and equations for Vi and Vf found in Refs.1–5.

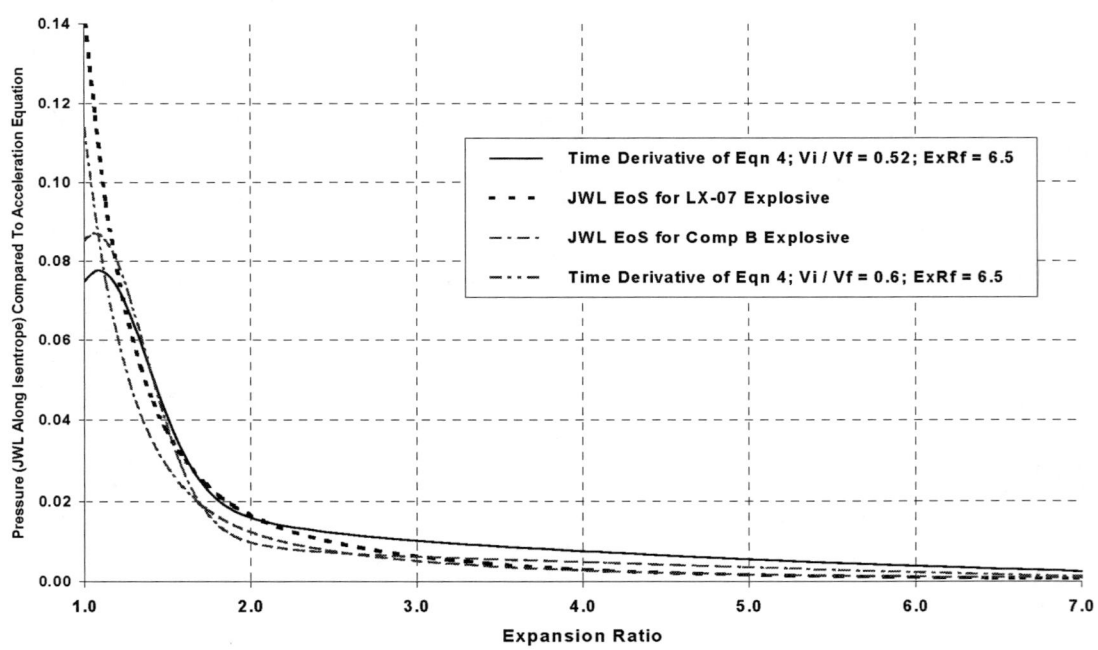

**Figure 2.** Comparison of JWL calculated pressures to the time derivative of Eqn.4 for an expanding cylinder.

Figure 2 shows that the time derivative of Eqn.4 creates a smoothly varying equation comparable to a JWL equation-of-state (EoS) [9]. As shown in Fig.2, both are found to be comparable for military explosives such as LX-07. However, there are distinct differences. Firstly, the new equation only represents the gas-push 2nd stage propulsion cycle whereas the JWL model was formed to describe the entire acceleration provided by the explosive. Thus, the JWL EoS must initially employ a high gas pressure in order to play "catch-up" since the JWL method assumes an initial velocity of zero rather than recognizing an initial free-surface velocity. The JWL EoS pressure also drops more quickly to a negligible pressure well before the time ExR reaches 6.5 whereas the time derivative of Eqn.4 continues a decaying pressure until the expansion ratio at which the final "steady-state" velocity is achieved in experiments.

## CONCLUSIONS

This paper has shown that the gas-push propulsion stage of detonation-driven propulsion by widely different explosives – cast, pressed plastic-bonded, liquid, and aluminized – can be represented on a common basis. This paper also has presented a new equation modeling gas-push propulsion in a manner consistent with previous propulsion concepts and modeling.

The data from "standard" cylinder tests and "scaled" cylinder tests used to construct Eqn.4 may have introduced a weakness. The data represent use of cylindrical geometry with very few different ratios of $t_{cyl}$ / $R_{ex}$ – a ratio shown to significantly affect Vi/Vf [2,5]. Thus, it would be beneficial to expand the available data through new experiments during which $t_{cyl}$ and $R_{ex}$ are varied beyond historically used values. Data should also be taken at as large an ExRf as possible.

## ACKNOWLEDGEMENTS

The author wishes to thank Dr. C.A. Weickert for encouraging him to continue this research over the past decade.

## REFERENCES

1. Backofen, J.E. and Weickert, C. 2000, "Initial Free-Surface Velocities Imparted by Grazing Detonation Waves", in *Shock Compression of Condensed Matter* – 1999, M.D. Furnish, L.C. Chhabildas, and R.S. Hixon, eds., American Institute of Physics., Part 2, pp. 919 – 922

2. Backofen, J.E. and Weickert, C.A. 2002, "Effect of an Inert Material's Thickness and Properties on the Ratio of Energies Imparted by a Detonation's 1st and 2nd Propulsion Stages", in *Shock Compression of Condensed Matter* – 2001, M.D. Furnish, N.N. Thadhani, and Y. Horie, eds., American Institute of Physics, pp. 954 – 957

3. Kennedy, J.E. 1998, "The Gurney Model of Explosive Output for Driving Metal", Chapter 7 in *Explosives Effects and Applications*, J.A. Zukas and W.P. Walters, eds., Springer-Verlag, New York, pp. 221 – 257

4. Backofen, J.E. 2003, "Confirmation of the Effects of Cylinder Wall Thickness and Material Properties on Measurement of an Explosive's Gurney Velocity", BRIGS Note 03-1, BRIGS Co., Oak Hill, VA, 9 February

5. Backofen, J.E. 2003, "Detonation-Driven Propulsion – Additional Comparisons Between Initial Free-Surface Velocity and Final "Steady-State" Velocity Measurements", BRIGS Note 03-5, BRIGS Co., Oak Hill, VA, 7 December

6. Kury, J.W., et al., 1965, "Metal Acceleration by Chemical Explosives", *Proc. 4th Symp. (International) Detonation*, White Oak, MD, October 12-15, pp. 3 - 13

7. Finger, M., et al., 1970, "Metal Acceleration by Composite Explosives", *Proc. 5th Symp. (International) Detonation*, Pasadena, CA, August 18-21, pp.137 - 149

8. Lee, E., et al., 1985, "The Motion of Thin Metal Walls and the Equation of State of Detonation Products", *Proc. 8th Symp. (International) Detonation*, Albuquerque, New Mexico, July 15-19, pp. 613 - 624

9. Hornberg, H. 1986, "Determination of Fume State Parameters from Expansion Measurements of Metal Tubes", *Propellants, Explosives, Pyrotechnics*, Vol.11, pp 23 - 31

CP845, *Shock Compression of Condensed Matter - 2005*,
edited by M. D. Furnish, M. Elert, T. P. Russell, and C. T. White

# EXPERIMENTAL METHOD TO DETERMINE THE DETONATION CHARACTERISTICS OF A VERY NON-IDEAL HIGH EXPLOSIVE

## G. Baudin, C. Le Gallic, F. Davoine, and P. Bouinot

*Centre d'Etudes de Gramat, French Ministry of Defense, 46500 – Gramat, France*

**Abstract.** Common experimental configurations used to determine HE detonation velocity-curvature are right circular cylinders detonated in air. The steadily propagating detonation front is curved and its velocity depends upon the diameter of the cylinder. This configuration requires several experiments with different diameters and sufficiently long cylinders to assume a steadily propagating detonation front. This last hypothesis is practically not achieved for non-ideal HE using reasonably long cylinders. To elude this problem, a special explosive device called "logosphere", developed by CEA, has been adapted to non ideal HE. It provides a well define spherically diverging detonation wave and allows measurements of the detonation velocity-curvature relationship by means of piezoelectric pins without any perturbation. VISAR and IDL diagnostics record the material velocities at the rear surface of the explosive through transparent windows. The particle velocity values are used to determine the curved detonation states using the detonation velocity-acceleration-curvature model of Louis Brun.

**Keywords:** Detonation wave, non ideal high explosive.
**PACS:** 47.40.-x, 47.40.Nm.

## INTRODUCTION

Detonation characteristics and ballistic performance of highly non-ideal high explosives (HE) strongly depend upon the geometric configuration used (scale effect).

The detonation velocity-curvature relationship is known as an intrinsic characteristic of steadily propagating detonation front in a high explosive. It allows the determination of the reaction rate just behind the detonation front, which is an important data for modeling. Common experimental configuration used to determine detonation velocity-curvature relationship is a right circular cylinder of explosive detonating in air. The detonation wave propagating in such a charge reaches a steady state after a transient phase of detonation buildup and unsteady regime. Hence in that case, the length of right circular cylinder must be large enough to obtain the steady state detonation wave. Its front is curved, and its velocity strongly depends upon the diameter of the cylinder. The curvature is generally measured on the axis of the right cylinder. This experimental configuration requires several experiments with different diameters and sufficiently long cylinders to assume a steadily propagating detonation front.

This last hypothesis is practically not achieved for highly non-ideal HE using reasonably long cylinders. To elude this problem, a special explosive device called "logosphere", developed by CEA [1], has been adapted to the considered Al/RDX/AP/binder and HMX/NTO/HTPB HE.

## "LOGOSPHERE" EXPERIMENT

The explosive charge is machined in such way that the meridian of the outer boundary is a logarithmic spiral, Fig. 1, which constant angle θ between the spherical detonation front and the explosive's meridian is small enough to avoid any perturbation on detonation propagation.

**Figure 1.** Explosive device called "logosphere"

This experimental configuration provides a well-defined spherically diverging detonation wave and allows measurements of the detonation velocity-curvature relationship by means of piezoelectric pins without any perturbation to the spherical geometry, using the quasi-steady state propagation approximation. It is clear that the spherically diverging detonation wave is intrinsically unsteady and that the characteristic of such detonation waves would be a detonation velocity-acceleration-curvature relationship. Fig. 2 represents the relationships obtained for three HE : HTPB-HMX-NTO 16-12-72, HTPB-RDX-AP-Al 12-20-43-25 and HTPB-RDX-AP-Al 14-50-24-12.

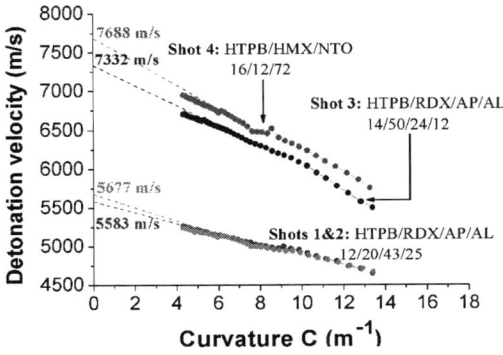

**Figure 2.** Detonation velocity-curvature relationships for three HE

Other instrumentation associated to this configuration is VISAR and DLI (Doppler Laser Interferometer) diagnostics to record the shock material velocities at the surface of the explosive through PMMA transparent windows (Fig. 3).

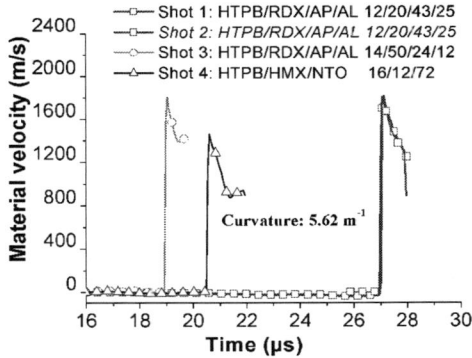

**Figure 3.** Material velocities recorded for a 5.62 m$^{-1}$ curvature.

## DETERMINATION OF THE DETONATION CHARACTERISTICS AND REACTION RATE JUST BEHIND THE FRONT

Knowing both detonation velocity and Hugoniot curve of the transparent windows, the measured particle velocities, Fig; 3, are used to determine the detonation states pattern and the reaction rate just behind the detonation front using the velocity-acceleration-curvature model of detonation developed by Louis Brun [2] and the classical quasi-steady detonation approach.

In this model the detonation is considered as a sonic shock and combustion wave followed by chemical reactions. It is an extension of the Chapman-Jouguet model to curved detonation waves. The velocity-acceleration-curvature relationship may be then described by

$$\frac{dD_n}{dt} = 2\Phi^2(D_n)[K(D_n) - C] \tag{1}$$

$$\Phi^2(D_n) = \frac{D_n^2}{3\frac{\gamma+1}{\gamma} - 2\frac{d\log\gamma}{d\log D_n}} \tag{2}$$

$$K(D_n) = \frac{\gamma+1}{2}\frac{\overset{\rho}{\sigma}.d\overset{\rho}{\lambda}/dt}{D_n} \tag{3}$$

where the polytropic coefficient $\gamma=\gamma(D_n)$. The detonation characteristics are given by equations similar to the Chapman-Jouguet model, i.e.

$$\frac{v_J}{v_0} = \frac{\gamma_J}{\gamma_J+1}, \quad P_J = \frac{\rho_0 D_n^2}{\gamma_J+1} \qquad \text{(4) and (5)}$$

$$u_{nJ} = \frac{D_n}{\gamma_J+1}, \quad \frac{c_J}{D_n} = \frac{\gamma_J}{\gamma_J+1} \qquad \text{(6) and (7)}$$

The classical quasi-steady detonation hypothesis leads to the detonation velocity-curvature relationship $K(D_n)=C$. A similar model can been found in the "Physics of Explosion" book of Andreev and al [3].

The material velocity is measured through PMMA windows. Therefore, the determination of the material velocity of the curved detonation wave before its interaction with the PMMA windows, needs several additional hypothesis:

- A quasi one-dimensional flow with a frozen sound velocity (slow variation of the reaction coordinate behind the sonic detonation front).
- A quasi-steady detonation wave (slow variation of the detonation velocity), hypothesis used to calculate the detonation velocity from the piezoelectric pins times.
- The acoustic approximation to determine the detonation products state just after the detonation wave interaction with the window; such an assumption has been proposed by Zababakhin [4].
- A locally slow variation of the polytropic index $\gamma$ so that pressure can be locally approached by the relationship $P=A\rho^\gamma$.

For quasi-steady detonation wave,

$$\sigma\frac{d\lambda}{dt} = \frac{2D_n C}{\gamma+1} \qquad (8)$$

In the acoustic approximation, the Riemann invariant is $u_2 + \dfrac{d\ln\rho}{d\ln c}c_2 = u_J + \dfrac{d\ln\rho}{d\ln c}c_J$ (Fig. 4). The relationship $P=A\rho^\gamma$ gives

$$\frac{P_2}{P_J} = \left(\frac{\rho_2}{\rho_J}\right)^\gamma = \left(\frac{c_2}{c_J}\right)^{\frac{2\gamma}{\gamma-1}} \qquad (9)$$

and the Riemann invariant becomes

$$u_2 + \frac{2c_2}{\gamma-1} = u_J + \frac{2c_J}{\gamma-1} \qquad (10)$$

where $c^2 = \left(\dfrac{\partial P}{\partial \rho}\right)_{S,\lambda} = \gamma A \rho^{\gamma-1}$.

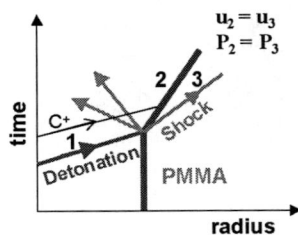

**Figure 4.** Flow-chart of the detonation interaction with the PMMA windows.

The P-u states 2 and 3 obtained after the detonation-PMMA interaction are equal (Fig. 4). Zababakhin [4] demonstrates the following relationship

$$P_2 = \frac{\rho_0 D_n^2}{\gamma+1}\left[\frac{\gamma^2-1}{2\gamma}\left(\frac{3\gamma-1}{\gamma^2-1} - \frac{u_2}{D_n}\right)\right]^{\frac{2\gamma}{\gamma-1}} \qquad (11)$$

connecting the states 1 and 2. Using a linear shock Hugoniot relationship for PMMA,

$$P_2 = \rho_{0F} D_F u_2 = \rho_{0F}\left(c_{0F} + s u_2\right)u_2$$

The knowledge of $P_2$, $u_2$ and $D_n$ allows the calculation of the polytropic index $\gamma$, and the complete thermodynamic state 1 of the detonation before its interaction with PMMA windows. Thus, the choice of an equation of state for the detonation products and inert HE mixing allows the determination of the reaction rate.

## RESULTS AND DISCUSSION

The results obtained for the three high explosives are given in Fig. 5 and 6. $\Delta u/c = (u_2-u_J)/c_J$

gives the validity of the acoustic approximation ($\Delta u/c < 9\%$). Fig. 5 represents the polytropic index and the products of the thermicity by the reaction rate.

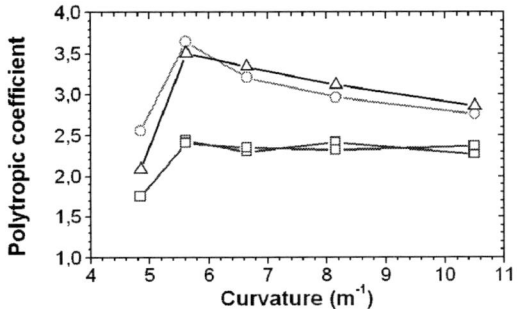

**Figure 5.** Polytropic versus curvature - squares: RDX-AP-Al-HTPB 20-43-25-12, circles: RDX-AP-Al-HTPB 50-24-12-14, triangles: HMX-NTO-HTPB 12-72-16.

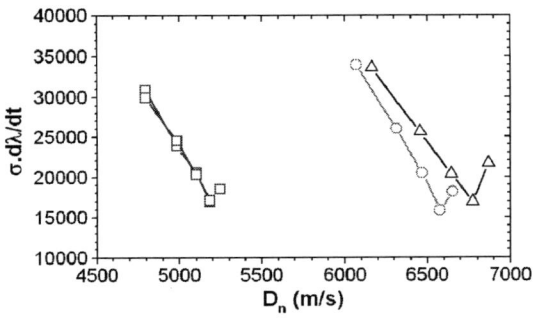

**Figure 6.** Polytropic index and $\sigma \dfrac{d\lambda}{dt}$ versus curvature and detonation velocity - same legend than Fig. 5.

To determine the reacted mass fraction of HE and the reaction rate just behind the detonation front, it is necessary to postulate an equation of state (EOS) for the mixing of detonation products and inert high explosives. As an illustration, the simple $P = \dfrac{(\gamma(v)-1)(e+\lambda Q)}{v}$ EOS was used to calculate the reacted mass fraction $\lambda$ and the reaction rate $d\lambda/dt$ at the detonation front according to the detonation velocity-curvature relationship.

The reacted constituents' fractions at the detonation front are given Fig. 7. For RDX-AP-Al-HTPB 20-43-25-12, whole of the RDX and only a fraction of AP react at the detonation front. The chemical reactions behind the front run with a characteristic time in the range 22-33 µs. Such data are useful for detonation modeling of such a non-ideal high explosive.

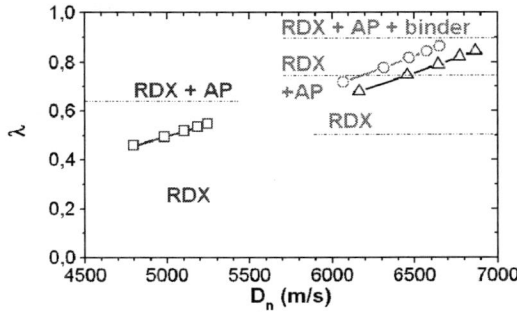

**Figure 7.** Reaction coordinate at the detonation front - same legend than Fig. 5.

## CONCLUSIONS

The special explosive device "logosphere" adapted to highly non-ideal HE and the method developed here provide data on the detonation states and the reaction rate just behind the detonation front. The perspective of this work is an application to more representative equation of state and the coupling of the curved detonation wave model with an hydrocode.

## REFERENCES

1. Aveillé, J., Baconin, J., Carion, N., Zoé, J., "Experimental study of spherically diverging detonation waves," 8th Symposium on Detonation, pp. 151-156, 1985.
2. Brun, L., "Un nouveau modèle macroscopique de la détonation non soutenue dans les explosifs condensés," 3rd International Symposium on High Dynamic Pressure, La Grande Motte, pp. 103-107, 1989.
3. Andreev, S.G., Babkin, A.V., et al, "Physics of Explosions," tome 1 (in Russian), Phyzmatlit, Moscow, 2002.
4. Zababakhin, E.I., "Some problems of the gasdynamics of explosions," RFNC-VNIITF Publishing House Snezhinsk, 2001.

CP845, *Shock Compression of Condensed Matter - 2005,*
edited by M. D. Furnish, M. Elert, T. P. Russell, and C. T. White

# ADVANCES IN THE UNDERSTANDING OF THE LARGE-SCALE GAP TEST

## S. J. Burley[1], N. K. Bourne[1], V. Fung[2], R. Hollands[2], J. C. F. Millett[3], A. M. Milne[4] and A. Wood [4]

[1]*University of Manchester, PO Box 88, Manchester, M60 1QD, UK*
[2]*BAE Systems Land Systems, Glascoed, Usk, Monmouthshire, NP15 1XL, UK*
[3]*Defence Academy of the UK, Cranfield University, Shrivenham, SN6 8LA, UK*
[4]*Fluid Gravity Engineering Ltd, 83 Market Street, St Andrews, Fife, KY16 9NX, UK*

**Abstract.** Knowledge of the sensitivity of high explosives to shock is important to avoid unwanted detonations in service. The large-scale gap test (LSGT) is used in the UK as one of the key qualification tests for energetic materials. This geometry consists of a donor charge, a (poly)methyl methacrylate (PMMA) attenuator (or gap) and a test or acceptor charge. The gap thickness is varied until 50% of test acceptors are detonated. In this work the shock to detonation behaviour of a UK secondary explosive formulation was measured. Manganin pressure gauges were embedded at various longitudinal positions and radii in acceptor charges of varying lengths, and the charges were subjected to varying gap output pulses, characterized in earlier work. For longer acceptor lengths and higher gap output pressures, detonation was observed. Curvature was also measured. The test configurations were modelled using a reactive model derived from parallel work done using plate impact. The predictions of the hydrocode and the results of the experiments are compared.

**Keywords:** detonation, sensitivity, shock, manganin
**PACS:** 47.40, 62.50

## INTRODUCTION

There are several tests to measure the sensitivity of munitions to external thermal, electrical, and mechanical stimuli. One event that can cause unwanted detonation in a plastic-bonded explosive (PBX) is impact, that causes an inert shock to start running.

The presence of pores within a PBX can lead to local high temperatures, from flow and impact associated with jetting, and/or by compression of included gas [1]. These high temperatures can start local chemical reactions, which may transit to detonation, depending on confinement.

The most controlled investigation of the shock to detonation transition uses planar input waves.

However, the equipment required is specialised, and in practise explosive shock sensitivity is measured using methods such as the large-scale gap test (LSGT). A standard donor charge drives a shock through various thicknesses of attenuator, or gap, into samples, or acceptors, of the explosive. The lengths of the donor and acceptor charges are fixed (at 50.8 and 139.7 mm in the UK specification) respectively. The means of determining detonation in the LSGT is the deformation of a 10 mm thick, steel witness plate.

The aim of the work described here was to investigate the run to detonation of an explosive in conditions similar to the LSGT. The wave output from the gap is curved, and pressures at large radii are lower than at the centre. Measurements of run to detonation under these conditions will aid understanding of both the behaviour of the acceptor explosive, and the LSGT itself.

The tests were modelled in two-dimensions using the multimaterial Eulerian hydrocode EDEN [2]. The two-dimensional geometry makes the solution more sensitive to the physical and chemical models employed. Some measurements are compared to numerical predictions below to validate the models and to illuminate processes occurring in the test.

## MATERIALS

Both the donor and the acceptor charges were UK PBXs produced by BAE Systems Land Systems. The donor was given the designation ROWANEX 3601 and contained 60% by mass of the explosive triamino-trinitro-benzene (TATB) and 35% by mass of cyclotrimethylene trinitramine (RDX) crystals in a 5% wax binder. The acceptor charge used was given the designation RF-38-22 and consists of 88% RDX in an HTPB based binder. A typical microstructure is shown in Fig. 1.

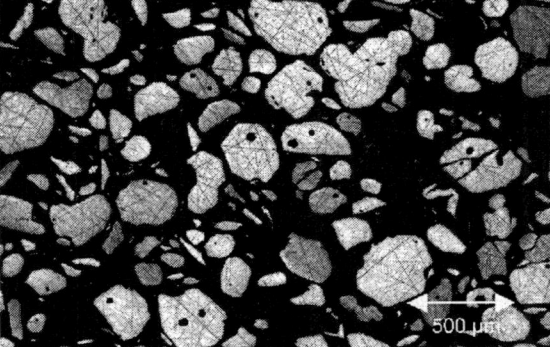

**Figure 1** Optical micrograph of RF-38-22.

## EXPERIMENTAL PROCEDURE

The experimental configuration is shown in Fig. 2, and consisted of a square cylinder donor charge, a PMMA gap, the acceptor charge, and an instrumented target. All components were 50 mm in diameter, and aligned within a suitable cardboard tube. Except for the acceptor charge length and the lack of confinement, the test configuration is consistent with the large-scale gap test described in EMTAP 22 [3].

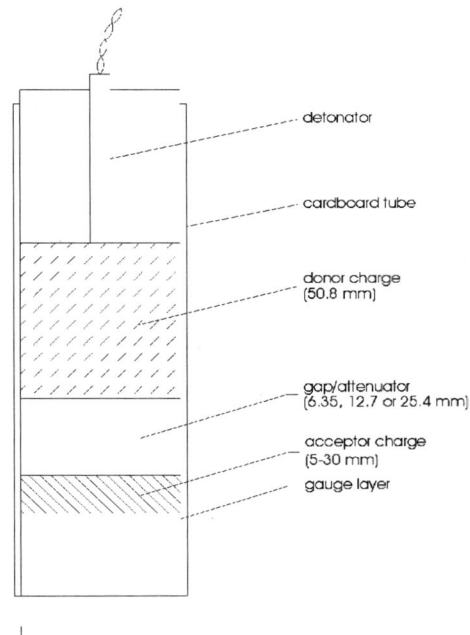

**Figure 2.** Experimental configuration. Connections to the manganin gauges are not shown.

The instrumented target consisted of several manganin foil gauges (Micro-Measurements type C-880113-B) sandwiched between 1.5 mm and 25 mm thick 50 mm diameter PMMA cylinders. The gauge calibration was taken from Rosenberg [4] giving an accuracy of ±2%. The gauges were energised using pulse power supplies. Data was recorded using a Tektronics TDS 5104 digital storage oscilloscope. The donor charge was initiated by a L2A1 detonator connected to a Reynolds fireset. Accurate timing was ensured by the use of a multichannel timing device giving separate trigger signals to the fireset, pulse power supplies, and storage oscilloscopes.

The on-axis pressure pulse at the output from the gap was measured in previous work [5, 6], see Table 2.

**Table 2 Output pressures for several gap lengths.**

| gap length (mm) | Pressure (GPa) |
|---|---|
| 6.35 | 9 ±1 |
| 12.7 | 7.8 ±0.8 |
| 25.4 | 5.4 ±0.5 |

## NUMERICAL MODELLING

EDEN makes use of a grain-burning model, in which the surface area available for burning is determined by the initial grain size distribution [7]. Thus, the effects of grain size on the shock to detonation transition can be numerically investigated unlike most continuum models which offer no insight into the underlying behaviour.

The acceptor was modelled using a two material grain burning model with a Vieille burn law

$$\frac{dR}{dt} = a\left(\frac{p}{p_0}\right)^n, \tag{1}$$

where $R$ is the grain radius and $a$, $p_0$, and $n$ are model parameters. In this work, $a$ and $p_0$ were fixed, and the initial value of $R$ and $n$ were selected by the best fit to experimental data.

## RESULTS

### Experimental

Fig. 3 shows the on-axis pressure for several acceptor lengths at an input pressure of 5.4 GPa. Note that the separation in time of the curves is arbitrary, and aids visibility of the data.

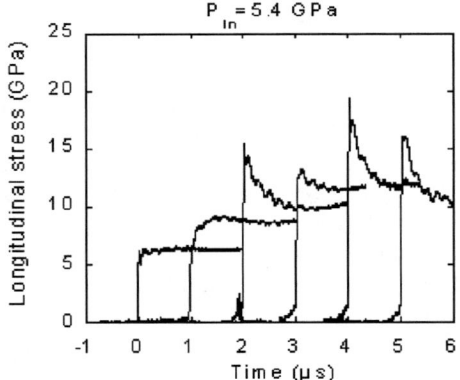

**Figure 3.** Pressure histories for acceptor lengths of 5.05, 6.40, 8.20, 9.85, 12.55, and 30.85 mm (with arbitrary offsets of 0, 1, 2, 3, 4, and 5 μs respectively).

There is a general trend in Fig. 3 towards higher pressures with longer run lengths. Initially

the pressure is close to the input pressure and although the pulse appears square in shape, the gauge only records reliably for 1 μs before 2D curvature perturbs the signal at the gauge [6]. The pulse develops into a Taylor wave as detonation occurs. However, there is some variability in the results.

Measurements were also made of the on-axis pressure for several acceptor lengths and for input pressure of 7.8 GPa and 9 GPa. At these input pressures the growth of reaction was faster, as expected.

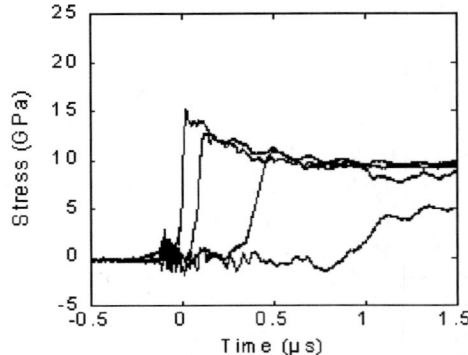

**Figure 4.** Pressure histories at several radii.

Fig. 4 shows the pressure history at radii of 0, 5.4, 11.8 and 18 mm for an input pressure of 5.4 GPa and an acceptor length of 8.2 mm. The pressure history at a radius of 18 mm differs from the on-axis measurement in several ways: the pressure is only about half the magnitude; the rate of rise in pressure is much lower; and there is a delay of approximately 1 μs. There is gradual increase in delay, and decrease in pressure achieved, as one moves from the centre to the edge of the cylinder.

### Numerical modelling

A large number of simulations were performed to model the field trials. Processing of the PBX reduces the mean grain size from that of the original RDX [7] and a mean grain size of 150 μm was chosen by fitting.

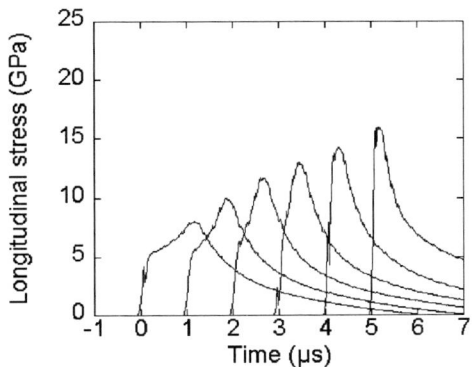

**Figure 5.** Calculated pressure histories for acceptor lengths of 5.05, 6.40, 8.20, 9.85, 12.55, and 30.85 mm.

Fig. 5 is a compilation of the on-axis pressure histories from several simulations and can be directly compared to the data of Fig. 3. The agreement for the initial pressure is reasonable for acceptor lengths of 5.05, 6.4, 9.85, and 30.85 mm (offsets of 0, 1, 3, and 5 µs respectively), but is less good for acceptor lengths of 8.20 and 12.55 mm (offsets of 2 and 4 µs respectively). While all the calculations show a steady growth towards steady detonation with distance, the experiments do not. Since the data of Fig. 3 is a composite of separate experiments we plan to investigate round to round variability in future tests.

The predictions of the off-axis measurements such as those shown in Fig. 4 also showed some variability. In keeping with Fig. 5, good agreement was only achieved in some tests. This work will be reported more extensively elsewhere.

## CONCLUSIONS

Measurements of the shock to detonation transition in a PBX have been successfully performed. The experimental configuration was similar to the UK qualification large-scale gap test [3].

Some variability in the results was seen. It is not clear if this is due to the materials used, experimental technique, or a feature of the LSGT.

No measurements of a detonation failure due to a small input pressure were made. This may be the subject of further work.

The grain burning model built into EDEN has predicted run to detonation and curvature, with reasonable agreement with experiment, by using a single value for the grain size. This was consistent with the original RDX size distribution. However, consistency in the experiments leads to difficulties in precisely defining parameters in the model.

Further work is planned to better understand the variability observed. Further improvements to the experiment include probes to monitor the progress of the reaction front through the acceptor. It is also hoped to study the shock response of further materials in this test configuration.

## ACKNOWLEDGEMENTS

We thank John Bradford, Alan Watkins, Gwyn Sayce, Colin Cook and Ken Mugleston of BAE Systems Land Systems Glascoed, and Gary Cooper of Cranfield University for their support with the experimental programme. This work was funded by Drs Ron Moss and Peter Eickhoff of DOSG in support of EMTAP 22.

## REFERENCES

1. Bourne, N. K. and A. M. Milne, "On Cavity Collapse and Subsequent Ignition", presented at the 12th International Detonation Symposium, San Diego, CA, 2002.
2. Milne, A. M., "EDEN Users' Manual", Fluid Gravity Engineering, St. Andrews, Fife, UK, 2004.
3. Eickhoff, P., "Manual for EMTAP Test Number 22, June 1978." (personal communication, 2004).
4. Rosenberg, Z., ,D. Yaziv, and Y. Partom, "Calibration of Foil-Like Manganin Gauges in Planar Shock Wave Experiments", J. Appl. Phys. 51, 3702-3705, 1980.
5. Bourne, N. K., S. J. Burley, G. A. Cooper, V. Fung, and R. Hollands, "Re-Calibration of the UK Large-scale Gap Test", Propellants, Explosives, Pyrotechnics 30, 196-198, 2005.
6. Bourne, N. K., A. M. Milne, and R. A. Biers, "Measurement of the Pressure Pulse from a Detonating Explosive", J. Phys. D Appl. Phys. 38, 1984-1988, 2005.
7. Bourne, N. K. and A. M. Milne, "Shock to Detonation Transition in a Plastic Bonded Explosive", J. Appl. Phys. 95, 2379-2385, 2004.

CP845, *Shock Compression of Condensed Matter - 2005*,
edited by M. D. Furnish, M. Elert, T. P. Russell, and C. T. White
2006 American Institute of Physics 0-7354-0341-4/06/$23.00

# ATMOSPHERIC EFFECTS ON THE COMBUSTION OF DETONATING ALUMINIZED EXPLOSIVES

## Joel R. Carney[1], J. Scott Miller[1,2], Jared C. Gump[1], and G. I. Pangilinan[1]

*[1]Research and Technology Department, Indian Head Division, Naval Surface Warfare Center, Indian Head, MD 20640*
*[2]Present address: Qualis Corporation / EM50, NASA Marshall Space Flight Center, Huntsville, AL 35812*

**Abstract.** The detonation and subsequent combustion of aluminized explosive formulations depend heavily on the oxidation reactions of aluminum. Fuel-rich formulations require oxygen from an external source (nominally an oxygen-containing atmosphere or detonation products) to burn the fuel to completion. Dynamic spectroscopic measurements are made for an aluminized explosive (PBXIH-135) to investigate the effect of changing atmospheres on the combustion properties of aluminum. The explosive formulation is tested under normal atmospheric conditions and in an atmosphere of nitrogen. Light emission (from 350-550 nm) from the explosive event is collected in a spectrometer and dispersed temporally in a streak camera. Aluminum emission (centered at 396 nm) is commonly observed in each atmosphere although the emission persists longer in nitrogen. Aluminum nitride (AlN) is observed as an intermediate in the oxidation of aluminum when oxygen is removed from the atmosphere. New, nitrogen-containing species (near 387 and 418 nm) also arise in the nitrogen atmosphere experiments. A slower, less intense aluminum monoxide (AlO) emission observed in the nitrogen atmosphere may correspond to the slower oxidation reactions of aluminum and detonation products ($CO_2$ and $H_2O$). The peak assignments and global kinetics of each species are presented and the implications of these results on atmospheric effects are discussed.

**Keywords:** Spectroscopy, detonation, combustion, aluminum.
**PACS:** 32.30.Jc, 39.30.+w, 82.33.Vx.

## INTRODUCTION

The full capability of fuel-rich explosive formulations has not been reached due to the non-efficient use of excess fuel in the post-detonation combustion. The harsh conditions inherent in such explosive events inhibit the advancement of this understanding though experimentation. The present study uses the light emission from an explosive event to provide some insight into the reactivity of selected fuels in the combustion phase. Experiments were performed to compare the reactivity of explosives with and without the presence of oxygen in the surrounding atmosphere.

Aluminum (Al) is a common fuel used in current fuel-rich explosive formulations. The energy release resulting from the complete oxidation of Al is significantly higher than hydrocarbons when compared by volume. However, the hot fuel ejected during a detonation must encounter sufficient oxidizing species to react it to completion. It must also react in a short time scale to support shock front propagation.

Experiments were conducted to investigate the dynamic response of post-detonation combustion products and intermediates with and without the presence of external oxygen to support combustion reactions. The results show the dramatic effect that atmospheric oxygen can play in the post-detonation combustion of each explosive.

## METHODS

Specific details describing the experimentation used can be found elsewhere. [1, 2] A brief description follows. Experiments were conducted at the Sigmund J. Jacobs Detonation Science Facility at the Naval Surface Warfare Center, Indian Head Division. An 85-liter steel chamber was used to provide atmospheric control over the tests. For tests in nitrogen, the chamber was equipped with polycarbonate panels for optical access and a Mylar sheet at one end to relieve pressure after detonation.

One-inch right cylindrical charges (nominally 20g) of PBXIH-135 were confined radially by a ¼" thick steel sleeve and mounted horizontally in a v-block inside the chamber. A light collection assembly was mounted eight inches from the exit surface of the charge. This was comprised of an optical fiber (f/4) and a 1" diameter, 50 mm focal length, plano-convex lens separated by 50 mm within a lens tube. The optical configuration created a cylindrical light collection volume that was ½" in diameter, extending from the lens to the output surface of the charge.

The optical fiber connected to the lens assembly transmitted the light from the explosive event to a remote spectrometer and streak camera for time-resolved spectroscopic measurements. The spectrometer had a range of 350 to 550 nm, covering the strong emission peaks of aluminum and aluminum monoxide.

A shock velocity gauge, described previously, [3] was used to measure the velocity of the propagating front 175 mm downstream of the explosive charge. Two optical fibers were mounted to a post, offset a measured distance from one another. The surface of the illuminated fibers faced the oncoming shock front from the explosive. The arrival of the propagating shock front could be determined by the sharp decrease in reflected light signal (indicating an increase in density near the fiber tip) from the surface of the fibers as a function of time. An accurate shock velocity (+/- 0.1 mm/$\mu$s) could be determined by dividing the physical separation of the fibers by the observed arrival time difference.

Digital framing camera images were recorded to supplement the previously described measurements with time and spatial information. The camera (Imacon 200, DRS Hadland Ltd.) recorded 16 individual images with a 10 ns exposure time and interframe separations of 6 $\mu$s. A Xenon flashlamp was used to backlight the event in a shadowgraphy configuration.

## RESULTS

Streak spectroscopy images from the experiments in air (a) and nitrogen (b) are presented in Fig. 1. The vertical axis represents the streak time (streak time is relative to the detonator trigger), while the horizontal axis marks the wavelength range. Aside from the strong emission peak centered at 396 nm (representing Al emission), the two images were very different. The known emission peaks of aluminum monoxide (AlO) between 450 and 550 nm were not a predominant feature when oxygen is removed from the atmosphere.

Horizontal cross sections of the images presented in Fig. 1 are provided in Fig. 2. These spectra represent the emission intensity of each respective experiment at a streak delay time of 11 $\mu$s. The intensity of the Al emission at 396 nm was normalized for comparison. In air, the spectrum (Fig. 2, blue) exhibited strong Al and AlO emission peaks. However, the AlO peaks were not visible in a nitrogen atmosphere at this particular time delay (Fig. 2, red) and new peaks arose near 387 and 418 nm, flanking the strong Al emission. A strong feature at 508 nm was also present in the nitrogen atmosphere, likely due to aluminum nitride (AlN).

The area under each spectroscopic feature of interest was extracted to quantify the time-dependent emission intensity. A baseline was drawn above the broad background. The intensity was integrated at each time step within the area bound by the peak and the baseline. Fig. 3 presents three such traces. The blue and black lines represent AlO emission in the air and nitrogen test, respectively. AlO emission from the nitrogen test was much weaker and spread out over a longer time scale. AlN emission from the nitrogen test (red line) follows a similar trend to the AlO emission in the air test.

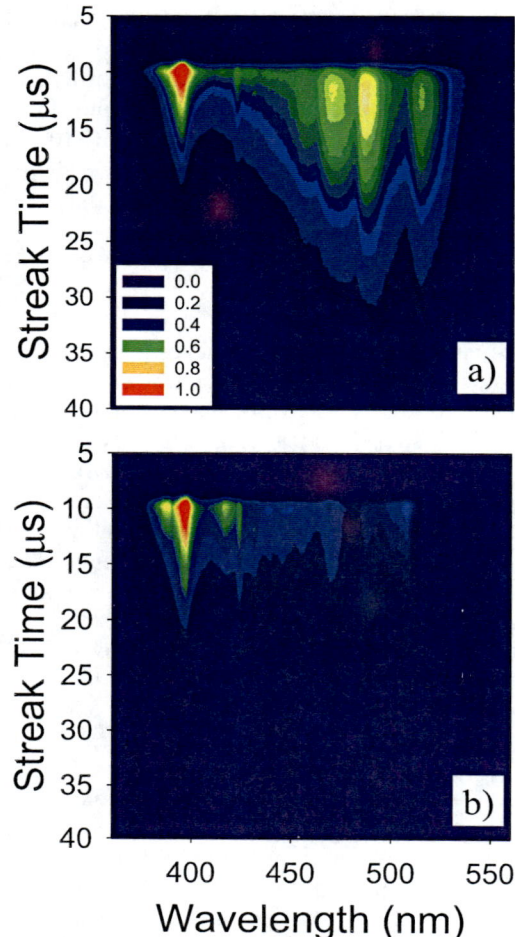

**FIGURE 1:** Streak spectroscopy images of the light emission following the detonation of PBXIH-135 in air (a) and in a nitrogen atmosphere (b). A color code for the scaled intensities of both images is included in (a).

The shock front velocity gauge placed 175 mm from the charge surface measured a shock velocity of 2.5 mm/µs for both tests. Digital framing images confirmed this propagation velocity. The images also showed that a majority of the light emission came from the region just behind the shock front. A comparison between two selected images is made in Fig. 4. At a streak delay of 22 µs, the shock and combustion front propagated a distance of approximately 55 mm. The position of the output surface of the energetic charge was at

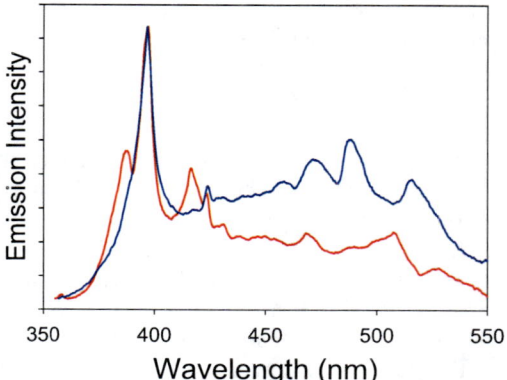

**FIGURE 2:** Single-track spectra at a streak delay of 11 µs, representing horizontal cross sections of the streak images in air (blue) and nitrogen (red).

the right edge of each image and the grid spacing was 8 mm (square edges).

## DISCUSSION

When oxygen was removed from the immediate atmosphere of a fuel-rich aluminized explosive like PBXIH-135, optical diagnostics observed a number of changes. There was a striking difference between the two spectroscopic images, signifying a change in both the predominant product channels and the dynamics of the emission channels. First, the primary intermediate channel for the early-time oxidation

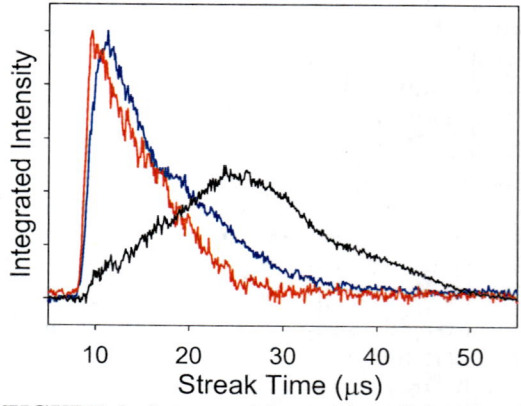

**FIGURE 3:** Integrated intensities of the chemical transients observed during the test in air (AlO/blue) and in a nitrogen atmosphere (AlN/red and AlO/black).

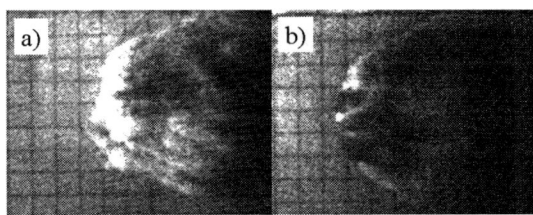

**FIGURE 4:** Shadowgraph images of shock and combustion fronts propagating from right to left. Images (a. Air, b. Nitrogen) were recorded 22 μs after the detonator trigger (exposure time of 10 ns).

of aluminum appeared to shift from AlO to AlN. Next, aluminum emission persisted for longer times. Therefore, the available hot aluminum was consumed much quicker when oxygen was present in the atmosphere.

Third, light emission from AlO transients in the combustion phase was significantly reduced and extended to longer timescales when external oxygen was removed. This slower AlO observation may be attributed to the slower oxidation reactions of Al with detonation products like $CO_2$ and $H_2O$.[4-7] This may be the source of the bimodal behavior [2] that we observed in our reacting front: "fast" AlO produced in the first 10 μs following the detonation can be attributed to oxidation by the external oxygen while "slow" AlO is produced by slower reactions with detonation products.

Lastly, spectral signatures at 387 and 418 nm were greatly enhanced when oxygen was removed from the atmosphere. These features matched the expected spectral position of CN emission (although the intensity match is not exact) and would imply that nitrogen plays a strong roll in the oxidation of carbon from the detonation products if oxygen is not available.

## CONCLUSIONS

Time resolved optical diagnostics have been used to show the profound effect that external oxygen has on the reactive pathways of fuel-rich aluminized explosives. The fast reactions (within the first 50 μs) following the detonation of these small charges display the effect. Undoubtedly, the longer-term reactions with aluminum and the turbulent environment following the detonation

will be effected by the changes in these "set-up" reactions.

## ACKNOWLEDGEMENTS

The authors would like to thank the Defense Threat Reduction Agency for funding this work (Advanced Energetics Program, Project Manager: William H. Wilson). We would also like to thank Professor Nick Glumac (University of Illinois UC), Dr. Von Whitley (NSWCIH), and Dr. Jeremy Monat (NSWCIH) for helpful discussions. Lastly, we would like to thank Mr. Robert N. Hay of Indian Head for his efforts as our energetics technician.

## REFERENCES

1. Carney, J. R., et al., "Time-resolved optical measurements of the post detonation combustion of aluminized explosives," Review of Scientific Instruments, Submitted for publication.
2. Miller, J. S. and Pangilinan, G. I., "Measurements of aluminum combustion in energetic formulations," in Shock Compression of Condensed Matter, 2003 (Furnish, M. D., et al.), 867.
3. Monat, J. E., et al., "Novel optical fiber-based gauge for measuring transient pressures," in Shock Compression of Condensed Matter, 2003 (Furnish, M. D., et al.), 1281.
4. Glumac, N., et al., "Temperature measurements of aluminum particles burning in carbon monoxide," Combustion Science and Technology, 177, 485, 2005.
5. Bucher, P., et al., "PLIF and ratiometric temperature measurements of aluminum particle combustion in $O_2$, $CO_2$, and $N_2O$ oxidizers, and comparison with model calculations," Proceedings of the Combustion Institute, 27, 2421, 1998.
6. Bucher, P., et al., "Condensed-phase species determinations about Al particles reacting in various oxidizers," Combustion and Flame, 117, 351, 1999.
7. Servaites, J., et al., "Ignition and combustion of aluminum particles in shocked H2O/O2/Ar and CO2/O2/Ar mixtures," Combustion and Flame, 125, 1040, 2001.

CP845, *Shock Compression of Condensed Matter - 2005*,
edited by M. D. Furnish, M. Elert, T. P. Russell, and C. T. White
© 2006 American Institute of Physics 0-7354-0341-4/06/$23.00

# THE ROLE OF BINDERS IN CONTROLLING THE COOK-OFF VIOLENCE OF HMX/HTPB COMPOSITIONS

## M. D. Cook[1], C. Stennett[2], P. J. Haskins[1], R. I. Briggs[1], A. D. Wood[1] and P. J Cheese[3]

[1]*QinetiQ, Fort Halstead, Sevenoaks, Kent TN14 7BP, UK*
[2]*RMCS, Shrivenham, Swindon, SN6 8LA, UK*
[3]*Defence Ordnance Safety Group, Walnut 2c #67, MoD Abbey Wood, Bristol, BS34 8JH, UK*

**Abstract.** There is a clear difference in cook-off vulnerability between highly-loaded pressed compositions such as LX-14 (pressed 95.5% HMX/4.5% binder), which yield violent responses, and cast compositions with low loadings, such as CPX 301 (85% RDX/15% HTPB), which yield relatively mild responses. These two classes of composition differ primarily in the quantity of binder, and in the manufacturing method used in production. An experimental study was conducted in an attempt to determine the filling proportion beyond which violent responses are observed. Here we describe a series of small-scale cook-off experiments which studied pressed compositions of 88%, 91%, 95% and 96% HMX, mixed with cured, cross-linked HTPB. The experiments used a novel glass-windowed test vehicle, instrumented internally with thermocouples. A trend of increasing event violence with increasing proportion of HMX was found, although in none of the experiments was mass reaction recorded. The results from these experiments are discussed.

**Keywords:** HMX, explosive, Cook-Off, violence
**PACS:** 47.70.Fw, 66.30Xj

## INTRODUCTION

Historical evidence has shown that certain pressed PBXs with high HMX content, e.g. LX-10 (95% HMX/5% Viton) and LX-14 (95.5% HMX/4.5% estane) respond violently to thermal stimuli. With this type of explosive, cook-off tests often result in deflagration and sometimes even detonation. By contrast, work on cast PBXs containing a relatively large proportion of a flexible binder, e.g. CPX 301 (85% RDX/15% HTPB), respond much less violently to thermal stimuli. These observations have led us to the question: what is the minimum binder proportion required to ensure low-violence cook-off responses?

To address this question we have carried out a series of cook-off experiments on a range of pressed HMX/HTPB compositions in which the HMX content has varied form 88% to 96%.

## EXPERIMENTAL PROCEDURE

The Small-Scale Glass Windowed Cook-Off (SSGWCO) test vehicle used in this study is a development of that reported previously [1]. This earlier work clearly showed that the capability to observe activity within the vehicle during heating was particularly useful. The new vehicle design includes improved pressure sealing capacity, to

~100 bar static pressure, and a more flexible instrumentation capability.

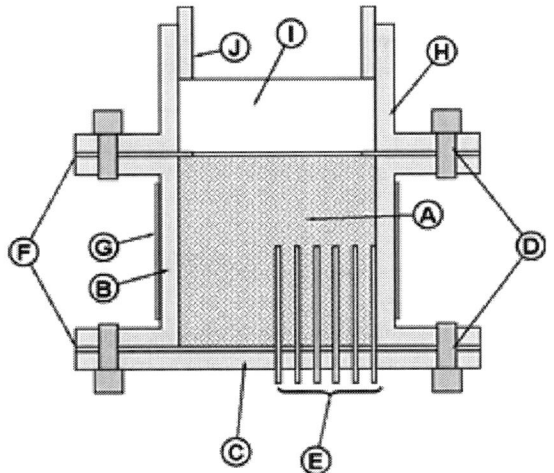

**FIGURE 1.** SSGWCO test vehicle cross-section

Fig. 1 shows a cross-sectional diagram of the SSGWCO test vehicle. The filling, A, is contained within a flanged mild steel tube, B, of internal dimensions 50mm x 50mm. The wall thickness is 5mm. A base plate, C, of 5mm thickness closes the lower end of the vehicle, and is clamped to the tube flange by six retaining bolts, D. Six K-type 1.5mm thermocouples, E, are fitted through the base plate so that their ends lie at the mid-point of the filling. The thermocouples are distributed evenly across the vehicle.

PTFE gaskets, F, are placed between the vehicle body and the end assemblies to provide gas sealing. The heating element, G, consisting of 1m of nickel-chrome heating tape, is wrapped around the tube body. The upper closure assembly, H, is clamped to the top flange of the tube with six retaining bolts. A glass window, I, is clamped to the top flange gasket using the window retaining ring, J.

The electrically heated Cook-off method has been described previously [1,2]. In this method, heating is achieved by a heating element wrapped around the outer surface of the test vehicle. The heating element consists of a 1m length of nickel-chrome heating tape, of 3.2mm width and 0.1mm thickness, and whose resistance is approximately $3\Omega$. The heating element is connected to the electrical supply through a variable transformer which allows control of the current applied, and hence control of the heating effect. In these experiments the current was controlled manually.

## HEATING PROFILES

Two heating rates were chosen for the experiments, namely: 30°C/hour and 600°C/hour. Heating was carried out at these approximate rates, to maintain a constant temperature profile across the filling, as measured on the 6 thermocouples. In the faster rate experiments the aim was to achieve 120°C variation in temperature from centre to edge at the time of ignition. For the slower rate experiments, the aim was to achieve 30°C temperature variation across the sample at the time of ignition.

## COMPOSITIONS

Five formulations were chosen for this study, all consisting of HMX and cross-linked HTPB in different proportions. These are shown in Table 1.

The HMX was type B, having a bimodal particle size distribution of 35% (vol.) in the size range 45-150μm, and 65% (vol.) of a size smaller than 45μm. The density of the crystals was 1.84 g/cc.

**TABLE 1.** Formulations used. Two pellets of each composition were made; the filling density for each is shown. 'V' suffix indicates pressing under vacuum to improve filling density; all other compositions were pressed at atmospheric pressure

| HMX | HTPB | % TMDs Pellet 1, Pellet2 |
|---|---|---|
| 88% | 12% | 94.0%, 96.9% |
| 91% | 9% | 94.7%, 94.7% |
| 95% | 5% | 94.8%, 94.5% |
| 96% | 4% | 91.4%, 91.4% |
| 96% (V) | 4% (V) | 94.7%, 98.6% |

The binder system in all the formulations consisted of HTPB, DOS and a mixture of MDI/IPDI isocyanate curing agent. After 24 hours partial curing the mixture was single-end pressed to a load of 127MPa at ambient temperature. The pellets were then inserted into the test vehicles, and cured for four days at ambient pressure and 60°C.

The filling densities, as shown in Table 1, are rather lower than would be expected for service use. This is probably a result of single-end pressing; isostatic pressing at elevated temperatures routinely yields filling densities >99% TMD.

The manufacture of the 88/12 compositions proved to be particularly difficult as this mix ratio is typically cast, rather than pressed, and remains semi-liquid even at room temperature. Partial curing of the pressing mixture gave a more workable material for pressing, but yielded finished pellets with variable filling density.

## RESULTS

A marked expansion of the filling was observed during the slow heating tests as the internal temperature reached 160-170°C. This coincided with a clear endotherm visible on the temperature recordings. This was attributed to the β-δ phase change in HMX [3]. δ-HMX occupies a greater volume than β-HMX, and therefore expansion of the material occurs at the transformation point. This expansion, beginning at a temperature of 165°C, was clearly visible in the slow heating rate experiments, but not so marked in those heated at the faster rate. At faster heating rates, much of the material remains at a temperature lower than that required for the β-δ transformation.

Table 2 summarises the experimental results. The violence of response is categorised into three divisions: A, B and C. Response A is characterised by a mild pressure failure of the vehicle, ejection of the sealing gaskets and bulging of the instrumentation plate to no more than 3mm. Response B is characterised by a medium pressure failure, with snapping of one or more retaining bolts, and with bulging of the instrumentation plate to approximately 6mm. Response C is the most violent, and is characterised by snapping of the retaining bolts, separation of one or both end closure assemblies, and bulging of the instrumentation plate to approximately 10mm. A graphical summary of the results is given in Fig. 2, which is a plot of average dent depth of the

SSGWCO vehicle base-plate and the percentage of binder.

**TABLE 2.** Experimental results summary. $T_{ign}$ is recorded at the surface of the explosive; $\Delta T$ is the temperature difference from centre to edge.

| Shot ID | HR | Fill | Response type | $T_{ign}$ °C | $\Delta T$ °C |
|---|---|---|---|---|---|
| 1 | Fast | 88/12 | A +burn | 230 | 128 |
| 2 | Fast | 91/9 | A +burn | 234 | 132 |
| 3 | Fast | 95/5 | A +burn | 239 | 111 |
| 4 | Fast | 96/4 | B | 229 | 106 |
| 5 | Fast | 96/4 | B | 227 | 112 |
| 6 | Slow | 88/12 | B + burn | 213 | 35 |
| 7 | Slow | 91/9 | B | 214 | 37 |
| 8 | Slow | 95/5 | B | 214 | 30 |
| 9 | Slow | 96/4 | C | 208 | 22 |
| 10 | Slow | 96/4 | C +burn | 216 | 42 |

There was no evidence of mass reaction of the filling in any of the experiments. In all cases, it was clear that, immediately after the pressure burst, a considerable portion of unreacted filling was present. In five experiments, this unreacted material was found both inside the test vehicle and scattered around the test chamber; in the other five, the remaining material burned away in a steady fashion following the pressure burst.

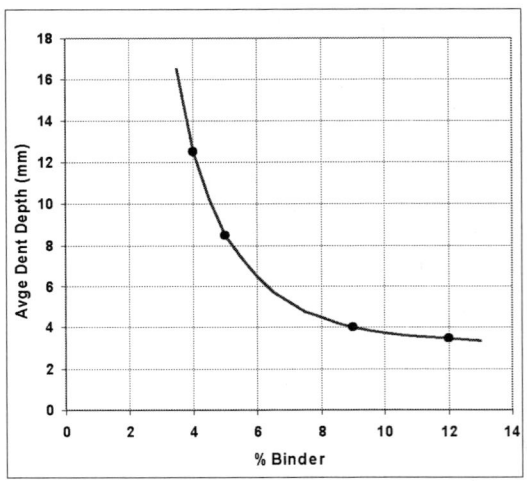

**FIGURE 2.** Plot of average dent of vehicle base-plate versus percentage binder.

## DISCUSSION

Although none of the events observed in these tests were particularly violent there is a clear trend of increasing violence with increase in HMX loading. Mass reaction could not however be ruled out under higher confinement than that afforded by the SSGWCO test vehicle. The simple criterion illustrated in Fig. 2 would suggest that there is a significant increase in reaction violence when the binder content is reduced from 8% to 6%. This may indicate the binder concentration limit beyond which non-violent cook-off responses cannot be guaranteed.

The behaviour at ignition is interesting. It is clear that, at the point of ignition, only a small proportion of the filling undergoes reaction. The remainder is ejected from the failed vehicle, or burns away in a steady fashion over a relatively long period. Previous experiments with RDX/TNT resulted in a similar ejection of unreacted material after the pressure burst [1], although in these cases post-event burning was not observed. Examination of the recovered PBX shows that no more than approximately 15g of PBX was consumed by the initial ignition.

The low filling density of all the compositions (~ 95% TMD) indicates that these compositions contained a number of small voids. It is well known that these should increase the sensitivity of the composition to initiation by mechanical means [4]. However, it is possible that by the time the ignition occurs any voids will have been closed either by simple thermal expansion or as a result of the $\beta$-$\delta$ phase transformation in HMX. Indeed, our results show no obvious sensitisation to the initial thermal explosion as we recovered damaged but unreacted pieces of PBX.

## CONCLUSIONS

The experiments reported here clearly show a general trend of increasing event violence with increasing HMX proportion. Although we have not been able to demonstrate a transition to violent mass reaction our results suggest that a binder concentration of 6-8% may be required to ensure benign cook-off responses.

We can interpret our results on the basis of the hypothesis that there are generally two possible stages in the cook-off reaction process. The first stage consists of the ignition of a small volume of the filling, driven purely by the external thermal stimulus. The size of this volume depends upon the temperature profile across the filling, and is therefore largely governed by the heating rate. At slower heating rates a greater volume of material ignites, and a more violent response is therefore generally observed.

Under certain conditions the initial ignition can give rise to propagation of reaction throughout the bulk material. The existence, or otherwise, of this second stage will depend on the mechanical-thermal stimuli produced by the ignition, and the sensitivity of the remaining bulk material. It appears that in our experiments the system does not undergo this secondary stage, quite possibly as a result of early failure of the confinement.

## REFERENCES

1. Stennett, C., Cook, M. D., Briggs, R. I., Haskins, P. J., Fellows, J., "Direct Observation of Cook-off Events Using a Novel Glass-Windowed Vehicle and Pipe Bombs", in *12th International Detonation Symposium*, San Diego, CA, 2002.
2. Cheese, P., Briggs, R. I., Fellows, J., Haskins, P. J., Cook, M. D., "Cook-off Tests on Secondary Explosives", in *10th International Detonation Symposium*, ONR 33300-5, 1998, pp 272-278.
3. Menikoff, R., Sewell, T. D., Combust. Theory Modelling **6**, 103-125 (2002).
4. Bowden, F. P., Yoffe, Y. D., *Initiation and Growth of Explosion in Liquids and Solids*, Cambridge University Press, 1952.

CP845, *Shock Compression of Condensed Matter - 2005,*
edited by M. D. Furnish, M. Elert, T. P. Russell, and C. T. White
© 2006 American Institute of Physics 0-7354-0341-4/06/$23.00

# THERMAL BEHAVIOR OF $Fe_2O_3$/Al THERMITE MIXTURES IN AIR AND VACUUM ENVIRONMENTS

## L. Durães[1,2], R. Santos[3], A. Correia[3], J. Campos[2], and A. Portugal[1]

[1]*Chemical Engineering Dept., Fac. of Sci. and Tech., Univ. of Coimbra, Pólo II, 3030-290 Coimbra, Portugal*
[2]*Laboratory of Energetics and Detonics, Av. da Univ. de Coimbra, 3150-277 Condeixa-a-Nova, Portugal*
[3]*Centro Tecnológico da Cerâmica e do Vidro, R. Cor. Veiga Simão, 3020-053 Coimbra, Portugal*

**Abstract.** In this work, the thermal behavior of $Fe_2O_3$/Al thermite mixtures, in air and vacuum, is studied. The individual reactants and three mixtures - stoichiometric and over aluminized - are tested, by Simultaneous Thermal Analysis (STA) and heating microscopy, with a heating rate of 10 °C/min. The STA results show that the presence of $O_2$ from air, or from residual air in vacuum, influenced the reaction scheme. The Al oxidation by this oxygen was extensive, making the thermite reaction with $Fe_2O_3$ unviable. There was also evidence of significant conversion of the $Fe_2O_3$ into $Fe_3O_4$, supporting the previous conclusion. So, the STA curves for the three mixtures were similar and displayed features of the individual reactants' curves. The heating microscopy images confirmed the STA conclusions, with one exception: the thermal explosion of the Al sample close to 550 °C. The absence of this phenomenon in STA results was explained by the limited amount of material used in each sample.

**Keywords:** Combustion, thermite, thermal behavior, iron oxide, aluminum.
**PACS:** 81.20.Ka, 81.70.Pg, 82.33.Vx.

## INTRODUCTION

Self-propagating high-temperature thermite reactions are commonly used as heat sources or to synthesize advanced materials. In the $Fe_2O_3$/Al classical system, often applied in pyrotechnics and explosive formulations, ~850 kJ/(mol of $Fe_2O_3$) are released, according to:

$$Fe_2O_3 + 2\,Al \;\rightarrow\; 2\,Fe + Al_2O_3 \qquad (1)$$

The reactive behavior of this system under heating is important to characterize its safety limits. Also, an understanding of the reaction scheme is crucial for improved combustion control. Some authors studied this system by thermal analysis, to determine the reaction scheme. Air [1,2], Ar [3,4] and He [2] were used as environments and the heating rates were usually 10 °C/min. Sarangi et al.

[1] have quantified the heats and kinetic parameters for the reaction, using several reactant ratios and heating rates (5-15 °C/min). Wang et al. and Mei et al. [2,4] also identified the intermediary products at 960 °C and 1060 °C. Analyzing all these authors' results, it can be concluded that 1-3 exothermic peaks may appear after the Al melting, depending on the experimental conditions. The proposed reactions to explain these peaks differ: thermite reaction, reaction of Al with $O_2$ from the environment, intermetallic reaction and/or partial reduction of $Fe_2O_3$ by Al to form lower iron oxides and hercynite.

The thermal behavior of stoichiometric and over aluminized $Fe_2O_3$/Al mixtures was studied using STA (DTA/TGA), in static air and vacuum, and heating microscopy. TGA curves are crucial for the assignment of the occurring phenomena. TGA was used in [1], but the end temperature of

1100 °C limited the conclusions. In this work, the end temperature was 1500 °C. Studies under vacuum were not reported before.

## EXPERIMENTAL PROCEDURE

Iron oxide powder (Fe$_2$O$_3$ Bayferrox 180 – Bayer), 97 % pure and $d_{50}$=1.6μm, and pyrotechnic coated aluminum powder (Al black 000 índia – Carob), 89.3 % pure and $d_{50}$=18.6μm, were mixed in stoichiometric and over aluminized ratios in a rotary mixer for 24 hours. The molar equivalence ratios – (O necessary to oxidize the existing Al to Al$_2$O$_3$)/(O present in Fe$_2$O$_3$ of the mixture) – were 1.00, 1.27 and 1.59 for T100, T127 and T159 mixtures, respectively. Corresponding Fe$_2$O$_3$/Al molar compositions were: 1:2, 1:2.54 and 1:3.19.

STA runs were carried out in a Netzsch STA 449 C Jupiter analyzer, from ~20 to 1500 °C and a heating rate of 10 °C/min, in static air or vacuum (10$^{-5}$ bar) environments. Samples of reactants and mixtures of ~7 mg, in bulk form, were analyzed in alumina pans. With the same heating rate, hand pressed cubic samples of reactants and mixtures were heated and filmed up to 1400 °C in a Leitz Wetzlar Germany heating microscope coupled to a Sony CCD-Iris video camera. To confirm the phases present at 1500 °C, a T100 sample was heated in air in a tube furnace, at 10 °C/min, held for 1 hour at 1500 °C and quenched in ambient air. The products were analyzed by X-ray diffraction (XRD), in powder form, with a Phillips PW 1710 diffractometer, at 40 kV and 30 mA, using Cu Kα radiation (λ=0.154 nm).

## RESULTS AND DISCUSSION

The STA results are presented in Figs. 1-4. Table 1 compiles the more important DTA peak temperatures, obtained by the tangent method. The changes up to peak A are due to secondary phenomena involving residual impurities. The overlapping of phenomena makes the interpretation of peaks difficult near the end temperature.

The STA curves for the three mixtures are comparable and retain the main characteristics of the individual reactants' curves, showing that the

thermite reaction may not have occurred. The involvement of O$_2$ from air, or from residual air in vacuum, in the reactions is notorious, considering the observed positive mass changes. Exothermic peak A is attributed to the oxidation of the aluminum particles surface by the O$_2$ from the environment, forming an alumina film. This peak was not reported in [1-4], but was observed in Al powders where the particles surface is not already fully oxidized (coated or kept in an inert gas) [5-8]. This peak was also detected in environments like He [7] and Ar [7,8]. Measures to avoid the Al oxidation by traces of O$_2$ in inert atmospheres are given in [6]. Endothermic peak B corresponds to the melting point of Al (660.4 °C, at P=1atm). The next stages of aluminum oxidation by O$_2$, corresponding to exothermic peaks C and F, are explained as follows [9,10]: the melted Al, confined in the alumina film, occupies a larger volume than solid Al (+6.4%) and its thermal expansion coefficient is 3.5 times higher than alumina; so, fissures open in the alumina shell and melted Al comes out and oxidizes, forming a thicker film and re-encapsulating the melted Al –

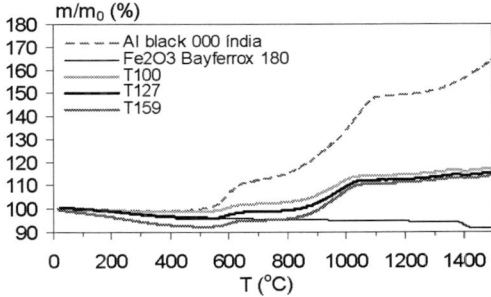

**Figure 1.** Mass changes obtained by STA in static air.

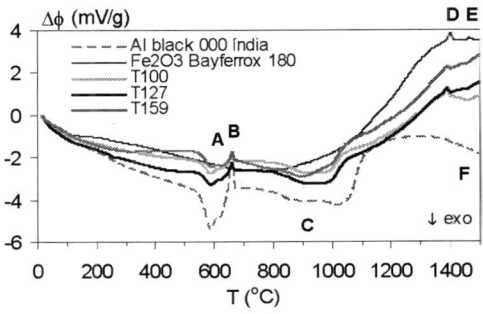

**Figure 2.** Energy changes obtained by STA in static air.

**TABLE 1.** DTA peak temperatures (°C). (Vd. Discussion for meaning of the codes and Figs. 2 and 4)

| | Al black | Fe₂O₃ Bayf. | T100 | T127 | T159 | Code |
|---|---|---|---|---|---|---|
| **A I R** | exo, 583.2 | - | exo, 585.1 | exo, 586.1 | exo, 587.7 | A |
| | endo, 658.0 | - | endo, 659.2 | endo, 659.5 | endo, 660.7 | B |
| | 2 exo, 909.7, 1018.8 | - | 2 exo, 897.3, 975.6 | 2 exo, 897.2, 978.8 | 2 exo, 889.5, 998.5 | C |
| | - | endo, 1400.3 | endo, 1389.7 | endo, 1389.1 | endo, 1387.9 | D |
| | - | endo, 1467.8 | masked | masked | endo, 1437.9 | E |
| | exo, begin 1391.5 | - | exo, 1461.9 | exo, 1457.1 | exo, 1455.1 | F |
| **V A C U U M** | exo, 589.7 | - | exo, 592.4 | exo, 594.0 | exo, 592.8 | A |
| | endo, 663.2 | - | endo, 665.5 | endo, 665.3 | endo, 662.8 | B |
| | 3 exo, ~750, 843.1, 1035.5 | - | 2 exo, 968.1, 1024.4 | 2 exo, 769.8, 998.9 | exo, 983.1 | C |
| | - | endo,1196.0 | endo,1097.5 | endo,1172.0 | endo,1136.3.0 | D |
| | - | endo, 1421.4 | endo, 1431.9 | 1 or 2 endo, 1399.7, 1454.4 masked | 3 endo, 1305.7, 1336.4, 1379.2 | E |
| | exo, 1441.8 | - | exo, 1235.2 | exo, ~1430 masked | exo, 1255.1 | F |

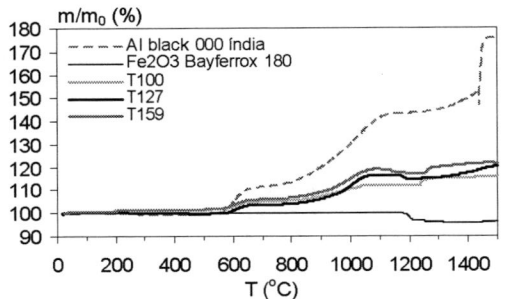

**Figure 3.** Mass changes obtained by STA in vacuum.

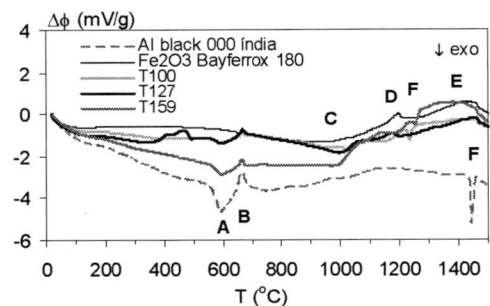

**Figure 4.** Energy changes obtained by STA in vacuum.

peak C; the remaining part of active Al only oxidizes at very high temperatures, by the same process – peak F. If a sufficiently high heating rate and sample mass was used, the destruction of the alumina film would be extensive, originating the particles' auto-heating and the samples' ignition. Chernenko et al. [11] maintain that the Al oxidation by the environmental $O_2$ is necessary to trigger the thermite. But the Al particles are separated by $Fe_2O_3$ (capacitance) and conditions to auto-heating and ignition of Al are difficult to achieve. Endothermic peaks D and E correspond, respectively, to the reduction of $Fe_2O_3$ to $Fe_3O_4$ and the partial reduction of $Fe_3O_4$ to FeO, with oxygen release, according to [12]. Higushi and Heerema [13] reported the first transition at 1375 °C, in atmospheric air, and near 1100 °C, at lower $O_2$ pressures, in agreement with the present results.

Comparing the results in air and vacuum, it can be stated: i) the phenomena involving the $O_2$ of the environment occurred at higher temperatures in vacuum, due to the lower availability of $O_2$; ii) the phenomena with oxygen release occurred at lower temperatures in vacuum, due to the pressure difference. Considering the TGA results for the thermite mixtures, the fraction of active Al that is oxidized to $Al_2O_3$ by the $O_2$ of the environment is within 71-91%. This range could shift to higher values if positive and negative mass changes were not overlapped for T>1000 °C. The XRD result of the quenching test and the matching ICDD database patterns are shown in Fig. 5. Due to rapid cooling, the products have low crystallinity and some present crystal lattice distortions (dislocations). The phases represented by symbols are overlapping and their peaks are minor. During cooling, the major part of $Fe_3O_4$ and FeO oxides re-oxidizes to $Fe_2O_3$ [13]. The presence of ε-Fe can be attributed to residual thermite or, most probably, to the FeO reduction near 1400 °C [12]. Phase "x" is not assigned, but may be a solid solution close to $AlFeO_3$ (30-0024). Therefore, the occurrence of the thermite reaction is unlikely, due to the large amounts of iron oxides detected.

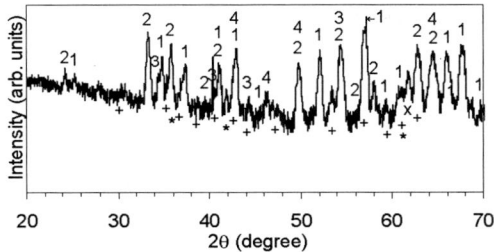

**Figure 5.** XRD pattern of products of quenched T100 combustion. 1–$\alpha$-Al$_2$O$_3$ (10-0173), 2–$\alpha$-Fe$_2$O$_3$ (84-0311), 3–FeO (49-1447), 4–$\epsilon$-Fe (34-0529), (+)–Fe$_3$O$_4$ (85-1436), (*)–FeO (06-0615).

The heating microscopy results confirm the previous discussion. The cubic samples' volume increased or decreased near the DTA temperatures where incorporation or release of oxygen occurred. Volume reductions due to sintering and increases due to Al dilation were also observable. In thermite mixtures, between 700 °C and 900 °C, little white points - alumina - appeared gradually on the cube surface. Only the Al sample behaved differently from STA. Its thermal explosion occurred close to 550 °C, in two combustion stages (vd. Fig. 6): i) at 545 °C, an orange color wave propagated slowly through the sample surface; ii) between 546 °C and 548 °C, a bright white color wave started at the center and propagated fast through all the sample volume. Gromov et al. reported [14] temperatures of 1400 °C and 2500 °C for these stages, respectively. In the STA experiments it did not happened, since the limited amount of material used in each sample favored the heat dissipation and the particles auto-heating was not achieved.

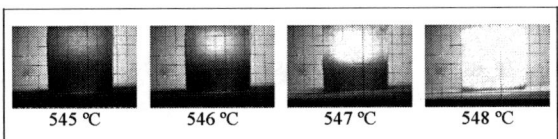

| 545 °C | 546 °C | 547 °C | 548 °C |

**Figure 6.** Thermal explosion of Al reactant observed with heating microscopy. Grid square = 0.5 mm$^2$.

## CONCLUSION

The O$_2$ present in the environment was the key factor of the reaction path for the Fe$_2$O$_3$/Al system during the heating. The Al oxidation by this oxygen is almost complete, even in vacuum, making the thermite reaction unviable. The high end temperature used in experiments and the TGA results were crucial to draw this conclusion. STA runs in high purity inert environment are under way, with careful procedures to avoid O$_2$ traces.

## REFERENCES

1. Sarangi, B. et al., Kinetics of Aluminothermic Reduction of MnO$_2$ and Fe$_2$O$_3$: a Thermoanalytical Investigation, ISIJ Int., 36 (9), pp.1135-1141, 1996.
2. Wang, S. et al., Analysis of the Aluminothermic Reaction Process, Key Eng. Mat., 224-226, pp.745-748, 2002.
3. Turetsky, A., and Young, G., New Insights in High Temperature Pyronol Chemistry, in Proc. 18th Int. Pyrot. Sem., pp.879-897, 1992.
4. Mei, J. et al., Mechanisms of the Aluminum-Iron Oxide Thermite Reaction, Scripta Mater., 41 (5), pp.541-548, 1999.
5. Jones, D. et al., Thermal Characterization of Passivated Nanometer-Sized Aluminum Powders, in Proc. 27th Int. Pyrot. Sem., pp.821-830, 2000.
6. Sandén, R., Characterization of Electro-Exploded Aluminum", in Proc. 29th Int. Annu. Conf. of ICT, pp.77.1-77.10, 1998.
7. Mench, M. M. et al., Propellant Burning Rate Enhancement and Thermal Behavior of Ultra-Fine Aluminum Powders, in Proc. 29th Int. Annu. Conf. of ICT, pp.30.1-30.15, 1998.
8. Botta, P. M. et al., Mechanochemical Synthesis of Hercynite, Mat. Chem. Phys., 76, pp.104-109, 2002.
9. Kashporov, L. Y. et al., Temperature of the Aluminium Based Mixtures Ignition by the Stationary Burning Wave, in Proc. 20th Int. Pyrot. Sem., pp.543-555, 1994.
10. Frolov, Y., Some Phenomenological Aspects of Heterogeneous Combustion of Condensed Systems, Propell. Explos. Pyrot., 25 (4), pp.161-167, 2000.
11. Chernenko, E. V. et al., Inflammability of Mixtures of Metal Oxides with Aluminum, Combust. Explo. Shock Waves, pp. 639-646, May 1989.
12. Wriedt, H., Fe-O (Iron-Oxigen), in Binary Alloy Phase Diagrams, vol. 2, 2nd ed., (Massalsky, T. et al., eds.), ASM Int., pp.1739-1744, 1990.
13. Higushi, K., and Heerema, R., Influence of Sintering Conditions on the Reduction Behaviour of Pure Hematite Compacts, Miner. Eng., 16, pp.463-477, 2003.
14. Gromov, A. et al., Kinetic and Thermodynamic Features of Combustion of Superfine Aluminum Powders in Air, in Proc. 34th Int. Annu. Conf. of ICT, pp.58.1-58.13, 2003.

CP845, *Shock Compression of Condensed Matter - 2005*,
edited by M. D. Furnish, M. Elert, T. P. Russell, and C. T. White
© 2006 American Institute of Physics 0-7354-0341-4/06/$23.00

# SHOCK WAVE INITIATION OF MIXTURE LIQUID EXPLOSIVES

## A.V. Fedorov, A.L. Mikhailov, D.V. Nazarov,
## S.A. Finyushin, A.V. Men'shikh, V.A. Davydov

*RFNC-VNIIEF, Sarov, 607190, Russia*

**Abstract.** We investigated initiation of liquid HE consisting of tetranitromethane (TNM) and nitrobenzene (NB). Smooth stable (when mass of NB<20%) and pulsing unstable detonation wave front was registered (20-50% NB). We registered shock wave, shock compressed explosive (SCE) detonation wave and normal detonation wave for unstable detonation front on different parts of the front. In case of normal and SCE detonation wave we registered parameters rise during 3-25 nsec until the start of chemical reaction. We consider it to be the induction period of thermal explosion inside detonation wave front.

**Keywords:** detonation wave structure, liquid tetranitromethane, particle velocity, Neumann spike.
**PACS:** 47.40.Nm, 62.50.+p, 82.33.Vx, 82.40.Fp.

## INTRODUCTION

Experimentally [1, 2], in homogeneous HE(s) it was observed that detonation wave front might be both rough with pulsation in nitromethane (NM) and smooth in tetranitromethane (TNM). It was found that shock wave (SW) starts to propagate as in inert material but then turns to detonation jump wise. Previously, we studied detonation wave structure in TNM-based liquid explosives [3-5]. Smooth stable front was observed in TNM and TNM/NM-46/54, while pulsed unstable front was recorded in TNM/NB-74/26. The goal of the present work was a more profound research of detonation wave (DW) structure in shock-initiated liquid explosives (like oxidizer/fuel). The research was concentrated on the following HE(s): tetranitromethane/nitrobenzene (TNM/NB)-95/5,90/10,85/15,80/20,74/26 and 50/50.

## EXPERIMENT

Experimentally, we recorded particle velocity of liquid explosive interface with LiF using Fabry-Perot laser interferometry technique [3-5]. Onto LiF single crystal surface in contact with HE it was deposited a ~1μm Al coating.

The internal diameter of liquid-HE-containing cells was 6-15 mm; while their length was 2…50 mm. Liquid HE was initiated by detonator. In several experiments between detonator and liquid HE it was placed a layer of highly sensitive HE based on plasticized PETN.

Aluminum foil with the thickness of 20 μm was inserted between PETN and liquid explosive. In all cases spherical detonation wave propagated in liquid HE(s). The method records the particle velocity profile at the interface HE/LiF - $U_{LiF}(t)$, which then was recalculated and HE states were defined.

## EXPERIMENTAL RESULTS

*TNM/NB compositions with up to 20% of NB:*

In all studied compositions (95/5; 95/10; 85/15; 80/20) detonated from detonator (the shock wave amplitude in liquid explosives was 17 GPa) stable detonation front was recorded with Neumann spike value close to that in TNM, i.e. $P_N$=21.6 GPa [3].

It is well known that shock wave critical pressure to initiate TNM is 8 GPa at the detonation pressure of $P_{C-J}$=16 GPa [2, 7]. Further, it will be shown that if NB content is more than 20%, the detonation wave front becomes unstable and pulsed. Probably, in the considered compositions NB content increase leads to rising critical pressure of initiating shock waves. Accordingly, in compliance with [8], detonation velocity and pressure rise as well. Fast energy release kinetic corresponds probably to compositions with up to 20% of NB, and slow energy release kinetic – to compositions with more then 20% of NB.

## TNM/NB-74/26:

The present composition is close to the stoichiometric one of 76.85/23.25 and has the density of $\rho=1.51$ g/cm$^3$, while possessing the maximum detonation velocity is D=7.5 km/sec of all TNM/NB liquid explosives. Its P-U diagram is shown below in fig.1.

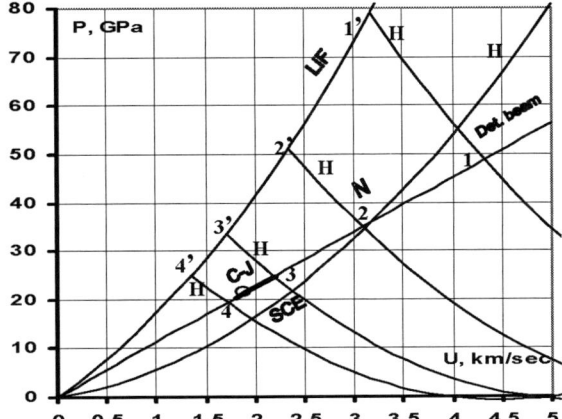

**Det.beam** - detonation beam; **H** - TNM/NB-74/26 shock adiabat; **LiF** - LiF shock adiabat; **C-J** - Chapman-Jouguet state; **N**– Neumann spike; **SCE**– range of shock-compressed HE states; **1,2,3,4** - wave amplitudes in TNM/NB-74/26; **1',2',3',4'** - LiF states recorded for 1,2,3,4 amplitudes.

**FIGURE 1**. TNM/NB-74/26 P-U diagram

In the considered detonator-initiated composition unstable pulsed mode was recorded. The $U_{LiF}(t)$ profiles recorded experimentally at the thickness of 2...30 mm in liquid explosives might be subdivided into three groups (fig.2):

(a)   Shock wave (fig.2.a): The pressure in the liquid explosive was 16...22 GPa ($U_{LiF}$=1.4...1.7). This corresponds to the shock-compressed HE region in fig.1.

(b)   Rising and gradually dropping front (fig.2b): During the first 3-9 nsec the $U_{LiF}$ rise was recorded that was followed by a gradual decrease. Experimentally, a smoothened spike was recorded. Particle velocity in LiF spike ranged within $U_{LiF}$=1.46...1.68 km/sec. This corresponded to pressures in the liquid explosive that are close to Chapman-Jouguet ones. Such front should be implemented in case of reaction breakdown and sharp parametric attenuation of SCE detonation wave spike (fig.2a).

(c)   Front with velocity rise and record break (fig.2c): In the wave forefront it was recorded a fast $U_{LiF}$ growth for the first 5...25 nsec before recording break. The recorded $U_{LiF}$ values ranged within $U_{LiF}$ =0.5...3.17 km/sec (fig.2c) at the moment of recording break. High particle velocities (more than 3.0 km/sec) might be explained by propagation of oblique detonation wave moving across SCE (see fig.5) [6, 7]. Maximum pressure in the considered liquid HE composition was P=55 GPa. Previously, similarly high pressures [2, 7] were recorded in SCE waves traveling across shock-compressed HE in nitroglycerine (P~50 GPa).

Experimentally, the above three structures were recorded randomly regardless of a particular front point hit by a laser beam.

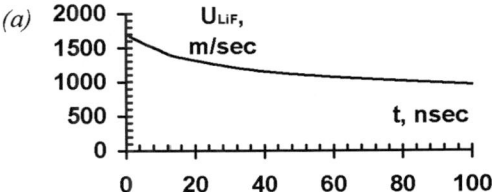

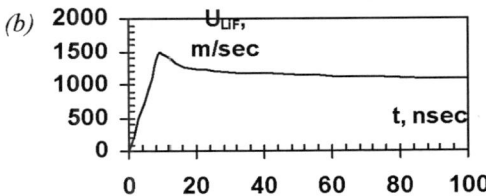

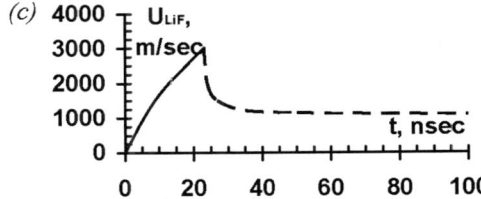

**FIGURE 2**. Recorded particle velocity typical profiles in TNM/NB-74/26 composition

In case of initiation from plasticized PETN with thickness 11 mm, a normal detonation wave was recorded in the liquid explosive h=50mm (fig.3c) with Neumann spike and Chapman-Jouget (C-J) state. Front rise duration till Neumann spike was $\tau$=5 nsec, while chemical reaction range from

Neumann spike till C-J state was $\Delta t \sim 13$nsec. Then, Taylor release wave was recorded at the profile.

Neumann spike pressure in the considered composition was $P_N=34$ GPa, while detonation pressure was $P_{CJ}=21.4$ GPa. Probably, the front rise of $\sim 5$nsec is the induction period as described in [7]. In fig.3a and 3b there are shown similar cases from thermal explosion theory, where the detonation wave induction period ranges within shock wave front limits. From fig.3 it is obvious that the recorded profile (fig.3c) coincides with the predicted one (fig.3a)

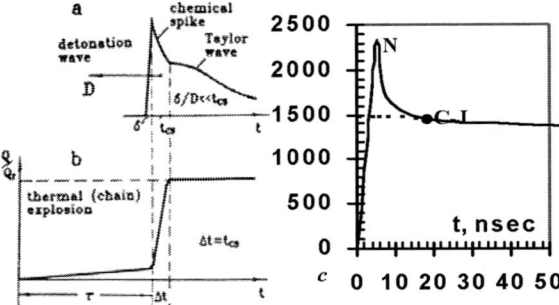

**FIGURE 3**. Similarities for explosive molecule transformation during detonation (***a***) and thermal or chain explosion (***b***) [7]; ***c***- recorded detonation wave profile in TNM/NB-74/26.

*TNM/NB-50/50*

Density of this mixture is $\rho=1.38$ g/cm$^3$, detonation velocity is D=6.5 km/sec. The thickness of the liquid HE charges was 5...30 mm.

There were recorded three $U_{LiF}$(t) profile types, which are shown in fig 4.

(a) <u>Front with spike</u>: in case of initiation by detonator or 3mm-thick plasticized PETN, in the considered composition it was recorded the average $U_{LiF}$ value of 2.62 km/sec in the spike that corresponds to the liquid explosion compressed at P=37 GPa. This value corresponds to SCE detonation wave. $P_{C-J}$ evaluated at n=3 polytrope index is $P_{C-J}=14.6$ GPa. Spike pressures drop fast down to P=15.5 GPa for $\sim 15$ nsec that is followed by smooth rate decrease in Taylor release wave. We think that pressure P=15.5 GPa corresponds to the pressure of $P_{C-J}$.

(b) <u>Rising and smoothly declining front</u>: It was recorded the profile with the velocity rise up to 450...600 m/sec for the initial 3...4 nsec that was followed by shock wave jump up to $U_{LiF}=2.1$ km/sec

smooth decline. This corresponds to Neumann spike of $P_N=26$ GPa $U_N=2.85$ km/sec (fig.4b). Probably, the velocity rise for the initial 3...4 nsec is the induction period.

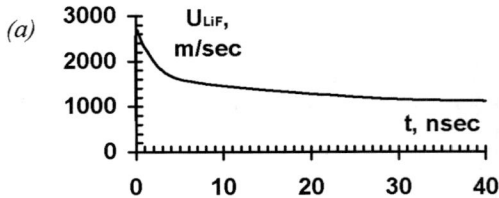

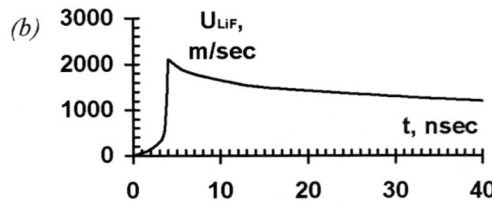

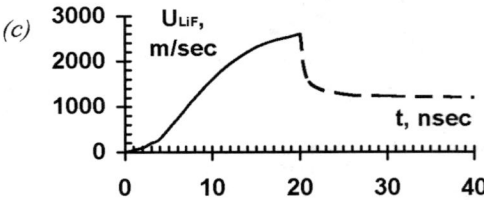

**FIGURE 4**. Recorded particle velocity profile types in TNM/NB-50/50.

(c) <u>Fastly rising front till recording break</u>: It was recorded a fast velocity rise in LiF up to 0.5...2.6 km/sec, where recording was broken. Maximum particle velocity $U_{LiF}$ at the HE/LiF interface was U=2.6 km/sec at the break point that corresponds to the pressure in the liquid explosive of 37 GPa and coincides with the maximum pressure for the structure in fig.4a.

## DISCUSSIONS AND CONCLUSIONS

Shock wave velocity in mixture TNM/NB-74/26 is in the range 5.4-6.1 km/sec (P=16-22 GPa), and maximum velocity of SCE detonation wave is D=9.05 km/sec (P=55 GPa), totally detonation wave velocity for overall pulsing front is 7.5 km/sec. A qualitative picture of emerging local explosions and pulsed detonation is shown in fig.5 from [2,7]. From fig.5 it is obvious that during SCE detonation wave motion through shock-compressed HE there should be implemented maximum step-wise pressure profiles (firstly the shock wave, then SCE detonation wave).

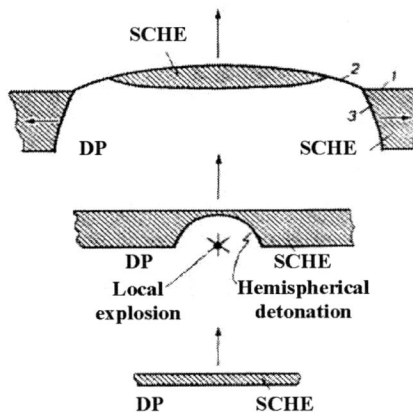

1-straight shock wave; 2- oblique detonation wave;
3- transversal wave with triple configurations;
DP-detonation products; SCHE-shock-compressed HE

**FIGURE 5**. Qualitative picture of pulsed detonation and emerging local explosions [7].

The recorded maximum velocities in LiF correspond to pressure amplitudes of P=55 GPa in the 74/26 composition and P=37 GPa in the 50/50 composition in the liquid explosives, which is 2.5 times higher than $P_{C-J}$ in the aforementioned compositions. Obviously, these values are maximum pressures, which may be implemented in oblique (2) wave in compliance with the schematic of fig.5. The recorded Neumann spike pressures practically coincide with the spikes computed based on shock adiabats (detonation beam cross-section with shock adiabat). With regard to the composition 74/26, the computational Neumann spike was $P_N$=36 GPa, while the experimentally recorded value was $P_N$=34 GPa. As for the composition 50/50, the experimental and computational Neumann spike was $P_N$=26 GPa. Shock adiabats of the recorded compositions were fitted using the additive method based on TNM and NB well known adiabats.

The following adiabats were obtained:
- for the composition 74/26: D=1.897 +1.767U;
- for the composition 50/50: D=2.207+1.51U.

In case of normal detonation waves we registered a rise of parameters in the front during 3-5 nsec. To our mind, this rise is a period of induction of thermal explosion inside the detonation wave front. The reasons for appearing of induction period are

following: in mixed liquid HE TNM detonates while NB does not; when big amount of NB in the mixture the last as inert component obviously greatly increases critical pressure of initiation and delays developing of chemical reaction, therefore detonation can not be developed according to ZND model.

That is why pulsing wave appears, where detonation is developed in SCE parts of detonation front. After self-ignition and beginning of chemical reaction excess oxygen is liberated from TNM and reacts with fuel (NB).

It is well known [7] that pulsed unstable detonation is characteristic for HE(s) with slow energy release kinetics. All gaseous mixtures and weak liquid explosives belong to such category. Reaction breakdown and shock compressed HE detonation occurs in random separate points and therefore the front is non-uniform and has a complicated 3D structure. Results of the present work are amazing and not understandable to the end. We don't understand why the registered velocity (pressure) (see fig.2-b,c and fig.4-b,c) begins smoothly growing from zero point? Although in all other experimental works the growing begins from shock wave pressures.

## REFERENCES

1. Campbelle, A.W., Davis W.B., Travis Y.R. // Phys. of Fluids, #4, v.498, 1961, c.118-122.
2. Dremin, A.N., Savrov S.D., Trofimov V.S., Shvedov K.K.//Detonation waves in condensed media//Moscow, Nauka, 1970 (Rus).
3. Fedorov, A.V. et al. // Proceedings of Shock Compression of Condensed Matter, Snowbird, 1999, pp.801-804.
4. Fedorov, A.V. // Proc. XII International Detonation Symposium. August 11-16, San Diego, 2002.
5. Fedorov, A.V. // Proceedings of Shock Compression of Condensed Matter. Atlanta, 2001, pp. 910-913.
6. Dremin, A.N.//Chemical physics // 1995, Vol4, i.12, pp 22-40. (Rus)
7. Dremin, A.N. Toward Detonation Theory. Springer. 1999.
8. Eremenko, L.T. et al // Proceedings of the conference "Chemical physics of combustion and explosion processes. Detonation" // Chernogolovka, 1977, pp.76-79. (Rus)
9. Dremin, A.N.//Fizika Goreniya i Vzryva (Rus), 2000, Vo. 36, i.6.

CP845, *Shock Compression of Condensed Matter - 2005,*
edited by M. D. Furnish, M. Elert, T. P. Russell, and C. T. White
© 2006 American Institute of Physics 0-7354-0341-4/06/$23.00

# STUDY OF DETONATION WAVE STRUCTURE IN SOLID AND LIQUID TETRANITROMETHANE (TNM)

## A.V. Fedorov, A.L. Mikhailov, D.V. Nazarov, S.A. Finyushin, A.V. Men'shikh, V.A. Davydov, T.A. Govorunova.

*RFNC-VNIIEF, Sarov, 607190, Russia*

**Abstract.** Investigations of detonation front structure and parameters in solid and liquid tetranitromethane were done using Doppler Fabry-Perot velocimeter. We recorded the particle velocity of explosion products, braking on the HE/window interface. Smooth front of the detonation wave and concave negative-going particle velocity profile were recorded for liquid TNM. The experimental records indicate that because of solid TNM heterogeneity flow, turbulization occurs behind detonation wave front what appears in the form of velocity fluctuations on the U(t) profile.

**Key words:** detonation wave structure, liquid tetranitromethane, particle velocity, Neumann spike.
**PACS:** 47.40.Nm, 62.50.+p, 82.33.Vx, 82.40.Fp.

## INTRODUCTION

It is well known that detonation wave structure in solid heterogeneous and liquid homogeneous HE(s) is different. Both smooth detonation front with laminated flow and unstable pulsed front with turbulent flow behind detonation wave front are typical for liquid explosives. Previously, we investigated heterogeneous and liquid homogeneous HE(s) using laser interferometer with high time resolution [1, 2]. Similar methods were used by other authors as well for detonation wave structural studies [3-12]. It is of interest to research wave structure of one and the same HE, when it is in solid and/or liquid initial state. Tetranitromethane (TNM) was chosen as testing substance, whose physical and chemical properties are described in [11-14].

## EXPERIMENT

TNM's detonation wave structure was investigated using Fabry-Perot method and looking through LiF single crystal window onto the interface clad with 0.5...1.5 μm Al coating [1]. Experimental layout is shown schematically below in fig.1. Laser radiation was focused on reflecting surface into the focal spot of ~100...200 μm with short-focus lens.

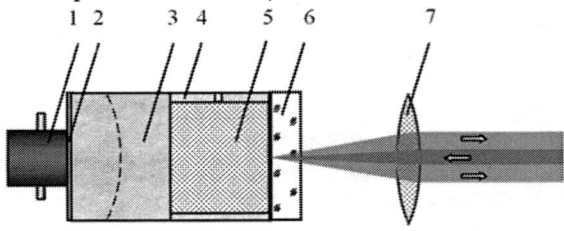

1- detonator;
2- plasticized PETN, Ø30×2 mm;
3- TNT/RDX-3/7, Ø30×30 mm;
4- plexiglas cell Ø30×30 mm;
5- TNM, Ø26×30 mm;
6- LiF with 0.5-1.5 μm Al coating;
7- focusing lens

**FIGURE 1.** Experimental set-up.

TNM was initiated by an detonator. In several experiments highly sensitive PETN-based HE interjacent layer, or hexogen-based HE charge, was inserted between TNM and detonator. Different initiation methods allowed varying pressure drop gradient of the wave, which initiated TNM.

Spherical expanding wave propagated in TNM. When the DW reached the TNM/LiF interface coated with aluminum, LiF was loaded and the shock wave was reflected into the interjacent HE. The interface velocity profile reproduced that of the interjacent HE. Particle velocities obtained in LiF were recalculated to define corresponding values for the test HE.

The internal diameter of the TNM-containing cell was 30 mm, while its thickness was 10-15mm. TNM solidification temperature was $14.2^0$C. TNM was pored into the cells and solidified before the experiment at the temperature of $0...5^0$C. The view of the cell with solid TNM is shown in fig.2.

**FIGURE 2.** Test assembly with solid TNM.

Crystal density of solid TNM is 2.00 g/cm$^3$. TNM density during solidification was measured based on volume changes. During solidification TNM density increased from $\rho$=1.64 g/cm$^3$ to $\rho$=1.82 g/cm$^3$. Voids were formed inside TNM during volume shrinking, their sizes according to microscopic analyses could be up to 100 μm. Solid TNM density is 91% of maximum crystalline one.

## EXPERIMENTAL RESULTS

*Liquid TNM:*

In experiments with liquid TNM we recorded a smoothly concave decreasing LiF U(t) interface particle velocity profile with Neumann spike (fig.3).

From fig.3 it is obvious that detonation in liquid TNM developed in a stable mode with Neumann spike. Stable detonation development in all experiments was also confirmed by the coincidence of recorded Neumann spikes in LiF ($U_{LiF}$=1.6 mm/μsec) for all charge lengths (10...30mm).

Profile coincidence testified the smoothness of the detonation wave front in liquid TNM. Even in works by Kormer in 1960-s [15] it was shown that mirror-like detonation front exists in liquid TNM, i.e.

non-uniformity sizes are less than the light wavelength.

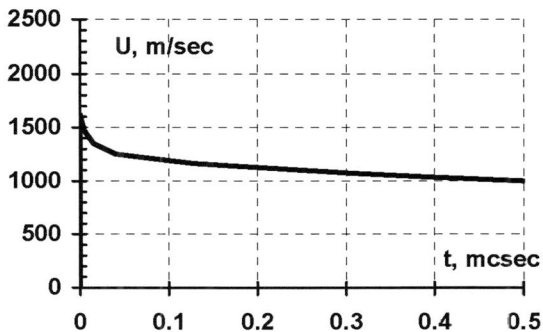

**FIGURE 3.** Particle velocity profile in liquid TNM (b).

*Solid TNM:*

Three experimental series were staged. In the first series in case of detonation from detonator (pressure in TNM was 17 GPa), there was no detonation at test thicknesses. In the second experimental series TNM was initiated by PETN-based HE charges with the thickness of 4...11 mm; and it was recorded transition to detonation with sloping forefront. For the initial 40-110 nsec the particle velocities were the following: $U_{LiF}$=1.43...1.54 km/sec (fig.4).

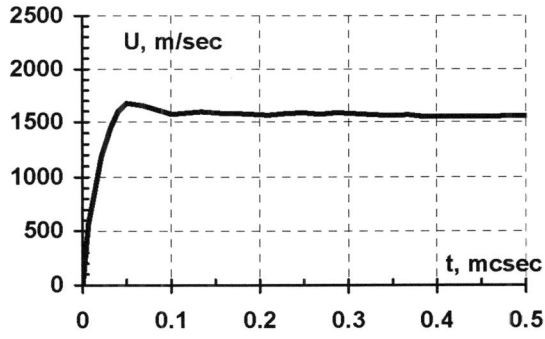

**FIGURE 4.** Transition to detonation mode in solid TNM.

In the aforementioned experiments it was recorded the initial detonation initiation stage with chemical spike. Probably, small HE thickness leads to a considerable delay of solid TNM initiation. However, after the wave traveling through the TNM thickness of 30mm, there was not yet steady-state detonation wave.

In the third experimental series, when TNM was initiated by a primer Ø30×30 mm from the composition TNT/RDX-3/7, in case of TNM thickness of 30 mm, it was formed the detonation wave with chemical spike. The spike value in solid TNM ($\rho$=1.82 g/cm$^3$) was P=28.7 GPa and U=2.27 km/sec. Besides, there were recorded oscillations at the profile U(t) that were connected with heterogeneous nature of solid TNM. Pulsation size behind detonation wave front ranged within 30-200 m/sec, while the pulsation period was 30-180 nsec (fig.5).

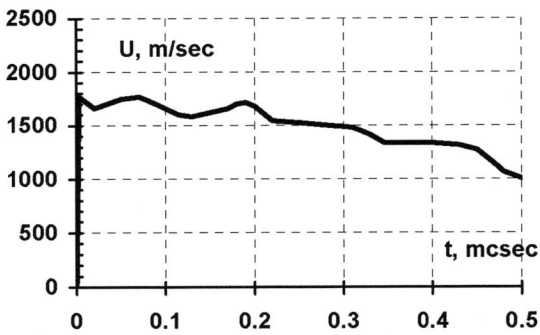

**FIGURE 5**. Particle velocity pulsation profile in solid TNM.

The pulsations were caused by formation of pores inside the HE during solidification. As is shown in [1, 2], inside the pores there are formed micro jets, which move before detonation wave and penetrate into the initial HE casing flow turbulence. This is the main reason for pulsation behind the detonation wave front.

Also micro jets are the main reason for "hot spot" formation inside HE. Many researchers consider that adiabatic compression of gas occurs inside pores and consequent self ignition ("hot spot"). We think that really micro jets appear inside pores, cracks, gaps, which velocities may be considerably higher than detonation velocity due to effects of cumulation like hollow charge. They penetrate into unreacted HE and not only perturb the detonation front and result to oscillations on particle velocity profiles, but start the chemical reaction in separated places ("hot spot") even before arrival of the main detonation wave.

In fig.6 it is shown an integral plot of particle velocities in liquid and solid TNM, and velocity pulsation is obvious.

So, in TNM it was recorded a significant particle velocity profile change during the transition from homogeneous to heterogeneous state.

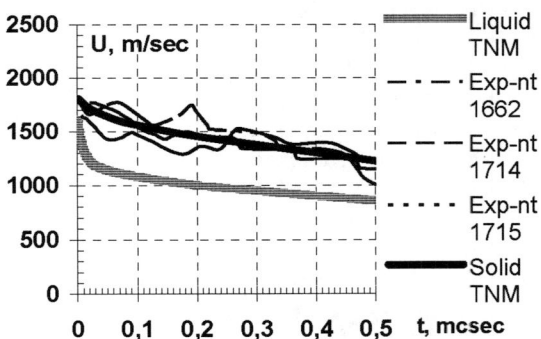

**FIGURE 6**. Particle velocity profile in TNM.

In several experiment it was fixed a recording break at the moment of the detonation wave coming to the HE/LiF interface. The break is caused either by the interface turning due to time jitter of the wave incidence onto the interface, or microstructure destruction of thin Al coating on LiF at the moment of the detonation wave outflow. Arguably, this might be attributed to heterogeneous nature of solid TNM and pores inside it.

## DISCUSSIONS AND CONCLUSIONS

In the experiments with liquid TNM it was always recorded Neumann spike and smoothly concave dropping profile U(t), which prove the smoothness of the detonation wave forefront in the liquid explosive. Chapman-Jouget state in TNM was $U_J$=1.53 mm/$\mu$sec, $P_J$=16 GPa. Neumann spike in liquid TNM was P=21.6 GPa, U=2.07 mm/$\mu$sec, which 1.36 times higher than Chapman-Jouget state.

In experiments with weak imitation of solid TNM by detonator there was no detonation at the considered thicknesses.

When PETN-based plasticized HE charge was used for initiation, at the thickness ranging within 4-11 mm it was recorded the mode of transition from shock to detonation wave.

In case of initiation by hexogen-based HE (TNT/RDX-3/7) a normal detonation wave is formed.

Due to heterogeneous nature of solid TNM the flow becomes turbulent behind the detonation wave front that is manifested in characteristic velocity oscillations at U(t) profile. The oscillation amplitude was $\Delta$U=20...220 m/sec, while the duration was $\Delta$t=40...180 nsec.

966

## ACKNOWLEDGEMENTS

Authors thank V.S.Sasik, A.A.Frolov, A.R.Gavrish, E.V.Filinov, I.A.Vidashov, G.B.Krassovsky for their help in fulfilling this work and valuable discussions.

## REFERENCES

1. Fedorov A.V., Menshikh A.V., Yagodin N.B. //Detonation front structure in heterogeneous HE(s) // Chemical Physics, 1999, Vo.18, i.11, pp.64-68 (Rus).
2. A.V. Fedorov Shock Compression of Condensed Matter. Atlanta, Part Two, 2001, pp. 910-913.
3. Leroy, Green G., Tarver C., Erskine D. Proccedings of 9th Symposium on Detonation. Portland, Oregon. 1989.
4. Gustavsen R.L., Sheffield S.A., Alkon R.R., Detonation wave profiles in HMX based explosives, pp. 739-742 in Shock Compression of Condensed Matter - 1997, S. C. Schmidt, D.P. Dandekar and J. L. Forbes (eds), AIP Press, 1998.
5. Utkin A.V., Kanel G.I.// Fizika Goreniya I Vzryva 1989, Vo.25, i.5 (Rus).
6. Utkin A.V., Malyarenko S.I., Kanel G.I./ Detonation Symposium// Suzdal, 1989 (Rus)
7. Tarver C.M., Kury J.W., Breithaupt R.D. J. Appl. Phys. 1997. 82(8). P. 3771.
8. Utkin A.V., Kanel G.I.// The 8th Symposium on Combustion and Detonation// Chernogolovka, 1986 (Rus).
9. Erskine D.J., Tarver C.M., Green L.G. Shock waves in condensed matter. 1989. P. 717.
10. Kury J.W., Breithaupt R.D. Proceedings of IX Symposium on detonation. USA. 1989. P. 1378.
11. Campbelle A.W., Davis W.B., Travis Y.R. Phys. of Fluids, 4, 498, 1961.
12. Dremin A.N., Savrov S.D., Trofimov V.S., Shvedov K.K.//Detonation waves in condensed media// Moscow, Nauka, 1970 (Rus).
13. C. Mader, Numerical modeling of detonations, Univ. California Press, Berkeley 1983.
14. Nesterenko D.A., Eremenko L.T.// Proceedings of the XXI International Pyrotechnical Seminar// 1995, pp. 627-637(Rus).
15. Kormer S.B.//Uspekhi Fizicheskikh Nauk // 1968, Vo.94, i.4, pp.641 (Rus)

CP845, *Shock Compression of Condensed Matter - 2005*,
edited by M. D. Furnish, M. Elert, T. P. Russell, and C. T. White
© 2006 American Institute of Physics 0-7354-0341-4/06/$23.00

# PROTON RADIOGRAPHY OBSERVATIONS OF THE FAILURE OF A DETONATION WAVE TO PROPAGATE TO THE END OF A CONICAL EXPLOSIVE CHARGE

## Eric N. Ferm [1], Fesseha Mariam [1], and LANSCE Proton Radiography Team [1]

[1] *Los Alamos National Laboratory, Los Alamos NM 87545*

**Abstract.** Failure diameter is a well-known property of explosive materials, being the critical diameter below which a steady detonation wave will not be able to support itself and ultimately fails to propagate. A detonation wave traveling down a uniform cylindrical charge larger than its critical diameter will reach a steady detonation velocity which is a function of the diameter of the explosive as well as other material properties, notably density and temperature. In this work, we use proton radiography to study the propagation of detonations down conical PBX 9502 charges beginning at diameters larger than failure diameter and ending at diameters much smaller than failure diameter. Experiments show cases where complete detonation of the cone occurs as well as cases where failure is observed significantly before the end of the cone and significant portions of the charge remain unreacted. Wave velocities and densities are obtained from the multiple image proton radiography experiments and compared with failure diameter effect curves for PBX 9502.

**Keywords:** Explosive behavior, Diameter Effect Curve, Failure Diameter.
**PACS:** 47.40.-x, 47.40.Nm, 82.40.Fp.

## INTRODUCTION

Failure diameters are obtained by firing a series of rate sticks of various diameters to determine the smallest diameter that exhibits a steady detonation velocity. The "Diameter Effect Curve" is obtained by plotting detonation velocity of the rate stick versus the charge diameter. Campbell and Engelke [1] studied the rate stick data for many heterogeneous explosives and found a fitting form for the diameter effect curves, which fits heterogeneous and homogeneous explosives quite well. Shown in Fig. 1 is the Diameter effect curve for PBX 9502 using the fit found in Campbell and Engelke [1]. Other methods of determining failure diameter have been examined by firing charges

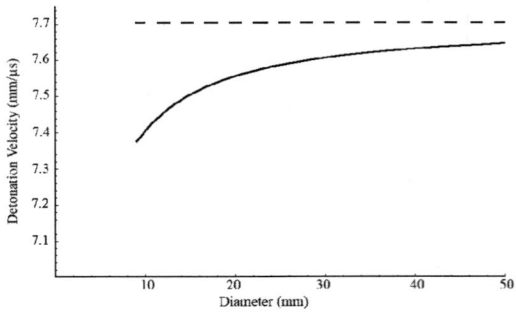

**Figure 1.** Diameter Effect Curve for PBX 9502 using the fit found in Campbell and Engelke [1]. The dashed line is the asymptote for large charges where a detonation velocity of 7.706 mm/μs is expected. Failure diameter is estimated at 9 mm with a detonation velocity of 7.37 mm/μs, 4% lower than the asymptote.

of slowly varying cross sections, monitoring the detonation with witness plates, and carefully measuring the velocity. Wedges, cones and prisms of explosives have been used to measure approximate diameter effect curves [2, 3, 4]. These techniques assume that the quasi-steady detonation velocity plotted against diameter is an approximation of the diameter effect curve. This assumption is clearly not true for a large angle cone where the detonation wave crosses the interfaces much faster than any signals can propagate into the center of the charge. For small cone angles the boundary condition on the edge becomes important to determining the detonation velocity. The question then becomes, "What is the lag in boundary effect influencing the detonation velocity of the charge?" or "What is the approach of the Quasi-steady solution to the diameter effect curve as a function of cone angle?"

## EXPERIMENTAL PROCEDURE

The proton radiographs for this work were obtained at the LANSCE Proton Radiography Facility. Using the accelerator's proton beam and magnetic optics, several image planes are created. The object to be radiographed is placed at an image location and at subsequent image locations scintillators emit light proportional to the transmitted proton beam. The image is recorded by cameras and the normalized transmission and path length images of the object in the beam are obtained by using the characterization of the entire imaging system. Beam pulses and cameras can be configured to take multiple images of a dynamic object with interframe times of greater than 0.7 μs. A more complete discussion of the radiography can be found in King et al [5].

The failure cone experiments were designed with the intent to show a range of possible explosive propagation behavior, to be reasonable to machine, and to minimize the amount of unburned explosive discovered after the shot in the containment system. The charges are described in Fig. 2.

Seventeen dynamic images of the 25-2.5 mm truncated-cone charge were taken, showing it detonating to nearly the end of the charge at least 22 mm past the 9 mm failure diameter position of the cone. Shown in Fig. 3 is two of the path-

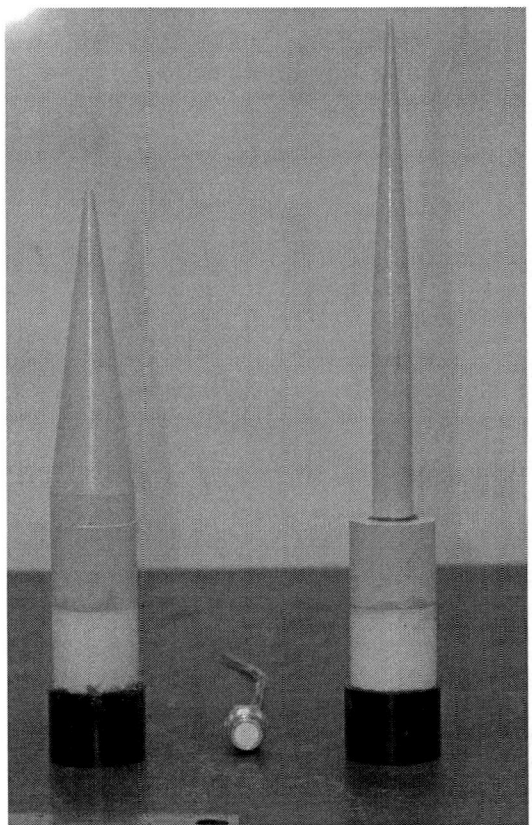

**Figure 2.** Failure cone charges used in Prad 0145 and 0147 experiments. On the left is a 25-mm failure cone and on the right is the 12-mm failure cone charge. The initiation and booster system for these charges consisted of a SE-1 detonator, 0.5" PBX 9407, 25 mm of PBX 9501 followed by 25.4 mm of PBX 9502, having a density of 1.894 g/cc. The conical charges were built from 100-mm long cylinders with a 10-mm cylindrical section followed by a 90-mm long truncated cone. This was intended to reduce concerns of glue joints in the test section. The 25 mm cone began at 25 mm and reduced to 2.5 mm, resulting in a reduction of 1 mm diameter for 4 mm of travel (7.1°). The 12 mm cone began at 12 mm and reduced to 2 mm, resulting in a reduction of 1 mm diameter for 9 mm of travel (3.2°). The 12-mm charge had an additional 12-mm-diameter 50-mm-long PBX 9502 charge preceding the cone to insure the wave velocity was steady and consistent within the 12 mm stick before entering the conical section.

length images from this experiment with the latest images showing little indication of unburned explosive. The wave velocity, however, fell significantly in the last few frames of the sequence.

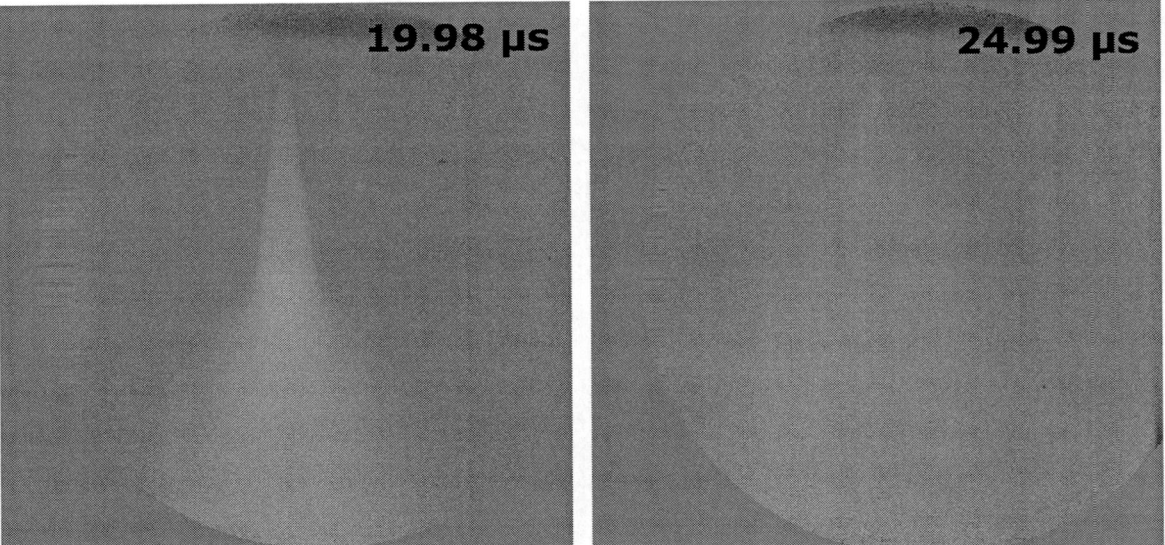

**FIGURE 3.** 25 mm Failure cone experiment, PRad 0145, detonated to the end of the cone. The tip does not seem to expand as quickly as other regions of the charge, but there is not strong indication of unburned explosive at the tip of the cone

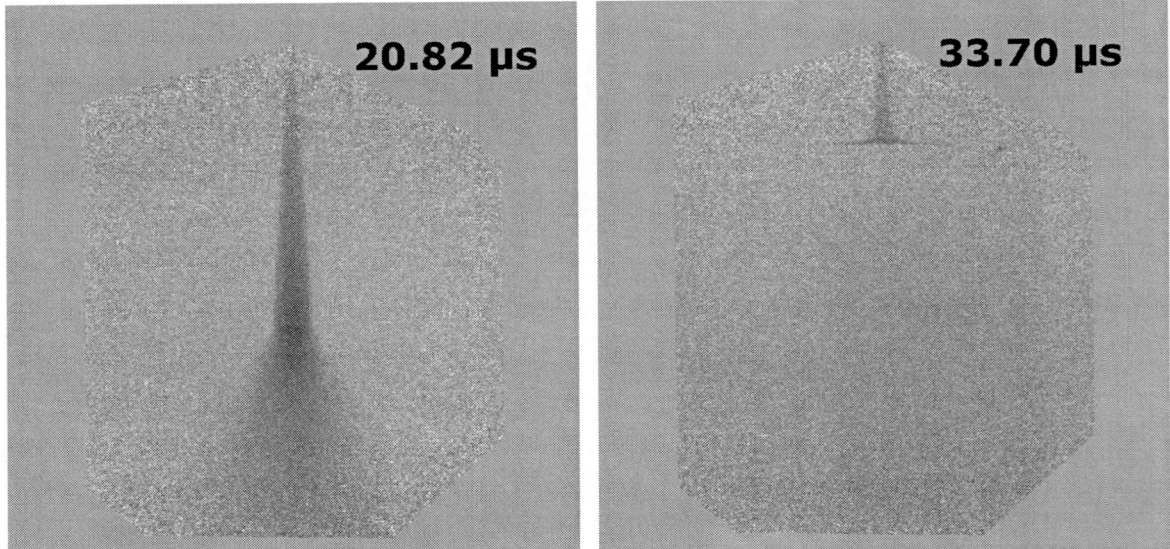

**FIGURE 4.** 12 mm Failure cone experiment PRad 0147 detonated 36 mm past the 9 mm failure diameter position of the cone to a diameter slightly less than 5.4 mm. The unburned explosive is visible in all the subsequent frames and no inert shocks were observed ahead of the material discontinuity between burned and unburned explosive.

In Fig. 4, two of the nine images from the 12 mm failure cone experiment are shown. In this case the charge detonated 36 mm past the 9 mm failure diameter position of the cone and ultimately failed at a diameter between 5.4 and 4.6 mm. The image shows regions where unburned explosive remains for over 4 µs.

Shown in Fig. 5 are the trajectories of the center of the detonation wave. Both experiments show an abrupt change to a much lower velocity. Since this velocity is much lower than shock velocity in the unreacted explosive, it is clear that any subsequent shock waves after the detonation failed were not observable in the radiography. The diameter effect curve indicates that a 25-mm diameter cylindrical charge would propagate at 7.59 mm/µs, while a 12-mm diameter charge would propagate at 7.46 mm/µs. It is interesting that in both cones, the detonation wave propagated at the same velocity as the initial diameter of the cone with no indication of velocity decay due to the changing diameter of the cone.

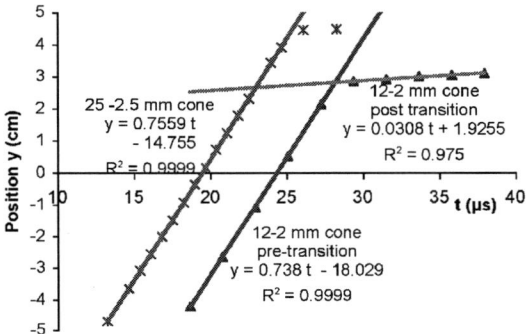

**Figure 5.** Detonation wave trajectories from the 25 - 2.5 mm and 12-2 mm diameter truncated cone experiments (× and ▲ respectively).

## CONCLUSIONS

The analysis of the detonation wave made from the radiographs presented here are based on the measurement of velocity along the center of the charge. The fact that the detonation velocity does not change until it abruptly changes in the 0.7 µs interframe time implies that the boundary condition is not influencing the wave velocity on the central streamlines in a continuous manner. The transverse wave velocity, which propagates the boundary information to the center of the charge, does not allow for the velocity to slowly lag the quasi-steady velocity, but once it arrives the detonation is quenched. The 33.7 µs image of Fig. 4 shows an interesting detail of unburned explosive feathering out into a thin layer at the point of failure. Insufficient resolution exists on the current radiographs to see the evolution of the outside edge of the experiment, but one possibility is that failure begins on the outside and works to the center, leaving a cone of unreacted material which has

been pushed into this thin layer by the explosive products behind it.

Further experiments are planned where the failure region is positioned into the higher resolution region of the radiograph, with the ability to resolve more details on the outer portions of the cone. Measurement of the detonation velocity on the outside of the charge and better imaging of the unburned region will illuminate the process by which failure occurs in conical charges.

## ACKNOWLEDGEMENTS

Larry Hill and Charles Mader are also interested in the topics of conical and wedge charges. Our interaction through the High Explosive Working Group at LANL as well as through personnel communications has been encouraging and challenging.

Failure cones are challenging to make and without the cooperation of our machinist, Paul Chapman, they would not have been possible. Jim Faulkner designed and procured shot hardware and helped in fielding of the shots.

This work was supported by the Department of Energy under contract W-7405-ENG-36 to the University of California and we gratefully acknowledge their support.

## REFERENCES

1. Campbell, A. W., and Engelke, R., "The Diameter Effect in High-Density Heterogeneous Explosives", The Sixth International Symposium on Detonation, White Oak, MD, 1976, pp. 642-652.
2. Urizar, Manuel J., Peterson, Suzanne W., and Smith, Louis C., "Detonation Sensitivity Tests", LASL Report, LA-7193-MS, 1978.
3. Ramsay, John B.,"Effect of Confinement on Failure in 95 TATB/5 KEL-F", The Eighth International Symposium on Detonation, Albuquerque, New Mexico, 1985, pp. 372-379.
4. Asay B. W, and McAfee, J. M., "Temperature Effects on Failure Thickness and the DDT in PBX 9502", The Tenth International Symposium on Detonation, Boston, MA, 1993, pp. 485-489.
5. King, N. S. P., Ables, E., Adams, K., et al., Nuclear Instruments & Methods in Physics Research A **424**, 84 (1999).

CP845, *Shock Compression of Condensed Matter - 2005*,
edited by M. D. Furnish, M. Elert, T. P. Russell, and C. T. White
© 2006 American Institute of Physics 0-7354-0341-4/06/$23.00

# CRITICAL CONDITIONS FOR IGNITION OF ALUMINUM PARTICLES IN CYLINDRICAL EXPLOSIVE CHARGES

David L. Frost[1], Samuel Goroshin[1], Jeff Levine[1], Robert Ripley[2], & Fan Zhang[3]

[1]*McGill Univ., Mech. Eng. Dep't., 817 Sherbrooke St. W., Montreal, Quebec, Canada H3A 2K6*
[2]*Martec Ltd., 1888 Brunswick St. Suite 400, Halifax, Nova Scotia, Canada B3J 3J8*
[3]*DRDC - Suffield, PO Box 4000, Stn Main, Medicine Hat, Alberta, Canada T1A 8K6*

**Abstract.** The critical conditions for the ignition of spherical aluminium particles dispersed during the detonation of long cylindrical explosive charges have been investigated experimentally. The charges consist of packed beds of aluminium particles, ranging in size from 3 – 114 μm in diameter, and saturated with sensitized liquid nitromethane (NM). The ignition conditions depend on both the charge and particle diameters with the most reactive particles corresponding to an intermediate size (~54 μm dia). With increasing charge diameter, three particle reaction regimes are observed: i) sub-critical (no particle reaction), ii) near-critical (reaction at isolated spots, or radial bands or rings), and iii) super-critical (continuous reaction of the particle cloud). To monitor the onset of aluminum oxidation, visible radiation from the charge is recorded, through a slit, with a line spectrometer.

**Keywords:** Aluminum particle ignition, combustion, metalized explosive
**PACS:** 47.40.-x, 47.40.Nm.

## INTRODUCTION

The effect of metallic particle additives on the detonation properties of liquid explosives has been investigated by a number of researchers [1-3]. Typically, adding particles reduces the detonation velocity and pressure due to heat and momentum transfer to the particles, although the critical charge diameter for detonation failure may decrease or increase, depending on the competing effects of the sensitization due to the formation of hot-spots and desensitization due to the dilution of the explosive by the added mass.

For a liquid explosive charge containing metallic particles, a second critical diameter can be identified: the critical charge diameter for which the particles ignite and react within the combustion products (CDPI). Above the CDPI, the residence time of the particles in the hot detonation products is sufficient to overcome the quenching effect of the unsteady expansion of the products such that the particles react. The CDPI will depend on the particle size and morphology as well as the particle material and oxidizing gases present. It should be noted that due to the high relative velocity between the particles and the hot combustion product gases, the particle ignition/reaction processes will be very different from the classical problem of metal particle ignition and reaction in a quiescent gas studied extensively since the 1960's.

The existence of a CDPI was observed in an earlier study [4] for cylindrical charges contained magnesium particles saturated with sensitized NM, and found to be a strong function of particle size. In the present work, the earlier experiments have been extended to the case of aluminum particles to determine the influence of particle material on the onset of particle reaction. Interpretation of the results is assisted with computations with a multiphase model.

## EXPERIMENTAL

The charges consisted of 122-cm long glass tubes, with inner diameters ranging from 9 to 74 mm. The charges contained packed beds of spherical aluminum particles (Valimet, CA) ranging from 3 – 114 µm in diameter, and saturated with NM sensitized with 10% by weight Triethylamine. The detonation was initiated in a 10-cm high section of liquid at the top of the charge with a detonator and a 10-g C4 booster.

The optical diagnostics used in the present investigation consisted of a high-speed Phantom VII video camera, a 3-color pyrometer and a line spectrometer. To spatially resolve the light emission from the detonation front and combustion products, the pyrometer and spectrometer viewed a segment of the charge, 70 cm from the top of the charge, through a horizontal 2.5-cm slit constructed with black Bristol-board and located 60 cm from the charge.

Light was collected for the pyrometer and line spectrometer using two conventional 35-mm cameras, located 60 m from the charge. A description of the pyrometer and calibration procedure has been previously published [5]. The line spectrometer differs from a conventional broad-band spectrometer in that the light emission from only 3 wavelength bands was recorded. For the line spectrometer, the filter wavelengths were chosen to monitor the emission from gaseous aluminum oxide AlO which exhibits a characteristic emission spectrum. In particular, two of the filters were set at 450 nm and 568 nm to monitor the intensity of the continuous background spectrum. The third narrowband filter wavelength was chosen to be 488 nm which is close to the most intense AlO band (486.6 nm). By using a linear interpolation, the background continuous spectrum at 488 nm can be estimated ($I_2$) and compared with the peak intensity measured at 488 nm ($I_1 + I_2$). The ratio $I_1/I_2$ then gives a measure of the relative intensity of the AlO band and can be used to monitor the presence of aluminum oxidation.

## RESULTS

Tests were carried out with the following different Valimet aluminum particle size designations (with corresponding Microtrac particle distribution for the batch used): H-2 (3.0±1.5µm); H-5 (8±4µm); H-10 (13±10µm); H-15 (20±10µm); H-30 (36±16µm); H-50 (54±21µm); H-60 (63±21µm); and H-95 (114±40µm). For the smallest particles (H-2), charges with a diameter of 41 mm id or less failed to detonate. For the largest particles (H-95), charges of 19 mm id or larger detonated, but the dispersed aluminum particles never ignited. For example, Fig. 1 shows the inert dispersal of H-95 particles from a 19-mm dia charge.

**FIGURE 1.** Detonation of 19-mm dia charge and dispersal of H-95 Al particles; 114 µs between frames.

For H-50 particles, particle reaction was evident with two morphologies: i) a near-critical regime in which discontinuous particle reaction occurs in the conical dispersed particle cloud at isolated regions (in the shape of spots, bands, or rings, perhaps associated with instabilities that develop at the particle cloud surface) which grow and eventually coalesce, or ii) a super-critical regime in which continuous reaction of the particle cloud occurs. These two regimes are illustrated in Fig. 2.

A summary of the dependence of the particle reaction behavior on charge and particle size is shown in Fig. 3. The boundary between the sub-critical (no reaction) and near-critical regimes is a U-shaped curve. In this figure, the H-2 (3 µm) and H-5 (8 µm) particles appear slightly more reactive than the larger H-10 and H-15 particles; this may be due to the fact that the particle mass fraction in the charge differs: 73±1% for H-2 and H-5 in comparison with 77±1% for all the larger particles.

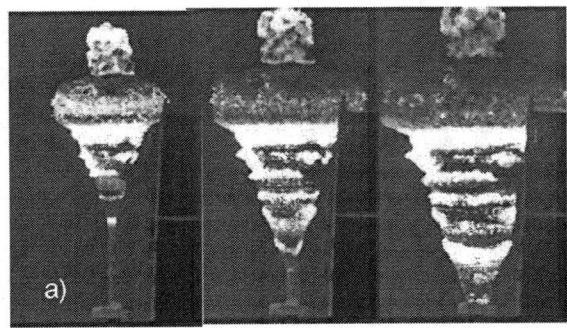

**FIGURE 2.** Particle reaction morphology for charges with H-50 particles: a) discontinuous reaction bands (or rings) for 34 mm dia tube; time between frames 57 μs, b) continuous particle reaction zone for 74 mm dia tube; time between frames 80 μs.

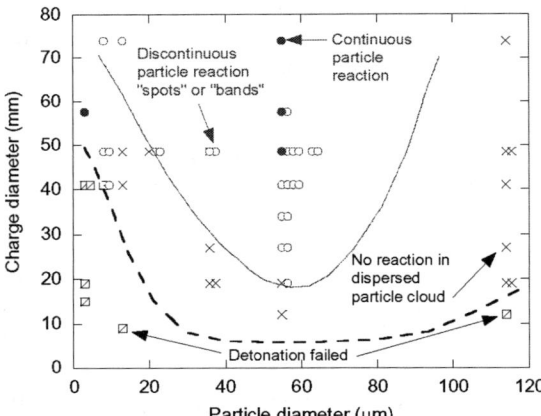

**FIGURE 3.** Dependence of particle reactivity on charge and particle diameter.

For the near-critical case of discontinuous particle reaction, there is some evidence from the line spectrometry that even in the "dark" zones,

there is some aluminum oxidation taking place, which may be obscured by soot at the periphery of the expansion cone. For example, Fig. 4 shows the pyrometer intensity and ratio $I_1/I_2$ for H-60 particles in a 48.5 mm id tube in which the particle reaction is discontinuous.

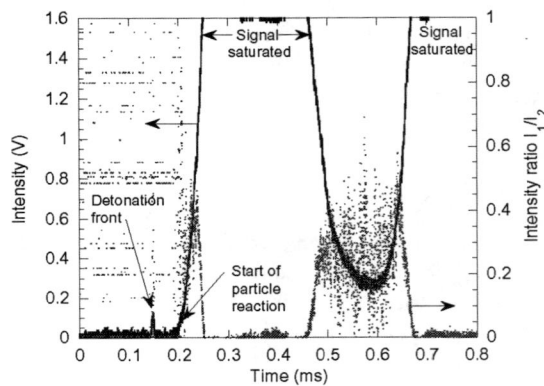

**FIGURE 4.** Light intensity and intensity ratio $I_1/I_2$ for 48.5 mm dia charge containing H-60 particles.

At a time of 0.15 ms, the detonation moves past the slit, then 50 μs later (corresponding to a distance of 250 mm behind the detonation which moves at a velocity ~ 5 mm/μs) intense particle reaction is visible. When a "dark" zone moves in front of the slit about 0.3 ms later, the overall light intensity drops, but the $I_1/I_2$ ratio at this point is about 0.3 which indicates the presence of gaseous aluminum oxides.

## DISCUSSION

For the largest particles (114 μm), the residence time in the hot combustion products is not sufficient to induce particle reaction. However, the susceptibility for particle reaction is not solely dependent on particle size, given that an intermediate particle size reacted with the smallest tube diameter. Hence, the fluid mechanics of the particle dispersal process and subsequent mixing with the surrounding air must also be relevant to the particle reaction dynamics. The initiation of discontinuous particle reaction sites appears to be correlated with perturbations that develop on the particle cone surface (see Fig. 2a) which would promote mixing with the nearby air which remains relatively hot due to the shock passage.

Results from the multiphase code calculations are shown in Figs. 5 and 6 for 54 µm particles in a 74 mm dia charge. Fig. 5 shows that the highest fluid temperatures are attained immediately behind the detonation front and behind the conical blast wave that is transmitted into the surrounding air. The expansion cooling of the products behind the detonation front is also evident.

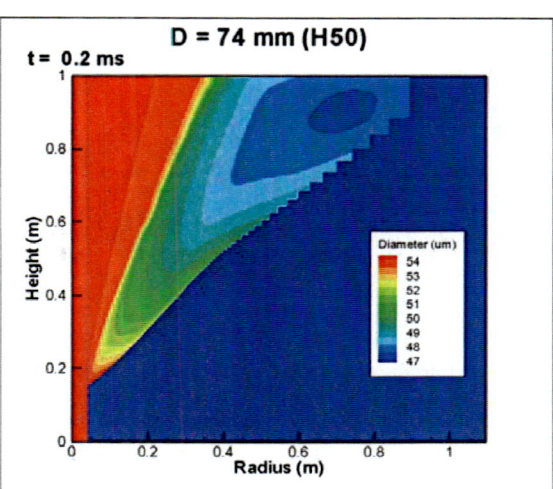

FIGURE 6. Variation in particle size due to burning during detonation of 74-mm dia charge with 54 µm particles. The particles in the central region remain unburnt.

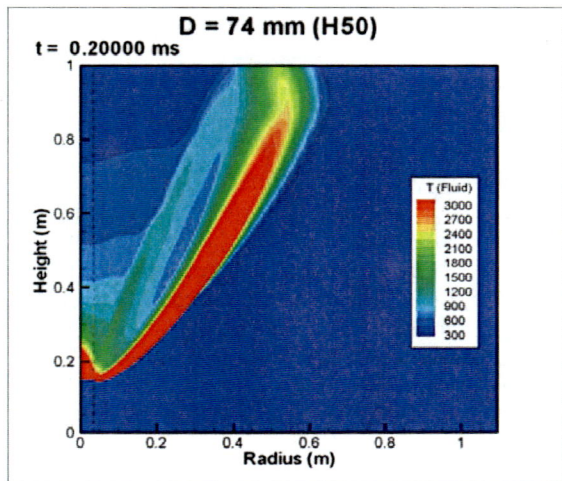

FIGURE 5. Fluid temperature during detonation of 1.1-m long, 74-mm dia charge containing H-50 (54 µm) particles. The detonation front, moving from top to bottom, is near the bottom of the charge at the time shown (0.2 ms).

Fig. 6 shows the variation of particle size due to the reaction of the particles. When the particles reach a threshold temperature (the Al melting point of 933 K is used), reaction of the Al particles is initiated. Note from Fig. 6, that the burning particles are concentrated near the outer region of the conical particle cloud (some of which have penetrated the shock front), where the particle number density is low and the fluid temperature is high, whereas the particles in the central core remain unburnt. Similar calculations with a 19-mm dia charge exhibit negligible particle reaction due to the more rapid expansion cooling of the combustion products, consistent with experimental observations.

## ACKNOWLEDGEMENTS

The authors gratefully acknowledge C. Ornthanalai, S. Janidlo, M. Cairns, F. Jouot, and the Field Operation Section for their assistance in the performance of the experimental trials and R. Lynde and S. Trebble for high-speed photography support. This work was funded partially by the Advanced Energetics Program of DTRA.

## REFERENCES

1. Kato, Y. and Brochet, C., Proc. 6th Symp. on Detonation, 124, 1976.
2. Lee, J.J., Frost, D.L., Lee, J.H.S., and Dremin, A., Shock Waves 5, **115**, 1995.
3. Haskins, P.J., Cook, M.D., and Briggs, R.I., Proc. 12th APS Topical meeting on Shock Compression of Condensed Matter, 890, 2001.
4. Frost, D. L., Zhang, F., Murray, S., and McCahan, S., Proc. 12th Symp. on Detonation, San Diego, 2002.
5. Goroshin, S., Frost, D. L., Levine, J., Yoshinaka, A., and Zhang, F., to appear in J. of Propellants, Explosives and Pyrotechnics, 2005.

CP845, *Shock Compression of Condensed Matter - 2005,*
edited by M. D. Furnish, M. Elert, T. P. Russell, and C. T. White
© 2006 American Institute of Physics 0-7354-0341-4/06/$23.00

# EXPLOSIVELY DRIVEN COMBUSTION OF SHOCK-DISPERSED FUELS

## P. Neuwald

*Fraunhofer-Institut für Kurzzeitdynamik, Ernst-Mach-Institut, Eckerstr. 4, 79104 Freiburg, Germany*

**Abstract.** The paper presents small-scale experiments with 1-g charges that explore the topic of post-detonation energy release due to the combustion of explosively dispersed fuels in the ambient air. To this end we have designed a new prototype small-scale charge, called Shock-Dispersed Fuel (SDF) charge. It consists of a lightweight, small paper cylinder filled with about one gram of a combustible powder (e.g., flake aluminum) surrounding a spherical PETN booster of 0.5 g. We have tested the SDF charges in a number of different environments, realized as closed steel vessels of simple geometry (barometric bombs). Three of the bombs vary in volume (6.6 l, 21.2 l and 40.5 l), while their aspect ratio L/D is kept constant at about 1. Two further bombs are comparable to the smallest bomb in volume (6.3 l), but provide different aspect ratios: L/D = 4.6 and 12.5. In addition, we have also performed tests in a tunnel-model with an L/D = 37.5. Our basic goal is to assess the performance of the charges by means of the combustion-related pressure built-up. Thus we contrast experiments on SDF charges in air with tests in nitrogen, to inhibit combustion, and with tests on conventional charges. Experiments and theoretical estimates on the expected overpressure allow one to formulate various indicators of the combustion effectiveness. For SDF charges these indicate that the combustion effectiveness decreases with increasing volume of the barometric bomb, and also with increasing aspect ratio at constant volume. This bears importance to the performance of SDF charges in tunnel environments. The performance losses reflect – at least in part – geometry-specific constraints on the mixing between fuel and air.

**Keywords:** After-burning, combustion, TNT, flake aluminum.
**PACS:** 47.40.-x, 47.40.Nm, 82.40.Fp.

## BACKGROUND: A BRIEF HISTORY OF EMI SMALL-SCALE EXPERIMENTS

When the Fraunhofer Institute for High Speed Dynamics (Ernst-Mach-Institut) was founded in 1959, shock wave propagation and reflection phenomena were among the institute's central research topics. The Experimental Fluid Dynamics Group contributed numerous fundamental studies, which were mainly performed in gas-driven shock tubes [1]. This experimental approach changed to some extent, when we were asked in the early eighties whether laboratory scale experiments on the blast wave phenomena were feasible. In response the institute developed a method to reliably detonate small PETN charges with masses ranging from 0.2 gram to 1.5 gram. Initially these charges were utilized to study height-of-burst effects; later on they became a means to investigate the blast loading of structures and buildings at model scales ranging from 1:10 to 1:200 [2]. In the late nineties we again adopted new aspects to our research; a change reflecting that the interest of the blast wave community was shifting from the loads caused by ideal explosives towards additional effects from the after-burning of fuel-rich explosives (e.g., TNT).

Currently the deliberate exploitation of such a post-detonation energy release is widely discussed; a topic of common interest to both the design specialist of energetic materials and the blast load analyst.

## SMALL-SCALE CHARGES

The basis of the two charge types discussed in this paper are custom-made spherical small-scale PETN-charges. They consist of nearly pure PETN at a density of about 1 g/cm³. Two electrical ignition wires are embedded. A thin resistance wire at the top bridges the small gap between these wires. Care is taken that the resistance wire is located in the center of the explosive sphere. A high voltage discharge explodes the bridge wire. This explosion creates an over-driven detonation through the charge.

Since PETN exhibits negligible after-burning EMI has designed a composite PETN/TNT charge to enhance the after-burning effects. A spherical PETN charge of 0.5 gram constitutes the core of the composite charge. Around this core an outer shell of TNT is built up. Its density is also around 1 g/cm³. Shell masses can be varied; our standard is a mass of 1 gram.

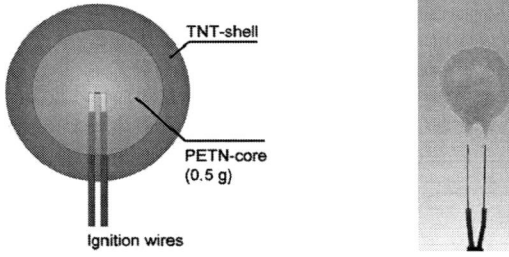

**Figure 1.** Schematic and photograph of the composite PETN/TNT charge.

In addition, we have designed shock-dispersed fuel (SDF) charges to study the post-detonation combustion of various solid fuels. In their standard form they consist of a lightweight paper cylinder with a height and diameter of 14 mm. Again a spherical PETN charge of 0.5 g is inserted into the center of the cylinder. It acts as the dispersing booster. The remaining volume (1.6 cm³) of the cylinder is filled with the fuel to be dispersed, usually in the form of a loosely packed powder.

For larger amounts of fuel we used a larger paper cylinder with a diameter and height of 17 mm.

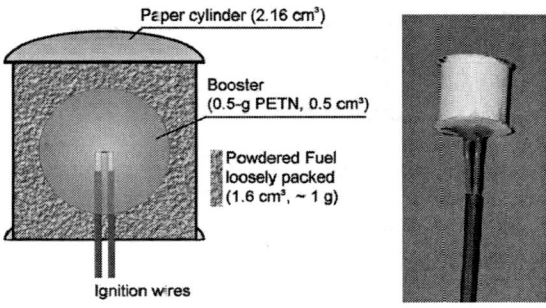

**Figure 2.** Schematic and photograph of the standard shock-dispersed fuel (SDF) charge.

We have tested a variety of fuels [3]; this paper however will focus on flake aluminum, which consistently gave the fastest and most complete fuel consumption. The flakes were rather thin (about 1 to 2 μm), while their lateral dimension varied between 20 to 200 μm. The standard SDF held 1 gram of the flake aluminum.

**Figure 3.** SEM image of the flake aluminum used as SDF fill.

**TABLE 1.** Thermodynamic properties of PETN, TNT and Al.

|      | $\Delta H_{Det}$ [kJ/g] | $\Delta H_{Comb}$ [kJ/g] | air consumption [liter/g] |
|------|------|------|------|
| PETN | 6.24 | 8.20 | 0.40 |
| TNT  | 4.61 | 15.04 | 2.75 |
| Al   | —    | 30.95 | 3.35 |

## EXPERIMENTS IN BAROMETRIC BOMBS

Roughly we can subdivide the energy release from non-ideal explosives into three phases: the detonation of the condensed explosive, anaerobic reactions in the hot, pressurized detonation products and combustion with ambient air. Since the combustion phase is typically much slower than the initial detonation, the most reliable effect of the combustion is the built-up of quasi-steady over-pressure in a closed vessel. Thus we performed our experiments in five different *barometric bomb calorimeters*: closed steel vessels of simple geometry, instrumented with a number of piezo-electric and piezo-resistive pressure gages mounted flush into the walls.

**TABLE 2.** Geometry of the barometric bombs.

|   | volume[l] | L/D | L[mm] | D[mm] | cross-section |
|---|---|---|---|---|---|
| A | 6.6 | 1.05 | 210 | 200 | circular |
| B | 21.2 | 1.00 | 300 | 300 | circular |
| C | 40.5 | 1.03 | 379 | 369 | circular |
| D | 6.3 | 4.63 | 555 | 120 | circular |
| E | 6.3 | 12.50 | 1000 | 80 | square |

### Experimental Procedure and Data Analysis

The charges were detonated at the center of the barometric bombs, which were filled with either air at ambient conditions or with nitrogen to inhibit the combustion phase. For each charge type and each barometric bomb at least five identical tests were performed. The pressure gages moni-tored the system for about 40 ms, yielding 8 to 12 time-resolved pressure histories at various locations. These pressure histories are initially dominated by the strong primary blast front and its multiple reflections at the vessel walls. In addition, the combustion of the fuel increases the quasi-steady pressure in the vessel. At least for fast reacting fuels the maximum quasi-steady overpressure is attained within just a couple of blast reflection cycles. To extract the quasi-steady part of the overpressure history we thus had to low-pass filter the data at a cut-off frequency low enough to eliminate the wave structure from the signal (0.5 kHz or lower, depending on the geometry of the barometric bomb). The smoothed curves were inspected for their maximum, which

was identified with the final quasi-steady overpressure attained in the test. Presented here are the mean results from all available gages, averaged over the number of test repetitions.

### Assessment of the Combustion Performance

The experiments were supplemented by thermodynamic equilibrium calculations, assuming an adiabatic explosion of the barometric bomb's content at constant volume (CVE). If the atmosphere in the barometric bomb is set to nitrogen, the result of the calculation is an estimate of the quasi-steady overpressure due to the initial detonation. Setting the atmosphere to air yields an estimate of the maximally possible overpressure for the case of complete combustion of the fuel.

The CVE calculations were used to define two performance characteristics for the charges [4].

The first is a pressure ratio or relative performance given by:

$$\Pi = \frac{p_{Exp/Air}}{p_{Ref/N2}} \qquad (1)$$

where the reference value is the overpressure from the detonation of a conventional charge without energy release due to after-burning. Here we have based the reference value on the CVE calculations for the composite PETN/TNT charge in nitrogen. Please note that the maximum possible relative performance (predicted by CVE calculations) for a given charge is not a constant, but depends on the volume of the barometric bomb.

The second characteristic, the combustion effectiveness or completeness (in terms of the quasi-steady overpressure) is defined as the ratio of the experimentally observed excess pressure from combustion to its theoretically predicted counter-part:

$$\delta = \frac{p_{Exp/Air} - p_{CVE/N2}}{p_{CVE/Air} - p_{CVE/N2}} \qquad (2)$$

The combustion effectiveness $\delta$ is related to the mass-fraction of fuel actually burned [6].

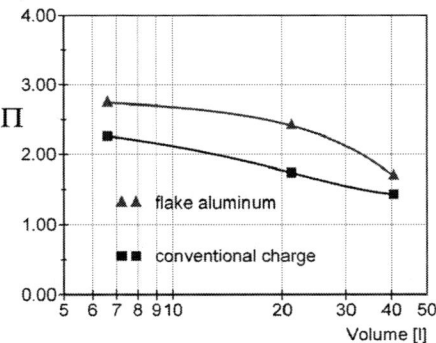

**Figure 4.** Relative performance of shock-dispersed flake aluminum and of the composite PETN/TNT charge as a function of barometric bomb volume.

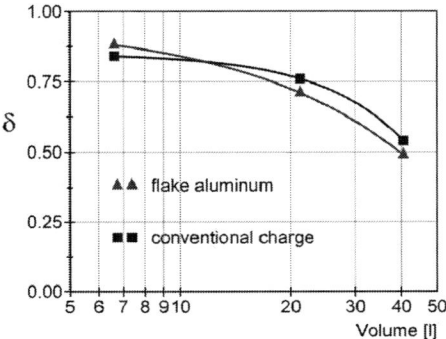

**Figure 5.** Combustion completeness of shock-dispersed flake aluminum and of the composite PETN/TNT charge as a function of barometric bomb volume.

### Results: Dependence on Volume

The experimentally observed performance characteristics as a function of barometric bomb volume are depicted in Figures 4 and 5. At 6.6 liter and 21.2 liter a very significant portion of the overpressure is due to combustion; at 40.5 liter the yield factor is still about 1.5 to 1.8. In terms of the quasi-steady overpressure 1 gram flake aluminum has a larger relative performance than 1 gram TNT. The combustion efficiency is very similar for TNT and for the flake aluminum. For 6.6 liter it is around $0.85 - 0.88$, for 21.2 liter around 0.75, dropping to around 0.5 for 40.5 liter. The decrease of performance with increasing volume reflects that in smaller volumes the blast wave reflections stir up the detonation products early and enhance the mixing with the ambient air more efficiently.

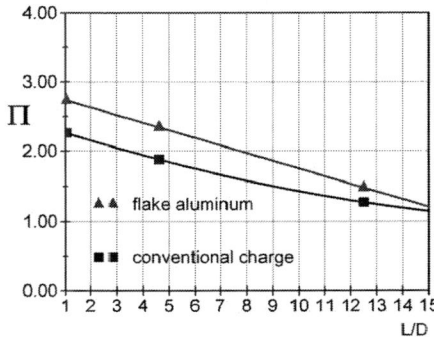

**Figure 6.** Relative performance of shock-dispersed flake aluminum and of the composite PETN/TNT charge as a function of the aspect ratio L/D.

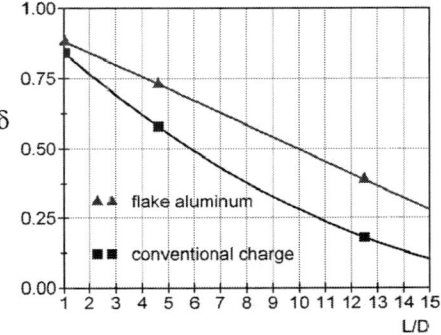

**Figure 7.** Combustion completeness of shock-dispersed flake aluminum and of the composite PETN/TNT charge as a function of the aspect ratio L/D.

### Results: Dependence on Aspect Ratio L/D

A more dramatic decrease of the combustion performance was observed for an increase of the aspect ratio L/D (Figs. 6 and 7). Even though the volume is near to that of optimum performance in vessels with an aspect ratio around 1, the combustion efficiency of shock-dispersed flake aluminum drops down to 0.4 for an aspect ratio of 12.5. The combustion efficiency of the TNT products is affected even more (around 0.2 for L/D = 12.5). In long and narrow geometries the blast propagation soon becomes essentially one-dimensional. Thus the wave reflection cycle depends on the length of the barometric bomb and is of comparatively low frequency and hence becomes inefficient in promoting the mixing. Similarly the detonation products soon fill a section of the barometric bomb and the mixing surface becomes constrained to virtually twice the cross-section of the bomb. In effect, the combustion products soon separate the remaining fuel from the ambient air, rendering further combustion extremely slow.

In addition, the barometric bomb is not an adiabatic system. Heat transfer to the walls occurs at the expense of interior temperature. This effect can be stronger in more narrow geometries.

The better performance of flake aluminum in comparison to the TNT products might be due to the fact that the aluminum particles can be accelerated to velocities in excess of the expansion of the detonation products cloud. Thus they can be expelled in to the ambient air, where they burn, a process that one might call "ballistic" mixing.

The dependence on the aspect ratio bears importance to the performance of after-burning and the combustion of shock-dispersed fuels in tunnel environments, which will be inspected in the next section.

## EXPERIMENTS IN A LONG TUNNEL SECTION

In addition to the test described above we also performed experiments in geometries resembling long, closed tunnel sections. The cross-section of these tunnel sections were squares of 8 cm x 8 cm, corresponding to the barometric bomb E, the length of the tunnel section could be varied from 1m to 6 m. Here we will report about some experiments in a 3-m long section, closed on both ends, with the charges being detonated near one end of the tunnel.

Since it takes about 8 ms for the primary blast wave to travel down the tunnel and back again, low-pass filtering of the pressure signals would have to be at an overly low cut-off frequency. Thus it is not easy to establish a reliable value for the quasi-steady overpressure that builds up in the tunnel section.

Instead, we monitored the pressure signals from side-on pressure gages along the tunnel wall. Figure 8 shows the peak overpressure of the primary blast wave down the tunnel length. The lowest curve holds for a spherical PETN charge of 0.5 gram. The sudden jump at x = 650 mm occurs when the reflection at the closed tunnel front

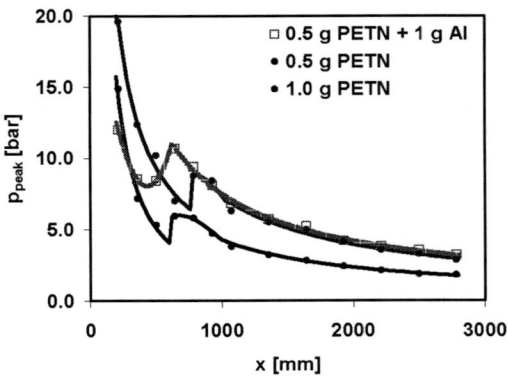

**Figure 8.** Side-on peak overpressure of the primary blast front for three cases: spherical charge of 0.5 g PETN, spherical charge of 1 g PETN and SDF charge with 1 g flake aluminum. Charges located at x = 80 mm, tunnel diameter 80 mm, tunnel length 3000 mm.

coalesces with the primary blast wave propagating down the tunnel. Close to the charge location an SDF charge filled with 1 gram flake aluminum generates comparable peak overpressures. However, for locations beyond x = 360 mm the peak overpressure from the SDF charge exceeds that from the 0.5-g PETN charge and corresponds to the peak from a 1-g spherical PETN charge. This demonstrates that initially the combustion of the flake aluminum actually couples energy into the blast front, though no further enhancement of the peak overpressure is observed at larger distances. This is reflected in the pressure loads at the tunnel end wall. In terms of reflected peak overpressure the SDF charge was found to be equivalent to a PETN charge of about 1 gram. In terms of the overpressure impulse the equivalent amount of PETN is slightly larger, namely about 1.2 gram.

To study the dynamics in the tunnel environment somewhat further, we mounted a number of photo-diodes into the tunnel wall. 22 mm long apertures, 2 mm in diameter, constrained their field of view to a narrow line normal to the tunnel axis. In addition, we have designed a simple point probe for the electro-conductivity of the detonation products cloud [5], which was installed in various tests at different locations along the tunnel axis. The probe voltage is a non-linear function of the electro-conductivity and around 0.7 V for a conductivity of $10^{-3}$ S/m, around 4.2 V for $10^{-2}$ S/m and around 8.4 V for 0.1 S/m. The probe is saturated at a voltage of 9.4 V, which is attained for conductivities in the order of 1 S/m and above.

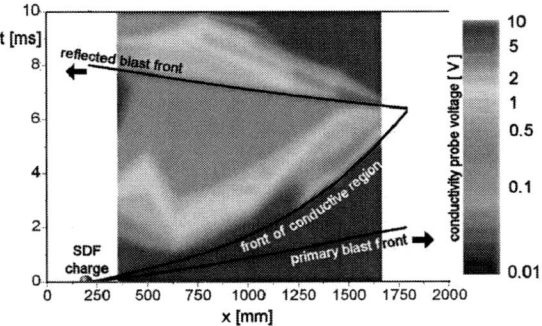

**Figure 9.** Approximate wave diagram for the electro-conductive region evolving from the detonation of an aluminum-filled SDF charge. Conductivity in Volt at conductivity probe. Tunnel length 3000 mm, charge location x = 214 mm.

Figures 9 depicts a wave diagram constructed from the conductivity measurements. Please note that the full length of the tunnel is 3000 mm, while the figure only shows the region from x = 0 to x = 2000 mm. The wave diagram shows the primary blast front, the blast reflection at the tunnel end and the conductive region. It is apparent how the expansion of the conductive region rapidly slows down, so that the conductive front falls back behind the primary blast front. At about 3.5 ms the primary blast front is reflected at the tunnel end-wall and returns back up the tunnel. Another 2.5 ms later the reflected wave impinges onto the conductive region at about x = 1800 mm and sweeps it backwards. The region is recompressed and the conductivity reinforced. The same phenomena become apparent in a the corresponding wave diagram for the luminous region, constructed from the photo-diode measurements. The general appearance of the luminous region and the conductive region is quite similar; the conductive region however has a better defined, steeply rising front. If we identify the regions of intense luminance or conductivity with regions of possible combustion activity, the increasing gap to the blast front adds plausibility to the fact that combustion-related energy reinforces the blast only in the initial stages, i.e., when the gap is still small.

**SUMMARY**

Small-scale experiments as described above are a versatile means for parametric studies on basic aspects of the combustion-related energy release. Since they are easily repeated, they also provide some insight into the r.m.s. variations of the results, which can be significant in any scenario involving turbulent flow fields. Thus the experiments are also well suited for comparison to numerical simulations addressing the problem of explosively driven combustion [7, 8].

**REFERENCES**

1. Reichenbach, H., "In the footsteps of Ernst-Mach – A historical review of shock wave research at the Ernst-Mach-Institut", Shock Waves 2, pp. 65-79, 1992

2. Reichenbach, H., Neuwald, P., Kuhl, A. L., "Role of Precision Laboratory Experiments in the Understanding of Large-Scale Blast Phenomena", Proc. 17th Int. Symp. Military Aspects of Blast and Shock, Las Vegas, 2002

3. Neuwald, P., Reichenbach, H., Kuhl, A. L., "Shock-Dispersed-Fuel Charges – Combustion in Chambers and Tunnels", Proc. 34th Int. ICT-Conference: Energetic Materials – Reactions of Propellants, Explosives and Pyrotechnics, Karlsruhe, 2003

4. Neuwald, P., Reichenbach, H., Kuhl, A. L., "Shock-Dispersed Flake Aluminum – Performance in Environments of Different Geometries", Proc. 36th Int. ICT-Conference: Energetic Materials – Performance and Safety, Karlsruhe, 2005

5. Neuwald, P., Reichenbach, H., Kuhl, A. L., "Combustion of Shock-Dispersed Flake Aluminum in a Long Tunnel Section", to be published in: Proc. 20th Int. Colloq. on the Dynamics of Explosions and Reactive Systems, Montreal, 2005

6. Kuhl, A. L., "Thermodynamics of Combustion of TNT Products in a Chamber", Proc. 5th Int. Seminar on Flame Structure, Novosibirsk, 2005

7. Bell, J. B. et al., "Numerical Simulation of the Combustion of PETN/TNT Products with Air in Closed Chambers", to be published in: Proc. 20th Int. Colloq. on the Dynamics of Explosions and Reactive Systems, Montreal, 2005

8. Khasainov, B. et al., "Model of Non-premixed Combustion of Aluminium–Air Mixtures", to be published in: Proc. 14th APS Conf. Shock Compression of Condensed Matter, Baltimore, 2005

CP845, *Shock Compression of Condensed Matter - 2005,*
edited by M. D. Furnish, M. Elert, T. P. Russell, and C. T. White
© 2006 American Institute of Physics 0-7354-0341-4/06/$23.00

# SEMICONDUCTOR MODEL OF DETONATION

## K.F. Grebenkin

*Russian Federal Nuclear Center- Institute of Technical Physics,
Snezhinsk, Chelyabinsk region, 456770, Russia.*

**Abstract.** A review of a semiconductor model of detonation is presented. The model is based on the assumption that energy is transferred from the hot spots by electron thermal conductivity. Evaluation of the crystalline TATB band gap is given, and recent measurements of electric conductivity in the shock compressed TATB are discussed. Estimations of TATB reaction time and the rate of the burning wave propagation from the hot spots are given. Some experiments where the effects of conductivity, caused by electron excitation in shock compressed crystalline explosives, could become apparent, are discussed.

**Keywords:** Detonation, TATB, band gap, hot spots, thermal conductivity.
**PACS:** 47.40.-x, 62.50.+p, 72.80.Le.

## INTRODUCTION

A concept of the hot spot mechanism of heterogeneous explosives detonation initiation results in a conclusion that the macrokinetic rate of the HE decomposition and the release of energy is proportional to the rate of the burning wave propagation from the hot spots – D. The last may be estimated as follows [1]

$$D \approx \sqrt{\chi_m / \tau_m},$$

where $\chi_m$ and $\tau_m$ are the thermal diffusivity coefficient, and the HE reaction time at a temperature close to that of the explosion products, respectively.

There is no direct experimental data on the thermal conductivity and chemical reaction rates for the burning wave propagation from the hot spots. Therefore, realistic values of the thermal conductivity coefficient, as well as the HE reaction time in the burning wave, are required to develop a physical model of the detonation initiation.

Several years ago the semiconductor model of detonation [2,3] was suggested. This model is based

on the assumption that energy in the burning wave, propagating from the hot spots, is transferred by electron thermal conductivity.

A review on the semiconductor model of detonation is given in the report.

### Band gap of TATB molecular crystal

When discussing the possibility of electron energy transfer from hot spots, a question concerning the crystalline explosives band gap should first be clarified. In papers [2,3] the band gap was considered an effective parameter of the temperature-based macrokinetic rate equation that was written as

$$\frac{dF}{dt} \sim D \sim \sqrt{\chi(T)} \sim \exp\left(-\frac{\varepsilon_g}{4kT}\right), \qquad (1)$$

where $F$ is the burn fraction of HE, $k$ is the Boltzman constant, and $\varepsilon_g$ is the effective band gap.

A value near 1.7 eV was chosen [2,3] to reproduce the results of the experiments on detonation initiation by shock waves in the hydrocode calculations.

Later, the TATB molecular crystal band gap was evaluated by means of the density functional method as $\varepsilon_g = (3 \pm 1)$ eV [4] at normal conditions, with a reduction to $\sim (1.5 - 2)$ eV for the high pressures typical for the shock wave initiation.

These estimates of the TATB band gap have been confirmed by other authors [5,6].

### Reaction time in the burning wave

There is no experimental data on HE reaction times at the burning wave conditions (2000-4000 K, 10 – 30 GPa). Extrapolation of the reaction rates measured at lower pressures and temperatures may result in severe errors. Molecular dynamics calculations [7] have shown that the reaction time in the burning wave is on the order of about ten picoseconds for TATB. This estimate was used when computer modeling the rate of the burning wave propagation from the hot spots.

### Measurements of the shock compressed TATB electro conductivity.

To verify the hypothesis that the shock wave transforms crystalline TATB into a semiconductor state it was suggested [3,7] the electro conductivity of the shock compressed HE be measured. TATB is a good insulator at normal conditions. However, it was predicted [7], that its electro conductivity may increase up to $\sigma \sim (10^2 - 10^4)$ (om*m)$^{-1}$ at near-threshold values of the shock compression pressure of 10 - 15 GPa.

The effect of many orders of magnitude increase in TATB electro conductivity after the shock wave action has been confirmed in [8].

The experimental records (see fig. 1) were interpreted as 4 stages of the electro conductivity growth process: 1 – increase of crystalline HE electro conductivity immediately after the shock wave enters the sample, 2 – increase of a volume fraction of the highly conductive detonation products due to growing hot spots, 3 – slow increase of electro conductivity due to a rise in temperature caused by the growth of carbon clusters in the explosion products, 4 – decrease of the electro conductivity due to sample unloading.

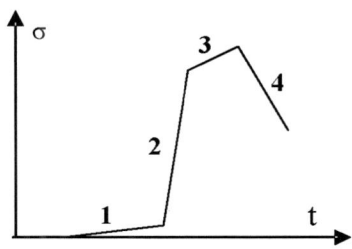

**Figure. 1.** Scheme of the experimental record of the shock compressed TATB electro conductivity dependence on time.

Typical values of the specific electro conductivity at the first stage were about 1 (om*m)$^{-1}$ at 10 – 15 GPa. Thus, TATB compression by shock waves results in an electro conductivity increase up to a level typical of semiconductors.

### The burning wave propagation rate.

The rate of the burning wave propagation from the hot spots can be evaluated from its obvious relation with the detonation initiation delay time, T, and the average distance between the hot spots, δ, as

$$T \approx \delta / D.$$

For typical values $\delta \approx 10$ mkm and $T \approx 0.1 - 1.0$ mks it gives $D \approx 10 - 100$ m/s.

Calculations of the rate of the burning wave propagation from the hot spots in the shock compressed TATB [9] were carried out based on the experimental data [8] that was used for evaluation of the electron thermal conductivity. Rate values near 10 - 100 m/sec were obtained in the calculations. This agrees with the above mentioned values expected from the detonation delay times.

A strong dependence of the burning wave rate on the crystalline TATB temperature beyond the hot spots was observed in the calculations. This may mean that the well-known experimental dependence of the detonation initiation delay time on the initiating shock wave intensity (Pop-plot) is a consequence of the strong dependence of the electron thermal conductivity of the shock compressed crystalline explosive on the temperature.

Certainly, experimental data on the microscopic processes occurring in the HE molecular crystals, including information on the kinetics of excitation and recombination of electrons as well as their mobility, the rate of electron and phonons relaxation etc., is needed in order to refine the modeling of the burning wave rate and its dependence on the initiating shock wave intensity.

## Interpretation of some experiments

There are some experiments that can be interpreted based on the concept of electron excitation to the conductivity zone of a crystalline HE. For example, shock wave compressed RDX was studied [10], and the conclusion was made that the shock wave transforms RDX into a semiconductor state.

Another example [11] is the visible light absorption observed after the shock wave coming through a PETN monocrystal, which is transparent for visible light at normal conditions. This may be a result of the shock induced excitation of electrons into the conductivity zone, as the visible light absorption observed in the experiments [11] may be caused by the intraband transitions of the conductivity electrons.

The high burning wave rate in the compressed HMX, measured experimentally [12], and its strong dependence on pressure, may be a consequence of the band gap narrowing caused by pressure. This effect must result in an increase of the electron concentration in the conductivity zone. Hence, the electron thermal conductivity will grow and, as a result, the burning wave rate will increase with rising pressure. Based on such an interpretation, it is possible to evaluate the HMX band gap pressure coefficient from the relation (1) as follows

$$\frac{d\varepsilon_g}{dP} \approx -4kT \cdot \frac{d\ln D}{dP}. \qquad (2)$$

This gives a value near - 0.08 eV/GPa for pressures over 5 GPa. This estimation is close to the result of calculation of the TATB band gap dependence on pressure [5]. It is proposed that TATB, HMX, and other HEs with nitro groups have similar zone structures [4].

By the way, the lower burning wave rate in TATB [13], compared to HMX at the same high pressure, could be explained by a lower TATB

explosion products temperature. This implies lower electron thermal conductivity in TATB and, according to the relation (1), a lower macrokinetic burning rate. So, the lower electron thermal conductivity due to the relatively low TATB explosion products temperature may be a fundamental reason why TATB is less sensitive than HMX or RDX.

Finally, let us discuss why the measured values of the specific electro conductivity of the shock compressed crystalline TATB [8] was much less than predicted [7]. This difference may be explained by the fact that in the estimations [7] the burning wave rate was determined by the electron thermal conductivity of the shock compressed bulk HE out of the hot spots. Actually, from the solution of the burning wave propagation in a media with non-linear thermal conductivity [1] it becomes clear that the burning wave rate is determined by a coefficient of thermal conductivity in the preheating zone, $\sigma_2$, having a temperature close to that of the explosion products, $T_2$. The conductivity in the preheating zone $\sigma_2$ may be estimated as

$$\sigma_2 = \sigma_1 \cdot \exp\left(-\frac{\varepsilon_g \cdot (T_1 - T_2)}{2kT_1 \cdot T_2}\right). \qquad (3)$$

Given that the specific electro conductivity of the bulk TATB beyond the hot spots was measured in the experiments as $\sigma_1 \sim 1$ $(om*m)^{-1}$, and $\varepsilon_g$=1.5 eV at $T_1$=600 - 700 K, Eq. (3) gives $\sigma_2$ $\sim 10^4$ $(om*m)^{-1}$, just the very thing that was expected in [7]. Based on the Videman-Frantz law, the electron thermal conductivity of crystalline TATB in the preheating zone can be evaluated as

$$\lambda_e \approx L \cdot T_2 \cdot \sigma_2 \approx 0.6 \, Wt / m \cdot K$$

where L – is the Lorentz constant. The phonon thermal conductivity is an order of magnitude less than $\lambda_e$ at such temperatures.

## CONCLUSIONS

The idea that electron thermal conductivity may play an important role in the detonation processes is not new. Previously it was discussed as relating to the process of energy transfer from the shock wave front into the non disturbed crystalline HE [14,15]. A specific concept for heterogeneous HE reaction initiation in the hot spots and the forthcoming burning wave propagation was not considered in the earlier works at all.

The principal feature of the semiconductor model of detonation is that the electron thermal conductivity is considered to be a function of the energy transfer from the hot spots. Electron transport may determine the rate of the burning wave propagation from hot spots. According to the relation (1) the rate of the burning wave is directly related to the macrokinetical rate of chemical reactions in heterogeneous explosives.

Some arguments supporting the semiconductor model have been presented in the given report. These include evaluation of the crystalline TATB band gap, its electro conductivity measurement after the shock wave action, and burning wave rate calculations. The model may open a way to explanation of some of the experimental data for the first time, including the reason for the low sensitivity of TATB. Based on the presented concept, we have suggested a temperature-based physical model of detonation for LX-17, the TATB-based composition [16].

That is why, we believe the hypothesis that electrons may transfer the energy from the hot spots is worth further detailed study.

The most urgent need is to ultimately clarify if the crystalline TATB band gaps can be reduced up to 1.5 – 2.0 ev at the detonation pressures. This could be performed by measuring the temperature dependence of TATB electro conductivity at heating up to 200 – 300 C and the static compression up to 10 - 20 GPa.

The main aim of the given report was to initiate such experiments.

## REFERENCES

1. Krishenik, P.M., Shkadinsky, K.G., "High-Temperature Thermal Front with Non-Linear Thermal Conductivity". Russian Academy of Science. Doklady (in Rus.). V. 392, no. 6, p. 761, 2003.
2. Grebenkin, K.F., "Physical Model of Detonation Initiation in Pressed Crystalline High Explosives". Technical Physics Letters. V. 24, p. 789, 1998.
3. Grebenkin, K.F., " Crystalline High Explosives as Semiconductors". Proceeding of Chelyabinsk Scientific Center (in Rus.). no. 2, p. 1, 1998.
http://www.sci.urc.ac.ru/news/1998_2/2-2-1.pdf
4. Grebenkin, K.F., Kutepov, A.L., "Band Gap Estimation for Triaminotrinitrobenzene Molecular Crystal by Means of Density Functional Method". Semiconductors. V. 34, p. 1161, 2000.
5. Wu, C.J., Yang, L.H., Fried, L.E., Quenneville, J., Martinez, T.J. "On the Role of Pressure Induced Metallization in Energetic Materials: Electronic Structure of Solid TATB Under Uniaxial Compression". Phys. Rev. B67, p. 235101, 2003.
6. Manaa, R. "Shear-induced Metallization of Triaminotrinitrobenzene Crystals". Chem. Phys. Lett. V. 83, no. 7, p. 1352, 2003.
7. Grebenkin, K.F., Zherebtsov, A.L., Kutepov, A.L. Popova, V.V. "On Experimental Verification of the Semiconductor Model of Detonation". Technical Physics. V. 47, no. 11, p. 1458, 2002.
8. Gorshkov, M.M., Grebenkin, K.F., Zaikin, V.T. et al. "Predetonation Conductivity of a TATB-Based Explosive". Technical Physics Letters. v. 30, no. 8, p. 631, 2004.
9. Grebenkin, K.F., Zherebtsov, A.L., Popova, V.V., Taranik, M.T. "Evaluation of the Burning Wave Propagation Rate from the Hot Spots in Detonating TATB". Proc. of V-th Khariton Scientific Talks, Sarov, Russia. 2003 , p. 189.
10. Chambers, G.P., Lee, R.G., Oxby, T.J. et al. "Electromagnetic Properties of Pre-Detonating Explosives". In Shock Compression of Condensed Matter-2001. Edited by Furish M.D., Thadhani N.N. and Horie Y. AIP. 2002, p. 894.
11. Gruzdkov, Y.A., Gupta, Y.M., Dick, J.J. "Time-resolved absorption spectroscopy in shocked PETN single crystals". In Shock Compression of Condensed Matter – 1999. Edited by Furish M.D., Thadhani N.N. and Horie Y. AIP. 2000. p. 929.
12. Esposito, A.P. et al. "Reaction Propagation Rates in HMX at High Pressure". Propellants, Explosives, Pyrotechnics. V. 28, no. 2, p. 83, 2003.
13. Foltz, M.F. Pressure Dependence of the Reaction Propagation Rate of TATB at High Pressure. Propellants, Explosives, Pyrotechnics. V. 18, p. 210, 1993.
14. Cook, M.A. "The Science of Industrial Explosives". Salt Lake City. 1974.
15. Williams, F.E. "Electronic States of Solid Explosives and Their Probable Role in Detonations" Adv. Chem. Phys. 21, 289, 1971.
16. Grebenkin, K.F., Zherebtsov, A.L., V.V., Taranik, M.T. et al. "Temperature-Based Macrokinetic of Shock Wave Initiation of Detonation". Abstr. of VII-th Khariton Scientific Talks, Sarov, Russia, p. 18, 2005.

CP845, *Shock Compression of Condensed Matter - 2005*,
edited by M. D. Furnish, M. Elert, T. P. Russell, and C. T. White
© 2006 American Institute of Physics 0-7354-0341-4/06/$23.00

# DETONATION WAVE PROFILE IN PBX 9501

## Ralph Menikoff

*Theoretical Division, MS-B214, Los Alamos National Laboratory, Los Alamos, NM 87545*

**Abstract.** Measurements of a CJ-detonation wave in PBX 9501 with a VISAR technique have shown a classical ZND profile for the reaction zone. This is compatible with one-dimensional simulations using realistic equations of state and an Arrhenius reaction rate fit to available data from other experiments. Moreover, the reaction zone width is less than the average grain size in the PBX. In contrast to initiation, which requires hot spots, the reaction rate from the bulk shock temperature is sufficiently high for propagating a detonation wave. This raises questions with burn models used for both ignition and propagation of detonation waves.

**Keywords:** PBX 9501, reaction zone profile, CJ detonation wave, ZND model
**PACS:** 47.40.-x,47.70.Fw

## INTRODUCTION

A planar detonation wave can be promptly initiated with a projectile from a gas gun. Several experimentalist have measured the wave profile using a VISAR technique. Here we study the underdriven or CJ-detonation wave profile in PBX 9501 — an HMX based plastic-bonded explosives pressed to within 1.5 % theoretical maximum denstiy — from the experiments by Gustavsen, Sheffield and Alcon [1, 2]. The experiments used two VISARs with different fringe constants in order to determine velocity jump at shock front. The estimated timing resolution is between 1 and 3 ns; the better resolution when velocity jump corresponds to an integer number of fringes.

The VISAR record displays an abrupt rise within the time resolution. Thus, despite the heterogeneities in the PBX and the laser spot size of many grains, the wave front is a shock. Moreover, the shape of the profile corresponds to a classical ZND reaction zone. We note that similar results have been found by Fedorov [3]. The measured ZND profile motivates us to run one-dimensional simulations with an Arrhenius reaction rate to compare with the VISAR data.

We begin in the next section with a discussion of equations of state for both the reactants and the prod-ucts of PBX 9501. Since chemical reaction rates are temperature sensitive, a complete EOS is needed. A key parameter for the shock temperature is the specific heat. Next we discuss the Arrhenius rate parameters. Some parameter sets for HMX in the literature and commonly used in simulations yield a rate at the von Neumann spike which is unreasonably large by several orders of magnitude. The next section describes the simulations. Simulated VISAR data is compatible with the experimental data. The reaction zone width is substantial less than the average grain size. Thus, in contrast to initiation, which requires hot spots, the reaction zone of a propagating detonation wave is dominated by the reaction rate from the bulk temperature.

## EQUATIONS OF STATE

Data for the unreacted PBX 9501 Hugoniot and several fits are shown in fig. 1. The data up to 15 GPa is compatible with several linear fits in the literature. Extrapolated to detonation velocity (D=8.8 km/s) gives a large difference for the particle velocity (2.8 to 3.9 km/s), and would have a large effect on the von Neumann spike pressure and temperature. An equation of state fit to HMX isothermal compression data to 27 GPa [4, 5] is compatible with Hugoniot data, including high pressure (40 GPa) single crystal HMX data. This is the basis for a complete EOS described in [6]. It uses the specific heat determined from molecular dynamics simulations [7].

In the range of interest for the reaction zone profile ($T = 2000$ to $3000$ K) $C_V \approx 2.0 \times 10^{-3}$ (MJ/kg)/K. This is compatible with the pseudo-classical limit for lattice vibrations; $C_V = (3N - N_H)R/M$, where $N$ is the number of atoms per molecules, $N_H$ is the number of hydrogen atoms, $M$ is the molecular weight, and $R$ is the gas constant. For HMX ($C_4N_8O_8H_8$), $N = 28$, $N_H = 8$ and $M = 0.296$ kg/mole gives $C_V = 2.1 \times 10^{-3}$ (MJ/kg)/K. We note that this is larger than published data [8, §5.3, p. 112] which extends only up to $\beta$–$\delta$ transition temperature; $C_P = 1.57 \times 10^{-3}$ (MJ/kg)/K at $T = 450$ K.

For the equation of state of the reaction products a SESAME table [9] is used. The products EOS is fit to data on overdriven detonations and release isentropes in PBX 9501 [10, 11]. The thermal part of the EOS is based on the assumptions that the CJ temperature is 3000 K and the specific heat is 0.5 cal/g ($2.07 \times 10^{-3}$ (MJ/kg)/K).

From the equations of state, impedance matches with the VISAR windows (PMMA and LiF) can be used to estimate the end states of the reaction zone; von Neumann spike and CJ states. This provides a consistency check on both experiments and simulations. The graphical solution to the impedance match is shown in fig. 2.

## REACTION RATE

We assume a first order Arrhenius rate; $(1 - \lambda)k\exp(-T_a/T)$. For PBX 9501 we use an activation temperature $T_a = 17922$ K and multiplier $k = 5.6 \times 10^5 \, \mu s^{-1}$ based on the "global rate" of Henson *et al.* [12]. The temperature in the ZND profile — based on the EOS of PBX 9501 — varies from 2100 K at the von Neumann spike to 3000 K at the CJ state. In this temperature range, the inverse reaction rate varies from 1 to 10 ns.

A set of Arrhenius parameters commonly used for simulations is based on differential scanning calorimetry experiments of Rogers [13]; see also [8, §5.7, p. 113]. The reaction rate for these two sets of parameters is shown in fig. 3. We note that the rates differ by several orders of magnitude.

Henson's rate and Rogers' rate cross at 470 K. We note that Rogers' calorimetry experiments covered a narrow temperature range about the melting temperature of HMX; from 544 K to 558 K [13, fig. 11]. Rogers' rate would give a sub ps reaction time for a CJ detonation. Since the time for a detonation wave to cross a unit cell in an HMX crystal is 0.1 ps, Rogers' rate is unphysically large. In contrast, Henson's rate is compatible with the 3 high pressure data points for single crystal HMX; experiments by Craig reported in [14, p. 1065] and [15, p. 218].

## REACTION ZONE PROFILE

Simulations have been run with the $Am^{ri}t_a$ environment of James Quirk [16, 17] using an adaptive mesh algorithm in order to resolve fully the reaction profile. Simulations are initialized with a steady ZND profile and use a piston boundary condition. The piston is given a ramp velocity to simulate a Taylor wave following the CJ state. The velocity profile immediately before the detonation wave impacts the VISAR window is shown in fig. 4. The reaction zone width is 24 $\mu$m at 90 % burn fraction, substantially less than the average grain diameter (140 $\mu$m) in PBX 9501.

We note that a propagating detonation wave is subject to a pulsating instability; see [18, chpt. 6A] and references therein. This is a generic property of an Arrhenius reaction rate. The calculations are over a sufficiently short distance of run that the instability does not have time to develop. Presumably the instability would be ameliorated by hot spots and three-dimensional effects.

Comparison of simulated VISAR profiles with experimental data is shown in fig. 5. The simulated profiles are consistent with the impedance match, fig. 2, for the equations of state being used. Moreover, they are compatible with the experiments using a PMMA window. The von Neumann spike is clipped in the VISAR experiment with a LiF window. This is due to the time resolution. For the fringe constants used, the PMMA experiments have a 1 ns resolution while the LiF experiments have a 2-3 ns resolution.

Similar experiments with an HMX based PBX and LiF window were performed by Fedorov. With 1 ns time resolution, the VISAR record [3, fig. 2a] shows a von Neumann spike amplitude close to the value in the simulation, fig. 5. Thus in order to resolve the von Neumann spike, 1 ns or better time resolution is needed. We note that the reaction time in Fedorov's experiment is nearly twice as large as shown in fig. 5. This is due to initiating the PBX with a detonator rather than a flyer plate. A detonator results in a curved detonation wave and a detonation velocity lower than the CJ speed for a planar wave. This in turn lowers the temperature behind the von Neumann spike and due to the sensitivity of the Arrhenius rate can have a large effect on the reaction time.

Finally, we note that similar experiments have been done with PBX 9501, PBX 9404, EDC 37. These PBXs have similar high percentage of HMX and low porosity. But they have different binders and initiation sensitivity. Nevertheless, their reaction zone profiles are nearly the same [1, fig. 2]. This is consistent with the reaction rate being dominated by the bulk shock temperature.

Most burn models for coarse resolution engineering simulations use pressure dependent reaction rate motivated by homogenization or a volume average over small region containing many hot spots. Since the underlying reaction mechanism for propagating detonation waves appears to be different, this raises the question as to why burn models work as well as claimed when applied to applications involving the curvature effect and corner turning.

## ACKNOWLEDGMENTS

This work was carried out under the auspices of the U. S. Dept. of Energy at LANL under contract W-7405-ENG-36. The author thanks Rick Gustavsen for providing VISAR data files from the detonation wave profile experiments.

## REFERENCES

1.  Gustavsen, R. L., Sheffield, S. A., and Alcon, R. R., "Detonation Wave Profiles in HMX Based Explosives," in *Shock Compression of Condensed Matter – 1997*, 1998, pp. 739–742.
2.  Gustavsen, R. L., Sheffield, S. A., and Alcon, R. R., "Progress in Measuring Detonation Wave Profiles in PBX9501," in *Eleventh (International) Symposium on Detonation*, 1998.
3.  Fedorov, A. V., "Detonation Wave Structure in Liquid Homogeneous, Solid Heterogeneous and Agatized HE," in *Twelfth (International) Symposium on Detonation*, 2002.
4.  Yoo, C., and Cynn, H., *J. Chem. Phys.*, **111**, 10229–10235 (1999).
5.  Menikoff, R., and Sewell, T. D., *High Pressure Research*, **21**, 121–138 (2001).
6.  Menikoff, R., and Sewell, T. D., "Complete equation of state for beta-HMX and implications for initiation," in *Shock Compression of Condensed Matter – 2003*, 2004, pp. 157–160.
7.  Goddard, W. A., Meiron, D. I., Ortiz, M., Shepherd, J. E., and Pool, J., Annual technical report, Tech. Rep. 032, Center for Simulation of Dynamic Response in Materials, Calif. Inst. of Tech. (1998), http://www.cacr.caltech.edu/ASAP/publications/cit-asci-tr/cit-asci-tr032.pdf.
8.  Gibbs, T. R., and Popolato, A., editors, *LASL Explosive Property Data*, Univ. of Calif. Press, 1980.
9.  Shaw, M. S. (2004), private communication.
10. Hixson, R. S., Shaw, M. S., Fritz, J. N., Vorthman, J. N., and Anderson, W. W., *J. Appl. Phys.*, **88**, 6287–6293 (2000).
11. Fritz, J. N., Hixson, R. S., Shaw, M. S., Morris, C. E., and McQueen, R. G., *J. Appl. Phys.*, **80**, 6129–6149 (1996).
12. Henson, B. F., Asay, B. W., Smilowitz, L. B., and Dickson, P. M., "Ignition Chemistry in HMX from Thermal Explosion to Detonation," in *Shock Compression of Condensed Matter – 2001*, 2002, pp. 1069–1072.
13. Rogers, R. N., *Thermochimica Acta*, **3**, 437–447 (1972).
14. Campbell, A. W., and Travis, J. R., "The Shock Desensitization of PBX-9404 and Composition B-3," in *Eighth Symposium (International) on Detonation*, 1986, pp. 1057–1068.
15. Mader, C. L., *Numerical Modeling of Explosives and Propellants*, CRC Press, 1998, second edn.
16. Quirk, J. J., "Amrita - A Computational Facility for CFD Modelling," in *29th Computational Fluid Dynamics*, VKI Lecture Series, von Karmen Institute, 1998, chap. 4, URL http://www.amrita-ebook.org/pdf/vki/cfd29/jjq/vki:cfd29::jjq_l1.pdf.
17. Quirk, J. J., "AMR_sol: Design Principles and Practice," in *29th Computational Fluid Dynamics*, VKI Lecture Series, von Karmen Institute, 1998, chap. 5, URL http://www.amrita-ebook.org/pdf/vki/cfd29/jjq/vki:cfd29::jjq_l2.pdf.
18. Fickett, W., and Davis, W. C., *Detonation*, Univ. of Calif. Press, 1979.
19. Sheffield, S. A., Gustavsen, R. L., Alcon, R. R., Robbins, D. L., and Stahl, D. B., "High Pressure Hugo-

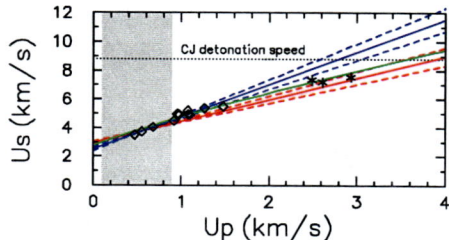

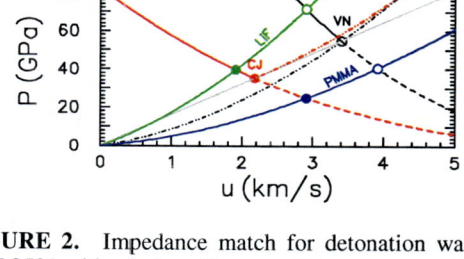

**FIGURE 1.** Unreacted Hugoniot for PBX 9501. Red and blue lines are from [8, §7.3, p. 116]; dashed lines are error bars and gray region is domain of fit. Green curve is fit to isothermal data [4, 5]. Black dotted line is CJ detonation velocity (8.8 km/s). Diamonds are data points from [19] and stars are single crystal HMX data from [20, p. 595].

**FIGURE 2.** Impedance match for detonation wave in PBX 9501 with window. Green and blue curves are Hugoniot loci for LiF and PMMA, respectively. Black and red curves are for reactants and products, respectively. Gray is Rayleigh line corresponding to CJ detonation velocity. Labels VN and CJ denote von Neumann spike and Chapman-Jouguet state, respectively. Open circles are match from VN spike and solid circles are match from CJ state.

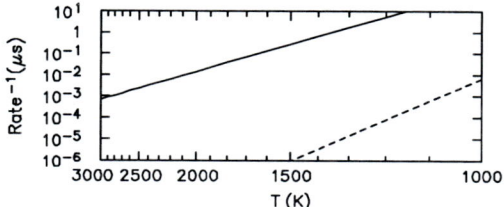

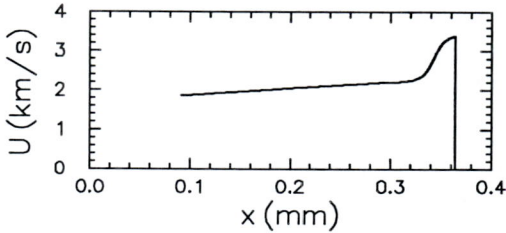

**FIGURE 3.** Inverse reaction rate vs temperature. Temperature is plotted on inverse scale. Dashed curve uses Arrhenius parameters in [8, §5.7, p. 113] ($T_a = 26522$ K, $k = 5.0 \times 10^{13} \, \mu s^{-1}$) and solid curve based on [12] ($T_a = 17922$ K, $k = 5.6 \times 10^5 \, \mu s^{-1}$).

**FIGURE 4.** Wave profile immediately before detonation wave impacts window.

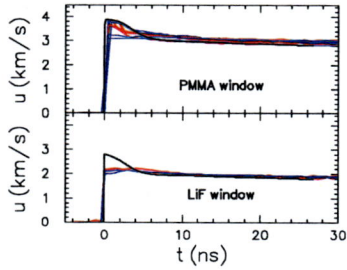

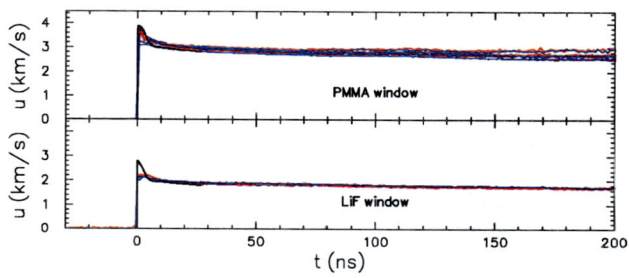

**FIGURE 5.** Comparison with VISAR data from [1, 2]. Top figures are for PMMA window and bottom are for LiF window. Left and right figures are on 30 and 200 ns time scale, respectively. Red and blue curves are experiments and black is simulations. VISAR used two laser beams with different fringe constants per experiment. Experiments varied drive pressure for initiation and the length of PBX sample.

niot and Reaction Rate Measurements in PBX9501," in *Shock Compression of Condensed Matter – 2003*, 2004, pp. 1033–1036.

20. Marsh, S. P., editor, *LASL Shock Hugoniot Data*, Univ. of Calif. Press, 1980.

CP845, *Shock Compression of Condensed Matter - 2005*,
edited by M. D. Furnish, M. Elert, T. P. Russell, and C. T. White
© 2006 American Institute of Physics 0-7354-0341-4/06/$23.00

# EVALUATION OF ALUMINUM PARTICIPATION IN THE DEVELOPMENT OF REACTIVE WAVES IN SHOCK COMPRESSED HMX

**R. J. Pahl, W. M. Trott, S. Snedigar, and J. N. Castañeda**

*Sandia National Laboratories*\*, *Albuquerque NM 87185*

**Abstract.** A series of gas gun tests has been performed to examine contributions to energy release from micron-sized and nanometric aluminum powder added to sieved (212-300μm) HMX. In the absence of added metal, 4-mm-thick, low-density (64-68% of theoretical maximum density) pressings of the sieved HMX respond to modest shock loading by developing distinctive reactive waves that exhibit both temporal and mesoscale spatial fluctuations. Parallel tests have been performed on samples containing 10% (by mass) aluminum in two particle sizes: 2-μm and 123-nm mean particle diameter, respectively. The finely dispersed aluminum initially suppresses wave growth from HMX reactions; however, after a visible induction period, the added metal drives rapid increases in the transmitted wave particle velocity. Wave profile variations as a function of the aluminum particle diameter are discussed.

**Keywords:** Shock Initiation, aluminum, reactive wave
**PACS:** 47.40.-x, 47.40.Nm, 82.40.Fp.

## INTRODUCTION

There has been a long standing interest in adding micron and nano-sized Al particulate to formulations to improve performance. This addition of Al affects detonation properties as has been shown previously[1]. The nature and magnitude of these effects will most likely be determined by understanding the timing of Al energy release in relation to the passing of the initial reaction wave[2].

We recently have conducted several experiments with HMX/Al mixtures to probe the timing of Al energy release. The gas-gun studies reported in this paper were motivated in part by earlier experimental and theoretical work on low-density (~65% TMD) pressings of sugar and HMX[3,4]. In particular, mesoscale modeling reveals that rapid deformation occurs at material contact points in such pressings subjected to impact. Localization effects first occur at these contact points, and subsequently, plastic flow into the interstitial pores produces localized regions of sustained elevated temperature. Both stress and temperature fields display a substantial degree of large amplitude fluctuations. Wave profiles generated by these materials can exhibit distinct ordered wave structures that are distributed over multiple grain dimensions and often coherent temporally. Velocity interferometry provides a useful diagnostic for relating measured wave structures to a well-characterized sample microstructure.

Adapting the previous experimental design to include HMX samples containing finely dispersed Al offers a number of possible advantages in unraveling the complex role of this metal in the

---
\*Sandia is a multiprogram laboratory operated by Sandia Corporation, a Lockheed Martin Company, for the United States Department of Energy 's National Nuclear Security Administration under Contract DE-AC04-94AL85000.

material energetics. When efficiently mixed, added micron-sized or nanometric aluminum powder tends to coat the explosive grains, remaining in intimate contact during sample pressing. This morphology promotes rapid exposure of the metal additive to both material deformation and sustained elevated temperatures under shock loading. The former process is likely useful in stripping surface oxide from the metal while the latter condition may be expected to promote aluminum chemistry. Finally, the use of low-density samples (with minimal grain fracture during pressing) is helpful in defining a material geometry that can be reasonably approximated in mesoscale numerical simulations.

## EXPERIMENTAL SETUP AND DIAGNOSTICS

A schematic diagram of the target design used in the present work is shown in Fig. 1. The experimental assembly consists of a Kel-F impactor and sample cup. A PMMA window is used to confine the porous HMX or HMX/Al bed in the sample cup. The target fixture is designed to accommodate simultaneous measurements using three diagnostics: (1) a fiber-coupled, single-point VISAR, (2) a line-imaging optically recording velocity interferometer system (ORVIS), and (3) a fiber-coupled streak spectrometer. The first two diagnostics utilize a 0.225-mm-thick buffer layer of Kapton and a reflective coating to maintain adequate return light throughout the test. The third diagnostic required direct optical access to the sample material. These conditions were met in each fixture by boring a 25-mm-diameter, 0.250-mm-thick cylindrical recess into the window and repolishing the resulting surface. The buffer was positioned in this cavity and affixed to the PMMA surface in the usual manner.

The HMX was taken from a 212-300 μm sieve cut from Grade B, class 1 HMX produced by Holston. The two Al particle sizes used in these tests were 2 μm (H-2 produced by Valimet) and 123 nm (produced by Technanogy). Both had a spherical morphology. The aluminized targets contained 10% aluminum powder by mass. To ensure thorough dispersion of the Al powder, the

Al was first mechanically mixed with the HMX powder and then ultrasonically agitated in hexane for several minutes. SEM images of pellets pressed from this mixture have shown adequate dispersal of Al with minimal agglomeration. The pure HMX targets were pressed to 64% and 68% TMD. In all HMX/Al test cases, the targets were pressed to a nominal density of 68% TMD. The density difference between Al and $Al_2O_3$ and the particle mass fraction of the oxide were accounted for in the target density calculation.

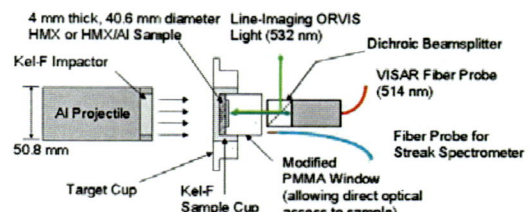

**Figure 1.** Diagram of target with diagnostics

Details of the VISAR and line-imaging ORVIS are available elsewhere[4,5]. The streak camera based spectrometer consists of a fiber coupled Acton 300 0.3 m spectrometer coupled to a Hadland 500 streak camera. The purpose of this diagnostic was to attempt the detection of AlO emission, signaling the combustion of Al.

## DATA AND ANALYSIS

The results discussed in this paper describe the significant variation in sample response that was observed over a fairly narrow range of impact velocity (0.40-0.51 km/s). VISAR records reveal many of the essential differences, as shown in Figs. 2 and 3. The response of the two different densities of pure HMX (cf. Fig. 2) represent "baseline" cases in which the 68% TMD and 64% TMD pressings match the porosity and mass content of the explosive, respectively, in the aluminized samples. In all cases, the transmitted wave profiles exhibit an initial ramp-up followed by a steadily increasing particle velocity driven by explosive energy release. As expected from long-established trends as a function of pressing density, profiles from the 68% TMD material reflect a somewhat lower sensitivity to shock initiation and early reaction growth. In particular, the velocity profile for 68% TMD HMX under impact at 0.425 km/s (Fig. 2b, bottom trace) exhibits both substantially lower amplitude and a longer propagation time through the 4-mm-thick sample as compared to the lower density HMX under nearly identical impact conditions (Fig. 2a, top trace). Significant oscillations in the particle velocity also occur; these fluctuations are somewhat more prominent in the 64% TMD material. These features are likely driven by shock reverberations in material with incomplete pore collapse; however, reactive growth from initial "hot spots" also seems to play a role since the observed oscillations are much more prominent than those seen in inert surrogate materials (e.g., sugar) under equivalent impact conditions[4].

As illustrated in Fig. 3, the addition of Al results in dramatic changes in the observed VISAR records. Clearly, the addition of finely dispersed Al initially suppresses the energy release corresponding to HMX decomposition. For example, Figs. 2b and 3 allow a direct comparison of response at impact velocities near 0.45 km/s. Under these conditions, the aluminized materials exhibit very slow reaction growth early on whereas the pure 68% TMD HMX is quickly driven to vigorous reaction. The wave transit time for the HMX/123-nm Al sample is ~1 μs longer than that for the pure HMX. This reaction suppression likely arises from a complex combination of physical factors including differences in shock impedance and particle size distribution as well as a probable reaction "bottleneck" due to the aluminum coating of the individual grains. For the two aluminized materials, initial energy release occurs faster in the case of the HMX/2-μm Al, perhaps as a result of a less uniform aluminum coating (comprised of >1000x fewer particles for a given mass percent).

The VISAR records for the aluminized samples indicate a significant pull back or quenching of the reactive wave build-up. After an induction period, rapid build-up to explosion then occurs, most likely fueled in part by Al combustion. The quenching period may correspond to Al undergoing phase change. The induction time is shorter at higher particle velocities, consistent with faster energy release leading to a shorter time needed for phase change. It can also be seen that the smaller particle size results in a shorter induction time, most likely the result of the

larger surface area to volume ratio of the smaller particles.

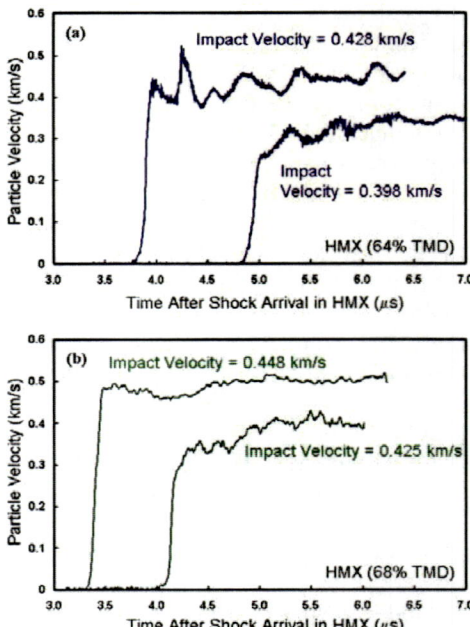

**Figure 2.** Comparison of the VISAR traces for a) 64% TMD HMX and b) 68% TMD HMX targets

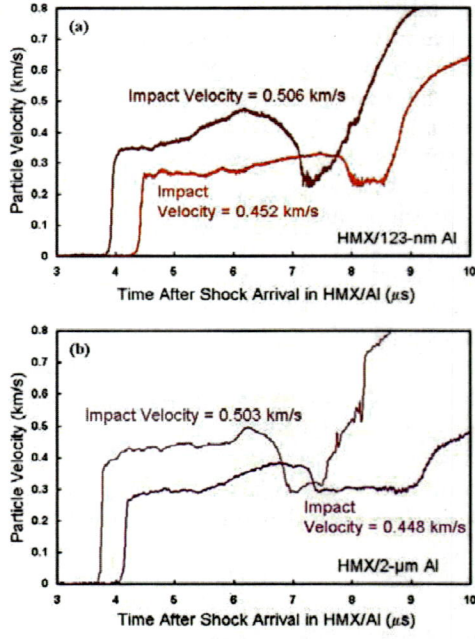

**Figure 3.** VISAR data from a) HMX/123-nm Al and b) HMX/2-μm Al targets

For the conditions explored in this study, the utility of the ORVIS diagnostic was limited by a 4-μs upper limit in the data recording time of our instrumental setup. In particular, we were unable to record profiles from the aluminized samples that captured the complete history of wave phenomena. The available results do suggest that significant spatial as well as temporal fluctuations are present in the transmitted waves. Figure 4 shows a typical spatially resolved velocity profile. Particle velocities are consistent with the single-point VISAR measurements (cf. Fig. 3a, top trace).

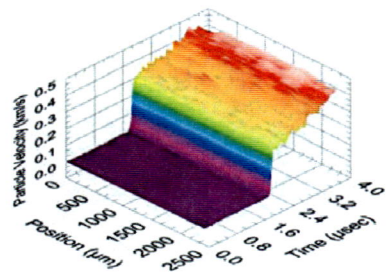

**Figure 4.** Line-Imaging ORVIS data from HMX/123-nm Al sample at impact velocity of 0.5 km/s

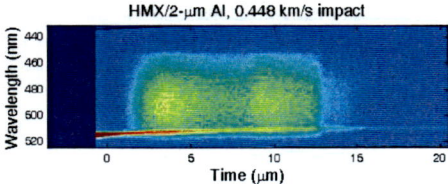

**Figure 5.** Streak Spectroscopy data from HMX/2-μm Al sample at impact velocity of ~0.45 km/s

It was hoped that spectroscopic data, in particular AlO emission signatures, would indicate if and when Al combustion is occurring in reference to the induction period seen in the VISAR data. A typical streak image is shown in Fig. 5. As this image clearly shows, no spectral features were seen outside of the VISAR laser probe. The absence of AlO or any other spectral features is most likely due to the high pressures and densities in the particle bed. AlO emissions are most likely broadened and quenched to a degree where they cannot be seen within the sensitivity range of this diagnostic. It is, however, noticeable that light emission from the surface of the bed decreases slightly after reaching peak intensity at around 4 μs, and then increases around 9 μs. This

corresponds to the induction time and subsequent increase in particle velocity, thus giving further indication that some quenching is taking place.

## CONCLUSIONS AND SUMMARY

The data shown here indicates that under low density, moderate loading conditions, aluminum reaction begins a few microseconds after reaction of HMX. The initial energy release by HMX creates a high temperature, high pressure environment where the Al particulate undergoes a phase change before reacting and contributing energy to the developing detonation front. This proposed phase change in Al is evidenced by a pull back and induction time in particle velocity traces.

## ACKNOWLEDGEMENTS

We would like to acknowledge John Liwski for his work in conducting the light gas gun operations and Mel Baer and Anita Renlund for their insight on the data resulting from these tests.

## REFERENCES

1. Gilev, S. D., and Trubachev, A. M., "Detonation Properties and Electrical Conductivity of Explosive-Metal Additive Mixtures" in Combustion, Explosion and Shock Waves, v38, No2, pp. 219-234 (2002).
2. Victorov, S. B., "The Effect of $Al_2O_3$ Phase Transitions on Detonation Properties of Aluminized Explosives," in Proc. 12th Int. Detonation Symp. Aug. 11-16, 2002
3. Baer, M. R., et al., "Computational Modeling of Heterogeneous Reactive Materials at the Mesoscale," in Shock Compression of Condensed Matter, 1999 (M.D. Furnish, L.C. Chhabildas, R.S. Hixson, eds.), part I, pp. 27-33.
4. Baer, M. R. and Trott, W. M., "Theoretical and Experimental Mesoscale Studies of Impact-Loaded Granular Explosive and Simulant Materials," in Proc. 12th Int. Detonation Symp. Aug. 11-16, 2002
5. Trott, W. M., et al., "Dispersive Velocity Measurements in Heterogeneous Materials," Sandia National Laboratories, SAND2000-3082, December 2000.

CP845, *Shock Compression of Condensed Matter - 2005,*
edited by M. D. Furnish, M. Elert, T. P. Russell, and C. T. White
© 2006 American Institute of Physics 0-7354-0341-4/06/$23.00

# COMPARISON OF FAILURE THICKNESS AND CRITICAL DIAMETER OF NITROMETHANE

## Oren E. Petel and Andrew J. Higgins

*McGill University, Department of Mechanical Engineering, Montreal, Quebec, H3A 2K6, Canada*

**Abstract.** The critical diameter and failure thickness of both neat liquid nitromethane and a 65% nitromethane/35% nitroethane blend confined by aluminum are determined experimentally. A comparison of these two parameters provides insight into the failure mechanism of detonation in these explosives. If the failure of detonation in a critical charge diameter (or thickness) experiment is due to reaction quenching resulting from expansion losses (wave curvature), then it is expected that the failure thickness should be half the value of the critical diameter. The critical diameter and failure thickness of neat nitromethane confined in aluminum are found to be 2.5 mm and 0.75 mm respectively for a temperature range of $26 \pm 1^{\circ}$C. The critical diameter and failure thickness of the 65NM/35NE blend confined in aluminum are found to be 6.2 mm and 1.7 mm respectively for a temperature range of $28 \pm 1^{\circ}$C. The ratio of critical diameter to failure thickness for these experiments is found to lie between 3:1 and 4:1 rather than 2:1 as expected from wave curvature theory. By comparing the experimentally determined values of critical diameter and thickness for the test explosives and examining the failure patterns recovered on witness plates, a mechanism of propagation in thin rectangular channels is proposed based on complex wave interactions.

**Keywords:** Critical diameter, failure thickness, nitromethane, nitroethane, detonation.
**PACS:** 47.40.-x, 47.40.Nm.

## INTRODUCTION

The comparison of the failure thickness and the critical diameter of an explosive can provide insight into the failure mechanism of detonations. Detonation waves in homogeneous condensed explosives are known to fail due to lateral expansion losses which result in detonation front curvature and reaction quenching. Considering the curvature of a detonation wave in a two-dimensional channel (curvature in one-dimension) and that of a detonation in a tube of circular cross-section (axisymmetric curvature), one would expect the failure diameter of the detonation in the tube to be twice the failure thickness of the channel.

This suggested relationship for the two charge geometries has previously been confirmed for PBX 9502 [1] by a comparison of failure thickness obtained in wedge tests to critical diameter results for the same explosive [2].

In the present study, a direct comparison of the critical diameter and failure thickness of two explosives is made. Although this comparison has previously been made for PBX 9502 [1], this comparison of failure thickness to critical diameter has not been made for liquid explosives.

## EXPERIMENTAL DETAILS

The present study involved two experimental setups: one to measure the critical diameter and the other to measure the failure thickness of an explosive. The set of test explosives used in this series of experiments was neat nitromethane (NM) and a blend of nitromethane and nitroethane (NE). The blend consisted of 65NM/35NE by weight.

The critical diameter of each of the test explosives was measured with aluminum 6061 T6 confinement. The walls of the aluminum tubing were approximately 3 mm thick. In the experiment, a detonation was first initiated in a larger steel tube to establish a steady detonation, which then transitioned into the smaller test section, as shown in Fig. 1a. The test section was at least 10 tube diameters in length for all experiments, to ensure the detonation was not still overdriven from the transition to the smaller test section.

Fig. 2a shows a rendered schematic and cross-sectional view of the experiment used to measure the failure thickness of the explosive. The experiment consisted of four sections: a line wave generator (50 ns planarity), a booster, a "stepped" section (to eliminate overdriven effects from booster), and the test section. The first two sections used sensitized NM (5% DETA/w) as the explosive, due to the fact that the test explosives did not have the required sensitivity for our detonators. A 13 μm Mylar™ membrane separated the two explosives. The test section was 76 mm in width and 130 mm in length with varying thickness. The booster section was 25 mm in length and had a thickness of 13 mm. The "stepped" section was 25 mm in length and 6 mm in thickness.

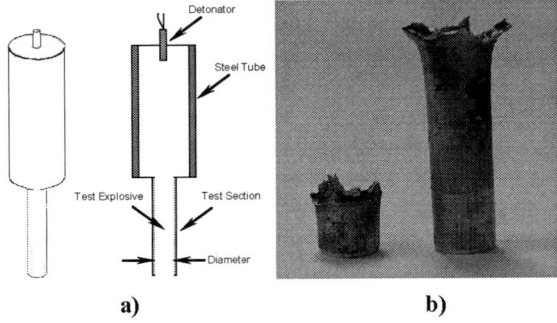

**Figure 1. a)** Rendered schematic and cross-sectional view of the setup for critical diameter measurements. **b)** Recovered tubes in which detonation failed.

The main structure of the experiment was made from polypropylene and the explosive was confined by aluminum in the test section. One of the plates confining the explosive in the test section was intentionally made thick (typically 10 mm) and acted as a witness plate. The second aluminum plate providing confinement in the test section was 3 mm thick. The thickness of the test section was altered by changing the thickness of the witness

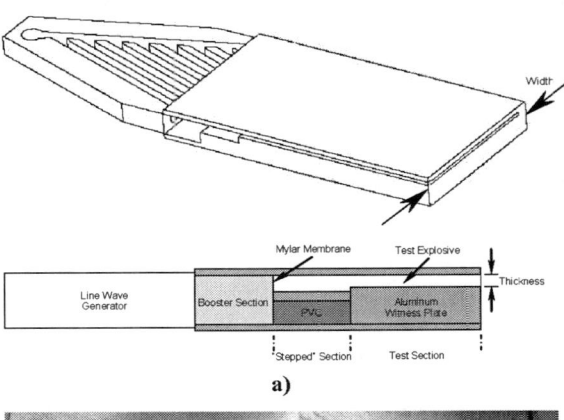

a)

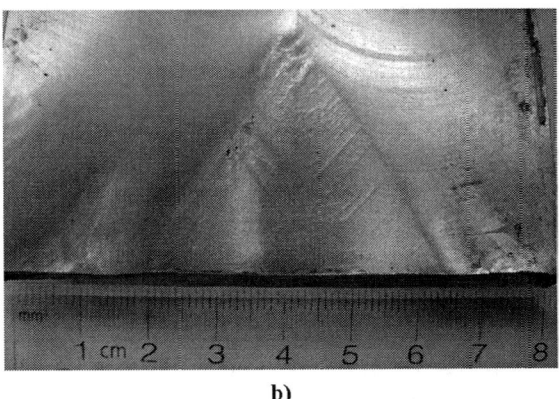

b)

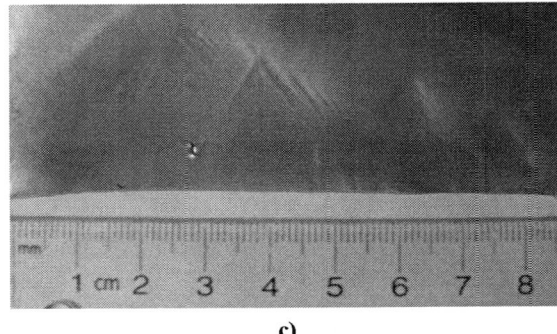

c)

**Figure 2. a)** Rendered schematic and cross-sectional view of the experimental setup for failure thickness measurements. **b)** Witness plate for a "Go". **c)** Witness plate for a "No Go".

plate. The experiment was designed with a test section aspect ratio in the 40:1 range (width : thickness).

## RESULTS

The results of the critical diameter experiments were determined by the recovery of the test sections, indicating detonation failure, as seen in Fig. 1b. The results of the failure thickness experiments were determined by the imprint left in the recovered witness plates (Fig. 2b and 2c). Although the test section had a large aspect ratio, the lateral confinement was polypropylene. The effects of lateral boundary expansion eventually overtook the test section, causing failure of the detonation wave due to lateral curvature. This typically occurred 20 channel heights into the test section, by which distance it was clear if the detonation was able to propagate along the centerline. The detonation was determined to successfully propagate in the test section if it could propagate 10 channel heights along the channel centerline without failing. Successful propagations are denoted as "Go's", while failures are called "No Go's". Fig. 2b shows an example witness plate of a "Go", while Fig. 2c shows the picture of a "No Go".

The critical diameter of neat NM confined in aluminum was determined to be $2.35 \pm 0.15$ mm, while its failure thickness was approximately $0.75 \pm 0.1$ mm (Fig 3a). The temperature of the neat NM for these experiments was $26 \pm 1°C$.

The critical diameter of the NM/NE blend confined in aluminum was determined to be $6.2 \pm 0.2$ mm, while its failure thickness was approximately $1.7 \pm 0.1$ mm (Fig. 3b). The temperature of the NM/NE blend for these experiments was $28 \pm 1°C$.

## DISCUSSION

From wave curvature considerations, the critical diameter of an explosive is expected to be twice its failure thickness. The experimental results from the present study for two different liquid explosives indicate that this is not the case. The ratio of the critical diameter to the failure thickness

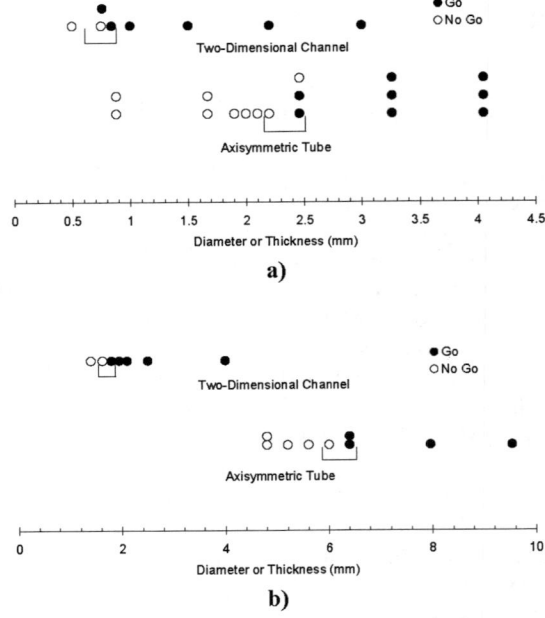

**Figure 3.** **a)** Results for neat NM. **b)** Results for 65NM/35NE blend.

for neat NM and the NM/NE blend was found to be $3.2 \pm 0.6$ and $3.6 \pm 0.4$ respectively. Since these results differ significantly from the theoretical predictions, it is important to consider any sources of error in the experiment.

The first consideration should be the effect of a non-planar detonation (along the width) entering the two-dimensional (rectangular) test section. Since detonation wave failure is considered to be due to wave curvature in one dimension only, any further wave curvature in the other dimension would cause an increase in the failure thickness. Since the failure thickness is smaller than expected, it is not due to the planarity of the detonation wave. Also, as discussed above, the charge was designed so that the detonation in the test section was not overdriven by the booster charge or the transition to the smaller test section area.

Although the experimental ratios of critical diameter to failure thickness are not in agreement with theoretical expectations based on front curvature, these results can be explained by an extension of Dremin's failure theory [3] to explosive slab geometries of high aspect ratio. Dremin et al. [4] observed that detonation failure in tubes filled with nitromethane is due to dark waves, which are pockets of no-reaction emanating from the confining tube walls. Dremin hypothesizes that a detonation in a tube will fail when a dark wave generated at the confining wall is able to propagate to the axis of symmetry prior to being overtaken laterally by a reinitiated detonation front along the tube wall.

From the recovered witness plates (Fig. 2b and 2c), there are many apparent zones of detonation failure and reinitiation in the two-dimensional rectangular channels. In the case of the "No Go" results, failure waves are observed to propagate outward in both directions from the centerline of the charge, leaving triangular regions of no detonation. When these regions merge, the detonation fails completely. In the case of "Go" results, failure waves are also observed to propagate outward from the center of the charge, but are closely followed by waves of detonation reinitiation, leaving thin strips of failure. The reinitiation waves likely originate from neighbouring sections of the charge that have not failed. This mechanism provides a means to continue propagating detonation that would be absent in a cylindrical charge. In a cylindrical charge at critical diameter, it is unlikely that the wave can spontaneously reinitiate.

This interpretation implies that the aspect ratio is a parameter in the experiment, since contributions from other regions of the front will have a role in laterally reinitiating zones of failure. Reinitation could also possibly be caused by the interactions of decoupled shocks from failed regions of the detonation front. Indeed, this effect has been observed in gas phase detonations in highly irregular mixtures. In experiments examining gas-phase detonation failure in rectangular channels, a dependence on channel aspect ratio has been found [5].

The relation between critical tube diameter and critical slot height has also been investigated in gas phase detonations by Benedict et al. [6]. Their study found that the ratio is approximately 4:1 with the critical tube diameter being $13 \lambda$ ($\lambda$ is the detonation cell size) and the critical slot height in the limit of high aspect ratio converges to about $3.25 \lambda$. This result is in quantitative agreement with the results of the present study.

## CONCLUSION

The critical diameter and failure thickness of two explosives in aluminum confinement have been determined experimentally. The relationship between these two measured parameters has been discussed and a possible failure and propagation mechanism has been proposed. The results point to a more complicated failure mechanism in a slab geometry than curvature in the detonation front due to mass divergence. The proposed mechanism involves transverse wave interactions allowing the detonation to propagate in much thinner slab thickness in spite of the appearance of many detonation failure zones.

## ACKNOWLEDGEMENTS

The authors would like to thank Cyrus Foster for assistance in conducting the experiments.

## REFERENCES

1. Ramsay, J.B., *Proc. 8th Symp. (Int) on Det.*, pp. 372-379, 1985.
2. Campbell, A.W., *Propellants Explosives and Pyrotechnics*, v. 9, pp. 183-187, 1984.
3. Dremin, A.N., *Towards Detonation Theory*, Springer, New York, 1999.
4. Dremin, A.N., Rozanov, O.K., and Trofimov, V.S., *Combust. Flame*, v. 7, pp. 153-162, 1963.
5. Radulescu, M.I, Ph.D. Dissertation, McGill University, 2003.
6. Benedick, W.B., Knystautas, R., and Lee, J.H.S., *Dynamics of Shock Waves, Explosions, and Detonations*, AIAA Progress in Astronautics and Aeronautics, v. 94, pp. 546-555, 1983.

CP845, *Shock Compression of Condensed Matter - 2005,*
edited by M. D. Furnish, M. Elert, T. P. Russell, and C. T. White
© 2006 American Institute of Physics 0-7354-0341-4/06/$23.00

# EFFECT OF SHOCK PRECOMPRESSION ON THE CRITICAL DIAMETER OF LIQUID EXPLOSIVES

**Oren E. Petel[1], Andrew J. Higgins[1], Akio C. Yoshinaka[2] and Fan Zhang[2]**

[1]*McGill University, Department of Mechanical Engineering, Montreal, Quebec, H3A 2K6, Canada*
[2]*Defence R&D Canada—Suffield, Medicine Hat, Alberta, T1A 8K6, Canada*

**Abstract.** The critical diameter of both ambient and shock-precompressed liquid nitromethane confined in PVC tubing are measured experimentally. The experiment was conducted for both amine sensitized and neat NM. In the precompression experiments, the explosive is compressed by a strong shock wave generated by a donor explosive and reflected from a high impedance anvil prior to being detonated by a secondary event. The pressures reached in the test sections prior to detonation propagation was approximately 7 and 8 GPa for amine sensitized and neat NM respectively. The results demonstrated a 30% - 65% decrease in the critical diameter for the shock-compressed explosives. This critical diameter decrease is observed despite a significant decrease in the predicted Von Neumann temperature of the detonation in the precompressed explosive. The results are discussed in the context of theoretical predictions based on thermal ignition theory and previous critical diameter measurements.

**Keywords:** Critical diameter, nitromethane, superdetonation, shock precompression.
**PACS:** 47.40.-x, 47.40.Nm.

## INTRODUCTION

The critical diameter of an explosive has previously been shown to be highly dependant on the initial state of the explosive [1]. Small variations in initial temperature cause drastic changes in the critical diameter of nitromethane (NM) and other liquid explosives (NG, TNT) [1-3]. These changes in critical diameter have been attributed to the temperature sensitivity of the reaction kinetics in the detonation front.

In the present study, the explosive will be shock compressed and subsequently a detonation will propagate through the explosive at this shocked state. Past studies have focused on the effects of shock compression on the initiation of reaction in NM [4]. This study looks at the propagation of a detonation through a previously shocked explosive.

## EXPERIMENTAL DETAILS

Two sets of experiments were conducted in this study. The first used sensitized NM (3% DETA by weight) and the second used neat NM. The experimental design was based on the experiments conducted by Petel et al. [5] and can be seen in Fig. 1. In all experiments, the donor and test explosive are the same explosive.

In the experiments involving sensitized NM, the donor explosive was 75 mm in diameter and 100 mm in height. The donor capsule was positioned on a PVC attenuator disk that was 165 mm diameter and 19 mm in height. In contact with the attenuating disk was a 152 mm internal diameter PVC test capsule, 25 mm in height. The test capsule had a transparent PVC tube running across its diameter, the diameter of which was varied in order to determine the critical diameter.

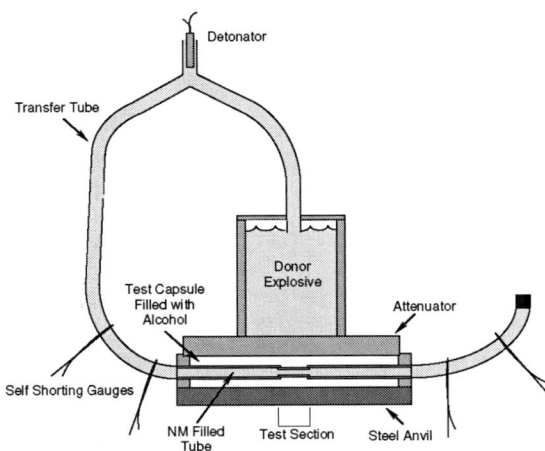

**Figure 1**. Cross-sectional schematic of the experimental setup used.

The section with the varied diameter was considered to be the test section. The test capsule surrounding the PVC tube was filled with denatured alcohol, which was used for its transparency under shock compression and acted as a hydrostatic bath on the time scale of the experiments. The test capsule was bounded on the bottom by a steel anvil. There was a 1.6 mm slit in the steel anvil, which was also filled with denatured alcohol and provided optical access to the experimental test section.

The setup of the experiments involving neat NM had the same components as the setup described above, only the scale was larger. For these experiments the donor explosive was 150 mm in diameter and 225 mm in height. The PVC attenuator disk had a 206 mm diameter and was 19 mm in height. The PVC test capsule was 32 mm in height, 203 mm in diameter and was filled with water instead of alcohol. Once again the diameter of the tube crossing the test section was varied to find the critical diameter. The steel anvil at the base of the test capsule was 50 mm thick. There was no slit in this steel anvil.

The experiments began with the initiation of a detonation in the tube filled with the explosive. The initial tube branched into two separate tubes. The first tube entered the donor capsule, initiating the donor explosive. The detonation in the donor explosive transmitted a shock wave into the attenuating disk which then entered the test capsule, compressing the PVC tube containing the test explosive. The shock wave then reflected from the

high impedance steel anvil, further compressing the test section.

Meanwhile, the detonation in the transfer tube propagated the full length of the tube and entered the test section as the shock reflection from the steel anvil occurred. The detonation propagated into the test section which remained at the highly compressed state. The dimensions of the charges were determined by the requirement that rarefaction waves from the sides of the capsule or bottom of the anvil do not penetrate the test section on the timescale of the experiment.

For the sensitized NM experiments, data was collected with streak photography as well as self-shorting gauges to record the detonation as it enters and exits the test section. The experiments with the neat NM used only the self-shorting gauges to determine the results of the experiment. The results of the experiments were categorized as either a "Go" or "No Go". A "Go" consists of an experiment in which the detonation propagated through the entire test section and exits the test capsule. A "No Go" is considered to be the case in which the detonation failed to propagate through the entire test capsule.

Control tests were also performed to determine the critical diameter of NM in the same tubing under ambient conditions. The control tests were conducted at a temperature of $25 \pm 1\ ^{\circ}C$

Experiments were also conducted with manganin strain gauges (Dynasen model MN4-50-EK) in order to estimate the post-shock pressure in the test explosive. In this set of experiments, the PVC test tube was removed and a manganin gauge was suspended on a Mylar bridge at the same height as the center of test section.

## RESULTS

The results from these experiments are presented in Fig. 2, showing the Go/No Go results as a function of the tube diameter in the test section. In Fig. 2a, there is evidence of a decrease in the critical diameter of sensitized NM due to shock precompression. The critical diameter of the test explosive at ambient conditions is approximately 2.4 mm. The critical diameter of the same test explosive while being compressed to a final

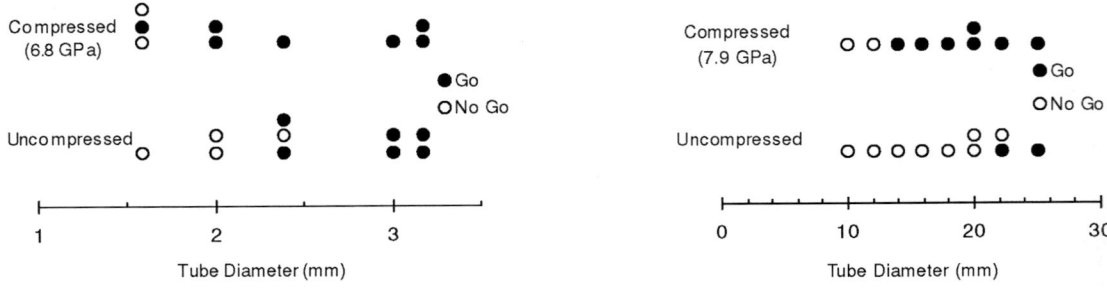

**Figure 2. a)** Results for NM + 3% DETA/wt experiments. **b)** Results for neat NM experiments.

pressure of approximately 6.8 GPa lies between 1.6 and 2 mm.

The results of the experiments with neat NM as the test explosive are shown in Fig. 2b. These results also show evidence of a decrease in critical diameter due to shock precompression of the test explosive. The control experiments conducted with the test explosive at ambient conditions found a critical diameter of roughly 22 mm, in good agreement with published values of $d_c$ of NM in weak confinement [1]. The critical diameter of the same explosive under shock precompression was between 12 and 14 mm. The final pressure reached in these tests was 7.9 GPa.

It is important to note that all tube diameters given in Fig. 2 are the nominal sizes prior to compression. In fact, due to shock compression, the tube cross-section becomes elliptical, with the minor axis being approximately 35 % smaller than the original diameter. Thus the difference in critical diameter between compressed and uncompressed NM in these experiments is even greater than stated above.

## DISCUSSION

To interpret these results, it is necessary to consider the effect of the shock precompression on the explosive. The passage of the shock through the test explosive raises the pressure, temperature and density of the explosive. Each of these state variables has a competing influence on the critical diameter of the explosive. Also, since the explosive is now confined at high pressure, the expansion of the confinement should be less severe, which would cause a decrease in the critical diameter. The fact that the density of the explosive is higher should

also decrease the critical diameter from energy density considerations. Although all these factors play a role in determining the effect of precompression on the critical diameter of the explosive, we expect the dominant effect should be due to changes in the temperature of the reaction zone of the detonation according to Arrhenius kinetics.

To understand the influence of shock precompression on the reaction zone of the detonation, the von Neumann spike will be examined. This calculation will be performed using the NM EOS from Winey et al. [6] to approximate the post-shock temperatures. It is assumed that this EOS is valid for sensitized nitromethane as well. From the manganin gauge measurements, the first shock wave and the reflected shock bring the explosive to pressures of 2.8 GPa and 6.8 GPa respectively and temperatures of 604 K and 734 K respectively.

The detonation velocity of the sensitized NM is approximately 6 km/s, and the detonation velocity of the precompressed sensitized NM is approximately 7.5 km/s as measured by Petel et al. [5] for the same precompression conditions. Assuming that the $U_s$-$u_p$ Hugoniot can be extrapolated linearly to these detonation velocities, the state at the von Neumann spike can be approximated using the NM EOS [6]. The calculations show that the von Neumann pressure and temperature are 17.8 GPa and 1740 K respectively for initially uncompressed NM. The calculation also shows that a 7.5 km/s detonation propagating in NM precompressed to 6.8 GPa and 734 K has a von Neumann pressure and temperature of 35.4 GPa and 1208 K respectively.

Since the reaction zone thickness is known to be extremely temperature dependent, such a large decrease in the von Neumann temperature (500°C) should result in a significantly longer reaction zone, in the precompressed case, thus increasing the critical diameter. From Arrhenius kinetics, assuming that the activation energy is the same for both detonations, the reaction zone of the precompressed detonation should be orders of magnitude longer than the uncompressed detonation. According to this estimate, the critical diameter should increase drasctically, however, just the opposite effect is seen. The reactions in the precompressed explosive are occuring at a much lower post-shock temperature than in the ambient explosive, but the precompressed explosives exhibit a smaller critical diameter. These results point to a more dominant role of pressure in the reaction zone at these extreme conditions. This result appears to contradict much of the literature which supports the thermal mechanism as being dominant in controlling detonation kinetics.

Recent molecular dynamics simulations have shown that at high pressures (compression factors between 2 and 3) it is possible to have proton transfer in nitromethane [7-9]. Engelke et al. [10] have shown that under high static pressure, nitromethane has an increased concentration of its aci ion form due to proton transfer. It has previously been shown that increased presence of the aci ion increases the reaction sensitivity of nitromethane [11] and could possibly explain the results obtained in this study. However, the role of pressure in the reaction mechanism of nitromethane has been widely discussed [4]. It was suggested that the reaction mechanism of nitromethane at high pressures could possibly involve a reaction precursor via an intermediate molecule [4, 12].

## CONCLUSION

The critical diameter of both amine sensitized and neat NM is observed to decrease under shock precompression. This experimental result is in contrast to the prediction that a detonation in the precompressed explosive has a significantly lower von Neumann temperature than for a detonation in the ambient explosive. That the critical diameter decreased under these conditions points to the possibility of a strong pressure dependency of the activation energy for chemical reactions at high

pressure. This study cannot single out one reaction mechanism over the other, but has shown that there is a definite pressure dependency in the reaction mechanism, whether it be through dissociation or through a reaction precursor.

## ACKNOWLEDGEMENTS

The authors would like to thank David Mack and Cyrus Foster for assistance in conducting the experiments. The dedicated assistance of the field support team of DRDC Suffield is gratefully acknowledged. This work was supported under Department of National Defence Contract W7702-04R026/001/EDM.

## REFERENCES

1. Campbell, A.W., Malin, M.E., Holland, T.E., *J. Appl. Phys.*, v 27, pp. 963, 1956.
2. Enig, J.W., Petrone, F.J., *Proc. 5th Symp. (Int) on Det.*, pp. 99-104, 1970.
3. Johansson, C.H., Persson, P.A., *Detonics of High Explosives*, Academic Press, London, 1970.
4. Winey, J.M., Gupta, Y.M., *J. Phys. Chem.*, v. 101, pp. 10733-10743, 1997.
5. Petel, O.E., Tanguay, V., Higgins, A.J., Yoshinaka, A.C., Zhang, F., *13th APS Topical Meeting: SCCM*, pp. 843-847, 2004.
6. Winey, J.M., Duvall, G.E., Knudson, M.D., Gupta, Y.M., *J. Chem. Phys.*, vol. 113, pp. 7492-7501, 2000.
7. Decker, S.A., Woo, T.K., Wei, D., and Zhang, F., *Proc. 12th Symp. (Int) on Det.*, 2002.
8. Margetis, D., Kaxiras, E., Elstner, M., Frauenheim, T., Manaa, M.R., *J. Chem. Phys.*, v. 117, pp. 788-799, 2002.
9. Manaa, M.R., Reed, E.J., Fried, L.E., Galli, G., Gygi, F., *J. Chem. Phys.*, v. 120, pp. 10146-10153, 2004.
10. Engelke, R., Schiferl, D., Storm, C.B., Earl, W.L., *J. Phys. Chem.*, v. 92, pp. 6815-6819, 1988.
11. Engelke, R., Earl, W.L., McMichael, R.C., *J. Chem. Phys.*, v. 84, pp. 142-146, 1986.
12. Bardo, R.D., *Proc. 9th Symp. (Int) on Det.*, pp. 235-245, 1989.

CP845, *Shock Compression of Condensed Matter - 2005,*
edited by M. D. Furnish, M. Elert, T. P. Russell, and C. T. White
© 2006 American Institute of Physics 0-7354-0341-4/06/$23.00

# MESO-SCALE PROBING OF CRZ STRUCTURE IN PBX: DW OSCILLATIONS FROM IGNITION UP TO FAILURE

## I. Plaksin, J. Direito, C. S. Coffey*, J. Campos, J. Ribeiro, R. Mendes and J. Kennedy**

*LEDAP – Laboratory of Energetics and Detonics, University of Coimbra, PORTUGAL*
*\*Naval Warfare Surface Center - Indian Head, MD, USA*
*\*\*Hazards & Explosives Research & Education, LLC, Santa Fe, NM, USA*

**Abstract.** Our latest contribution to the meso-scale study of PBX detonations is presented. We apply a 96-channel optical analyzer for simultaneous measurements of light irradiation from the detonation front surface, and the stress field induced by the chemical reaction zone in optical monitor. This paper addresses three major emerging topics: the dominant role of shear at the shock initiation of PBX, the origination of DW oscillations and related cellular structure of DF, and the relation between failure diameter and the size of detonation cells. New experimental evidence of the existence of a thermal precursor at SDT is presented and discussed.

**Keywords:** PBX, mesoscale, ignition, initiation: shear, failure diameter
**PACS:** 47.40.-x, 47.40.Rs

## INTRODUCTION

Application of a 96-channel fast optical analyzer allows for measurements of the mechanisms of shock ignition and micro-detonics with resolution in the scale of PBX' coarse explosive grains[1-5,7]. A large amount of registration channels allows parallel recording[1-7, 9] in the same experiment of both anisotropic processes – light irradiation from the surface of detonation front (DF), and the formation of a shock field caused by detonation flow in multi-layer optical monitor. From a combination of these nano-second resolution measurements we identify and quantify fluctuations in the chemical reaction zone (CRZ) at SDT, and at self-sustained detonation wave (DW) propagation in PBX, then analyze them as a function of HMX grain sizes, mass ratios, binder's nature (inert/energetic) and general DF curvature[1-7,9].

From the fine resolution of the CRZ's 3D structure and light irradiation patterns, we have obtained new insights in the scenario of shock initiation and detonation failure.

## TOWARDS A DOMINANT ROLE OF SHEAR IN INITIATION OF PBX-s BY mini-SWG

Conclusions about the dominant role of shear at initiation of PBX charges by a miniature SW Generator (SWG) was presented in our previous paper[2] (see also references). For better resolution of the shock ignition scenario, we performed a set of tests with acceptor charges – simulants of PBAN-128 and PBXN110, and with PETN/Epoxy 82/18. The applied experimental setup[2] and typical results are presented in Figure 1.

SW caused by mini-SWG is attenuated in a 230μm PMMA gap up to 9.5GPa. The induced shock field has a significant shear component[2] in the periphery zone $\theta \approx 60°$. It was clearly identified for all tested PBX's that intense reaction arises in zones of maximum shear, which confirms our previous conclusions.

We detected the origination of weak precursors (ramp profile) prior to the main shock front. Such predecessor events were theoretically predicted[8] ("heat precursor") and then were identified in our previous

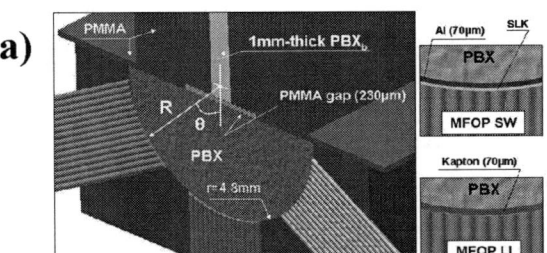

a)

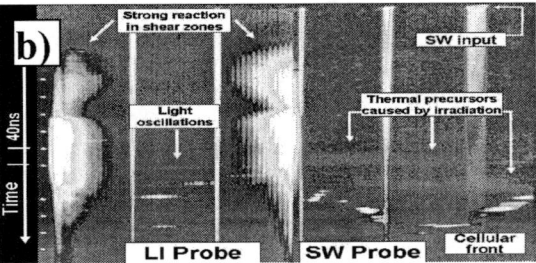

b)

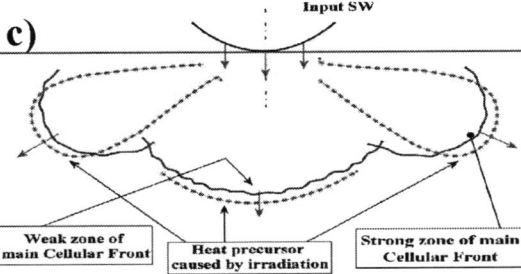

c)

**FIGURE 1.** Experiment with the PBAN-128 simulant (77 HMX $d_{50}$=20μm / 23 Epoxy); (a) Setup; (b) streak record; (c) Scheme of shock ignition. Two parallel Multi-Fiber Optical Probes (MFOP) provided simultaneous measurements of light irradiation (MFOP LI) and SW output (MFOP SW) in the same R-θ range.

work[9] ("Irradiation Impact"). We suggest that the heat precursor of the main shock is caused by thermal expansion/phase transitions of explosive crystals ahead of the main wave front through absorption of visible and infrared photons coming from the high temperature reaction zone. The main front has a cellular structure over its full surface. In the axial zone, visible reaction spots (indicated as localized light oscillations) rise behind the heat precursor. However, the reaction (associated with light irradiation) is significantly less intense in this zone than in the periphery one. Figure 1 (d) summarizes the shock ignition scenario in terms of identified cellular front and heat precursor.

## OSCILLATIONS AT SDT

Experiments[1-6] performed on a wide range of HMX-grain sizes, porosities and binders provide solid evidence that DF and CRZ, in self-sustained DW, have a cellular structure. Detonation cells, whose scale exceeds the HMX-grain mean size by a few times, result from the reaction localizations in CRZ due to the highly unstable kinetics of the shock reaction of coarse grains[1,7]. The formation of cells is due to a synergetic mechanism, with the multiple ejecta of overdriven micro-jets, originally called a "DF roughness". Their dominant role in the origination of a wide spectrum of oscillating instabilities in inertial confinement is presented in previous work[5].

The experimental setup shown in Figure 2 was applied for fine resolution measurements of oscillations at SDT in pore-free HMX/water samples subjected to a 15.4GPa shock pulse; results are shown in Figs 3 – 4.

Shock was induced in a PVC stack attenuator by detonation of a long (80x30x20mm) booster charge.

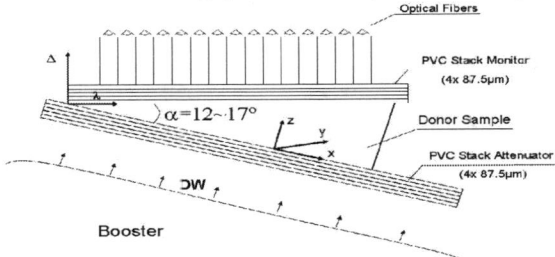

**FIGURE 2.** Experimental setup. Input (15.4GPa) shock was induced in a PVC stack attenuator by detonation of a long (80x30x20mm) booster charge.

**Figure 3.** Streak records for Figure 2 configuration.

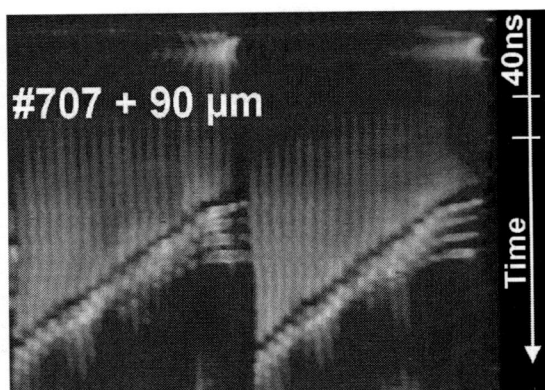

**FIGURE 4.** Streak records for Figure 2 configuration.

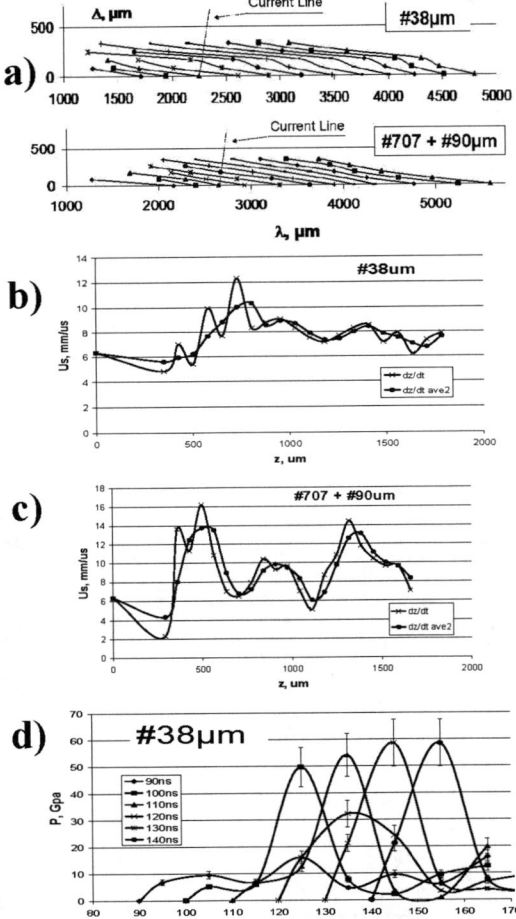

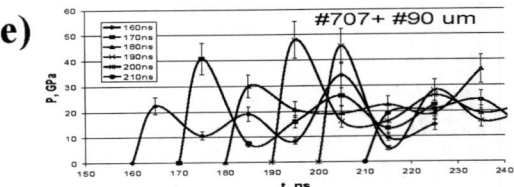

**FIGURE 5.** Experimental results: (a) 10ns-isochrones of SW front in PVC SM; (b) and (c): D-z diagrams of SDT; (d) and (e): fragments of 3D stress fields.

Fine resolution of SDT was carried out through registration of a z-t diagram of front output from the acceptor sample, and λ-t diagrams of SW motion in the PVC stack monitor (SM). Results obtained in the HMX(38-45μm)/Water with a mixture of grains 1/3 HMX(90-106μm) + 2/3 HMX (707-820μm) experiment are presented in Figure 5. Streak records of SW motion in the PVC SM allow determination of shock front isochrones and further measurements of both vector fields, $\mathbf{Us}(\lambda,\Delta)$ and stresses $\mathbf{P}(\lambda,\Delta)$, along the current lines. As it can be seen in Figure 5 (b and c), SDT is attended with the overdriven phase (1st oscillation) followed by a set of less intense and more regular oscillations. It's clearly indicated also in pattern of light intensity (see Figures 3 and 4). In the case of the fine grained "#38μm"-sample, the well pronounced ramp of the stress profiles, $\mathbf{P}(t)$, may be evidence that the effect of irradiation impact is stronger for this explosive than in case of coarse-grained "#707 + #90 μm" because the Compton scattering of photons is greater for fine particles.

Origination of oscillations at SDT can be schematically represented by the diagram shown in Figure 6.

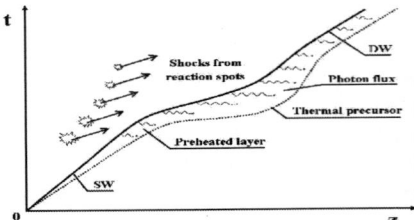

**FIGURE 6.** Diagram of SDT in terms of heat precursor.

## OSCILLATIONS UP TO DW FAILURE

Comprehensive studies of CRZ disruption were conducted on conical PBX charges[4, 6]. Tests (see Figure 7) were performed with eleven different PBX samples ($5° < \alpha < 10°$, $\rho_0 = 0.97$-$0.99$TMD, all confined by Kapton) distinguished by HMX/binder mass ratios (77/23, 82/18, 88.5/11.5, which simulate PBAN-128, PBXC-121, PBXN-110, respectively), HMX' grain sizes (narrow mono-modal fractions with $d_{50}$=20, 58, 77, 155, 195 and 508μm) and binder (HTPB, GAP, Epoxy).

Results showed a common regularity in oscillation patterns at disintegration of the CRZ structure. In all tested PBX-s, we found that in the last oscillation, the ratio between failure diameter, $\emptyset_f$, and wave length, $\lambda$, is constant: $\emptyset_f/\lambda=1.28\pm0.03$. The oscillation pattern of DW disintegration can be conceived as a process of sequential dieing out of elementary cells (one after another) followed by their separation from CRZ. At the latest oscillation, contraction of the detonation front surface is best matched to the loss, by CRZ, in one elementary cell of an area of $(\lambda/2)^2$. Hence it follows: 1 - at DW failure CRZ consists on number of $n = \pi(1.28)^2 \approx 5$ cells, and 2 - detonation critical diameter ($\emptyset_{crit}$) needs to be limited by 6 cells.

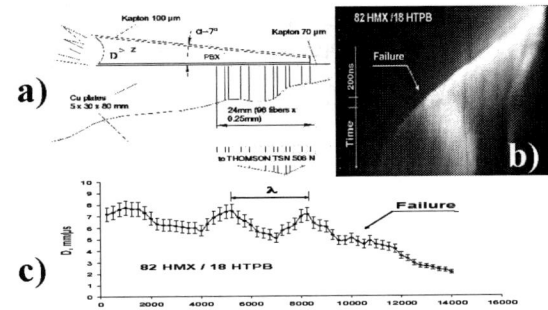

**FIGURE 7.** Conical charge test with PBXC-121 simul.; (a) setup; (b) streak record; (c) D-z diagram of DW decay.

## SUMMARY

We obtained new experimental evidence that the process of initiation and disintegration of the cellular DF runs in a progressive oscillating regime. DW transition from $\emptyset_{crit}$ to $\emptyset_f$ corresponds to contraction of a number of cells from 6 to 5. Origination of thermal precursor needs a comprehensive study.

## ACKNOWLEDGMENTS

This work was supported by the NSWC-Indian Head, under the contract n⁰ SC-IP 050303.

## REFERENCES

1. Plaksin, I. I., J. Campos, P. Simoes, A. Portugal, J. Ribeiro, R. Mendes and J. Gois. "Detonation Study of Energetic Micro-Samples". Proc., 12th Symp. (Int.) on Detonation, San Diego, CA, 11-16 Aug. 2002, p.p. 42-50.
2. Plaksin, I., J. Campos, R. Mendes, J. Ribeiro, J. Direito, D. Braga and C. S. Coffey. "Effect of Shear Stress in Shock Initiation of PBX". Shock Compression of Condensed Matter - 2003. AIP Conference Proceedings 706. Edited by M. D. Furnish, Y. M Gupta and J. W. Forbes, p.p.1013-1016.
3. Plaksin, I., , J. Campos, R. Mendes, P. Simoes, A. Portugal, J. Ribeiro and J. Gois. "Coarse Explosive Particles of PBX as Dominant Factor of Detonation Instability". Ibid, p.p. 887-890.
4. Plaksin, I., J. Campos, J. Direito, J. Ribeiro, R. Mendes and J. Gois. "Birth and Death of PBX's Detonation in Mezzo-Scale: Effect of Grain Sizes and Binder's Nature in Ignition and Failure of Detonation". Proc., 35th Int. Conf. of ICT on Energetic Materials. June 29 – July 2, 2004. Karlsruhe, FRG, paper 24.
5. Plaksin, I., J. Campos, J. Direito, J. Kennedy, S. Coffey, R. Mendes, J. Ribeiro, J. Gois."Micro-Ejecta from Detonation Front as Ignored Factor in Performance of PBX Detonation". Proc., 36th Int. Conf. of ICT on Energetic Materials. June 28 – July 1, 2005. Karlsruhe, FRG, paper 11.
6. Mendes, R., J. Ribeiro I. Plaksin, J. Direito and J. Campos. "Phenomenon of Detonation Wave Failure in PBX based on HMX with the Inert Binder". Ibid, paper 168.
7. Pedroso, L. M., I. Plaksin, P. Simoes, J. Direito, J. Campos and A. Portugal. "Triazine pre-Polymer Energetic Binder – Shock Behavior and Detonation Performance in Scale of HMX Grain Size». Ibid, paper169.
8. Zababakhin E. I. and Simonenko V. A. "Converging Wave in Heat-Conducting Gas". In Prikl. Mat. i Mekh., 1965, Vol. 29, Issue 2, pp. 334-336 (Russian).
9. Plaksin, I., J. Direito, J. Campos and S. Coffey. "Kinetics Diagrams at Shock Ignition of PBX and Non-Ideal Explosives". Int. Conf. New Models and Hydro-codes for Shock Wave Process, Univ. of Maryland, 16-21 May, 2004.

CP845, *Shock Compression of Condensed Matter - 2005*,
edited by M. D. Furnish, M. Elert, T. P. Russell, and C. T. White
© 2006 American Institute of Physics 0-7354-0341-4/06/$23.00

# THERMO-KINETIC STUDY OF CORE-SHELL NANOTHERMITES

## A. Prakash[1], A.V. McCormick[2] and M. R. Zachariah[1]

[1]*Department of Chemistry and Biochemistry, and Department of Mechanical Engineering, University of Maryland, College Park, MD 20742*
[2]*Department of Chemical Engineering and Material Science, University of Mineesota, Minneapolis, MN 55455*

**Abstract.** This article presents the formulation of a new nano-thermite mixture ($Al/KMnO_4$) for application in energetic materials. Reactivities of different thermite mixtures have been compared by a constant volume combustion experiment and the results indicate that the reactivity of the new formulation is two orders of magnitude greater than the traditional formulations. We also present a generic technique for synthesizing core-shell nanostructured composite particles as a means of controlling the reactivity and initiation. By coating a strong oxidizer nanoparticle ($KMnO_4$; ~150 nm) by a layer of mild oxidizer ($Fe_2O_3$; ~4-15 nm) the measured reactivity in terms of pressurization rate (psi/μs) could be varied by more than a factor of 10. The composite oxidizer particles were synthesized by a two-temperature aerosol process where the non-wetting interaction between the two components of the particle causes phase segregation into a core-shell structure. We also show that the characteristic reaction times and particle length scales can be used to measure characteristic diffusion rates which are indicative of the reactivity of a thermite nanocomposite.

**Keywords:** Nano-thermite, pyrotechnics, core-shell, pressurization rate.

## INTRODUCTION

Thermite metastable intermolecular composites (MIC) are a fascinating class of high-energy materials and have been a subject of extensive research over the past decade [1-3]. Such materials are composed of an intimate mixture of nanoparticles of two components - fuel and oxidizer. Since the goal is to enhance the reactivity, the use of nanoscale material reduces the mass transport limitations between the fuel and oxidizer and the reaction moves towards kinetic control [4]. Aluminum nanoparticles are invariably used as fuel, while there are a host of metal oxide nanoparticles that are used as oxidizer for the MIC. Thermodynamic calculations of adiabatic flame temperatures and reaction enthalpy help us choose the possible MIC components from a large number of possible nano-thermite combinations. Of the numerous thermodynamically possible thermite

formulations [5], the most widely used MICs are nano-Al (fuel) combined with $MoO_3$, $CuO$ and $Fe_2O_3$ oxidizers. We recently reported a new MIC formulation of nano-Al combined with $KMnO_4$ nanoparticles [6] the reactivity of which is about two orders of magnitude higher than the traditional formulations.

Although there has been considerable success in formulating new composites with greater energy release rates [1-3, 6], the subject of achieving a precise control over reactivity of nanothermites is an opportunity for further research. There have been some studies done on the size-dependent reactivity [7, 8] of nanoparticles as a means to control reactivity. In one of our earlier works [9], we reported a method of charge enhanced particle assembly to realize greater energy release rates.

A MIC mixture of $Al/KMnO_4$ although very reactive, would have very poor shelf life compared to the other traditional MIC combinations, the reason being the strong oxidizing nature of $KMnO_4$. In such a MIC, the oxidizer slowly converts all the aluminum to its oxide reducing the shelf life. In this contribution we report a novel technique to moderate the reactivity of nanoenergetic materials. We describe the synthesis of composite oxidizer nanoparticles with a core containing the strong oxidizer (potassium permanganate) and the shell of variable thickness of a relatively mild oxidizer (iron oxide). Using a composite oxidizer (potassium permanganate coated with Iron oxide) as described in this paper, we are able to tune the reactivity over a relatively large dynamic range by changing the thickness of the less reactive oxidizer. An additional benefit is that the structure has an improved shelf life, and also reduces the sensitivity of the nanoenergetic mixture to any unintended initiations.

## EXPERIMENTAL PROCEDURE

The composite oxidizer nanoparticles were synthesized by a new single-step, two-temperature aerosol spray-pyrolysis method, the schematic of which is shown in Figure 1. A key to the synthesis of the desired microstructure is to employ the difference in the characteristic temperatures of the two components. The difference between thermal decomposition temperature of iron nitrate (<100 °C) and the melting point of potassium permanganate [10] (~240 °C) in this case is the key to obtaining coated particles. An aqueous solution of Iron (III) nitrate nonahydrate $(Fe(NO_3)_3.9H_2O)$ and potassium permanganate $(KMnO_4)$ is sprayed into droplets using a collision-type nebulizer. The total salt precursor concentration is kept constant at about 2 wt %. The initial droplet size is about 1 μm in diameter (measured by a high-sensitivity laser aerosol spectrometer). The moisture from the aerosol is absorbed in silica-gel diffusion dryer. The aerosol is then passed through two tube furnaces, the first one maintained at about 120 °C (above decomposition temperature of iron nitrate), and the second at about 240 °C (~ melting point of permanganate). The composite particles are then collected on 0.6 μm DTTP filter manufactured by Millipore.

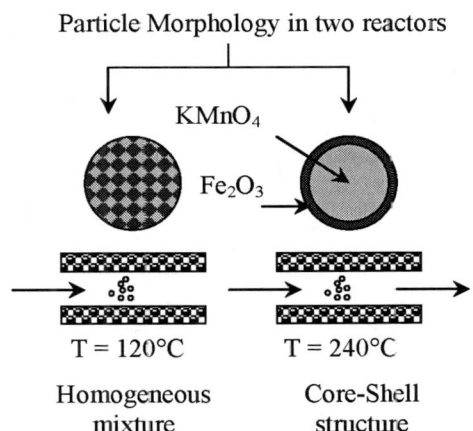

**FIGURE 1.** Schematic of aerosol experiment for synthesis of core-shell structured oxidizer nanocomposite.

In the first furnace iron nitrate decomposes to form iron oxide while the permanganate remains a solid. Under these conditions an intimate mixture of the two components exists. In the second furnace, as potassium permanganate melts, the solid iron oxide particles become more mobile in presence of a liquid like matrix. This increased mobility of the iron oxide particles helps them rearrange and aggregate towards the exterior of the particle. Presumably, the liquid like potassium permanganate molecules do not like to wet the surface of solid iron oxide and hence they phase segregate and pool in the center of the particle to minimize surface energy. This results in the formation of a composite nanoparticle of potassium permanganate coated with iron oxide.

## RESULTS AND DISCUSSION

A TEM (transmission electron microscope) image along with the elemental map of the composite nanoparticles is shown in Figure 2. The TEM image clearly shows core-shell structure particles. The nature of the core-shell structure is confirmed with the elemental map, which shows high iron intensity on the perimeter of the particles and high manganese content in the interior. Our experiments indicate that a certain minimum amount of iron nitrate precursor was necessary to successfully coat the potassium permanganate particles. We find that an iron nitrate to potassium permanganate ratio of at least 3:1 (by weight in the precursor) was necessary to coat the particles, and corresponds to a particle of calculated 86% by volume KMnO$_4$ (for a 150 nm composite particle with a ~ 4 nm iron oxide coating). The elemental map shown in Figure 2 is in agreement with our estimation of coating thickness. Using a smaller

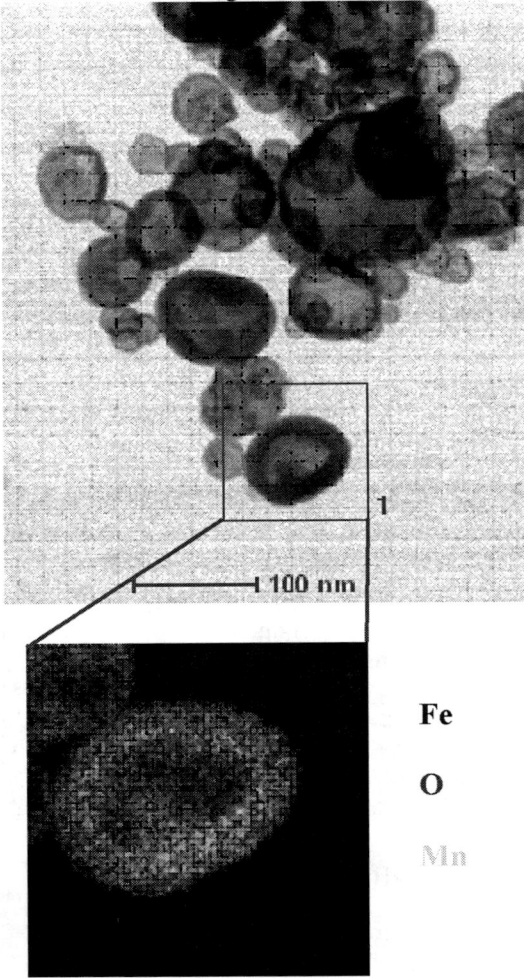

**FIGURE 2.** TEM image and elemental map showing the core-shell structure of the nanocomposite.

amount of iron nitrate in the precursor resulted in the formation of the particles that were uncoated or partially coated.

Having coated a very strong oxidizer with a very mild oxidizer, we measured the reactivity of

these composite oxidizer particles with aluminum to see if we could modulate the reactivity by changing the coating thickness. One approach to estimate the reactivity of MICs is to measure the pressurization rate during confined combustion. A fixed amount of MIC (25 mg) is ignited in a small volume (~13 cc) pressure vessel, and the pressure of the vessel is monitored as a function of time. The initial slope of the pressure rise is defined as the pressurization rate, and is reported in units of psi/μs. The more reactive the MIC is, the faster the pressure in the vessel rises and the higher is the pressurization rate. The details of the pressure vessel apparatus and measurement protocol can be found in a prior work [6].

Figure 3 shows the pressurization rate measured as a function of different stoichiometric proportions of fuel and oxidizer. The three

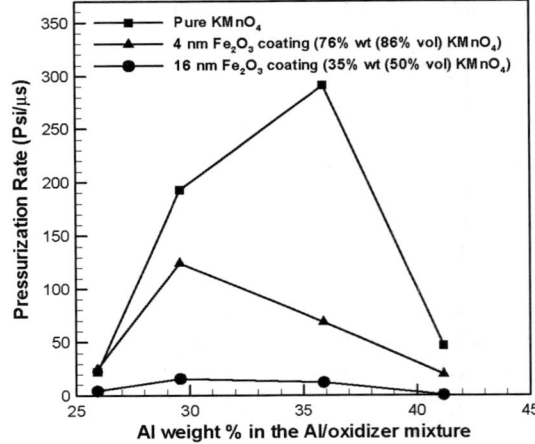

**FIGURE 3.** Pressurization rate for different composite MICs as a function of stoichiometry.

different curves show measurements for three oxidizers of different compositions. We notice that increasing the coating thickness of a weaker oxidizer steadily moderates the reactivity of the MIC. The maximum pressurization rate measured for Al/KMnO$_4$ MIC is about 290 psi/μs while that for Al/Fe$_2$O$_3$ is ~ 0.017 psi/μs. In principle, we should be able to tune the reactivity between these two limits by modifying the composition of the oxidizer. We also found that by simply mixing the two oxidizer nanoparticles one cannot control the reactivity. We believe that the diffusion of reaction species through a layer of weak oxidizer is the rate limiting step for the thermite reaction. In other words, diffusion rate of species through the weak oxidizer layer is very slow compared to diffusion through potassium permanganate. We calculate this characteristic diffusion rate by taking the ratio of square of the coating thickness to the characteristic reaction time obtained from the pressurization rate, as shown in Table 1. The size of the composite particle is ~150 nm. We see that characteristic diffusion rate of species through a layer of iron oxide is ~ 1e-11 $m^2$/s whereas, the diffusion rate through permanganate is ~ 1e-9 $m^2$/s. A constant

**TABLE 1.** Estimated diffusion rate of reaction species through oxidizer layer.

| $KMnO_4$ (vol %) | P-rate (psi/μs) | Coating thickness of iron oxide (nm) | Reaction Time (μs) | Diffusion coefficient ($m^2$/s) |
|---|---|---|---|---|
| 0 | 0.02 | 77 | 210 | 2.82e-11 |
| 49.5 | 16.03 | 16.2 | 12.91 | 2.04e-11 |
| 83 | 121.02 | 4.9 | 1.66 | 1.48e-11 |
| 86.7 | 124.58 | 3.8 | 1.60 | 8.79e-12 |
| 100 | 290.87 | 0 | 1.29 | 4.36e-9 |

diffusion rate of species through iron oxide obtained from four different experiments suggests that the once the reaction is initiated, it is the diffusion of species that limits the reaction rate, which in this case is diffusion through iron oxide.

## CONCLUSIONS

We have been able to create core-shell type composite oxidizer nanoparticles by taking advantage of the temperature difference between the decomposition temperature of the shell precursor (iron nitrate) and the melting point of the core (potassium permanganate). We have demonstrated that one could tune the reactivity of the nanoparticles by varying the coating thickness of the shell material.

## REFERENCES

1. Prakash, A.; McCormick, A. V.; Zachariah, M. R., "Aero-Sol-Gel Synthesis of Nanoporous Iron-Oxide Particles: A Potential Oxidizer for Nanoenergetic Materials," Chemistry of Materials, 16, 1466, 2004.

2. Granier, J. J.; Pantoya, M. L., "Laser Ignition of Nanocomposite Thermites," Combustion and Flame, 138, 373, 2004.
3. Tillotson, T. M.; Gash, A. E.; Simpson, R. L.; Hrubesh, L. W.; Satcher, J. H.; Poco, J. F., "Nanostructured Energetic Materials using Sol-Gel Methodologies," J. Non-Cryst. Solids, 285, 338, 2001.
4. Aumann, C. E.; Skofronik, G. L.; Martin, J. A., "Oxidation Behavior of Aluminum Nanopowders," J. Vac. Sci. Technol., B: Microelectron. Nanometer Struct-Process, Meas., Phenom., 13, 1178, 1995.
5. Fischer, S. H; Grubelich, M. C., "Theoretical Energy Release of Thermites, Intermetallics, and Combustible Metals," presented at the 24th Int. Pyrotechnics Seminar, Monterey, CA, July 1988.
6. Prakash, A.; McCormick, A. V.; Zachariah, M. R., "Synthesis and Reactivity of a Super-Reactive Metastable Intermolecular Composite (MIC) Formulation of Al/KMnO₄," Adv. Mater. , 17(7), 900 2005.
7. Granier, J. J; Pantoya, M. L., "The Effect of Size Distribution on Burn Rate in Nanocomposite Thermites: a Probability Density Function Study" Combustion Theory and Modeling, 8, 555, 2004.
8. Tomasi, R; Munir, Z. A., "Effect of Particle Size on the Reaction Wave Propagation in the Combustion Synthesis of Al2O3-ZrO2-Nb Composites, J. Am. Ceram Soc., 82, 1985, 1999.
9. Kim, S. H.; Zachariah, M. R., "Enhancing the Rate of Energy Release from NanoEnergetic Materials by Electrostatically Enhanced Assembly," Advanced Materials, 16(20), 1821, 2004.
10. Moghaddam, A. Z.; Rees, G. J., "Thermal Decomposition of Potassium Permanganate. Thermogravimetry, Differential Scanning Calorimetry and Hot-stage Microscopy Studies," Fuel, 1984, 63, 653.

CP845, *Shock Compression of Condensed Matter - 2005,*
edited by M. D. Furnish, M. Elert, T. P. Russell, and C. T. White
2006 American Institute of Physics 0-7354-0341-4/06/$23.00

# A STUDY OF THE EFFECT OF ELECTRICAL ENERGY INPUT ON DETONATION FAILURE IN WEDGES OF THE TATB-BASED EXPLOSIVE EDC35

## Darren Salisbury, Ron Winter and Lester Biddle

*AWE, Aldermaston, Reading, Berkshire, RG7 4PR, UK*

**Abstract.** Experiments have been conducted to investigate detonation failure in wedges of the Insensitive High Explosive EDC35 (95/5 TATB/Kel-F) with and without addition of external energy via an applied electrical field. The thickness of the plane wedges varied from 6 mm to 1 mm along a 100 mm length with a width of 50 mm. The wedge was initiated along a line at its thick edge. Streak photography was used to record the progression of the detonation wave from the thick end of the wedge and its subsequent failure towards the thin end of the wedge. Three experiments were conducted: 1) with no external electrical energy, 2) with input of 1.25 kJ from a 25 kV pulsed power source applied to electrodes mounted in contact with the 50 mm x 100 mm faces of the wedge and 3) with input of 5 kJ applied to the electrodes. Analysis of the streak records suggests that failure thickness was reduced.

**Keywords:** TATB, electrical energy, wedges, detonation failure
**PACS:** 47.40.-x

## INTRODUCTION

The concept of an explosive having a characteristic "failure diameter" is well established. This is defined as the smallest diameter cylinder along which a detonation wave will propagate. Similarly there is a failure thickness which is the thickness of the thinnest slab of explosive along which it is possible for a detonation wave to propagate. Ramsay (1) demonstrated that failure thickness in slab geometry is half the cylindrical failure diameter. Campbell and Engelke (2) reported failure diameters for various explosive compositions. HMX-based explosives such as PBX9501 exhibit failure diameters of the order of 1mm whereas TATB-based compositions typically have failure diameters of an order of magnitude larger. Ramsay (1) found that thin layers of confinement reduced the failure thickness of an explosive considerably. The reduction in failure thickness was found to increase with increasing impedance and thickness of the confining material.

Ramsay also showed that failure thickness decreased with increasing temperature.

It is conceivable that an ability to control the failure thickness of a given explosive could form the basis of an explosive switch or shutter. In such a device, detonation would only be able to propagate along a channel that is thinner than the ambient failure thickness, when either 1) the confinement is increased 2) the HE temperature is increased or 3) external energy is added to the reaction zone and detonation front. This paper describes a preliminary experimental study of the effect of added electrical energy on detonation failure.

It is known that the conductivity of the reaction zone in a detonating explosive is higher than that of the unreacted explosive or the detonation products. Time resolved measurements of this conductivity have been used as a means of studying the physics and chemistry of the detonation process (3). Furthermore, in principle, this distribution of conductivity allows the performance of the explosive to be enhanced by depositing external energy in the reaction zone. This is achieved by applying an electric field across the conducting reaction zone causing current to flow and dissipate energy via joule heating. A performance increase can be observed experimentally by an increase in

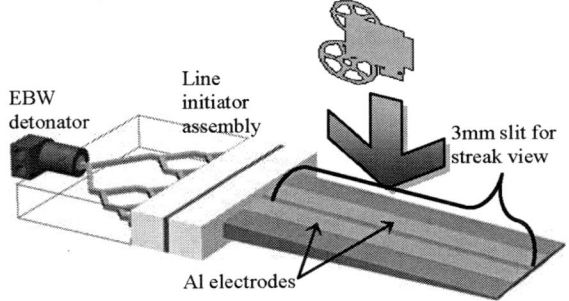

**FIGURE 1.** Schematic of EED experiment assembly

detonation velocity. Lee et al (4) studied detonation enhancement in 2 mm thick Primasheet placed between copper electrodes. Electrical energy delivered by a 5 kJ source increased the average and local detonation velocities by an estimated 3% and 10.4% respectively. The objective of the experimental study reported here is not to enhance performance over a sustained time frame but to assess the effect of external energy on detonation failure thickness.

## EXPERIMENTAL DESIGN

The insensitive high explosive EDC35 (TATB/Kel-F 95/5) was selected for this study. Hutchinson et al (5) confirmed that the EDC35 failure diameter of approximately 8mm is consistent with that measured in the near-identical composition PBX 9502 (6). Because of this relatively large failure diameter, it was judged that changes in failure behaviour would be easier to resolve experimentally.

The wedge geometry, as depicted in Fig. 1, simplified experimental assembly and allowed an optical streak diagnostic to be used. The thickness of the plane wedges varied from 6 mm to 1 mm along 100 mm length (defined here as the direction in which detonation propagates), and the width of the wedge was 50 mm. The wedge was initiated along a line at its thick edge. A Cordin 134 rotating mirror streak camera, running at 22 mm/µs was used to record the progression of the detonation

wave from the thick end of the wedge and its subsequent failure towards the thin end of the wedge. Aluminium electrodes, 11 mm wide and 0.25 mm thick were positioned in pairs separated by 3 mm on both large faces of the wedge. The 3 mm gap orientated parallel to the direction of detonation propagation is viewed using a streak camera. The electrodes are positioned in identical positions on both faces of the wedge to maintain symmetry. Analysis of the streak record gives a continuous measurement of detonation front position and consequently determination of the detonation failure point. Due to the geometry of the wedge a change in failure thickness of say 1mm, translates to a 20mm shift in failure position as viewed by the streak camera on the face of the wedge.

Not shown in Figure 1 are sheets of 50 µm Kapton polyimide film which were positioned on both sides of the wedge between the explosives surface and the Aluminium electrodes. The purpose of this insulating film was firstly to prevent possible tracking around the experimental assembly. Furthermore, as polyimide is known to breakdown at high pressures, the film in the assembly ensured the voltage across the explosive thickness held off until the arrival of the detonation wave. A low current insulation test showed that the 2 thicknesses of film held off 28 kV between the aluminium electrodes.

Pulsed power supplies designed for driving Xenon flash tubes were used as the energy source for this experimental series. Four channels were available for use, each comprising a 4 µF capacitor that is charged to 25 kV, theoretically giving 1.25 kJ of energy from each output.

The overall objective of this study was to demonstrate the concept of reducing the failure thickness of the explosive by increasing the energy of the detonation by electrical means. To achieve this the following three experiments were fired in this Electrically Enhanced Detonation (EED) study. EED2/1 was a control shot to record the detonation velocity history and failure without external energy applied. EED2/2: was fired with nominal 1.25 kJ energy, input at a time when the detonation wave had propagated a significant distance into the wedge, and EED2/3 was fired with nominal input of 5 kJ input just before detonation was established along the thick end of the wedge.

## RESULTS AND ANALYSIS

The three streak films were digitised using an optical scanner to enable data reduction via image analysis software. Approximately 50 optical density

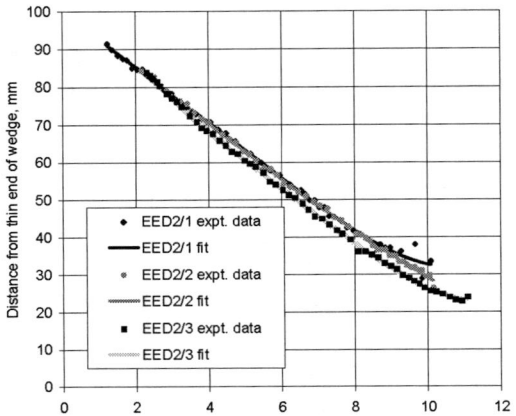

**FIGURE 2.** Detonation wave distance-time data and polynomial fits derived from streak records

column profiles were analysed to determine the location of the leading edge of the detonation wave. A fifth order polynomial function was fitted to these points to give a continuous position-time history. Figure 2 shows the detonation wave position histories and polynomial fits for the three experiments. Velocity-time profiles, shown in Fig. 3, were derived by differentiating the polynomial position histories. A static full-frame image at the beginning of the streak record enabled the distance axis to be scaled and its absolute position fixed. This enabled the velocity-time histories to be converted to velocity-HE thickness plots as shown in Fig. 4.

The data from the control experiment EED2/1 shows that the detonation wave propagates initially at an approximately constant velocity of around 7.3 mm/μs. Failure appears to start at a thickness of approximately 3.8 mm which is consistent with Ramsay's prism test data on unconfined PBX9502 (1). On this basis it can be concluded that the confining electrodes have had no effect on the failure thickness. Ramsay reports a reduction of failure width from 4mm to 2.2 mm in PBX9502 confined with 0.25 mm aluminium. In the experiments reported here the Kapton film between the EDC35 and the aluminium effectively acts like

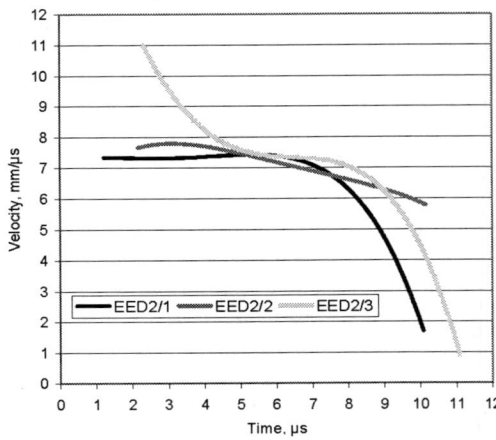

**FIGURE 3.** Detonation wave velocity-time histories

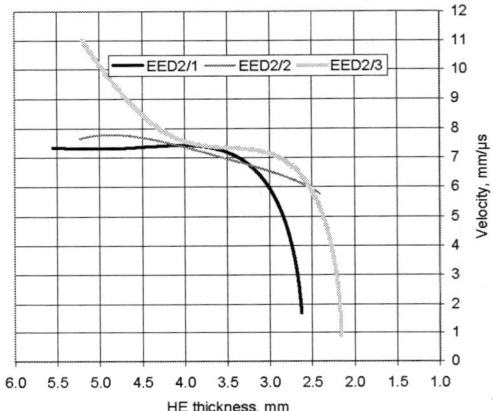

**FIGURE 4.** Derived wave velocity-explosive thickness plots

a free surface because of its low acoustic impedance.

The data from EED2/2 suggest that although the detonation velocity before the 4 mm thickness appears unsteady, (and indeed appears to start failing earlier than the control), the application of the external electric field at around 6.5 μs has retarded the failure process. A current monitor probe fielded in this experiment recorded a long rise-time of 7.5 μs to reach peak-current of 4.5 kA. Therefore in the timescale of the experiment only a fraction of the available energy could be delivered to the assembly.

In the third experiment, EED2/3, the electric field was applied before detonation was established in the wedge to ensure there would be enough time for the circuit to deliver the current and for the energy to feed forward to the detonation front. Also, to maximise the possibility of producing an observable effect, the amount of energy delivered to the HE was increased. This was achieved by connecting all 4 available channels of the power supplies in parallel across the explosive. As shown in Figs. 3 and 4 the velocity data obtained from this third experiment implies a very high detonation velocity at early time of up to 10 mm/μs which is not considered credible. The cause of this anomaly is not known but it could be due to light reflection or electrical arcing from the electrodes encroaching across the slit view. After 5 μs the data implies a detonation velocity more consistent with the previous two experiments. However this velocity is sustained for longer and failure appears to begin when the wave reaches approximately 3 mm thickness, compared to 3.8 mm in the control experiment. It is evident from the velocity histories plotted in Fig. 4 that the detonation wave propagation and failure behaviour are different for the three experiments.

Although the failure process does not lead to a sharp transition from detonation to shock, qualitatively the streak camera data suggest that the failure thickness *has* been reduced. However it is useful to quantify the effect in some way. A hydrocode calculation was performed incorporating a WBL model (7) for the EDC35 detonation that does not model detonation failure. Detonation velocity along the wedge was calculated to be a near-constant 7.35 mm/μs. Arbitrarily, failure thickness was defined as the wedge thickness at which the velocity reduced to 90% of the constant velocity derived from the hydrocode model. The derived failure thicknesses are shown in Table 1.

| Experiment | Applied energy | Detonation wave failure thickness, mm |
|---|---|---|
| EED2/1 | 0 | 3.2 |
| EED2/2 | 25 kV, 1.25 kJ (late in time) | 3.1 |
| EED2/3 | 25 kV, 5 kJ (early in time) | 2.7 |

**TABLE 1**. Detonation failure thickness arbitrarily defined as thickness at which velocity falls to 90% of WBL predicted velocity of 7.35 mm/μs.

## CONCLUSIONS

Further work is required to address issues of experimental reproducibility and to develop the electrical diagnostics. However, the results of this experimental study suggest that the failure width of an explosive can indeed be reduced by adding external energy via electrical means. We note that the energy required to reduce failure width may be relatively small if the timing of its input is precisely controlled to give an extra "kick" to a detonation wave which is on the point of failing.

It may be possible to use the phenomenon studied here to create an electrically controlled explosive switch which may have safety and other applications.

## ACKNOWLEDGEMENTS

The authors gratefully acknowledge Garry Reece and Aled Jones for their assistance in performing these experiments

## REFERENCES

1. Ramsay, J. B., "Effect of Confinement on Failure in 95 TATB / 5 Kel-F." 8th International Detonation Symposium, 1985
2. Campbell, A.W. and Engelke, R., "The Diameter Effect in High Density Heterogeneous Explosives.", 6th International Detonation Symposium, 1976
3. Tasker, D. G. and Lee, R. J., "The Measurement of Electrical Conductivity in Detonating Condensed Explosives." 9th International Detonation Symposium, 1989
4. Lee, J. et al, "Enhancement of Detonation Properties by Electrical Energy Input." Shock Compression of Condensed Matter, 1999
5. Hutchinson, C. D., Foan, G. C. W., Lawn, H. R. and Jones, A. J., " Initiation and Detonation Properties of the Insensitive High Explosive TATB/ Kel-F 900 95/5." 9th International Detonation Symposium, 1989
6. Campbell, A.W., "Diameter Effect and Failure Diameter of a TATB-based Explosive.", Propellants, Explosives and Pyrotechnics, Vol.9, 1984, pp183-187.
7. Lambourn, B. D. and Swift, D. C., "Application of Whitham's Shock Dynamics Theory to the Propagation of Divergent Detonation Waves", 9th International Detonation Symposium, 1989

CP845, *Shock Compression of Condensed Matter - 2005*,
edited by M. D. Furnish, M. Elert, T. P. Russell, and C. T. White
© 2006 American Institute of Physics 0-7354-0341-4/06/$23.00

# FLAME SPREAD ACROSS SURFACES OF PBX 9501

## S. F. Son[1] and H. L. Berghout[2]

[1]*Los Alamos National Laboratory, P.O. Box 1663, Los Alamos NM 87545*
[2]*Chemistry Dept., Weber State University, Ogden UT 84408*

**Abstract.** We report the results of flame-spread experiments of PBX 9501 (HMX-based explosive). The horizontal flame spread rate, $V_f$, for PBX 9501 is curve-fit with a power law function of pressure from 0.077 to 17.3 MPa, specifically, $S_f = 0.259 P^{0.538}$ (cm/s) where $P$ is the dimensionless pressure $p/p_0$ with $p_0 = 0.1$ MPa. $V_f$ is of the same order of magnitude as normal deflagration and varies nearly as the square root of pressure, as scaling estimates predict. In the vertical orientation, the flame propagation downward was observed to be slightly faster than horizontal flame spread presumably because of the melt layer flowing downward on the sample.

**Keywords:** PBX 9501, explosives, combustion, flame spread.
**PACS:** 82.33.Vx

## INTRODUCTION

Accidental ignition of bare explosive charges is a potential safety issue. A critical question is how fast a flame spreads across charge to engulf the entire charge. The answer to this question could affect fire suppression and evacuation planning. Although flame spread has been studied with some solid propellants (see Kumar and Kuo for a review[2]), it has not been considered for explosives, and in particular for nitramines that are important in both propellant and explosive formulations. Nitramines have received extensive study and detailed chemistry has been developed,[3] consequently data for a nearly pure material is of interest to modelers. Several reviews are available for flame spread with the emphasis on fuels in an oxidizing environment.[4,5]

Recent studies have focused on flame spread in cracks and gaps.[6] Flame spread across the surface of PBX 9501 represents the limiting case of an infinitely wide crack. The rate of flame spread across a two dimensional surface of PBX 9501 represents the limiting case of an infinitely wide crack. Flame spread is related closely to ignition, it can be viewed as a continuous ignition process occurring across a surface. Together with burning rate[1], and ignition[7], flame spread data can be used to calibrate and validate combustion modeling. Here we report on experiments of the flame-spread rate across the surface of horizontal and vertical flat samples of PBX 9501 at various pressure levels. Simple scaling arguments are made and compared to experimental results.

## EXPERIMENTAL PROCEDURE

The energetic material used in this work is PBX 9501, a high explosive composed of 95% HMX by weight in a plasticized, estane-based binder. Samples for these experiments are machined to the required shapes from PBX 9501 molding powder that has been hydrostatically pressed to 1.83 g/cm³. The surface flame-spread experiments use 4.0 × 2.0

× 0.2-cm PBX 9501 flats that are mounted either vertically or horizontally in an argon pressurized combustion vessel, or in atmospheric conditions. An electrically heated NiCr wire uniformly ignites an entire edge of the flat. For pressurized experiments, a mixture of ammonium perchlorate (AP) and 3,6-dihydrazino-s-tetrazine (DHT) is spread along one of the 2.0-cm edges of the flat, and uniformly ignites the entire edge of the sample. Flame spread across the sample is recorded by video in profile from a 4.0-cm edge of the sample and analyzed digitally. An Omega Model PX605-10KGI pressure transducer monitors the pressure in the combustion vessel. Tektronix Model TDS 460A and TDS 540A digital oscilloscopes capture the pressure sensor outputs for later storage and analysis.

**Table 1. Flame spread as a function of pressure**

| Pressure (MPa) | Orientation | Spread Rate | Normal Burn Rate[a] |
|---|---|---|---|
| 17.3 | Horizontal | 4.4 cm/s | 2.1 cm/s |
| 8.0 | Horizontal | 2.4 cm/s | 1.0 cm/s |
| 3.5 | Horizontal | 1.7 cm/s | 0.47 cm/s |
| 1.4 | Horizontal | 1.3 cm/s | 0.21 cm/s |
| 0.55 | Horizontal | 0.57 cm/s | 0.080 cm/s |
| 0.21 | Horizontal | 0.37 cm/s | 0.036 cm/s |
| 0.077 | Horizontal | 0.23 cm/s | 0.014 cm/s |
| 17.4 | Vertical | 5.4 cm/s | 2.1 cm/s |
| 1.5 | Vertical | 1.3 cm/s | 0.21 cm/s |

[a] From fit of Son, et al.[15]

## SURFACE FLAME SPREAD

Convective heat transfer that drives combustion in a slot is minimized in the case of flame spread across an open flat. All these flame-spread experiments were performed in quiescent conditions. Figure 1 is an image from a typical video record showing flame spread across the surface of PBX 9501 at 8.0 MPa. Arrows identify the flame front and the regressing explosive surface.

The flame is viewed edge-on as it progresses across the explosive surface. Table 1 contains flame spread rate results covering a range of pressures between 0.077 to 17.3 MPa for both horizontal flame spread and few vertical flame spread experiments. In the case of vertical flame spread,

sample ignition occurs at the top of the sample and the flame progresses downward. Table 1 also contains normal burn-rate data for PBX 9501 based on the work of Son, et al.[1] The normal burn-rate is also called the layer-by-layer rate. These data are plotted in Fig. 2 along with a fit of the flame-spread data. The horizontal flame spread rate in PBX 9501 over the range of pressures in our study appears to

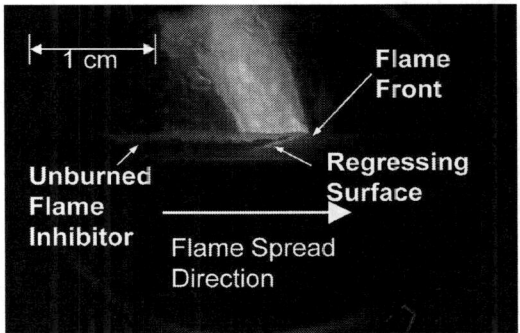

**FIGURE 1.** Image of flame spread across the surface of PBX 9501 at 8.0 MPa. The flame is moving from left to right in this view.

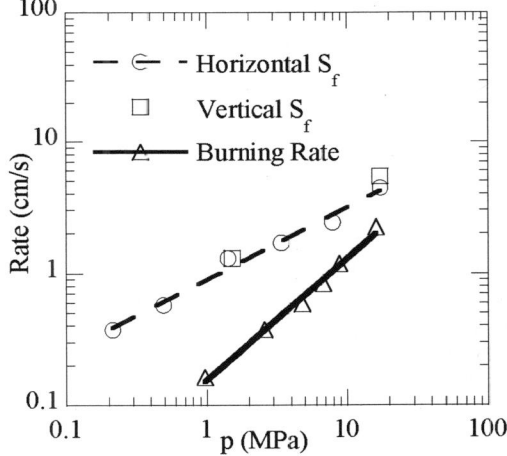

**Figure 2**. PBX 9501 surface flame spread rate. The dashed line is a fit of the horizontal flame spread data. Normal burn rate data for PBX 9501 for Son, et al.[1] is included for comparison.

obey a rate law of the form $S_f = 0.259 P^{0.538}$ (cm/s) where $P$ is the dimensionless experimental pressure defined as $P = p / p_0$ with $p_0 = 0.1$ MPa.

The pressure dependence of horizontal flame spread in PBX 9501 is weaker than that of the normal burn rate, which obeys a pressure law of the form $r_b = 0.018\,P^{0.924}$ (cm/s) .[1]    At 17.3 MPa, the surface flame spread rate is somewhat greater than twice the normal burn rate for PBX 9501. The normal burn rate for PBX 9501 decreases more quickly with decreasing pressure than does the surface flame spread rate. At 0.21 MPa, the surface flame spread rate is nearly ten times the normal burn rate. Combining the rate law for flame spread with the rate law for normal burning in PBX 9501 suggests that at pressures near 110 MPa the two rates should be about equal. At pressures above the limit where the flame spread rate and flame burn rate become equal, the flame spread rate and normal burning rate must remain equal since flame spread cannot logically progress at a slower rate than the normal burn rate of the material. However, normal burning would be unlikely at such high pressures due to reaction propagation into the very small voids and cracks that exist in all pressed PBX 9501 samples.[8]

The orientation of the explosive's surface produces the largest effect at high pressure where flame spreads across a vertical surface about 25% faster than it spreads across a horizontal surface. This phenomenon might be anticipated under some circumstances such as combustion of a fuel in an oxidizing atmosphere. In that case, heat-driven product-gas convection away from the reaction zone will tend to draw more reactant gas into the reaction zone. Since PBX 9501 is an energetic material burning in inert argon gas, flow of fresh argon to the reaction zone from the surrounding atmosphere is not necessary and could actually impede flame spread if convection interferes with heat transfer back to the PBX 9501 surface. During combustion of PBX 9501, the solid material melts before vaporizing. The main heat releasing reactions occur in the gas phase. The observed result is possibly due to the flow of melted PBX 9501 flowing down the face of the explosive. It is difficult to confirm from the video record if flow of the melt is actually occurring but a similar process has been observed in the combustion of aluminum metal in an oxygen rich atmosphere.[9]

## DISCUSSION

A very simple analysis was considered with the goal of determining what the pressure dependence of spread rate should be and how it compares to normal deflagration. We consider two zones, preheat and reaction zones. The reaction zone is of the thickness $\delta$. If we consider an energy balance of the preheat zone we can write,

$$\langle \rho \rangle \langle C \rangle S_f (T_2 - T_1) = \frac{\langle k \rangle (T_3 - T_2)}{\delta}, \quad (1)$$

where $\langle \ \rangle$ signifies some unspecified averaging of gas and solid phases. Here $\rho$ is density, $C$ is heat capacity, $S_f$ is the flame-spread rate and $k$ is the thermal conductivity. This averaging somewhat parallels the scaling analysis of Sirignano.[5] The reaction zone thickness scales as,

$$\delta \propto S_f \Big/ \left( \frac{d\lambda}{dt} \right), \quad (2)$$

where $\lambda$ is a reaction progress variable. Using Eqn. 2 and assuming a simple Arrhenius rate law yields, we find,

$$S_f^2 \propto \frac{k(T_3 - T_2)}{\langle \rho \rangle \langle C \rangle (T_2 - T_1)} \lambda^n p^{n-1} A e^{E/RT}. \quad (3)$$

where $A$ is a prefactor constant, $n$ is the reaction order, and $E$ is activation energy. This result is of the same order of magnitude as the result obtained from a similar analysis of normal deflagration. Considering only the terms that may have a strong dependence on pressure, we see that the flame spread rate dependence on pressure is,

$S_f \propto \sqrt{p^{n-1} / \langle \rho \rangle}$. The density is some weighted average across phases and temperatures. Assuming the ideal gas law, the gas phase weighting results in, $S_f \propto p^{(n-2)/2}$. For second-order reactions (the most common case) no pressure dependence is predicted, $S_f \propto p^0$. This result is the same as for gas phase deflagration.[10] Physically this limit could represent a very reactive gas above a solid. The condensed phase limit is likely more realistic because a mass-weighting average would be approximately equal to the solid density since $\rho_c >> \rho_g$. For the solid weighting limit we see the following dependence, $S_f \propto p^{(n-1)/2}$. Again, for 2nd order kinetics, we see that $S_f \propto p^{1/2}$, comparing well with the measured dependency here. For normal deflagration, following similar scaling arguments, we obtain $r_b \propto p^{n/2}$, which is also obtained by high activation energy asymptotics.[11] Normal burning rates of energetic materials scale linearly with pressure, as the above expression predicts for 2nd order reactions. Indeed, most energetic materials exhibit near linear dependence with pressure, as does PBX 9501.[1,12] Assuming temperature independence of the gas phase kinetics, a pressure exponent less than one results for 2nd order reactions for HMX, matching deflagration results somewhat better.

Other analyses, also predict a pressure dependence of flame spread.[2] Fernandez and Williams[13] predict a $p^{0.8}$ dependence considering a somewhat different problem. Propellant experiments with a composite propellant (very heterogeneous) show an approximately linear pressure dependence at pressures below atmospheric.[14] These experiments match the assumptions of the simple scaling analysis here more closely than other experimental results available in the literature.

Flame spread on the surface of relatively homogeneous material under quiescent conditions has been experimentally studied. Open-surface flame spread experiments that investigate the limiting case of and infinitely wide crack show that the flame spread rate for PBX 9501 as a function of pressure from 0.077 to 17.3 MPa obeys $S_f = 0.259 P^{0.538}$ (cm/s). This is consistent with scaling analysis that predicts a pressure dependence of $S_f \propto p^{(n-1)/2}$, which becomes $S_f \propto p^{0.5}$ for second order overall reactions in the gas phase. A few experiments were oriented vertically and flame spread downward proceeded measurably faster perhaps due to melt flowing. Future studies should include more detail analysis and modeling of the flame spread.

## ACKNOWLEDGEMENTS

We acknowledge the support of Los Alamos National Laboratory, under contract W-7405-ENG-36. In particular, we acknowledge the support of Dr. Deanne Idar and the High Explosives Science Project.

## REFERENCES

1 Son, S. F., Berghout, H. L., Bolme, C. A. et al., Proceedings of the Combustion Institute 28(pt.1), 919 (2000).

2 Kumar, M. and Kuo, K. K., Prog Astronaut Aeronaut 90, 305 (1984).

3 Davidson, J. E. and Beckstead, M. W., presented at the 26th Symposium (International) on Combustion/The Combustion Institute, 1996.

4 Fernandez-Pello, A. C. Combustion science and technology 39 (1-6), 119 (1984); A. C. Fernandez-Pello and T. Hirano, Combustion science and technology 32 (1-4), 1 (1983); Indreks Wichman, (1992); F. A. Williams, presented at the Symp (Int) on Combust, 16th, MIT; Aug 15-20 1976; Cambridge, MA, USA, 1976.

5 Sirignano, W. A., Combustion Science and Technology 6, 95 (1972).

6 Dickson, P. M., Fugard, C. S., Henson, B. F., et al., presented at the Detonation Symposium 11th; Snowmass, CO (United States, 1998 (unpublished); H. L. Berghout, S. F. Son, C. B. Skidmore et al., Thermochimica Acta 384 (1-2/SISI), 261 (2002); H. L. Berghout, S. F. Son, and B. W. Asay, Proceedings of the Combustion Institute 28(pt.1), 911 (2000).

7 Ali, A. N., Son, S. F., Asay, B. W. et al., Combustion Science and Technology 175, 1551 (2003); A.N. Ali, S.F. Son, B.W. Asay et al., Journal of Applied Physics 97 (6), 1 (2005).

8 Maienschein, J. L. and Chandler, J. B., presented at the to appear in the Proceedings of the Eleventh Symposium (International) of Detonation, Snow Mass, CO, 1998 (unpublished).

9 Yeh, C. L., Johnson, D. K., Kuo, K. K., et al., Combust. Sci. and Tech. 137, 195 (1998).

10 Kuo, K. K. The Principles of Combustion, 2005.

11 Williams, F. A. AIAA J 11 (9), 1328 (1973).

12 Ward, M.J., Son, S.F., and Brewster, M.Q., Combustion and Flame 114 (3-4), 556 (1998).

13 Fernandez-Pello, A., and Williams, F. A., presented at the Symp on Combust, 15th Int, Proc; Aug 25-31 1974; Tokyo, Japan, 1974.

14 McAlevy, R. F., and Magee, R. S., presented at the Combustion Inst, 12th Symp (Int) on Combustion; July 14-20 1968; Poitiers, France, 1968; R. F. McAlevy, R. S. Magee, J. A. Wrubel et al., AIAA Journal 5 (2), 265 (1967).

CP845, *Shock Compression of Condensed Matter - 2005,*
edited by M. D. Furnish, M. Elert, T. P. Russell, and C. T. White
© 2006 American Institute of Physics 0-7354-0341-4/06/$23.00

# EXPERIMENTAL VALIDATION OF DETONATION SHOCK DYNAMICS IN CONDENSED EXPLOSIVES

## D. Scott Stewart*, David E. Lambert[†], Sunhee Yoo* and Bradley L. Wescott**

*Mechanical and Industrial Engineering, University of Illinois, Urbana, IL 61801*
[†]*Air Force Research Laboratory, Munitions Directorate, Eglin AFB, 32542*
**Theoretical and Applied Mechanics, University of Illinois, Urbana, IL 61801*

**Abstract.** Experiments on the HMX-based, condensed explosive PBX-9501 were carried out to validate a reduced asymptotically derived description of detonation shock dynamics (DSD) where it is assumed that the normal detonation shock speed is determined by the total shock curvature. The passover experiment has an embedded lead disk in a right circular cylindrical charge of PBX-9501 and is initiated from the bottom. A range of dynamically changing states, with both divergent (convex) and converging (concave) shock shapes are realized as the detonation passes over the disk. The time of arrival of the detonation shock at the top surface of the charge is recorded and compared against the DSD simulation and a separate multi-material simulation (DNS). A new wide-ranging equation of state (EOS) and rate law is used to describe the explosive and is employed in both the theoretical (DSD) calculations and the multi-material simulations. The experiment, theory and simulation are found to be in excellent agreement.

**Keywords:** Detonation shock dynamics, PBX-9501, experiment, validation

## INTRODUCTION

For most applications of condensed explosives, the length of the reaction zone that powers the lead detonation shock is a fraction of a millimeter or smaller. The typical ratio of the device size to the reaction zone is typically $O(1000)$ or larger. The asymptotic theory of detonation shock dynamics (DSD), is a theory that assumes that the radius of curvature of the shock is large compared to the length of the reaction zone that supports the detonation and accounts for change to the detonation shock speed due to shock curvature during quasi-steady evolution. The present short account describes a validation of the reduced theory by experiment and by direct simulation. A full account of this work and complete references are found in [1], and [2].

The simplest DSD theory derives the result that total curvature $\kappa = \kappa_1 + \kappa_2$, is a function of the normal detonation shock velocity, $D_n$, written as

$$\kappa = F(D_n), \qquad (1)$$

with the property $F(D_{CJ}) = 0$, where $D_{CJ}$ is the Chapman-Jouguet velocity. To derive (1) from theory, one assumes that the explosive is described by an EOS and rate law of the general form $e(p, v, \lambda)$ and $r(p, v, \lambda)$, and that the Euler equations hold. One solves for the quasi-steady detonation structure in the shock-attached frame. The equations for the structure contain both the normal detonation speed $D_n$ and the total curvature $\kappa$. Since it is assumed that the detonation structure passes through a sonic plane near the end of the reaction zone, the value of $D_n$ is not independent of $\kappa$ and their relationship is determined as a nonlinear eigenvalue. Since the normal shock, curvature relation is dependent on the explosive's equation of state and reaction rate law, experiments serve as a powerful constraint on the allowable forms for $e(p, v, \lambda)$ and $r(p, v, \lambda)$, as well as a check on the theory.

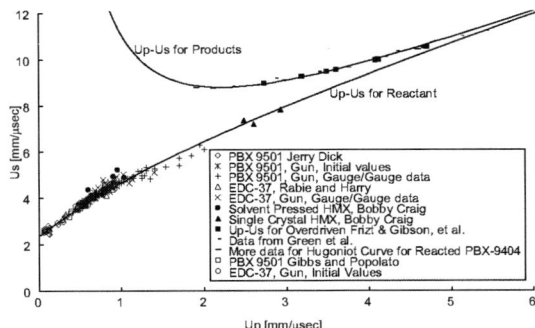

**FIGURE 1.** PBX 9501 $U_p - U_s$ Hugoniots with experimental data shown

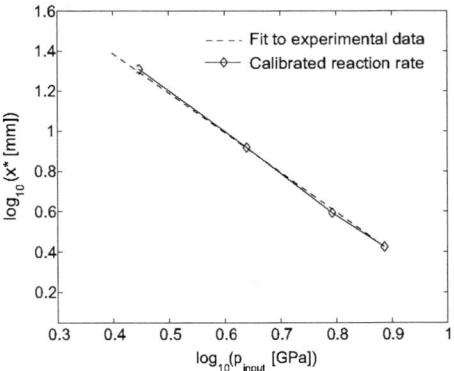

**FIGURE 2.** "Pop"-plot, run to detonation distance versus input shock pressure ($P_{input}$) for PBX 9501 from experiments and direct simulation results.

## DESCRIPTION OF THE EXPLOSIVE: WIDE RANGING EOS AND RATE LAW

Space does not permit recitation of the detailed forms used for the equation of state or the calibration procedure used to assign the the model parameters for the wide ranging rate law for PBX-9501; full details are given in [1]. The methods used in the calibration are also described in detail in [2]. Here we provide a brief summary. Davis developed a wide ranging equation of state for detonation products whose form was chosen to accurately describe the behavior of adiabatic $\gamma$ (dimensionless sound speed) and Grüneisen gamma, $\Gamma$. Davis also developed a similar reactants equations of state fit to PBX-9404, 9501. Stewart, Yoo and Davis in the 12-th Detonation Symposium, proposed a modification of Davis' reactant equation of state and introduced a closure model to develop a mixture EOS that includes the reaction progress variable of the form $e(p,v,\lambda)$ and that uses the standard rules for a binary mixture of reactants and products, where $\lambda$ is the mass fraction of the products and is the reaction progress variable. The equation of state parameters were fit to the shock Hugoniot data for both reactants and products. Fig. 1 shows a plot of the particle velocity, shock velocity ($U_P, U_S$) Hugoniot (top curve) calculated from the products EOS, and Hugoniot (bottom curve) calculated from the reactants EOS, compared with experiment. The experiment data shown was was compiled by R.Gustavsen, LANL. The calibration considers work done by expanding gases and the temperatures of the reactants and products.

A rate law is proposed for PBX-9501 with a single-term, fractional depletion, pressure dependent reaction rate of the form

$$r(p,v,\lambda) = k(1-\lambda)^v \left( \frac{P}{P_{CJ}} \right)^N . \qquad (2)$$

Shock to detonation data ("Pop"-plot) and detonation shock speed curvature data, is used to calibrate the parameters of the rate law. Hull's experimental data suggests the $D_n$, $\kappa$ relation is linear near the $CJ$ point. The depletion exponent $v$ is picked primarily to match Hull's data, see Fig. 3. The pressure exponent $N$ and rate constant $k$ are adjusted to match the shock initiation data. One-dimensional, reverse impact simulations were carried out using the specified EOS and rate law to match the published experimental data for PBX-9501, and the results are shown in Fig. 2. The calibrated rate law parameters were found to be $p_{CJ} = 36.3$ GPa, $k = 110 \mu sec^{-1}$, $N = 3.5$, $v = 0.93$.

Figure 3 shows the detonation velocity curvature relation calculated using this EOS/rate law pair described by asymptotic theory. This $D_n - \kappa$ relation is the shock motion rule that is used to compute the shock motion according to the reduced DSD description. Hull's $D_n$, $\kappa$ experimental data is also shown.

## THE PASSOVER EXPERIMENT

The experimental set up is shown in Fig. 4. The PBX-9501 explosive (white material) has a disk of

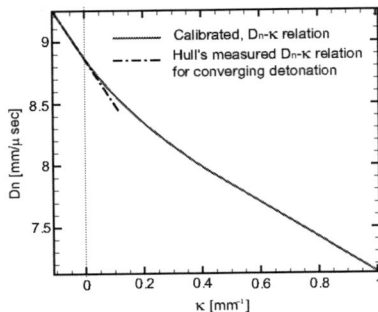

**FIGURE 3.** The $D_n, \kappa$ relation for PBX-9501 calculated from the wide-ranging EOS and rate law model

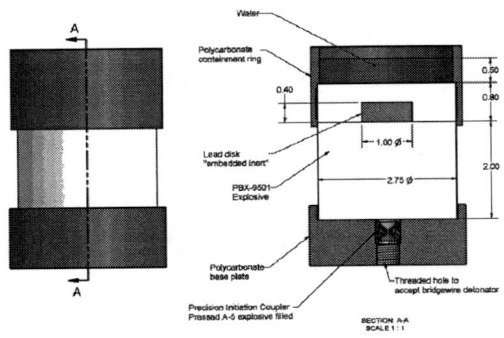

**FIGURE 4.** Assembly sketch for DSD validation experiment

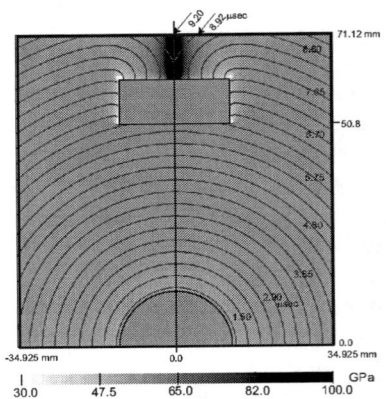

**FIGURE 5.** DSD simulation of the axisymmetric passover experiment. The grey-scale shows the shock pressure in GPa when a shock passes a point $(x,y)$ in the explosive.

pure lead (grey object) embedded along the central axis. A detailed description of the experiment is found in [1]. The experiment transforms a single, quasi-steady, convex hemispherical shock into a shock with a high concavity at the central implosion axis. PBX-9501 was chosen to test previous experimental DSD-characterization from Hull's work. Pure lead was selected as the inert for its high shock-impedance and well-characterized shock Hugoniot properties. The charge is initiated at the bottom of the PBX-9501 charge. The detonation shock front propagates as a simply connected surface with convex, positive curvature hemisphere for 50-mm and then encounters the lead disk within the top piece of PBX-9501. The shock speed in the inert lead is much lower than the detonation velocity in the PBX-9501 and a diffraction event occurs as the detonation sweeps about the disk and encompasses it. Water sits atop the charge and extinguishes the reactive shock as it transmits into the water. Light from the explosive/water plane records the time-of-arrival of the detonation shock through a single 150-micron slit aperture plate as captured by a Cordin 132A camera. Four (4) passover experiments were conducted with identical hardware. Figure 5 (as computed from DSD-theory) illustrates the subsequent shock motion in the experiment.

## COMPARISON OF DSD, NUMERICAL SIMULATIONS (DNS) AND EXPERIMENT

Two different types of simulations were carried out to compare with the time of arrival results obtained by the passover experiments. The first uses the reduced DSD-model defined by the $D_n, \kappa$ shock motion rule, shown in Fig. 1, subject to inert angle confinement boundary conditions. The lead disk is described by a Mie-Gruneisen $(U_p, U_s)$ EOS $e(p,v)$. Shock polar analysis for the DSD confinement derives the angles at the PBX-9501/lead interface (the interior angle between the shock and interface normals) as $\omega = 35^o\,(sonic)$ and $66^o\,(subsonic)$. The initial shock in the DSD simulation was a hemi-sphere of radius 5 $mm$ centered at the bottom of the charge.

The multi-material numerical simulation (DNS) was carried out with the wide-ranging EOS and rate law for the explosive and the Mie-Gruneisen EOS for the lead. An outflow boundary condition is used at the lateral boundaries. The multi-material simulation code combines two high-order solvers, a high-order total variation diminishing (TVD) solver for the Euler equations and a level-set solver to move the material interface that separates explosive and lead. The DNS simulation uses an initial condition of a hemispherical hot spot of radius $R = 5\ mm$, centered at the bottom. Figure 6 shows a comparison of the DSD and DNS simulation, to show that they give consistent results (i.e. that the shocks overlap).

## CONCLUSION

Figure 7 is a composite of TOA records for the passover experiments that includes experiments, DSD simulations, DNS simulations and the ideal Huygens construction. The Huygens construction assumes that the detonation propagates at a constant normal shock speed of $D_{cj} = 8.86\ mm/\mu sec$ and does not account for the slowing of the wave due to curvature effects. The excellent level of agreement, both qualitative and quantitative, between experiment, DSD and DNS is encouraging because it indicates that one can use the wide ranging EOS/rate law and the corresponding DSD description effectively to model real explosives and predict complex dynamic behaviors.

## ACKNOWLEDGMENTS

This was work supported by the US Air Force Research Laboratory, Munitions Directorate, US Air Force Office of Scientific Research and US Department of Energy, Los Alamos National Laboratory.

## REFERENCES

1. Lambert, D. E., Stewart, D. S., Yoo, S. and Wescott, B. L. , Experimental Validation of Detonation Shock Dynamics in Condensed Explosives, to appear in the *Journal of Fluid Mechanics*, Los Alamos Report number LA-UR-05-0169 (2005).
2. Wescott, B. L., Stewart, D. S., and Davis, W. C., Equation of state and reaction rate for condensed-phase explosives., to appear in *J. Applied Physics.*, (2005)

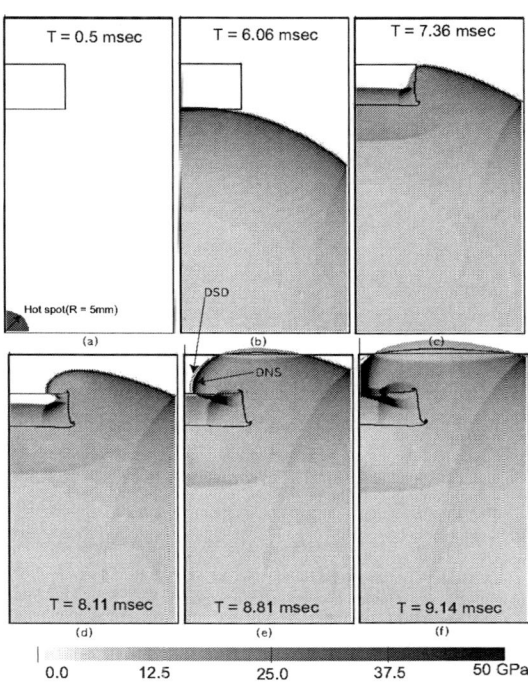

**FIGURE 6.** Comparison of DSD and DNS simulation

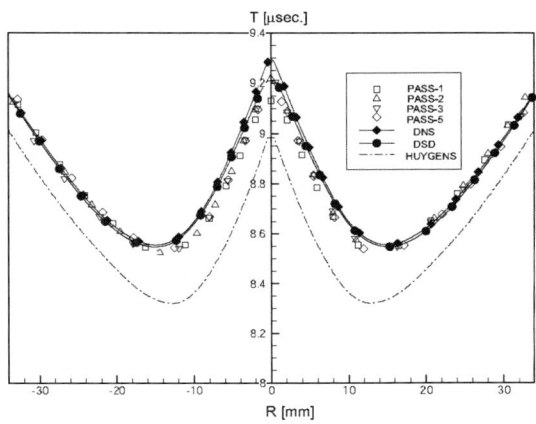

**FIGURE 7.** TOA comparison of experiments, DNS, and DSD simulations at the top of the charge

CP845, *Shock Compression of Condensed Matter - 2005*,
edited by M. D. Furnish, M. Elert, T. P. Russell, and C. T. White
© 2006 American Institute of Physics 0-7354-0341-4/06/$23.00

# NANO-ALUMINUM REACTION WITH NITROGEN IN THE BURN FRONT OF OXYGEN-FREE ENERGETIC MATERIALS

**B. C. Tappan, S. F. Son, D. S. Moore**

*P.O. Box 1663, MS C920, Los Alamos National Laboratory, Los Alamos, NM 87545*

**Abstract.** Nano-particulate aluminum metal was added to the high nitrogen energetic material triaminoguanidium azotetrazolate (TAGzT) in order to determine the effects on decomposition behavior. Standard safety testing (sensitivity to impact, spark and friction) are reported and show that the addition of nano-Al actually decreases the sensitivity of the pure TAGzT. Thermo-equilibrium calculations (Cheetah) indicate that the all of the Al reacts to form AlN in TAGzT decomposition, and the calculated specific impulses are reported. T-Jump/FTIR spectroscopy was performed on the neat TAGzT. Emission spectra were collected to determine the temperature of AlN formation in combustion. Burning rates were also collected, and the effects of nano-Al on rates are discussed.

**Keywords:** High-nitrogen, explosives, nano-aluminum, decomposition
**PACS:** 82.33.Vx, 82.33.Ln

## INTRODUCTION

Nano-particulate aluminum metal (nano-Al) has received a large amount of attention in the field of energetic materials due to the extremely fast time scale of reaction, leading to greater kinetic control of reaction rates as opposed to mass transport control. One of the primary areas involving nano-Al research is in metastable intermolecular composite (MIC) materials, which are thermite systems consisting of nano-Al and a metal oxide, and which show dramatic increases in reaction rates over conventional thermites with micron particle sizes [2]. Work has been done to investigate the effect of nano-Al in traditional propellant formulations, showing successful increases in performance in ammonium nitrate based materials that typically do not yield the theoretical energy release. However, to date there has been no investigation of the contribution of nano-Al on the burning of oxygen-free high-nitrogen energetic materials. Because of the highly reactive nature of nano-Al, the reaction to form

AlN is shown to be possible, whereas temperatures needed to react conventional micron sized Al are not reached with the high nitrogen materials.

## BACKGROUND

High-nitrogen energetic materials, often free of oxygen, derive their energy from a high positive heat of formation rather than oxidation of hydrogen and carbon bound on the same molecule, as is the case with traditional explosives. These materials are valuable as gas generants or propellants because a relatively high theoretical specific impulse can be achieved despite the relatively low decomposition temperature. However, certain applications call for higher reaction temperatures, but still require a high fuel density, such as in air-breathing rocket motors. The addition of aluminum metal to increase reaction temperature is one way to achieve this goal, but this has only recently become an viable option with the advent of nano-particule aluminum, which will react at considerably lower temperatures than micron-sized aluminum, thus enabling heat release from the formation of aluminum nitride [2]. While the energy released in the formation of AlN is only 28.5% of that of $Al_2O_3$ ($\Delta_f H_{AlN} = 317.4$ kJmol$^{-1}$ and $\Delta_f H_{Al2O3} = 1675.7$ kJmol$^{-1}$), a roughly 40% increase in calculated flame temperature (Cheetah) is achieved. In addtion, as an oxidizing environment is introduced, $Al_2O_3$ may form from the AlN. The compound triaminoguanidium azotetrazolate (TAGzT) is particularly interesting because of its exceptionally high heat of formation, fast burning rate and high molar and volumetric gas production [1,3]. Because TAGzT contains no oxygen, the chemistry that dominates the ignition and combustion is more centered on reactions in the condensed phase rather than oxidation chemistry in the gas phase. The burning rate of the neat material and with 25 wt% 38 nm Al was studied at various pressures of inert gas. T-jump/FTIR spectroscopy was utilized to provide a "snap shot" of the condensed phase reactions involved in the pathway of decomposition for pure TAGzT. TAGzT has an exceptionally high burning rate, in fact, at the higher-pressure range, the burning rate for TAGzT is in the neighborhood of the fastest measured CHNO material, DAATO$_{3.5}$ and the high nitrogen material, BTATz [3]. Emission spectroscopy was utilized to determine the temperature of the burning TAGzT/Al, and efforts were made to observe the AlN emission bands, albeit without success.

## EXPERIMENTAL PROCEDURE

**Material.** The TAGzT used for these experiments was synthesized by the NSWC at Indian Head and used as received. The synthesis and physical properties of TAGzT were first reported by Tremblay, and later by Hiskey et al. [1]. The material is a yellow, needle-like crystalline solid having a theoretical maximum density of 1.60 g/cm$^3$, decomposes at 195°C and has a heat of formation of +257 kcal/mol [1]. The impact sensitivity ($H_{50}$ – height at which there is a 50% probably of initiation with 2 kg weight) for pure TAGzT is 25 cm, and friction sensitivity (BAM) is 10 kg, whereas the 75% TAGzT/ 25% Al

has an $H_{50}$ of 28.5 cm and a friction sensitivity of 14.6 kg. Both pass spark sensitivity testing.

**Burning Rate.** Cylindrical pellets 6.3 mm x 6.4 mm of TAGzT and 75% TAGzT 25% 38 nm Al were burned in a 2L stainless steel vessel under a constant pressure of argon from 2 - 70 atm. A thin film of acrylic was used to prevent the flame front from spreading down the pellet sides after ignition using a resistively heated wire. The combustion event was filmed at 30 fps and 250 fps using a Canon XL1 3CCD Digital Video Camcorder and Red Lake MotionScope PCI 8000S high-speed-video system, respectively. Pressure was monitored with an Omega Model PX605-10KGI static pressure transducer. Optical records were analyzed using commercially available computer graphics software to obtain the burning rate data. Typical images are shown in Fig. 1.

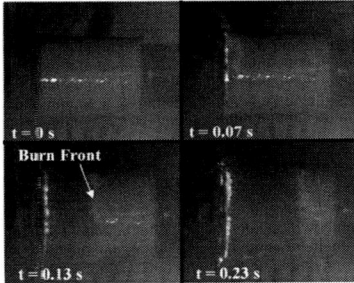

**Figure 1.** Frame sequence of burning TAGzT at 200 psig, filmed at 30 fps. The flame front is moving from left to right, and the orange luminosity on the left of the pellet is the o-ring of the pellet fixture being heated by the hot exhaust gasses.

**Flash Pyrolysis/FTIR Spectroscopy.** T-Jump/FTIR spectroscopy was performed on the neat TAGzT and this technique has previously been described in detail [3]. Briefly, approximately 200 μg of sample was thinly spread onto the center of a Pt ribbon filament, which is very rapidly heated (2000°C/sec) to a constant, predetermined set temperature using a high-gain, fast-response power supply. This filament was inserted into a 25 cm$^3$ cell with ZnSe windows to allow the diagnostic IR beam to pass a few millimeters above the surface of the filament to obtain gaseous product spectra in near real time as they evolve.

**Emission Spectroscopy.** Time-resolved emission

spectra were recorded using a Princeton Instruments IRY-700 intensified gated diode array (1024 pixels of 25 μm x 2.5 mm size; only the center 740 intensified pixels record useful information) in the image plane of an Acton 320i spectrometer with 150 lines/mm grating blazed at 300 nm. Wavelength calibration was performed using an Hg penlamp (Oriel). The spectral response of the entire optical train from sample to detector was calibrated using a NIST traceable standard irradiance source. The temperature was estimated by fitting the response-corrected emission spectrum to Planck curves at constant emissivity. Figure 2 shows a still frame of a burning rate/emission spectrum measurement of 75% TAGzT/25% nano-Al.

**Figure 2.** Burning rate measurement of 75% TAGzT 25% Al showing highly luminous flame front from which emission spectra were collected.

## RESULTS AND DISCUSSION

Certain features of high-nitrogen materials are immediately apparent from all the data. The fact that much of the decomposition occurs in the condensed phase suggests that burning rates can be faster at lower pressure, while having a lower pressure dependence (compared to classical materials such as HMX, which have pressure exponents approaching 1) because solid phase reactions are relatively independent of pressure. Figure 3 shows the burning rate measurement of neat TAGzT and 75% TAGzT/ 25% nano-Al. The pressure exponent is essentially unaffected by the addition of nano-Al, changing from 0.672 to 0.742. This small difference, however, is probably still less than it appears, as inspection of the graph shows a wider scatter in the data for the TAGzT/Al mixture, leading to a slightly high pressure exponent value.

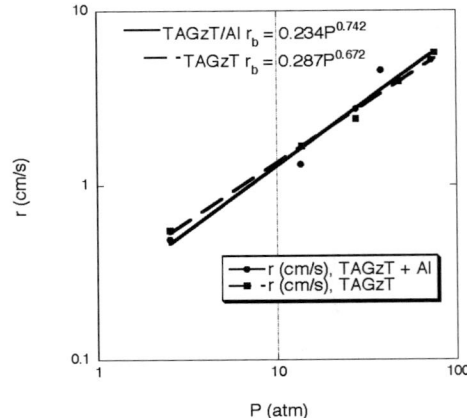

**Figure 3.** Burning rate of TAGzT (blue squares) and 75% TAGzT 25% Al (black circles).

The measurements made in the burning rate experiments and T-jump/FTIR spectroscopy indicate that the decomposition and the ignition behavior of TAGzT are dominated by condensed phase reactions, which helps to explain the remarkably high burning rates at low pressures. The reaction as determined by the t-jump/FTIR spectroscopy is given in equation 1, and the expected gas-phase reaction is given in equation 2 (calculated products for the 75% TAGzT/ 25% Al are given in equation 3).

$$TAGzT \rightarrow 5.08\ NH_3 + 4.00\ HCN$$
$$+ ?NH_2CN + 6.46\ N_2 \qquad (1)$$

$$5.08\ NH_3 + 4.00\ HCN \rightarrow 4.50\ N_2$$
$$+ 5.65\ H_2 + 2.53\ C + 1.47\ CH_4 \qquad (2)$$

$$75\%\ TAGzT/25\%\ Al \rightarrow 1.55\ H_2 + 1.52\ N_2$$
$$+ 0.62\ C + 0.82\ AlN \qquad (3)$$

Cheetah calculations considering a rocket chamber at 68 atm and 1 atm exhaust pressure indicate the formation of 11.00 $N_2$, 6.06 $H_2$, 1.47 $CH_4$ and 2.53 C, on a mole product per mole explosive basis. Assuming the condensed phase products seen in the T-jump react to produce the products seen in the Cheetah calculation leads to equation 2. Equation 2 is exothermic by -416 kJ/mol, which when added to the heat of reaction in equation 1, which is exothermic by −770 kJ/mol, yields a net heat of reaction of −1186 kJ/mol. In accordance, the heat of formation of TAGzT is 1076 kJ/mol or about 10% less than the net heat of reaction. Equations 1 and 2 suggest that roughly 65% of the reaction energy is released in the condensed phase, which is concurrent with the assumption based on the slope of the burning rate.

Figure 4 shows the temperature measurement for the 75% TAGzT/ 25% Al, which yielded a temperature of 1750 ± 100 °C. Due to the strong thermal background, the AlN bands at 507.8 and 527.6 nm are not immediately apparent, however it is believed that AlN is formed based on the measured temperature, as it is the only route for the Al reaction, and the energy dispersive scattering (EDS) pattern (Fig. 5) indicates a large amount of bound nitrogen.

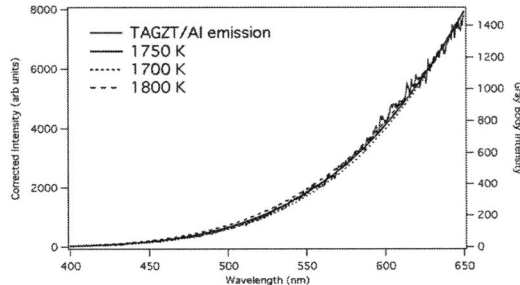

**Figure 4.** Temperature measurement from emission of deflagrating TAGzT and Al (see text for details). The best fit is for a temperature of 1750 ± 100 °C.

Cheetah rocket calculations at the same pressure as our measurement (20 atm) indicate a chamber temperature of 2200 °C and an exhaust temperature of 1657 °C, which puts our measurement in the calculated range, considering heat losses. The calculated $I_{sp}$ at 20 atm chamber pressure and 1 atm exhaust for the 75% TAGzT/ 25% Al is 204.1 seconds, were as the for the same conditions pure TAGzT is 188.8 seconds.

## CONCLUSIONS

Despite non-observance of a clearly defined emission band for AlN, from the analysis of combustion products, measured temperature of combustion, and the fact that no other decomposition route is available, it is indicted that AlN is formed in the combustion of TAGzT and nano-Al. The temperature measurement matches well with Cheetah calculations.

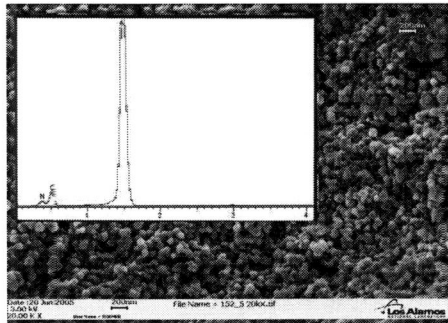

**Figure 5.** Scanning electron micrograph with energy dispersive scattering plot (inset) of AlN collected after burning rate measurement. The formed AlN is very similar in size to the original 38 nm Al.

## ACKNOWLEDGEMENTS

This work was funded by the Defense Threat Reduction Agency's Advanced Energetics Program and B.C.T.'s Agnew National Security Postdoctoral Fellowship. The Los Alamos National Laboratory is operated by the University of California for the U.S. Department of Energy under Contract W-7405-ENG-36.

## REFERENCES

1. (a) Tremblay, M., *Can. J. Chem.* **1964**, 42, 1157. Hiskey, M.A., Goldman, N., Stine, J. R., *J. Energetic Materials*, **1998**, *16*, 119.
2. Aumann, C.E., Skofronick, G. L., and Martin, J. A., Oxidation Behavior of Aluminum Nanopowders, *J. Vac. Sci. Technol., B: Microelectron. Nanometer Struct.-Process., Meas., Phenom.* **1995**, *13*, 1178.
3. Ali, A.N., Son, S.F., Hiskey, M.A., Naud, D.L.,. *J. Prop. and Power*, **2004**, *20*, 120.
4. Simmons, J.D. and McDonald, J.K., "The Emission Spectrum of AlN" *J. Mol. Spectro.* **1972**, *41*, 584.
5. (a) Brill, T.B., Brush, P.J., James, K.J., Shepherd, J.E., and Pfeiffer, K.J., *Appl. Spectrosc.*, **1992**, *46*, 900. (b) Shepherd, J.E., and Brill, T.B., *Tenth Int. Deton. Symp.*, Naval Surface Warfare Center, White Oak, MD, 1993, p.849. (c) Thynell, S.T., Gongwer, P.E., and Brill, T.B., *J. Propuls. Power*, **1996**, *12*, 993.

CP845, *Shock Compression of Condensed Matter - 2005*,
edited by M. D. Furnish, M. Elert, T. P. Russell, and C. T. White
© 2006 American Institute of Physics 0-7354-0341-4/06/$23.00

# DETONATION REACTION ZONES IN CONDENSED EXPLOSIVES

## Craig M. Tarver

*Lawrence Livermore National Laboratory*
*P.O. Box 808, L-282, Livermore, CA 94551*

**Abstract.** Experimental measurements using nanosecond time resolved embedded gauges and laser interferometric techniques, combined with Non-Equilibrium Zeldovich – von Neumann – Doring (NEZND) theory and Ignition and Growth reactive flow hydrodynamic modeling, have revealed the average pressure/particle velocity states attained in reaction zones of self-sustaining detonation waves in several solid and liquid explosives. The time durations of these reaction zone processes are discussed for explosives based on pentaerythritol tetranitrate (PETN), nitromethane, octahydro-1,3,5,7-tetranitro-1,3,5,7-tetrazocine (HMX), triaminitrinitrobenzene(TATB) and trinitrotoluene (TNT).

**Keywords:** Explosives, detonation, reaction zone
**PACS:** 47.40.-x, 82.40.Fp

## INTRODUCTION

It is essential to know the pressure/particle velocity history in the reaction zone of a detonating condensed phase explosive to determine the momentum produced in surrounding materials. The Non-Equilibrium Zeldovich - von Neumann – Doring (NEZND) theory was developed to identify the non-equilibrium chemical processes that precede and follow exothermic chemical energy release within the reaction zones of self-sustaining detonation waves (1-10). Previously, the chemical energy released was merely treated as a heat of reaction in the conservation of energy equation in the Chapman-Jouguet (C-J) (11,12), Zeldovich – von Neumann-Doring (ZND) (13-15), and curved detonation wave front theories (16). NEZND theory has explained many experimentally determined detonation wave properties including: the induction time delays for the onset of chemical reaction; the rapid rates of chain reactions that form the reaction product molecules; the de-excitation rates of the initially highly vibrationally excited products; the feedback mechanism that allows the chemical energy to sustain the leading shock wave front; and the establishment of the three-dimensional lead shock wave front structure of all detonation waves.

Along with an understanding of these physical processes, it is also necessary to have a practical reactive flow model that can be used to predict shock initiation and detonation wave propagation in one-, two-, and three-dimensional hydrodynamic computer codes. The Ignition and Growth model (17) has been very successful in this regard. It was formulated using compression and pressure dependent reaction rate laws and calibrated to available experimental data. Nanosecond time resolved data has been obtained using: embedded pressure gauges (18); embedded particle velocity gauges (19); electrical conductivity probes (20); and laser interferometric techniques, such as VISAR (21) and Fabry-Perot (22). One-dimensional detonation wave reaction zone profiles for nitromethane, PETN, HMX, TATB and TNT are discussed in this paper.

## NEZND THEORY OF DETONATION

Figure 1 illustrates the various processes that occur in the NEZND model of detonation in condensed explosives. At the head of every detonation wave is a three-dimensional Mach stem shock wave front. Behind the shock front, the phonon modes are excited, followed by multi-phonon excitation of the low frequency vibrational (doorway) modes and then excitation of the higher frequency modes by multi-phonon up-pumping and internal vibrational energy redistribution (IVR)(23). Energy equilibration has been studied in shocked liquid and solid explosives by Dlott et al. (24) and Fayer et al. (25). Only after the explosive molecules become vibrationally excited can chemical reactions begin. Recently McGrane et al. (26) showed using laser generated shocks that induction times of tens of picoseconds were required for the dissociation of nitro groups in poly (vinyl nitrate) films.

The induction time for the initial endothermic bond breaking reaction can be calculated using high pressure - high temperature transition state theory. The reaction rate constant K is given by:

$$K = (kT/h) \, e^{-S} \sum_{i=0}^{s-1} (E/RT)^i \, e^{-E/RT} / i! \qquad (1)$$

where T is the equilibrated temperature, E is the activation energy, k, h, and R are Boltzmann's, Planck's, and the gas constant, respectively, and s is the number of neighboring vibrational modes interacting with the transition state. Reaction rate constants have been calculated for detonating solids and liquids using Eq. (1) with realistic equations of state and values of s (10).

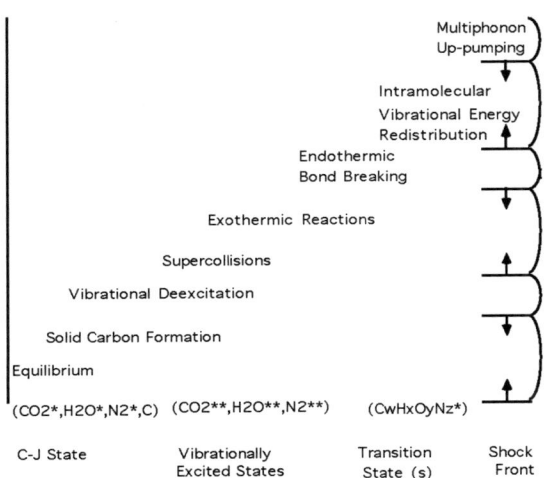

FIGURE 1. The Non-Equilibrium Zeldovich - von Neumann - Doring (NEZND) model of detonation

Following the induction and endothermic initial bond breaking processes, exothermic chain reactions follow in which reaction product gases are formed in highly vibrationally excited states (2). These excited products either undergo reactive collisions with surrounding explosive molecules or non-reactive collisions with their neighbors. Some non-reactive collisions are "super-collisions"(27) in which transfer several quanta of vibrational energy. Since reaction rates increase rapidly with each quanta of vibrational energy available, reactive collisions dominate and the main chemical reactions are extremely fast. Once the chain reactions are completed, the rest of the reaction zone is dominated by vibrational de-excitation of the gaseous product molecules and solid carbon formation. These processes control the approach to the equilibrium C-J state and the average reaction zone length. The relative lengths of these major time dependent processes determine the average pressure - time profiles of the detonation zones discussed below.

## PETN DETONATION REACTION ZONES

In a Fabry-Perot study of PETN detonation, Tarver et al. (28) found that all of the experimental data could be accurately calculated using a C-J model of the PETN detonation wave. Since the Fabry-Perot system used in that study has a five to ten nanosecond time resolution, this implies that the reaction zone length in detonating PETN is less than 10 ns or 80 μm. PETN is well-oxidized and forms little or no solid carbon in its reaction products (20). Thus the formation and vibrational de-excitation of its gaseous products can proceed very rapidly without the interference of collisions with developing carbon clusters. The calculated induction time for the initial PETN decomposition reaction is a few tenths of a nanosecond (10), which is close to the experimental induction times for "super" detonation observed in PETN pressed nearly to theoretical maximum density (TMD) (28). Thus determination of the PETN detonation reaction zone profile requires the use of subnanosecond techniques, such as new VISAR's, that are becoming available (29).

## NITROMETHANE REACTION ZONES

The liquid nitromethane detonation reaction zone has been studied by several techniques with the most recent and fastest time resolution being the "home made" VISAR studies of Sheffield et al. (29).

Sheffield et al. measured a von Neumann spike state of approximately 20 GPa that lasted for 1 to 3 ns, in good agreement with calculated estimates (4,10). A rapid decrease in particle velocity was then measured which lasted for about 7 ns, followed by a slower decrease that lasted for approximately 50 ns. These particle velocity results correlate well with the electrical conductivity probe measurements of Hayes (20) that show that carbon formation begins 1 ns or so behind the detonation wave front and increases to a maximum at about 5 ns, which lasts for at least 20 ns. Thus the nitromethane detonation wave profile appears to show an induction time of 1 to 3 ns, rapid exothermic reaction for perhaps 7 ns, and product equilibration and solid carbon formation that lasts another 20 to 50 ns. The exact location of the C-J state at which the reaction zone ends and the rarefaction wave begins is difficult to determine, because the slope differences in particle velocity or pressure values near the C-J state are very small.

## HMX DETONATION REACTION ZONES

It has long been known that the main chemical energy release in detonating HMX-based explosives must occur within 20 ns to explain their detonation velocity versus charge diameter data (30). More recent VISAR measurements of Gustavsen et al. (31) on detonating PBX 9501 (95% HMS, 2.5% BDNPA/F and 2.5% Estane) show a rapid decrease in particle velocity for about 20 ns followed by a slower rate of decrease for several tens of ns. The measured peak pressure of approximately 40 GPa agrees well with theoretical predictions and the unreacted equation of state in the Ignition and Growth model, which was calibrated to embedded gauge and Fabry-Perot data. Figure 2 shows the experimental detonation velocity versus inverse cylindrical charge radius data (32) for PBX 9404 (94% HMX, 3% nitrocellulose, and 3% chloro ethyl phosphate) compared to Ignition and Growth reactive flow calculations which assume that 90% of the chemical energy is released within 20 ns followed by the remaining 10% being released in 80 more ns. The main energy release and vibrational equilibration in detonating HMX takes about 20 ns and solid carbon cluster formation takes another 60 to 80 ns.

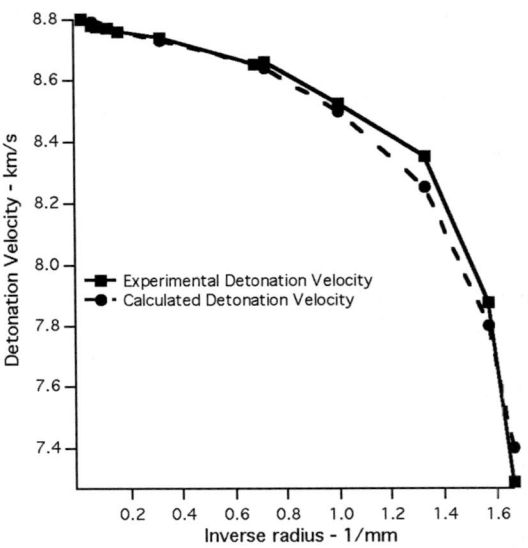

**FIGURE 2.** Experimental and calculated detonation velocity versus inverse radius curves for PBX 9404

## TATB DETONATION REACTION ZONES

TATB-based explosives are oxygen poor and thus produce a great deal of solid carbon in their detonation products. The main energy release occurs over approximately 60 to 80 ns and the overall reaction zone length is about 2.5 to 3 mm or 300 to 400 ns. Several embedded gauge and laser interferometry studies have shown these general features (33). Ignition and Growth modeling of LX-17 (92.5 % TATB and 7.5 % Kel-F) and PBX 9502 (95% TATB and 5 % Kel-F) detonation has accurately calculated a great deal of one-, two- and three-dimensional experimental data using a 80% energy release over the first 60 ns followed the remaining 20% released over another 240 ns (34,35). The experimentally measured von Neumann spike state for LX-17 is 33.5 GPa, which agrees closely with extrapolated unreacted shock Hugoniot data and the unreacted equation of state used in the Ignition and Growth model. The C-J state is not readily apparent from the experimental records, but a C-J pressure of 27 GPa for LX-17 can be used to predict all existing detonation and overdriven data.

## TNT DETONATION REACTION ZONES

Like TATB, TNT is under oxidized and produces a great of solid carbon. Kury et al. (36) modeled laser interferometric data on detonating and overdriven TNT using: a von Neumann spike pressure of 25 GPa; a fast reaction of 90% of the TNT in 80 ns, a slow reaction of the remaining 10% in another 200 ns; and a C-J pressure of 19 GPa. Electrical conductivity measurements in TNT (20) showed that the conductivity increased for over 100 ns behind the shock front. Thus the detonation reaction zone profile of TNT is very similar to that of TATB.

## DISCUSSION

One-dimensional detonation reaction zone profiles have been measured by several ns resolution techniques and modeled using Ignition and Growth. Experimental resolution of the 3D shock front structures of condensed phase explosive detonation waves is now possible (37). Detonation reaction zone temperature measurements are required to build reactive flow models based on Arrhenius kinetics.

## ACKNOWLEDGMENTS

This work was performed under the auspices of the U.S. Department of Energy by the University of California, Lawrence Livermore National Laboratory under Contract No.W-7405-ENG-48.

## REFERENCES

1. Tarver, C. M., "*On the Chemical Energy Release in Self-Sustaining Detonation Waves in Gaseous and Condensed Explosives,*" Ph. D. thesis, The Johns Hopkins University, Baltimore, MD, 1973.
2. Tarver, C. M. *Comb. Flame* **46**, 111-133 (1982).
3. Tarver, C. M., *Comb. Flame* **46**, 135-155 (1982).
4. Tarver, C. M., *Comb. Flame* **46**, 157-179 (1982).
5. Tarver, C. M., Fried, L. E., Ruggerio, A. J., and Calef, D. F., *Tenth International Detonation Symposium*, Office of Naval Research ONR 33395-12, Boston, MA, 1993, p. 3-11.
6. Tarver, C. M., in *Shock Compression of Condensed Matter-1997*, S. C. Schmidt, D. P. Dandekar, and J. W. Forbes, eds., AIP Press, 1998, pp. 301-304.
7. Tarver, C. M., *J. Phys. Chem. A* **101**, 4845–4851 (1997).
8. Tarver, C. M., in *Shock Compression of Condensed Matter – 1999*, M. D. Furnish, L. C. Chhabildas, and R. S. Hixson, eds., AIP, 2000, pp. 873-877.
9. Tarver, C. M., in *Shock Compression of Condensed Matter – 2001*, N. Thadhani, Y. Horie, and M. Furnish, eds., AIP Press, pp. 42-49.
10. Tarver, C. M., "What is a Shock Wave to an Explosive Molecule?" in *High Pressure Shock Compression of Solids VI*, Y. Horie, L. Davison, and N. N. Thadhani, ed., Springer-Verlug, New York, 2003, pp. 323-340.
11. Chapman, D. L., *Phil. Mag.* **213**,5,47,90 (1899).
12. Jouguet, E., *J. Appl. Math.* **70**,6,1,347 (1904).
13. Zeldovich, Y. B., *J. Exper. Theor. Phys. (USSR)* **10**, 542 (1940).
14. Von Neumann, J., *Office of Science Research and Development, Report No.* 549 (1942).
15. Doring, W., *Am. Physik* **43**, 421 (1943).
16. Wood, W. W. and Kirkwood, J. G., *J. Chem. Phys.* **29**, 957 (1958).
17. Tarver, C. M., Hallquist, J., and Erickson, L. M., *Eighth Symposium (International) on Detonation*, NSWC MP86-194, Albuquerque, NM, 1985, pp. 951-961.
18. Tarver, C. M., Parker, N. L., Palmer, H. G., Hayes, B. and Erickson, L. M., *J. Energetic Materials* **1**, 213-250 (1983).
19. Hayes, B. and Tarver, C. M., *Seventh Symposium (International) on Detonation*, NSWC 82-334, Annapolis, MD, 1981, pp. 1029-1039.
20. Hayes, B., *Fourth Symposium (International) on Detonation*, ACR-126, White Oak, MD, 1965, pp. 595-601.
21. Sheffield, S. A., Bloomquist, D. D., and Tarver, C. M., *J. Chem. Phys.* **80**, 3831-3844 (1984).
22. Tarver C. M., Tao, W. C., and Lee, C. G., *Prop., Explosives, Pyrotech.* **21**, 238-246 (1996).
23. Weston, Jr., R. E. and Flynn, G. W. *Ann. Rev. Phys. Chem.* **43**, 559-592 (1993).
24. Hong, X., Chen, S., and Dlott, D. D., *J. Phys. Chem.* **99**, 9102-9109 (1995).
25. Holmes, W., Francis, R. S., and Fayer, M. D., *J. Chem. Phys.* **110**, 3576-3583 (1999).
26. McGrane, S.D., Moore, D. S., and Funk. D. J., *J. Phys. Chem. A* **108**, 9342-9347 (2004).
27. Bernshtein, V. and Oref, I., *J. Phys. Chem.* **100**, 9738-9758 (1996).
28. Tarver, C. M., Breithaupt, R. D., and Kury, J. W., *J. Appl. Phys.* **81**, 7193–7202 (1997).
29. Sheffield, S. A., Engelke, R., Alcon, R. R., Gustavsen, R. L., Robbins, D. L., Stahl. D. B., Stacy, H. L. and Whitehead, M. C., *Twelfth International Detonation Symposium*, Office of Naval Research ONR 333-05-2, San Diego, CA, 2002, pp. 159-166.
30. Green, L. G. and James, E. Jr., *Fourth Symposium (International) on Detonation*, ACR-126, White Oak, MD, 1965, pp. 86-91.
31. Gustavsen, R. L., Sheffield, S. A., and Alcon, R., *Eleventh International Detonation Symposium*, ONR 33300-5, Aspen, CO, 1998, pp. 821-827.
32. Campbell, A. W. and Engelke, R., *Sixth Symposium (International) on Detonation*, ACR-221, Coronado, CA, 1976, pp. 642-652.
33. Tarver C. M., Kury, J. W., and Breithaupt, R. D., *J. Appl. Phys.* **82**, 3771–3782 (1997).
34. Tarver, C. M. and McGuire, E. M., *Twelfth International Detonation Symposium*, ONR 333-05-2, San Diego, CA, 2002, pp. 641-649.
35. Tarver, C. M., *Propellants, Explosives, Pyrotechnics* **30**, 109-116 (2005).
36. Kury, J. W., Breithaupt, R. D., and Tarver, C. M., *Shock Waves* **9**, 227-237 (1999).
37. Plaksin, I., Campos, J., Simoes, P., Portugul, A., Riberio, J., Mendes, R., and Gois, J., *Twelfth International Detonation Symposium*, ONR 333-05-2, San Diego, CA, 2002, pp. 650-658.

CHAPTER XV

# EXPLOSIVE AND INITIATION STUDIES

CP845, Shock Compression of Condensed Matter - 2005,
edited by M. D. Furnish, M. Elert, T. P. Russell, and C. T. White
© 2006 American Institute of Physics 0-7354-0341-4/06/$23.00

# HOT SPOTS FROM DISLOCATION PILE-UP AVALANCHES

## R.W. Armstrong[1] and W.R. Grise[2]

[1]Center for Energetic Concepts Development, University of Maryland, College Park, MD 20742
[2]Department of Industrial Technology, Morehead State University, Morehead, KY 40351

**Abstract.** The model of localized adiabatic heating associated with release of a dislocation pile-up avalanche is described and re-evaluated. The model supplies a fundamental explanation of shear banding behavior in metal and non-metal systems. Now, a dislocation dynamics description is provided for more realistic assessment of the hot spot heating. Such localized heating effect was over-estimated in the earlier work, in part, to show the dramatic enhancement of the work rate, and corresponding temperature build-up, potentially occurring in the initial pile-up release, say, at achievement of the critical dislocation mechanics-based stress intensity for cleavage. Proposed applications are to potentially brittle metal, ionic, and energetic material systems.

**Keywords:** Dislocations, shear flows, dynamical loading, energy conversion
**PACS:** 61.72Ff, Hh, Lh; 47.20Ft; 83.50.-v; 84.60.-h

## INTRODUCTION

The dislocation pile-up model is shown in Fig. 1.

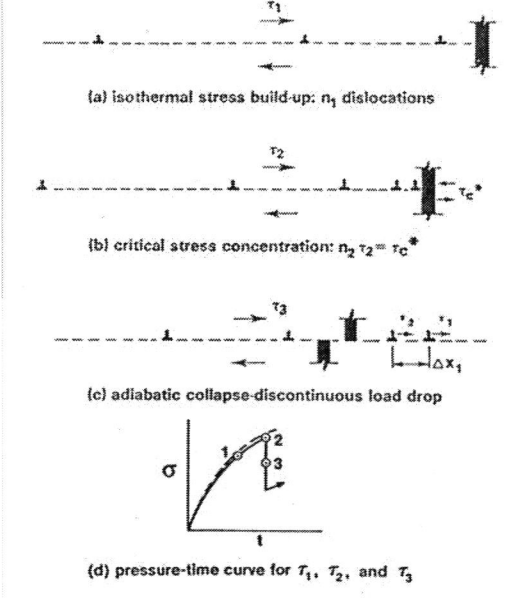

(a) isothermal stress build-up: $n_1$ dislocations

(b) critical stress concentration: $n_2 \tau_2 = \tau_c^*$

(c) adiabatic collapse-discontinuous load drop

(d) pressure-time curve for $\tau_1$, $\tau_2$, and $\tau_3$

**FIGURE 1.** Stages of dislocation pile-up release [1].

## AVALANCHE CHARACTERISTICS

Two important aspects of the avalanche-assisted enhancement of the local material plastic work rate are derived from the critical condition:

$$n\,(\tau_a - \tau_o) = \tau_c^{\,*}$$

for which n is the number of (free) pile-up dislocations, $\tau_a$ is the applied shear component of stress, $\tau_o$ is the lattice friction stress resisting individual dislocation movement, and $\tau_c^{\,*}$ is the critical component of shear stress. First, substitution of the linear dependence of n on effective stress and slip diameter gives a microstructural stress intensity, $k_s$, evaluated at the (highest) crack nucleation limit as $\pi G b^{1/2}/4\alpha$, where G is the shear modulus, b the dislocation Burgers vector and $\alpha = 2(1-\nu)/(2-\nu)$, with $\nu$ being Poisson's ratio [2]. Thus, n has its largest value at this $\tau_c^{\,*}$. Secondly, at sudden pile-up release, the first now free dislocation is driven by the effective stress, $(n - 1)\,(\tau_a - \tau_o)$, and the one behind by $(n - 2)\,(\tau_a - \tau_0)$, and so on [2]. The combined result is an appreciably enhanced work rate with greatest potential temperature rise.

The temperature rises for such dislocation avalanches were over-estimated by the relations

$$\Delta T \leq [k_s \ell^{1/2} v/16\pi K] \ln [2K/c^* vb]$$

or

$$\Delta T > [k_s \ell^{1/2}/16\pi] [2v/c^* bK]^{1/2}$$

dependent on whether $[2K/c^* vb] > 1.0$, or $< 1.0$, respectively [1]. The material constants for metals and ionic solids fit the first condition and those for molecular energetic materials fit the second condition. Substitution of a thermally-activated dislocation velocity for $v$

$$v = v_o \exp[-(G_o - \int bA \, d\tau_{th})/kT]$$

led [3], then, with $A = W_o/b\tau_{th}$ and $\tau_{th}$ proportional to an exponential dependence on the drop-weight height for 50% probability of initiation, $H_{50}$, to prediction of a log-log relationship for $H_{50}$ versus $\ell^{-1/2}$.

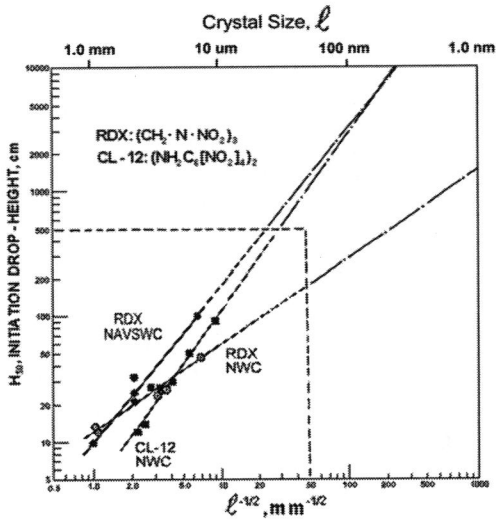

**FIGURE 2.** $H_{50}$ vs $\ell^{-1/2}$ for impacted crystals.

In the equation for $v$, $G_o$ is the Gibbs free energy for dislocation activation in the absence of a thermal component of stress, $\tau_{th}$, $A$ is dislocation activation area, and $k$ is Boltzmann's constant. Figure 2 gives reasonable confirmation of the predicted behavior

measured for RDX, $([CH_2 N NO_2]_3)$, and CL-12, $([NH_2 C_6 \{NO_2\}_4]_2)$.

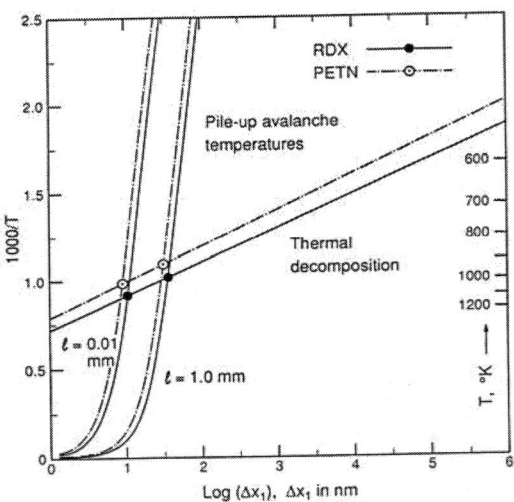

**FIGURE 3.** Pile-up and explosion temperatures.

The pile-up predictions have been compared with thermal explosion predictions for RDX and PETN, $(C [CH_2 OH]_4)$ [3,4].

In Figure 3, the thermal explosion temperatures themselves follow an Arrhenius law that carries through the analysis to give a reciprocal dependence of the critical temperature on the logarithm of the hot spot size, $\Delta x_1$. The pile-up temperatures are shown for two crystal sizes that may be seen from the comparison of curve-and-line intersections to give a higher required temperature for initiation of thermal decomposition for smaller crystal sizes [4]. Furthermore, the easier initiation of PETN compared to RDX, at the same crystal sizes, is seen to occur because of the lower thermal explosion temperature for PETN, that is interpreted to result because of the lesser stability of the PETN molecule compared to RDX.

The relative brittleness of RDX and related crystal structures may be assessed in one way in terms of a cleavage susceptibility index $(\gamma/Gb)^{1/2} = 0.066$ for RDX [5] and 0.070 for PETN; values $< 0.29$ are indicative of brittleness in metals. The index compares the ease of cracking with the difficulty of generating dislocations. A further comparative relationship of plastic flow to cracking is shown below on an indentation hardness stress-strain basis [6] in Figure 4.

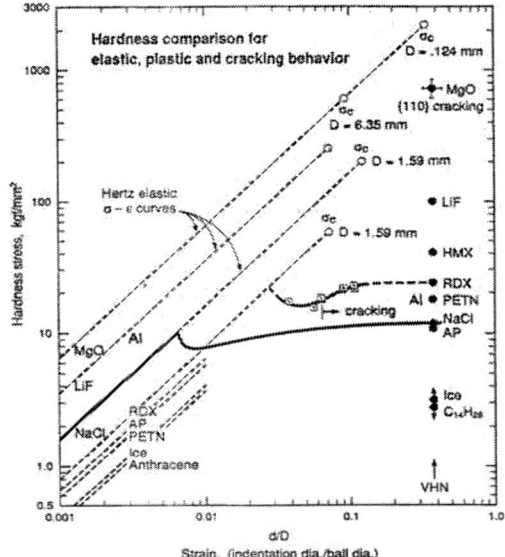

**FIGURE 4.** Elastic/plastic/cracking hardnesses.

In the Figure, with Al recently added [7], the hardness stress is the equivalent mean pressure on a (steel) ball indenter and the effective strain is the contact diameter, d, divided by the ball diameter, D. Vickers (diamond pyramid) hardness numbers, VHN, are plotted at (d/D) = 0.375. The elastic unloading doesn't alter d for a plastic indentation. The main point here, however, is to note that the hardness stress for RDX is ~3 times lower than the hardness stress needed elastically, $\sigma_c$, for cracking at the same ball size. The ratio of hardness stresses provides an estimate of the number of dislocations needed plastically to reach the cracking stress.

The new consideration then is the extent to which the analytic dislocation pile-up equations

elastic/plastic/cracking basis for assessing the been made for various types of pile-up configurations [8] and the perhaps surprising result of applicability at small numbers leads to the possibility of illustrating the proposed avalanching effect in a numerical model of such a breakthrough.

A pioneering numerical model description of pile-up release dynamics was given by Gerstle and Dvorak [9] for the hypothesized case of a relatively weak obstacle and employing small dislocation numbers. In the model, the obstacle resistance of a grain boundary was represented by a narrow region requiring a higher viscosity than the grain interior. Thus, the piled-up dislocations at the single-ended slip band tip were held up until forced through the boundary region by others following behind. An exponential dependence of the dislocation velocity on the effective shear stress was employed with constants fitted to the grain size dependent yielding of steel. Figure 5 provides an example result for a pile-up of 17 dislocations in which $x_j$ is the position of the j'th dislocation counted from the lead position and $(t/t_y)$ is the relative time scale determined by the time for the lead dislocation to pass through the obstacle. In the Figure, the dashed "s" curve is the average positional movement for all of the dislocations.

for dislocation number, pile-up length, and effective shear stress might be applicable at small dislocation numbers. Such comparison has

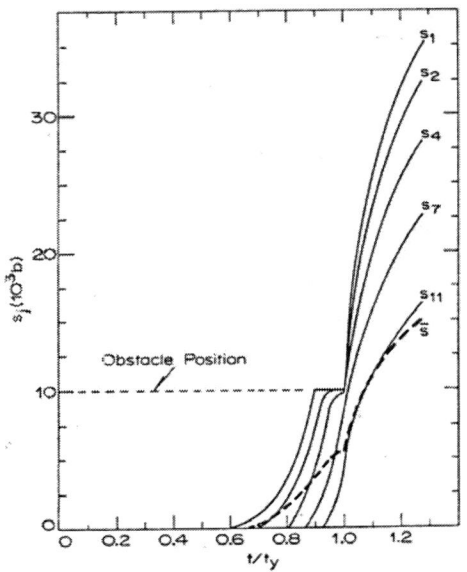

**Figure 5.** Dislocation pile-up releases [9].

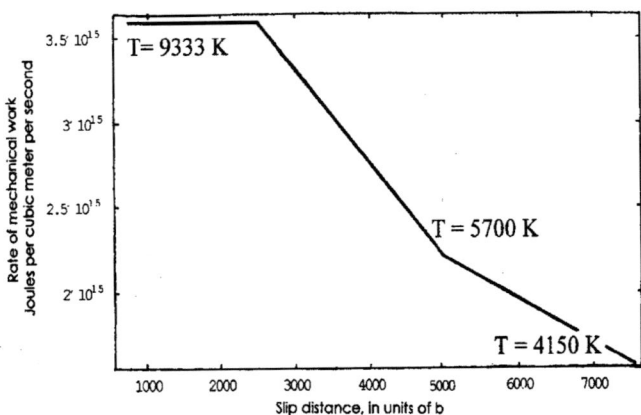

**Figure 6**. Plastic work rate and temperature rise for a modeled dislocation avalanche in iron [11].

Attention is directed in Figure 5 to the speed at which the lead dislocations are released. Even for this case of a release (obstacle) stress only just greater than 3 times the effective applied stress, the lead dislocation is seen to move initially at greater than 100 times the average dislocation velocity leading up to the obstacle.

Taylor and Quinney [10] are generally credited with the experimental observation, made at large material straining, that most of the plastic work goes into heating the deformed material. In the present case modeled after Gerstle and Dvorak and without loss or creation of additional dislocations in a slip length, $\ell$, containing sixteen dislocations, the work rate, that is assumed to be confined within the slip band thickness for the released dislocations, is expressed [11] as

$$\Delta W/\Delta t = (1/\ell) \sum \tau_{i,\,eff}\; v_i$$

in which the sum is over all dislocations and the effective stresses are evaluated at each i'th dislocation with its corresponding velocity at the time $t_i$. Figure 6 shows evaluation of the work rate achieved over micron distances. The temperatures at each position are computed for the total conversion of the plastic work. Though still relatively high, the temperatures are lower than those previously overestimated for pile-up release at cracking [2] and, for which, the dislocation shear wave speed, more than 100

times greater than for the lead dislocation here, had been employed for the released dislocation velocities.

## REFERENCES

1. Armstrong, R.W., Coffey, C.S. and Elban, W.L., Acta Metall. **30**, 2111 (1982).
2. Armstrong, R.W., and Elban, W.L., Mater. Sci. Eng. A **122**, L1 (1989).
3. Armstrong, R.W., Coffey, C.S., DeVost, V.F., and Elban, W.L., J. Appl. Phys. **68**, 979 (1990).
4. Armstrong, R.W. Ammon, H.L., Elban, W.L., and Tsai, D.H., Thermochim. Acta, **384**, 303 (2002).
5. Armstrong, R.W., and Elban, W.L., Mater. Sci. Eng. A **111**, 35 (1989).
6. Armstrong, R.W., and Elban, W.L., *Dislocations in Solids* edited by F.R. N. Nabarro and J.P. Hirth, Elsevier Sci. Publ., Oxford, U.K., 2004, **12**, p. 403.
7. Armstrong, R.W., and Elban, W.L., Mater. Sci. Tech., in print.
8. Armstrong, R.W., Mater. Sci. Eng. A, in print.
9. Gerstle, F.P., and Dvorak, G.J., Philos. Mag. **29**, 1337; Ibid., 1347 (1974).
10. Taylor, G.I., and Quinney, H., Proc. Roy. Soc., Lond., **A143**, 307 (1934); Zerilli, F.J., and Armstrong, R.W., *Shock Compression of Condensed Matter – 1997*, edited by Schmidt, S., Dandakar, D. and Forbes, J.W., Amer. Inst. Phys., N.Y., 1998, CP429, p. 215.
11. Grise, W.R., *Dislocation Pile-Ups and Their Role in Nanosized Crystal Hotspots*, NRC/AFOSR SFFP Report, Eglin AFB, FL. 2003.

CP845, *Shock Compression of Condensed Matter - 2005,*
edited by M. D. Furnish, M. Elert, T. P. Russell, and C. T. White
© 2006 American Institute of Physics 0-7354-0341-4/06/$23.00

# ELECTROMAGNETIC RADIATION FROM THE DETONATION OF METAL ENCASED EXPLOSIVES

## W. T. Brown[1], M. F. Schmidt[1], P. T. Dzwilewski[2], and T. M. Samaras[2]

[1]*Applied Research Associates, Inc, Capital Area Division Alexandria VA 22314*
[2]*Applied Research Associates, Inc, Rocky Mountain Division Littleton CO 80127*

**Abstract.** Electromagnetic radiation accompanying the detonation of chemical explosives was first reported in 1954. Such emissions result from detonations of both bare and cased explosives. However, the dominant wavelengths of emissions from these two types of explosions generally differ by as much as three or four orders of magnitude. We present results of far-field and near-field experimental measurements of electric fields emitted by metal encased and show that metal fracture is the dominant mechanism leading to these emissions. Additionally, we present results of computational analysis of explosive fracture of steel cylinders performed to investigate the correlation between the time-dependent fragment size distribution and the pattern of electromagnetic emissions.

**Keywords:** Electromagnetic radiation, explosives, fracture.
**PACS:** 41.20Jb, 62.20Mk, 52.50.+p, 79.90.+b

## INTRODUCTION

In a short 1954 article in Nature, Kolsky [1] first described electromagnetic (EM) emissions from the detonation of conventional explosives[1]. Subsequent studies examined these emissions, experimentally and theoretically [e.g. 2 and 3]. However, the literature on this topic contains conflicting information about the nature of these emissions. Much of this discrepancy arises because comparisons are made among diverse experimental configurations. Data were obtained at various frequency ranges and at diverse distances from the explosions. Some data were obtained over an ultra-broadband; others were obtained in narrow frequency bands. A recent publication [4] addresses some of the differences associated with

detonation environment; however, it is beyond the scope of this paper to discuss these differences. We consider only the distinctions between free-field detonations of bare and metal encased explosives when the shock wave, detonation products or case fragments do not interact with the ground or other surfaces during the measurement times. We also compare emissions from explosives encased in cylinders with emissions from impact fracture of cylinders, and find excellent temporal and spectral agreement between these two types of emissions.

## EXPERIMENTAL PROCEDURE

We performed a series of explosive experiments that included extensive near-field measurements to investigate the detonation environment. All experiments were conducted using right circular cylinders of two sizes: those with 114 mm diameters and 3.81mm wall thickness, and those with diameters of 153 mm and 5.08 mm wall thickness. We also performed two experiments

---

[1] Soviet literature references a 1940 publication on this topic: Ivanov, , A. G., "Seismic-electrical Effects of Second Kind", Transactions of USSR Academy of Sciences. Geography and Geophysics Series. (Russian), 5, p 699, 1940; however we have not yet located this publication.

using bare cylindrical explosive charges. We tested cylinders in pre-scored and un-scored configurations (Fig. 1). The pre-scored cylinders allowed us to control the fragment size and to more closely examine the processes associated with individual fragments. We used two scoring patterns: grid and diamond. To examine the effects of material properties, the un-scored cylinders were fabricated with either 1026 mild steel or 4140 heat-treated steel. In two experiments, the explosives were bare charges, with no case of any type.

**Figure 1.** Three types of steel cylinders used to investigate explosively generated EM emissions.

We designed experiments to examine details of processes occurring in the near-field of detonating explosives. Results were used to assist in developing a more complete picture of the mechanisms responsible for the EM emissions during detonations. We focused on measuring processes that influence the EM environment: electric charge of the case fragments and of the detonation products, degree of ionization of the detonation products and voltage distribution of the detonation products. Additionally, we obtained video images of the detonations at two framing rates (100,000 frames/sec and 750, 000 frames/sec) along with remote temperature and pressure measurements.

EM emissions resulting from the detonations were recorded using EG&G asymptotic conical dipole (ACD) sensors in series with baluns, connected with low-loss coaxial cables to LeCroy high-speed digital oscilloscopes. Configurations were optimized to obtain data in the range of 125 MHz to 2 GHz at approximately 25 m from the explosions. Prior to each experiment, we performed periodic measurements of the EM background to reduce the possibly of observing signals unrelated to the detonations. We performed a total of twenty-six such experiments.

## RESULTS AND DISCUSSION

Fig. 2 shows a pair of Rugowski coil placed near an explosive-filled cylinder (on left) to determine fragment charge; Langmuir and voltage probes were used in similar configurations. Details of these near-field physical measurements are not given here; we summarize key results in Table 1. There has been some speculation in the literature [4] that spark discharges between the detonation products and metal fragments are responsible for such emissions. We conclude that measured values are not consistent with spark discharge processes.

**FIGURE 2.** A typical experimental setup for near-field detonation measurements. The arrow shows the path a pre-scored fragment follows through dual Rugowski coils.

**TABLE 1.** Summary of electrical property measurements

| Instrument | Measurement | Mean Value |
|---|---|---|
| Rugowski coils | Fragment charge | -2.5 pico-Coul |
| Langmuir probes | Detonation gas charge density | $2 \times 10^{19}/m^3$ |
| Voltage probes | Detonation gas voltage | -4.5 V |

In this paper, we concentrate on results of our measurements of EM emissions in the far-field. We used two pairs of D-dot sensors to obtain vertical and horizontal components of the electric displacement vector; the pairs of sensors were separated by ninety degrees at 25 m from the detonation point. Both components display the same characteristic features; however, the details differ. When viewed on a time scale that includes the full detonation event ($\sim$200 - 250 $\mu$s), the emissions occur as a series of random spikes, of varying amplitude. Examination of such emissions at finer time resolution shows that each "spike" consists of numerous dipole-like oscillations with durations of about 20-30 ns.

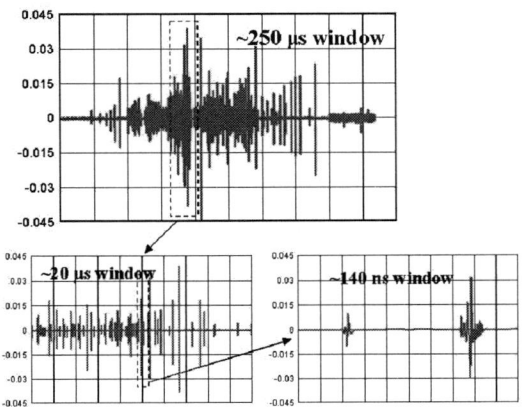

**Figure 3.** Typical EM emission from the detonation of a cased explosive. An event of $\sim$ 250 $\mu$s, when viewed in a time window of $\sim$140 ns includes many dipole-like oscillations of 20-30 ns duration.

As shown in Figure 3, these pulses are typically separated by approximately 40-50 ns. Although the general patterns are repeatable, details are random;

these events are stochastic. We have used Mott fragment distribution statistics [5, 6] as a basis for analyzing the statistics of the EM emissions. For explosive fragmentation of a cylinder of diameter (d) and wall thickness (T), the average fragment size is given by:

$$L_0 = C \cdot T^{5/6} d^{1/3} (1 + T/d) \qquad (1)$$

where C is a constant defining coupling between metal and explosive. The number of fragments with a size greater than or equal to L is given by

$$N_f(L) = N \cdot \exp(-L/L_0) \qquad (2)$$

Curran [7] and Grady [8] also show that one can define the fragment shape through a parameter $\eta$ that describes the ratio of the sides of a rectangular fragment. If $L_\theta$ is the length in the circumferential ($\theta$) direction, then the axial length is given by $L_z = \eta \cdot L_\theta$ where,

$$\eta = 1 + \left[ \left( \frac{\dot{\varepsilon}_\theta}{\dot{\varepsilon}_z} \right)^{2/3} - 1 \right] \cdot e^{-nD/C_g} \qquad (3)$$

Here the subscripts refer to the corresponding components of strain rate. D is detonation velocity, $c_g$ is the crack velocity of the metal, n is a constant. The components of the electric fields are given by

$$E_i = A \sin(\omega t)[1 - \exp(-t/\tau)] \qquad (4)$$

where the subscript refers to either the axial (z) or circumferential ($\theta$) component. Plots of the number of pulses of a given peak amplitude follows the same statistics as the fragment size distribution, which suggests a relationship between fragmentation and EM emissions. Misra [8] found that quasi-static fracture of metals leads to EM emissions; Molotskii [9] developed a theory to describe the Misra effect, in which the amplitude of the emission has a power-law dependence on yield strength. Further, we find that each component depends on the case thickness (T) and the fragment length (L); when applied to our data we find that

$$A_{\theta(z)} \propto L_{z(\theta)} \cdot T \cdot \left[ \frac{Y(\dot{\varepsilon})}{\sigma} \right]^5 . \qquad (5)$$

Note that the amplitude of the circumferential component of the E-field depends on the surface area of the fragment along the axial direction, and vice versa.

The frequency increases with tensile strength of the metal [9]; here the strength depends on strain-rate,

$$\omega = \omega_0 \left( \frac{Y(\dot{\varepsilon})}{\sigma} \right)^2 . \qquad (6)$$

By assuming logarithmic dependence on strain rate

$$Y = Y_0 \left[ 1 + \ln(\dot{\varepsilon}/\dot{\varepsilon}_0) \right] \qquad (7)$$

we find good agreement with our explosion data. We also find no emissions from bare explosives in the range of 125 MHz to 2 GHz.

## CONCLUSIONS

Analysis of EM emissions from detonations of metal encased explosives provides strong evidence that fracture of the metallic case is the dominant cause of these emissions. The data agree with Molotskii's model of the Misra effect. Additional work [11] shows that similar emissions occur during impacts. The absence of emissions from bare explosives adds credence to this argument.

## ACKNOWLEDGEMENTS

Funding provided by the Defense Threat Reduction Agency under Contract DTRA01-0033-SC. Discussions with D. E. Grady were valuable.

## REFERENCES

1. Kolsky, H. "Electromagnetic Waves Emitted on Detonation of Explosives, Nature, 173, p77, January 1954.
2. Anderson, V. H. and Lang, C. L. "Electromagnetic Radiation from Detonation of Solid Explosives", J. Appl. Phys., Vol. 36, No. 4, p1494, April 1965.
3. Hays, B. "The Detonation Electric Effect", J. Appl. Phys. Vol. 38, No. 2, p507, February 1967.
4. Brown, W. T. "Electromagnetic Emissions from Chemical Explosions: A Literature Review, DTRA Report RT-2001-01-002 (SRF-151), November 2001.
5. Mott, N. F. "A Theory of the Fragmentation of Shells and Bombs", British Ministry of Supply Report A. C. 4035, 1943.
6. Grady, D. E. and Kipp, M. E. "Geometric Statistics and Dynamic Fragmentation", J. Appl. Phys., Vol. 58, 1210-1222, 1985.
7. Curran, D. R. "Simple Fragment Size and Shape Distribution Formulae for Explosively Fragmenting Munitions", Int. J. Impact Engng. Vol. 20, 197-208, 1997.
8. Misra, A. "Electromagnetic Effects at Metallic Fracture", Nature. 254, 133-134, 1975.
9. Molotskii, M. I. "Dislocation Mechanism for the Misra Effect", Sov. Tech. Phys. Lett. 6(1), 22-23, 1980.
10. Grady, D. E. "Models and Analysis tools for Fragmenting Munitions" Report AEA Project 151.
11. Brown, W. T., Schmidt, M., and Calahan, K., "Electromagnetic Radiation From The High-Strain-Rate Fracture Of Mild Carbon Steel", to be published.

CP845, *Shock Compression of Condensed Matter - 2005,*
edited by M. D. Furnish, M. Elert, T. P. Russell, and C. T. White
© 2006 American Institute of Physics 0-7354-0341-4/06/$23.00

# EXPERIMENTAL STUDY OF GRIT PARTICLE ENHANCEMENT IN NON-SHOCK IGNITION

**Richard V. Browning[1], Paul D. Peterson[2], Edward L. Roemer[2], Michael R. Oldenborg[3], Darla G. Thompson[2] and Racci Deluca[2]**

*[1]127 Piedra Loop, Los Alamos NM 87544*
*[2]Los Alamos National Laboratory, Los Alamos NM 87545*
*[3]Dept. of Aerospace Engineering, University of Colorado at Boulder, Boulder, CO 80303*

**Abstract.** The drop weight impact test is the most commonly used configuration for evaluating sensitivity of explosives to non-shock ignition. Although developed 60 years ago and widely used both as a material compression test and as a test bed for understanding the ignition process itself, little is known about the flow mechanisms or involvement of grit particles as sensitizing agents. In this paper, we present the results of a series of experiments designed to study the flow mechanisms and events leading up to ignition. The experimental configuration used involves two pellet sizes, 3 and 5 mm in diameter, tested with three conditions: (1) smooth steel anvils, (2) standard flint sandpaper, and (3) shed grit particles loaded between the steel anvils and the pellet faces. Diagnostics include optical micrographs, and scanning electron micrographs. Un-reacted samples show a variety of morphologies, including what appear to be quenched reaction sites, even at very low drop heights. Quasi-static crushing experiments were also done to quantify load-time histories.

**Keywords:** PBX 9501, Bruceton drop weight impact machine, Type 12, Type 12b, grit, sandpaper
**PACS:** 62.50.+p, 82.3.Vx, 82.40.Fp

## INTRODUCTION

The drop weight impact test, as developed at the Bruceton Naval Research Laboratory 60 years ago [1] is still one of the standard tests used to characterize the handling safety of energetic materials. In developing the test, the goal was to create a characterization test that would order materials by impact sensitivity in a pre-conceived sequence. The original work focused on various anvil arrangements, starting with Type 1, Type 2, up through the Type 12 and Type 12b configurations used today. Although the basic machine, with its 2.5 or 5.0 kg weight, striker and anvil configuration remained essentially unchanged through the different configurations, the detailed arrangement of the sample between the faces of the striker and anvil was varied extensively.

Today, the most commonly used configurations for solids are the Type 12 flat faced striker and anvil with sandpaper under the sample and the Type 12b that omits the sandpaper. However, the interaction of the sandpaper with the sample is not well understood, and sometimes sensitizes things but in other situations appears to desensitize the sample [3].

Our original goal was to use Type 12 and Type 12b configurations to evaluate the enhancement caused by the presence of grit as modeled in a previous paper [2]. But as we investigated the test, it became obvious how little was really known about the detailed ignition mechanics of the experiment. Chaudhri and Field have studied gas compression ignition [4]. In another study Field used transparent anvils and high speed cameras to study the appearance of gas products from ignition events, and characterized the thermal conditions with clever diagnostics; however, the pressures involved and details of the flow are still not well understood [5]. Others have used the drop weight impact test as a convenient tool for studying the fast chemical reactions occurring in ignition events [6]. Here we report on our initial studies of grit involvement in the drop weight test and discuss evidence of possible mechanisms for ignition in these situations.

## EXPERIMENTAL PROCEDURE

Pellets with 48 mg PBX 9501 were pressed into two different cylindrical shapes, 5mm D by 1.35 mm H (L/d=0.3) and 3 mm D by 3.8 mm H (L/d=1.3. A series of drops were then done, one at a low height of 10 cm where a no-go event is expected, followed by a few drops at higher heights to obtain at least one go and one no-go event. Three different conditions were evaluated: bare anvils as in the Type 12b configuration, flat sandpaper underneath the pellet as in Type 12, and shed grit particles underneath the sample.

These configurations differ from the standard test as done at Los Alamos in that normally a powder sample of 30 mg is placed in a pre-dimpled cup of sandpaper. It is known [8] that the method of placing the powder in the cup and the pre-placement of the striker on top of the pile of powder can influence the outcome of the experiment. The pressed pellets provide a more repeatable sample and allow for tests with loose grit particles.

Quasi-static compression tests were done on samples of these pellets for quantitative information on the compression of the pellets. Load deflection data is recorded, and for several samples the loading process was interrupted and the samples photographed at a sequence of states.

Finally a conventional evaluation of the 50% drop height was done for the pellets. This was done for the bare anvil configuration and we plan on doing the sandpaper and grit particle arrangements as well.

## RESULTS AND DISCUSSION

Optical and electron microscope images were taken of the remains of the un-reacted samples. Some images are shown in Figs. 1–4.

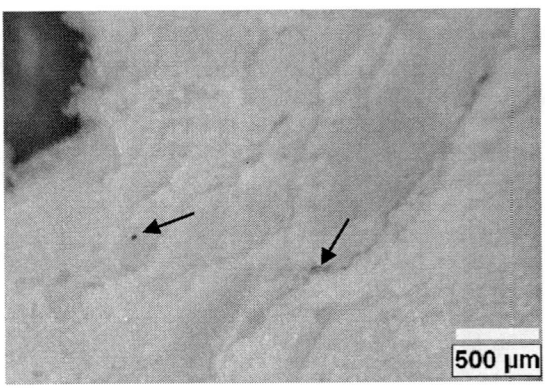

**Figure 1.** Sample without grit, drop height of 57 cm, shows fissures that could trap gases, and a small black speck towards left of center that could be a reaction site. The dark area at upper left is a green paper background supporting the sample. In color, a small area of red or pink is visible along the lower edge of the green.

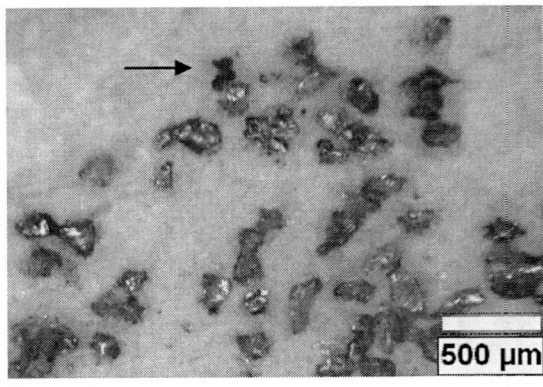

**Figure 2.** This sample was covered with garnet shed grit particles, drop height at 10 cm. Note the spiral-like dark region just above top center grit particle – possibly a quenched ignition site.

A variety of features are evident, including colored specs, black regions, and fissures. Depending on light reflections, some evidence of twinning of the HMX crystals is seen. In one case (Figure 4) a metal chip was seen on the top (striker side) of a recovered sample. Scratches were also visible on both the striker and bottom anvil for all three test configurations in this study. We had hoped to see metal chips in these experiments but were surprised by the location. It became obvious later that the pellets are reduced to a very thin disk when fully compressed, at which point the grit particles and larger HMX particles span the distance between the anvils.

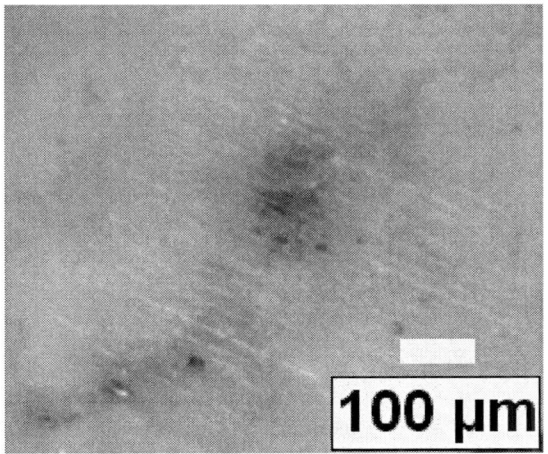

**Figure 3.** This image from another grit covered sample shows light striations from possible twinning over darkened areas that might be reaction sites. Drop height was 40.5 cm.

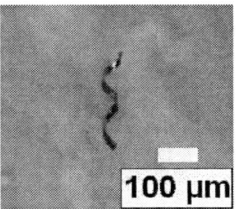 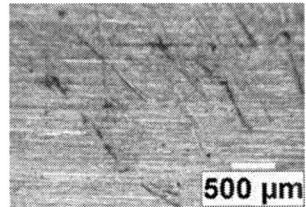

**Figure 4.** Left -- Spiral-shaped metal shaving recovered from the top (striker side) of a no-go sample. Right – Scratches on the bottom anvil after Type 12b test. Similar scratches were visible on the striker.

A few quasi-static compression tests were also done to quantify the energy flow into the sample. The potential energy in the drop weight is sufficient to heat the sample to approximately the melting point, but this is not enough to cause significant reaction in the HMX on the time scales of the impact event. The load deflection results done on the Instron indicate that very little energy

is actually absorbed by the samples during the initial crushing event (see Figure 5). Although the Instron tests are done at relatively low rates, we do not expect substantial changes in the nominal crushing stress of the samples at strain rates achieved in the drop weight impact machine. The measurements of the compressed pellet diameter also allow us to estimate the actual pressures seen in the pellets. Pressure estimates were based on preliminary numerical simulations of the complete drop weight impact machine and the sample. In these simulations the pressures are initially very low as the sample begins to crush, but builds up rather suddenly at the end of the compression stroke to values over 1000 MPa (10 kb).

Under these assumptions the pressures turned out to be rather substantial — 400 MPa (4 kbar), at the 30 kN load limit of the Instron. Clearly, the drop weight impact machine is capable of generating much higher pressures, although we have not been able to quantify the actual value. Values in the 1000-2000 MPa (10-20 kbar) range are likely, limited only by the yield strength of the anvil.

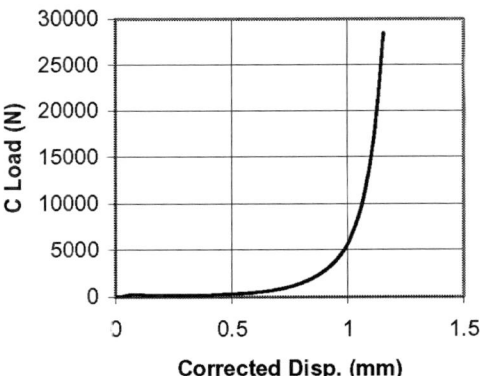

**Figure 5.** Load-deflection, after correction for stiffness of testing machine, of a 5 mm diameter by 1.35 mm high pellet. This indicates a pellet height of only 150-200 microns at the load capacity of the machine. Measurements made after unloading show the wafers at about 300 microns. The initial peak load is barely visible in this chart at 0.1 mm of displacement.

The 50% drop height values show substantial differences between the shorter and longer pellets as summarized in Table 1. These results are unexpected in that the energy absorbed in the initial crushing of the sample is very small compared to the total energy in the drop weight.

**Table 1.** Standard "Bruceton up/down method" drop heights for PBX 9501 pellets and reference materials. The PBX 9501 samples were tested without sandpaper (Type 12 B test).

| Sample | 50% Height (cm) | σ Log Units |
|---|---|---|
| Short Pellets | 58.9 | 0.07 |
| Tall Pellets | 91.4 | * |
| TNT Standard | 244.7 | 0.04 |
| PETN Standard | 14.0 | 0.10 |
| HMX Standard | 25.9 | 0.08 |

\* The tall pellets gave a very wide range of heights, from 64.0 to 143.0 cm, so sigma is out of range on the chart.

## CONCLUSIONS

Our current concept of the drop weight ignition sequence involves a small amount of energy deposited during the initial crushing of the sample, whether a pellet or pile of powder. When the sample becomes thin enough that individual crystals can span the gap between the anvils, then a higher lateral restraining force develops, allowing for the development of high pressures and leading to high temperatures caused by adiabatic heating of the materials. The Gruneisen gamma is the critical material constant in this process, and surprisingly little is known about gamma for the polymeric materials used as binder, or even the explosive materials themselves.

Clearly, future work is needed to obtain better values for the Gruneisen gamma, on time scales of 10 micro-sec to 10 ms. We also plan to do well resolved calculations of the dynamics of the drop weight machine to study how the shape and size of the anvil, striker and drop weight interact during the deformation of the pellet.

The importance of grit particles on ignition itself is still not clearly defined. They can have an effect, but in the drop weight impact configuration, the interaction of grit particles is subtle. When adiabatic compression can serve as the ignition mechanism, then the added ignition sources caused by having a bed of grit particles on one face might

be important in some situations. However, the simple mechanical interaction of the grit particles as additional drag sources as the anvils get close together might be the more likely situation. On the other hand, grits might cause ignition events at such low drop heights that adiabatic compression simply doesn't generate enough temperature to cause ignition.

## ACKNOWLEDGEMENTS

We are grateful for continued support from the ASC program, Tom Dey manager, for the modeling work done on this project. Dennis Montoya and Ken Laintz provided experimental support with the drop weight impact machine.

## REFERENCES

1. H. Dean Mallory, Ed., "The Development of Impact Sensitivity Tests at the Explosives Research Laboratory Bruceton, Pennsylvania During the Years 1941-1945," NAVORD Report 4236, March 1956

2. Browning, R.V., Peterson, P.D., Roemer, E.L., Scammon, R.J., "Grit Particle Enhanced Non-Shock Ignition of Explosives," in Shock Compression of Condensed Matter, 2003, (M.D. Furnish, Y.M. Gupta, and J.W.Forbes, eds.), part II, pp. 921-924.

3. Gibbs, T.R., Popolato, A., "LASL Explosive Property Data," U. California Press, Berkeley 1980, pp 446-453

4. Chaudhri, M.M., Field, J.E., "The Role of Rapidly Compressed Gas Pockets in the Initiation of Condensed Explosives," Proc. Royal Soc. London, A(1974) 340, pp. 113-128

5. Field, J.E., et.al., "Hot-Spot Ignition Mechanisms for Explosives and Propellants," Phil. Trans. R. Soc. Lond. A (1992) 339, pp. 269-283

6. Buntain, G.A., et.al., "Decomposition of Energetic Materials on the Drop-Weight-Impact Machine," Ninth Detonation Symposium, pp. 1037-1043

CP845, *Shock Compression of Condensed Matter - 2005*,
edited by M. D. Furnish, M. Elert, T. P. Russell, and C. T. White
2006 American Institute of Physics 0-7354-0341-4/06/$23.00

# MEASUREMENT OF IGNITION AND REACTION PARAMETERS IN NON-IDEAL ENERGETIC MATERIALS

## E. J. Cart, R. H. Granholm, V. S. Joshi, H. W. Sandusky, and R. J. Lee

*Naval Surface Warfare Center, Indian Head Division, Indian Head, MD 20640*

**Abstract.** Two small-scale tests were performed to measure ignition and reaction parameters in non-ideal energetic materials. Hydrocode modeling underway will determine the effectiveness of this approach. The time to reaction and the ignition conditions are derived from the newly developed hybrid Hopkinson bar experiments, whereas the growth criteria are based on the recently developed small-scale shock reactivity test (SSRT). The hybrid Hopkinson bar test simultaneously measures the mechanical behavior and ignition conditions of explosives. The reactivity test measures the potential of a material to be an explosive regardless of its sensitivity, thus avoiding the problem of scale, inherent in most small-scale explosive tests.

**Keywords**: non-ideal explosives, Hopkinson Bar, Small-scale, Ignition and Growth models
**PACS**: 81.70.Bt, 82.40.Fp

## INTRODUCTION

A hybrid Hopkinson bar[1] and the small-scale reactivity test (SSRT)[2] are being used to investigate time to reaction and growth, respectively. The ultimate goal is to offer a means of determining a first order approximation of sensitivity for new formulations containing ingredients that are not yet readily available. Miller has successfully approximated ignition and growth parameters using large-scale gap tests and wave curvature measurements.[3] The approach here is to obtain similar results from much smaller scale tests.

For time to reaction, an earlier study by Davis had shown promise in modeling impact sensitivity by correlating drop weight test data to a critical energy fluence defined by $P^2\tau$, where P and $\tau$ are the magnitude and duration of the stress in the sample, respectively.[4]

The hybrid Hopkinson bar combines aspects of two standard tests configurations. The split Hopkinson pressure bar (SHPB) is used to measure the mechanical behavior of materials at high strain rates [5], whereas, the drop-weight impact test is used to measure the sensitivity of explosives to impact. In the SHPB a greater fraction of the elastic wave energy travels to the transmitter bar. This leaves less energy deposited into the sample, decreasing the likelihood of ignition, therefore making this test unsuitable for studying ignition. The hybrid set-up eliminates this problem. The hybrid Hopkinson bar was chosen for these ignition studies in lieu of the drop weight test because its input loading can be better defined.

The second test of interest is the SSRT, which was developed by Sandusky and Granholm [6]. It explores explosives at their earliest stage of scale-up, using a gram or less, making it convenient for laboratory use. The measurement of shock reactivity is possible without having to obtain shock-to-detonation transition. The SSRT, in which explosive components are approximately the diameter of a detonator, is briefly described in a companion paper [6].

The approach is to perform these studies on PBXN-111, which is a well-characterized cast-cured explosive. When completed, the modeling will be validated from known data on this explosive. In addition to it being well characterized, PBXN-111 is attractive because its basic formulation is similar to many non-ideal explosives and propellants, containing a crystalline nitramine explosive (RDX or HMX), ammonium perchlorate (AP), powdered aluminum, and a plastic binder system.

## EXPERIMENTAL METHODS

### Hybrid Hopkinson Bar

A schematic of the set-up is shown in Fig. 1. The transmitter bar from the SHPB is replaced by a fixed anvil. Two relatively short 2.54 cm

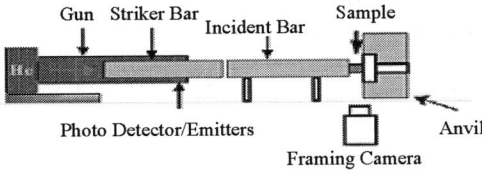

**Figure 1.** Hybrid Hopkinson bar set-up.

diameter by 30.5 cm long hardened steel bars replace the striker and incident bars of the SHPB, or the weight and striker respectively, in the drop-weight impact test. Compressed gas drives the striker at velocities in excess of 20 m/s, impacting the incident bar, causing high strain rate compression of the sample.

Impact velocity is measured by two pairs of high-speed (~1ns rise-time) photo detector/emitter sensors placed 5.08 cm apart at the end of the gun barrel. The output of the second sensor is connected to a trigger for the camera. A customized high-speed framing camera is used to view the deformation of the explosive sample as well as any light from ignition if it occurs. Line scan imaging was used for capturing the strain data and for initial analysis, as shown in Fig. 2. Rather than full-field views, only a single row of 1280 pixels was recorded, as shown by the line on Fig. 2. This camera captures 13,200 lines at 2 μs each, thus creating a streak image for a total of 26.4 μs.

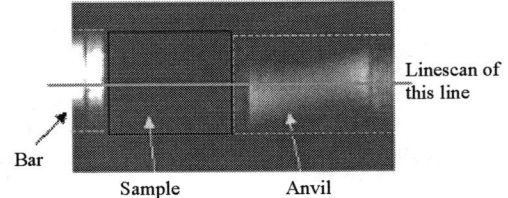

**Figure 2.** Initial full-screen image of the set-up. Line shows 1280 x 1 linescan of moving marker.

### Small-Scale Shock Reactivity Tests

See "Prompt Reaction of Aluminum in Detonation Explosives" by Sandusky and Granholm (this meeting) for the SSRT arrangement. A small sample is placed in a hole in a steel containment block. The sample length is less than the diameter of the detonator so that the entire sample receives a strong shock. At the bottom of the set-up is a soft aluminum witness block. The depth of the dent in the witness block indicates the reaction growth occurring within a microsecond timeframe from the shock loading. The relatively short sample length limits shock attenuation, thus shock reactivity is possible without having to obtain shock-to-detonation transition.

The SSRT was characterized with samples where HMX was the only energetic ingredient. Making comparisons with explosives having low detonation pressure, such as PBXN-111, required dilution of the HMX samples. These were blends of Class 1 HMX and melamine powders in various proportions, but always totaling 70 %vol. For samples with <50% volume (%vol) of Class 1 HMX, shock reaction occurs without transiting to detonation, which is referred to as the range of lower reaction.

## RESULTS

### Hybrid Hopkinson Bar

Two streak images obtained from the camera are shown in Fig. 3. The bands of light on either side of the dark sample region were produced by placing reflectors on the edges of

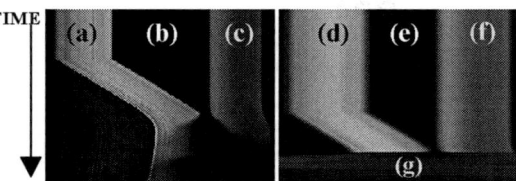

TIME

**Figure 3.** Multi-frame (linescan) images when the striker compresses the material. Thinner sample sees ignition. 3(a,d) Striker moving into the sample. 3(b) First test sample 0.64 mm thick. 3(c,f) Anvil. 3(e) Second test sample 0.33 cm. 3(g) Black line indicating ignition.

the incident bar and the anvil. Measurements can be made from either edge of the band. The difference in distance between the two light bands show sample compression in time. The cylindrical PBXN-111 sample is 0.73 cm diameter by 0.64 cm thick. After compressing the sample to a minimum, the bar then rebounded as observed from the back edge of the reflected incident band. The sample fractures out and obscures the leading edges of the bands but no reaction is observed. The second image of Fig. 3 illustrates a second test where the sample was thinner (0.33cm thick) which has the effect of increasing the stress on the sample. The striker was maintained at a relatively similar velocity. The material experiences ignition at the start of the black line at the bottom of the image when the reaction products block the light to the camera.

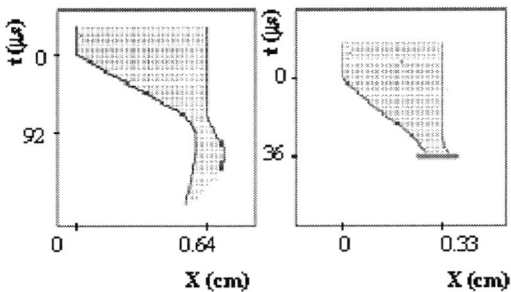

**Figure 4.** Pixel vs. time profiles generated from streak images in Fig. 3.

An edge detection technique was used to produce the pixel vs. time profiles seen in Fig. 4. Using this input the strain, $\varepsilon_v$, and strain rates, $d\varepsilon_v/dt$, can be calculated assuming a constant volume,

$$d\varepsilon_v/dt = (dh/dt)/h, \qquad (1)$$

where $dh/dt$ is the change in height of the sample and $h$ is the thickness of the sample. We can calculate the rate of energy deposition, $dW/dt$ with the equation

$$dW/dt = \sigma_v (d\varepsilon_v/dt). \qquad (2)$$

A load cell can be added to directly measure stress, $\sigma_v$. Different striker velocities and sample thickness can be used to vary energy and energy rate combinations to be plotted along the ignition threshold curve, which is shown in Fig. 5. The markers on the graph represent targeted experimental range of values.

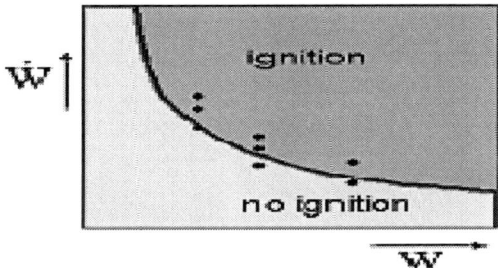

**Figure 5.** Ignition threshold curve.

### Small-Scale Shock Reactivity Tests

A fit through the data from HMX/ melamine/HTPB samples is illustrated in Fig. 6 along with the actual dent depths for some energetic materials whose major ingredient is AP. Those samples without HMX were assigned an equivalent HMX %vol, (%vol$_{equiv}$ ), assuming that equal values of detonation pressure ($P_{CJ}$) resulted in the same dent depth. If the sample contained RDX, %vol$_{equiv}$ was the %vol RDX multiplied by the 0.88 ratio of the $P_{CJ}$ for RDX relative to HMX, as if the RDX is the only component that reacts promptly to a strong shock. For the AP-based samples without any nitramine, %vol$_{equiv}$ was the ratio of their $P_{CJ}$, usually calculated with the Cheetah thermochemical code, with the $P_{CJ}$ for HMX. While other schemes could be used for %vol$_{equiv}$, those described at least provide a means of comparing similar samples.

Along with PBXN-111 are shown the results of two other related samples. One has HMX substituted for the RDX, resulting in a dent within the ±0.1 mm repeatability for the test. The other has somewhat more RDX, thus the higher %vol$_{equiv}$; but unlike PBXN-111, the RDX is predominantly a fine particle size that reduces its shock reactivity, as shown by the reduced dent. The samples of AP, AP/HTPB, and AP/Al all contained 60 %vol of 200 μm AP. Filling the voids in porous AP with HTPB results in a more stoichiometrically balanced mixture, with a higher calculated P$_{CJ}$ and estimated %vol$_{equiv}$, but the shock reactivity is reduced. Without any binder, adding Al to AP increases the dent, more so in the 90/10 blend than in the 80/20 blend. When including an energetic binder to make an AP/Al mix detonable, the dent is comparable to that for the porous bed of 80/20 AP/Al.

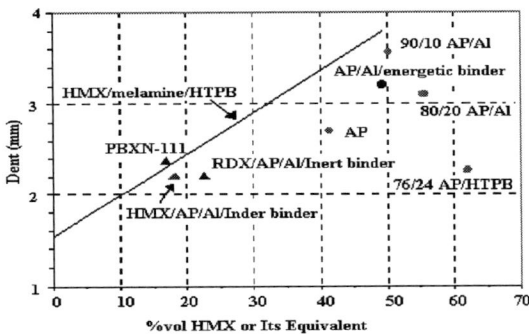

**Figure 6.** SSRT data in the range of lower reaction.

## SUMMARY

A hybrid Hopkinson bar has been developed and illustrated how it can be used to explore ignition phenomena. Also, the small-scale shock reactivity test has been used to provide data on rate of energy release. The results from these methods will be combined to obtain a first estimate for ignition and growth parameters for future use in developing reaction models. This approach is attractive because it only requires a small amount of material.

## FUTURE WORK

The hybrid Hopkinson bar test series will be completed to obtain additional data points for the ignition threshold curve. Results from this test and the SSRT will be modeled using the DYNA-2D hydrocode to explore useful parameters that may be obtained for PBXN-111 in a Lee-Tarver ignition and growth model. The resulting reaction model will be validated with existing large-scale data for PBXN-111.

## ACKNOWLEDGEMENTS

This work was funded by the Technology Improvement Program at NSWC-Indian Head Division.

## REFERENCES

1. Joshi, V.S. and Guirguis, R.H., "A Hybrid Drop Weight-Hopkinson Bar Test to Characterize the Reactive and Mechanical Properties of Explosives," NAVSEA Indian Head Division, 22$^{nd}$ JANNAF PSHS meeting, June 13-17, 2005, Charleston, SC, in print.
2. Sandusky, H.W., Granholm, R.H., and Bohl, D.G., "Small-Scale Shock Reactivity Test," IHTR 2701, NAVSEA Indian Head Division, to be printed.
3. Miller, P.J, "A Simplified Method for Determining Reactive Rate Parameters for Reaction Ignition and Growth in Explosives," Naval Air Warfare Center, China Lake, CA, Materials Research Society Symposium: Volume 418, Decomposition, Combustion, and Detonation Chemistry of Energetic Materials, Nov. 27-30, 1995, Boston, MA, in print.
4. Davis, J.J., "Characterization of Plastic Deformation and Chemical Reaction in Titanium-Polytetrafluoroethylene Mixture," American University, UMI Dissertation Services, 1997.
5. Gray, G. T., "Classical split Hopkinson bar techniques" in ASM Handbook, 10th edition, vol. 8, pp. 462-476, 2000.
6. Sandusky, H.W. and Granholm, R.H., "Prompt Reaction of Aluminum in Detonation Explosives," this meeting.

CP845, *Shock Compression of Condensed Matter - 2005,*
edited by M. D. Furnish, M. Elert, T. P. Russell, and C. T. White
© 2006 American Institute of Physics 0-7354-0341-4/06/$23.00

# LX-04 VIOLENCE MEASUREMENTS-STEVEN TESTS IMPACTED BY PROJECTILES SHOT FROM A HOWITZER GUN

## Steven K. Chidester, Kevin S. Vandersall, Lori L. Switzer, and Craig M. Tarver

*Lawrence Livermore National Laboratory*
*Livermore, CA 94550*

**Abstract.** Characterization of the reaction violence of LX-04 explosive (85% HMX and 15% Viton A by weight) was obtained from Steven Impact Tests performed above the reaction initiation threshold. A 155 mm Howitzer propellant driven gas gun was used to accelerate the Steven Test projectiles in the range of approximately 170-300 m/s to react (ignite) the LX-04 explosive. Blast overpressure gauges, acoustic microphones, and high-speed photography characterized the level of high explosive reaction violence. A detonation in this velocity range was not observed and when comparing these results (and the Susan test results) with that of other HMX based explosives, LX-04 has a more gradual reaction violence slope as the impact velocity increases. The high binder content (15%) of the LX-04 explosive is believed to be the key factor to the lower level of violence.

**Keywords:** Steven Impact Test, Explosive safety, LX-04, HMX explosives, ignition threshold
**PACS:** 82.33.Vx, 82.40.Fp

## INTRODUCTION

In general, the Steven Impact Test is a safety test involving high explosive (HE) targets impacted at increasingly greater velocities with projectiles until you get a "GO" (reaction). For the most part, these reactions involve a burning or deflagration process in lieu of a full-scale detonation. Naturally, the lowest velocity where you get a "GO" is the "reaction threshold" and typically involves several experiments to determine. Performing experiments above the "reaction threshold" as presented in this work can also act to characterize the level of violence observed in the reaction. Both the "reaction threshold" and violence level data can be utilized in various hydrodynamic reactive flow models for safety predictions that may not be directly tested.

Research on the Steven Test has been performed at Lawrence Livermore National Laboratory [1-7] as well as a modified version of this test at Los Alamos National Laboratory [8-10]. Converting overpressure transit data from impact to equivalent point source energy dates back to the Susan Test [11] and is discussed in a prior publication [12].

The Steven Impact Test results to date have increased the fundamental knowledge and practical predictions of impact safety hazards for confined and unconfined explosive charges. As discussed in the prior publications [1-10], friction, shear, and strain are the main contributing mechanisms to reaction although continuing research is still investigating these individual areas and combinations of mechanisms.

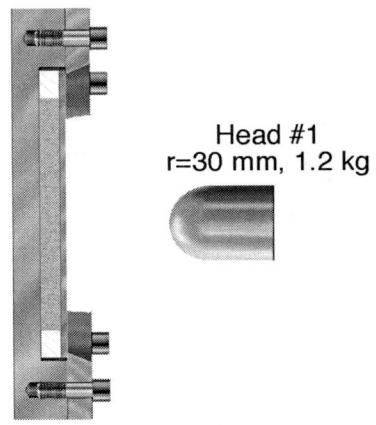

Head #1
r=30 mm, 1.2 kg

**FIGURE 1.** Schematic diagram of the standard Steven Impact Test arrangement used in this work.

## EXPERIMENTAL PROCEDURE

For these experiments, a 155 mm diameter smooth bore Howitzer gun located at LLNL Site 300, bunker 850 was utilized to fire five projectiles with various velocities at Steven Impact Test targets on an outdoor firing table. The experimental geometry of the Steven Impact Test target is shown in Fig. 1. The steel projectile head (see Fig. 1) is attached to a sabot body that is accelerated via a propellant charge into the target. External blast overpressure gauges were placed around the target at a 3.05 m standoff for direct comparison to the Susan test data [11].

This work was performed using LX-04 (85% HMX, 15% Viton A) energetic material samples to determine if the gradual reaction violence slope continues as the impact velocity increases. Normally a 76 mm diameter smooth bore light gas gun is used for these tests, but the Howitzer gun was used instead due to a higher velocity capability.

As shown in Fig. 1, the projectile consists of a hemispherical 30.05 mm radius steel head having a mass of 1.2 kg. The test projectile is accelerated into a 110 mm diameter by 12.85 mm thick explosive charge confined by a 3.18 mm thick steel plate on the impact face, a 19.05 mm thick steel plate on the rear surface, and 26.7 mm thick steel side confinement. A Teflon ring around the explosive provides radial confinement. Blast overpressure gauges, microphones, and high-speed photography characterized the level of reaction violence.

## RESULTS/DISCUSSION

The tabulated results for this work are shown in Table 1. Included are details about the experiment number, impact velocity, and reaction violence. All of these tests were performed at ambient temperature (20°C) using LX-04 as the target material with 277-month stockpile age and had a sample density of 1.863 g/cm$^3$. The reaction violence was obtained by converting the measured over-pressures to grams of TNT equivalent reaction [12].

The test results were video taped with fast framing cameras. Two of the frames from a test are presented in Figures 2 and 3. The grid in the background of the movie frames is used as a conformation of the projectile velocity. The visual evidence obtained by the movies helps provide a rough correlation of the over-pressure gauge results. The reaction observed is clearly not a detonation and this is confirmed by the over-pressure gauge results. Intentional detonations with TNT and other explosives were reported previously [3].

**TABLE 1.** Summary of LX-04 (sample density 1.863 g/cm$^3$) high velocity Steven Test results performed at ambient temperature (20°C).

| EXPT | PROJECTILE VELOCITY m/s, (ft/s) | VIOLENCE–TNT EQUIVALENT (g) |
|------|------|------|
| WRL158 | 173, (569) | 160 |
| WRL157 | 231, (759) | 154 |
| WRL156 | 250, (820) | 180 |
| WRL155 | 293, (961) | 185 |
| WRL154 | 298, (979) | 175 |

**FIGURE 2.** Frame showing projectile prior to projectile impact.

**FIGURE 3.** Frame showing reaction violence after projectile impact.

Figure 4 presents a comparison of results for the Steven Impact Test (dashed lines) and the Susan Impact test (solid lines) regarding the reaction violence in the form of blast overpressure related to a TNT equivalent as a function of the projectile velocity. All five tests with the Howitzer gun with velocities ranging from 173 to 298 m/s had about the same reaction violence with no noticeable trend.

There is good reaction violence correlation between the Steven Tests and the Susan Tests in order of decreasing explosive reaction violence PBX 9404, LX-10, PBX 9501, and LX-04. The five recent Steven Tests reported here with LX-

04 do not follow the gradually increasing violence trend of the Susan Test. A possible explanation for this is that in the Susan test the explosive sample may remain under inertial confinement between the Susan projectile and the impact barrier longer than does the explosive in the Steven test geometry. This longer confinement duration could allow the ignited LX-04 to react gradually more violently in the Susan Test as the velocity increases. This postulated explanation could be tested in future three-dimensional hydrodynamic modeling and experiments using increased confinement of the LX-04 Steven Test charges.

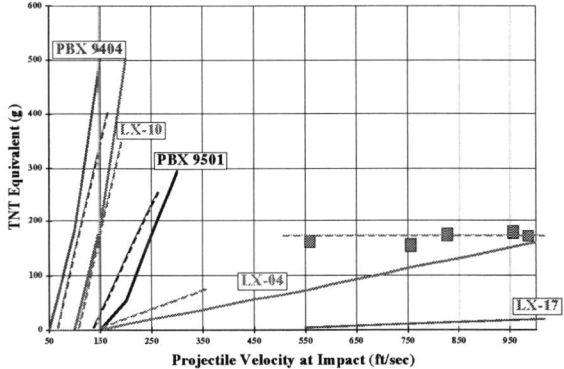

**FIGURE 4.** Comparison of results for the Steven Impact Test (dashed lines) and the Susan Impact test comparing violence in the form of blast overpressure related to a TNT equivalent as a function of the projectile velocity.

## SUMMARY AND FUTURE WORK

Steven Test targets containing LX-04 explosive samples were impacted at velocities up to 298 m/s. In comparing these results to those for LX-04 from the Susan Test, the LX-04 reaction violence did not gradually continue to increase as previously predicted. All five tests with the Howitzer gun with velocities ranging from 173 to 298 m/s had about the same reaction violence with no noticeable trend although an asymptote may have been reached.

Future work is planned to incorporate these Steven Test reaction violence results into HE reactive flow computer models for LX-04 that can then be used to make reaction violence predictions for other impact scenarios and for LX-04 in other geometries.

## ACKNOWLEDGEMENTS

The following members of the 155 mm gun crew at Site 300 (Bunker 850) are thanked for their hard work: Karen Luis, Tommy Rambur, Jim Browning, and Anthony Regalado. In addition, this work would not have been possible without the unwavering management and funding provided by Ron Streit. This work was performed under the auspices of the U. S. Department of Energy by the University of California, Lawrence Livermore National Laboratory under Contract No. W-7405-Eng-48.

## REFERENCES

1. Chidester, S.K., Green, L.G., and Lee, C.G., "A Frictional Work Predictive Method for the Initiation of Solid High Explosives from Low Pressure Impacts," Tenth International Detonation Symposium, ONR 33395-12, Boston, MA 1993, pp. 785-792.
2. Chidester, S. K., Tarver, C. M., and Lee, C. G., "Impact Ignition of New and Aged Solid Explosives," Shock Compression of Condensed Matter-1997, edited by S.C. Schmidt et. al., AIP Conference Proceedings 429, AIP Press, New York, 1998, pp. 707-710.
3. Chidester, S. K., Tarver, C. M., and Garza, R., " Low Amplitude Impact Testing and Analysis of Pristine and Aged Solid High Explosives," Eleventh (International) Symposium on Detonation, ONR 33300-5, Arlington, VA, 1998, pp. 93-100.
4. Chidester, S. K., Tarver, C. M., DePiero, A. H., and Garza, R. G., "Single and Multiple Impact of New and Aged High Explosives in the Steven Impact Test," Shock Compression of Condensed Matter-1999, M.D. Furnish, L. C. Chhabildas, and R. S. Hixson, eds., AIP Conference Proceedings 505, New York, 2000, P. 663-666.
5. Niles, A. M., Garcia, F., Greenwood, D. W., Forbes, J. W., Tarver, C. M., Chidester, S. K., Garza, R. G., and Switzer, L. L., "Measurement of Low Level Explosives Reaction in Gauged Multi-dimensional Steven Impact Tests," Shock Compression of Condensed Matter-2001, Furnish, M. D., Thadhani, N. N., and Horie, Y, eds. CP-620, AIP Press, New York, (2002).
6. Vandersall, K.S., Chidester, S. K., Forbes, J. W., Garcia, F., Greenwood, D. W., Switzer, L. L., and Tarver, C. M., "Experimental and Modeling Studies of Crush, Puncture, and Perforation Scenarios in the Steven Impact Test," Proceedings of the 12th International Detonation Symposium, San Diego, CA, August, 2002, pp.131-139.
7. Switzer, L. L., Vandersall, K. S., Chidester, S. K., Greenwood, D. W., and Tarver, C. M., "Threshold Studies of Heated HMX-Based Energetic Material Targets Using the Steven Impact Test," Shock Compression of Condensed Matter-2003, edited by M.D. Furnish, Y.M. Gupta, and J.W. Forbes, pp. 1045-1048, 2004.
8. Idar, D. J., Lucht, R. A., Straight, J. W., Scammon, R. J., Browning, R. V., Middleditch J., Dienes, J. K., Skidmore, C. B., and Buntain, G. A., "Low Amplitude Insult Project: PBX9501 High Explosive Violent Reaction Experiments," Eleventh International Detonation Symposium, Aspen, CO, 1998, pp. 101-110.
9. Scammon, R. J., Browning, R. V., Middleditch, J., Dienes, J. K. Haverman, K. S., and Bennett, J. G., "Low Amplitude Insult Project: Structural Analysis and Prediction of Low Order Reaction," Eleventh International Detonation Symposium, Aspen, CO, 1998, pp. 111-118.
10. Browning, R. V., "Microstructural model of mechanical initiation of energetic materials," Shock Compression of Condensed Matter-1995, S. C. Schmidt and W. C. Tao, eds, AIP Press, New York, 1996, p. 405-408.
11. Dobratz, B.M. and Crawford, P.C., LLNL Explosives Handbook, Lawrence Livermore National Laboratory Report UCRL-52997 change 2, 1985.
12. Green, L.G. and Dorough, G.D., "Further studies on the Ignition of Explosives, Fourth Symposium (International) on Detonation, October 12-15, 1965, White Oak, MD, pp. 477-486.

CP845, *Shock Compression of Condensed Matter - 2005,*
edited by M. D. Furnish, M. Elert, T. P. Russell, and C. T. White
© 2006 American Institute of Physics 0-7354-0341-4/06/$23.00

# LINKS BETWEEN THE MORPHOLOGY OF RDX CRYSTALS AND THEIR SHOCK SENSITIVITY

## H. Czerski*, M. W. Greenaway*, W.G. Proud* and J. E. Field*

*PCS Group, Cavendish Laboroatory, Madingley Road, Cambridge CB3 0DS, UK*

**Abstract.** It has been known for some time that batches of the secondary explosive RDX from different manufacturers show significant variation in their shock sensitivity. No clear correlation between shock sensitivity and either chemical composition or morphology has been identified. As yet no comprehensive study has been reported covering RDX from different manufacturers, different production methods, and different particle sizes. In this work we use a range of techniques to study the morphology of RDX grains more closely and to assess which hotspot mechanisms might be dominant. Crystals were characterised using mercury porosimetry, environmental scanning electron microscopy (ESEM) and optical microscopy. This range of methods yields quantitative data on internal void size and number and surface porosity. Shock sensitivity is quantified using small-scale gap tests and these demonstrate clear differences in sensitivity between batches from different manufacturers. The samples used are from three manufacturers, produced by both the Woolwich and Bachmann processes, and of two different sizes, so a comprehensive study of how morphology might affect hotspot formation is possible.

**Keywords:** RDX, morphology, gap test, shock sensitivity
**PACS:** 47.40.-x, 47.40.Nm

## INTRODUCTION

There is considerable interest the topic of RDX shock sensitivity because of the desire for insensitive munitions. In recent years many studies have been carried out to look for links between sensitivity and morphology. Variations in shock sensitivity between different batches as great as 50% have been measured. Differences in chemical composition have been ruled out as an explanation and so the current focus is relating morphology to possible hotspot mechanisms. Many aspects of the morphology have been suggested as crucial factors in hotspot generation, including surface roughness and voids [1, 2], internal defect population [3, 4],and dislocation motion [5, 6].

In this study we investigate shock sensitivity differences between unpressed granular beds of each batch. Our sample set comprises seven different batches, including samples from three manufacturers and in two size classes, produced by both Bachmann and Woolwich processes. The class 5 samples used here are labelled 5a–d and the class 1 samples are 1a–c. In order to examine both surface and internal morphology, we have used Environmental Scanning Electron Microscopy (ESEM), optical microscopy and mercury porosimetry. Shock sensitivity was assessed using a small scale gap test.

## SHOCK SENSITIVITY

The experimental design of the small scale gap test [7] is shown in figure 1. The shock pressure decayed by a factor of approximately two for each additional 2mm in gap thickness. The detonators used produced a shock of the order of 15 GPa.

In order to test the RDX crystals in the as-supplied state, the charges were not pressed, but were poured into the confinements in small increments

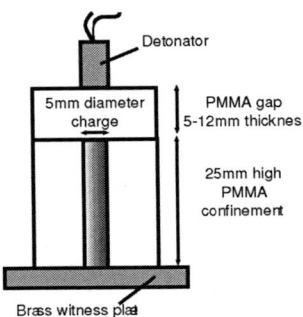

**FIGURE 1.** The design of the small-scale gap test

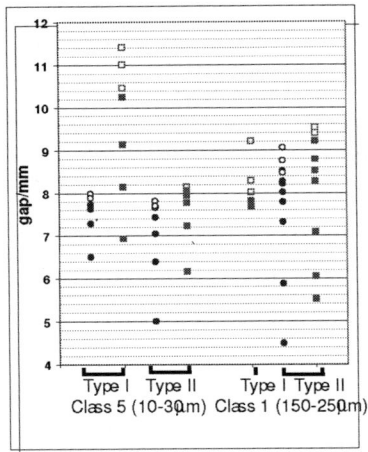

**FIGURE 2.** Results of gap tests on two size classes. Open symbols represent "no go" events and filled symbols represent "go" events.

and tapped. The differences in final density were only 1–2% for the medium size class. For the smallest size class, tapped densities for different batches ranged from 32% to 49% of the theoretical maximum density. Repeating some of the sensitivity tests with samples pressed to 50% showed no change in sensitivity so we are confident that testing unpressed samples gives the best representation of the as-supplied crystal, without the complications introduced by damage during pressing. The results for both size classes are shown in figure 2.

There are significant differences in shock sensitivity, most notably between the two class 5, type I samples (5a) and (5b). This suggests that at least some of

the observed differences in the shock sensitivity of cast materials may be due to crystal features, even before any binder-crystal interaction.

## MORPHOLOGY

Morphological crystal features can be studied using a range of techniques, each appropriate for a different scale and a different type of feature. For this work, optical microscopy, mercury porosimetry and environmental scanning electron microscopy were chosen. Bowden and Yoffe [8] identified the size of a critical hotspot at 0.1–10 $\mu$m so we expect that any morphological features leading to critical hotspots would be approximately this size.

### Optical Microscopy

Samples in index-matched fluid were examined and photographed. Typical crystals are shown in figure 3. Black areas on the photographs arise from the refractive index mismatch between the crystal and the contents of its internal closed voids. This allows the identification of internal voids with diameters of 2 $\mu$m and above. For the larger size class, each of 10 crystals were inspected and the numbers of voids in different size classes were counted. The size classes used were < 5 $\mu$m, 5–15 $\mu$m, 15–50 $\mu$m and > 50 $\mu$m. For the smaller crystals, the small size meant that size classes were not appropriate and instead the total number of voids in each crystal large enough to be resolved was counted. Most voids were almost spherical in shape.

A summary of the results for the quantitative analysis of voids is shown in tables 1 and 2. For the smallest size class, the most sensitive crystals have only 1 or 2 voids per crystal, although a few crystals with many voids (up to 20) skew the average.

For both size classes, we observed that the most sensitive crystals contained the fewest internal voids. This suggests that either void content is irrelevant here, or that internal voids may act to suppress shock-to-detonation transition in this situation. In view of the large volume of literature correlating void content with sensitivity, it seems that for RDX in these gap tests, factors other than void content dominate.

**TABLE 1.** Optical microscopy results. Total number of voids compared with critical gap for the smallest size class

| Sample label | (5a) | (5b) | (5c) | (5d) |
|---|---|---|---|---|
| Crit. gap/mm | 7.5 | 7.8 | 8.1 | 10.3 |
| Av. no. of voids | 3.1 | 4.1 | 0.1 | 1.5 |
| Mode void no. | 3 | 2 | 0 | 0 |
| Mean crystal size/$\mu$m | 29 | 27 | 12 | 16 |

**TABLE 2.** Void data from optical microscopy the larger size class

| Sample label | (1a) | (1b) | (1c) |
|---|---|---|---|
| Crit. gap/mm | 7.9 | 8.5 | 9.3 |
| Av. no. of voids | 35 | 21 | 9 |
| Av. crystal size/$\mu$m | 237 | 177 | 165 |

### Environmental Scanning Electron microscopy

Environmental Scanning Electron Microscopy (ESEM) can be used to examine surfaces with sub-micron resolution but the electron beam damages RDX. At low resolution, the beam power per unit area is small and damage is insignificant on the timescale required for imaging. However, if an attempt is made to focus on scales of a few tens of microns, the surface is quickly damaged and becomes porous. Damage can be identified by zooming out and looking for the edges of the closely scrutinised area. ESEM is useful for inspecting surface features of 10 $\mu$m or larger but can only give approximate particle shapes for the smaller crystals. The crystals of 1a are generally rounded but ESEM shows that the many smaller crystals adhering to the surfaces are angular. 1a and 1b both have irregular steps on the surface with a scale of a few microns. The surfaces of 1c are much smoother, with a few pits of around 5 $\mu$m in size. No observed surface feature correlated with sensitivity.

### Mercury porosimetry

Mercury porosimetry is used to assess surface voids and roughness down to 0.01 $\mu$m in size. The surface tension of mercury means that high pressure is needed to force it into small voids or cracks. This allows measurement of how the cumulative pore area increases as pore diameter decreases. The cumulative area curves are shown in figure 4. Samples from batches in the same size class are similar but there are differences in the detail. The dotted line in each case shows the more sensitive material. For the larger size class the more sensitive material has fewer voids in the micron size range, but the converse is true for the smaller size class.

### DISCUSSION

If the dominant mechanism for hotspot formation was the collapse of internal voids, we would have expected to see that more sensitive crystals have more voids. However, for both size classes, the more voids there were in a crystal, the less sensitive it was. We conclude that the collapse of voids above 2 $\mu$m in size is not the dominant mechanism for hotspot formation in these samples. Smaller internal voids may exist but detecting them was beyond the scope of this work. Another possible mechanism for hotspot formation is friction, a surface effect. ESEM images of the larger size class showed no correlation between observable surface features and sensitivity. Mercury porosimetry suggested that 1c had more surface features than 1b on the micron size scale, and that 1b had more surface features than the sub-micron size scale.

Mercury porosimetry on the smaller size class failed to show any distinguishing features that might explain the significant difference in sensitivity between 5b and the other three samples. The mercury porosimetry provides data about inter-granular voids as well as surfaces features. For the smaller size class, the intergranular voids are expected to be on the scale of 10 $\mu$m and a difference is seen between 5a and 5b. Sample 5a has significantly lower porosity on this size scale than 5b and is also less sensitive. Without data from the other class 5 samples, it is difficult to draw any conclusion from this.

The correlations seen by many authors with different aspects of the crystal morphology may be an indication of an underlying cause for the shock sensitivity differences which is below the detection level of the techniques used. However, any such feature might require conditions to form which might also cause larger scale features, and this could explain

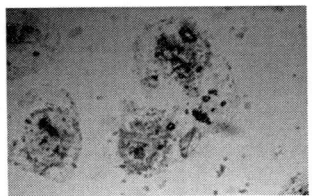 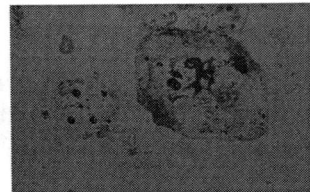 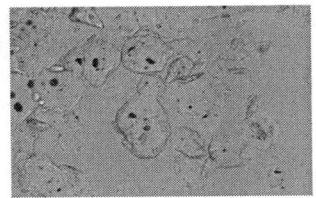

**FIGURE 3.** Typical crystals of each class 1 batch. They are shown in order from lowest to highest sensitivity: 1a, 1b, 1c.

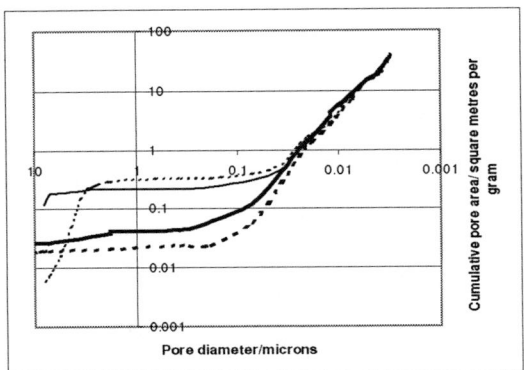

**FIGURE 4.** Mercury Porosimetry results. The thick lines are class 1 samples and the thin lines are class 5. The dashed line is the more sensitive sample in each case.

previously observed correlations.

## CONCLUSIONS

There were differences of up to 50% in shock sensivity between batches of RDX from different manufacturers which could not be explained by particle size and density variations. This implies that some feature of the crystals themselves is responsible for the difference.

There are several possible hotspot formation mechanisms, for example void collapse and friction. Crystals have many morphological features and each feature has the potential to affect a different mechanism. By isolating the dominant morphological features associated with shock sensitivity, it is possible to infer the dominant hotspot mechanism. The lack of correlations seen in this work between internal voids greater than 2 $\mu$m in size and sensitivity suggests that the hotspot mechanisms

associated with larger voids are not dominant here. No significant differences were detected in the surface features which might account for sensitivity variation because of external void collapse or friction. The detection of internal crystal features less than 2 $\mu$m in size and more detailed surface analysis may provide more information about the dominant hotspot mechanism in this system.

## ACKNOWLEDGMENTS

The authors would like to thank [dstl]and AWE Aldermaston for financial support for this project.

## REFERENCES

1.  Khasainov B. A., Ermolaev B. S., Presles H. N., and Vidal P., Shock Waves 7, 89-105 (1997).
2.  v. d. Heijden A. E. D. M., Bouma R. H. B., and v. d. Steen A. C., Propellants, explosives, pyrotechnics 29, 304-313 (2004).
3.  Borne L., Patedoye J.-C., and Spyckerelle C., Propellants, explosives, pyrotechnics 24, 255-259 (1999).
4.  Baillou F., Dartyge J. M. , Spyckerelle C., and Mala J. , 10th Symposium on Detonation, 816-823 (1993?).
5.  Armstrong R. W., Coffey C. S., DeVost V. F., and Elban W. L., J. Appl. Phys. 68 (1990).
6.  Halfpenny P. J., Roberts K. J., and Sherwood J. N., Philosophical Magazine A 53, 531-542 (1986).
7.  Chakravarty A., Gifford M.J., Greenaway M. W., Proud W. G., and Field J. E., SCCM, 1007-1010 (2001).
8.  Bowden F.P., Yoffe A. D. "The Initiation and Growth of explosion in liquids and solids", Cambridge University Press, 1952

CP845, *Shock Compression of Condensed Matter - 2005*,
edited by M. D. Furnish, M. Elert, T. P. Russell, and C. T. White
© 2006 American Institute of Physics 0-7354-0341-4/06/$23.00

# FRICTIONAL HEATING AND IGNITION OF ENERGETIC MATERIALS

## P.M. Dickson, G.R. Parker, L.B. Smilowitz, J.M. Zucker & B.W Asay

*Los Alamos National Laboratory, Los Alamos, NM 87545.*

**Abstract.** For many years, powder friction tests have been an integral part of explosives sensitivity and safety testing. More recently, oblique impact tests have been used in the hazard assessment of monolithic charges. However, these tests are simply threshold tests for reaction, and relatively little work has been done to try to examine the processes that lead to frictional heating and ignition of energetic materials. We report the results from a series of experiments in which energetic materials are subjected to frictional heating under closely-controlled conditions (normal load, sliding speed, grit quantity and composition, substrate). The response of the energetic material and grit, if present, is observed by optical and infrared high-speed photography to determine the nature of the interactions between the test material, grit and substrate, and the mechanisms by which the energetic material may be heated to ignition.

**Keywords:** Friction, frictional heating, hot spots, HMX, PBX 9501.

## INTRODUCTION

The effect of grit properties on the frictional ignition of powder explosives was first investigated by Bowden and Gurton [1]. In a series of elegant experiments they demonstrated that during the frictional interaction of two surfaces in the presence of grit, enhanced heating occurs at grit particles, and that the maximum temperatures achieved at such hot spots are dependent on the melting point of the grit. Grit only sensitizes explosive if the melting point of the grit is higher than the ignition temperature of the explosive. The results are relatively insensitive to grit hardness. Dyer and Taylor [2] extended this work to examine the frictional interaction between pressed or cast explosives and various surfaces. When rubbed against a metal file or another piece of explosive, no ignitions were observed. With sand bonded to steel, ignition was enhanced by increasing the normal load or sliding speed, but the biggest factor was the presence of loose grit. They concluded that rubbing an explosive against a rough surface in the absence of grit is generally not effective in causing ignition, which most readily occurred at hot spots produced by grit-on-grit or grit-on-substrate interactions.

Frictional ignition of explosives is of considerable concern from a safety point of view, and has been implicated in a number of accidents involving the handling of explosives, but little work has been done to investigate the details of the process. Of particular interest is the precise nature of the interactions between the grit, explosive and surface during such an impact. The experiments referred to above inferred mechanisms, but did not permit direct observation of these interactions. The experiments described here are an attempt to make such observations with the plastic-bonded explosive PBX 9501 (95% HMX, 2.5% estane, 2.5% BDNPA/BDNPF) using transparent substrates and high-speed visible and infrared photography. The objectives are to observe the response of the grit during the impact (fracture, translation and rotation), and to measure hot spot size, temperature and interaction (if any). Experimental variables are grit composition, size, morphology, and distribution, together with sliding speed, normal load and duration of contact. A pulsed copper vapour laser is used to provide illumination for visible imaging, and a Nd:YAG laser is used to probe the HMX $\beta - \delta$ phase change by measuring second harmonic generation [3].

## EXPERIMENTAL PROCEDURE

Bowden and Gurton [1], and Dyer and Taylor [2], used a technique whereby the explosive was pre-loaded against the substrate, which was then rapidly moved. In these tests we use an alternative approach in which the substrate, a rotating disk of toughened glass or vitreous alumina (sapphire), is moving relative to the sample before contact is made. The intention is to make this test more representative of a glancing impact. Alumina is the substrate of choice for the purpose of infrared imaging as it is quite transparent at the wavelengths of interest, but it has the drawback of a very high thermal conductivity compared to glass, and tends to quench frictional hot spots. Figure 1 shows a schematic of the apparatus. The rotating disk is 152 mm in diameter and 19 mm thick. A 10 mm diameter disk of PBX is placed in the sample holder, which is then brought into contact with the spinning disk by the pneumatic actuator. Sliding speed is determined by the rotational speed of the disk (present range 2.5 – 15 m s$^{-1}$), and normal force is controlled by the pneumatic pressure (0 – 450 N). The duration of the contact is determined by the control signal to the actuator, with a practical minimum of around 0.1 s. Time-resolved records of both normal and frictional force are obtained from force transducers. Rotational speed is measured by laser reflection from a stripe-coded strip around the edge of the rotating disk. Single or multiple grit particles are placed in advance on the explosive sample, and may be embedded or loose.

High-speed visible photography is achieved using a Vision Research Phantom high-speed digital video camera synchronized to the copper vapour laser. Infrared photography is performed with a Santa Barbara Focal Plane IR camera, which is calibrated to enable temperatures to be estimated to +/- 20 ˚C. The cameras view the event from above through the rotating substrate.

Several different grit materials have been tried, including spherical silica (400 μm and 800 μm particles) and zirconium (200 μm), but the results given are for ordinary sand, which is commonly defined as 60 – 2000 μm $\beta$ quartz (SiO$_2$).

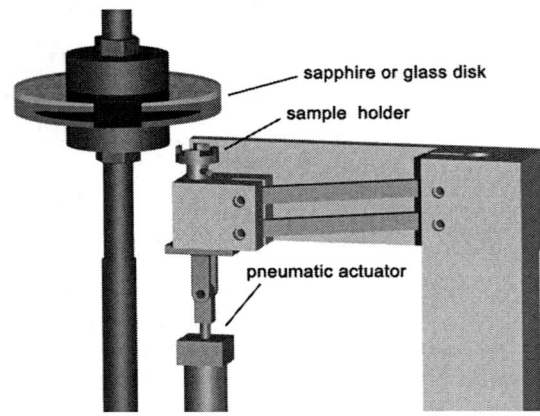

**FIGURE 1.** Schematic view of friction apparatus.

## RESULTS AND DISCUSSION

Figure 2 shows a typical record of normal force, lateral force and calculated coefficient of friction during a 0.5 s contact with a sample of PBX 9501 and several grains of sand at a sliding speed of approximately 14 m s$^{-1}$. The oscillation in the normal force, also reflected in the lateral force, is due to a very slight wobble in the rotation of the disk. The frequency of oscillation is equal to the angular frequency of rotation.

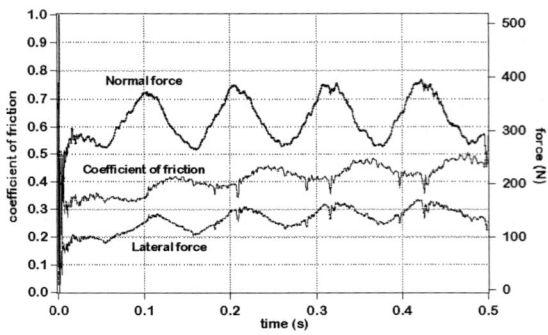

**FIGURE 2.** Normal force, lateral force & coefficient of friction.

1058

During the impact, the grit is pushed into the explosive sample and so the normal and lateral forces represent a combination of those arising due to the explosive and the grit. The calculated coefficient of friction is a little lower than for a simple sand-on-glass or sand-on-alumina interaction.

Figure 3 shows a sequence of visible and infrared images from a 0.5 s contact between PBX 9501 and sand particles (0.5 – 1 mm). The direction of motion of the rotating disk as viewed relative to the sample is upwards. The first pair of images is at impact, and it is apparent that substantial heating occurs very soon after contact is made. The sand particles do not appear to translate much relative to the sample (less than one particle diameter), but they are pushed into the sample and in some cases rotate slightly in the process. Subsequently, some noticeable pulverization of the particles does occur. Softening and melting of the binder is evident around some of the particles, and migration of liquid binder probably leads to some lubrication at the interface. Figure 4 shows the temperature distribution across one of the grit particles, indicating that while the explosive surface is being heated directly, these particles do constitute significant hot spots.

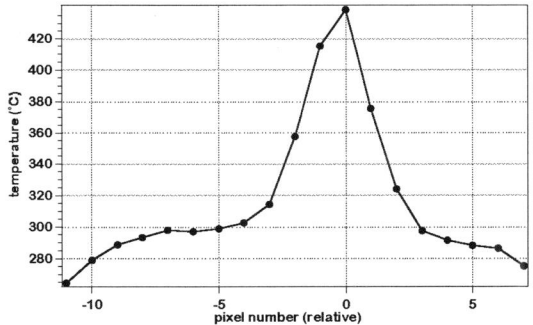

**FIGURE 4.** Estimated temperatures across a grit particle.

This sample did not undergo a propagating reaction, but subsequent inspection by plane-polarized light microscopy (figure 5) showed that decomposition had occurred in the vicinity of the embedded grit particles, with widespread melting at the surface.

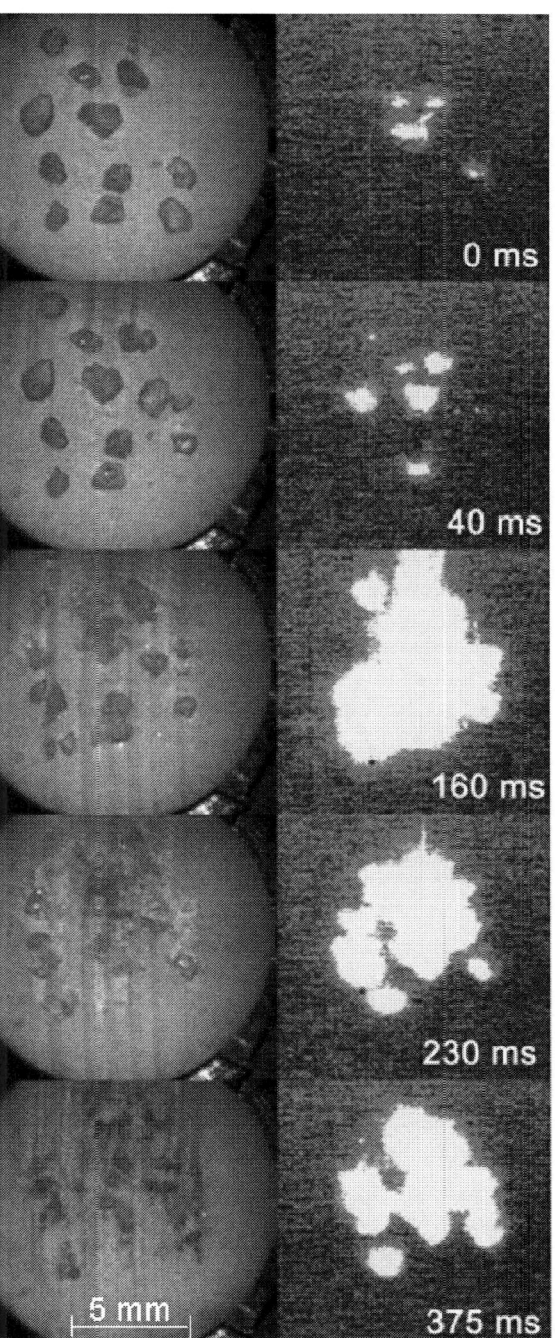

**FIGURE 3.** Visible and infrared images during an impact.

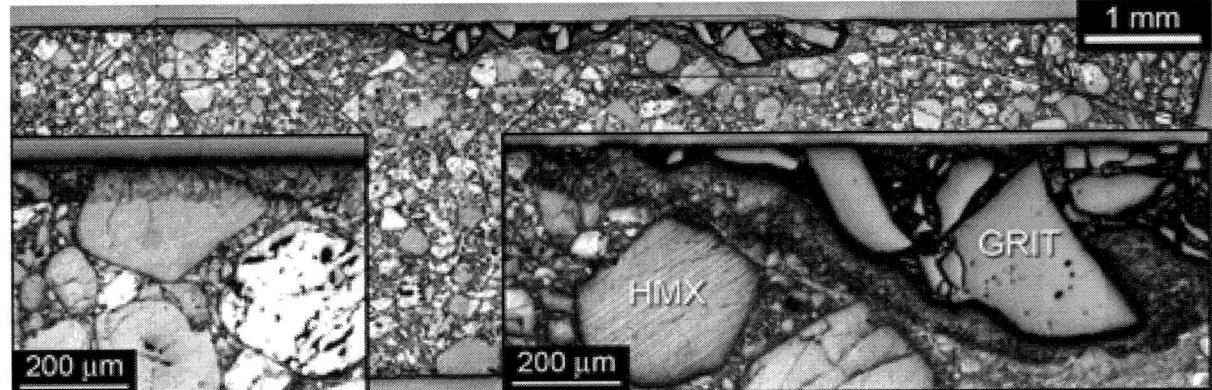

**FIGURE 5.** Visible light image of interface .

Examination of the surface by probing with a 1064 nm laser indicated that HMX crystals across much of the surface had undergone the β – δ phase change, which occurs at around 170˚C. Figure 6 is a compilation of data by Henson [4], showing the time-to-ignition for HMX as a function of temperature. At temperatures in excess of 400˚C HMX ignites in a few tenths of a second, which is consistent with the observed behaviour.

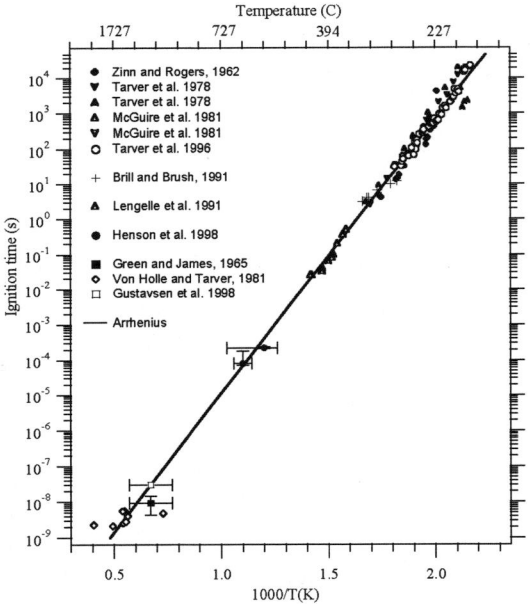

**FIGURE 6.** Ignition times for HMX / PBX 9501 as a function of temperature.

## CONCLUSIONS

Direct observations of the frictional interaction of explosive, grit and substrate have been performed, showing that under the conditions of these experiments, the grit tends to embed in the explosive and is preferentially frictionally heated. Temperatures sufficient to achieve ignition in times of the order of $10^{-1}$ s are rapidly reached, and the onset of decomposition has been observed.

## ACKNOWLEDGEMENTS

Funding for this work was provided by the DOE Surety Program. The authors thank Dr P.M. Howe for valuable discussions and suggestions on this subject.

## REFERENCES

1. Bowden, F. P. & Gurton, O. A., 1948, Nature, Lond., **162**, 654.
2. Dyer, A. S, & Taylor, J. W., 1970, Initiation of detonation by friction on a high-explosive charge, In Proc. 5th Int. Symp. on Detonation, Pasadena, CA, pp. 291–300. Arlington, VA: Office of Naval Research.
3. Henson, B. F., Smilowitz, L. B., Asay, B. W, & Dickson, P. M., The β – δ phase transition in the energetic nitramine octahydro-1,3,5,7-tetranitro-1,3,5,7-tetrazocine, 2002, J. Chem. Phys. **117**, 3789.
4. Henson, B. F., Asay, B. W, Smilowitz, L. B., & Dickson, P. M., Ignition chemistry in HMX from thermal explosion to detonation, In Shock Compression of Condensed Matter—2001 (ed. M. D. Furnish, N. N. Thadhani & Y. Horie), AIP Conference Proceedings, vol. 620, pp. 1069–1072.

CP845, *Shock Compression of Condensed Matter - 2005*,
edited by M. D. Furnish, M. Elert, T. P. Russell, and C. T. White
© 2006 American Institute of Physics 0-7354-0341-4/06/$23.00

# THERMAL COOK-OFF EXPERIMENTS OF THE HMX BASED HIGH EXPLOSIVE LX-04 TO CHARACTERIZE VIOLENCE WITH VARYING CONFINEMENT

## Frank Garcia[1], Kevin S. Vandersall[1], Jerry W. Forbes[2], Craig M. Tarver[1], and Daniel Greenwood[1]

[1]*Energetic Materials Center, Lawrence Livermore National Laboratory, Livermore, CA 94550*
[2]*Center for Energetic Concepts Development, University of Maryland, College Park, MD 20742*

**Abstract.** Thermal cook-off experiments were carried out using LX-04 explosive (85% HMX and 15% Viton by weight) with different levels of confinement to characterize the effect of confinement on the reaction violence. These experiments involved heating a porous LX-04 sample in a stainless steel container with varying container end plate thickness and assembly bolt diameter to control overall confinement. As expected, detonation did not occur and reducing the overall confinement lowered the reaction violence. This is consistent with modeling results that predict that a lower confinement will act to lower the cook-off pressure and thus the overall burn rate which lowers the overall violence. These results suggest that controlling the overall system confinement can modify the relative safety in a given scenario.

**Keywords:** Thermal explosion, cook-off, LX-04, HMX explosives, violence, confinement
**PACS:** 82.33.Vx, 47.70.Fw

## INTRODUCTION

Energetic materials can react violently when heated, resulting in a thermal explosion, or commonly called "cook-off." This situation generally poses two hazards, the initial thermal explosion event and any resulting fragments that may initiate a reaction in any adjacent energetic materials in close proximity. A better understanding is needed for safe handling, transportation, and storage of explosive devices. Experiments that measure the violence with energetic materials of known size, confinement, and heating rate are used to measure the relative safety of different scenarios and also calibrate computer models. These computer models can then be utilized to run a number of scenarios that are not as easily tested. In this work, the

confinement of a previous experiment (12 mm cover thickness) is reduced by two levels (6 mm and 3 mm cover thickness) and tested to observe the effects of confinement changes. These are referred to as the Thermal EXplosion Test (TEXT) experiments and the two additional experiments are compared with each other and the initial experiment.

Often times, thermal explosion events are generalized into "fast cook-off" or those occurring in minutes to a few hours and "slow cook-off" or those occurring over many hours or days. The experiments discussed here use a relatively quick heating rate of ~5.7°C/min until 170°C, then 1°C/min until explosion which means the experiment will take place over a few hours (i.e. "fast cook-off"). Comparing these tests to experiments with slow heating rates of ~1°C/hour [1], which translates to experiments occurring over a few days can reveal different results. The combined results of these two extremes help understand the effect on heating rate. Modeling of these "fast" and "slow" cook-off examples will aid in understanding how heating rates affect the overall violence.

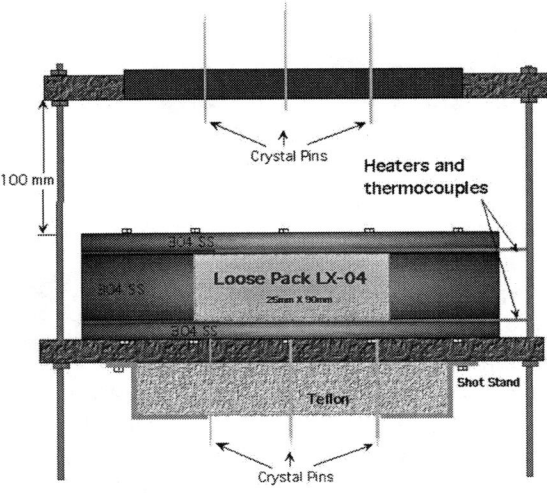

**FIGURE 1.** Diagram of the assembly for TEXT XI and TEXT XII experiments.

## EXPERIMENTAL PROCEDURE

Two thermal explosion experiments were performed using LX-04 donor charges confined in 304 stainless steel cased donor assemblies with decreasing confinement from a prior test (TEXT X). These experiments, TEXT XI and XII used a porous LX-04 donor (55.6% TMD) and a pin assembly at a 10 cm standoff. The experiment assembly for TEXT XI and XII is shown in Figure 1.

Heating of the donor occurred at a rate of 5.7°C a minute until the thermocouples located next to the nichrome heater foils recorded 170°C and soaked for about 30 minutes. Then the heating rate in the heater package was set at 1°C/min until the explosive thermally reacted. Thermocouples were also placed on the case

exterior to measure the thermal lag of temperatures from the heaters to the outer case. The assembled experiments were placed inside a large steel expendable cylinder before firing to protect the firing chamber walls.

Several experiments have been previously performed with LX-04 explosives in this same series. Experiments TEXT VIII and TEXT IX [2] in addition to TEXT X [3] were detailed in prior publications and will not be fully described here, but will be referred to in the following sections for comparison and discussion. Extensive details about a prior series on PBX9501 are also indicated more specifically elsewhere [4].

## RESULTS/DISCUSSION

The donor assembly in TEXT XI and XII cooked off as expected at a 232°C heater temperature (216°C exterior temperature) and 244°C heater temperature (227°C exterior temperature), respectively. The higher temperature recorded in TEXT XII is attributed to release of pressure in the donor as the thin cover plates buckled due to pressure buildup allowing gases to escape. The thermocouple outputs recorded for TEXT XII are included in Figure 2 for reference.

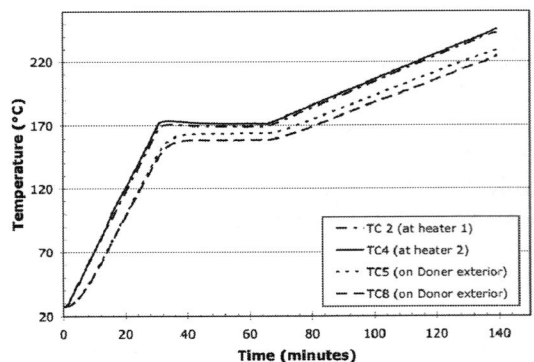

**FIGURE 2.** Thermocouple outputs recorded for experiment TEXT XII. Not shown are thermocouples 2 and 3 (similar to thermocouples 2 and 4, respectively) and thermocouples 6 and 7 (both similar to thermocouple 5).

In TEXT XI, the cook-off was not violent enough to trigger the scopes from the pin signals so a velocity was not measured. In TEXT XII, the thin plates deformed instead of breaking the bolts and throwing a plate. The violence was found to decrease with decreasing confinement. The assembly hardware used and cook-off temperatures as discussed above are shown in Table 1 for reference. Prior TEXT experiments had 12 mm thick top and bottom plates with a 3 mm thick copper sealing plate attached by 9.5 mm diameter Grade 8 assembly bolts. The follow up tests used 6 and 3 mm thick plates with 1.5 mm thick copper sealing plates for TEXT XI and TEXT XII respectively, both with 6.4 mm diameter bolts.

**TABLE 1.** Assembly bolt diameter, cover plate thickness, and temperatures near the heaters and on donor exterior for LX-04 TEXT experiments.

| EXPT | BOLT DIA. | COVER THICKNESS | TEMP. AT COOKOFF |
|------|-----------|-----------------|------------------|
| TEXT X | 9.5 mm | 12 mm | - / 232 ° C |
| TEXT XI | 6.4 mm | 6 mm | 216 / 232 ° C |
| TEXT XII | 6.4 mm | 3 mm | 227 / 244 ° C |

Photographs showing the before and after images of experiments TEXT XI and XII are shown in Figures 3 and 4. The charring seen in Figure 3 (b) is the result of the acrylic base holding the experiment catching on fire shortly after the experiment which makes it look more violent than actuality. The fact that the plate impact did not trigger the adjacent pins below or the acceptor pin array 100 mm above, seems to reveal a rather low plate velocity and thus lower violence. This is backed up by the standard video (images not shown due to graininess) where the cook-off event could be discerned even at the low frame rate of approximately 1 frame per second. In TEXT XII, the reason for the pins not triggering is obvious due to the plates becoming bulged and venting high-pressure gases and not resulting in a flying plate.

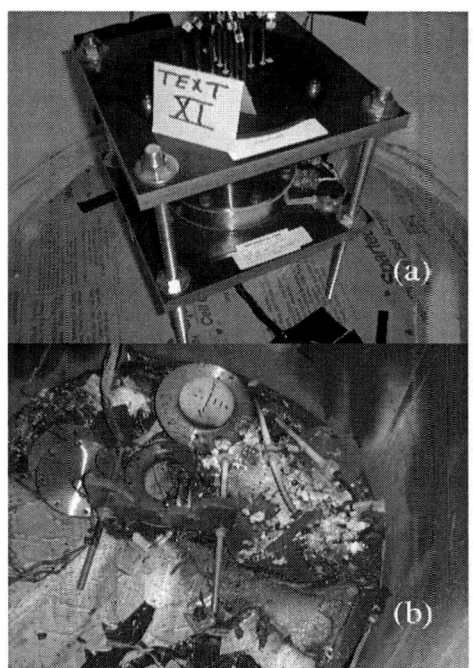

**FIGURE 3.** (a) photograph of TEXT XI experiment before and (b) after thermal explosion.

**FIGURE 4.** (a) photograph of TEXT XII experiment before and (b) after thermal explosion. Note the bulged cover plate and in tact bolts holding assembly together.

These results are consistent with the previous LX-04 cook-off experiments. TEXT VIII had a full density donor during cook-off and did not have significant violence to even blow off the donor charge cover plates. This was the reason for going to the porous donor in TEXT IX, which sent a ramp wave into the Teflon acceptor in contact with donor top plate with peak pressures of 0.8 GPa and a decaying ramp wave. The porosity plays an essential part in allowing the LX-04 to accumulate appreciable violence during thermal explosion.

In TEXT X, a porous LX-04 donor assembly was cooked off and the assembly plate was accelerated into a nearby heated (100°C) acceptor at a 10 cm standoff with a final velocity of 350 m/s. The carbon resistor pressure gauge results (without temperature corrections) showed ramp waves with peak pressures of 0.7 GPa and rise times of ~2$\mu$s. The ramp pressure wave decays very rapidly (i.e. does not build to detonation) as it moves through the acceptor charge and the rise times become more dispersed.

Modeling was performed in the prior TEXT experiments on the Donor cook-off and the modeling parameters were adjusted to obtain the correct measured Donor plate velocity. Because a measured plate velocity was not obtained in these experiments, this modeling was not performed. However, the modeling appears to agree with the result of a lower violence because it predicted that a slower burn rate would result in a lower pressure inside the Donor, and thus a lower overall violence (i.e. lower plate velocity).

## SUMMARY

Two experiments, TEXT XI and XII, involving cook-off of a confined porous LX-04 explosive were performed with decreasing confinement from a prior test (TEXT X). In TEXT XI, the cook-off was not violent enough to trigger the scopes from the pin signals so a velocity was not measured. In TEXT XII, the thin plates deformed instead of breaking the bolts and throwing a plate. These results showed that the violence was found to decrease with decreasing confinement. The prior modeling performed agrees with this result since it predicted that with lower confinement, the burn rate observed and thus pressure pushing the donor confinement plates is less resulting in a reduced violence.

## ACKNOWLEDGEMENTS

Jerry Dow at LLNL is generously thanked for funding this work. Assistance by Rich Villafana, Steve Kenitzer, and Gary Steinhour is gratefully acknowledged. This work was performed under the auspices of the U. S. Department of Energy by the University of California, Lawrence Livermore National Laboratory under Contract No. W-7405-Eng-48.

## REFERENCES

1. Wardell, J. F. and Maienschein, J. L., "The Scaled Thermal Explosion Experiment," Proceedings of the 12th International Detonation Symposium, San Diego, CA, August, 2002, pp. 384-393.
2. Forbes, J. W., Garcia, F., Tarver, C. M., Urtiew, P. A., Greenwood, D. W., and Vandersall, K. S., "Pressure Wave Measurements During Thermal Explosion of HMX-based High Explosives," Proceedings of the 12th International Detonation Symposium, San Diego, CA, August, 2002, pp.837-845.
3. Garcia, F., Vandersall, K., Forbes, J., Tarver, C., Greenwood, D., "Pressure Wave Measurements Resulting From Thermal Cook-off of the HMX Based High Explosive LX-04," Shock Compression of Condensed Matter-2003, edited by M.D. Furnish, Y.M. Gupta, and J.W. Forbes, pp. 947-950, 2004.
4. Garcia, F., Forbes, J. W., Tarver, C. M., Urtiew, P. A., Greenwood, D. W., and Vandersall, K. S., "Pressure Wave Measurement Resulting from Thermal Cook-off of the HMX Based Explosive LX-04," Shock Compression of Condensed Matter-2001, Furnish, M. D., Thadhani, N. N., and Horie, Y, eds. CP-620, AIP Press, New York, 2002, p. 882.

CP845, *Shock Compression of Condensed Matter - 2005*,
edited by M. D. Furnish, M. Elert, T. P. Russell, and C. T. White
© 2006 American Institute of Physics 0-7354-0341-4/06/$23.00

# ONE-DIMENSIONAL SHOCK AND DETONATION CHARACTERIZATION OF ULTRAFINE HEXANITROSTILBENE

## S. G. Goveas[1], J. C. F. Millett[2], N. K. Bourne[3], I. Knapp[2]

[1] *Atomic Weapons Establishment, Aldermaston, Reading, Berkshire, RG7 4PR, United Kingdom.*
[2] *Defence Academy of the United Kingdom, Cranfield University, Shrivenham,*
*Swindon, SN6 8LA, United Kingdom.*
[3] *University of Manchester, P.O. Box 88, Sackville Street, Manchester, M60 1QD, United Kingdom.*

**Abstract**. A series of plate impact experiments was performed, using a single-stage gas gun, on die-pressed, high density (92 % theoretical maximum) samples of ultrafine hexanitrostilbene (HNS). This enabled investigation of the inert shock response and subsequent detonation of the material. Shock magnitudes up to *ca.* 6 GPa were investigated by varying the flyer and target plate materials, and impact velocities. In each case, the shock length was chosen to be longer than the pellet (*ca.* 3 mm). Shock wave profiles and transit times were diagnosed using embedded miniature (1 mm$^2$) manganin stress gauges placed at the front and rear of the shock assemblies. The results have been interrogated to establish the non-reactive Hugoniot of the HNS and deduce information on its run-to-detonation. Analysis of measured stresses and calculated pressures suggests that pressed HNS possesses little strength behind the shock front. These and other features are discussed and compared with existing data.

**Keywords:** HNS, hexanitrostilbene, Hugoniot, shock, detonation, reaction.
**PACS:** 47.40.Nm, 62.50.+p, 82.33.Vx

## INTRODUCTION

Hexanitrostilbene[i] (HNS) is a thermally stable secondary explosive. As such, it has found many uses in both the military and industry, since its discovery in 1964 by Shipp [1]: for example, detonators, detonation cord, aircraft escape systems, demolition line cutters and charges for deep oil wells (where temperatures can exceed 200°C).

There exists a significant body of work investigating and quantifying the chemical, physical, shock and detonation properties of this material. However, little information has been published on the shock response and run-to-detonation characteristics of ultrafine HNS (UF-HNS) compacts. This paper presents the results and

analysis of one-dimensional plate impact experiments to establish the non-reactive Hugoniot of high density, die-pressed samples of UF-HNS, and first steps towards characterizing run-to-detonation.

Comparisons are made with similar research, into the shock response of different HNS powders, by Sheffield *et al* (plate impact with quartz gauges) [2], Davies *et al* (plate impact with piezoresistive gauges) [3] and Roth (explosive plane-wave lenses and streak photography) [4]. The evidence suggests that the bulk Hugoniot for HNS compacts, at high density, is not highly dependent on particle size.

In addition, the close correlation observed between measured stresses and calculated hydrodynamic pressures illustrates that pressed UF-HNS samples possess little strength behind the shock front.

---

[i] 1,1-(1,2-ethenediyl)bis-(2,4,6-trinitrobenzene)

## EXPERIMENTAL

All plate impact experiments were performed using the 6 m long, 50 mm bore single stage gas gun at the Defence Academy of the United Kingdom. Impactor plates were manufactured from 5 mm lapped copper and tungsten discs mounted onto acetal (polyoxymethylene) sabots. Impact velocity was measured to an accuracy of 0.5 % using a sequential pin-shorting method and tilt was fixed to be less than 1 mrad by means of a precision machined end piece to the gun barrel.

Pellets of UF-HNS (Specific Surface Area, SSA of 8.3 $m^2$ $g^{-1}$) were die-pressed to a diameter of 12.93 mm and density of 1.6 g $cm^{-3}$ (92 % theoretical maximum). A standard pellet length of *ca.* 3 mm was chosen to ensure the shock length was longer than the pellet, in all cases.

Miniature manganin stress gauges (MicroMeasurements type C-951213-C), insulated with 50 μm mylar, were fixed at the front and rear of each UF-HNS pellet using epoxy adhesive: this gauge type has a small active area of 1 $mm^2$ to reduce susceptibility to lateral releases from the edge of the pellet. The first gauge was supported upon the front of the pellet with a cover plate of 1 mm copper. In order to protect the second gauge, it was mounted between 1.8 mm of polymethyl-methacrylate (PMMA), in contact with the rear surface of the pellet, and a further 12 mm of PMMA. The specimen configuration and gauge placements are illustrated in Fig. 1.

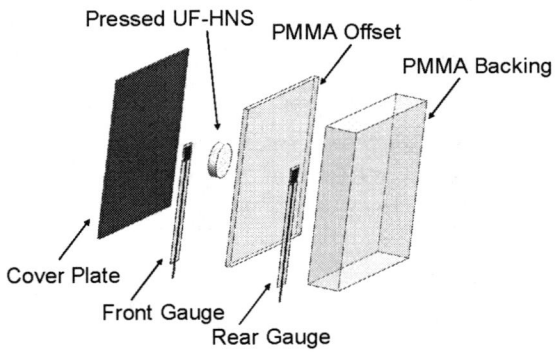

**Figure 1**. Specimen and gauge configuration.

Gauge records were converted to stress-time traces using the methods of Rosenberg et al [5].

## RESULTS AND DISCUSSION

Impact velocities were in the range of 840 - 1075 m $s^{-1}$ corresponding to induced (front face) stresses of 3.9 - 8.9 GPa. Typical examples of stress-time traces are presented in Fig. 2.

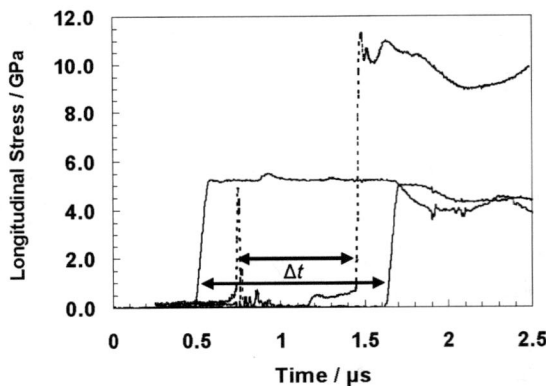

**Figure 2**. Stress-time traces for 2 experiments: front gauge on left, rear gauge on right. The solid lines relate to a copper impact at 1002 m $s^{-1}$ (non-reactive response) and the dashed lines to a tungsten impact at 1075 m $s^{-1}$ (reactive response).

### Non-Reactive Response

For impacts generating stresses up to 5.7 GPa, the gauge traces showed a rapid rise in signal with a relatively flat top. The front gauge gives the input stress profile. The rear trace is a good indication of the shape of the stress pulse through the sample, as there is a close impedance match between the PMMA, epoxy adhesive and gauge backing, and so the rise time is limited only by the thickness of the manganin gauge element. There was no evidence of a break in slope that would indicate the presence of a Hugoniot elastic limit (Fig. 2).

The temporal spacing ($\Delta t$) between the traces for the front and rear of each specimen, in combination with the known separation of the gauges, has been used to obtain shock velocity ($U_s$). The particle velocity ($u_p$) was determined using impedance matching techniques. These characteristics have been plotted in Fig. 3.

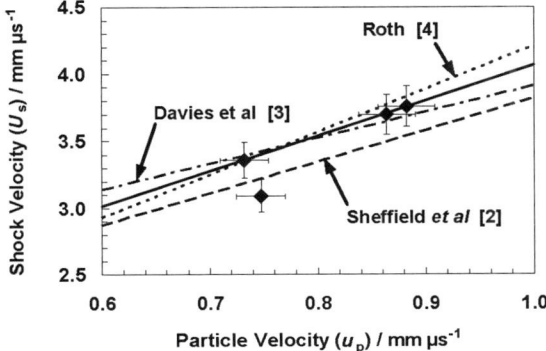

**Figure 3**. Shock Hugoniot of pressed UF-HNS in $U_s$ - $u_p$ space (solid line), compared to those of other pressed HNS samples (broken lines) [2, 3, 4].

A straight line has been fitted to the data set assuming the standard linear Hugoniot relationship between shock and particle velocities ($U_s=c_0+Su_p$): one point was disregarded due to severe ringing on the front gauge record. The fitted Hugoniot compares favourably with those determined for other HNS powders by a variety of means [2, 3, 4]: illustrated in Fig. 3 and listed in Table 1.

It is interesting to observe that this agreement is despite clear differences in the ranges of shock stress tested, the techniques used and the specific surface areas (SSA), suggesting that the Hugoniot is not highly dependent on the HNS particle characteristics (e.g. size distribution).

**Table 1**. Comparison of HNS Hugoniots

| Type & SSA / $m^2$ $g^{-1}$ | Density / g $cm^{-3}$ | Hugoniot / $U_s$ | Stress / GPa |
|---|---|---|---|
| UF-HNS 8.3 | 1.60 | $1.43 + 2.63\,u_p$ | 3.9 - 5.7 |
| HNS I [2] 1.5 - 2.0 | 1.58 | $1.45 + 2.37\,u_p$ * | 0.5 - 1.7 |
| HNS II [3] 0.3 - 0.7 | 1.58 | $1.98 + 1.93\,u_p$ | 0.4 - 3.5 |
| HNS [4] 0.27 | 1.57 | $1.00 + 3.21\,u_p$ | 1.2 - 5.2 |

* Calculated from measurements in Reference 2.

The equation of the fitted line has been used, in conjunction with the sample density ($\rho_0$), to determine hydrodynamic pressure ($P_{HD}$) through the relationship:

$$P_{HD} = \rho_0 U_s u_p \qquad (1)$$

It should be noted that this does not take into account the elastic response or strength of the material and, hence, is a measure of the hydrodynamic response. However, comparison with the measured stresses from the gauges is a useful exercise, giving insight into the materials response to shock loading. The results are presented in Fig. 4.

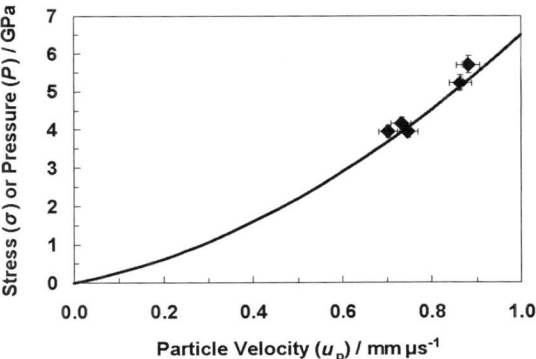

**Figure 4**. Shock Hugoniot of pressed UF-HNS in $\sigma_x$ - $u_p$ space. The curve is according to Equation 1, using the linear fit in Fig. 3 and Table 1.

Clearly, the agreement between the calculated hydrodynamic pressure (Equation 1) and the directly measured (non-reactive) stress is good. This further supports the linear relationship between shock velocity and particle velocity.

The longitudinal stress ($\sigma_x$) generated during shock loading can be expressed as a function of hydrostatic pressure ($P_{HS}$) and sample shear strength ($\tau$) as follows:

$$\sigma_x = P_{HS} + \frac{4}{3}\tau \qquad (2)$$

It is acknowledged that hydrodynamic pressure will vary slightly from hydrostatic pressure, which is generally calculated from extrapolation of ambient pressure bulk moduli data, and so strength cannot be determined from a combination of Equations 1 and 2. Nevertheless, it can be seen that failure to take into account the shear strength response of a material to shock loading may lead to a degree of disagreement between the calculated hydrodynamic pressure and the measured stress. Fig. 4 shows excellent agreement between these characteristics, indicating that the shear strength of the pressed UF-HNS in this stress regime is low. Furthermore, divergence does not occur, suggesting that the shear strength is near constant with increasing shock stress. This is consistent with expectations for pressed samples of crystalline explosive powder without a binder.

## Reactive Response

One experiment appears to have resulted in exothermic chemical reaction: the rear face longitudinal stress was measured in excess of 11 GPa (Fig. 2), yet the input stress was expected to be 8.9 GPa.

Unfortunately, the front gauge shorted and so there is no accurate measure of input stress. However, an interesting feature can be seen on the associated rear gauge trace, which provides insight of the reactive shock. The small increase in the gauge signal at 1.2 $\mu$s (around 0.3 $\mu$s before the shock reaches the gauge) is thought to mark the arrival of the shock at the UF-HNS/PMMA interface: the signal being generated by the slight piezoelectric property of the PMMA. This indicates an average shock velocity of 6.4 mm $\mu$s$^{-1}$ in the UF-HNS, which is close to the detonation velocity for HNS (at the same density) of 6.8 mm $\mu$s$^{-1}$ published by Lee et al [6] and Kilmer [7]. Therefore, it is deduced that prompt detonation was achieved. Furthermore, if this value for detonation velocity is assumed appropriate, the run-to-detonation distance can be calculated to be of the order of 0.3 mm.

## CONCLUSIONS

The non-reactive shock Hugoniot has been determined for die-pressed, high density (92 % theoretical maximum) samples of ultrafine hexanitrostilbene (UF-HNS). This has been compared to measured Hugoniots, from published sources, for a range of other HNS powders. Agreement was good, indicating that the shock response of such samples are not dominated by particle characteristics (e.g. size distribution).

Comparison of measured shock stresses with those calculated, from shock and particle velocities, has demonstrated that the pressed UF-HNS has little and near constant shear strength for shocks between 3.9 and 5.7 GPa. There was no evidence of chemical reaction in this regime, but explosive decomposition was observed following a shock of greater amplitude, 8.9 GPa. Interpretation of the stress gauge records and comparison with published HNS detonation properties suggest that this was prompt detonation.

## ACKNOWLEDGMENTS

The authors would like to thank Jim Clements of Cranfield University for his valuable technical assistance in preparing the UF-HNS samples.

## REFERENCES

1. Shipp, K. G., *Reactions of α-Substituted Polynitrotoluenes. I. Synthesis of 2,2',4,4',6,6'-Hexanitrostilbene*, J. Org. Chem, **29**(9), 2620, 1964.
2. Sheffield, S. A., Mitchell, D. E. & Hayes, D. B., *The Equation of State and Chemical Kinetics for HNS Explosive*, Proc. Sixth Symposium (Int.) on Detonation, pp. 748-754, Arlington, Virginia: Office of Naval Research, 1976.
3. Davies, F. W., Shrader, J. E., Simmerschied, A. B. & Riley, J. F., *The Equation of State and Shock Initiation of HNS II*, Proc. Sixth Symposium (Int.) on Detonation, pp. 740-747, Arlington, Virginia: Office of Naval Research, 1976.
4. Roth, J., *Shock Sensitivity and Shock Hugoniots of High-Density Granular Explosives*, Proc. Fifth Symposium (Int.) on Detonation, pp. 219-230, Arlington, Virginia: Office of Naval Research, 1970.
5. Rosenburg, Z., Yaziv, D. & Partom, Y., *Calibration of Foil-Like Manganin Gauges in Planar Shock Wave Experiments*, J. Appl. Phys, 51(7), 3702, 1980.
6. Lee, E. L., Walton, J. R. & Kramer, P. E., *Equation of State for the Detonation Products of Hexanitrostilbene at Various Charge Densities*, Lawrence Livermore National Laboratories, Livermore, California, UCID-17134, 1976.
7. Kilmer, E. E., *Heat-Resistant Explosives for Space Applications*, J. Spacecr. Rockets, 5(10), 1216, 1968, cited in Lawrence Livermore Handbook of Explosives (ed. Dobratz, J.), 1982.

CP845, *Shock Compression of Condensed Matter - 2005*,
edited by M. D. Furnish, M. Elert, T. P. Russell, and C. T. White
2006 American Institute of Physics 0-7354-0341-4/06/$23.00

# COMPARISON OF REACTION KINETICS OF I-RDX® AND RDX AT HIGH PRESSURE

## Jared C. Gump[*] and Suhithi M. Peiris

*Naval Surface Warfare Center - Indian Head Division, Indian Head, MD 20640*
*\* Author to whom correspondence should be addressed*

**Abstract.** Reactions and kinetics of the new less-sensitive form of RDX developed by Eurenco (known as I-RDX®) need to be compared to standard RDX, especially at shock pressures and temperatures. To evaluate the effect of pressure samples of I-RDX® and standard RDX were compressed to various static pressures in anvil cells. To evaluate the effect of temperature samples were initiated with different fluences of a 5ns pulsed Nd:YAG laser. As a measure of global reaction rate, changes in transmittance through the samples were monitored during reaction. Both RDX and I-RDX® were initially transparent under pressure, becoming opaque soon after initiation and then clear as the final gaseous products are formed. A comparison of the global reaction times obtained under various pressure and fluence conditions for I-RDX® and RDX samples are presented.

**Keywords:** RDX, kinetics, spectroscopy, gem anvil cell
**PACS:** 82.30.Lp, 07.35.+k

## INTRODUCTION

Due to the close proximity of personnel and ammunition on military ships, the Navy is very interested in decreasing the sensitivity of its energetic ingredients. RDX is a common ingredient in many munitions. Eurenco has developed a less sensitive form of RDX known as I-RDX® that has shown reduced shock sensitivity in Naval Ordnance Laboratory (NOL) Large Scale Gap Tests. [1-3] Previously Kuntz et al. investigated the laser initiation of standard RDX. [4] They reported the minimum laser fluence required to elicit a reaction from crystalline as well as powdered RDX as a function of pressure.

The goal of this study is to compare the reaction kinetics of I-RDX® from Eurenco and standard RDX from Holston, which usually contains approximately 7% HMX as a chemical impurity. I-RDX® is speculated to contain fewer chemical impurities and perhaps fewer defects in its crystal lattice. The question addressed in this study is

whether I-RDX® exhibits the same reactions and reaction rates as the standard material widely used in munitions. Further, if the crystals do have less impurities and defects, does compression affect both materials similarly, or does reaction behavior change with static high pressure? The appearance of any difference in reaction rates would provide insight into the reaction growth of the less sensitive material.

## MATERIALS AND METHOD

Laser initiation experiments were performed with a 5ns pulsed Nd:YAG laser set at 532nm wavelength. Samples of I-RDX® obtained from Eurenco and standard RDX with approximately 7% HMX impurity were loaded into gem anvil cells (GACs) with cubic zirconia gems. The gasket material was 127 micron thick Inconel 600 drilled with a circular sample chamber of 150 microns in diameter. The samples were compressed in the GAC to pressures between 0.5 and 3 GPa. To avoid side reactions or other interactions with the sample, no pressure medium was used. Hence the experiments were performed on samples under non-hydrostatic pressure. The pressure inside each cell was determined by including a small sphere of ruby (on the order of 50 μm in diameter) in the gasket hole with the sample and using the ruby fluorescence method. [5] Once under a pressure of about 0.8 GPa both RDX and I-RDX® become transparent.

The experimental arrangement to measure time-resolved absorption is shown in Fig. 1. Loaded cells were placed at the focal position of the laser. A single laser pulse was used to initiate the sample. This laser-initiation varies from traditional shock initiation due to the higher temperatures achieved (laser ~20,000K, shock ~4,000K). The reaction was backlit using the focused light of a pulsed xenon flashlamp with duration of approximately 17 μs. Light transmitted through the cell was sent through a 532 nm bandpass filter to reduce the intensity of the direct laser light, and then collected into a fiber optic cable. This light is then passed through a spectrometer to achieve spatial or wavelength dispersion, a streak camera for temporal dispersion, and on to a CCD detector that measures its intensity. The spectrometer was centered at 490 nm and had a range of about 100 nm. The streak camera was set to collect light at a rate of 1 μs/mm for a total time of 24.5 μs. The dependence on laser fluence was investigated by varying the energy per pulse of the laser.

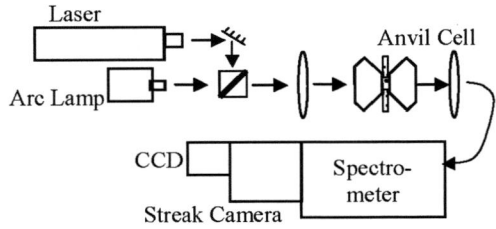

**FIGURE 1**. Experimental setup showing laser and flashlamp light being sent through the GAC into the detection system.

Before each experiment an initial CCD image was obtained of only the light that is transmitted from the flashlamp through the pressurized sample cell. This transmitted image is considered a reference intensity $I_o(T,\lambda)$. Then the sample was initiated with the laser (at time T=0) and a transmitted image was recorded during reaction to yield $I(T,\lambda)$. The "change in absorbance" ($\Delta A$) was calculated with respect to the reference intensity, as shown in equation (1).

$$\Delta A(T,\lambda) = \log_{10} [I_o(T,\lambda) / I(T,\lambda)] \quad (1)$$

Each CCD image is binned into ~0.5μs time intervals, resulting in a series of transmitted intensity versus wavelength graphs at various times. Then the $\Delta A$ intensity within 200-pixels around a particular wavelength was averaged to obtain $\Delta A$ versus time, at that wavelength. The latter is called a Reaction Profile based on the hypothesis that absorbance is proportional to the quantity of RDX that the light is transmitted through (similar to Beer's law).

For all experiments, a broadband emission was seen during the laser pulse, which corresponds to interaction between the laser and the sample/cell. This emission lasted for about one microsecond, so any results occurring within that time were masked.

## RESULTS

To study laser-initiated reaction kinetics, over 15 samples each of RDX and I-RDX® were compressed to various initial pressures between 0.97 – 2.77 GPa and laser-initiated with a laser pulse with fluence between 700 – 2800 J/cm². Powdered samples of both RDX and I-RDX® became transparent only above ~0.8 GPa. Since these experiments track changes in transmittance through the samples, pressures below 0.8 GPa were not investigated. Above about 2 GPa the samples reached maximum absorbance within the first microsecond of initiation. That is, higher-pressure samples reacted so fast that their maximum absorbance occurred within the broadband emission produced by the laser. Therefore, reaction kinetic studies above 2 GPa pressure did not produce quantitative data.

Considering the laser fluence range studied (~700-2800 J/cm²), at the lowest fluence, a few of the I-RDX® samples had small amounts of unreacted material remaining in the cell. In contrast, all the material of every RDX sample always reacted completely for the reported range of fluence. That is, in I-RDX® at low laser fluence, reaction did not propagate through the entire sample area, instead only the local region around the laser pulse reacted completely.

After the first microsecond of light generated upon laser-initiation, CCD images of transmitted light showed an initial darkening, followed by subsequent higher transmission or brightening (see Fig. 2). The amount of time required to reach the maximum darkness decreased as the initial sample pressure was increased.

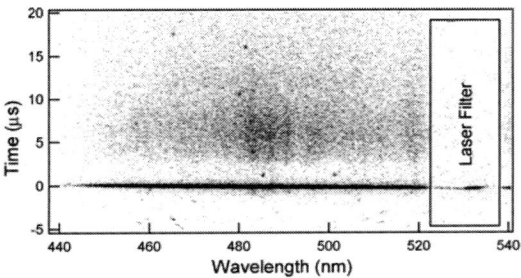

**FIGURE 2**. A typical CCD image of transmitted light. The laser flash defines time (T) zero. During the first few microseconds transmitted light is reduced followed by an increase as the reaction goes to completion.

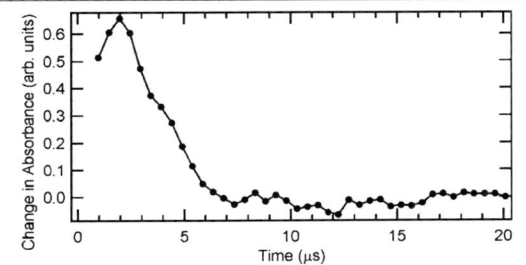

**FIGURE 3**. A typical reaction profile (change in absorbance) plotted against time from initiation.

When this transmitted light is used to calculate change in absorbance, the initial darkening results in a maximum absorbance. Change in absorbance was then used, as described in the Materials and Methods section, to produce a Reaction Profile. Figure 3 shows a typical reaction profile. The y-axis corresponds to the change in absorbance and the x-axis is time in microseconds. The zero point in time represents the arrival of the laser pulse. This figure shows the change in absorbance initially increasing with time, reaching a maximum and then decreasing. As mentioned previously, the first

microsecond of data is masked by the broadband emission occurring during the laser pulse.

## DISCUSSION

The remnant of some un-reacted I-RDX® sample in the anvil cell when initiated at low fluence suggests that fluence can affect reaction propagation. The samples were initiated in a localized region (laser spot diameter ~0.10mm, sample diameter ~0.15mm) and the reaction diffused outward. If the fluence was low enough, the remnant of un-reacted I-RDX® suggests that reaction propagation was arrested before the entire sample was consumed. There were no instances of un-reacted material remaining when RDX samples were initiated at any fluence within our reported range.

As shown in Figure 3, reactions occurred with an initial increase followed by a decrease in $\Delta A$. This reaction profile can be explained by the same two-step reaction process previously proposed by Brill et al. [6] Under pressure in the GAC, both RDX and I-RDX® appear transparent. When the laser initiates reaction, the first step is formation of dark intermediate species. These dark intermediate products absorb or scatter the incident flashlamp light. As the sample continues to react, transparent gaseous products are formed. Eventually all the sample converts to gas and the cell is once again transparent.

In comparing RDX with I-RDX®, the main difference between the two materials is that I-RDX® is less sensitive to stimulus such as friction, shock, etc. than RDX. [1-3] Therefore, it is assumed here that the important part of the reaction of these materials is the first step or the formation of the dark intermediates. Hence, in this paper we use the time to maximum absorbance (hereto referred to as $\Delta T$) as the criteria for comparison of these two materials.

Within the fluence range studied, $\Delta T$ for either RDX or I-RDX® appeared to be independent of fluence. Figure 4 shows the time to maximum absorbance versus fluence for samples at a pressure of ~1 GPa. There appeared to be no appreciable differences in time of the first reaction step to initiating laser fluence from RDX as opposed to I-RDX®. Error bars in $\Delta T$ reflect the 0.5 µs time resolution.

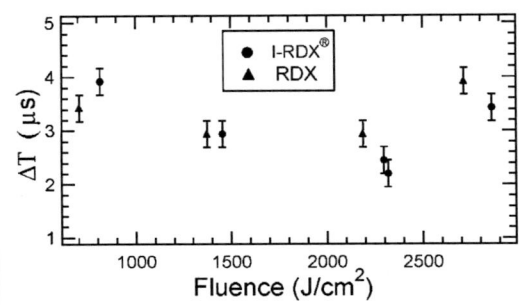

**FIGURE 4**. Fluence dependence of the first reaction step ($\Delta T$) at ~1 GPa pressure.

For both RDX and I-RDX® the time to maximum absorbance decreases with increasing pressure. Figure 5 shows a plot of $\Delta T$ versus pressure for both RDX and I-RDX®. The fluence varies slightly from point to point in Figure 5, but as discussed above, changes in fluence does not seem to influence reaction time. At pressures of 2 GPa and greater, the reaction occurs so fast that absorbance was maximized in less than a single microsecond. This time regime is dominated by broadband emission from laser exposure, so above 2 GPa, exact $\Delta T$ could not be determined, as indicated by arrowheads in Figure 5.

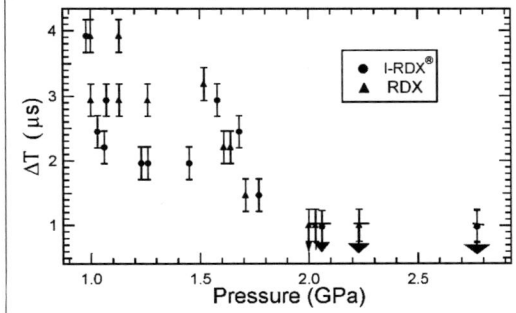

**FIGURE 5**. Pressure dependence of $\Delta T$. At pressures above 2 GPa absorbance maxima occur within 1 $\mu$s.

The decrease in $\Delta T$ with pressure is likely due to the same effect that led to an increase in the rate of thermal decomposition in the work of Miller, et al. [7] According to reference 7, the bimolecular decomposition mechanism of alpha RDX (the room temperature stable form used in this study) has a negative volume of activation. Therefore, any decrease in volume caused by compression favors the reaction rate of RDX, similar to the effect shown in Figure 5.

The only difference between RDX and I-RDX® highlighted from this study is that reaction did not propagate through the entire sample in a few of the I-RDX® samples when initiated at low fluence. Though once initiated, neither fluence changes nor pressure changes produced any significant difference in the time to production of intermediate species. These observations indicate that it takes more laser fluence (or energy) to initiate I-RDX® in such a way as to efficiently propagate reaction than it would take to initiate efficient reaction in RDX. However, once reaction is initiated, the reaction rates of the two forms of RDX are indistinguishable at pressures investigated in this study.

## CONCLUSIONS

These experiments show that once initiation occurs, there is little difference between the reaction rates of RDX and I-RDX® in the pressure and fluence ranges studied. However, when initiated at low laser fluence some I-RDX® samples had un-reacted material remaining while the reaction propagated through the RDX samples leaving no un-reacted material. This suggests that it takes more stimulus to initiate I-RDX® to react efficiently than it would take to initiate RDX.

## REFERENCES

1. Doherty, R.M. and D.S. Watt. Reduced Sensitivity RDX, Round Robin Program. in Insensitive Munitions & Energetic Materials Technology Symposium. 2004. San Francisco.
2. Freche, A., J. Aviles, L. Donnio, and C. Spyckerelle. Insensitive RDX (I-RDX). in Insensitive Munitions and Energetic Materials Symposium. 2000. San Antonio, TX.
3. Watt, D.S. and R.M. Doherty. Reduced Sensitivity RDX - Where are we? in 35th Annual Conference of ICT. 2004. Karlsruhe, Germany.
4. Kunz, A.B., M.M. Kuklja, T.R. Botcher, and T.P. Russell, Initiation of chemistry in molecular solids by processes involving electronic excited states. Thermochimica Acta. **384**: p. 279-284 2002.
5. Peirmarini, G.J., S. Block, J.D. Barnett, and R.A. Forman, Calibration of the pressure dependence of the R1 ruby fluorescence line to 195 kbar. Journal of Applied Physics. **46**(6): p. 2774-2780 1975.
6. Brill, T.B., H. Arisawa, P.J. Brush, P.E. Gongwer, and G.K. Williams, Surface Chemistry of Burning Explosives and Propellants. Journal of Physical Chemistry. **99**: p. 1384-1392 1995.
7. Miller, P.J., S. Block, and G.J. Piermarini, Effects of Pressure on the Thermal Decomposition Kinetics, Chemical Reactivity and Phase Behavior of RDX. Combustion and Flame. **83**: p. 174-184 1991.

CP845, *Shock Compression of Condensed Matter - 2005*,
edited by M. D. Furnish, M. Elert, T. P. Russell, and C. T. White
2006 American Institute of Physics 0-7354-0341-4/06/$23.00

# LAGRANGIAN ANALYSIS OF VELOCITY GAUGE DATA TO DETERMINE REACTION RATE HISTORIES IN EDC37

## C A Handley

*AWE, Aldermaston, Reading, RG7 4PR, U.K.*

**Abstract.** The Lagrangian analysis technique was applied to EDC37, an HMX-based explosive. The method was tested against an analytic model before being used to analyse particle-velocity-gauge data from two sustained-shock gas-gun experiments. This work provides evidence that the first stages of reaction in EDC37 are endothermic, as well as indicating that reaction-rate histories in explosives are bell-shaped.
**Keywords:** Explosions, Shock Wave Effects, Particle Velocity Analysis, Reaction Kinetics
**PACS:** 62.50.+p, 82.33.Vx

## INTRODUCTION

In contrast to experimental observations of the shock initiation of explosives [1], particle-velocity histories calculated using the Lee-Tarver reaction-rate model [2] have a distinctive concave-upwards shape. In his simple reactive-hydrodynamics model CIM, Lambourn [3] showed that a bell-shaped reaction-rate history can at least qualitatively match experimental particle-velocity histories.

In this investigation, a Lagrangian analysis [4-9] methodology was used to analyse the results of shock initiation experiments in EDC37 (a plastic-bonded, HMX-based explosive) [1]. Pressure, volume and energy data were extracted from polynomial fits to the experimental particle-velocity gauge traces. These were combined with an equation of state (EOS) to generate reaction-rate histories.

## LAGRANGIAN ANALYSIS

The Lagrangian analysis technique is based on the one-dimensional equations of the conservation of momentum, mass and internal energy. In integrated form, these are:

$$p(h,t) = p_{sh}(h_{sh},t) - \rho_0 \int_{h_{sh}}^{h} \left( \frac{\partial u}{\partial t} \right)_h dh \quad (1)$$

$$v(h,t) = v_{sh}(h,t_{sh}) + \frac{1}{\rho_0} \int_{t_{sh}}^{t} \left( \frac{\partial u}{\partial h} \right)_t dt \quad (2)$$

$$e(h,t) = e_{sh}(h,t_{sh}) - \frac{1}{\rho_0} \int_{t_{sh}}^{t} p \left( \frac{\partial u}{\partial h} \right)_t dt \quad (3)$$

where $h$ and $t$ are the Lagrangian position and time respectively, $u$ is the particle velocity, $p$ the pressure, $v$ the specific volume and $e$ the specific internal energy. $h$ can be related to the Eulerian distance coordinate by $\partial h = \frac{\rho}{\rho_0} \partial x$. $p_{sh}(h_{sh},t)$ is the shock pressure at the current location of the shock front $h_{sh}$ (the starting point of the integration) at time $t$. $v_{sh}(h,t_{sh})$ and $e_{sh}(h,t_{sh})$ are the specific volume and specific internal energy respectively, at the time the shock arrived $t_{sh}$ at the Lagrangian position $h$.

The essence of the Lagrangian analysis method is that, if a fit can be made to experimental particle-velocity gauge data so that $u(h,t)$ is known, the derivatives of $u$ can be integrated using equations (1), (2) and (3) to determine $p(h,t)$, $v(h,t)$ and $e(h,t)$. Then, by assuming an EOS for the partially reacted explosive, a reaction-rate $\lambda(h,t)$ can be determined.

### Fitting process

Polynomial relations, like those described by Maw [9], can be used to produce an accurate fit to particle-velocity gauge traces for individual experiments.

Here, a scaling technique is employed. Define $\mu$ and $\tau$ as a scaled velocity and time respectively, using the relations:

$$\mu = \frac{u - u_{sh}}{u_p - u_{sh}} \quad \text{and} \quad \tau = \frac{t - t_{sh}}{t_p - t_{sh}}$$

where $u$ and $t$ are particle velocity and time respectively, with $u_{sh}$ and $t_{sh}$ being their values at the shock. $u_p$ is the peak particle velocity reached at a given location, with $t_p$ being the time of the peak. A polynomial relation, $\mu = a\tau^4 + b\tau^3 + c\tau^2 + d\tau + e$, can be used to describe the shape of particle-velocity gauge traces. Two different polynomial fits were used, one for $t \leq t_p$ and one for $t > t_p$. Polynomial relations in $h$ were used as parameters in the calculation of $a$, $b$, $c$, $d$ and $e$, and also to specify $t_{sh}(h)$, $t_p(h)$, $u_{sh}(h)$ and $u_p(h)$.

Values for $p_{sh}(h)$, $v_{sh}(h)$ and $e_{sh}(h)$, the pressure, specific volume and specific energy at the shock for a given location $h$, can be determined from the particle velocity at the shock, $u_{sh}$, and the Hugoniot for the explosive, $U_s = c_0 + su_{sh}$, using the Rankine Hugoniot relations.

## APPLICATION

### Analytic solution

CIM [3] is a simple reactive-hydrodynamic model that is an exact analytic solution of the partial differential equations of motion in a reactive fluid. As a test of the method, CIM's analytic expression for $u(h, t)$ was used for Lagrangian analysis. The resulting $p$, $v$ and $\dot{\lambda}$ histories were compared to the analytic expressions, showing excellent agreement and providing confidence that the numerical schemes are of sufficient accuracy.

### Fit to the analytic solution

A smooth set of particle-velocity histories, which at least qualitatively resemble the experimental records for EDC37 gas-gun shot 1159 [1], were produced using the CIM model. The fitting process described above was used to derive an approximate $u(h, t)$ for Lagrangian analysis, providing a more ad-

vanced test of the method than the purely-analytic solution.

The polynomial fit to the CIM particle-velocity traces is shown in Fig. 1, along side a comparison of the $\dot{\lambda}$ histories produced by Lagrangian analysis with those coming directly from the model. The $\dot{\lambda}$ histories were calculated using the same EOS.

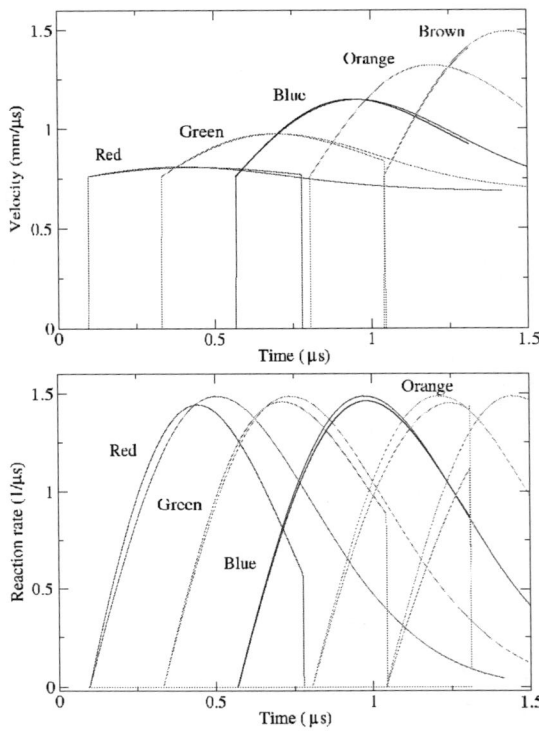

**FIGURE 1.** Above: Analytic particle-velocity histories from the CIM model (long traces) and the polynomial fit (short traces). Below: Analytic reaction-rate histories from the CIM model (long traces) and those from Lagrangian analysis (short traces).

Inspection of Fig. 1 and similar plots of $p$, $v$ and $e$ shows that, although the fit to the simulated particle-velocity histories is excellent, only approximate matches to the $p$, $v$, $e$ and $\dot{\lambda}$ traces are obtained. The first trace (in red) seems to be a particularly poor match. This is because the Lagrangian analysis technique is based on integrating the derivatives of the fitted particle-velocity field with respect to $h$ and $t$ but, in the region of the red gauge, there is little information regarding the derivative $\partial u / \partial h$, nor how the derivative $\partial u / \partial t$ varies with $h$.

This test of the Lagrangian analysis method

proves that we can have some confidence in the results for $p$, $v$, $e$ and $\dot{\lambda}$. However, the poor fit of the red gauge serves as a reminder that the results are only as good as the particle-velocity fit employed and, in regions where the derivatives of $u$, $\partial u/\partial h$ and $\partial u/\partial t$, are poorly known, the Lagrangian analysis method is not reliable.

### EDC37 experiments

The Lagrangian analysis technique was applied to fits of the experimental particle-velocity traces for a number of EDC37 gas-gun shots. The results of two of these, shots 1160 and 1159 [1], are presented in Fig. 2 and 3.

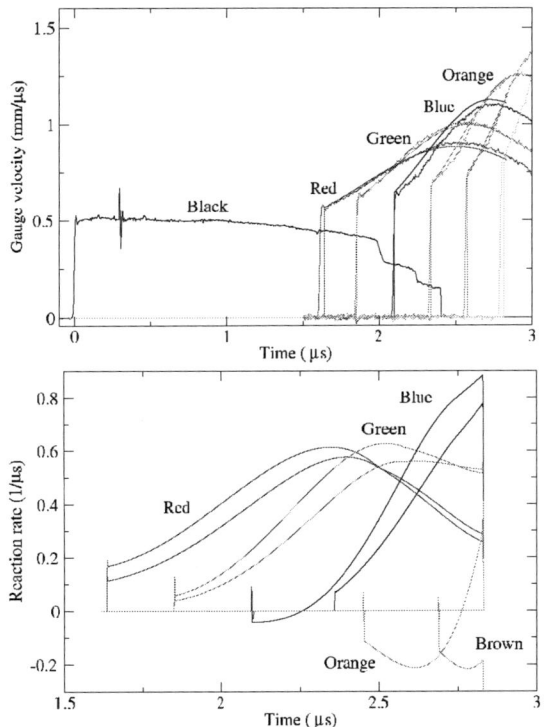

FIGURE 2. Above: Experimental particle velocity traces for shot 1160 and a polynomial fit to that data. Below: Reaction-rate histories determined by Lagrangian analysis, using the ISE V and ISE P equations of state.

The particle-velocity fit is good for both experiments, although it was only considered to be accurate within a certain range. Outside this range, the velocity was set to zero, which explains the discontinuities

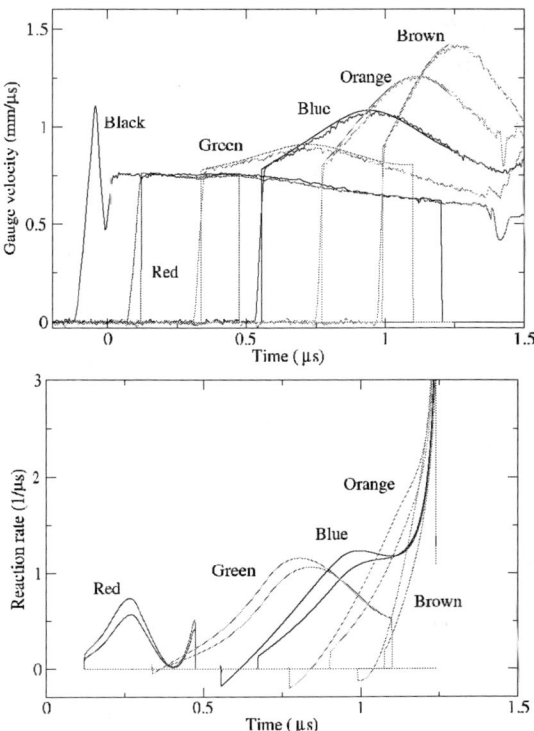

FIGURE 3. Above: Experimental particle velocity traces for shot 1159 and a polynomial fit to that data. Below: Reaction-rate histories determined by Lagrangian analysis, using the ISE V and ISE P equations of state.

in the reaction-rate traces.

The two reaction rates in Fig. 2 and 3 correspond to two different equations of state: ISE V and ISE P [10]. Both these use a finite strain EOS [11] for the solid unreacted explosive and a JWL for the gaseous products. ISE V assumes that the solid and gaseous components occupy equal specific volumes, whereas ISE P assumes pressure equilibrium. Two separate EOS were used to demonstrate the differences in $\dot{\lambda}$ caused by changes in EOS. There is no guarantee that either of these EOS is 'right' for EDC37.

The reaction rates in Fig. 2 and 3 give an indication of the shape of reaction-rate histories that might be used successfully in future reaction-rate models for explosives.

Conclusions which are not based on the particular EOS that has been chosen can be drawn by considering the pressure-volume curves traced out by each gauge. $p(v)$ traces for shot 1160 are shown in Fig. 4, along with the Hugoniot and a theoretical estimate for the principal isentrope, based on an assumed form for Gruneisen gamma.

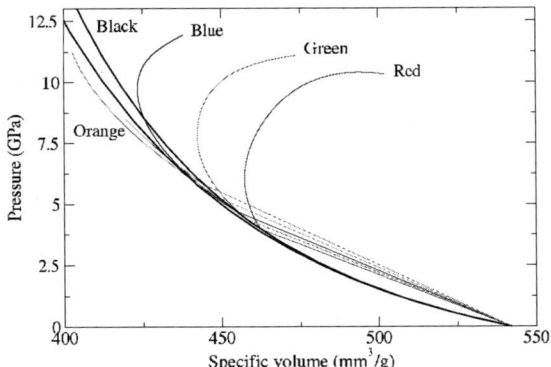

**FIGURE 4.** $p(v)$ traces determined by Lagrangian analysis for shot 1160. The black curves are the principal Hugoniot (upper curve) and principal isentrope (lower curve).

The $p(v)$ traces fall below the Hugoniot for both shots; for shot 1160, they even fall below the isentrope. This observation seems to be relatively robust: a number of reasonable modifications to the $u(h,t)$ fit, as well as changes of the Hugoniot parameters on which the $p(v)$ traces and principal isentrope are based, were made. These showed that, whilst the trajectories of the $p(v)$ traces were visibly affected by the modifications, they still fell below the principal isentrope.

Exothermic reactions release energy, so their reaction products have higher specific entropy than their reactants. A mixture of unreacted explosive and reaction products will therefore lie above the principal isentrope in $p, v$ space, for exothermic reactions. In contrast, for shot 1160 the $p(v)$ traces fall below the principal isentrope, which may indicate that the first stages of reaction in EDC37 are endothermic. This conclusion would be consistent with commonly-believed theories of the reaction chemistry of HMX-based explosives [12].

## CONCLUSIONS

Lagrangian analysis is based on an imperfect fit to the experimental particle-velocity gauges and derived reaction rates are based on a particular EOS. Nevertheless, two conclusions may be drawn from this work:

Lagrangian analysis suggests that the appropriate reaction rate to use for the modelling of EDC37 experiments is bell-shaped, as illustrated in Fig. 2 and 3.

Pressure-volume curves are found to pass below the principal Hugoniot, and even below the principal isentrope, providing evidence that the first stages of reaction in EDC37 are endothermic.

## ACKNOWLEDGEMENTS

The author is endebted to Gustavsen et al., for providing me with their experimental data.

## REFERENCES

1. Gustavsen, R. L., Sheffield, S. A., Alcon, R. R., Hill, L. G., Winter, R. E., Salisbury, D. A., and Taylor, P., *Shock Compression of Condensed Matter-1999*, AIP Conference Proceedings 505, New York, 2000, pp. 879-882.
2. Lee, E. L., and Tarver, C. M., *Phys. Fluids*, **23**(12), 2362-2372 (1980).
3. Lambourn, B. D., *Shock Compression of Condensed Matter-2003*, AIP Conference Proceedings 706, New York, 2004, pp. 367-370.
4. Fowles, R., and Williams, R. F., *Journal of Applied Physics*, **41**(1), 360-363 (1970).
5. Seaman, L., *Journal of Applied Physics*, **45**(10), 4303-4314 (1974).
6. Cowperthwaite, M., and Rosenberg, J. T., *Proceedings of the Seventh Symposium (International) on Detonation*, 1981, pp. 1072-1083.
7. Vantine, H. C., Rainsberger, R. B., Curtis, W. D., Lee, R. S., Cowperthwaite, M., and Rosenberg, J. T., *Proceedings of the Seventh Symposium (International) on Detonation*, 1982, pp. 466-477.
8. Forest, C. A., *Shock Compression of Condensed Matter-1989*, Elsevier Science Publications B.V., 1990, pp. 189-192.
9. Maw, J. R., *Shock Compression of Condensed Matter-2001*, AIP Conference Proceedings 620, New York, 2002, pp. 1027-1030.
10. Cowperthwaite, M., *Proceedings of the Seventh Symposium (International) on Detonation*, 1981, pp. 498-505.
11. Jeanloz, R., *Journal of Geophysical Research*, **94**, 5873-5886 (1989).
12. Tarver, C. M., Chidester, S. K., and Nichols III, A. L., *J. Phys. Chem.*, **100**, 5794-5799 (1996).

CP845, *Shock Compression of Condensed Matter - 2005,*
edited by M. D. Furnish, M. Elert, T. P. Russell, and C. T. White
© 2006 American Institute of Physics 0-7354-0341-4/06/$23.00

# MEASUREMENT OF TEMPERATURE AND IGNITION TIME DURING FAST COMPRESSION AND FLOW IN PBX 9501

## B. F. Henson, L. Smilowitz, J. Romero, B. W Asay and P. M. Dickson

*Chemistry and Dynamic Experimentation Divisions, Los Alamos National Laboratory, Los Alamos NM 87545*

**Abstract.** We have made radiometric temperature measurements on a microsecond time scale during the compression and flow of PBX 9501 subsequent to impact. A cylindrical sample was fired into a sapphire window normal to the cylinder axis at velocities on the order of several hundred meters per second. Cylindrically symmetric flow resulted which led to the classic circular ignition pattern at the outermost radial distance from the center at times of a few to tens of microseconds and temperatures on the order of 1000 degrees Celsius. We also observed a difference in ignition pattern for samples of beta or delta PBX 9501. We report the times and temperatures of ignition and relate them to our model of PBX 9501 decomposition kinetics. We also discuss these results in the context of various other methods of thermally and mechanically heating PBX 9501 and note the invariance of the decomposition kinetics of HMX to the method of heating.

**Keywords:** Non-shock initiation, decomposition kinetics, PBX 9501.
**PACS:** 82.33.Vx, 82.30.Lp.

## INTRODUCTION

We have demonstrated in previous work that the thermal decomposition and ignition of explosive formulations based on HMX (octahydro-1,3,5,7-tetranitro-1,3,5,7-tetrazocine) exhibit a linear relationship between the logarithm of the ignition time, $\ln(t)$, and the inverse temperature when viewed over a broad range in temperature.[1] We have also shown that heating by a variety of different means including direct heating, shear and friction, laser, and even strong shock all lead to an ignition time indicated by the temperature induced in the material.[1,2] We have also shown that a relatively simple kinetic scheme which captures the essential steps in the decomposition from solid to gas reproduces the simple linear relationship of $\ln(t)$ to inverse temperature.[3,4]

The impact of these experimental and modeling studies has been twofold. The first has been the determination of rate constants which govern the decomposition to a sufficient accuracy that the underlying physical mechanisms they represent could be determined. For instance the identification of the rate constants due to the solid $\beta$–$\delta$ phase transition, and subsequent sublimation and vaporization processes have been identified as key elements of the mechanism of decomposition. It is the analysis of ignition and decomposition phenomena over such a broad range of temperature that has made such accurate determinations possible. As a result the study of decomposition in these materials has moved far past such initial confusions as the kinetic compensation effect discussed by Brill.[5] The second impact of these studies has been to demonstrate the considerable simplification afforded by the single relationship of $\ln(t)$ to inverse temperature. In the past the modeling of ignition response to thermal and the different mechanical stimuli have been treated as separate and isolated problems. We have shown with this body of work that a single model of thermal response will suffice to calculate ignition in any scenario, provided the temperatures induced in the material by the stimuli can be measured or calculated. This is a dramatic simplification of the general problem of predictive modeling in HMX based explosive formulations.

In this paper we present measurements of temperature and ignition time in another regime of response for the HMX based formulation PBX 9501.[6] We present radiometric measurements of temperature during compression and flow of PBX 9501 subsequent to impact on a sapphire substrate. We show that measured temperatures and ignition times in these experiments are consistent with the suite of data and modeling presented previously. The measurements extend the validity of the HMX decomposition model and represent the highest temperature non-shock ignition data obtained to date.

## EXPERIMENTAL PROCEDURE

We used the light gas gun developed at Los Alamos for these experiments.[7] PBX 9501 samples (right circular cylinders, 0.73 mm thick x 3.54 mm diameter) are mounted on a 1.66 g teflon projectile and fired at a 20 mm thick sapphire window in order to allow optical and IR access to the sample during impact. Impact velocity was measured with an accuracy of 10% over the range 170 to 350 m/s. The impact results in a weak shock (~1 GPa). The dimensions of the experiment are such that lateral release in the sample is fast compared to the observed compression and flow of material in the experiment. Upon impact the sample is compressed and material flows laterally in the field of view with cylindrical symmetry. The projectile and window assembly are shown in Fig. 1.

Visible images of the post ignition burning in these experiments were obtained with a fast Cordin framing camera using the light generated by the burning sample. We observe two distinct modes of propagation of ignition in these experiments depending upon the HMX polymorph used as the sample. The observed propagation of ignition for samples of β-HMX PBX 9501 is consistent with the classic idea of ignition in such a flow field, where propagation begins in the high shear region

at the furthest extremity of the flow. This is illustrated in the top frame of Fig. 2.

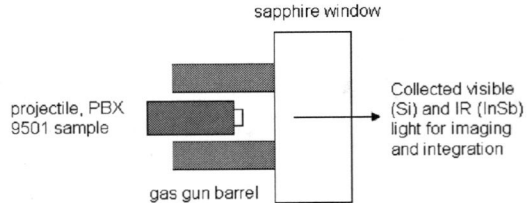

**Figure 1.** Schematice of the experimental apparatus. A teflon projectile carries a PBX 9501 cylindrical sample to impact a sapphire mirror. Impact velocities were on the order of 170 – 350 m/s. Visible and IR light were imaged and integrated through the sapphire window subsequent to impact.

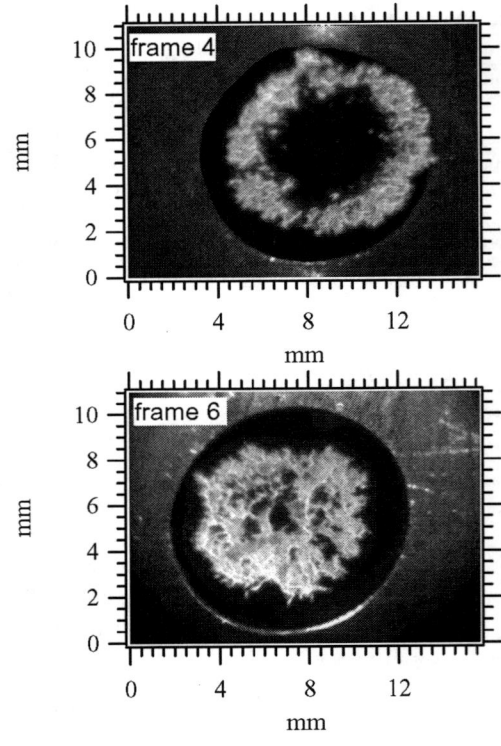

**Figure 2.** Top frame: Visible self light image of burning β-HMX subsequent to impact. Bottom frame: Visible self light image of burning δ-HMX subsequent to impact. Both are superimposed on a barrel image. Both images were acquired approximately 10 μs after impact with a 1 μs integration time.

In contrast to this behavior samples of δ-HMX PBX 9501 ignited over the full field of view, either by the same mechanism, greatly accelerated, or by a different mechanism of heating. These differences in propagation have been discussed previously,[8] and in this paper we concentrate on the ignition step as measured by radiometric temperature in the center of the flowing samples.

In the analysis described here we measure the IR radiant power from the surface in order to calculate surface temperature. We use 8 elements of an InSb detector ($\lambda$ = 4 to 5 μm). The elements are arranged in a linear array and are .08 mm on a side. We imaged the center of the sample with a SiGe lens and approximately 1:1 magnification, yielding an integrated surface area of 0.0064 mm$^2$ per element. The rise time of the amplification electronics used in these experiments was 0.5 μs. The calibration of the detectors with standard black body sources and the calibration of IR emission from the surface of PBX 9501 as a function of temperature have been described elsewhere.[9]

## RESULTS AND DISCUSSION

An example of the calculated IR temperature and optical intensity obtained at the sample center upon impact by δ-HMX PBX 9501 is shown in Fig. 3. The optical intensity was recorded by Si photodiode imaging 1 mm$^2$ at the sample center. Impact is recorded by the initial flash in the optical intensity. The rise in temperature from the delta sample is initially faster than our resolution, exhibits a plateau at an isothermal boundary condition, and then rises quickly, simultaneously with the optical intensity. We define the final rise in temperature and optical intensity as ignition. Analysis of these data involves plotting the recorded ignition time, from temperature rise to ignition, as a function of the plateau temperature.

An example of an experiment using a β-HMX PBX 9501 sample is shown in Fig. 4. As with the propagation of ignition, the temperature records indicate a different mechanism for beta and delta HMX samples. In the β-HMX samples the heating is approximately linear in time subsequent to impact, with a reproducible dip just prior to the rise in optical intensity denoting ignition. This behavior

is quantitativley consistent with the results of weak shock experiments reported by Tarver and von Holle,[10] and interpreted by us in a previous paper.[1]

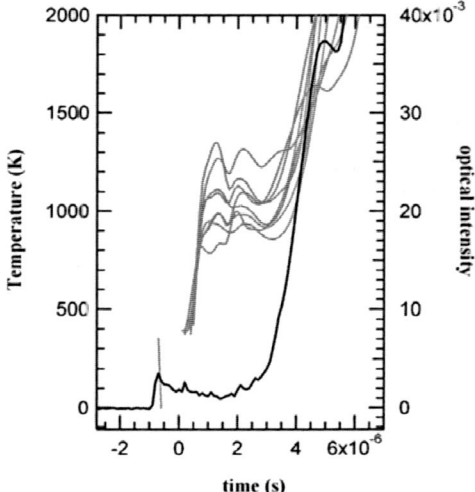

**FIGURE 3.** Calculated temperature (grey lines) and integrated optical intensity (black line) for an example experiment on a δ-HMX PBX 9501 sample.

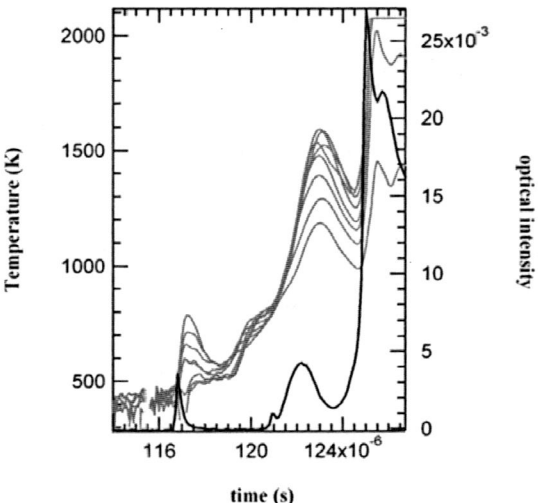

**FIGURE 4.** Calculated temperature (grey lines) and integrated optical intensity (black line) for an example experiment on a β-HMX PBX 9501 sample.

Here we adopt the analysis of Ref. [1] and plot the inverse of the observed linear heat rate as a function of the inverse peak temperature.

A compilation of ignition data is shown in Fig. 5 as the result of the flow experiments reported here (blue squares, δ-HMX, red squares, β-HMX) and from thermal explosion[11,12] (filled and open circles) laser ignition[13] (open squares) friction and shear ignition[2] (open diamonds) shock to detonation[10] (open triangles) and steady detonation (filled and open inverted triangles).[14,15]

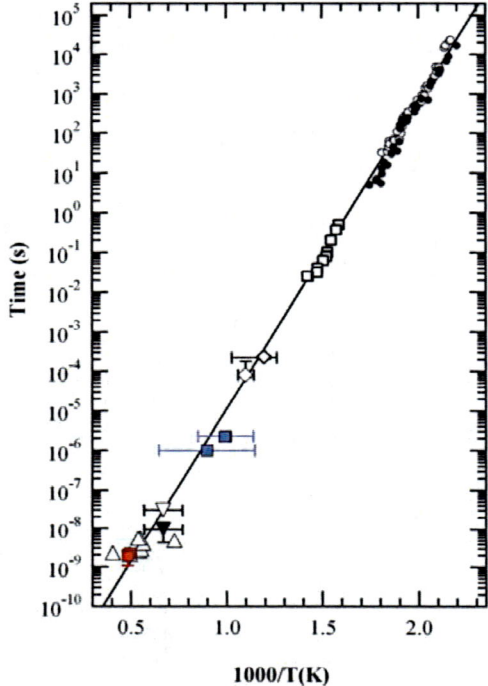

**Figure 5.** Compilation of ignition data (see text). The results of this work are shown as the blue squares (δ-HMX samples) and red squares (β-HMX samples).

## CONCLUSIONS

Even given the large error bars associated with the δ-HMX experiments the measured temperature and time to ignition reported here are consistent with previous behavior exhibited by HMX in this and other laboratories. The δ-HMX experiments further serve to provide data in a region of temperature lacking in experiments.

The principle result of this work is therefore further evidence of the range of applicability of a single thermal response model for HMX based materials. In addition, the idea that the ignition response of HMX based materials depends not on the mechanical stimuli but on the resulting generation of temperature is further confirmed.

## ACKNOWLEDGEMENTS

This work was supported by the US Department of Energy under contract number W-7405-ENG-36.

## REFERENCES

1. Henson, B. F., L. Smilowitz, B. W Asay, P. M. Dickson and P. M. Howe, 12th International Symposium on Detonation, p. 987 (2002).
2. Henson, B. F., B. W Asay, P. M. Dickson, C. Fugard and D. J. Funk, 11th International Symposium on Detonation, p. 325 (1998).
3. Henson, B. F., L. Smilowitz, B. W Asay, P. M. Dickson, in Shock Compression of Condensed Matter, 2001 (M.D. Furnish, N. N. Thadhani and Y. Horie, eds.), part II, p. 1069.
4. Henson, B. F., L. Smilowitz, B. W Asay, P. M. Dickson and D. K. Zerkle, JANNAF 22nd Propulsion Systems Hazards Subcommitte Meeting, Charleston SC (2005).
5. Brill, T. B., P. E. Gongwer and G. K. Williams, J. Phys. Chem. **98**, 12242 (1994).
6. PBX 9501 is composed of a bimodal distribution of HMX crystals (~ 120 and 30 μm) bonded with Estane and the eutectic mixture of bis(2,2-dinitropropyl) acetal and bis(2,2-dinitropropyl) formal.
7. Laabs, G. W., D. J. Funk and B. W Asay, Rev. Sci. Inst. **67**, 195 (1996).
8. Asay, B. W, B. F Henson, L. Smilowitz and P. M. Dickson, J. Energ. Mat. **21**, 223 (2003).
9. Henson, B. F., D. J. Funk, P. M. Dickson, C. Fugard and B. W Asay, in Shock Compression of Condensed Matter, 1997 (S. C. Schmidt, D. P. Dandekar and J. W. Forbes, eds.), p. 805.
10. Von Holle, W. G.and C. M. Tarver, 7th International Symposium on Detonation, p. 993 (1981).
11. Tarver, C. M. and T. D. Tran, Comb. and Flame, **137**, 50 (2004).
12. Tarver, C. M., S. K. Chidester and A. L. Nichols, J. Phys. Chem. **100**, 5794 (1996).
13. Lengelle, G., A. Bizot, J. Duterque and J.- C. Amiot, Rech. Aerosp. **2**, 1 (1991).
14. Green, L. G. and E. James Jr., 4th International Symposium on Detonation, p. 86 (1965).
15. Gustavsen, R. L., S. A. Sheffield, and R. R. Alcon, 11th International Symposium on Detonation, p. 451 (1998).

CP845, *Shock Compression of Condensed Matter - 2005*,
edited by M. D. Furnish, M. Elert, T. P. Russell, and C. T. White
2006 American Institute of Physics 0-7354-0341-4/06/$23.00

# ESTIMATION OF TWO-DIMENSIONAL INITIATION THRESHOLDS FROM POP PLOT DATA

## H.R.James

*AWE, Aldermaston, Reading, Berkshire RG74PR UK.*

**Abstract.** There is a close correspondence between the Pop Plot run distance and the run distance at the two-dimensional initiation threshold for a given explosive. This is in contrast to the differences between the Pop Plot distance and the run distance at the one-dimensional initiation threshold. The reasons for the two-dimensional correspondence are explored and estimates of the two-dimensional threshold, obtained from what is a one-dimensional measurement derived from a sustained shock (the Pop Plot), are compared with experimental results for a wide range of conventional explosive types (primary, melt-cast and plastic-bonded).

**Keywords:** Explosives, shock initiation, reaction growth.
**PACS:** 82.40.Fp, 82.33.Vx, 62.50.+p.

## INTRODUCTION

In shock initiation studies involving heterogeneous explosives, it has been found [1] that in one-dimensional flyer impacts the run-to-detonation (RTD) increases as the flyer thickness is reduced (at constant impact velocity) towards the initiation threshold. It will be shown that this is not the case for two-dimensional projectiles such as flat or round-nosed rods, or spheres. For these projectiles, as the diameter is reduced towards the threshold at a constant impact velocity, the run distance remains close to that of the Pop Plot [2]. Only if the projectile diameter approaches the failure diameter of the explosive is there an increased RTD. It will be argued that the change in behaviour results from differences in the pattern of pressure release at the position of the relative thresholds. This leads to differences in the cooling of the chemical reactions and hence the alteration of the subsequent reaction growth.

The identification of the boundary of sustained pulses that define the Pop Plot is derived. This identification, coupled both with the observation that the RTD from the two-dimensional threshold corresponds to the Pop Plot, and the knowledge of the shock structure at this threshold [3], leads to a relationship between the Pop Plot and the 2D initiation threshold criterion. It should be emphasised that the increasing RTD near the one-dimensional threshold precludes this type of relationship for flyer impacts.

## RELEASE WAVES AT ONE- AND TWO-DIMENSIONAL THRESHOLDS

The one-dimensional flyer threshold has long been associated with the critical energy concept [4]. In this criterion the amount of energy involved in initiation, and deposited into the explosive by the impact, is assumed to be bounded by the arrival of the rarefaction originating from the rear of the flyer, at the flyer/explosive interface (the "rearward" rarefaction). To make the resulting parameter $(E_C)$ which defines this threshold correspond to the observed 2D threshold for the same explosive, this duration had to be identified with the time to maximum non-divergent shock volume in the explosive [3,5]. This does not change the time at which the 1D impact triggers initiation, but places the 2D threshold at the time when the head of the peripheral rarefaction has traversed about a third of the projectile radius.

Figure 1 shows a 2D Eulerian hydrocode simulation of a projectile, whose geometry lies at the boundary of both 1D and 2D behaviour, impacting an explosive. No explosive reaction has been modelled, and the time chosen to display the internal energy contours corresponds to both thresholds. It can be seen that the release due to the rearward rarefaction is far steeper than that induced by the peripheral rarefaction. Analysis of a point in the high explosive (HE) near the interface shows it would undergo a rate of drop in hydrodynamic internal energy with time, which is an order of magnitude greater for a 1D impact than at the 2D threshold which corresponds to the same impact velocity.

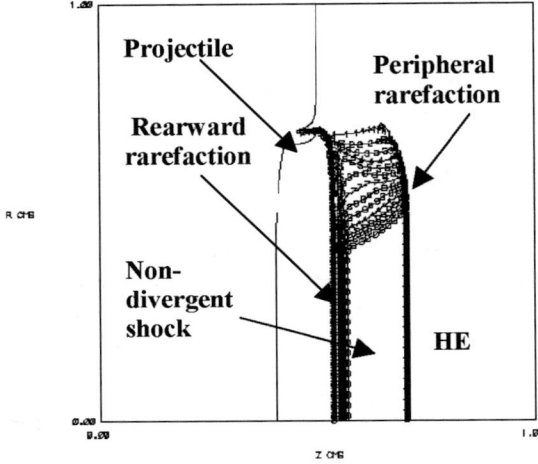

**Figure 1.** Internal energy contours in the HE showing the different patterns of release at the initiation threshold in 1D and 2D impacts.

The difference in the rarefaction gradient is both due to the differences in equations-of-state of the two materials, and the greater path length that must be travelled by the peripheral release wave to reach the threshold position. The rearward rarefaction only has to travel the thickness of the flyer. The peripheral rarefaction has been travelling for the time taken for the shock to cross the flyer thickness, plus the return time of the release wave, ie to first order it is in existence for approximately double the time of the rearward rarefaction. This gives additional time for the head of the release to separate from the slower moving elements in the tail.

The rate of release of chemical energy triggered by the shock stimulus then has to compete with the rate

of reduction of that stimulus due to the rarefaction. Obviously the 1D impact will have a far greater effect on the subsequent reaction growth than for the 2D projectile (see figures 2 and 3).

The "Effective Diameter" used in figure 3 results from the reduction in the diameter of the non-divergent portion of the transmitted shock as it crosses a barrier.

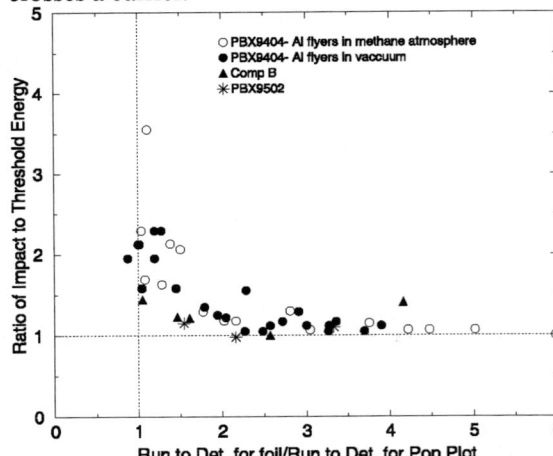

**Figure 2.** Increase in run distance as 1D projectiles approach the initiation threshold [6-8]. The points shown go from threshold impacts (with long RTD) to impacts well in excess of the threshold energy where the RTD reduces to the value of the Pop Plot.

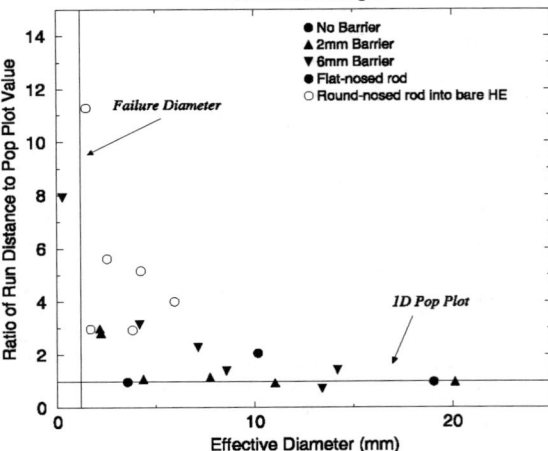

**Figure 3.** 2D Projectiles all at (or just above) the initiation threshold, but increases in run distance are only seen for projectile diameters near HE failure diameter [9].

This has the effect of making the stimulus appear to come from a projectile with a smaller initial diameter ($D_E$) than the actual diameter $D_0$. The relationship

$$D_E = D_0 - \frac{2d}{W}\left(C^2 - \{W - U\}^2\right)^{1/2} \qquad (1)$$

was derived in [10], where d is the barrier thickness; and W,C and U are the shock, sound and particle velocities in the shocked region of the barrier. Spheres or round-nosed rods have an effective diameter of $D_0/3$ on impacting bare HE, as only part of their diameter is involved in forming a region that approximates to a non-divergent shock [11]. Areas of HE that contain divergent regions of shock are assumed to play no part in forming the initiation conditions.

The data in figure 3 [9] had to be manipulated to give run distances from the transit times that are recorded in this reference. This involved assuming that the initial shock in the HE did not grow significantly until the onset of detonation occurred, ie the shock effectively ran through an inert material. In PBX9404 this is not a bad assumption, as is illustrated in figure 4. Here the increase of shock velocity with position was recorded in a wedge test. As can be seen, the major acceleration of the shock only occurs over a very short distance towards the end of the run to detonation. The majority of the run occurs close to the initial shock value.

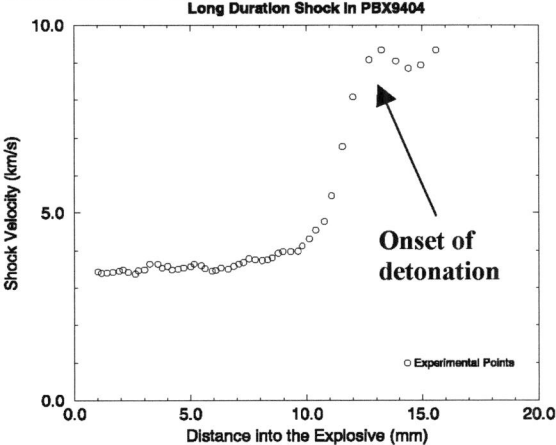

**Figure 4.** Acceleration of the shock front during growth to detonation in PBX9404 [12].

### RELATIONSHIP OF THE REARWARD RAREFACTION TO THE POP PLOT

By treating the explosive as behaving as an inert material over the RTD distance, it is relatively straightforward to calculate the position in the HE that a release wave, originating from the rear of a flyer, will catch the shock propagating into the explosive. Conversely, still using the inert approximation, the flyer thickness (s) required to allow the rarefaction to catch the shock at the detonation position can be found: -

$$s = X\frac{(U_H + C_H - W_H)(W_F + V)C_F}{W_H C_H (W_F + U_H + C_F)} \qquad (2)$$

where X is the RTD position in the HE, subscript F refers to flyer conditions, H to HE conditions and V is the impact velocity.

Comparison of s with the flyer thicknesses that define the 1D initiation threshold for PBX9404 reveals the value of s to be some 1.4-2.7 times the threshold thickness. The impact energy transmitted into the explosive, which is responsible for initiation, is proportional to the flyer thickness. From figure 2 it can be seen that the ratio of impact to threshold energy required for the RTD to be reduced to the value from the Pop Plot is approximately the same as the ratio of s to threshold thickness. Hence the onset of the Pop Plot distance occurs, to first order, at the point where the rearward rarefaction catches the shock just as it transits to detonation. It should be noted that the Pop Plot gives the minimum RTD distance for a given impact pressure. Also for thinner flyers between the 1D threshold position and the Pop Plot RTD, the rarefaction overtakes the shock before detonation occurs, causing the RTD to increase.

### PERIPHERAL RAREFACTIONS AND THE 2D THRESHOLD

By analogy with the 1D situation, the radial position of the head of the peripheral release wave at the instant of detonation defines the shock radius corresponding to the sustained shock boundary. Before that radius (but after the threshold condition is satisfied) the low rarefaction gradient could also ensure no RTD increase, in contrast to the 1D case. If, however, detonation has not occurred by the time the release has reached the centre-line then considerable cooling of the reaction would be expected due to the cylindrically convergent rarefaction. Since, away from the HE failure diameter, no RTD increase is observed for 2D threshold impacts, this implies that either the sustained shock boundary is spatially close to the initiation position, or the rarefaction gradient has little effect on the reaction. More probably there is a combination of the two effects, although the current first-order theory is predicated on the sustained shock position. The RTD distance is given by the Pop Plot relationship

$$\ln X = A_1 + A_2 \ln P \qquad (3)$$

where P is the initial impact pressure, and $A_1$ and $A_2$ are constants for a given explosive. The radial motion of the rarefaction can be found by studying figure 5.

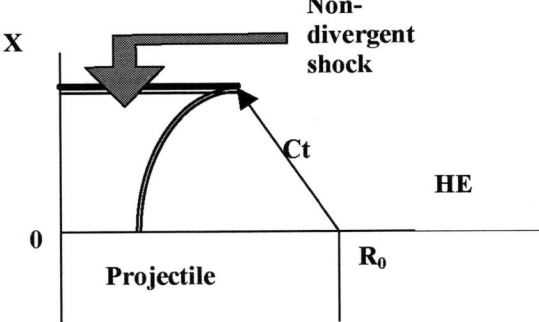

**Figure 5.** Shock structure due to peripheral rarefaction in axes moving with the interface. Structure is reflected about the OX line.

It is assumed the sustained shock boundary lies in the vicinity of the centre-line. Consequently X (the uncompressed distance into the HE) is of the order of

$$X = \frac{D_0 W_H}{2\sqrt{C_H^2 - \left(W_H - U_H\right)^2}} \qquad (4).$$

Combining (3) and (4) eliminates X and allows $D_0$ to be associated with the impact velocity (V) – assuming the hugoniots for the projectile and non-reactive HE are known. The 2D threshold can then be expressed in terms of $V^2D_0$ [13,16]. Results are shown in Table 1.

**TABLE 1.** Comparisons between theory and experiment.

| HE | $A_1$ | $A_2$ | $V^2D_0$ (expt.) $mm^3\mu s^{-2}$ | $V^2D_0$ (theory) $mm^3\mu s^{-2}$ |
|---|---|---|---|---|
| PBX-9404 | -5.04 | -1.37 [14] | 4.35 [9] | 5.39 |
| Comp. B | -4.38 | -1.50 [14] | 10.76 [16] | 15.00 |
| Tetryl | -5.86 | -1.24 [15] | 2.14 [16] | 3.62 |

## DISCUSSION AND CONCLUSIONS

The results indicate that the theory successfully follows the type of explosive, but the higher threshold value implies detonation occurs sometime after the rarefaction reaches the centre-line. The lack of RTD-increase probably results from a combination of the detonation being reasonably close to this event, and the low rarefaction gradient.

The theory needs a factor of 0.6-0.8 over the wide range of HE types studied, to bring it into line with experiment. With this factor it could be used to give an approximate estimate of the initiation threshold for explosives which only possess a Pop Plot.

## REFERENCES

1. Green LG, Nidick EJ & Walker FE, *Critical Shock Initiation Energy of PBX9404, A New Approach*, LLNL Report UCRL-51522 (1974).
2. Ramsay JB & Popolato A, *4th Symposium (International) on Detonation*, p233, Silver Spring MD, 12-15 Oct (1965).
3. James HR, *Propellants, Explosives, Pyrotechnics* **13**, 35 (1988).
4. Walker FE & Wasley RJ, *Explosivstoffe* **17**, 9 (1969).
5. James HR, *Propellants, Explosives, Pyrotechnics* **21**, 8 (1996).
6. Gittings EF, *4th Symposium (International) on Detonation*, p373, Silver Spring MD, 12-15 Oct (1965).
7. Belanger C & Matte Y, *Shock Sensitivity of CX-84A and Composition B to Flyer Plate Test*, Valcartier, Canada, DREV Report M-2642/83 (1983).
8. Dick JJ, *J.Energetic Materials*, **5**, 267 (1987).
9. Bahl KL, Vantine HC & Weingart RC, *7th Symposium (International) on Detonation*, p325, Annapolis MD, 16-19 June (1981).
10. Cook MD, Haskins PJ & James HR, *9th Symposium (International) on Detonation*, p1441, Portland OR, 27 Aug-1 Sep (1981).
11. James HR & Hewitt DB, *Propellants, Explosives, Pyrotechnics* **14**, 223 (1989).
12. Dick JJ, *On the Nature of the Buildup to Detonation in Solid High Explosives During Plane Shock Initiation*, LANL Report LA-UR-80-329 (1980).
13. Held M, *Explosivstoffe* **16**, 98 (1968).
14. Mader CL, *Numerical Modeling of Detonations*, University of California Press, (1979).
15. Gibbs TR & Popolato (eds.), *LASL Explosive Property Data*, University of California Press, (1979).
16. Slade DC & Dewey J, *High Order Initiation of Two Military Explosives by Projectile Impact*, BRL Report 1021 (1957).

CP845, *Shock Compression of Condensed Matter - 2005*,
edited by M. D. Furnish, M. Elert, T. P. Russell, and C. T. White
© 2006 American Institute of Physics 0-7354-0341-4/06/$23.00

# OBSERVATION OF SHOCK INITIATION PROCESS IN GAP TEST

**S. Kubota[1], Y. Ogata[1], Y. Wada[1], K. Katoh[1],
T. Saburi[1], M. Yoshida[1] and K. Nagayama[2]**

[1] *Research Center for Explosion Safety, National Institute of Advanced Industrial
Science and Technology, 16-1 Onogawa, Tsukuba, Ibaraki, 305-8569, Japan*
[2] *Department of Aeronautics and Astronautics, Faculty of Engineering,
Kyushu University, Fukuoka, 812-8581, Japan*

**Abstract.** We have conducted the experiments for shock sensitivity of high energetic materials by gap test. The set up of gap test have been improved to observe the shock initiation phenomena in acceptor charge by high-speed video. The length of gap material was varied to observe the reaction process under various situations. The luminescence at the surface of acceptor holder was used Go/Nogo decision. The distance from the gap end to the luminescence area increases with increasing in gap length. In the critical gap length in which the sympathetic detonation does not occur, the remarkable decomposition of acceptor charge was observed as gas expansion.

**Keywords:** Gap test, explosive, sympathetic detonation, high speed video, SDT
**PACS:** S 47.40.-x, 47.40.Nm.

## INTRODUCTION

A gap test has been utilized to estimate the sensitivity for high energetic materials. The set up of this test is simple and judgment of the results is easy, but the test includes much important information on the reaction process of the energetic materials. Therefore many researchers have been combined the gap test and optical measurement to make clear the initiation phenomena [1,2].

We have developed the hydrodynamic code to estimate various explosion phenomena [3,4]. One of the most important phenomena for risk assessment of the energetic materials is the sympathetic detonation of high explosives. The gap test is suitable for checking the reliability of calculation of the sympathetic detonation or for obtaining the parameters of initiation model based on those data with high accuracy. The gap tests have been observed by high speed video to collect the data on initiation phenomena. We shall present the results of a series of the gap test and discuss the aspects on the reaction of the acceptor charge.

## EXPERIMENTAL SET UP

The gap test configuration is illustrated in Fig. 1. Both of the donor and the acceptor charge is RDX based explosive which have 7100 m/s detonation velocity. Those charges are filled up PMMA pipe with 26 mm of inner diameter and 2 mm thickness. The height of donor is 50 mm and that of acceptor is 40 mm. The large and the small size PMMA plates are set as gap material. By using the large size plate we can take the photography of acceptor from the side view without disturbance of gas expansion. Small size plate is to adjust the gap length. The steel block with 35 mm diameter was used as witness block. The gap length was varied from 10-30 mm. High speed video made by Shimadzu Corporation Japan was set its maximum framing rate 1,000,000 f/s.

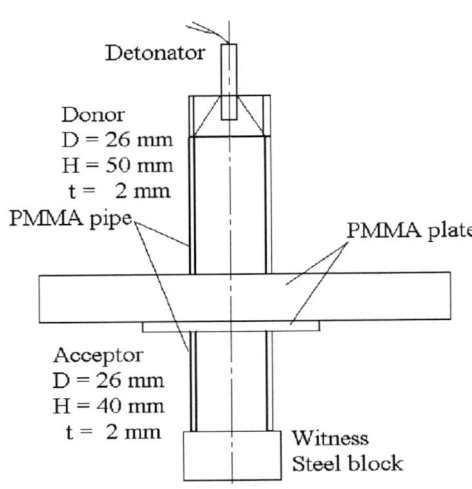

FIGURE 1. Gap test experimental configuration

FIGURE 2. Recovered witness steel block in the case of the gap length is 22mm or less

## RESULTS AND DISCUSSION

Since the luminescence at the surface of acceptor holder based on the reaction of acceptor charge, it was used Go/Nogo decision. The results can be roughly divided two conditions. First one is that the luminescence at the surface of acceptor can be confirmed before the shock wave arrived at the steel block. In such cases we judge the condition is 'Go'. Second one there is no luminescence in the acceptor during gap test observation. When the gap lengths are 22 mm or less, sympathetic detonation occurs in acceptor. Fig.2. is the witness steel block recovered under such conditions, and considerable damage was given by reaction of the acceptor charge. Fig.3 presents the high speed photographs of acceptor part obtained by high speed video. All cases are corresponds to the luminescence is confirmed in acceptor. When the 10 mm gap length, luminescence can be confirmed from the interface of the gap material and the acceptor. Detonation occurs with no measurable delay at the interface. While in the case of 20 mm gap length the acceptor is bright near the interface, but be bright from the halfway in the acceptor. The distance from the interface of the gap and the acceptor to luminescence area increases with increasing in gap length.

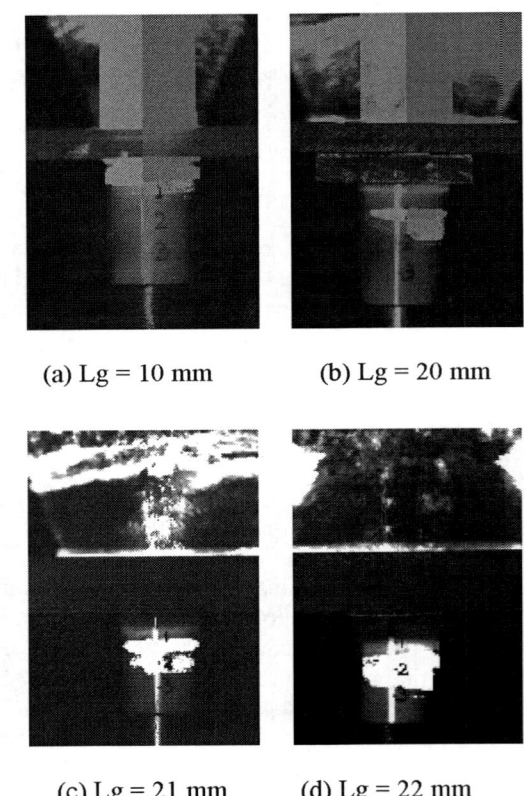

(a) Lg = 10 mm     (b) Lg = 20 mm

(c) Lg = 21 mm     (d) Lg = 22 mm

FIGURE 3. The high speed photographs of acceptor part in gap test. The gap lengths are 22 mm or less, the sympathetic detonation occurs in acceptor. Lg ; Gap length

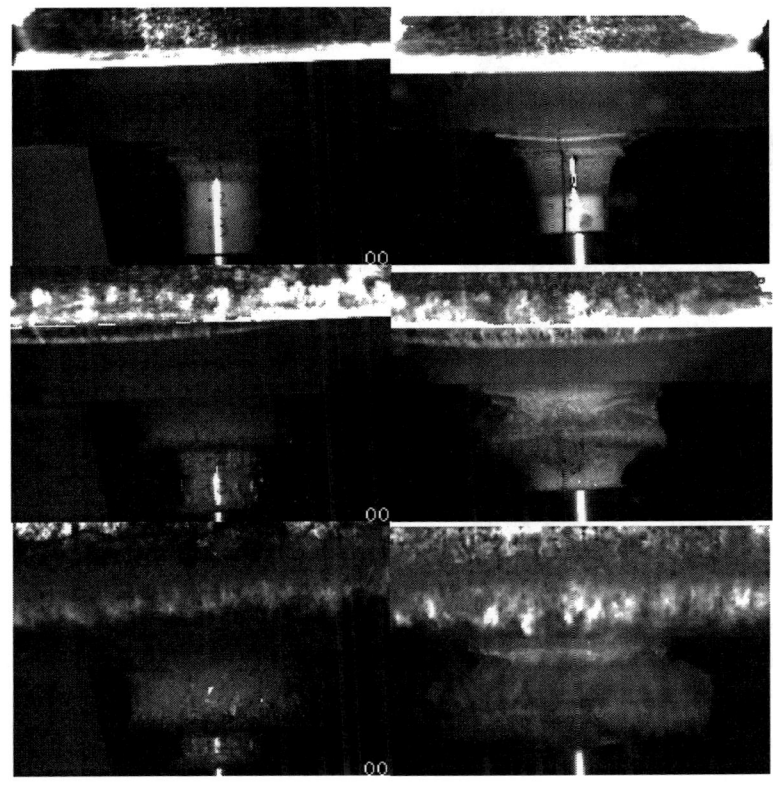

(a) Lg = 26 mm          (b) Lg = 23 mm

FIGURE 4. The high speed photographs of acceptor part in gap test. Lg ; Gap length
The gap lengths are 26 mm and 23 mm, the sympathetic detonation does not occur.
The times of each frame are 35, 55 and 75 microseconds from the initiation

The propagation velocities of luminescence area on the acceptor surface were estimated. In the case of 10 mm gap length, the velocity is about 7500 m/s, while in the case of 20 mm it is about 10,000 m/s. It is thought that those results are influenced by the geometric effect, the temporal response of the shape of the shock front in the acceptor during SDT process. Under the conditions that the gap length is relatively long, input shock wave to the acceptor has curved front, and the pressure of the shock front becomes high near the central axis of the acceptor. The shock wave is increasingly accelerated near the central axis under the influence of the decomposition of explosive. Consequently, the curvature radius of the shock front with the reaction becomes still smaller. In such cases, the propagation velocity of the luminescence area observed on the acceptor surface is influenced of the reaction wave from the axis direction. In an appearance, it becomes faster than the steady state detonation velocity. While if the gap length is short and the front configuration of the input shock wave is closely plane geometry, the SDT occurs simultaneously in the same cross section of acceptor. Therefore, observed propagation velocity is almost the same as the steady state detonation. When the gap length was 22 mm, the experiment was conducted 3 times, and the same result was obtained on each cases. SDT did not occur in acceptor when the gap lengths are 23 mm or long. The photographs in Fig.4 were obtained under such conditions.

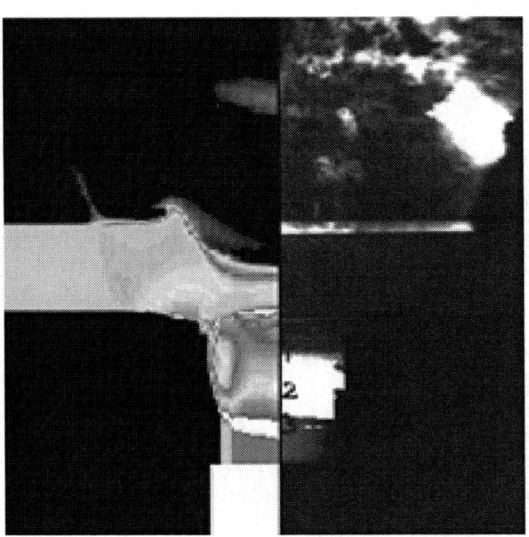

**FIGURE 5.** Comparison of the numerical result and the high speed photograph of gap test; left hand side calculated density distribution. Lg : 21 mm

Although the SDT did not occur, the remarkable decomposition of acceptor could be seen as gas expansion in Fig. 4 (b). The steel block had no damage in this case. The experiments with 23 mm gap were done three times and had the same results each other. Therefore the 23 mm gap corresponds to critical gap in which the sympathetic detonation does not occur. In addition, when there is no decomposition, the shock speed in explosive could be estimated. The density change accompanying shock wave propagation was observed as the contrast change on high speed photographs. Based on those photos the average velocity of the shock was 2500 m/s in the case of 30 mm gap length.

The numerical simulations for a series of gap test have been performed to establish reliable code for various explosion phenomena and the determination method for the parameters of the initiation model. CIP method has been adopted to reduce numerical diffusion caused by calculation of advection term in Euler equations [5,6]. To solve the initiation process of energetic material, ignition and growth model has been employed [7]. Fig. 5 shows the comparison of the numerical simulation and the experiment. From this result, it can be considered that our numerical code can be well reproduced the shock initiation phenomena.

## SUMMARY

The high-speed photography of the gap test was conducted. Both of the donor and the acceptor charge is RDX based explosive. The height of donor is 50 mm and acceptor is 40 mm with 26 mm inner diameter. The Gap material is PMMA. The luminescence at the surface of the acceptor holder was used Go/Nogo decision. When the gap lengths are 22 mm or less, the sympathetic detonation occurs in the acceptor. The distance from the gap end to the luminescence area increases with increasing in gap length. In the critical gap length in which the sympathetic detonation does not occur, there in no damage on the steel block but the remarkable decomposition was observed by high speed video.

## REFERENCES

1. Sultanoff M. et al., "Shock induced sympathetic detonation in solid explosive charges", third symposium (international) on Detonation, 1960, pp.520-533.
2. Bouma, R.H.B. and van der Steen A. C., "Experimental determination of shock and detonation wave profiles during shock to detonation transition for different initiation modes", eleventh symposium (international) on Detonation, 1998, pp. 640-646.
3. Kubota, S et al., "Simulation of sympathetic detonation by a CIP Eulerian code", Computational Ballistics II, 2005 pp.107-114.
4. Liu, Z., Kubota, S., Otsuki, M., Yoshimura, K., Okada, K., Nakayama, Y., Yoshida, M., Fujiwara, S., "Development of a 2D Eulerian code for reactive shock analysis using CIP scheme. Journal of the Japan explosives society, 63(5), pp.264-270, 2002.
5. Takewaki, H., Nishiguch, A., Yabe, T., "Cubic interpolated pseudo-particle method (CIP) for solving hyperbolic-type equations". Journal of computational physics, 61, pp.261-268, 1985.
6. Yabe, T., "A universal cubic interpolation solver for compressible and incompressible fluids" Shock Waves, pp.187-195 1991.
7. Lee, E.L. and Tarver, C.M., "Phenomenological model of shock initiation in heterogeneous explosives" Phys. Fluids, 23(12), pp. 2362-2372, 1980.

CP845, *Shock Compression of Condensed Matter - 2005,*
edited by M. D. Furnish, M. Elert, T. P. Russell, and C. T. White
© 2006 American Institute of Physics 0-7354-0341-4/06/$23.00

# SHOCKS AND DISCONTINUITIES IN PARTICLE METHODS

## L. D. Libersky[1] and P. W. Randles[2]

[1]*Los Alamos National Laboratory, Los Alamos, NM 87545*
[2]*Defense Threat Reduction Agency, Kirtland, NM 87117*

**Abstract.** We briefly examine the propagation of strong shocks across density discontinuities as computed with two meshfree particle codes: Smoothed Particle Hydrodynamics and Dual Particle Dynamics. Density ratios considered are typical of those that appear in explosive-metal interactions. The work is motivated by the desire to eliminate numerical artifacts from degrading solutions of problems involving explosive initiation and spall. We observe that incorporating aspects of the Riemann solution into the SPH pair-wise interpolation scheme improves the results significantly. DPD results are superior.

**Keywords:** SPH, DPD, Particle, Meshfree, Simulation, Continuum, Code, Hydrodynamics, Shocks.
**PACS:** 02.70.Ns, 02.60.Cb

## INTRODUCTION

Smoothed Particle Hydrodynamics (SPH) is a conceptually simple and robust continuum method [1,2] that uses no fixed background spatial mesh - only Lagrange points (particles) with variable connectivity and kernel interpolation. SPH does an adequate job of treating shocks [3] in the continuum but leaves troublesome remnants of shock passage across density discontinuities, e.g., explosive–metal interfaces. Typically, the error is large enough to cause concern for problems where rate laws drive essential physics, such as burn rates for HE and damage evolution for fracture. The source of the error is traced to the SPH interpolations requiring smooth fields. If we do not adhere to this "rule" then we should expect numerical error. Furthermore, since SPH is a colocational scheme (all field variables carried on all particles) the error generated will persist as a free or zero-energy mode. A possible "fix" for this problem would be to smooth the density across interfaces, but this would play havoc with solid equations of state; it is also not practical in higher dimensions. Therefore, interactions between particles representing different materials (densities) will require that some explicit appeal be made to contact boundary conditions. The great strength of SPH is its robustness and conceptual simplicity owing to its pair-wise interaction scheme which eliminates geometrical complications. If possible, it would seem expedient introduce contact conditions in like manner, in keeping with the spirit of SPH. We will compare results of such an approach with Dual Particle Dynamics (DPD), a more recent particle method [4], in which gradient estimates and boundary conditions are formulated for enhanced accuracy and stability.

## FORMULATION

One of the traditional forms of SPH for hydrodynamic flow is:

$$\frac{d\rho_i}{dt} = \sum_j \frac{m_j \rho_i}{\rho_j} \left( \vec{U}_i - \vec{U}_j \right) \cdot \vec{\nabla}_i W_{ij} \qquad (1)$$

$$\frac{d\vec{U}_i}{dt} = -\sum_j \frac{m_j}{\rho_i \rho_j} \left( P_i + P_j \right) \vec{\nabla}_i W_{ij} \qquad (2)$$

$$\frac{dE_i}{dt} = \sum_j \frac{m_j \left( P + P_j \right)}{2\rho_i \rho_j} \left( \vec{U}_i - \vec{U}_j \right) \cdot \vec{\nabla}_i W_{ij} \qquad (3)$$

where $\rho$, $U$, $E$, $P$, $m$, $W$, are the density, velocity, specific internal energy, pressure, mass, and the interpolating kernel respectively. The kernel is a function of the particle coordinates ($r$) and smoothing length ($h$). Parshikov [5] has written these equations in the frame of the radial between particles "i" and "j"

$$\frac{d\rho_i}{dt} = -\sum_j \frac{m_j \rho_i}{\rho_j h} \left( U_i^R - U_j^R \right) W_{ij}' \qquad (4)$$

$$\frac{d\vec{U}_i}{dt} = -\sum_j \frac{m_j \left( P_i + P_j \right)}{\rho_i \rho_j h} W_{ij}' \frac{\vec{P}_j - \vec{P}_i}{\left| \vec{P}_j - \vec{P}_i \right|} \qquad (5)$$

$$\frac{dE_i}{dt} = -\sum_j \frac{m_j \left( P_i + P_j \right)}{2\rho_i \rho_j h} \left( U_i^R - U_j^R \right) W_{ij}' \qquad (6)$$

where

$$U^R = \vec{U} \frac{\vec{P}_j - \vec{P}_i}{h \left| \vec{P}_j - \vec{P}_i \right|} \qquad (7a)$$

and

$$\nabla_i W_{ij} = W_{ij}' \frac{\vec{P}_i - \vec{P}_j}{h \left| \vec{P}_i - \vec{P}_j \right|} \qquad (7b)$$

and then brings in the influence of a Riemann solution (superscript star) using the substitutions

$$\tfrac{1}{2} \left( U_i^R - U_j^R \right) \rightarrow U_{ij}^{*R} \qquad (8a)$$

$$\tfrac{1}{2} \left( P_i - P_j \right) \rightarrow P_{ij}^{*} \qquad (8b)$$

giving

$$\frac{d\rho_i}{dt} = -2\sum_j \frac{m_j \rho_i}{\rho_j h} \left( U_i^R - U_{ij}^{*R} \right) W_{ij}' \qquad (9)$$

$$\frac{d\vec{U}_i}{dt} = 2\sum_j \frac{m_j P_{ij}^{*}}{\rho_i \rho_j h} W_{ij}' \frac{\vec{P}_j - \vec{P}_i}{\left| \vec{P}_j - \vec{P}_i \right|} \qquad (10)$$

$$\frac{dE_i}{dt} = -2\sum_j \frac{m_j P_{ij}^{*}}{2\rho_i \rho_j h} \left( U_i^R - U_{ij}^{*R} \right) W_{ij}' \qquad (11)$$

The hydrodynamic Riemann solutions for the velocity and pressure appearing in (8) are

$$U_{ij}^{*R} = \frac{U_j^R \rho_j C_j + U_i^R \rho_i C_i - P_j + P_i}{\rho_j C_j + \rho_i C_i} \qquad (12)$$

$$P_{ij}^{*R} = \frac{P_j \rho_i C_i + P_i \rho_j C_j - \rho_i C_i \rho_j C_j \left( U_j^R - U_i^R \right)}{\rho_j C_j + \rho_i C_i} \qquad (13)$$

Parshikov, who applied his technique to all particle interactions, offered concern about the relatively high numerical viscosity associated with the method. We believe that his approach may provide significant improvement to SPH if applied only between particles of different kinds (normal densities) and the SPH solution using Monaghan's artificial viscosity [3] retained in the interior. We will test this hypothesis for two hydrodynamic shock problems in one-dimension.

**Step down in density**

Consider a 200 Kb shock in steel (generated by steel on steel impact) traversing a steel-HMX boundary. Such problems are encountered in shock initiation events where impacts on covered HE are involved. The density contrast for this case is on the order of 10/2. Pressure profiles for the SPH (Eq's 1-3) solution is shown in Fig. 1a and the Parshikov modified SPH solution (Eq's 9-11) in Fig. 1b. The boundary-modified solution shows significant improvement over standard SPH results where no attempt is made to account for the discontinuity in density. Fig. 1c shows the DPD result where no oscillations are observed at the boundary and the lead shock is much sharper.

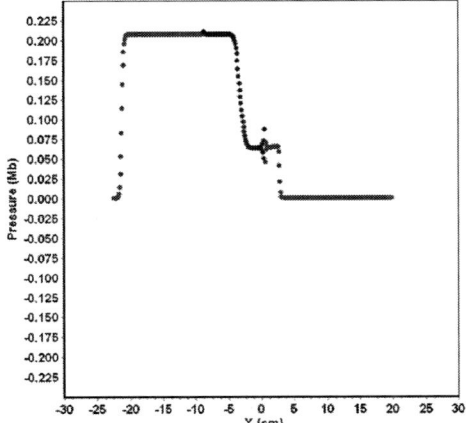

**Figure 1a**. Shock over density step-down, SPH.

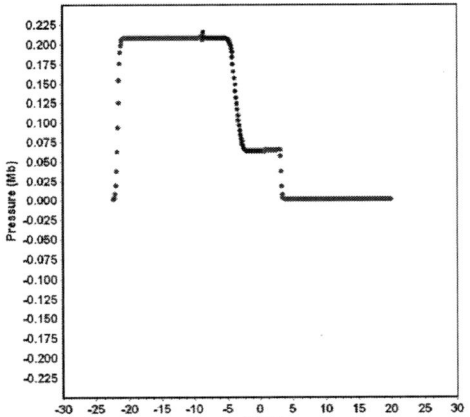

**Figure 1b**. Shock over density step-down, mod SPH.

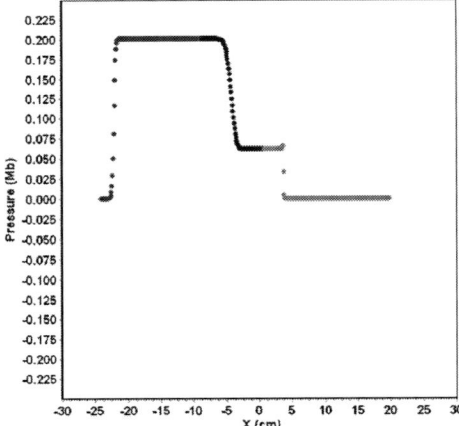

**Figure 1c**. Shock over density step-down, DPD.

## Step up in density

Consider a 100 Kb shock in HMX traversing a HMX-steel boundary. Such problems are encountered during warhead explosions where detonation waves encounter metal surroundings. The density contrast for this case is on the order of 2/8. Pressure profiles for the SPH solution is shown in Fig. 2a and the Parshikov modified SPH solution in Fig. 2b. The boundary-modified solution shows significant improvement over standard SPH results. Fig. 2c shows the DPD result where, again, no oscillations are observed at the boundary and the lead shock is much sharper.

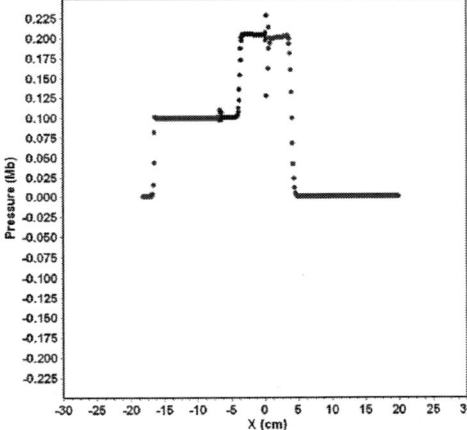

**Figure 2a**. Shock over density step-up, SPH.

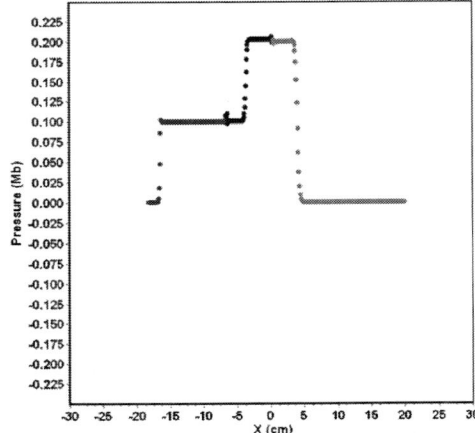

**Figure 2b**. Shock over density step-up, mod SPH.

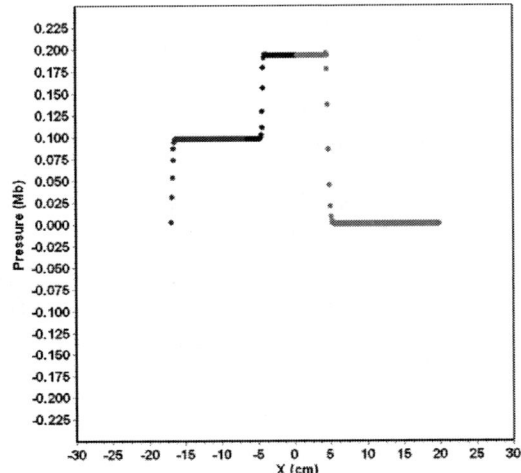

**Figure 2c**. Shock over density step-up, DPD.

## CONCLUSIONS

We have performed simple one-dimensional simulations with SPH, modified SPH with Riemann solver at the contact boundaries, and DPD. These numerical techniques are all Lagrange meshfree particle methods. The goal was to examine the ability of the different methods to handle the propagation of strong shocks across density discontinuities that are typical of those encountered in many problems involving high explosives and metals. Since the problems simulated are only one-dimensional, they do not interrogate the boundary conditions adequately, but they do give direction as to the possible relative merits of the techniques. We observe that application of Parshikov's technique of a Riemann-modified SPH at contact boundaries has merit and that further development to include the full stress tensor and higher dimensions seems justified. This technique retains the robustness and simplicity of the SPH pair-wise point integration scheme, as well as the inaccuracies of such a paradigm. Contact conditions in DPD give superior results, but are more difficult to implement in higher dimensions as geometrical considerations are involved. We have implemented contact boundaries in DPD in two dimensions [4] using Riemann-constrained linear moving-least squares interpolations and see encouraging results.

## REFERENCES

1.  Gingold, R.A. and Monaghan, J.J., "Smoothed Particle Hydrodynamics: Theory and Application to Non-Spherical Stars", Mon. Not. Roy. Astron. Soc., 181, 375, 1977..
2.  Lucy, L.B., "A Numerical Approach to the Testing of Fusion Process", The Astron. J., 82, 0103, 1977.
3.  Monaghan, J.J., On the Problem of Penetration in Particle Methods", J. Comput. Phys. 82, 1, 1989.
4.  Randles, P.W. and Libersky, L.D., "Boundary Conditions for a Dual Particle Method", Computers and Structures, 83, 1476, 2005.
5.  Parshikov, A.N. and Medin, S.A., "Smoothed Particle Hydrodynamics Using Interparticle Contact Algorithms", J. Comp. Phys, 180, 358, 2002.

CP845, Shock Compression of Condensed Matter - 2005,
edited by M. D. Furnish, M. Elert, T. P. Russell, and C. T. White
© 2006 American Institute of Physics 0-7354-0341-4/06/$23.00

# MEASUREMENTS OF THE DDT PROCESS IN EXPLODING BRIDGEWIRE DETONATORS

## Eric S. Martin[*], Keith A. Thomas[*], Steven A. Clarke[*], James E. Kennedy[†], D. Scott Stewart[‡]

[*]Los Alamos National Laboratory, DX-1, Los Alamos, NM 87545
[†]Hazards & Explosives Research & Education, LLC, Santa Fe, NM 87505
[‡]Mechanical and Industrial Engineering, University of Illinois, Urbana, IL 61801

**Abstract.** The deflagration-to-detonation transition (DDT) of low density (0.88 g/cc) PETN during exploding bridgewire (EBW) initiation has been studied using laser interferometry and streak photography. Cutback experiments using VISAR have confirmed a 1.0 mm run-distance to detonation in this low density PETN powder. In a detonation system using a combination of low and high density powders, an apparent center of initiation (COI) analysis of streak data has yielded a surprisingly similar result. This data suggested that a compaction of low density powder to near theoretical maximum density (TMD) may occur before the onset of detonation, which is consistent with work done previously [2-4]. These experiments show this is not the case and COI analysis reveals a non-ideal initial propagation front. Additionally, data show that although function time increases significantly with decreasing firing voltage, the apparent COI changes very little. This indicates that the detonation criterion is not dependent upon the rate of deflagration, but on a volume of material that must be burned in a confined space to create the critical pressure needed at the compaction front.

**Keywords:** DDT, PETN, exploding bridgewire, detonation, initiation
**PACS:** 82.33.Vx

## INTRODUCTION

Recent efforts have focused on characterizing the initiation properties of low density PETN. Interferometry measurements have been made in the past which reveal a distinct run-distance of about 1 mm in 50% TMD PETN [1]. Previous COI analysis on detonators using low density PETN as the initial pressing has shown the apparent center to be about the same location. Given the ideality of PETN detonation and the combination of low and high density powders in the detonators, there must be another explanation for this coincidence.

The evidence suggested a compaction front may be forming before the onset of detonation, which would be consistent with work done on other low density powders [2-4]. Therefore, streak camera analysis and accurate function time measurements were used to further characterize this initiation process, revealing new information about exploding bridgewire initiation of PETN.

## EXPERIMENTAL PROCEDURE

Expanding upon previous work by Kennedy [1], VISAR measurements were made on cutbacks at 4 different heights and two firing voltages. The cylinder heights used for the cutbacks were 0.655, 0.914, 1.181, and 1.450 mm. These 4 mm diameter cylinders were loaded with PETN to a target density of 0.88 g/cm$^3$ on top of a 1.5 mil diameter by 40 mil bridgewire. When using cylinders shorter than 0.655 mm, the target density could not be met reliably. The PETN used was recently measured by Fisher subsieve permeametry to be 12,000 cm$^2$/g.

The fireset used in all of these experiments uses a 0.5 uF capacitor and an N-MCT solid-state switch. The solid-state switch gives a jitter time near 2-3 ns and allows us to make accurate relative time measurements.

The interferometry system itself is a dual-leg Valyn VISAR system and differs from the system used by Kennedy [1]. Using this new setup resulted in limited useful data beyond early time velocity measurements, likely due to a weak return signal.

Streak camera measurements were made on the bare cutback pellets as well as fully assembled detonators. These detonators use a 2.5 mm by 4 mm diameter initial pressing of 0.88 g/cm$^3$ PETN and have a larger diameter, 1.0 mm high, 1.65 g/cm$^3$ PETN output pellet, as shown in Fig. 1. The PETN in this detonator uses a smaller specific surface of 3,650 cm$^2$/g. Additionally, the bridgewire is only 20 mils long. A shock-fluorescing salt, aluminum fluosilicate, was applied to the outermost 5-mil aluminum cup to capture shock arrival times across the surface with the streak camera. All measurements were made parallel to the bridgewire. These fully assembled detonators were fired at a variety of voltages such that comparisons of function time and COI could be made.

A center of initiation (COI) analysis technique was applied to all streak camera data. Given some constant detonation velocity from the observed surface, a Huygen's reconstruction is applied to the measured detonation wave across a limited fitting range to estimate where that wave appeared to have originated from. This point is a quantifiable and comparable reduction of complex waveform data and is described solely by an X and Y coordinate and the detonation velocity used. Point A in Fig. 1 exemplifies the COI location when observing the detonator surface. The ideas behind this type of analysis can be traced back to the optics of explosives by Busco [5] and have been embellished by others [6].

## RESULTS AND DISCUSSION

The velocity measurements obtained with the VISAR corresponds very well the the expected C-J

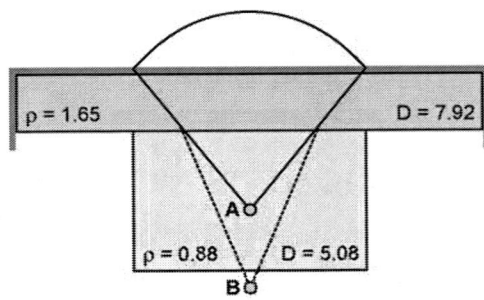

**Figure 1.** Point A is the calculated COI from the detonator surface. Point B is the refracted COI due to the lower detonation velocity in the initial pressing.

pressure of ~7 GPa for 50% TMD PETN. At both firing voltages of 1500 V and 2500 V, there is an obvious DDT transition between 0.914 mm and 1.181 mm above the bridgewire. Since the few VISAR measurements made did not improve upon those done by Kennedy [1], they are not shown.

Detonators using low density PETN in the initial pressing have a calculated COI of 1.00±0.08 mm above the bridgewire. This puts the COI in about the same place as that measured by VISAR. However, COI analysis assumes a constant detonation velocity through the entire system. The system observed uses two different pressing densities in the initial pressing and in the output pellet. When an appropriate correction for refraction is made to the COI using Snell's Law, a physically impossible COI is calculated. (Fig. 1) Based upon previous work by Luebcke [2], and McAfee [4], it was thought that a higher density plug may be forming and then detonating in the

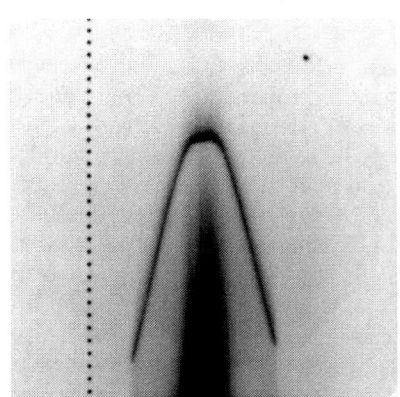

**Figure 2.** Streak of 0.655 mm cutback at 2500 volts.

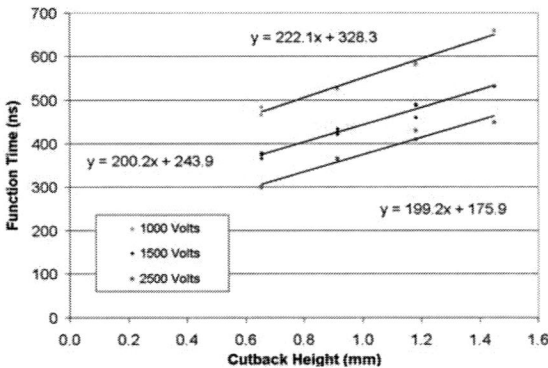

**Figure 3.** Streak camera measurements of function time.

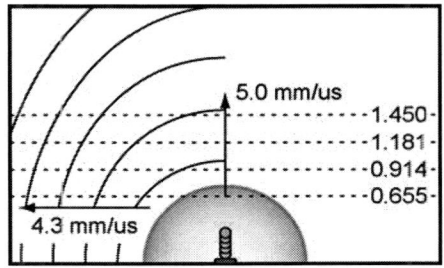

**Figure 5.** Small migration of the COI is likely due to differences in detonation velocities.

initial pressing before detonation. To answer this question, streak camera measurements were made in the cutbacks in order to get an idea of early time detonation wave formation.

A sample streak image of the shortest cutback at 0.655 mm is shown in Fig. 2. Note the nose of the breakout is fairly ragged indicating the burning front is not smooth. A COI analysis could not be made on the shortest cutbacks because of this. The edges of the breakout clearly show a constant detonation of about 4.3 mm/us. This compares very well with the measurements made by Blackburn and Reithel across a clear bridged header [7].

The shock arrival times for the cutbacks at various voltages are shown in Fig. 3. The inverse of the slope yields a velocity of 5.0 mm/us on axis with a hard-fire, with a velocity of 4.2 mm/us when fired near threshold. This corresponds very well to the expected velocity of 5.08 mm/us for 0.88 g/cm³

PETN [8]. These data indicate that at the range of heights examined in the cutbacks, there is no change in density of the initial pressing prior to its detonation.

COI calculations are shown in Fig. 4. At constant voltage, the COI increases slightly for increasing heights of cutback. This could be due to the difference of detonation velocities on axis versus the velocity perpendicular to the axis. At this scale, curvature in the detonation front may be affecting velocity or losses might be propagating from the bridged header. Either way, this results in an elliptical waveshape forming in the initial pressing, causing the COI to migrate slightly upwards when observing increasing heights. This is shown pictorially in Fig. 5.

Before interpreting the differences between threshold and hard-fire voltages, Fig. 6 should be considered. The function time curve shown is typical for an EBW detonator. The right side of the

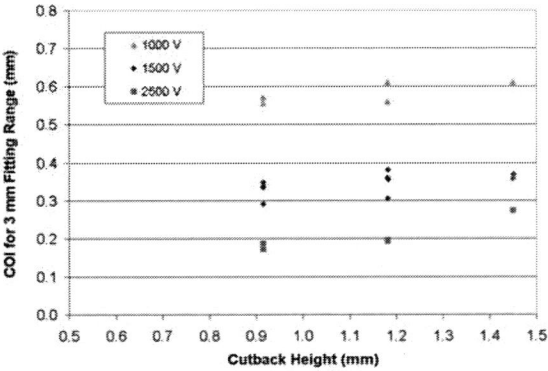

**Figure 4.** Apparent COI calculations for the cutbacks.

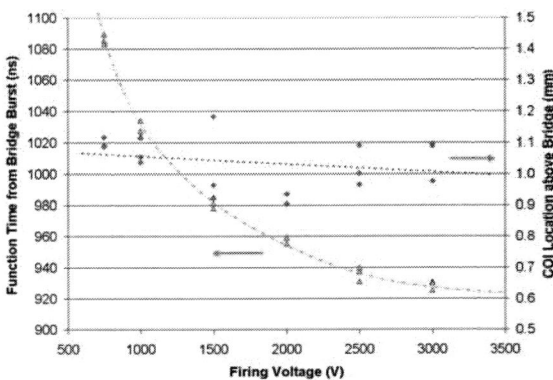

**Figure 6.** The detonator wave shape changes very little for a wide range of firing voltages.

graph is the calculated COI scaled to the function time by the detonation velocity used in the calculation. Clearly, while the function time changes about 200 ns from threshold to hard-fire, the average COI changes less than 0.1 mm. The difference seen at early time nearly disappears in the full detonator assembly. The significance of this should not be understated. Across a wide range of voltages, the output of the detonator remains nearly the same. This lack of variation indicates that initiation of low density PETN in exploding bridgewires may be more a function of material burned in a confined space, rather than a rate of deflagration.

The surprisingly larger range of COI values calculated in the initial pressing from Fig. 5 is likely due to deviations in the initial burn front. It is entirely feasible that the amount of energy delivered to the bridgewire affects the shape of that burn front. Fig. 7 depicts how these detonation fronts may look based upon COI calculations at the 1.450 mm surface. The difference in COI could be due to a flatter initial burn front at 2500 V, perhaps indicating a larger number of hot spots around the bridgewire. Further experimental and modeling work will be pursued to address some of the questions raised here.

## CONCLUSIONS

New VISAR shots confirm a run-distance of 1 mm at 1500 and 2500 volts. Streak records indicate there is little change in the accepted 50% TMD PETN detonation velocity throughout the range of pressing lengths examined. Accurate time measurements with a low jitter fireset verify a 5.0 mm/us velocity. If there is a higher density plug

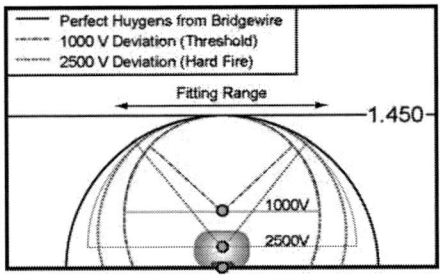

**Figure 7.** Shown is a depiction of how a non-ideal wave propagation could yield the range of COIs calculated.

forming, it must be in a radius less than the smallest pressing of 0.655 mm.

While observing distinct differences in wave shape due to firing voltage at early time in the cutbacks, those variations are minimized in the full detonation train. The lack of variation for a wide range of function times suggests that initiation may be more a function of a critical volume of material burned, rather than the rate of deflagration.

## ACKNOWLEDGEMENTS

This work was supported by the U.S. Department of Energy, Los Alamos National Laboratory. Special thanks to Michael Martinez and Dennis Jaramillo for their hard work executing these experiments.

## REFERENCES

1.  Kennedy, J.E. et al., "Mechanisms of Exploding Bridgewire and Direct Laser Initiation of Low-Density PETN," Proc., 29th Intl. Pyrotechnics Seminar, pp. 781-785, July 2002
2.  Luebcke, P.E., Dickson, P.M., and Field, J.E., "Deflagration-to-Detonation Transition in Granular Pentaerythritol Tetranitrate," J. Appl. Phys., 79 (7), pp. 3499-3503, 1996
3.  Gifford, M.J., Luebcke, P.E., and Field, J.E., "A New Mechanism for Deflagration-to-Detonation in Porous Granular Explosives," J. Appl. Phys., 86 (3), pp. 1749-1753, 1999
4.  McAfee, J.M., Asay, B.W., Campbell, W., and Ramsay, J.B., "Deflagration to Detonation Transition in Granular HMX," Ninth (Int.) Detonation Symposium Vol 1, pp. 265-279, 1989
5.  Busco, M., "Optical Properties of Detonation Waves (Optics of Explosives)," Proceedings of the Fifth (Int.) Detonation Symposium, pp. 513-522, 1970
6.  Hill, Larry and Forest, Charles, "SE-1 Detonator Characterization Tests," Los Alamos Memo, DX-10-95-207, 1995
7.  Blackburn, James and Reithel, Robert, "Exploding Wire Detonators: Sweeping-Image Photographs of the Exploding Bridgewire Initiation of PETN," Exploding Wires Vol. 3, W.G. Chace and H. Moore, eds., Plenum Press, N.Y. , pp. 153-173, 1964
8.  Hornig, H.C., Lee, E.L., Finger, M., and Kurrle, K.E., "Equation of State of Detonation Products," in Proc. 5th Symp. (Int.) on Detonation, Office of Naval Research, Washington, DC, ACR-184, pp. 503-512, 1970

CP845, *Shock Compression of Condensed Matter - 2005*,
edited by M. D. Furnish, M. Elert, T. P. Russell, and C. T. White
2006 American Institute of Physics 0-7354-0341-4/06/$23.00

# IMPACT INITIATION OF RODS OF PRESSED POLYTETRAFLUOROETHYLENE (PTFE) AND ALUMINUM POWDERS

## Willis Mock, Jr. and William H. Holt

*Naval Surface Warfare Center, Dahlgren Division, Dahlgren, VA 22448-5100*

**Abstract.** A gas gun has been used to investigate the shock initiation of rods of a mixture of 74 wt% PTFE and 26 wt% aluminum powders. The rods were sabot-launched into 4340 steel anvils at impact velocities ranging from 104 to 963 m/s. A framing camera was used to observe the time sequence of events. At low velocity, no initiation occurred. Above an initiation threshold, the initiation time dropped abruptly from 56 $\mu$s just above threshold to 4 $\mu$s at the highest impact velocity. Several high velocity experiments were performed for pure PTFE material for comparison with the PTFE/Al rods.

**Keywords:** PTFE/Al, PTFE, shock initiation, gas gun, initiation threshold, framing camera
**PACS:** 47.40.-x, 47.40.Nm, 61.41.+e, 82.40.Fp

## INTRODUCTION

The purpose of the present study is to investigate the impact initiation of solid rods of a composite mixture of PTFE and aluminum powders. The average initial particle sizes were 28 $\mu$m and 9 $\mu$m for the PTFE and Al, respectively. The stoichiometric mixture of 74 wt% PTFE and 26 wt% Al was chosen. Aluminum powder was selected due to its large heat of reaction when forming $AlF_3$ with PTFE [1]. The powders were pressed and sintered into a billet with a theoretical maximum density of 2.27 g/cm$^3$ [2], and then machined into 50.8 mm long by 7.59 mm diameter rods [3] for the present experiments. Figure 1 is a photomicrograph of unshocked PTFE/Al material. In addition, several high velocity experiments were performed for pure PTFE rod material [4] for comparison with the PTFE/Al rod experiments.

## EXPERIMENTAL

A 40 mm bore gas gun was used to perform the impact experiments [5]. A series of eleven experiments was performed, nine with PTFE/Al rods and two with pure PTFE rods. For all of the experiments except one, the rod was impacted directly onto a 4340 steel anvil. The anvils (50

mm diameter x 25.4 mm thick) had a hardness of RC 38-40. The impact surface of the anvils was lapped to a final finish with 9 micron lapping film. Figure 2 is a schematic for a direct impact experiment. A rod was secured by epoxy into a 10 mm deep recess in the front of a 6061-T6 aluminum sabot for launch. The target assembly consisted of the steel anvil, two polycarbonate end pieces (not labeled in Fig. 2), and a surrounding polymethylmethacrylate (PMMA) tube (76 mm inside diameter x 6.4 mm thick). The barrel and target assembly were evacuated to 50-100 mTorr vacuum before the shot. Sabot velocity was measured just before impact to an accuracy of approximately 1% with three charged pins (not shown in Fig.2) in the side of the barrel.

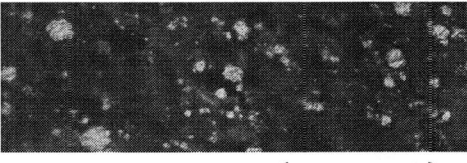

**Figure 1.** Photomicrograph of unshocked PTFE/Al material. The Al particles (light regions) are imbedded in the PTFE matrix (dark regions).

For the direct impact configuration, the sabot is in contact with the back of the PTFE/Al rod for the entire time after rod impact occurs. To ensure that this does not alter the rod initiation time, one impact experiment was performed for the reverse ballistic configuration shown in Fig. 3. For this configuration, the back of the PTFE/Al rod is free. A 4340 steel anvil (35.6 mm diameter x 25.4 mm thick) was launched into the rod that was held in position by a small amount of epoxy on the end of a 1.6 mm diameter steel support rod.

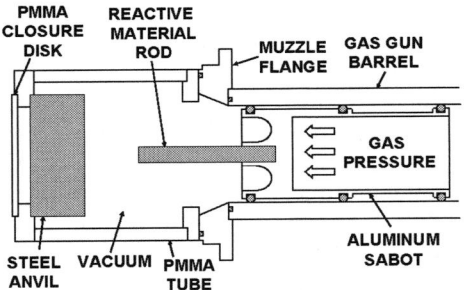

**Figure 2.** Direct impact configuration.

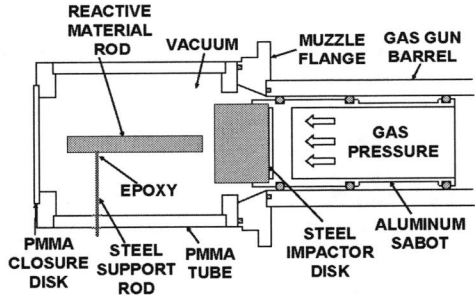

**Figure 3.** Reverse ballistic configuration.

A 68 frame DSR Hadland framing camera [6] with a frame rate up to 500,000 frames/s was used to observe the time sequence of events. The camera controlled a high intensity flash lamp that illuminated the target region during impact. The camera, which was triggered with a sabot velocity pin signal, was controlled from a PC through a fiber optic cable. The camera exposure time was 250 ns for all the experiments. The time between frames was 8, 4, or 2 μs depending upon the impact velocity.

## ANALYSIS AND DISCUSSION

Table 1 summarizes the experimental results. The impact stress range was 4.5-63.3 kbar for the PTFE/Al rods. No initiation occurred (as indicated by no observed light output) for the 4.5 kbar

**Table 1.** Summary of experimental results. Impact stresses calculated from data in References 7-12.

| Experiment No. | Material | Impact Velocity (m/s) | Calculated Impact Stress (kbar) | Observation of First Light After Impact (μs) |
|---|---|---|---|---|
| 1 | PTFE/Al | 104 | 4.5 | below threshold |
| 2 | PTFE/Al | 172 | 7.8 | 56 |
| 3 | PTFE/Al | 207 | 9.6 | 32 |
| 4 | PTFE/Al | 222 | 10.4 | 28 |
| 5 | PTFE/Al | 303 | 15.0 | 18 |
| 6 | PTFE/Al | 466 | 25.1 | 12 |
| 7 | PTFE/Al | 614 | 35.3 | 8 |
| 8 | PTFE/Al | 765 | 46.0 | 6 |
| 9 | PTFE/Al | 963 | 63.3 | 4 |
| 10 | PTFE | 464 | 21.9 | - |
| 11 | PTFE | 773 | 42.0 | - |

impact stress experiment. For the higher impact stresses, initiation was observed which decreased from 56 to 4 μs as the impact stress increased. First light was observed at discrete points, suggesting that the light may be occurring at cracks in the material. It is not known if first light is due to heating or reaction effects in the rod material.

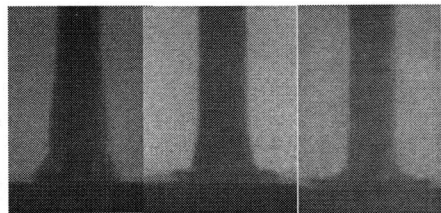

**Figure 4.** Shape of PTFE/Al rod at observation of first light for the 7.8, 10.4, and 35.3 kbar experiments (left to right in figure). The rod moves downward and impacts the steel anvil. First light, due to its faintness, is not visible in these small frames. It was taken as the first indication of light in a frame if it was visible in at least several consecutive frames or grew in brightness.

Figure 4 shows the shape of the PTFE/Al rod for three selected experiments at the observation of first light. For the 7.8 kbar experiment, considerable rod deformation can be observed away from the impact surface, suggesting that rod strength effects are important. For the 10.4 kbar impact stress, less deformation is observed away from the impact surface, and for the 35.3 kbar experiment rod deformation occurs only at the impact surface. Figure 5 shows the evolution of rod deformation and emitted light for the 10.4 kbar experiment. This sequence of photographs shows

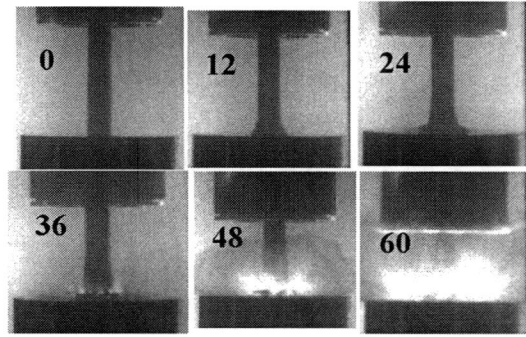

**Figure 5.** Evolution of PTFE/Al rod impacting steel anvil. Impact stress is 10.4 kbar. Impact occurs at 0 μs. The time after impact in μs is given in each frame. A cloud begins to form at 36 μs. At 60 μs light emission is occurring from the cloud impact with the sabot.

rod shape at three times before initiation (0, 12 and 24 μs) and at three times after initiation (36, 48, and 60 μs). First light was observed at 28 μs after impact, and grows from discrete points into a single glowing source at the impact interface. An expansion cloud forms after initiation and travels from the impact interface towards the sabot at an estimated velocity of about 1 km/s. The observation that the expansion cloud glows on impact with the sabot suggests that it consists partly of particles of original material that react on impact with the sabot.

Figure 6 is a plot of first light after impact versus calculated impact stress for the PTFE/Al rod experiments. The dashed vertical line at 7.3 kbar is the threshold impact stress. This value is the average stress for two impact experiments in which no initiation occurred [13]. In addition, the data point represented by the square just above threshold (7.5 kbar, 70 μs) is the average of several experiments [13]. Figure 6 indicates that for these experiments no initiation of the PTFE/Al rods occurs if the impact stress is below 7.3 kbar. Just above threshold the initiation time drops abruptly from about 70 μs to about 20 μs at 15 kbar. It then decreases more slowly to 4 μs at 63.3 kbar. The nine data points above threshold were fitted with the curve $T(\sigma-7.3)^{0.5} = 48$, where T is in μs and σ is in kbar.

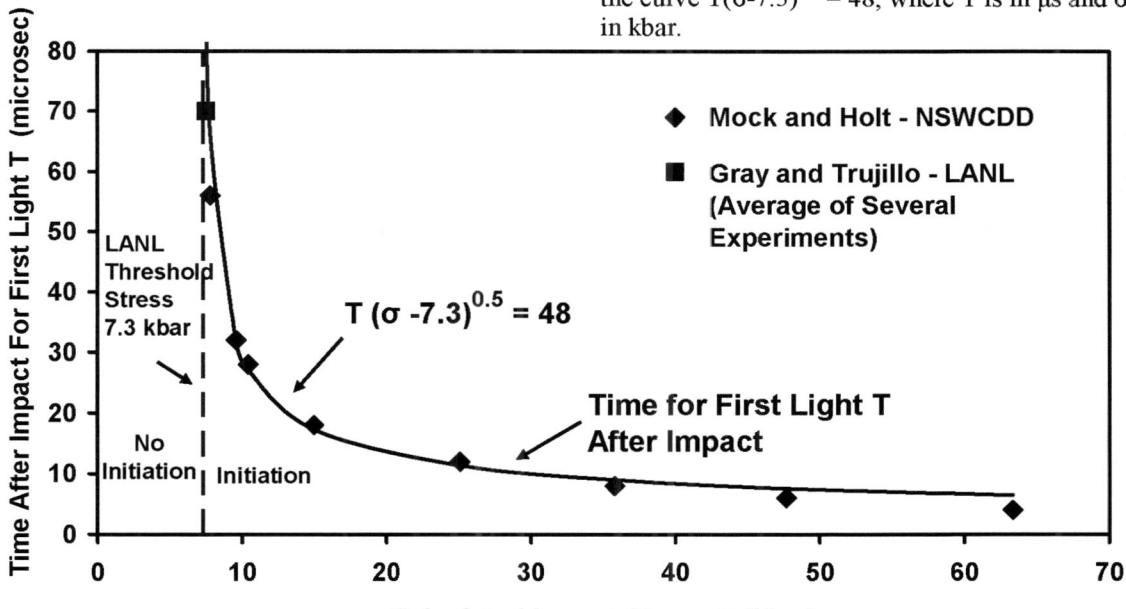

**Figure 6.** Time after impact for first light.

The experimental point at 9.6 kbar and 32 μs was determined using the reverse ballistic configuration. As can be observed from Fig. 6, this point falls on the curve with the other data points that were obtained using the direct impact configuration. This agreement indicates that the observation of first light was not influenced by launching the rods at the steel anvil with a sabot. Another indication is that the back of the PTFE/Al rod could be observed until it was obscured by light at 72 μs after impact. During this time the back of the rod did not move, which suggests that if an impact compression wave did propagate to the back of the rod, it did not cause the back surface to move during the time of interest.

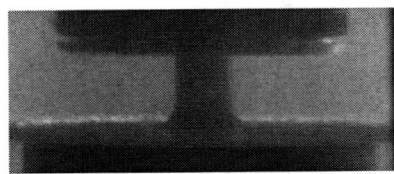

**Figure 7.** Observation of points of light on the top edge of the mushroomed portion of the PTFE rod at 30 μs after impact for the 42.0 kbar impact experiment. The sabot with PTFE rod moves downward and impacts the steel anvil shown at the bottom in the figure. Note that the mushroomed PTFE material extends beyond the 35.6 mm diameter steel anvil.

The pure PTFE material for Experiments 10 and 11 in Table 1 showed more extensive radial flow without obvious brittle fracture compared with the PTFE/Al material. Small points of light were observed on the edge of the mushroomed portion of the PTFE rods beginning at about 36 and 20 μs after impact, for the 21.9 and 42.0 kbar impact experiments, respectively. This suggests heating or dissociation of the mushroomed rod material, or the impact of the mushroomed material with the inside of the PMMA tube. (The distance from the edge of the steel anvil to the inside surface of the PMMA tube is 13 mm.) Figure 7 shows this phenomenon for the 42.0 kbar experiment. Approximately 60% of the rod is mushroomed. It is not known if this light phenomenon would occur if the rod were not being carried by the sabot.

**ACKNOWLEDGEMENT**

This work was funded by Judah Goldwasser, Program Officer for Undersea Warheads and Energetic Materials, of the Office of Naval Research.

**REFERENCES**

1. CRC Handbook of Chemistry and Physics, 75th Edition, D. R. Lide, Editor-in-Chief, CRC Press, 1995, p. 5-5.
2. ATK Alliant TechSystems, ATK Thiokol Propulsion, Brigham City, UT.
3. DE Technologies Inc., 3620 Horizon Dr., King of Prussia, PA.
4. Part ZTR-05, Small Parts, Inc.,Miami Lakes, FL.
5. W. Mock, Jr. and W. H. Holt, Report NSWC/DL TR-3473, Naval Surface Weapons Center, Dahlgren, VA, July 1976.
6. DRS Imaging , 138 Bauer Dr., Oakland, NJ.
7. PTFE/Al Hugoniot $\sigma = 43.3u + 44.3u^2$ with $\sigma$ in kbar and u in mm/μs provided by W. D. Reinhart and L. C. Chhabildas, Sandia National Laboratories, Albuquerque, NM (private communication).
8. The calculated 4340 steel Hugoniot (with RC40 hardness) is $\sigma = 461u$ for $u \leq 0.0426$ and $\sigma = 4.42 + 352u + 105u^2$ for $u \geq 0.0426$. The 4340 steel Hugoniot was calculated using: the measured HEL of 19.6 kbar [9]; the U-u relationship U = 4.578 + 1.33u, density 7.81 gm/cm$^3$, the 801 kbar shear modulus [10]; the 1655 kbar bulk modulus and Poisson's ratio 0.287 [11]. The PTFE Hugoniot was calculated using U = 1.682 + 1.819u [12], and the measured PTFE density 2.17 g/cm$^3$.
9. O. E. Jones, F. W. Neilson, and W. B. Benedict, J. Appl. Phys. **33**, 3224 (1962).
10. D. J. Steinberg, Report UCRL-MA-106439, Lawrence Livermore National Laboratory, Livermore, CA, February 1996.
11. Metals Handbook, Desk Edition, edited by H. E. Boyer and T. L. Gall, American Society of Metals, 1985, p. 2-16.
12. R. G. McQueen, S. P. Marsh, J. W. Carter, J. N. Fritz, and W. J. Carter, in *High-Velocity Impact Phenomena*, edited by Ray Kinslow, Academic Press, 1970, p. 371.
13. Experiments with PTFE/Al rods from the same batch used for the present experiments were performed by G. T. Gray III and C. P. Trujillo, Los Alamos National Laboratory, Los Alamos, NM (private communication).

CP845, *Shock Compression of Condensed Matter - 2005*,
edited by M. D. Furnish, M. Elert, T. P. Russell, and C. T. White
© 2006 American Institute of Physics 0-7354-0341-4/06/$23.00

# UNDERSTANDING THE MECHANISMS LEADING TO GAS PERMEATION IN THERMALLY DAMAGED PBX 9501

## G. R. Parker, P. M. Dickson, B. W Asay, L. B. Smilowitz, B. F. Henson and W. L. Perry

*P.O. Box 1663, MS C920, Los Alamos National Laboratory, Los Alamos, NM 87545*

**Abstract.** We present data that indicate that thermally damaged PBX 9501 is substantially more permeable than the pristine material and that this may have a significant effect on the pre-ignition slow cook-off process, as well as the post-ignition flame spread process. Experiments indicate that the mechanism responsible for the formation of interconnected matrix porosity is likely dominated by nitroplasticizer decomposition in the early stages of the permeability evolution followed by secondary, slower HMX decomposition.

**Keywords:** Thermal damage, explosives, permeability, cook-off
**PACS:** 82.33.Vx, 82.33.Ln

## INTRODUCTION

Understanding cook-off behavior of explosives is important and *a priori* prediction through numerical modeling is highly desirable. In the past, a considerable amount of experimental and modeling work has been done to investigate the processes involved in the slow cook-off of PBX 9501 (comprised by % weight; 95% HMX crystals, 2.5% Estane 5703, 1.25% BDNPA nitroplasticizer and 1.25% BDNPF nitroplasticizer). However, in much of the early work, little attention was paid to the possible effects of gas transport within the system. The work contained in this article addresses gas transport through measurement of Darcy flow to obtain the specific permeability of PBX 9501 damaged at elevated temperatures.

## BACKGROUND

Effective reduced Arrhenius chemistry schemes have been used to predict time-to-explosion [1,2], and more recently modified [3] to allow for prediction of reaction location, which is recognized

as an important parameter in determining reaction violence, in cook-off models for PBX-9501.

This led to the following scheme for HMX decomposition:

$$HMX_\beta \leftrightarrow HMX_\delta \qquad (R1)$$
$$HMX_\beta + HMX_\delta \leftrightarrow HMX_\delta + HMX_\delta \qquad (R2)$$
$$HMX_\delta \rightarrow GAS \qquad (R3)$$
$$HMX_\delta + GAS \rightarrow GAS\ PRODUCTS \qquad (R4),$$

where the gaseous products formed in R3 react auto-catalytically with remaining HMX to accelerate HMX decomposition in R4. R4 is an empirical approximation of the late exothermic chemistry responsible for thermal runaway during slow cook-off.

This scheme produced good agreement when used to model the LANL radial cook-off tests. However, these tests are confined but unsealed, and significant questions have been raised by other researchers regarding the effects of gas permeation. A number of tests have apparently indicated that gas-tight systems behave differently to unsealed systems, especially under slow cook-off conditions,

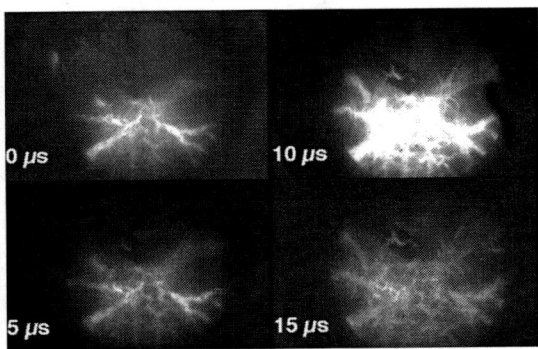

**Figure 1.** Post-ignition flame spread in thermally damaged PBX 9501, demonstrating flame penetration and convective burning [9].

which would suggest that the permeation of gas through the plastic bonded explosive (PBX) affects the response, either by transferring heat (modifying heat transport rates) or reactive species (modifying reaction rates), or both. Specifically, HMX decomposition products will be entrapped within a matrix of binder that is also decomposing at its own rate. If these gaseous species are able to permeate away from the R4 point of origin, through voids opened by binder decomposition, then the auto-catalysis will be reduced and decomposition rate will decrease as well. At the same time, these same permeating gases will accelerate HMX decomposition in surrounding material into which the gases permeate. This "reaction spreading" phenomenon has significant implications for modeling ignition location and violence of slow cook-off, as demonstrated by Zerkle et al. [4].

In addition to the pre-ignition influence that gas permeation may have, there are other post-ignition implications. For a material to be permeable, it must have a network of interconnected porosity. Therefore knowing a material to be permeable allows one to infer that it also has gas-flow pathways. While pristine PBX 9501 has generally been regarded as approximately impermeable, it is well known that its porosity, and thus possibly its permeability, increase as physical and chemical changes occur at elevated temperatures [5]. Experimental results indicate that flame spread into defects in HE is dependant on both defect aperture and pressure [6-8]. As a confined explosive burns and pressure increases, a critical pressure may be reached at which the flames can penetrate the bulk

and transition the burn from laminar to convective. As observed by Dickson et al. [9] in their mechanically confined cook-off (MCCO) experiments, thermally damaged PBX 9501 can develop morphology conducive to significant flame spread (Fig. 1), increased reaction rate, and ultimately increased violence. Extending this hypothesis, it is reasonable to speculate that deflagration-to-detonation transition (DDT) phenomena may occur in larger, mechanically- or inertially-confined masses of damaged PBX 9501.

## EXPERIMENTAL

A permeameter was constructed that permitted quantitative dynamic measurement of gas-flow by recording the pressure decay rate of a constant volume gas reservoir as gas permeated through a specimen and parametric study was performed to measure the permeability of thermally damaged PBX 9501 [10, 11].

Samples mounted in pristine condition were subsequently heated to 185˚C, pressurized and held at temperature for periods of 2-20 hours. Upon heating, pressure would begin to drop at an increasing rate for approximately 2-3 hours, after which there was an observed exponential pressure decay. Data from this regime, where pressure decay rate is quasi steady, were used to calculate permeability of radially confined PBX 9501 undergoing thermal damage (Fig. 2). We found that permeability increases initially by 3 orders of magnitude to $3 \times 10^{-16}$ m$^2$ after 2–3 hours at 185˚C (Fig. 3). During the quasi-steady regime, permeability continues to rise, though at a lower rate. This indicates a secondary mechanism of long-duration damage.

Another series of experiments was performed to address concerns that the interface between two machined sections may introduce a preferential gas-flow pathway. Pristine samples of PBX 9501 were machined into half-cylinder sections and were then potted together so that the flat faces were touching. One of the samples in this series was potted with thermocouples (125 $\mu$m bead) taped between the faces of the midplane section (Fig. 4).

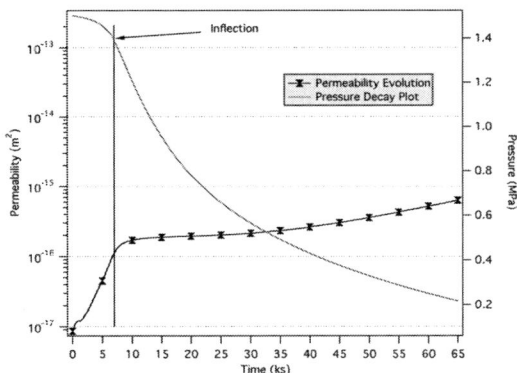

**Figure 2.** Typical pressure decay curve and accompanying permeability history for PBX 9501 damaged and soaked at 185°C.

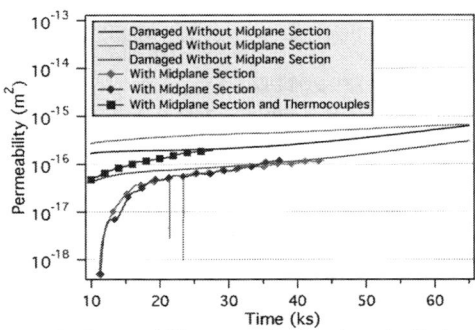

**Figure 3.** Permeability records showing similarity between non-sectioned and midplane-section samples.

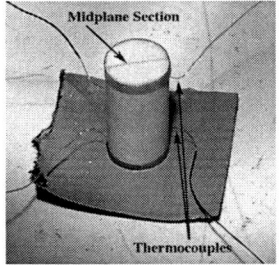

**Figure 4.** Preparation of a midplane-sectioned permeability specimen (before potting in specimen holder).

The other two samples were potted without thermocouples between the faces. Results indicate a midplane interface had a significant effect initially (before heating), but little-to-no effect on the overall permeability measurement at elevated temperatures, where permeabilities were close to those of non-sectioned samples (Fig. 3). It would appear likely that the compressive effect of the volume expansion due to the HMX $\beta$ - $\delta$ phase transition and normal thermal expansion, coupled with binder melt, was sufficient enough to "seal" the interface to the extent that it was no longer a preferential flowpath and that the observed permeability therefore resulted from matrix flow.

## DISCUSSION

Upon examination of the pressure decay curves from experiments where the specimen had an applied pressure gradient throughout the entire heating profile, there is typically a point of inflection where the curve transitions from an increasing to decreasing rate (Fig. 2). The inflection point separates the permeability evolution into two regimes: the first 2–3 hours where permeability increases by 3 orders of magnitudes and a quasi-steady regime for the remainder of the history, which typically ran for 15 additional hours.

During the first 3 hours at 185°C, a network of interconnected porosity throughout the specimen was established progressively allowing gas to permeate at an increasing rate indicating a mechanism of solid-to-gas decomposition at work. Upon examination of Thermal Gravimetric Analysis (TGA) of HMX alongside the binder used in PBX 9501, it is clear the nitroplasticizer (BDNPA-F) is the only constituent of PBX 9501 that would decompose rapidly enough, at these temperatures, to explain the observed permeability (Fig. 5).

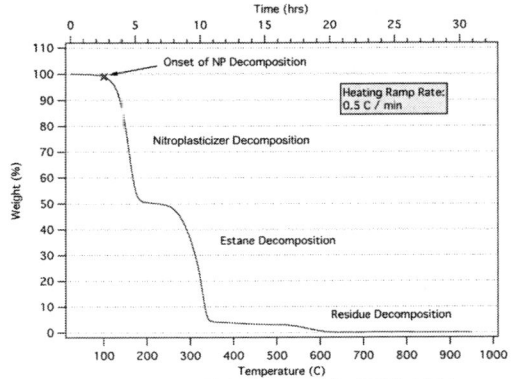

**Figure 5.** TGA of binder used in PBX 9501.

Above 120°C, nitroplasticizer decomposition is well underway and over the duration of 3 hours most of it will have decomposed leaving voids in its place. Decomposition nucleation sites would be ubiquitous throughout the bulk of the PBX leading to countless bubbles. As decomposition continues, pressure from the evolving gases would cause the bubbles to grow and join, ultimately forming the interconnected network. Once nitroplasticizer decomposition neared completion, the rate of permeability increase would slow and eventually become quasi-steady.

After the initial permeability regime, decomposing HMX is the only component remaining in the PBX that could lead to further development of a permeable porosity network. As demonstrated by Behrens [12], HMX crystals decompose from their surfaces inward, leaving a highly porous residue layer. Over time the slower HMX surface regression would contribute to the overall extent of the porosity network, thus also permeability of the PBX.

## CONCLUSIONS

We have presented data that indicate that thermally damaged PBX 9501 is substantially more permeable than the pristine material and that this may have a significant effect on the pre-ignition slow cook-off process, as well as the post-ignition flame spread process.

## ACKNOWLEDGEMENTS

This work was funded by Los Alamos HE Science and Surety Programs. We thank Lloyd Davis and Mary Sandstrom for TGA.

## REFERENCES

1. McGuire, R. R. & Tarver, C. M., "Chemical decomposition models for the thermal explosion of confined HMX, TATB, RDX, and TNT explosives". In *Proc. 7th Int. Symp. on Detonation, Annapolis, MD*, **1981**. pp. 56–64. Arlington, VA: Office of Naval Research.
2. Tarver, C. M., McGuire, R. R., Lee, E. L., Wrenn, E. W. & Brein, K. R., "The thermal decomposition of explosives with full containment in one-dimensional geometries". In *Proc. 17th Int. Symp. on Combustion, Pittsburgh, PA*, **1978**. p. 1407. Pittsburgh, PA: The Combustion Institute.
3. Dickson, P. M., Asay, B. W, Henson, B. F., Fugard, C. S. & Wong, J., "Measurement of phase change and thermal decomposition kinetics using the Los Alamos radial cookoff test". In *Shock Compression of Condensed Matter—1999*. AIP Conference Proceedings, vol. 505, pp. 837–840. Bethlehem, PA: American Institute of Physics.
4. Zerkle, D. K. and Luck, L. B., "Modeling Cook-off of PBX 9501 with Porous Flow and Contact Resistance", *39th Joint Army Navy NASA Air Force (JANNAF) Combustion Meeting*, Colorado Springs, CO, USA, 1-5 December **2003**.
5. Parker, G., Peterson, P., Asay, B., Dickson, P., Perry, W., Henson, B., Smilowitz, L. & Oldenborg, M., " Examination of Morphological Changes that Affect Gas Permeation Through Thermally Damaged Explosives," Propellants, Explosives and Pyrotechnics, 29 **2004** No. 5.
6. Son, S. F., "The Combustion of Explosives", *AIP Conference Proceedings: Shock Compression of Condensed Matter*, Atlanta, GA, USA, 24-29 June **2001**, 1059-1064.
7. Berghout, H. L., Son, S. F., Slidmore, C. B., Idar, D. J., and Asay, B. W, "Combustion of Damaged PBX 9501 Explosive," *Therm. Acta* **2002**, *384*, 261-277.
8. Berghout, H. L., Son, S. F. and Asay, B. W, "Measurement of Convective Burn Rates in Gaps of PBX 9501". *AIP Conference Proceedings: Shock Compression of Condensed Matter*, Snowbird, UT, USA, 27 June – 2 July **1999**.
9. Dickson, P.M., Asay, B.W, Henson, B.F. & Smilwitz, L.B., "Thermal cookoff response of confined PBX 9501," Proceedings of the Royal Society London, A, 460, 2052, **2004**.
10. Parker, G. R., Asay, B.W., Dickson, P.M., Henson, B.F. and Smilowitz, L.B., "Effect of Thermal Damage on the Permeability of PBX 9501," *Shock Compression of Condensed Matter, American Physical Society Topical Conference*, Portland, OR, July 20-25, **2003**.
11. Asay, B. W, Parker, G., Dickson, P., Henson, B. & Smilowitz, L., "Dynamic Measurement of the Permeability of an Explosive Undergoing Thermal Damage," Journal of Energetic Materials, 22(1), **2004**.
12. Behrens, R., "Thermal Decomposition of HMX: Morphological and Chemical Changes Induced at Slow Decomposition Rates," In *Proc. 12th Int. Symp. on Detonation, San Diego, CA*, **2002**. Arlington, VA: Office of Naval Research.

CP845, *Shock Compression of Condensed Matter - 2005*,
edited by M. D. Furnish, M. Elert, T. P. Russell, and C. T. White
2006 American Institute of Physics 0-7354-0341-4/06/$23.00

# PROMPT REACTION OF ALUMINUM IN DETONATING EXPLOSIVES

## H. W. Sandusky and R. H. Granholm

*NAVSEA Indian Head Division, Indian Head MD 20640*

**Abstract.** The potential of aluminum (Al) reaction to boost detonation energy has been studied for decades, most recently spurred by the availability of nanometer-sized particles. A literature review is consistent with results from the small-scale shock reactivity test (SSRT). In this test, <1/2-g samples in confinement are shock loaded on one end, and the output at the other end dents a soft witness block. For samples in which 0.3 g of cyclotetramethylenetetranitramine (HMX) was mixed with 8 μm Al, the deepest dent occurred at 15% Al. When ammonium perchlorate (AP) was mixed with the same Al, the increased dents were consistent with changes in detonation velocity previously reported on similar mixtures. One outcome of this study is a new interpretation for the participation of Al in large scale gap tests on plastic-bonded explosives, which was discussed by Bernecker at this meeting in 1987.

**Keywords:** Shock reaction, aluminum, gap tests
**PACS:** 82.33.Vx, 82.40.Fp.

## INTRODUCTION

The consensus in the literature is that Al at least partially reacts within the detonation zone of pure explosives, increasing both detonation velocity (D) and pressure, but is delayed in explosives with binders. Al reaction forms a solid product, thereby reducing the moles of gas products, but has such high thermal energy that it contributes to pressure by heating the gas products.

Price et al. [1] used an equation attributed to Kamlet for D that is proportional to the ¼ power of the product of gas moles/mass of explosive, heat of reaction, and mass fraction of gas products. This assumes reaction occurs within the detonation zone, which is referred to as ideal conditions. Figure 11 of reference 1 shows for beds of AP and Al packed at 55% of theoretical maximum density (TMD) that D peaks at ~15 %weight (%wt) Al according to the equation and at ~10 %wt in experiments on 95/5, 90/10, and 85/15 mixtures of 7-9 μm AP and 7 μm Al. Measurements were made at various charge diameters and extrapolated to infinite diameter so that they approached the ideal conditions in the calculations. The figure also shows that the measurements for D are less than the calculations as %wt Al increases. In similar measurements [2] for trinitrotoluene (TNT) mixtures at TMD, there was no effect for up to 32.2 %wt Al with large particle size (δ), <149 μm, whereas adding NaCl depresses D. This indicates that the increased energy from reacting some of the Al compensates for the dilution by the rest.

Tao et al. [3] studied the prompt reaction of both pentaerythritol tetranitrate (PETN) and TNT when mixed with 5 to 20 %wt of either 5 or 18 μm spherical Al. With PETN, both particle sizes of Al reacted within 1.5 μs, increasing the prompt energy release by 18 to 22% compared to PETN alone. With TNT, 5 to 10 %wt of 5 μm Al reacts promptly, while 10 %wt of 18 μm Al may not.

Including binders, however, inhibits Al reaction. Finger et al. [4] showed in cylinder expansion tests on mixtures of 85/15 HMX/Viton with up to 15 %volume (%vol) of 5 μm Al that Al reaction began contributing to wall energy 4 μs after the detonation zone. In the plate dent tests of Smith [5], replacing some of the Kel-F with 2 to 20 μm atomized Al or 1 to 3 μm Al in mixtures with cyclotrimethylenetrinitramine (RDX) did not affect dent depth over the range of 65 to 92 %vol RDX. At ~65 %vol RDX, a sample with flake (400 mesh) Al had a somewhat deeper dent, while a sample with an inert mix of LiF and BaF$_2$ had a somewhat shallower dent.

Nanometer-sized Al did not increase its reactivity in the detonation zone. Brousseau et al. [6] measured D for TNT, Composition B, HMX-based plastic-bonded explosives (PBXs), and ammonium nitrate/fuel oil (ANFO) with both micron-sized and nanometric Al. D was increased when nanometric Al was mixed with TNT but declined when mixed in Comp B, PBXs (even those with energetic binders), and ANFO. Lefrancois et al. [7] showed in cylinder tests no significant change in D for mixtures of RDX, AP, Al, and wax with 100 nm versus 5 μm Al.

## SMALL-SCALE EXPERIMENTS

The small-scale shock reactivity test (SSRT) measures the propensity of energetic materials to react within a microsecond timeframe from a strong shock, even for samples whose size is smaller than their critical diameter ($d_c$) for propagating steady detonation. The arrangement shown in Fig. 1 has a mild steel block with a central hole in which fits a RISI RP-80 detonator, an optional polymethyl methacrylate (PMMA) shock attenuator, and a 6.35-mm long sample. The downstream end of the sample contacts a soft Al witness block, which is dented by the combined shock from the detonator and sample reaction. All of the tests described in this report were without a PMMA gap so that the full output of the detonator is applied to the sample, thereby simulating a small volume of explosive within a detonating charge. The SSRT was characterized with various HMX mixtures, which provided a basis for comparison with results from a range of explosives [8].

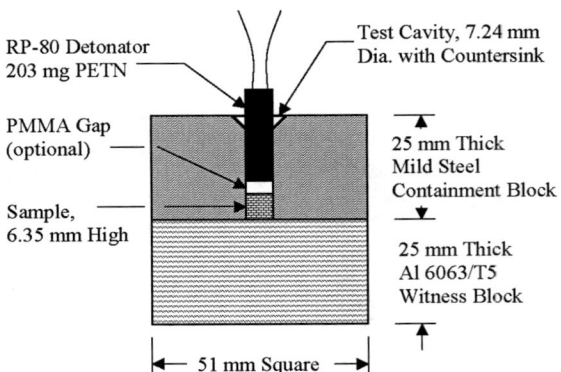

**Figure 1.** SSRT test arrangement.

Various amounts of Valimet H-5 Al with δ = 8 μm and inerts were blended into samples that were consistently 60 %vol of Class 1 HMX (~298 mg) or 200 μm AP (~308 mg). Most samples were porous beds (without liquid) so that the hot HMX or AP reaction products had direct contact with the additive. These results are plotted in Fig. 2 relative to %wt of additive in the total solid (additive plus HMX or AP), which does not include the liquid for hand mixes.

When adding Al to porous HMX, the dent increased to a peak at 15 %wt Al and then slowly decreased, but was still greater at 25 than at 10 %wt Al. An opposite but much smaller effect occurred from adding melamine and glass beads to porous HMX, and when HMX/glass blends were mixed with hydroxy-terminated polybutadiene (HTPB) to fill the voids. For these various inert additions, the dent decreased by the same amount as the solid additive increased.

HMX/HTPB without any Al produced a slightly higher dent than the HMX by itself; although, the difference is within the ±0.1 mm repeatability for the test. With a 90/10 HMX/Al blend hand mixed with HTPB, the dent depth decreased by the same amount as when blending inerts with HMX. When a 80/20 HMX/Al blend was mixed with HTPB, the dent depth no longer decreased as for the inert additives, with or without HTPB, but was about the same as with no added Al. This indicates a small amount of Al reaction compensating for a primarily inert effect. The increased amount of Al in this mix is offset by less

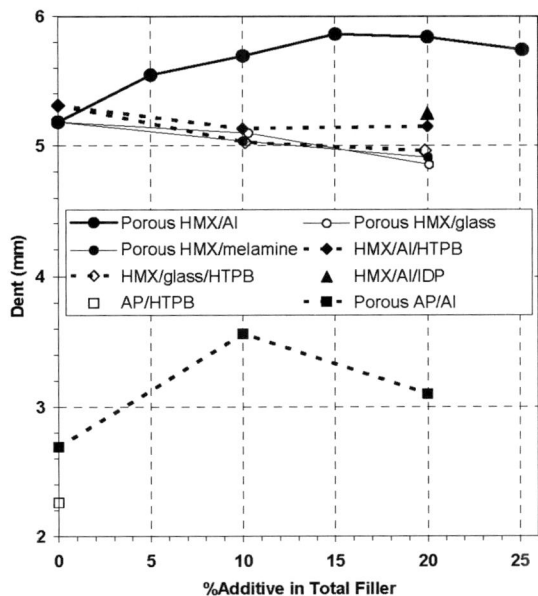

**FIGURE 2.** Effects of adding Al versus inerts and fuels to 60 %vol HMX or to 60 %vol AP.

HTPB. Since the 29 %vol HTPB is about the lower limit for a successful hand mix and poor mixing could promote Al reaction, the HTPB was replaced with the lower viscosity isodecyl pelargonate (IDP), a common plasticizer for HTPB. The 0.1 mm deeper dent with IDP is a small effect from probably other than mixing.

The dents from AP and its mixtures were much less than those from HMX, but still significantly more than the 1.70 mm depth from inert samples [8]. Unlike the more shock reactive HMX, filling the pores in AP with HTPB reduced its dent; but, adding Al to porous AP increased the dent, similar to the effect Price et al. [1] had observed on D.

## CONTRIBUTION OF AL REACTION IN LARGER SCALE TESTS

Bernecker studied the contribution of Al to shock reactivity by comparing measurements from PBXW-108, which is RDX in a plasticized HTPB binder, and PBXN-109, in which Al replaces some of the RDX. Unconfined samples ranging from 38.1 to 50.8 mm in diameter were shock loaded by

a large-scale gap test (LSGT) donor. Streak camera measurements of run distance to detonation (x) were compared [9,10] with those from wedge tests. Data from Figure 3 of reference 9 were scaled and redrawn below, also as Figure 3. The wedge test measurement of x at a given shock pressure in the explosive ($P_E$) for both compositions is always less than that in the gap test due to the absence of rarefactions in the timeframe of the measurements. Also, x is measured on the lateral surface of the sample in a gap test, which may be somewhat beyond the internal location for the onset of detonation, whereas x in the wedge test is equivalent to an internal location for the onset of detonation.

At the higher shock pressures for each composition, the slope of 1/x versus $P_E$ from gap tests is comparable to that slope from wedge tests. The PBXW-108 data have a somewhat higher slope than that for PBXN-109 at these shock pressures, meaning that an increase in shock pressure reduces the run distance to detonation more so in PBXW-108. At the lower shock pressures, $P_E$ <6.0 GPa, the PBXN-109 gap test data have a higher slope, now the same as for PBXW-108. This slope break for the PBXN-109 does not appear in the wedge test data.

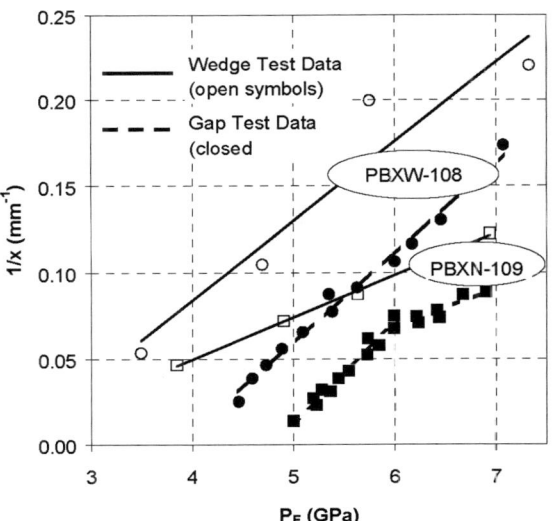

**Figure 3.** Run distance to detonation in gap and wedge tests for PBXW-108 and PBXN-109 from Figure 3 of Reference 9.

Bernecker made the following conclusion [9] about Al reaction in PBXN-109. "In the PMMA arrangement [LSGT], the aluminum apparently only reacts above 6.0 GPa. Moreover, in this pressure range the data appear to indicate that the participation of aluminum slows down the RDX kinetics." PBXN-109 has similar proportions of ingredients as the SSRT samples in Fig. 2 that were 80/20 HMX/Al blends mixed with HTPB or IDP. Since these samples indicate little Al reaction and HMX has the same oxygen balance as RDX, little of the Al probably reacts in PBXN-109 at high $P_E$. Thus, an alternative interpretation of Bernecker's data is that there is little Al reaction at high $P_E$ and that the reduced amount of RDX in PBXN-109 relative to PBXW-108 results in less slope (lower shock reactivity) in the $1/x$ versus $P_E$ plane.

In the pressure range <6.0 GPa, Bernecker addressed [10] the effect of lateral rarefactions on the PBXN-109 gap test. "The presence of two linear regions may be associated, in part, with the radial pressure profiles provided by the 2-D [two-dimensional] shock loading of the pentolite/PMMA system." Less RDX in PBXN-109 requires a larger $d_c$, and so shock buildup is more quickly quenched by rarefactions, as shown by the higher slope in the lower range of $P_E$. At these lower pressures for shock initiation, Al reaction is even less likely.

## CONCLUSIONS

Adding Al enhanced the reaction of porous HMX and AP in the SSRT, peaking in the range of 10 to 15 %wt. When adding a liquid to HMX/Al mixtures, there was little if any enhancement from the Al. These results are in agreement with the literature, indicating that the SSRT can be useful for studying variations in the major ingredients of formulations.

Bernecker's gap test data for RDX-based compositions with and without Al can be interpreted differently assuming that shock initiation is simply dependent on the RDX concentration and that the Al does not contribute. The Al may not even contribute at the higher shock pressures associated with the detonation wave.

## REFERENCES

1. Price, D., Clairmont, A. R., Jr., and Erkman, J. O., "Explosive Behavior of Aluminized Ammonium Perchlorate," Combustion and Flame, Vol. 20, pp. 389-400, 1973.
2. Price, D., "Aluminized Organic Explosives," NOLTR 72-62, Naval Ordnance Laboratory, 8 Jun 1972.
3. Tao, W. C., Tarver, C. M., Kury, J. W., Lee, C. G., and Ornellas, D. L., "Understanding Composite Explosive Energetics: IV. Reactive Flow Modeling of Aluminum Reaction Kinetics in PETN and TNT Using Normalized Product Equation of State," Proceedings of Tenth International Detonation Symposium, ONR 33395-12, Office of Naval Research, pp. 628-636, 1993.
4. Finger, M., Hornig, H. C., Lee, E. L., and Kury, J. W., "Metal Acceleration by Composite Explosives," Proceedings of Fifth Symposium (International) on Detonation, ACR-184, Office of Naval Research, pp. 137-151, 1970.
5. Smith, L. C., "On Brissance, and a Plate-Denting Test for the Estimation of Detonation Pressure, " Explosivstoffe, Nr. 5, 1967, pp. 106-110, and Nr. 6, pp. 130-134, 1967.
6. Brousseau, P., Dorsett, H. E., Cliff, M. D., and Anderson, C. J., "Detonation Properties of Explosives Containing Nanometric Aluminum Powder," Proceedings of Twelfth International Detonation Symposium, San Diego, CA, 11-16 Aug 2002, to be printed.
7. Lefrancois, A., Baudin, G., Le Gallic, C., Boyce, P., and Coudoing, J-P, "Nanometric Aluminum Powder Influence on the Detonation Efficiency of Explosives," Proceedings of Twelfth International Detonation Symposium, San Diego, CA, 11-16 Aug 2002, to be printed.
8. Sandusky, H. W., Granholm, R. H., and Bohl, D. G., "Small-Scale Shock Reactivity Test," in Proceedings of the 40th JANNAF Combustion Subcommittee Meeting, to be printed.
9. Bernecker, R. R., Clairmont, A. R., Jr., Sandusky, H. W., and Smith, M. S., "Participation of Aluminum in Two-Dimensional Shock Initiation Experiments," in Shock Compression of Condensed Matter 1987, (S.C. Schmidt, N.C. Holmes, eds.), pp. 573-576.
10. Bernecker, R. R. and Clairmont, A. R., Jr., "Shock Initiation Studies of Cast, Damaged, and Granulated PBXs," in Proceedings of Tenth Symposium (International) on Detonation, ONR 33395-12, Office of Naval Research, pp. 499-506, 1993.

CP845, *Shock Compression of Condensed Matter - 2005,*
edited by M. D. Furnish, M. Elert, T. P. Russell, and C. T. White
© 2006 American Institute of Physics 0-7354-0341-4/06/$23.00

# TOWARD A NEW PARADIGM FOR REACTIVE FLOW MODELING

## R. G. Schmitt

*Sandia National Laboratories P.O. Box 8500, Albuquerque, NM 87185-0836*

**Abstract.** Traditional reactive flow modeling provides a computational representation of shock initiation of energetic materials. Most reactive flow models require ad hoc assumptions to obtain robust simulations, assumptions that result from partitioning energy and volume change between constituents in a reactive mixture. For example, most models assume pressure and/or temperature equilibrium for the mixture. Many mechanical insults to energetic materials violate these approximations. Careful analysis is required to ensure that the model assumptions and limitations are not exceeded. One limitation is that the shock to detonation transition is replicated only for strong planar shocks. Many models require different parameters to match data from thin pulse, ramp wave, or multidimensional loading, an approach that fails for complex loading. To accurately simulate reaction under non-planar shock impact scenarios a new formalism is required. The continuum mixture theory developed by Baer and Nunziato is used to eliminate ad hoc assumptions and limitations of current reactive flow models. This modeling paradigm represents the multiphase nature of reacting condensed/gas mixtures. Comparisons between simulations and data are presented.

**Keywords:** Shock initiation, SDT, reactive flow, CTH, multiphase, Baer-Nunziato.
**PACS:** 47.40.-x, 82.40.fp, 47.70.-n

## INTRODUCTION

Shock initiation behavior is one of the most important characteristics with regards to safety, performance, and hazards of energetic materials. There is a long history of experimental, theoretical, and numerical investigations into shock initiation [1,2,3,4]. The traditional numerical method for analyzing shock to detonation transition (SDT) is called reactive flow modeling. The mathematical equations are represented by Euler equations with an additional reaction rate expression representing the conversion of condensed-phase energetic material to reaction products. For condensed phased energetic materials the problem lies in the construction of complex equation of state (EOS) surfaces to represent the unreacted and fully reacted materials. Analytic and tabular EOS's

provide necessary ingredients for simulating detonation physics problems. Complex EOS representations and mixture assumptions create mathematical closure problems for the governing equations. Many reactive flow models solve these problems by assuming that the unreacted and fully reacted materials have the same pressure and temperature [5]. This is nonphysical because even though pressure is communicated at acoustic velocities it is not in equilibrium within the reaction zone and temperature equilibrates by much slower diffusion processes and is definitely not in equilibrium. Based on thermochemical arguments the concept of hot-spots also indicates that temperature is not truly in equilibrium within reacting condensed-phase materials.

It is recognized that the SDT behavior of energetic materials under strong planar shock loading is adequately modeled using any number of reactive flow models. A variety of experiments to characterize SDT response have been developed. These include wedge tests, plate impact, embedded gauge, gap tests, and more recently isentropic loading. Tarver has repeatedly demonstrated the utility of embedded pressure gauges using ignition and growth models developed at Lawerence Livermore National Laboratories [5]. Researchers working at Los Alamos National Laboratory have developed embedded particle velocity gauges with shock tracking that provide the highest quality integrated measurements of SDT [6]. Some reactive flow models begin to fail as they are applied to this suite of experiments. Often the models require nonunique or multiple sets of parameters to match the experimental results or fail to adequately match data associated with critical loading conditions such as thin-pulse loading. This is the anticipated outcome when using phenomenologically based models. It is important to recognize the limitations of these models.

The goals of the model presented here are to eliminate the ad hoc rules associated with partial reaction states and to develop the thermodynamically consistent mathematical framework necessary to address complex combustion phenomena. The complex combustion processes include but are not limited to enhanced surface area burning, volumetric burning, and nondetonative reactive wave phenomena. All of these combustion processes can result in significant energy release. There is a long history of application of continuum mixture theory to reactive flow modeling at Sandia National Laboratories. The formalism chosen here was first presented in Baer and Nunziato [7]. Baer-Nunziato was demonstrated to simulate accurately deflagration-to-detonation transition (DDT) in granular explosives. This paper discusses the application of the theory to near theoretical maximum density high explosive PBX 9501 undergoing SDT.

## THEORETICAL MODEL

The continuum mixture theory is represented by conservation equations for each phase in Lagrangian form, see [7] for details. Each phase is represented by governing equations for conservation of mass, momentum, and energy. In addition, an evolutionary equation for the volume fraction is necessary for mathematical closure of the equation set. The conservation of mass for phase 'a' is given by:

$$\rho'_a = -\rho_a \nabla \cdot v_a + c_a^+ \qquad (1)$$

The superscript '+' identifies the interphase source term for mass. The variable $\rho$ is the partial mass density, $v$ is the velocity. The governing equations for conservation of momentum and energy also contain source terms. This mathematical model allows each phase in a chemically reacting mixture to exist at its own thermodynamic state. The phases in the mixture then exchange mass, momentum and energy based upon physical mechanisms such as viscous stress relaxation, heat transfer, and drag. The component conservation equations when summed over all components yield the usual set of conservation equations for the mixture.

A key aspect of the multiphase mixture formulation is the specification of the interphase source terms. It has been demonstrated that the multiphase mixture formulation yields a system of equations that is completely hyperbolic with a set of eigenvalues and eigenvectors that are all real [8]. This implies that the equations can be solved by standard numerical methods for hyperbolic systems without appealing to special algorithms to stabilize nonphysical solutions. Functional forms for the interphase source terms are determined by appealing to the second law of thermodynamics to place physically meaningful limiting conditions upon the admissible properties of the these terms. This procedure has been demonstrated as a useful technique that guarantees by construction that the second law is **NOT** violated. Also the grouping of terms to define the interphase source terms often yields mathematical and physical insight into the nature of the governing equations. Equations for the interphase source terms are available in the literature [7]. The specific equation for the mass source term is:

$$c_s^+ = -\gamma_s \left[ (1-\lambda)\tau_s^{-1} + \lambda \langle S/V \rangle \dot{r} \right] \qquad (2)$$

where $\tau_s$ is the characteristic time scale for energy localization and is reserved for future development; $\lambda$ is the gas-phase induction parameter, the surface to volume ratio <S/V> is multipled by the "linear" burn rate $\dot{r}$. The effective linear burn rate is typically represented by the empirical pressure-based expression, rate=$AP^n$. The surface to volume ratio is represented by the form factor $(1-F)^x F^y$ where F is the mass fraction or the volume fraction. Additional relationships are under development. The mixture theory applied to the volumetric source term has the flexibility of using the thermodynamic states for the condensed, gas or mixture properties. The future direction for these rate expressions is to incorporate temperature-based chemistry and energy localization mechanisms. This model is being developed in the Sandia National Laboratories Eulerian shock physics code, CTH [9].

## RESULTS AND DISCUSSION

The standard technique for determining the sensitivity of explosives is the wedge test. This experiment yields the pop-plot for the material at a given initial state [10]. While this point data is necessary for analysis of SDT it is not sufficient to evaluate the reaction rate expression or the mixture assumptions. Additional data from gap tests and flyer plate impacts provide insight into the thermochemical processes, but embedded particle velocity and pressure gauges provide the most stringent tests for reactive flow modeling. The embedded particle velocity gauge data from Gustavsen et al. [6] provide information concerning the transition to detonation within PBX 9501 subject to planar shock waves between 3 and 5 GPa. In addition, shock tracking data clearly illustrate the location and timing of the SDT process.

Two flyer plate impact experiments from [6] are used to evaluate the multiphase reactive flow model. The impact of Vistal at 816 m/s results in an impact stress of 5.21 GPa. The fully dense unreacted Hugoniot for PBX 9501 and the shock properties of Vistal form [6] are used. A Sesame table is used for the detonation product EOS. The distance to transition is approximately 5.6 mm. The shock tracker data (thick line) is cross plotted with the numerically generated Xt diagram of the transition in Fig. 1. As shown, the model replicates the initial shock speed, amplitude, reactive wave growth and transition to detonation accurately. The

particle velocity at 10 embedded gauge locations is cross plotted with the numerical simulation in Fig. 2. As shown, the model reproduces the particle velocity gauge data.

The impact of vistal at 653 m/s results in an impact stress of 3.9 GPa. The distance to transition is approximately 8.8 mm. The shock tracker data is cross plotted with the numerical simulation of the transition in Fig. 3. The model replicates the initial shock speed, amplitude, reactive wave growth and

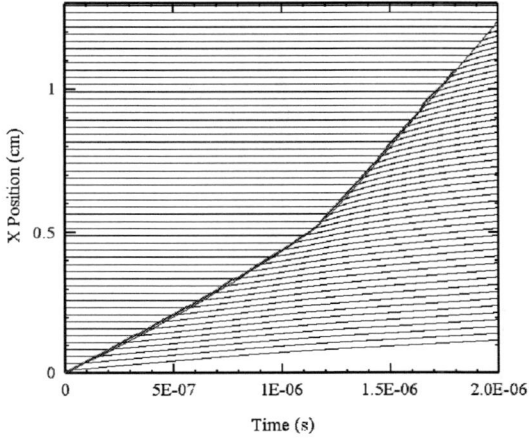

**Figure 1.** The shock tracker data for an impact pressure of 5.21 GPa from [6] is cross plotted with the numerical simulations using the multiphase reactive flow model.

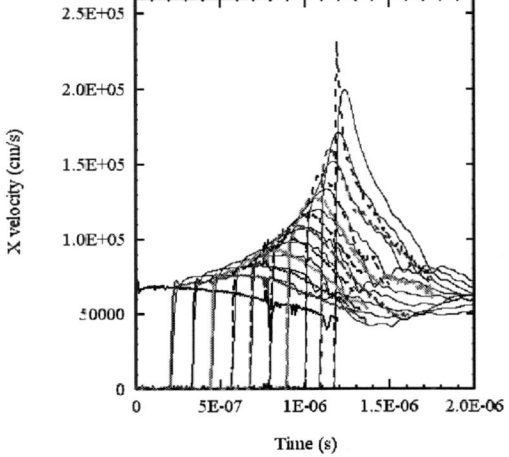

**Figure 2.** The embedded particle velocity gauge data for an impact pressure of 5.21 GPa from [6] is cross plotted with the numerical simulations using the multiphase reactive flow model.

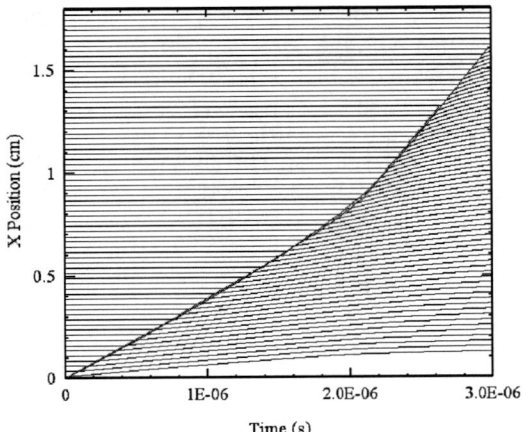

**Figure 3.** The shock tracker data for an impact pressure of 3.9 GPa from [6] is cross plotted with the numerical simulations using the multiphase reactive flow model.

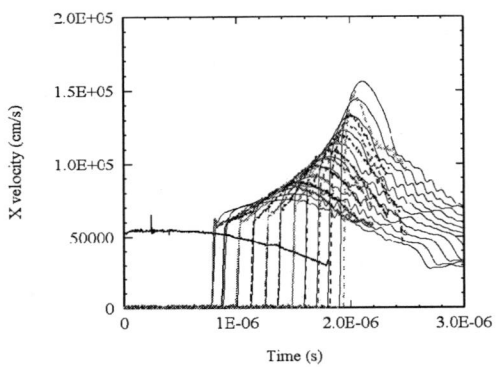

**Figure 4.** The embedded particle velocity gauge data for an impact pressure of 3.9 GPa from [6] is cross plotted with the numerical simulations using the multiphase reactive flow model.

transition to detonation accurately as shown. The particle velocity at 11 embedded gauge locations is cross plotted in Fig. 4. As shown, the model reproduces the particle velocity gauge data.

## CONCLUSIONS

The Baer-Nunziato continuum mixture theory is undergoing development for application to multidimensional SDT and fragment impact modeling. The model is demonstrated to match the pop-plot and particle velocity embedded gauge data for PBX 9501. The preliminary evaluation is that the current reaction rate forms are adequate to simulate the details of the transition to detonation

as presented in this paper. However, more loading conditions especially at lower stress levels need to be evaluated. The remaining data from [6] needs to be simulated before more advanced forms for reaction rate parameters can be developed.

## ACKNOWLEDGEMENTS

The author gratefully acknowledges many fruitful discussions with Mel Baer and Marlin Kipp of Sandia National Laboratories. Sandia is a multiprogram laboratory operated by Sandia Corporation, a Lockheed Martin Company, for the United States Department of Energy's National Nuclear Security Administration under contract DE-AC04-94AL85000.

## REFERENCES

1. Bowden, F. P. and Yoffe, A.D., *Initiation and Growth of Explosions in Liquids and Solids*, Cambridge University Press, Cambridge, Mass. 1952.
2. Bowden, F. P. and Yoffe, A.D., *Fast Reactions in Solids*, Cambridge University, Cambridge Butterworth Scientific Publications, London 1958.
3. Mader, C. L. *Numerical Modeling of Detonation*, University of California press, 1979.
4. International Detonation Symposia, 1951-2002. Editions 1-12.
5. Lee, E. L. and Tarver, C. M., "Phenomenological Model of Shock Initiation in Heterogenoeus Explosives," Phys. Fluids 23, pp. 2362-2372, 1980.
6. Gustavsen, R. L., Sheffield, S. A., Alcon, R. R., and Hill, L. G., "Shock Initiation of New and Aged PBX 9501 Measured with Embedded Electromagnetic Particle Velocity Gauges," Los Alamos National Laboratories report LA-13634-MS, Sept. 1999.
7. Baer, M. R. and Nunziato, J.W., "A Two-Phase Mixture Theory for the Deflagration-to-Detonation Transition (DDT) in Reactive Granular Materials," Int. J. Multiphase Flow, Vol. 12, no. 6, pp. 861-889, 1986.
8. Embid, P. F. and Baer, M. R., "Mathematical Analysis of A Two-Phase Model for Reactive Granular Material," Sandia National Laboratories Report SAND88-3302, 1989.
9. McGlaun, J. M., Thompson, S. L., Elrick, M. G., "CTH – A Three-Dimensional Shock-Wave Physics Code," International Journal of Impact Engineering, Vol. 10, pp. 351-360, 1990.
10. Gibbs, T. R. and Popolato, A., *LASL Explosive Property Data*, University of California Press, Berkely, CA,1980.

CP845, *Shock Compression of Condensed Matter - 2005*,
edited by M. D. Furnish, M. Elert, T. P. Russell, and C. T. White
2006 American Institute of Physics 0-7354-0341-4/06/$23.00

# THE EFFECT OF PRECURSOR SHOCKS ON THE GROWTH TO DETONATION OF EDC37

## S. S. Sorber and R. E. Winter

*Hydrodynamics Department, AWE, Aldermaston, Reading, UK, RG7 4PR.*

**Abstract.** The response of the HMX-based explosive EDC37 to shock loading has been studied using electromagnetic particle-velocity gauges. One of the aims of the work was to determine the effect of a relatively weak pre-shock on the growth to detonation of a following, stronger, shock. Ideally this requires a comparison between the response of the sample when shocked by a simple sustained shock with that generated by a shock of the same amplitude, but preceded by a weaker pre-shock. Although our limited results do not allow a direct comparison, a normalisation technique has been developed which allows us to interpolate the growth-of-reaction curves for any chosen input shock. Comparison then allows the effective origin of the main shock in the pre-shock experiments, that is the plane at which the reaction starts to grow, to be located. It is found that this growth origin is located slightly before the plane at which the main, and the slower-moving pre-cursor shocks, coalescence. The distance between the effective growth origin of the main shock and the coalescence point depends on the stress of the pre-cursor shock.

**Keywords:** Shock Initiation, EDC37, single shock, double shock, predictive
**PACS:** 62.50.+p

## INTRODUCTION

Experiments in which the response of EDC 37 to single and double shocks has been studied using embedded velocity gauges have been reported in previous papers [1-3]. The wider programme aim is to develop a predictive capability for the response of EDC37 to complex shocks. The work described within this paper examines single and double shock data with the aim of determining patterns or trends in the data which will allow the response of EDC37 to double shocks to be predicted.

## NORMALISATION OF SUSTAINED SINGLE SHOCK RESULTS

The normalisation process requires the six single shock experimental results to be shown in one plot. To make this "easier on the eye", the individual gauge histories for each shot shown in Fig.1 have been reduced to two points.

As illustrated in Fig.1 the arrival of the peak of the initial shock at each gauge location can be depicted as a point in particle velocity, ($u_p$) versus time (t)

space where t is the time at which the shock front arrives at the gauge. A similar procedure can be followed for the peak of the reactive wave.

The curves generated by these points for Experiment 1159 are shown in Fig.2. The vertical line shows the run distance for this experiment as determined by the shock tracker gauge records [3].

Salisbury et al [2] plotted the growth curves for each velocity gauge experiment in a normalised form.

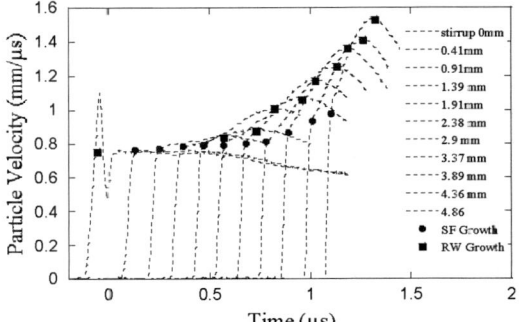

**FIGURE 1.** Typical particle velocity vs. time records for a sustained shock (Shot 1159).

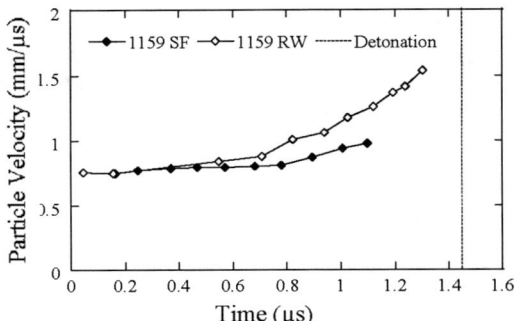

**FIGURE 2.** Shot 1159: curves showing the growth of the shock front and the reactive wave.

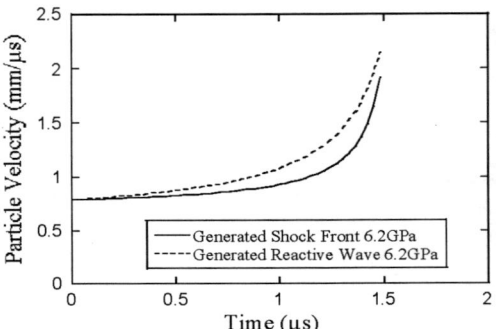

**FIGURE 4.** Generated growth curves for a notional input shock of initial pressure 6.2GPa.

In the normalised depiction particle velocity is plotted as $(u_p - u_p^{in})/(u_p^{CJ} - u_p^{in})$, where $u_p^{in}$ is the input particle velocity at the first gauge and $u_{CJ}$ is the Chapman-Jouget particle velocity. Time is plotted as $t/t_{run}$, where $t_{run}$ is the run to detonation time.

In this paper results from a fifth experiment are added and the normalisation technique is taken further by fitting an equation through the normalised points. The curve is constrained to pass through the points (0,0) and (1,1), the latter representing detonation.

Figure 3 shows normalised particle velocity versus normalised time plotted for the experiments with input shocks of 2.69GPa, 3.52GPa, 4.1GPa, 4.91GPa, 5.92GPa and 10.80GPa. The data showing the growth of the shock wave and reactive wave was fitted by the hyperbolic equation:

$$u_p^{Normalised} = \frac{At_N}{(1+A-t_N)}, \text{ Where } t_N = \frac{t}{t_{det}} \quad (1)$$

A = 0.051479 for the shock front growth
A = 0.128628 for the reactive wave growth

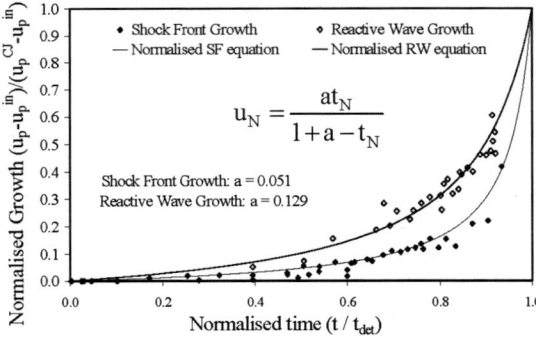

**FIGURE 3.** Growth data for six sustained shock experiments plus analytic fits.

As shown in Fig.4 the normalised curves allow the growth plots corresponding to any chosen input pressure to be generated.

## EFFECT OF PRE-CURSOR SHOCKS

Particle velocity gauge results for two of the double shock plate impact experiments are shown in Fig. 5 and Fig. 6.

In each experiment a weak "pre" shock precedes a stronger main shock. For the lowest precursor stress, shots 1175 and 1176, shown in Fig.5, it is clear that the pre-shock displays no increase in amplitude with time indicating that no significant reaction is occurring. This has been termed an "unreactive" pre-shock. The time at which the waves coalesced and the run-to-detonation conditions for this configuration were determined from the plot of wave arrival times [3] in Fig.7.

In the next pair of experiments, 1194 and 1195, shown in Fig.6, the effect of significant reaction behind the pre-shock was investigated by increasing the impact velocity and the pulse length between the two shocks, relative to shots 1175/76.

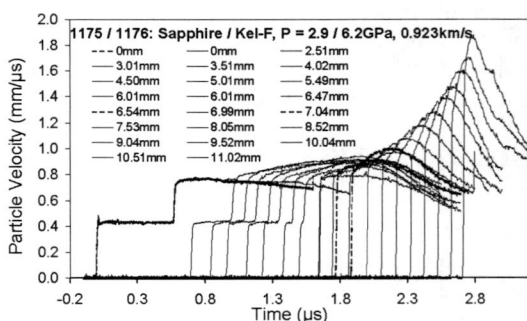

**FIGURE 5.** Particle velocity records for an unreactive pre-shock

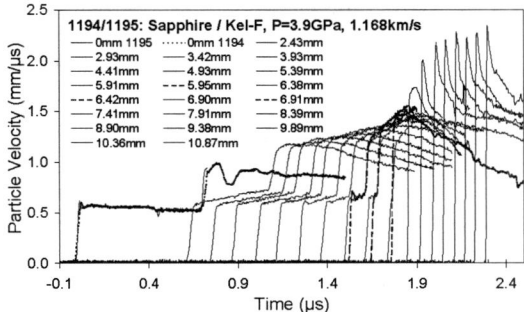

**FIGURE 6.** Particle velocity records for a reactive pre-shock.

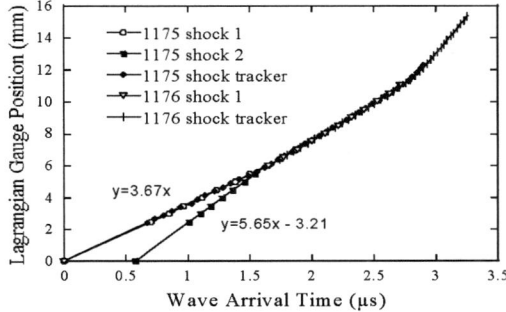

**FIGURE 7.** Plot of wave arrival times for an unreactive preshock, showing coalescence and turn over to detonation.

This allowed the reaction from the pre-shock to develop before coalescence. The rising particle velocity behind the pre-shock and the increase of shock amplitude with depth in Fig.6 shows that the wave is causing the EDC37 to react. The waves coalesce at 7.0 mm and 1.80μs.

To determine the effect the pre-shock has on the reaction generated by the main shock, it is necessary to compare the reaction generated by the main shock with that which would be generated by a single shock having the same amplitude.

In Fig.8 the shock front and reactive wave growth for 1175/1176, in which the main shock pressure was 6.2GPa, are shown together with the growth curves generated using equation (1) for a 6.2GPa input single shock.

In Fig.9 the growth curves have been shifted in time to the right until the generated shock front curve overlays the experimental shock front points. This allows the effective time origin of the main shock in the double shock experiment to be determined.

It is seen from Fig.9 that the effective time origin of the main shock in the double shock experiment is at 1.28μs. The effective origin of the main shock may be compared with the time at which the two

waves in the double shock experiments coalesce. The coalescence time, as judged by examination of the shock tracker records [3] is 1.59 μs and is shown by the dashed vertical line on the graph.

As shown in Figs.10 and 11 this method may also be applied to the double shock experiments, 1194/95, in which the pre-shock was reactive. In this case the main shock growth starts 0.4μs before coalescence, slightly earlier than the 0.31μs observed with the un-reactive pre-shock.

## DISCUSSION

In seeking an explanation for the effects described above a number of factors must be considered. One of these factors is shock desensitisation. It has long been considered that a pre-shock can desensitise a heterogeneous explosive to a following main, stronger, shock (see, for example reference 4). Later research by Campbell and Travis [5] concluded that a finite time was needed for a pre-shock to become effective in desensitising the explosive.

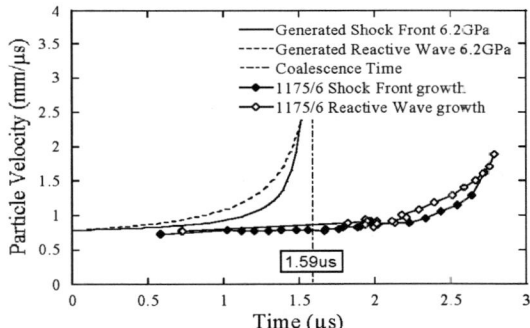

**FIGURE 8.** Comparison between growth curves of a 6.2GPa sustained shock and a 6.2GPa main shock preceded by a 2.9GPa precursor shock.

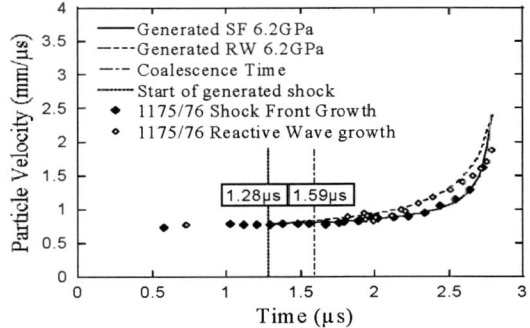

**FIGURE 9.** The sustained shock growth curves have been shifted to overlay the shock with the precursor growth curves. With a reactive preshock the main shock starts to grow 0.31μs before coalescence.

One explanation for this effect is that the pre-shock deactivates the hot spots which form the reaction points for the main shock but that it takes time to do this. Our observation that the main shock starts to grow before it coalesces with the pre-cursor shock appears to support Campbell and Travis's desensitisation concept. A further factor to be taken into account in analysing the effect of a pre shock on the growth of a following main shock is that shocking a material to a given pressure by two (or more) shocks will generate a lower temperature than if the material was shocked to the same pressure by a single shock. For example, back-of-envelope calculations show that shocking EDC37 to a pressure of 6.2GPa via a precursor shock of 2.94GPa gives a temperature of 126 °C whereas a single shock at the same pressure gives a temperature of 196 °C. Although this goes some way towards explaining why a double shock is less efficient than a single shock at detonating an explosive, it does not explain why growth in our experiments appears to start before coalescence.

## CONCLUSIONS

The growth curves from sustained shock experiments fired with different impact pressures can be normalised to give a single curve that can then be used to generate a growth curve for any chosen input shock. These "generated" shocks have allowed the effect of a weaker pre-shock on the growth of a following main shock to be isolated.

Comparing the growth of a single shock with that of a shock of the same amplitude but preceded by an un-reactive pre-shock reveals that the growth to detonation of the main shock begins to grow at a point slightly before the faster-moving main shock catches the pre-shock. This observation is consistent with the concept of a desensitisation time as described by Campbell and Travis [5].

When the pre-shock is reactive, the time between the main shock growth origin and the coalescence point increases relative to that seen in the unreactive shock case. This suggests that a reactive shock is less effective at desensitising the explosive.

We envisage that the descriptions presented in this paper will be used to validate reaction models and help to develop a capability to predict the response of EDC37 and other HMX-based explosives to complex shocks.

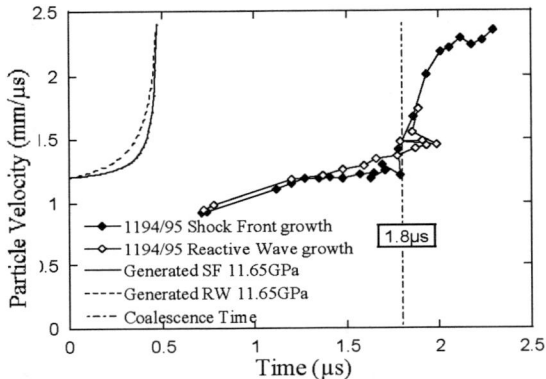

**FIGURE 10.** Comparison between growth curves for a 11.65GPa sustained shock and a 8.5GPa shock preceded by a 3.9GPa precursor shock.

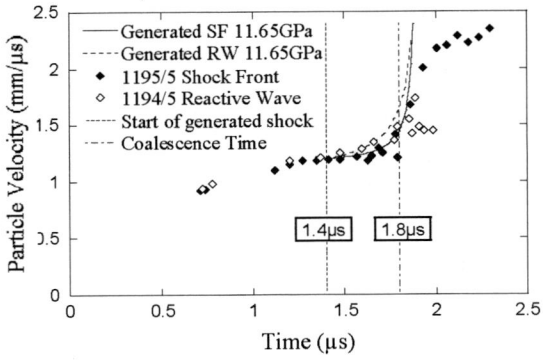

**FIGURE 11** The growth curves for the sustained shock have been shifted to overlay the growth curves for the shock with the precursor. It is seen that with a reactive preshock the main shock starts to grow 0.4µs before coalescence.

### REFERENCES

1. Gustavesen R. L., Sheffield S. A., Alcon R. R., Hill L. G., Winter R. E., Salisbury D. A., and Taylor P., *in Shock Compression of Condensed Matter-1999*, AIP New York, 1999, p879.
2. Salisbury D A, Winter R E, Taylor P, Gustavsen R L, Sheffield S A, and Alcon R. R., *Proceedings of the Twelfth Symposium (International) on Detonation*, San Diego, 2002, forthcoming publication.
3. Gustavsen R. L., Sheffield S. A., Alcon R. R., Winter R. E., Taylor P. and Salisbury D. A., *in Shock Compression of Condensed Matter-2001*, AIP New York, 2001, p999.
4. Plant. J, Phys Bull 23, p203-207 (1972).
5. Campbell A. W. and J. R. Travis, *Proceedings of the Eighth Symposium (International) on Detonation*, 1985, p1057.

CP845, *Shock Compression of Condensed Matter - 2005,*
edited by M. D. Furnish, M. Elert, T. P. Russell, and C. T. White
© 2006 American Institute of Physics 0-7354-0341-4/06/$23.00

# SHOCK COMPRESSION OF SOLID WITH VOIDS BY GRIDLESS LAGRANGIAN SPH

**Katsumi Tanaka**

*National Institute of Advanced Industrial Science and Technology, Tsukuba Ibaraki, 305-8568 Japan*

**Abstract.** The mechanism of formation of local hot spots with a single or multiple voids has been studied numerically by Smoothed Particle Hydrodynamics (SPH). Temperature increase was not enough to initiate emulsion explosives for incident shock pressure lower than 10GPa in the case of single void. It is important for formation of local hot spot to consider complex physical interaction and chemistry in multiple voids.

**Keywords:** SPH, local hot spot, shock initiation, emulsion
**PACS:** 47.40.-x, 47.40.Nm.

## INTRODUCTION

Microscopic studies of mechanism of the shock initiation by local hot spots in heterogeneous explosives have been developed extensively. Earlier Eulerian studies for the shock propagation through an air bubble in water have shown a hot spot with locally high temperature and high pressure [1]. Numerical studies of collapsing bubble in nitromethane by PIC (Particles in Cell) and Eulerian hydrodynamics in early 1960s have shown the mechanism of local hot spots [2]. Recent study by free Lagrangian computation for shock waves propagating through water including an air bubble show the shock interaction and adiabatic heating of air bubble [3]. Recent development of SPH gives useful results on the local hot spots by collapse of bubble or cavities. SPH is gridless Lagrangian. The main advantage of SPH is integral scheme that can avoid the instability by mesh tangling and distortion which usually occur in Lagrangian finite difference or finite element method. SPH can be applied to high velocity impact phenomena, jetting in large cavity of shaped charge and other problem accompanying large deformation. Details of the SPH technique is described elsewhere[4] .

## NUMERICAL

The two dimensional hydrodynamic equations of continuity, momentum and energy can be solved using F-CEL2D which is the VOF (Volume of Fraction) scheme applicable for multiple materials. SPH uses kernel approximation based on randomly distributed particles. Hydrodynamic equation can be discretized by kernel function $W$ as follows.

$$< f(x) > = \int_{-2h}^{2h} f(x')W(x-x',h)dx'$$

$$< \nabla \cdot f(x) > = \int_{-2h}^{2h} f(x')\nabla \cdot W(x-x',h)dx'$$

Kernel function $W$ is defined by

$$\int_{-2h}^{2h} W(x-x',h)dx' = 1$$

$$W(x-x',h) = 0 \ , \ |x-x'| \geq 2h \ ,$$

where $h$ denotes the smooth length which corresponds to a mesh size for the Lagrangian

computation by finite difference. SPH has usual disadvantage of Lagrangian methods that SPH analysis for large density or velocity difference is unstable typically in case of multiple material problem. Numerical technique for the definition of boundary conditions for rigid boundary or cylindrical symmetry condition is required.
The Mie-Gruneisen equation for condensed materials and the revised Kihara-Hikita (KHT) equation for detonation products are applied. Arrhenius reaction rate law is included for thermal reaction. Air is assumed polytropic gas.

## RESULTS AND DISCUSSION

First, computational results by SPH are compared with F-CEL2D. SPH can trace the time variation of specified material easily comparing with usual Eulerian technique but its resolution is relatively lower than finite difference.

### Square void in ammonium nitrate

Shock compression of a square void or air bubble in solid ammonium nitrate is studied. Numerical results by SPH and Eulerian computation by F-CEL are well compared as shown in Fig.1 where tracking particles are used to trace the material movement. A void or an air bubble of 0.5mm thick and 2mm in width is assumed for SPH and F-CEL2D respectively. Two hundred thousand particles or meshes are used in this calculation. Ammonium nitrate is assumed non-reactive solid. Linear relation for Hugoniot between shock velocity $U_s$ and particle velocity $u_p$, $U_s$(km/s)=2.5 +1.5$u_p$ is used with Mie-Gruneisen constant of 0.7. For ammonium nitrate, initial density and the Gruneisen constant are 1.725g/cm$^3$ and 0.7 respectively. Reaction in ammonium nitrate is not considered. Piston condition to SPH and mass applied piston to F-CEL2D is used as boundary condition corresponding to incident shock pressure of 10GPa.

Although ammonium nitrate is insensitive, local hot spot in a solid by shock compression is not enough to induce detonation initiation in this case. However, temperature of air bubble which is isentropicaly compressed is quite high as shown in

earlier numerical works [1,3]. Air bubble gives little effect to shock propagation and void collapse.

### Spherical void in emulsion-explosive matrix

The emulsion explosive is a typical commercial explosive for rock blasting, mining and quarrying. In the emulsion explosive ammonium nitrate is solved in water with emulsified oil. The water-in-oil emulsion explosive is mixed with micro hollow glass or porous solid. Emulsion explosive considered in this work is composed of ammonium nitrate 72wt%, sodium nitrate 10%, water 12% and oil 6%. Detonation velocity of emulsion explosive without void is 7400m/s, pressure is 18.4GPa and temperature is 1147K by C-J calculation. Volumetric fraction of void is assumed to be 30% which corresponds to detonation velocity of 5700m/s, C-J pressure of 8.94GPa and temperature of 1686K. Ammonium nitrate based explosives generally have low detonation temperature.

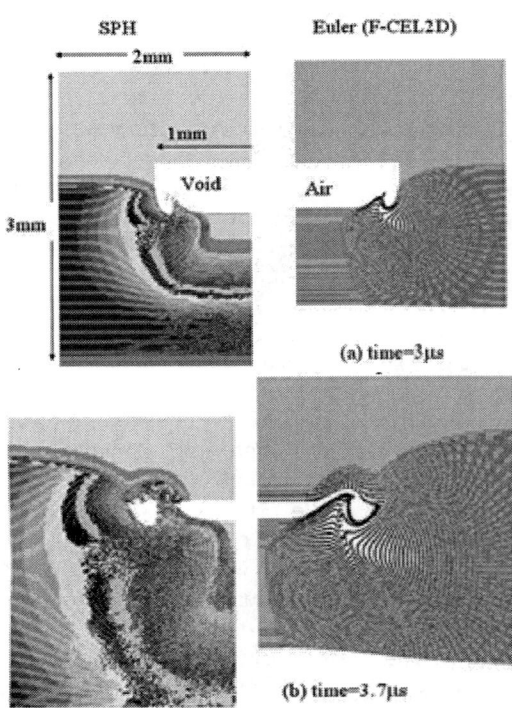

**Figure 1.** Comparison of shock compression of a void by SPH and an air bubble by Euler calculation. Left figures are pressure contours by SPH. Right figures

show material movement by F-CEL2D traced by using tracking particles.

Numerical results by SPH for shock compression of a spherical void in emulsion matrix including thermal reaction are shown in Fig. 2. Incident shock pressure is 10GPa which is nearly C-J pressure. Temperature increase around void is low to initiate explosive mixture as shown in Fig.3. It suggests that possible amount of direct initiation by local hot spot generated in a single void or a bubble in explosive is small, because rarefaction wave from free surface boundary of void reduce quickly local hot temperature generated by impact of jet. Numerical results show the jet and shock Mach reflections during cavity collapse. Effect of glass microballoon is shown in Fig.4. Temperature increase by a jet can be seen only in a glass microballoon.

## Multiple voids in emulsion-explosive matrix

Local hot spot for the case of a single void is not enough to show whole process of detonation in heterogeneous explosives. A numerical study for the case of multiple void in emulsion explosive mixture is shown in Fig.5 where cavity jet impacts and Mach reflections can be seen through a few voids. Although it is still difficult to study the whole microscopic mechanism of shock initiation and detonation propagation, the physically complex interaction is important.

.

## CONCLUSIONS

SPH results for void collapse by shock compression give similar results as computed by Eulerian hydrodynamic code. Shock compression of a single void shows small temperature increase which is insufficient to enhance shock initiation. The artificial void with glass microballoon in emulsified water-in-oil explosive gives small temperature rise but it generates large shock interaction and disturbance. Mechanism of detonation initiation for heterogeneous explosives should be studied based on reliable reaction chemistry including not only a bubble collapse but also shear bands, friction and other effects.

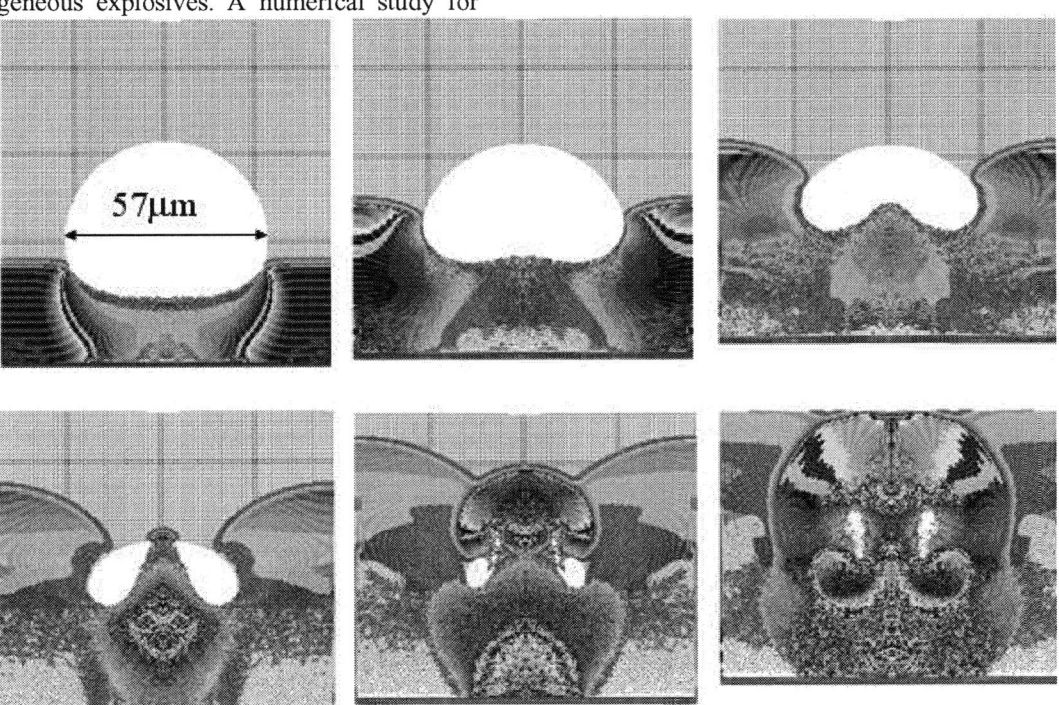

**FIGURE 2.** Shock cavity interaction in water-in oil emulsion explosive mixture to incident pressure of 10GPA with a time interval of 5ns right to left and up to down.

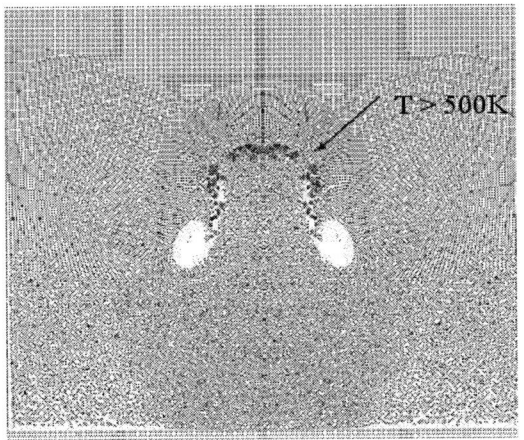

**FIGURE 3.** Temperature higher than 500K (dark points) at 30ns in Fig.2..

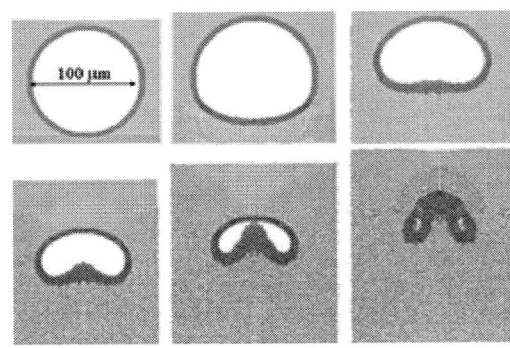

**FIGURE 4.** Collapsing glass microballoon in emulsion explosive mixture with time interval of 100ns.

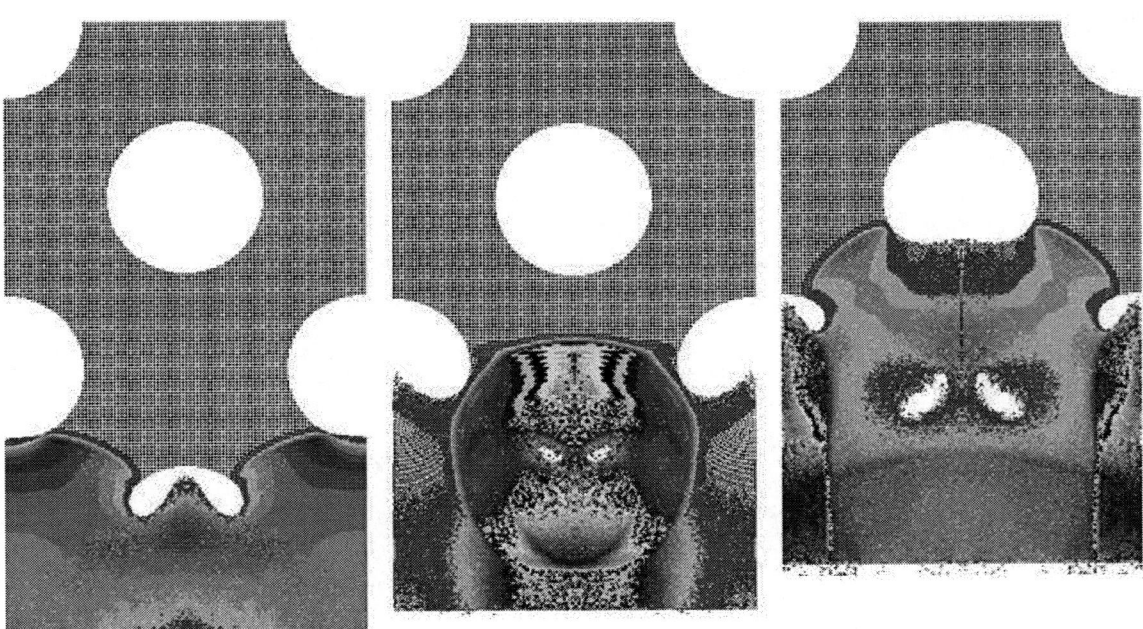

**Figure 5.** Shock interaction and jet formation in multiple voids for time interval of 5ns.

## REFERENCES

1. Tanaka, K., "Numerical analysis of shock compression of heterogeneous medium including air bubbles", 21st Int. Symp on Shock Waves, paper 6210 ,1997.
2. Mader, C. L. ,"Numerical Modeling of Detonation", Univ. California Press, 1979
3. Ball, G. J., *et a*l. "Shock induced collapse of a cylindrical air cavity in water ;a Free-Lagrange simulation", Shock Waves 10, 265, 2000
4. Liu, G. R. and Liu, M. B. "Smoothed Particle Hydrodynamics", World Scientific, 2003

CP845, *Shock Compression of Condensed Matter - 2005*,
edited by M. D. Furnish, M. Elert, T. P. Russell, and C. T. White
© 2006 American Institute of Physics 0-7354-0341-4/06/$23.00

# PROCESSING, APPLICATION AND CHARACTERIZATION OF (ULTRA)FINE AND NANOMETRIC MATERIALS IN ENERGETIC COMPOSITIONS

**A. E. D. M. van der Heijden[1], R. H. B. Bouma[1], E. P. Carton[1], M. Martinez Pacheco[2], B. Meuken[1], R. Webb[1], and J. F. Zevenbergen[1]**

*[1] TNO Defence, Security and Safety, P.O. Box 45, 2280 AA Rijswijk, The Netherlands*
*[2] The Netherlands Institute for Metals Research, Delft University of Technology, The Netherlands*

**Abstract.** The energetic materials research at TNO Defence, Security and Safety, The Netherlands is focusing at the development and characterization of explosives (insensitive munitions), gun/rocket propellants and pyrotechnic compositions and their ingredients. The application of reactive, (ultra)fine and nanometric materials in these compositions has gained increased interest over the past few years. Current research topics focus on the processing, application and characterization of (1) (ultra)fine energetic crystals and composite nano-clusters in plastic bonded explosives, (2) metastable intermolecular composites (MICs) and (3) self-propagating high-temperature synthesis (SHS). In this paper these topics will be highlighted in more detail.

**Keywords:** Nanocomposites, MICs, SHS
**PACS:** 61.46.+w, 81.05.Mh, 82.33.Vx, 82.40.Fp.

## INTRODUCTION

It is well-known that parameters like mean size, shape, purity and crystallographic structure of solid particles generally affect their physico-chemical properties. From a powder technology viewpoint flowability, dust formation, bulk density, agglomeration, mechanical strength, caking, hygroscopicity, etc. can be mentioned as important parameters which may vary depending on the particle properties. For energetic materials also properties like impact and friction sensitivity, susceptibility towards an electrostatic discharge, decomposition, reactivity towards other chemicals, internal crystal quality (purity, inclusions, defect content), burning rate etc. play an important role. These different properties can be effectively used in order to tailor the processing and/or product properties of the energetic formulations containing these energetic particles. In the more recent past nanosized particles became commercially available and were introduced in energetic formulations, since it was assumed that because of their distinctly different particulate properties as compared to their macroscopic counterparts, they would also lead to significant changes in the properties of the energetic formulations in which they are applied.

This paper provides several examples of the work carried out at TNO Defence, Security and Safety on processing, characterization and application of (ultra)fine and nanometric materials in energetic formulations [1].

## EXPERIMENTAL PROCEDURE

### (Ultra)fine energetic crystals and composite nanoclusters

Several co-crystallization methods have been investigated and optimized with the aim to obtain nanocomposite materials consisting of nano-aluminium (n-Al) and the explosive material RDX (cyclotrimethylene trinitramine, $C_3H_6N_6O_6$). For the experiments n-Al from Nanotechnologies Inc. was used, with a mean size of 40.7 nm. The final products were characterized (SEM, thermal analysis, ageing and hazard tests) and one of the co-crystallization methods was successfully scaled up for the preparation of cured, HTPB-based plastic bonded explosives (PBXs), partly containing co-crystallized RDX/n-Al nano-composites with the objective to assess the influence of these nanocomposites on the detonative properties of the PBX. Hereto the PBXs were subjected to shock initiation tests using the Mega Ampere Pulser (MAP) set-up. During this test a flyer plate is accelerated to impact with the PBX sample. The critical flyer velocity $v_{critical}$ is determined at which the sample initiates ('go/no go' level). Details on the MAP set-up can be found elsewhere [2].

## Metastable intermolecular compounds

Welding and incendiary mixtures of e.g. fine aluminium powder with a metallic oxide (e.g. $Fe_2O_3$) – generally referred to as thermites – yield an intense heat when ignited. Thermites are attractive energetic materials because they have a high energy density, show highly exothermic reactions and a high combustion temperature. Nonetheless, the application of thermite materials has been limited because of the relative slow energy release rate compared to other energetic materials. Metastable Intermolecular Composite (MIC) materials are comprised of a mixture of oxidizer and fuel with particle sizes in the nanometer range. One important characteristic of MICs is the fact that the energy release rate can be tailored by varying the size of the ingredients. The nano-scale proximity of the reactants minimizes distances over which the reactants must diffuse in order to react, resulting in dramatically increased rates compared to conventional powder blends. Drawbacks are the sensitivity towards spark ignition and their ageing properties [3-5].

Applications of MICs are e.g. the replacement of Pb-containing igniters [6].

The susceptibility of several MICs and conventional thermites towards an electrostatic discharge has been investigated at TNO by using a specially designed ESD set-up [7]. By including a resistor bank, the spark duration produced by the discharge circuit can be varied from a few μs to ms, since it is known that some materials are more susceptible to initiation by a short, but powerful spark (e.g. explosives), while other materials require a stimulus with a longer duration (e.g. pyrotechnics).

## Self-propagating high temperature synthesis

Self-propagating high-temperature synthesis (SHS), also known as solid-flame combustion or combustion synthesis, is a gasless combustion process. SHS is a cost-effective method for producing high-purity refractory compounds and advanced ceramics, including functionally gradient composite materials [8]. The basis of the reaction synthesis relies on the ability of gasless and highly exothermic reactions to be self-sustaining.

SHS has assumed significance for the production of intermetallics, ceramics and cermets, because it is a very rapid processing technique which avoids complex furnaces [9]. One of its drawbacks is the high porosity of the final product (typically 50% of the Theoretical Maximum Density). Therefore, a subsequent densification step is required, though this is often hard to achieve in ceramic composite materials, because they are highly deformation-resistant. However, when the reaction product is still hot, one may take advantage of its ductile behaviour and apply techniques as hot (quasi-isostatic) pressing, hot rolling or shock waves to eliminate or reduce porosity.

Powder metallurgical arcing contact materials are composite materials used in medium and high current applications such as circuit breakers and switchgears [10]. The homogeneity of their micro-structure is essential for the physical characteristics and the switching performance of the electrical contacts. The main requirements for an arcing contact electrical material include high electrical conductivity, elevated melting point and high hardness. Recently $TiB_2$-based cermets in a Cu or

Al metallic phase for electrical contact applications have been successfully produced via SHS and quasi-isostatic pressing (QIP) at TNO [11]. As starting materials high-purity (99.9%) powders of amorphous B (1 μm), Ti (45 μm), Cu (63 μm) and Al powder (1.2 μm) have been used in the tests.

## RESULTS AND DISCUSSION

### (Ultra)fine energetic crystals and composite nanoclusters

· Several examples of co-crystallized RDX/n-Al composites are shown in Fig. 1. Generally, the size of the more or less rounded particles ranged from several tens of microns to submicron [1]. The coating of the n-Al with RDX was not ideal, although at least a part of the n-Al particles appeared to be covered by RDX, as was shown by X-ray micro-analysis, taken from the centre of the largest RDX/n-Al composite shown in Fig. 1(a). This clearly shows that Al is present, next to C, N and O resulting from RDX. As no n-Al particles or clusters can be observed at the surface of this RDX particle – contrary to Fig. 1(b) – it is concluded that the n-Al is inside the RDX crystal.

The impact and friction sensitivity data shows that RDX mixed with n-Al (40.7 nm) is more friction sensitive than the same RDX mixed with 6 μm Al (80 vs. 160 N). The co-crystallized RDX/n-Al is even more sensitive (60 N). Probably this is related to a more intimate contact between the RDX and n-Al particles in the co-crystallized sample compared to the physical blends. Clearly, the presence of (n-)Al sensitizes the RDX, since a blank measurement (no Al present) yields a friction sensitivity of 240 N. A re-crystallized RDX sample without (n-)Al showed the lowest sensitivity towards a frictional stimulus (360 N). For all samples tested, the impact sensitivity did not change (all 7.5 Nm).

Thermal analysis tests (TG/DTA, DSC) were performed in order to assess differences in reactivity and decomposition characteristics of RDX/n-Al compositions. The results obtained with co-crystallized RDX/n-Al and a physical blend of recrystallized RDX and n-Al indicated a lowering of the melting and decomposition temperatures of

RDX compared to what is generally found for (coarse) RDX. The results indicate that the recrystallized RDX consists of a range of particle sizes, the smaller ones revealing properties similar to nanometric RDX and the larger ones showing features of 'standard' RDX. The lower melting and decomposition temperatures of nanometric RDX can be attributed to the fact that as a result of the high surface area of the nanometric particles (large surface to volume ratio), the bulk properties become governed by surface properties. This is substantiated by other literature data [12] and quantum mechanical calculations [13].

*(a)*

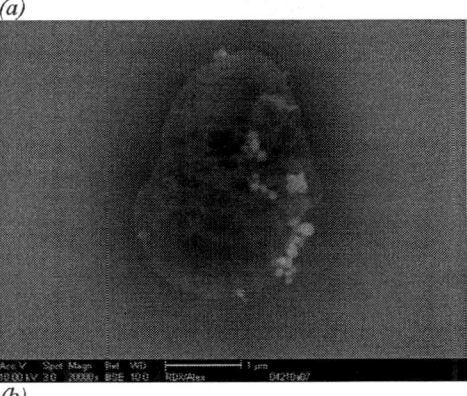

*(b)*

**Figure 1.** Examples of co-crystallized RDX/n-Al composites. In (b) n-Al particles are attached to an RDX particle.

Plastic bonded explosives (PBXs) were prepared according to the compositions shown in Table 1. In order to prevent problems with PBX processing (increased viscosities when using small particles), the fine particle content was limited to ~18 wt%. The results of the shock initiation (MAP) tests are

**Table 1.** PBX batches prepared on 250 g scale (contents in wt%). Solid load is 76.0 wt% for all formulations.

| PBX | Coarse RDX | Fine RDX incl. n-Al [a] | Fine RDX | n-Al [a] | Al (6 μm) | HTPB |
|---|---|---|---|---|---|---|
| RU-172 | 57.9 | – | 14.5 | – | 3.6 | 24.0 |
| RU-173 | 57.9 | – | 14.5 | 3.6 | – | 24.0 |
| RU-174 | 57.9 | 18.1 (co-cryst. with n-Al) | – | – | – | 24.0 |

[a] n-Al from Technanogy, mean size 40.7 nm, $Al_2O_3$ content 56.6 wt%.

listed in Table 2. The PBX containing conventional Al (6 μm) is less shock sensitive than the PBXs containing n-Al (higher $v_{critical}$). No difference was observed between the PBX with co-crystallized RDX/n-Al and the physical blend of RDX and n-Al. The detonation velocity is the same for all three PBXs. The difference between the reference and the n-Al containing PBXs may well be related to the difference in the overall active Al content in the 6 μm Al (~99 wt%) and the n-Al (56.6 wt%). However, it is practically impossible to replace Al with n-Al while keeping all other properties constant (solid load, RDX content, active Al content, etc.).

**Table 2.** Shock initiation results of three PBX formulations by flyer impact (125 μm kapton); $v_{critical}$ = critical flyer impact velocity; $V_{det}$ = measured detonation velocity.

| PBX | $v_{critical}$, km/s | $V_{det}$, km/s |
|---|---|---|
| RU-172 | 3.7 ± 0.2 | 7.3 ± 0.2 |
| RU-173 | 3.1 ± 0.2 | 7.3 ± 0.2 |
| RU-174 | 3.1 ± 0.2 | 7.2 ± 0.2 |

Thermodynamic calculations using a kinetic module in Cheetah 2.0, on composition PBXN109, have addressed the influence of Al reaction rates on $V_{det}$, detonation pressure ($p_{det}$) and temperature [14]. Although a high Al reaction rate leads to a fast heat release, $V_{det}$ and $p_{det}$ may be negatively influenced because of the fact that the reaction of Al mainly yields solid reaction products ($Al_2O_3$), thereby taking away oxygen from the gaseous detonation products. This may therefore explain some of the results of (other) Al-containing PBXs in which conventional Al has been replaced by n-Al.

## Metastable intermolecular compounds

Table 3 shows the threshold ESD energies, i.e. the minimum energy for the ignition of indicated samples. The value $\frac{1}{2}CV^2$ is the energy stored on the capacitor. When possible the value of $E_{spark}$, the energy deposited in the spark through the sample, is given as well. With no series resistance in the discharge circuit, short duration pulses of the order of μs are created. With a 10 kΩ resistance, the pulse duration will be on the order of 0.1-1 ms, depending on the capacitor value.

The order of ESD sensitivities is more or less the same looking either at the capacitor energy in a short and long duration pulse, or at $E_{spark}$. Samples #1, #4 and #6 are the most sensitive, and #7 is the least sensitive. Actually #6 is so sensitive that the reproducibility in sensitivity threshold is difficult to measure and/or the ESD equipment has difficulty in reproducibly generating a low $E_{spark}$.

The ESD sensitivity depends on the mean size of the reactants used. This is evident by comparing samples #7 and #8, containing Al and n-Al respectively, and previous measurements presented on various mixtures of n-Al and (n-)$MoO_3$ [7]. Sample #4, though consisting of micronsized particles of Al and $Bi_2O_3$, is rather spark sensitive, indicating that a nano-version of this composition might initiate even more easily.

Samples #1 and #2 have the same stoichiometry of n-Al and n-$Fe_2O_3$, but use a slightly different mean particle size of Al. Assuming that reactivity increases with decreasing particle size, sample #2 was expected to be more sensitive than #1, but the opposite is observed. This may well be related to the higher $Al_2O_3$ content of the n-Al used in #2.

Another interesting phenomenon is the influence of sample stoichiometry on ESD sensitivity: the stoichiometric mixture of B and Ti (sample #6) is more reactive than the non-stoichiometric one (#5).

MICs can also be rather impact and friction sensitive, as is illustrated by the following examples: the impact sensitivity of a mixture containing n-Al, $MoO_3$, NC and IDP

densities of the $TiB_2$-Al cermets are high (max. 98% of the theoretical mean density, TMD), but those of the Cu-based cermets are too low (57-67% TMD). Numerical simulations with ABAQUS have

**Table 3.** ESD sensitivities of several MICs compared to conventional thermites.

| Sample | Sample ingredients | Specifications | Resistance, k$\Omega$ | 1/2 $CV^2$ ($E_{spark}$[a]), J |
|--------|--------------------|----------------|------------------------|--------------------------------|
| #1 | 2 n-Al + n-$Fe_2O_3$ | n-Al 100 nm ($Al_2O_3$ content 31.9 wt%); n-$Fe_2O_3$ 90 nm | 0<br>10 | 0.026<br>1.8 |
| #2 | 2 n-Al + n-$Fe_2O_3$ | n-Al 40.7 nm ($Al_2O_3$ content 56.6 wt%); n-$Fe_2O_3$ 90 nm | 0<br>10 | 0.29 ($3.4 \cdot 10^{-3}$)<br>4.4 |
| #3 | C + Ti | C, Ti < 10 μm | 0<br>10 | 0.40 ($1.3 \cdot 10^{-3}$)<br>16.8 |
| #4 | 2 Al + $Bi_2O_3$ | Al < 10 μm; $Bi_2O_3$ < 6 μm | 0<br>10 | 0.051 ($\sim 2 \cdot 10^{-3}$)<br>0.50 |
| #5 | B + Ti | B < 1 μm; Ti < 10 μm; (non-stoichiometric ratio) | 0<br>10 | 0.40 ($2.3 \cdot 10^{-3}$)<br>6.4 |
| #6 | 2 B + Ti | B < 1 μm; Ti < 10 μm (stoichiometric ratio) | 0<br>10 | 0.13 ($0.6 \cdot 10^{-3}$)<br>0.01 |
| #7 | 2 Al + Ti | Al, Ti < 10 μm | 0<br>10 | 1.3 ($13.8 \cdot 10^{-3}$)<br>no initiation |
| #8 | 2 n-Al + Ti | n-Al 40.7 nm ($Al_2O_3$ content 56.6 wt%); Ti < 10 μm | 0<br>10 | 0.58 ($6.2 \cdot 10^{-3}$)<br>16.8 |

[a] True spark energy is measured from the voltage and current through the sample. At the moment sensors permit reliable measurements only at spark energies above ~ 1 mJ and at relatively short duration pulses (so only at 0 k$\Omega$).

(65.0/24.4/5.3/5.3 wt%) was > 50 Nm, whereas for composition with n-Al, n-$MoO_3$ and VitonA (65.0/24.4/10.6 wt%) this was 4 Nm; the friction sensitivity in both cases was found to be < 5 N. These hazard properties show that MICs can be extremely sensitive towards impact, friction and spark, and should be handled only if proper safety measures have been applied.

A potential application of MICs is as energetic fragments which ignite upon impact with a target [15]. Currently tests are planned with pressed MIC and conventional thermite samples (selected from table 3) fired onto a target with a 30 mm caliber canon, as well as experiments to determine the deformation sensitivity of the MICs using a so-called ballistic impact chamber (BIC) test set-up [16], which has recently been constructed at TNO.

### Self-propagating high temperature synthesis

SHS experiments have been carried out with $TiB_2$ in an Al and Cu metallic phase. The relative

been performed, modelling the cooling of the $TiB_2$ cermet after the SHS reaction in order to optimize the QIP conditions for reaching higher densities of the $TiB_2$-Cu cermets. The simulations show that the Cu-based cermet cools down sufficiently slowly, allowing an extension of the time-window during which the load is applied with the objective to increase the density by squeezing molten Cu into the open pores in the ceramic phase without cracking or breaking the ceramic [17].

In the near future the focus of this research will shift to the kinetics of an SHS reaction, i.e. assessment of the combustion velocity as a function of particle size and reactants ratio.

### CONCLUSIONS

This paper illustrates that the availability of (ultra)fine and nanomaterials provides new tools to tailor the properties of energetic formulations. Their use, however, may also introduce drawbacks, such as an increased susceptibility to impact, friction and spark, requiring adequate safety measures. Another disadvantage is related to potential difficulties in the processing of castable compositions like PBXs and solid composite propellants when introducing nanomaterials, since these will increase the viscosity of the mixture. This should be able to be circumvented by using an extrusion rather than a casting technique. Furthermore, the combination of experiments and numerical modelling using dedicated software tools, has proven to be a successful approach in developing a better understanding of the phenomena playing a role when including (ultra)fine and/or nanomaterials in energetic formulations. Obviously, the application of nanomaterials is not limited to the energetic formulations discussed in this paper, but may also comprise rocket and gun propellants. TNO is involved in these developments as well.

## ACKNOWLEDGEMENTS

The authors acknowledge financial support from EOARD (contract F61775-02-C4093) and The Netherlands Institute for Metals Research (partner agreement MCX.02130).

## REFERENCES

1. Van der Heijden, A.E.D.M., Bouma, R.H.B., Van der Steen, A.C., and Fischer, H.R., "Application and characterization of nanomaterials in energetic compositions", Proc. Materials Research Society Vol. 800, 2004, pAA5.6
2. Prinse, W.C., Van Esveld, R.J., Oostdam, R., Van Rooijen, M.P., and Bouma, R.H.B., Proc. 23rd International Congress on High-Speed Photography and Photonics, 20-25 September 1998, Moscow, Russia
3. Kearns, M., "Development and applications of ultrafine aluminium powders", Materials Science and Engineering A 375-377, 2004, 120-126
4. Jones, D.E.G., Turcotte, R., Kwok, Q.S.M., Vachon, M., Guindon, L., Lepage, D., and Gertsman, V.Y., "Thermal characterization of coated aluminum nanopowders", CERL report 2004-25 (OP-J), NATAS Conference, Williamsburg, VA, USA, 2004
5. Walter, K.C., Aumann, C.E., Carpenter, R.D., O'Neill, E.H., and Pesiri, D.R., "Energetic materials development at Technanogy Materials Development", Proc. Materials Research Society Vol. 800, 2004, pAA1.3
6. Son, S.F., Hiskey, M.A., Naud, D.L., Busse, J.R., and Asay, B.W., Lead-free electric matches, Proc. 29th Intl. Pyrotechnics Seminar, Westminster, CO, USA, 2002, p871
7. Martinez Pacheco, M., Bouma, R.H.B., and Katgerman, L., "Electrostatic discharge initiation of Ti+C mixtures and the thermite Al+MoO$_3$", 35th Intl. Conf. of ICT, Karlsruhe, Germany, 2004
8. Khina, B.B., and Loban, D.N., "Modeling SHS in porous systems using cellular automata approach" National Academy of Sciences, 2000, pp.412-418
9. Holt, J.B., "Self-propagating, high-temperature synthesis", Engineered Materials Handbook, vol.4 Ceramics and Glasses, pp.227-231
10. Schrott, O., "Preparation of WC/Ag contact materials with different homogeneity", Struers Journal of Materialography, 2003
11. Martinez Pacheco, M., Bouma, R.H.B., Carton, E.P., Stuivinga, M., and Katgerman, L., "Synthesis of electrical contact materials via combustion synthesis reactions", NIMR Conference Building Bridges in Metallurgy, 2004, poster nr.10-1
12. Pivkina, A., Frolov, Y., Zavyalov, S., Ul'yanova, P., and Schoonman, J., "Thermal properties of nanosized energetic materials", (see: http://www.udman.ru/nano/11694F19.htm)
13. Kuklja, M.M., "Thermal decomposition of solid cyclotrimethylene trinitramine", J. Phys. Chem. B, 105 (2001) 10159
14. Verbeek, H.J., and Bouma, R.H.B., "Kinetic thermodynamic calculations for composition PBXN-109", presented at the Gordon Conference, Energetic Materials, Tilton NH, USA, June 2002
15. Jones, J.W., "Energy dense explosives", US Patent US 6,679,960 B2, Jan. 20, 2004
16. Namkung, J., and Coffey, C.S., "Plastic deformation rate and initiation of crystalline explosives", Shock Compression of Condensed Matter – 2001, ed. Furnish, M.D., Thadhani, N.N., and Horie, Y., American Institute of Physics, pp1003-1006, 2002
17. Martinez Pacheco, M., Bouma, R.H.B., and Katgerman, L., "Experiments and numerical modelling of gasless combustion processes", 36th Intl. Conf. of ICT, Karlsruhe, Germany, 2005

CP845, *Shock Compression of Condensed Matter - 2005*,
edited by M. D. Furnish, M. Elert, T. P. Russell, and C. T. White
© 2006 American Institute of Physics 0-7354-0341-4/06/$23.00

# SHOCK INITIATION EXPERIMENTS ON PBX9501 EXPLOSIVE AT 150°C FOR IGNITION AND GROWTH MODELING

## Kevin S. Vandersall, Craig M. Tarver, Frank Garcia, and Paul A. Urtiew

*Energetic Materials Center*
*Lawrence Livermore National Laboratory*
*Livermore, CA 94550*

**Abstract.** Shock initiation experiments on the explosive PBX9501 (95% HMX, 2.5% estane, and 2.5% nitroplasticizer by weight) were performed at 150°C to obtain in-situ pressure gauge data and Ignition and Growth modeling parameters. A 101 mm diameter propellant driven gas gun was utilized to initiate the PBX9501 explosive with manganin piezoresistive pressure gauge packages placed between sample slices. The run-distance-to-detonation points on the Pop-plot for these experiments showed agreement with previously published data and Ignition and Growth modeling parameters were obtained with a good fit to the experimental data. This parameter set will allow accurate code predictions to be calculated for safety scenarios involving PBX9501 explosives at temperatures close to 150°C.

**Keywords:** Explosive, PBX9501, shock to detonation transition, ignition and growth
**PACS:** 82.33.Vx, 82.40.Fp

## INTRODUCTION

Interest exists in studying safety to shock impact of HMX (octahydro-1,3,5,7-tetranitro-1,3,5,7-tetrazocine) based explosives such as the commonly used PBX 9501 (95% HMX, 2.5% estane, and 2.5% BDNPA-F nitroplasticizer by weight), especially at elevated temperatures where the relative sensitivity to shock increases. Prior studies on PBX 9501 include wedge tests [1], embedded particle velocity gauges [2-4], VISAR at low input shock pressures [5,6], and embedded manganin gauges [7] at both ambient and elevated temperature. Another HMX based explosive, LX-04 (85% HMX, 15% Viton) has also been studied extensively at elevated temperatures [7,8]. In this paper, the shock sensitivity of PBX 9501 at 150°C was measured using embedded manganin pressure gauges.

## EXPERIMENTAL PROCEDURE

Shock initiation experiments were performed on the explosive PBX 9501 using the 101 mm diameter propellant driven gas gun at Lawrence Livermore National Laboratory (LLNL). Figure 1 shows a description of a typical experiment. The projectile consisted of a polycarbonate sabot with a 6061-T6 Aluminum flyer plate on the impact surface. As seen in Figure 1, the target includes buffer plates in contact with the high explosive at both the front and rear of the assembly to hold the material in place and sandwich the nichrome heater foils. The explosive was in the form of thin disks (with starting density approximately 1.82 g/cm$^3$) with gauge packages inserted in between with the total explosive thickness being 20 mm. The manganin piezoresistive foil pressure gauges placed within the explosive sample were "armored" with sheets of Teflon insulation on each side of the gauge. Manganin is a copper-manganese alloy that changes electrical resistance with pressure (i.e. piezoresistive). Also used were PZT Crystal pins to measure the projectile velocity and tilt (planarity of impact). During the experiment, oscilloscopes measure change of voltage as result of resistance change in the gauges which were then converted to pressure using the hysteresis corrected calibration curve published elsewhere [9,10].

From the data of the shock arrival times of the gauge locations, a plot of distance vs. time ("x-t plot") is constructed with the slope of the plotted lines yielding the shock velocities with two lines apparent, a line for the un-reacted state as it reacts and a line representing the detonation velocity. The intersection of these two lines is taken as the "run-distance-to-detonation," which is then plotted on the "Pop-Plot" showing the run-distance-to-detonation as a function of the input pressure in log-log space.

## REACTIVE FLOW MODELING

The Ignition and Growth reactive flow model [11] uses two Jones-Wilkins-Lee (JWL) equations of state, one for the un-reacted explosive and another one for the reaction products, in the form:

$$p = Ae^{-R_1V} + Be^{-R_2V} + \omega C_V T/V \quad (1)$$

where p is pressure in Megabars, V is relative volume, T is temperature, $\omega$ is the Gruneisen coefficient, $C_V$ is the average heat capacity, and A, B, $R_1$ and $R_2$ are constants. The equations of state are fitted to the available shock Hugoniot data. Table 1 contains the modeling parameters and reaction rate constants for these experiments. The reaction rate equation is:

$$dF/dt = \underbrace{I(1-F)^b(\rho/\rho_0 - 1 - a)^x}_{0<F<F_{Igmax}} +$$

$$\underbrace{G_1(1-F)^c F^d p^y}_{0<F<F_{G_1max}} + \underbrace{G_2(1-F)^e F^g p^z}_{F_{G_2min}<F<1} \quad (2)$$

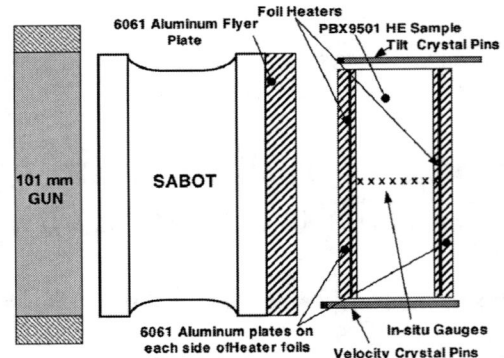

**FIGURE 1.** Typical description of a shock initiation experiment.

**Table 1.** Ignition and Growth modeling parameters.

| UNREACTED JWL | PRODUCT JWL |
|---|---|
| A=7320 Mbar | A=16.689 Mbar |
| B=-0.065278 Mbar | B=0.5969 Mbar |
| $R_1$=14.1 | $R_1$=5.9 |
| $R_2$=1.41 | $R_2$=2.1 |
| $\omega$=0.8867 | $\omega$=0.450 |
| $C_V$=2.7806x10$^{-5}$ Mbar/K | $C_V$=1.0x10$^{-5}$ Mbar/K |
| $T_0$ = 423°K | $E_0$=0.102 Mbar |
| Shear Modulus=0.0354 Mbar | - |
| Yield Strength=0.002 Mbar | - |
| $\rho_0$=1.762 g/cm$^3$ | - |

| REACTION RATES | |
|---|---|
| a=0 | x=4.0 |
| b=0.667 | y=1.0 |
| c=0.667 | z=2.0 |
| d=0.277 | $F_{igmax}$=0.3 |
| e=0.333 | $F_{G1max}$=0.5 |
| g=1.0 | $F_{G2min}$ =0.5 |
| I=1.4 x 10$^{11}$ $\mu s^{-1}$ | $G_1$=190 Mbar$^{-2}\mu s^{-1}$ |
| - | $G_2$=400 Mbar$^{-2}\mu s^{-1}$ |

where F is the fraction reacted, t is time in $\mu s$, $\rho$ is the current density in g/cm$^3$, $\rho_0$ is the initial density (calculated based on thermal expansion data), p is pressure in Mbars, and I, $G_1$, $G_2$, a, b, c, d, e, g, x, y, and z are constants. This reaction rate law models the three stages of reaction generally observed during shock initiation of solid explosives. Table 2 details the Gruneisen parameters used.

**Table 2.** Gruneisen parameters for inert materials.

| INERT | $\rho_0$ (g/cc) | C (km/s) | $S_1$ | $S_2$ | $S_3$ | $\gamma_0$ | a |
|---|---|---|---|---|---|---|---|
| 6061-T6 Al | 2.703 | 5.24 | 1.4 | 0.0 | 0.0 | 1.97 | 0.48 |
| Teflon | 2.15 | 1.68 | 1.123 | 3.98 | -5.8 | 0.59 | 0.0 |

## RESULTS/DISCUSSION

Table 3 contains the experimental flyer velocities, impact pressures, and run distances to detonation for the two PBX 9501 shots performed at 150°C.

**Table 3.** PBX9501 gun experiments at 150°C.

| SHOT | IMPACT VELOCITY | INPUT PRESSURE | RUN TO DET |
|---|---|---|---|
| 4663 | 0.72 km/s | 3.3 GPa | 7.1 mm |
| 4664 | 0.54 km/s | 2.3 GPa | 11.2 mm |

The resulting data points are plotted on the Pop-plot as shown in Figure 2. The data from this work are plotted as filled circles and agree well with the previous data by LANL at 150°C [1] shown in the open circles. The other data from the same work are also plotted for reference to compare this data to other test temperatures. It is unclear at this time all of the factors (density changes at temperature, thermal energy, etc.) and to what amounts attribute to this sensitivity increase.

The in-situ gauge records are shown in Figures 3 and 5 for experiments 4663 and 4664 respectively. An increase in pressure can be observed as the shock progresses through and reacts the explosive material until a full detonation is observed. Ignition and Growth reactive flow modeling results are shown in Figures 4 and 6 in the form of simulated gauge records. They simulate the experimental records in Figures 2 and 4 respectively. From

comparing these records a good agreement can be seen.

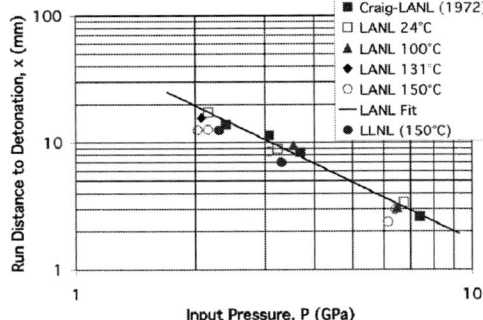

**FIGURE 2.** Pop-Plot comparing the data from this work with that of previous experiments.

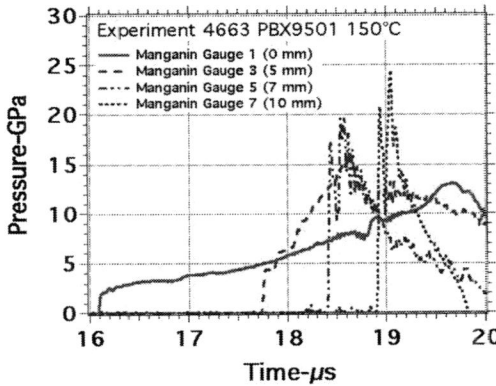

**FIGURE 3.** Experimental pressure histories for 150°C PBX 9501 impacted by an aluminum flyer plate at 720 m/s.

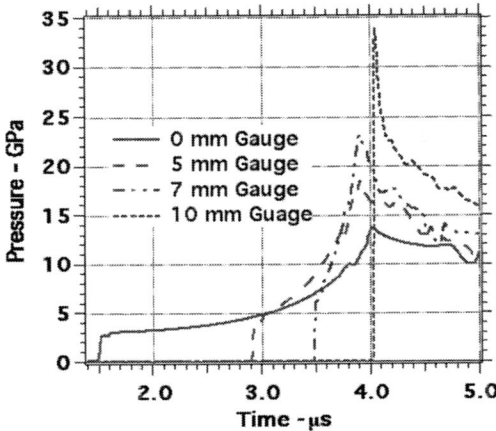

**FIGURE 4.** Calculated pressure histories for 150°C PBX 9501 impacted by an aluminum flyer plate at 720 m/s.

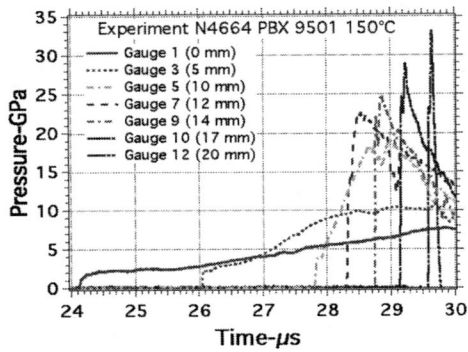

**FIGURE 5.** Experimental pressure histories for 150°C PBX 9501 impacted by an aluminum flyer plate at 540 m/s.

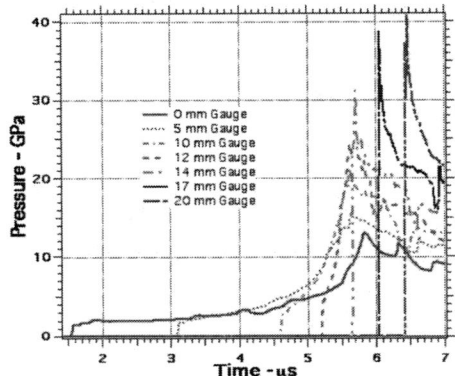

**FIGURE 6.** Calculated pressure histories for 150°C PBX 9501 impacted by an aluminum flyer plate at 540 m/s.

## SUMMARY

Shock initiation experiments on the explosive PBX9501 (95% HMX, 2.5% estane, and 2.5% nitroplasticizer by weight) were performed at 150°C to obtain in-situ pressure gauge data and Ignition and Growth modeling parameters. The run-distance-to-detonation points on the Pop-plot for these experiments showed agreement with previously published data and Ignition and Growth modeling parameters were obtained with a good fit to the experimental data.

## ACKNOWLEDGEMENTS

The High Explosives Response program provided funding for this research. Special thanks go to the 101 mm gun crew in the High Explosives Application Facility (HEAF) including Rich Villafana, Steve Kenitzer, and Gary Steinhour. This work was performed under the auspices of the U. S. Department of Energy by the University of California, Lawrence Livermore National Laboratory under Contract No. W-7405-Eng-48.

## REFERENCES

1. LASL Explosive Property Data, Terry R. Gibbs and Alphonse Popolato, Editors, University of California Press, pp. 353-358, 1980.
2. Sheffield, S. A., Gustavsen, R. L., Hill, L. G., and Alcon, R. R., *Eleventh International Detonation Symposium*, ONR 33300-5, Snommass, CO, 1998, pp. 451-458.
3. Sheffield, S. A., Gustavsen, R. L., and Alcon, R. R., Shock Compression of Condensed Matter-1999, M. D. Furnish, L. C. Chhabildas, and R. S. Hixson, eds., AIP Conference Proceedings 505, Snowbird, UT, 1999, pp. 1043-1048.
4. Gustavsen, R. L., Sheffield, S. A., Alcon, R. R., and Hill, L. G., "Shock Initiation of New and Aged PBX 9501," Proceedings of the 12th International Symposium on Detonation, San Diego, CA, August, 2002, pp. 530-537.
5. Dick, J. J., Shock Compression of Condensed Matter-1999, Furnish, M. D Chhabildas, L. C., and Hixson, R. S.,eds., AIP Conference Proceedings 505, Snowbird, UT, 1999, pp. 683-686.
6. Dick, J, J., Martinez, A. R., and Hixson, R. S., *Eleventh International Detonation Symposium*, ONR 33300-5, Snommass, CO, 1998, pp. 317-324.
7. Tarver, C. M., Forbes, J. W., Urtiew, P. A., Garcia, F., "Shock Sensitivity of LX-04 at 150°C," Shock Compression of Condensed Matter-1999, pp.891-894.
8. Urtiew, P. A., Forbes, J. W., Tarver, C. M., Vandersall, K. S., Garcia, F., Greenwood, D. W., Hsu, P. C., and Maienschein, J. L., "Shock Sensitivity of LX-04 with Delta Phase HMX at Elevated Temperatures," Shock Compression of Condensed Matter - 2003, pp. 1053-1056.
9. Vantine, H.C., Erickson, L.M. and Janzen, J., "Hysteresis-Corrected Calibration of Manganin under Shock Loading", J. Appl. Phys., **51** (4), April 1980.
10. Vantine H., Chan J., Erickson L. M., Janzen J., Lee R. and Weingart R. C., "Precision Stress Measurements in Severe Shock-Wave Environments with Low Impedance Manganin Gauges," Rev. Sci. Instr., **51**. pp. 116-122 (1980).
11. Tarver, C. M., Hallquist, J. O., and Erikson, L. M., "Modeling Short Pulse Duration Shock Initiation of Solid Explosives," *Eighth Symposium (International) on Detonation*, Naval Surface Weapons Center NSWC MP86-194, Albuquerque, NM, 1985, pp. 951-961.

CP845, *Shock Compression of Condensed Matter - 2005*,
edited by M. D. Furnish, M. Elert, T. P. Russell, and C. T. White
2006 American Institute of Physics 0-7354-0341-4/06/$23.00

# VIBRATIONAL SPECTROSCOPIC STUDIES OF REDUCED-SENSITIVITY RDX UNDER STATIC COMPRESSION

## Chak P. Wong[*] and Jared C. Gump

*Naval Surface Warfare Center - Indian Head Division, Indian Head, MD 20640*
*\* Author to whom correspondence should be addressed*

**Abstract.** Explosive formulations with reduced-sensitivity RDX showed reduced shock sensitivity using Naval Ordnance Laboratory (NOL) Large Scale Gap Test, compared with similar formulations using standard RDX. Molecular processes responsible for the reduction of sensitivity are unknown and are crucial for formulation development. Vibrational spectroscopy at static high pressure may shed light on the mechanisms responsible for the reduced shock sensitivity as shown by the NOL Large Scale Gap Test. I-RDX®, a form of reduced- sensitivity RDX was subjected to static compression at ambient temperature in a Merrill-Bassett sapphire cell from ambient to about 6 GPa. The spectroscopic techniques used were Raman and Fourier-Transform IR (FTIR). The pressure dependence of the Raman mode frequencies of I-RDX® was determined and compared with that of standard RDX. The behavior of I-RDX® near the pressure at which standard RDX, at ambient temperature, undergoes a phase transition from the $\alpha$ to the $\gamma$ polymorph is presented.

**Keywords:** Vibrational spectroscopy, RDX, Raman, IR.
**PACS:** 78.30.-j, 07.35.+k.

## INTRODUCTION

Many explosives use RDX as an ingredient. Its structural properties have great influence on the sensitivity and safe handling of munitions. Various kinds of reduced-sensitivity RDX have been discovered [1] that yielded reduced shock sensitivity when incorporated in formulations. The exact physical reasons for this reduction are not known, although microscopy [2] and Nuclear Quadrupole Resonance (NQR) [3] have shown that this type of RDX has fewer defects. We use vibrational spectroscopy to characterize both a shock insensitive form of RDX, I-RDX®, and a standard shock sensitive form of RDX, called standard RDX. The results will also be compared with Raman [4] and IR [5] studies under compression reported for RDX.

We have performed vibrational studies of I-RDX® under static compression using Raman and IR spectroscopy. The main objective of this work is to determine pressure-induced changes in Raman and IR spectra of I-RDX®.

## MATERIALS AND METHOD

Powdered samples of I-RDX® supplied by SNPE (~ 0.1% HMX), France and standard RDX with ~ 7 % HMX from Holston, USA were used for our studies. This amount of HMX is within the impurity range for RDX supplied by this company. Samples were loaded in a Merrill-Bassett Gem Anvil Cell, with sapphire anvils. The gasket was 0.127 mm thick Inconel 600 alloy with a 0.250 mm gasket hole. Raman spectra were collected with a Spex 1403 spectrometer of 0.85 m focal length coupled to a liquid nitrogen-cooled CCD. Back scattering geometry was used for spectra collection and the 514.5 nm line of an argon ion laser was used for sample excitation. Raman frequencies from 50 - 3100 cm$^{-1}$ were obtained. Infrared absorption spectra were collected with a Thermo Nicolet Nexus

870 FTIR spectrometer from 650 to 4000 cm⁻¹. Due to absorption of IR radiation by sapphire, the useful spectral region was reduced to 1800-4000 cm⁻¹. Pressure was determined by monitoring the pressure-dependent fluorescence of a ruby chip in the cell. Non-hydrostatic loading was used for all the samples studied. No pressure medium was used to avoid medium - sample interaction. All studies were at ambient temperature. Raman spectra were collected from ambient to ~ 6 GPa and IR spectra from ambient to ~5 GPa.

### RESULTS

Raman and IR spectra of I-RDX® under static compression are shown in Figures 1-4. I-RDX® spectral lines were identified to be RDX lines by comparison with known RDX Raman and IR studies [6]. Mode assignments were those reported in the literature [6]. Only RDX lines were observed for I-RDX®. No HMX lines were observed, since the trace amount of HMX was below the detectable limits of our spectrometers. The absence of HMX lines was consistent with the fact that I-RDX® has essentially no HMX [7]. For comparison we have verified with Raman and IR spectroscopy that the standard RDX sample contained HMX as an impurity. In the following we will focus on I-RDX® results. Due to the limitation of space only spectra in a few spectral regions are shown.

Figure 1 shows the Raman spectra of lattice modes and some of the lowest energy internal modes as a function of pressure. The lines at ~76 and 116 cm⁻¹ under ambient conditions are laser plasma lines. The lattice modes are at ~75, 87 and 105 cm⁻¹ at ambient. The internal modes are the N-NO₂ torsion and C-N-C bending with torsion modes [6]. Figure 2 shows the Raman spectra of internal modes of higher energy. These included stretching ring modes and other modes [6]. The CH₂ internal stretching modes of I-RDX® [6] are shown in Figures 3 and 4 observed with Raman and IR spectroscopy respectively.

The IR spectra in Figure 4 show the pressure dependence of I-RDX®, as well as one standard RDX spectrum at 0.22 GPa for comparison. Standard RDX spectra under compression up to 5 GPa were also obtained. The only significant difference between the compression of standard RDX versus I-RDX® was the appearance of two HMX impurity lines in the standard RDX spectra. The pressure-induced peak shifts for both samples were reversible, as seen for I-RDX® in the top two spectra in Figure 4.

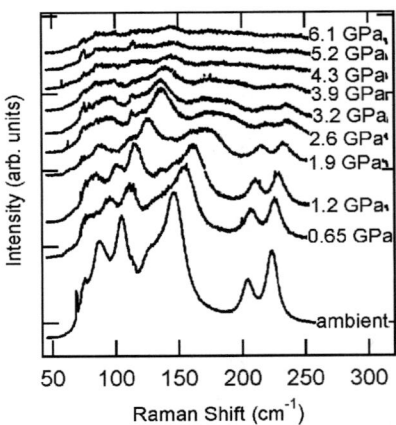

**Figure 1.** Pressure dependence of I-RDX® Raman spectra showing lattice modes and low energy internal modes.

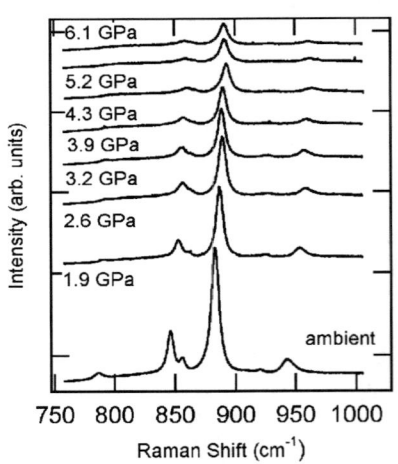

**Figure 2.** Pressure dependence of I-RDX® Raman spectra showing higher-energy internal modes.

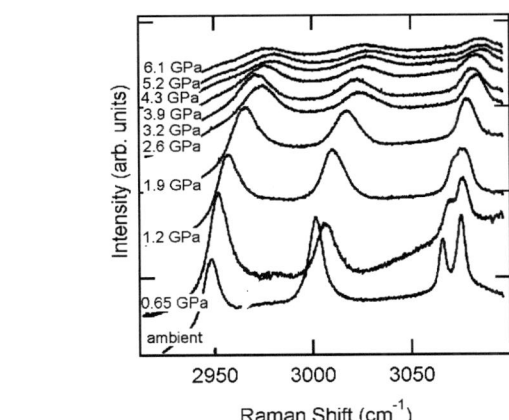

**Figure 3.** Pressure dependence of I-RDX® Raman spectra showing $CH_2$ internal modes.

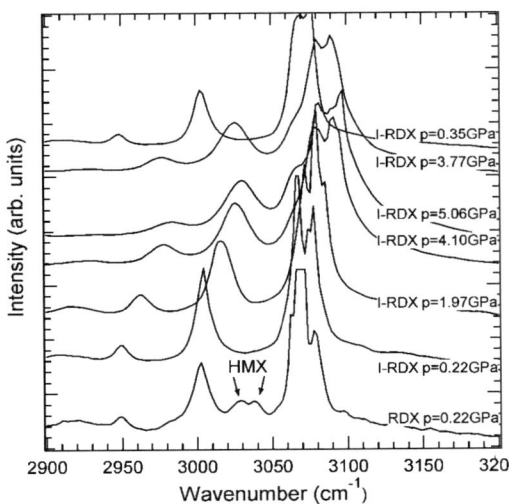

**Figure 4.** Pressure dependence of IR spectra of $CH_2$ internal modes for I-RDX®. The bottom spectrum is a standard RDX spectrum shown for comparison with I-RDX® spectra. The top two spectra were taken during decompression. Flat tops on some peaks result from detector saturation.

## DISCUSSION

Figures 1-4 show that I-RDX® Raman and IR modes increase in wavenumbers as pressure was increased. The Raman and IR lines also decreased in intensity and broadened with increasing pressure.

Peak positions were obtained by fitting with Gaussian waveforms. From the peak position versus pressure graphs, assuming linear dependence on pressure, slopes were generated for data below ~ 3 GPa and compared with those reported in reference 4 which reported RDX studies to a pressure of 3 GPa. The slopes are the pressure coefficients, $dv/dP$, where $v$ is the Raman frequency and P is the pressure.

Table 1 shows the Raman pressure coefficients of I-RDX® and standard RDX in a pressure range from ambient to ~3.0 GPa with maximum uncertainties of ~ 0.12 and ~ 0.08 cm$^{-1}$/ kbar for the lattice modes and internal modes respectively. The pressure coefficients for lattice modes and internal modes are similar between both types of RDX. However, the lattice mode pressure coefficients of I-RDX® and standard RDX obtained in our study were larger than those reported in [4], although the internal mode pressure coefficients reported in that work were similar to those we report here. The reason for this difference in the lattice mode pressure coefficients is not known, but lattice modes are more sensitive to molecular environments.

Figure 5 shows the dependence of peak positions with pressure for a few vibrational modes. It is observed that the Raman data is linearly continuous to ~ 4 GPa but the slopes decrease at higher pressures. This change in slope at ~ 4 GPa clearly corresponds to the $\alpha - \gamma$ transformation in RDX as observed in other work [5,8]. The IR data show similar trends upon compression up to the maximum pressure achieved. It is unclear whether fluctuation in the Raman data from 2-3 GPa is all associated with the phase change or if some unidentified source of uncertainty is also contributing.

Because the trace amount of HMX in I-RDX® is beyond the detectable limits of our instruments, the relationship between HMX and pressure coefficients has not been investigated. In particular it is not clear whether the absence of HMX was the reason for the reduced sensitivity in formulations. It is not certain whether variation in HMX content contributed to the smaller lattice mode Raman pressure coefficients reported in the literature [4] since the sample defect or impurity levels were not described.

**Table 1**. Raman Pressure Coefficients of I-RDX ®
and RDX*

| ν (cm⁻¹) | dv/dP(cm⁻¹)/ kbar | Assignment |
|---|---|---|
| 75(68) | 0.77,<u>0.6</u>,(0.35) | Lattice Mode |
| 87.1(84) | 1.3,<u>1.2</u>,(0.31) | Lattice Mode |
| 104.7(102) | 1.2,<u>1.3</u>,(0.69) | Lattice Mode |
| 883.4(881) | 0.25,<u>0.22</u>,(0.16) | Ring stretch |
| (1209) | (0.38) | Ring stretch |
| 1214.9 | 0.28,<u>0.23</u> | Ring |
| (1565) | (0.2) | $NO_2$ a stretch |
| 1571.1 | 0.17,<u>0.17</u> | $NO_2$ a stretch |
| (1586) | (0.03) | $NO_2$ a stretch |
| 1592.7 | 0.08,<u>~0</u> | $NO_2$ a stretch |
| (1592) | (0.22) | $NO_2$ a stretch |
| 1596.4 | 0.22,<u>0.15</u> | $NO_2$ a stretch |
| (2938) | (1.05) | $CH_2$ stretch |
| 2947.8 | 1.0,<u>0.84</u> | $CH_2$ stretch |
| (2990) | (1.0) | $CH_2$ stretch |
| 3001.1 | 0.85,<u>0.78</u> | $CH_2$ stretch |
| (3055) | (0.1) | $CH_2$ stretch |
| 3065.9 | 0.64,<u>0.63</u> | $CH_2$ stretch |
| 3075.2 | 0.28,<u>0.24</u> | $CH_2$ stretch |

\* Literature values of RDX in parenthesis [4];
  standard RDX values are underlined.

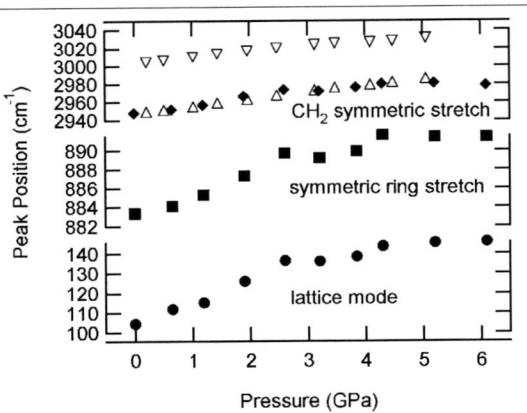

**Figure 5**. Peak position dependence on pressure for
Raman and IR spectra of several modes for I-
RDX®. (Raman ◆ ■ ● ; IR ▽ Δ )

# CONCLUSIONS

We have observed firstly that for the ambient phase the experimentally determined Raman lattice mode pressure coefficients of I-RDX® and standard RDX were similar. But they were larger than those reported in the literature. However, the internal mode pressure coefficients were similar in these two materials and also similar to those reported in the literature. Secondly, above a critical pressure the pressure coefficients of all the modes changed abruptly indicating a possible phase transition. This pressure is very similar to the pressure which induces the α to γ phase transition in RDX. Our results show that although defect concentrations in I-RDX® and standard RDX influence the shock sensitivity of formulations, they have not affected the pressure coefficients, although defects may have affected other spectral characteristics such as line width and intensity.

# REFERENCES

1. Doherty, R. M. and Watt, D. S., 2004 Insensitive Munitions & Energetic Materials Technology Symposium, Nov. 15-17, 2004, San Francisco.
2. Van der Heijden, A. E. D. M.,Bouma, R. H. B. and Van der Steen A.C., *Propellants, Explosives Pyrotechnics* 29, 304 (2004).
3. Caulder, S. M., Buess, M.L., Garroway, A.N. and Miller, P.J., Proceedings of Shock Compression of Condensed Matter- 2003, AIP, CP706, 929 (2004).
4. Owens, F.J. and Iqbal, Z. Chem. Phys. 74, 4242 (1981).
5. Goto, N., Yamawaki, H., Tonokura, K., Wakabayashi, K., Yoshida, M. and Koshi, M., *Material Science Forum* Vols. 465-466, 189 (2004).
6. Iqbal, Z., Suryanarayanan, K., Bulusa, S., and Autera, J.R., U.S. NTIS, AD-752899 (1972).
7. Spyckerelle, C., Freche, A. and Eck, G., 2004 Insensitive Munitions & Energetic Materials Technology Symposium, Nov. 15-17, 2004, San Francisco.
8. Baer, B.J., Oxley, J. and Nicol, M., *High Pressure Research* 2, 99 (1990).

CP845, *Shock Compression of Condensed Matter - 2005,*
edited by M. D. Furnish, M. Elert, T. P. Russell, and C. T. White
© 2006 American Institute of Physics 0-7354-0341-4/06/$23.00

# UK MARGINAL INITIATION CHARACTERISATION TEST (MICT) FOR HIGH EXPLOSIVES

**Mark Wright, Peter Williams, John Richardson, Elizabeth Edmonds, Andrew Jones and Rodney Drake**

*Atomic Weapons Establishment, Aldermaston, Reading, RG7 4PR, United Kingdom.*

**Abstract.** The UK Marginal Initiation Characterisation Test (MICT) has been developed to cover two areas of research: as an initial screening test for candidate energetic materials; and as a technique for characterising the effects of composition changes, ageing and temperature on detonation spreading performance. Tests have been undertaken on UF-TATB, UF-TATB/HMX and a highly loaded HMX PBX. The initiation system has been tuned to deliver a suitable pressure/time profile to marginally initiate each of these compositions. Subsequently, a study of sensitivity and divergence as a function of charge attributes (e.g. composition and density) has been completed. In addition, first stage hydrocode modelling of the MIC Test has been completed using CTH. The modelling output has highlighted the relationship between the prominent features in the results and the confinement around the explosive sample.

**Keywords:** MICT, Divergence, Screening Test, Sensitivity, UF-TATB, HMX
**PACS:** 82.33.Vx, 82.40.Fp

## INTRODUCTION

The Atomic Weapons Establishment (AWE) has identified the requirement for a small-scale screening and performance test to aid characterisation and development of existing and novel explosives. This requirement is focussed on the need to study incipient changes within explosive materials due to ageing, environmental, compositional and manufacturing variations using a reproducible small-scale test. The Marginal Initiation Characterisation Test (MICT) has been developed to fulfil this requirement, which was based on the Floret test, introduced by J Kennedy *et al.* at LANL [1-4] and later adapted by T Tran *et al.* at LLNL [5, 6].

## MICT EXPERIMENT

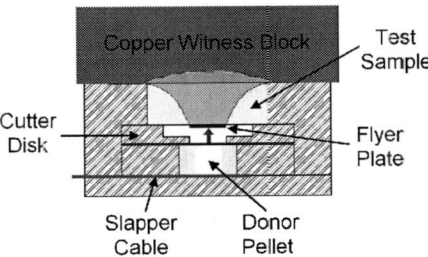

**Figure 1. MICT Schematic Diagram**

The MICT has been designed to study the effects of density, age, composition and temperature variations on detonation spreading within a 12.6-mm diameter explosive sample. A schematic representation of the MICT is shown in Figure 1.

The test uses a pre-defined explosively driven flyer plate to impact the material being characterised. The flyer parameters (e.g. diameter) are chosen to marginally initiate the test sample, such that the run-to-detonation distance is comparable to the length of the pellet. Any small variations in the run-to-detonation region will affect the extent of reaction (detonating fraction) within the sample pellet. The effect of variations in the composition and environment can then be studied through the comparative analysis of the dent profiles created in a copper witness block.

In most of the experiments the donor explosive was a 6-mm diameter by 3-mm thick 1.62-g/cm$^3$ PETN pellet, which was detonated by an electrical slapper system. This in turn accelerated a 0.125-mm thick aluminium flyer through a 'cutter hole' of the predetermined size to impact into the acceptor pellet being tested (12.6-mm diameter by 4-mm thick). The resulting dent in the copper witness block was profiled using either a Verdict Tricom 300 QCT Co-ordinate Measurement Machine or, more recently, a Uniscan OSP100A Optical Surface Profiler.

## RESULTS AND INTERPRETATION

Experiments have been undertaken on a highly loaded HMX PBX (91% with 8% K10 and 1% NC), UF-TATB/HMX (72.5/25% with 2.5% HTPB binder), UF-TATB and an inert crystalline material. These materials were chosen to provide a range of explosive compositions to commission the test. The experiments on the inert acceptor pellets were undertaken to assess the contribution of the PETN donor to the dent profile.

### Spot Size and Performance Study

A series of firings is required for a new material to optimise the flyer characteristics to ensure marginal initiation. An initial study on the suitability of the MICT to characterise the HMX PBX was undertaken using a range of flyer diameters, from 5-mm down to 1-mm. A similar study was also undertaken on the UF-TATB/HMX, using flyer diameters between 3-mm and 6-mm.

Figure 2 shows the 2-D cross-sectional area from each of the dent profiles (at a density of 1.8-g/cm$^3$) plotted against the flyer diameter used. Additionally the results from the inert acceptor experiments have been included to define the 'no-reaction' level, below which the profile is attributed to the PETN donor pellet alone.

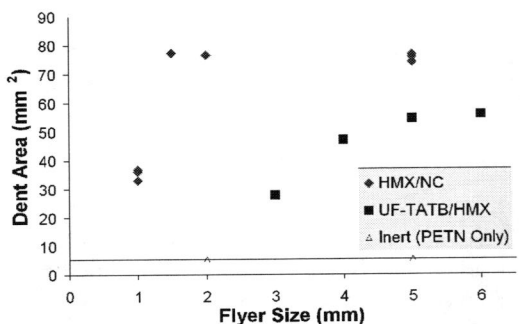

**Figure 2: Spot Size and Relative Performance at a Pellet Density 1.8 g/cm$^3$**

The results show that the HMX PBX does undergo a marginal response with a 1-mm flyer, but this is part of a very abrupt change from full detonation at 1.5-mm to a predicted failure at under 1-mm. There was concern that this abrupt detonation-to-failure transition would lead to a large variability in the results. However, a flyer size of 1-mm was chosen for an initial density study of the HMX PBX to investigate this further.

UF-TATB/HMX showed a much more gradual transition from detonation to failure than the HMX PBX. It was also noted from the asymptote of the graph at large diameters that the sample was close to fully detonating at a flyer diameter of 6-mm. These precursor firings suggested that a flyer size of 4-mm would be suitable for the initial characterisation of this material in the MICT.

Figure 2 also highlights the flexibility of the MICT, such that it can be used to rank the output pressures of different explosive compositions by studying the prompt detonation asymptote at large flyer diameters.

## UF-TATB Density Experiments

All of the UF-TATB experiments have been undertaken with a 5-mm aluminium flyer. The choice for this was based on collaborations with LLNL and LANL and knowledge of the approximate spot size of the UF-TATB. However, a spot size study for this material at a density of 1.8-g/cm$^3$ is planned for the near future to corroborate this choice.

The effect of density on the detonation spreading performance in UF-TATB is shown in Figure 3. The 2-D profiles through the deepest part of the witness block show an increased dent depth and lateral spreading as the density is lowered, suggesting that due to a reduced run distance a greater fraction of the pellet is detonating. These results highlight the increased impact sensitivity and improved divergence of UF-TATB at lower densities.

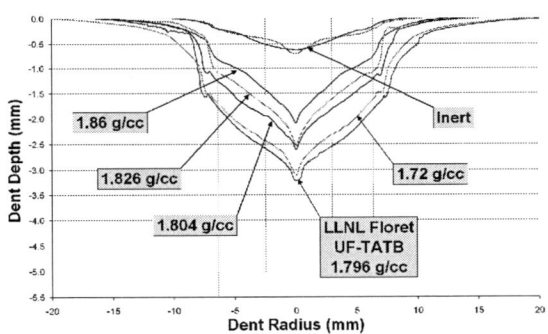

**Figure 3. MICT UF-TATB Witness Block Dent Profiles**

A profile from a Floret test undertaken by T Tran *et al.* at LLNL on UF-TATB is also shown in Figure 3. The LLNL result shows a very similar profile to the UK experiments with, however, a slightly deeper dent depth for a comparable density. Differences in the test configurations and fabrication of the US and UK UF-TATB pellets account for this variation.

## UF-TATB/HMX Density Experiments

The effect of altering the pellet density on the detonation spreading performance in UF-TATB/HMX is shown in Figure 4. As with the UF-TATB experiments, there is a general increase in dent depth as the density is lowered and a greater fraction of the pellet detonates. However, unlike the UF-TATB results there is little change in the width of the profiles as the density is decreased. This is likely to be due to the percentage of HMX in the composition that promotes good divergence at all of the densities tested.

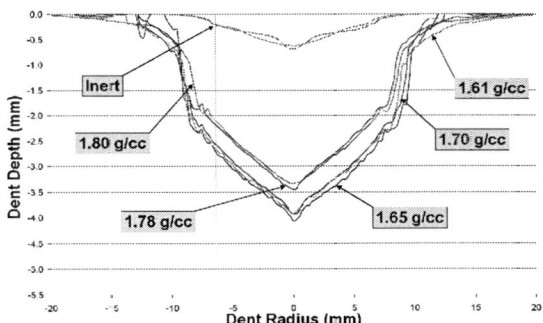

**Figure 4. MICT UF-TATB/HMX Witness Block Dent Profiles**

A further series of experiments are planned using a 3-mm flyer due to concerns that the lower densities (under 1.7-g/cm$^3$) were characteristic of a prompt detonation and were no longer undergoing a marginal response.

## HMX PBX Density Experiments

The initial results showed a similar trend to the UF-TATB/HMX profiles, where the dent depth is deeper as the pellet density is lowered due to the increased sensitivity yielding a higher detonating fraction. Also the profiles displayed almost no change in lateral spreading characteristics throughout the tests, highlighting the strong divergence of HMX based compositions across the density range tested.

However, some of the firings produced results that, while still showing the same general trends, introduced a large variability in the dent depth relationship with density. As mentioned earlier, this was highlighted as a concern due to the abrupt detonation-to-failure region displayed in the spot size study. A further series of tests is planned to investigate the possible reasons behind the variation in these results.

Figure 5 shows the dent area plotted against pellet density for all of the compositions reported. This graph highlights the general trend of increased sensitivity at lower pellet density displayed by all of the compositions. This also highlights the marked variability in the HMX PBX results compared to the other compositions tested.

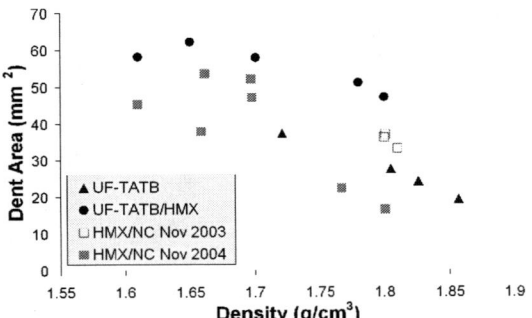

**Figure 5: Effect of Density of Sensitivity**

## HYDROCODE MODELLING

Initial hydrocode modelling has been completed using CTH (Version Jan01). The results suggest that the prominent features seen in the experimental dent profiles could be explained by shock interactions with the confinement disks. The interaction causes regions of intense pressure that are likely to affect the resulting profiles in the copper witness block. Further development of the modelling code and the materials database is required to gain a more quantitative analysis of the MICT results.

## SUMMARY

A test has been developed to detect small changes in explosives by measuring the detonation spreading performance of the reaction wave under marginal initiation conditions. Explosives of current interest to AWE with different output pressures and divergence characteristics have been studied to bracket the experimental conditions required for marginal initiation.

The results have shown an increased sensitivity for flyer impact with lower pellet density in all of the compositions tested. The

magnitude of this effect between the different materials tested seems to be consistent. However, the UF-TATB based compositions displayed reduced divergence performance at higher pellet densities.

## ACKNOWLEDGEMENTS

The authors would like to thank Tri Tran (LLNL) and Jim Kennedy (LANL) for all their help and advice throughout this work; and the invaluable help and support from the people within the Technical Operations Group and Explosives Technology Facility at AWE.

## REFERENCES

1. Lee K-Y, Kennedy J E, Hill L G, Spontarelli T, Stine J R and Kerley G I, *Synthesis, Detonation Spreading and Reaction Rate Modelling of Fine TATB*, International Detonation Symposium **11th**, 362 - 370, 1998.
2. Kennedy J E, Plaksin I, Thomas K A, Martin E S, Lee K-Y, Akinci A, Asay B W, Campos J and Direito J, *Instrumented Floret Tests of Detonation Spreading*, Shock Compression of Condensed Matter **13th**, 1500 - 1503, 2003.
3. Kennedy J E, Lee K-Y, Son S F, Martin E S, Asay B W and Skidmore C B, *Second-Harmonic Generation and the Shock Sensitivity of TATB*, Shock Compression of Condensed Matter **CP505**, 711 - 714, 1999.
4. Kennedy J E, Lee K-Y, Spontarelli T and Stine J R, *Detonation Spreading in Fine TATBs*, International Pyrotechnics Seminar **24th**, 743 - 748, 1998.
5. Tran T D, Simpson R L, Pagoria P F, Hoffman D M, Weber S, Cunningham B, Lee R, Hanks R, Chau H H, Hodgins R and Cutting J L, *Updates on LLM-105 Development as an Insensitive High Explosive Booster Material*, UCRL-Press **145524**, 2001.
6. Tran T D, Pagoria P F, Hoffman D M, Cunningham B, Simpson R L, Lee R S and Cutting J L, *Small-Scale Safety and Performance Characterisation of New Plastic Bonded Explosives Containing LLM-105*, International Detonation Symposium **12th**, 440 - 447, 2002.

CP845, *Shock Compression of Condensed Matter - 2005*,
edited by M. D. Furnish, M. Elert, T. P. Russell, and C. T. White
© 2006 American Institute of Physics 0-7354-0341-4/06/$23.00

# INITIATION OF DETONATION IN MULTIPLE SHOCK-COMPRESSED LIQUID EXPLOSIVES

## A. C. Yoshinaka[1], F. Zhang[1], O. E. Petel[2] and A. J. Higgins[2]

[1]*Defence R&D Canada Suffield, Medicine Hat, Alberta, T1A 8K6, Canada*
[2]*McGill University, Department of Mechanical Engineering, Montreal, Quebec, H3A 2K6, Canada*

**Abstract.** Initiation and resulting propagation of detonation via multiple shock reverberations between two high impedance plates has been investigated in amine-sensitized nitromethane. Experiments were designed so that the first reflected shock strength was below the critical value for initiation found previously. Luminosity combined with a distinct pressure hump indicated onset of reaction and successful initiation after double or triple shock reflection off the bottom plate. Final temperature estimates for double or triple shock reflection immediately before initiation lie between 700-720 K, consistent with those found previously for both incident and singly reflected shock initiation.

**Keywords:** Detonation, initiation, pre-compression, temperature, nitromethane.
**PACS:** 47.40.-x, 47.40.Nm.

## INTRODUCTION

The initiation of homogeneous liquid explosives by planar shock waves has been studied extensively. There exists, however, disagreements over the mechanism of the initiation process. The classic view suggests that initiation is governed by thermal decomposition after shock compression leading to the formation of a "super-detonation" propagating in the explosive originally compressed by the incident shock and eventually overtaking the latter [1-3]. In contrast to thermal decomposition behind the shock, Dremin [4] and Walker et al. [5] suggested that detonation ignition is controlled by the excitation of kinetic motion degrees of freedom within the shock front as it propagates into the explosive.

In order to further investigate the mechanism of initiation in homogeneous explosives, others have performed experiments whereby a reflected shock was used as a means to bring the explosive to a thermodynamic state off of the principle Hugoniot prior to initiation. Presles et al. [6] attempted to initiate pure nitromethane in this fashion using aluminum as the reflective surface. As no detonation was observed, it was concluded that the reflected shock was insufficiently strong. Higgins et al. [7,8] conducted incident and reflected initiation experiments in amine-sensitized nitromethane using steel as a reflective surface or anvil. They found the critical reflected shock pressure to reach at least 1.4 times that of the critical incident shock value. The temperatures for incident and reflected initiation, however, are equal thus suggesting that temperature seems to be the controlling variable in initiating detonation, independent of the shock path used.

To further examine the limit of thermal decomposition as the mechanism for detonation initiation at higher degrees of pre-compression, the present study examines the results from attempts at initiating detonation in a homogenous liquid explosive through multiple shock reflections between two high impedance anvils.

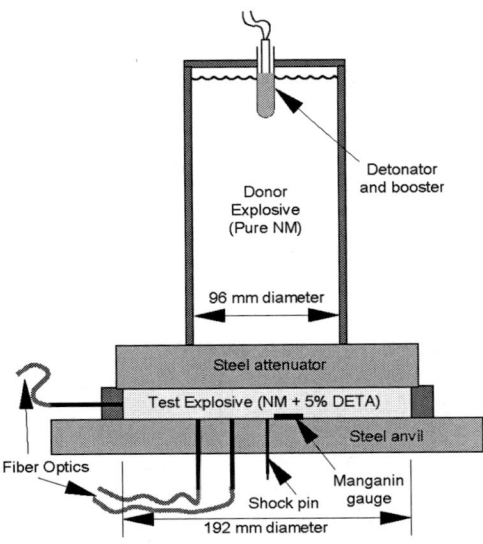

**Figure 1.** Schematic of experimental setup

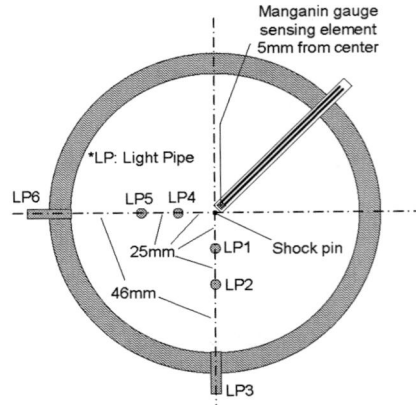

**Figure 2.** Receptor instrumentation layout

## EXPERIMENTAL DETAILS

The experiments are conducted using a configuration similar to that used previously [7,8], as shown in Fig. 1. A donor charge consisting of pure nitromethane (NM) is detonated and sends a shock wave through a steel attenuator. The shock then transmits into the receptor containing the test explosive, before reflecting on a steel anvil instrumented with light pipes and a manganin gauge (Dynasen model MN4-50-EK) as shown in Fig. 2. A thin Mylar sheet (0.25 mm thick) of size only slightly larger then the gauge had to be inserted between it and the metal anvil for electrical insulation. Gauge error was estimated at ±5% over the range of pressures used in these tests.

The gauge is located 5 mm from the center, whereas anvil mounted light pipes are positioned at radii of 25 and 50 mm. The latter consist of polished acrylic rods 32 mm long by 6.35 mm in diameter. As the reverberating shock does not reach the receptor inner diameter during the time of interest, side-on light pipes like those used previously [7,8] (essentially a bare fiber looking through a 80 μm thick Mylar window) observe the event through the wall surrounding the test charge. Two arrays of light pipes spaced 90 degrees apart are used in order to confirm initiation in the charge center. Luminosity from each light pipe is transmitted through a 1 mm acrylic fiber into an amplified photodiode circuit (Thorlabs model PDA55).

Thicknesses of 22.3-25.5 mm for the attenuator, 6.2 mm for the receptor and 12.7-25.4 mm for the anvil are chosen such that multiple reverberation of a single shock within the receptor is possible on the time scale of the experiment. The test explosive consists of commercial grade NM blended with 5% diethylenetriamine (DETA) sensitizer; the same prototypical mixture used previously [7,8] for incident and reflected initiation studies. The uncompressed liquid explosive has a measured detonation velocity of 6.0 km/s. Initial liquid temperature is maintained between 23-30°C, and since the room temperature critical diameter for NM+5%DETA is less than 1 mm, the current receptor thickness should be sufficient to propagate a steady detonation.

## RESULTS AND DISCUSSION

Results from an experiment using a 22.3 mm attenuator with a 12.7 mm anvil are presented in Fig. 3. Zero time corresponds to the arrival of the first incident shock on the anvil. The double shock pressure rise to 4.9 GPa on first reflection is a result of using Mylar insulation between the gauge and the anvil. The pressure history at the anvil surface shows two plateaus, indicating two reflections, before rising to a peak value of 11.37 GPa at 4 μsec. This time also coincides with the appearance of luminosity as observed though the wall mounted (side-on) light pipes. It is

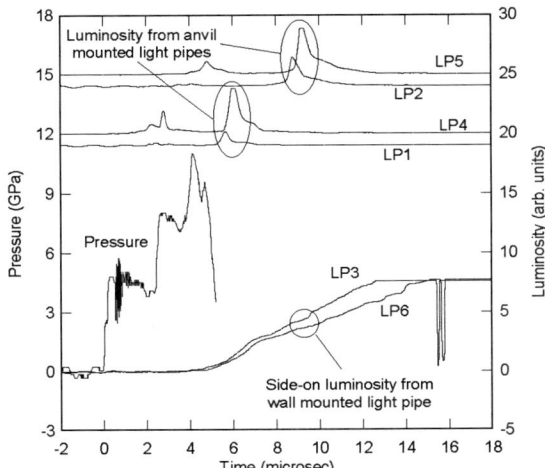

Thicknesses: 22.3mm steel attenuator / 6.2mm receptor / 12.7mm steel anvil

**Figure 3.** Pressure and luminosity records for the 22.3 mm thick steel attenuator

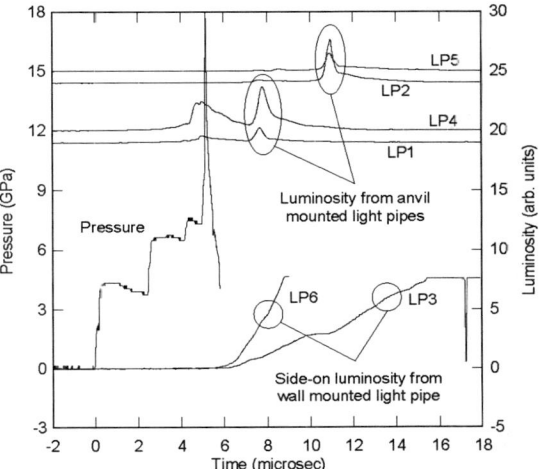

Thicknesses: 25.5mm steel attenuator / 6.2mm receptor / 25.4mm steel anvil

**Figure 4.** Pressure and luminosity records for the 25.5 mm thick steel attenuator

important to keep in mind that manganin gauges deployed in these experiments have a design pressure slightly in excess of 10 GPa. Thus, they cannot be expected to record pressure quantitatively after initiation. The earlier luminosity observed from the anvil mounted (end-on) light pipes, between 1-3 μsec and 3-5 μsec at 25 and 50 mm from the center respectively, is associated with shocked air at the anvil-air interface underneath the charge. Simplified shock trajectory estimates as well as LS-DYNA simulations further support evidence of shocked air luminosity at those times.

Stronger and sharper light signals from these end-on light pipes are observed after the final pressure jump, where peak-to-peak time of arrival between light pipes gives an average velocity of 8.0 km/sec in both radial directions. The side-on light pipe shows a momentary drop in luminosity at 15.5 μsec as a result of the detonation reaching the wall. Increase in luminosity shortly thereafter results from the shock traversing the Mylar window and emerging into air. Average velocity based on arrival time between the 50 mm end-on radial light pipe and the wall is 7.1 km/sec in both radial directions. Detonation decay in the radial direction results from lower sample pre-compression closer to the receptor wall. Both pressure jump and average velocities confirm that

detonation successfully initiated within the receptor.

Fig. 4 shows results of an attempt at achieving more reverberations during pre-compression using both a thicker attenuator and anvil of 25.5 and 25.4 mm respectively. The thicker anvil resulted in a delayed shock arrival at the anvil-air interface as evidenced by the first luminosity peak at 5 μsec. This time also coincides to when the incident shock emerges underneath the anvil at the light pipe locations. Both pressure gauge and wall mounted light probes indicate initiation of detonation after three shock reflections. Finally, average detonation velocities between light probes at 25-50 mm and 50-96 mm (wall) are 7.8 km/sec and 6.8 km/sec respectively for both radial directions.

Successful initiation requires the application of a shock of sufficient strength and duration, which in turn will be affected by charge boundary conditions. The foregoing apparatus, however, is geometrically similar to that used in a previous work [8], so a comparison can be made in order to examine the influence of compression path and pressure on the initiation process. The Winey et al [9] equation of state, which calculates temperature in pure nitromethane under multiple shock reverberations, is used assuming that the small concentration of DETA present in the test mixture has negligible effects in the unreacted NM temperature changes. Since the experiments

**Table 1.** Shock pressures and temperatures during pre-compression

| Initiation technique | Pressure (GPa) | Estimated temperature (K) |
|---|---|---|
| Incident shock initiation | 4.3[8] | 720 |
| Single reflected shock initiation* | | |
| Incident | 2.5-2.8[8] | 576-602 |
| Reflected | 6.0-6.5[8] | 701-724 |
| Double reflected shock initiation* | | |
| 1st incident | 2.02** | 532 |
| 1st reflected | 4.80 | 652 |
| 2nd incident | 6.29** | 685 |
| 2nd reflected | 7.89 | 715 |
| Triple reflected shock initiation* | | |
| 1st incident | 1.84** | 515 |
| 1st reflected | 4.35 | 632 |
| 2nd incident | 5.65** | 664 |
| 2nd reflected | 6.64 | 683 |
| 3rd incident | | |
| 3rd reflected | 7.49 | |

*From a steel anvil          **Calculated

detailed here only had one pressure gauge mounted on the steel surface, the NM shock pressure immediately before reflection off the anvil is calculated using LS-DYNA. Conditions similar to those found in experiments of Figs. 3 and 4 are simulated and results are presented in Table 1. For the case with a double reflection, a final temperature of 715 K is obtained. For the triple reflection pre-compression test, however, calculated pressures beyond the second reflection were not successful at reproducing the measured values. Although final temperature cannot be estimated, experimental final shock pressure is found to be slightly lower then that for the double reflection case, which should result in a lower final temperature.

## CONCLUSIONS

An estimated temperature of 700-720 K seems to be reasonable for the double or triple shock reflection leading to a successful initiation of detonation. This temperature value is consistent with that obtained previously from the incident and singly reflected shock initiation tests. This suggests that temperature is a dominant variable for initiation at the studied shock pre-compression levels, regardless of the difference in shock path and pressure magnitude.

## ACKNOWLEDGEMENTS

The authors would like to thank Luc Légaré, Kiril Mudri and Mark Churcher for their assistance in conducting the experiments. The dedicated assistance of the Field Operations Services section of DRDC Suffield is gratefully acknowledged.

## REFERENCES

1. Chaiken, R.F., J. Chem. Phys., Vol. 33, pp. 760-761, 1960.
2. Campbell, A.W., Davis, W.C., and Travis, J.R., Phys. Fluids, Vol. 4, pp. 498-510, 1961.
3. Sheffield, S.A., Engelke, R., and Alcon, R.R., 9th Symp.(International) on Det., 1989, pp. 39-49.
4. Dremin, A. N., Phil. Trans. R. Soc. Lond. A 339:355-364, 1995.
5. Walker, F.E., and Wasley, R.J., Combust. Flame, Vol. 15, pp. 233-246, 1970.
6. Presles, H.N., Fisson, F., and Brochet, C., Acta Astronautica, Vol. 7, pp. 1361-1371, 1980.
7. Higgins, A.J., Jetté, F.X., Yoshinaka, A.C., Lee, J.H.S., and Zhang, F., APS Topical Conference on Shock Compression of Condensed Matter, 2001, pp. 1023-1026.
8. Higgins, A.J., Jetté, F.X., Yoshinaka, A. and Zhang, F., 12th Symp. (International) on Det., 2002.
9. Winey, J.M., Duvall, G.E., Knudson, M.D., and Gupta, Y.M., J. Chem. Phys., Vol. 113, pp. 7492-7501, 2000.

CHAPTER XVI

# SHOCK-INDUCED MODIFICATIONS AND MATERIALS SYNTHESIS

CP845, *Shock Compression of Condensed Matter - 2005,*
edited by M. D. Furnish, M. Elert, T. P. Russell, and C. T. White
© 2006 American Institute of Physics 0-7354-0341-4/06/$23.00

# DEFECT SUBSTRUCTURES IN PLATE IMPACTED AND LASER SHOCKED MONOCRYSTALLINE COPPER

**Bu Yang Cao[1], Marc A. Meyers[1], David H. Lassila[2], Matt S. Schneider[1], Yong Bo Xu[3], Daniel H. Kalantar[2], Bruce A. Remington[2]**

[1]*MATS, Univ. of California, San Diego, 9500 Gilman Dr.La Jolla, CA 92093-0411 USA*
[2]*Lawrence Livermore National Laboratory, Livermore, CA 94550 USA*
[3]*Chinese Academy of Sciences, Inst. of Metal, Shenyang, Liao Ning 110016 China*

**Abstract.** Monocrystalline copper samples with orientations of [001] and [221] were shocked at pressures ranging from 20 GPa to 60 GPa using two techniques: direct drive lasers and explosively driven flyer plates. The pulse duration for these techniques differed substantially: 40 ns for the laser experiments at 0.5 mm into the sample and 1.1 ~1.4 μs for the flyer-plate experiments at 5 mm into the sample. The residual microstructures were dependent on orientation, pressure, and shocking method. For the flyer-plate experiments, the longer pulse duration allow shock-generated defects to reorganize into lower energy configurations. Calculations show that the post shock cooling for laser shock is $10^3 \sim 10^4$ faster than that of the plate-impact shock, propitiating recovery and recrystallization conditions for the latter. At the higher pressure level extensive recrystallization was observed in the plate-impact samples. An effect to contribute significantly to the recrystallization is the existence of micro-shearbands, which increase the local temperature.

**Keywords:** laser, shock compression, plate impact, shear localization in copper, shock waves, explosives
**PACS: 62.50.+p**

## INTRODUCTION

Flyer-plate impact and laser shock are two typical loading methods employed in shock-recovery experiments. Significant differences in the residual microstructure in monocrystalline copper shocked by these two methods have been observed. The objective of this paper is to demonstrate the differences of the residual microstructures are to a large extent due to how the heat generated inside the samples during shock is extracted. Post-shock recovery and recrystallization processes dominate the residual microstructures, if the time interval and temperature are sufficient. The unique advantage of laser shock compression

over plate impact, namely, the rapid post-shock cooling, is discussed.

## EXPERIMENTAL PROCEDURE

Explosively driven flyer plates and direct drive lasers produce different shock pulses. For the plate impact experiments reported herein, the duration of the pulse at a depth of 5 mm from the impact interface was in the 1.1—1.4 μs range. The triangular shape laser shock pulse duration is 40 ns at energy around 300 J, which produces an initial pressure of approximately 60 GPa. The facilities used for plate impact and laser shock have been described in previous papers [1,2]. Monocrystalline coppers with orientations of <001> and <221> were shock-compressed by both laser (at ambient temperature) and plate impact (at 88 K) from 20 GPa to 60 GPa.

## EXPERIMENTAL RESULTS

The microstructures are characterized by stacking faults for both the 30-40 GPa plate impacted and laser shocked <100> samples. The average spacing between stacking faults is 230 and 450 nm for the laser shocked samples and 180 and 220 nm for the plate impacted sample (Fig. 1(a)). It shows the two sets of stacking faults as the traces of [$\bar{2}20$] and [220] orientations in the (001) plane. Four stacking fault variants *viz* the (11$\bar{1}$)1/6[112], (111)1/6[$\bar{1}\,\bar{1}\,2$], ($\bar{1}$11)1/6[$1\bar{1}2$], and ($1\bar{1}1$)1/6[$\bar{1}12$] are observed in 40 GPa laser shocked samples. The stacking faults are similar to the ones observed by Murr [3]. It should also be noted that, in the 30 GPa plate impacted <100> samples, we observed isolated recrystallization as well as localized deformation bands. These were absent for the laser shocked specimens.

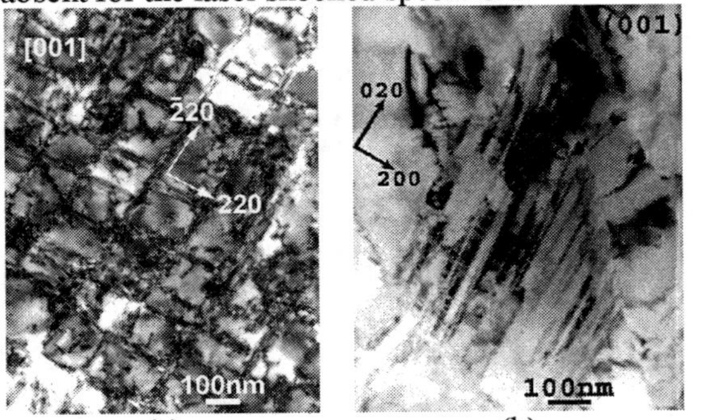

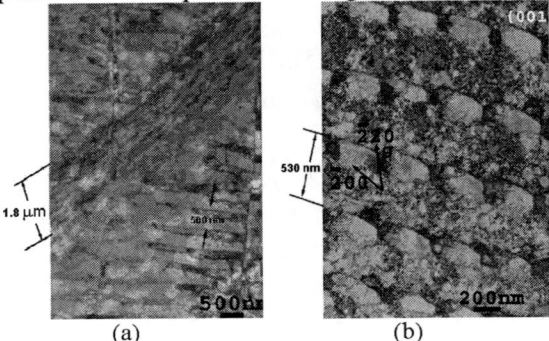

**Figure 1.** (a) Stacking faults in 30 GPa plate impacted <100> sample; (b) Micro-bands in 30 GPa plate impacted <221> samples.

The substructure of the plate impacted <221> sample shocked at 30 GPa contains micro-bands, whose morphologies vary through this sample. Some large bands, shown in the left part of Fig. 1 (b), have a width around 120 ~130 nm. Micro-bands with a width of 20~30 nm were found within these large bands. Huang and Gray [4] proposed a model to explain the formation of micro-bands, based on the development of coarse slip bands. The laser shocked <221> samples are characterized by a greater density of twins than bands. Although

some bands with width of 100 ~ 200 nm were observed very similar to those big bands in 30 GPa plate impacted samples, twins with ($1\bar{1}1$) habit plane were more prevalent throughout the sample.

**Figure 2.** TEM for 57 GPa plate impacted <100> copper samples: (a) overview of the sample (x10K); (b) dislocation circles.

At 55-60 GPa, Micro-twins occur in both plate impacted and laser shocked <100> samples. There are micro-twins with ($\bar{1}11$) as habit plane in plate impacted samples. The sizes for micro-twins vary from 80 nm to 180 nm. For the laser-shocked samples, there are two sets of micro-twins along [$2\bar{2}0$] and [$\bar{2}\,2\,0$].

For the 57 GPa plate impacted samples, there are deformation bands, slip bands, recrystallized regions and dislocation tangles in addition to micro-twins. Fig. 2 (a) shows a deformation band of approximately 1.8 μm width traversing the specimen. The appearance of these stacking faults is different from the ones shown in Fig. 1(a). There is evidence for recovery processes within them. These broad bands are absent after laser shock because of the much smaller time. Indeed, the shock velocity is approximately 5.6 mm/μs. A duration of 1.4 μs can generate heterogeneities extending over a few mm. On the other hand, laser shock, with duration of only 2ns, is much more restricted in its ability to generate inhomogeneities. In Fig. 2 (b), regular dislocation cell arrays can be seen. Between two arrays, there are dislocation tangles and in some places the density of dislocation is very high. Mughrabi and Ungár [5] found some dislocation cell structures very similar to our observations, but they are quite unlike the cells observed by other investigators (e.g., Johari and Thomas [6]). The distances between the repeated structures in both Fig. 2 (a) and (b) have the same width of around 500 nm. The periodicity of the features of Fig. 2(a) is remarkable. It is speculated that these features are due to the recovered stacking-fault arrays seen in Fig. 2 (b). The major difference between the laser shocked samples and plate impacted samples in 55-60 GPa regime is the presence of fully recrystallized regions in the latter.

The <221> samples plate impacted at 57 GPa were full of large recrystallized grains (Fig. 3(a)). Annealing twins grow in the recrystallized grains. In 60 GPa laser shocked <221> samples, there is a high density of dislocation, as shown in Fig. 3(b). These dislocations are tangled and some bands were formed as a result of heavy dislocation density. A few of deformation twins were also found in this sample.

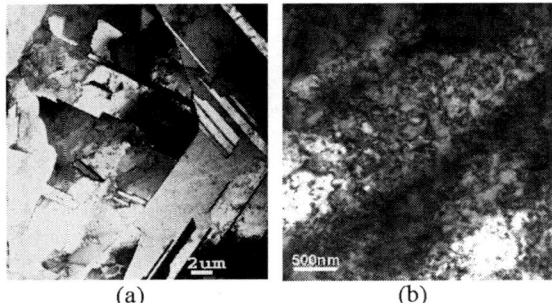

(a)                               (b)

**Figure 3.** (a) TEM showing annealing twins and recrystallized grains in 57 GPa plate impacted <221> sample; (b) Dislocation structures in 60 GPa laser shocked <221> samples.

## ANALYSIS

It is important to notice that laser and plate-impact shocks have different wave shapes and duration times because this likely results in very different effects on the heat generated during the shock and the heat transfers afterwards. Based on the progress of the shock pulse and its decay, the residual temperatures immediately after shock can be calculated [7] (Fig. 4). To calculate the heat transfer after shock, a semi-infinite heat transfer model was adopted [8]. The following assumptions were made: 1) Conduction is one-dimensional; 2) Copper sample is a semi-infinite medium; 3)

Temperature profiles at time t=0 are shown in Fig. 4 .

Fig. 5 shows the change of temperature with time for 30 GPa plate impacted samples. The maximum temperature (at surface) changes from approximately 160 K to 100 K during a period of 1000 s. For 57 GPa (Fig.6), the temperature changes from 360 K to 140 K during this same time period. This period of time should be sufficient to induce some microstructural changes

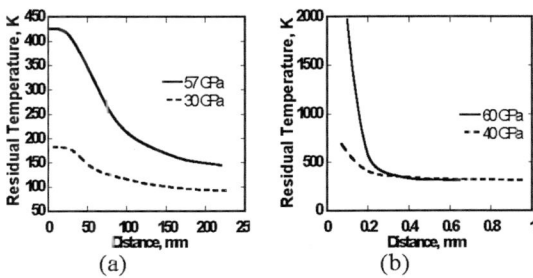

(a)                               (b)

**Figure 4.** Residual temperature inside the sample immediately after shock: (a) plate-impact shock; (b) laser shock.

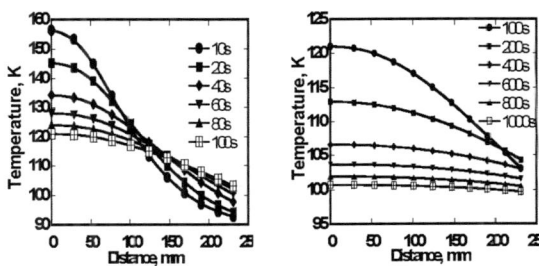

**Figure 5.** Temperature change for copper plate impacted at 30 GPa.

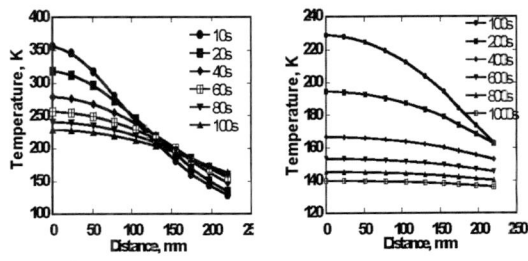

**Figure 6.** Temperature change for copper plate-impacted at 57 GPa.

inside the samples. For laser shock, the region which is affected by the temperature rise is much shorter (up to 1mm). The temperature excursions in laser shocked samples are shown in Fig. 7.Based on these analyses, a qualitative comparison of the plate impact and laser shock can be estimated. The temperature decays in the laser shocked sample are $10^3 \sim 10^4$ faster than those in the plate impacted sample. These results explain why, although the peak pressures of laser shock are much higher than those of impact (resulting in higher residual temperatures), the post-shock microstructures in plate impact samples show a greater effect of post shock thermal excursion.

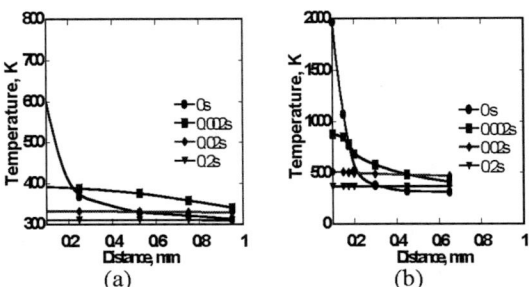

(a)　　　　　　　(b)

**Figure 7.** Temperature change in laser shocked copper: (a) at 200J (40 GPa); (b) at 300J (60 GPa).

TEM observations confirm the presence of localized regions of concentrated shear. The plastic deformation in these regions substantially exceeds those predicted from uniaxial strain, and one can expect local fluctuations in temperature. The temperature rise in the shear localization areas can be calculated as [1]:

$$\Delta T_d = \frac{\beta}{\rho C_p} \int_{\varepsilon_0}^{\varepsilon_1} \sigma d\varepsilon \qquad (1)$$

where $C_p$ is the heat capacity, and $\beta$ is the Taylor factor (0.9-1.0 here). We use the Johnson-Cook [9] equation strength of the material $\sigma$. The temperature change due to the plastic deformation is expressed as:

$$\frac{T - T_r}{T_m - T_r} = 1 - \exp[\frac{-0.9(1 + C\log\frac{\dot{\varepsilon}}{\dot{\varepsilon}_0})}{\rho C_p(T_m - T_r)} \times (\sigma_0\varepsilon + \frac{B\varepsilon^{n+1}}{n+1})] \ (2)$$

Where, $T_r = 90$ K, $T_m = 1356$K, B = 53.7 MPa, C = 0.026, $\sigma_0$ = 330 MPa (the value for shock hardened copper), n = 0.56, m = 1.04. There is considerable local heat generation around heavily deformed areas (such as deformation bands) as show in Fig. 8. These regions can act as initiation sites for post-shock recrystallization.

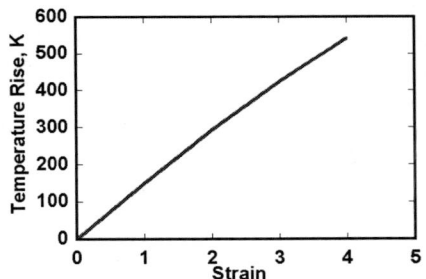

**Figure 8:** Temperature rise due to plastic deformation.

## ACKNOWLEDGEMENTS

Supported by DOE Grants DEFG0398DP00212 and DEFG0300SF2202.

## REFERENCES

1. D. H. Lassila, T. Shen, B. Y. Cao, and M. A. Meyers, Metal. and Mat. Trans., 35A (2004) 2729-2739.
2. M. A. Meyers, F. Gregori, B. K. Kad, M. S. Shheider, D. H. Kalantar, B. A. Remington, G. Ravichandran, T. Boehly, J. S. Wark, Acta. Metall. 51 (2003)1211-1228.
3. L. E. Murr, in: M. A. Meyers and L. E. Murr (Eds.), Shock Waves and High-Strain-Rate Phenomena in Metals, Plenum, NY, 1981, pp. 607-673.
4. J. C. Huang and G. T. Gray III, Acta metall. 37. No. 12, (1989) 3335-3347.
5. H. Mughrabi, T. Ungár, W. Kienle, and M. Wilkens, Phil. Mag. A, 53 (1986) 793-813.
6. O. Johari and G. Thomas, Acta. Metall. 12 (1964) 1153-1159.
7. M. A. Meyers, Dynamic Behavior of Materials, John Wiley and Sons, Inc, New York, 1994.
8. F. Kreith and M. S. Bohn, Principles of Heat Transfer, Brooks/Cole, CA, 2000.
9. G. R. Johnson and W. H. Cook, Proc. 7th Int. Symp. On Ballistics, ADPA, the Netherlands, 1983.

CP845, *Shock Compression of Condensed Matter - 2005*,
edited by M. D. Furnish, M. Elert, T. P. Russell, and C. T. White
© 2006 American Institute of Physics 0-7354-0341-4/06/$23.00

# THE INFLUENCE OF INTERSTITIAL OXYGEN ON THE ALPHA TO OMEGA PHASE TRANSITION IN TITANIUM AND ZIRCONIUM

**E. Cerreta[1], G.T. Gray III[1], A.C. Lawson[1], C.E. Morris[1], R.S. Hixson[2], and P.A. Rigg[2]**

[1]*MST-8, Los Alamos National Laboratory, Los Alamos New Mexico 87545*
[2]*DX-2, Los Alamos National Laboratory, Los Alamos New Mexico 87545*

**Abstract.** The pressure for the $\alpha$ to $\omega$ phase transition was investigated for two grades of titanium and three grades of zirconium. A series of shock experiments were conducted from 5 to 35GPa and revealed that the pressure for the phase transition increases with increasing interstitial oxygen content and is completely suppressed in low purity materials. For the high purity Ti and Zr in this study, the pressure for the phase transition occurred at 10.4 and 7.1GPa, respectively and no reverse transformation was observed upon unloading. Increasing the oxygen content increases the number of octahedral sites occupied; this is postulated to increase the pressure for the phase transition. Neutron diffraction was utilized to quantify the volume fraction of metastable $\omega$ phase and to characterize the microstructures within the high purity, shocked, and "soft" recovered specimens

**Keywords:** Shock wave, Metastable phases, Solid-solid phase transformation, Impurity concentration
**PACS:** 62.50.+p, 64.60.My, 64.70.Kb, 68.37.Lp, 61.72.Ss, 68.18.Jk

## INTRODUCTION

Ti and Zr have been studied extensively under high pressures and the crystallography and morphology of the high-pressure omega phase is well documented [1-5]. Specifically, the $\alpha$ to $\omega$ phase transformation has been characterized and modeled as a function of temperature, pressure, and even alloying additions for Ti and, more recently, for Zr [6-10]. Although, the response of Ti and Zr to dynamic loading continues to be examined, little experimental data exists concerning the influence of chemistry, in particular interstitial content, on the detailed structure/property relationships of hcp alloys subjected to shock loading.

At ambient temperature and pressure, pure Ti and Zr are stable as the $\alpha$-phase, a hexagonal closed packed (hcp) phase [11]. At high pressures the $\omega$-phase, which has a simple hcp structure, is stabilized in both Ti and Zr [6, 7]. This martensitic phase transformation can occur under shock, quasi-static, or hydrostatic loading conditions, however the pressures for the phase transition under shock loading conditions are typically higher than those observed under static-loading conditions [8]. Unlike the $\alpha$-$\varepsilon$ transition in pure Fe, the $\alpha$ to $\omega$ transformation in Ti and Zr exhibits a large hysteresis that is responsible for retention of the metastable high-pressure $\omega$-phase to atmospheric pressure [5]. The $\omega$-phase is morphologically similar to the $\omega$-phase formed in as-quenched $\beta$-phase alloys based on Ti and Zr.

Recently, the effects of interstitials on the $\alpha$ to $\omega$-phase transition and the effects of shock hardening in hcp metals and alloys, have received increased interest [2, 10, 12]. While the crystallography of transformation has been well examined, the influence of interstitial content on

**TABLE 1.** Chemistries of the Ti and Zr Alloys in wt% PPM

| Material | O | C | N | Al | V | Fe |
|---|---|---|---|---|---|---|
| High Purity Ti | 360 | 60 | 10 | 4 | 3 | 5 |
| A-70 Ti | 3700 | 170 | 240 | - | - | 1,800 |
| High Purity Zr | <50 | 22 | <20 | <20 | <50 | <50 |
| Low Purity Zr I | 390 | 70 | 15 | <20 | <25 | 125 |
| Low Purity Zr II | 1,200 | 270 | 80 | - | - | 2,400 |

the phase change has received less attention. The purpose of the present study is to systematically investigate the effects of interstitial oxygen on the α to ω-phase transition in Ti and Zr.

## EXPERIMENTAL PROCEDURE

This investigation was performed on two grades of Ti and three grades of Zr, each with varying levels of interstitial oxygen. Table 1 gives the chemistries for each of the 5 materials and processing details are provided elsewhere [9, 12]. To investigate the influence of alloy content on ω-phase formation, wave profile and shock recovery experiments were conducted on each of the materials as a function of shock pressure. All shock recovery experiments were performed on an 80mm single stage launcher utilizing a shock assembly consisting solely of Ti or Zr and the wave profiles were measured with a VISAR built at Los Alamos [13]. The precision of the wave velocity measurements is estimated to be approximately 1% in particle velocity. Photomultiplier circuits were utilized that had a 1ns rise time.

Optical and transmission electron microscopy (TEM) as well as neutron and X-ray diffraction were performed to characterize the as-received, as-annealed, and shocked materials. TEM foils were observed in a JEOL 2000EX at 200kV and a Phillips CM30 at 300kV. Neutron diffraction experiments were performed on the High Intensity Powder Diffraction (HIPD) instrument at the Los Alamos Neutron Scattering Center (LANSCE). Bulk specimens were examined without special preparation because the neutron penetration is large enough to sample the entire specimen.

## RESULTS AND DISCUSSION

The shock wave profile experiments conducted on Ti with two different impurity concentrations were performed to quantify the effects of

interstitials on the pressure of the α to ω-phase transition. The high-purity Ti was shocked at peak shock pressures ranging as high as 15.6 GPa and the transition pressure was identified as 10.4 GPa. The shock-wave profiles measured using VISAR at 6.4 and 15.6 GPa are given in Fig. 1. From this figure the α to ω-phase transition can be clearly identified in the 15.6 GPa shock but is absent in the 6.4 GPa wave profile. The lack of a rarefaction shock in the 15.6 GPa case suggests that no reverse phase transformation occurred. In the A-70 Ti, the phase transformation was completely suppressed for peak pressures up to 35 GPa.

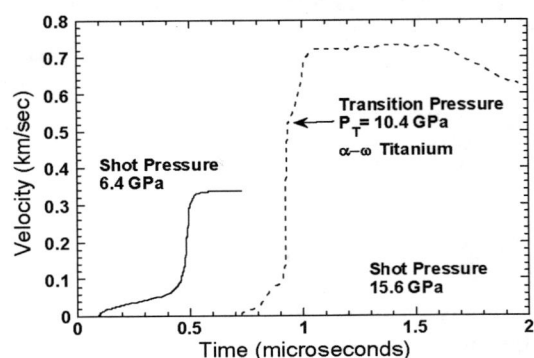

**FIGURE 1.** VISAR traces for the high purity Ti material at 6.4 and 15.6 GPa

Shock wave profile experiments were also conducted on the three types of zirconium with increasing impurity concentration to quantify the effects of interstitial content on the pressure for the α-ω phase transition. The VISAR wave profiles for each of the materials are given in Fig. 2. From this figure, the α-ω phase transition for the high purity and low purity I material can be identified as 7.1 and 8.3 GPa, respectively. Clearly, the pressure for phase transformation increases with increasing impurity content. Again, the lack of a rarefaction shock indicates that no reverse transition occurred. The phase transition in the low purity II material was completely suppressed within the peak pressures examined.

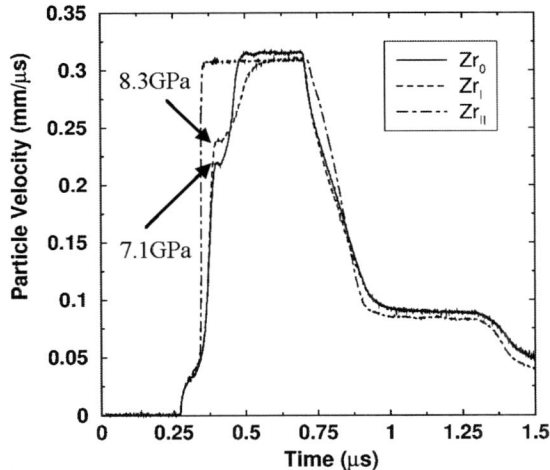

**FIGURE 2.** VISAR traces for the each of the three zirconium materials investigated: high purity ($Zr_0$), the low purity I ($Zr_I$), low purity II ($Zr_{II}$) materials, respectively

Rietveld analysis of the neutron diffraction data was used to quantify volume fractions of the $\alpha$ and the $\omega$ phases in the recovered shock prestrained. In the case of the Ti, both the high purity and the A-70 were shocked to 5 and 11GPa. A volume fraction of 28% $\omega$-phase with lattice parameters $a$ = 0.4614nm and $c$ = 0.2832nm was measured in the 11 GPa shock recovered high purity titanium specimen. Metastable $\omega$-phase was detected in neither the high purity Ti specimen shocked to 5GPa nor the A-70 Ti at both pressures. In the case of the zirconium specimens, neutron diffraction was only performed on high purity Zr specimens shocked to pressures just above (8GPa) and just below (5.8GPa) the $\alpha$ to $\omega$-phase transition pressure. No $\omega$-phase was observed in the 5.8GPa shocked zirconium, but almost 40% by volume of the Zr preshocked to 8GPa was retained metastable $\omega$-phase.

The pressure to induce the $\alpha$-$\omega$ phase transition was observed in the VISAR measurements to increase with increasing oxygen concentration and is eventually inhibited at high enough impurity levels. Oxygen, which has an atomic radius of 0.65Å, sits in the octahedral sites rather than the tetrahedral sites of $\alpha$-phase zirconium and titanium. The occupation of octahedral sites by oxygen leaves little free space within the hexagonal $\alpha$-phase zirconium prism and actually creates a lattice microstrain in titanium, which was measured by neutron diffraction to be 0.68 and 0.52 for the high purity and A-70 Ti utilized in this study, respectively [9, 14].

The influence of interstitials on the crystallography of the phase transformation must be considered. In the high purity zirconium, the orientation relationship observed between the $\alpha$ and the $\omega$-phase is $[\bar{1}010]_\alpha//[\bar{2}11\bar{3}]_\omega$ and $(0001)_\alpha//(\bar{1}101)_\omega$[12]. The transformation occurs in two steps [4]. The first step is a lattice shear parallel to the $[\bar{1}010]_\alpha$ directions, which translates atoms along the $[\bar{1}2\bar{1}0]_\alpha$ direction to positions along the $[\bar{1}101]_\omega$ direction. The second step is a subtler atom shift on the $(0001)_\alpha$ planes. Atoms in the $(0001)$ planes shift in $[01\bar{1}0]$ directions, but atoms in adjacent planes shift in the opposite sense to effectively collapse closely spaced $(01\bar{1}0)$ planes. It is the first step of this transformation that is hindered by an increased interstitial content. In specimens with a high oxygen content, many of the octahedral sites are occupied. In Fig. 4, as zirconium atoms are sheared in a direction parallel to the $[\bar{1}010]_\alpha$ direction, the necessary space within the lattice to accommodate the shearing atoms is occupied and the required stress for shear deformation is likely to increase. Furthermore, within this transformation, the $\alpha$-phase octahedral sites 1 and 3 labeled in Fig. 4 are not translated into the correct position in the $\omega$-phase. This lack of registry causes the $\alpha$ to $\omega$-phase transformation to be less favorable as the oxygen concentration increases.

A similar phenomenon occurs in the Ti, where the observed orientation relationship is $(0001)_\alpha$ // $(\bar{1}2\bar{1}0)_\omega$ and $[\bar{1}\bar{1}20]_\alpha//[0001]_\omega$. Here, when the $\alpha$-phase transforms to the $\omega$-phase, there is a coordinated set of atom shifts. The first shift involves row 1 in Fig. 5 moving in the $[\bar{1}2\bar{1}0]$ direction with a magnitude of 0.81Å and then in the

[$10\bar{1}0$] direction with a magnitude 0.224Å [3]. The first shift occurs in the same manner for three neighboring atomic rows and in the opposite direction for the next three, and so on. The second shift occurs in the opposite sense for successive (0001) planes. This transforms the ($1\bar{2}10$) α plane into the (0001) ω plane.

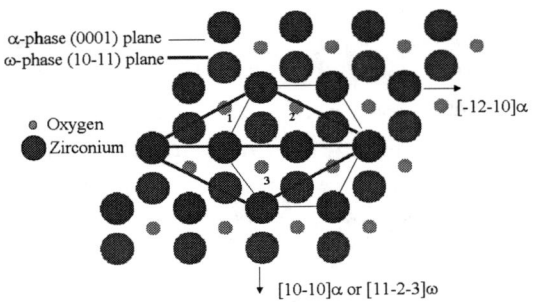

**FIGURE 3**. Schematic of α to ω-phase transformation for the [$\bar{1}010$]α//[$\bar{2}11\bar{3}$]ω orientation relationship: the location of interstitials during the shear in the [$\bar{1}010$]α direction .

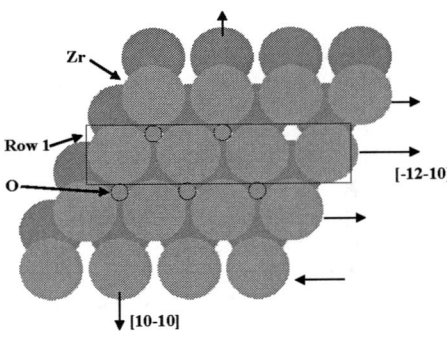

**FIGURE 4.** Basal planes in α-phase zirconium with some octahedral sites filled with oxygen atoms. Directions for the atom shifts for the α-ω phase transition are also given

The details of this phase transformation are given elsewhere [3, 4]. Shown in Fig. 5, as octahedral sites fill, this transformation is likely to become more difficult because the oxygen atoms occupy the necessary space within the lattice required to accommodate the atom movements

during the α-ω phase transformation. Additionally, just as was observed in the first orientation relationship, here the atom shifts within the α-phase do not shift the interstitial atoms into the correct interstitial site within the ω-phase and registry is not retained.

## CONCLUSIONS

As the concentration of large interstitial elements like oxygen increases (smaller atoms such as hydrogen are likely to occupy tetrahedral sites), it becomes increasingly more difficult for the lattice to accommodate atom shifts necessary for the phase transformation and higher pressures are required to induce the phase transformation. At high enough interstitial levels the transformation is likely to be suppressed completely. This is observed in the A-70 titanium and the lowest purity zirconium, $Zr_{II}$.

## ACKNOWLEDGEMENTS

This work has been performed under the auspices of the United States Department of Energy

## REFERENCES

1. Greeff, C., D. Trinkle, and R. Albers, JAP, 2001. **90**(5): p. 2221-6.
2. Hennig, R.G., et al., Nat. Mater., 2005: p. 129-134.
3. Rabinkin, A., M. Talianker, and O. Botstein, Acta Metall., 1981. **29**: p. 691-698.
4. Song, S.G. and G.T. Gray, *Phil. Mag. A*, 1995. **71**(2): p. 275-290.
5. Usikov, M.P. and V.A. Zilbershtein, Phys. Stat. Sol. A, 1973. **19**: p. 53-58.
6. Vohra, Y.K., et al., Acta Metall., 1980. **28**: p. 683-685.
7. Jamieson, J.C., Science, 1963. **140**: p. 72-73.
8. Sikka, S.K., Y.K. Vohra, and R. Chidambaram, Prog. Mat. Sci., 1982. **27**: p. 245-311.
9. Gray, G.T., C.E. Morris, and A.C. Lawson. in *Titanium '92*. 1992. San Diego, Ca: TMS.
10. Trinkle, D., et al., Phys. Rev. Lett., 2003. **91**(2): p. 025701-4.
11. Partridge, P.G., Metal. Rev., 1967. **12**: p. 169-194.
12. Cerreta, E., et al., Acta Mater., 2005. **53**: p. 1751-1758.
13. Hemsing, W., Rev. Sci. Inst., 1979. **50**(1): p. 73-78.
14. Conrad, H., Prog. Mat. Sci., 1981. **26**: p. 123-403.

CP845, *Shock Compression of Condensed Matter - 2005*,
edited by M. D. Furnish, M. Elert, T. P. Russell, and C. T. White
© 2006 American Institute of Physics 0-7354-0341-4/06/$23.00

# INVESTIGATION OF SHOCK-INDUCED REACTIONS IN A NI+AL POWDER MIXTURE

## D.E. Eakins, N.N. Thadhani

*School of Materials Science and Engineering.*
*771 Ferst Dr., Love Manufacturing Bldg.*
*Georgia Institute of Technology, Atlanta GA 30332*

**Abstract.** The shock-compression and reaction response of equi-volumetric micron-scale (~50-60% dense) spherical nickel and aluminum powder mixtures is investigated in the range of the calculated crush-up pressure (P = 0.4 GPa) and up to 6 GPa. Time resolved stress measurements (using PVDF gauges) coupled with VISAR data is used to determine the shock states. Evidence of reaction or lack thereof is inferred by comparing the measured states with calculated Hugoniot state of reaction products based on the ballotechnic model proposed by Bennett and Horie, (Shock Waves 4:127-136). Post-impact microstructural analysis of recovered material and comparison of calculated and measured product states is used to establish the criterion for reaction occurring in the shock or post-shock states.

**Keywords:** Nickel, aluminum, shock-induced reactions, powders
**PACS:** 62.50.+p, 82.40.Fp

## INTRODUCTION

The initiation of reaction shortly behind the shock front in powder mixtures has been the subject of many research efforts [1-11]. The mechanism(s) responsible for so-called shock-induced reactions however are poorly understood and remain the source of debate. Following the work on silicide forming powder mixtures (e.g. Ti+Si, Nb+Si, Mo+Si) [9], it has been suggested that reaction initiation is controlled by the mechanisms of short time-scale mixing, and is most encouraged in systems in which the crush-up behavior of reactive components is comparable.

The Ni+Al system can be considered a hard/soft mixture, where the difference in bulk static yield strength between reactants is in excess of 100 MPa. When composed of spherical particles, the system is one in which early densification is carried preferentially by the aluminum [12]. Filling in of the interparticle spaces eliminates the void volume necessary for localized Ni plastic flow, reducing the overall extent of mechanically-driven mixing. The question posed is then can reactant configuration be manipulated to encourage shock-induced reactions in an otherwise dissimilarly deforming powder mixture.

The dependence of crush-up behavior on powder pre-treatment has been shown in prior work on NiTi [11]. It is proposed that similar effects can be instrumented through careful selection of component configuration (e.g. particle morphology, packing density, etc.). The highly exothermic Ni+Al intermetallic forming system ($\Delta H_R = -1.38\,\text{kJ}/g$) is ideal to study the relationship between starting powder configuration and reaction initiation response.

The overall goal of the current research is to address this problem in the Ni+Al system through the investigation of varying starting powder configurations, including shape and size of particles. Time-resolved Hugoniot measurements will be used to provide evidence of inert-deviant behavior occurring at the high-pressure state, while a description of the crush-up mechanics will be obtained through post-shock microstructure characterization of recovered materials.

## EXPERIMENTAL PROCEDURE

The first series of experiments were conducted on micron-scale (-325 mesh) spherical nickel and aluminum powders mixed at a 1:1 volumetric ratio, and tested within the range of stresses from the crush-up to 6 GPa. The target assembly consisted of an OFHC Cu containment ring fastened to a Cu driver, within which the powders were statically pressed to an initial density of ~60%, shown schematically in Fig. 1. A transparent fused silica backer plate served as the VISAR window. Piezoelectric stress gauge packages were bonded to the sample-side surfaces of the driver and backer plates, and comprised of Bauer-type PVDF gauges sandwiched between 0.001" Teflon insulation. To avoid electrical interference of the gauges from the sample, a thin (150 nm) evaporated coating of Al separated the packages from the specimen. Data reduction was performed using the PlotData software developed by Sandia National Labs [13].

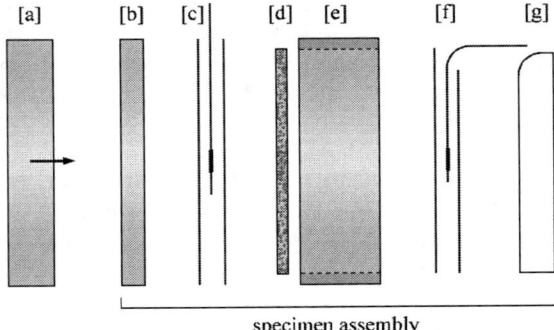

specimen assembly

**Figure 1.** Schematic of the powder specimen assembly used in the Hugoniot measurement experiments. [a] Cu flyer, [b] Cu driver, [c] input gauge package, [d] pressed Ni+Al powder specimen, [e] Cu containment ring, [f] propagated gauge package, [g] rear VISAR window. The powder layer was approximately 3.85 mm thick, 50.8 mm in diameter.

An 80-mm diameter helium driven single-stage gas gun was used to perform impact experiments in the range of 250 to 1000 m/s. Measurement of the input stress pulse, combined with time-of-travel between the input and propagated gauge package, allowed determination of the $P$, $U_s$ shock state. Additional measurements of the equilibrated particle velocity $U_p$, between the specimen and window material were also made in several of the experiments.

Several theoretical models were used as the basis for inferring the occurrence of reaction at the high-pressure state. The inert response of the shocked powder was calculated by the mixture theory developed by McQueen [14], while limits for the complete formation of plausible reaction products were determined using the ballotechnic model [5, 7].

A soft-catching technique allowed partial recovery of the shocked specimen for post-shock characterization of the microstructure. Optical and scanning electron microscopy, coupled with stereological measurement techniques, provided quantitative descriptions of the microstructure environment in specimens recovered from 250, 618, and 998 m/s impact experiments.

## RESULTS AND DISCUSSION

The results of time-resolved Hugoniot measurement experiments on the first series of Ni+Al powder mixtures are collected in Table 1. Italicized values of input stress are the result of calculation based upon the measured equation of state.

**Table 1.** Hugoniot measurement experimental results

| Shot No. | Velocity (mm/µs) | Input Stress (GPa) | Shock Velocity (mm/µs) |
|---|---|---|---|
| 0407 | 0.4753 | *1.671* | 1.065 |
| 0412 | 0.8037 | *3.597* | 1.475 |
| 0414 | 1.0170 | *5.119* | 1.828 |
| 0506 | 0.9408 | 2.850 | 1.367 |
| 0513 | 0.9220 | 4.868 | 1.790 |
| 0514 | 0.9980 | 5.602 | 1.892 |
| 0520 | 0.2504 | 0.503 | 0.625 |

The Hugoniot of the mixture in $U_s$-$P$ space up to 6 GPa is shown in Fig. 2. For convenience,

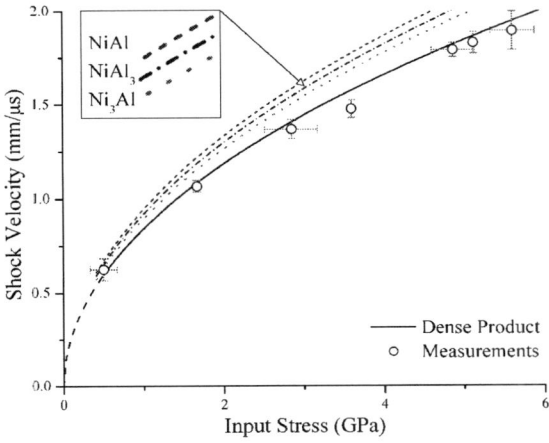

**Figure 2.** Ni+Al $U_s$-$P$ Hugoniot up to 6 GPa. Measured shock states appear to follow the Hugoniot of the inert dense product calculated through mixture theory. Error bars correspond to measurement uncertainties.

reference to shock compressibility curves follow the naming convention used previously by Bennett and Horie [7]. The solid curve corresponds to the inert behavior of the dense product calculated following McQueen's mixture theory. In the case of no reaction, experimental data should lie along this curve, provided the model is appropriate for the mixture. Deviations from this curve are indicative of events such as reaction product formation occurring shortly behind the shock front, as established in prior work on Ti+Si and Ni+Ti systems [9, 11].

Limits for shock-induced reaction were calculated using the ballotechnic model, which performs a constant pressure adjustment of the dense product Hugoniot in proportion to the product heat of formation. Since mixing is not expected to be complete at the pressure range tested, formation of both nickel and aluminum-rich reaction products was considered. The reaction product curves were first constructed in $P$-$V$ space, and subsequently transferred to $U_s$-$P$ space through the jump conditions. The equiatomic NiAl, NiAl$_3$, and Ni$_3$Al phases are shown in Fig. 2. It should be emphasized that the curves differ from the Hugoniots of the solid aluminides, and are rather the signature of reaction product for complete formation of any particular phase.

Superposition of the experimental results in Fig. 2 reveal good agreement with the dense inert product Hugoniot, indicating the absence of detectable shock-induced reaction within the pressure range explored.

Scanning electron micrographs imaged from specimens recovered from the low, intermediate, and high velocity experiments are shown in Fig. 3. At 250 m/s, preferential deformation of the softer Al phase was observed, although regions of retained porosity suggested incomplete densification. The particle boundaries were easily discernable in the Al and Ni cluster interiors, indicating lack of mixing between particles irrespective of species. Similar characteristics were witnessed in the specimen tested at 618 m/s, with

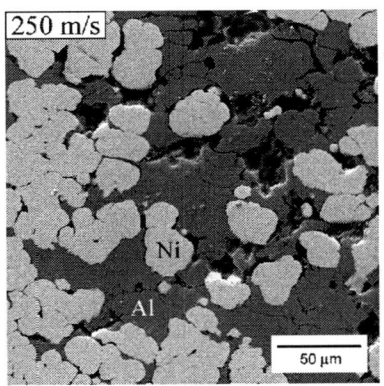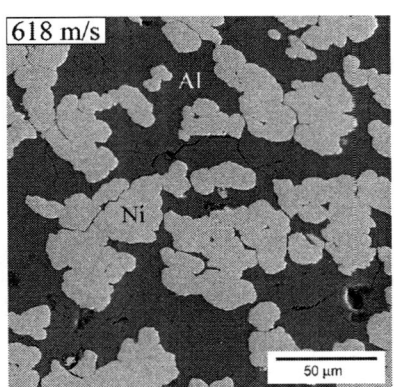

**Figure 3.** Micrographs revealing the microstructure of specimens recovered from experiments conducted at 250, 618, and 998 m/s. Evidence of mixing and the formation of the NiAl$_3$ reaction product was observed only at the highest velocity.

clustering of nickel and filling in of the void space by the heavily deformed Al.

Examination of the 998 m/s recovered specimen yielded a markedly different microstructure. In addition to the deformation processes noted previously, mixing between the parent Ni and Al phases resulting in localized formation of the $NiAl_3$ product phase was observed. Inspection of the unreacted aluminum phase indicated no evidence of Al-Al particle boundaries due to its melting, suggesting reaction proceeded through Ni dissolution in molten Al. Nearly 8% by area of the microstructure consisted of the $NiAl_3$ intermetallic.

## CONCLUSIONS

The first series of experiments investigating the role of powder configuration on shock-induced reactions in Ni+Al mixtures has been performed using time-resolved Hugoniot measurements in the range of crush-up to 6 GPa. Based upon the ballotechnic model, no evidence of shock-induced reaction is observed. However, post-shock characterization revealed presence of melted and resolidified Al and a limited amount of the $NiAl_3$ phase formed via dissolution of Ni in Al and reprecipitation. Although small deviations from inert behavior resulting from the influence of only 8% reaction are challenging to distinguish from measurement error, melting of the parent Al tends to suggest reaction product formation in time scales of thermal equilibration and not the microsecond duration of the high-pressure state.

The difficulty in initiating shock-induced reactions at the high-pressure state in mixture of Ni and Al powders of spherical morphology appears to be due to the hard/soft powder characteristics and limited mixing between dissimilar species. Future work is planned on mixtures incorporating flake Ni, as well as nano scale Ni and Al.

## ACKNOWLEDGEMENTS

Funding was provided by an AFOSR/MURI Grant No. F49620-02-1-0382 and through a National Defense Science Engineering Graduate Fellowship awarded to Dan Eakins.

## REFERENCES

1. Batsanov, S. S., Doronin, G. S. and Klochkov, S. V., "Synthesis Reactions Behind Shock Fronts", Combusion, Explosions, and Shock Waves, 22, 765, 1987.
2. Horie, Y. and Kipp, M. E., "Modeling of shock-induced chemical reactions in powder mixtures", Journal of Applied Physics, 63, 5718, 1988.
3. Bennett, L. S., Sorrell, F. Y., Simonsen, I. K., Horie, Y. and Iyer, K. R., "Ultrafast chemical reactions between nickel and aluminum powders during shock loading", Applied Physics Letters, 61, 520, 1992.
4. Song, I. and Thadhani, N. N., "Shock-Induced Chemical Reactions and Synthesis of Nickel Aluminides", Metallurgical and Materials Transactions A, 23A, 41, 1992.
5. Graham, R. A., Anderson, M. U., Horie, Y., You, S.-K. and Holman, G. T., "Pressure Measurements in Chemically Reacting Powder Mixtures with the Bauer Piezoelectric Polymer Gauge", Shock Waves, 3, 79, 1993.
6. Iyer, K. R., Bennett, L. S., Sorrell, F. Y. and Horie, Y., "Solid state chemical reactions at the shock front", 1337, Year.
7. Bennett, L. S. and Horie, Y., "Shock-Induced Inorganic Reactions and Condensed Phase Detonations", Shock Waves, 4, 127, 1994.
8. Thadhani, N. N., "Shock-Induced and Shock-Assisted Solid-State Chemical Reactions in Powder Mixtures", Journal of Applied Physics, 76, 2129, 1994.
9. Thadhani, N. N., Graham, R. A., Royal, T., Dunbar, E., Anderson, M. U. and Holman, G. T., "Shock-induced chemical reactions in titanium-silicon powder mixtures of different morphologies: time-resolved pressure measurements and materials analysis", 82, 1113, 1997.
10. Vandersall, K. and Thadhani, N. N., "Inverstigation of "shock-induced" and "shock-assisted" Chemical Reactions in Mo + 2Si Powder Mixtures", Metallurgical and Materials Transactions A, 34A, 15, 2003.
11. Xu, X. and Thadhani, N. N., "Investigation of shock-induced reaction behavior of as-blended and ball-milled Ni+Ti powder mixtures using time-resolved stress measurements", Journal of Applied Physics, 96, 2000, 2004.
12. Dunbar, E., Thadhani, N. N. and Graham, R. A., "High-pressure shock activation and mixing of nickel-aluminum powder mixtures", Journal of Materials Science, 28, 2903, 1993.
13. PlotData, Ver. 2.0, Sandia National Laboratories, Albuquerque.
14. McQueen, R. G., Marsh, S. P., Taylor, J. W., Fritz, J. N. and Carter, W. J., "The Equation of State of Solids from Shock Wave Studies, High Velocity Impact Phenomena", Academic Press, 1970.

CP845, *Shock Compression of Condensed Matter - 2005*,
edited by M. D. Furnish, M. Elert, T. P. Russell, and C. T. White
© 2006 American Institute of Physics 0-7354-0341-4/06/$23.00

# SHOCK COMPRESSION OF FePt AND FePt/Fe$_3$Pt NANOPARTICLES: EXCHANGE-COUPLED NANOCOMPOSITE MAGNETS

## Z. Q. Jin[1,2], J. Li[1], N. N. Thadhani[1], Z. L. Wang[1], T. Vedantam[2], J. P. Liu[2]

[1] *School of Materials Science and Engineering, Georgia Institute of Technology, Atlanta, Georgia 30332*
[2] *Department of Physics, University of Texas at Arlington, Arlington, Texas 76019*

**Abstract.** The shock-compression response of partially-ordered 10-20 nm size FePt and FePt/Fe$_3$Pt nanocomposite particles has been studied in this work. The chemically synthesized and annealed nanoparticle powders were packed into three-capsule plate-impact fixtures at ~45% density and shock consolidated using a gas-gun at impact velocities of 500 and 750 m/s. The compacts were recovered as disc-shaped bulk magnets, with up to ~90% higher density than the initial packing density. Transmission electron microscopy analysis revealed plastic deformation and flow of nanoparticles in the process of void annihilation and densification. High-resolution imaging revealed complete retention of nano-size of particles in the dense magnets. Shock compression of the FePt nanoparticles also resulted in an order-to-disorder phase transition from fct-to-fcc structure. However, annealing at temperatures around 700°C caused complete reversal and formation of ordered fct structure. The resulting magnetic property measurements showed energy product $(BH)_{max}$ up to 14 MGOe and coercivity $H_c$ up to 14.6 kOe, which are higher than the values measured for the starting nanoparticles.
**Keywords**: nanocomposite magnets, exchange coupling, Fe-Pt, nanoparticle deformation
**PACS**: 64.70.Nd; 75.60.-d, 75.50.Ww; 75.50.Bb, 75.30.Et

## INTRODUCTION

FePt alloys have contributed to the development of permanent magnets due to their high magneto-crystalline anisotropy and chemical stability. FePt magnets can be synthesized by metallurgical methods and deposition techniques. Typical 3-D FePt cast alloys usually have very low magnetic properties. Deposition techniques yield maximum energy products of 45 MGOe, but these can only be used to obtain 2-D structures, such as coatings or thin films [1]. In recent years chemical synthesis of 1-D mono-dispersed FePt nanoparticles, has aroused significant interest in high-performance permanent magnets, particularly if exchange-coupling can be attained with the soft (Fe$_3$Pt) phase of nano-scale domain size [2]. Nanoparticles can now be synthesized with well-controlled, tailored morphologies (diverse shapes and narrow size distribution), and compositional uniformity. Chemically-synthesized FePt nanoparticles have a disordered fcc structure, which does not show hard magnetic properties. However, its transition to ordered fct phase at elevated temperature, promotes large magnetocrystalline anisotropy needed for permanent magnets. Fabrication of bulk (*3-D*) magnets from ultrafine *1-D* nanoparticles, requires densification without loss of nano-scale structure. Shock compression has the potential to retain nano-structure in consolidated compacts starting with powders of nano-sized grains [3]. Here, we report on shock-compression response of FePt and FePt/Fe$_3$Pt nanoparticles to form bulk (3-D) permanent magnets with retained nanostructure.

## EXPERIMENTAL PROCEDURE

FePt nanoparticles were chemically synthesized by polyol reduction of Pt acetylacetonate and thermal decomposition of $Fe(CO)_5$ in an environment consisting of oleic acid, oleyl amine and long-chain 1,2 hexadecanediol [2]. Thermal decomposition of $Fe(CO)_5$ was utilized to prepare $FePt/Fe_3Pt$ nanocomposites. Subsequent annealing at 400 °C in a forming gas was used to remove the surfactant from the particle surface. The samples were then cold-pressed into steel capsules and shock compressed at impact speeds of 500 and 750 m/s using a single-stage plate-impact gas gun [5], to produce ~10 mm diameter by 1 mm thick discs. The shock-compaction fixture was designed with an air gap surrounding the capsules to minimize radial wave-focusing effects and allow the loading history to be dominated by lower-pressure planar waves. AUTODYN-2D simulations were performed to predict the shock-loading conditions employing the P-α densification model. The recovered compacts were characterized by density measurements, X-ray diffraction (XRD) using Cu $K_\alpha$ radiation, and transmission electron microscopy (TEM). Magnetic properties were measured using a superconducting quantum interference device (SQUID) with a maximum applied field of 70 kOe.

## RESULTS AND DISCUSSION

The phase structure of FePt samples as examined with XRD analysis is shown in Fig. 1. The chemically-synthesized and annealed (at 400 °C) sample shows a disordered fcc structure as the dominant phase (Fig. 1a), along with traces of the chemically ordered fct structure, although previous studies have reported that annealing above 550°C is necessary to obtain a high degree of ordering of pure FePt nanoparticles [6], in the absence of elements, such as Ag, Cu, Sn, Pb, Sb and Bi [9]. The shock-compacted samples, show no significant difference in structure, as revealed by XRD traces of both impact and rear surfaces, indicating a uniform shock compression response throughout the entire disc-shaped bulk magnet, since the radial wave focusing effect has been minimized due to an air gap placed around powder-containing capsule.

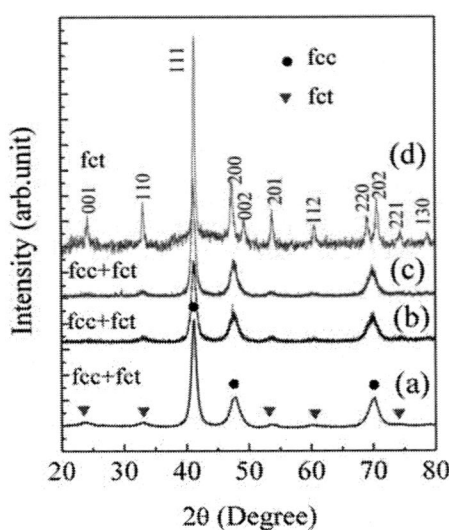

Fig 1. XRD patterns of FePt samples. (a) The chemically synthesized sample was annealed at 400 °C prior to shock compaction. (b) and (c) Rear and impact surfaces of shock compacted samples, respectively. (d) Shock compacted sample annealed at 700 °C.

Further analysis of XRD patterns also revealed the reduction of peak intensity of fct structure, due to rereversal of ordered (fct) to disordered (fcc) phase due to shock-wave induced rearrangement of Fe and Pt atoms. After annealing the shock compacted sample at 700 °C, complete ordering to the fct phase is observed, as shown by the XRD trace in Fig. 1(d). The XRD peak intensity ratios also indicate isotropic magnetic characteristics of shock consolidated and annealed bulk magnet. Similar effect has been observed for both FePt and $FePr/Fe_3Pt$ shock consolidated at 500 and 700 m/s.

TEM images comparing the (a,b) as-synthesized and annealed (to remove surfactant); and (c,d) shock-compacted samples, are shown in Fig. 2. The as-synthesized spherical particles (Fig. 2 (a,b)) show 10-20 nm size particles of uniform size distribution. Following shock compaction, the nanometric grain size appears to be fully retained, with some grains of even smaller (<10 nm) size, as illustrated in the TEM image in Fig. 2 (c,d)). Retention of nano-structure in shocked compacts is similar to that observed in our work on exchange-coupled $Pr_2Fe_{14}B/\alpha$-Fe nanocomposite magnets made from powders of nano-scale grain size [5].

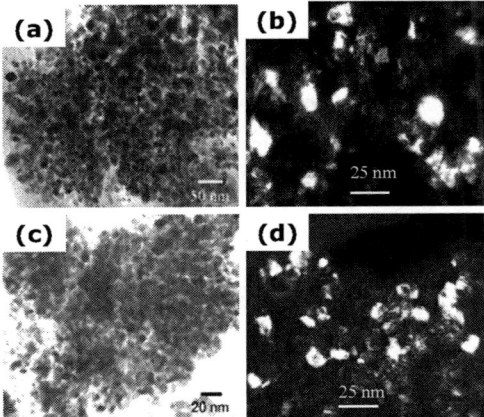

Fig. 2. TEM images comparing (a,b) starting (as-synthesized/annealed) nanoparticles; and (c,d) shock-consolidated compacts.

High-resolution TEM observation of interparticle regions revealed clear evidence of plastic deformation of the nano-size FePt particles during dynamic shock compaction. Figure 3 shows the TEM image illustrating the severe plastic deformation of the nanoparticle in an attempt to fill the void space. The plastic deformation and flow processes [12] lead to densification to ~78% TMD in the case of compacts made at 500 m/s and ~84% TMD in compacts made at 750 m/s. EDX analysis also revealed the presence of disordered fcc and ordered $L1_0$ phases, consistent with results of XRD analysis. The disordered phase formed during shock compaction, fully reverts back to the ordered fct structure upon annealed at 700°C.

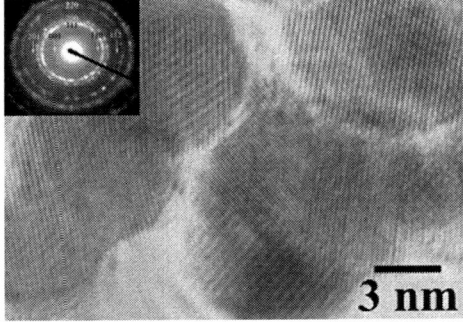

Fig. 3. High-resolution TEM image illustrating severe plastic deformation of FePt nanparticle in an attempt to fill voids

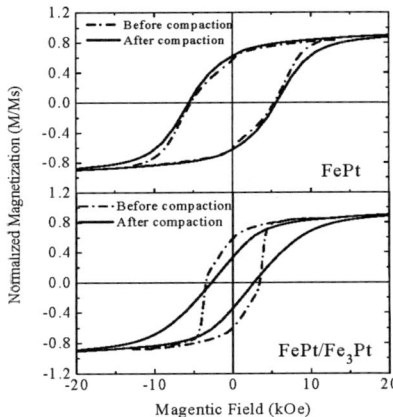

Fig. 4. Hysterisis loops of FePt and FePt/Fe$_3$Pt samples, before and after compaction.

Results of magnetic property measurements are shown in Fig. 4, which compares the hysterisis loops of FePt and FePt/Fe$_3$Pt samples, before and after compaction. Both types of samples show hard magnetic behavior. The hysterisis loop of the FePt compact is smooth, and has better squareness possibly due to more enhanced interparticle interaction. The heterogeneous structure of the FePt/Fe$_3$Pt nanocomposite leads to reduction in loop squareness. However, the hysterisis loop squareness and overall magnetic properties are significantly enhanced upon post-shock annealing. Figure 5 shows the hysteresis loops for the shock compacted FePt and FePt/Fe$_3$Pt samples upon annealing at different temperatures. While both as-compacted samples show hard magnetic behavior and coercivity of several kOe, verifying existence of fct phase, annealing at temperatures of 400° to 600°C results in completion of ordering process and drastic increase in coercivity, remanence, loop squareness and maximum energy product for both samples. Maximun energy products of 14 and 9 MGOe are obtained for FePt and FePt/Fe$_3$Pt nanomaterials, and their respective coercivities are 14.6 and 13 KOe. Improved magnetic properties achieved in the shock consolidated/annealed compacts are due to nanostructure retention and exchange coupling between FePt (hard) and Fe$_3$Pt (soft) phases, and also amongst the hard FePt nanoparticles, as evidenced by high remanence ratios ($M_r/M_s$~0.65).

1159

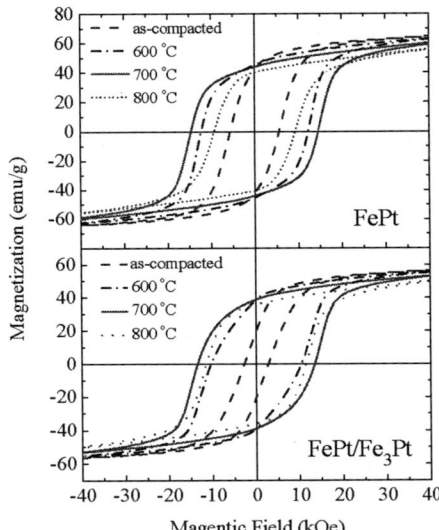

Fig. 5. Hysteresis loop of shock compacted samples and the samples annealed at different temperature for 1 hour.

Magnetic property measurements performed on samples obtained from several regions of the same compact, revealed uniformity in magnetic properties, which indicates the uniform shock pressure and density distribution throughout the 1-mm thick disks produced by shock compaction.

## CONCLUSIONS

Shock compression of FePt and FePt/Fe$_3$Pt nanoparticles has been investigated. The size retention of ultrafine particles was observed with change in particles morphology, which results in the enhanced densification. Heat treatment improved the hard magnetic properties of bulk magnets up to 14 MGOe. The high performance allows this material to be used as permanent magnets, where excellent corrosion and demagnetization resistance are necessary.

## ACKNOWLEDGEMENTS
This work was supported by US DoD/DARPA through ARO under grant DAAD19-03-1-0038 and by ONR/MURI under grant N00014-05-1-0497.

## REFERENCES

[1]  Liu, J. P., Luo, C. P., Liu, Y., and Sellmyer, D. J., *Appl. Phys. Lett.* **72**, 483 (1998).
[2]  Sun, S., Murray, C. B., Weller, D., Folks, L., and Moser, A., *Science*, **287**, 1989 (2000).
[3]  Thadhani, N. N., Graham, R. A., Royal, T., Dunbar, E., Anderson, M. U., and Holman, T. T., *J. Appl. Phys.* **82**, 1113 (1997).
[4]  Shkuratov, S. I., Talantsev, E. F., Dickens, J. C., Kristiansen, M., and Baird, J., *Appl. Phys. Lett.* **82**, 1248 (2003).
[5]  Jin, Z. Q. Chen, K. H., Li, J., Zeng, H., Cheng, S. F., Liu, J. P., Wang, Z. L., and Thadhani, N. N., *Acta Mater.* **52**, 2147 (2004).
[6]  Watanabe, Y., Kimura, N., Hono, K., Yasuda, K., and Sakurai, T., *J. Magn. Magn. Mater.* **170**, 289 (1997).
[7]  Watanabe, M., Masumoto, T., Ping, D. H., and Hono, K., *Appl. Phys. Lett.* **76**, 3971 (2000).
[8]  Sun, S., Fullerton, E. E., Weller, E., and Murray, C. B., *IEEE Trans. Magn.* **37**, 1239 (2001).
[9]  Kang, S. S., Harrell, J. W., and Nikles, D. E., *Nano Lett.* 2, 1033 (2002).
[10] Maeda, T., Kai, T., Kikitsu, A., Nagase, T., and Akiyama, J. I., *Appl. Phys. Lett.* **80**, 2147 (2002).
[11] Kitakami, O., Shimada, Y., Oikawa, K., Daimon, H., and Fukamichi, K., *Appl. Phys. Lett.* **78**, 1104 (1991).
[12] Elkins, K., Li, D. R., Poudyal, N., Nandwana, V., Jin, Z. Q., Chen, K. H., and Liu, J. P., *J. Phys. D : Appl. Phys.* **38**, 2306 (2005).

CP845, *Shock Compression of Condensed Matter - 2005*,
edited by M. D. Furnish, M. Elert, T. P. Russell, and C. T. White
© 2006 American Institute of Physics 0-7354-0341-4/06/$23.00

# FEATURES OF THE SHOCK AND DETONATION WAVES IN CYLINDRICAL EXPLOSIVE COMPACTION

## J. B. Ribeiro, R. L. Mendes, I. Ye. Plaksin, and J. A. Campos

*Mechanical Engineering Department, Faculty of Sciences and Technology, University of Coimbra,
Pinhal de Marrocos, Pólo II, 3030-201 Coimbra, Portugal*

**Abstract.** Despite of the significant amount of the work that is being done in the field of the explosive consolidation, a considerable lack of time and spatial resolved experimental data is being feel by the people working in the area; moreover, many of the attempts made to overcome this problem were done in conditions far way from the ones used in real consolidation experiments. To fill this gap is necessary to perform the characterization of both, the consolidation and the detonation waves, in conditions close to the ones used in real experiments and with a spatial resolution approaching the characteristic size of the powder. Using an experimental technique developed at our laboratory, based on the utilization of a 64 channels optical fiber strip, connected to an electronic streak camera, spatial and temporal resolved details of the compaction and detonation waves front shape and pressure, were obtained and are presented. The results refer to a cylindrical configuration set-up and alumina powder, two values of the initial density, two characteristic sizes of the powder particles and two powder container materials.

**Keywords:** explosive consolidation, alumina, shock wave, consolidation wave.
**PACS:** 81.20.Ev, 81.05.Je, 81.40.Vw, 62.50.+p.

## INTRODUCTION

Explosive consolidation of powders of metals, ceramics, polymers and their mixtures is deserving a considerable amount of attention by the scientists working in the areas of shock wave physics and material science in the last decades. The interest for this subject have grown-up recently with the advent of the nanomaterial science because it is believed that this process will preserve the potentially enhanced properties of the bulk nanocrystalline materials by hindering the undesirable grain growth associated to conventional sintering processes. However, despite of this renewed interest on the subject and despite of all theoretical, numerical and experimental (mainly sample recover analysis) work done, there is a considerable lack of real time experimental information about the shape of the consolidation wave [CW], with suitable temporal and spatial resolutions, obtained in conditions close to the ones used in real consolidation experiments. This is particular true for the case of the cylindrical configuration, one of the most used and one of the most interesting for consolidation of powders. In this specific configuration the loss of energy during the compaction is compensated with the convergence of the wave. When these two effects are well balanced the shape of the CW is expected to be conical and it is supposed to lead to cylindrical samples of consolidated powder with homogeneous density throughout its entire radius. However, even for this situation, in which the two effects are well balanced, problems are waited to occur in the center of the sample due to accumulation of a finite quantity of energy in an approaching to zero volume. What really happens for this region of the sample is being nowadays depicted from the analysis of the recovered samples, because the main diagnostic technique, x-ray photography, used for these experiments does not have enough resolution to offer clear information. The clarification of the role of the powder container in the consolidation process is another point that needs clarification. The dependence of the final properties of the consolidated samples on its material, and on its thickness, is recognized but explanations for that dependence are not clear. Finally, the intrinsic oscillating behavior of the shock wave propagating process, that is believed to occur for all materials, and was already described with detail for some explosives and foams, it is also expected to be found in this process [1]. The magnitude and the importance of those oscillations are just now starting to be analyzed in the consolidation experiments [1].

In this paper we are presenting an original experimental methodology able to generate data that are believed to contribute for the clarification of some of the points stated before.

## EXPERIMENTS

### Experimental set-up

As it was already referred the experimental characterization of the consolidation wave was done for the well known cylindrical configuration [2]. The main diagnostic element of the set-up is a strip of 64 independent optical fibers, each one of them with 250 µm of diameter, that is connected to an electronic streak camera, THOMSON TSN 506 N, without any intermediate optics. This strip was divided in two parts, one with 28 fibers, used for the CW monitorizarion and the other, with 36 fibers, for the detonation wave [DW] characterization. The part of the fiber used for the CW monitorisation is inserted between the two half parts of the bottom plug of the powder container, which has a conical top surface, and covered with a stack of Kapton layers (each layer with 125 µm of thickness). The other part of the fibers is cover with an aluminum foil, in which, at knows positions, are open "window slits" for light entrance, and placed within the explosive charge oriented along the powder container wall. A

schematic representation of the described set-up can be seen in Fig. 1.

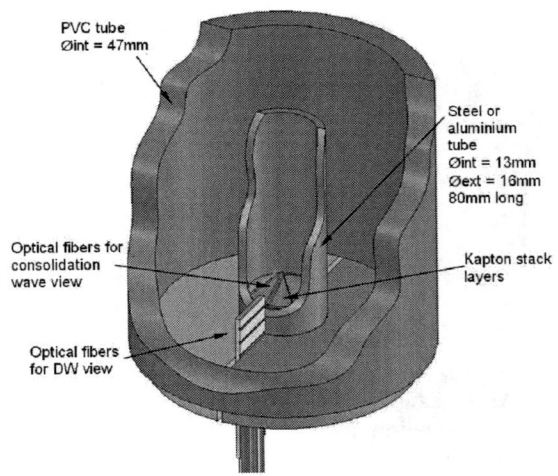

**Figure 1.** Schematic representation of the experimental set-up

### Materials

Two different alumina powders have been used in these experiments. In the Table 1 are resumed their most important characteristics.

**TABLE 1.** Characteristics of the powders

| Origin | Effective density, $g/cm^3$ | $Al_2O_3$ content, % | $d_{50}$, µm | Crystallite size, nm |
|--------|-----------------------------|----------------------|--------------|----------------------|
| Sulzer | 3.90 | > 99.7 | 22.5 | - |
| Lab. pilot plant | 3.85 | > 99.8 | 3.5 | 70 |

For all the experiments the used explosive was an ammonium nitrate aqueous emulsion sensiblized with 5% w/w of perlite hollow microspheres. The main characteristics of this explosive are D = 4.60 mm/µs and density = 1.10 $g/cm^3$.

### Experimental work plan

Beyond the experimental identification of the characteristics features of the consolidation process, one of our objectives is to understand how those features change with typical parameters of the consolidation process, like the initial density of the powder samples, the characteristic size of the powder particles and the nature of the material of the powder container. So an experimental work plan involving the consideration of four different experimental situations and five experiments was depicted and followed. The main characteristics of each one of those experiments are presented in Table 2.

**TABLE 2.** Experiments main characteristics

| Experiment | Powder $d_{50}$, $\mu m$ | $\rho_{00}$, $g/cm^3$ | Powder container material | $\rho_{final}$[1], $g/cm^3$ |
|---|---|---|---|---|
| #1 | 22.5 | 2.17 | Steel | 3.16 |
| #2 | 22.5 | 2.29 | Steel | 3.38 |
| #3 | 22.5 | 2.20 | Aluminum | 3.19 |
| #4 | 3.5 | 2.54 | Steel | hole |
| #5 | 3.5 | 1.22 | Steel | lost |

## RESULTS

### Typical results and data presentation

Streak records like the one shown in the Fig. 2 are the typical results of the performed experiments. Two different areas can be identified in those streak records; one referring to the curvature of the DW near the powder container wall, and the other referring to the interaction of the CW with the conical shaped bottom plug. From the streak records like this one, because the position of the fibers is known, and it is assumed that both the detonation and the CW are moving at constant velocity (det. velocity), is possible to obtain the real shape of the referred waves. Results of that treatment are shown in the graphic of the Fig. 3.

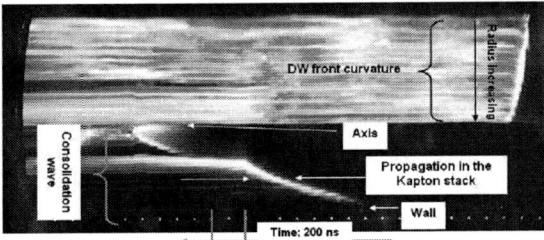

**Figure 2.** Typical streak record obtained with the experimental set up shown in Fig. 1. Experimental situation #3

Once obtained the real shape of the CW, it is straightforward the determination of its local velocity, using a procedure similar to what was used by Carton [3]. These results are not shown.

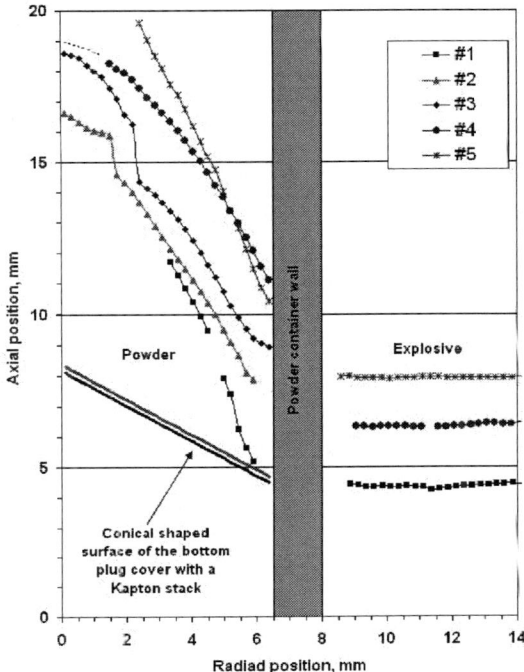

**Figure 3.** Consolidation and detonation wave shapes. DW and CW have been artificial displaced for not overlaying.

The evaluation of the pressure is much more cumbersome. In fact the interaction of the CW with the Kapton stack generates on this a shock wave not parallel to the surface and which is changing slope while the CW propagates toward the center. A real representation of that process, made with data from experiment #2, is shown in the Fig. 4. The evaluation of the pressure by the impedance matching technique, will obey to the determination of $U_{SKapton}$, for which, as it can be seen from the Fig. 4, will be necessary to know the inclination of the shock wave and the two tangential velocities $U_{Stan1}$ and $U_{Stan2}$. The results obtained for the experiment #2 are shown in Fig. 5.

---

[1] Evaluated based on diameter differences.

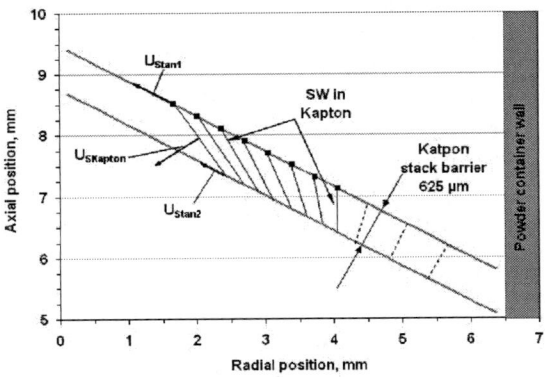

**Figure 4.** Shock wave propagation in the Kapton stack for the experiment #2

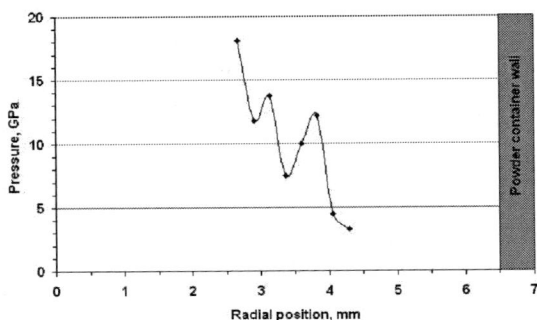

**Figure 5.** Consolidation wave pressure, for the experimental situation #2, as a function of the radial position.

This concludes the presentation of the kind of data that can be taken from the streak records obtained with the experimental set-up described before. The determination of the shock velocity and the pressure in the powder with such detail as reached here, together with a recovered sample analysis, is expected to allow a complete description of the consolidation phenomena.

### Results discussion

Leaving out the parametric analysis of the results for the moment, due to the lack of space, we would like to emphasize on this point the obtained consolidation wave shape profiles (*vd.* Fig. 3). The detail with which they were obtained allows the observation of a non monotonous propagation of the CW towards the center, and at least for two of them, (exp. #2 and #3) one big discontinuity in the central zone. We have tried to find the explanation

for this discontinuity coupling this information with the analysis of the recovered samples, which can be observed, for those two experiments, in the Fig. 6, but possible explanations are not obvious. Crack formation and/or the existence of a secondary wave, during the propagation toward the center, are possibilities that need to be explored.

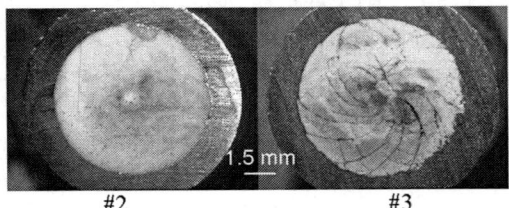

**Figure 6.** Photos of the recovered samples of experiments #2 and #3 near the base, where the monitorization of the consolidation wave was made.

### CONCLUSIONS

A new time and space detailed experimental technique for CW monitorization was presented. Data obtained with such technique allow a complete characterization of the wave. The results obtained for the particular case of the alumina show a non monotonous behavior and significant discontinuities which causes need to be further investigated.

### ACKNOWLEDGEMENTS

Funding was provided by FCT - Portuguese Foundation for Science and Technology, through ADAI under the contract POCI/EME/55398/2004. The authors thank to Susana Ferreira and ICEMS for the photos of the recovered samples.

### REFERENCES

1. Plaksin, I., et al. "Novelties in physics of explosive welding and powder compaction" Journal de Physique IV, Vol. 110 (September 2003), pp. 797-802.
2. Pruemmer, R. A., "Latest results in the explosive compaction of metal & ceramic powders and their mixtures", Fourth Int. Conf. of the Center for High Energy Forming, Estes Part, Co, 9-13th July, 1973.
3. Carton, E. P., et al., "Dynamic compaction of powders by an oblique detonation wave in the cylindrical configuration" Journal of Applied Physics, vol. 81, no. 7, pp. 3038-3045, 1997.

CP845, *Shock Compression of Condensed Matter - 2005*,
edited by M. D. Furnish, M. Elert, T. P. Russell, and C. T. White
© 2006 American Institute of Physics 0-7354-0341-4/06/$23.00

# SHOCK-INDUCED CHEMISTRY IN POLYDIMETHYLSILOXANE

**Robert Sander, Norm Blais, Ray Engelke, Dana Dattelbaum,
Steve Sheffield, Rhonda McInroy.**

*MS.J567, Los Alamos National Laboratory, Los Alamos NM 87545*

**Abstract.** Polydimethylsiloxane (PDMS) is a common silicone polymer. Understanding its decomposition product distribution is required for calculating its equation of state under shock conditions. We have detonated small samples of HMX explosive in contact with the polymer in a high vacuum chamber and used a time-of-flight mass spectrometer to analyze the chemical products of PDMS decomposition. We have used the computer code CTH to model the time history of pressure and temperature in the sample. The time scale of a few nanoseconds in these experiments generates products that are significantly different than the equilibrium products observed in thermal pyrolysis experiments. The mass spectrum under shock conditions predominately shows monomers to heptamers of dimethylsiloxane, and the n-mers minus methyl groups. We have compared spectra of high molecular weight liquid PDMS, crosslinked solid PDMS, and silica filled solid PDMS.

**Keywords**: polydimethylsiloxane, decomposition, shock, mass spectrometer
**PACS**:82.40.Fp, 47.40.-x

## INTRODUCTION

Detonations of high explosives can cause shockwaves in nearby materials. The Pdv compression work of the shock can transiently heat the materials and cause chemical reactions such as decomposition. Because the high temperatures and pressures are so transient, the reaction products may be different from the equilibrium products observed in slow pyrolysis experiments. Because plastics and polymers are common structural materials it is of interest to examine their shock-induced chemistry. Polymeric materials are also of interest because the decomposition into shorter chain material, or even monomers, is driven by the high temperatures, while the high pressure might be expected to preserve the highly linked compact molecular structure. Polydimethylsiloxane (PDMS) is a common engineering polymer that is available as a liquid medium-chain-length

polymer, and also as a long chain solid, and more rigid crosslinked plastic, and with inorganic filler. Its structure is:
$[-O-Si(CH_3)_2-]_n$ where n can range up to a few hundred, and the monomer has a weight of 74 AMU. Essentially, it has the linear backbone of quartz surrounded with organic hydrocarbon arms.

## EXPERIMENTAL

The apparatus used for these measurements has been previously described in detail.[1-3] It is a large vacuum chamber, with several stages of differential pumping between the region of the detonation and an ultra-high vacuum time-of-flight mass spectrometer. The mass spectrometer is pulsed to extract ions every 12 μs. This allows us to obtain many mass spectra of the chemical products from the gas flow without wall reactions, before the buildup of gas in the chamber produces a Mach stem and shuts off the flow.    A schematic of the apparatus is shown in Fig. 1.

The carousel shown in Fig.1 allows us to mount several shots inside the vacuum without having to return to atmospheric pressure. The mass spectrum is produced by ionizing the neutral decomposition products in a 90 eV. electron beam.  There is an inherent difficulty in electron impact ionization, because the electron beam can also cause decomposition. Because the detonation products have a high speed, the sideways extraction of the ions in the time-of-flight mss spectrometer requires two separate detectors: an upstream detector for low speed products, and a downstream detector for faster or higher mass products.

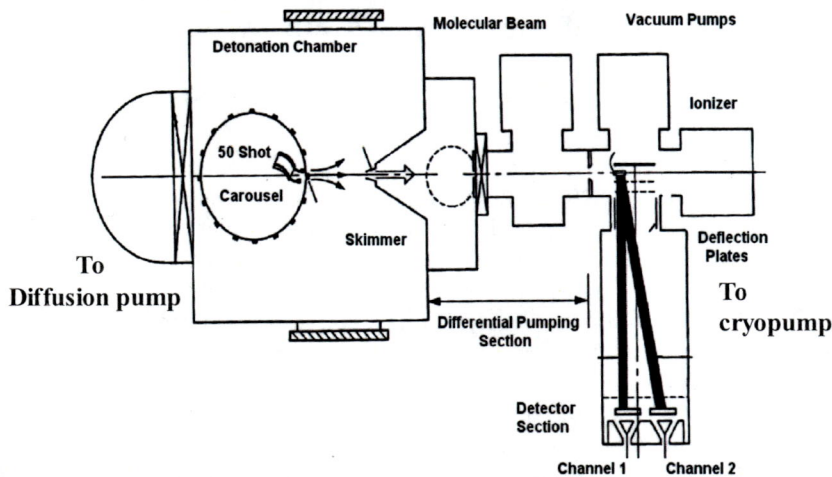

**Figure 1.** A schematic view of the apparatus. The flight path from detonation region to the ionizer of the mass spectrometer is  1.29 meters.    A mass spectrum from 0 to 300 AMU is produced by pulsing the ionizer every 12 μsec.

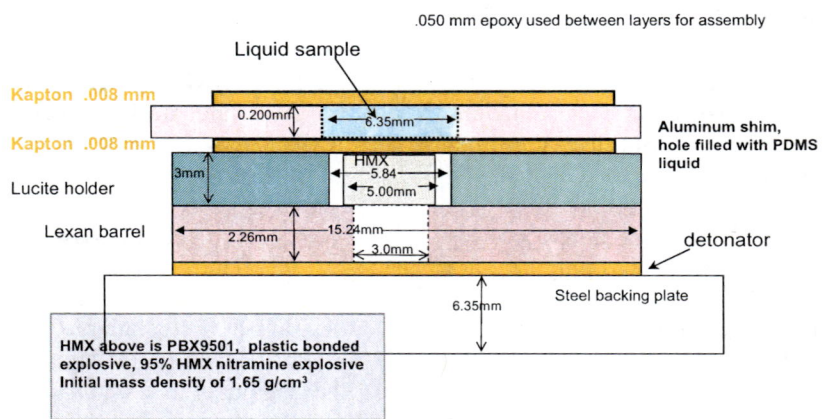

**Figure 2.** A schematic of the assembled shot, showing how the shock wave from the explosive is coupled to the liquid polymer sample.

A schematic view of the assembled shot is shown in Fig. 2. The solid polymer samples were easier than liquids to assemble adjacent to the explosive. The liquid samples were coupled to the explosive and contained in the mass spectrometer vacuum with thin Kapton foils. The detonator is a slapper, a necked-down strip line of copper coated Kapton pulsed with ca. 300 Joules from a capacitive discharge unit. The slapper accelerates down the Lexan barrel initiating the HMX pellet.

There are two types of data that can be obtained: 1) the mass spectrum of the chemical decomposition products, and 2) the arrival time history (with 12 μs resolution) of a given mass peak at the ionizer. Figure 3 shows the lower end of a mass spectrum near the peak of the arrival time distribution. The three peaks corresponding to monomer (mass 74), dimer and trimer are shown. The trimer minus a methyl group is also seen. As expected, the upstream detector shows the lighter fragments, and the downstream detector shows the heavier ones. The time-of-flight mass spectrometer starts a new scan after mass 300. One might have expected the thermodynamically most stable product, silicon carbide, which is seen in slow pyrolysis.

Seeing the lowest three n-mers in this scan suggests that the trimer may be formed and then break up to give the dimer and monomer. The trimer could be a six membered ring structure. Cyclic ring structures have been previously suggested in conventional mass spectra of silicones.[4]

We see similar low n-mer fragments in the mass spectrum of low molecular wt. PDMS vaporized at 150 °C. So we cannot say with certainty whether a trimeric form is produced by the shock or by the ionization or by some combination of the two sequential processes.

By looking at earlier scans which have higher velocity products, with different deflection voltages to allow focusing of higher masses into the detectors we can see the higher mass fragments. In Fig. 4 is data showing the downstream detector as a solid line, and the upstream detector as a dashed line. The lighter masses impact the upstream detector while the heavier masses arrive at the downstream detector. Peaks up to the heptamer are seen. Some of the peaks show smaller satellites at a mass corresponding to loss of one to three methyl groups (15 mass units).

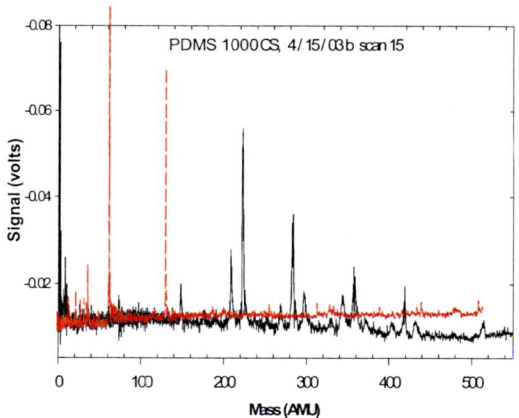

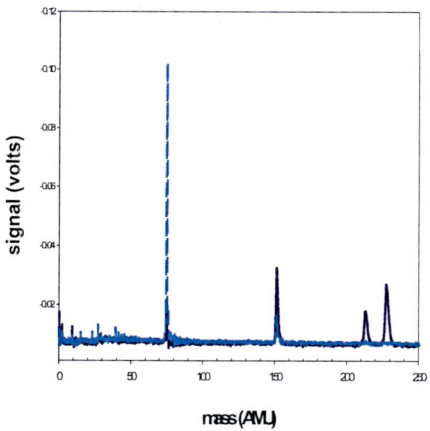

**Figure 3.** This shows a mass scan from a shot of 1000 centistokes viscosity liquid PDMS with molecular weight 16457, mass scan 21, or 252 μsec after the detonator was fired. The downstream detector data is the dashed line, and the upstream detector is the solid line. The data is smoothed with a 5 point running average

**Figure 4.** This shows mass scan 15 of an earlier shot of 1000 centistokes PDMS. The unshocked material contains about 220 monomer units. The monomer and dimer peaks are seen on the upstream detector (dashed) and the tetramer through heptamer are seen on the downstream detector (solid). Data is smoothed with a 5 point running average.

The spectrum looks quite different from the spectrum we observe of a vapor phase PDMS sample, and also from literature spectra [4] Perhaps the shock is producing vibrationally hot molecules that fragment more easily (loss of methyl) during the ionization process. The decomposition of PDMS may be affected by the very short time that it is exposed to high pressures and temperatures. For that reason we used a two-dimensional hydrodynamic code, CTH,[5] to calculate the pressure history at a series of points inside of the liquid volume. In Fig. 5 we show results for four Lagrangian points spaced at 66μm successively deeper in the PDMS cell. Even for the mass point with the longest exposure to high pressure, the particle only experiences high-pressure for about 30 ns.

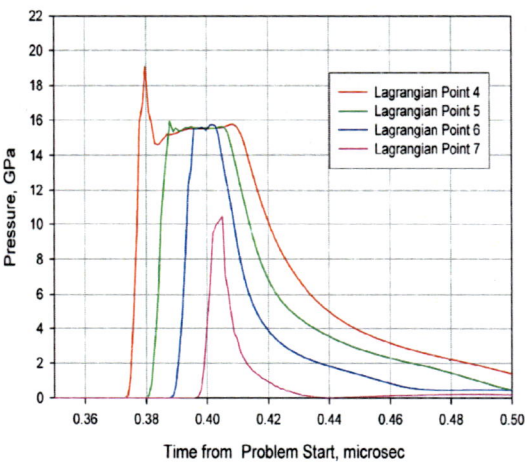

**Figure 5.** CTH calculation of the time history of the pressure in the shocked PDMS for the configuration of Fig. 2.

The temperatures associated with the pressures shown in Fig. 5 can be estimated using CTH. They range from 1500 K to 2700 K for the four points. The decay time of the temperature is slightly longer than for the pressure.

## CONCLUSIONS

We have measured the mass spectrum of PDMS shocked to 16 GPa by a small charge of HMX explosive under free expansion conditions. We observe n-mers of dimethylsiloxane ranging from n=1 to 7. In addition, n-mers missing 1 to 3 methyl groups are observed. We take this to imply that under the conditions of the current experiments there is a only limited amount of decomposition, and that the lowest energy thermodynamic products do not have time to form.

## ACKNOWLEDGEMENTS

This work is supported by the U.S. Department of Energy under contract number W-7405-ENG-36

## REFERENCES

1. Blais, N.C., Fry, H.A., Greiner, N.R., "Apparatus for the Mass Spectrometric Analysis of Detonation Products Quenched by Adiabatic Free Expansion," Rev. Sci. Instr. **64** 174, (1993).
2. Blais, N.C., Engelke, R., Sheffield, S.A., "Mass Spectroscopic Study of the Chemical Reaction Zone in Detonating Liquid Nitromethane," J.Phys.Chem. **101** 8285-8295, (1997).
3. Engelke, R. Blais, N.C., Sheffield, S.A. Sander, R.K. "Production of a Chemically- Bound Dimer of 2,4,6-TNT by Transient High Pressure," J. Phys. Chem. **A105**, 6955-6964, (2001).
4. Moore, J.A." Mass Spectrometry" in "The Analytical Chemistry of Silicones" pp 421-470, A. Lee Smith ed., John Wiley, N.Y. (1991).
5. McGlaun, M.C., Kemtyk, L.N., Elrich, M.G., Thompson, S. L., "A Brief Description of the Three-dimensional Shock Wave Physics Code CTH," Sandia report, Sand89-0607,(1989).

CP845, *Shock Compression of Condensed Matter - 2005*,
edited by M. D. Furnish, M. Elert, T. P. Russell, and C. T. White
© 2006 American Institute of Physics 0-7354-0341-4/06/$23.00

# LONGITUDINAL SHOCK WAVE DEPOLARIZATION OF Pb($Zr_{52}Ti_{48}$)$O_3$ POLYCRYSTALLINE FERROELECTRICS AND THEIR UTILIZATION IN EXPLOSIVE PULSED POWER

**Sergey I. Shkuratov[1], Evgueni F. Talantsev[1], Jason Baird[2], Henryk Temkin[3], Larry L. Altgilbers[4], and Allen H. Stults[5]**

[1]*Loki Incorporated, Rolla, MO 65409, U.S.A.*
[2]*Loki Incorporated and University of Missouri - Rolla, Rolla, MO 65409, U.S.A.*
[3]*Department of Electrical Engineering, Texas Tech University, Lubbock, TX 79409, U.S.A.*
[4]*U.S. Army Space and Missile Defense Command, Huntsville, AL 35807, U.S.A.*
[5]*U.S. Army Research, Development and Engineering Command, Huntsville, AL 35898, U.S.A.*

**Abstract.** A poled lead zirconate titanate Pb($Zr_{52}Ti_{48}$)$O_3$ (PZT) polycrystalline piezoelectric ceramic energy-carrying element of a compact explosive-driven power generator was subjected to a longitudinal explosive shock wave (the wave front traveled along the polarization vector $P_0$). The shock compression of the element at pressures of 1.5-3.8 GPa caused almost complete depolarization of the sample. Shock wave velocity in the PZT was determined to be 3.94 ± 0.27 km/s. The electric charge stored in a ferroelectric, due to its remnant polarization, is released during a short time interval and can be transformed into pulsed power. Compact explosive-driven power sources utilizing longitudinal shock wave depolarization of PZT elements of 0.35 to 3.3 $cm^3$ volume are capable of producing pulses of high voltage, with amplitudes up to 22 kV, and up to 350 kW peak power.

**Keywords:** shock compression of solids, shock depolarization, ferroelectric materials, explosive pulsed power
**PACS:** 62.50.+p; 77.80.Bh

## INTRODUCTION

A wide range of modern devices relies on the energy chemically stored in high explosives, propellants, metastable intermolecular composites, and high-energy-density nanocomposites. Explosive-driven electric generators, considered the most effective modern compact sources of pulsed power, are one class of such devices [1].

The design and performance of recently developed autonomous pulsed power sources utilizing the electromagnetic energy stored in ferroelectric materials was described previously [2]. Compact explosive-driven generators based on shock wave depolarization of ferroelectric energy-carrying elements have been demonstrated to have reliable and controllable electrical operation [2]. This paper presents the results of an experimental investigation of the depolarization of ferroelectric energy-carrying elements within compact ferroelectric generators (FEGs) under the action of shock waves generated by the detonation of high explosive charges.

## EXPERIMENTAL PROCEDURE

The test objects were commercial polycrystalline lead zirconate titanate

Pb($Zr_{52}Ti_{48}$)$O_3$ (PZT) piezoelectric ceramic disks (supplied by EDO Corp.). Their parameters are as follows: density $\rho_0$ = 7.5 x $10^3$ kg/m$^3$, dielectric constant $\varepsilon$ = 1300, Curie temperature 320° C, Young's modulus 7.8 $10^{10}$ N/m$^2$, piezoelectric constant $d_{33}$ = 295 x $10^{-12}$ C/N, and piezoelectric constant $g_{33}$ = 25 x $10^{-3}$ m$^2$/C.

A schematic diagram of the experimental setup is shown in Fig. 1. A shock wave was initiated at the front face of the PZT disk by a light aluminum impactor (flyer plate) accelerated to high velocity by the detonation of desensitized RDX high explosives (HE, with Chapman-Jouguet state pressure of 22.36 GPa, and detonation velocity of 8.1 km/s). In the devices, a ferroelectric disk was mounted on a copper backplate that provided mechanical impedance matching to minimize reflection of stress waves when they reached the rear face of the PZT disk. Silver contact plates were deposited on both faces of the PZT disk. The overall dimensions of the shock wave ferroelectric generators used in the experimental series did not exceed 50 mm. A detailed description of these devices can be found in [2].

The PZT module depolarization current and generated voltage waveforms were monitored with commercial current and voltage probes. A Model 411Pearson Electronics current monitor was used to measure the pulse current, and the voltage pulses were monitored using a Tektronix P6015A high voltage probe.

In this series of experiments, PZT disks were poled parallel to their short axes to their full remnant polarization values, $P_0$ = 30 µC/cm$^2$. The flyer

plates provided longitudinal impacts on the ferroelectric bodies so that the shock waves traveled in a direction parallel to the polarization vector, $P_0$. Before flyer plate impact in a test, the electric field in the ferroelectric sample is equal to zero because the dipole moment of the sample, $P_0$, obtained during the poling procedure is compensated for by surface charges. When an impact shock depolarizes the ferroelectric disk, free charges in the volume of the disk are redistributed. An electric field then exists in the PZT and a pulsed electric potential (electromotive force, or EMF) appears on the metallic contact plates of the ferroelectric module until a new equilibrium state is reached.

The pulsed EMF causes a pulse of electric current, $I(t)$, to flow in the electrical circuit. Integration of the $I(t)$ waveform from 0 to $t$ gives the momentary value of the electric charge, $\Delta Q(t)$, released in the circuit during explosive operation of the FEG:

$$\Delta Q(t) = \int_0^t I(t) \cdot dt. \qquad (1)$$

This charge is equivalent to the electric charge released by the ferroelectric energy-carrying element during shock depolarization.

We performed a series of experiments with PZT disks of four sizes: diameter $D$ = 26 mm and thickness $h$ = 0.65 mm, $D$ = 27 mm and $h$ = 2.1 mm, $D$ = 25 mm and $h$ = 5.1 mm, $D$ = 25 mm and $h$ = 6.5 mm.

### RESULTS AND DISCUSSION

A typical waveform of the current pulse generated by an FEG containing a PZT disk of $D$ = 27 mm/$h$ = 2.1 mm is shown in Fig. 2. The load resistance and inductance were $R_L$(100 kHz) = 0.2 Ω and $L_L$(100 kHz) = 0.53 µH, respectively. The current pulse amplitude was $I(t)_{max}$ = 213 A, with full width at half maximum (FWHM) of 0.5 µs.

Figure 2 also shows the evolution of the electric charge, $\Delta Q(t)$, released in the electrical circuit of the generator during shock wave action. The maximum charge released in the circuit in this experiment was $\Delta Q(13.2$ µs$)_{depol}$ = 157 µC.

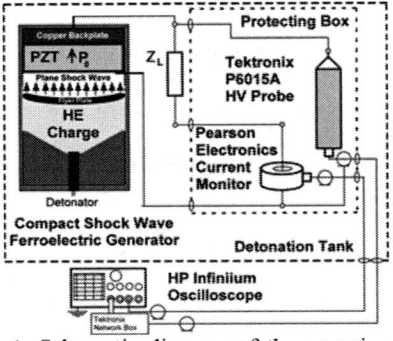

**Figure 1.** Schematic diagram of the experimental setup for investigation of the depolarization of PZT energy-carrying elements in FEGs.

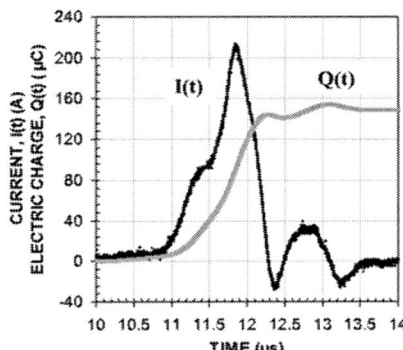

**Figure 2.** A typical waveform of the current pulse (black) generated by an FEG containing a PZT disk ($D$ = 27 mm/h = 2.1 mm), and the corresponding electric charge, $\Delta Q(t)$, (gray) released due to the shock wave depolarization of the disk.

The average value of the total electric charge released from PZT disks of this size under shock wave action in seven experiments of this series was $\Delta Q_{depol\ aver}$ = 168 ±17 μC. This result was obtained with FEGs loaded with 12 to 18 g of HE. During experimentation, reduction of the HE mass below 12 g resulted in decreasing $\Delta Q_{depol}$.

The initial electric charge, $Q_0$, formed by the poling procedure and stored in the PZT energy-carrying elements can be determined as follows:

$$Q_0 = P_0 \cdot A, \qquad (2)$$

where $P_0$ is the remnant polarization of the ferroelectric sample and $A$ is its area. Accordingly, PZT disks with $P_0$ = 30 μC/cm$^2$ and $A$ = 5.7 cm$^2$ have $Q_0$ = 171 μC.

Based on our experimental results, the electric charge released by PZT disks under shock wave action, $\Delta Q_{depol}$, is nearly equal to $Q_0$. This is direct evidence of practically complete depolarization, $\Delta Q_{depol} / Q_0$ = 0.98, of the PZT due to shock wave compression. Therefore, the physical effect of complete shock wave depolarization of the PZT ferroelectrics was detected experimentally.

Shock compression of materials results in simultaneous increase in the temperature of the material and in mechanical compression of the crystal lattice. Therefore, the depolarization of the PZT ferroelectric sample may be due to the 180-degree switching of existing domains, to nucleation and growth of new ferroelectric domains, or to a ferroelectric-to-paraelectric phase transition.

In the following manner, we calculated the shock pressures, $P_{SW}$, required to produce the experimentally-detected shock wave depolarization of PZT. Assuming a perfectly elastic impact of an aluminum flyer plate of infinite diameter with an PZT element, also of infinite diameter, and assuming no plastic or fluidic behavior in either material at the moment of impact, the pressure acting on the front face of the ferroelectric disk, $P_{SW}$, can be estimated using the following equation [3]:

$$P_{SW} = (m \cdot 2 \cdot s)/(\tau^2 \cdot A_{FP}), \qquad (3)$$

where $m$ is the mass of the aluminum flyer plate, $A_{FP}$ is the flyer plate area, $s$ is the gap between the flyer plate and the ferroelectric energy-carrying element (acceleration path), and $\tau$ is the flight time of the flyer plate preceding the impact.

To determine the flight time of the flyer plate, a series of experiments was performed with generators in which the shock wave in the PZT disk was initiated by the direct action of explosive detonation (i.e., without a flyer plate). The flight time of the flyer plate, $\tau$ = 5.1 ± 0.2 μs, was determined from the shift in the time scale of the voltage pulses generated by the FEGs with a flyer plate (Fig. 1) versus FEGs utilizing the direct action of a detonation shock wave. This value is in good agreement with that obtained in another series of experiments performed with generators having transparent Lexan® bodies, in which the free motion of the flyer plate was recorded using a high-speed Cordin 010-A framing camera.

Substituting the flyer plate mass $m$ = 5.1 g, acceleration gap $s$ = 0.5 cm, flyer plate area $A_{FP}$ = 5 cm$^2$, and $\tau$ = 5.1 ± 0.2 μs into Eq. (3) gives the pressure at impact of the flyer plate on the front face of the ferroelectric element, $P_{SW}$ = 3.8 ± 0.3 GPa. This value is an upper bound, since the real impact situation will produce plastic behavior in the flyer and since the component diameters are not infinite. Relaxation waves from free surfaces and energy expended in the material by permanent deformation subtract from the pressure available at impact. In fact, experimental results have shown the flyer plate to have "splashed" on the PZT surface, which typically shows little or no indication of deformation.

Exploring further the electrical output obtainable from compact explosive-driven FEGs,

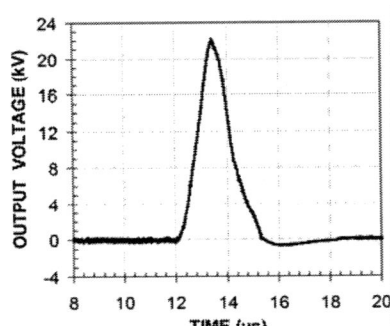

**Figure 3.** Waveform of the pulsed EMF produced by a shock wave ferroelectric generator containing a PZT disk of $D = 26$ mm/$h = 6.5$ mm.

several designs of high-voltage and high-power FEGs have been studied. A typical waveform of an EMF pulse produced by a high-voltage FEG containing a PZT disk of $D = 26$ mm/$h = 6.5$ mm is shown in Fig. 3. The EMF pulse amplitude was $U_g(t)_{max} = 22.0$ kV with FWHM of 1.1 μs.

The load impedance in these high-voltage experiments was 100 MΩ, therefore the current in the electrical circuit of the generator was negligibly small (less than $3 \times 10^{-4}$ A) and there was practically no interference with the electrical current flowing through the PZT module during shock wave induced depolarization. Moreover, transition processes in the electrical circuit had no effect on the EMF pulse waveform generated by the PZT disk. In this mode of electrical operation, the increase in the EMF pulse from zero to its maximum value was the direct result of the depolarization of the ferroelectric energy-carrying element due to shock wave action. The EMF pulse rise time corresponded to the shock front propagation time through the PZT disk thickness, $h$. Therefore, the velocity of the shock wave front could be determined by utilizing the following relationship:

$$U_S = h/\tau_f \qquad (4)$$

where $\tau_f$ is the time of increase of the EMF pulse from zero to its maximum value. Accordingly, the shock wave velocity in the PZT was determined to be $U_S = 3.94 \pm 0.27$ km/s.

The basic equation for shock wave pressure in condensed matter [3],

$$P_{SW} = \rho_0 \cdot U_S \cdot U_P \qquad (5)$$

allows one to obtain the pressure in a shock-compressed body (here $\rho_0$ is the density of the material before shock action and $U_P$ is the particle velocity). The particle velocity in the PZT samples, corresponding to the shock wave velocity determined from the experimental data and Eq. (4), above, can be found from the Hugoniot for Pb(Zr$_{52}$Ti$_{48}$)O$_3$ [4]; $U_P = 0.050 \pm 0.004$ km/s. The pre-shocked density of Pb(Zr$_{52}$Ti$_{48}$)O$_3$ is $7.5 \times 10^3$ kg/m$^3$. Substituting these values for $Us$, $U_P$ and $\rho_0$ into Eq. (5) results in $P_{SW} = 1.5 \pm 0.2$ GPa. The estimations of the pressure in the bulk of PZT (1.5 GPa) and at the PZT/flyer plate interface (3.8 GPa) upon impact are the lower and upper bounds, respectively, of the pressure generated in the ferroelectric modules.

In the high-power mode (the load resistance and inductance were $R_L$(100 kHz) = 0.24 Ω and $L_L$(100 kHz) = 0.7 μH), FEGs containing a PZT disk of $D = 26$ mm/$h = 0.65$ mm generated high power pulses of amplitude up to $W(t)_{max} = 350$ kW with FWHM of 0.1 μs.

## SUMMARY

The effect of complete shock wave depolarization of Pb(Zr$_{52}$Ti$_{48}$)O$_3$ ferroelectrics under shock pressure $P_{SW} = 1.5 - 3.8$ GPa was detected experimentally. Miniature primary power sources (PZT unit volume 0.35 to 3.3 cm$^3$) based on this effect are capable of producing pulses of high voltage, with amplitudes up to 22 kV and peak powers up to 350 kW.

## REFERENCES

1. Altgilbers, L.L. et al, "Magnetocumulative Generators" (Springer-Verlag, 2000).
2. Shkuratov, S.I., et al, "Compact High-Voltage Generator of Primary Power Based on Shock Wave Depolarization of Lead Zirconate Titanate Piezoelectric Ceramics", Rev. Sci. Instruments, vol. 75, pp. 2766-2769, 2004.
3. Trunin, R.F., "Shock Compression of Condensed Materials" (Cambridge University Press, 1998).
4. Reynolds, C.E. and Seya, G.E., "Two-Wave Shock Structures in the Ferroelectric Ceramics Barium Titanate and Lead Zirconate Titanate", J. Appl. Phys., vol. 33, pp. 2234-2241, 1962.

CP845, *Shock Compression of Condensed Matter - 2005*,
edited by M. D. Furnish, M. Elert, T. P. Russell, and C. T. White
© 2006 American Institute of Physics 0-7354-0341-4/06/$23.00

# SHOCK-WAVE SYNTHESIS AND HPHT SINTERING
# OF CUBIC SILICON NITRIDE

## A. S. Yunoshev[1], V. V. Silvestrov[1], A. A. Kalinin[2], and Yu. N. Pal'yanov[2]

[1]*Lavrentyev Institute of Hydrodynamics, Lavrentyev Av., 15, Novosibirsk 630090*
[2]*Institute of Mineralogy and Petrography, Koptyug Av., 3, Novosibirsk 630090*

**Abstract.** We have used the shock-wave method to synthesize the high-pressure cubic phase of silicon nitride. To consolidate the received nano-powder into the bulk sample, the special high-quality HPHT apparatus was used. As a result, we have obtained bulk samples of cubic silicon nitride up to 6 mm in size. The Vickers microhardness of this superhard material equals 30-50 GPa, which significantly exceeds the data in other investigations.

**Keywords:** Shock synthesis, cubic silicon nitride, superhard material.
**PACS:** 62.25.+g, 62.50.+p, 64.70.Kb.

## INTRODUCTION

An increased interest in the synthesis of new superhard materials has been observed over the last few years. Note a prediction of the existence of crystal carbon nitride with a proposed Vickers microhardness of HV=70-95 GPa (comparable to that of diamond) [1], and the synthesis of cubic silicon nitride $c$-$Si_3N_4$ [2] and boron carbonitride [3], with HV=37-43 and 65-80 GPa, respectively. These materials are synthesized at high static pressures of 15-30 GPa and temperatures up to 3000°C as bulk samples, no more than 1-2 mm in size using a diamond anvil cell or multi-anvil apparatus.

The purpose of this work is the synthesis of cubic silicon nitride in an amount sufficient for study of its physical-mechanical properties. This material was first synthesized in 1999 using diamond anvil cell and laser heating methods. Synthesis of cubic silicon nitride from its polymorphic modifications (α- and β- phase) requires pressures up to 30 GPa and temperatures of about 2800 K. In this way, a 15 micron sample was prepared.

Later, the c-phase was obtained by shock-wave methods using flyers accelerated by a propellant gun [5]. The initial sample was a mixture of sub-micron $\beta$-$Si_3N_4$ powder and a copper powder. At a shock pressure of 19 GPa and calculated temperatures of ~3000 K, the occurrence of a new phase was registered. The yield of cubic silicon nitride at 60 GPa and 2400 K achieved about 100 %. The nano-grain size of c-$Si_3N_4$ was 10-50 nm. Little is known about compacting or sintering a nano-powder into a bulk sample.

To synthesize the $c$-$Si_3N_4$, we used a shock-wave technique with high explosive loading. To achieve a consolidation of the received nano-powder into the bulk sample a pressure-free split sphere apparatus was used, capable of static pressures up to 7 GPa and temperatures to 2500°C. This apparatus allows us to maintain these stable pressure-temperature parameters for several hours up to several weeks, and is used to grow large artificial diamonds [6].

## EXPERIMENTAL PROCEDURE

The starting matter for synthesis was a micron-size $\beta$-$Si_3N_4$ powder, or its mixture with copper powder. The sample was pressed into a copper container to a density of 60-80 % of the theoretical density under vacuum. The volumetric part of copper was changed from 0 up to 80 %. The initial thickness of the sample was 2-3 mm. The shock was generated by high explosive (HE) placed on the container, or by the impact of a duralumin flyer 5 or 8 mm in thickness explosively accelerated to 5.3 or 3.4 km/s, accordingly.

After the experiment, the container was opened, and the copper was removed by washing the product in nitric acid. Then the residue was washed out in water and dried. To analyze the synthesized matter we used X-ray diffraction (XRD) structural analysis with a $CuK_\alpha$ emission line with a wavelength of 1.5418 Å.

For dynamic loading of pure hexagonal silicon nitride in the pressure range of 40-75 GPa, no phase transition was observed if the recovery container was still sealed after the shot. The negative factor here is the high temperatures, resulting in annealing of the new phase at rarefaction.

For the residual temperature to be lower, the silicon nitride powder was mixed with copper powder. At a volume fraction of copper greater than 60 %, sample porosity of about 1.43, and shock pressure in the container cover of 53 GPa, the yield of the new cubic phase was nearly 100 %.

nitride for one shot, is shown in Fig. 1. The feature of this scheme is the use of two charges of explosives. The first charge (cast TNT/RDX 50/50) accelerates the flyer up to 3.4 km/s. The second lateral charge (loose ammonite or RDX) was used to decrease an intensive radial deformation of the container and to prevent the formation of a central hole in the upper cover container. The second charge was initiated by first one by six pieces of detonation cord of necessary length. The application of the second charge allows the inner volume of the container to enlarge by a factor of five.

Manganin gauges were used to measure shock pressures near the sample. In the first shot, the gauge was placed inside of the sample, and in the second shot, it was placed at the distance of 3 mm below the sample. Shock pressure-time profiles for the explosively driven impact are shown in Fig. 2. The measurements show that the shock pressure in the sample has not reached the shock intensity in the upper cover of the container and has not exceeded 35 GPa. This corroborates the observation in [7] that the shock-induced phase transition of $\beta$-$Si_3N_4$ to $c$-$Si_3N_4$ occurs only at pressures above 36 GPa. The phase transition may be completed with a longer shock pulse duration and a pressure close to the threshold pressure of phase transformation.

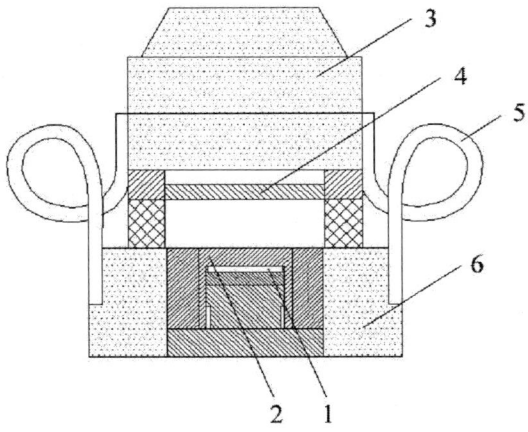

**Figure 1.** Experimental assembly: 1 – sample; 2 – container; 3 – main HE charge; 4 – flyer plate; 5 – detonating cord; 6 – lateral HE charge.

The experimental assembly, which provides a synthesis of about up to two grams of cubic silicon

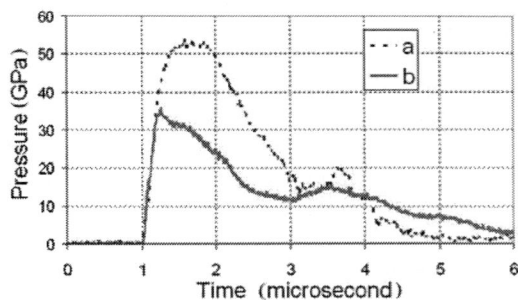

**Figure 2.** Pressure-time profiles for explosively driven impact of 3.4-km/s by duralumin plate 8 mm in thick: under the upper cover of copper container (a), and 3-mm below the sample of pressed copper/silicon nitride mixture (b).

## RESULTS OF SHOCK SYNTHESIS

Figure 3 shows the X-ray spectra of the initial hexagonal and the synthesized cubic silicon nitride.

The new material displays three broad lines with maximums at $2\theta = 32.2$, $38.2$, and $46.5°$. These lines are most intense in the interval $2\theta = 20{-}50°$, and their positions to within $0.5°$ agrees with the data for shock synthesized cubic silicon nitride with a spinel structure reported in [5]. The cubic phase is estimated to be present in an amount no less than 95 vol. %.

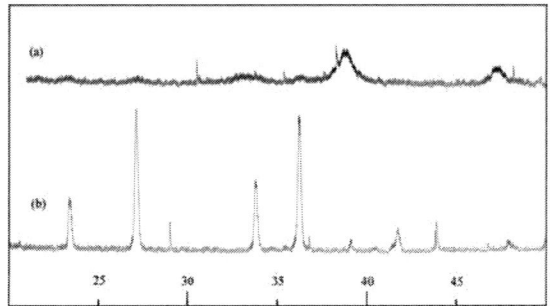

**Figure 3.** XRD spectra of the synthesized cubic silicon nitride (a) and an initial $\beta - Si_3 N_4$ (b).

We also used transmission electron microscopy (TEM) to study our product. TEM-pictures show the presence of both crystalline and amorphous phases. Local chemical analysis of the amorphous phase gave an average atomic composition close to $Si_3N_4$. Therefore this product contains crystalline and amorphous phases of silicon nitride. More than 95% of the crystalline phase is the $c$-$Si_3N_4$ phase and the rest is the $\beta$-$Si_3N_4$ phase. The amount of amorphous phase is hard to estimate but we suppose that it is not more then 20 %. We can not separate the phases. The presence of an intermediate amorphous phase may result in the formation of a stronger bulk sample because of the difficulties in the compaction of pure hard materials. Superhard nanocrystallites embedded in an amorphous matrix are currently the most promising concept for the synthesis of novel superhard materials [3].

Thus, the synthesized $c$-$Si_3N_4$ is an ultrafine powder with a size of the coherent-scattering region of 10-20 nm, which is about one tenth of the size of the corresponding region for the initial $\beta$-phase (0.1-0.15 µm). Subsequently, both explosive compaction and high-pressure high-temperature (HPHT) sintering were used to obtain the bulk samples.

## HPHT SINTERING AND MICROHARDNESS

Before sintering, the specimens were pressed to a density of about 60 % of the grain density, and were wrapped in 10 µm in thick platinum foil. The HPHT sintering of $\beta$-$Si_3N_4$ and $c$-$Si_3N_4$ powders was realized at a static pressure of 5-6 GPa and a temperature of 1100°C over the course of 5 hours. As a result, bulk specimens 3-6 mm in diameter and 2.5-3 mm in height are obtained (Fig. 4). Density of the specimens was determined by the Archimedean method.

**Figure 4.** Sintering bulk sample of cubic silicon nitride (scale factor – 1 mm).

The Vickers microhardness of the bulk samples was measured using an indentation force from 2 to 0.3 N. The indentation diagonal size ranged from 3 to 12 microns. At each force, loading was carried out for 5-10 measurements. The results are given in Fig. 5. For both specimens the dependence of microhardness on indentation load is observed.

The bulk sample of $\beta$-$Si_3N_4$ was tested first. Its density is 3.4±0.1 g/cc; its Vickers microhardness is 20±2 GPa and almost independent of the indentation load. It points to strong bonding between grains. Good mechanical quality of the sample means the HPHT technique may be used to consolidate superhard ceramic powder into bulk samples.

Density of the sintered bulk sample of a $c$-$Si_3N_4$ is 3.6±0.1 g/cc , significantly lower than the calculated value of 4 g/cc [4]. XRD analysis of the sintered sample shows the appearance up to 20 % of $\beta$-$Si_3N_4$. According to [5], cubic silicon nitride synthesized by shock-waves is stable up to a temperature of 1370°C. Therefore, the appearance of $\beta$-$Si_3N_4$ the sintered sample is probably caused by transition of a part of an amorphous phase to a hexagonal one under HPHT conditions.

Despite the content of $\beta$-$Si_3N_4$ in sintered cubic silicon nitride sample, its microhardness is high and exceeds 30 GPa at indentation loads less than 2 N, and reaches 55±13 GPa at an indentation force of 0.3 N. Microhardness at 1 N is two times greater than the values listed in [8] at the same load. Unfortunately, we can not measure the microhardness at indentation loads less than 0.3 N.

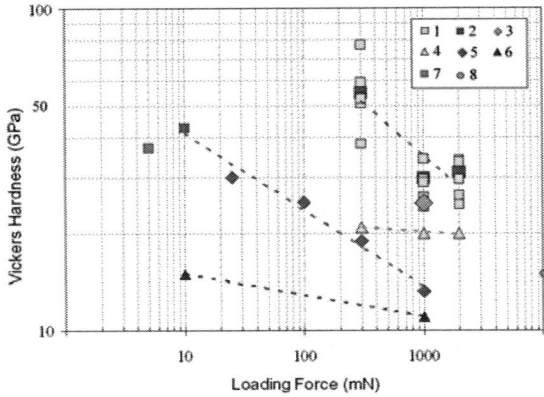

**Figure 5.** Microhardness of different phases of silicon nitride. Our data: 1-3 – $c$-$Si_3N_4$, 4 – $\beta$-$Si_3N_4$; 1 – separate measurements, 2-4 – average values. Other authors: $c$-$Si_3N_4$ – 5 [8], 7 [2]; $\beta$-$Si_3N_4$ – 6 [8], 8 [2].

We also have tried to compact the $c$-$Si_3N_4$ powders by explosion. A sample with an abundance of cracks was produced. In this case the appearance of the β-phase was not observed. Its microhardness reached HV=27±2 GPa at the indentation load of 1 N (Fig. 5, point 2).

## CONCLUSIONS

At shock pressure of 35 GPa in a mixture of copper and $\beta$-$Si_3N_4$ powders almost the whole of silicon nitride transforms to a cubic phase.

The explosive synthesis method allows us to receive up to 1.5 grams of the cubic phase of silicon nitride in one shot in laboratory conditions.

HPHT sintering allows us to produce specimens of nano-crystal silicon nitride up to 6 mm in

size with a microhardness much greater than those obtained in previous experiments.

## ACKNOWLEDGEMENTS

Funding was providing by Program of Presidium of RAS under project #11/1, by the Siberian Division of RAS under Integration Grant #29, and by RFBR under grant #03-03-33174. The authors thank Pr. A.A. Deribas for helpful discussions and T.S. Teslenko for providing the XRD spectra.

## REFERENCES

1. Liu, A. Y., Cohen, M. L., "Prediction of new low compressibility solids," Science, vol. 245, pp. 841-842, 1989.
2. Zerr A., Miehe G., et al., "Synthesis of cubic silicon nitrid," Nature, vol. 400, pp. 340-342, 1999.
3. Zhao, Y., He D.W., et al., "Superhard B-C-N materials synthesized in nanostructured bulks," J. Mater. Res., vol. 17, no. 12, pp. 3139-3145, 2002.
4. Sekine, T., Hongliang He, et al., "Shock-induced transformation of $\beta$-$Si_3N_4$ to a high-pressure cubic-spinel phase," Appl. Phys. Lett., vol. 76, no. 25, pp. 3706-3708, 2000.
5. Sekine, T., Mitsuhashi, T., "High-temperature metastability of cubic spinel $Si_3N_4$," Appl. Phys. Lett., vol. 79, no. 17, pp. 2719-2721, 2001.
6. Pal'yanov Yu.N., Sokol A.G., Borzdov Yu.N., Khokhryakov A.F., Gusev V.A., Sobolev N.V., "Synthesis and Characterization of Diamond Single Crystals up to 4 Carats," Translations (Doklady) of the Russian Academy of Science/Earth Science Sections, Vol. 355A, No. 6, pp. 856-858, 1997.
7. Hongliang He, T. Sekine, T. Kobayashi, H. Hirosaki "Shock-induced phase transition of β-$Si_3N_4$ to c-$Si_3N_4$," Phys. Rev. B, vol. 62, No. 17, pp 11412-11417, 2000.
8. Tanaka I., Oba F., et al., "Hardness of cubic silicon nitride," J. Mater. Res., vol. 17, no. 4, pp. 731-733, 2002.

CP845, *Shock Compression of Condensed Matter - 2005*,
edited by M. D. Furnish, M. Elert, T. P. Russell, and C. T. White

# NUMERICAL SIMULATION OF SUPERFAST SHOCK-INDUCED CHEMICAL REACTION IN TITANIUM – SILICON MIXTURE

## S. A. Zelepugin[1], V. B. Nikulichev[2], and O. V. Ivanova[3]

[1]*Dept. for Structural Macrokinetics, Tomsk Scientific Centre, SD RAS, Tomsk, 634021 Russia*
[2]*Kyrgyz-Russian Slavonic University, Bishkek, 720000 Kyrgyzstan*
[3]*Tomsk State University, Tomsk, 634050 Russia*

**Abstract.** A phenomenological zeroth-order kinetic model for computations of shock-induced solid-state chemical reactions in porous mixtures is proposed. In the model a porous mixture is considered as a continuous medium whose thermomechanical properties are determined at each time step depending on mass fractions of the components. The kinetic relationships are characterized by a constant rate of chemical transformation under shock wave loading. The heat release due to chemical transformation is introduced in the energy equation. The effect of the dispersity of the mixture components on the reaction rate is taken into account by varying the constants that enter the kinetic model. The results of the numerical computations for porous Ti-Si mixture reflect the fact that the process can be divided into several stages (dynamic compaction, shock-wave propagation, reaction of synthesis). It is shown that an increase in the chemical-reaction rate can give rise to flow regimes in which the unloading wave almost stops.

**Keywords:** Shock waves, chemical reaction, numerical simulation.
**PACS:** 47.40.Nm, 82.40.Fr, 02.60.Cb

## INTRODUCTION

At present, the macrokinetics of solid-state chemical reactions under dynamic loading is an important line of fundamental and applied investigations in extreme chemistry. Solid-state reactions of synthesis in porous mixtures generated by shock compression occur under high pressure and strain rates and are accompanied by rapid release of a large amount of thermal energy. This leads to a considerable increase in temperature, which usually exceeds the melting points of the mixture components. Special features of the behavior of reacting mixtures of inorganic substances were studied, for example, in [1-4] by different experimental methods. The aim of the present paper is to study numerically the special features of superfast exothermic chemical reactions in porous mixtures under shock wave loading. A mixture is treated as a continuous medium whose thermomechanical properties are determined at each time step depending on mass fractions of the components. The heat release due to chemical transformation is introduced in the energy equation. Numerical computations were carried out by the finite element method.

## FORMULATION OF THE PROBLEM

To simulate numerically the processes of high velocity impact loading, we use a model of a damaged medium characterized by the presence of microcavities (pores and cracks). The total volume of the medium comprises the undamaged part and microcavities of zero density. The level of damage of the medium is characterized by the specific volume of pores $V_f$. The system of equations governing the nonstationary, adiabatic (for both elastic and plastic deformations) motion of a compressible medium with allowance for the evolution of microdamages and chemical transformations comprises the continuity equation, the equation of motion, the energy equation [5-8].

The system includes the kinetic equation of shock-induced solid-state chemical reaction:

$$\frac{d\eta}{dt} = \begin{cases} 0, & \text{if } \eta = 1 \text{ or if} \\ & (T < T_\eta \text{ and } p < p_\eta), \\ f(p_\eta), & \text{if } \eta < 1 \text{ and if} \\ & (T \geq T_\eta \text{ or } p \geq p_\eta), \end{cases}$$

$$f(p_\eta) = \begin{cases} K_0, & \text{if } p < p_\eta, \\ K_p K_0, & \text{if } p \geq p_\eta, \end{cases}$$

Here $T$ is the temperature, $T_\eta$, $p_\eta$, $K_p$, $K_0$ are constants, $p$ is the average pressure, $\eta$ is the conversion ratio.

In modeling chemical reactions under shock wave loading, we use the zeroth-order kinetic relation, characterized by a constant rate of chemical transformations [9]. The constant of the reacting mixture $K_0$ is a structurally dependent quantity and is determined primarily by the component particle size. The value of $K_0$ increases with increase of dispersity of the mixture components.

To construct the equation of state for the mixture, we use experimental shock adiabats of its components. The density, specific volume, specific energy, and Grüneisen constant of the mixture were calculated by the relationships:

$$\frac{1}{\rho} = V = \sum m_n V_n , \qquad E = \sum m_n E_n ,$$

$$\frac{V}{\gamma} = \sum m_n \left(\frac{V}{\gamma}\right)_n ,$$

where $m_n$ is the mass fraction of the $n$th component.

## NUMERICAL RESULTS

Interaction of an impactor (1) with a diameter of 90 mm and a height of 10 mm with a recovery ampoule (2) with a diameter of 100 mm and a

height of 50 mm is modeled. The investigated specimen (3) was 2 mm in height and 20 mm in diameter and represented the pressed porous mixture in a proportion of 46 mass fraction titanium and 54 mass fraction silicon that corresponds to a stoichiometric reaction with synthesis of titanium disilicide ($TiSi_2$). Initial porosity of the specimen was of 40%. The specimen was placed between a cover of the ampoule with height of 7 mm and a steel internal electrode (5) with diameter of 20 mm and height of 16 mm. Numerical computations were carried out for impact velocity of 1800 - 2200 m/s.

Figure 1 shows chronogram for an initial impact velocity of 2200 m/s, which characterize the moment of dynamics of deformation of the interacting bodies.

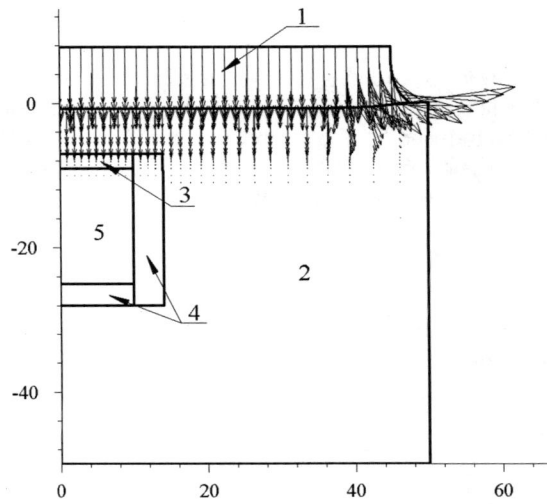

**Figure 1.** Computed configuration of interacting bodies for initial impact velocity 2200 m/s and time 0.5 μs.

A superfast chemical reaction in the specimen occurs when the critical value $p_\eta = 23$ GPa or $T_\eta = 1693$ K is exceeded. The enthalpy of the chemical reaction $\Delta H$ was 1730 kJ/kg and $K_p = 2$. The results presented in Fig. 1 correspond to $K_0 = 173$ GJ/(kg sec).

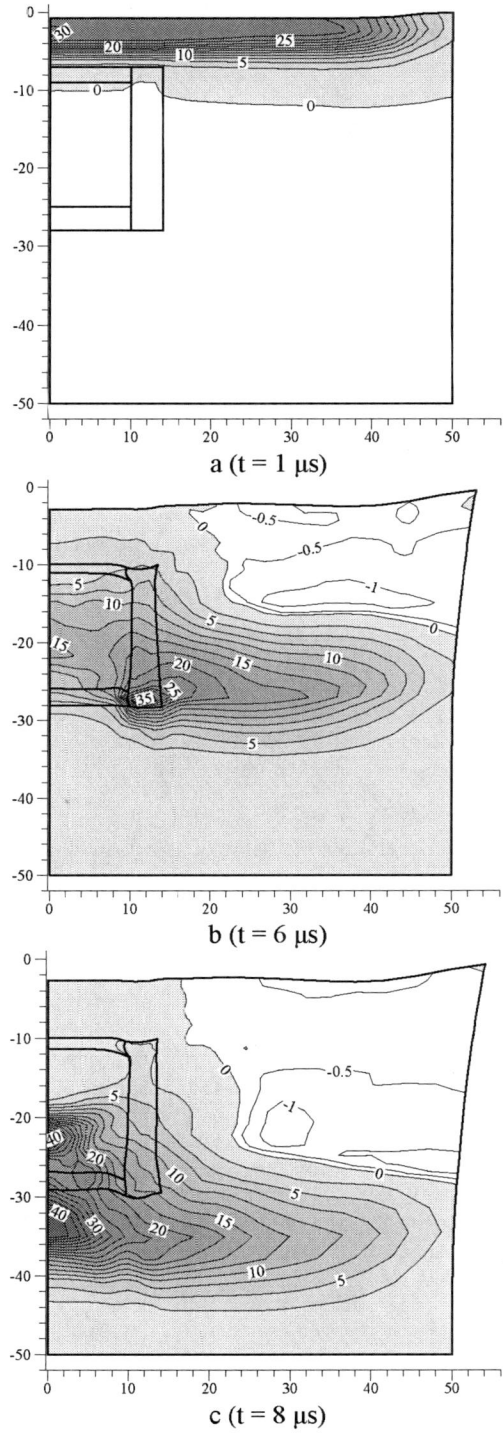

a (t = 1 μs)

b (t = 6 μs)

c (t = 8 μs)

**Figure 2.** Isobars in radial cross-section of the ampoule.

In Fig. 2 isobars (GPa) in radial cross-section of the ampoule for different moment of time are presented for $v_0$ = 2200 m/s.

At the time 1 μs, we observe a flat shock-wave front in the lid of the ampoule. At the same time, the unloading waves propagating from the free surfaces of the projectile and the ampoule reduce the level of pressure in the shock-wave front, with the pressure in the specimen still maintained relatively high up to the time 9 μs. A special feature of the wave process is the formation of a local extremum of compression observed in the bottom corner section of the insulating cup (4 in Fig. 1) at the time 6 μs (Fig. 2b). By the time 8 μs, the compression wave arrives at the axis of symmetry and splits into two waves moving upwards and downwards along the axis of symmetry.

An analysis of the temperature distributions in the ampoule shows that the highest temperatures are observed within the specimen and the lateral wall of the insulating cup. In the specimen, the increase in temperature is basically caused by the chemical reaction, while in the insulating cup it results from shear deformation.

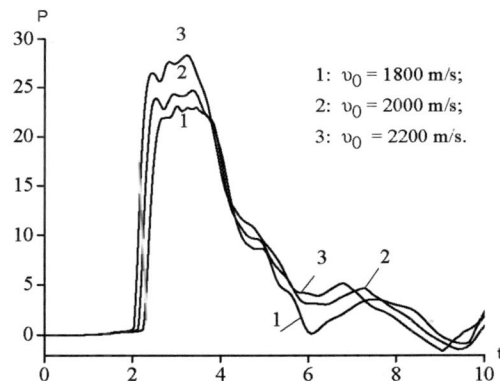

1: $v_0$ = 1800 m/s;
2: $v_0$ = 2000 m/s;
3: $v_0$ = 2200 m/s.

**Figure 3.** Pressure (GPa) versus time (μs) in the computational cell of the simulated specimen.

Fig. 3 illustrates that for impact velocity of 1800 m/s the maximum of pressure in the selected element is 22.5 GPa, for $v_0$ = 2000 m/s - 23.5 GPa, for $v_0$ = 2200 m/s - 27 GPa. The history of changes of specific volume of pores on fig. 4 shows that process of collapse of voids takes place in the given area of the sample approximately after 1 μs. After 9 μs the growth of pores is observed under action of a

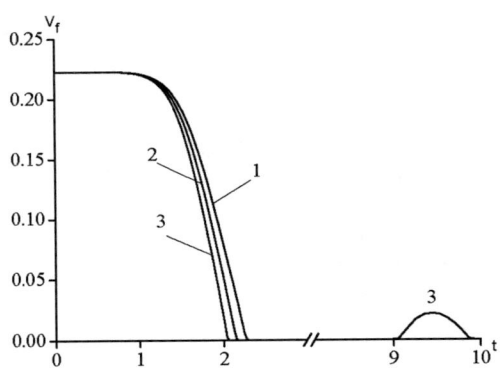

**Figure 4.** Specific pore volume ($cm^3/kg$) versus time ($\mu s$) in the computational cell of the simulated specimen.

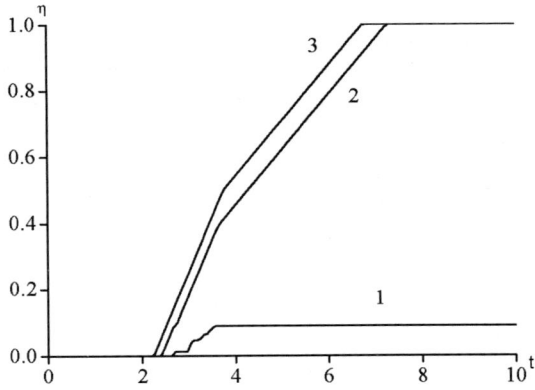

**Figure 5.** Conversion ratio versus time ($\mu s$) in the computational cell of the simulated specimen.rarefaction wave during some interval of the process.

Fig. 5 shows the conversion ratio versus time in the computational cell of the specimen. For 1800 m/s, the level of chemical transformation in the computational cell is low, while for the velocities 2000 and 2200 m/s the reaction in the computational cell is fully completed.

## CONCLUSIONS

Numerically with use of phenomenological model the processes of chemical transformations in porous mixtures under shock wave loading are investigated. It has been demonstrated that a local extremum of pressure is formed in the recovery ampoule used for electrical measurements, which could affect the measurement accuracy in physical experiments. It has been established that in the conditions under study it is the high level of pressure that initiates the chemical reaction in Ti-Si mixture, though the pressure criterion is operative within a limited period of time. It has been pointed out that temperature is of crucial importance for the completeness of chemical transformation of the mixture. The chemical reaction of synthesis proceeds completely if during the action of high dynamic pressure the temperature has enough time to exceed its critical value.

## ACKNOWLEDGEMENTS

This work was supported by the Russian Foundation for Basic Research under Grants 03-01-00122, 05-03-98001.

## REFERENCES

1. Boslough, M.B., and Graham, R.A., "Submicrosecond shock-induced chemical reactions in solids: first real-time observations," Chem. Phys. Lett., vol. 121, no. 5-6, pp. 446-452, 1985.
2. Horie, Yu., and Kipp, M., "Modeling of shock-induced chemical reactions in powder mixtures," J. Appl. Phys., vol. 63, no. 12, pp. 5718-5727, 1988.
3. Gogulja, M.F., Voskobojnikov, I.M., Dolgoborodov, A.Yu., etc., Khimicheskaja Fizika, vol. 11, no. 2, pp. 244-247, 1992.
4. Thadhani, N.N., "Shock-induced chemical reactions and synthesis of materials," Progress in Materials Science, vol. 37, pp. 117-226, 1993.
5. Gorelskii, V.A., Zelepugin, S.A., Nabatov, S.S., Nikulichev, V.B., Khimicheskaja Fizika, vol. 18, no. 5, pp. 102–107, 1999.
6. Gorelskii, V.A., Zelepugin, S.A., Kim, V.V., Smolin, A.Yu., Khimicheskaja Fizika, vol. 19, no. 2, pp. 27-31, 2000.
7. Zelepugin, S.A., and Nikulichev, V.B., Combustion, Explosion, and Shock Waves, vol. 36, no. 6, pp. 845-850, 2000.
8. Gust, W.H., "High impact deformation of metal cylinders at elevated temperatures," J. Appl. Phys., vol. 53, no. 5, pp. 3566-3575, 1982.
9. Gryadunov, A.N., Shteinberg, A.S., and Dobler, E.A., Doklady, vol. 321, no. 5, pp. 1009-1014, 1991.

CHAPTER XVII

# INSTRUMENTATION

CP845, *Shock Compression of Condensed Matter - 2005,*
edited by M. D. Furnish, M. Elert, T. P. Russell, and C. T. White
© 2006 American Institute of Physics 0-7354-0341-4/06/$23.00

# PIEZOELECTRIC POLYMER SHOCK GAUGES

## F. Bauer

*Institut Franco-Allemand de Recherches de Saint-Louis, (ISL), 68300 Saint-Louis, France*

**Abstract.** The science and technology of piezoelectric materials has long been dominated by the availability of specific materials with particular properties. Piezoelectric PVDF (Poly(vinylidene fluoride) polymer and copolymers of PVDF with trifluoroethylene have shown to have the potential for new shock-wave sensors. Since 1981 through 1995, the piezoelectric response of PVDF was studied in a cooperative effort with François Bauer of ISL, France, R.A. Graham of Sandia National Laboratories and L.M. Lee of Ktech Corporation, Albuquerque. Among the known piezoelectric polymers, the PVDF plays an important role in measuring mechanical and physical state of matter under shock loading. The present paper presents the history of the development of the PVDF gauge. After 24 years of research in this area, main relevant results and data obtained are summarized as well as some original applications of the PVDF gauges.

**Keywords:** Polyvinylidene fluoride, piezoelectric polymer, shock gauge, experimental techniques.
**PACS:** 77.22Ej, 77.65Ly, 71.20Rv, 07.50Hp.

## INTRODUCTION

Piezoelectric materials are widely used as active elements in stress gauges to provide nanosecond, time-resolved stress measurements of rapid impulsive stress pulses produced by impact, explosion or rapid deposition of radiation. In the earliest work the gauges used crystalline sensors made of either x-cut quartz or various cuts of lithium niobate with thicknesses of many millimeters. The wave transit times through such sensors range from many tens of nanoseconds to a few microseconds. The upper response limits of the crystalline sensors are limited by either, or both, mechanical or electrical properties: dynamic yielding of the sensors or dielectric breakdown due to the large internal fields produced by the piezoelectric effect.

Piezoelectricity in polyvinylidene fluoride (PVDF) was discovered by Kawai in 1969 [1], and existence of ferroelectricity in PVDF was confirmed by Kepler in 1978 [2].

In 1981, Bauer [3] called attention to the strong, and well characterized electrical signals observed from shock-compressed, piezoelectric films of the polymer PVDF. Early work [3,5], which explored the behavior of PVDF for high pressure applications, led to recognition of the need for highly reproducible properties for PVDF. In particular, for applications and materials studies requiring high quality and reproducibility, individual samples must be produced with well defined electrical properties in a process, which can control amplitude, duration and history of the electric field. In 1982, we developed this new poling technology [3,4] to process PVDF such that its physical properties exhibit reproducibility approaching that of piezoelectric single crystals.

An international cooperative study between Institut Franco-Allemand de Recherches de Saint-Louis (ISL) and Sandia National Laboratories to develop the PVDF gauge, has shown that 25 µm thick PVDF film can be reliably used in a wide range of precise stress and stress-rate measurements [6,7].

The main purpose of this paper is to present the history of the development of the PVDF gauge. After 24 years of research in this area, main relevant results and data obtained are summarized as well as some original applications of the PVDF gauges. References have also been collected for further information.

### FIRST INVESTIGATIONS ON PVDF

In 1982 [4,5], a novel experimental arrangement was designed and built for measuring directly the polarization of PVDF. The ISL poling system [4], allows us to adjust, in real time, the predetermined remanent polarization as well as the maximum displacement current measured at the coercive field for an individual PVDF gauge. With the cyclic poling process, reproducible remanent polarizations as large as 9 $\mu C/cm^2$ can be routinely achieved in commercial processes. Each PVDF gauge is carefully prepared from polymeric film which is typically 25 $\mu m$ in thickness.

First investigation in 1982 conducted under uni-axial compression has shown that the electrical charge released in a short-circuit was a continuous function of the pressure applied up to 0.9 GPa, Fig. 1.

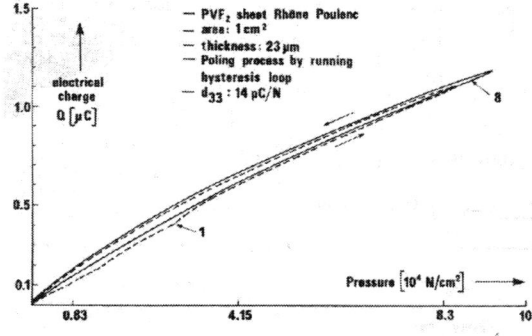

Fig. 4 Electrical charge versus static
**Figure 1.** Electrical charge versus static compression.

It was clear that the piezoelectric polymer gauge has substantial properties.

In looking at these results and the first records of the voltage measured on a resistive load [5], carried out under shock loading, R.A. Graham realized at the same time that PVDF could be a strong candidate for shock gauge development. A cooperative program between the Institut Franco-Allemand de Recherches de Saint-Louis (ISL), R.A. Graham, R.P. Reed, M.U. Anderson, L.M. Moore of Sandia National Laboratories, L.M. Lee of the Ktech Corporation of Albuquerque, New Mexico, was established to study the response of piezoelectric polymer stress gauges to shock loading.

### PIEZOELECTRIC PVDF SHOCK RESPONSE

Studies of the response of ISL piezoelectric polymer stress gauges began in 1982 at ISL and in 1984 at Sandia National Laboratories. For these studies, standard impact loading experiments were developed for our ISL powder gun and for the Sandia 25 meter compressed gas gun [6,7,15]. It should be borne in mind, that, as shown on Fig. 2, the gauge element was placed directly on the impact face of a target either z-cut sapphire or z-cut quartz. In the impact experiment the symmetry conditions for identical impactor and sample materials require that precisely one half of the velocity at impact be imparted to the sample. With appropriate attention to detail, the velocity at impact can be determined to an accuracy and precision of 0.5%. The pressure resulting from the velocity imparted to the sample is the product of

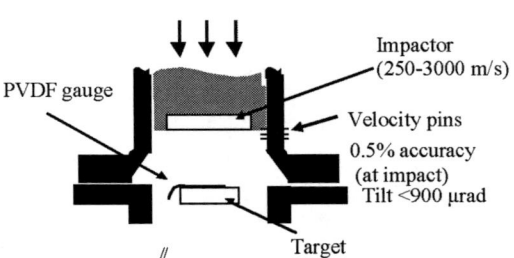

**Figure 2.** The piezoelectric response of PVDF film is studied with the impact of standard materials with samples placed on the impact surface of the target.

the velocity, the density and wave speed. With the use of single crystal materials of high elastic limits, pressures can be routinely achieved up to about 10 GPa with uncertainties of about 1%. For stresses greater than 12 GPa, copper and aluminum were used [12,15]. When used in the electrical current mode, the gauge produces a signal which is directly dependent of the rate-of-change of stress. The piezoelectric charge associated with a particular peak pressure was observed upon integration of the current pulse. Techniques describing the electrical current measurement and gauge assembly are described in references [ 6,7,15,19].

## RELEVANT RESULTS

The works of ISL, Sandia and Ktech have led to determine, for both 1 mm$^2$ and 9 mm$^2$ PVDF gauges, the piezoelectric response versus stress, based on experimental and computed data.

It has been demonstrated that the released charge is a continuous function of the shock pressure, Fig. 3, and that the piezoelectric response is independent of the loading path [15,19].

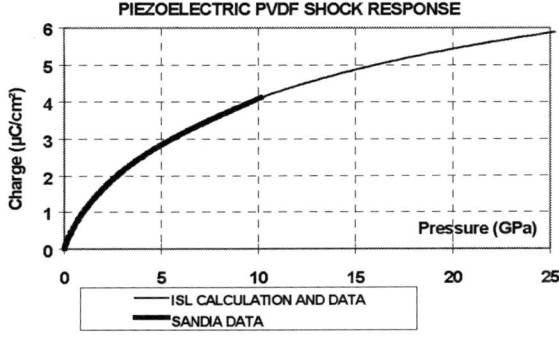

**Figure 3.** The electrical charge observed at various peak pressures is shown for PVDF. The experimental and computed data are on the same curve.

We have [19] established that the experimental electrical charge Q released by the piezoelectric PVDF film could be correlated with the global one-dimensional true strain, $\varepsilon$, of PVDF. The following relationships [19,20] have been found to fit well the dependence of the electrical charge Q versus true strain $\varepsilon$:

- for Q < 0.8288 $\mu$C/cm$^2$:

$$Q = 1.88 \cdot \ln (1 + 10. \, \varepsilon) + 0.0047; \quad (1)$$

- for Q > 0.8288 $\mu$C/cm$^2$:

$$Q = 12.16. \, \varepsilon + 0.16; \quad (2)$$

where $\varepsilon = \ln [(v(t)/v(t_0)]$, v = PVDF specific volume and t = time.

## FERROLECTRIC COPOLYMERS

P(VDF-TrFE) copolymers exhibit tailorable ferroelectric, piezoelectric and structural properties that may be superior to those of PVDF for some shock gauge applications. G. Samara and F. Bauer [17] have investigated these properties as functions of static and dynamic pressure. The temperature-hydrostatic pressure phase diagram has been determined for P(VDF$_{0.77}$-TrFE$_{0.23}$) and PVDF. following established procedures. The important features are (with increasing temperature) a prominent molecular relaxation process centered around $T_\beta$=270K, a ferroelectric transition ($T_C$) and the melting transition ($T_m$). All of these transitions shift to higher temperatures with increasing pressure, as well as for PVDF. The strong increase in $T_C$ and $T_m$ with pressure and the relaxational responses has provided a rational explanation for why it is possible to use these polymers as piezoelectric shock gauges to high shock pressures and accompanying high temperature.

## RELEVANT APPLICATIONS

The development of these gauges has brought significant advances in the diagnostic and understanding of shock wave phenomena as well as some original application of PVDF. Let us recall here some applications of the PVDF gauges

L. Moore investigated the capability of PVDF for measurements of the shock output of specific detonators [12,14].

J. P. Romain et al. [16] demonstrated that piezoelectric copolymer gauges could be used for short shock profiles induced by pulse laser in thin Al or Cu targets.

Initial airblast measurements with PVDF have been reported by Biele [18]. Reproducible pressure measurements of detonation blast pressures were obtained using this new PVDF transducer [20].

PVDF gauges have been utilized successfully for Hugoniot measurements of electrical shielded porous explosives [19]. Shock pressure profiles have been obtained with PVDF gauges in situ porous H.E. in a detonation regime [20].

Ballistic pressure gauges have been also developed for pyrotechnic combustion diagnostics.

## CONCLUSIONS

The piezoelectric response of PVDF under shock loading has been recalled and summarized. The research studies resulted from a 24 year cooperative effort of ISL, Sandia National Laboratories and Ktech Corporation.

## REFERENCES

1. Kawai, H., "Piezoelectricity of poly(vinylidene fluoride)", Japan J. Appl. Phys., vol. 8, pp. 975-981, 1969.
2. Kepler, R.G., "Saturation remanent polarization of poly(vinilydene fluoride)", Org. Coatings Plast. Chem., vol. 38, pp.706-708, 1978.
3. Bauer, F., "Behavior of Ferroelectric Ceramics and PVF$_2$ Polymers under Shock Loading", in Shock Compression of Condensed Matter, 1981 (W.J. Nellis, L. Seaman, R.A. Graham, eds), pp. 251-266.
4. Bauer. F. French Patent 822102S (1982), U.S. patents 4611260, (1986), and 4684337, (1987).
5. Bauer, F., "PVF$_2$ Polymers: Ferroelectric Polarization and Piezoelectric Properties under Dynamic Pressure and Shock wave action", Ferroelectrics, vol. 49, n° 49, pp. 231- 239, 1983.
6. Lee, L. M., Williams, W. D., Graham, R. A., Bauer, F., „Studies of the Bauer Piezoelectric Polymer Gauge (PVF$_2$) under Impact Loading", in Shock Compression of Condensed Matter, 1985 (Y.M. Gupta, eds), Plenum Press, pp. 497-502.
7. Bauer, F., "Ferroelectric Properties and Shock Response of a poled PVF$_2$ Polymer and of VF$_2$/C$_2$F H Copolymers" in Shock Compression of Condensed Matter, 1985 (Y.M. Gupta, eds), Plenum Press, pp. 483-496.
8. Reed, R. P., Greenwoll, J. I., "The PVF$_2$ Piezoelectric Polymer Shock Stress Sensor Signal Conditionnig and Analysis for Field test Application", Sandia report, SAND87-0154, November 1988.
9. Reed, R. P., Graham, R. A., Moore, L. M., Lee, L. M., Fogelson, D. J., Bauer, F., "The Sandia Standard for PVDF Shock Sensors", in Shock Compression of Condensed Matter, 1989 (S.C. Schmidt, J.N. Johnson, L.W. Davison, eds.), pp. 825-828.
10. Lee, L. M., Hyndman, D. A., Reed, R. P., Bauer,F., "PVDF Applications in Shock Measurements". in Shock Compression of Condensed Matter, 1989 (S.C. Schmidt, J.N. Johnson, L.W. Davison, eds.), pp. 821-824.
11. Soulard, L., Bauer, F., "Applications of Standardized PVDF Shock Gauges for Shock Pressure Measurements in Explosives", in Shock Compression of Condensed Matter, 1989 (S.C. Schmidt, J.N. Johnson, L.W. Davison, eds.), pp. 817-820.
12. Moore, L.M., Graham, R. A., "Response of Standardized PVDF Piezoelectric Polymer Gauges to Direct Shock Pressures Between 8 and 32 GPa", in Shock Compression of Condensed Matter, 1989 (S.C. Schmidt, J.N. Johnson, L.W. Davison, eds.), pp. 817-820.
13. Marlin, P. L., Penazzi, L., Bensoussan, Ph., "Laser-Induced Shock Waves Measurements in Aluminum with PVDF and Quartz Stress Gauges", in Shock Compression of Condensed Matter, 1989 (S.C. Schmidt, J.N. Johnson, L.W. Davison, eds.), pp. 837-840.
14. Moore, L., Graham, R. A., Reed, R. P., Lee, L. M., Bauer, F., Warren T. W., "Standardized Piezoelectric Polymer (PVDF) Gauge for Detonator Response Measurements", 3$^{rd}$ Symposium International Hautes Pressions dynamiques (CEA eds.), Paris, 1989.
15. Graham, R. A., "Solids under High Pressure Shock Compression", New York: Springer Verlag, pp 103-113, 1993.
16. Romain,.J. P. Bauer, F., Zagouri, D., Boustie, M., "Measurements of Laser induced Shock Pressures using PVDF Gauges", in Shock Compression of Condensed Matter, 1993 (S.C. Schmidt, J.W. Shaner, G.A. Samara, M. Ross eds.), pp. 1915-1918.
17. Samara, G. A., and Bauer, F., "The Role of Hich Pressure in the Study and Applications of the Ferroelectric Polymer Polyvinylidene Fluoride and its Copolymers", Ferroelectrics, vol. 171, n° 1-4, pp 299-311, 1995 and references therein.
18. Biele, J., "Reflectometric Detection of Shock Wave Propagation within a Concrete Wall", in Shock Compression of Condensed Matter, 1999 (M.D. Furnish, L.C. Chhabildas, R.S. Hixson, eds.), part II, pp. 1029-1032.
19. Bauer, F., "PVDF Shock Sensors: Applications to polar Materials and High Explosives", IEEE Transactions on Ultrasonics, Ferroelectrics, and Frequency Control, vol. 47, pp. 1448-1458, 2000.
20. Bauer, F., "PVDF gauge piezoelectric response under two-stage light gas gun impact loading", in Shock Compression of Condensed Matter, 2001 (M.D. Furnish, N.N. Thadhani, Y. Horie, eds.), part II, pp. 1149-1152.
21. Bauer, F., "PVDF Shock Compression Sensors in Shock Wave Physics", in Shock Compression of Condensed Matter, 2003 (M.D. Furnish, Y.M. Gupta, J.W. Forbes, eds.), part II, pp. 1121-1124.

CP845, *Shock Compression of Condensed Matter - 2005,*
edited by M. D. Furnish, M. Elert, T. P. Russell, and C. T. White
2006 American Institute of Physics 0-7354-0341-4/06/$23.00

# CALCULATING THE RESISTANCE OF LATERAL MANGANIN GAUGES IN A STEEL MATRIX.

## E. J. Harris and R. E. Winter

*AWE, Aldermaston, Reading, Berks, RG7 4PR, UK*

**Abstract.** High resolution computer simulations of manganin gauges mounted to measure lateral stresses in a steel matrix subject to nominally 1-D shock have been run using the AWE Eulerian code Shamrock. The components of the stress and strain in each of the cells in the manganin element have been obtained from the code. Piezoresistive and plastic strain coefficients derived by previous workers have been used to compute the resistivity of each cell in the gauge and, ultimately, the total resistance change of the gauge. The predicted resistance changes are compared with existing experimental data.

**Keywords**: Stress gauges, lateral gauges, manganin gauges
**PACS**: 62.50; 07.05.Tp

## INTRODUCTION

Feng and Gupta [1] and Feng, Gupta and Wong [2] used a Lagrangian hydrocode to model a manganin gauge mounted laterally in a polycrystalline silicon carbide matrix shocked by a copper flyer to particle velocities ranging from 150 m/s to 1300 m/s. Stresses and strains in the deforming manganin were derived and, using piezo-resistive and electrical strain coefficients, the change in resistivity in each calculational cell of the manganin gauges was computed. A manganin equation of state allowed the change of resistance of the *cell* due to its thinning to be computed. The resistance of the *gauge* was calculated by summing the resistances of the cells in parallel.

This paper describes work in which the basic method described above has been used to analyse a series of experiments reported by Hammond et al [3]. In Hammond's experiments Micro Measurements encapsulated "T" Gauges, Type J2M-SS-580SF-025, were mounted in a lateral configuration between layers of polymer protective padding of various thicknesses within an EN3B

mild steel target. The impact velocity for all of the experiments was 400 m/s.

## BASIC THEORY

In the experiments the length of the gauge (in the z direction) is 15 mm but in the simulations we assume that the gauge is infinitely long and its length does not change during the time of interest. The resistance of a cell, R, of resistivity, $\rho$, length L and cross-sectional area, A, is given by

$$R = \frac{\rho L}{A}$$

The resistance *change* of a cell, $\frac{\Delta R}{R_0}$, resulting from changes in resistivity and cross-section of $\Delta\rho$ and $\Delta A$ respectively is given by:

$$\frac{\Delta R}{R_0} = \frac{\Delta\rho}{\rho_0} - \frac{\Delta A}{A_0} \qquad (1)$$

The resistivity change of a cell of the manganin conductor is given by the relation:

$$\frac{\Delta\rho}{\rho_0} = \pi_{11}\sigma_z + \pi_{12}\sigma_x + \pi_{12}\sigma_y + \eta\gamma^p \qquad (2)$$

where $\sigma_x$, $\sigma_y$, and $\sigma_z$ are the stresses in the x, y and z directions respectively. Following Feng and Gupta [1] the equivalent plastic strain, $\gamma^p$, is defined as equal to $\sqrt{I_2'^p}$, the second invariant of the plastic strain tensor. $\pi_{11}$ and $\pi_{12}$ are piezo-resistive coefficients and $\eta$ is the electrical strain hardening coefficient. Feng, Gupta and Wong [2] determined the piezoresistive coefficients of manganin (in GPa$^{-1}$) to be:

$$\pi_{11} = 0.01121 - 0.00009556\, p_g$$
$$\pi_{12} = 0.004108$$

where $p_g$ is the pressure in the cell in GPa, and the strain coefficient as:

$$\eta\gamma^p = (0.56 - 5.729\gamma^p)\gamma^p \quad \text{if} \quad \gamma^p \leq 0.043 \quad \text{or}$$
$$\eta\gamma^p = 0.01352 \quad \text{if} \quad \gamma^p > 0.043.$$

The resistance change of the cell is obtained by combining Eqns. 1 and 2:

$$\frac{\Delta R}{R_0} = \pi_{11}\sigma_z + \pi_{12}\sigma_x + \pi_{12}\sigma_y + \eta\gamma^p - \frac{\Delta A}{A_0}$$

## SIMULATIONS

In Hammond's experiments 50 mm diameter copper flyers impacted mild steel targets consisting of two 25 mm x 25 mm x 50 mm EN3B steel blocks. As shown in Fig. 1 only part of the experimental region, chosen to avoid reflections from the extremities of the problem, was modelled in the simulations. Simulations were run using the AWE ALE Code Shamrock.

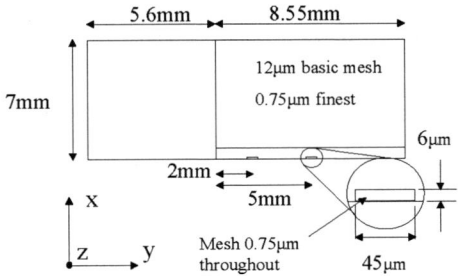

**FIGURE 1.** Shamrock Set-up.

The equations of state for copper and steel took the following form:

$$p = \frac{A_1\mu + \tilde{A}_2\mu^2 + (B_0 + B_1\mu + B_2\mu^2)\varepsilon + (C_0 + C_1\mu)\varepsilon^2}{\varepsilon + \varepsilon_0}$$

where $\tilde{A}_2 = A_2$ when $\mu \geq 0$ or $\tilde{A}_2 = A*_2$ when $\mu < 0$

$$\mu = \frac{v_{ref}}{v} - 1, \quad \varepsilon = \frac{E}{v_{ref}} \quad \text{and} \quad \varepsilon_0 = \frac{E_0}{v_{ref}}.$$

V is specific volume in cc/g and E is internal energy in Mbcc/g. Constants are given in Table 1.

The following EoS was used for the PMMA:

$$p = A_1\mu + \tilde{A}_2\mu^2 + \varepsilon(B_0 + B_1\mu)$$

where $\tilde{A}_2 = A_2$ if $\mu \leq 0$ or $\tilde{A}_2 = A_2*$ if $\mu > 0$

and A1=0.06032; A2=0.1361; A2*=-0.1361; B0=0.75 and B1=4.8932.

We assume that the PMMA has zero strength.

**TABLE 1. EoS Parameters for Steel and Copper**

|  |  | Steel | Copper |
|---|---|---|---|
| $v_{ref}$ | cc/g | 0.1266 | 0.112 |
| A$_1$ |  | 1.6311 | 1.178 |
| A$_2$ |  | 2.109 | 1.147 |
| A*$_2$ |  | -2.109 | -1.147 |
| B$_0$ |  | 1.679 | 1.724 |
| B$_1$ |  | 2.706 | 1.015 |
| B$_2$ |  | 5.743 | 6.000 |
| C$_0$ |  | 0.4181 | 0.4667 |
| C$_1$ |  | 0.5374 | 0.560 |
| E$_0$ | Mbcc/g | 0.1266 | 0.0949 |
| G | GPa | 54.7 | 47.7 |
| Y | GPa | 0.12 | 0.7 |

The manganin EoS below was based on shock velocity particle velocity data published by Keough [4].

$$p = p_r(v) + \frac{\Gamma(v)}{v}(E - E_r(v))$$

where

$$p_r(v) = \frac{a^2(v_0 - v)}{(v_0 - b(v_0 - v))^2} \quad \text{and} \quad E_r(v) = \frac{1}{2}p_r(v)(v_0 - v)$$

and a=3.7 mm/$\mu$s ; b=2.28 and $\Gamma(v)=2$. Values of Yield strength (0.75 GPa) and Rigidity modulus (43.5 GPa) were taken from Ref. 6.

1188

## RESISTIVITY MAPS

Figure 2 shows the effect of varying the strength of the steel matrix on the resistivity of a gauge positioned 2 mm from the impact face in a polymer layer of half thickness 172 µm. It is seen that as the strength of the matrix is increased both the distortion of the gauge and its average resistivity decrease.

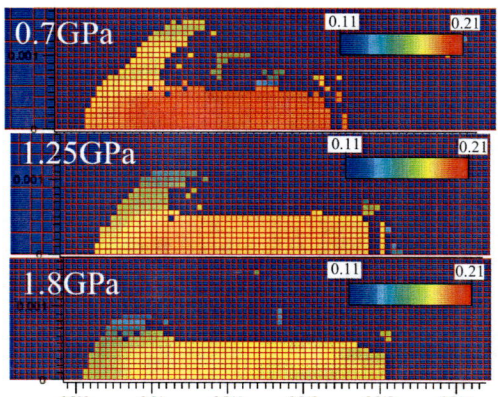

**FIGURE 2.** Resistivity maps for 2 mm gauge in 172 µm polymer run with three different strength in the steel matrix. Note that, although the maps are shown vertically aligned, in fact the gauges in the lower strength calculation are displace significantly more to the right than the high strength calculations.

## RESISTANCE CHANGE

Once the resistivity of each cell in the gauge has been computed the resistance of the gauge can be calculated.

Quantities for cells which initially contain manganin are assigned the superscript i and quantities for meshes which contain manganin at time, t, are assigned the superscript j. The initial area of the Eulerian meshes is $A_0$. Note that, in general, some meshes will be mixed initially and at time, t and, further, that the mesh areas may change due to refinement during the course of the calculation. The areas of the manganin in the $i^{th}$ and $j^{th}$ meshes are designated $A^i$ and $A^j$ respectively, The resistivity of the manganin is $\rho_0$ initially and $\rho^j$ at time t. The fraction of the mesh volume

occupied by manganin initially and at time t are termed $V_{frac}^i$ and $V_{frac}^j$ respectively. The total resistance of the gauge is R and the resistances of the $i^{th}$ and $j^{th}$ cells are $R^i$ and $R^j$.

The following equations can be written:

$$A^i = A_0 F_{frac}^{\;i} \qquad (3)$$

$$A^j = A_0 F_{frac}^{\;j} \qquad (4)$$

Using Eqn. 1:

$$R_0 = \frac{1}{\sum_i \dfrac{1}{R^i}} = \frac{1}{\sum_i \dfrac{A^i}{\rho_0 L}}$$

$$R = \frac{1}{\sum_j \dfrac{1}{R^j}} = \frac{1}{\sum_j \dfrac{A^j}{\rho^j L}}$$

where L is the length of the gauge.

The relative resistance change of the gauge, $\dfrac{\Delta R}{R_0}$, is given by

$$\frac{\Delta R}{R_0} = \frac{R}{R_0} - 1$$

$$= \frac{\sum_i \dfrac{A^i}{\rho_0 L}}{\sum_j \dfrac{A^j}{\rho^j L}} - 1 = \frac{\sum_i A^i}{\sum_j \dfrac{\rho_0}{\rho^j} A^j} - 1$$

$$= \frac{\sum_i A^i}{\sum_j \dfrac{A^j}{\left(1 + \dfrac{\Delta \rho^j}{\rho_0}\right)}} - 1$$

which, incorporating Eqns. 3 and 4 gives:

$$\frac{\Delta R}{R_0} = \frac{\sum_i V_{frac}^{\;i}}{\sum_j \dfrac{V_{frac}^{\;j}}{\left(1 + \dfrac{\Delta \rho^j}{\rho_0}\right)}} - 1$$

## COMPARISON BETWEEN CALCULATION AND EXPERIMENT.

Figure 3 shows calculated resistances for gauges in polymer layers of three different thicknesses. The shear strength of the steel was set at 1.8 GPa [5] in these calculations. Note that each of the three profiles eventually becomes horizontal and that the three profiles converge to the same level. This implies that, in the three cases, the same, constant, y stress is eventually achieved in the steel and the polymer. However, in the experimental records, displayed in Fig. 4, the traces tend to rise at late time and show no signs of converging. A comparison between measured and calculated resistance is shown for 22 μm and 272 μm gauges in Fig. 5. It is seen that the initial jumps in resistance are well matched but that the experimental and calculated curve diverge at late time. A possible explanation for the late time discrepancy is that the manganin conductors are actually 15 mm long (in the z direction) but are modelled as infinitely long. Consequently at later times z stresses in the manganin in the real gauge will be increased by the polymer pressure exerted on the ends of the gauge (the x, y faces).

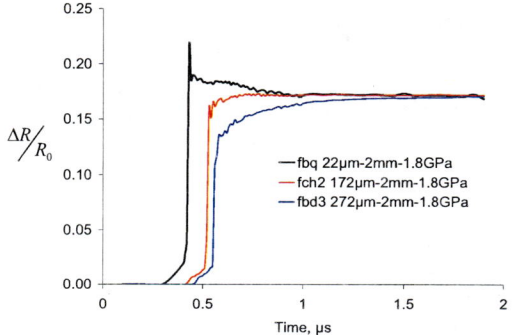

**FIGURE 3.** Computed resistance changes for gauges in polymer layers of half-thickness 22 μm, 172 μm and 272 μm.

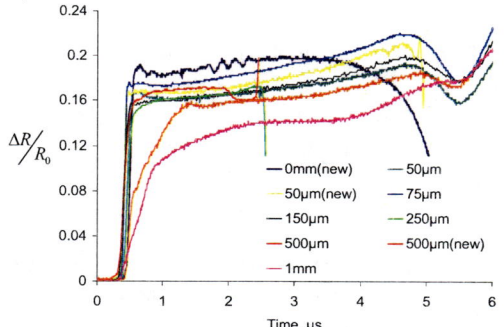

**FIGURE 4**. Resistance changes measured by Hammond for a gauge positioned 2 mm from the impact surface. Unlike the calculations the profiles do not converge at late time.

## CONCLUSIONS

The AWE Eulerian Code SHAMROCK has been used to compute the resistance change of Micro Measurements encapsulated "T" Gauges, Type J2M-SS-580SF-025 in a EN3B steel matrix. Maps have been generated illustrating the distribution of resistivity throughout the shocked gauge. An algorithm is presented for computing resistance change in a Eulerian code. The results suggest that the shear strength of EN3B used in Hammond's [1] experiments was 1.8 GPa. discrepancies between the shapes of the computed and measured profiles at late time may be due to the fact that the actual gauges have a finite length.

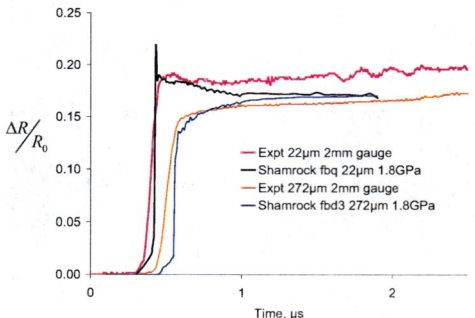

**FIGURE 5.** Comparison between calculated and measured resistance for a gauge in 22 μm and 272 μm polymer layers.

## REFERENCES

1. Hammond, R. I., Church, P. D., Grief, A., Proud, W. G. and Field, J. E., *Shock Compression of Condensed Matter*, APS, 1125, (2003).
2. Feng, R. and Gupta, Y. M., *J Appl. Phys.* 83 (2), 747, (1998).
3. Feng, R., Gupta, Y. M. and Wong, M. K. W., J. *Appl. Phys.* 82 (6), 2845, (1997).
4. Keough, D. D., "Pressure Transducer for Measuring Shock Wave Profiles", SRI Report DASA-1414, November 1, (1963).
5. Winter, R. E., Harris, E. J. and Hammond, R. I., "Measurement of strength of EN3B mild steel using lateral gauges", this meeting.
6. Rosenberg, Z., Yaziv, D. and Partom, Y., *J Appl. Phys.* 51, 3702, (1980).

CP845, *Shock Compression of Condensed Matter - 2005*
edited by M. D. Furnish, M. Elert, T. P. Russell, and C. T. White
© 2006 American Institute of Physics 0-7354-0341-4/06/$23.00

# VALIDATION OF A MULTITECHNICAL DEVICE AIMED TO REACH THE TEMPERATURE OF A MATERIAL UNDER SHOCK LOADING

## B. Legrand, E. Blanco, and E. Martinez

*Commissariat à l'Energie Atomique, Centre Ile-de-France, BP12, 91680 Bruyères-le-Châtel, France*
*(bruno.legrand@cea.fr)*

**Abstract.** The temperature is one of the parameters which occurs in the equations of state, to describe the behaviour of materials submitted to an intense shock (tens of GPa). A multispectral pyrometer is used to determine the temperature. This diagnostic is well known on classical experimental set-up, but it allows to obtain only one temperature measurement. The aim of this work is to validate a device able to realize two simultaneous measurements of temperature, coupled with chronometry and velocity measurements, to improve our knowledge of the behaviour and of the temperature of a material under shock loading. It was necessary to resize and to characterize the existing diagnostics and taking into account the duration of the phenomena (700 ns), as well as mechanical and luminous disturbances generated by the shock. The true temperature of the shocked material is obtained from a radiance measurement at short wavelength. So the pyrometer was reconfigured. This assembly was validated on bismuth samples. The results obtained enable to consider new prospects, like the addition of a reflectivity measurement, in order to reduce uncertainty on the temperature.

**Keywords:** Pyrometry, temperature, emissivity, shock, bismuth, plate impact.
**PACS:** 07.20.Ka, 62.50.+p, 78.47.+p.

## INTRODUCTION

To simulate numerically the behaviour of the matter under shock, it is interesting to well know the thermodynamic parameters in order to determine the best equation of state (EOS) of the materials. The temperature is a significant parameter to differentiate the EOS, but in the case of shock-compressed metals its measurement remains difficult. The Hugoniot temperature is not measurable experimentally, only free surface or interface temperature measurement (using transparent anvil) are possible. Taking into account the brevity of the phenomena (lower than 1 μs) and destructive aspect of the experiments, the optical pyrometry seems to be the most adapted

measurement method to determine the temperature of materials shocked loading. This technique appeared, at the C.E.A., twelve years ago. In 1993, M. Mondot develops a pyrometer with three channels, to measure the free surface temperature of a sample [1]. Measurement was done under a secondary vacuum level, to free itself from the luminous phenomena of ionization of the air. In 1997, E. Blanco takes measurement through a lithium fluoride window and extends the range of the pyrometer to the visible spectrum in order to measure higher temperatures [2]. The number of analysis channels passes to eight. In 2002, D. Partouche improves the experimental building used by Blanco, to make it less sensitive to the luminous disturbances. He also realized some experiments where the temperature measurement and reflectivity measurement are coupled [3]. Finally, B. Legrand validates a new set-up, based on the use of a 60 mm diameter launcher which allows to measure two interface temperatures and shock velocity during the same experiment [4].

Classically, the Hugoniot temperature $T_H$ is determined from several experiments which are dedicated to measure free surface temperature $T_{FS}$ or interface temperature using window anvil ($P_{I1}$ or $P_{I2}$) (see Fig.1).

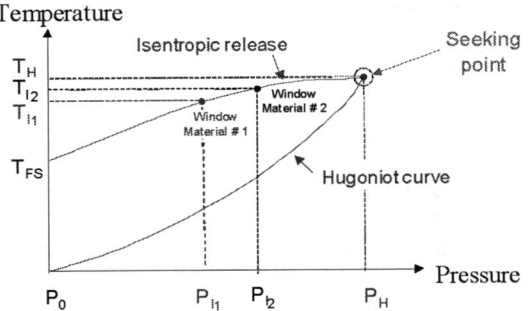

P$_I$, T$_I$ : Pressure and Temperature at the Interface.
T$_{FS}$ : Temperature of Free Surface.

**Figure 1.** Schematic diagram of the scientific objective.

From theses results, the isentropic release of the material is rebuilt. The goal is to find the coordinates of the required point ($P_H$, $T_H$). Until now, the standard set-up allowed to acquire only one temperature measurement. Consequently, two experiments were needed to determine the Hugoniot temperature. The mains difficulties were, first to obtain the same shock pressure on the two shots and secondly, to have the same properties of the studied materials. The convolution of theses parameters generates some uncertainty of $T_H$ determination.

The objective of this work is to design and validate an experimental device intended to determine, under shock, the Hugoniot temperature of a material, from two simultaneous measurements of interface temperature ($P_{I1}$ and $P_{I2}$). Coupling to other techniques of measurement this device will improve the accuracy of the system. The principal restriction concerning the

set-up size is governed by the use of a 60 mm diameter powder-gun.

## EXPERIMENTAL PROCEDURE

The experiments of optical pyrometry on metals are limited to a measurement on the surface, because they are opaque (optical thickness lower than the wavelength). So during the shock, it is only possible to measure the radiance emitted by the surface of the sample. The spectral distributions of radiance of the real sources have the same aspect as those of the black body, but emitted energies are always lower, at equal temperature and wavelength. In the relation (1), we note that the monochromatic radiance measured $L_\lambda(\lambda, T)$ is proportional to the blackbody radiance $L_\lambda^0(\lambda, T)$, obtained during the pyrometer calibration step and the sample emissivity $\varepsilon_\lambda$:

$$L_\lambda(\lambda,T) = \varepsilon_\lambda \cdot L_\lambda^0(\lambda,T), \qquad (1)$$

with the blackbody radiance gives by the Planck formula:

$$L_\lambda^0(\lambda,T) = \frac{C_1 \cdot \lambda^{-5}}{exp\left(\dfrac{C_2}{\lambda \cdot T}\right) - 1}, \qquad (2)$$

where $C_1$ and $C_2$ are linked to the fundamentals physics constants.

The emissivity value is defined between 0 and 1. For a considered material it depends of its temperature $T$, the wavelength $\lambda$, the emission direction $\theta$ and for opaque material of the surface state.

At thermal equilibrium, the emissivity of opaque sample depends only of the reflectivity $\rho_\lambda$:

$$\varepsilon_\lambda = 1 - \rho_\lambda, \qquad (3)$$

In this case, it is easy to find in the literature some results in static condition. But there is no information during dynamic shock, so it is necessary to take hypothesis on the emissivity value.

A window anvil is glue to the sample (see Fig.2), in order to measure the emitted radiation along the isentropic release ($P_{I1}$ or $P_{I2}$).

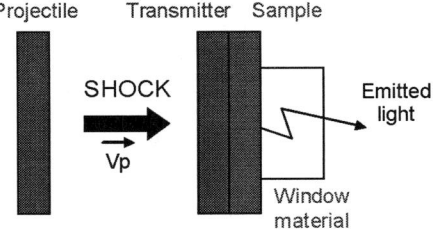

**Figure 2.** Principle of optical pyrometry measurement.

The emitted light from the interface is collected by the optical head of the pyrometer. So the window material must remain transparent under shock.

The pyrometer configuration allows to work on a large wavelength range, from 0.4 µm to 3.5 µm. Four kinds of detectors are used: photomultiplier in the visible range, Ge in the near IR, HgCdTe and InSb in the middle IR.

## Set-up description

The sample material used to validate the multi diagnostic set-up is bismuth considering the lot of temperature experimental results [2] [3], as well as a theoretical studies [5].

For anvil materials, we have chosen lithium fluoride (LiF) and sapphire ($Al_2O_3$). The LiF remain transparent under shock until 160 GPa [6]. The $Al_2O_3$, remain transparent at less until 50 GPa. But it emits light when it is shocked [7].

The duration of the measurement is limited by the size of the window. We choose an anvil with 10 mm in thickness and 15 mm in diameter. With theses dimensions, an observation time between 400 ns to 1 µs is obtained in our pressure range.

The proposed experimental set-up (see Fig.3) allows to obtain simultaneous:
- two temperature measurements at different sample-window interface.
- shock wave celerity, from 8 chronometrical probes.
- the particle velocity at a LiF-sample interface, from Doppler Laser Interferometry (D.L.I.).

- target velocity, from D.L.I. and 9 chronometrical probes.

The set-up supported four samples.

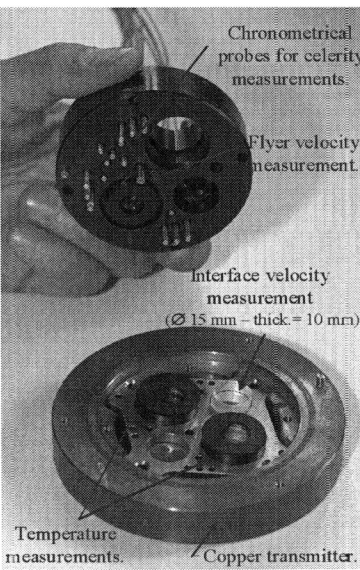

**Figure 3.** Photography of the set-up.

Comparatively to the classical device [3], where only one temperature and the projectile velocity were measured, this new set-up offer the advantage to study simultaneously four samples under the same dynamic solicitation. This set-up offers significant possibilities of experimentations. For example, the comparison between two different materials or the influence of the glue thickness can be made.

## RESULTS AND DISCUSSION

Results are given in Table 1. The interface temperature is calculated assuming the dynamic emissivity value is between 0.1 and 1.

The first experiment (T101) was realized with the classical device [3]. The signals obtained are considered as reference for the new set-up. The interface temperature was measured across a LiF window. The pyrometer was used in its initial configuration with 1x8 channels of analysis.

**TABLE 1.** Experimental data on bismuth with standard (T101) and multidiagnostic set-up (T105 & T107).

| Shot | $V_p$, m/s | $U_{Bi/LiF}$, m/s | $P_{Bi/LiF}$, Gpa | $P_H$, GPa | $T_{LiF}$, K | $T_{Al2O3}$, K |
|------|-----------|-------------------|-------------------|------------|--------------|----------------|
| T 101 | $1562 \pm 10$ | - | $19.49 \pm 0.93$ | $28.01 \pm 1.36$ | $1769 \pm 164$ | - |
| T 105 | $1507 \pm 38$ | $1079 \pm 11$ | $18.69 \pm 0.89$ | $26.62 \pm 1.77$ | $1635 \pm 140$ $1663 \pm 144$ | - |
| T 107 | $1438 \pm 11$ | $1026 \pm 8$ | $17.70 \pm 0.85$ | $24.92 \pm 1.20$ | $1572 \pm 130$ | $2280 \pm 270$ |

The shot T105 is dedicated, primary, to verify the size device component and secondly to validate simultaneous measurement. Two interface temperature measurements using LiF anvils were performed in order to verify the consistency of the results. Moreover we verify that the particle velocity measurement, realized by DLI, do not disturb the pyrometric measures. The pyrometer was reconfigured in 2x4 channels of analysis. The two pyrometric channels were used in the same configuration. Filters as well as the sensitivities were comparable.

The T107 experiment validates the multidiagnostic set-up in its final configuration, with two different materials windows: lithium fluoride and sapphire.

The interface temperatures measurements acquired during these experiments are consistent with the previous results and theoretical values. They validate our multidiagnostic device.

## CONCLUSIONS

We have validated a multidiagnostic device which allowed to measure several physical parameters. The simultaneous measurement of two interfaces temperatures under the same shock loading is a significant improvement for physical determination of $T_H$. The uncertainty of theses temperature measurements is less than 10 % around 1500 K. One limitation of the accuracy is the knowledge of the dynamic emissivity. Our future objective is to integrate an emissivity measurement on the device. The dimensioning of this set-up is compatible with projectile velocity below 1600 m.s$^{-1}$. The resizing of a set-up based on the use of a double-stage powder gun (48 mm diameter, maximum target velocity: 3500 m.s$^{-1}$) will permit to measure interface temperature at higher pressure.

## REFERENCES

1. Mondot, M., "La température des matériaux sous choc", Thèse de doctorat en sciences physiques, CNAM Paris, 1993.
2. Blanco, E., "Température et émissivité des matériaux sous choc – Etude expérimentale par pyrométrie optique à travers un matériau fenêtre" Thèse de doctorat en énergétique, University of Paris X – Nanterre, 1997.
3. Partouche-Sebban, D., "Développement des techniques ultra-rapides de pyrométrie optique pour l'étude du comportement des métaux sous choc et application à la caractérisation du diagramme de phase du Bismuth dans la gamme 10-20 GPa", Thèse de doctorat en sciences, University of Paris XI, 2002.
4. Legrand, B., "Conception et validation d'un dispositif multitechique destiné à accéder à la température sous choc des matériaux", Mémoire d'ingénieur, CNAM Paris, 2004.
5. Wetta, N., Pelissier, J.L., "A model-potential approach for bismuth (II). Behaviour under shock loading", Physica A, vol. 289, p. 479-497, 2001.
6. Wise, J.L., Chhabildas, L.C., "Laser interferometer measurements of refractive index in shock compressed materials", Shock waves in condensed matter, 1985, New York: Plenum 1986.
7. Hare, D.E., Holmes, N.C., "Shock-wave-induced optical emission from sapphire in the stress range 12 to 45 GPa: Images and spectra", Phys. Rev. B 66, 014108, 2002.

CP845, *Shock Compression of Condensed Matter - 2005*,
edited by M. D. Furnish, M. Elert, T. P. Russell, and C. T. White
2006 American Institute of Physics 0-7354-0341-4/06/$23.00

# STUDY OF SENSITIVITY AND REPEATABILITY OF PIEZOELECTRIC SENSORS

## C. E. Lloyd, M. W. Greenaway and W. G. Proud

*Physics and Chemistry of Solids Group, Cavendish Laboratory, Cambridge, CB3 0HE, UK*

**Abstract.** The sensitivity and repeatability of stress and density measurements obtained using commercially available piezoelectric probes have been studied for high-velocity ($> 450$ m s$^{-1}$) gas gun-driven spray experiments. The probes used are Dynasen Piezopins, in which the sensor element is a small (($0.4 \pm 0.05$) mm thick, ($1.2 \pm 0.1$) mm diameter) PZT (Lead Zirconate Titanate) disk. The probe gives an output voltage V(t) proportional to the time derivative of the force normal to the poled axis of the PZT. The stress level is obtained using the time-integrated voltage. It is assumed that there is complete momentum transfer between the spray and the Piezopin, therefore the spray density can be found from the stress level. The spray is produced by accelerating aluminum powder (10 μm grain size) in a gas gun. Spray density measurements are compared with values measured from x-ray images, and stress measurements are compared with extrapolated values from the spray densities obtained from the x-ray images.

**Keywords:** Piezoelectric devices, stress measurement, density measurement
**PACS:** 85.50.-n, 07.10.Lw, 06.30.Dr

## INTRODUCTION

When a shock wave arrives at a free surface, small particles can be ejected and a spray is produced. Such sprays have hydrodynamic properties and may travel at velocities of a few kilometers per second [1-3]. In the last two decades, the dynamics of ejection has attracted the attention of both experimental and theoretical physicists [1-6]. Intricate techniques including interferometry [2-4], radiography [4] and holographic techniques [5] have been used to study the stability of shock loaded surfaces. However, the need to develop a more compact diagnostic tool to measure time-dependent stresses and densities of shock-induced sprays has turned the focus on piezoelectric probe technology.

Manufacturers claim that their piezoelectric probes have rise times of a few tens of nanoseconds, and their resilience to initial stages of impact suggests that they could be a useful diagnostic tool for the study of sprays. However, relatively little quantitative data has been published on their output performance with fast moving sprays [7]. This study concentrates on characterizing the output of the CA-1136 Piezopin that is commercialized by Dynasen, Inc. Particular attention is given to evaluating its sensitivity and repeatability.

The CA-1136 Piezopin consists of a small disk (1.2 mm diameter, 0.4 mm thick) PZT-5A element housed between a copper target and brass support, insulated by a Kynar jacket and mounted in a brass tube. The potential difference that is developed across the PZT element during impact can be recorded using an oscilloscope terminated at 50 Ω.

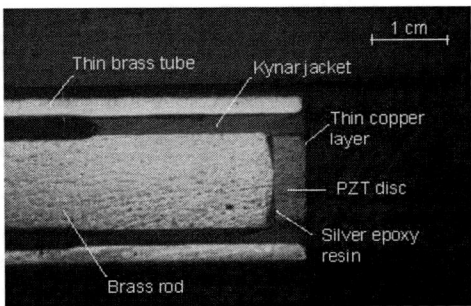

**Figure 1.** Photograph of cross-section of the CA-1136 Piezopin, commercialized by Dynasen, Inc.

When there is one-dimensional stress equilibrium in the PZT, the polarization $p_z$ evolved across its $z$-axis when subjected to stress $\sigma_{zz}$ is

$$p_z(t) = d_{33}\sigma_{zz}(t), \qquad (1)$$

where $d_{33}$ is the piezoelectric charge coefficient for the PZT along the poled $z$-axis. For PZT-5A, $d_{33} = 374$ pC N$^{-1}$.

Since the probe arrangement is largely resistive in nature when terminated at $Z = 50$ $\Omega$, the stress $\sigma_{zz}(t)$ can be related to the time-integrated output voltage $V(t)$ of the probe,

$$\sigma_z(t) = \frac{\int V(t)dt}{ZAd_{33}}, \qquad (2)$$

where $A$ is the area of the PZT element. For impacting material that transfers all its momentum to the Piezopin, the density $\rho(t)$ can be related to the stress $\sigma_{zz}(t)$ and impacting velocity $u$,

$$\rho(t) = \frac{\sigma_{zz}(t)}{u^2} = \frac{\int V(t)dt}{ZAd_{33}u^2}. \qquad (3)$$

Since the ultimate aim of characterizing the response of the Piezopin is to obtain accurate stress and density measurements of ejecta spray, the performance of the probe in a well-controlled spray environment should be investigated. The gas gun-driven spray technique detailed in this study was chosen over traditional shock-induced spray ejecta experiments to assess the performance of the Piezopins. The gas gun technique allows greater control over the particle size distribution, the velocity and approximate diameter of the spray. Complimentary diagnostics such as high-speed photography and x-ray imaging can also be used in the experimental arrangement.

## EXPERIMENTAL PROCEDURE

A spray consisting of aluminum particles (mean diameter 10 μm) was produced by accelerating a thin-walled, powder-filled polycarbonate sabot down a 25 mm diameter gun barrel. A laser and photodiode arrangement was placed at the end of the gun barrel as a velocity meter. A sabot stripper at the end of the barrel stopped the sabot and allowed the powder to travel onwards, thus producing a spray. A Piezopin was placed a few centimeters away from the end of the barrel, and sampled the spray. The voltage response was recorded on a 500 MHz oscilloscope.

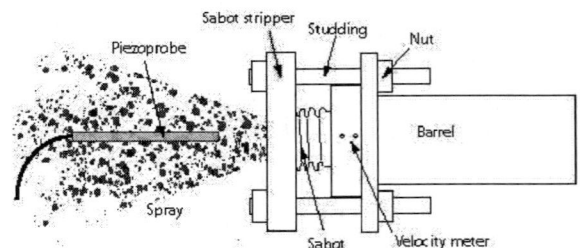

**Figure 2.** Schematic of the experimental arrangement.

A DRS Hadland Ultra-8 high-speed camera and an x-ray unit that produced polychromatic x-ray pulses of 30 ns FWHM were used to obtain independent density measurements of the spray. Agfa Curix UV-L Medical x-ray film encased in Agfa x-ray cassettes with imaging screens were used to obtain x-ray images of the spray.

The calibration for the relation between the intensity measured by x-ray film and the density of the material through which the beams traversed was achieved by imaging wedges of aluminum powder. The mass of powder per unit area was known for the whole wedge, since the dimensions of the wedge and the packing density of the powder were known. Therefore by obtaining an x-ray image of the wedge, each intensity level on the x-ray image could then be converted to its corresponding mass per unit area. This was done for each piece of film used (see Fig. 3). Information regarding the dimensions of the spray, such as its thickness, was obtained from the high-speed camera images. Spray density estimates for given areas of spray imaged on the x-ray film could then be obtained by dividing the equivalent mass per unit area by the thickness of the spray at that point.

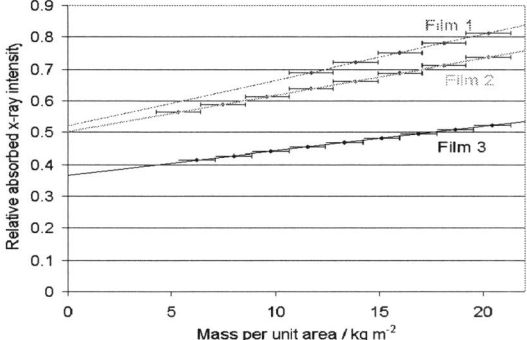

**Figure 3.** Calibration graph relating the relative absorbed x-ray intensity to the equivalent mass per unit area for aluminum powder wedges of known density.

In all the Piezopin experiments, the probes were positioned such that their central axes were aligned with that of the gun barrel. The spray was therefore assumed to impact the Piezopins along the direction of poling of PZT. Equations (2) and (3) could therefore be used to obtain stress and density values from the voltage signals from the probes. It should be noted that x-ray images could not be obtained in the same spray experiments as when the Piezopin were used, due to the intense electromagnetic pulse generated by the x-ray unit.

The impact velocity of the spray $u$ was required to obtain the density of the spray. The velocity meter at the end of the barrel measured the sabot velocity, and therefore did not give an accurate spray velocity. The spray velocity could however be obtained from the x-ray images and photographic images of the spray. Another method was to sample the spray using two Piezopins of a known distance apart along the $z$-axis. This technique had the added convenience of obtaining

quantitative data of how the spray profile changed with time. It also proved to be a more straightforward and reliable method of measuring velocity than the photographic and x-ray techniques, since it did not require trigger synchronization.

## RESULTS AND DISCUSSION

High speed photographs of the aluminum spray was found to be diffuse and diverged at an angle of 28° from the barrel axis.

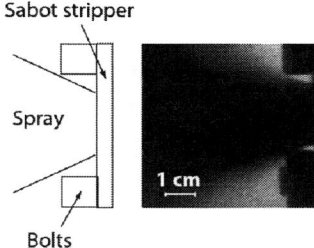

**Figure 4.** High-speed image of an aluminum spray produced by the gas gun arrangement.

An x-ray image of aluminum spray taken at the same instant as the frame shown in Fig. 4 is shown in Fig. 5. Only central parts of the spray absorbed sufficient x-radiation to show contrast on the x-ray film. The maximum spray density detected by the x-ray image was 120 kg m$^{-3}$ (4.4% theoretical maximum density).

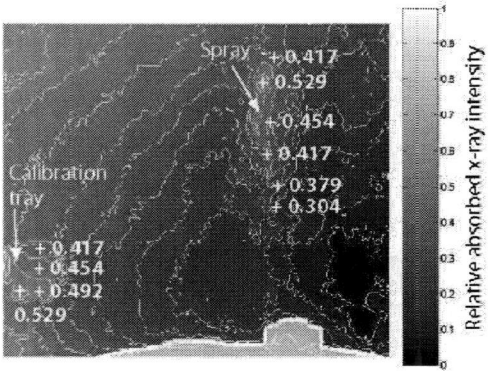

**Figure 5.** X-ray image of an aluminum spray accelerated by a gas gun. The contours indicate regions of constant intensity and the relative absorbed x-ray intensity for various points in the spray are labeled. The spray velocity is 675 m s$^{-1}$.

Using equation (3) and its assumptions, an object in the path of a spray of density 120 kg m$^{-3}$ travelling at a velocity of 675 m s$^{-1}$ would experience a stress of ~55 MPa.

Stress and density traces were obtained by the Piezopins in a series of four experiments. Fig. 6 shows typical Piezopin density traces as a function of time. Two Piezopins were placed at a known distance apart in this particular experiment. The probe furthest from the barrel detected a slightly lower density and stress than the probe in front. This is consistent with the divergence of the spray shown in Fig. 4. The velocity measured from the time delay between the detection of the spray was similar to the velocities measured using the photographic and x-ray techniques. This demonstrates that the Piezopins are suitable as time of arrival sensors.

All the Piezopin traces were of similar shapes, although peak stresses ranged from 450 MPa to 1025 MPa. Peak densities varied from 1880 kg m$^{-3}$ to 4300 kg m$^{-3}$. The stress values measured by the Piezopins were typically ~10 times higher and the density measurements were ~20 times higher than the values measured by the x-ray technique.

The density for pure aluminum is ~2700 kg m$^{-3}$; the densities and therefore the stresses measured by the Piezopins are overestimates. The x-ray technique, with its measurement of lower stress and density values, gave more credible values. Several reasons could

account for the Piezopin's unexpectedly high response. Damage inflicted onto the probe in the early stages of spray detection is the obvious cause. It is also known that the constitutive relations governing the PZT response become non-linear if subjected to high stresses; this is not accounted for in the relations outlined in equations (1) to (4) [8].

## CONCLUSIONS

Dynasen Piezopins gave reproducible measurements for aluminum sprays. If their increased pressure and density response is taken into account, they offer a compact and relatively simple method of studying the behavior of spray ejecta. This research is on-going and forms part of a larger study.

## ACKNOWLEDGEMENTS

The authors are grateful to AWE Aldermaston, UK for its financial support. EPSRC is thanked for grants in support of the high-speed camera system.

## REFERENCES

1. Speight, C. S., Harper, L., Smeeton, V. S., *Rev Sci Instrum* **60** (12), pp. 3802-3808, 1989.
2. Asay, J. R., Mix, L. P., Perry, F. C., *App Phys Lett* **29**(5), pp. 284-287, 1976.
3. Asay, J. R., Baker, L. M., J App Phys **45**(6), pp. 2540-2546, 1974.
4. Asay, J. R., Trucano, T. G., *Int J Impact Engng* **10**, pp. 35-50, 1990.
5. Sorenson, D. S., Minich, R. W., Romero, J. L., Tunnell, T. W., Malone, R. M., *J App Phys* **92**(10), pp. 8530-5836, 2002.
6. Chen, J., Jing, F. Q., Zhang, J. L., Chen, D. Q., Wang, J. H., *J Phys-Condens Mat* **14**, pp. 10833-10837, 2002.
7. Charest, J. A., Mace, J. L., *Shock Compression of Condensed Matter-2001*, (M.D. Furnish, N.N. Thadhani, Y. Horie, eds.), part II, pp. 1153-1156, 2002.
8. Hall, D. A., *J Mater Sci* **36**(19), pp. 4575-4601, 2002.

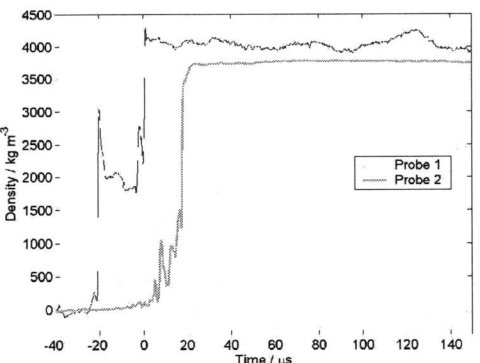

**Figure 6.** Densities measured by two Piezopins for aluminum spray. The Piezopins were placed (9.2 ± 0.1) mm apart, the time delay between the spray detection in the two traces was (18.8 ± 1.0) μs, therefore the spray velocity was (489 ± 27) m s$^{-1}$.

CP845, *Shock Compression of Condensed Matter - 2005*,
edited by M. D. Furnish, M. Elert, T. P. Russell, and C. T. White
2006 American Institute of Physics 0-7354-0341-4/06/$23.00

# TEMPORAL PROFILES OF EXPLOSIVELY-GENERATED PRESSURES IN SOLIDS MEASURED BY AN OPTICAL FIBER-BASED GAUGE

## J. E. Monat, J. R. Carney, V. H. Whitley, and G. I. Pangilinan

*Research and Technology Department, Indian Head Division,*
*Naval Surface Warfare Center, Indian Head, MD 20640*

**Abstract.** A new gauge is being developed and calibrated to measure pressures generated from explosives as a function of time in solids where traditional gauges are not applicable. A laser-pumped optical fiber-based gauge was embedded in Modified Gap Test cylinders. The gauge responded to pressure and showed other features likely due to additional interfaces beyond the fiber tip, including the shock front. Since the temporal profile of the pressure is not well known, hydrocode and ray-tracing modeling are being used to understand the results. The gauge shows promise for time-resolving explosively-generated shock pressures in solids.

**Keywords:** Pressure, shock, optical fiber, laser, refractive index
**PACS:** 47.40.-x, 47.40.Nm, 62.50.+p, 07.60.Vg.

## INTRODUCTION

Testing small amounts of energetic materials reduces scale-up time and costs. Small-scale tests require gauges that are not affected by electrical interference from a nearby detonator as piezo-resistive and -electric gauges are[1]. VISAR[2] requires a reflective surface, which may not be possible in a given test. Ruby fluorescence spectroscopy can measure dynamic pressures [1], but the required streak camera increases complexity and cost.

A gauge designed to fulfill small-scale testing requirements is being developed. The GRIP (Gauge using Refractive Index for Pressure) uses an inexpensive photodiode and oscilloscope for detection, and has been demonstrated in underwater tests[3]. To extend its use to solids, the Modified Gap Test (MGT) has been used with two application methods. In both cases the GRIP worked well as a time-of-arrival gauge, but hydrocode modeling is needed for pressure calibration. The challenges of each method have been identified.

## MATERIALS AND METHODS

### GRIP gauge calibration

The optoelectronic setup (Fig. 1) uses a 532 nm continuous-wave laser and has been described previously [3].

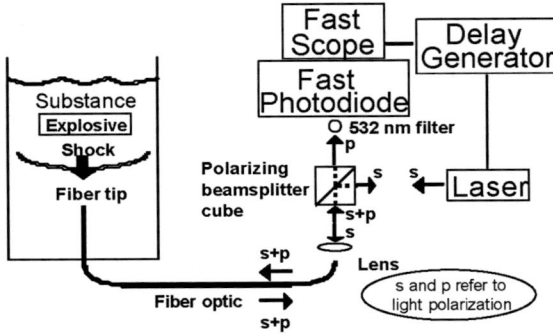

**Figure 1.** GRIP optoelectronic setup.

Conversion of measured signal to pressure proceeds through refractive indices:

$$S = W\phi\left[\left(\frac{n_{subst}(t) - n_{fiber}}{n_{subst}(t) + n_{fiber}}\right)^2 + s_0\right] \quad (1)$$

where $S$ is the photodiode signal (V), $W$ (mW) is the laser power entering the optical fiber, $\phi$ (V/mW) is the overall conversion efficiency of light input into the fiber to voltage produced by the detector, $n_{subst}(t)$ is the index of refraction of the substance at the tip/substance interface vs. time, $n_{fiber}$ is the optical fiber core refractive index, and $s_0$ is the scattering fraction of the optical system. Fitting $\phi$ and $s_0$ to static reflection data provided a calibration.

## Energetics experiments

A more complete report of the experimental results herein has been made[4]. In the Sigmund Jacobs Detonation Science Facility, the MGT (RP80 detonator, 50.8 mm diameter x 50.8 mm tall pentolite charge) was performed for several pentolite-GRIP gap distances, at which the peak pressure is known[5-7]. To allow light to pass straight through the PMMA cylinder (the gap material), flats were machined on opposing sides as in previous reports[7]. Care was taken to remove air bubbles in front of the GRIP, particularly in the adhesive used to affix the GRIP assembly to the

PMMA. An Imacon 200 framing camera provided visual records.

In one GRIP method, a thin layer (~100 μm) of transparent colorless Dow Corning RTV silicone was allowed to dry on the tip of an optical fiber. A previous report[8] suggested that this layer alone would comprise the sensing medium ($n_{subst}$ in Eqn. 1), but static tests showed reflection from other interfaces, which was undesirable.

A boot of fluid (BOF) GRIP method was also tested using 4:1 methanol:ethanol (commonly used in diamond anvil cells up to 10 GPa) as the sensing medium. A layer of black paint at one end of the polyethylene boot, below the fluid, succeeded in reducing reflections from other interfaces.

## RESULTS AND DISCUSSION

### Experimental data

Fig. 2 shows two silicone-on-tip shots. The dashed vertical lines are the shock arrival times according to the framing camera, which demonstrate good agreement. The GRIP signals show an increase upon shock arrival, meaning that another interface than fiber tip/silicone, likely silicone/adhesive, was the main signal source, thus complicating the analysis.

GRIP signal data from four boot of fluid shots are shown in Fig. 3. Each showed a rapid decrease in signal when the shock arrives (set as time zero). A pre-shock increase in signal was noticed and will be addressed in the next section.

While the MGT is a simple, well-studied test, insertion of materials of different impedances modifies the expected pressures. Thus hydrocode calculations of the relevant pressures must be performed to interpret gauge results.

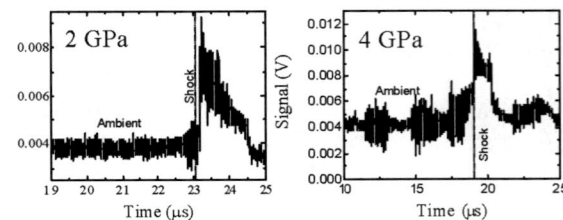

**Figure 2.** Silicone on tip GRIP data; nominal gap test pressures are labeled.

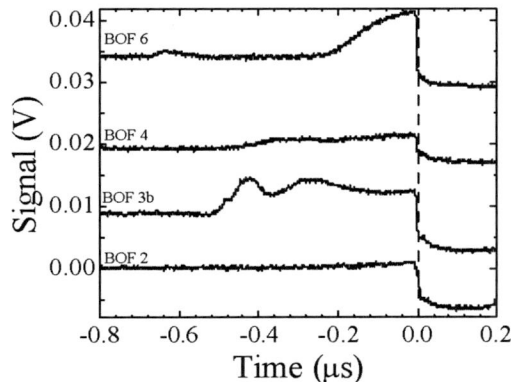

**Figure 3.** Boot of fluid GRIP data; nominal gap test pressures (in GPa) are labeled above each trace.

### Ray-tracing modeling

Because the GRIP gauge is based on the reflection of light back through an optical fiber, any laser light reflecting off other interfaces may affect the signal. Laser light reflecting off the approaching shock front seemed the most likely explanation for the pre-shock increase in signal in the boot of fluid tests (Fig. 3). To test this hypothesis, simple static experiments were performed where the shock front was simulated by an oil-water interface which the GRIP was moved towards.

The GRIP signal as a function of position relative to the interface is shown in Figure 4. The baseline value, corresponding to reflection from the fiber tip-oil surface only, was obtained until the GRIP came to within a few mm of the oil-water interface. Near the interface, the signal increased to a maximum before decreasing precipitously at the interface itself. This is in accord with our hypothesis.

To quantify the effect, a simple two-dimensional model of reflection off an interface was developed. The fiber tip was viewed as a series of point sources of light with the shock front as a flat surface approaching the tip. The shock front, though actually approximately spherical, may be approximated as flat with respect to the 400 μm diameter fiber tip: the gaps tested herein are greater than 25.4 mm; and a sphere 25.4 mm in radius deviates from planarity over a 400 μm chord by

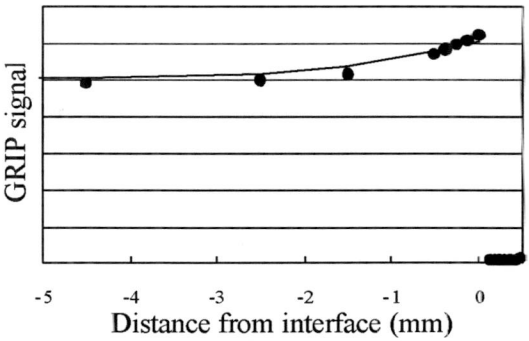

**Figure 4.** Points: GRIP signal when approaching and passing through (at 0 mm) an oil-water interface. Line: fit to model of light reflecting off interface.

only 9.4 μm. The angular distribution of each point source was modeled as that of an optical fiber at far field, where the intensity peaks parallel to the fiber's axis and approaches zero at the critical angle[9]. Only photons reflecting back within the fiber core diameter were assumed to be detected.

Fitting this model to the static data gave a reasonable fit, Figure 4, though the model does show a signal rise sooner than is observed. Before fitting to the dynamic data in Figure 3, the effect of the black paint had to be considered, Figure 5. Before the shock has reached the paint, essentially all the laser light is absorbed by the paint so the shock has no effect. While the shock is within the paint layer, light may be partially absorbed by the paint, reflected off the shock, then absorbed further by re-traversing the paint, whereupon it can re-couple into the fiber and affect the GRIP signal. Once the shock has moved through the paint, the paint has no attenuating effect on reflection off the shock.

As a first approximation, Beer's law was used to model the absorption of light by the paint:

$$A = al(t) = a[d(t) - f] \qquad (2)$$

where $A$ is the absorption, $a$ is the absorptivity per unit length (cm$^{-1}$), $l$ is the depth of paint (cm) in front of the shock, $d$ is the distance from the tip to the shock, and $f$ is the depth of the fluid between the fiber tip and the beginning of the paint. Since it

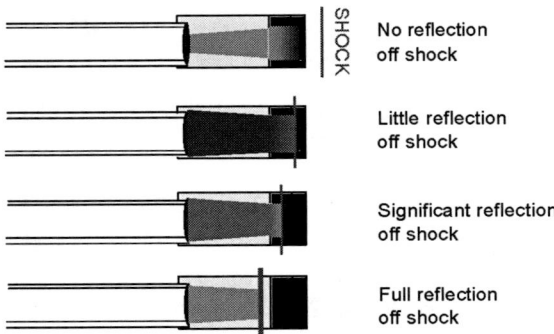

**Figure 5.** Effect of black paint on reflection off shock wave.

was difficult to measure the paint and fluid depths, $f$ was fitted to the data. The $d(t)$ was determined by setting the rapid decrease in GRIP signal as $d=0$ and using the nominal shock velocity in the gap test, 3.84 mm/μs for a gap corresponding to 6 GPa.

The model fit the dynamic data well (Figure 6) and gave $f = 0.5$ mm and $a = 1.3$ mm$^{-1}$ (meaning that 1 mm of paint would absorb 95% of the laser light), both physically reasonable values. We thus conclude that reflection off the shock front is responsible for the pre-shock signal peak observed.

## CONCLUSIONS

The Gauge using Refractive Index for Pressure (GRIP) was applied to Modified Gap Tests. Both

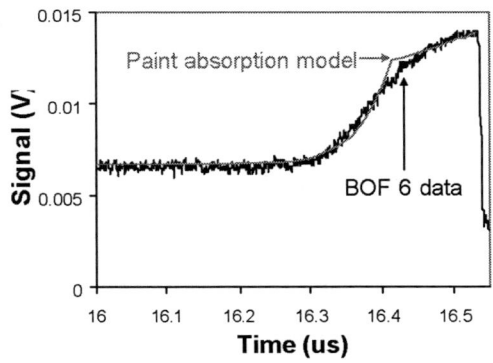

**Figure 6.** Fit of model of reflection off shock front to dynamic data (boot of fluid, nominally 6 GPa pressure).

the silicone-on-tip and boot-of-fluid applications functioned as time-of-arrival gauges. Challenges with using these as pressure gauges were identified, including the need for hydrocode modeling. A pre-shock increase in signal in the boot-of-fluid tests was hypothesized to be from reflection from the approaching shock front. Static experiments where a GRIP approached an oil-water interface and simple ray-tracing modeling demonstrated the plausibility of this explanation.

## ACKNOWLEDGEMENTS

We thank Dr. Jared Gump for static pressure tests, Gary Johns for static GRIP measurements, and Robert Hay for technical assistance. Funding was provided by NAVSEA Indian Head In-House Laboratory Independent Research program and Lawrence Livermore National Laboratory (Vlad Georgevich).

## REFERENCES

1. Pangilinan, G. I., Russell, T. P., Baer, M. R., Namkung, J., Chambers, P., *Appl. Phys. Lett.,* **77,** 684, 2000.
2. Sheffield, S. A., Bloomquist, D. D., Tarver, C. M., *J. Chem. Phys.,* **80,** 3831, 1984.
3. Monat, J. E., Carney, J. R., Pangilinan, G. I., *Shock Compression of Condensed Matter-2003,* 1281, 2003.
4. Monat, J. E., Carney, J. R., Whitley, V. H., Pangilinan, G. I., *J. Japan Explosives Soc.,* submitted for publication.
5. Erkman, J. O., Edwards, D. J., A. R. Clairmont, J., Price, D., Calibration of the NOL Large Scale Gap Test; Hugoniot Data for Polymethyl Methacrylate, Naval Ordnance Laboratory, White Oak, NOLTR 73-15, 1973.
6. Liddiard, T. P., Forbes, J. W., A Summary Report of the Modified Gap Test and the Underwater Sensitivity Test, Naval Surface Warfare Center, Dahlgren, NSWC TR 86-350, 1987.
7. Liddiard, T. P., Jr., Price, D., Recalibration of the Standard Card-Gap Test, Naval Ordnance Laboratory, White Oak, NSWC TR 65-43, 1965.
8. Staudenraus, J., Eisenmenger, W., *Fortschritte der Akustik-DAGA '92,* 301, 1992.
9. http://www.tpub.com/neets/tm/109-9.htm, Integrated Publishing, accessed October 29, 2004.

CP845, *Shock Compression of Condensed Matter - 2005*,
edited by M. D. Furnish, M. Elert, T. P. Russell, and C. T. White
© 2006 American Institute of Physics 0-7354-0341-4/06/$23.00

# THE EFFECT OF SAMPLE ROUGHNESS AND PLANARITY ON GAUGE RESPONSE TIMES

**W.G. Proud, J. Wang and D.L.A. Cross**

*PCS, Cavendish Laboratory, Madingley Road, Cambridge, CB3 0HE. UK.*

**Abstract.** This study presents a simple analytical approach for determining the significance of sample planarity on its response under extreme pressure. Plate impact experiments are carried out on Copper and PMMA targets of varying roughness and the resulting shock wave are measured with manganin stress gauges. For PMMA, Velocity Interferometer System for Any Reflector (VISAR) and high-speed photography are also used. Results show that the transit times and final stresses reached for both materials are consistent with expected values calculated from the Hugoniot. Samples angled at one degree give an increased rise time consistent with the additional distance traversed by the shock in reaching the gauge.

**Keywords:** Gauge repsonse, Experimental Procedure, Hugoniot.
**PACS:** 62.50.+p, 61.43.Gt, 07.35.+k

## INTRODUCTION

One commonly used diagnostic, in many shock studies, is the manganin gauge [1,2]. These gauges are piezoresistive and have been widely used in a variety of material types e.g. ceramics, metals and glasses [ *e.g.*3-5]. As in all experimental studies, it is important to use a robust technique and pay close attention to factors such as sample planarity, alignment and surface finish. However, when the time resolution of the gauge is taken into account, a limit exists beyond which improved sample preparation will not produce better information. This is relevant as some samples such as composites are not amenable to producing flat surfaces.

In this study, two well-characterised materials, copper (Cu101) and polymethylmethacrylate (PMMA, Perspex, ICI) are used. The gauges are micro-measurements LMSS210FD-050 option SP60. The diagnostics also included VISAR (Valyn) and, where appropriate, high-speed photography using an Ultranac FS501 camera.

By studying the response of well-characterised materials with a range of surface finishes and varying sample alignment, it is possible to check these effects on the gauge response. No attempt is made to reduce the impedance mismatch at the impact interface due to surface roughness; all impacts are conducted in a rough vacuum of 1 mbar.

The response time of a manganin gauge mounted, in a slow-setting epoxy, between plates of a sample material of differing impedance to the gauge package is approximately 200 ns. If the material is impedance matched to the gauge package e.g. PMMA, the response time is much lower, 20 ns. This difference can be explained by the stress having to rise and equilibrate in the whole gauge package in the first case and the stress equilibrating in the thin manganin element of the gauge in the second. In these experiments, all back-surface gauges are mounted in PMMA, have a nominal time resolution of 30 ns.

## EXPERIMENTAL PROCEDURE

All impacts were conducted in the plate impact facility in the Cavendish Laboratory, Cambridge. The target materials were all machined and polished to be flat and parallel to within 10 μm over the full plate width. The surface was finished with a fine grade emery paper. Having attained this degree of flatness the plate destined to be the impact plate was placed in a milling machine and a series of square cross section grooves cut into the impact surface.

In all cases the first plate, where the grooves are locate, was 2.2 mm thick, the second plate 4.9 mm thicl and the rear PMMA plate was 18.0 mm thick. The front plates varied from smooth mirror-like finish, rough front plates (grooves 0.25mm wide, 0.25 deep and 1mm apart) and very rough front plates (0.5x0.5x1mm). For each material, both smooth and very rough samples were impacted at 250, 500 and 900m s⁻¹. Further samples were deliberately misaligned by one-degree angle at low velocity. The impact geometry is shown in figure1.

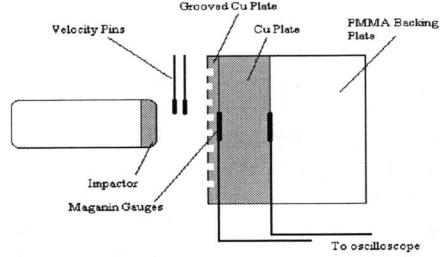

**Figure 1.** Schematic of target and impactor

For PMMA, the sides of the sample to allow high-speed photography. A 1cmx1cm square of 25 μm thick brass shim was placed next to the rear gauge, acting as a reflector for the VISAR. The measured velocities were compared with the stress curves obtained from the gauges.

## RESULTS

In figure 2, the gauge traces from copper of a rough-faced sample and a smooth-faced sample are compared, under low velocity impact. In the front gauge of the rough-face sample, ramping is seen,

however, after 5 mm more travel, no ramping is seen in the higher time resolution, rear gauge. Upon impact, the material flows to fill the grooves, figure 3 illustrates this for a very rough Cu plate. The incoming projectile is travelling at 0.25 mm μs⁻¹ so that in 1μs, approximately the ramp time, it travels half the depth of the 0.5mm grooves. To a first approximation after 1μs, a smooth plate has effectively been formed. A sharp rise in stress follows the ramp with the ultimate stress being effectively identical to that of the smooth-faced sample.

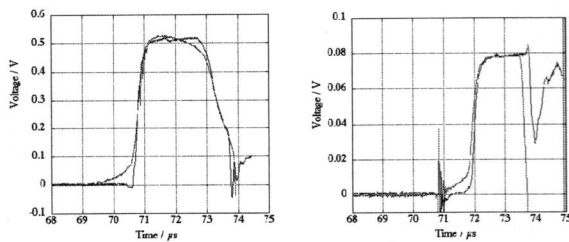

**Figure 2.** Comparison of the raw data from impact, at 250 m s⁻¹ on smooth and rough-faced copper targets. Left, comparison of front gauges, the trace from the rough sample ramps: right, rear gauges, very little difference in the traces.

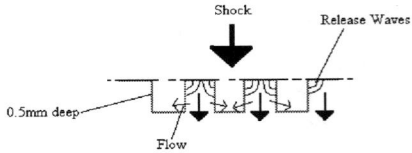

**Figure 3.** Schematic of the processes on a very rough Cu faced target under low velocity impact

For 0.5mm deep grooves, using the same logic, a ramp time of 500 ns would be expected. A value of 400 ns, was seen. While the argument presented here is a simplified explanation of a complex phenomenon but does allow a guide in assessing which of the smaller features of a gauge trace can be assigned to surface finish.

If the ramping repsonse of the front gauge traces, for rough and very rough faced samples, are superimposed, figure 4, they follow a similar form indicating a similarity in the flow process. Again the ultimate stress reached in the loading cycle was the same in both samples, the rear gauge traces were identical.

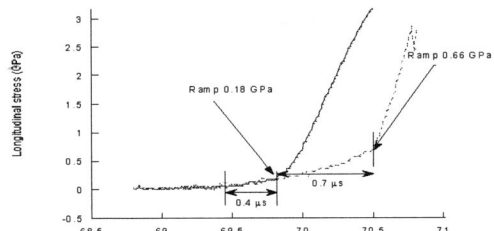

**Figure 4**. Superposition of Cu ramps. Impact at 250 m s$^{-1}$, 0.25 mm grooved and 0.50 mm groove (dotted line)

At 900 m s$^{-1}$, no ramping is seen, with a very rough-surface target: the correpsonding traces from all targets are identical. The impact velocity is so high, the distances involved so small that, given the gauge time resolution, only a single step is seen.

For PMMA samples the trend is similar. Here the time resolution of both gauges is ~30 ns. The estimation of the rise time outlined above applies but with the less ductile nature of PMMA in these conditions there is no ramp, instead a step occurs, followed by a slight decay then a second rise to the stress predicted by the Hugoniot (figure 5).

VISAR traces, obtained from the plane of the rear gauge, for the PMMA samples show exactly the same features, as seen in the gauge traces, implying the gauges are seeing a target response and not features due to gauge effects.

In figure 6, images from two high-speed sequences are shown, in both the field of view is ~50 mm x 25 mm. The shock travels from left to right. The two vertical lines, towards the outer edge of the image in 6(a) are the gauge locations. Between these two lines the shock front is visible. The width of the shock front agrees with the rise time seen in the gauge traces. The width of the rising edge of the shock front is seen to reduce with distance. Also some roughness can be seen in the wave front due ot the surface finish. Later in the sequence this roughness caused by the individual grooves reduces as the wavelets coalesce into a single sharp front.

Figure 6(b) shows the shock propagating from an 250 ms$^{-1}$ impact with an initial 1 degree angle between target and impactor. The angled shock wave can be seen sweeping through the gauge locations, this time only the shock and the second gauge location (vertical line) can be seen in the image. The angle of the shock wave is ~10 degrees

as expected given the difference in the velocity of the impactor compared to the shock speed in PMMA.

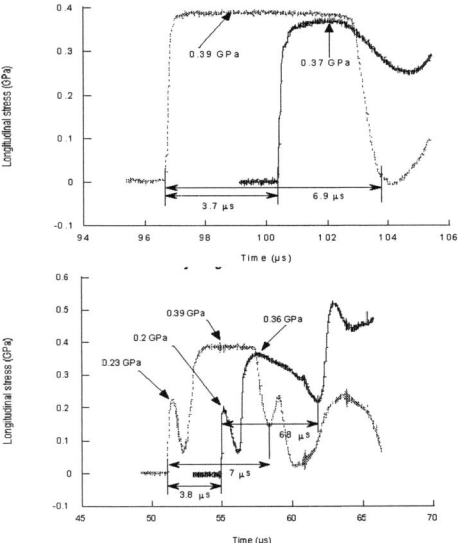

**Figure 5** Comparison of gauge response on smooth (top) and very rough (bottom) PMMA targets at 250 m s$^{-1}$.

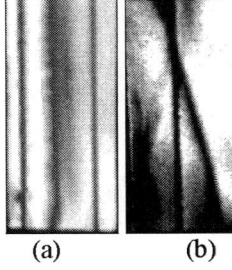

(a)       (b)

**Figure 6**. Images from a high-speed sequence of impacts on PMMA. (a) for a very rough impact surface and (b) for angled impact on smooth faced target.

The maximum recommended degree of misalignment of samples between samples and impactor is less than one milliradian. In this experimental series some samples were accurately misaligned by 1 degree. This should produce a strongly ramping signal. The effect of the angle on a smooth copper target and that of a rough PMMA target are shown in figure 7(a). As can be seen there is a noticeable difference. In the case of copper the angled gauge gives a higher response 7(b). Over the series of angled impacts with both rough- and smooth-faced samples a variety of response levels were found but all are within 20% of the level seen with smooth-faced and accurately aligned samples.

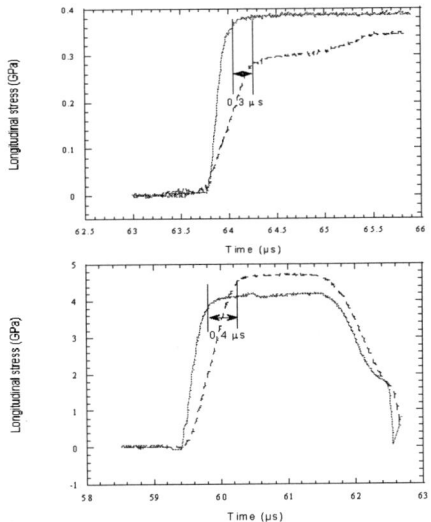

**Figure 7.** The effect of angled impact on smooth-faced PMMA (top) and smooth-faced copper (bottom)

## CONCLUSIONS

Experiments were carried out on Copper and PMMA, analysis of the traces from piezoresistive gauges showed that the transit times and final stresses reached for both materials were consistent with expected values irrespective of grooves in the impact surface. This view was further supported by results from VISAR and high-speed photography in the case of PMMA.

Ramping of the shock wave was seen at low velocity with grooved samples. Superposition of the ramps supports the view that the same process took place irrespective of groove thickness. The width of the ramp, in the gauge traces, can be estimated from knowledge of the surface roughness and the impact velocity.

For the angled impacts, the increase in rise time can be quantified by considering the tilt of the sample and calculating the extra distance traversed by the shock to reach the gauge. The value of the stress gained while not as consistent as that seen with rough surfaces was still relatively close.

Obviously some features of the shock process may be lost in the blurring of the gauge response, e.g. the Hugoniot elastic level (HEL) level due to sample roughness and misalignment. In terms of alignment and surface finish the arrangements would disgrace even the least capable experimentalist, yet the manganin gauges used still give consistent results, especially with regard to roughness. This is useful for those who are studying samples which are impossible or hazardous to polish to a high-degree of accuracy or where alignment of the gauges is difficult, or those with an interest in the collapse processes of pores under shock loading.

## ACKNOWLEDGEMENTS

ESPRC is thanked for their support of the high-speed imaging systems. William Proud is QinetiQ Senior Research Fellow at the Cavendish Laboratory, Cambridge. Prof. J.E. Field provided invaluable help during this experimental series.

## REFERENCES

1. Rosenberg, Z., D. Yaziv, et al. (1980). "Calibration of foil-like manganin gauges in planar shock wave experiments." J. Appl. Phys. 51: 3702-3705.
2. Partom, Y., D. Yaziv, et al. (1981). "Theoretical account for the response of manganin gauges." J. Appl. Phys. 52: 4610-4616.
3. Brar, N. S., S. J. Bless, et al. (1992). Response of shock-loaded AlN ceramics determined with in-material manganin gauges. Shock-Wave and High-Strain-Rate Phenomena in Materials. M. A. Meyers, L. E. Murr and K. P. Staudhammer. New York, Marcel-Dekker: 1023-1030.
4. Church, P., R. Townsley, et al. (2000). Consideration of stress gauges in the modelling of plate impacts. Shock Compression of Condensed Matter - 1999. M. D. Furnish, L. C. Chhabildas and R. S. Hixson. Melville, New York, American Institute of Physics: 1083-1086.
5. Rosenberg, Z., Y. Meybar, et al. (1981). "Measurement of the Hugoniot curve of Ti-6Al-4V with commercial manganin gauges." J. Phys. D: Appl. Phys. 14: 261-266.

CP845, *Shock Compression of Condensed Matter - 2005*,
edited by M. D. Furnish, M. Elert, T. P. Russell, and C. T. White
© 2006 American Institute of Physics 0-7354-0341-4/06/$23.00

# CALIBRATION OF COMMERCIAL GAUGES OF VARYING GEOMETRY TO MEASURE THE LATERAL COMPONENT OF STRESS

## Z. Rosenberg, N.K. Bourne[*], J.C.F. Millett[+]

*RAFAEL, P.O. Box 2250, Haifa, Israel.*

[*]*University of Manchester, Sackville Street, Manchester, M60 1QD. United Kingdom.*

[+]*Defence Academy of the United Kingdom, Cranfield University, Shrivenham, Swindon, SN6 8LA. United Kingdom.*

**Abstract.** A series of experiments have been performed which show that the response of manganin gauges in lateral orientation is dependent upon the gauge geometry. Below a lateral stress of *ca.* 3.5 GPa T-shaped gauges have a lower change in resistance than their more common grid shaped counterparts. However, above this level, T and grid gauges have a common response. A preliminary analysis indicates that the T-gauges are behaving more like an embedded wire gauge in this stress regime. Such behaviour must be accounted for when using these gauges, especially in low impedance materials shocked to low stress levels.

**Keywords**: Shock, stress gauge, gauge geometry, lateral stress, gauge calibration
**PACS**: 62.50

## INTRODUCTION

The direct measurement of lateral stresses with piezoresistive gauges has been the focus of intense research over the past twenty years. This has resulted in a better understanding of the response of the gauges to various loading conditions [1-4]. In addition to these calibration efforts, a lot of work has been done on various materials (including metals [5, 6], ceramics [7], glasses [8, 9] and polymers [10]) in order to map their dynamic strengths under one-dimensional loading conditions. This is achieved by monitoring the longitudinal ($\sigma_x$) and lateral ($\sigma_y$) stresses, and assigning the difference between the two to the compressive strengths that exist behind the shock front at high-pressures. In the

works cited above, it was shown that the yield strength of the gauge material, as well as its geometry (foil, wire) and orientation, strongly influence the relative resistance change. Thus, care must be taken in calibrating a given piezoresistive transducer before one can use it to monitor longitudinal and lateral stresses in a shock loaded specimen.

In the work presented here we report on several plate impact experiments, which were performed with two versions of commercial manganin gauges, which differ in their geometry. We found that although the two gauges were made from the same alloy, their response under lateral stress conditions was quite different in the low stress range (0 – 3.5 GPa) while it coincided at higher stresses. A possible explanation will be given for this observation.

## EXPERIMENTAL

Plate impact experiments were performed using a 50 mm bore, 5 m single stage gas gun [11]. The gauges which we compared are the 48 $\Omega$, grid-like gauge (MicroMeasurements C-951213-C) and the 25 $\Omega$, T-shaped gauge (MicroMeasurements J2M-SS-580SF-025) shown in Fig. 1.

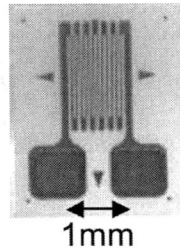

1mm

a.    Grid gauge

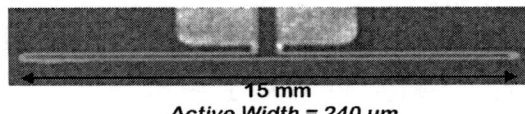

15 mm
**Active Width = 240 µm**

b.    T-gauge

**Figure 1.** Manganin stress gauges used in this investigation.

The fact that they are made from the same alloy is evident by their identical calibration curves for longitudinal stress measurements [2]. The two types of gauge were embedded in polymethyl-methacrylate (PMMA) dural (aluminium alloy 6082-T6) and soda-lime glass in order to measure their relative resistance changes under identical loading conditions. In the PMMA experiments, a 20 mm copper anvil was fixed to the back of the target to allow the stress to ring up. Various details on gauge placement, velocity, tilt measurement and data reduction are given in previous papers [5, 11, 12]. Impactor materials were polyoxymethylene (POM), dural and copper, in the impact velocity range 197 to 621 m s$^{-1}$. A schematic of the target assembly and gauge placement is presented in Fig. 2.

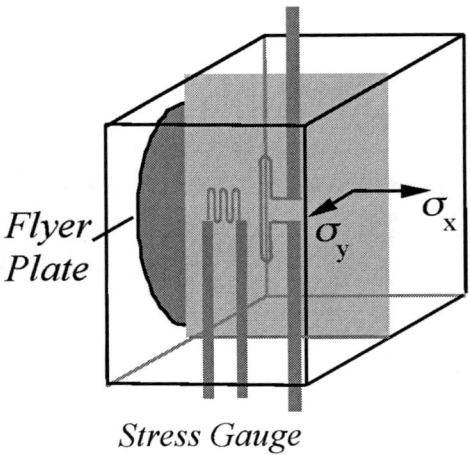

Stress Gauge

**Figure 2.** A schematic diagram showing specimen configuration and gauge placement.

## RESULTS AND DISCUSSION

In Fig.s 3 and 4 results are presented for low and high velocity shots for gauges in PMMA. 20 mm of copper was fixed to the back of the target assemblies. Therefore, as the stress reverberated between the copper flyer and anvil, it would increase in a series of steps, and thus differences between the gauge responses could be observed in a single shot. The traces have been zeroed at the same time to aid interpretation.

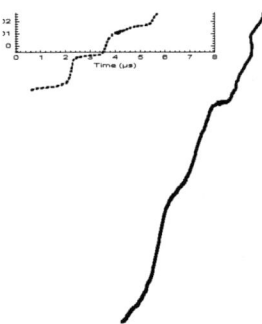

**Figure 3.** Comparison of grid and T gauge traces in lateral orientation in 6 mm PMMA. 20 mm of copper was used as an anvil, and the flyer was 20 mm copper at 197 m s$^{-1}$.

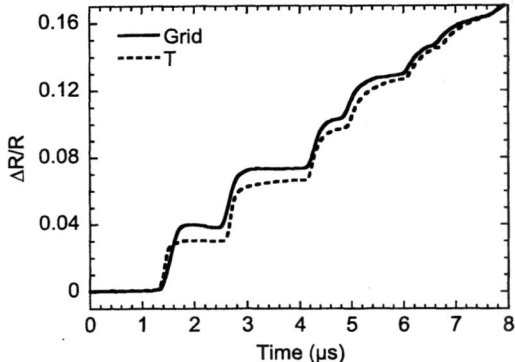

**Figure 4.** Comparison of grid and T gauge traces in lateral orientation in 6 mm PMMA. 20 mm of copper was used as an anvil, and the flyer was 20 mm copper at 390 m s⁻¹.

One can clearly see that the resulting relative resistance changes ($\Delta R/R$) are quite different in the low stress experiment (maximum $\sigma_x$ = 3.7 GPa), while for the higher stress experiment (maximum $\sigma_x$ = 7.6 GPa), the two traces clearly converge at the higher ring up stresses. This behaviour was observed in all our shots on PMMA and we found that the convergence lateral stress was *ca.* 3.5 GPa ($\Delta R/R$ *ca.* 0.1). The extraction of the lateral stress in all our experiments was done using the calibration curves for the grid gauge [5, 12], which have been checked thoroughly over a number of years. Thus, the main "surprise" in our experiments is that the T-gauge responds differently in the lower stress region. This can be seen in Fig. 5 where all our data are presented in terms of $(\Delta R/R)_{grid}$ versus $(\Delta R/R)_T$.

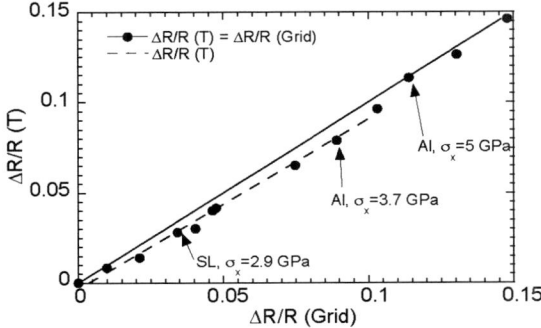

**Figure 5.** Response of the T gauge plotted as a function of the response of the grid gauge in lateral orientation.

We have pointed out additional data points were we compared the response of grid and T-gauges in dural (Al) and soda-lime glass (SL) under single shock conditions. Two lines are drawn through the data. The solid line represents an equal response of the two gauges, whilst the dotted line follows the actual response of the T-gauge. One can clearly see that the T-gauge data is offset by a certain amount from that of the grid gauge in the low stress region, while it coincides with it at higher stresses. We have included the data from some experiments on soda-lime glass and dural which show similar trends.

A possible explanation for this phenomenon can be gained by examining the calibration curve for the lateral grid gauge shown in Fig. 6 [13].

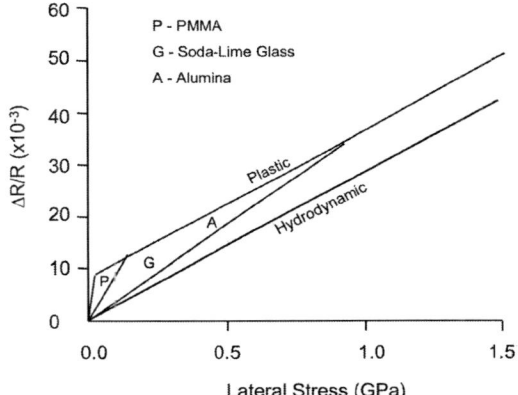

**Figure 6.** Calibration curve for the manganin grid gauge in lateral orientation.

This curve, in terms of $\Delta R/R$ versus $\sigma_Y$, consists of two parts; a straight line at lower stresses which represents the elastic response of the gauge and a curved part for the higher stresses where the gauge is beyond its yield point. The intersection of the two parts is dependent on the specimen's properties (impedance). The "plastic" part of the calibration curve is parallel to a hydrodynamic curve, which is obtained for a gauge that has a circular cross section (wire configuration). For this geometry, the stresses inside the gauge elements are different than those for a foil gauge as explained elsewhere [14]. These differences in the stresses within the gauges result in a different hydrodynamic pressure inside the gauge material, which will cause a difference in the resulting relative resistance changes. From earlier work [14], we know that the stress tensors and the pressures inside wire and foil gauges are given (for lateral stress configuration) by,

$$\sigma_{ij}^{wire} = \begin{pmatrix} \sigma_y & & \\ & \sigma_y & \\ & & \sigma_y \end{pmatrix}; P_{wire} = \sigma_y$$

$$\sigma_{ij}^{foil} = \begin{pmatrix} \sigma_y + Y_g & & 0 \\ & \sigma_y & \\ 0 & & \sigma_y \end{pmatrix}; P_{foil} = \sigma_y + \frac{1}{3}Y_g$$

where $Y_g$ is the yield stress of the gauge. Thus the difference between the pressures inside the two gauge geometries amounts to $1/3\ Y_g$, which for our gauge is about 0.25 GPa.

These observations can explain the difference between the two ranges of the T gauge response. In the lower stress range it behaves as a wire gauge with the pressure inside the gauge equal to the lateral stress in the specimen. With increasing stress (and strain), the T gauge behaves like the grid gauge (foil) with the same calibration curve for both geometries.

## CONCLUSIONS

Over the past years, we have been involved in extensive work characterising the response of manganin gauges to shock loading. This has indicated a dependence upon several factors including gauge geometry, when used to measure the lateral component of stress. In this series of experiments, we have shown that at low lateral stresses (below 3.5 GPa), a T-shaped gauge has a lower change in resistance than a grid gauge when shocked under identical loading conditions. A simple analysis suggests that in these circumstances, the T-gauge behaves as though it were an embedded wire gauge. Failure to take this into account can lead to significant errors (of order 0.25 GPa) in the lateral stress measurement and corresponding calculation of materials' properties. This error will be large at low stresses, but will diminish as stress increases. This is of particular significance in low impedance materials such as polymers. Further work is in progress to clarify this issue.

## ACKNOWLEDGMENTS.

We would like to thank Matt Eatwell, Ivan Knapp and Yann Meziere of Cranfield University for performing the shock loading experiments discussed in this paper.

## REFERENCES.

1. Gupta, S. C. and Gupta, Y. M., *J. Appl. Phys.*, **57**, 2464 (1985)
2. Rosenberg, Z., Yaziv, D. and Partom, Y., *J. Appl. Phys.*, **51**, 3702 (1980)
3. Chartagnac, P. F., *J. Appl. Phys.*, **53**, 948 (1982)
4. Rosenberg, Z., Partom, Y. and Yaziv, D., *J. Appl. Phys.*, **52**, 755 (1981)
5. Millett, J. C. F., Bourne, N. K. and Rosenberg, Z., *J. Phys. D. Applied Physics*, **29**, 2466 (1996)
6. Gray, G. T., Bourne, N. K. and Millett, J. C. F., *J. Appl. Phys.*, **94**, 6430 (2003)
7. Millett, J. C. F. and Bourne, N. K., *J. Mater. Sci.*, **36**, 3409 (2001)
8. Brar, N. S. and Bless, S. J., *High Pressure Research*, **10**, 773 (1992)
9. Bourne, N. K., Millett, J. C. F. and Rosenberg, Z., *J.Appl. Phys.*, **81**, 6670 (1997)
10. Millett, J. C. F., Bourne, N. K. and Gray III, G. T., *J. Phys. D. Applied Physics*, **37**, 942 (2004)
11. Bourne, N. K., *Meas. Sci. Technol.*, **14**, 273 (2003)
12. Rosenberg, Z. and Partom., Y., *J. Appl. Phys.*, **58**, 3072 (1985)
13. Rosenberg, Z. and Brar, N. S., *J. Appl. Phys.*, **77**, 1443 (1995)
14. Rosenberg, Z. and Partom, Y., *J. Appl. Phys.*, **57**, 5084 (1985)

CP845, *Shock Compression of Condensed Matter - 2005*,
edited by M. D. Furnish, M. Elert, T. P. Russell, and C. T. White
© 2006 American Institute of Physics 0-7354-0341-4/06/$23.00

# FAST INTERNAL TEMPERATURE MEASUREMENTS IN PBX9501 THERMAL EXPLOSIONS

**L. Smilowitz[1], B. F. Henson[1], M.M. Sandstrom[1], B. W Asay[2], D. M. Oschwald[2], J.J. Romero[1] and A.M. Novak[2]**

*[1]Chemistry, Los Alamos National Laboratory, Los Alamos NM 87545*
*[2]Dynamic Experimentation, Los Alamos National Laboratory, Los Alamos NM 87545*

**Abstract.** We have made spatially and temporally resolved temperature measurements internal to a thermal explosion in PBX9501, which is a plastic bonded explosive composed of 95% HMX and 2.5% estane mixed with 2.5% nitroplasticizer (BDNPA/F). In order to study the evolution of ignition in a thermally treated piece of explosive, we have pushed the time resolution of several different temperature diagnostics. In this paper, we will discuss the details of the time response of these diagnostics including temperature uncertainties. The temperature measurements are made both by thermocouples with corrections applied to compensate for the thermocouple response time and with optical pyrometry. An additional goal of adding high energy radiography diagnostics to future experiments has motivated an effort to synchronize thermal explosions to an external clock. In this paper, I discuss our current capabilities for controlling and measuring the development of an ignition within a piece of heated PBX9501.

**Keywords:** Thermal explosion, PBX 9501, optical pyrometry, laser ignition
**PACS:** 07.20.–n, 82.33.V, 78.20.Ci

## INTRODUCTION

The radial thermal explosion experiment was designed to control ignition location[1-3]. Our goal for this series of experiments was to study the transition from heating in place to ignition propagation through heated HE. To this end, we modified our thermocouple based temperature measurements for faster time response and added fiber optics for pyrometric temperature measurements. A future goal for these experiments is to add a high energy radiation imaging technique such as x-ray or proton radiography to get information on morphology prior to ignition and burn propagation, and coupling to metal walls post-ignition. These techniques require synchronization of the thermal explosion with the time clock of the radiation diagnostic. To this end, we have worked out a laser ignition technique using an additional

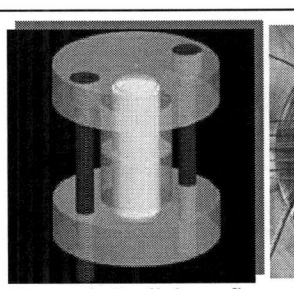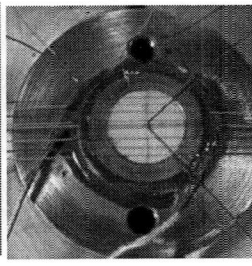

Figure 1: Radial configuration and midplane diagnostics. The aluminum casing on the shot configuration is shown with partial transparency so that the midplane location is visible.

fiber optic embedded in the explosive to provide a laser pulse at the central ignition volume. In this paper, I will discuss the capabilities and limitations of the different temperature measurement techniques and present initial results on synchronizing a thermal runaway event.

## EXPERIMENTAL SETUP

The radial experiment is shown in figure 1. The left panel shows the outer aluminum casing with partial transparency so that the internal cylinders of PBX 9501 are also visible. The right panel shows the diagnostics placed at the midplane of the HE. These radials are conducted with half inch diameter half inch tall halves held together by the aluminum casing. There is one joint in the aluminum casing which is sealed with a combination of RTV and JB Weld to minimize gas loss. Success in sealing the experiments is judged by the consistency of the time to ignition from shot to shot and agreement with time to ignition as a function of temperature measured by others[1-4]. These shots are heated from the outer aluminum casing using resistive heating wire. They are designed with enough ullage at the ends to allow for the difference in thermal expansion between the HE and the aluminum and for the volume expanding $\beta-\delta$ phase change. The heat profile used is a heat to 70C to balance the halves, followed by a 5C/minute ramp to 178C and a soak at this temperature for 20 minutes to allow the $\beta-\delta$ phase transition to complete and the material to re-

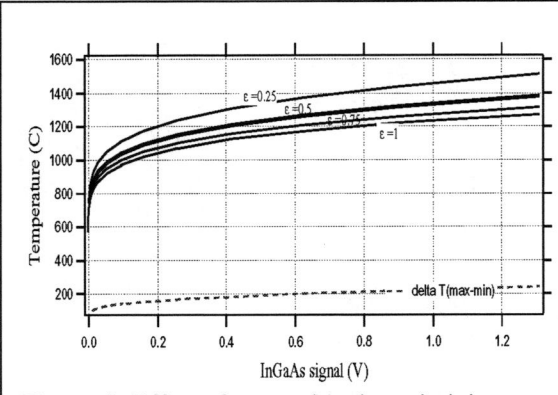

**Figure 2:** Effect of uncertainty in emissivity on calibrated temperature from fiber optic.

equilibrate at the boundary temperature. After 20 minutes, there is a final ramp of 5C/min to 205C and the boundary is held at this temperature until ignition (for example, see Figure 4). During this soak, the PBX 9501 begins generating heat and the aluminum boundary becomes a heat sink driving ignition to the spot furthest from the walls to the center of the piece. The midplane of the piece is outfitted with an array of temperature sensing diagnostics (Fig 1) including type K thermocouples and 200 micron core silica fiber optics.

## DIAGNOSTICS

The thermocouple voltage outputs are measured directly using a Tektronix 500MHz scope as an analog to digital converter with a 20 microsecond time base. Our laser heating measurements of these thermocouples in air and in water give us a response time of between 1.5 ms and 10 ms depending on the thermal conductivity of the surrounding medium[5-7]. We use this range to deconvolve the temperature measurements in the ignition volume and back out actual HE temperatures. The fiber optic output is coupled to an InGaAs amplified photodiode with a response time of 100ns. The gain setting and fiber diameter chosen give us a minimum temperature sensitivity of 800K. We calibrate the fiber output assuming an emissivity of the thermally damaged PBX 9501 of 0.625 ± 0.375. This uncertainty in emissivity correlates to a temperature uncertainty of ± 100K for peak temperatures measured at 1500K as shown above. The emissivity of pristine PBX 9501 at room temperature is 0.7. The emissivity certainly changes with the thermal decomposition, but the range of 0.25 to 1.0 is chosen to bracket the emissivity between that of a metal and that of a perfect blackbody. The impact of this scale of temperature uncertainty is illustrated in Figure 3. While a 100K uncertainty at low temperatures would limit their utility in pinning down a reaction mechanism, an uncertainty of 100K in the measured temperature regime of 1500K still provides a constraint on reaction rate. Figure 3 shows the effects of this temperature uncertainty on constraining the reaction mechanism on an Arrhenius-type log time versus inverse temperature plot. The temperature error bars are shown as the

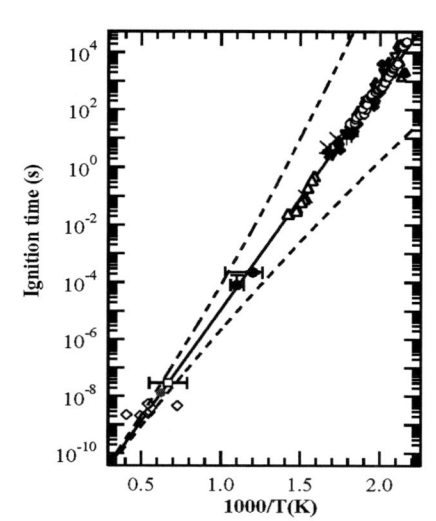

**Figure3:** k(T). HMX ignition time vs inverse temperature. Solid line is a single arrhenius rate, the data are a collection of data taken from different experiments reported in the literature[3]. The dashed lines are calculations using the 100K temperature uncertainty. The red dot shows the position for 1600K

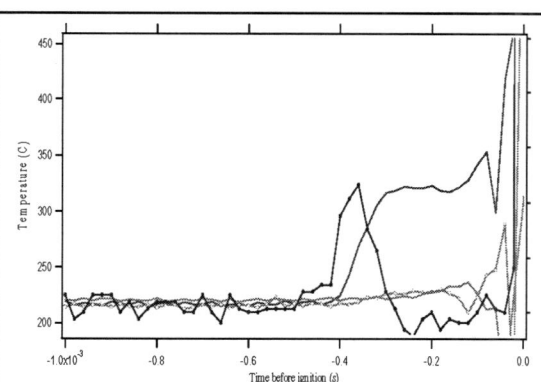

**Figure 4:** Laser synchronization. Black line shows laser signal. Red line is the laser heating on the thermocouple nearest the fiber. Time zero is ignition breakout seen as final rise in all thermocouple traces.

dashed line above and below the data line. The red point is the measured high temperature.

## LASER SYNCHRONIZATION

Both our experimental and modeling results have pointed to an exponentially steep spatial gradient in temperature analogous to the exponentially steep temporal gradient set up at the ignition volume[4,7]. To gain more insight into the evolution of the central ignition volume, we would like to perform proton radiography to image through the experiment at times just prior to and subsequent to the break out of ignition. The difficulty with applying a high energy imaging diagnostic such as proton or x-ray radiography to a thermal explosion is the inability to pretrigger these diagnostics. They require the shot to be synchronized to the radiation source. In order to utilize these diagnostics, we have worked on a laser synchronization method designed to minimally perturb the evolution to self-ignition in a thermal explosion while allowing synchronization within the time window of the proton beam. Initial work using a 5-7 nanosecond Nd:YAG did not allow us to couple enough light into the HE due to the fiber damage threshold limitations. One solution would be to use a larger diameter fiber, but this both makes the fibers more difficult to handle due to worse bend performance, and makes the fiber a larger perturbation. Instead, we opted for a microsecond pulse train to allow us to couple in larger total energies. We have chosen a trigger

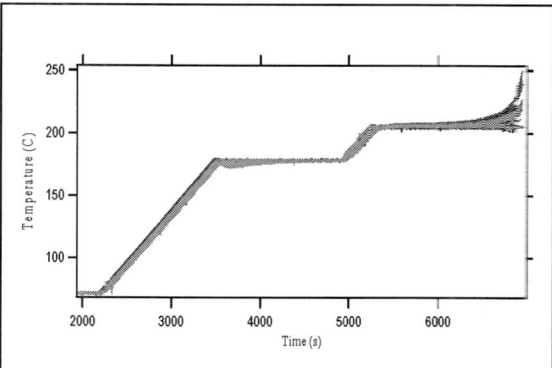

**Figure 5:** Comparison of a selfignition profile with two laser ignitions. Red is self ignited, green and blue are laser ignited.

temperature of 220C which is approximately one minute prior to the self ignition time of the explosive (see Fig. 5). Once we see this temperature on the central thermocouples, we use an SRS DG535 pulse generator to provide a TTL pulse to the flashlamps of a Spectra-Physics Indi Laser run in free run mode. This produces a series of pulses approximately 160microseconds long with each pulse in the train being approximately 100ns duration and a total energy coupled through the fiber or 25 mJ. The tradeoff here is between the desire not to perturb the thermal explosion and the need to laser trigger before the shot autoignites. The laser is meant to provide a temperature jump at a small volume near the central ignition location just before it would have autoignited in order to synchronize the ignition breakout time with diagnostics. We have successfully synchronized 3 thermal explosions using 25mJ pulses igniting the explosions when central temperatures reached 220C (approximately 1-2 minutes before self ignition would have occurred). The ignition delay ranged from 250microseconds to 360microseconds with one source of uncertainty here caused by the drift in laser power (25mJ-29mJ) coupled in to the fiber. Figure 4 shows the timing of the laser pulse, shown in black, and the subsequent heating of the nearest thermocouple, red, with time zero being the ignition time.

## DISCUSSION

The goal of our laser ignition work was to allow the HE to self heat until a time as close to ignition as we could reliably reach. When we allow the radials to run all the way to self-ignition, we typically see self heating above the boundary temperature begin at the HE nearest the outer walls and then propagate inward as the thermal excursion increases above the boundary and the boundary continues to sink the HE temperature. This period of HE self heating above the boundary lasts for about 28 minutes with the final hundred seconds exhibiting an exponentially steep rise in temperature at the center most thermocouples. Maximum temperatures typically observed at the central ignition range from 230C to 250C depending on the exact location relative to ignition in both space and time. The laser pulse provides a

temperature jump at a small volume near the central ignition location prior to autoignition in order to synchronize the ignition breakout time with diagnostics. We use 220C as the temperature fiducial for providing a laser pulse. Figure 5 compares the full temperature profiles for 3 shots- one of which is allowed to autoignite and the other two triggered manually as described above. The steep spatial gradients set up in this shot configuration mean that the burn propagation through the HE should be the same for the laser ignited shot as it would have been had the central volume been allowed to continue the last minute to self ignition.

Future work includes mapping out the ignition delay/ central temperature phase space to understand the tradeoff between making the experiment more robust (igniting further in time before self-ignition would occur) and less perturbing from the self-ignition trajectory of the material.

## ACKNOWLEDGEMENTS

Funding was provided by the High Explosive Sciences and HE Surety Programs at Los Alamos National Laboratory

## REFERENCES

1. Dickson, P.M., B.W. Asay, B.F. Henson, C.S. Fugard, J. Wong, AIP Conference Proc. SCCM-1999, ed. M.D. Furnish et al. AIP Conf. Proc.505, NY, pp. 837-840.
2. Kaneshige, M.J., A.M Renlund, R Schmitt, R Erikson, AIP Conference Proc. SCCM; 2004; v.706, p.351-354
3. Henson, B.F., B.W. Asay, L. Smilowitz,P.M.Dickson, AIP conf Proc. SCCM 2001, ed. MD Furnish et al, AIP Conf. 620, pp1069-1072.
4. Henson, B.F., L. Smilowitz , B.W. Asay,P.M. Dickson, D. Zerkle, 40th JANNAF Combustion Subcommittee Meeting, Charleston, SC (2005)
5. Asay, B.W., S.F. Son, P.M. Dickson, L.B. Smilowitz, B.F. Henson, Propellants Explosives Pyrotechnics; Jun 2005; v.30, no.3, p.199-208
6. Kardos, P.W., Chem Eng., 1977; v.84, p79-83
7. Smilowitz, L., B.F. Henson, M.M. Sandstrom, B.W. Asay, D. Zerkle J.J. Romero, A.M.Novak, P.M. Dickson, D.M. Oschwald, 40th JANNAF Combustion Subcommittee Meeting, Charleston, SC (2005)

CP845, *Shock Compression of Condensed Matter - 2005,*
edited by M. D. Furnish, M. Elert, T. P. Russell, and C. T. White
2006 American Institute of Physics 0-7354-0341-4/06/$23.00

# STRESS-INDUCED RESISTIVITY OF PADDED LATERAL STRESS GAUGES

## R E Winter and E J Harris

*Hydrodynamics Department, AWE, Aldermaston, Reading, Berkshire, RG7 4PR, UK*

**Abstract.** In principle, the strength of a material subjected to a plane shock can be estimated by measuring the stress developed in a plane perpendicular to the shock front. High resolution hydrocode calculations of manganin gauges mounted laterally in an elastic / perfectly plastic steel matrix have been conducted. The simulations, run with various thicknesses of strength-less polymer mounting layer, show how the stresses and strains at different positions in the gauge element vary with time. Time plots of the individual stresses in the gauge reveal that, in the post-transient phase, the three residual orthogonal stresses in the gauge are surprisingly similar, indicating near-hydrodynamic conditions. Making the simplification that the gauges are free of shear stress allows a relationship between the gauge resistivity and the pressure in the polymer layer in which the gauge is mounted to be derived. This approximate analysis gives resistivities which are a close match to those derived from simulations run on a series of configurations.

**Keywords**: Stress gauges, lateral gauges, manganin gauges
**PACS**: 62.50; 07.05.Tp

## INTRODUCTION

The process of deducing the shear strength of a material from the measured resistance change of shock-loaded lateral gauges may be divided into two main stages. The first stage, and the main subject of this paper, is to deduce the stress-strain state of the gauge, (as distinct from the stress state in the sample), from its measured resistance change. In this paper the analysis is simplified by assuming that the polymer layer in which the manganin element is mounted behaves as a fluid. With this proviso the problem reduces to deriving the pressure in the polymer from the resistance change of the gauge. The second part of the unfolding procedure, which will be addressed in a future paper, is to use the pressure in the polymer to deduce the stress in regions of the sample which are far enough away from the gauge to be unaffected by it (termed here the "far-field stress").

As indicated in Fig. 1 the gauge is envisaged as divided into cells which consist of thin fibres of conductor oriented along the z direction. It is assumed that the length, L, of the gauge does not change.

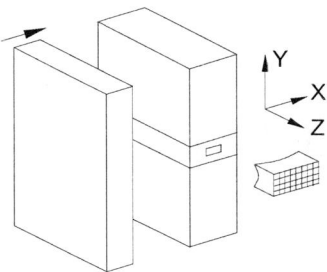

**FIGURE 1** A schematic of the lateral gauge configuration.

Following Feng and Gupta [1] the total resistivity change of a cell due to changes in the stress field is given by:

$$\frac{\Delta\rho^{stress}}{\rho_0} = \pi_{11}\sigma_z + \pi_{12}\sigma_x + \pi_{12}\sigma_y \qquad (1)$$

where $\rho^{stress}$ is resistivity, $\sigma_x, \sigma_y$ and $\sigma_z$ are x, y and z stresses and $\pi_{11}$ and $\pi_{12}$ are the piezoresistive coefficients.

It is evident from Eqn. 1 that, even if the contribution made to resistivity by stress is known, there is insufficient information to derive the individual orthogonal stresses. However it is possible that in an actual gauge the stresses will be inter-related in a known way, thus permitting any component stress in the gauge to be derived from a knowledge of resistivity change.

The purposes of this paper are to build a picture of the processes within a lateral manganin gauge by running hydrocode simulations of a range of configurations and thereby to assess whether the stresses in the gauge, and therefore, from Eqn. 1, the stress-dependant resistivity, can be simply related to the pressure in the polymer in which the gauge is mounted.

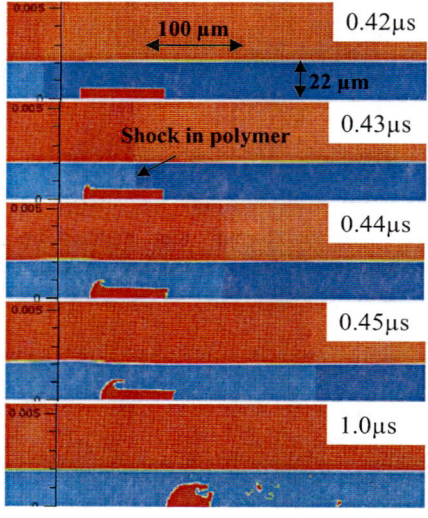

**FIGURE 2** Shamrock sequence showing the deformation of a 2mm/22μm gauge. The gauge continues to deform after the shock wave has passed.

## SIMULATIONS

The configuration is based on that used in an experimental study conducted by Hammond et al, [2]. The AWE Automatic Mesh Refinement (AMR) Eulerian code Shamrock was used to model manganin wires of width 45μm and half-thickness

6μm, mounted in polymer layers of different thicknesses. In the notation used in this paper gauges mounted 2mm and 5mm from the impact surface in a 172μm half-thickness polymer layer (for example) are referred to as "2mm/172μm gauges" and "5mm/172μm gauges" respectively. The shear strength assigned to the steel matrix was also varied. In all of the simulations, shocks were generated by a copper plate impacting at 400m/s.

The time sequence of density plots in Fig. 2 shows the distortion of a 2mm/22μm gauge. In the first frame, at 0.42μs from the instant of impact, the shock wave generated in the polymer layer by the impact is seen approaching the gauge. By 0.43μs the wave is just over half way along the gauge and severe "mushrooming" distortion of the front edge of the gauge is evident. The "petals" at the head of the mushroom continue to distort after the shock wave passes through.

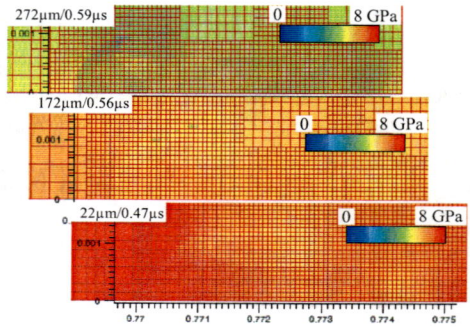

**FIGURE 3** Pressure maps for gauges in 272μm, 172μm and 22μm layers. Pictures are aligned vertically to show the relative horizontal displacement of the gauges. The gauge outlines are in the same positions as in figure 4 but are difficult to distinguish because the pressures in the gauge and the polymer are similar.

Figures 3 and 4 show the effect of polymer thickness on the calculated deformation. The times of the three frames in each of these figures are chosen to show the state of the gauges 0.04 μs after the shock in the polymer layer reaches the centre of the gauge.

The pressure maps in Fig. 3 show that, for all three thicknesses, the pressure in the polymer is very similar to the pressure in the gauge suggesting that the pressure in the gauge can be regarded as equal to the pressure in the polymer. Further, the average pressure is higher when the gauge is mounted in a thinner polymer layer.

The depiction shown in Fig. 4 uses Eqn. 1 to map the calculated contribution made to the resistivity by stress. Note that the resistivity in the centre of the gauge appears fairly typical of the whole gauge. It is seen that resistivity is higher for gauges in thinner polymer.

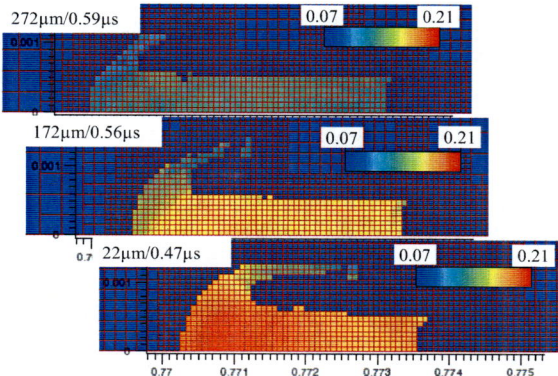

**FIGURE 4.** Maps of stress-induced resistivity for 3 polymer thicknesses.

## SIMPLIFYING THE STRESS STATE

Plotting the stress state of a cell in the centre of the gauge provides an estimate of the stress state of the whole gauge.

Figure 5 shows Shamrock results for a cell near the centre of a 2mm/172μm gauge run with 1.8GPa strength in the steel. Pressure, ($p^{gauge}$), and x, y and z stresses are shown. It is seen that, in the first ~0.1μs, the x stress is significantly greater than the y or z stresses. At later times, however, the traces become parallel, indicating residual, constant, shear stresses in the gauge. Knowledge of the three orthogonal stresses in the cell, together with the piezo-resistive coefficients allows the contribution that the stress state of the cell makes to its resistivity to be calculated. The following piezo-resistive coefficients of manganin, (in GPa$^{-1}$), were determined by Feng, Gupta and Wong [3].

$$\pi_{11} = 0.01121 - 0.00009556 p^{gauge}$$
$$\pi_{12} = 0.004108 \,.$$

In Fig. 6 the trace marked "$\pi_{12}\sigma_x + \pi_{12}\sigma_y + \pi_{11}\sigma_z$" shows the resistivity change obtained in the 2mm/172μm gauge by using the stresses determined by SHAMROCK.

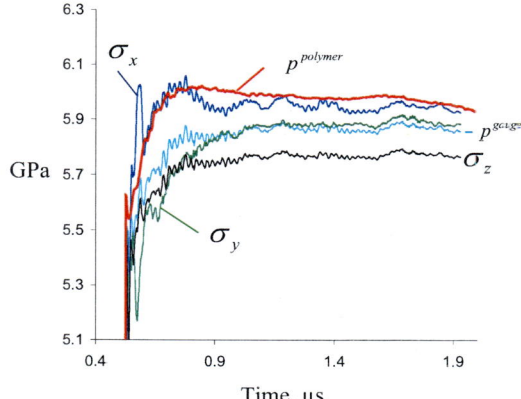

**FIGURE 5.** Pressure and x, y and z stress profile in the centre of a 2mm/172μm gauge run with matrix strength of 1.8GPa shown for comparison with a pressure profile from a polymer-only calculation.

The observation that, after the first few tenths of a microsecond, the x, y and z stresses seemed very similar to each other, and therefore, to the pressure in the cell, suggests that the derived resistance would not change significantly if it was assumed that the stress in the cell was hydrodynamic. In other words if:

$$\sigma_x = \sigma_y = \sigma_z = p^{gauge} = \frac{\sigma_x + \sigma_y + \sigma_z}{3} \,.$$

This approximation allows the pressure in the cell to be linked to its pressure-dependant resistivity, $\frac{\Delta\rho^{approx}}{\rho_0}$. From Eqn. (1) the contribution that stress makes to the resistivity of the cell becomes:

$$\frac{\Delta\rho^{approx}}{\rho_0} = \pi_{11} p^{gauge} + 2\pi_{12} p^{gauge} \qquad (2)$$

The line in Fig. 6, labelled "$(\pi_{11} + 2\pi_{12})p^{gauge}$", shows the resistivity obtained by using Eqn. (2). It is seen that with the 2mm/172μm gauge the hydrodynamic approximation overestimates the resistivity change by ~0.0006, which is ~0.5% of the resistance change caused by stress at a loading pressure of 6.5GPa.

The ideas presented above may be taken a stage further. Calculations have been run with the gauges absent and pressure profiles obtained at the positions corresponding to the gauge locations. These are much quicker than calculations run with the gauge included. The "polymer" trace in Fig. 5 shows the result of a 2mm/172μm polymer-only calculation. It is seen that the pressure in the polymer, ( $p^{polymer}$ ), is higher than that in the gauge.

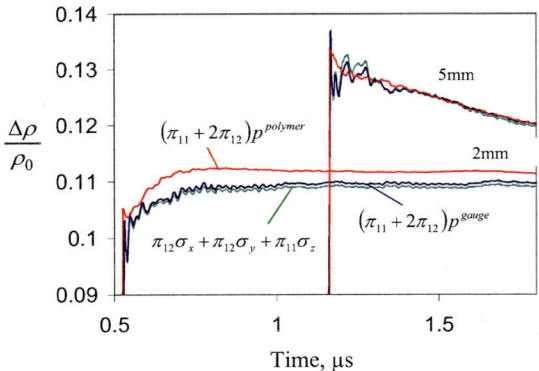

**FIGURE 6.** Resistivity changes for 2mm/172μm and 5mm/172μm gauges.

The line labelled " $(\pi_{11}+2\pi_{12})p^{polymer}$ " in Fig. 6 shows the resistivity change obtained by assuming that all the stresses in the gauge are equal to the *polymer* pressure at the corresponding point in a polymer-only calculation. In this case the "polymer" approximation overestimates the resistivity change by ~0.002. At a loading pressure of 6.5GPa this corresponds to an overestimate of ~2%.

The analysis described above for the 2mm/172μm gauge has been repeated for eight other gauge configurations and different matrix strengths. In general it was found that as the thickness of the polymer layer was increased the stresses became more widely spread and averaging the stresses introduces a bigger error. For example with the 2mm/272μm gauge setting the stresses equal to the gauge and polymer pressure overestimates the resistivity change by 0.0015 and 0.005 respectively. With gauges in polymer thinner than 172μm the approximation becomes more accurate.

Based on the limited number of configurations run to date we conclude that Eqn. 2 provides a reasonably accurate estimate of the resistivity change caused by the stress in a lateral manganin gauge during the period after the transient stresses have died away.

## CONCLUSIONS

Simulations enable the evolving stress/strain distribution in a lateral gauge to be determined and it is found that the stresses and strains in the manganin element depend on factors such as the thickness of the polymer and the strength of the matrix. Further it is found that during the few tenths of a microsecond from the time at which the shock in the polymer reaches the gauge, the gauge is subject to varying deviatoric stresses. On the other hand, at later times a stable pattern of residual deviatoric stresses becomes established.

An obvious simplification is to assume that the stress state in the gauge is free of deviatoric stresses, with a time dependant pressure equal to the pressure that would have been generated at the gauge position in a configuration with the polymer layer but without the gauge. It has been shown that for eight gauge configurations simulated, the average error in resistivity change associated with making this simplification would be ~0.3%. Therefore, over the range of pressures studied, we consider that Eqn. 2 provides a good method of relating the stress component of the resistivity of manganin in a lateral gauge to the pressure in the polymer in which the gauge is mounted.

## REFERENCES

1. Feng R and Gupta Y M  1998, "Determination of lateral stresses in shocked solids: Simplified analysis of piezoresistance gauge data", *J Appl. Phys.* 83 (2), pp 747-753
2. Hammond R I, Church P D, Grief A, Proud W G, and Field J E, "Dependence of measured lateral stress on thickness of protective padding around gauge," pp. 1125-1128 in M. D. Furnish, Y. M. Gupta and J. W. Forbes (eds), *Shock Compression of Condensed Matter* – 2003, AIP Press, 2004.
3. Feng R,. Gupta Y M, and Wong M K W 1997, "Dynamic analysis of the response of lateral piezoresistance gauges in shocked ceramics" *J. Appl. Phys.* 82 (6), pp 2845-2855

CP845, *Shock Compression of Condensed Matter - 2005*,
edited by M. D. Furnish, M. Elert, T. P. Russell, and C. T. White
© 2006 American Institute of Physics 0-7354-0341-4/06/$23.00

# AN AUTOMATED TEST BED FOR VISAR PROBE CHARACTERIZATION

## T. R. Salyer*, N. S. Khalsa* and L. G. Hill*

*Los Alamos National Laboratory, Los Alamos, New Mexico 87545 USA*

**Abstract.** Accurate characterization of VISAR probes is helpful for their effective fielding on a given experiment. Much stands to be gained through optimal placement and choice of probe as well as optimal target surface preparation. Revelations through a series of dynamic shots can be time consuming, expensive, and inefficient. An automated system to measure probe illumination and return characteristics independent of the VISAR helps to alleviate these problems. Motion of a target reflector is simulated via linear traverses and a rotation stage. Laser illumination is provided to yield a probe response measured with sensitive optical power detectors. A beam profiler is used for 2-D analysis of the illumination spot over the full range of target travel. Furthermore, the whole system is automated through LabVIEW software control. A proposed standardized probe test consists of the 1-D axial response, sensitivity to target angle, sensitivity to target surface material and preparation technique, and the illumination beam characteristics. As the community recognizes the need for more specialized probes, such a tool enables the rapid development of new designs as well as the cataloging of current ones.

**Keywords:** VISAR, probe, characterization
**PACS:** 42.82.Bq, 42.87.-d, 07.60.-j

## INTRODUCTION

For multi-dimensional VISAR problems, fielding a probe to accommodate a long range-of-motion target with simultaneous tilt and possible surface degradation can be especially difficult. To improve data collection, a number of probe factors must be optimized including the type, the standoff, the angle to the target, and the surface preparation of the target itself. In practice this process is largely trial and error, and can be cumbersome if optimizations are incrementally reached shot after shot.

For targets whose reflectivity does not change drastically during a given shot, such as annealed pure copper tubes used for HE cylinder tests [1, 2], probe performance is mostly dictated by geometrical aspects of the target motion. Such geometrical optimizations are especially important for tests with poor reflectivity such as LANL HE sandwich tests [3, 4] with tantalum confinement.

To assist, an automated system has been developed to determine probe performance for a given target motion and surface preparation, both of which can be tailored to mimic actual shot parameters. The information is collected efficiently, so probe fielding can be optimized prior to a shot, nondestructively. With a full barrage of probe characterization tests (including illumination beam profiles), the anticipated response of the probe to any target motion or surface finish can be estimated with confidence.

For target surfaces that degrade due to processes such as material spall or stretching, probe response estimations based on pure geometrical target motion will vary from actual shot data, but may still indicate best-case performance in instances where the surface reflectivity degrades mostly upon initial shock. Furthermore, dynamic surface reflectivity changes can be quantified fairly accurately through the analysis of differences between the pre-shot probe tests and the final shot results.

## AUTOMATED TEST BED

The probe test bed has the capability of simulating complex wall motions like those encountered during multi-dimensional VISAR tests. The user facility requires minimal effort to accomplish a full probe characterization, and is automated through the use of customized LabVIEW control software.

The system is a compact desktop instrument. Primary hardware includes a laptop with controlling LabVIEW software, a sensitive optical power meter with twin detectors, a three-axis motion controller with linear and rotational actuators, a laser with power supply, a laser safety shutter with controller, and an optics breadboard with miscellaneous opto-mechanical parts under a light tight enclosure.

Details of the optics breadboard are shown in Figure 1. Light from a non-stabilized 1 mW green (543.5 nm) polarized HeNe laser is launched into a 50/125 $\mu$m glass multimode fiber that leads to a 90/10 fiber intensity splitter. The smaller portion of split light is passed to a silicon photodetector for monitoring fluctuations in the laser light intensity. The larger portion of light is sent to the probe, which is located in the inner chamber, separate from the stray light of the fiber beam launch. Light from the probe impinges on a target that is mounted to a set of electronically actuated translation and rotation stages. Two translation stages affect motion in the plane of the breadboard, and a single rotation stage tilts the target about the axis normal to the breadboard. In this manner, a complex target motion can be simulated, such as that from a cylinder test. The light returned from the probe is then passed through a fiber to another silicon photodetector for measuring the probe light return.

The LabVIEW software was developed to enable straightforward measurements of the probe light return. From the main LabVIEW control panel, the entire experiment can be controlled and set up for automated data acquisition. A target trajectory file is first loaded into the system, and then plotted for path verification and real-time target tracking. Next, a simple initialization process sets up the motion controller and adjusts settings on the optical power meter specific to the experimental parameters. Once the experiment begins, the system automatically traverses the target along its trajectory, averaging probe return and beam intensity monitor data at each point. Real-time data from each photodetector is plotted along with a

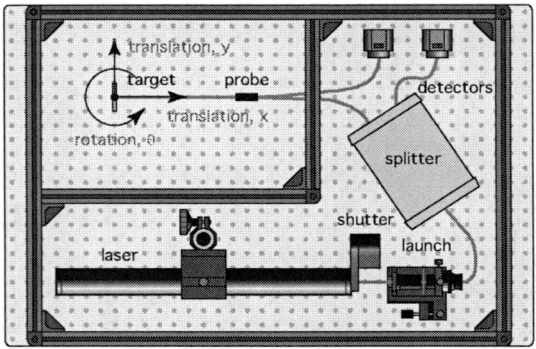

**FIGURE 1.** Optics breadboard schematic.

scaled ratio that indicates the response of the probe after the laser power fluctuations have been removed. Note that the experiment can be paused at any time. When the trajectory is complete, the results may be saved.

Additional control panels are used for adjusting the settings which govern the automated process. One control panel is for manually jogging the motion controllers. This is useful during the experimental setup phase when the probe is being aligned, or for close inspection of a particular area along the trajectory. Another panel is used to manually change any of the settings on the optical power meter, or to control the beam shutter or manually zero the detectors.

The last feature of the test bed is a commercial beam profiler, which is used to determine the illumination beam characteristics. The profiler consists of a sensitive camera placed in the path of the beam, and separate control software to acquire and analyze the data for determining beam profile parameters. The profiler can be translated along the path of the illumination beam for a full 3-D beam analysis over the region of interest.

## STANDARDIZED PROBE TEST

In general, probe performance is determined by the following criteria: the depth-of-field, the integral of light returned over the full range of target motion, the dynamic range of the returned light intensity, how well the return light matches the VISAR numerical aperture, and the quality/size of the illumination spot. The primary performance characteristic of any given VISAR probe is its light return efficiency, though the illumination spot qualities have a large effect on multi-dimensional VISAR problems where the initial shock moves transversely across the spot. Also, the minimum allowable spot size can be intimately linked to small-scale target surface features, thus supporting the optimization of the probe/target combination as a whole.

To effectively compare different types of probes, a standardized probe test has been proposed. The full test consists of the 1-D axial response, the sensitivity to target angle, the sensitivity to target surface material and preparation technique, and the illumination beam characteristics. Based on these criteria, the overall performance of a given probe can be well characterized. Gathering a large set of probe results will yield a complete catalog, thus allowing the most appropriate design to be selected when tailoring VISAR diagnostics for any given experiment.

Note that an ideal probe must perform well in all aspects, and must be ideally matched to the experiment under consideration. Though most probes fail to perform well in all aspects of performance, some are certainly better suited for particular applications. In some experiments, light collection efficiency may be most desired, whereas in others the depth-of-field may be more important. Typically, there are trade-offs between these features.

## PROBE CHARACTERIZATION

To illustrate the capabilities of the test bed, a commercial Valyn probe with a 30 mm target focus (#FOP-1000) was characterized with the proposed standardized probe test. Figure 2 depicts the 1-D axial response of the probe as well as the sensitivity to target angle using diffuse aluminum foil as a standard surface for targets out to 60 mm. The peak response off the standard surface is approximately 18%, and the probe goes blind with the target positioned within roughly 20 mm of the probe. A blockage of the returned light due to the gradient index (GRIN) illumination lens causes this blindness. Also note that the measured internal back reflections are insignificantly small. The probe sensitivity is halved at roughly 5 degrees of target tilt for the chosen standard surface, but will vary with differing target diffuseness.

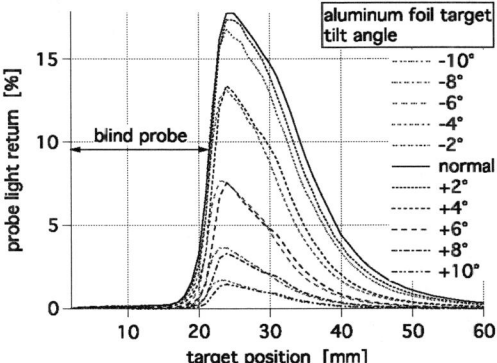

**FIGURE 2.** Sensitivity to target angle.

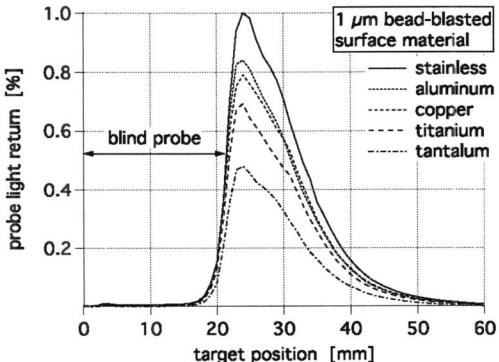

**FIGURE 3.** Sensitivity to target material.

The next tests determine the sensitivity of the probe to various target materials and surface preparations. Figure 3 shows the probe sensitivity to several typical target materials, each prepared with the same 1 $\mu$m silica bead-blast. Tantalum and titanium surfaces have very poor reflectivity as expected. Unfortunately, materials like tantalum are frequently necessary for HE confinement tests [3, 4].

Figure 4 depicts the response difference between two different surface preparations of aluminum. One surface is the standard dull foil, while the other was prepared with a 1 $\mu$m silica bead-blast. The specific alloys should be inconsequential to the results. Clearly the bead-blasted surface is more diffuse, but as a result will have better sensitivity to tilted targets. The ratio of peak responses is 21.2, which indicates the large role of surface preparation.

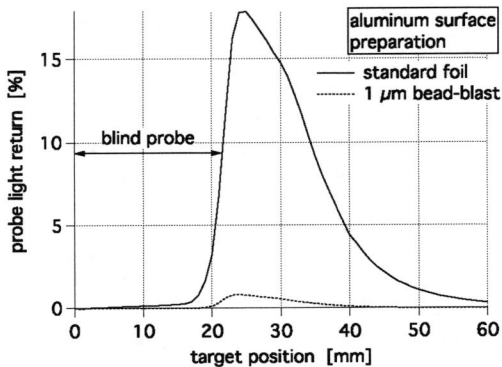

**FIGURE 4.** Sensitivity to target surface preparation.

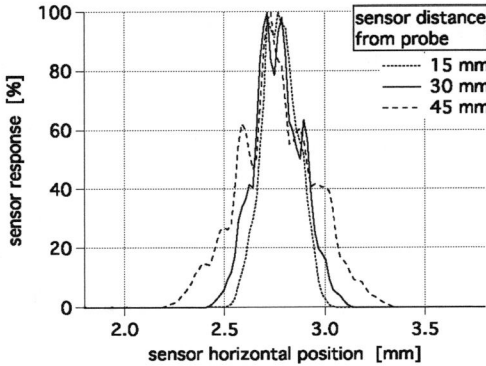

**FIGURE 5.** Horizontal profile through beam center.

A measure of the illumination beam characteristics is required to complete the probe characterization. Figures 5 and 6 depict a three-position beam profile at distances of 15, 30, and 45 mm from the front surface of the probe. Figure 5 shows the horizontal profile through the center, while Figure 6 shows the vertical profile through the center. Clearly the illumination spot size expands with increasing target distance. The average Gaussian beam width at the $1/e^2$ (13.5%) intensity mark is measured as approximately 432 $\mu$m at the design target location of 30 mm, though the output beam is not well collimated or circular. The approximate full-angle beam divergence (averaged from both horizontal and vertical beam data) is 7 mrad. Also note the clear beam nonuniformities due to the characteristic speckle pattern exiting the illumination fiber.

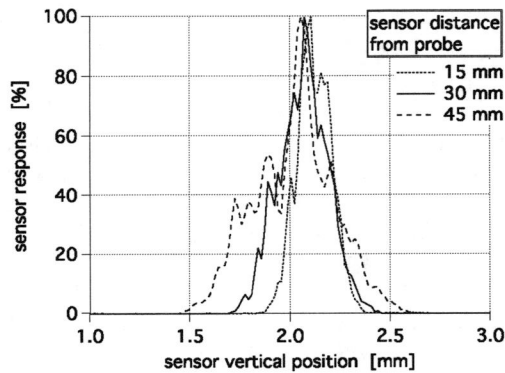

**FIGURE 6.** Vertical profile through beam center.

## CONCLUSIONS

The VISAR probe test bed is an accurate, convenient, low-cost method of probe characterization. For a specific experiment, the response to a given target motion may be reliably predicted from the standardized probe test provided that the target reflectivity stays fairly constant. Further optimization may be possible with a specific trajectory simulation on the test bed, and with tailored target surface preparation. For targets whose surface reflectivity changes dynamically, these changes may be quantified through the comparison of shot data and test bed results.

## ACKNOWLEDGMENTS

The authors gratefully acknowledge support from the United States Department of Energy.

## REFERENCES

1. Catanach, R., Hill, L., Harry, H., Aragon, E., and Murk, D., Cylinder test specification, Tech. Rep. LA-13643-MS, Los Alamos National Laboratory (1999).
2. Davis, L. L., and Hill, L. G., "ANFO Cylinder Tests," in *Shock Compression of Condensed Matter*, American Physical Society, 2001, pp. 165–168.
3. Hill, L. G., "Development of the LANL Sandwich Test," in *Shock Compression of Condensed Matter*, American Physical Society, 2001, pp. 149–152.
4. Hill, L. G., 9501 sandwich data, Memo. DX-2:02-119, Los Alamos National Laboratory (2002).

CP845, *Shock Compression of Condensed Matter - 2005*,
edited by M. D. Furnish, M. Elert, T. P. Russell, and C. T. White
© 2006 American Institute of Physics 0-7354-0341-4/06/$23.00

# A NEW SPIN ON AN OLD TECHNOLOGY: PIEZOELECTRIC EJECTA DIAGNOSTICS FOR SHOCK ENVIRONMENTS

W. S. Vogan*, W. W. Anderson[†], M. Grover**, N. S. P. King[‡], S. K. Lamoreaux[‡], K. B. Morley[‡], P. A. Rigg[†], G. D. Stevens**, W. D. Turley** and W. T. Buttler[‡]

*DX-3, MS-P940, LANL, Los Alamos, NM 87545
[†]DX-2, MS-P952, LANL, Los Alamos, NM 87545
**Bechtel Nevada, Special Technologies Laboratory, Santa Barbara, CA 93117
[‡]P-23, MS-H803, LANL, Los Alamos, NM 87545

**Abstract.** In our investigation of ejecta, or metal particulate emitted from a surface subjected to shock-loaded conditions, we have developed a shock experiment suitable for testing new ideas in piezoelectric mass and impact detectors. High-explosive (HE) shock loading of tin targets subjected to various machined and compressed finishes results in significant trends in ejecta characteristics of interest such as areal density and velocity. Our enhanced piezoelectric diagnostic, "piezo-pins" modified for shock mitigation, have proven levels of robustness and reliability suitable for effective operation in these ejecta milieux. These field tests address questions about ejecta production from surfaces of interest; experimental results are discussed and compared with those from complementary diagnostics such as x-ray and optical attenuation visualization techniques.
**Keywords:** Shock physics, piezoelectric diagnostics, ejecta.
**PACS:** 62.50.+p, 77.65.-j

## INTRODUCTION

Among other diagnostics such as the Asay foil [1], x-radiography attenuation and optical transmission techniques [2], piezoelectric crystals are being used to quantitatively measure *ejecta*, or metal particulate ejected from a shocked surface. Crystals mounted on the ends of copper wires to make "pins" lend themselves well to these measurements due to their compactness. To determine their value as an effective and reliable ejecta diagnostic, we compared three types of "piezo-pin": crystal type PZT-5A (a form of lead zirconate titanate, $PbZr_xTi_{1-x}O_3$), crystal type LN (lithium niobate, or $LiNbO_3$), and crystal type LN with buffers of foam and titanium foil affixed. PZT-5A, also known as Navy type II, is favored for its high sensitivity to applied stress, whereas LN is a more robust piezoelectric material due to its higher Curie temperature (1400 K, versus PZT's 620 K).

The foam and Ti shields were used to ensure retention of impacting particulate and to prevent perforation of the crystal, mitigating the shock. For calculation of the incident stress and density measured by the crystal, the operation of the crystal itself was described by the first order phenomenological piezoelectric effect [3]; other assumptions made in the calculations are outlined in Ref. [4].

## EXPERIMENTAL METHODS

A complete description of the experimental methods is given in [4]; a sketch of the experimental apparatus is shown in Figure 1. The experiments were carried out in a high explosive (HE) chamber at Bechtel Nevada's Special Technologies Laboratory in Santa Barbara, CA. The HE drive (9501, 2 g, 12.7 mm ci-

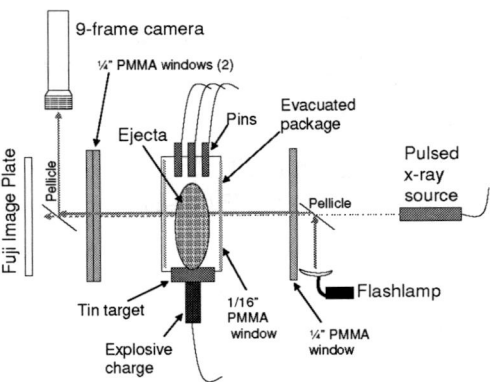

**FIGURE 1.** A sketch of the experimental configuration is shown. The x-ray and optical beams pass through the PMMA windows of the HE chamber, traversing the optically clear plastic windows of the package that the chamber contains.

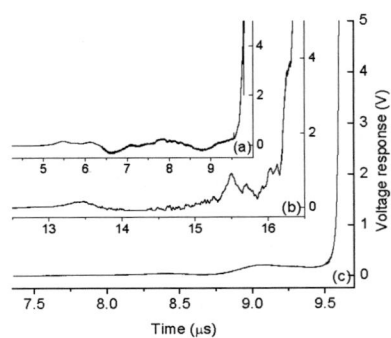

**FIGURE 2.** Center pin responses for selected shots: targets of $d_{gr}$ (a) 18, (b) 7 and (c) 1 $\mu$m. The height of the pin above the target, $h$, was gauged from 18 to 25 mm.

ameter and thickness), backed by an RP-80 detonator, was press-fit to the back of the target. Target surface preparation consisted of a "fly-cut" machined finish, effectively a series of "saw-tooth" grooves of depth $d_{gr}$ ranging from 1 to 24 $\mu$m. In order to ascertain ejecta density levels, piezoelectric sensors were used. The "piezo-pins," procured from Dynasen, Inc. [5], consisted of a 1.27 mm diameter crystal of 0.5 mm thickness mounted via silver epoxy to the end of a 25.4 mm long brass rod of the same diameter; the assembly is insulated with Kynar and encased in brass, with a thin layer of copper foil atop the crystal itself to complete the circuit. The pin crystals used were PZT-5A (lead zirconate titanate, Navy Type II) and LN (lithium niobate, $y + 36^o$-cut) of nominal sensitivities 400 and 24 pC/N. A so-called "shielded" pin was fabricated by affixing titanium foil of 0.25 mm thickness and 2.35 mm diameter to the pin's end via silver epoxy, and a "cap" of polyurethane foam [6] of the same diameter to the foil via the same; the foam was of 1/3 or 2/3 mm thickness. The thickness of the foil was chosen in an attempt to mitigate additional shock propagation effects from the foil's edge which would complicate the crystal's primary signal; the foam was used for the purpose of retaining the impacting ejecta, since the calculations of stress and density assume an inelastic impact, as well as for mitigating shock-induced trauma to the piezoelectric

crystal.

Results from non-viable pins displaying early cessation of voltage response (indicating premature failure of the piezoelectric crystal) or negative voltage response (possibly indicative of off-axis impact of particulate [7]) were not used to calculate ejecta densities.

## RESULTS AND DISCUSSION

Representative pin responses are shown in Figure 2. For a given pin voltage response, the time of surface impact is typically determined at a point 1/3 to 1/2 of the trace's base-to-peak amplitude for the primary peak subsequent to the ejecta signal. This time and the initial height above the target, $h$, are used to estimate the free surface velocity $u_{fs}$ [8], ranging from 1.83 to 1.91 mm/$\mu$s. The pressure in the target just prior to shock breakout was calculated from free surface velocity measurements to be 260 - 280 kbar [9]. At this shock unloading pressure, the tin is believed to be partially melted as the pressure unloads at the free surface [10].

The time of arrival of the fastest impacting particulate is observed at the first deviation of the signal from the baseline, and used to determine the velocity of the fastest particulate, $u_{ej}$. Figure 3 shows $u_{ej}$ as a function of surface finish, as well as the ratio $u_{ej}/u_{fs}$. The values of $u_{ej}$ determined from framing

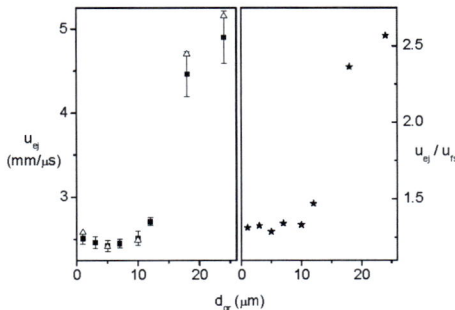

**FIGURE 3.** The peak ejecta velocities $u_{ej}$ averaged over all pins are shown as a function of target groove depth (solid squares); the error bars depict standard deviation of $u_{ej}$ for the pin results. Also shown are $u_{ej}$ calculated from selected shadowgraphy data (hollow triangles). On the right sector of the graph, the ratio of the pin-generated $u_{ej}$ to $u_{fs}$ for each shot is plotted (solid stars).

camera data are shown with good comparison to the pin-generated data.

The most direct method of comparison between piezo-pin and x-ray data is via calculated densities; Figure 4 shows a comparison of densities, as a function of $u$, calculated from x-ray images and from pin responses for selected targets. Differences in the x-ray and piezo-pin density values are due in part to the fact that the range of step wedge areal densities available for comparison to transmission in these shots was $\sim$ 18 - 150 mg/cm$^2$, necessitating some amount of extrapolation in extrema of the density profile. Nevertheless, the general correlation of the piezo-pin and x-ray results is striking and speaks to the validation of the pins as a quantitative diagnostic.

The data from Figure 4 were taken from shielded pins only; a comparison of shielded and unshielded pin performance is shown in Figure 5. A measure of mass accumulated on the piezoelectrc crystal is the areal density, $\rho_A = dm/dA$, expressed in Figure 5 in units of mg/cm$^2$ as a function of $d_{gr}$, and determined at the point of free surface impact, $u_{fs}$. The calculation of $\rho_A$ involves a double integration of the voltage data represented in Figure 2. From the initial comparison in Figure 4, we have confidence that the shielded pins perform well in comparison to x-ray radiogra-

phy data. The data in Figure 5 provide a single-point comparison to demonstrate that the unshielded pins of both PZT and LN type prove unreliable, particularly in high ejecta fields.

## CONCLUSIONS

We have measured characteristics of ejecta from shocked Sn targets with a breakout pressure of 260 - 280 kbar, and have made the following observations: The velocity of the free surface, as far as the surface arrival is coherent, is less than 2 mm/$\mu$s, and varies slightly with respect to surface finish. Detectable peak ejecta velocity varies according to surface groove characteristics. Values for $u_{ej}$ and density derived from optical shadowgraphy and x-ray images correlate well with those calculated from pin responses, confirming those results as well as the viability of the pins as density diagnostics (for a more complete discussion of this correlation, see Ref. [4]). Finally, the pins shielded with foam and Ti foil proved to be consistently viable; LN pins were shown to be more consistent in reliability than were PZT pins. Our tests of new concepts in shock mitigation have hence proven successful in their purpose, comparing well with x-ray and shadowgraphy diagnostics.

## ACKNOWLEDGMENTS

This work was supported by the United States Department of Energy. We gratefully acknowledge the contributions by members of the Los Alamos Dynamic Experimentation, Physics, Applied Physics and Materials Science and Technology Divisions and of Bechtel Nevada's Special Technologies Laboratory, citing in particular R. S. Hixson (LANL DX-2), R. D. Fulton (LANL P-23), B. K. Park (LANL P-23), J. E. Hammerberg (LANL X-7), P. T. Reardon (LANL MST-7), F. P. Garcia (LANL MST-7), L. R. Veeser (BN/STL) and G. Macrum (BN/STL).

## REFERENCES

1. Asay, J. R., "Ejection of material from shocked surfaces,' Appl. Phys. Lett., 29, 284, 1976.

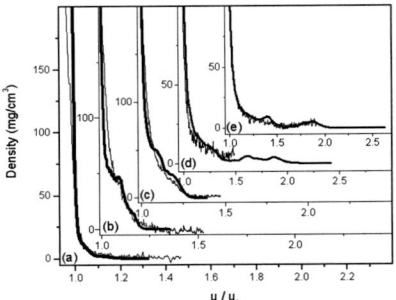

**FIGURE 4.** X-ray (thin lines) and pin-generated (thick lines) density calculations for selected targets are shown as a function of velocity normalized to $u_{fs}$: targets of $d_{gr}$ (a) 1; (b) 5; (c) 10; (d) 18; (e) 24 $\mu$m. The densities are shown as a function of $u = h/t$ normalized to the nominal free surface velocity, $u_{fs}$.

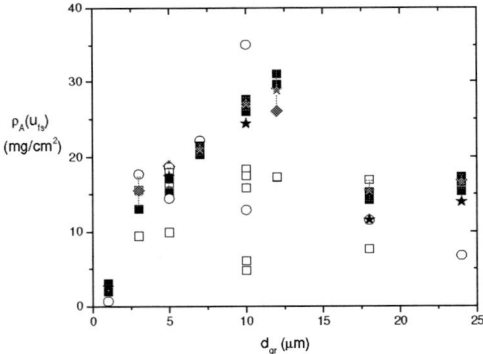

**FIGURE 5.** The areal density at the point of surface impact, $\rho_A(u_{fs})$, is shown for all pins. The solid squares show the areal densities calculated from shielded LN data; the hollow squares represent unshielded LN data; the hollow circles represent unshielded PZT data; and the solid black stars represent data derived from x-ray images. The averages from viable pin data are shown by solid gray stars with standard deviations. It is obvious that the best tracking of the x-ray derived data is given by the shielded LN pins.

2. Buttler, W. T., et al., "Quantitative measurement of ejecta from shocked surface: A comparison of several techniques," LA-UR-02-5211.
3. Speight, C. S., Harper, L., and Smeeton, V. S., "Piezoelectric probe for the detection of shock-induced spray and spall," Rev. Sci. Instrum., 60, 3802, 1989.
4. Vogan, W. S., Anderson, W. W., Grover, M., Hammerberg, J. E., King, N. S. P., Lamoreaux, S. K., Macrum, G., Morley, K. B., Rigg, P. A., Stevens, G. D., Turley, W. D., Veeser, L. R., and Buttler, W. T., "Piezoelectric characterization of ejecta from shocked tin surfaces," J. Appl. Phys., 98, 113508, 2005.
5. www.dynasen.com.
6. General Plastics Manufacturing Co., "Last-A-Foam" FR-7130 of 0.5 g/cm³ density and 83.2 $\mu$m closed cell diameter.
7. Pajewski, W., "Piezoelectric probe response to shock waves," Proceedings of Vibration Problems, 14, 383, 1973.
8. The free surface velocity, $u_{fs}$, is represented by the unloading time to the time of arrival of the free surface at the position of the probe; $u_{fs}$ so determined thus constitutes an upper bound on the asymptotic velocity of the free surface.
9. Results of IMPACT, a locally developed program for calculating equation-of-state quantities from the Hugoniot. For an impact parameter, a range of 1.8 to 1.9 mm/$\mu s$ in the free surface velocity $u_{fs}$ was used; this determines the cited range in break-out pressure.
10. Mabire, C., and Heil, P. L., "Shock induced polymorphic transition and melting of tin," Shock Compression of Condensed Matter, No. 505, 93, 1999.
11. Asay, J. R., and Bertholf, L. D., "A model for estimating the effects of surface roughness on mass ejection from shocked materials," SAND78-1256, 1978.
12. Cheret, R., Chapron, P., Elias, P., and Martineau, J., "Mass ejection from the free surface of shock-loaded metallic samples," Shock Waves in Condensed Matter, 651, 1985.

CHAPTER XVIII

# EXPERIMENTAL TECHNIQUES

CP845, *Shock Compression of Condensed Matter - 2005*,
edited by M. D. Furnish, M. Elert, T. P. Russell, and C. T. White
© 2006 American Institute of Physics 0-7354-0341-4/06/$23.00

# INFLUENCE OF SHOCK WAVE MEASUREMENT TECHNIQUE ON THE DETERMINATION OF HUGONIOT STATES

## C.S. Alexander, T.J. Vogler, W.D. Reinhart, D.E. Grady, M.E. Kipp, and L.C. Chhabildas

*Sandia National Laboratories\*, P.O. Box 5800, Albuquerque, NM 87185-1131*

**Abstract.** In theory, a shock wave traveling through a material gives rise to a well defined Hugoniot state. However, in practice, the measurement technique used to probe the shocked state imparts on the data a unique set of experimental artifacts which can affect interpretation of this data. Two commonly used methods for acquiring shock wave data, VISAR and inclined-mirror measurements are examined to determine the effects of the measurement technique on the final Hugoniot determination. Recent plate impact experiments on the ceramic silicon carbide are used to calibrate a one-dimensional computer model, which is then used to simulate experimental VISAR and inclined mirror data. The results, which highlight potential pitfalls in interpretation of experimental data, will be discussed and solutions to the discrepancies will be proposed. Further, this work is extended to include ceramics that undergo phase transitions.

**Keywords:** shock waves, VISAR, inclined mirror, silicon carbide, WONDY.
**PACS:** 62.50.+p, 07.35.+k.

## INTRODUCTION

The common methods used to acquire shock data from impact experiments can be classified as either *in-situ* or free surface techniques [1,2]. Experiments utilizing pins or inclined mirror (IM) [3,4] approaches record data at a free surface while embedded thin-film gauges and VISAR (velocity interferometer systems for any reflector) [5,6] mimic conditions internal to the sample. For the purposes of this study, IM and VISAR techniques will be used as representative examples of each class of experiments.

Each approach has a unique set of advantages and disadvantages. For example, inclined mirror

techniques benefit from fast detector response times and since all data is recorded on a single record, there is no need to synchronize instrumentation. However, wave reflections from the free surface can be a problem. VISAR systems minimize this issue through the use of impedance matched windows but suffer from the integration time delay. The focus of this study is to examine the effects of these unique characteristics on the resultant analysis.

The issue addressed in this paper is reflected shock waves. For a single shock there would be no reflections and either approach would have little complication. However, in materials such as ceramics which tend to have higher yield strength there are typically larger elastic wave precursors. In these cases, reflections originating from the elastic precursor at the free surface or window interface result in potentially significant delay of

\*Sandia is a multiprogram laboratory operated by Sandia Corporation, a Lockheed Martin Company, for the United States Department of Energy's National Nuclear Security Administration under Contract DE-AC04-94AL85000.

the plastic shock and perturbation of the data which complicates the determination of the shock arrival.

While many materials do not generate these discrepancies, recent studies on high strength ceramics $B_4C$ [7], SiC [6], and AlN [8] have shown discrepancies between IM and VISAR results. This study will examine the influence of the measurement technique as a function of material strength in the range typical of ceramics. Since the differences tend to originate with the determination of shock velocity via the shock time of arrival (TOA), this aspect of the experiments will be scrutinized.

## COMPUTER MODEL

The computer simulations in this study utilize WONDY, a one-dimensional Lagrangian finite-difference wave propagation code [9]. Two material models were examined. An elastic-perfectly plastic (EPP) model wherein the material yields completely at a single yield strength was used for simplicity in understanding some of the basic effects of reflected waves. In order to more accurately reproduce real experimental results, an anisotropic work-hardening (AWH) model was used. The specific material parameters were adjusted within the AWH model to give a close approximation of experimental results of symmetric impact of the ceramic SiC [6] as shown in Fig. 1.

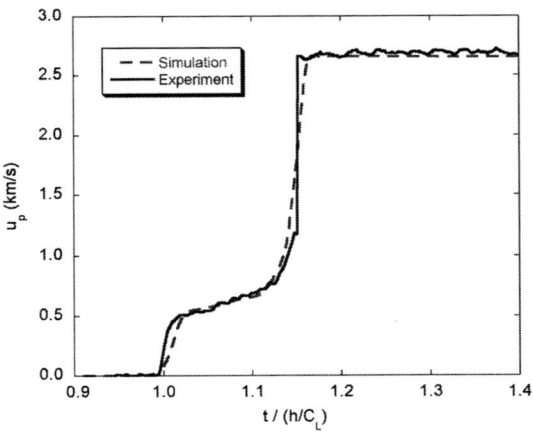

**Figure 1.** Comparison of the AWH model (dashed) to experimental symmetric impact data on SiC (solid) [6].

Regardless of the material model used, the geometry of the problem was for a 6mm thick impactor hitting a 12mm thick target at impact velocities of 0.5 – 6.0 km/s. In order to simulate experimental measurement conditions the target was backed with either 12mm of target material (a virtual *in-situ* condition), 12mm of LiF (VISAR condition), or a free surface (inclined mirror condition). In all cases, the point of measurement was the interface between the target and the backing material. The extremely thick impactor and target are used to expand the time scale over which interactions between waves occur.

## METHOD OF ANALYSIS

In order to allow for comparison of the various techniques, a well-defined procedure was used to analyze the data for each case. These procedures were chosen to best approximate actual experimental data analysis techniques.

*In-situ* data is collected as particle velocity as a function of time. TOA of the plastic wave is determined by the time at which the velocity jumps. The virtual *in-situ* gauge is used to determine the "true" TOA since the elastic-plastic wave structure is not perturbed in this case. Error is then determined as a percentage of the *in-situ* value for the other measurement techniques according to the relation:

$$\% \text{ Error} = \frac{\text{TOA}_{exp} - \text{TOA}_{is}}{\text{TOA}_{is}} \times 100 \quad (1)$$

where $\text{TOA}_{is}$ is the virtual *in-situ* TOA and $\text{TOA}_{exp}$ is the TOA for an experimental condition.

Due to reflections, there is not a single jump associated with the plastic wave arrival at the measurement interface in the VISAR data. In this case, TOA is determined as the midpoint of the last discrete jump prior to the plateau associated with the Hugoniot state.

Inclined mirror data is collected as position as a function of time. TOA of the plastic wave is determined from the location of a kink in the data. The precise location of the kink is found from taking the intersection of linear fits to the data above and below the kink. In cases where intermediate slopes, arising from reflections, could be distinguished, the latest arriving kink was used to find the shock TOA and the last intermediate slope was used for the linear fit below the kink. In cases where no clear linear data was present below the kink, a best fit was used below the kink.

## RESULTS AND DISCUSSION

The effect of material yield strength (Y) is important relative to the maximum stress on a sample ($\sigma_{max}$) during an experiment. Accordingly, results will be presented as a function of $Y/\sigma_{max}$. $Y/\sigma_{max}$ will first be varied by holding $\sigma_{max}$ constant by using a steady impact velocity while varying Y in the computational model. This condition allows for easier comparison of the effects of changes in $Y/\sigma_{max}$. It is more realistic however to have a fixed material strength and vary the stress levels. This will later be done by varying the impact velocity while holding the material model properties constant.

Simulated data is shown in Fig. 2 for both experimental configurations. Impact conditions were such that the peak stress on the sample was fixed at 42 GPa. In general, the EPP model yields particle velocity or position profiles with discrete features, while the AWH model tends to produce more rounded curves. In the limiting cases of both large and small values of Y, the data approaches a single wave structure; hence, for $Y/\sigma_{max}$ approaching zero or one, the error is expected to be small.

Relative error is shown in Fig. 3 as a function of $Y/\sigma_{max}$ and includes information from simulations omitted from Fig. 2 for clarity. As expected, the error is small for $Y/\sigma_{max}$ near zero. The AWH model data shows error decreasing as $Y/\sigma_{max}$ approaches one and similar behavior is expected for the EPP model only at higher values of $Y/\sigma_{max}$ than shown. While the magnitude of the error is comparable for both experimental configurations using the EPP model due to the sharpness of the features, there is considerably larger error for the IM setup using the more realistic AWH model resulting from increased rounding of the curves. This error reaches a

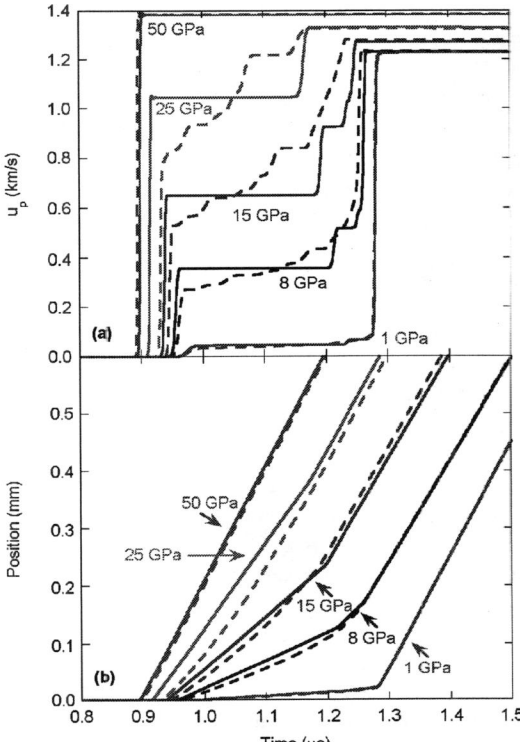

**Figure 2.** Simulated results for (a) VISAR and (b) IM experiments using both the EPP (solid curves) and AWH (dashed curves) models where the yield strength (indicated) is varied at a fixed stress of 42 GPa.

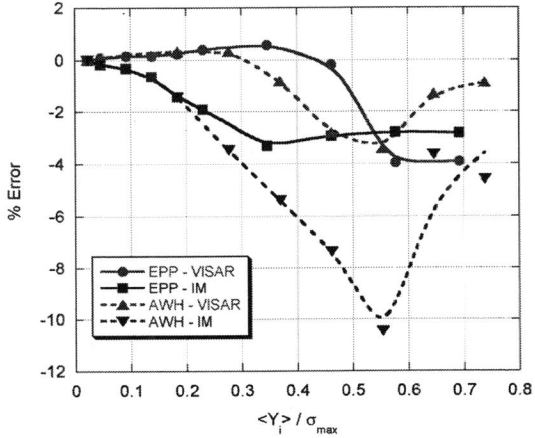

**Figure 3.** Error in determination of the shock TOA at fixed stress based on the "true" value from a virtual *in-situ* gauge. Solid points correspond to the EPP model while hollow points are for the AWH model. Lines shown are guides to the eye.

maximum magnitude near $Y/\sigma_{max} = 0.55$, which corresponds to a yield strength of approximately 23 GPa.

As mentioned previously, it is more realistic to have fixed material yield strength and vary the stress through the impact conditions. This was done using the AWH model similarly to the data of Fig. 2 except the average yield strength was fixed at 16GPa, a reasonable value for SiC [6], while the impact velocity was varied to give peak stresses from 17 to 117 GPa. The resulting errors are shown in Fig. 4. Similar behavior is seen as in Fig. 3 except for a slight increase in the magnitude of the error and a minor shift in the maximum error magnitude to higher values of $Y/\sigma_{max}$.

SiC is not the only material to show these types of effects. Aluminum nitride (AlN), another ceramic material, has been studied previously [10] using similar techniques. The situation is further complicated in AlN by the presence of a structural phase transformation at about 20 GPa. This results in a three wave structure in shock experiments where the stress exceeds this level. The risetime of the three wave structure results in increased wave interactions thereby yielding a greater potential for error. Using a realistic model of AlN, where $Y/\sigma_{max}$ is 0.18, errors of +1% were observed for the VISAR and inclined mirror configurations at the second wave arrival. These values are consistent to those determined using SiC as a model with a two

wave structure. However, at the third wave arrival, the errors were +6% and -10% for the VISAR and inclined mirror configurations respectively. These values are much larger than for the second wave arrival and are expected to increase further for materials with $Y/\sigma_{max}$ closer to 0.5. Thus, the potential for very large errors exists regardless of experimental technique.

## CONCLUSIONS

Characteristics of the stress wave pulse that can have a significant effect on the techniques used to determine its time of arrival have been summarized. *In-situ* or free surface techniques can be used to accurately determine the shock states of materials provided precursory waves which alter the wave profiles are taken into consideration. While these effects are negligible for many systems, in the case of stiffer ceramic materials caution must be used when analyzing the experimental data.

## REFERENCES

1   Graham, R.A. and Asay, J.R., *High Temp.- High Press.* **10**, 355 (1978).
2   Chhabildas, L.C. and Graham, A., *AMD (Symposia Series) (American Society of Mechanical Engineers, Applied Mechanics Division)* **83**, 1 (1987).
3   Ahrens, T.J., Gust, W.H., and Royce, E.B., *J. Appl. Phys.* **39**, 4610 (1968).
4   Sekine, T. and Kobayashi, T., *Phys. Rev. B* **55**, 8034 (1997).
5   Barker, L.M. and Hollenbach, R.E., *J. Appl. Phys.* **43**, 4669 (1992).
6   Vogler, T.J., Reinhart, W.D., Chhabildas, L.C., and Dandekar, D.P., *submitted to J. Appl. Phys.* (2005).
7   Vogler, T.J., Reinhart, W.D., and Chhabildas, L.C., *J. Appl. Phys.* **95** (8), 4173 (2004).
8   Kipp, M.E. and Grady, D.E., *J. de Physique IV* **4-C8**, 249 (1994).
9   Kipp, M.E. and Lawrence, R.J., WONDY V - A one-dimensional finite-difference wave propagation code, Sandia Report, SAND81-0930, April 1982.
10  Grady, D.E. and Kipp, M.E., unpublished work

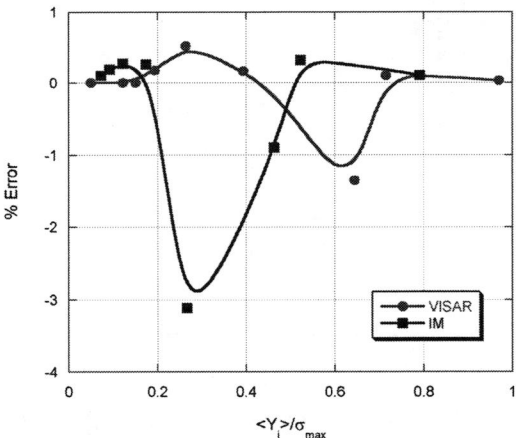

**Figure 4.** Error in determination of the plastic wave TOA using the AWH model with a fixed average yield strength of 16 GPa. Lines shown are guides to the eye.

CP845, *Shock Compression of Condensed Matter - 2005*,
edited by M. D. Furnish, M. Elert, T. P. Russell, and C. T. White
2006 American Institute of Physics 0-7354-0341-4/06/$23.00

# LIMITATIONS OF THE HOPKINSON PRESSURE BAR FOR HIGH-FREQUENCY MEASUREMENTS

## Richard G. Ames[1]

[1]*Naval Surface Warfare Center, Dahlgren Division, Code G22, Building 221*
*17320 Dahlgren Rd*
*Dahlgren, VA 22448*

**Abstract.** The Hopkinson Pressure Bar is a measurement tool that has been used for a variety of applications, including measurement of high-strain-rate loads in material test specimens, characterization of blast fields produced by high explosives, and impulse measurements for a variety of ballistic test specimens. The technique is generally used for integrated measurements, including impulse or total strain energy. However, the Hopkinson Pressure Bar has also been used to measure the time-history of a variety of dynamic loads. The limitations on this type of measurement are strict, however, and generally require bars of extremely small diameter for most measurements of practical interest. In particular, dynamic loads with significant energy in the higher frequencies can become severely distorted as they are propagated down the bar. This paper provides a review of the relevant theory behind the Hopkinson Pressure Bar technique and derives a relationship that provides a practical limitation on the use of the technique for high-frequency measurements.

**Keywords:** Hopkinson Pressure Bar, Frequency Response, High-Strain Rate, Dispersion
**PACS:** 81.70.-q, 81.70.Bt, 62.65.+k

## NOMENCLATURE

| | |
|---|---|
| $c$ | = frequency-dependent phase velocity |
| $c_0$ | = zero-frequency wave speed |
| $C$ | = magnitude of decaying load |
| $E$ | = modulus of elasticity of bar |
| $f$ | = temporal frequency, $f = \omega/2\pi$ |
| $k$ | = radius of gyration |
| $P$ | = load applied to end of bar |
| $r$ | = radius of bar |
| $s$ | = Laplace transform variable |
| $t$ | = temporal coordinate |
| $T$ | = period, $T = \Lambda/c$ |
| $u$ | = longitudinal displacement of bar |
| $x$ | = spatial coordinate |
| $\gamma$ | = spatial angular frequency, $\gamma = 2\pi/\Lambda$ |
| $\Lambda$ | = wavelength (in the bar) |
| $\rho$ | = density of bar material |
| $\sigma$ | = stress component along x-axis |
| $\omega$ | = temporal angular frequency, $\omega = 2\pi/T$ |

## PROBLEM DESCRIPTION

The analysis presented here considers a semi-infinite bar of uniform circular cross section with radius r. This bar is constructed of a homogeneous material with density $\rho$, modulus E, and Poisson ratio $\nu$. The effects of reflection from a free end are not present as a result of the semi-infinite assumption.

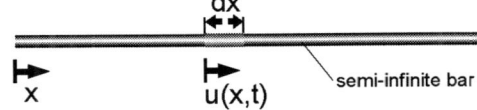

The spatial and temporal coordinates are x and t, respectively, and the longitudinal displacement u is allowed to vary in both coordinates. Linear elasticity is assumed and body forces are neglected for all of the analyses presented here.

The objective of this analysis is to determine the manner in which the time-history of a displacement at x = 0 is transmitted down the length of the bar. To this end, three approaches are presented: the classical theory which results in a simple wave-function solution, a modified theory that accounts for the effects of lateral dispersion (due to Love, ref. 3), and a comparison to the exact solution provided by Pochhammer.

## CLASSICAL THEORY

Under the additional assumption that the stress is uniform across the cross section of the bar and neglecting the effects of lateral inertia, it can be shown that the governing equation of motion for the bar reduces to the following form (e.g. Ref. 2):

$$\frac{\partial^2 u}{\partial t^2} = \frac{E}{\rho}\frac{\partial^2 u}{\partial x^2} \qquad (1)$$

This equation is the classical wave equation and, as such, has the D'Alembert solution

$$u(x,t) = f(x - c_0 t) + g(x + c_0 t)$$

where the wave speed is

$$c_0 = \sqrt{\frac{E}{\rho}}$$

It is apparent, then, that under the assumptions of a uniform stress distribution and zero lateral dispersion, a displacement u(0,t) produced at x = 0 will be propagated down the bar at a speed of $c_0$ and without distortion.

## THE IMPROVED THEORY DUE TO LOVE

For certain types of loads, the effects of lateral dispersion are significant. Because momentum and energy must be conserved down the length of the bar, the dispersion effects will, in general, have only a minor effect on integrated effects such as impulse or total strain energy. However, the displacement waveform that is produced at x = 0 can be drastically altered as it is propagated down the length of the bar if lateral dispersion effects are present.

Love (Ref. 3) presented an analysis that included the effects of lateral dispersion through the Poisson effect. Love presented his derivation under energy considerations (solving Hamilton's equation using appropriate expressions for the kinetic and potential energies) and derived the following equation of motion:

$$\frac{\partial^2 u}{\partial t^2} = c_0^2 \frac{\partial^2 u}{\partial x^2} + \upsilon^2 k^2 \frac{\partial^4 u}{\partial t^2 \partial x^2} \qquad (2)$$

where $\upsilon$ is the Poisson ratio for the bar material and k is the radius of gyration of the bar cross-section (for a circular cross-section, $k^2 = \frac{1}{2}r^2$). This equation is different from the classical wave equation by the introduction of the second term on the right hand side; it is this term that arises from the inclusion of the lateral strain effects in the energy balance.

In order to examine the effects of lateral dispersion on an arbitrary displacement-time history applied to the free end, this analysis considers an arbitrary harmonic function of the form

$$u(x,t) = Ae^{i(\gamma x + \omega t)} \qquad (3)$$

where A is an arbitrary amplitude and $\gamma$ and $\omega$ are the spatial and temporal angular frequencies, respectively, of the function.

This approach is appropriate for the analysis presented here due to the fact that Equation (2) is a linear partial differential equation. As such, the principle of superposition applies to any collection of particular solutions that satisfy Equation (2); specifically, sums of arbitrary functions given by Equation (3). By considering any function U(x,t) as the sum of its Fourier components, the results derived for the arbitrary harmonic function given in Equation (3) can be extended to the function U(x,t).

In particular, this analysis will show that as the frequency $\omega$ of the harmonic function is increased, the effects of the lateral motion become more pronounced and the waveform deviates from the dispersion-free propagation predicted by classical theory. Therefore, the high-frequency Fourier components of U(x,t) will become distorted as the waveform is propagated down the bar.

Substituting Equation (3) into Equation (2) and defining the phase-dependent velocity c as

$$c = \frac{\Lambda}{T}$$

where T is the period of the harmonic disturbance and $\Lambda$ is the wavelength of that disturbance in the bar, also using the relations

$$\omega = \frac{2\pi}{T} \qquad \gamma = \frac{2\pi}{\Lambda}$$

the following relationship can be derived:

$$\frac{c}{c_0} = \frac{1}{\left[1 + \dfrac{4\pi^2 \upsilon^2 r^2}{2\Lambda^2}\right]^{\frac{1}{2}}} \qquad (4)$$

Equation (4) shows a number of important relationships. First, as the wavelengths of the disturbance become very long relative to the radius of the bar, the frequency-dependent phase velocity approaches the material sound speed $c_0$. In this case, the long-wavelength (low-frequency) disturbance would be propagated down the bar with very little dispersion.

However, for disturbances which have shorter wavelengths, the phase velocity deviates from the nominal sound speed. This result has implications for disturbances that contain a significant amount of energy in the higher frequencies: these components will be propagated down the bar at a speed slower than the sound speed and the signal will become distorted.

## COMPARISON TO THE EXACT THEORY OF POCHHAMMER

Pochhammer (ref. 6) provided a generalized theory of the vibrations of a cylindrical rod. His theory is considered exact to the extent that the longitudinal displacement and stress are allowed to vary over the cross-section of the bar and the radial displacement is given by the generalized Hooke's law.

The resulting equations of motion are complicated and are not presented here. Of relevance to the discussion here is the manner in which these equations predict the transmission of displacements down the length of the bar. These results, along with those given by the classical theory and Equation (4) are presented below in Figure 1. The data for the exact theory due to Pochhammer are taken from Ref. 5, Table 11.1 (page 406).

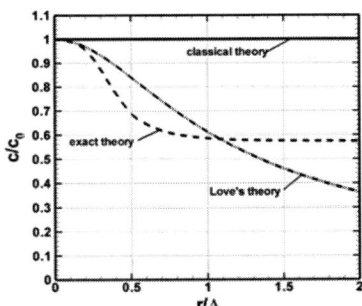

**Figure 1**: Comparison of normalized phase velocities predicted by the three theories presented here.

The plot above shows the predicted relationships between the normalized phase velocity $c/c_0$ and the normalized bar diameter $r/\Lambda$. As stated above, for very small bar diameters (or very long wavelengths), the phase velocity equals that predicted by the classical theory. As the bar diameter becomes large (or the wavelengths become small), the predicted phase velocities begin to diverge from the classical theory.

The three theories agree to within about 5% for normalized bar diameters of 0.2 or less. As a result, wavelengths of more than 5 times the bar diameter can be expected to traverse the bar with relatively little distortion (i.e., the classical theory applies). For shorter wavelengths (or larger bar diameters), the differences in phase velocity become substantial and decrease by approximately 40% for wavelengths on the order of the bar diameter.

## THE LIMITS OF DISTORTION-FREE WAVE PROPAGATION

Figure 1 shows that Love's theory and the exact theory are in good agreement for normalized phase velocities in the range of 0.95 – 1.00. This fact allows the use of Equation (4) in setting practical limitations for the use of the pressure bar technique. For example, if the range of normalized phase velocities is restricted to the range of agreement between Love's theory and the exact theory, Equation (4) can be written to require

$$\frac{c}{c_0} > 0.95$$

In terms of the properties of the bar and the wavelength of the displacement, this reduces to

$$r < \frac{0.465\Lambda}{2\pi\upsilon} \qquad (5)$$

Inequality (5) presents an extremely useful result: it provides a practical, closed-form solution for the maximum bar diameter that will transmit an essentially distortion-free signal of wavelength $\Lambda$ down the length of the bar.

Because this result limits the range of phase velocities to those that approximate the nominal (zero-frequency) sound speed, Inequality (5) can be rewritten in terms of the frequency of the displacement function using

$$\Lambda = \frac{2\pi c}{\omega} \approx \frac{2\pi c_0}{\omega} = \frac{c_0}{f}$$

This yields the result in terms of the displacement function frequency:

$$r < \frac{0.465 c_0}{2\pi\upsilon f} \qquad (6)$$

Using the material properties for steel ($\nu = 0.29$ and $c_0 = 5060$ m/s), Inequality (6) shows that bar diameters in the range of practical use, 10 mm – 100 mm, provide distortion-free signal propagation for frequencies of less than 10 – 100 kHz. While this bandwidth limitation would likely not affect integrated measures such as impulse or total strain energy, the actual time history of the waveform being propagated down the bar is likely to be distorted for most measurements of practical interest.

In particular, high explosive blast measurements generally require bandwidths greater than 1 MHz in order to accurately capture the time history of the loading. Based on the results presented above, using the Hopkinson Pressure Bar for this type of measurement would require a bar diameter on the order of a few millimeters. Useful measurements can still be made with a bandwidth limitation of 100 kHz; this bandwidth would require a bar of approximately 10 mm or less.

## CONCLUSIONS

Conclusions drawn from the analysis presented in this paper include the following:

1. Use of a steel Hopkinson Pressure Bar to characterize the temporal evolution of air shock loads requires bars of 10 mm diameter or less.
2. The short-wavelength components of a load applied to a Hopkinson Pressure Bar are distorted when the bar diameter is more than 5 times the shortest wavelengths.
3. Inequality (6) provides a practical limitation on bar diameters for a given maximum loading frequency

## REFERENCES

1. Kolsky, H., *Stress Waves in Solids*, Dover Publications, 2003
2. Graff, K.F., *Wave Motion in Elastic Solids*, Dover Publications, 1992
3. Love, A.E., *A Mathematical Treatise on the Theory of Elasticity*, Dover Publications, 1944
4. DuChateau, P and Zachmann, D., *Applied Partial Differential Equations*, Harper and Row, 1989
5. Davies, R.M., "A Critical Study of the Hopkinson Pressure Bar", *Phil. Trans. R. Soc.* #240 (1948), p 375-457
6. Pochhammer, L, *J. reine angew. Math.* #81 (1876)
7. Gere, Timoshenko, *Mechanics of Materials*, PWS Publishing Company, 1990

CP845, *Shock Compression of Condensed Matter - 2005,*
edited by M. D. Furnish, M. Elert, T. P. Russell, and C. T. White
© 2006 American Institute of Physics 0-7354-0341-4/06/$23.00

# INTERMEDIATE AND HIGH STRAIN-RATE TESTING OF SOFT MATERIALS

## Simon P. Anderson*, Elisavet Palamidi* and John J. Harrigan*

*School of Mechanical, Aerospace and Civil Eng., University of Manchester, Manchester M60 1QD, UK*

**Abstract.** Strain-gauged bars are often employed as load cells for direct impact testing of materials and are incorporated within the split Hopkinson pressure bar (SHPB). Low impedance bars (e.g., magnesium or polymer bars) are desirable when testing soft specimens such as various energetic materials and cellular solids. However, due to the rheological properties of polymer bars, wave dispersion and attenuation occurs. For relatively large diameter bars and high frequency waves, geometrical wave dispersion due to radial inertia also occurs. In this paper the spectral finite element method (SFEM) is applied to the SHPB to obtain the stress-strain curves of specimens under investigation by an inverse analysis. The method presented makes use of a higher-order rod approximation, applicable to viscoelastic bars, that accounts for wave dispersion. To demonstrate the technique experimental results for balsa wood using both magnesium alloy and PMMA bars are provided.

**Keywords:** Spectral elements, wave propagation, wave separation, inverse analysis, Hopkinson bar
**PACS:** 02.70.Hm, 02.30.Zz, 07.05.Kf

## INTRODUCTION

The split Hopkinson Pressure bar (SHPB) is a well known apparatus for determining the mechanical properties of materials under high strain-rate loading. Conventional SHPB analysis requires a clear distinction between waves travelling in opposite directions in the bars, and therefore strain measurements are typically made at the centre of the bars. The maximum strain that a specimen undergoes in a test is determined by the duration of the incident stress wave which is in turn dependent upon the length and material properties of the striker bar. If it is required to increase this duration to obtain higher strains then the length of the input bar must also be increased in order to avoid the occurrence of wave superposition at the strain gauge. Alternatively, a method of wave separation can be employed to deduce the forward and backward travelling pulses by taking additional measurements along the bar [1, 2].

In this paper the non-linear stress-strain curve of a test specimen is determined by an inverse analysis method that makes use of at least two strain records from the input bar and one from the output bar. In order to 'deconvolve' the stress and strain histories of the specimen from the measured strain histories, the transfer functions of the bars require determining. This is achieved using the spectral finite element method (SFEM) [3], which is a frequency domain matrix method with a similar formulation to the finite element method (FEM). The significant advantage of the SFEM over the FEM is that the exact frequency response of the element is described at the nodes irrespective of its length. As a frequency domain method it is an ideal tool for the solution of dynamic inverse problems since the system transfer functions are readily determined. Moreover, the limitations on the length of the bars and the test duration are removed since wave superposition is accounted for in the transfer functions.

## SPECTRAL FINITE ROD ELEMENT

The frequency domain load-displacement relationship of the SFEM can be expressed as [3]

$$\{\hat{P}(\omega_n)\} = [\hat{K}(\omega_n)]\{\hat{U}(\omega_n)\}, \quad n = 0, 1, \ldots, N_f - 1, \tag{1}$$

where $\omega$ is circular frequency, hat denotes a frequency domain term, and $N_f$ is the number of discrete frequencies that Eq. (1) is solved at. For an axial rod element (e.g., the input bar in Fig. 1), the vector of applied loads $\{\hat{P}\} = \{\hat{f}_1, \hat{f}_2\}^T$, the vector of nodal displacements $\{\hat{U}\} = \{\hat{u}_1, \hat{u}_2\}^T$, and the frequency domain dynamic stiffness matrix is given by

$$[\hat{K}] = \frac{EA\gamma}{1 - e^{-2\gamma L}} \begin{bmatrix} 1 + e^{-2\gamma L} & -2e^{-\gamma L} \\ -2e^{-\gamma L} & 1 + e^{-2\gamma L} \end{bmatrix}, \tag{2}$$

where the dependency of terms on frequency is understood and omitted for brevity. In Eq. (2), $EA$ is the axial rigidity, $L$ is the length of the element, and $\gamma = \gamma(\omega)$ is the propagation coefficient for axial waves in the rod. This can be determined either experimentally [4] or analytically using, for example, the elementary wave equation, however the latter does not take geometrical wave dispersion into consideration and may therefore become too inaccurate with large diameter bars and high frequency disturbances. In this study a four-mode rod equation is used which is comparable in accuracy to the exact solution of the Pochhammer equation [5]. Given the nodal displacements, the displacement at a distance $x$ from node 1 within the rod is given by

$$\hat{u}(x) = \hat{N}_1(x)\hat{u}_1 + \hat{N}_2(x)\hat{u}_2, \tag{3}$$

where the displacement shape functions are given by

$$\hat{N}_1(x) = \frac{e^{-\gamma x} - e^{-\gamma(2L-x)}}{1 - e^{-2\gamma L}},$$

$$\hat{N}_2(x) = \frac{e^{-\gamma(L-x)} - e^{-\gamma(L+x)}}{1 - e^{-2\gamma L}}.$$

## APPLICATION TO A HOPKINSON BAR

The SFEM can be applied to the SHPB setup in order to determine the stresses on the proximal and distal ends of a specimen and the average strain that it is subject to. This is achieved by modelling the input

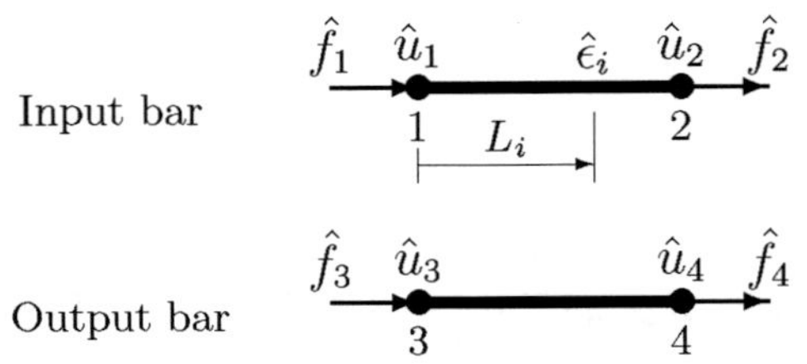

**FIGURE 1.** Spectral finite rod elements used to model Hopkinson pressure bars.

and output bars as two separate spectral rod elements with externally applied nodal forces as shown in Fig. 1. $f_1$ on the input bar refers to the impact force history due to the striker bar. $f_2$ is the force exerted on the end of the input bar by the specimen, and is therefore equal to the force on the proximal end of the specimen. Similarly, $f_3$ is the force on the distal end of the specimen and $f_4$ is the end force on the free end of the output bar which is essentially equal to zero. Considering the input bar first, the strain response, $\varepsilon_i$, at position $i$ along the bar can be represented as the superposition of multiple input forces, $f_j$, located at $j$, such that

$$\hat{\varepsilon}_i = \sum_j \hat{h}_{ij} \hat{f}_j, \tag{4}$$

where $\hat{h}_{ij}$ is the transfer function that relates the force at $j$ to the strain response at $i$. For example, for $j = 1$ (i.e., the first node), the transfer function is determined by

$$\hat{h}_{i1} = \begin{bmatrix} \hat{N}'_1(L_i) & \hat{N}'_2(L_i) \end{bmatrix} [\hat{K}]^{-1} \begin{Bmatrix} 1 \\ 0 \end{Bmatrix}, \tag{5}$$

where prime denotes differentiation with respect to $x$. For two nodal forces being reconstructed from multiple strain responses, the problem can be expressed as

$$\begin{pmatrix} \hat{\varepsilon}_1 \\ \hat{\varepsilon}_2 \\ \vdots \end{pmatrix} = \begin{pmatrix} \hat{h}_{11} & \hat{h}_{12} \\ \hat{h}_{21} & \hat{h}_{22} \\ \vdots & \vdots \end{pmatrix} \begin{Bmatrix} \hat{f}_1 \\ \hat{f}_2 \end{Bmatrix}, \tag{6}$$

where $\hat{f}_1$ and $\hat{f}_2$ are solved using a least squares method. Once the end forces are known, the end displacements are simply obtained through Eq. (1). A similar method is then applied to the output bar where only one strain measurement is required since there is only one force history to be reconstructed,

1238

namely, $\hat{f}_3$. The average strain in a specimen of width $b$ is then given by, $\varepsilon_s = (u_2 - u_3)/b$.

## EXPERIMENTAL RESULTS

Results from two experimental tests carried out on along the grain balsa wood specimens are presented in this section; the first employed metallic bars and the second used polymer bars. A detailed investigation into the quasi-static and dynamic response of balsa wood at similar strain rates to this study was carried out by Vural and Ravichandran [6].

The first test used 23 mm diameter magnesium alloy AZ31B bars with a nominal length of 1 m. Three strain gauges were instrumented on the input bar at distances of 374 mm, 492 mm and 701 mm from the striker end, and two gauges on the output bar at distances of 374 mm and 701 mm from the specimen end. The propagation coefficient of the bars was determined analytically using the four-mode rod equation [5].

Fig. 2 shows two of the five strain responses that were measured in the test. Fig. 3 shows two of the nodal force histories, $f_2$ and $f_3$, that were computed using Eq. (6). The perturbations towards the end of the histories are due to an ill-conditioned coefficient matrix, that is typical of inverse problems, magnifying experimental noise. The nodal displacement histories on the proximal and distal ends of the specimen were subsequently determined using Eq. (1). From these time histories the average stress-strain curve of the specimen was readily constructed and is shown in Fig. 4. The length of the striker bar was not sufficient to cause densification during propagation of the first pulse, however this was achieved during reloading of the second pulse up to a strain of approximately 0.8, while the plateau stress was observed to be nearly constant at 8.5 MPa.

The second test used 20 mm diameter PMMA bars. The input bar had a length of 1 m and was instrumented with two strain gauges at 500 mm and 922 mm from the striker end. The output bar had a length of 842 mm and had one strain gauge located at 74 mm from the specimen end. The propagation coefficient was determined experimentally following reference [4]. The Poisson's ratio was obtained by simultaneously taking axial and hoop strain measurements at the same position along the bar during a cal-

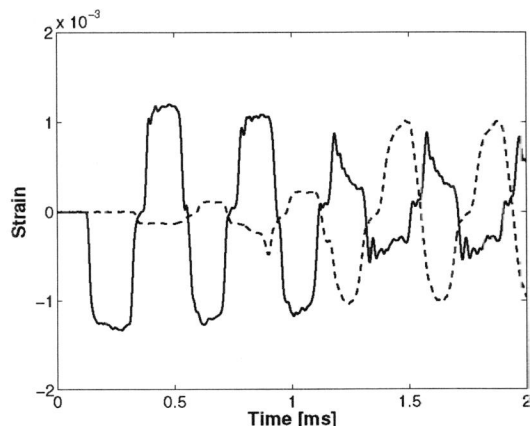

**FIGURE 2.** Strain histories from SHPB test using magnesium alloy bars (374 mm on input bar, solid line; 374 mm on output bar, dashed line).

ibration impact test. The results indicate that, up to a frequency of 30 kHz, the Poisson's ratio increases from 0.32 to 0.38. The frequency dependent complex elastic modulus, required by Eq. (2), was computed from the experimental propagation coefficient and Poisson's ratio using the rearranged four-mode rod equation.

Fig. 5 shows the three measured strain histories that were used to compute the stress-strain curve shown in Fig. 6. Looking at the strain records, wave

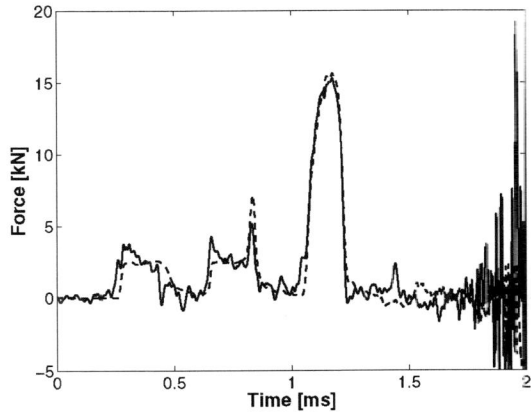

**FIGURE 3.** Reconstructed force histories on the proximal ($f_2$, solid line) and distal ($f_3$, dashed line) ends of the specimen.

1239

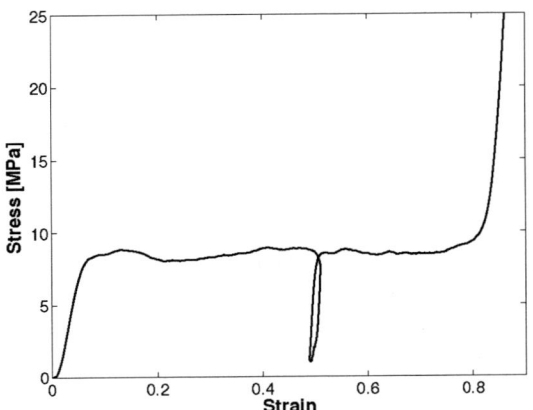

**FIGURE 4.** Stress-strain curve for along the grain balsa (specimen density, 94.5 kgm$^{-3}$; test strain rate, ~2500 s$^{-1}$).

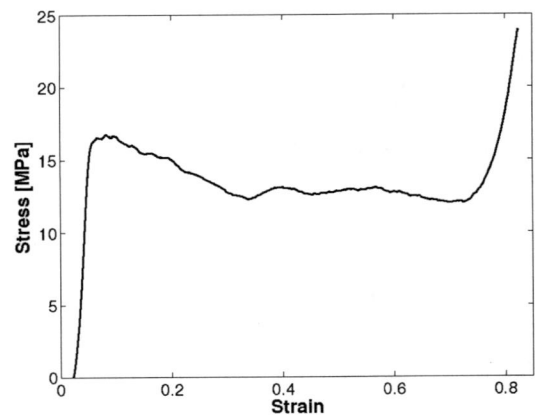

**FIGURE 6.** Stress-strain curve for along the grain balsa (specimen density, 142.8 kgm$^{-3}$; test strain rate, ~2400 s$^{-1}$).

superposition appears to occur at all three gauge positions, including the gauge at the centre of the input bar. This can be a problem with viscoelastic bars since material wave dispersion occurs. Without the use of longer bars, it would be necessary to separate the forward and backward travelling waves. However, this is effectively included in the transfer function determined from the SFEM.

**CONCLUSION**

The SFEM has been applied to the SHPB to obtain the transfer functions of the input and output bars. From this, the stress-strain curve of a test specimen can be determined by an inverse analysis using strain response measurements. As wave superposition may occur at the strain gauges, there is no limitation on the length of the bars or the duration of the test.

**ACKNOWLEDGEMENTS**

The authors are grateful to ONR Global for their support through the STEP program.

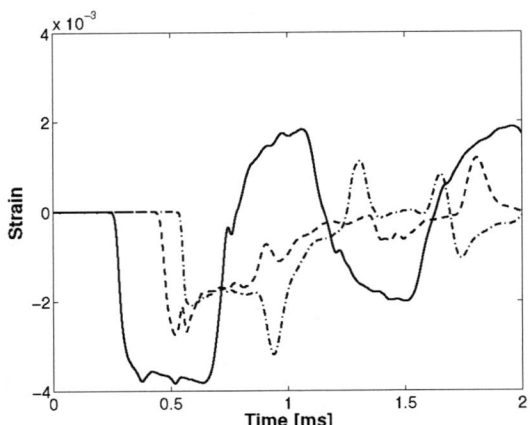

**FIGURE 5.** Strain histories from SHPB test using PMMA bars (500 mm on input bar, solid line; 922 mm on input bar, dashed line; 74 mm on output bar, dash-dot line).

**REFERENCES**

1. Lundberg, B., and Henchoz, A., *Experimental Mechanics*, **17(6)**, 213–218 (1977).
2. Bacon, C., *International Journal of Impact Engineering*, **22(1)**, 55–69 (1999).
3. Doyle, J. F., *Wave Propagation in Structures*, Springer–Verlag, New York, 1997, 2nd edn.
4. Bacon, C., *Experimental Mechanics*, **38(4)**, 242–249 (1998).
5. Anderson, S. P., *Journal of Sound and Vibration*, **290**, 290–308 (2006).
6. Vural, M., and Ravichandran, G., *International Journal of Solids and Structures*, **40(9)**, 2147–2170 (2003).

CP845, *Shock Compression of Condensed Matter - 2005,*
edited by M. D. Furnish, M. Elert, T. P. Russell, and C. T. White
© 2006 American Institute of Physics 0-7354-0341-4/06/$23.00

# MORE ON THE STRENGTH OF MATERIALS
# UNDER HIGH SHOCK PRESSURES

**Y. Ashuach, Z. Rosenberg, E. Dekel and A. Ginzberg**

*RAFAEL, P.O. Box 2250, Haifa, Israel*

**Abstract.** We present a relatively simple and direct technique that is based on simultaneously measuring two stress histories, in a plate impact experiment with a structured target plate. The basic idea is to measure the stresses on the Hugoniot and release path simultaneously and to extract the strength under the high pressure from the difference between the two. Thus, the target is composed from two halves, one of which is a thick Plexiglas plate in which a manganin gauge is embedded some 2-4 mm from the impact face. The other half consists of the specimen backed by a thick Plexiglas, with another gauge at the interface or at some distance into the plastic. The flyer material is a thick specimen disc. The two stress records correspond to the direct impact of the specimen on Plexiglas, resulting in a point on the Hugoniot of the specimen, and the second to a point on the unloading of the specimen. Results are presented for different aluminum alloys, steels and an alumina.

**Keywords:** Shock waves, stress gauges, dynamic strength.
**PACS:** 62.50

## INTRODUCTION

One of the more important issues in the field of the dynamic response of solids concerns the accurate determination of their strength at the high pressures and temperatures existing behind strong shock waves. Several experimental techniques have been developed in order to measure these strengths, all of which rely on the structure of particle-velocity or stress histories inside the shocked specimens. These structures result from the various elastic and plastic shock and release waves which sweep the specimen in a typical 1D experiment (see [1-4] for example). The most rigorous technique for determining strengths at high shock stresses is the self consistent technique of Asay and Lipkin [1] which is based on

monitoring the particle velocities under reshock and release from a given state. This technique has been also applied successfully recently to strong ceramics [5-6].

The purpose of the work presented here was to use manganin stress gauges to determine directly the difference between the shock Hugoniot and the release curves of the specimen, from which a quantitative estimate can be obtained for its dynamic strength. The basic idea is to use a split target in a 1D experiment with two gauges embedded in a well defined soft material (Plexiglas for example). One of the gauges is embedded between two Plexiglas discs while the other is placed at the specimen-Plexiglas interface. The flyer is made of the specimen material.

This arrangement results in a direct measurement on the Hugoniot of the specimen, through its direct impact on the Plexiglas half of the target, and a point on its release curve intersecting the Hugoniot of Plexiglas, from the interface gauge. Comparing the two stress amplitudes one can have a measure for the strength of the material, through the distance between the Hugoniot and the release curves. Some initial results on various aluminum alloys, steel and alumina specimens will be described and discussed.

## EXPERIMENTAL CONFIGURATION

The experimental configuration is shown schematically in Fig. 1. Basically, this is a very simple way to obtain data on shock and release in a single shot. We used commercial manganin gauges (manufactured by Micro-Measurements) which were calibrated in our lab. As we are only interested in the shock waves traversing Plexiglas on both sides of the target, no need for hysteresis corrections were necessary to analyze the data. Fig. 2 shows schematically the elasto-plastic behavior of loading and unloading a specimen in a 1D experiment and the resulting intersections of these curves with the Hugoniot of Plexiglas. Clearly, the higher the strength of a specimen, a larger separation between these points is expected. All the experiments described here were performed with our 64mm gas gun using the standard techniques to measure impact velocity, embed the gauges and reduce the data.

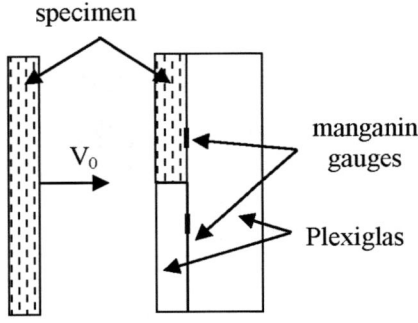

**FIGURE 1.** Schematic description of the experimental configuration.

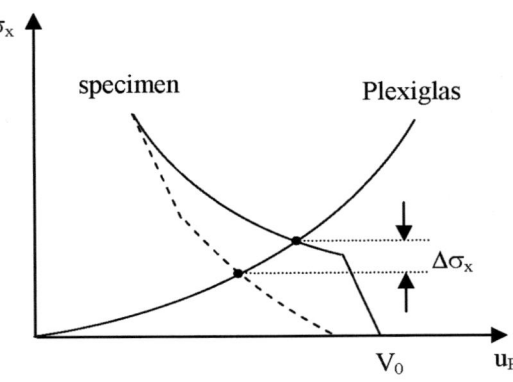

**FIGURE 2.** The elasto-plastic behavior of 1D loading and unloading of a specimen.

## RESULTS AND DISCUSSION

The first experiments were performed with aluminum specimens (2024-T351, 6061-T6 and 7075-T6). These are relatively similar materials having yield strengths in the range of 0.3-0.5GPa with the 7075 alloy somewhat stronger than the other two. Fig. 3 shows the resulting gauge records of this experiment (impact velocity of 405m/s). The stress difference in this experiment (0.05GPa) was intermediate to that obtained for the 6061 alloy (0.04GPa) and the 7075 alloy (0.06GPa) at the same impact velocity. In order to quantify

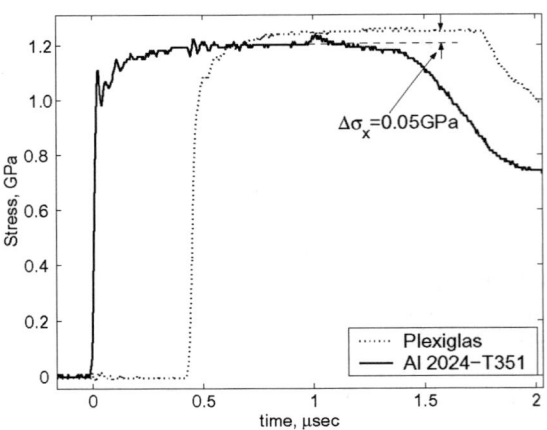

**FIGURE 3.** Gauge records of the experiment with aluminum 2024-T351 specimen.

these differences we performed several 1D simulations of these experiments using the strength of the material as a parameter. The resulting simulations agreed with the experimented records for all three alloys. Thus, our main conclusion is that the technique is sensitive enough to measure even such relatively small strengths and that the gauges are accurate and reproducible to follow these trends.

The next experiments were performed with stainless steel (304L, impact velocity of 685m/s) and a carbon steel (1020, impact velocity of 588m/s). Figs. 4-5 show the results for these materials from which we find an appreciable difference between the two pairs. We expect that the stress difference will be higher for the 304L because of the high strain hardening property of this material. The HEL for the 1020 shot is clearly evidenced in the gauge record (Fig. 4).

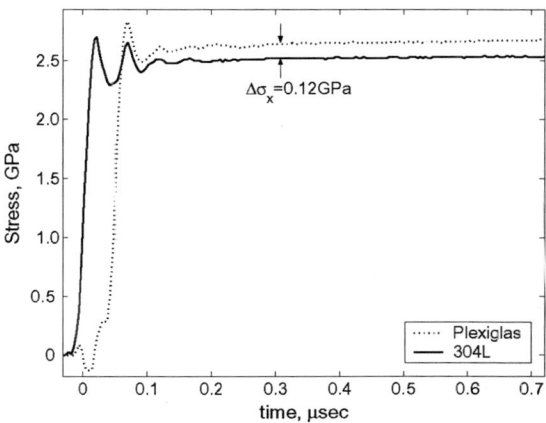

**FIGURE 5.** Gauge records of the experiment with 304L stainless steel specimen.

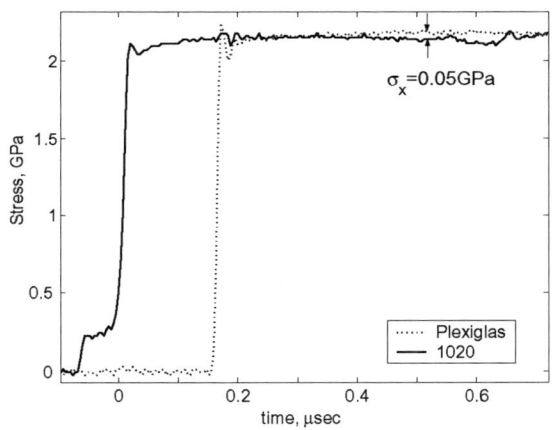

**FIGURE 4.** Gauge records of the experiment with 1020 steel specimen (impact velocity of 588m/s).

Fig. 6 shows the results of 1D numerical simulations which we performed for these experiments. Material parameters for both steel alloys and Plexiglas were taken from published data. The strength of the material (von-Mises yield criterion) was varied as a parameter and the output result was given in term of the stress difference between the two gauges. We did not perform a sensitivity study with other material properties in order to check possible influence on the resulting strength values. This will be done in the near future. From this figure we find that the strength of the 1020 steel at the high stress conditions remains around 0.5GPa, while that of the 304L increases to 1.4GPa. This is quite a high value which is probably the result of the strain rate and the strain hardening effects on this alloy.

Another experiment with 1020 steel was performed at a higher velocity (693m/s) in order to obtain an impact stress above the $\alpha \rightarrow \varepsilon$ phase transition. We expected the release curve from that point to be much farther from the Hugoniot because of the volume change involved in the transition. This should have resulted in a much larger difference between the two stress records, as we clearly obtained in this experiment (Fig. 7).

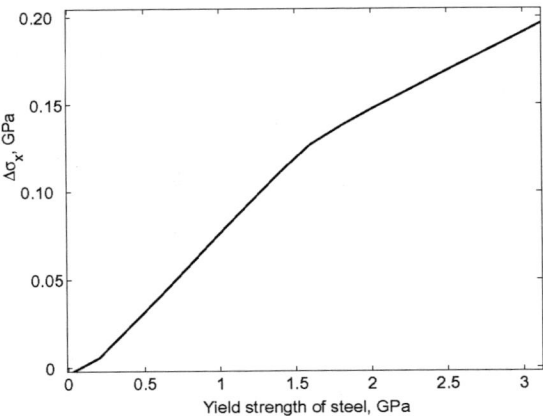

**FIGURE 6.** 1D Numerical simulation results describing the stress difference between the two gauges as a function of steel strength.

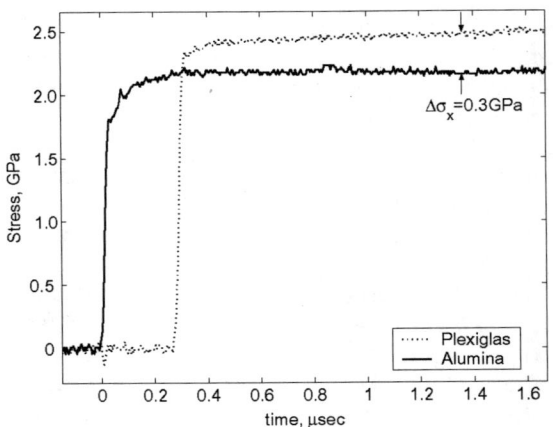

**FIGURE 8.** Gauge record of the experiment with an alumina specimen.

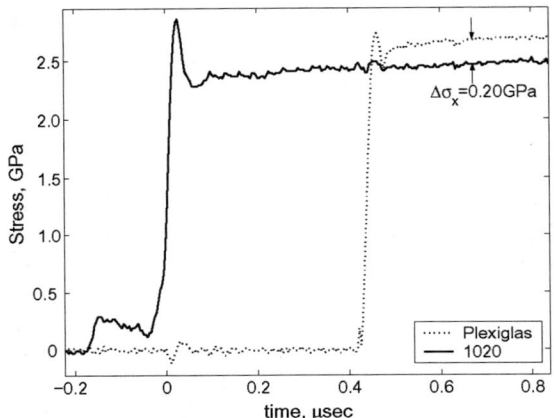

**FIGURE 7.** Gauge records of the experiment with 1020 steel specimen (impact velocity of 693m/s).

Finally, Fig. 8 shows the resulting records from an experiment with an alumina specimen (AD85, impact velocity was 680m/s, stress level above the HEL). It is clearly seen that the difference between the two records is much higher than in all the previous cases as the alumina is much stronger. Thus, the technique we propose here will be especially suitable for high strength solids like armor ceramics and strong steels.

## CONCLUDING REMARKS

We suggest a relatively simple technique, based on manganin stress gauges, to monitor the strength of shock loaded materials directly. The idea is to place two gauges behind a target that is composed of two halves, and compare the stress levels as monitored by the gauge. Several examples are given to demonstrate the usefulness of the technique.

## REFERENCES

1. J. R. Asay and J. Lipkin, *J. App. Phys.*, **49**, 4242 (1978).
2. J. R. Asay and G. I. Kerley, *Int. J. Impact Eng.*, **5**, 69 (1985).
3. J. R. Asay et al., *Shock Waves in Condensed Matter - 1985*, ed. Y. M. Gupta, p.145.
4. Z. Rosenberg, Y. Partom and D. Yaziv, *J. App. Phys.*, **56**, 143 (1984).
5. D. P. Dandekar, W. D. Reinhart and L. C. Chhabildas, *J. Phys. IV France*, **110**, 827 (2003).
6. W. D. Reinhart and L. C. Chhabildas, *Shock Compression in Condensed Matter - 2003*, eds. M. D. Furnish, Y. M. Gupta and J. W. Forbes, p.759.

CP845, *Shock Compression of Condensed Matter - 2005*,
edited by M. D. Furnish, M. Elert, T. P. Russell, and C. T. White
© 2006 American Institute of Physics 0-7354-0341-4/06/$23.00

# TEMPERATURE CONTROLLER SYSTEM FOR GAS GUN TARGETS

## S. M. Bucholtz[1], R. J. Gehr[1], T. D. Rupp[1], S. A. Sheffield[2], D. L. Robbins[2]

[1]*Honeywell Federal Manufacturing & Technologies, Los Alamos, NM 87544\**
[2]*Los Alamos National Laboratory, Los Alamos, NM 87545*

**Abstract.** A temperature controller system capable of heating and cooling gas gun targets over the range -75°C to +120°C was designed and tested. The system uses cold nitrogen gas from a liquid nitrogen Dewar for cooling and compressed air for heating. Two gas flow heaters control the gas temperature for both heating and cooling. One heater controls the temperature of the target mounting plate and the other the temperature of a copper tubing coil surrounding the target. Each heater is separately adjustable, so the target material will achieve a uniform temperature throughout its volume. A magnetic gauge membrane with integrated thermocouples was developed to measure the internal temperature of the target. Using this system, multiple magnetic gauge shock experiments, including equation-of-state measurements and shock initiation of high explosives, can be performed over a range of initial temperatures. Successful heating and cooling tests were completed on Teflon samples.

**Keywords:** Experimental techniques, temperature, shock.
**PACS:** 83.85.-c, 73.43.Fj, 43.40.Yq.

## INTRODUCTION

The initial temperature of a material is an important variable in shock compression experiments. [1-4] Relatively small temperature changes can significantly affect the material properties, including the Hugoniot, strength, and elastic-plastic properties. Additionally, temperature changes on plastic bonded explosives (PBX) not only affect the explosives-binder strength and Hugoniot, but also the shock initiation sensitivity. [5] Thus, it is highly desirable to be able to accurately control the initial temperature of a target material used in a shock physics experiment.

The goal of this work is to develop a heating/cooling system for use with the Los Alamos National Laboratory (LANL) DX-2 single- and two-stage gas guns. This temperature controller needs to be capable of controlling the heating/cooling of a gas gun target to a predetermined temperature within a four hour timeframe, permitting an experiment to be completed in a single day. The system needs to be compatible with the current target mounting and alignment systems and with the standard diagnostics, including LANL's embedded magnetic gauges. In addition it needs to be low cost, as many of the components are destroyed in the experiment.

Typical gas gun targets that have been studied include polymers, high explosives, and foams. All of these materials have poor thermal conductivity, suggesting that conductive heating through the back surface of the sample would be too inefficient to achieve a uniform temperature throughout the sample. Therefore the temperature controller design includes both conductive and radiative heating of the target.

In this paper we describe the design of the controller system, the modeling which was used to determine our design feasibility, and also the heating and cooling tests that were performed.

## TEMPERATURE CONTROLLER DESIGN

The design concept utilized in the temperature controller system is to enclose the target sample in a

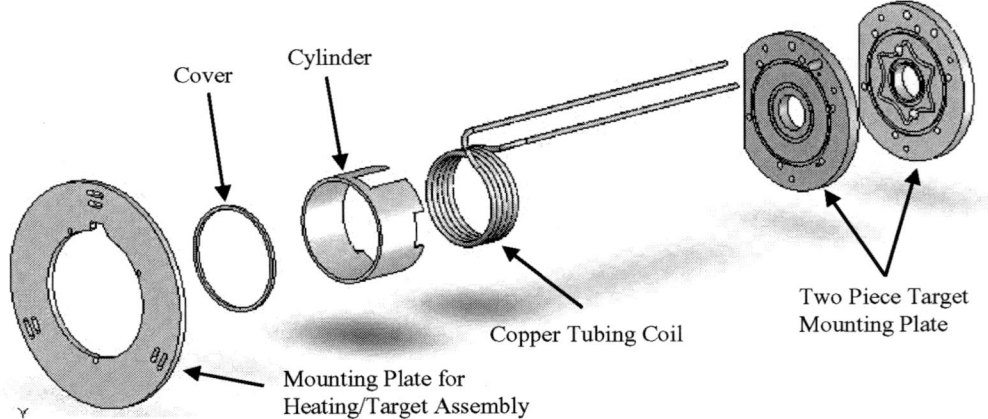

**Figure 1.** The temperature controller system assembly. The target mounting plate, copper coil, and cover create an enclosed environment to control the target sample temperature.

small space and to accurately control the temperature of the walls of that space. To this end, the temperature controller system consists of a mounting plate, coil and cylinder, cover plate, gas flow heaters, flow meters, and tubing. (See Fig. 1.) The target sample is attached to the mounting plate and is surrounded by the coil and the cover plate. The mounting plate and the coil receive separate gas flows whose flow rates and temperatures are controlled with the flow meters and gas flow heaters, respectively. To cool the sample, cold nitrogen gas from a liquid nitrogen Dewar is used; to heat the sample, compressed air is used.

The mounting plate is an assembly consisting of two pieces of aluminum sealed together by o-rings. A star-shaped groove for gas flow is milled in the area between the pieces, creating efficient heat transfer between the gas and the plate. (See Fig. 1.) The target sample is mounted to the plate using an adhesive, chosen both for compatibility with the sample material and for the desired sample temperature.

The coil is simply a piece of 3/16" outer-diameter soft-annealed copper tubing wrapped into a three-turn coil using a mandrel. The inner diameter of the coil is slightly greater than three inches, which allows the target to be mounted and the projectile to pass unimpeded. This design allows for a two-mechanism heat transfer. Conductive heating comes from the mounting plate to which the target is attached. Radiative heating comes from the coil of copper tubing surrounding the target.

The cover is an aluminum ring covered with aluminum foil and insulation. It is attached to the cylinder with a piece of zinc wire, which is used due to its low melting temperature. Immediately before the gun is fired, a piece of nichrome wire is heated to melt the zinc wire and drop the cover.

## COMPUTER MODELING

Once a basic design of the temperature controller was completed, computer modeling was used to determine the feasibility of the design. The ABAQUS [6] finite element analysis code was utilized by Honeywell FM&T's modeling and simulation division. Two-dimensional cylindrical representations of the temperature controller were input into the code. The mounting plate, the cylinder, and cover were modeled as solid aluminum pieces. The coil was modeled as a solid block. A 1" thick, 2" diameter piece of PBX-9501 was used as the target sample. Simulations were performed in both vacuum and air for both heating with +65°C gas flow and cooling with -50°C gas flow. All models were run both with and without the cover in place.

The results of the finite element analysis models were consistent. For both heating and cooling runs in an air environment, the sample failed to reach uniform temperature. The variations were between 10°C and 30°C across the sample. This indicated that our design would not work in an air environment. In a vacuum environment, however, the samples showed a significantly higher uniformity. The variations were between 4°C to 5°C across the sample.

The conclusions drawn from these simulations were as follows. First, a vacuum environment leads to a more uniform sample temperature. Second, the mounting plate temperature must be approximately the same as the desired sample temperature. Third, the coil temperature must exceed the desired sample temperature by several degrees. Finally, the models were inclusive as to whether or not the cover would make a difference in sample temperature or temperature uniformity. Overall, the simulations suggested that the design should meet the system specifications.

## EXPERIMENTAL METHOD

A Teflon sample 2 inches in diameter and 1 inch thick was used for the temperature control tests. Three thermocouples were inserted into holes bored into the sample: one at the sample center; one ½" deep and ¼" up from the mounting plate; and one ¼" deep and ¾" up from the mounting plate. In addition, once the sample was attached to the mounting plate, thermocouples were glued to the front and back faces, providing a total of five temperature readings at a number of sample locations.

Thermocouples were also attached to the mounting plate and the copper coil to provide feedback on their temperatures relative to the sample temperature. Data collection was accomplished using an InstruNet 100B direct to sensor data acquisition unit [7] communicating with a PC. Adjustments to the thermocouple offsets were determined using a hand held thermocouple reader and a bundle of reference junctions immersed in an ice water bath.

The gas flow heaters were controlled using variable transformers (variacs). These variacs were initially set to approximately the correct voltage level for the desired sample temperature and left in that position until the mounting plate and copper coil approached this temperature. The variacs were then actively adjusted to stabilize the mounting plate and coil at a temperature within a few degrees centigrade of the desired temperature. As the sample temperature approached its target level, final adjustments to the variacs were made to stabilize the sample.

## EXPERIMENTAL RESULTS

### Heating Tests

The goals of the heating tests were to reach 50°C, 100°C and 150°C in fewer than four hours with less than +/- 2°C variation across the sample. Both the 50°C and 100°C tests were successful; the 150°C test, discussed below, encountered a problem. In the 50°C test, the Teflon sample reached temperature without overshooting in about 100 minutes. (See Fig. 2.) It was then maintained at this temperature for 70 minutes with only small fluctuations. The mean temperature of the sample was 48.5°C, with a spread of less than +/-1.5°C. In the 100°C test, the sample reached temperature with some overshoot in about 120 minutes. This temperature was then maintained for 60 minutes. (See Fig. 2.) The mean temperature of the sample was 100.7°C, with a spread of less than +/-1.5°C.

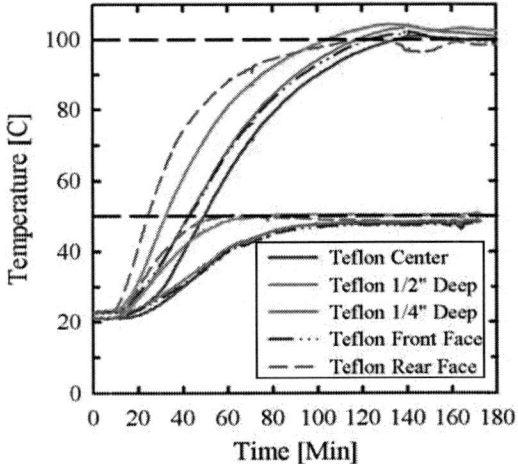

**Figure 2.** Results from the heating tests to +50°C and +100°C. The five lines represent data from the five thermocouples in and on the sample. The temperature spread for each test is less than +/-1.5°C.

During the 150°C test, the temperature of the rear face of the sample increased in conjunction with the mounting plate as expected, but the temperature of the rest of the sample lagged behind unexpectedly. After 180 minutes, the rear face was at temperature, but the front face was more than 60°C cooler. It was therefore concluded that a problem had arisen. The test was aborted and the target chamber reopened. It was discovered that the cover plate had dropped off, leaving the sample open to the chamber. Controlled tests of the zinc wires using weights and a hot plate proved that the wires softened at temperatures over 130ºC, sufficient to allow the cover drop off. Therefore a modification of the cover plate mounting system is necessary to achieve temperatures over approximately 120ºC. This test demonstrated that the cover is critical for temperature uniformity.

### Cooling Tests

The goals of the cooling tests were to reach -25ºC and -50ºC in fewer than four hours with less than +/- 2ºC variation across the sample. In the -25ºC test, the Teflon sample reached temperature in about 130 minutes, (Fig. 3) and was then maintained at this temperature for 40 minutes with only small fluctuations. The mean temperature of the sample was -24.6ºC, with a spread of +/-2.7ºC. This spread is greater than expected, and slightly out of specification. Experiments involving the gas flow rates and tubing insulation are underway in an effort to correct this issue.

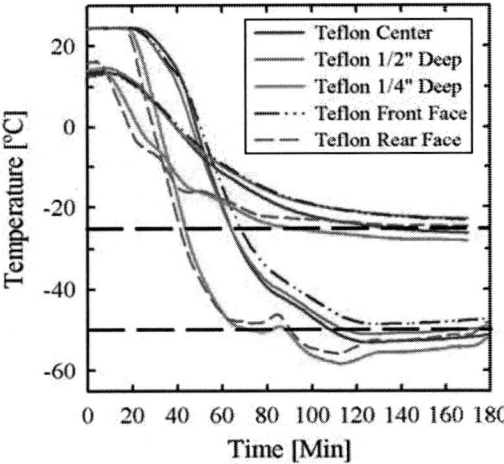

**Figure 3.** Results from the cooling tests to −25ºC and −50 ºC. The five lines represent data from the five thermocouples in and on the sample. The temperature spreads are +/-2.7ºC and +/-3.2ºC, respectively.

In the -50ºC test, the sample reached temperature in about 130 minutes and was then maintained at temperature for 50 minutes. (See Fig. 3.) The mean temperature of the sample was -51.2ºC, with a spread of +/-3.2ºC.

## CONCLUSIONS

A temperature controller system for gas gun targets has been designed to accurately control the temperature of low thermal conductivity samples (such as explosives and polymers). The system has been successfully used to control a gas gun sample target to the desired temperature with minimal overshoot/undershoot. It is flexible in that it can either heat or cool the sample and can work with either the single- or two-stage gas guns at the LANL DX-2 facility. Tests on a simulated target reached the following temperatures and variations: +48.5ºC +/-1.5ºC, +100.7ºC +/-1.5ºC, -24.6ºC +/-2.7ºC, -51.2ºC +/-3.2ºC. The cold temperature variations are higher than hoped, but minor modifications to the cooling system should decrease the temperature spread.

## ACKNOWLEDGMENTS

This work is supported by the NNSA Enhanced Surveillance Campaign through contract DE-ACO4-01AL66850. The authors would like to acknowledge Pete Chavez and Joe Lloyd for their invaluable assistance at the LANL gas gun facilities. We would also like to thank Jim Mahoney and the modeling team at KCP for the finite element analysis models.

## REFERENCES

1. Nellis, W. J. and Mitchell, A. C., "Shock compression of liquid argon, nitrogen, and oxygen to 90 GPa (900 kbar)," J. Chem. Phys. **73**, 6137-6145 (1980).
2. Mitchell, A. C. and Nellis, W. J., "Equation of state and electrical conductivity of water and ammonia shocked to the 100 GPa (1 Mbar) pressure range," J. Chem. Phys. **76**, 6273-6281 (1982).
3. Urtiew, P. A., et al., "Shock Initiation of LX-17 as a Function of Its Initial Temperature," in The Ninth Symposium (International) on Detonation, Portland, OR (1989) pp. 112-122.
4. Stahl, D. B., et al., in Shock Compression of Condensed Matter, 1999, (M.D. Furnish, L.C. Chhabildas, R.S. Hixson, eds.), pp. 1087-1090.
5. Dallman J. C. and Wackerle J., "Temperature-Dependent Shock Initiation of TATB-Based High Explosives," in The Tenth International Detonation Symposium, Boston, MA (1993) pp. 130-138.
6. ABAQUS, A suite of general and special purpose analysis products, Version 6.4, ABAQUS, Inc.
7. GW Instruments, Inc. Website www.instrunet.com

CP845, *Shock Compression of Condensed Matter - 2005*,
edited by M. D. Furnish, M. Elert, T. P. Russell, and C. T. White

# EXPLOSIVE FORMING OF AEROSPACE COMPONENTS

## E.P. Carton, M. Stuivinga, H.J. Verbeek

*TNO Defence, Security and Safety, P.O. Box 45, 2280AA Rijswijk, The Netherlands*

**Abstract** Results are presented of the development of explosive forming technology for metal sheets and plates. Explosive forming is labor intensive, but requires only single-sided tooling it can be used economically for small series of hard to deform metals, like nickel, titanium and aluminum alloys, that are generally used in aerospace applications. As the alloys can be formed in their hardened (tempered) condition, the formed components do not need a heat-treatment afterwards (at which deformations generally occur). Plate velocity calculations and measurements can be used for engineering calculations. Examples are given of aerospace parts of aluminum and nickel alloys for process development and the fabrication of demonstrators.

**Keywords:** Dynamic plastic deformation, Explosive forming, Gurney-theory.
**PACS:** 81.20.Hy, 81.40.Vw

## INTRODUCTION

Explosive forming is a high-energy-rate plastic deformation process for metal plates and tubes in which an explosive charge is used as energy source. It has been developed about fifty years ago and has been used in the seventies of the last century for several aerospace components such as the forming of large aluminum alloy sheets into dome segments for the Saturn rockets in the USA [1]. Nowadays, the process is only applied by a small number of companies and research institutes. An overview of this technology and its users has recently been published [2].

Only single sided dies are required which is a large benefit considering the high machining costs for the manufacture of dies with complicated shapes. Also no large machines or special power supply are required, which enable the forming of large and thick-walled metal plates. However, the clamping of metal plates and the handling of explosive charges make it a rather labor intensive activity which drives up the recurring costs. The combination of its modest start-up costs and the labor intensive

character make this forming process only profitable for the fabrication of single pieces and small series.

## THE FORMING PROCESS

Explosives form a powerful, cheap and highly reproducible source of energy. Contrary to other forming methods it does not require a machine. This is considered as an advantage since the finite size and strength of a machine intrinsically limit the products in size and thickness that can be formed. As opposed to that, an explosive charge can be adjusted to any scale strength and thickness of the metal to be formed.

The explosive charge can be concentrated in one spot or distributed such that the energy release is spread over the metal evenly (using detonation cords). Usually, the explosive is not in direct contact with the plate, but situated under water which is a good shock wave transmitting medium since it is a rather incompressible and pore free space filler.

The plate is placed over a die and clamped at the edges. The space between the plate and the die may be vacumized before the forming process in order to prevent back pressure (and local metal oxidation due to high gas temperatures) from compressed air.

Usually the die and explosive charge are lowered in a water tank in which the forming process takes place. However, for small parts it is easier to place a water filled bag on top of the die and detonate the explosive charge placed in the water at the required stand-off.

At the moment of detonation a shock wave is generated in the water which propagates in all directions with a speed depending on the shock intensity (typically about 1600 meter per second a few centimeters away from the high explosive charge). When the shock wave arrives at the metal surface it is reflected back into the water and the metal obtains momentum and starts to move into the die cavity. The velocity of the metal depends on the intensity of the shock wave in the water and the mass of the plate, but is generally on the order of tens of meter per second. Due to the much higher velocity of the shock wave, the metal starts to move practically as one body. As the water can only generate stress perpendicular to the metal surface, the metal is also accelerated only in this direction. Due to the constraints generated by the die and the clamping of the metal, the periphery of the metal can not move together with the rest of the plate and the plate will start to bend over the radius of the edges of the die. This bending motion is then moving through the metal plate with its own velocity.

Depending on the amount of clamping the plate will experience an in-plane tensile stress and the metal will stretch (resulting in plate thinning) or the plate will start to pull in metal from below the clamps. Normally both plate reactions occur at the same time. The metal will be slowed down as more and more of its kinetic energy is transferred to plastic deformation of the plate.

When all kinetic energy is used before the metal has touched the die surface, the metal is free-formed and will need additional (explosive) forming steps.

When the shock wave accelerated metal hits the die surface it will be formed according to the shape of the die and the required form is obtained.

## GURNEY THEORY

A simple and first order estimate of the *maximal* plate velocity can be obtained using the Gurney theory [3]. This theory (based on energy balance) is capable of calculating the maximal velocity of explosively accelerated metal plates, spheres and tubes. Apart from the empirically determined specific energy of the explosive used (the so-called Gurney energy of an explosive type, $E_g$), only the mass ratio between the explosive (C) and the metal part ($M_P$) is needed. The original theory derifed in [3] applies to a cylinder completely filled with explosive material. Due to the presence of the water the theory has to be slightly adjusted, resulting in equation 1. The plate velocity then also depends on the mass of the water and on the inner (a) and outer (b) radius of the water (see figure 1). Depending on the standoff distance and the form in which the explosive has been applied underwater (concentrated or as a linear charge), the shocked water will have the form (and volume) of a sphere or cylinder. For example, we will calculate the maximal velocity of an expanding aluminum cylinder with a radius ($b_0$) of 90 mm, a wall thickness of 2.5 mm and a height of 250 mm. The cylinder is filled with water and a detonation cord is positioned on its axis of symmetry (12 g/m PETN, radius $a_0$ of 4 mm). First, the Gurney velocity of PETN should be found; from literature one finds $\sqrt{2E_{g,PETN}}$ equals 2930 m/s [3]. Then the mass of the plate and the shocked water should be calculated. However, for thin plates and regular stand-off distances (>50 mm), the mass of the shocked water greatly exceeds the plate mass and an upper bound of the plate velocity can be found using only the shocked water mass. In our case the metal cylinder is free formed (die less) and completely surrounds the explosive charge. Therefore, the efficiency of the forming process will be rather high. The detonating cord generates a cylindrical shock front in the water, with a velocity distribution that relates to 1/r. The highest water velocity occurs at the explosion products-water interface. The water in the cylinder has a volume of 6.4 liter and therefore weights 6.4 kg. The mass of the aluminum cylinder $M_P$ is about 1 kg. The mass of the explosive charge is only 3 gram (0.25 m * 12 g/m).

$$V_P = \sqrt{2E_g \left( \frac{M_P}{C} + \frac{M_W}{C} \frac{2b^2}{b^2 - a^2} \ln\left(\frac{b}{a}\right) \right)^{-1/2}} \quad (1)$$

Using equation (1), initially a maximal plate velocity of 34 m/s is obtained. Upon expansion of the cylinder the kinetic energy distribution within the water further accelerates the plate. At a radial expension of 10 mm the plate velocity is 43 m/s.

The initial velocity of the plate is close to the 30 m/s that has been measured experimentally using strain gauges on explosively expanded cylinders, and laser-triangular plate position measuring methods.

The laser positioners are based on the measurement of the position of the laser light reflection on the (moving) plate surface.

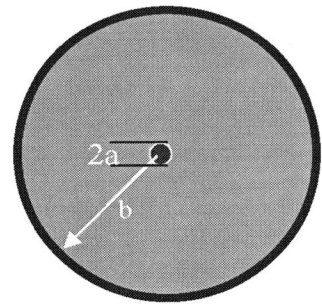

**FIGURE 1.** Schematic set-up for explosive forming a water filled metal cylinder (mass M) with a detonation cord (mass E) at a stand-off (R).

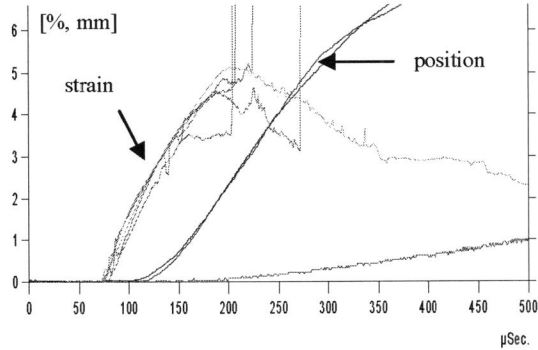

**FIGURE 2.** Result of strain and plate position measurements using strain gauges and laser reflection.

In Figure 2 the results of the measurements are presented. All three strain gauges give the same strain and strain-rate information, and two laser triangular positioners also agree. The third positioner had a wrong setting (integrating) resulting in a too low measuring response. From the slope of the lines (strain and position versus time) one can determine the strain-rate (375 s⁻¹) and plate velocity (30 m/s). The calculated plate velocity is not far off but indeed an upper bound velocity. The measurements do not show a further acceleration of the metal plate upon cylinder expansion. This is probably due to energy escaping the water cylinder system. This is not surprising considering the small length to diameter ratio of the cylinder used here.

Once the (upper bound) maximal plate velocity is known, one can estimate the strain-rate and the kinetic energy of the plate. Knowing the strain required for the metal preform to obtain the required (die) form and the yield stress of the metal, the required energy for the forming can be compared to the available kinetic energy of the metal. In many cases the final form can only be obtained by explosively forming the product in at least two forming steps, e.g. a forming and a calibration shot.

**EXAMPLES**

Research has been done on explosive forming segments of an AA2024 stiffening ring for the main engine frame of ESA's space vehicle Ariane 5. The standard fabrication route involves hammering using several (lead) dies where the alloy is in the annealed condition. This has to be followed by a heat treatment to the high strength aged condition (T3).

By explosive forming this alloy can be formed directly in the T3 condition and only a single sided die is needed. An explosively formed ring segment together with the starting metal plate is shown in Figure 3.

A second example of the explosive forming of aerospace components is the forming of a gas mixer for gas turbine engines. This product is typically made by welding together numerous hot formed segments (each containing one fold). The double curved product can, however, be made out of just one flat plate by explosive forming, see Figure 4. The starting material is a circular disk which is first simply bended to contain the number of folds that are required, see

Figure 4(a). This preformed plate is then positioned in a die that has the double curved form. By explosive forming using an axially positioned under water explosive charge, the final shape is obtained, see Figure 4(b). This fabrication method is quite simple and a lot of weld lines (as well as their inspection) are avoided.

## CONCLUSIONS

Explosive forming is a batch process for small series of complex parts of hard to form materials. The forming in more then one process step makes it rather labor and costs intensive, thereby reducing the production capacity. However, only single-sided dies are used, which will reduce the initial production costs.

This forming method is capable of fabrication of complicated double curved products with the metal in the aged (hardened) condition. This avoids heat treatments and welding of several small parts after their forming and therefore may lead to better and less expensive fabrication of products.

For engineering calculations of the forming process, the plate velocity can be estimated using the Gurney-theory. Here one has to incorporate the mass of water as well as its velocity distribution.

(a)

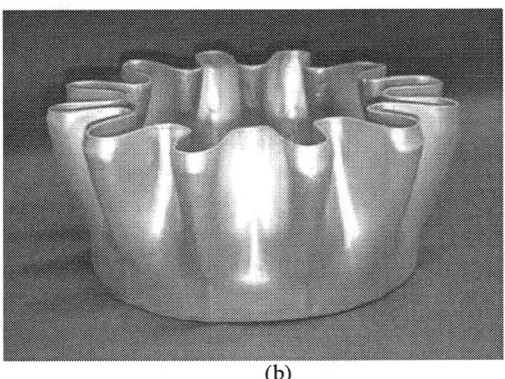
(b)

**FIGURE 4.** Gas mixer for gas turbine exhaust made by explosive forming (b) using a preformed disk (a). (Courtesy of Exploform BV)

This has been supported by strain and plate velocity measurements that were obtained using (resistance) strain gauges and laser-triangular positioners, respectively.

### ACKNOWLEDGEMENT

We would like to thank Dutch Space BV (Leiden, The Netherlands) for her active involvement in the explosive forming research on the Ariane 5 ring segments. This part of research was financially supported by the NIVR and the Dutch Ministry of Economic Affairs

### REFERENCES

1. Enzra, A.A., Principles and practice of explosive metalworking, Garden City Press Limited, London, 1973.
2. Mynors D.J., and Zhang B., Journal of Materials Processing Technology 125-126, pp. 1-25, 2002.
3. Meyers, M.A., Dynamic Behavior of Materials, John Wiley & sons, 1994.

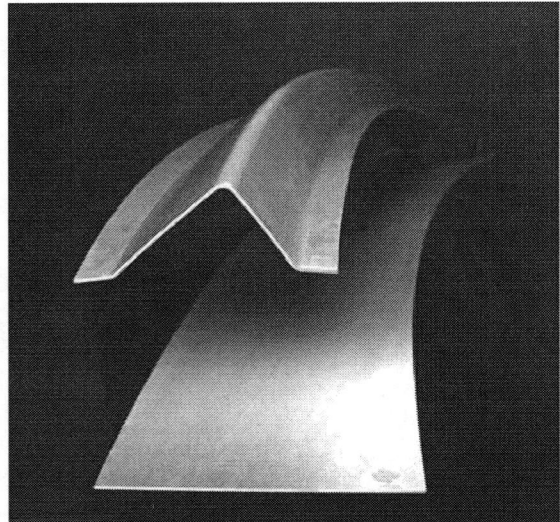

**FIGURE 3.** Starting material (AA2024-T3) and explosively formed ring segment.

CP845, *Shock Compression of Condensed Matter - 2005*,
edited by M. D. Furnish, M. Elert, T. P. Russell, and C. T. White
© 2006 American Institute of Physics 0-7354-0341-4/06/$23.00

# X-RAY DIFFRACTION STUDIES OF THE STRUCTURES OF DYNAMICALLY COMPRESSED BE, AL, LIF, KCI, AND SIO₂

## L.A. Egorov, A.l. Barenboim, V.V. Mokhova, A.I.Samoilov

*Russian Federal Nuclear Center All-Russia Research Institute of Experimental Physics*
*607190, Sarov, Nizhni Novgorod region, Russia*

**Abstract.** Results from an X-ray diffraction structural study of crystals under shock-wave compression are presented. An analysis of the date leads to the conclusion that at the shock front, the crystals undergo universal structural changes over the entire range of pressures studied. X-ray diffraction patterns suggest that the relaxation process of structural change is related to a change in the state of the crystal electron subsystem responsible for the chemical bonding in substance. The process of structural change at shock front can be divided into two stages: 1) uniaxial compression of the structure, the compression direction coincides with the direction of shock-wave propagation; 2) transformation of the unstable, uniaxially compressed, initial structure to a structure determined by the degree (depth) of relaxation of the electron subsystem responsible for the chemical binding of the ionic component of the crystal. X-ray diffraction patterns of a number of crystals at both stages of the process of structural change are presented.

**Keywords:** X-ray diffraction, dynamical compressed crystals, relaxation.
**PACS:** 62.50.+p, 61.10.Nz

## INTRODUCTION

Over the last two decades we have implemented a program of dynamically compressed crystal studies using x-ray diffraction analyses of the structures realized behind a shock front. We obtained the first x-ray diffraction data for restructuring of the material subjected rapid deformation. Polycrystalline samples of beryllium and aluminum, as well as single crystal samples of lithium fluoride, were studied at pressures higher than their Hugoniot elastic limits (HEL's), single crystal samples of α-quartz and lithium fluoride were studied at pressure below the HEL's, and single crystal samples of potassium chloride were studied at pressures above the phase transformation point.

## EXPERIMENTAL RESULTS

The parameters of the structures realized were determined with the relative error of measurement of the interplanar spacing of not worse than 0.5%. The recorded diffraction patterns of crystalline samples undergoing rapid deformation demonstrate the universal process of structural material reorganization. Outside the shock front, the transformation process slows down, and this allows us to observe the spectrum of the "frozen" structural states. As a rule, among them there is equilibrium structure state, whose parameters correspond to the pressure and temperature for the material studied; other states are nonequilibrium structures.

The diffraction patterns imply that a relaxation process of substance restructuring connects changes of a crystal's electron subsystem that chemically bond matter. The process of substance restructuring can be divided into two stages. The first stage is uniaxial compression of the sample structure; the direction of the compression coincides with the shock front propagation direction. One can use the assumption that such structure deformation moves the sample structure (and substance) into the "transition" state analogous to H.Eyring`s activated complex [1].

Apparently the relaxation of the crystal`s electron subsystem occurs during this stage. The second stage converts the unstable structural "transition" state into states which are determined by a degree (a depth) of the electron subsystem relaxation. The x-ray diffraction patterns of the "transition" states of the dynamically compressed crystals are presented in [2-5].

The examples are of structures of NaCl and Si samples uniaxial compressed with the direction of the compression coincides with the direction of the shock front propagation are displayed there. In the same way the x-ray diffraction patterns for the number of dynamically compressed crystals are presented in [6-9]. The examples of the structural states generated by second stage of the substance restructuring process are displayed there.

Details of the experimental procedure are described elsewhere [8]. The examples of the x-ray diffraction patterns of the dynamically compressed crystals are presented on Fig. 1 and Fig. 2. These patterns from [8,9], display the spectra of the structural states arising behind a shock front. No diffraction patterns display the "transition" structural state; the apparent reason for this is using

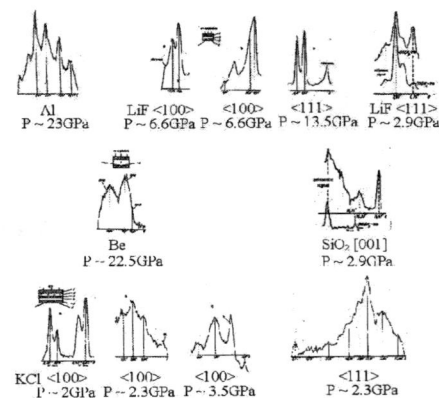

**Figure 2.** Optical density profiles of photographs

of the nonoptimal experimental geometry for registration such states.

The x-ray diffraction pattern of a KCl sample is presented in Fig. 3, clearly displays the peculiarities of the structural transformation while the substance relaxes. The experimental geometry differs from the ordinary geometry by the orientation of the anode of the x-ray tube. This difference becomes clearer when the experimental geometries presented on Fig. 1 and Fig. 3 are compared. The single crystals of KCl sample were dynamically compressed along a crystalline direction <100> under the pressure .~ 0.2 GPa.

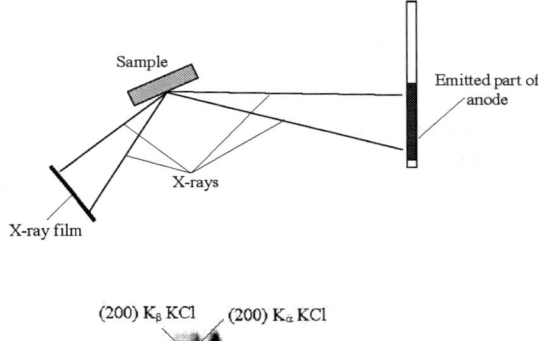

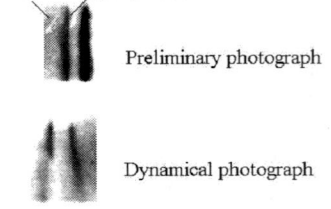

**Figure 3**. Photographs of the shock compressed crystal KCl

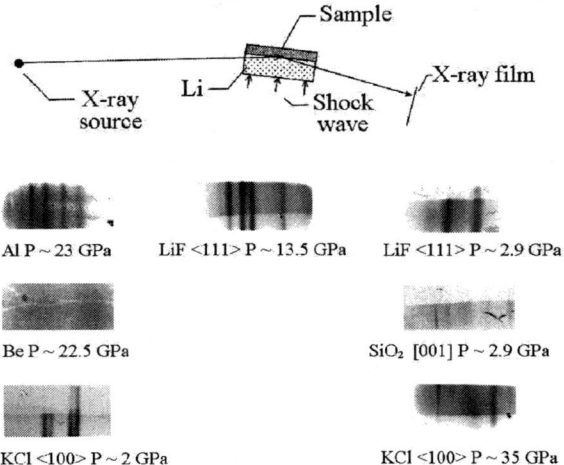

**Figure 1**.Diffraction photographs of number of shock compressed crystals. Exposure time was ~250 ns.

The preliminary photograph registers two components (Kα and Kß x-rays) of the radiation of the molybdenum anode. The angular distance between them equals the difference of the Bragg`s diffraction angles ΘKα and ΘKβ for Kα and Kß rays accordingly emitted from different points of the anode that satisfy for the crystal. The distance (l ) between the diffraction reflection on X-ray film is l = R tg (ΘKα – ΘKβ ), R – a distance between the simple and X-ray film. In the ordinary using experimental geometry for x-ray spot anode the angle distance between these rays is twice as large: l = R tg 2 (ΘKα – ΘKβ )..

The dynamically photograph shows the diffraction pattern formed by two superposed patterns. One pattern is identical to the diffraction pattern of the preliminary photograph with little increase of the distance between Kδ and Kß x-rays, the other characterizes itself the twice as large angle distance between them. The complete diffraction pattern can be understood by assuming that the layer of substance investigated disintegrated into two structural states. One is uniaxial compression state of the crystal structure - "transition" state. Other structural state can be related to the part of the investigated layer of the simple transformed by second stage of the restructuring process. This state has the peculiarity of x-ray scattering characteristic for polycrystalline states: there is the dilatation of the profile of x-ray reflection and the angle distance between Kδ and Kß x-rays is the twice as large of the difference of its Bragg`s diffraction angles. Disintegration of the part of the "transition" state takes place, and a small crystallites little slightly disorientated (for ~ 5°) relatively to the direction of shock front propogation were appeared. Apparently this is the characteristic peculiarity of the structural states generated by second stage of the substance restructuring process. This part of the complete diffraction pattern was generated automatically by the brightest spot of anode with the ordinary experimental geometry.

The nonequilibrium structures presented in Figures 1 and 2 have the unexpected packing symmetry. So, for cubic symmetry crystals the unit cell of the Bravais lattice is distorted: the angles between the edges different from $90^0$ ; hexagonal singony crystals have the nonequilibrium ratio of the hexagonal prism height to the base edge.

Formation such structures cannot be understood on the basis of existing ideas physical processes occurring with rapid deformation of a material. The concepts of dislocation kinetics, which are the basis for these ideas, do not incorporate process which changes the packing symmetry of the atoms. Apparently, the representation of transformation of chemical bonds in the condensed matter under dynamical loads must be modified.

## ACKNOWLEDGEMENTS

Funding was provided by the ISTC, under Project ISTC #2514. The authors thank Dr. L.C. Chhabildas for valuable comments.

## REFERENCES

1. Eyring H., "The Activated Complex in Chemical Reactions" J..Chem.Phys. 1935, N1, pp.107-108
2. Jamet F., Bauer F., "Analyse Radiocristallgraphique de la Deformation de la Structure Cristalline Chlorure de Sodium Soumise a une Compreeion par de Choc", Comportement des Milieux Denses Sous Hautes Preeions Dynamiques, Paris, 1978, pp.409-421.
3. Müller F., Schulte E., "Shock Wave Compression of NaCl single Crystals observed by Flash X-ray Diffraction" Z. Naturforsch, 1978, **33**, pp. 918-923.
4. Wark J.s., Whitlock R.R., Hauer A., Swain J.E. and Solone P.J. "Shock launching in silicon studied with use of pulsed X-ray Diffraction", J.Physical Review B Rapid. Comm., 1987, **35**, N17, pp..9391-9394.
5. Zaretsky E. "Pulse X-ray Diffraction study of Shock-Compressed NaCl", J. Phys.IV, France, 1997, **7**, pp.329-334.
6. Johnson Q, Mitchell A.C., Evens L., "X-ray Diffraction Study of crystalls undergoing shock wave compression", J. Appl. Phys. Letters, 1972, **21**, N1, pp. 29-30
7. Johnson Q. and Mitchell A.C., "First X-ray Diffraction Evidence for a Phase Transition during Shock Wave Compression", J. Phys.Review Letters, 1972, **29**, N20, pp.1369-1371.
8. Egorov L.A., Barenboim A.l., Makeev N.G., Mokhova V.V., and Rumyantsev V.G., "X-ray diffraction studies of the structures of dynamically compressed Be, Al, LiF, KCl, and Fe+ 3%Si", JETP 1993, 76(1), pp. 73-81
9. Egorov L.A., A.l. Barenboim L.A., Mokhova V.V., and Samoilov A.I. "X-ray diffraction measurements for structural parameters for dynamically

compressed SiO$_2$, Si, and LiF under pressures below elastic Hugoniot limit" J.Khimich. P., **14**, N 2-3, 1995, 100-105.

CP845, *Shock Compression of Condensed Matter - 2005*,
edited by M. D. Furnish, M. Elert, T. P. Russell, and C. T. White
2006 American Institute of Physics 0-7354-0341-4/06/$23.00

# SMALL-SCALE SHOCK REACTIVITY AND INTERNAL BLAST TEST

## R. H. Granholm and H. W. Sandusky

*Naval Surface Warfare Center, Indian Head, MD 20640*

**Abstract.** Explosives react from a strong shock, even in quantities too small for detonation. The potential for a new material to be an explosive can be evaluated from this shock reactivity. The recently developed small-scale shock reactivity test (SSRT) uses very high confinement to allow prompt reactions to occur in less than half-gram samples well below critical diameter. Early and late-time reactions are simultaneously measured from a single sample subjected to the output from an RP-80 detonator. Prompt reactions are quantified by a dent in a soft aluminum witness block, while later reactions, such as from fuel/air combustion, are measured by recording blast pressure. Internal blast quasi-static pressure is obtained by confining the sample apparatus within a three-liter chamber. Late-time reaction effects of plastics, and results from HMX, HMX/Aluminum, and a plastic-bonded explosive (PBX) are reported.

**Keywords:** Small-scale, shock reactivity, internal blast, Cheetah, heat of combustion.
**PACS:** 82.33.Vx, 82.40.Fp.

## INTRODUCTION

Recent interest in internal blast, or explosion in confined space, has generated data on a variety of explosives tested in large-scale chambers up to 180 cubic meters in volume, with charge weights up to 23 kg [1]. The Small-Scale Shock Reactivity and Internal Blast Test (SSBT) adds an internal blast measurement to the previously developed SSRT by conducting the test within a small chamber and recording the pressure of the reaction products. In the SSRT [2], a 7.2 mm diameter by 6.4 mm high sample is radially confined in a steel block. On top of the sample is an RP-80 detonator and under it is a soft aluminum witness block. The detonator shock and any prompt reaction of the $< \frac{1}{2}$ g sample dents the witness block. These reaction products then vent into the SSBT chamber for observations of late-time reactions. The total explosive (detonator plus sample) mass to chamber volume is similar to that in the large-scale tests.

## EXPERIMENTAL ARRANGEMENT

The setup consists of a 3-liter explosion-proof electrical junction box for enclosing the SSRT apparatus, as shown in Fig. 1.

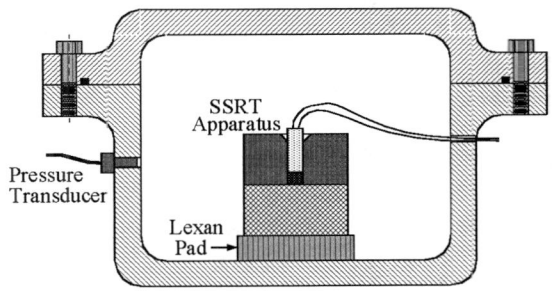

**Figure 1.** SSRT apparatus inside 3-liter chamber.

The chamber is cast aluminum with internal dimensions 15x15x12.5 cm, with 1.3 cm wall thickness (# AXJ664-N4, made by Akron Electric, Inc.), and scales closely to the large-scale chamber. Rectangular geometry is presumed to enhance air-mixing when compressive waves rebound from corners. The chamber is sealed except for a 5 mm hole for detonator leads and a small gas port for flushing with nitrogen in some tests to show reactions of just the explosive without any additional oxygen. In that case the chamber is purged with nitrogen for 15 minutes at 15 liters/min. A pressure gage is centrally mounted in one chamber wall.

As will be shown, plastic or any excess fuel can contribute to quasi-static pressure, so tape or other combustibles were eliminated from the chamber, which was carefully cleaned before each shot. The Lexan pad beneath the SSRT apparatus and the Teflon insulation on the detonator leads were recovered intact from all tests.

### Instrumentation

Chamber pressure is measured with a Kulite XTE-190 piezo-resistive transducer having a response time of a few microseconds. The gage is powered by a 9V battery and recorded without amplification on a Nicolet Integra 40 oscilloscope, with a 5 ms pretrigger in all tests. Pressure profiles from the SSBT are similar to those from large-scale chamber tests (Fig. 2).

Following reaction of the sample, pressure waves reach the chamber walls and produce multiple reflections which appear as high-frequency ringing at the beginning of the recorded pressure traces. Since the chamber is essentially sealed, a quasi-static pressure develops and then decays as the product gases cool and slowly vent from the detonator lead hole. In the SSBT, pressure decays by 80 - 90% in 0.5 s. The method of selecting peak quasi-static pressure shown

in Fig. 2 was approximated in the current study without smoothed fits of the data.

## RESULTS AND DISCUSSION

### Detonator Without Sample

The 0.204 g of explosive in the detonator is often more than half the mass of the sample. The special bare output Teledyne RISI RP-80 used in the SSRT has PETN initiator and output pellets in a Delrin sleeve, without the usual aluminum cup. The header is a glass-filled phenolic plastic (~35% phenolic) weighing 0.75 g. As shown by the difference between air and $N_2$ atmospheres in Fig. 3, the plastic contributed so much internal blast that the sleeve was switched to brass for all subsequent tests. The plastic in the header reacts more slowly than the plastic sleeve because it is in contact with only the low-density initiating pellet. This slow reaction is seen in the brass-sleeved RP-80 trace in Fig. 3, where peak quasi-static pressure is not reached until 45 ms. The Delrin-sleeved detonator reaches its peak more quickly, due to direct contact with the output pellet, and the higher temperature and pressure in the chamber.

In nitrogen atmosphere, both sleeve types gave the same pressure. In air, peak pressures increased by 3.5 and 6 times, respectively, for brass and Delrin sleeve detonators. The air-filled 3-liter chamber (~2900 cc free space with the SSRT apparatus) contains 0.80 g oxygen which could consume 0.32 g of phenolic plastic, or 0.75 g Delrin. In nitrogen, the detonators reach peak quasi-static pressure rapidly, in about one millisecond. In air, there is again a rapid initial rise which then increases more slowly as the remaining

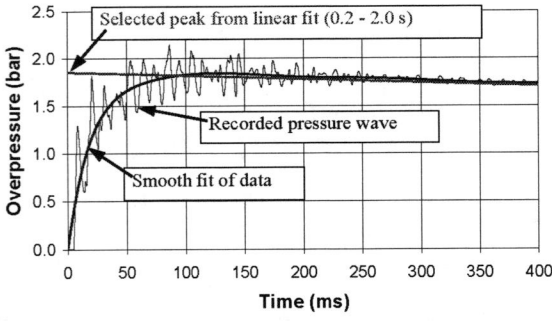

**FIGURE 2.** Example pressure profile from large-scale (180 m³) chamber test (with permission, R. J. Lee [3]).

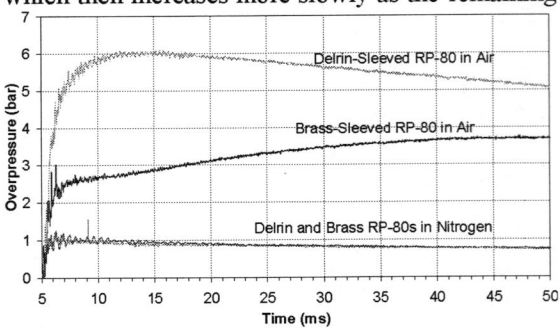

**FIGURE 3.** Detonator-only pressure traces in 3-liter chamber.

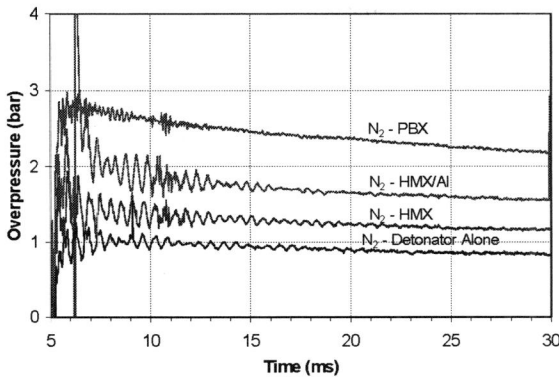

**Figure 4.** Pressure traces in 3-liter chamber for tests in $N_2$.

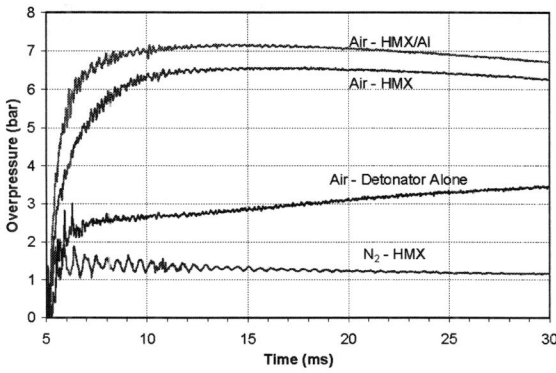

**Figure 5.** Pressure traces in 3-liter chamber for tests in air, including one test in $N_2$ copied from Fig. 4.

fuel in the explosive, and plastic in the detonator, mixes with air and burns.

### Tests with Samples

Three sample materials were tested: 0.3 g of HMX Class 1 powder, alone and with 0.075 g Valimet H-5 atomized aluminum (8 μ), and a cast-cured plastic-bonded explosive (PBX) containing ammonium perchlorate, aluminum, RDX, and binder.

Pressure profiles of the three samples in nitrogen are shown in Fig. 4, along with a detonator-alone shot. Only a relatively small pressure rise is seen, going from detonator alone, to HMX, to HMX/Al. The PBX gave a higher pressure than the HMX/Al mainly due to a larger sample mass.

The HMX and HMX/Al samples in air are shown in Fig. 5, along with the detonator-alone trace. The HMX-in-$N_2$ trace from Fig. 4 is reproduced in Fig. 5 to provide a common reference for comparison between the two figures, and shows the dramatic difference between the $N_2$ and air experiments. Adding 0.3 g of HMX approximately doubles the detonator-alone pressure and increases the initial rate (dp/dt). Adding 0.075 g Al to the HMX gives another ~10% pressure increase.

### Theoretical Quasi-Static Pressure

The Cheetah 3.0 thermochemical code [4] was used to calculate heats of reaction and peak quasi-static overpressures, using the BKWS product library and constant volume explosion. Heat of reaction and generated gas volume are two factors contributing to quasi-static pressure, but in these experiments the pressure correlates with reaction heat because the generated gas volume is much smaller than the air in the chamber: 3 g air compared to ~ 0.5 g explosive. Thus the dominant source of pressure is heating the existing air. In internal blast the theoretical maximum reaction heat is the heat of combustion in air or the heat of detonation in its absence. Calculated pressure correlates with reaction heat, whether in air or in nitrogen, as seen in Fig. 6.

### Discussion

Measured and calculated data are summarized in Table 1 and plotted in Fig. 6. The efficiency of the samples in generating quasi-static pressure is the percent of theoretical pressure achieved. Tests in nitrogen all had efficiencies below 100%, and so appear below the calculated curves in Fig.6. In air, the measured values are higher than theoretical because of the contribution of plastic in the detonator, which is not included in the calculated values. The HMX/Al measured value in air is

**TABLE 1. Calculated and measured values. Calculations include detonator but not detonator plastic.**

| Sample Material | Detonator | Atm. | Sample Mass (mg) | Heat of Reaction* (total calories in sample+detonator) | Peak Quasi-Static Pressure (bar) | | |
|---|---|---|---|---|---|---|---|
| | | | | | Calculated (Cheetah) | Measured | %measured/ calculated |
| Detonator Alone | Delrin | air | - | 370 | 2.0 | 6.5 | 310 |
| Detonator Alone | Brass | air | - | 422 | 2.3 | 2.5(3.8at45ms) | 110(160) |
| Detonator Alone | Delrin | $N_2$ | - | 295 | 1.7 | 1.0 | 59 |
| Detonator Alone | Brass | $N_2$ | - | 280 | 1.6 | 1.0 | 64 |
| HMX 60 % TMD | Brass | $N_2$ | 298.4 | 692 | 3.4 | 1.4 | 44 |
| HMX/Al 80/20 | Brass | $N_2$ | 374.0 | 965 | 4.7 | 2.1 | 44 |
| HMX 60 % TMD | Brass | air | 298.1 | 1053 | 5.3 | 7.1 | 139 |
| HMX/Al 80/20 | Brass | air | 371.8 | 1604 | 7.1 | 7.6 | 111 |
| PBX | Brass | $N_2$ | 445.8 | 1103 | 5.0 | 2.8 | 57 |

● Cheetah: heat of combustion for in-air tests, heat of detonation for in-nitrogen; BKWS product library.

dropping back towards the calculated line as the amount of oxygen remaining in the chamber diminishes.

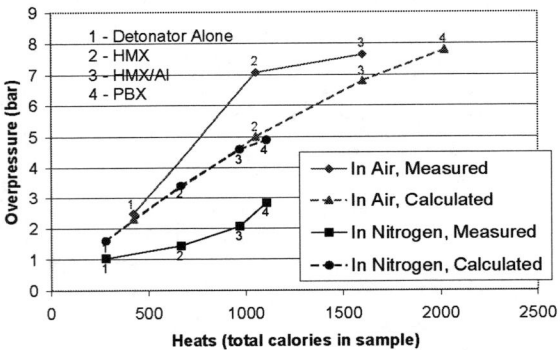

**Figure 6.** Measured and predicted quasi-static pressure vs heats of combustion (in-air tests) and heats of detonation (in-$N_2$ tests) in the 3-liter chamber. All tests with brass-sleeved detonators.

## Conclusions

We have devised a small-scale test useful for studying internal blast. High confinement allows reaction in small samples well below critical diameter. The effect of excess fuel, inside or outside the explosive, is clearly seen, and affects the correlation of quasi-static pressure with heat of combustion.

## ACKNOWLEDGEMENTS

Funding was provided by the NAVSEA Indian Head Division Core Research Program. We thank Bill Lawrence and Richard Lee for helpful discussions on internal blast.

## REFERENCES

1. Lee, R., Chang, J., Lawrence, G., Cart, E., Mychajlonka, K., and Chernoff, M., "Thermobaric ACTD Payload Program Parametric Study of Internal Blast," NSWC report IHTR 2648, to be printed.
2. Sandusky, H. W., Granholm, R. H., Bohl, D. G., "Small-Scale Shock Reactivity Test," NSWC report IHTR 2701, July, 2005.
3. Lee, R. J., Lawrence, G. W., Chernoff, M. P., Cart, E. J., Chang, J. C., and Mychajlonka, K. L., "Parametric Study of Internal Blast for Explosives," Proceedings of the 40th JANNAF Combustion Subcommittee Meeting, June, 2005.
4. Fried, L. E., Howard, W. M., Souers, P.C., Vitello, P. A., Cheetah 3.0 thermochemical code, Energetic Materials Center, LLNL.

CP845, *Shock Compression of Condensed Matter - 2005*,
edited by M. D. Furnish, M. Elert, T. P. Russell, and C. T. White
2006 American Institute of Physics 0-7354-0341-4/06/$23.00

# MEASUREMENT OF STRENGTH OF EN3B MILD STEEL USING LATERAL GAUGES

## R I Hammond[1][*], R E Winter[2] and E J Harris[2]

[1] *PCS, Cavendish Laboratory, Madingley Road, Cambridge, CB3 0HE, UK*
[2] *AWE, Aldermaston, Reading, Berks, RG7 4PR, UK*

**Abstract.** Hammond and co-workers, (SCCM, 2003, p1125), measured the resistance change of lateral manganin gauges mounted in mild steel samples. All of the experiments were conducted at an impact velocity of 400m/s giving a shock pressure of ~7GPa. The results provide data from which the strength of the sample material at a shock pressure of ~7GPa can be derived. The analysis is simplified by assuming that the polymer in which the gauge is mounted has no strength. First, calibration curves obtained by previous workers for longitudinal gauges were used to estimate the pressure in the polymer surrounding each gauge corresponding to its measured resistance change. Hydrocode simulations were then used to determine the material strength that matched the observed polymer pressures at the gauge positions. An elastic-perfectly plastic model was assumed for both the sample and the manganin.

**Keywords**: Stress gauges, lateral gauges, manganin gauges
**PACS**: 62.50; 07.05.Tp

## INTRODUCTION

Millett, Bourne and Rosenberg (1), who used lateral gauges to measure the shear stress of copper, iron and mild steel, discovered that adding a 25μm Mylar sheet on each side of the gauge extended the survival time of the gauge, especially at high pressures. In 2003 Hammond et al (2) explored this idea further by using the configuration shown in Fig. 1 to measure the resistance change of lateral gauges cladded with different thicknesses of protective padding. Figure 1 also defines the co-ordinate system used in this paper.

The purpose of the work reported in this paper is to determine the flow strength of EN3B which is compatible with Hammond et al's results.

## EXPERIMENTS

Micro Measurements encapsulated "T" Gauges, Type J2M-SS-580SF-025 were used. The nominal impact velocity for all of the experiments was 400m/s, chosen to generate a longitudinal stress at which it was judged the gauges would easily survive even without protective padding.

The manganin elements of each gauge were encapsulated in a polyimide sheath of total thickness 45μm. Experiments were fired with a bare gauge and with additional polymer padding of 50μm, 75μm, 150μm, 250μm, 500μm and 1000μm on each side of the gauge.

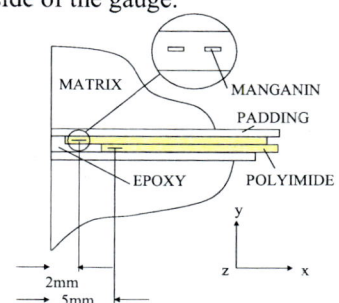

**FIGURE 1.** Schematic of the experimental set-up Gauges were overlapped as shown.

Hammond's results showed that the shapes of the profiles depend on the thickness of the polymer

---

[*] Now at TWI Ltd., Granta Park, Great Abington, Cambridgeshire, CB1 6AL, UK

layer. In the thinly padded experiments a sharp rise in resistance is followed by a fall. Then the resistance rises slowly until it falls again due to release from the back of the flyer. As the padding increases in thickness the rise time increases and the peak amplitude reached tends to decrease. The data from the gauges positioned 5mm from the impact surface showed that, in general, the gauge at 5mm records a higher initial jump in resistance than does the gauge at 2mm.

## ANALYTICAL APPROACH

Following an approach similar to that described by Feng and Gupta (3) and Feng et. al, (4) we have run high resolution hydrocode simulations of lateral manganin gauges in a steel matrix. Simulations in which the gauge element is included, are described in Refs. 5 and 6. We hope that these calculations will eventually yield a relation between measured resistance change and the shear strength of the matrix. However in this paper we use a less laborious method which comprises a combination of forward and reverse analysis

We have simplified the problem of interpreting lateral gauge data by following Chartagnac (7) in assuming that the polymeric material in which the gauges are mounted has no strength and, further, that the length of the gauge does not change. With these assumptions the problem of deriving the shear strength of the material from the measured resistance change of the gauges reduces to the two parts discussed in the next two sections.

## CALCULATING THE PRESSURE IN THE POLYMER

The first part of the analysis is to deduce the pressure in the polymer mount from the resistance change of the lateral gauge.

Several workers (for example Barsis et al (8)) have noted the distinction between 1D and 2D configurations in longitudinally-mounted gauges. The 1D configuration applies when the manganin element of the gauges has the form of a flat sheet in a plane parallel to the shock front. In this case strains in the y and z directions, (defined in Fig. 1), are zero and the deformation is deemed to be 1D. Alternatively, if the gauge consists of thin wires, the deformation in the x and y directions is finite and the deformation is deemed to be 2D. As noted by Chartagnac (7), when the polymer mount is

assumed to behave as a fluid, the pressure/resistance relation for the *lateral* gauges should be the same as that for gauges mounted in the *2D longitudinal* configuration. Therefore past calibration for 2D longitudinal gauges, either measured directly or derived from 1D data, may be used.

Calculations in which encapsulated T gauges are modelled in both longitudinal and lateral configurations provide some justification for applying the calibration obtained with 2D longitudinal gauges to lateral gauges.

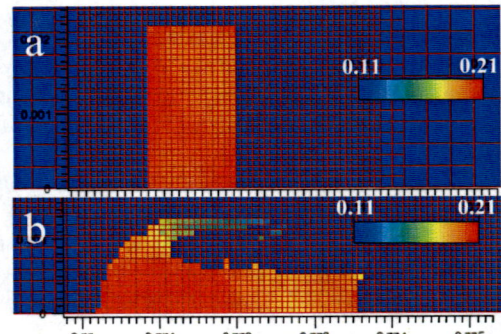

**FIGURE 2**. Maps of Resistivity change, (defined as $\Delta\rho/\rho_0$) for (a) a longitudinal and (b) a lateral gauge.

Figures 2 (a) and (b) shows for comparison computed *resistivity,($\rho$)*, changes for a 12μm x 45μm manganin conductor mounted in a polymer layer of total thickness 45μm, in longitudinal and lateral configurations. In both cases the centre point of the gauge is 2 mm from the impact surface and the times are chosen to show the situation 0.04μs after the initial shock has passed the centre of the gauge. The total calculated resistance changes of the longitudinal and lateral gauges were 0.2011 and 0.1960 and the calculated average pressures in the gauge were 6.91GPa and 6.88GPa respectively. The pressure coefficients of resistivity, (K), at these two sampling points, defined as $\Delta R / pR_0$, are 0.0291GPa$^{-1}$ and 0.0284GPa$^{-1}$ for the longitudinal and transverse gauges respectively. The calculated *resistance* change of the longitudinal gauge is slightly higher than that of the lateral gauge despite the fact that the lateral gauge obviously suffers more plastic flow. The reason for this is that the stresses, particularly the z stress that acts along the length dimension of the gauge, is significantly greater with the longitudinal than with the lateral orientation.

Barsis et al (8) found that, over the pressure range 3 to 8 GPa K is equal to 0.0286GPa-1 for 2D loading of manganin gauges. The fact that this value lies between the values derived from the two calculations run using piezo-resistive and plastic strain hardening coefficients from Refs. 3 and 4 gives confidence in its applicability to our configuration. Therefore it is used in this paper to convert the measured resistance change in Hammond's experiments, to pressure in the polymer mount.

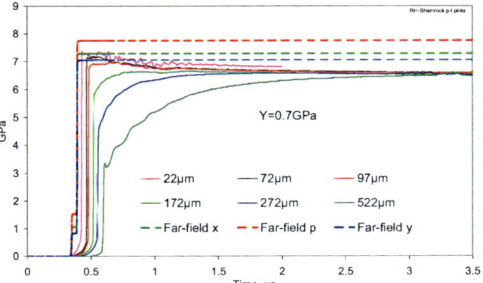

**FIGURE 3**. Shamrock calculations for the 2mm gauge.

## CALCULATING THE SHEAR STRENGTH

We have shown in Refs 5 and 6 that, for the purposes of relating polymer pressure to far field stress, calculations run without the manganin element present may be used.

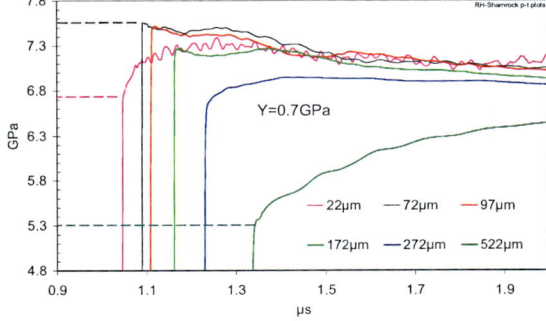

**FIGURE 4.** Shamrock calculations for the 5mm gauge. The dashed lines show how the "first jump" was measured.

Figure 3 show results from a set of Shamrock calculations run with layers of strengthless PMMA of a range of thicknesses corresponding to the padding thicknesses in Hammond's experiments. The shear strength assigned to the steel was 0.7GPa. Pressure profiles for a position 2mm from the impact face are shown. Also displayed for comparison are the computed far field pressure and

x and y stresses. Figure 4 shows similar data for a position 5 mm from the impact surface. Figure 6 illustrates that the shear strength that best fits the experimental data can be found by running calculations with different assumed shear strengths in the steel. Comparing the simulations to the data for a polymer layer of half-thickness 72μm suggests that the data is best matched by setting the shear strength of the steel to ~1.8GPa.

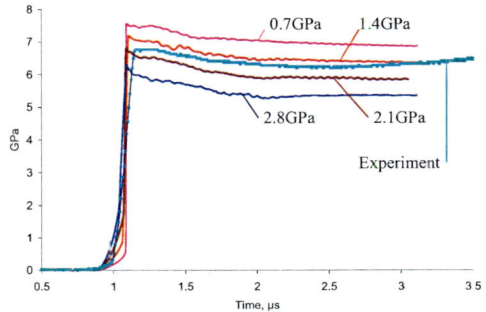

**FIGURE 5.** Calculations for the 2mm/72μm gauge compared with experiment. Comparison suggests that assigning a strength of ~1.8GPa to the steel gives the best match to the data.

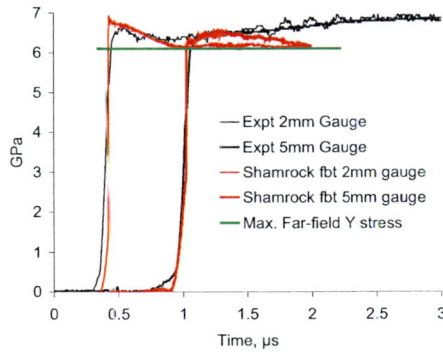

**FIGURE 6.** Experiment vs Shamrock for a polymer layer of half-thickness 22μm. A shear strength of 1.8GPa was used in the calculation. The green line shows the computed maximum far-field y stress.

Comparisons between experiment and calculations, run with 1.8GPa shear strength in the steel, are shown for polymer thicknesses of 22μm, 97μm and 272μm respectively in figures 6 to 8. It is seen that the closeness of the match between experiment and calculation depends on the polymer thickness. However in all cases the relative amplitudes of the 2mm and 5mm profiles are similar for calculation and experiment.

We have attempted to add objectivity to the experimental/calculational comparison by estimating the pressure of the "first jump" as shown, for example, in Fig. 5. This was repeated for the 2mm and 5mm gauges for all polymer thicknesses and the 2mm and 5mm first jumps were averaged. The averaged first jumps for experiment and for simulations run with a shear strength of 1.8 GPa, are shown in Fig. 9. This diagram suggests to us that 1.8GPa provides the optimum match to Hammond's data.

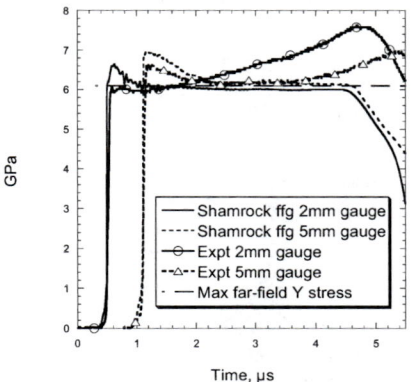

FIGURE 7. Experiment vs Shamrock for a polymer layer of half-thickness 97µm. A shear strength of 1.8GPa was used in the calculation.

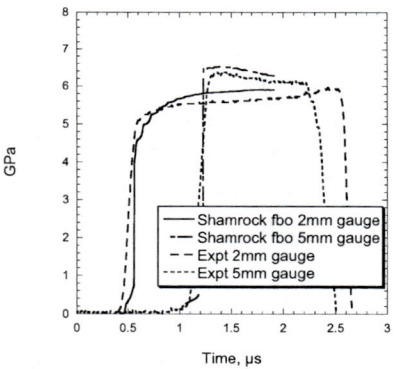

FIGURE 8. Experiment vs Shamrock for a polymer layer of half-thickness 272µm. The match to experiment has been improved by using a shear strength of 1.8GPa in the calculation.

## CONCLUSIONS

Hammond's experiments have been analysed using a two stage process. In the first, reverse analysis, stage the pressure in the polymer mounting layer is derived from the measured resistance change by using an existing calibration curve for 2D longitudinal manganin gauges. Then the relationship between the shear strength of the steel and the pressure generated in the polymer layer was determined, for the set of experimental configurations, by iterating using the AWE hydrocode, Shamrock.

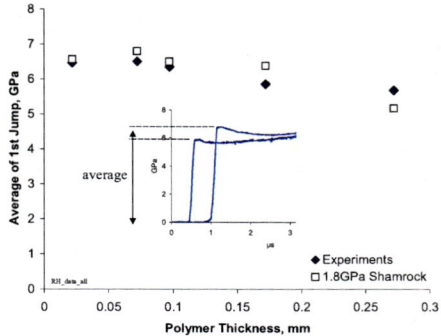

FIGURE 9. Average of 2mm and 5mm first jumps vs. polymer half-thickness.

The work illustrates that meaningful measurements of shear strength can still be made in those situations where gauges need to be padded.

The shear strength of EN3B mild steel shocked to a longitudinal stress of 7.1GPa was found to be 1.8GPa.

## REFERENCES

1.  Millett, J. F. C., Bourne, N. K. and Rosenberg, Z, J Phys. D. 29, 2466, (1996).
2.  Hammond, R. I., Church, P. D., Grief, A., Proud, W. G. and Field, J. E., APS SCCM, 2003
3.  Feng, R. and Gupta, Y. M., J Appl. Phys. 83 (2), p747, (1998)
4.  Feng, R., Gupta, Y. M. and Wong, M. K. W., J. Appl. Phys. 82 (6), 2845, (1997)
5.  Harris, E. J. and Winter, R. E., "Calculating the resistance of lateral gauges in a steel matrix", this meeting.
6.  Winter, R. E. and Harris, E. J., "Simulation of Lateral Stress Gauges", this meeting.
7.  Chartagnac, P. F., J. Appl. Phys 53(2), p948, (1981).
8.  Barsis, E., Williams, E. and Skoog, C., J Appl. Phys 41, p5155, (1970).

CP845, *Shock Compression of Condensed Matter - 2005*,
edited by M. D. Furnish, M. Elert, T. P. Russell, and C. T. White
© 2006 American Institute of Physics 0-7354-0341-4/06/$23.00

# SHOCK COMPRESSION SPECTROSCOPY WITH HIGH TIME AND SPACE RESOLUTION

## Wentao Huang[1], James E. Patterson[2], Alexei Lagutchev[1] and Dana D. Dlott[1*]

[1]*School of Chemical Sciences, University of Illinois at Urbana-Champaign,
600 S. Mathews Ave., Urbana, IL 61801, USA*
[2]*Present address: Institute for Shock Physics, Washington State University, Pullman, WA 99164-2816*

**Abstract.** Femtosecond laser-driven shock compression experiments are described using nonlinear coherent vibrational sum-frequency generation spectroscopy (SFG) probing of molecular materials. SFG selectively monitors molecular groups at surfaces and interfaces, providing a high degree of spatial resolution. In initial experiments a self-assembled monolayer of long-chain alkane molecules is studied, where SFG sees only the methyl ($-CH_3$) head groups. The plane of methyl groups is just 1.5Å thick. Shock-induced bending of the chain and shock-induced rotations around carbon-carbon bonds are observed. Possible future directions are discussed.

**Keywords**: laser-driven shock, femtosecond spectroscopy, molecular monolayers
**PACS**: 62.50.+p, 78.47.+p, 68.35.Ja, 78.30.-j

## INTRODUCTION

In this paper we describe a technology that lets us measure the effects of shock compression in molecules with the time and space resolution (femtoseconds and angstroms) needed to make a close connection to atomistic simulations. Femtosecond shock compression techniques have improved a great deal recently, and temporal resolution at the shock front of a few picoseconds has been obtained [2,4-6]. Probing steep shock fronts with spatial resolution on the order of angstroms is a significant problem that has not been overcome until now. In shock breakout experiments on metals and semiconductors, spatial resolution of a few nanometers has been obtained since light pulses probe only a thin skin layer at the surface [4,5], but this will not work for dielectric slabs of molecular materials. We use a nonlinear optical technique termed "vibrational sum-frequency generation" spectroscopy (SFG) [7].

With SFG it is possible to obtain spatial resolution on the order of a single atomic diameter [8].

Here we describe the SFG method and its application to shock compression measurements on molecular monolayers of long-chain hydrocarbons. Then we briefly discuss some future directions.

## EXPERIMENTAL

In SFG, infrared (IR) light that is resonant with molecular vibrational transitions is mixed with visible light to generate a coherent sum frequency signal via the second-order nonlinear susceptibility $\chi^{(2)}$. In centrosymmetric media $\chi^{(2)}$ vanishes, so at the interface between two such media, SFG selectively probes molecular vibrations at the interface and ignores the vastly greater number of vibrations in the bulk media. For instance as shown in Fig. 1, we can probe self-assembled monolayers (SAMs) of octadecyl thiol (ODT) on the gold {111} surface having the structure,

---

* Corresponding author, email dlott@scs.uiuc.edu.

(gold surface)-S-$(CH_2)_{17}$-$CH_3$

In the geometry shown in Fig. 1, where the methyl head groups of the SAM are in contact with a centrosymmetric liquid, deuterated ethylene glycol, SFG sees only transitions of the methyl head group, a layer just 1.5Å thick [6]. We probe the methyl C-H stretching transitions $\nu_{CH}$ near 3000 cm$^{-1}$ using SFG with *ppp* polarization, which measures the instantaneous methyl tilt angle $\theta$ (Fig. 1b) via the ratio of $\nu_s$ to $\nu_{as}$ intensities Fig. (1c) [1].

Spectra are acquired with the broadband multiplex [9] SFG set up in Fig. 2. The IR is a broadband fs pulse, the visible is a narrowband ps pulse, and a spectrograph and CCD detect all three CH-stretch transitions simultaneously.

In our cited works [2,3,6], we used 170 fs, 115 µJ pulses at 2 J/cm$^2$ to drive a shock through a layered assembly consisting of Cr, Ni and then Au, which resulted in pressures of 1-1.6 GPa in the SAM layer. We had a narrow range of achievable shock pressures due to nonlinear effects [10] occurring when the laser pulse passed through a 1.6 mm thick glass substrate. At high optical fields, filamentation and optical breakdown occur in the glass, distorting the pulse and limiting the energy that reaches the shock generation layers.

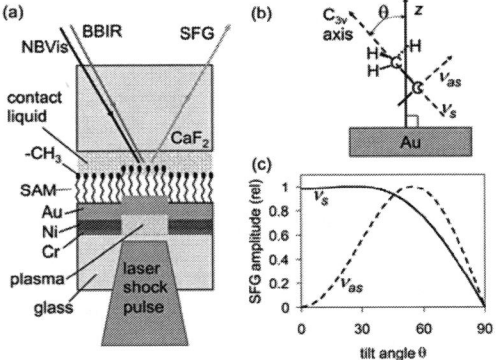

**Figure 1**. (a) One element of a target array for shock compression of self-assembled monolayers (SAMs). Vibrational sum-frequency generation (SFG) probes CH stretch transitions of the terminal methyl –CH$_3$ groups. (b) The methyl tilt angle is $\theta$. (c) Calculated SFG intensities for $\nu_s$ and $\nu_{as}$ CH-stretch transitions as a function of $\theta$. Reproduced from ref. [2], copyright APS.

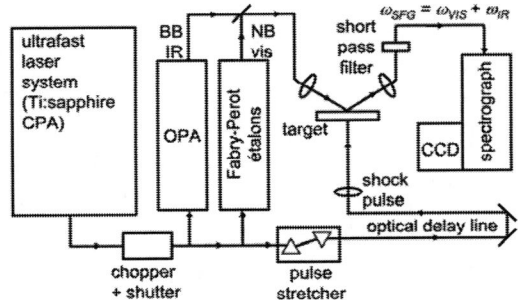

**Figure 2**. Schematic of laser apparatus. OPA = optical parametric amplifier.

**Figure 3**. Images of Au surface after shock breakout. The laser pulse has a Gaussian radial profile with a diameter (1/e$^2$) of 200 µm. With 100 fs pulses, the pulse is distorted passing through the glass substrate. The distortion is negligible with 840 fs pulses. At higher pulse energies the shock diameter is greater and more gold is blown away (dark areas).

Below we will report new data as a function of shock strength, using stretched pulses. Since the shock front risetime is $t_r \sim 3$ ps, the time resolution is hardly degraded using pulses up to 1 ps. Using a prism pulse stretcher we varied the pulse duration and looked at the breakout at a free Au surface, as shown in Fig. 3. Shock breakout drives an Au plug off the surface. We try to make this plug as smooth and large as possible. With 110 fs pulses above 100 µJ, the plug shows the effects of filamentation. With 840 fs pulses there is little distortion and much more pulse energy is converted to shock.

# SHOCK COMPRESSION OF MONOLAYERS

Figure 4 shows data on two SAMS that have either an even number (ODT) or an odd number (pentadecane thiol, PDT) of carbon atoms. Because the carbon chain has an all-*trans* zigzag structure, the equilibrium methyl tilt angle is larger in PDT (60°) than in ODT (25°) [11]. Figures 4a and 4c show SFG spectra. The broad part is a nonresonant signal from Au and its shape tracks the spectrum of the broad-band IR. The molecular transitions are the sharper dips [9]. In the more tilted PDT, the $\nu_{as}$ transition is relatively larger, as shown in Fig. 1a.

Figures 4b and 4d show data with a 1-1.6 GPa shock that has a steep <3 ps onset and an ~15 ps decay. In the figure we plot a vibrational response function [3] normalized so that each transition ranges from 0-1. Behind the shock front, both PDT and ODT lose intensity in both $\nu_s$ and $\nu_{as}$. Figure 1c shows this indicates a large methyl tilt angle close to 90°.

Figure 5 shows shock propagation data [3]. The Au layer thickness was increased in 10 nm increments, which increases the arrival time in ~3 ps increments. The slope in Fig. 5b gives a velocity in Au of 2.8 ± 0.3 km/s, consistent with the small-amplitude shock velocity[12] in Au of 3.07 km/s. The best-fit line intercepts the x-axis near $t$ = 32 ps. Small-amplitude shock propagation through 25 nm of Cr at 5.2 km/s and 100 nm of Ni at 4.6 km/s

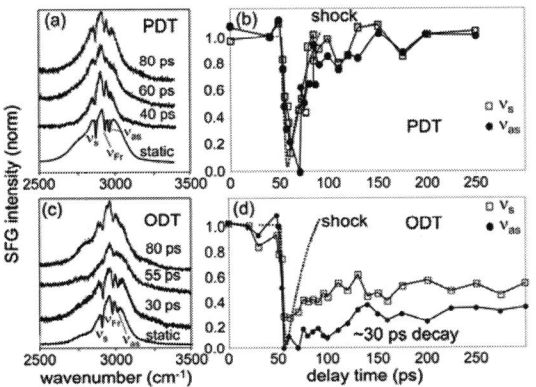

**Figure 4.** (a) and (c) SFG spectra of PDT (15-carbon) and ODT (18-carbon) monolayers. (b), (d). Time-dependent vibrational response functions from SFG for -CH$_3$ transitions shocked with 120 μJ pulses. Reproduced from ref. [2], copyright APS.

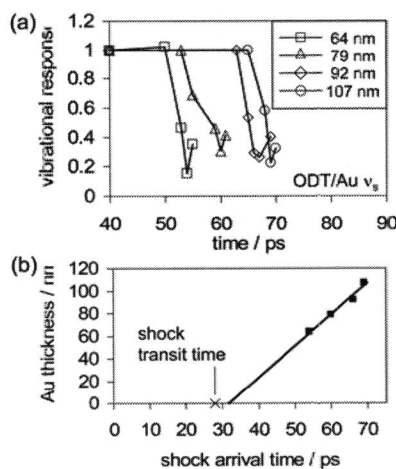

**Figure 5.** The Au layer thickness was increased in ~10 nm increments. (a) The shock front arrival time is determined from the SFG signal from the $\nu_s$ transition of ODT. (b) The slope gives a shock velocity in Au of 2.8 ± 0.3 nm/ps. The intercept results from an ~27 ps transit time through Cr and Ni plus an ~5 ps shock build up. Reproduced from ref. [3], copyright ACS.

should require 27 ps. The remaining 5 (±1) ps represents the shock build-up time[13].

With PDT, shock compression is entirely elastic. The SFG signal recovers instantaneously as the shock unloads. ODT evidences a long-lasting loss of SFG signal. The PDT results confirm that the shock does not decompose the SAM, so the ODT result is attributed to shock-generation of SFG-invisible high-tilt methyl group conformations of ODT. At longer times ODT shows a $\nu_s/\nu_{as}$ ratio greater than before the shock, consistent with more upright structures. Thus at least three ODT conformations need to be considered [1].

We have explained these results [1,2] using molecular simulations of uniaxial isothermal compression. An infinite wall is lowered onto a periodic lattice of SAM molecules as the structure is varied to minimize energy. This does not incorporate dynamical effects of the shock front, so the simulation needs higher pressures to obtain the same molecular deformation as a real shock [1]. With PDT, as shown in Figs. 6a and 6b, uniaxial compression causes both methyl and whole-chain bending. The cold compression curve for PDT is

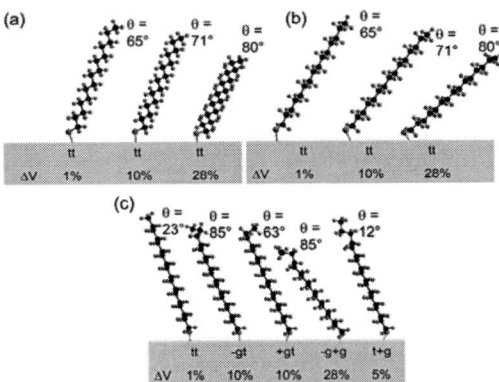

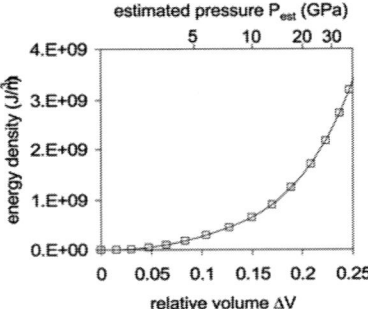

**Figure 6**. Models of PDT and ODT on Au. (a), (b) PDT from two perspectives. With increasing compression the methyl tilts but the all *trans* (tt) conformer remains dominant. (c) All-*trans* (tt) ODT under compression can generate *gauche* defects at the first and second dihedral (tg, gt or gg). When the shock unloads, gg quickly relaxes to the more upright tg. Reproduced from ref. [1], copyright ACS.

**Figure 7**. Calculated isothermal planar compression curves for periodic lattices of PDT on Au. Only the tt conformer is stable in this pressure range. Reproduced from ref. [1], copyright ACS.

shown in Fig. 7. Shock compression of PDT is entirely elastic in this pressure regime.

Because of the more upright methyl groups, ODT is quite different from PDT. Uniaxial compression causes ODT to deform via rotation around C-C bonds near the methyl head group. The original all-*trans* (t) configuration can be converted to *gauche* (g) configurations. *Gauche* chains are shorter and fatter, so shock compression favors *gauche* over *trans*. The configurations that are

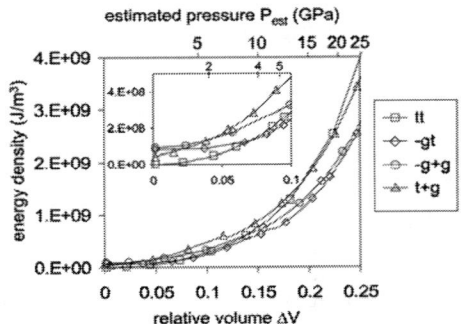

**Figure 8**. Calculated isothermal planar compression curves for periodic lattices of ODT/Au conformers. Several conformers have similar stabilities at pressures of a few GPa. Reproduced from ref. [1], copyright ACS.

energetically accessible in our shock experiments involve rotations at the first or second dihedrals only. These configurations shown in Fig. 6c are denoted tt, gg, tg and gt (gt means gtt...t, etc.). Figure 8 shows cold compression curves for these conformers. Based on these results, we explain the ODT shock data in Fig. 4d as follows. When the shock front arrives, it can create both single *gauche* gt and double *gauche* gg, which are comparable in energy and have high tilt angles making them almost SFG-invisible. When the shock unloads the gt molecules spontaneously reconvert back to tt and the gg molecules convert to tg which is nearly upright [1].

These experiments demonstrate that we can measure the methyl tilt angle in real time behind a shock front and relate changes in tilt to changes in molecular structure.

## NEW RESULTS ON MONOLAYERS

Using the 840 fs pulses we have measured the pressure-dependent response of the ODT SAM. Figure 9 shows ODT data using 50 μJ and 150 μJ pulses (120 μJ in 170 fs was used earlier). At the lower pressure, we see recurrences with an ~50 ps period, attributed to shock reverberations between the Au-SAM and Cr-glass interfaces (see Fig. 1a). At higher pressures there is more plasma and less solid matter at the Cr-glass interface so there is more damping of the reverberations.

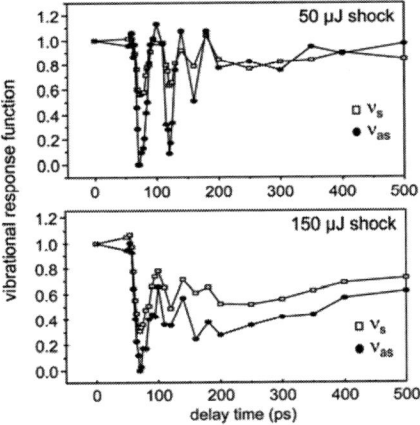

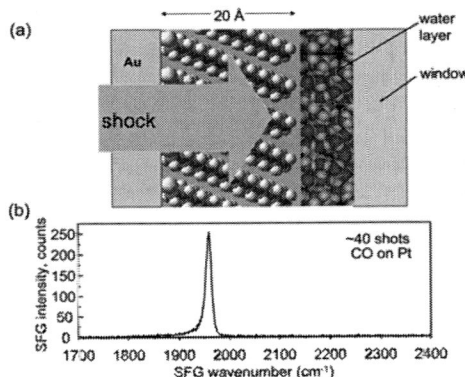

**Figure 9**. ODT monolayers shocked at lower and higher pressures show recurrences from shock reverberations between the Au/ODT and Cr/glass interfaces.

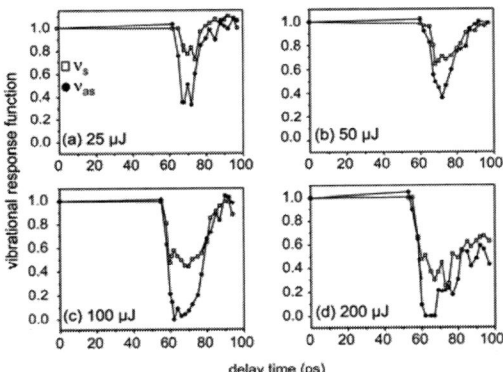

**Figure 10**. Short time behavior of SFG signals from ODT shocked with 840 fs duration pulses at the indicated energies.

A closer look at the short-time behavior of ODT as a function of shock strength is shown in Fig. 10. After the shock arrives the $v_{as}/v_s$ ratio decreases but the ratio depends little on the shock strength. A roughly constant ratio is expected from Fig. 1c in the $\theta > 60°$ tilt range, suggesting that even the weaker shocks tilt the methyl groups past 60°. Increasing the shock strength slows down the SFG signal recovery. The faster recovery with weaker shocks suggests that the weaker shocks produce a more elastic response.

**Figure 11.** (a) Schematic of experiment to measure dynamic shock compression of a thin water layer on the surface of a SAM. (b) Intense SFG spectrum from a monolayer of CO on Pt obtained with 40 laser shots.

## FUTURE DIRECTIONS

The sample arrangement shown in Fig. 1 is a versatile platform for studying shock compression with high time and space resolution. The technology of monolayer engineering is highly developed. Quite a few different SAM-forming compounds are available having the same chain structure but different head groups such as $-CO_2H$, $-OH$, $NH_3$, and $-C_6H_5$ (phenyl). We are also interested in studying shock compression of water layers a few molecules thick. Such water layers can be adsorbed on the surface of the SAM, as in Fig. 11a. Sending a shock through the SAM, causing the head group to tilt might be expected to disrupt the water hydrogen bonding, and the SFG spectrum is a sensitive probe of hydrogen bonding [14]. Recently the Gupta group [15] showed that shock freezing of water is extremely sensitive to nucleation processes at the window surface. Silica windows that are presumably OH-terminated promoted freezing whereas sapphire did not. Our arrangement would allow us to study water in contact with a variety of tailored hydrophilic and hydrophobic surfaces.

The spectra in this paper were obtained by averaging 5000 laser shots. It would be helpful to obtain monolayer spectra with a single shot. Preliminary indications are that this is possible, based on our measurements of a CO monolayer on a Pt surface [16]. Its SFG spectrum is shown in Fig.

11b. The spectrum is sharp, strong and intense with a signal-to-noise ratio of about 200:1 averaging just 40 laser shots. The ultimate time resolution of such measurements is determined as followed. In SFG the IR pulse coherently excites the molecular dipoles, and the visible pulse scatters from the coherent polarization. The polarization free-induction decay occurs with time constant $T_2$, and SFG signals are emitted only during this process. The value of $T_2$ for CO can be estimated from the spectral width $\Delta v = 14$ cm$^{-1}$ using $\Delta v = (\pi c T_2)^{-1}$ to be about 750 fs.

## ACKNOWLEDGEMENTS

This material is based on work supported by the U.S. Department of Energy, Division of Materials Sciences under Award No. DEFG02-91ER45439, through the Frederick Seitz Materials Research Laboratory at the University of Illinois at Urbana-Champaign and through the Stewardship Sciences Academic Alliance Program from the Carnegie-DOE Alliance Center under grant number DE-FC03-03NA00144. Additional support was provided by the US Air Force Office of Scientific Research under award number F49620-03-1-0032, and by the US Army Research Office under award number W911NF-05-1-0345.

## REFERENCES

1. Patterson, J. E. and Dlott, D. D., Ultrafast shock compression of self-assembled monolayers: a molecular picture, *J. Phys. Chem. B* **109**, 5045-5054 (2005).
2. Patterson, J., Lagutchev, A. S., Huang, W., and Dlott, D. D., Ultrafast dynamics of shock compression of molecular monolayers, *Phys. Rev. Lett.* **94**, 015501 (2005).
3. Lagutchev, A. S., Patterson, J. E., Huang, W., and Dlott, D. D., Ultrafast Dynamics of Self-Assembled Monolayers Under Shock Compression: Effects of molecular and substrate structure, *J. Phys. Chem. B* **109**, 5033-5044 (2005).
4. Funk, D. J., Moore, D. S., and Gahagan, K. T., Ultrafast measurement of the optical properties of aluminum during shock-wave breakout, *Phys. Rev. B.* **64**, 115114 (2001).
5. Gahagan, K. T., Moore, D. S., Funk, D. J., Rabie, R. L., Buelow, S. J., and Nicholson, J. W., Measurement of shock wave rise times in metal thin films, *Phys. Rev. Lett.* **85**, 3205-3208 (2000).
6. Patterson, J., Lagutchev, A. S., and Dlott, D. D., Shock compression of molecules with 1.5 angstrom resolution, *AIP Confer. Proc.* **706**, 1299-1302 (2004).
7. Zhu, X. D., Suhr, H., and Shen, Y. R., Surface vibrational spectroscopy by infrared-visible sum frequency generation, *Phys. Rev. B* **35**, 3047-3050 (1987).
8. Bain, C. D., Davies, P. B., Ong, T. H., and Ward, R. N., Quantitative analysis of monolayer composition by sum-frequency vibrational spectroscopy, *Langmuir* **7**, 1563-1566 (1991).
9. Richter, L. J., Petralli-Mallow, T. P., and Stephenson, J. P., Vibrationally resolved sum-frequency generation with broad-bandwidth infrared pulses, *Opt. Lett.* **23**, 1594-1596 (1998).
10. Moore, D. S., Gahagan, K. T., Reho, J. H., Funk, D. J., Buelow, S. J., Rabie, R. L. et al., Ultrafast nonlinear optical method for generation of planar shocks, *Appl. Phys. Lett.* **78**, 40-42 (2000).
11. Nishi, N., Hobara, D., Yamamoto, M., and Kakiuchi, T., Chain-length-dependent change in the structure of self-assembled monolayers of n-alkanethiols on Au(111) probed by broad-bandwidth sum frequency generation spectroscopy, *J. Chem. Phys.* **118**, 1904-1911 (2003).
12. Marsh, S. P., *LASL Shock Hugoniot Data,* University of California Press, Berkeley, CA, 1980.
13. Schoen, P. E. and Campillo, A. J., Characteristics of compressional shocks resulting from picosecond heating of confined foils, *Appl. Phys. Lett.* **45**, 1049-1051 (1984).
14. Richmond, G. L., Molecular bonding and interactions at aqueous surfaces as probed by vibrational sum frequency spectroscopy, *Chem. Rev.* **102**, 2693-2724 (2002).
15. Dolan, D. H. and Gupta, Y. M., Nanosecond freezing of water under multiple shock wave compression: Optical transmission and imaging measurements, *J. Chem. Phys.* **121**, 9050-9057 (2004).
16. Lu, G. Q., Lagutchev, A., Dlott, D. D., and Wieckowski, A., Quantitative vibrational sum-frequency generation spectroscopy of thin layer electrochemistry: CO on a Pt electrode, *Surf. Sci.* **585**, 3-16 (2005).

CP845, *Shock Compression of Condensed Matter - 2005*,
edited by M. D. Furnish, M. Elert, T. P. Russell, and C. T. White
© 2006 American Institute of Physics 0-7354-0341-4/06/$23.00

# SANS AND CONTRAST VARIATION MEASUREMENT OF THE DIFFERENT CONTRIBUTIONS TO THE TOTAL SURFACE AREA IN PBX 9501 AS A FUNCTION OF PRESSING INTENSITY

## Joseph. T. Mang and Rex. P. Hjelm

*Los Alamos National Laboratory, Los Alamos, NM 87545*

**Abstract.** We have used small-angle neutron scattering (SANS) in conjunction with the method of *contrast variation* to measure the surface area ($S_{HB}$, $S_{HV}$, and $S_{BV}$) associated with the three interfaces (HMX-binder (HB), HMX-voids (HV) and binder-voids (BV)) in pressed pellets of PBX 9501 (95% HMX and 5% binder, by weight). These interfaces are of interest as they may influence the transmission of microstresses and *hot spot* formation under shock conditions. Because of the difficulty in making measurements, little is known about the microstructure of pressed PBX 9501 parts and thus how it is affected by processing. Here, we explore the effect of varying the pressing intensity on the PBX 9501 microstructure and in particular, how the three interfaces (HB, HV and BV) are affected. Disk-shaped samples of PBX 9501 were die-pressed with applied pressures ranging between 5,000 and 29,000 psi. SANS measurements were performed on 4-5 pellets at each pressure. Analysis of the SANS data indicates systematic changes in $S_{HB}$, $S_{HV}$, and $S_{BV}$ with applied pressure.

**Keywords:** PBX 9501, pressing parameters, microstructure
**PACS:** 61.12.Ex, 81.40.-z

## INTRODUCTION

The high explosive (HE) material, PBX 9501, is a composite, consisting of a crystalline high explosive (95 wt %) and a polymeric binder (5 wt %, Estane + nitroplasticizer). Because of its composite nature, and the structure of its primary components, the microstructure of PBX 9501 is highly complex. Like other HE materials, PBX 9501 possesses both naturally occurring and process-related defects (cracks, voids, etc.). Such features are known to have the potential to dramatically affect the ignition sensitivity and performance of energetic materials. While these features can be characterized in the raw material, little is known about how they are affected during processing.

As each structural component undoubtedly has a role in the overall safety and performance of high explosives, our objective, is to understand the structural details of PBX9501 and relate these details to data on the performance and safety of these materials. Here, we explore the effect of pressing intensity on the high explosive

microstructure, particularly how the interfacial regions are affected by the variation of pressing intensity.

A pressed piece of PBX 9501 gives rise to three unique interfacial regions between the components: HMX-binder (*HB*), HMX-voids (*HV*) and binder-voids (*BV*). These interfaces are of interest because they can influence the transmission of microstresses through the material under shock conditions [1]. Figures 1 and 2 show optical images of the PBX 9501 microstructure. The white and grey structures in Fig. 1 are HMX crystals whereas the dark regions are (dirty) binder and void regions. Also visible in Fig. 1 are interfaces between HMX and voids. Fig. 2 shows an expanded view of the so called dirty binder region of the microstructure. In this case, the image has been processed and the interfaces between HMX (grey) and binder (black) can be seen clearly.

In the preparation of PBX 9501, variables such as pressing intensity, number of pressing cycles, dwell time, or rest time (between cycles) are often varied in order to achieve nominal density. Although bulk density is met, differences in pressing conditions can lead to variations in microstructure (particularly the interfacial regions) between samples.

We have used SANS and *contrast variation* [2], which takes advantage of the difference in which deuterium and hydrogen scatter neutrons, to distinguish scattering arising from the different interfaces (*HB*, *HV*, and *BV*) in the PBX 9501 microstructure. By exploiting swelling capabilities developed through concurrent studies of the PBX 9501 binder alone [3], we have made a quantitative measure of the surface area of the interfacial regions and its evolution with pressing intensities. Such knowledge will aid in the development of full-scale constitutive models for PBX 9501.

## SANS AND CONTRAST VARIATION

The scattered intensity, I(Q), observed in a SANS experiment is directly related to the structure of the sample through the squared Fourier transform of the scattering length density, $\rho(r)$. Scattering derives from fluctuations in $\rho(r)$, which reflects microscale structure in the sample density, chemical and isotopic composition. Scattering intensity is measured in absolute units of differential cross section per unit volume, I(Q) (cm$^{-1}$) as a function of scattering vector, **Q**, of magnitude Q = $(4\pi/\lambda)\sin\theta$, where $\lambda$ is the wavelength of the incident neutron and $\theta$ is half of the scattering angle.

For the current system, the scattering regime probed is such that QR >> 1, where R is the average size of a particle. The observed scattering then arises from the interfaces between the particle and surrounding media and follows a negative power-law, I(Q) = $I_o Q^{-p}$, where $I_o$ and $p$ are constants. For the special case of smooth particle interfaces, $p = 4$ and, $I_o = 2\pi S \Delta\rho^2$, where $S$ is the surface area per unit volume. This *Porod* approximation allows for *in situ* measurements of surface area.

The scattering signal from PBX 9501 arises primarily from three interfaces. In the Porod region, the scattered neutron intensity for this three phase system is given by:

$$I(Q) = \frac{2\pi}{Q^4}\left[\Delta\rho_{HB}^2 S_{HB} + \Delta\rho_{HV}^2 S_{HV} + \Delta\rho_{BV}^2 S_{BV}\right] = \frac{2\pi}{Q^4} I_p, \quad (1)$$

where $\Delta\rho_{ij}$ and $S_{ij}$ are the scattering length density

contrast and surface area per unit volume, respectively, between phases i and j. The use of the

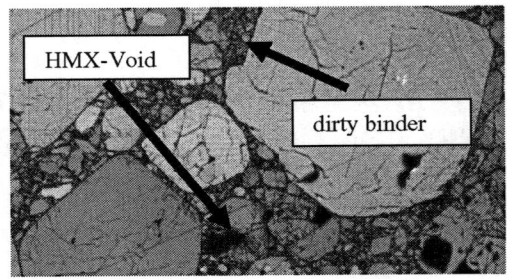

**Figure 1.** Optical microscopy image of PBX 9501, indicating an HMX-Void interface and a region of so called *dirty binder*.

contrast variation technique enables the scattering arising from the different interfaces to be separated. For the current studies, the samples are immersed in a solvent that swells the binder and fills the voids. The scattering length density of the binder regions then is a volume-weighted average of the scattering length density of the binder ($\rho_B$) and the solvent ($\rho_{sol}$). The scattering length density of the voids is equal to the solvent. Making these substitutions in Eq. 1, we see that $I_p$ shows a quadratic dependence on the scattering length density of the solvent:

$$I_p(\rho_{sol}) = A\rho_{sol}^2 + B\rho_{sol} + C \quad (2)$$

The parameters A, B and C contain cross terms, but are related to the different surface areas; the solution of Eq. 2 yields the individual surface areas.

## EXPERIMENTAL

To evaluate the effect of pressing intensity on the microstructure of PBX 9501, disk-shaped samples (1-1.5 mm thick, 9.5 mm diameter) were die-pressed with applied pressures ranging between 5,000 and 29,000 psi. The die was heated to 90 °C and the samples underwent two intensifications (30 s each) with a 25 s relaxation period in between. With the exception of the sample prepared at 5000 psi, there was no significant variation in the average pellet density over the range of samples.

SANS measurements were made in conjunction with the method of contrast variation on the Low-Q Diffractometer (LQD) of the Lujan

Center at Los Alamos National Laboratory (LANL). The individual samples were placed in cells having pathlengths of 1 or 2 mm with fused silica windows. The samples were then immersed in various toluene mixtures containing different fractions of perdeuterated toluene. The toluene mixtures infuse the voids that are external to the HMX crystals and swell the binder. The difference in neutron scattering length between hydrogen (-3.7 fm) and deuterium (6.7 fm), leads to changes in

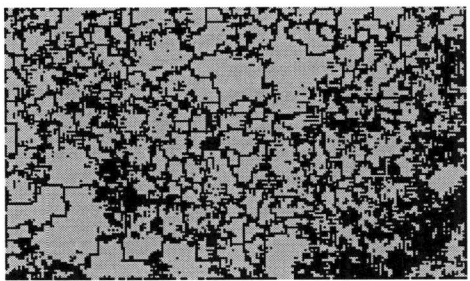

**Figure 2.** Image analysis of a dirty binder region in PBX 9501 microstructure, displaying HMX-Binder interfaces. Black regions represent binder and grey regions represent HMX.

$\rho(r)$ across the interfaces, resulting in a variation of the $\Delta\rho_{ij}$ in Eq (1). Samples were swelled for ~24 hours before SANS measurements were made at which time the swelling reached equilibrium. Toluene was used because of the limited solubility of HMX and our experience with swelling of Estane [3]. We have found that while Estane swells in toluene, it still retains a significant fraction of its mechanical integrity. So, the choice of toluene as a swelling agent limits the amount of perturbation of the original PBX 9501 microstructure.

## RESULTS

SANS (corrected for solvent and empty cell scattering), measured at different contrasts, for samples prepared with an applied pressure of 10 kpsi, are shown in Fig 3. As seen in the figure, the intensity of the scattering changes with the percent deuteration of the solvent. These changes indicate that the samples are thoroughly swelled and are related to the relative contributions of the surface area of the three interfaces to the total surface area. The solid lines in the figure are the result of fits to the Porod law (Eq. 1), and yield $I_p$ for the different

contrasts. $I_p$ is plotted in Fig. 4 as a function of the fraction, $f$, of deuterated toluene used to swell the binder.

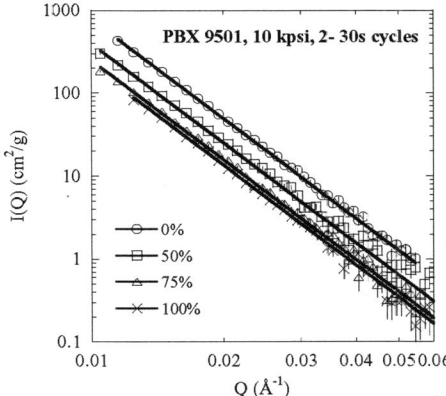

**Figure 3.** SANS as a function of the momentum transfer, Q, for different solvent deuteration levels for PBX 9501 pressed at 10 kpsi. The solid lines are fits to the Porod law which yield $I_p$.

The solid line in Fig. 4 is the result of a fit to Eq. 2. The surface areas determined from the fit are $S_{HB} = 1.20 \pm 0.01$ m$^2$/g, $S_{HV} = 0.19 \pm 0.01$ m$^2$/g and $S_{BV} = 0.04 \pm 0.01$ m$^2$/g. Similar analysis was performed on the SANS data from the remaining samples. A plot of the measured surface area, at the three interfaces, as a function of the pressing intensity is shown in Fig. 5. As we see from the figure, the surface area at the HMX-Binder interface increases with the pressing intensity up to 15 kpsi, above which the surface area is seen to decrease. This behavior suggests that as the pressing intensity is increased to 15 kpsi, more damaged is induced to the microstructure, in the form of broken or cracked HMX crystals that become filled with binder. At higher pressing intensities, this process appears to be less efficient as noted by the decline in $S_{HB}$. This result is consistent with previous SANS studies of neat HMX and recent optical microscopy studies of similarly prepared samples [2,4].

The surface area measured at the HMX-Void interface ($S_{HV}$) shows an initial decrease, followed by an increase above 15 kpsi. This is consistent with the $S_{HB}$ results and again suggests that at lower pressures, pressing-induced damage to the HMX crystals is filled-in with binder. At higher pressures though, more damaged is induced than

1273

can be accommodated by the binder. No clear trend can be seen with increasing pressure at the Binder-Void interface. This may be a result of the processes described above. However, we anticipate that this interface would be the most perturbed by the swelling process.

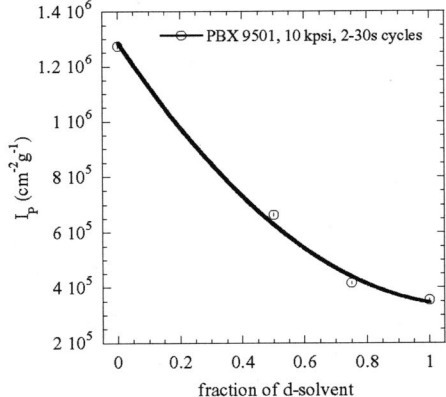

**Figure 4.** $I_p$ values for the 10 kpsi sample obtained from Porod analysis. The solid line in the figure is the result of a fit to a Eq. 2 and yields the various surface area contributions.

## SUMMARY AND FUTURE WORK

SANS and contrast variation measurements were made on samples of PBX 9501 that were prepared with applied pressures ranging from 5–29 kpsi. The samples were swollen with different mixtures of deuterated and non-deuterated toluene in order to vary the overall scattering length density contrast. Analysis of the data has confirmed that we can separate the scattering signal arising from the three interfaces of interest. The associated surface areas ($S_{HB}$, $S_{HV}$, and $S_{BV}$) were measured and found to be dependent upon the pressing intensity.

Our results will help determine how differences in pressing conditions can lead to variations in the microstructure of high explosive systems. The ability to measure $S_{HB}$, $S_{HV}$, and $S_{BV}$ in a pressed sample is novel for these systems and may aid in the understanding of how microstresses are propagated through HE samples under shock.

For our future studies, PBX 9501 will be formulated with different fractions of deuterated binder, which will eliminate any perturbation of the microstructure caused by swelling.

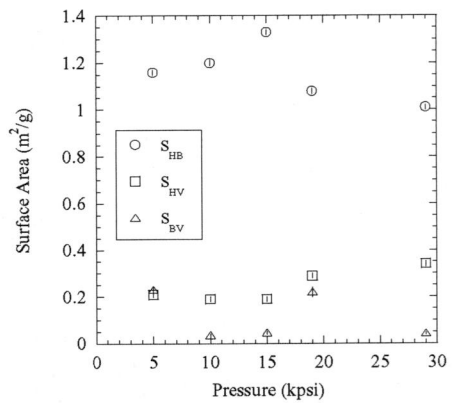

**Figure 5.** The various contributions to the total surface area as a function of the applied pressure.

## ACKNOWLEDGMENTS

This work was performed under the auspices of the US Department of Energy at Los Alamos National Laboratory, operated by the University of California under Contract No. W-7405-ENG-36. The authors wish to thank Paul D. Peterson for providing the optical images of PBX 9501. This work benefited from the use of the SANS instrument, LQD at the Manuel Lujan, Jr. Neutron scattering Center of the Los Alamos National laboratory, supported by the DOE office of Basic Energy Sciences.

## REFERENCES

1. Roessig, K. M. and Foster, J. C., "Experimental Simulation of Dynamic Stress Bridging in Plastic Bonded Explosives", in Shock Compression of Condensed Matter, 2001 (M.D. Furnish, N.N. Thadhani, Y. Horie, eds.), part II, pp. 829-832.
2. Mang, J. T., Skidmore, C. B., Hjelm, R. P. and Howe, P. M., "Application of Small-Angle Neutron Scattering to the Study of Porosity in Energetic Materials", J. Mater. Res. 15, 1199, 2000.
3. Mang, J. T., Hjelm, R. P., Orler, E. B., Wrobleski, D. A., "Distribution and Polymer Domain Composition of a Solvent-Swollen Segmented Polyurethane by Small-Angle Neutron Scattering", PMSE Preprint, 2005.
4. Peterson, P. D., Fletcher, M. A. and Roemer, R. L., "Inluence of Pressing Intensity on the Microstructure of PBX 9501", J. of Energetic Materials 21, 247, 2003.

CP845, *Shock Compression of Condensed Matter - 2005,*
edited by M. D. Furnish, M. Elert, T. P. Russell, and C. T. White
© 2006 American Institute of Physics 0-7354-0341-4/06/$23.00

# THREE-DIMENSIONAL DYNAMIC DEFORMATION MEASUREMENTS USING STEREOSCOPIC IMAGING AND DIGITAL SPECKLE PHOTOGRAPHY

## H. J. Prentice and W. G. Proud

*PCS Group, Cavendish Laboratory, University of Cambridge, Madingley Road, Cambridge CB3 0HE,
United Kingdom*

**Abstract.** A technique has been developed to determine experimentally the three-dimensional displacement field on the rear surface of a dynamically deforming plate. The technique combines speckle analysis with stereoscopy, using a modified angular-lens method: this incorporates split-frame photography and a simple method by which the effective lens separation can be adjusted and calibrated *in situ*. Whilst several analytical models exist to predict deformation in extended or semi-infinite targets, the non-trivial nature of the wave interactions complicates the generation and development of analytical models for targets of finite depth. By interrogating specimens experimentally to acquire three-dimensional strain data points, both analytical and numerical model predictions can be verified more rigorously. The technique is applied to the quasi-static deformation of a rubber sheet and dynamically to Mild Steel sheets of various thicknesses.

**Keywords:** Stereoscopic imaging, speckle photography, plate bulging.
**PACS:** 42.30.Ms, 91.10.Lh, 62.20.Fe.

## INTRODUCTION

The bulging and penetration of a plate has been traditionally measured using the fine grid technique [1]: by applying a regular grid to the surface of the plate and using high-speed photography, the distortion caused by an impacting projectile can be measured. Measurements can be made directly from the grid or more accurate measurements can be made using the automated fine grid technique, which tracks the co-ordinates of the grid and determines the displacement between frames.

Digital Speckle Photography (DSP) [2] provides an alternative method for measuring the in-plane displacements. This pattern-recognition technique can be implemented by applying a random pattern of speckles onto the material surface, and using image correlation to track sub-regions of the material surface as it deforms. By further utilising a stereoscopic system [3, 4], in addition to the measurement of out-of-plane displacements, the accuracy of in-plane displacement measurements is increased, producing a three-dimensional image of plate bulging. These data can then be used to interrogate model simulations of plate penetrations.

Use of speckle photography combined with stereoscopic lensing systems to determine three-dimensional shape and deformation has been achieved with two CCD cameras [3, 4].

The logistic and synchronisation problems associated with the use of two cameras, in particular for the investigation of high-speed events, has been previously recognised and an optical wedge [5] has been used in order to perform split-frame photography [6] using the translated lens method. However, this approach does not overcome a fundamental problem with the translated lens system in which the requirement of overlapping fields in paired images results in loss of image resolution for the area of interest. The new method presented in this paper addresses this issue using an optically simple solution and the angular lens stereo system.

## STEREOSCOPIC MIRROR METHOD

The angular lens configuration is such that the object, lens and image planes are inclined. The accuracy to which out-of-plane displacements may be determined increases with lens separation owing to increased image disparity. However, as the lens separation, and hence inclination, increases, the images produced with this system are increasingly triangulated. Such distortions to the images decrease the correlation between the speckle patterns, thus imposing an upper bound for lens separation. In addition, the lens separation is also limited by the scale of the deformation under examination; large deformations can occlude the comparative areas of interest from each view-point.

The co-ordinates of each lens with respect to the object are usually determined via a calibration experiment in which either the object [3] or the camera [4] is mounted upon an accurate translation stage and displaced a small distance. This is not practical if the object is to be impacted, the camera is large or if the system is to be portable.

By introducing a system of mirrors and a single camera, we can easily modify the lens separation to optimize accuracy with an acceptable amount of triangulation, calibrate the system in almost any environment, and ensure that the image pairs are synchronised in time. Fig. 1 illustrates the mirror assembly and ray paths. The mirrors result in two images, formed side-by-side, which can be captured by a camera. The effective angular lens positions are illustrated by the equivalent images, $I'$, that would have been formed in the absence of mirrors. The first mirrors encountered by rays are free to rotate so as to shift the position of the images either closer or further apart as desired. The displacement field measured for a small rotation of the side-mirrors can be used to calibrate the system and determine the effective lens co-ordinates with respect to the object.

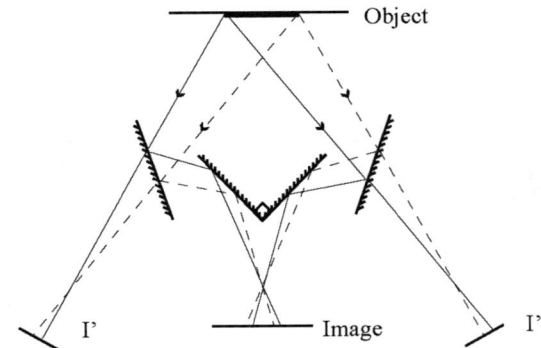

**Figure 1.** Mirror Method. The scale of the assembly depends upon the feature under investigation; the side mirrors can be rotated and translated to alter the effective lens positions, $I'$.

## EXPERIMENTAL AND RESULTS

### Quasi-static Deformation

A quasi-static deformation was investigated in order to verify the success of this mirror method in extracting accurate three-dimensional information.

A rubber membrane marked with a random speckle pattern (20 mm x 30 mm) was pulled taut about a cylindrical shell with diameter $(65 \pm 1)$ mm. A drum-stick was mounted upon a calibrated translation stage (with accuracy $\pm 0.1$ mm) and aligned to move along the cylinder's axis and perpendicular to the rubber membrane's plane. A bulge was formed in the membrane by moving the drum-stick a controlled distance along the axis. The mirror method was used to view the rubber sheet stereoscopically and a DRS Hadland Ultra-8 digital camera was used to obtain both stereoscopic images, and photographs of the bulge profile. Fig. 2 shows a typical frame.

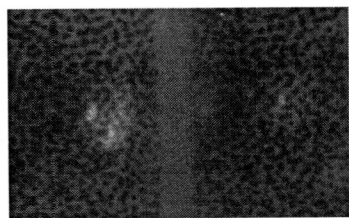

**Figure 2**. View of typical speckled split-frame.

Changes in a speckle pattern are determined by applying a Digital Image Cross Correlation algorithm (DICC) [7]. The distortion in each split-frame image caused by the optical components and system misalignments can be calibrated by correlating each stereoscopic image with a direct image of the speckle pattern in order to adjust the optics to reduce relative translation, rotation or magnification of the images and to account for any systematic errors.

The relative in-plane displacements between images are determined by applying DICC between split-frame image pairs and between reference and deformed images. These displacements are used to determine the topography of the surface.

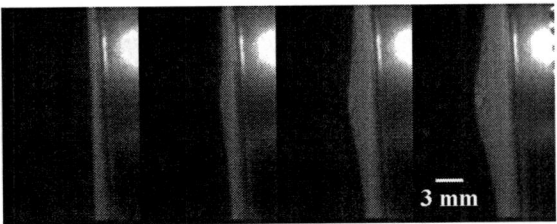

**Figure 3**. Photographs of rubber surface profile. The drum-stick is moving from right to left. The four successive frames show the profile at depths of penetration of 0 mm, 1 mm, 2 mm and 3 mm.

Fig. 3 shows the successive bulges in the membrane. The profiles were measured directly by examining contours of intensity. A number of sources of error contribute to the uncertainties in the measurement of the profile. Most importantly, these include errors arising from the pixilated quality of the images which causes some blurring, the presence of dark speckles on the surface which reduce the brightness of the rubber surface at the points of interest, and misalignments of the membrane plane along the lens axis. The profiles

for each depth of bulge, both measured directly (noisy profile) and reconstructed from stereoscopic measurements, are shown in Fig. 4. The set of stereoscopic photographs were obtained before the set of profile photographs.

Whilst the peak maxima are not at exactly 1 mm, 2 mm or 3 mm, owing to the errors associated with moving the drum-stick along the bulge axis, it is clear that the profile reconstructed from data obtained stereoscopically is in good agreement with the profile measured directly from the photograph. This includes the subtle asymmetries due to variation in tension across the membrane and friction between the drum-stick and the rubber. The relative smoothness of the reconstructed profile is in part by virtue of its lower sampling frequency. The differences between the direct and reconstructed profiles at the side of the bulge are attributed to a loss of tension in the rubber between the two sets of photographs.

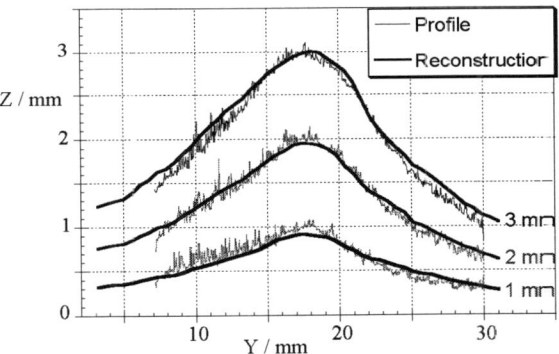

**Figure 4.** Matching the profile measured directly from photographs with that reconstructed from stereoscopic measurements. The fluctuations in the direct profile measurements are approximately ± 0.1 mm. The error in the reconstructed profile points is approximately ± 0.05 mm. The Y axis represents the position on the membrane with arbitrary zero position.

## Dynamic Metal Plate Bulging

Bulging of a thin Mild Steel plate upon normal impact by a stainless steel sphere (diameter 13 mm) was investigated for varying plate thickness. The sphere was fired at $(70.0 \pm 0.8)$ m/s in air using a small gas-gun. The velocity on exit from the gun barrel was measured using a laser light-gate. The rear of the plate was photographed stereoscopically during the first 30 μs after impact using a DRS Hadland Ultra-8 digital camera. The plate was sufficiently wide that boundary reflections would not interfere with the region of interest for this duration; however, this made photography of the side profile of the bulging plate very inaccurate and the results could not be verified in the same way as for the quasi-static rubber membrane deformation. Target rear-faces were painted white before applying a black speckle pattern in order to maximise contrast. Two flashes were employed in order to reduce the effect of differing regions of shadow caused by the bulge in each image which impedes successful speckle pattern correlation results.

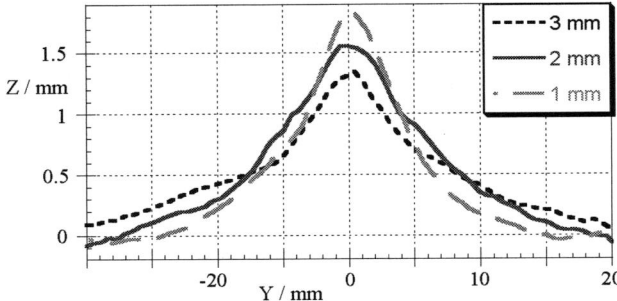

**Figure 5.** Profiles taken from three-dimensional reconstructions of the plate rear surface at $(30 \pm 1)$ μs after impact. The experimental uncertainty in the results is $\pm 0.05$ mm.

Fig. 5 illustrates the profiles at $(30 \pm 1)$ μs after impact for Mild Steel plates with thicknesses, 1.0 mm, 2.0 mm and 2.8 mm. The general deformation appears to be more localised for thinner plates, with a larger peak displacement.

Post-mortem examination of the plates display circular shoulders with approximate radius $(5 \pm 1)$ mm; this is indicates the onset of failure. The dynamic results hint at the early stages of the formation of this shoulder; however, without examining this feature in the three-dimensional data, it cannot be conclusively distinguished from noise and is likely to be a consequence of the finite sub-image size used in the DICC algorithm.

## CONCLUSIONS

Stereoscopic speckle photography is a non-invasive and data rich method, applicable to a wide range of length-scales and time-scales. The modifications to the angular-lens system include the use of a single camera, rather than two. In addition to the non-invasive calibration method offered and ease of alignment, the single camera has further advantages in terms of cost, convenience and space, and guarantees temporally synchronised images from different viewpoints.

This non-invasive stereoscopic mirror technique has been shown to successfully reconstruct object surfaces and has the potential to provide detailed material strain and deformation data in three dimensions. It has applications in a large range of experimental scenarios, including high strain rate experiments, owing to its simple implementation and the data-rich results yielded. Such results are vital for comparison with numerical and analytical model predictions.

## ACKNOWLEDGEMENTS

The authors would like to thank J. P. Curtis and N. J. Lynch at QinetiQ and EPSRC for their support of this research.

## REFERENCES

1. Goldrein, H. T., Palmer, S. J. P. and Huntley, J. M., *Opt. Lasers Eng.* **23**(5), 305-318, 1995.
2. Sirohi, R. S., Ed., *Speckle Metrology*, Marcel Dekker, New York, 1993.
3. Synnergren, P., *Opt. Eng.* **36**, 2302-2310, 1997.
4. Helm, J.D., Sutton M.A. and McNeill, S.R., Proc. SPIE **2350**, 32-45, 1994.
5. Gorham, D. A., *J. Phys.E: Sci. Instrum.* **15**, 562-564, 1982.
6. Prentice, H. J., et al., in Shock Compression of Condensed Matter, 2003.
7. Sjodahl, M., Benckert, L. R., *J. Appl. Opt.* **32**, 2278-2284, 1993.

CP845, *Shock Compression of Condensed Matter - 2005*,
edited by M. D. Furnish, M. Elert, T. P. Russell, and C. T. White
© 2006 American Institute of Physics 0-7354-0341-4/06/$23.00

# X-RAY MICROTOMOGRAPHY OF SUGAR AND HMX GRANULAR BEDS UNDERGOING COMPACTION

## M. W. Greenaway[1], P. R. Laity[2] and V. Pelikan[3]

[1]*PCS Group, Cavendish Laboratory, Madingley Road, Cambridge CB3 0HE, United Kingdom.*
[2]*Department of Materials Science and Metallurgy, Univ. of Cambridge, New Museums Site, Pembroke Street,*
*Cambridge CB2 3QZ, United Kingdom.*
[3]*Institute of Energetic Materials, Univ. of Pardubice, CZ-532 10, Pardubice, Czech Republic.*

**Abstract.** Granular beds are an important simulant of damaged PBXs which have developed porosity. Recent developments in X-ray microtomography have provided us with the ability to resolve energetic crystals contained within a polymer matrix or granular bed. Although electron microscopy offers better spatial resolution, it yields little information beyond the surface. Methods to look inside a granular bed, PBX or energetic crystal involving polishing or optical microscopy have enjoyed only limited success. The information now available using X-ray microtomography surpasses these methods, as will be shown in this paper. Two-dimensional slices through a sample are obtained and can be reconstructed to form a three-dimensional image of the entire bed. Slices through energetic crystals reveal the presence of intragranular pores. Granular beds of sugar and HMX have been subjected to compaction experiments; fracture, bed rearrangement, changes in porosity and other affects are clearly visible and quantifiable. It is hoped the extent of bed fracture will make it possible to approximate the extent of energy dissipation due to material fracture over the different compaction regimes.

**Keywords:** Microtomography, compaction, granular, HMX, sugar.
**PACS:** 45.70.Cc, 87.59.Fm.

## INTRODUCTION

The quasi-static compaction of granular beds of energetic material is important to several areas of explosives research and development. A compaction law describes the dependence of intergranular stress on porosity during compaction. Such a law is necessary to feed continuum models of the deflagration-to-detonation transition (DDT) such as that under development at the Department of Applied Mathematics and Theoretical Physics, University of Cambridge (U.K.). For a more thorough background refer to [1-5]. This is not a comprehensive list of references but highlights some of the key work in the field.

Traditionally a compaction law has been derived from the quasi-static experiments conducted at the Naval Surface Weapons Center (NSWC), Silver Spring (U.S.A.) during the 1980s [6-8]. However, recent research has shown the possibility of conducing these experiments at higher strain-rates [9,10]. In this more recent version of the test, a Split Hopkinson Pressure Bar is used to drive a well-defined compaction wave through the sample and measurements of the input, transmitted and reflected stress pulses enables calculation of the sample porosity and intergranular stress as a function of time.

In this paper, we report the results of quasi-static compaction experiments (similar but more simplistic to those of NSWC) and show X-ray microtomographs resolving individual crystals and pores inside beds. The primary aim of these experiments was the development of the X-ray tomography technique rather than the compaction experiment itself. Experiments were conducted with two types (differing grain size distribution) of sugar and HMX.

## EXPERIMENTAL PROCEDURE

Compaction was achieved in these experiments using an Instron 5567 materials testing machine. The granular bed is held in a small polymethylmethacrylate (PMMA) tube, supported in a larger steel confinement. The PMMA is included because of its partial transparency to X-radiation. The PMMA mount has outside and inside diameters of 18 mm and 6 mm respectively and the inner surface was polished to aid movement.

A schematic of the compaction apparatus is given in Fig. 1.

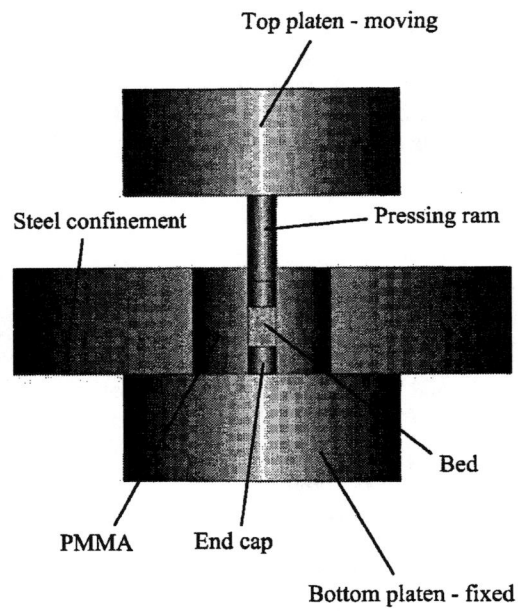

**Figure 1.** Schematic of the loading apparatus.

The load-cell (30 kN) is attached to the top platen and measures the force applied to the bed. It

is recognized that owing to the nature of our sample and apparatus, this is not identical to the transmitted force. The porosity of the bed is calculated from the measured platen displacement. Again, this is not entirely accurate because the system is not perfectly rigid and does give under high load (known as the "system compliance"). However, these inaccuracies were considered acceptable for these experiments where the primary aim was to develop the X-ray microtomography technique.

### Granular materials

Cyclotetramethylene tetranitramine (HMX) is a familiar energetic material widely available to the explosives community in a variety of grain size distributions. In these experiments, we used U.K. HMX Types A and C. Crude particle size analysis indicated that Type A has a mean size of circa 1000 µm and Type C circa 25 µm.

As a comparison, regular sugar was subjected to the same tests. Sieved caster and icing sugar with respective grains sizes 212-300 µm and < 63 µm were chosen.

### X-ray microtomography

Our design enabled the PMMA cylinder to be removed from the outer steel confinement and placed in the X-ray machine. Friction and the two end caps hold the bed in place. The steel confinement only acts to give the whole system mechanical strength during loading.

The sample rotates between the X-ray source and detector with images taken at increments of rotation angle. Two-dimensional reconstructions of the sample are made computationally from the hundreds of intensity patterns obtained.

PMMA is quite transparent to our X-rays but does reduce the image quality. The dimensions of our PMMA mount meant the X-rays had to traverse 12 mm of PMMA.

### RESULTS AND DISCUSSION

The compaction data obtained from beds of HMX are given in Figures 2. The equivalent data from sugar is not presented. For all materials, a number of different runs were made where the experiment was stopped at a different maximum solid volume fraction. The traces show a period of unloading and some hysteresis.

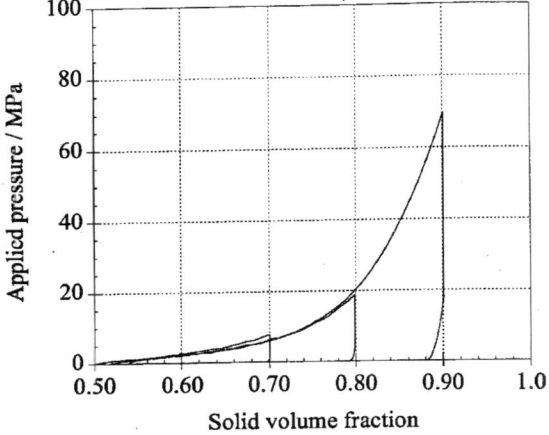

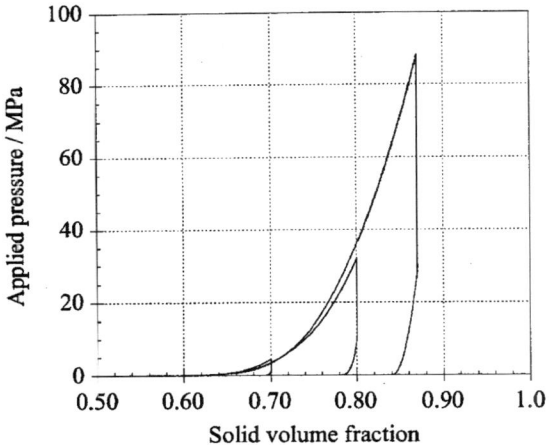

**Figure 2.** Compaction data for HMX A (top) and HMX C (bottom).

Given the rudimentary nature of our apparatus, the agreement between different runs is quite good. The unloading is interesting; the pressure drops off very drastically once loading has stopped and the implications of this should interest those producing pressed pellets for research or ordnance. It is also interesting that wide spread grain fracture can be noticed during loading through localized deviations from the otherwise smooth loading trace. These do not reproduce clearly in Fig. 2.

## X-ray microtomography

At the time of going to print, only images of uncompacted beds were available for publication. An example X-ray microtomograph of a bed of HMX A is given in Fig. 3. This bed is held in a PMMA confinement and the asymmetric nature of this outer region is a consequence of a slight offset in the alignment of source, sample and detector. The wall of PMMA has been partially cropped to maximize magnification and resolution inside the bed.

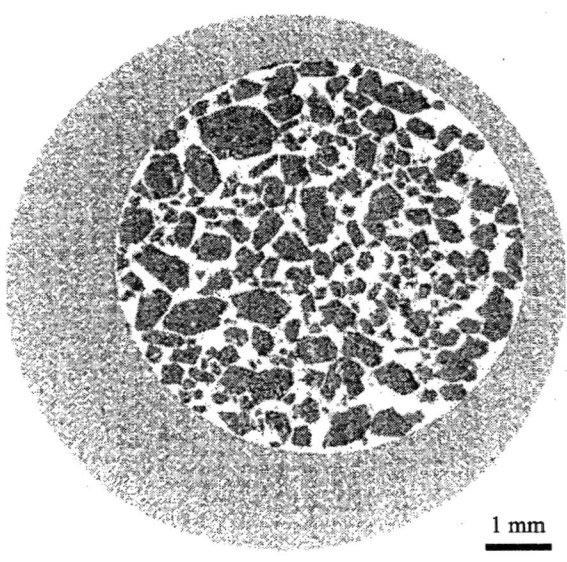

**Figure 3.** X-ray microtomograph of a free-poured (uncompacted) bed of HMX A.

Fig. 3 shows the presence of intragranular (inside the crystal) pores and cracked grains. The influence of intragranular pores on the shock sensitivity of granular explosives is particularly topical. See the paper by Czerski *et al* in these proceedings for further discussions on the subject. We can begin to resolve features circa 10 μm in this image and certainly many pores between circa 20 and 400 μm are visible in each image obtained from this bed (of which Fig. 3 is only an example).

Microtomographs of single crystals of RDX obtained clearly resolve internal pores above 10 μm in diameter. An example of this is given in Figure 4.

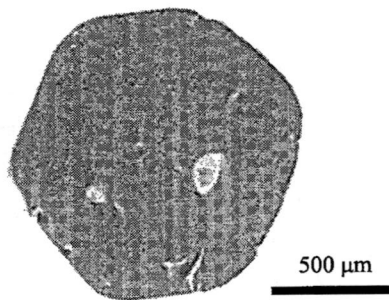

500 μm

**Figure 4.** An X-ray microtomograph of a single RDX crystal.

Some on going work focuses on quantifying these intragranular pores. The role of pores between grains as ignition hot-spot sites is well-known, however, the importance of intragranular pores is a matter of some debate. It is hoped images like these will help shed some light on these issues in the near future.

## CONCLUSIONS

Similar experiments by Elban & Chiarito [8] showed that the transmitted force is circa 60% the applied force for samples like to ours. The applied force does not account for losses, elasticity and yield at the PMMA walls. A more rigorous analysis such as that developed by Kuo *et al* [11] is necessary to better deduce the compaction law.

However, our primary aim was to investigate the capabilities of X-ray microtomography to yield quantitative information on the physical processes occurring inside the bed. The images obtained clearly show the grain outline, intragranular pores, intergranular pores and grain cracking. Computational techniques to quantify many of these properties are under development and this research is on-going.

## ACKNOWLEDGEMENTS

This research was funded jointly by EPSRC and dstl. The authors would like to thank Dr. W. G. Proud, Dr. N. Nikiforakis and Prof. J. E. Field for overseeing the project and Dr. C. A. Lowe for many useful discussions. This paper is dedicated to the late Josef Pelikan.

## REFERENCES

1. Baer, M.R. and Nunziato, J.W., Int. J. Multiphase Flow 12, pp.861 (1986).
2. Sheffield, S.A. *et al.*, in *High-pressure shock compression of solids IV*, edited by L. Davison, Y. Horie, and M. Shahinpoor (Springer-Verlag, New York, 1997).
3. Gonthier, K.A. *et al.*, in *Shock Compression of Condensed Matter - 1997*, edited by S.C. Schmidt, D.P. Dandekar, and J.W. Forbes (American Institute of Physics, Woodbury, New York, 1998), pp. 289.
4. Menikoff, R. and Kober, E., in *Shock Compression of Condensed Matter - 1999*, edited by M.D. Furnish, L.C. Chhabildas, and R.S. Hixson (American Institute of Physics, Melville, New York, 2000), pp. 397.
5. Lowe, C.A. and Greenaway, M.W., J. Appl. Phys. (accepted July 2005).
6. Coyne Jr., P.J. and Elban, W.L., in *Shock Waves in Condensed Matter - 1983*, edited by J.R. Asay, R.A. Graham, and G.K. Straub (North-Holland, Amsterdam, 1984), pp. 147.
7. Sandusky, H.W. *et al.*, in *Shock Waves in Condensed Matter - 1983*, edited by J.R. Asay, R.A. Graham, and G.K. Straub (North-Holland, Amsterdam, 1984), pp. 567.
8. Elban, W.L. and Chiarito, M.A., Powder Technol. 46, pp.181 (1986).
9. Roessig, K. and Hiermaier, S., in *Energetic Materials: Structure and Properties* (DWS Werbeagentur und Verlag GmBH, Karlsruhe, 2004), paper 176.
10. Greenaway, M.W., J. Appl. Phys. 97(9), article no. 093521 (2005).
11. Kuo, K.K. *et al.*, in *Proc. 16th JANNAF Combustion Meeting Vol. 1 Chemical Propulsion Information Agency Publ.* 308 (1979), pp. 559.

CP845, *Shock Compression of Condensed Matter - 2005*,
edited by M. D. Furnish, M. Elert, T. P. Russell, and C. T. White
© 2006 American Institute of Physics 0-7354-0341-4/06/$23.00

# INVESTIGATION OF EJECTA PRODUCTION IN TIN USING PLATE IMPACT EXPERIMENTS

## P. A. Rigg*, W. W. Anderson*, R. T. Olson*, W. T. Buttler* and R. S. Hixson*

*Los Alamos National Laboratory, Los Alamos, NM 87545*

**Abstract.** Experiments to investigate ejecta production in shocked tin have been performed using plate impact facilities at Los Alamos National Laboratory. Three primary diagnostics – piezoelectric pins, Asay foils, and low energy X-ray radiography – were fielded simultaneously in an attempt to quantify the amount of ejecta produced in tin as the shock wave breaks out of the free surface. Results will be presented comparing and contrasting all three diagnostics methods. Advantages and disadvantages of each method will be discussed.

**Keywords:** tin, ejecta, plate impact
**PACS:** 62.50.+p, 79.90.+b, 42.25.Hz, 41.50.+h

## INTRODUCTION

The reflection of a shock wave from the free surface of a solid can lead to the ejection of material from that surface under the right conditions [1]. The form that this ejected material takes can range from a fine spray to the formation of jets depending on the material, surface roughness, and the loading/unloading conditions at the free surface [2]. Understanding the mechanisms leading to these phenomena, collectively referred to as ejecta production, is important for developing robust, physics-based models for the next generation of hydrocodes. The pioneering work by Asay, et. al. [3] demonstrated that the areal density of ejecta produced from a shocked free surface could be measured by suspending a thin foil above the free surface and recording the motion of this foil using velocity interferometry.

Despite a long standing interest in this problem, our understanding of ejecta production needs significant improvement due in part to a lack of quantitative experimental data. The experiments described here are an attempt to obtain quantitative data of ejecta production using plate impact facilities at Los Alamos National Laboratory. Three independent diagnostics methods, including the Asay foil technique mentioned above, were used in this series of experiments to validate results and determine the strengths and weaknesses of each technique.

## EXPERIMENTAL METHOD

Polycrystalline tin disks approximately 38 mm in diameter by several mm thick were shock-compressed by impacting them with a solid metal flyer accelerated by a 50 mm light-gas gun or 40 mm powder-driven gun. Plate impact experiments produce well-characterized, one-dimensional loading and unloading of the tin free surface and thus provide a controlled environment in which to test and understand new diagnostics techniques. Small grooves were cut on the free surface side of most samples to enhance ejecta production at the time of shock breakout. The detail of these grooves is given in the Experimental Results Section. The powder gun experiments were performed at high enough stresses to produce ejecta without the grooved surface.

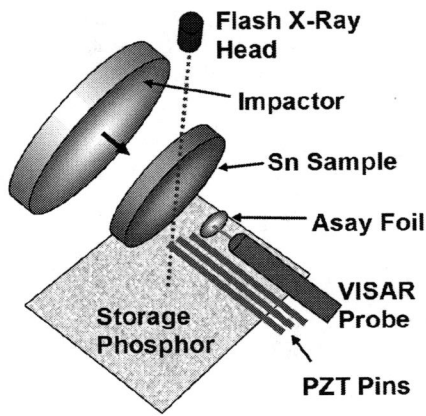

**FIGURE 1.** Representation of the experimental configuration used.

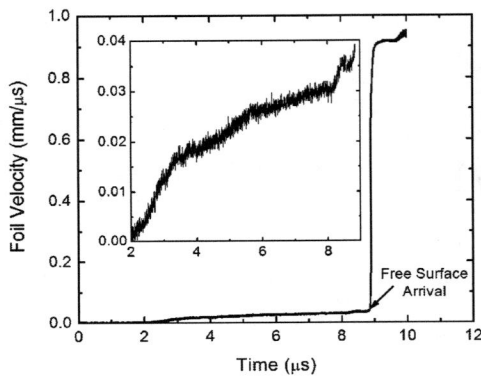

**FIGURE 2.** Asay foil velocity record from a typical experiment. Inset shows the data used to calculate the areal density.

Three diagnostics techniques were used to quantify the amount of ejecta produced from the tin samples: Asay foils, piezoelectric pins, and low energy X-ray radiography. Figure 1 shows a typical experimental configuration with the Asay foil above the center of the target. The Asay foil technique employed here uses a standard push-pull VISAR [4] to monitor the motion of a thin tungsten foil suspended above the target as it is struck by ejecta particles [3]. The measured foil velocity is then converted to an areal density of ejecta versus time using the principles of conservation of mass and momentum (see next section for details). Piezoelectric pins produce a current when ejecta particles stagnate against the surface. This current is proportional to the striking force through a constant piezoelectric coefficient for the active element of the pin. Again using conservation of mass and momentum, the measured voltage (into 50Ω) is converted to an areal density. For the work described here, either lead zirconium titanate (PZT) or lithium niobate (LiNbO$_3$) pins produced by Dynasen were used. Finally, X-ray radiography images were obtained during the ejecta process and converted to areal density values for comparison to the Asay foil and piezoelectric pin results.

Analysis of the Asay foil data was performed using two assumptions; first, that all ejecta were generated at the free surface simultaneously and second, that all collisions of the ejecta particles with the foil were inelastic. The total mass of ejecta that accumulates on the foil is calculated using conservation of momentum

$$m_f u_f + m_e u_e = (m_f + m_e)(u_f + \Delta u_f) \qquad (1)$$

where $e$ and $f$ indicate ejecta and foil values for the mass, $m$ and particle velocity, $u$, respectively and $\Delta u_f$ is the change in foil velocity measured with the VISAR probe. The areal density is then calculated from the peak velocity of the foil just before collision of the free surface with the foil (see Fig 2) and the exposed area of the foil.

To analyze the piezoelectric pin data, the signal measured from each pin is assumed to be directly proportional to the time derivative of the applied force to the active area of the pin. For the LiNbO$_3$ pins the proportionality factor was assumed to be 24 pC/N whereas the value used for the PZT pin analysis was 400 pC/N. The active area of all pins used here was 1.27 mm$^2$. Again, the areal density was calculated using conservation of momentum assuming that all ejecta was generated simultaneously and that all collisions with the pins were inelastic. A detailed explanation of the pin analysis is presented in these proceedings [5].

X-ray radiographs were analyzed by first converting the recorded pixel intensities to a corresponding tin thickness using the radiographic values of a tin step wedge placed in the beam path, but away from the ejecta field. Next, by assuming total symmetry of the ejecta field (a reasonable assumption for plate impact experiments), the total areal density of ejecta at the Asay foil or piezoelectric pin can be calculated by integrating along a line extending from that diagnostic to the target free surface. However, for the most accurate comparison, the radiograph itself must be obtained after ejecta production begins, but before the first ejecta impact the Asay foil or pins. The more spread out the ejecta field, i.e. the closer that first ejecta is to the foil or pin, the more accurate the measurement. Therefore, this puts tight constraints on the timing of the X-ray output with respect to the ejecta motion. In some cases, a test shot to obtain the timing was necessary.

## EXPERIMENTAL RESULTS

Five experiments were conducted to quantify the areal density of ejecta produced from tin under shock loading/unloading conditions and relevant parameters are summarized in Table 1. The first three experiments were performed using the 50 mm gas gun with tungsten flyers impacting grooved tin targets at roughly 600 m/s. For all experiments the grooves were 35 to 45 $\mu$m deep with a 60° dihedral and spaced 235 $\mu$m apart. The Asay foil assembly consisted of a ~55 $\mu$m thick tantalum foil affixed to the end of a 0.375 in. outer diameter brass tube. This tube served as an anchor point for the foil above the center of the target as well as a holder for the VISAR probe focused at the back of the foil. In Experiment 1, the foil was securely attached to the end of the tube whereas in all subsequent experiments, the foil was lightly tacked to the inside of an end cap allowing the foil to freely move toward the VISAR probe; both configurations had an exposed area of 46.6 mm$^2$.

The results of the first three experiments are summarized in Fig 3. Relatively good agreement was obtained for the areal density between the Asay foil results and radiography results. However, the areal density was approximately a factor of two higher using the PZT pin results. Two assumptions were made to explain the discrepancy: First, because the pins were closer to the sample edge, they may be subjected to ejecta generated at the plastic target plate/sample interface. For Experiments 1 and 2, the Asay foil was always placed over the center of the sample and radiography would be much less sensitive to this ejecta if it primarily consisted of plastic particles. Second, since the pins have a much smaller cross section than the Asay foil and the X-rays, they

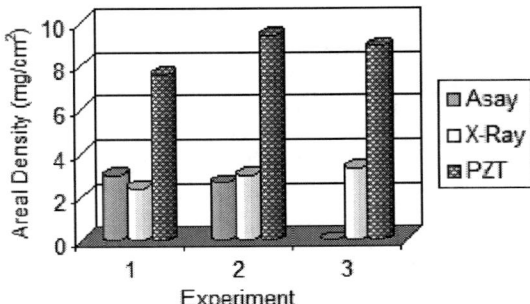

**FIGURE 3.** Summary of the results obtained for the three low pressure experiments performed using the 50 mm gas gun.

are affected negatively by jetting produced by the grooves on the tin surface. Experiment 3 was fielded without the Asay foil and with several pins located near the center without resulting in better agreement between the pin and radiography data.

To address the second possibility for the discrepancy, two higher stress experiments – using the 40 mm powder gun – were designed to cross the melt boundary in tin upon release at the free surface, producing enough ejecta to be diagnosed without using a grooved surface. Figure 4 shows the radiograph obtained during Experiment 4 with the tin step wedge used for calibration shown in the upper portion of the picture. This radiograph clearly shows a large

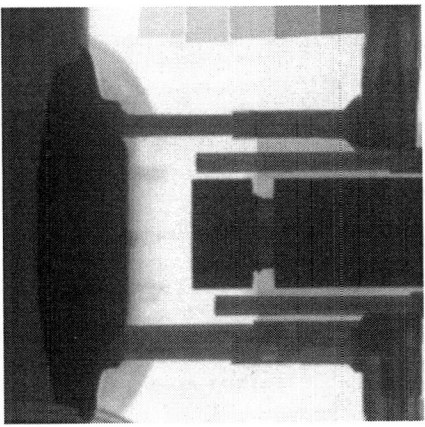

**FIGURE 4.** Radiograph obtained during Experiment 4 showing an annulus of material generated at the edges of the sample.

**TABLE 1.** Summary of ejecta experiments performed to date

| Experiment | Shot Number | Impactor | Impactor Velocity | Surface Finish | Asay Foil Height | Piezoelectric Pins | X-rays |
|---|---|---|---|---|---|---|---|
| 1 | 56-02-08 | Tungsten | 600 m/s | Grooved | 12 mm | PZT | Yes |
| 2 | 56-02-15 | Tungsten | 608 m/s | Grooved | 12 mm | PZT | Yes |
| 3 | 56-02-30 | Tungsten | 605 m/s | Grooved | N/A | PZT | Yes |
| 4 | 04-03-13 | Copper | 1367 m/s | 32 $\mu$-in. | 12 mm | PZT/LiNbO$_3$ | Yes |
| 5 | 04-03-14 | Copper | 1550 m/s | 32 $\mu$-in. | 12 mm | PZT/LiNbO$_3$ | Yes |

amount of ejecta being generated in the vicinity of the target plate/sample interface as previously suspected. Figure 5 shows the areal densities obtained from all three diagnostics techniques with the results from Experiment 5 qualitatively similar. Slightly better agreement is obtained between the foil and pins, particularly the LiNbO$_3$ pins suggesting at least that grooved targets are not appropriate for comparing these diagnostics. However, the X-ray results were clearly affected by the annulus of material generated at the edges of the sample and this will be addressed in future experiments.

## CONCLUSIONS

The experiments presented here were designed to directly compare three diagnostics techniques for quantifying the amount of ejecta produced from shocked tin samples. The low pressure experiments showed good agreement between results generated

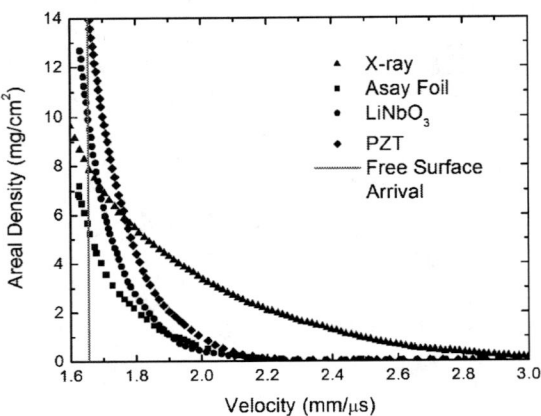

**FIGURE 5.** Areal densities calculated from Experiment 4 for all three diagnostics techniques.

from the Asay foil data and X-ray radiography data, but did not agree with the PZT pin data leading us to conclude that grooved targets are not ideal for comparing these techniques. The high pressure experiments showed better agreement between the pins (both PZT and LiNbO$_3$ pins were used) and Asay foil, but the X-ray results were tainted by the large amounts of ejecta produced at the sample edges. The results from this study, however, show promise for all three diagnostics techniques and future experimental work in this area is planned.

## ACKNOWLEDGMENTS

The authors gratefully acknowledge the assistance of M. Byers in conducting these experiments. This manuscript has been authored by employees of the University of California, operator of the Los Alamos National Laboratory under Contract No. W-7405-ENG-36 with the U.S. Department of Energy. Accordingly, the U.S. Government retains an irrevocable, nonexclusive, royalty-free license to publish, translate, reproduce, use, or dispose of the published form of the work and to authorize others to do the same for U.S. Government purposes.

## REFERENCES

1. Walsh, J. M., Shreffler, R. G., and Willig, F. J., *J. Appl. Phys.*, **24**, 349 (1953).
2. Sorenson, D. S., Minich, R. W., et al., *J. Appl. Phys.*, **92**, 5830 (2002).
3. Asay, J. R., Mix, L. P., and Perry, F. P., *Appl. Phys. Lett.*, **29**, 284 (1976).
4. Hemsing, W. F., *Rev. Sci. Inst.*, **50**, 73 (1979).
5. Vogan, W., Anderson, W., et al., "A New Spin on an Old Technology: Piezoelectric Ejecta Diagnostics for Shock Environments," in *these proceedings*.

CP845, *Shock Compression of Condensed Matter - 2005*,
edited by M. D. Furnish, M. Elert, T. P. Russell, and C. T. White

# MEGAVOLT ELECTRON BEAMS FOR ULTRAFAST TIME-RESOLVED ELECTRON DIFFRACTION

**F.M. Rudakov[1], J. B. Hastings[2], D. H. Dowell[2], J. F. Schmerge[2], and P. M. Weber[1].**

[1]*Department of Chemistry, Brown University, Providence, RI 02912*
[2]*SLAC, Stanford University, Menlo Park, CA 94025*

**Abstract.** We investigate the potential of using rf-guns as electron sources for relativistic, ultrafast time-resolved pump-probe electron diffraction experiments. We explore the feasibility of the experiment by simulating the electron trajectories using instrumental design parameters of the Gun Test Facility (GTF) at the Stanford Linear Accelerator (SLAC), and modeling the scattering event using relativistic differential scattering cross sections. The simulations, done for a 1500 nm thick aluminum foil, suggest that single-shot diffraction patterns can be obtained with electron pulses containing $\sim 10^7$ electrons, and that a time resolution approaching hundred femtoseconds is possible.

**Keywords:** Ultrafast electron diffraction, rf-gun, relativistic electron scattering.
**PACS:** 34.80.Bm. 61.14.Hg

## I. INTRODUCTION

In spectroscopy, the time-resolution of ultrafast time-resolved pump-probe experiments is limited only by the duration of the laser pulses, which is now in the regime of tens of femtoseconds. In contrast, time-resolved diffraction experiments, useful to determine time-dependent structural dynamics, has remained at time scales near 1 ps. Electron diffraction is particularly promising, since large cross sections allow the study of surface phenomena, the bulk structures of thin foils and membranes, as well as molecular structural investigations of gas phase samples.[1] Unfortunately, the space-charge interactions of the electrons within a pulse, and the initial kinetic energy distribution with which the electrons are generated, have made it difficult to obtain pulses much shorter than 1 ps.[2,3,4]

Approaches to improve the time resolution of pump-probe electron diffraction can involve the use of fewer electrons per pulse, requiring a longer data acquisition time to obtain the necessary signal to noise ratio.[5] Alternatively, it is possible to increase the electric field inside the electron gun, while reducing the flight distance between the gun and the target.[6] Both tend to reduce the time of flight of the electron pulse, thereby giving the electron pulse less time to spread. Even so, there may be limits to that approach, as the maximum DC electric field is 12 MV/m,[7] and the highest pulsed field is on the order of 25 MV/m.[8]

In the present work we explore the use of MeV electron beams, generated in a gigahertz rf-gun, for ultrafast time-resolved pump-probe diffraction. Advantages of this approach include:

1) Electron energies of 5-6 MeV are easily accessible with rf-guns. Intense, well collimated electron beams can be generated; rf electron guns are currently the brightest electron sources available.[9]

2) The electric fields are very high, up to 100 MV/m,[10] so that the electrons reach relativistic speeds within a few millimeters. Once they reach such speeds, the Coulomb repulsion stops playing an important role in pulse broadening, because, to change the speed of a relativistic electron, a much large momentum is required than that for a non-relativistic one. In addition, the short time of travel to the target gives little time for pulse broadening.

3) The AC field used in rf-guns can be timed so as to provide compression of the electron bunch.[11]

The combination of these advantages suggests that electron pulses generated by rf-guns can contain many electrons, while maintaining the desired ultrashort pulse durations. This gives rise to the possibility of getting complete diffraction patterns within a single electron pulse. Most important for the study of shock phenomena, the total (elastic plus inelastic) interaction of relativistic electrons with matter is smaller than that of non-relativistic ones, thus allowing for thicker samples. Finally, the short wavelengths of relativistic electrons implies very small scattering angles. As a result, the diffraction detector can be placed at a significant distance from the target.

The great potential of MeV ultrafast time-resolved electron diffraction (MeV-UED) serves as a motivation to explore the applicability of rf-guns to electron diffraction experiments. We are interested in optimal electron gun parameters, and ask about the time resolution that might be achievable for electron pulses with different amounts of charge. In addition, we investigate whether it would be possible to obtain reasonable quality electron diffraction patterns with a single shot, and how the time resolution of the experiment depends on the pulse duration of the laser that generates the electron pulse.

This paper is organized as follows: In section II we describe the test setup, as well as the computational tools to model the electron pulse propagation. Section III concerns the relativistic scattering cross sections and the computational method used to model the diffraction process. The results of our investigations are delineated in section IV.

## II. EXPERIMENTAL ARRANGEMENT

Our model studies are based on the experimental layout of the gun test facility (GTF) at SSRL, shown schematically in fig 1. This instrument was developed to study the brightness of electron beams from the RF photocathode gun for the Linac Coherent Light Source (LCLS) project, and is capable of producing the low divergence, short pulse electron bunches needed for UED. The GTF beam line configuration consists of a 1.6 cell photocathode rf gun and magnetic solenoid, followed by a 3-meter long linac, both operating at the s-band radiofrequency of 2.856 GHz. For the present simulations, the Linac is off, and serves as a drift tube for the electrons. The gun rapidly accelerates photoelectrons in a 110 MV/m field to an energy of 5 MeV, preserving the short pulse lengths and transverse beam quality. Downstream of the linac are a quadrupole lens doublet and a view screen. The copper photocathode of the RF gun is illuminated by the 1 mm diameter beam of a quadrupled Nd:glass laser, at 263 nm, which has a pulse duration of 2 ps. The distance between the cathode and the target is 0.7 meters, and the length of the linac region is 4 m. The solenoid is used to focus the beam to compensate for transversal pulse broadening caused by the rf-gun.

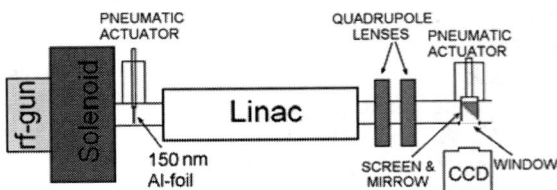

**Figure 1.** Layout of the beam line that serves as a basis for the computational studies.

The generation and propagation of the electron pulse was simulated using the General Particle Tracer program.[12] The program calculates trajectories in 6N phase-space, where the position $x$

and momentum $\mathbf{p} = \gamma m \mathbf{v}$ are used as coordinates. The equations of motion for a particle $i$ are

$$\frac{\partial p_i}{\partial t} = F_i \quad \text{and} \quad \frac{\partial x_i}{\partial t} = \frac{p_i c}{\sqrt{p_i^2 + m_i^2 c^2}} \quad (1)$$

respectively, where $F_i = q\left(E_i + [v_i \times B_i]\right)$, and $F_i$, $E_i$ and $B_i$ are the force, the electric field and the magnetic field acting on a particle. To reduce the computational demands, the program groups electrons into 'particles' that have no internal structure. Thus the total number of particles is less than the number of electrons, even though the total pulse has the correct mass and charge. The overall simulation procedure can be divided into three steps:

1) Electron emission and propagation to the target.

2) Electron scattering from the Al-foil.

3) Propagation in an external field free region towards the screen.

We used 832,195 particles in our simulations. The initial 6N particle coordinates were distributed such that the z-coordinate (propagation direction) is zero at the time of particle origination, and that the RMS beam radius is 0.43 mm. The velocities are distributed normally in all three directions, with a standard deviation of 0.001 c (0.25 eV). To explore the effect of laser pulse duration on the electron pulse, we performed calculations with different Gaussian distributions. The GPT model calculations were previously benchmarked against experiments.

The particle trajectories are propagated to the target under the influence of the electromagnetic field of the rf-gun, the solenoid, and the space charge field of the electron pulse. At the position of the target the electrons are scattered by the sample. Details relating to the simulation of this scattering event are described next.

### III. RELATIVISTIC ELECTRON SCATTERING

We calculate the differential cross-sections of Aluminum atoms following the procedure described by Salvat.[13] The Mott-formula describes the differential scattering cross section of an electron by a point charge $Z$, in the first Born approximation, as

$$\frac{d\sigma}{d\Omega_Z} = \frac{4Z^2}{s^4} \frac{1 - \beta^2 \sin^2(\theta/2)}{1 - \beta^2} \quad (2)$$

where $s = 2p \cdot \sin(\theta/2)$ is the magnitude of the momentum transfer, with $p$ the electron's momentum. As usual, $\beta$ is the electron velocity, expressed in units of the speed of light. All quantities are in atomic units ($e = m_e = \hbar = 1$).

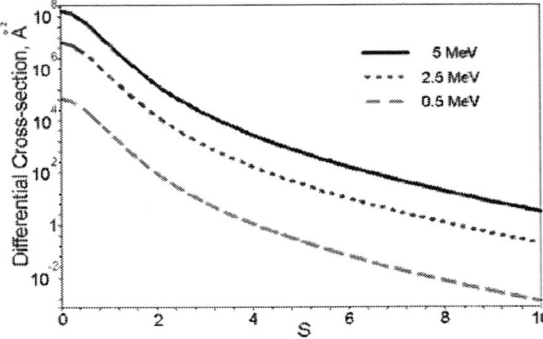

**Figure 2.** Differential scattering cross-sections of 0.5, 2.5 and 5 MeV electrons by aluminum atoms, as a function of the magnitude of the momentum transfer, $s$.

The electrons surrounding a nucleus create an additional scattering potential, which is accounted for by including a screening factor $F(s)$:

$$\frac{d\sigma}{d\Omega_{atom}} = [1 - F(s)]^2 \frac{d\sigma}{d\Omega_Z} \quad (3)$$

$F(s)$ is normalized such that $F(0) = 1$. It is straightforward to calculate the screening factor using electron density distributions derived from atomic structure calculations. For the present discussion the analytical approximations given by Salvat et al.[14] suffice.

Figure 2 shows the differential scattering cross sections for relativistic electrons with energies of 0.5 MeV, 2.5 MeV and 5 MeV. The curves reveal a dramatic increase of the differential cross sections as the electron energy is increased. As concerns the observation of a signal, this increase is, however, balanced by the smaller scattering angles for faster electrons, and the concomitant smaller area of the diffraction pattern. As a result of this balance the total scattering cross-section, calculated as

$$\sigma_{total} = 2 \cdot \pi \int_0^\pi \frac{d\sigma}{d\Omega} \sin(\varphi) d\varphi \ , \qquad (4)$$

remains independent of the electron energy once relativistic speeds are reached. For aluminum atoms and 5 MeV electrons the total cross-section is 0.02624 a.u.$^2$, or 0.00737 Å$^2$.

The interaction of electrons with the aluminum foil was simulated as follows: First we calculated the total number of electrons scattered. Then, knowing how many particles were single-scattered, we randomly picked a certain number of them and changed the directions of their velocity vectors so as to simulate electron scattering from the atom.

The number of particles that was scattered was calculated using the Poisson distribution

$$P(n) = e^{t/l} \frac{\left(\frac{t}{l}\right)^n}{n!} \ , \qquad (5)$$

where $t$ is the thickness of the sample, $n$ the number of scattering events, $l$ the mean free path for electrons in aluminum that can be calculated as $l = \frac{1}{\sigma_{total} N}$. $N$ is the number of atoms per unit volume. The normalized Poisson distribution for 1500 nm thick aluminum is plotted in figure 3. It is apparent that while multiple scattering is not rare, even in the 1500 nm foil single scattering events predominate.

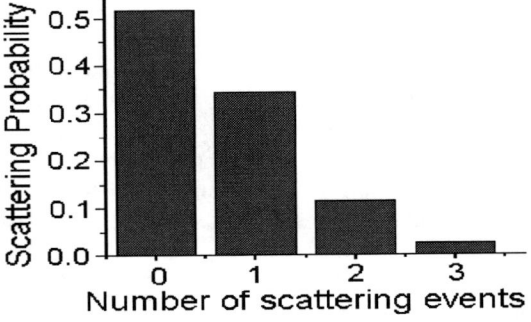

**Figure 3.** Histogram showing the probability of an electron to pass the sample without scattering, and with single, double, and triple scattering.

A single scattering event was simulated as follows: For each particle chosen to be scattered, a momentum transfer vector $s$ was generated (figure 4). $s$ is directed such that $|p'| = |p + s| = |p|$ (the scattering is presumed to be elastic, i.e. the electron energy remains constant). The magnitude of the vector $s$ was randomly picked with a distribution given by the function of the intensity of scattering $I(s)$. $I(s)$ was calculated in the kinematic approximation [15] and is represented by a sum of delta-functions at positions given by Braggs' law. The angle $\varphi$ determines the scattering direction and is uniformly distributed form 0 to $2\pi$.

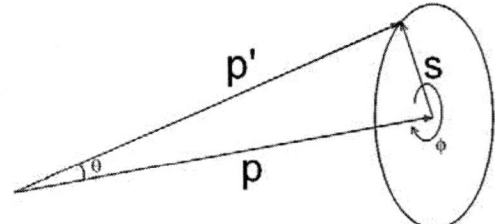

**Figure 4.** Geometry of the scattering simulation. The initial electron momentum $p$ is changed to $p'$, the momentum of the scattered electron. As the scattering event is elastic, $|p| = |p'|$. $\theta$ is the scattering angle, and $s = 2p \cdot sin(\theta/2)$ is the momentum transfer vector. The angle $\varphi$ determines the direction of scattering.

After scattering from the target foil, the electrons move in an external-field free region and experience only the space charge arising from other electrons of the pulse. To simulate these trajectories we again used GPT. The 6N phase coordinates obtained from the scattering simulations were used as the starting conditions for the GPT program.

To find optimal experimental conditions, all calculations were performed repeatedly with varying input parameters.

## IV. RESULTS AND DISCUSSION

Calculations were performed for several electron pulse charges. In the present section we present results for 5 MeV electron beams with a range of charges in the picoCoulomb range. This appears to be at the borderline where space-charge effects just start to affect the experimental observations. For each pulse charge the current through the solenoid was optimized to provide the best possible resolution of diffraction patterns. The required currents were slightly dependent on the charges.

The patterns for 5 pC and 2 pC pulses are shown in figure 5. Figure 6 shows the intensity as a function of $s$ for the 2 pC pulse. Several points are important. First we note that the diffraction patterns are well defined, implying that patterns can be obtained with individual electron pulses. In an experiment, the efficiency for detecting an electron may well be less than 100%. On the other hand, each point in figure 5 represents 15 electrons for the 2 pC pulse, and 37.5 electrons for the 5 pC pulse. Thus, even non-ideal experimental conditions should still allow for well-resolved and observable single-shot diffraction patterns.

Secondly we address the resolution of the diffraction patterns. Both patterns of figure 5 show clearly the diffraction signature of the aluminum. However, while the pattern with the 2 pC pulse is limited by the divergence of the electron beam, the pattern with the 5 pC pulse shows additional wash-out, presumably arising from the space-charge interactions within the pulse. Thus, at the currently employed RF gun geometry and focusing conditions, it appears that pulses with up to several picoCoulomb of charge can be employed without undue effects from space charge interactions.

One method to improve the spatial resolution for the high-charge pulse is to insert a collimator in the beam line, which minimizes both beam divergence and diameter. Separately, a reduction of the laser beam diameter at the photocathode could also improve the resolution.

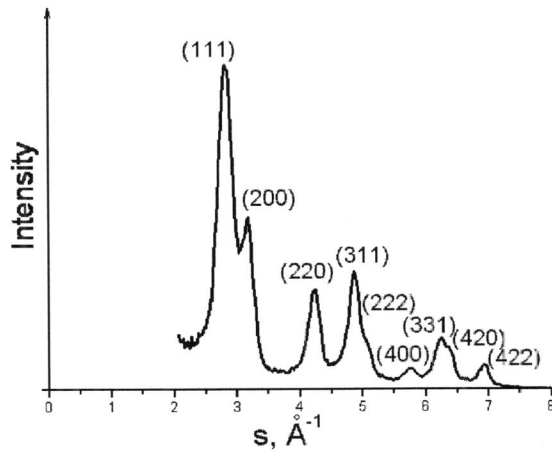

**Figure 6.** The intensity of scattering, *I(s)*, as calculated from the diffraction image shown in figure 5b.

The duration of the electron pulse at the point of the target is an important determinant for the time-resolution that is achievable in a pump-probe diffraction experiment. For the 2 pC pulse, at the target, we find that the pulse duration is 310 fs. Clearly, the RF gun achieves a pulse compression of more than a factor of 6, when compared with the duration of the laser pulse employed (2 ps). To explore the pulse durations achieved when generating the electron pulse with shorter laser pulses, we repeated the calculations with different laser pulse durations. As expected, figure 7 shows that the smaller charge content pulse provides a better time resolution. Additionally, at any charge, a tremendous reduction of the electron pulse duration can be achieved by employing shorter laser pulses for ejecting the electron. Off-the-shelf laser systems can nowadays produce pulse durations less than 50 fs. The combination of such lasers with current RF gun technology should thus be able to provide a time resolution for MeV-UED experiments approaching 100 fs.

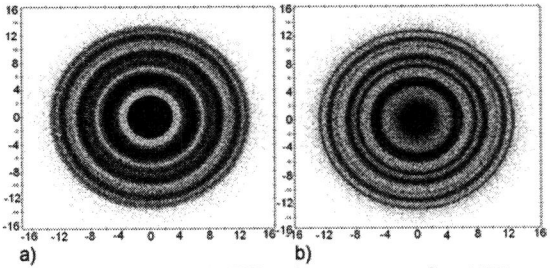

**Figure 5.** Calculated diffraction pattern of a 1500 nm aluminum foil, for a) 5pC electron pulse, b) 2 pC electron pulse. Both images were obtained with optimal focusing conditions.

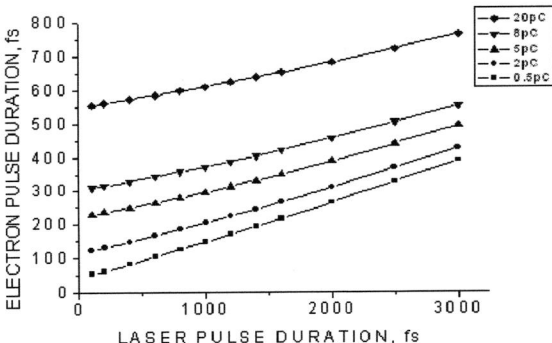

**Figure 7.** Electron pulse duration at the target as a function of the laser pulse duration, for pulses with 2 pC and 5 pC of total charge.

## V. CONCLUSIONS

Time resolved pump-probe electron diffraction with MeV electron beams is a technique of great promise for the study of transient phenomena in both bulk materials and rarified gases. Relativistic electrons are advantageous because of the greatly reduced space-charge interactions, and because they have a large penetration depth. For certain applications, the small scattering angles of short wavelength electrons offers the opportunity to place gun and detector instrumentation at large distances from the target.

We have shown that the space-charge limitations of relativistic electrons is so greatly reduced that it is possible to pack a large enough charge within a sub-picosecond electron pulse that it is possible to obtain a diffraction pattern in a single shot. Pulse compression within the RF gun counters temporal broadening effects, so that a time resolution approaching 100 fs appears possible. Initial test experiments, which are currently in progress, support these conclusions.

## ACKNOWLEDGEMENTS

This research was supported by the Chemical Sciences, Geosciences and Biosciences Division, Office of Basic Energy Sciences, Department of Energy, grant number DE-FG02-03ER15452, and by the Army Research Office under contract No. DAAD19-03-1-0140.

## REFERENCES

1. Hargittai, I. and Hargittai, M.; "Stereochemical Applications of Gas-Phase Electron Diffraction" Verlag Chemie, New York, 1988.
2. Srinivasan, R., Lobastov, V. A., Ruan, C.-Y. and Zewail, A. H., "Ultrafast Electron Diffraction (UED)" Helvetica Chimica Acta 86 (2003) 1763-1837.
3. Dudek, R. C. and Weber, P. M., "Ultrafast diffraction imaging of the electrocyclic ring-opening reaction of 1,3-cyclohexadiene," Journal of Physical Chemistry A, 105, 4167-4171 (2001).
4. Siwick, B. J., Dwyer, J. R., Jordan, R. E., Miller, R.J. D. "An Atomic –Level View of Melting Using Femtosecond Electron Diffraction," Science 302 (2003) 1382-1385.
5. Williamson, J. C., Cao, J., Ihee, H., Frey, H., Zewail, A. H., "Clocking transient chemical changes by ultrafast electron diffraction," Nature, (1997), 386, 159-162.
6. Siwick, B. J., Dwyer, J. R., Jordan, R. E., Miller, R.J. D., J. Appl. Phys. 92, 1643 2002; Siwick, B. J., Dwyer, J. R., Jordan, R. E., Miller, R.J. D., J. Appl. Phys. 94, 807 2003.
7. Park, H., Nie, S., Wang, X., Clinite, R., and Cao, J., "Optical Control of Coherent Lattice Motions Probed by Femtosecond Electron Diffraction" J. Phys. Chem. 109 (2005), 13854-13856.
8 Gallant, P. *et al.,* Rev. Sci. Instrum 71, 3627 (2002)
9. King, W. E., Campbell, G. H., Frank, A., Reed, B., Schmerge, J. F., Siwick, B. J., Stuart, B. C., Weber, P. M., "Ultrafast electron microscopy in materials science, biology, and chemistry," J. App. Phys. 97, 111101 (2005).
10. Wuensch, W., in *Proceedings of EPAC 2002* (European Physical Society Interdivisional Group on Accelerators and CERM, Geneva, 2002), pp. 134-138.
11. Wangler, T., RF-linear accelerators (Wiley series in beam physics and accelerator technology) ISBN: 0471168149
12. General Particle Tracer, www.pulsar.nl.
13. Salvat, B. F., "Elastic scattering of fast electrons and positrons by atoms," Physical Review A 43(1), 578.
14. Salvat, B. F., Martinez, J.D., Mayol, R., Parellada, J. "Analytical Dirac-Hartree-Fock-Slater screening functions for atoms (Z=1-92)," Physical Review A 36(2), 467.
15. Li, X. Z. "JECP/PCED – a computer program for simulation of polycrystalline electron diffraction pattern and phase identification," *Ultramicroscopy 99* (2004) 257-261.

CP845, *Shock Compression of Condensed Matter - 2005*,
edited by M. D. Furnish, M. Elert, T. P. Russell, and C. T. White
© 2006 American Institute of Physics 0-7354-0341-4/06/$23.00

# MEASUREMENTS OF STRAIN PROPAGATION IN HOPKINSON BAR SPECIMENS

## C.R. Siviour and W.G. Proud

*PCS Group, Cavendish Laboratory, Cambridge, CB3 0HE, UK,*

**Abstract.** The research presented continues the development of speckle metrology as a tool for making novel measurements using the split Hopkinson pressure bar (SHPB). Here, speckle is used to observe and quantify strain propagation in long specimens, by measuring the deformation field in the specimen as a function of time as an elastic wave passed through it. By combining these measurements with calculations of the stress in the specimen it is hoped to measure a high strain rate equivalent to the elastic modulus: the relationship between stress and strain in a propagating wave. The experiment outlined is a first step in this development.

**Keywords:** Hopkinson bar, Speckle metrology, Elastic wave propagation
**PACS:** 62.20.Fe

## INTRODUCTION

The split Hopkinson pressure bar [1] is a well-established technique for the evaluation of the mechanical properties of materials at strain rates between 100 and 10,000 s$^{-1}$. In particular, it is able to provide accurate measurements of the yield and flow stresses of homogeneously behaving materials over these rates. Unfortunately, all measurements produce a single value for the whole specimen, and is effectively an average property for that specimen; for specimens that do not deform homogeneously, such as those of materials that undergo strain localization, such measurements may not be an adequate reflection of material behavior.

A major drawback of high strain rate measurement techniques is that the finite speed of sound in the specimen means that an appreciable amount of deformation is accumulated during the time taken for a stress wave to travel from one end of the specimen to the other. This means that the specimen is not in mechanical equilibrium in the early stages of any experiment. Therefore, it is

generally accepted by experimentalists that it is not possible to calculate a high strain rate elastic modulus from Hopkinson bar data. Indeed, it may be argued that it is not possible to define either a modulus or a strain rate for a system that is not in mechanical equilibrium. However, it would be useful to be able to measure the relationship between stress and strain in the propagating waves.

Overall, it would be of great interest to be able to measure displacement fields in the specimen, and to be able to use these fields to relate stress and strain in the propagating wave. The latter may be possible if the usual, short, Hopkinson bar specimen is replaced by a longer rod of material. Furthermore, these measurements will assist with examination and development of models of the Hopkinson bar system.

In the SHPB system loading is provided by a travelling compressive stress pulse. When this reaches the first bar-specimen interface it splits into reflected and transmitted components. The transmitted portion travels to the second interface, where it again splits. A stress wave then reverberates within the specimen, contributing to

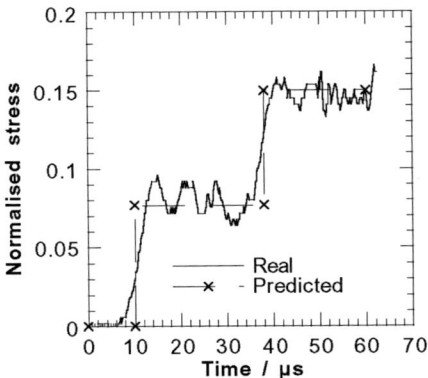

**Figure 1.** Comparison of the real and predicted output bar stresses, normalized with respect to the input bar stress, for a specimen of polycarbonate in magnesium alloy Hopkinson bars. The specimen diameter and length were 4.11 and 18.2 mm respectively. The time between stress increases agrees with the time taken for one full wave oscillation at 1.4 mm $\mu s^{-1}$ (26 $\mu s$).

the reflected and transmitted pulses in the bars in a stepwise fashion. As long as the specimen remains elastic the amount of reflection and transmission at the boundaries is governed by the standard elastic coefficients [2]. If the specimen is made longer than the rise time of the incident pulse, the stepwise increase can be seen in the transmitted bar, Fig. 1. The experiments described in this paper will evaluate the evoution of stress and strain in such a specimen as a function of time.

Experiments were performed on Bayer Makrolon® 2805 polycarbonate. Specimens were machined from injection moulded cylinders to a length of approximately 20 mm and a range of diameters between 5 mm and 10 mm. The experiments were performed in a Magnesium Hopkinson bar system, instrumented with semi-conductor strain gauges, which have a high sensitivity to strain.

## SPECKLE MEASUREMENTS OF SPECIMEN STRAIN

Speckle metrology [3] is a technique for measuring displacements in photographts by following the evolution with time of a random speckle pattern in the picture. The experimental procedure used to produce these images has been described in other publications [4,5,6] In this case, the specimens were speckled with black ink and, being transparent, were back lit. The exposure

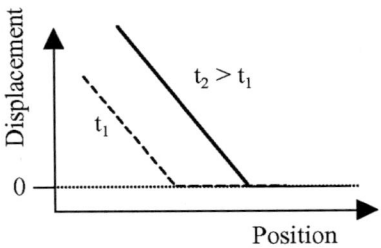

**Figure 2.** Expected displacement-relationship as a stress wave passes through a rod. The gradient behind the wavefront gives the strain.

times were 1 $\mu s$, and the interframe times 2 $\mu s$, 8 frames were taken.

The algorithm used to calculate displacements from the speckle photographs was developed by Sjödahl and co-workers [7.8.9]. The comparison between images was performed on the Fast Fourier Transform (FFT) of the images. Using the FFT allows the displacement resolution to be as good as 0.01 of a pixel on the digitized image.

In these experiments, speckle metrology was used to make measurements of $x$-displacement as a function of $x$-position on a series of photographs taken during the deformation, where $x$ is defined as the direction along the specimen axis. Speckle allows accurate calculations of displacement by following the evolution of a random pattern on the specimen surface. In principle, the use of areas of pattern, rather than single points, and the use of calculations in the Fourier, rather than spatial, domain, can allow displacement accuracies and sensitivities of 0.01 pixels on a digitized image (i.e. the displacement accuracy is greater than the resolution of the camera). In the case of a stress wave passing through a specimen, the shape of the displacement-position curve should be as shown in Fig. 2. Before the wave arrives at a point, the displacement is zero. Behind the wavefront, after a short transition region determined by the rise time of the stress pulse, the strain is constant, and is given by the gradient of the displacement-position graph in this region. With time, the gradient should remain the same, but the wavefront moves along the specimen at the acoustic wavespeed. This is in agreement with the measurement presented in figure 1; the time between subsequent steps is consistent with a wavespeed of 1.4 mm $\mu s^{-1}$. This is the expected wavespeed, $c = \sqrt{E/\rho}$,

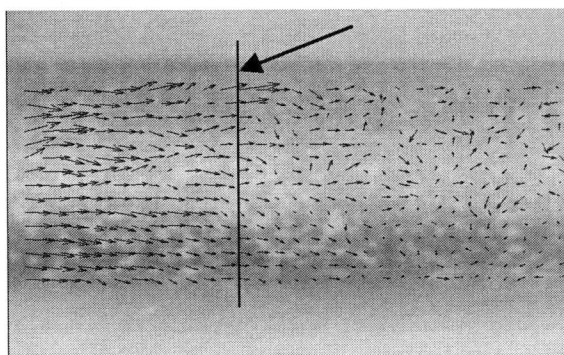

**Figure 3.** Displacement quiver plot for a stress wave passing through a specimen of polycarbonate. Not all of the specimen is visible in this image. The image is in negative for clarity.

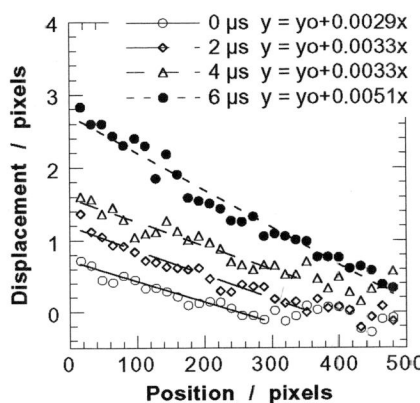

**Figure 4.** Plot of displacement against position along a polycarbonate specimen, at four different times during loading. The horizontal scale is 53 pixels mm$^{-1}$, so the movement of the wavefront agrees with a wavespeed of 1.4 mm μs$^{-1}$.

for a wave in a thin rod. The longitudinal sound speed in polycarbonate is $u_l = 2.2$ mm μs$^{-1}$, which is higher because of restoring shear forces.

Fig. 3 shows a displacement quiver plot for a specimen of polycarbonate. This specimen was 5.0 mm in diameter and 18.2 mm long. The boundary between displaced and undisplaced material is marked with a vertical line (arrowed). In Fig. 4 the displacement-position plots for 4 such photographs taken during a single experiment are given. The value given at each $x$-position is the average over all of the quivers at that value of $x$. The exact wavefront is difficult to define, but the strained region moves through the specimen at a rate consistent with a speed of 1.4 mm μs$^{-1}$. The strain behind the wavefront is approximately 3 mε in the first 3 frames, and 5 mε in the 4$^{th}$. Because the strain measurement is effectively averaged over a large number of displacement measurements it is very robust; in particular it is insensitive to the choice of wavefront position.

The values calculated are interesting; it is established that behind a propagating wave, the strain is given by $\varepsilon = v/u$, where $v$ is the particle velocity at the loading end of the specimen, and $u$ is the relevant acoustic wavespeed. In this case, $v$ is the same as the speed of the end of the input bar, and can be calculated from the reflected wave. In fact, it was found that if an acoustic speed of 2.2 mm μs$^{-1}$ was used, the strain calculated in this experiment was 3.3 mε, and if 1.4 mm μs was used the strain was 5.2 mε, in agreement with the two measured strains.

## CALCULATION OF SPECIMEN STRESS

Consideration of the force conservation requirement when a compressive wave interacts with a boundary shows that the wave in the second material is the same as the sum of the incident *and* reflected waves in the first. This means that the stress wave in the transmitted Hopkinson bar is not representative of the forward propagating wave in the specimen. Instead, the difference in magnitude of the incident and reflected waves, measured in the input bar, must be used. This can be divided by the specimen area to give the stress in the specimen.

In the elastice regime the stress waves propagated are small, even for a strong material such as polycarbonate. In order to maximise the accuracy, large diameter (10 mm) specimens were used. Currently, these experiments have not been performed in conjunction with speckle measurements, however strain estimates can be obtained from the particle velocity and wavespeed, as described above. Fig. 5 shows the specimen stress and particle velocity for a typical experiment. From the data presented in Fig. 5, the strain in the specimen during the first wave is either 4.4 mε or 2.7 mε, using 1.4 and 2.2 mm μs$^{-1}$ respectively. This gives the ratio between stress and strain as 3900 and 6000 MPa. Five repetitions of the experiment gave values of 3750 ± 150 and 5850 ± 150 MPa. The lower of these values compares well to the bulk modulus of the

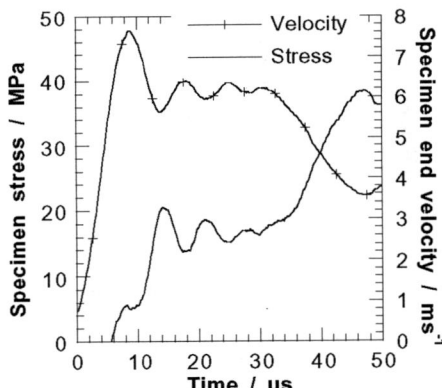

**Figure 5** Stress and specimen end velocity as a function of time for a polycarbonate specimen.

polycarbonate used: 3800 MPa.

## CONCLUSIONS

Using speckle metrology, it has been possible to measure the propagation of a strain wave through a Hopkinson bar specimen, and the strain caused by this wave. These measurements rely on the ability of the technique to produce high resolution displacement measurements. Fig. 3 shows that sub pixel accuracy is a requirement for such measurements.

Both the optical measurements, and those of transmitted stress, show that the strain wave passes through the specimen at 1.4 mm $\mu s^{-1}$. However, the initial strain in the specimen, when compared to the particle velocity at the end of the specimen, is actually that associated with the bulk longitudinal sound speed, 2.2 mm $\mu s^{-1}$. This only changes later to that associated with a sound speed of 1.4 mm $\mu s^{-1}$. This effect requires further investigation, but may be due to lateral movement of stress waves within the specimen.

Initial measurements of the ratio of stress and strain in a propagating wave indicate that they may be related by the bulk modulus. This may have implications for Hopkinson bar testing itself, which uses the strain on the surface of the bar to calculate the stress propagating through it; however, it should be noted that for many metals the bulk modulus and the Young's modulus are very similar. In any case, most laboratories calibrate theirs strain gauges dynamically 'in-situ'.

Further investigation will focus on making speckle and stress measurements in the same experiments, and producing more accurate calculations of the relationship between stress and strain.

## ACKNOWLEDGEMENTS

The authors would like to strongly acknowledge the assistance of DM Williamson and SG Grantham in performing the speckle analysis on these experiments. The software used was supplied by M Sjödahl of the University of Luleå, Sweden. CR Siviour would like to thank EPSRC, [dstl] and the Worshipful Company of Leathersellers for supporting his work. The high-speed camera used in this research is also supported by EPSRC. WG Proud is a QinetiQ Senior Research Fellow; we thank QinetiQ for their support. Finally, we would like to thank JE Field for his advice and encouragement.

## REFERENCES

1. Gray III, G.T., "Classic split-Hopkinson pressure bar testing", in ASM Handbook. Vol. 8: Mechanical Testing and Evaluation, 2000 (H. Kuhn and D. Medlin, Eds) pp. 462-476.
2. Briscoe, B.J. and R.W. Nosker, "The influence of interfacial friction on the deformation of high density polyethylene in a split Hopkinson pressure bar", Wear, 95, pp. 241-262, 1984.
3. Sjödahl, M., "Digital Speckle Photography", in Digital Speckle Pattern Interferometry and Related Techniques, 2001 (P.K. Rastogi, Ed.) pp. 337- 362.
4. Grantham, S.G., et al., "Speckle measurements of sample deformation in the split Hopkinson pressure bar", J. Phys. IV France, 110, pp. 405-410, 2003.
5. Grantham, S.G., et al., "High-strain rate Brazilian testing of an explosive simulant using speckle metrology", Meas. Sci. Technol., 15, pp. 1867-1870, 2004.
6. Siviour, C.R., et al. "Application of Speckle Metrology to the split Hopkinson pressure bar", In preparation.
7. Sjödahl, M. and L.R. Benckert, "Electronic speckle photography: Analysis of an algorithm giving the displacement with subpixel accuracy", Appl. Opt., 32 pp2278-2284, 1993.
8. Sjödahl, M., "Electronic speckle photography: Increased accuracy by non-integral pixel shifting", Appl. Opt., 33, pp. 6667-6673, 1994.
9. Sjödahl, M., "Accuracy in electronic speckle photography", Appl. Opt., 36, pp. 2875-2885, 1997.

CP845, *Shock Compression of Condensed Matter - 2005*,
edited by M. D. Furnish, M. Elert, T. P. Russell, and C. T. White
© 2006 American Institute of Physics 0-7354-0341-4/06/$23.00

# LASER-INDUCED MACH WAVES
# FOR ULTRA-HIGH-PRESSURE EXPERIMENTS

## Damian C. Swift* and Christian R. Ruiz*

*P-24, MS-E526, Los Alamos National Laboratory, Los Alamos, NM 87545*

**Abstract.** Laser-driven experiments are a principal technique for inducing pressures in the terapascal regime and higher. However, when high irradiance laser light interacts with matter, it generates fast electrons and x-rays, which may heat material ahead of hydrodynamic loading waves such as shocks. This preheat limits the scope for investigating properties of initially cold material and potentially reduces the accuracy of measurements. A new configuration for laser experiments is proposed, using convergence and irregular reflection of shocks to induce high pressures without such high laser irradiances. Related Mach wave generators have been developed previously for high-explosive drive; the design considerations for laser-driven Mach wave generators are typically dictated by constraints on the laser pulse duration and differ from high-explosive systems. Relations are presented between the pressures achievable with different variants of the laser drive technique and different combinations of materials in the Mach-interaction region. The prospects for isentropic compression using this type of experiment are discussed.
**Keywords:** shock, high energy density physics, Mach wave, laser ablation
**PACS:** 47.40.+x, 07.35.+k, 47.40.Nm, 52.38.Mf, 52.35.Tc

## INTRODUCTION

There are many needs for high pressure experiments, including investigations of the equation of state (EOS) and strength under shock and isentropic loading. Laser-driven experiments are desirable in a number of respects, e.g. the extremely high pressures which can be attained, the flexibility of loading (ease of changing the energy and the temporal shape of the pulse), the concentration of energy, which often reduces 'collateral momentum' (e.g. sabots from a gas gun, high pressure reaction products from high explosives) and so makes experiments easier to conduct and samples easier to recover. However, there are complications with laser experiments, including the onset of preheat at drive pressures over a few hundred gigapascals, and the difficulty of ensuring a drive pressure which is spatially uniform and adequately characterized in time (e.g. constant, for a shock).

Here we discuss designs for experiments which use shock dynamics to ameliorate some of the problems inherent in laser drives for high pressure applications. The basic idea is that a shock wave can be smoothed and increased in pressure, so high pressures can be reached without necessarily requiring laser irradiances high enough to induce preheating, and spatial variations can also be reduced. The relevant concepts have been applied for many years on a larger scale in systems driven by high explosives. Analogous designs are discussed for use with lasers, and some variants are suggested which would be uniquely suited for lasers. Simulation results are presented for initial trial experiments.

## SHOCK DYNAMICS AND MACH REFLECTION

Consider a solid cone, initially at some constant pressure (e.g. zero), when an elevated pressure is applied

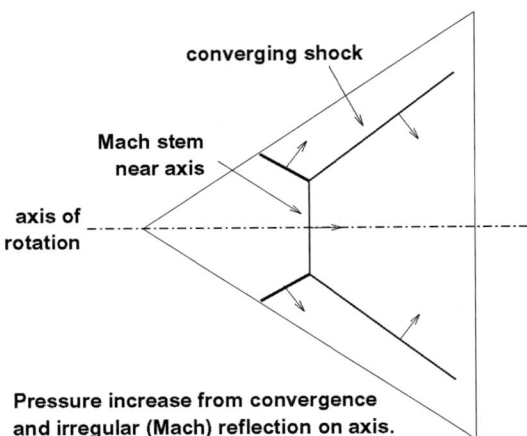

converging shock

Mach stem
near axis

axis of
rotation

Pressure increase from convergence
and irregular (Mach) reflection on axis.

**FIGURE 1.** Schematic of Mach wave generation in a solid cone.

to the conical surface. The applied pressure induces a conically-converging shock. Convergence increases the pressure as the shock propagates inward. Convergence also acts to smooth spatial variations in the shock: regions which lag have a larger negative curvature, thus locally greater convergence; the pressure increases and the shock accelerates to reduce the lag. Conversely, regions which are ahead of the rest of the shock tend to slow down. These smoothing effects are described by a hyperbolic equation, so deviations from symmetry typically cause underdamped oscillations toward it. It is preferable to start with a reasonably symmetric drive.

The cone axis acts as a rigid boundary: waves in a material with linear response would reflect and double the pressure; real EOS are nonlinear and the pressure is generally higher. At appropriate angles of incidence, nonlinear reflection results in the formation of a Mach stem [1] – a disk-shaped shock of higher pressure – which grows radially outward as it propagates along the axis. (Fig. 1.)

## EXPLOSIVE GENERATORS

Several groups have presented designs for Mach wave generators driven by the detonation of chemical explosives. There are design variants in the initiation of the explosive, which changes the phase speed of the detonation over the surface of the cone. The simplest design uses a solid cone, embedded in a cylinder of explosive [2]. Energy transfer from the detonating explosive can be optimized by imploding a hollow cone onto a coaxial solid cone; the implosion process accumulates mechanical work as the reaction products expand, which can easily induce a shock depositing more internal energy than from the shock transmitted directly from the detonation wave. One design has induced shocks of 250 GPa in the PMMA inner cone, which is transmitted as ~500 GPa in a Cu sample [3]. The pressure can be further increased if used to accelerate a flyer which then impacts a stationary target, A smaller system (0.5 kg of the HMX-based explosive EDC29), with a simpler initiation system to induce a ring initiation at the apex end, has been developed and tested, with sample pressures ~100 GPa [4].

## CONCEPTS FOR LASER-DRIVEN MACH WAVE GENERATORS

Laser-driven Mach wave generators can be envisioned with several different configurations. Most designs possible with detonating explosives have a laser-driven analog. Starting with the same basic platform of a cone, the pressure may be induced by laser ablation over the outer surface (Fig. 2). The pressure drive may also be induced in other ways, such as with a laser-heated hohlraum: laser energy is used to induce a thermal radiation field – typically soft x-ray – in a hohlraum, and the thermal radiation induces the shock over the cone; this scheme has the advantage that the radiation field is spatially smoother than the original laser beams.

The laser energy may also be used to generate pressure through confined plasma ablation [5]. The cone would be enclosed in a medium transparent to the laser light, e.g. sapphire. Confining the ablation plasma greatly increases the efficiency of energy transfer to the cone, similarly to the way in which laser-driven flyers operate.

Regardless of the way in which the pressure is applied to the outside of the cone, designs with a solid cone could be replaced by a hollow cone imploded onto a solid inner cone. This variant is appropriate if the pressure can be sustained for a time comparable with the time to accelerate the hollow cone to maximum speed.

An intriguing possibility for laser drives is that the

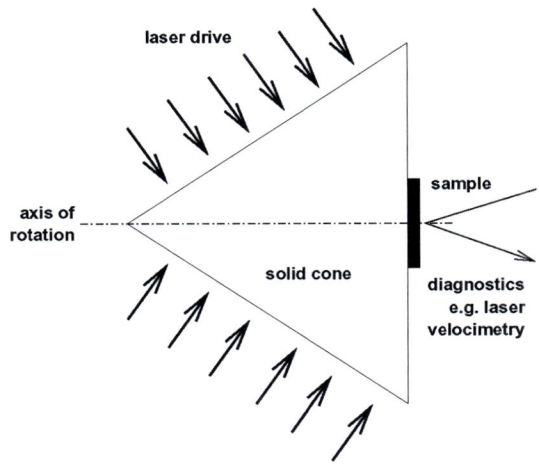

laser drive

axis of
rotation

sample

solid cone

diagnostics
e.g. laser
velocimetry

**FIGURE 2.** Schematic of Mach wave generation in a solid cone driven by laser ablation.

Mach wave could be induced in a planar target, by spatial variation of the laser irradiance or temporal variation in the arrival of different portions of the drive pulse over the surface. Specifically, if the drive beam has a ring of high irradiance around a central region of low irradiance, or if the central portion of the beam arrives later, a Mach wave may be induced.

Other degrees of freedom include the precise profile of the cone (a curved cross-section could be used) and the laser irradiance history.

## APPLICATIONS OF MACH WAVE GENERATORS

The obvious application of Mach wave generators is to shock loading experiments – or shock and release – to study the properties of the sample material. Ideally, the cone material would be chosen to have a well-characterized EOS, and the Mach wave generator would be reproducible and characterized carefully in advance. The sample would be attached to the flat base of the convertor cone in which the Mach wave is induced. Standard diagnostics include time-of-arrival (e.g. by change in reflectivity) and velocimetry of the free surface or surface in contact with a release window. Release windows are less likely to be useful at terapascal-scale pressures. The Mach wave configurations would be compatible with some of the more exotic diagnostics demonstrated on laser-driven experiments, including x-ray diffraction, x-ray radiography, emission spectroscopy, Raman spectroscopy, and polarization-dependent reflectivity.

Material placed on the base of the cone can be chosen to vaporize on shock and release by the Mach wave. If allowed to expand, the vapor would drive a compression wave into a sample spaced some distance from the Mach cone; this is analogous to the method of generating quasi-isentropic compression by vaporizing a layer of material directly with the laser [6], but should allow higher pressures to be reached without laser-induced preheat. Building on the isentropic compression concept, the expanding vapor can be used to accelerate a flyer shocklessly. This flyer can then be impacted with a sample to induce a shock wave. This scheme is closely analogous to the explosive-driven scheme in which pressures around 5 TPa were induced in the sample [3].

## SCOPING CALCULATIONS

Prototype design calculations were made, to allow design variants to be tested on lasers such as TRIDENT and OMEGA.

The baseline design was a cone 10 mm high with a base 10 mm in diameter, made of solid Cu. A representative time scale for low pressure experiments is the time for a weak shock to propagate along the axis: a couple of microseconds. Applying the drive pressure for a time of this order requires a 'long' laser pulse, hence only a relatively low irradiance is possible. Confined ablation is assumed to increase the energy efficiency. At some irradiance, damage to the transparent substrate would limit the energy transmission. A confined ablation pressure $\sim$10 GPa has been demonstrated in TRIDENT experiments with a sapphire substrate.

Hydrodynamic simulations were performed using a 2D Eulerian hydrocode, representing the drive pressure as a constant pressure void. Meshes were 0.1 mm square. With the baseline design, a narrow Mach wave was generated, the pressure increasing to of order 100 GPa near the base of the cone. (A 'transmissive' boundary was used over the base of the cone; this was not perfectly transmissive, and also reflected a shock wave. Thus the pressures calculated

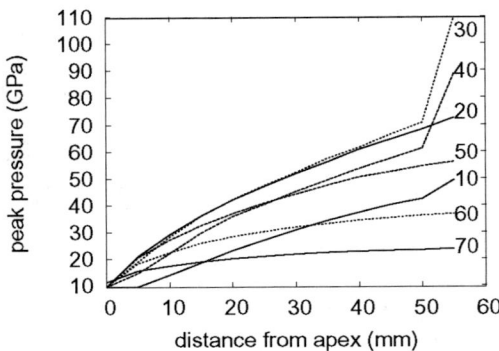

**FIGURE 3.** Pressure as a function of distance from the cone apex, for different cone half-angles (in degrees). The cone was Cu, with a sustained 10 GPa drive.

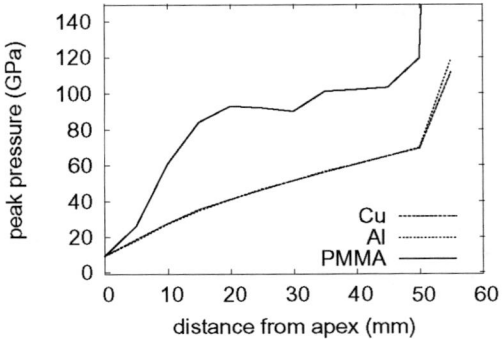

**FIGURE 4.** Pressure as a function of distance from the apex, for different cone materials. The cone had a half-angle of 28.8° – the optimum value for Cu – and had a sustained 10 GPa drive.

at the base were higher than 100 GPa.) Simulations were performed varying the height of the cone. The maximum pressure in Cu was achieved for a cone angle of 28.8° (Fig. 3).

The sensitivity to cone material was explored. Cu and Al gave similar Mach wave pressures; poly(methylmethacrylate) (PMMA) was predicted to give a significantly greater pressure (Fig. 4). The equation of state of PMMA may not be accurate at these pressures, so the predictions should be viewed with caution, but it seems likely that PMMA is a more appropriate material to use. Because of the impedance mismatch, low-impedance cone materials would induce an even greater pressure in the sample than would high-impedance materials.

Based on the confined ablation experiments performed at the TRIDENT laser, it should be possible to sustain the pressure over a cone of this area for ~50 to 100 ns, which should be ample for the Mach wave to form. The Mach pressure increases faster than linearly with the applied pressure, so it would be advantageous to use a higher drive pressure if possible. It is likely that confining windows can be pushed to somewhat higher pressures, but free ablation would be necessary for drives over a few tens of gigapascals.

## CONCLUSIONS

Several designs are possible for laser-driven Mach wave generators. These offer potentially great flexi-

bility in pressure and applications. It should be possible to perform meaningful initial scoping experiments with a laser of the energy and capability of TRIDENT, and thereafter to stage up to much higher pressures using OMEGA and NIF.

## ACKNOWLEDGMENTS

This work was performed under the auspices of the U.S. Department of Energy under contract # W-7405-ENG-36.

## REFERENCES

1. Harlow, F. H. and Amsden, A. A., Los Alamos National Laboratory report LA-4700, 1971.
2. Bushman, A. V., Kanel', G. I., Ni, A. L., and Fortov, V. E., *Intense Dynamic Loading of Condensed Matter*, Taylor and Francis, London, 1993.
3. Aveillé, J. and Protat, J. C., in Proc. 4$^e$ Symphosium (*sic*) International Hautes Pressions Dynamiques, Tours, France, 5-9 June 1995, Commissariat à l'Énergie Atomique, Paris, 1995).
4. Swift, D. C. and Edwards, R. J., unpublished, 1996.
5. Colvin, J. D., Ault, E. R., King, W. E., and Zimmerman, I. H., Phys. Plasmas, **10**, 7, 2940, 2003.
6. J. Edwards, K.T. Lorenz, B.A. Remington, S. Pollaine, et al, Phys. Rev. Lett., **92**, 075002, 2004.

CHAPTER XIX

# ISENTROPIC COMPRESSION EXPERIMENTS

CP845, *Shock Compression of Condensed Matter - 2005*,
edited by M. D. Furnish, M. Elert, T. P. Russell, and C. T. White
© 2006 American Institute of Physics 0-7354-0341-4/06/$23.00

# JUMP CONDITIONS FOR NONSTEADY WAVES

## William W. Anderson

*Los Alamos National Laboratory, Los Alamos NM 87545*

**Abstract.** The usual integral forms of the Rankine-Hugoniot jump conditions are valid only for steady waves that conserve flux. However, many waves encountered in practice are not steady. Here, more general forms of the jump conditions are presented. These can be used, with certain assumptions, for waves that do not conserve flux. The insights gained from these more general forms provide guidance for practical analysis of measured wave profiles. Using simple approximations, analytic expressions can be obtained that allow the evaluation of the validity of assumptions about use of the Rankine-Hugoniot equations for analysis.

**Keywords:** Nonsteady waves.
**PACS:** 62.50.+p, 47.40.Nm.

## INTRODUCTION

Nonsteady waves, i.e., waves with evolving profiles, are commonly encountered in dynamic experiments. Over the past several decades, analysis methods for such waves have been developed and refined [1-3], based on the Lagrangian analysis approach of Fowles and Williams [1]. These analyses assume that wave profiles are measured at several positions in a sample, which is usually not the case. Commonly, the wave profile, if measured at all, is measured at a single location. The usual approach in such cases is to apply the integral forms of the Rankine-Hugoniot conservation relations as an acceptable approximation. However, the Rankine-Hugoniot equations as commonly expressed are valid only for steady waves and their use introduces errors into the analysis of nonsteady waves. In this paper, I examine the forms of the conservation relations for nonsteady wave profiles and consider situations in which, with certain assumptions, the analysis of nonsteady waves in typical shock compression experiments can be taken beyond the use of the Rankine-Hugoniot equations. Consideration of the deviations of the conservation equations due to nonsteadiness also allows estimates of the amount of error that is introduced by application of the usual forms of the Rankine-Hugoniot equations.

## GENERALIZED JUMP CONDITIONS

The unique property of a steady wave is that the profile is constant, meaning that the characteristics are parallel in *x-t* space. One consequence of this is that mass flux through any portion of the wave at a given Lagrangian position is constant. In the case of a nonsteady wave, this condition is not met and we must appeal to the stronger requirement that total mass be conserved for any portion of the wave traversing a given distance in Lagrangian coordinates. Similar considerations apply to the conservation of momentum and energy. In the current treatment, it is assumed that the wave profile can be measured at one Lagrangian position and that a reasonable estimate of the wave profile can be had at a second position. An example might be a release wave generated by a strong shock at a low-impedance interface, in which the wave profile can be measured at some point in the sample and the wave profile is assumed to be a discontinuity at the interface. Two assumptions are required: that the wave characteristics, although not parallel, are straight, and that differences in phase velocities between different quantities can be ignored.

## Conservation of mass

The total mass $m_0$ traversed by a wave of unit surface area between two chosen Lagrangian positions $x_a = 0$ and $x_b$ (Fig. 1) is constant for all characteristics and is defined by the initial density $\rho_0$, the velocity $U_0$ of the leading edge, and the time $t_0$ for the leading edge to travel the distance $x_b$:

$$m_0 = \rho_0 U_0 t_0 \tag{1}$$

To obtain the density change induced by passage of the wave, we must abandon Lagrangian coordinates. If we choose the initial state as the rest state for Eulerian coordinates, then for any chosen mass velocity $u_1 \leq u_f$ (where $u_f$ is the final mass velocity achieved), the Eulerian propagation distance associated with the appropriate characteristic is

$$y(u_1) = (U_0 - u_1)t_0 + \int_0^{u_1} (u - u_1)\left(\left(\frac{dt}{du}\right)_{x_b} - \left(\frac{dt}{du}\right)_{x_a}\right) du \tag{2}$$

where the derivatives are obtained from the "known" wave profiles at positions $a$ and $b$. Since the mass traversed must still be $m_0$, then the density is required to be

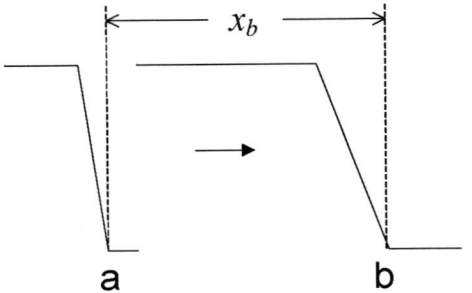

**Figure 1.** Schematic of a nonsteady wave traversing the region between two Lagrangian points $a$ and $b$ separated by distance $x_b$.

$$\rho_1 = \frac{m_0}{y(u_1)} \tag{3}$$

Hence, the density at mass velocity $u_1$ is obtained from (1)-(3):

$$\frac{\rho_1}{\rho_0} = U_0 t_0 \left[ (U_0 - u_1)t_0 + \int_0^{u_1} (u - u_1)\delta t' du \right]^{-1} \tag{4}$$

where

$$\delta t' = \left(\frac{dt}{du}\right)_b - \left(\frac{dt}{du}\right)_a \tag{5}$$

This expression applies to any point in the wave profile and can be used to obtain the jump resulting from the entire wave at location $b$ if $u_1$ is replaced by the final particle speed $u_f$ achieved by the material and if the wave is not increasing in amplitude as it propagates. Note that, when the derivatives are equal (*i.e.*, the wave is steady), the integral vanishes and (4) reduces to the usual form of the Rankine-Hugoniot equation for conservation of mass. It should also be noted that the assumption of straight characteristics means that the result in eq. (4), as well as those presented below, is spatially invariant, i.e., the value of the density (and stress and energy, as given below) for a given mass velocity is the same at all locations.

## Newton's Second Law

Here, we must assume, for lack of information, that the phase velocities of stress and particle velocity are equal. Euler's equation in Lagrangian coordinates gives

$$\rho_0 \left(\frac{du}{dt}\right)_x + \left(\frac{d\sigma}{dx}\right)_t = 0 \tag{6}$$

which reduces to

$$\left(\frac{\partial u}{\partial t}\right)_y = -\left(\frac{\partial u}{\partial y}\right)_t \left({}^E U + u\right) \tag{7}$$

1304

where $^{E}U$ is the Eulerian speed associated with the appropriate characteristic. Then, (7) can be rearranged to give

$$\frac{d\sigma}{du} = \rho\,^{E}U = \rho_0\,^{L}U \tag{8}$$

The expression in (8) can be integrated, noting that $^{L}U = U_0 t_0/t_1$, to give

$$\sigma_1 - \sigma_0 = \rho_0 U_0 t_0 \int_0^{u_1} \left( t_0 + \int_0^u \delta t' du' \right)^{-1} du \tag{9}$$

Again, we note that in the steady wave case, (9) reduces to the usual Rankine-Hugoniot equation.

## Conservation of Energy

The energy conservation equation in Lagrangian coordinates is

$$\left( \frac{\partial E}{\partial t} \right)_x + \left( \frac{\sigma}{\rho_0} \right) \left( \frac{\partial u}{\partial x} \right)_t = 0 \tag{10}$$

This expression, using (4) and (9), can be rearranged to give

$$dE_b = \left( \frac{\sigma}{\rho_0} \right) \frac{du}{dt} \bigg|_b \frac{t(u_1)}{U_0 t_0} \frac{dt}{du} \bigg|_b du \tag{11}$$

If we simplify this expression and substitute in earlier results, we get

$$E_1 - E_0 = \int_0^{u_1} \left[ \frac{\sigma_0}{\rho_0 U_0 t_0} + \int_0^u \left( t_0 + \int_0^{u'} \delta t' du'' \right)^{-1} du' \right] \times \left( t_0 + \int_0^u \delta t' du' \right) du \tag{12}$$

As with (4) and (9), the steady wave condition results in this expression reducing to the usual Rankin-Hugoniot equation for energy conservation.

## Limitations

The foregoing discussion describes the expressions that can be used for analyzing nonsteady waves in cases where the wave profile at two points can be measured or estimated. In practice, there is a significant shortcoming to the treatment as it currently exists, in that there is no satisfactory treatment of interfaces, which usually exist in real experiments. Interfaces, except those between materials with similar impedances, reflect waves, causing wave interactions that violate the assumption of straight characteristics. The most easily implemented, but rather unsatisfactory, approach would be to assume that the effects of the wave interactions on the straightness of the characteristics can be ignored. More complicated approaches could be implemented for specific experimental arrangements, but exhaustive discussion of interface effects is beyond the present discussion.

## EVALUATING THE RANKINE-HUGONIOT TREATMENT

Aside from actual analysis of nonsteady waves, the expressions presented here provide an assessment of the degree to which application of the usual forms of the Rankine-Hugoniot equations results in errors when determining the changes in state variables caused by passage of a nonsteady wave. As noted, (4), (9), and (12) reduce to the traditional forms of the Rankine-Hugoniot equations when the derivatives $(\partial t/\partial u)_x$ at positions $a$ and $b$ are equal. We can thus take the difference with the nonsteady wave case and obtain an estimate of the fractional errors in $\rho$, $\sigma$, and $E$.

For (4), the fractional difference is given by

$$\frac{\delta\rho}{\rho_f} = \frac{1}{(U_0 - u_f)t_0} \int_0^{u_f} (u_f - u)\delta t' du \tag{13}$$

If we approximate the derivatives by their average values,

$$\bar{t}' = \frac{\Delta t}{\Delta u} \approx t' \tag{14}$$

then we can estimate the fractional error as

$$\frac{\delta \rho}{\rho_f} \approx \frac{1}{2t_0} \frac{u_f^2}{U_0 - u_f} \delta \bar{t}' \tag{15}$$

One consequence of (15) that holds if the wave profile is symmetric is that selection of the midpoint of the profile as the wave arrival time allows the Rankine-Hugoniot equation to give a density jump identical to that given by (4). In a similar manner, for (9) and (12) and making the approximation presented in (13), we get

$$\frac{\delta \sigma}{\sigma_1 - \sigma_0} \approx \frac{\delta \bar{t}' u_f}{\delta \bar{t}' u_f - t_0} \tag{16}$$

and

$$\frac{\delta E}{E_f - E_0} \approx \frac{\sigma_0}{2\rho_0 U_0} \frac{\delta \bar{t}'}{t_0} u_f^2 + \frac{\delta \bar{t}'}{t_0} u_f$$

$$+ \left( \frac{\delta \bar{t}'}{t_0} u_1 + \frac{1}{2} u_f^2 \right) \ln \left( 1 + \frac{\delta \bar{t}'}{t_0} u_f \right) \tag{17}$$

It should be noted that the approximation in (14) will fail if the quantity $\delta' u_f$ is greater than $t_0$.

One might wish to attempt a similar sort of argument to assess how close a given loading path is to isentropic. However, knowledge of the detailed $P$-$V$ path of the isentrope is required prior to performing such an assessment. Since this information is usually not available, such an assessment usually cannot be performed.

## SUMMARY

In this paper, I have examined an approach to nonsteady wave analysis that is applicable to situations where reasonable knowledge of the wave profiles at two Lagrangian positions can be had.

Several significant shortcomings remain in practice and this approach is not intended to replace more rigorous methods that have been developed. From a practical standpoint, the most difficult problem may be the bending of characteristics that results from interactions with an interface, such as is common in experiments using VISAR or other optical diagnostics. Such interactions violate the assumption that characteristics are straight. The simplest approach is to ignore this effect, but more sophisticated approaches are probably warranted in most cases.

Although shortcomings exist, consideration of the expressions presented here does have immediate utility in evaluating the validity of using the simple Rankine-Hugoniot equations to estimate the stress, density, and energy jumps due to a nonsteady wave. A simple estimate of the validity of the Rankine-Hugoniot equations for a given situation can often be obtained using averages of the mass velocity time derivatives at the two positions being considered.

## ACKNOWLEDGEMENTS

This work was supported by the US DOE under contract W-7405-ENG-36.

## REFERENCES

1. Fowles, R., and Williams, R. F., "Plane Stress Wave Propagation in Solids", J. of Applied Physics 41, 360, 1970.
2. Cowperthwaite, M., and Williams, R. F., "Determination of Constitutive Relationships with Multiple Gauges in Nondivergent Waves", J. of Applied Physics 42, 456, 1971.
3. Aidun, J. B., and Gupta, Y. M., "Analysis of Lagrangian gauge measurements of simple and nonsimple plane waves", J. of Applied Physics 69, 6998, 1991.

CP845, *Shock Compression of Condensed Matter - 2005*,
edited by M. D. Furnish, M. Elert, T. P. Russell, and C. T. White
© 2006 American Institute of Physics 0-7354-0341-4/06/$23.00

# ISENTROPIC COMPRESSION EXPERIMENTS FOR MESOSCALE STUDIES OF ENERGETIC COMPOSITES

## M. R. Baer[1], C. A. Hall[1], R. L. Gustavsen[2], D. E. Hooks[2] and S. A. Sheffield[2]

*[1]Sandia National Laboratories, Albuquerque, NM, 87185*
*[2] Los Alamos National Laboratory, Los Alamos NM 87545*

**Abstract.** New experimental diagnostics and computational modeling provide an unprecedented means for improving the understanding of energetic material behavior at the mesoscale (grain or crystal ensemble levels). This study focuses on the determination of appropriate constitutive and EOS property data of the constituents of an energetic composite at high stress and moderate strain-rate states. The Sandia Z accelerator is used to determine the mechanical response of energetic composites via isentropic ramp wave compression loading. In this paper we describe an energy source method in CTH that models ramp loading for the analysis of ICE experiments. This approach is applied to design experimental configurations to probe the constituent response of PBX 9501 subjected to ~40 Kbar ramp load over 300 ns duration. Multiple VISAR are used to determine the averaged response of the composite material in comparison to the individual constituents including the effects of anisotropy of HMX crystals and the interactions of fine crystallites with binder material.

**Keywords:** Isentropic compression, Z accelerator, Explosives, CTH modeling.
**PACS:** 62.50.+p, 62.20.Fe.

## INTRODUCTION

Mesoscale simulations of energetic material composites have revealed that shock loading produces wave fields that are unsteady and three-dimensional [1]. The constituents of an energetic composite strongly interact in localized regions to control the sensitivity of initiation and sustained reaction. To model this behavior appropriate constitutive and EOS property data at high stress-strain states are needed.

An isentropic compression technique (ICE) has been developed at Sandia National Laboratories using the Sandia Z accelerator to provide ramped loading on multiple samples to high stress states over the duration of several hundred ns [2]. A comparative response of the various constituents of an energetic material composite can then be readily assessed. Of particular interest is the determination of the material interactions at states that are near to the threshold of reaction.

This work applies this technique to an energetic composite, PBX 9501, consisting of 95% HMX, 2.5% Estane and 2.5% Plasticizer BDNPA-F. A bimodal distribution of HMX crystallites is used to create an energetic composite with a high density whereby fine HMX crystallites are mixed with the polymeric binder material (known as "dirty binder") [3]. Of most interest is determining the effect of fine crystallites on the overall mechanical response of the energetic material.

## EXPERIMENTAL CONFIGURATION

In this experiment, fielded as Shot Z1251 at the Sandia Z Facility, various constituents of PBX 9501 were subjected to an isentropic ramp load to ~40 Kbar over 300 ns. At this loading condition it was believed that no reactive behavior was induced in the energetic material. Hence, the mechanical response could be assessed without the ambiguities of reactive behavior.

Sixteen samples were loaded simultaneously in this ICE configuration as illustrated in Figure 1. In this design, four 0.5 mm thick Al panels were mounted with the various constituents of PBX 9501. One of these panels included samples of the polymeric binder HTBP; the response of this material is not discussed here.

All samples were mounted with NaCl or PMMA windows and VISAR particle velocity measurements were made at these interfaces. Each panel included a measurement of the back surface of the panel whereby backward integration [4] could be used to define the drive conditions.

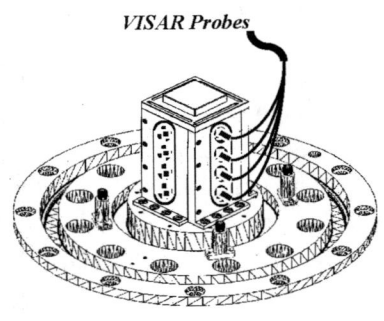
VISAR Probes

**FIGURE 1.** ICE configuration with four panels containing a total of sixteen samples. Drive measurement on each panel is not shown.

## DRIVE BOUNDARY CONDITION

In designing the experimental configuration and assessing the response of the isentropic loading, a method was devised to replicate ramp wave loading conditions in CTH Eulerian shock physics analysis. To mimic a Lagrangian stress boundary condition, an energy source was defined whereby energy is deposited in a thin ideal gas layer adjacent to the drive surface [5].

The equations of motion for a perfect ideal gas are solved for a thin layer that expands in thickness, $X(t)$, due to a volumetric energy source, Q. Using the ideal gas law and a constant isentropic index, $\gamma$, the time variation of Q is defined to replicate the stress and particle velocity at the drive surface. Combining the mass and energy conservation equations leads to the following:

$$Q = \frac{1}{(\gamma-1)X^{\gamma}} \frac{d}{dt}\left(X^{\gamma}p\right) \qquad (1)$$

where $X(t) = X_0 + \int_0^t u_p(\tau)d\tau$.

Backward integration from the measured plane [4] determines the stress and particle velocity states at the drive surface and a tabular volumetric source is defined for CTH analysis. Lagrangian calculations have also been performed using WONDY [6] to verify that the energy source method accurately reproduces the drive stress boundary condition.

## RESULTS AND DISCUSSION

Figure 2 displays an overlay of particle velocities of four samples of PBX 9501 with varied thickness. CTH calculations have been included whereby the PBX 9501 is modeled as a Mie-Grüneisen material without any material strength effects. The undisturbed density is given as $\rho_0$=1.837 g/cc, the shock Hugoniot represented as: $U_s$= 2.4 + 2.4 $u_p$ and the Grüneisen constant is $\Gamma_0$=1.14 [7].

In all of these four profiles, the loading appears to be isentropic with no shock formation. At the foot of each profile is a dispersive region that grows in extent with increasing sample thickness. This dispersive behavior is likely due to the mesoscopic effects of the binder interacting with the crystallites in the energetic composite. As the loading increases, the constituents appear to consolidate and the response becomes entirely hydrodynamic as suggested in the CTH calculations.

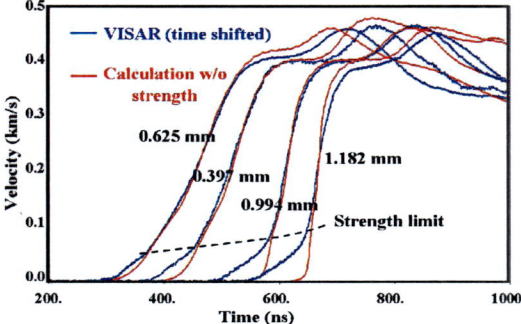

FIGURE 2. Particle velocity profiles in PBX 9501 of varied sample thickness (in mm). CTH calculations without any mesoscale effects or material strength effects are included.

Figure 3 displays a comparison of the HMX crystals with two orientations. Anisotropy effects are evident and a larger elastic precursor is observed along the (010) direction. Relaxation effects are also seen for each orientation. CTH calculations are overlaid using shock Hugoniot data and a perfectly plastic material strength model. Estimated material yield strength values are approximately a factor of two greater than those suggested from impact experiments [8].

Another panel contained samples of the Estane/NP and layers of "dirty binder". Figure 4 shows a comparison of a CTH calculation of the Estane material modeled using a liquid Hugoniot [9] that was recalibrated using LANL impact data [7]. In this material, the input ramp loading produces a shock; hence, the material response is not entirely isentropic. Nonetheless, the shock development and post response is well represented by a pure hydrodynamic model without any strength effects.

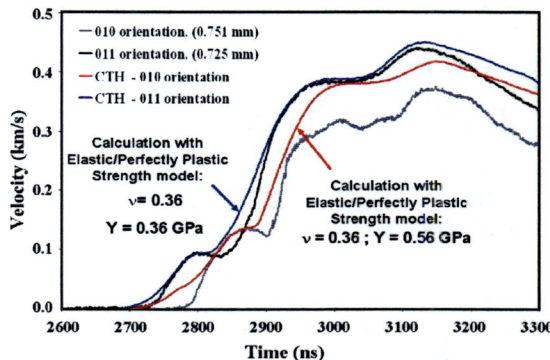

FIGURE 3. Particle velocity profiles in HMX single crystals at the interface of an (100) NaCl window.

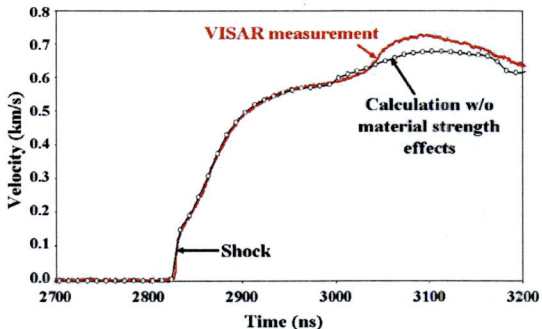

FIGURE 4. Measured and calculated particle velocity profiles in Estane/NP sample.

The difference in material behavior between the pure Estane/NP and the "dirty binder" is shown in Figure 5. An 80/20 mixture (by weight) of fine HMX crystallites and Estane/NP is represented as the "dirty" binder composition. A marked difference in response is seen in this comparison. Clearly, the fine crystallites cause wave dispersion at the front. Furthermore, shock formation did not occur in the "dirty" binder.

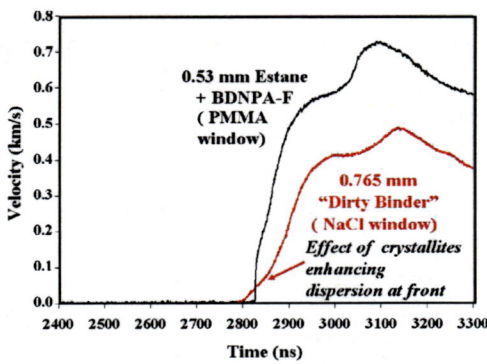

**FIGURE 5.** Measured particle velocity profiles in PBX 9501 "dirty" binder and Estane/BP without crystallites

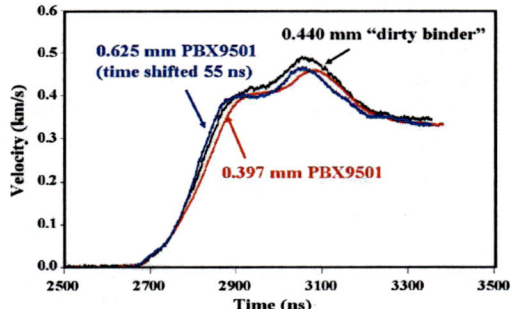

**FIGURE 6.** Measured particle velocities in two sections of PBX 9501 compared to a 80/20 "Dirty" binder sample.

Figure 6 displays a comparison of PBX 9501 to a similar thickness sample of "dirty" binder. Remarkably, the material behavior of the PBX 9501 is nearly identical to that of the binder which suggests that the overall response of the energetic composite is dominated by the interactions of the "dirty" binder materials. Furthermore, this suggests that the fine HMX crystallites in the PBX9501 aggregate might have an important role in mechanical-induced reactive behavior.

## CONCLUSION

This study has applied isentropic loading conditions to unravel the response of a typical energetic composite. Various constituents are simultaneously loaded and comparative response has suggested that fine crystallites mixed with polymeric binder constituents may have a significant role in controlling the mechanical behavior near the onset of reaction. Future work will address similar configurations at loading conditions that trigger reactive behavior.

## ACKNOWLEDGEMENTS
Sandia is a multiprogram laboratory operated by Sandia Corporation, a Lockheed Martin Company, for the US dept of Energy's NNSA under contract DE-AC04-94AL85000.

## REFERENCES

1. Baer, M. R. and Trott, W. M., "Mesoscale Descriptions of Shock-Loaded Heterogeneous Porous Materials", in Shock Compression of Condensed Matter, 2001 (M.D. Furnish, N. N. Thadhani, Y. Horie, eds.), part I, pp. 713-716.
2. Hall, C. A., et al, "Experimental Configuration for Isentropic Compression of Solids Using Pulsed Magnetic Loading", Rev. Scientific. Instrumentation. **72**, 9, 2001, pp. 3567-3595.
3. Skidmore, C. B., et al, "The Evolution of Microstructural Changes in Pressed HMX Explosives", 11th International Detonation Symposium, 1998, pp 556-564.
4. Hayes, D. B., "Backward Integration of the Equations of Motion to Correct for Free Surface Perturbations," Sandia National Laboratories Rpt SAND2001-1440, 2001.
5. Schmitt, R. G [personal communication].
6. Kipp, M. E. and Lawrence, R. J., "WONDY V – A One-Dimensional Finite-Difference Wave Propagation Code", Sandia National Laboratories Rpt: SAND81-0930, 1982.
7. Gustavsen, R. L, Sheffield, S. A., Alcon, R. R., and Hill, L. G. "Shock Initiation of New and Aged PBX9501 Measured with Embedded Electromagnetic Particle Velocity Gauges, LA-13634-MS, 199.
8. Menikoff, R. and Sewell, T. D., "Constitutive Properties of HMX Needed for Mesoscale Simulation", Combustion Theory Modeling, 6, 2002, pp. 103-125.
9. Woolfolk, R. W, Cowperthwaite, M. and Shaw, R., Thermochimica Acta, 5, 1973, p. 409.

CP845, *Shock Compression of Condensed Matter - 2005*,
edited by M. D. Furnish, M. Elert, T. P. Russell, and C. T. White
© 2006 American Institute of Physics 0-7354-0341-4/06/$23.00

# A STUDY OF POLYMER MATERIALS SUBJECTED TO ISENTROPIC COMPRESSION LOADING

## C. A. Hall[1], M. R. Baer[1], R. L. Gustavsen[2], D. E. Hooks[2], E. Bruce Orler[2], D. M. Dattelbaum[2], S. A. Sheffield[2] and G. T. Sutherland[3]

[1]*Sandia National Laboratories, Albuquerque, NM, 87185*
[2] *Los Alamos National Laboratory, Los Alamos NM 87545*
[3] *Naval Surface Warfare Center, Indian Head, MD, 20695*

**Abstract.** This work applies a ramped, quasi-isentropic compression loading technique (ICE) to investigate the mechanical behavior of polymers that are often used in energetic composites. The focus of this effort is the determination of appropriate constitutive and EOS property data at high stress states and moderate strain rates that is needed for detailed mesoscale modeling. Several thicknesses of samples were subjected to a ramp load of ~45 Kbar over 500 ns duration using the Sandia Z-machine. Profiles of transmitted ramp waves were measured at window interfaces using conventional VISAR. Shock physics analysis is then used to determine the nonlinear material response of the binder materials.

**Keywords:** Isentropic compression, Z accelerator, polymer binders, CTH modeling.
**PACS:** 62.50.+p, 62.20.Fe.

## INTRODUCTION

Polymeric binders such as Estane, Teflon, Kel F and HTPB are often used in energetic materials to bond polycrystalline and/or metallic constituents together as a composite. Previous isentropic loading work has suggested that binder materials play an important role in the mechanical behavior of an energetic composite particular at the threshold to reaction [1]. This study combines an experimental study with analysis to assist in unraveling the mechanical response of these materials when subjected to isentropic compression loading [2]. Although numerous planar impact studies of these materials have been documented, this work applies isentropic ramp loading to provide additional time-dependent property

information using simultaneously loaded multiple targets. A comparative response of a variety of materials is evaluated by subjecting these materials to identical loading conditions. Of particular interest is the determination of appropriate constitutive data for binder constituents that is needed for detailed mesoscale modeling of energetic composites [3].

## EXPERIMENTAL CONFIGURATION

Shot Z1405, fielded at the Sandia Z Facility, isentropically loaded multiple samples of polymers Estane, Teflon, Kel F and HTPB. These materials were subjected to a ramp load of 45 Kbar over 500 ns. This loading condition was purposely selected to overcome shock development in these soft materials as was observed in Z test 1251.

Sixteen samples mounted on 0.6 mm thick aluminum "drive plates" were simultaneously loaded. Panel 1 contained four samples of PBX 9501 binder (Estane/NP) of four thicknesses ranging from 0.324 to 0.597 mm. A second panel had four samples of C7-Teflon between 0.22 to 0.45 mm thick. Four samples of HTPB, ranging between 0.383 mm to 0.597 mm, were mounted on a third panel and the fourth panel contained two samples of Kel-F 800 (0.376 mm, 0.720 mm) and two samples of Kel-F 81 (0.47 mm, 0.855 mm).

All panels used LiF (100) windows and VISAR particle velocity measurements were made at the sample/window interfaces. Each panel also included a "drive measurement" of the back surface at the LiF/panel location whereby backward integration [4] provides an estimate of the drive conditions.

## RESULTS AND DISCUSSION

To circumvent shock development in these soft materials, the ramp loading condition of Z1251 was modified with an additional 200 ns rise time. This loading rate decreased significantly in the lower stress portion of the drive. Figure 1 displays the drive stress-time history as determined from the backward integration using the measured particle velocities at the back surface. Analysis of the Z data was conducted using an energy source method in CTH [1]. All calculations are based on historical EOS models.

Figure 2 displays an overlay of such calculations for various thicknesses of PBX 9501 binder with the experimental data. The simulations included no strength effects and the following parameters: $\rho_o$=1.27 g/cc, $U_s$ = 1.7+3.45 $u_p$-0.265 $u_p^2/c_0$ mm/$\mu$s, $c_0$= 1.7 mm/$\mu$s, $\Gamma_0$=1.14, [5].

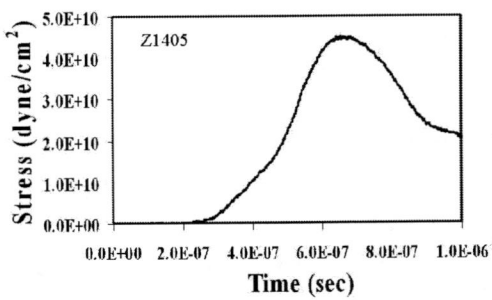

**FIGURE 1.** Stress history at the drive surface.

From this comparison it is seen that at higher stresses the material response appears to be more nonlinear than anticipated. Since no shock formation was observed further analysis using backward integration method [4] may yield additional insight to this behavior. The pronounced inflection at a velocity of 2.5E04 cm/s has not yet been fully explained.

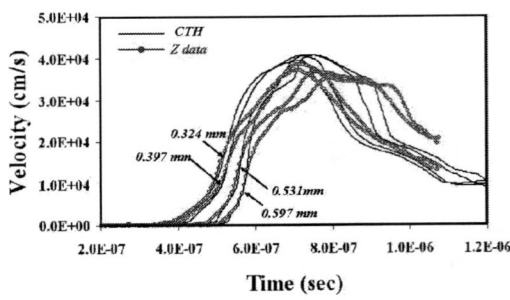

**FIGURE 2.** Particle velocities in the four samples of PBX9501 binder subjected to Z1405 ramp load. CTH calculations are overlaid with the VISAR measurements.

In a prior Z test (Z1251) HTPB samples were too thick for that particular time-dependent loading condition and shocks developed. Figure 3 displays the particle velocities observed for the panel mounted with HTPB on Z test 1405, with corresponding CTH calculations using EOS data of Millet, *et. al.* [6]. The HTPB had a density of $\rho_o$=0.93 g/cc and the material response was estimated using $U_s = 1.53+2.84\ u_p$ mm/μs. At the lower stresses the measured and predicted response are in reasonable agreement. However, at the higher stresses the HTPB appears to soften and eventually a shock forms upon interaction with the LiF window. It is hoped that sufficient shockless acceleration data exists for backward integration studies to help in unraveling this nonlinear behavior.

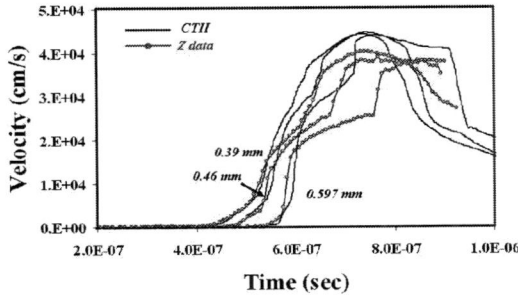

**FIGURE 3.** Particle velocity profiles in HTPB of varied sample thickness (in mm). CTH calculations without material strength effects are included.

Kel-F 800 is a binder used in TATB explosives and has many similarities to Kel-F 81 and Teflon; hence, the interest here is determining a comparative material response [7]. Figure 4 displays the measured response of the two types of Kel-F subjected to simultaneously ramp loading. Significant differences of material behavior were observed in the < 0.5 mm thick samples. At the 0.855 mm thickness, shock development in the Kel-F 800 sample was observed.

Figure 5 shows a comparison of CTH calculations for Kel-F modeled with a single element Maxwell Viscoelastic description (CTH library EOS data). Although the overall material

behavior appears to be well represented, there exist uncertain differences in behavior at the low stresses.

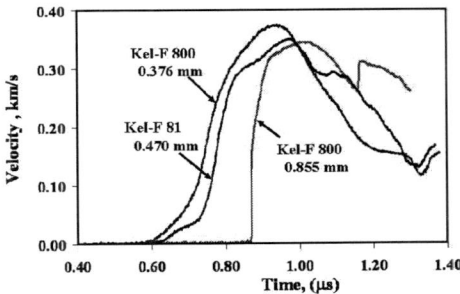

**FIGURE 4.** Particle velocity profiles in Kel-F at the interface of an (100) LiF window.

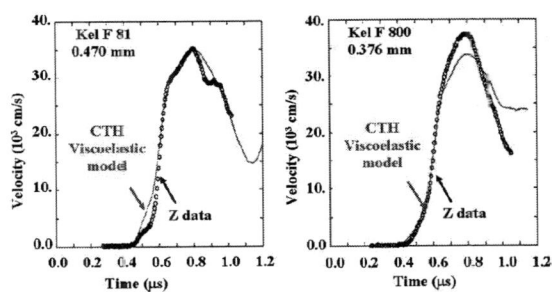

**FIGURE 5.** Measured and calculated particle velocity profiles in the two samples of Kel-F.

In the remaining panel, four samples of C7-Teflon were subjected to the Z1405 ramp loading. Prior impact studies have suggested that a phase transformation occurs near 5 Kbar [8]. Isentropic loading is of interest since subtle phase changes can be observed that would otherwise be over-driven during shock loading. Figure 6 displays the measured particle velocity indicating that shockless acceleration was achieved in all of these samples. The precursor at the acceleration wave front is well behaved and steepens with increasing thickness.

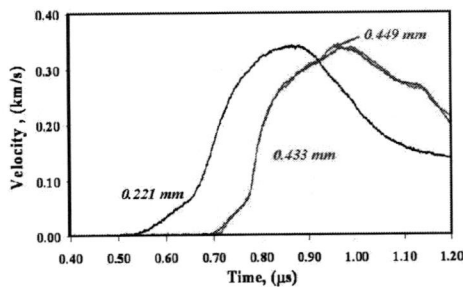

**FIGURE 6.** Measured particle velocity profiles in samples of Teflon.

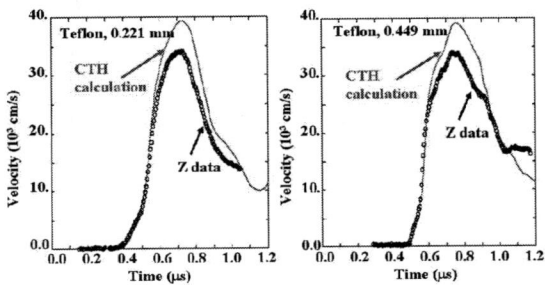

**FIGURE 7.** CTH simulation of Z1405 ramp loading of two Teflon samples.

In modeling the ramp loading in Teflon, prior shock Hugoniot data [9] was used. Figure 7 is a comparison of the CTH calculations with the Z data which shows that the response is well defined by the modeling without including any effect of a phase transformation.

## CONCLUSION

This study has applied isentropic loading conditions to unravel the comparative response of a various polymers that are used in energetic composites. Historical EOS and constitutive data has been used in analysis of this data and some differences in response are observed. In most of these materials identical ramp loading produced shockless acceleration and a more refined analysis of material behavior is possible using the Hayes backward integration technique [4]. Future work will address this data analysis.

### ACKNOWLEDGEMENTS
Sandia is a multiprogram laboratory operated by Sandia Corporation, a Lockheed Martin Company, for the US dept of Energy's NNSA under contract DE-AC04-94AL85000. Sponsorship of the NSWC effort was provided by the NSWC Core Research Program.

### REFERENCES

1. Baer, M. R., et al, "Isentropic Compression Experiments for Mesoscale Studies of Energetic Composites", this conference.
2. Hall, C. A., et al, "Experimental Configuration for Isentropic Compression of Solids Using Pulsed Magnetic Loading", Rev. Scientific. Instrumentation. 72, 9, 2001, pp. 3567-3595.
3. Baer, M. R. and Trott, W. M., "Mesoscale Descriptions of Shock-Loaded Heterogeneous Porous Materials", in Shock Compression of Condensed Matter, 2001 (M.D. Furnish, N. N. Thadhani, Y. Horie, eds.), part I, pp. 713-716.
4. Hayes, D. B., "Backward Integration of the Equations of Motion to Correct for Free Surface Perturbations," Sandia National Laboratories Rpt SAND2001-1440, 2001.
5. Gustavsen, R. L, Sheffield, S. A., Alcon, R. R., and Hill, L. G. "Shock Initiation of New and Aged PBX9501 Measured with Embedded Electromagnetic Particle Velocity Gauges, LA-13634-MS, 199.
6. Millett, J.C.F., Bourne, N.K., and Akhavan, J. "The Response of Hydroxyl-terminated Polybutadiene to One-Dimensional Shock Loading,", J. of Applied Physics, 95, no.9, pp 4722-4727.
7. Sheffield, S. A., et al.,"Dynamic Response of Fluorinated Semi-Crystalline Polymers: Kel-F 81 and Kel-F 800", 11th Sym. on Plasticity, Hawaii, 2005.
8. Champion, A. R., "Shock Compression of Teflon from 2.5 to 25 Kbar – Evidence for a Shock-Induced Transition", J. Applied Physics, 42, p 5546, 1971.
9. Carter, W. J. and Marsh, S., "Hugoniot Equation of State of Polymers", Los Alamos Rpt. LA-13006-MS, 1995.

CP845, *Shock Compression of Condensed Matter - 2005*,
edited by M. D. Furnish, M. Elert, T. P. Russell, and C. T. White
© 2006 American Institute of Physics 0-7354-0341-4/06/$23.00

# ISENTROPIC COMPRESSION DATA ON LX-04 EXPLOSIVE AT 150°C USING THE Z ACCELERATOR

## David E. Hare[1], Kevin S. Vandersall[1], Frank Garcia[1], Jean-Paul Davis[2], Clint Hall[2] and Jerry W. Forbes[3]

[1]*Lawrence Livermore National Laboratory, Livermore, CA 94550*
[2]*Sandia National Laboratory, Albuquerque, NM 87185*
[3]*Center for Energetic Concepts Development, University of Maryland, Department of Mechanical Engineering, College Park, MD 20742*

**Abstract.** Isentropic compression data was collected on LX-04 explosive (85% HMX and 15% Viton by weight) at 150°C using the Sandia National Laboratories Z accelerator facility. A ramp compression wave was applied to the explosive samples mounted on aluminum panels with VISAR interferometry measuring the sample and backing window interface velocity. Heating was obtained by wrapping band heaters around a thermal mass attached to each panel and temperatures were recorded by thermocouples at several locations on the panel. This work will outline the methods used, discuss the VISAR interface velocities, and present the preliminary data obtained on heated LX-04. These results demonstrate the ability to perform experiments on preheated samples to obtain isentrope data.

**Keywords:** Ramp compression, Z machine, HMX based explosive, LX-04, equations of state
**PACS:** 81.05.Lg, 82.33.Vx, 82.40.Vx, 82.40.Fp, 64.30.+t, 62.50.+p

## INTRODUCTION

The Z accelerator facility at Sandia National Laboratories (SNL) offers a viable tool for performing "Isentropic Compression Experiments" (ICE) [1-3]. This is accomplished by using a large-amplitude, short-duration current pulse delivered into a metal panel "floor" which backs the sample and creates a ramp compression wave from the magnetic pressure.

The ICE technique has been useful in studying the isentrope (and associated Hugoniot) of high explosives (HE). LX-04, an HMX based explosive, with 85% HMX and 15% Viton [4] has been studied in previous ICE experiments [5-7]. Prior works have also investigated the heated shock response and equation of state LX-04 [8-9]. In this work, a pre-heating technique was utilized to heat ICE samples and load them to investigate the technique and hopefully obtain usable data. The procedure utilized and results obtained will be discussed with suggestions for future work and improvements on subsequent experiments.

## PROCEDURE

LX-04 samples (85% HMX, 15% Viton-A) with thickness of 200-700 $\mu$m and 6 mm diameter were mounted onto 6061 aluminum panel assemblies. Figure 1 shows a photograph of the West panel showing the 3 measurement locations on the right and cavity location on the left for placement of the copper thermal mass that gets surrounded by a band heater. A block can be seen above the sample locations, which is covering the sample thermocouple locations. Thermocouples were also placed on the thermal mass near the band heaters.

Each panel contained 2 LX-04 samples backed by 6 mm diameter by 3 mm thick NaCl (100) windows. The aluminum floor thickness below each sample was approximately 1 mm. Table 1 contains the sample thickness and window information on each panel. The NaCl windows were used because they have a good acoustic impedance match to the LX-04 and the experiment pressure was expected below the NaCl 25 GPa phase transition. VISAR laser interferometry [10] was used to measure the window interface and panel free surface velocities.

All 4 of the panels similar to that pictured in Figure 1 were assembled in a "square short" arrangement with each panel making up a side of the square. Note that in Figure 1, the panel is laying sideways with the top being toward the right and the bottom at the left. This created an anode insert opening of 26 mm x 26 mm which was placed around a 20 mm by 20 mm cathode stalk to create a 3 mm gap all the way around. After assembly of the target was complete, the four panels were preheated to approximately 150°C and the Z machine was fired.

**TABLE 1.** Summary of sample thickness measurements and window information on panels with LX-04

| DETAIL | NORTH PANEL | SOUTH PANEL | WEST PANEL |
|---|---|---|---|
| Top sample t | 192 $\mu$m | 399 $\mu$m | 602 $\mu$m |
| Top window | NaCl | NaCl | NaCl |
| Middle (no sample) | NaCl | NaCl | NaCl |
| Bottom sample t | 293 $\mu$m | 504 $\mu$m | 708 $\mu$m |
| Bottom Window | NaCl | NaCl | NaCl |

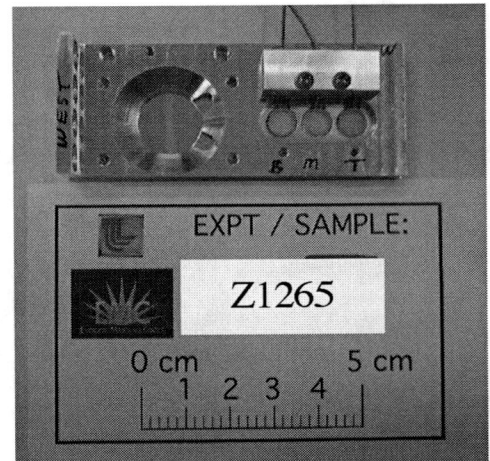

**FIGURE 1.** Photograph of samples and windows mounted on the West panel for experiment Z1265.

## RESULTS/DISCUSSION

The results for the experiment include the heating temperature profile, current profile from the machine, and velocity histories of the drive measurements and panels. Figure 2 displays the thermocouple readings from the South panel. All panels showed a similar heating profile with a temperature variation of approximately 2-5°C from panel to panel. Note that there was a slight temperature overshoot from the desired 150°C and the heaters were turned off prior to firing as shown by the peak followed by a dip at the end of the temperature record.

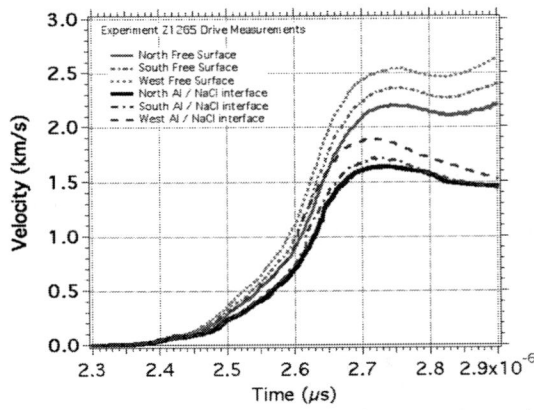

**FIGURE 2.** Temperature profile of the South panel.

The current pulse (average of 4 probes) achieved during the experiment is shown in Figure 3 and shows about a 15 MA peak pulse and a 400 ns duration pulse ramp rise time.

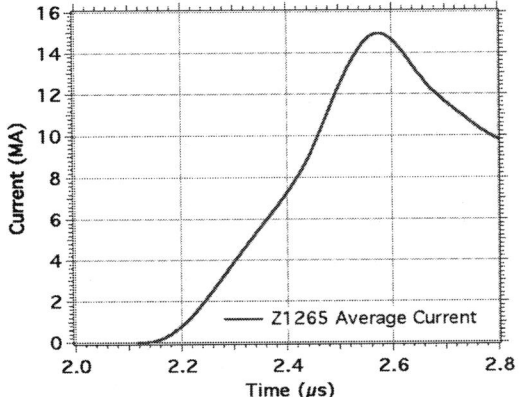

**FIGURE 3.** Current measurement for Z1265.

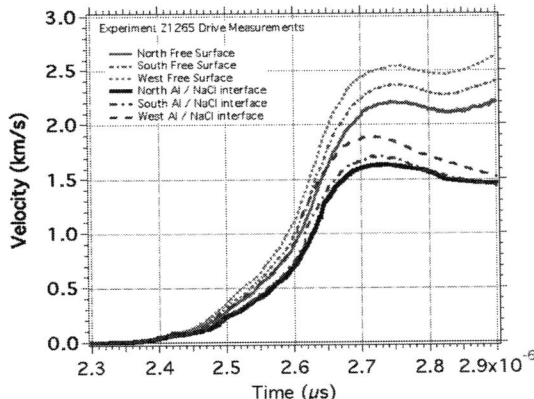

**FIGURE 4.** Drive measurements for Z1265 panels with LX-04 samples.

Figure 4 shows the drive measurement for all three panels with LX-04 samples. Note that the aluminum free surface velocity measurements show a higher velocity than the profiles for the Al/NaCl interface velocity due to the presence of the NaCl window. It can be noticed that the velocities are not consistent from panel to panel, but the relative ratios of free surface to Al/NaCl interface velocities on each panel correlate together. This appears to show that possibly the drive was not symmetric,

possibly due to slight expansion changes during the heating of the target assembly.

Figures 5, 6 and 7 show the VISAR traces for the North, South, and West panels with LX-04 samples respectively. As shown in Figure 4, the drive measurements on each panel are included in these plots. The North and South panels did yield a somewhat usable velocity profile for the thin samples, but the West panel did not. Note that in all cases, the thicker sample did not show a quality velocity trace. This may be due to loss of reflectively from the sample/window interface or the sample severely shocking up during the experiment resulting in several missed VISAR fringes.

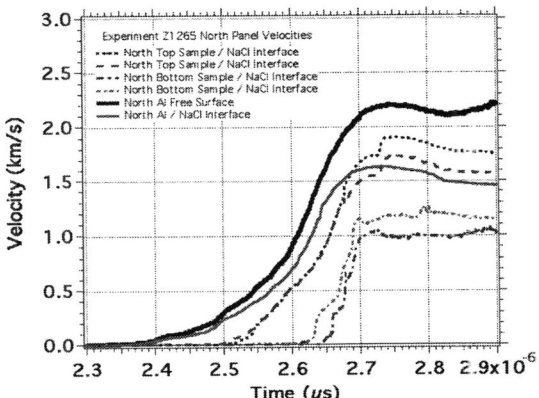

**FIGURE 5.** VISAR traces of North panel samples.

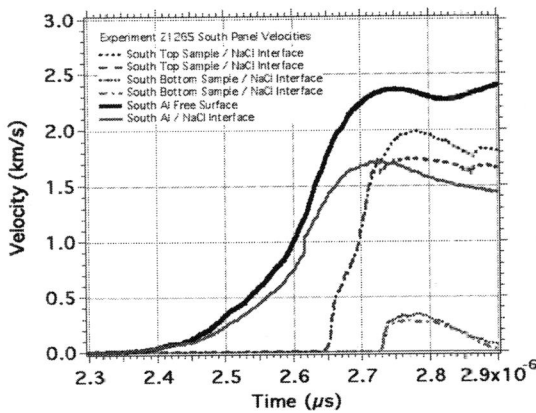

**FIGURE 6.** VISAR traces of South panel samples.

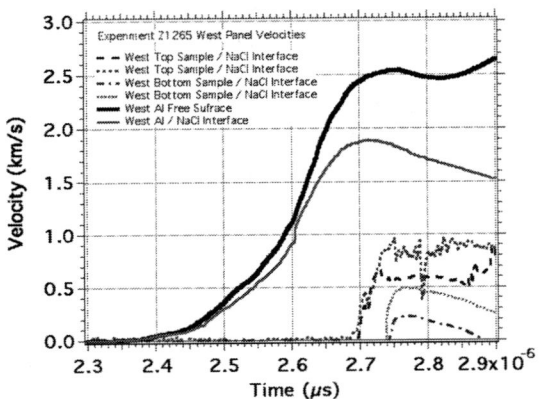

**FIGURE 7.** VISAR traces of West panel samples.

It may be possible to obtain isentrope data from the thin samples on the North and South panels, but the associated errors may be high due to the limited data.

## SUMMARY AND FUTURE WORK

An experiment using LX-04 samples pre-heated to approximately 150°C was performed on the Z accelerator at Sandia National Labs. This research demonstrates the ability to collect isentrope data on pre-heated samples. Window reflectivity at elevated temperature proved to be a problem for several of the window interfaces.

Future work is planned to perform further analysis on the data to investigate the possible calculation of an isentrope. It is unsure at this time whether an additional window and sample corrections will be needed due to the pre-heated nature of the experiment. Improvements to the pre-heating experiments including resolving any window reflectivity issues are in progress and future experiments are planned.

## ACKNOWLEDGEMENTS

This work would not have been possible without the assistance of LeRoy Green, Scott Humphery, Allen Elsholz and Tim Uphaus from LLNL, and from SNL the outstanding technical

staff of the Z-machine facility. This work was performed under the auspices of the U. S. Department of Energy by the University of California, Lawrence Livermore National Laboratory under Contract No. W-7405-Eng-48.

## REFERENCES

1. Hall, C.A., "Isentropic Compression Experiments on the Sandia Z Accelerator," Phys. Plasmas 7, 2069 (2000).

2. Reisman, D.B. et. al., "Magnetically driven isentropic compression experiments on the Z-accelerator," J. Appl. Phys. 89, 1625 (2001).

3. Reisman, D. B., Wolfer, W. G., Elsholz, A., "Isentropic compression of irradiated stainless steel on the Z accelerator", J. Appl. Phys. 93, 8592 (2003).

4. Owens, C., Nissen, A. and Souers, P. C., "LLNL Explosives Reference Guide," UCRL-WEB-145045 (2003).

5. Reisman, D. B. et al., "Isentropic Compression of LX-04 on the Z accelerator," Shock Compression of Condensed Matter – 2001, edited by M. D. Furnish, N. N. Thadhani, and Y. Horie, AIP press, 2002, pp. 849-852.

6. Reisman, D. B., Forbes, J. W., Tarver, C. M., Garcia, F, Hayes, D. B., Furnish, M. D., Dick, J. J., "Isentropic Compression of High Explosives with the Z Accelerator," Proceedings of the 12th International Detonation Symposium, San Diego, CA, August, 2002, pp. 343-348.

7. Hare, D. E., Reisman, D. B., Garcia, F., Green, L. G., Forbes, J. W., Furnish, M. D., Hall, C., Hickman, R. J., "The Isentrope of Unreacted LX-04 to 170 kbar," Shock Compression of Condensed Matter – 2003, edited by M. D. Furnish, Y. M. Gupta, and J. W. Forbes, AIP press, 2004, pp. 145-148.

8. Tarver, C. M., Forbes, J. W., Urtiew, P. A., Garcia, F., "Shock Sensitivity of LX-04 at 150°C," Shock Compression of Condensed Matter-1999, pp.891-894.

9. Urtiew, P. A., Forbes, J. W., Tarver, C. M., Vandersall, K. S., Garcia, F., Greenwood, D. W., Hsu, P. C., and Maienschein, J. L., "Shock Sensitivity of LX-04 with Delta Phase HMX at Elevated Temperatures," Shock Compression of Condensed Matter - 2003, pp. 1053-1056.

10. Barker, L. M., and Hollenbach, R. E., "Laser interferometer for measuring high velocities from any reflecting surface," J. Appl. Phys. 43, 4669-4675, (1972).

CP845, *Shock Compression of Condensed Matter - 2005*,
edited by M. D. Furnish, M. Elert, T. P. Russell, and C. T. White
© 2006 American Institute of Physics 0-7354-0341-4/06/$23.00

# DYNAMIC RESPONSE OF COPPER SUBJECTED TO QUASI-ISENTROPIC, GAS-GUN DRIVEN LOADING

**H. Jarmakani[1], J. M. Mc Naney[2], M. S. Schneider[1], D. Orlikowski[2], J. H. Nguyen[2], B. Kad[1], M. A. Meyers[1].**

[1]*Mechanical and Aerospace Engineering Dept, Materials Science Program, University of California, San Diego, La Jolla CA 92093 0418*
[2]*Lawrence Livermore National Laboratory, Livermore CA 94550*

**Abstract.** A transmission electron microscopy study of quasi-isentropic high-pressure loading (peak pressures between 18 GPa and 52 GPa) of polycrystalline and monocrystalline copper was carried out. Deformation mechanisms and defect substructures at different pressures were analyzed. Current evidence suggests a deformation substructure consisting of twinning at the higher pressures and heavily dislocated laths and dislocation cells at the intermediate and lower pressures, respectively. Evidence of stacking faults at the intermediate pressures was also found. Dislocation cell sizes decreased with increasing pressure and increased with distance away from the surface of impact.

**Keywords:** Isentropic compression, microstructural defects, dislocation activity, gas-gun loading.
**PACS:** 62.50.+p

## INTRODUCTION

Quasi-isentropic compression experiments (ICE) have been carried out since the early seventies. The main motivation behind such experiments was to simulate conditions occurring in the depths of planets. Entropy, the measure of the randomness of a system, does not change with depth in planets. Only temperature and pressure changes are experienced. As such, quasi-isentropic experiments come very close to replicating such conditions. Today, the interest behind these shockless experiments is focused on their ability to maintain the solid state of a material while it undergoes extreme pressures. The temperature rise during isentropic compression is significantly less severe than during shock compression. The solid state of a material can, thus, be retained, and an understanding and characterization of its response is possible.

Quasi-isentropic compression conditions can be achieved by various methods: gas-gun, laser, and magnetic loading. Early work on ICE with gas-gun by Lyzenga et al. [1] used a composite flyer plate with materials of increasing shock impedance away from the target material. Barker [2] placed powders of varying densities along a powder blanket and pressed the blanket to produce a pillow impactor having a smooth shock impedance profile. In the case of ICE via laser, McNaney et al. [3] used a shockless laser drive setup to compress and recover an Al alloy. A smoothly rising pressure pulse is generated by focusing a laser beam on a reservoir material (carbon foam), creating a plasma that "stretches out" through a vacuum and discharges onto the sample. In the case of magnetically driven experiments [4], the Z accelerator at Sandia National Labs (SNL) is capable of producing quasi-isentropic compression loading of solids using magnetic pulses. An advantage of this method is that a smoothly rising pressure profile can be generated without the initial spike at low pressures seen during impact experiments. Control over loading pressures and a rise time is also possible to meet experimental requirements [4].

In this work, monocrystalline and polycrystalline copper samples were quasi-isentropically loaded via gas-gun. Peak pressures obtained ranged between 18 GPa and 51 GPa.

## EXPERIMENTAL PROCEDURE

The two-stage gas gun and experimental set up for this work is located at LLNL. Functionally-graded material (FGM) impactors designed with increasing density profile or shock impedance, as depicted in Figure 1 below, were used to produce the smoothly rising pressure profiles [5].

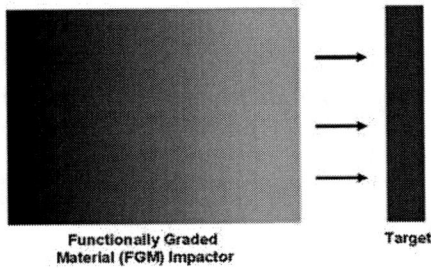

**Functionally Graded Material (FGM) Impactor**          **Target**

**Figure 1.** Illustration of FGM impactor hitting a target (darkness proportional to density).

The pressure profiles in Fig. 2 were obtained from simulations (CALE) carried out at LLNL. Five experiments, **A** (1700m/s), **B** (1260 m/s), **C** (730m/s), **D** (1760 m/s) and **E**, were carried out, with **A** experiencing the highest pressure of 51.5 GPa and **C** experiencing the lowest pressure of 17.7 GPa. Two distinct pressure profiles were attained, one having a hold-time of approximately 10 μsec and one having no hold time (or a "short pulse"). The as-received samples belonging to each batch were in the form of cylindrical specimens having an average diameter and thickness of 6mm and 3.6 mm, respectively.

Shock experiments are dictated by the following Swegle-Grady expression:

$$\dot{\varepsilon} = 7.84 \times 10^{-33} \times P^4 \qquad (1)$$

Where as the isentropic experiments have the following relationship:

$$\dot{\varepsilon} = 6.72 \times 10^{-9} \times P^{1.21} \qquad (2)$$

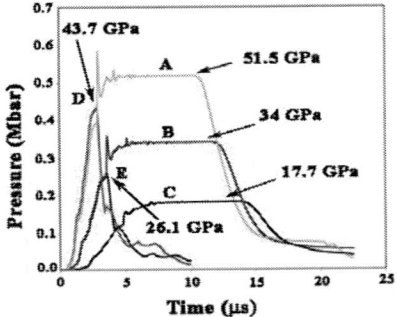

**Figure 2.** Pressure profiles of ICE experiments.

The figure below shows the strain rate versus pressure plot for these ICE experiments. Strain rates achieved during these ICE experiments are on the order of $10^4$/sec, $10^4$ to $10^5$ orders of magnitude lower than shock experiments.

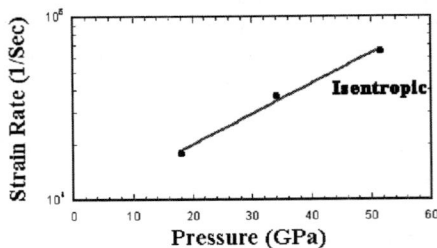

**Figure 3.** Strain Rate Vs. Pressure for ICE.

## Grain Size and Micro-Hardness Measurements

Monocrystalline and polycrystalline samples in each batch were identified by macro-etching the bottom surface of the samples. The average grain size of the polycrystalline samples was approximately 36 μm. Microhardness measurements were performed on a selected group of samples (**A, B** and **C**). The top surface was indented using a Vickers tip attached to a Leco: M-400-H1 microhardness machine. The load applied was 200 gF, with a hold time of 15 sec.

## TEM Sample Preparation

Cylindrical cuts having a diameter of 3mm through the center of each specimen were made by EDM. Four TEM foils were then sliced from each cylinder. The foils were electropolished and sent to Oak Ridge National Labs, where TEM was performed under the SHaRE program. TEM analysis was performed on monocrystalline samples of experiments **A**, **B** and **C** and polycrystalline samples of experiments **C** and **D**.

## RESULTS AND DISCUSSION

### Microhardness Results

The hardness value increased from 103 KgF/mm$^2$ to 111 KgF/mm$^2$ with increasing pressure from 18 GPa to 52 GPa. An interesting trend was noticed in the monocrystalline samples. Hardness increased from **A** to **B**, but then dramatically decreased in **C**. The plot below shows this phenomenon. It is hypothesized that this behavior could be due to orientation effects within the crystals or melting on the impact surface, causing a decrease in hardness.

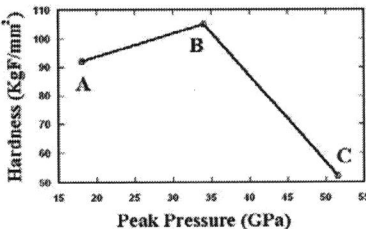

**Figure 4.** Hardness plot of monocrystalline samples.

### TEM Results
*Monocrystalline Samples*

A sample from **A**, 51.5 GPa, located at 97μm from the impact surface contained very clear twinned regions, as can be seen from Figure 5. Both small and large twins were observed. A diffraction pattern revealed twinning taking place at the [011] orientation. At 1.2 mm from the surface within the same sample, a large concentration of dislocation cells, with an average size of 0.154 μm was observed. At 1.7 mm into the sample, heavy dislocation activity was evident, with dislocation cells of 0.198 μm. Twinning was still evident as well.

A monocrystalline sample from **C**, 18 GPa, revealed mostly dislocation activity, with cell sizes ranging in size from 0.5 μm at 0.13 mm within the sample to 0.65 μm at 1.9 mm within the sample.

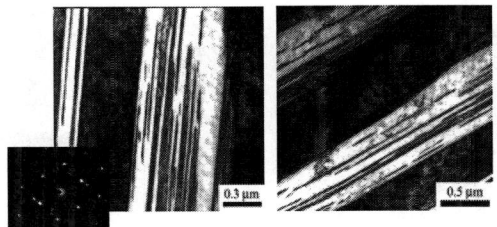

**Figure 5.** Twinned region at 97μm from the surface at a [011] orientation (experiment **A**).

The orientation of this sample and all other samples was determined to be [001]. An interesting feature within the top-most sample should be noted. Elongated cells were evident, but their sizes could not be discerned due to the limitations of the TEM pictures. It is not know whether these features are due to residual deformation or cratering introduced during specimen preparation. Figure 6 below shows typical dislocation activity found within this sample. The section is at 0.13 mm from the impact surface.

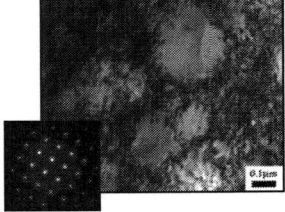

**Figure 6.** Dislocation cells at 0.13 mm form surface at [001] orientation (experiment **C**).

Analysis of a sample from **B** confirmed that the deformation at this intermediate pressure of 34 GPa was transitional, between **A** and **C**, as should be expected. Dislocation cells were mostly revealed at this pressure. Dislocation cell sizes at only two distances were determined, with sizes of 0.26 μm at 0.13 mm and 0.29μm at 0.74 mm.

*Polycrystalline Samples*

A polycrystalline sample from **D** revealed heavily dislocated lathes at 0.14 mm from the top surface, as seen in Figure 7 (a). Dislocation cells and elongated dislocation features, Figure 7 (b), were evident further into the sample, specifically 1.3 mm from the surface, and irregular dislocation activity was evident at 1.9 mm from the surface.

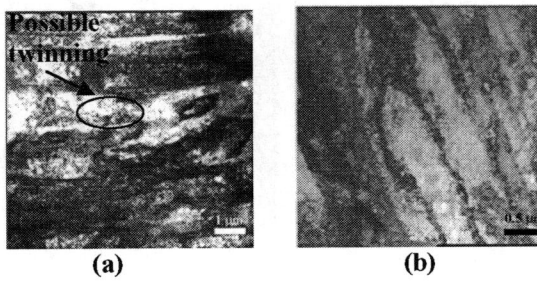

**(a)**                 **(b)**

**Figure 7.** (a) Heavily dislocated laths with possible twinning at 0.14 mm, (b) Elongated cells at 1.3 mm (experiment **D**).

The polycrystalline sample from **C** revealed more dislocation cells. An interesting feature was exposed, hinting at stacking faults and heavily dislocated laths. The distance from the impact surface is unknown since this sample was prepared at LLNL. The figure below reveals this feature.

**Figure 8.** Stacking faults and Dislocated Laths (from **C**).

## CONCLUSIONS

The deformation substructure is consistent with what is expected of impact treatment. In the case of the highest pressure, twinning in the monocrystalline sample was evident closest to the impact surface, and dislocation activity decreased with increasing distance from the impact surface for all experiments. This is consistent with work carried out by Schneider et al. [6] on laser shock of copper. It can be concluded that the twinning threshold lies between 34 GPa and 51 GPa. This is higher than in shock compression (~25 GPa) [6]. It is also noticed that the cell sizes achieved in ICE are larger than those of shock. In the case of the polycrystalline samples, heavily dislocated laths, regular, irregular and elongated dislocation features were evident. Evidence of stacking faults was even

noticeable at the low pressure of 17.7 GPa. The plot below summarizes the change in cell size with distance for the various pressures considered.

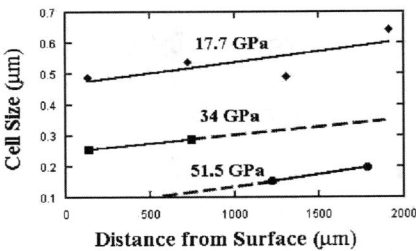

**Figure 9.** Cell-Size vs. Distance from impact surface.

## ACKNOWLEDGEMENTS

This work was performed under the auspices of the U.S. Department of Energy by University of California, Lawrence Livermore National Laboratory under Contract W-7405-Eng-48.

## REFERENCES

1. Lyzenga, G. A., and Ahrens, T. J., "One-Dimensional Isentropic Compression," in Shock Waves in Condensed Matter, 1981, W. J. Nellis, L. Seaman, and R. A. Graham, Eds., American Institute of Physics Conf. Proceedings No. 78 (1982) 231-235.
2. Barker, L. M., "High Pressure Quasi-Isentropic Impact Experiments", Shock Waves in Condensed Matter, 1984, J.R. Asay, R. A. Graham, and G. K. Straub, Eds., (Elsevier Sci. Pub., Amsterdam).
3. McNaney, J. M., Edwards, M. J., Becker R., Lorenz K. T., Remington B. A., "High Pressure, Laser Driven Deformation of an Aluminum Alloy," Met. Trans A, 35A, pp 265 (2004).
4. Hall, C. A. et al., "Experimental Configuration for Isentropic Compression of Solids using Pulsed Magnetic Loading", Review of Scientific Instruments, 72, 3587 (2001).
5. Nguyen, J.H. et al., "Specifically Prescribed Dynamic Thermodynamic Paths and Resolidification Experiments, Shock Compression of Condensed Matter, M. D. Furnish, L.C. Chhabildas, and R. S. Hixson, Eds., AIP Conf. Proc., Melville, New York (2004).
6. Schneider, M.S. et al, "Laser-Induced Shock Compression of Copper: Orientation and Pressure Decay Effects." Met Trans A, 35 A, 263 (2004).

CP845, *Shock Compression of Condensed Matter - 2005,*
edited by M. D. Furnish, M. Elert, T. P. Russell, and C. T. White
© 2006 American Institute of Physics 0-7354-0341-4/06/$23.00

# RESULTS FROM ISENTROPIC COMPRESSION EXPERIMENTS (ICE)

**D. G. Tasker[1], J. H. Goforth[1], H. Oona[1], P. A. Rigg[1], D. Dennis-Koller[1], J. King[1], D. Torres[1], D. Herrera[1], F. Sena[1], F. Abeyta[1], and L. Tabaka[1]**

*[1]Los Alamos National Laboratory, Los Alamos, NM 87545*

**Abstract.** We have developed high explosive pulsed power (HEPP) methods to obtain accurate isentropic EOS data with the isentropic compression experiment (ICE)[1][2]. In the HEPP-ICE experiment, fast rising current pulses (with risetimes from 400 to 600 ns) at current densities of many MA/cm, create continuous magnetic compression of materials to Mbar pressures. The response of materials to this isentropic loading, as determined with VISAR measurements of free surfaces, provides the required isentropic EOS. Experiments on copper will be presented here. The data are analyzed using conventional Lagrangian and Backward [3] techniques. In the present arrangement four samples can be studied at one time, but accurate EOS data can only be obtained from opposing sample pairs, because it is only these pairs that share the same magnetic fields.

**Keywords:** Isentropic compression, Equation of State, Copper.
**PACS:** 62.50.+p, 07.35.+k, 07.55.Db, 64.30.+t

## INTRODUCTION

Magnetic fields of up to 2.8 kT (28 MGauss) have been used to isentropically compress materials to >3 TPa [4][5]. However, until 1998 these pulsed experiments were only performed in cylindrical geometries. Then Asay [6] reported magnetic isentropic compression experiments (ICE), performed on the Sandia Laboratory's Z-machine, that yielded accurate isentropic equation of state (EOS) data at Mbar-pressures [7] in a planar geometry. The planar geometry has several advantages: the one-dimensional wave propagation into the sample under test greatly simplifies data analysis and therefore improves the accuracy of the technique; and it provides ready access to the sample, thus allowing a variety of diagnostic techniques to be used, most notably VISAR. We have demonstrated that the HEPP version of the ICE technique can also be used to obtain high accuracy isentropic EOS data at Mbar-pressures in a planar

geometry, and that the system is reliable, reproducible and predictable. With this system we may ultimately reach pressures of ~20 Mbar (2 TPa) [8]. We have performed experiments on OFHC copper and pure tungsten, but space limitations

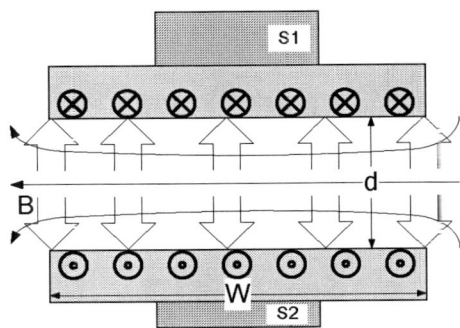

**Figure 1.** B-forces (block arrows) launch ramp waves towards samples S1 and S2. Circled dots and crosses indicate currents into and out of the paper.

restrict us to the results for copper. We compare techniques of data analysis and discuss some of the issues affecting accuracy.

## EXPERIMENTAL PROCEDURE

The physics of ICE and the methods of data recovery and analysis have been described previously [1][6]. In an ICE experiment two or more identical samples of different thicknesses are subjected to identical ramp wave loadings by magnetic forces that originate on the inside surfaces of a pair of parallel conductors, Figure 1. In the case of HEPP-ICE consider two equal and opposing currents (I) flowing in two flat and parallel conductors of equal width (W) and separated by a small distance (d); W and d are typically 12.7 mm and 0.5 mm. The magnetic pressure (B-force) on the inside surfaces is given by the vector product $\mathbf{P_B} = \mathbf{J} \times \mathbf{B}$, where J is the current per unit width. In the simplest case when W>>d, and J is uniform, the magnetic field on the inside surface of one conductor due the current in the opposing surface B = $\frac{1}{2}\mu_0$ J [9] and

$$P_B = \frac{1}{2}\mu_0 J^2$$

If J = 1 GA/m for example, then $P_B$ = 200$\pi$ GPa (6.28 Mbar). The time of application of these currents must be less than the time it takes a compression wave to traverse the thinnest sample and back again; up until then identical high pressure loads have been maintained in each sample. In the ICE experiments these current risetimes are typically 300 to 600 ns.

The velocities of the outer sample surfaces are measured with VISAR, either through windows or at vacuum interfaces. Using either Lagrangian or 'Backward' techniques (described later) the isentropic EOSs of the materials under test are obtained.

## EXPERIMENTAL DESIGN

Our prototype HEPP-ICE system [1] uses a compact explosive flux compression generator (FCG) [10], a 15 to 30-nH storage inductor, an explosively-formed fuse opening switch (EFF) [11], and explosively-actuated closing switches

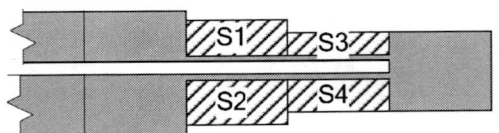

**Figure 2.** Load section. Current flows from connections on the left, on the inside surfaces of the slit, through the short circuit on the right and back again. The two sample pairs are S1/S2 and S3/S4.

[12]. It is capable of delivering 5 to 7 MA (dI/dt ~30 TA/s) into loads of 1 to 2 cm width with the required risetimes, producing ramp waves up to a 3 Mbar. During development of the system we systematically eliminated various problems, most notably to do with insulation, corona, and the closing switches. We now routinely obtain the desired current profiles on every shot.

One load section design is shown in Figure 2. It comprised a brass transmission line (not shown) converging from 30 cm to a width of 4 cm, and a copper load section tapering to 1.27 cm width. The conductors were separated by a 660-μm slit and insulated with Kapton sheets (not shown). Four stepped copper samples, each 12.7 mm wide, were mounted in the load section with thicknesses between 1.79 and 2.52 mm. VISARs measured particle velocities at the back face of each sample (which was evacuated to prevent air flash).

### Data analysis for OFHC copper

Figure 3 shows the raw VISAR data (top) and the compression isentrope (bottom), derived from Lagrangian analysis [13]. First the sound speed $C_L(u)$ is measured for each pair of wave profiles by dividing the known difference in sample thickness by the difference in arrival times of the wave at a given particle velocity u. The differential form of the momentum conservation equation is then used to calculate the change in pressure, d$\sigma$, for each step of u, du, going up the curve u(t), where $\rho_0$ is the initial density (at zero pressure), i.e., $d\sigma = \rho_0 c_L(u)du$. Hence we calculate continuous EOS relationships between pressure, sound speed, particle velocity, and density from 0 to the peak pressure. The EOS results for all four samples are shown in Figure 3, and there is agreement to within 0.2% of each other in pressure (the curves

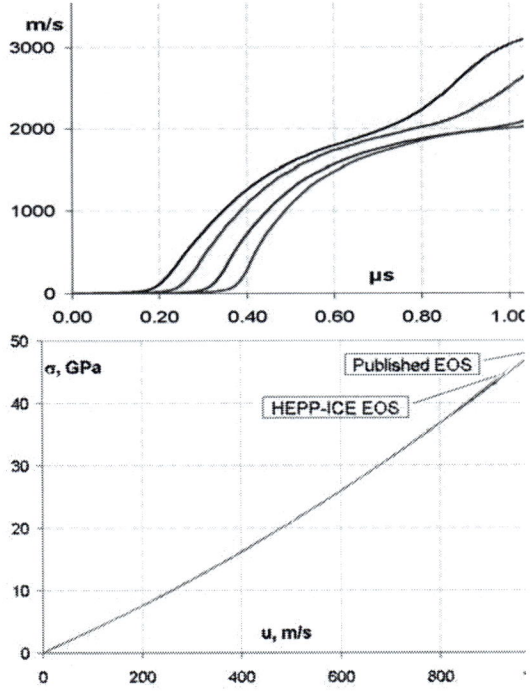

**Figure 3.** Top) Raw VISAR velocity data (m/s) vs. time (µs) for the four samples; Bottom) isentropic EOS results for each sample pair, pressure (GPa) vs. particle velocity (m/s).

lay on top of one another). Two published isentropes [14] are also shown, and they pass through the results (within 0.2%) up to 40 GPa.

One problem with the Lagrangian technique is that we must derive u from the measured free surface velocity $U_{fs}$. To a first approximation u is 1/2 $U_{fs}$, and then corrections can be made to adjust these data, e.g., [15]. But interactions of the reflected rarefactions at the free surface with oncoming ramp waves make accurate corrections difficult.

A better approach for ICE data analysis is the "Backward" technique [3]. Because there are no discontinuities in the ramp wave profiles, it is possible to calculate backwards in space from the rear faces of the samples to the inside surfaces driven by the magnetic loading, or to any convenient interface. The technique depends on the fact that the pressures (and particle velocities) at the inside surfaces are equal. If the calculations are performed with the correct EOS then the calculated pressure histories on these inside surfaces will match. First

an estimate of the compression isentrope in polynomial form. The polynomial is then adjusted in an iterative procedure until the differences between the two pressures (or the two particle velocities) are minimized. This procedure accurately accounts for the interactions of the reflected rarefactions with oncoming ramp waves.

## Backward Calculation Results

The inside surface pressures, Figure 4, were obtained with the Backward analysis using the same isentropes as before [14]. In this analysis we ignored the material strength of copper (280 MPa) which is small compared to the applied stresses. (In contrast, our results for tungsten – not shown here - show a material strength of ~3 GPa which cannot be ignored.) The agreement between each pair is within 0.2% in pressure, using the same isentrope, but the pressures for the two sets of samples differ significantly. The top pair of curves is for the two samples at the top of the load adjacent to the short circuit.

The difference is due to the fact that currents in parallel plate transmission lines are not perfectly

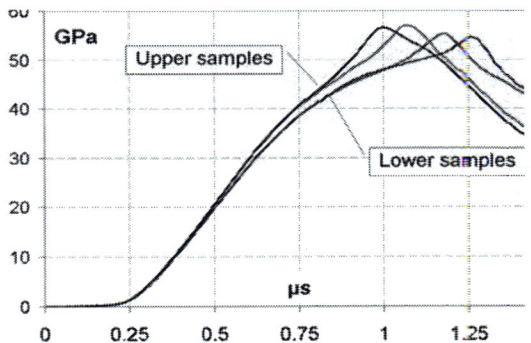

**Figure 4.** Results of backward calculations showing the apparent magnetic pressure at x = 0 for the two sets of samples, GPa vs. time (µs).

uniform across their width for rapidly changing currents. The electrical impedance to current flow is smaller at the edges than the center, so current initially flows preferentially along them; this reduces the magnetic pressure in the center. The tapered electrodes that connect to the load section also force the current density to be higher along the outer edges of the lower samples [2]. These dynamic processes, combined with skin effects, have no simple analytic description and can only be solved by 3-D computer hydrocodes with coupled MHD capabilities, see Goforth [8]. Fortunately we do not need to know the magnitude of magnetic fields for ICE. Opposing samples share the same B-field; the magnetic forces on them are exactly matched. The excellent agreement between the two sets of curves in the Backward calculation demonstrates that there is an exact match of the B-fields on opposing sample faces. The Backward technique has proven to be extremely useful not only in EOS data analysis, but also in understanding the operation of the HEPP-ICE system. For example, we have identified the effects of current non-uniformities, and other experimental artifacts.

## SUMMARY

We have demonstrated the HEPP version of the ICE technique can be used to obtain accurate isentropic EOS data at Mbar pressures in a planar geometry, and that the system is reliable, reproducible and predictable. The technique has be used to obtain accurate EOS data for OFHC copper, reported here, and tungsten up to 45 GPa, but can be extended to reach pressures of ~250 GPa (8 MGauss) with the prototype system and 1.7 TPa (20 MGauss) with an advanced system [8]. In the present arrangement four samples can be studied at one time, but accurate EOS data can only be obtained from opposing sample pairs, because it is only these pairs that share the same magnetic fields. In other words, only the sample-pairs S1/S2, S3/S4 in Figure 2 can be used; S1 and S3 or S4 cannot.

## ACKNOWLEDGEMENTS

The authors wish to acknowledge Jennifer McGuire and James Arellano for their outstanding hardware design and manufacturing support. Grateful thanks are due Dennis Hayes for invaluable help with the Backwards integration technique, and to Steve Sterbenz and Rendell Carver for their programmatic support and enthusiasm. This work was funded by the US Department of Energy.

## REFERENCES

[1] Tasker, D.G, et al., "Advances in Isentropic Compression Experiments (ICE) Using High Explosive Pulsed Power," Proc. APS Shock Compression of Condensed Matter (SCCM), 2003, p.1239.

[2] Tasker, D.G., et al., "Equation of State Experiments in Extreme Magnetic Fields," Tenth International Conference On Megagauss Magnetic Field Generation And Related Topics (Megagauss X), July 2004, Berlin, Germany, http://megagaussx.physik.hu-berlin.de/.

[3] Hayes, D.B., "Backward Integration of the Equations of Motion ... ," Sandia National Labs., SAND2001-1440, May 2001

[4] Fowler, C.M et al., "Nearly Isentropic Compression of Materials by Large Magnetic Fields : A Survey," Procs. Intl. Symp. On Intense Dynamic Loading and Its Effects, Chengdu, China, June 1992.

[5] Boyko, B.A., et al., Megagauss IIX, Oct 1998, Tallahassee, FL.

[6] Asay, J.R., Shock Comp. of Condensed Matter – 1999, Am. Inst. Phys., p.261.

[7] As the material strength of copper is small compared to the applied pressure we use the term "pressure" throughout. Strictly, the magnetic field effects are isotropic and are manifest as pressures whereas the material forces are manifest as stresses.

[8] Goforth, J.H., Megagauss IX, July 2002, Moscow-St. Petersburg, Russia, p. 137.

[9] All formulae are in SI units; permittivity of free space, $\mu_0 = 400\pi$ nH/m.

[10] D.J. Erickson, Proc. Fifth IEEE Pulsed Power Conference, June 10-12, 1985.

[11] J.H. Goforth, Proc. PPPS-2001, IEEE Pulsed Power Plasma Science conference, June 17-22, 2001, p.150.

[12] Tasker, D.G., at al., "High Current, Low Jitter, Explosive Closing Switches," Fifteenth IEEE International Pulsed Power Conference, Monterey, CA, June 2005.

[13] Aidun, J.B. and Gupta, Y.M., JAP, 43, 4669 (1972).

[14] Isentropes obtained from: Hugoniots in "Selected Hugoniots," Los Alamos Scientific Lab., May 1, 1969, LA-4169-MS; and Thermodynamic Los Alamos Sesame Tables 3336.

[15] Reisman, D.B., et al., "Magnetically-driven Isentropic Compression Experiments on the Z-accelerator," JAP, 89(3), p.1625, 1 Feb. 2001.

CHAPTER XX

# OPTICAL AND ELECTRICAL MEASUREMENTS

CP845, *Shock Compression of Condensed Matter - 2005*
edited by M. D. Furnish, M. Elert, T. P. Russell, and C. T. White

# SOME ASPECTS OF SHOCK-INDUCED RADIATION OF TRANSPARENT MEDIA AND ITS TRANSFORMATION WITH PRESSURE

## M. F. Gogulya and M. A. Brazhnikov

*Semenov Institute of Chemical Physics, Russian Academy of Sciences,
4, Kosygin Street, Moscow, 119991, Russia*

**Abstract.** Radiation of shocked compressed water, glycerol, and sulphur and its transformation with pressure has been studied. It was shown that radiation of shocked water loaded through different metal barriers (Al, Mg, and Cu) depended both on shock pressure (varied from 10.0 to 39.5 GPa) and the nature of the metal. It was proposed that radiation histories measured in transparent or semitransparent materials could be explained in terms of "contact" radiation and "volume" radiation. To verify this assumption there was measured radiation from glycerol shocked in pressure range 18.9-45.2 GPa.

**Keywords:** Shock compression: water, temperature, radiation, transparency.
**PACS:** 62.50.+p, 78.47.+p.

## INTRODUCTION

Various experimental techniques have been employed to determine shock temperature of water [1,2], but in this range up to 30 GPa, in which it is transparent, the temperature has not been measured previously. Radiation from shocked materials which retain their partial transparency is an important problem of basic research.

## EXPERIMENTAL RESULTS

Water was studied in the pressure range $p = 10.0$-$39.5$ GPa. Shock waves (SWs) were produced by impact of a plate accelerated by HE detonation products (DP) on a metal target or by expanding DP ("throwing" and "contact" set-ups of tests). The Al, Cu, and Mg targets with identical surface finish were used. Radiation was recorded by an optical pyrometer ($\lambda_{eff} = 420$ and $720$ nm) with a time resolution of ~0.02 μs. In Figs. 1 and 2, the results of tests with Al targets are shown. The radiation intensity is hereinafter given in units of temperature. The histories recorded at different pressures have features in common. First this is a sharp increase in the intensity at the instant when the SW arrives at the Al-water interface Peak of radiation within first tenth of microsecond is in conflict with what can be expected in the case of thermal radiation. See, e.g., Fig. 3, in which there is shown the radiation history for glass shocked at the same pressure as for water (curve 2 in Fig. 2). Another feature is that the temperatures are rather high (Fig. 1 and 2). However, an increase in shock pressure from 10.0 to 25.1 GPa does not lead to an adequate increase in the temperature; only to an increase in the duration of the peak radiation. Radiation detected within ~0.1-0.3 μs assumed to be of a non-thermal nature. In the tests, radiation was recorded until the shock arrived at the water-air interface, which was accompanied by a sharp increase in radiation of air. In the "throwing" set-up (Fig. 2), the signal rise is observed much earlier, than is expected and it drops to the "plateau" corresponding to 1500-1700 K. For the "throwing" tests, one can assume the presence of Al microspalls on the Al-water interface, which can react actively with water, leading to the observed increase in temperature. We performed tests with a Cu target at various pressures (Fig. 4), to verify this assumption. Radiation histories for the examined metals under almost the same shock-loading conditions (curve 2 in Fig. 4 and curve 3 in Fig. 2) clearly shows the role of the target material; for Cu target the secondary rise in radiation due to the entry of the SW into air is observed within the anticipated time (~1 μs).

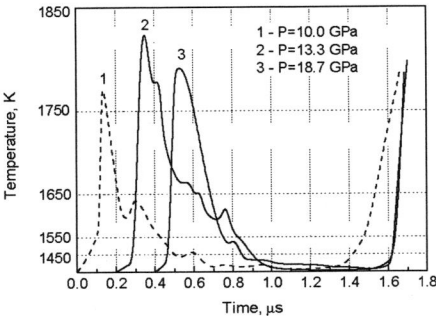

**Figure 1.** Water radiation histories *(λ$_{eff}$ = 720 nm)*: "contact" experiments; curves 2 and 3 are shifted in time by 0.2 and 0.4 μs respectively.

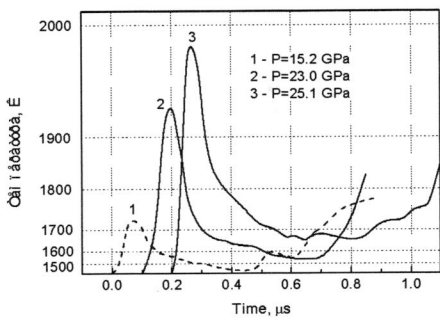

**Figure 2.** Water radiation histories (λ$_{eff}$ = 720 nm): the "throwing" set-up of experiments; curves 2 and 3 are shifted in time by 0.1 and 0.2 μs, respectively.

In the "throwing" set-up of tests with an Al target at *p* = 10.0-25.1 GPa, the secondary rise in radiation can be attributed to the water-Al reaction. The histories for tests with Al and Mg targets are shown in Fig. 5. In the case of more reactive Mg,

the radiation begins to grow when the SW still propagates over water and has not yet reached the interface.

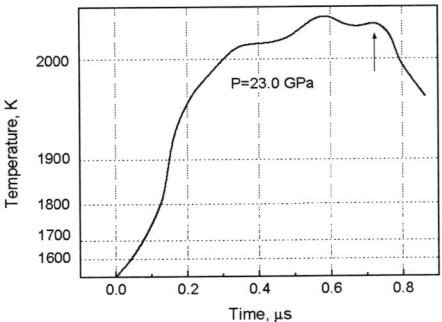

**Figure 3.** Kron-glass radiation history (λ$_{eff}$ = 720 nm). The "contact" set-up; the arrow shows the instant of arrival of the rarefaction.

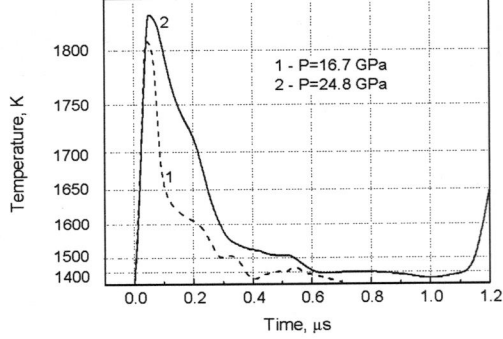

**Figure 4.** Water radiation histories (λ$_{eff}$ = 720 nm): the "throwing" set-up of experiments; a Cu target.

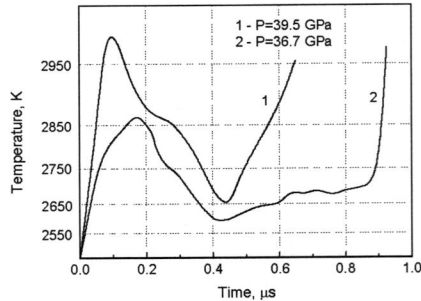

**Figure 5.** Water radiation histories (λ$_{eff}$ = 420 nm): the "throwing" set-up; 1- Mg target; 2 -Al target.

1330

Thus, water and active metals (Al and Mg) are capable of reacting in the range $p = 10\text{-}40$ GPa. Radiation of shocked water can be considered as the superposition of the "contact" luminosity due to processes at the target-water interface, the "volume" radiation emitted by a layer of shocked semitransparent liquid, and the radiation caused by chemical reactions. This conclusion is confirmed by the data given below.

## DISCUSSION

**"Contact" Radiation.** The role of contacting materials is illustrated by the test (see Fig. 6.) in which a SW sequentially entered Al, PMMA and glycerol. The radiation profile from water placed on an Al plate is also given. Both profiles were recorded at the same pressure approximately ($\sim 10$ GPa). The radiation of the "assembly" has a far more complicated nature. The question of whether the "contact" radiation occurs on the target surface or in the transparent material in the immediate proximity to the interface remains open. Consider data on shock-induced radiation of NaCl (see Fig. 7); we assumed that "contact" radiation arises in a layer of shocked material near the interface. The appreciably higher intensity measured for the pressed sample (0.97 TMD) was explained by its extremely high imperfection. In liquids; we believe, the radiation recorded at the instant of SW passage through the interface is localised in a layer near the interface.

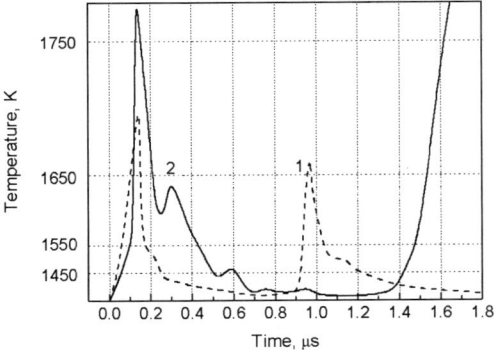

**Figure 6.** Radiation profiles ($\lambda_{eff} = 720$ nm) for SW propagation through an Al/PMMA/glycerol "assembly" (curve 1) and through water (curve 2).

The question is what are the imperfections in the liquids responsible for "contact" radiation. Real liquids are solutions of gases, liquids, and solids. The presence of such defects is typical of interfaces; therefore, the intensity of the "contact" radiation can depend also on the surface finish of the metal target. Shock-compression of liquids can produce conditions for formation of gas inclusions. Such "nuclei" of a new phase can serve as centres of non-thermal radiation.

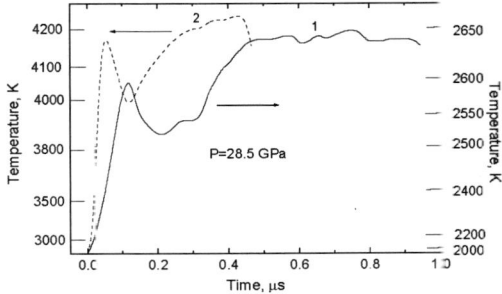

**Figure 7.** Radiation profiles ($\lambda_{eff} = 420$ nm) of shocked NaCl: 1 - pressed sample, 2 - single crystal; the Al target.

**"Volume" Radiation** Records of radiation of glycerol shocked in tests performed in the "throwing" set-up with Al targets are shown in Fig. 8. Radiation profile is transformed when the "contact" radiation is screened by the more powerful radiation generated in the volume as the pressure increases. At $p < 25$ GPa (Fig. 8a), the radiation is due primarily to processes near the interface and the volume radiation is negligible. Two-peak records were obtained at $p = 29\text{-}31$ GPa (Fig. 8b). It can be assumed that the second peak is due to an increase in the layer of shock-compressed glycerol. Over a time interval of 0.1-0.2 μs, the "contact" non-thermal radiation and the "volume" (possibly thermal) radiation are superimposed on each other. At higher pressures ($p = 45.2$ GPa; Fig. 8c), glycerol is apparently almost opaque and the first peak and the second rise in radiation merge together within the first 0.05 μs. The suggested description of the luminosity of shocked liquids in terms of "contact" radiation and "volume" can be applied to transparent or semitransparent media only. We examined radiation of shocked sulphur in the range 23.5-55.1 GPa. No "contact" radiation has been observed. Radiation histories of shocked sulphur can be considered as thermal. Temperatures in a direct SW are given in Fig. 9.

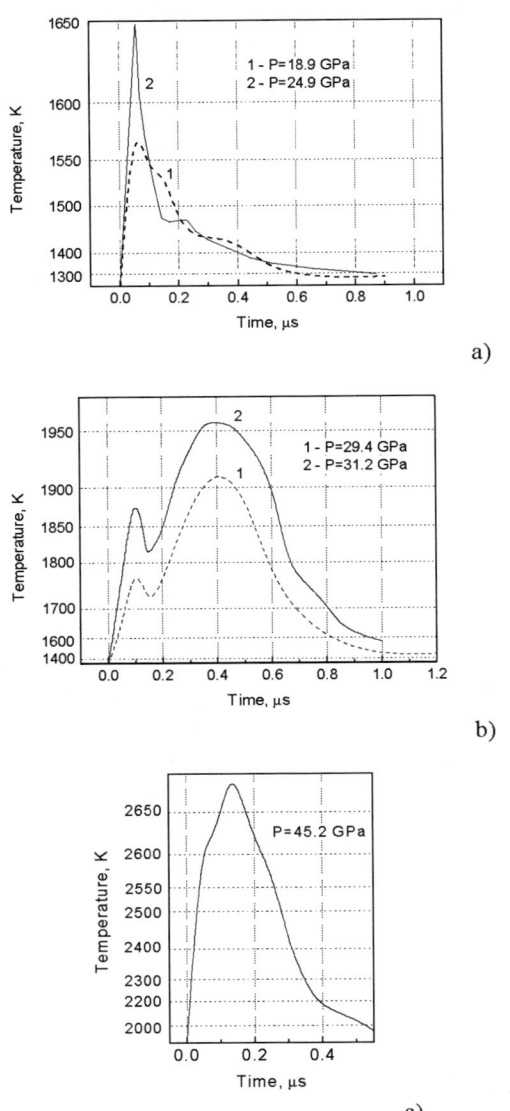

a)

b)

c)

**Figure 8.** Radiation profiles of shock-compressed glycerol ($\lambda_{eff}$ = 720 nm): the Al target .

Curve I was calculated for non-dissociated sulphur [3]. Curve II fits experimental data obtained for 4-mm thick uncovered samples (initial temperature 120-125 $^0$C); curve III – for 2-mm samples covered by "windows" (initial temperature 145-155 $^0$C). More detail explanation will be given

elsewhere [4]. Some changes in the structure of liquid sulphur possibly occur at p > 40 GPa.

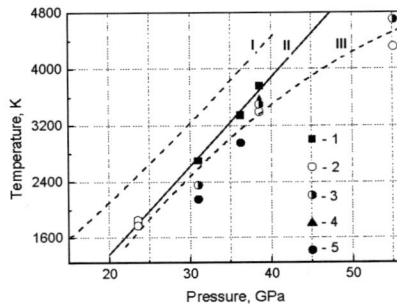

**Figure 9.** Temperature of liquid sulphur. Curves: I – calculated [3]; II, III – experimental for 4-mm and 2-mm thickness of liquid sulphur. Experimental points: 1 – without window (4 mm), other points with window 2 – LiF (2 mm), 3 – NaCl (2 mm), 4 – PMMA (2 mm), 5 – LiF (3 mm); sulphur thickness is given in brackets.

## CONCLUSIONS

A comparison of the records obtained for semitransparent shocked water, glycerol, NaCl shows that they are similar to some extent despite the difference between the materials. The non-thermal radiation generated by a SW should be treated as a result of the irreversible non-equilibrium nature of dynamic loading. There should be a certain general approach to explain the nature of shock-induced radiation in homogeneous media and its transformation with pressure.

## REFERENCES

1. Lyzenga, G. A., et al., "The temperature of shock-compressed water", J. Chemical Physics, vol. 76, no. 12, pp. 6282-6286, 1982.
2. Zeldovich, Ya. B., et al., "Study of the optical properties of transparent materials at hyperpessures", Doklady. Akaemii. Nauk SSSR, vol. 138, no. 6, pp. 1333-1336, 1961, (in Russian).
3. Voskoboinikov I.M. "The State of Sulphur behind Shock Wave Front", J. Combustion, Explosion, and Shock Waves, vol. 41, no 2, pp. 119-125, 2005, (in Russian).
4. To be published in J. Combustion, Explosion, and Shock Waves.

CP845, *Shock Compression of Condensed Matter - 2005*,
edited by M. D. Furnish, M. Elert, T. P. Russell, and C. T. White
© 2006 American Institute of Physics 0-7354-0341-4/06/$23.00

# DISPLACEMENT MAPS IN TAYLOR IMPACT USING SPECKLE RADIOGRAPHY

## S.G. Grantham[1], C.H. Braithwaite, W.G. Proud and D.M Williamson

*Physics and Chemistry of Solids, Cavendish Laboratory, Madingley Road, Cambridge CB3 0HE, United Kingdom*
*1. CHP Consulting Ltd, Augustine House, 6a Austin Friars, London EC2N 2HA*

**Absract**. A method was developed at the Cavendish Laboratory (Cambridge University) to determine the internal displacement characteristics of a polymer rod undergoing classic Taylor impact. Using a layer of lead filings on the central plane of the rod and Digital Speckle Radiography with flash x-rays it was possible to build up a dynamic displacement map. This technique can be used to derive material properties such as yield strength and provide data for model validation.

**Keywords:** Taylor impact, Speckle radiography
**PACS:** 62.50.+p, 06.60.Jn

## INTRODUCTION

The Taylor test [1], introduced in 1947, is a simple method of testing materials at high strains and high strain rates. In this paper we are concerned with the classic Taylor test rather than the more contemporary symmetric test [2].

The Taylor test provides suitable data for the validation of constitutive models, as well as determination of dynamic yield strengths [2]. With metals and other ductile materials the sample can be collected after the experiment and subjected to a variety of analytic methods (such as indentation hardness, crystallography and profilometry) to give these characteristics. With more brittle materials, such as polymers and glasses [3], this might not be possible due to extensive fragmentation of the sample and elastic recovery, as described by Briscoe and Hutchins [4]. Thus high-speed photography is employed to capture the impact as the deformation occurs, and before total fragmentation.

The disadvantage of this method is that while it is time resolved, features such as voiding and internal fracture may not be observed. In addition the process of determining small deformations is difficult and introduces inaccuracy.

A solution to this problem is the use of digital speckle radiography [5]. This technique allows the imaging of an internal plane within the sample and the determination of a displacement map of that plane. The measured displacement is a reproducible quantitative measure of deformation within the sample as it undergoes impact, and as such is superior to approaches based *solely* on photographic methods.

## EXPERIMENTAL

The research was carried out in two different methods to test the robustness of the technique. The speckle imaging technique was the same in both cases, the difference being in the triggering of the diagnostics. In both sets of experiments a small gas gun was used to fire rods made from a two-part polymer resin/hardener into a large mild steel block (flattened cylinder of radius 100 mm, length 300 mm and mass approx. 60 kg). The block, which was smooth on the impact surface, was fixed against an immovable barrier. This setup approximates to the semi-infinite perfectly rigid anvil required for the Taylor test.

A set of calibration experiments were conducted to determine the velocity variation of unseeded rods (mass 36.6 ± 0.3 g, diameter 19.6 mm and length 100 ± 0.4 mm) which was found to be 140 ± 9 m s⁻¹. The lead seeding used for the speckle pattern increased the weights of the rods to a range between 37.2 g and 40.0 g. Calibration data for the gun used, however, indicates that the velocities of the slightly heavier rods would not have been significantly different from the control experiments. In the control experiments a number of high-speed photographic sequences were taken (for example Figure 1), with the camera (Ultranac FS501) and flash triggered from a break trigger attached to the end of the barrel.

The speckle analysis requires two x-ray images, one of the rod before firing (the reference image) and one during the time it is impacting the target (the deformed image). In order to relate the time that the deformed image was taken to the impact of the projectile, a PVDF gauge was attached to the front of the metal target to act as an impact trigger.

In the second set of experiments the velocity and impact time were determined from a high-speed photographic sequence obtained simultaneously with the x-ray image. This was a much simpler system to operate, and was entirely triggered by the use of a break trigger. It also allowed the impact surface of the anvil to be free from any diagnostics. A delay generator was used to synchronise the system.

Flash x-rays of voltage 120 keV and flash width of 30 ns, were generated by a Scandiflash 150 keV set. The x-ray head was positioned in the same place every shot. The head was positioned to be over the impact face of the metal target to avoid any part of the impact being missed due to the cone-shaped nature of the x-ray emission. The film cassette was placed in the bottom of the vacuum chamber, below the impact point, and was kept horizontal at all times.

Preliminary experiments were carried out to ascertain which conditions were optimal for the technique to work. In particular this effort focused on whether the rod manufacturing process affected the results. In order to have a plane of lead filings in the centre of the rod it was necessary to make half

of the rod, allow it to set enough to support the lead (and it's own weight), and then make the second half. The high-speed photographic sequences showed that having a rod made in two parts did not significantly affect the fracture of the rod. Nor does the presence of the lead layer make a significant difference. Minimal rotation occurred in the system. Some rotation however can be accounted for in the processing algorithm.

## RESULTS

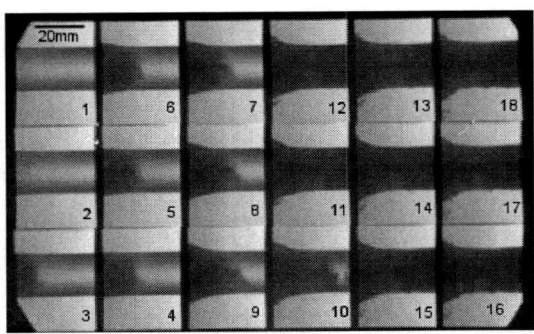

**Figure 1**. General photographic sequence of impacting rod, 8 μs interframe time and 2 μs exposure. The numbering shows the frame sequence.

The output data for this experiment was Polaroid photographs (Figure 1 shows a typical sequence) from the high-speed camera and developed x-ray film images.

The x-ray images were analysed using an algorithm due to Sjödahl [6,7]. This algorithm uses a cross correlation technique to match small sub images from the deformed image to the corresponding positions on the reference image. Sub-pixel accuracy is achieved by Fourier expanding the discrete data. For the purposes of this experiment the rear of the rod was used as the fiducial marker, as it was not deformed when the second, dynamic, x-ray image was obtained. This undeformed section allowed compensation for any rotation or translation that had occurred between the two images. Any effect of the initial plastic wave was too small to detect with the resolution of this technique.

The data can be presented as either a quiver plot (with arrows showing movement) or as a contour map. When a comparison is made between the high-speed sequences and the x-ray images two things should be taken into consideration. Firstly, the dark area of the photos does not correspond to the deformed section of the rod. The dark area actually represents the comminuted section of the rod. Hence some detectable deformation is expected in the x-ray images ahead of the corresponding black areas of the photographs. Secondly the photographs are taken at 90 degrees to the x-rays, and as such any non-rotationally symmetric features of the photographic images will not be seen in the x-ray images.

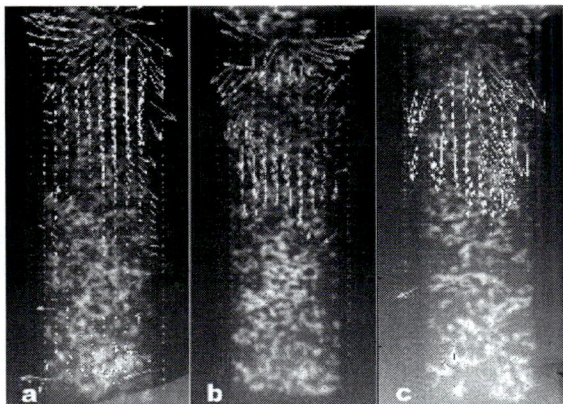

**Figure 2.** Analysed X-ray images at: a) 10 µs b) 20 µs c) 40 µs after impact (Arrows are magnified 15 times). Width of each rod is 19.6 mm

Figure 2 shows the results from the first set of experiments in a series of quiver plots. It can be seen that the general shape follows the one shown in Figure 1 (albeit rotated). There is a clear outwards bulge initially and then the front propagates down the length of the rod. In the later image times the rod fragments and is too damaged to allow for the algorithm to find correlation between sub images.

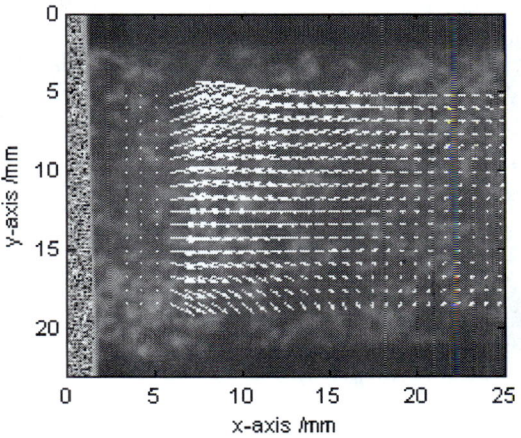

**Figure 3.** 15 ± 6 µs after impact at the left hand side of the image. Arrows are displacements in mm.

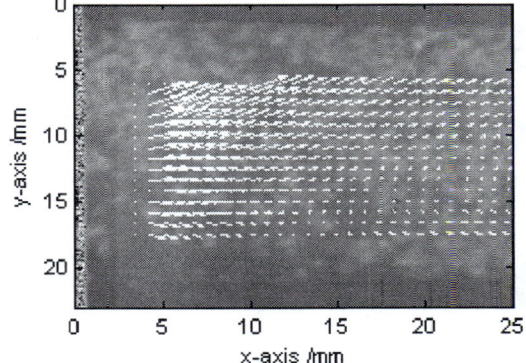

**Figure 4.** 17 ± 4 µs after impact at the left hand side of the image.

In the figures the deformation of the rod is imaged quite clearly. This is true of both the quiver plots (Figures 3 and 4) and, in a less quantitative way, the high-speed sequences (5 and 6). The sequences also show the asymmetric nature of fracture in brittle materials. This is also demonstrated very clearly in the quiver plots, with Figures 3 and 4 showing much greater deformation in the top half of the images. In Figure 2 it is noticeable that the displacements of adjacent areas can be quite different, again consistent with the nature of brittle fracture. It is interesting to note that the fractured area on the photograph is preceded by plastic deformation within the rod. In both cases the deformation of the rod is hard to determine from the high-speed photographic sequence, but is plainly visible on the quiver plots.

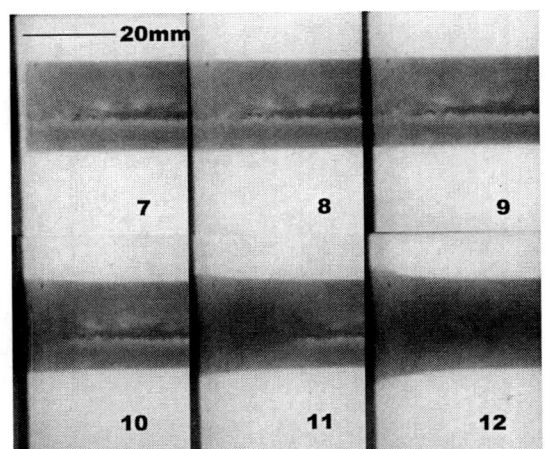

**Figure 5**. The photographic equivalent of Figure 3. Interframe times are 8 μs and exposure time is 2 μs.

Due to the perpendicular nature of the camera and x-ray set up it is not possible, without the aid of a rotationally symmetric impact to compare directly Figures 3 and 4 with Figures 5 and 6. However, the x-ray in Figure 3 corresponds to frame 10 in Figure 5 and the x-ray in Figure 4 corresponds to frame 11 in Figure 6.

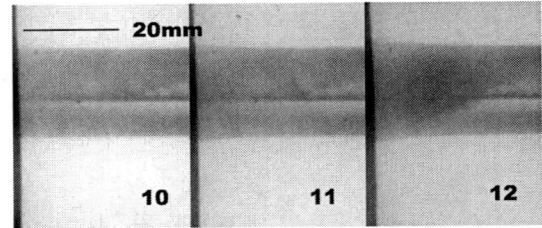

**Figure 6**. The photographic equivalent of Figure 4. Interframe times 8 μs and exposure time is 2 μs. The x-ray was taken in frame 11.

## DISCUSSION

The results of this investigation have clearly demonstrated that the technique presented is suitable for imaging the interior displacements of plastic rods undergoing conventional Taylor impact. This further demonstrated the subtlety and accuracy of the use of digital speckle radiography. Deformations that are difficult to detect using the high-speed photographic sequences are easily measured by this method. The fact that the algorithm generates quantitative data is also of

benefit, and is again advantageous over the photographic method.

This method can be applied in further investigations and in varied setups. For example a denser speckle pattern would provide increased resolution. Other setups might include mounting the rod and fiducial markers in an adjustable mount and fire a metal flier at the rod.

## ACKNOWLEDGEMENTS

H. Prentice and D. Chapman for discussions involving experimental method. M. Oberwittler and E. Stanistas for x-ray development. R. Flaxman, R. Marrah and D. Powell for technical assistance.

## REFERENCES

1. Taylor, G.I., *The use of flat ended projectiles for determining yield stress. 1: Theoretical considerations*. Proc. R. Soc. Lond. A, 1948. **194**: p. 289-299.
2. Erlich, D.C., D.A. Shockey, and L. Seaman, *Symmetric rod impact technique for dynamic yield determination*, in *Shock Waves in Condensed Matter – 1981*, W.J. Nellis, L. Seaman, and R.A. Graham, Editors. 1982, American Institute of Physics: New York. p. 402-406.
3. Willmott, G.R., and D.D. Radford, *Taylor impact of glass rods*. J. Appl. Phys., 2005. **97**: p. 93522.
4. Briscoe, B.J., and I.M. Hutchings, *Impact yielding of high density polyethylene*. Polymer, 1976. **17**: p. 1099-1102.
5. Synnergren, P., H.T. Goldrein, and W.G. Proud, *Application of digital speckle photography to flash X-ray studies of internal deformation fields in impact experiments*. Appl. Opt., 1999. **38**: p. 4030-4036.
6. Sjödahl, M., *Electronic speckle photography: Increased accuracy by non-integral pixel shifting*. Appl. Opt., 1994. **33**: p. 6667-6673.
7. Sjödahl, M., and L.R. Benckert, *Electronic speckle photography: Analysis of an algorithm giving the displacement with subpixel accuracy*. Appl. Opt., 1993. **32**: p. 2278-2284.

CP845, *Shock Compression of Condensed Matter - 2005*,
edited by M. D. Furnish, M. Elert, T. P. Russell, and C. T. White
© 2006 American Institute of Physics 0-7354-0341-4/06/$23.00

# DIAGNOSIS OF MBAR LASER PRODUCED SHOCKS IN TIN USING SHORT PULSE PROBES

**W. Grigsby[1], B. T. Bowes[1], D. A. Dalton[1], S. Bless[2], M. C. Downer[1], E. Taleff[3], J. Colvin[4], T. Ditmire[1]**

[1] *Texas Center for High Intensity Laser Science, The University of Texas at Austin,
1 University Station #C1510, Austin, TX 78712*
[2] *Institute for Advanced Technology, The University of Texas at Austin,
3925 West Braker Lane, Suite 400, Austin, TX 78759*
[3] *Department of Mechanical Engineering, The University of Texas at Austin,
1 University Station #C2200, Austin, TX 78712-0292*
[4] *Chemistry and Materials Science, Lawrence Livermore National Lab, Livermore, CA 94550 USA*

**Abstract.** We are studying shock induced melting using laser produced shock waves in tin foils. The diagnostics used for these studies include pump-probe reflectivity and interferometry using 40fs pulses. Rear surface expansion data from the interferometer suggests that we have reached pressures necessary to shock melt tin upon compression. We have seen no unambiguous change in the reflectivity data to date.

**Keywords:** Laser driven shock, time resolved, interferometry, melting, metals, tin
**PACS:** 42.55.-f, 62.20.-x, 62.60.-i, 78.20.-e

## INTRODUCTION

Understanding material behavior under extreme conditions is an important area of research in physics and material science. One method to study the behavior of materials under these conditions is to drive a strong shock wave through a material and watch its response. In many cases the material response is complicated by phase transitions such as lattice restructuring [1-3] and melting [1, 3-6]. To study these dynamics we are using lasers in high time resolution pump-probe experiments to develop a real time diagnostic on the phase of a shocked material. This technique enables probing of the entire phase history of a material as it shock compresses and releases.

Our work concentrated on characterizing laser produced shock waves in thin metal foils backed by a transparent window. We shocked the front surface of a foil and probed the rear surface at various delay times using linear reflectivity and rear surface interferometry. The reflectivity of the metal is used to help diagnose the state of the material in optical pump probe measurements, while interferometry is used to determine the shock parameters (i.e. particle velocity and shock velocity).

## EXPERIMENTAL DESCRIPTION

The laser that we used was a Ti:Sapphire chirped pulse amplification system, with a Gaussian shaped stretched pulse with a width of 600 ps FWHM. After amplification to ~1 J, the beam was split into two parts. Most of the energy was diverted into the shock driving beam (~700 mJ) which contained a delay leg that allowed the beam to arrive in the target chamber at the same time as the probe beam. The energy not contained in the shock beam traveled into the pulse compressor where it was compressed to ~40 fs before traveling to the target chamber. The short time duration of the probe allowed us to take time-resolved images of the rear surface using an ordinary CCD camera.

The experiment was performed inside of a vacuum chamber that allowed us to shock and probe without the worry of ionizing air near the focus of our beams (See Fig. 1). The shock driving laser was focused using an f/20 lens to a diameter of a few hundred microns FWHM, depending on the intensity for which we were aiming. This laser pulse drove an ablation-driven shock wave which traveled through the target [7]. For these experiments, our targets consisted of 4 μm of tin deposited onto the surface of a 50.8 mm diameter x 6 mm thick, (100) lithium fluoride window with a surface flatness of <= 0.3 μm. The target thickness was chosen so that the shock wave would not decay appreciably during propagation through the 4 μm of metal, while still allowing time for the shock to steepen up. The window was used as a support structure for the thin foil, and also as a medium to confine the rear surface of the target to prevent material ejecta. In order to shock and probe a fresh surface for each shot, an XYZ translation stage was used to raster the target between shots.

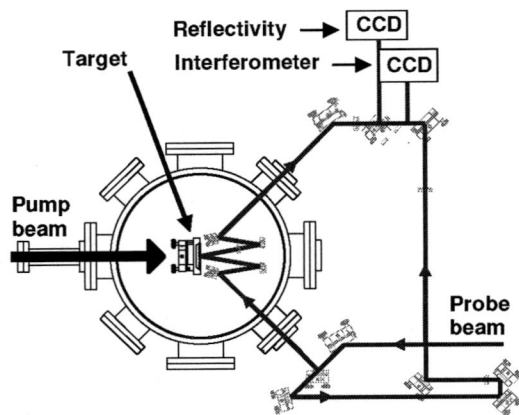

**FIGURE 1.** Diagram of experimental setup. The pump beam arrives from the left, and the probe beam arrives from the right.

**Diagnostics**

The main diagnostic used to determine shock parameters was a Mach-Zehnder style interferometer that probed through the back surface of the lithium fluoride and reflected off of the tin interface. To observe smooth and straight fringes with our probes, the surface flatness requirements were typically < 3 μm across the entire substrate.

One arm of the interferometer traveled into the vacuum chamber, reflected off of the target interface, and was then imaged onto the face of a CCD camera located outside of the chamber (See Fig. 1). The reference arm of the interferometer was located completely outside of the chamber and was adjusted to have equal imaging optics and equal optical delay to the probe arm. A combining beamsplitter was placed before the CCD, and adjustments were made to straighten and optimize the fringes on the camera. Before each shot was taken, a reference image was saved so that the change in surface position due to the shock wave could be measured accurately.

Data analysis was performed by Fourier analysis of the interferometric images, in a manner similar to Takeda, et al [8]. In this method, the image is first Fourier transformed, line by line. A super-Gaussian filter is then applied to the resulting spectrum to clean up the image. The primary background frequency is then subtracted by rotating the spectrum so this frequency lies at the origin. The new shifted spectrum is inverse Fourier transformed and the phase shift is extracted from the result. We extracted the distance $d$ that the surface expanded using,

$$d = \frac{\lambda \cdot \phi(x, y)}{4\pi n \cdot Cos(\theta)} \quad (1)$$

where $n$ is the index of refraction in the window, $\theta$ is the angle of the probe beam inside of the window (relative to the surface normal), and $\phi(x, y)$ is the unwrapped phase shift from the analysis of the interferograms.

We employed a second diagnostic that measured two-dimensional linear reflectivity. Our reflectivity data was obtained with the same probe pulse as in the interferometer, however a beamsplitter diverted a portion of the probe arm to a separate 16 bit CCD camera before the beam was

**TABLE 1.** Experimental Results Combined with Simulation Parameters

| Incident Laser Intensity (W/cm²) | Pressure From Measured $U_p$ Using the Hugoniot (±30kbar) | Laser Intensity Used in HYADES Simulation (W/cm²) | Pressure From HYADES Simulation (kbar) |
|---|---|---|---|
| $6 \times 10^{11}$ | 260 | $8 \times 10^{11}$ | 237 |
| $8 \times 10^{11}$ | 330 | $1.5 \times 10^{12}$ | 380 |
| $3 \times 10^{12}$ | 580 | $3 \times 10^{12}$ | 645 |
| $6 \times 10^{12}$ | 800 | $5 \times 10^{12}$ | 950 |
| $1 \times 10^{13}$ | N/A | N/A | N/A |

combined with the reference arm (See Fig. 1). This allowed us to obtain better reflectivity data without the added convolution of fringes in the interferometer. The reflectivity data was normalized to reference shots taken before the shock wave was driven into the target, so that we could more easily compare multiple shots with each other.

## RESULTS AND DISCUSSION

In these experiments, we used five different laser intensities to drive shock waves into our Sn/LiF targets. We measured the displacement of the interface as a function of probe delay for each of these five cases. The maximum displacement that we could probe was limited by the fringe size relative to the shock spot, and also by the contrast between adjacent fringes on the CCD. For example, interferometric data for the $1 \times 10^{13}$ W/cm² scan did not extend past 0.6 μm because of the small spot size on the CCD. For the other scans, where the fringes were distinguishable, we were able to observe up to a 2.1 μm expansion.

A line was fit to the linear portion of the expansion data, using a least-squares algorithm, to determine the particle velocity at shock breakout (see Fig. 2). The shock impedance mismatch between the tin and the lithium fluoride was accounted for so that the interface velocity could be converted into a particle velocity of the bulk tin. This particle velocity was then combined with Hugoniot data for tin to estimate the pressure observed at the interface within the target:

$$U_S = 2590 \text{ m/s} + 1.49 u_p \qquad (2)$$

$$P = \rho_o U_S u_p \qquad (3)$$

This pressure is compared with results from HYADES in Fig. 2 and in Table 1.

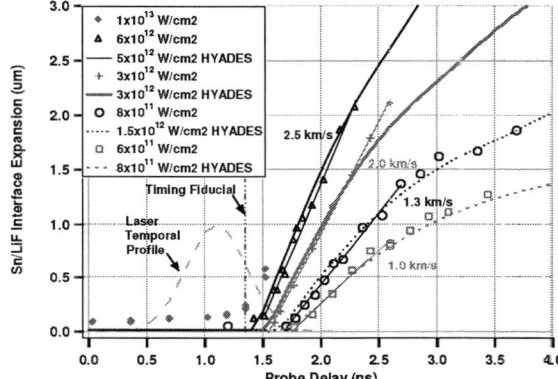

**FIGURE 2.** HYADES simulations (curves) are fit to four of the five Sn/LiF interface expansion data (symbols). The numbers printed next to the linear fits to the data (solid thin lines) are the slopes of the lines (i.e. interface velocities).

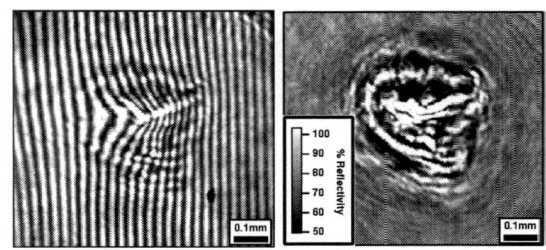

**FIGURE 3.** Interferometric and reflectivity data from a shock breakout. The laser intensity was $6 \times 10^{12}$ W/cm² and the probe delay was 450 ps after the timing fiducial.

Data obtained from the reflectivity diagnostic are not currently understood. They contain what look like interference rings (See Fig. 3) that may be due to a problem with the imaging system. More work must be done before any conclusions can be drawn from this data, and before it can be compared to Ref. [3].

## Simulations

Simulations of both the laser-target coupling and of the subsequent shock wave evolution in the target were run using HYADES, which is a one-dimensional radiation and hydrodynamics code developed by Jon Larsen [9]. Simulations were run with 150 mesh points spanning 4 μm of tin (feathered 5% for both surfaces) and 100 mesh points spanning 21 μm of lithium fluoride (feathered 5% from the front to the back). The EOSs used were SESAME #2160 and #7270 for tin and lithium fluoride, respectively. Results from these simulations were initially used to predict the approximate laser intensity necessary to shock melt tin at the rear surface of the target. These results agreed well with the initial LASNEX simulations that we performed. After the experiments were carried out, HYADES simulations were used to match the Sn/LiF interface expansion to measured data. The laser intensity parameter was slightly adjusted in the simulations until a match to the target expansion data was made. As Fig. 2 shows, the resulting simulations match the data very well. Additionally, the intensities necessary in three of the four simulations agree perfectly with the measured intensities, within error (See Table 1).

## CONCLUSIONS

We have shocked targets, consisting of 4 μm of tin coated on a (100) lithium fluoride substrate, with laser intensities ranging from $5 \times 10^{11}$ to $5 \times 10^{13}$ W/cm$^2$. The expansion of the rear surface of the tin due to the shock wave was measured as a function of probe delay using an interferometer. These data were compared with the known Hugoniot equations to estimate the pressure that the interface achieved. Preliminary HYADES simulations were able to match our expansion data very well, when the incident laser intensity was used as a variable parameter. Although we were able to shock at pressures above the melt pressure of tin, we have not yet seen a consistent enough change in the reflectivity of our target that we can attribute to shock melting.

## ACKNOWLEDGEMENTS

We gratefully acknowledge the useful discussions of Kim Budil, Alan Jankowski, Wayne King, and many others. This work was supported by Lawrence Livermore National Laboratories under Contract No. B541024 and by the U. S. Department of Energy, National Nuclear Security Administration under Contract No. DE-FC52-03NA-00156.

## REFERENCES

1.  Mabire, C. and Hereil, P.L., *Shock induced polymorphic transition and melting of tin up to 53 GPa (experimental study and modelling).* Journal De Physique Iv, 2000. **10**(P9): p. 749-754.
2.  Barker, L.M., *Shock-Wave Study Of Alpha - Epsilon Phase-Transition In Iron.* Bulletin Of The American Physical Society, 1975. **20**(1): p. 25-25.
3.  Swift, D.C., et al., *Dynamic response of materials on subnanosecond time scales, and beryllium properties for inertial confinement fusion.* 2005. **12**(5).
4.  Asay, J.R., *Experimental Determination Of Shock-Induced Melting In Aluminum.* Bulletin Of The American Physical Society, 1975. **20**(1): p. 20-20.
5.  Werdiger, M., et al., *Detecting of melting by changes of rear surface reflectivity in shocked compressed metals using an interferometric diagnostic method.* Laser and Particle Beams, 1999. **17**(3): p. 547-556.
6.  Elias, P., Chapron, P., and Laurent, B., *Detection of Melting in Release for a Shock-Loaded Tin Sample Using the Reflectivity Measurement Method.* Optics Communications, 1988. **66**(2-3): p. 100-106.
7.  Trainor, R.J., et al., *Ultrahigh-Pressure Laser-Driven Shock-Wave Experiments In Aluminum.* Physical Review Letters, 1979. **42**(17): p. 1154-1157.
8.  Takeda, M., Ina, H., and Kobayashi, S., *Fourier-Transform Method Of Fringe-Pattern Analysis For Computer-Based Topography And Interferometry.* Journal Of The Optical Society Of America, 1982. **72**(1): p. 156-160.
9.  Larsen, J.T. and Lane, S.M., *Hyades - a Plasma Hydrodynamics Code for Dense-Plasma Studies.* Journal of Quantitative Spectroscopy & Radiative Transfer, 1994. **51**(1-2): p. 179-186.

CP845, *Shock Compression of Condensed Matter - 2005*,
edited by M. D. Furnish, M. Elert, T. P. Russell, and C. T. White
© 2006 American Institute of Physics 0-7354-0341-4/06/$23.00

# LIQUID-SOLID PHASE TRANSITION OF BENZENE UNDER SHOCK COMPRESSION STUIDED BY TIME-RESOLVED NONLINEAR RAMAN SPECTROSCOPY

## K. G. Nakamura[1,2], A. Matsuda[2], and K. Kondo[1]

[1]*Materials and Structures Laboratory, Tokyo Institute of Technology, R3-10, 4259 Nagatsuta,*
*Yokohama 226-8503, Japan*
[2]*Institute for Molecular Science, Myodaiji, Okazaki 444-8585, Japan*

**Abstract.** The liquid-solid phase transition of benzene has been studied under laser-shock compression up to 4.2 GPa by using nanosecond time-resolved nonlinear Raman spectroscopy. A shock wave is generated by irradiation of 10-ns pulsed laser beam on the plasma confinement target and its pressure is estimated from the particle velocity, which is measured by a velocity interferometer. The ring-breathing mode shows blue shift under shock compression. Nanosecond time-resolved nonlinear Raman spectra show a rapid phase transition from liquid phase to solid phase under shock compression.

**Keywords:** Phase transition, benzene, Raman, nonlinear Raman spectroscopy
**PACS:** 78.30.-j, 62.50. +p.

## INTRODUCTION

It is essentially important to study the dynamics of materials under shock compression, in order to understand shock compression phenomena. Time-resolved spectroscopy is a well-suited tool to investigate vibrational and structural changes in molecules and crystals under shock compression. Nanosecond or picosecond time resolution can be achieved when we use a picosecond pulsed laser.

Benzene is one of the most widely studied compounds at high pressures because it is the most simple aromatic compound [1-6]. Schmidt *et al.* [1] and Kobayashi and Sekine [2] have obtained Raman spectra of benzene under shock compression up to 6 GPa. They observed blue shifts of the ring-breathing mode at high pressures. The shift in the shock compressed liquid benzene at pressures above 3 GPa agreed well with that of

solid benzene reported in the static compression experiment [3]. We suppose that the liquid-solid phase transition occurs under shock compression at pressures above 3 GPa. In this paper, we studied a phase transition of benzene by using nanosecond time-resolved nonlinear Raman spectroscopy: Stimulated Raman Scattering (SRS) and Coherent Anti-Stokes Raman Scattering (CARS).

## EXPERIMENTAL PROCEDURE

Figure 1 shows a schematic of the experimental setup for the CARS measurement. The fundamental light (1064 nm) of a nanosecond Q-switched Nd:YAG (yittrium aluminum garnet) laser was used for shock generation. The second-harmonic light (SHG: 532 nm) of a picosecond mode-locked Nd:YAG laser and a tunable light from an optical parametric generator (OPG) pumped by the third-harmonic light (THG: 355

nm) were used for CARS.   In the SRS                centered opposite the focus of the pump beam.

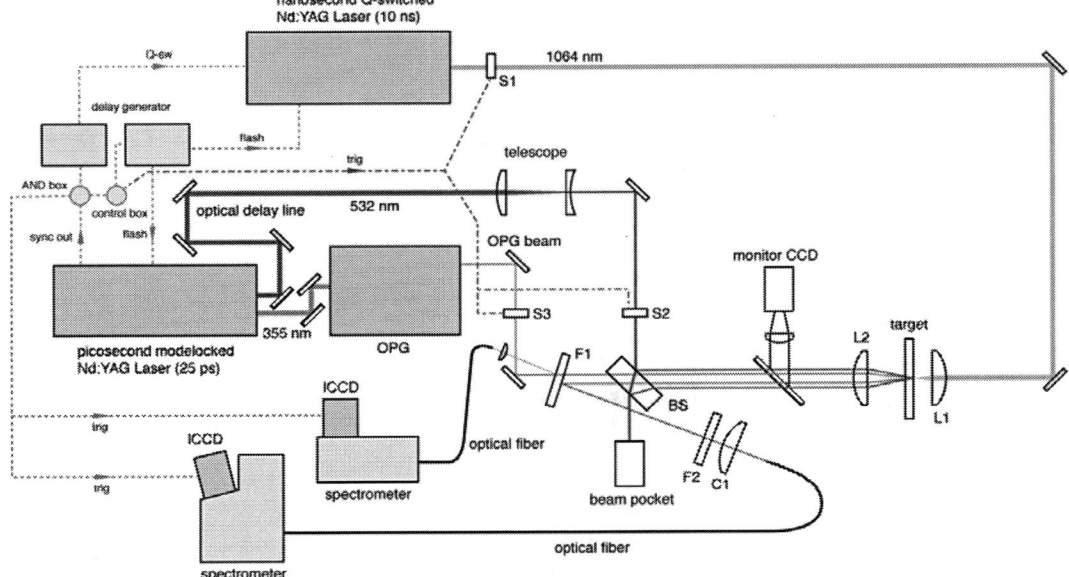

**Figure 1** A schematic of the experimental setup for CARS measurement under laser shock compression, S1-3: shutter, L1-2: lens, BS: beam splitter, F1: high pass filter, F2: notch filter.

measurement, only the SHG light was used for excitation of Raman scattering.

The target assembly was fabricated with a back-up float glass substrate (diameter of 15 mm, 3 mm thick), an aluminum foil (30 μm thick), a Teflon spacer (130 or 200 μm thick), and a cover float glass substrate (diameter of 15 mm, 150 μm thick). A thin polycarbonate debris shield is placed slightly displaced from the cover substrate to protect the optics from fragments of blow-off glass. The liquid sample was confined in the space surrounded by the aluminum foil, Teflon spacer and the cover glass substrate. Benzene (99.8 %) was obtained from Wako Pure Chemical Industries Ltd. By focusing the laser beam onto a spot on the aluminum, confined plasma is generated at the aluminum-glass interface, which drives a shock wave through the aluminum foil into the liquid benzene. Since the target is broken by a single shock event, the target is replaced every single shot.

Two SHG beams and the OPG beam were focused onto the target from the rear side with a diameter of 50 μm using an achromatic lens

The laser energy of the SHG light was 15 μJ and that of the OPG light renged from 0.3 to 10 μJ depending on the pressure. The reflection-type folded BOX CARS geometry [7,8] was used. The bandwidth of the OPG light was ranging from 20 to 60 cm$^{-1}$. The coherent anti-Stokes light was reflected at the benzene-aluminum interface travel backwards and introduced into an optical fiber, which connects to the spectrometer. The dispersed light was detected by the charge-coupled device (CCD) camera. The timing between the pump beam and probe beams was controlled by a delay and pulse generator (Stanford Research DG535). The timing jitter was within 2 ns. The spectrometer used for CARS measurement was reflection type (SOLAR TII) with a resolution of 1.7 cm$^{-1}$. On the other hand, the spectrometer used for SRS measurement was a transmission type (Kaiser) with resolution of 3 cm$^{-1}$.

Velocity of the interface between the aluminum and benzene was measured by using an ORVIS. In the ORVIS experiment (setup is not shown in Fig. 1), 532 nm light from continuous-wave diode laser was used. The ORVIS measurements were performed independently from the CARS and SRS measurements. The velocity is calibrated on the irradiated laser intensity. The shock pressure was estimated from the interface velocity, the Hugoniot equation of state, and the conservation equation. The shock wave profile inside the sample was calculated using one-dimensional fluid dynamic code.

## RESULTS AND DISCUSSION

At ambient pressure the ring-breathing mode has a frequency of 996 cm⁻¹. Nanosecond time-resolved SRS obtained under shock compression at 2.0 GPa showed a new high-frequency shifted peak, which is due to shock compressed benzene, at 1014 cm⁻¹ at delay time of 20 ns[5]. As the shock wave propagates through the liquid benzene, the intensity of the new high-frequency-shifted peak increased while the original peak decreased, because the volume of shock-compressed benzene increased with shock wave propagation inside the sample. The shift of the ring-breathing mode at 2 GPa agrees well with that obtained from the extrapolation of shifts obtained for liquid benzene using both static [1] and shock compressions [1,4].

In addition, the peak position of the new high-frequency-shifted peak did not change during shock wave propagation within 40 ns.

Figure 2 shows a typical example of nanosecond time-resolved SRS of benzene under shock compression at 4.2 GPa.[5] Under shock compression to 4.2 GPa, a new high-frequency shifted peak appears at 1027 cm⁻¹ at 12.5 ns. This new peak begins to shift towards lower frequency and reaches a stable state (1018 cm⁻¹) at about 30 ns. The frequency shift at delay time 12.5 ns corresponds to decrease of shock pressure, which is confirmed by ORVIS measurement. The intensity of the new peak increases as the delay time increases, and that of the original peak decreases. The shift ($\Delta\omega = 31$ cm⁻¹) at 12.5 ns is on extrapolation of reported frequency shift of liquid benzene and the shift ($\Delta\omega = 22$ cm⁻¹) at 30 ns agrees well with the reported value of the solid benzene. The change in shift may correspond to liquid-solid phase transition of benzene under shock compression.

In order to confirm this change, we performed the CARS experiment, which has much better spectral resolution. Figure 3 shows a typical example of nanosecond time-resolved CARS of benzene under shock compression at 4.2 GPa.

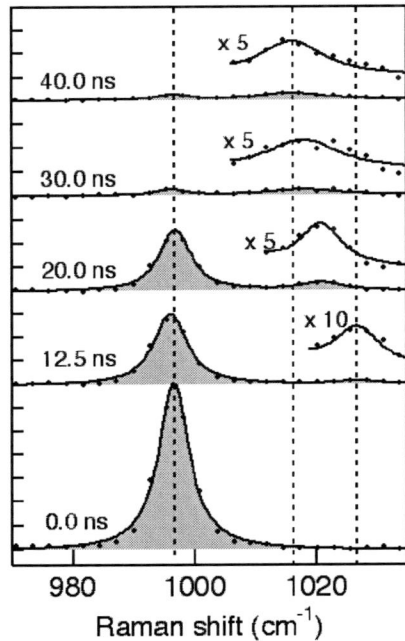

**Figure 2** Nanosecond time-resolved SRS of benzene under shock compression at 4.2 GPa.

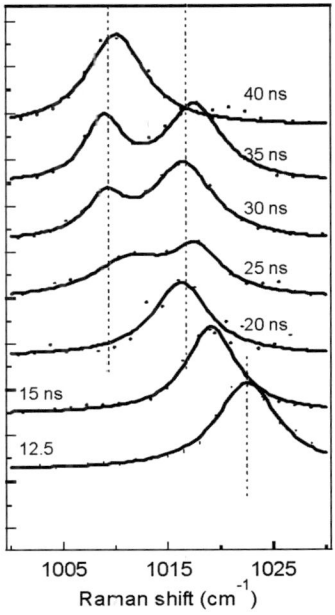

**Figure 3** Nanosecond time-resolved CARS of benzene under shock compression at 4.2 GPa.

Under shock compression, CARS measurement shows a new high-frequency shifted peak at 1022.5 cm$^{-1}$ at delay time of 12.5 ns. In the CARS measurements, the original ring-breathing mode peak and the new high-frequency shifted peak were detected separately, because the bandwidth of the OPG light is not wide enough to detect both peaks. This new peak begins to shift towards lower frequency up to delay time of 20 ns. At delay times between 25 and 35 ns, the peak splits into two peaks: a higher-wave-number peak (H-peak) at 1016.7 cm$^{-1}$ and a lower-wave-number peak (L-peak) at 1009.3 cm$^{-1}$. Relative intensity ratio of the L-peak to the H-peak increases as the delay time increases. The peak positions of the H-peak and the L-peak agree well with the peak positions for the liquid and solid benzene at 3.7 GPa, respectively. The frequency shift at delay time 12.5 ns corresponds to decrease of shock pressure, which is confirmed by ORVIS measurement and changes in frequency at delay time between 20 and 40 ns corresponds to liquid-solid phase transition.

## CONCLUSIONS

A liquid-solid phase transition of benzene has been studied under laser-shock compression up to 4.2 GPa by using nanosecond time-resolved nonlinear Raman spectroscopy. The ring-breathing mode shows blue shift under shock compression. Nanosecond time-resolved nonlinear Raman spectra show nanosecod phase transition from liquid phase to solid phase under shock compression within 20 ns.

## ACKNOWLEDGEMENTS

The authors thank M. Hasegawa for his help on constructing the experimental setup. This work was partially supported by Grant-in-Aid for Scientific Research #1477209 (Scientific Research on Priority Areas) from the Ministry of Education, Culture, Sports, Science and Technology (MEXT) of Japan.

## REFERENCES

1. Schmuidt, S. C., Moore, D. S., Schiferl, D., and Shaner, J., Phys. Rev. Lett., vol 50, pp. 661-664, 1983.
2. Kobayashi, T., and Sekine., T, Phys.l Rev. B, vol. 62, pp. 5281-5284, 2000.
3. Thiery, M. M., and Leger, J. M., J Chem. Phys., vol. 89, pp. 4255, 1988.
4. Matsuda, A., Nakamura, K. G., and Kondo, K., Phys. Rev. B, vol. 65, 174116 (4 pages), 2002.
5. Matsuda, A., Kondo, K., and Nakamura, K. G., Jpn. J. Appl. Phys., vol. 43, pp. L1614-L1616, 2004
6. Matsuda, A., Nagao, H., Nakamura, K. G., and Kondo, K., Chem. Phys. Lett., vol. 372, pp. 911-914, 2003.
7. Demtroder, W, "Laser Spectroscopy", Springer-Verlag, Berlin Hidelberg, 1996.
8. Levenson, M. D., and Kano, S. S., "Introduction to Nonlinear Laser Spectroscopy", Academic Press Inc., San Diego CA, 1988.

CP845, *Shock Compression of Condensed Matter - 2005*,
edited by M. D. Furnish, M. Elert, T. P. Russell, and C. T. White
© 2006 American Institute of Physics 0-7354-0341-4/06/$23.00

# PREDICTION OF COHERENT OPTICAL RADIATION FROM SHOCK WAVES IN POLARIZABLE CRYSTALS

## Evan J. Reed*[†], Marin Soljačić[†], Richard Gee* and J. D. Joannopoulos[†]

*Lawrence Livermore National Laboratory, Livermore, CA 94551*
[†]*Center for Materials Science and Engineering and Research Laboratory of Electronics, Massachusetts Institute of Technology, Cambridge, MA 02139*

**Abstract.** We predict that coherent electromagnetic radiation in the 1-100 THz frequency range can be generated in crystalline materials when subject to a shock wave or soliton-like propagating excitation. To our knowledge, this phenomenon represents a fundamentally new form of coherent optical radiation source that is distinct from lasers and free-electron lasers. General analytical theory and molecular dynamics simulations demonstrate coherence lengths on the order of mm (around 20 THz) and potentially greater. The emission frequencies are determined by the shock speed and the lattice constants of the crystal and can potentially be used to determine atomic-scale properties of the shocked material.
**Keywords:** shock wave, optical emission, molecular dynamics, sodium chloride,
**PACS:** 42.72.-g,47.40.Nm

## INTRODUCTION

The invention of lasers in 1958 made possible a staggeringly wide range of applications. The key characteristic of lasers that enables many of these applications is the fact that they are sources of *coherent* light. Almost 50 years later, very few distinct ways to generate coherent light have been realized. These include: "traditional" lasers based on stimulated emission and free-electron lasers, [1] each with its own unique practical advantages and disadvantages. This work presents what we believe is a new source of coherent optical radiation that is fundamentally distinct from lasers and free-electron lasers. We perform analytical theory and molecular dynamics simulations that predict that weak yet measurable *coherent* light can be observed emerging from a shocked polarizable crystal, typically in the range 1-100 THz. The *periodicity of the crystalline lattice* is the origin of the coherence of emitted radiation *rather than the "coherence" of the source generating the shock wave.*

This effect is predicted to be observable in a wide

variety of material systems under realizable shock wave conditions. In this work, we consider shock waves in NaCl. Some experiments on shocked single NaCl crystals have been reported. [2, 3] To our knowledge, coherent emission has never been observed because a shocked crystal is not an obvious system to discover (and hence look for) coherent radiation and the radiation is in a portion of the electromagnetic spectrum that is usually not observed in such experiments.

## ANALYTICAL THEORY

As a shock propagates through a polarizable crystal, a change in polarization can be induced which yields a time-dependent polarization current (even in materials with no static polarization). While it is not surprising that radiation should be emitted from the polarization currents induced by the shock, it is unexpected that this emission should be of a coherent nature. The coherent property of the radiation is the subject of this work. The frequencies of the polarization current are associated with the temporal period of the shock propagating through a single lattice unit of the crystal. The periodicity of the crystal lattice is the true origin of the coherent emission.

In this section we use an analytical approach to demonstrate the coherent nature of the emitted radiation. To represent the material, a polarizable element $P_n(t)$ that exists on each lattice point $n$ located at $x = na$ obeys the equation,

$$\frac{d^2 P_n(t)}{dt^2} =$$
$$\mu_n(t)E_n(t) - \Omega_n(t)^2 P_n(t) - f_n(t) - \gamma(t)\frac{dP_n(t)}{dt}. \quad (1)$$

Here, $\mu_n(t)$ is a polarizability-related parameter, $\Omega_n(t)$ is the resonant frequency of the $n^{\text{th}}$ polarizable element, and $\gamma(t)$ is an absorption term. The term $f_n(t)$ represents coupling to the shock wave and is a forcing term that generates shock-induced changes in polarization. $\Omega_n$ is the local transverse optical mode frequency ($\omega_T$) ranging from $10^{13}s^{-1}$ for phonons in ionic crystals to $10^{15}s^{-1}$ and higher for electronic excitations. Equation 1 can model many polarizable materials when combined with Maxwell's equations. In the case where $\Omega_n(t)$ and $\mu_n(t)$ are time-independent and $f_n = 0$, Equation 1 produces the usual polariton dispersion relation. [4]

To determine emission characteristics of this shocked polarizable material, we perform a symmetry analysis of the classical equations of motion of the system. There exists a time and space translational invariance of this system that gives rise to a Bloch-like property for the fields. Define a space and time translation operator $\hat{T}_m$ such that $\hat{T}_m g_n(t) \equiv g_{n-m}\left(t - m\frac{a}{v_s}\right)$. In the shock wave, suppose that the polarizable elements have the property that $\hat{T}_m \mu_n(t) = \mu_n(t)$, $\hat{T}_m \Omega_n(t) = \Omega_n(t)$, $\hat{T}_m f_n(t) = f_n(t)$, and $\hat{T}_m \gamma_n(t) = \gamma_n(t)$. Comparison of the fields in Equation 1 and Maxwell's equations with and without the application of $\hat{T}_m$ leads to the result that the electric field $E$ must be of the form,

$$E = \sum_k e^{ik(x-v_s t)} \sum_\ell E'_{k,\ell} e^{-2\pi i \ell \frac{v_s}{a} t} \quad (2)$$

where $\ell$ is an integer and $k$ is a wave vector and $H$ has a similar form. The Bloch-like property of the fields yields a condition on the radiation emitted by the shock wave. Possible frequencies in the fields in Eq. 2 are,

$$\omega_1 = k_1 v_s + 2\pi \ell \frac{v_s}{a} \quad (3)$$

where subscript 1 denotes the output radiation. Possible emission frequencies into the pre-shock and post-shock materials are those for which Eq. 3 and $\omega(k_1)$ for the pre-shock and post-shock materials have common solutions, respectively. The emission frequencies for $\ell \neq 0$ are highly anomalous since $\frac{2\pi}{a} >> k_1$. Since the lattice constant $a$ is typically several orders of magnitude smaller than the wavelength of optical light, these frequencies for $\ell \neq 0$ are larger than a typical Doppler shift ($k_1 v_s$) by several orders of magnitude. The confinement of the emitted radiation to discrete frequencies demonstrates the coherent nature in this model.

## MOLECULAR DYNAMICS SIMULATIONS

In this section, we numerically explore the light generated by a shocked polarizable material by performing molecular dynamics simulations of shock waves (see, for example, Ref. [5]) propagating through crystalline NaCl. Such commonly utilized simulations solve the classical equations of motion for atoms subject to an empirically-constructed interaction potential and incorporate thermal effects and deformation of the crystal lattice. In these calculations, planar shock waves are generated within 3D computational cells of perfectly crystalline atoms with cross section $17 \times 17$nm and length in the shock propagation direction ranging from 158nm to 235 nm (2-3 million atoms) at T=4.2K. Due to periodic boundary conditions in the directions transverse to the propagation direction, an infinitely planar shock propagates away from the constrained atoms into the the cell which is oriented along either the [111] or [100] directions of NaCl. We have performed simulations with computational cell cross-sections of up to $135 \times 135$ nm and obtained results in agreement with cross sections of $17 \times 17$ nm which indicates that the computational cell sizes are large enough that periodic boundary conditions do not play an artificial role in the behavior of the shocked material. Atomic interactions are treated using unit charge Coulomb interactions combined with Lennard-Jones interactions. These potentials are found to yield a lattice constant (5.64 Å) in agreement with the experimental value and to yield sound speeds that deviate from experimental measurements by 10-20%.

These molecular dynamics simulations do not explicitly solve Maxwell's equations for the electric and magnetic fields. However, since the wavelength of radiation emitted at frequencies considered here (longer than 10 $\mu m$) is much longer than the dimensions of the computational cell, it is expected that the total polarization current generated in the computational cell will be closely related to the generated electromagnetic radiation for frequencies above the phonon frequencies (above about 10 THz in this NaCl model) where the material has good transmission properties. The shock propagation direction component of the total electric polarization current is $J = \sum_i v_{z,i} q_i$ where $q_i$ is the charge and $v_{z,i}$ is the shock direction ($z$) component of the velocity of atom $i$. Emission characteristics are discussed in detail below.

Figure 1 shows results of about 30 ps duration simulations of shocks propagating in the [111] and [100] directions initiated with piston velocities of 200 m/s. This relatively small piston velocity generates a shock that applies a uniaxial strain of 0.03-0.04 to the post-shock material and increases the material temperature less than 1K. Figure 1 compares the shocked and unshocked Fourier transform of the shock propagation direction component of the total electric polarization current in the computational cell. Narrow peaks are observed in the shocked simulation that do not exist in the unshocked simulation. From the peak widths, the coherence length of the radiation emitted in vacuum is determined to be about 5mm and 3mm for the 16 THz, [111] peak and 22 THz, [100] peak, respectively. These lengths are comparable to those of some commonly used lasers. The coherence times are nearly Fourier transform-limited, suggesting that longer coherence lengths could be demonstrated by increasing the shock propagation time.

Equation 3 predicts that emission should occur in multiples of 5.4 THz in the [111] case since the periodic unit for the [111] direction in NaCl $a = 9.78$ and the shock speed observed in the simulation is $v_s = 5300 m/s$. The 16 THz and 32 THz peaks on the left of Figure 1 correspond to 3 and 6 times the fundamental frequency of 5.4 THz ($\ell = 3$ and $\ell = 6$ in Equation 3), in excellent agreement with theory (gray arrows). The 16 THz peak can be attributed

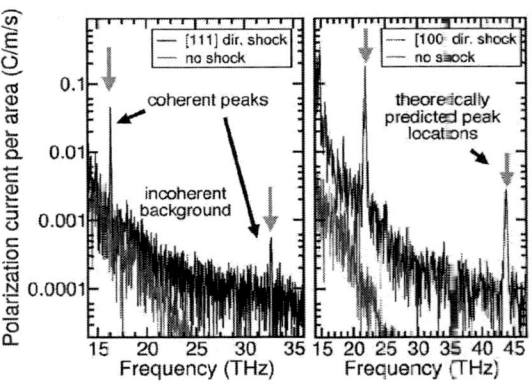

**FIGURE 1.** Fourier transform of the electric polarization surface current component in the shock propagation direction for molecular dynamics simulations of a shock propagating through NaCl in the [111] direction (left) and [100] direction (right). Narrow bandwidth, coherent peaks exist in the shocked simulations (black) that do not exist in the simulations without shocks. Gray arrows are emission frequencies predicted by Eq. 3. The coherence length for emission from the 16 THz peak in the [111] case is about 5 mm, comparable to that of some commonly used lasers. Thermal noise gives rise to an incoherent background.

to structure within the unit cell of distance $a = 3.26$ (i.e. $\ell = 1$ if $a = 3.26$ in Eq. 3) which is the distance between atomic lattice planes of like charge in the [111] direction (the NaCl crystal consists of alternating planes of positively and negatively charged atoms in the [111] direction.) Lattice planes are compressed as the shock propagates through the crystal, generating an alternating polarization current with a frequency associated with the rate at which the shock propagates through the lattice planes. If the shock speed is constant, the generated frequencies are constant and the coherence time of the emitted radiation is expected to be proportional to the time duration of the propagation of the shock wave. Equation 3 is also in good agreement with the [100] direction data where $a = 5.64$ and the observed shock speed is $v_s = 6200 m/s$ with peaks corresponding to the $\ell = 2$ and $\ell = 4$ cases. These peaks correspond to the distance between neighboring lattice planes in the crystal.

The power and spatial distribution of the emitted radiation can be simply estimated in the specific case where the shock front diameter is much less than the wavelength of the coherently emitted radiation. In this case, the propagating shock wave acts as a point source of radiation emitting a dipole-like distribution of radiation (peaked in the plane of the shock wave) with power,

$$p(\omega_0) = \frac{1}{4\pi\varepsilon_0} \frac{j_0^2 A^2 \omega_0^2}{3c^3} \qquad (4)$$

where $j_0$ is the polarization current per unit area on the shock front surface (plotted in Figure 1) and $A$ is the shock front area. For a $5\mu m$ diameter shock front (e.g., a laser-driven shock) and $j_0 = 0.18$C/s/m for the 22 THz peak on the right side of Figure 1, Equation 4 predicts the power coherently radiated at 22 THz is $3 \times 10^{-11}$ Watts while the shock propagates. Collecting and focusing this radiation can yield electric field amplitudes up to about 0.1 V/cm. We find that the power radiated can be increased by orders of magnitude when larger shock front areas and deviations from perfect planarity considered. [6]

The emission peaks studied in this work have frequencies that are up to two times higher than phonon frequencies. Observation of these peaks requires that there be spectral components of atom velocities at these higher frequencies around the shock front. Such high frequency components of atomic motion can be generated only if the shock front (or part of it) is very sharp. The natural shock front rise distances in the molecular dynamics simulations are several lattice planes. This effect represents a mechanism through which elastic shock wave rise times can potentially be experimentally measured; to our knowledge, such a measurement has never been made.

## CONCLUDING REMARKS

We have shown that a mechanical shock wave propagating through a crystalline polarizable material can produce coherent radiation. The key to this process is the periodicity of the crystal lattice. Suspected non-equilibrium (non-thermal) optical radiation in the visible regime has been observed from shocked dielectrics under some conditions, [7, 8, 9, 10] but this radiation appears to have a different origin than the coherent emission discussed here which occurs in a different part of the electromagnetic spectrum and results from the crystallinity of the lattice.

We note that it is well-known that coherent radiation at almost any frequency can be obtained from a *coherent* source using materials with a nonlinear optical response. This approach utilizes radiation from a coherent source to generate new coherent radiation. The coherent radiation mechanism presented in this work is fundamentally distinct from such nonlinear approaches in that coherence results from the crystal lattice rather than another coherent source. The optical nonlinearity mechanism works in amorphous materials while the shocked crystal mechanism does not.

## ACKNOWLEDGMENTS

We thank E. Ippen and F. Kaertner of MIT, J. Glownia, A. Taylor, and R. Averitt of LANL, and L. Fried and D. Hicks of LLNL for helpful discussions. This work was supported in part by the Materials Research Science and Engineering Center program of the National Science Foundation under Grant No. DMR-9400334. This work was performed in part under the auspices of the U.S. Department of Energy by University of California, Lawrence Livermore National Laboratory under Contract W-7405-Eng-48.

## REFERENCES

1. Sesseler, A. M., and Vaughan, D., *Am. Sci.*, **75**, 34 (1987).
2. Zaretsky, E., *J. App. Phys.*, **93**, 2496 (2003).
3. Muller, F., and Schulte, E., *Z. Naturforsch., A: Phys. Sci.*, **33**, 918 (1978).
4. Kittel, C., *Introduction to Solid State Physics*, John Wiley and Sons, New York, NY, 1996.
5. Bringa, E. M., Cazamias, J. U., Erhart, P., Stolken, J., Tanushev, N., Wirth, B. D., Rudd, R. E., and Caturla, M. J., *J. App. Phys.*, **96**, 3793 (2004).
6. Reed, E. J., Soljačić, M., and Joannopoulos, J. D., to be published.
7. Kromer, S. B., *Usp. Fiz. Nauk.*, **68**, 641 (1968).
8. Ahrens, T., Lyzenga, G., and Mitchell, A. C., "," in *High Pressure Research in Geophysics*, Manghnani Center for Academic Publication, Japan, 1982, p. 579.
9. Kondo, K., and Ahrens, T., *Phys. Chem. Miner.*, **9**, 173 (1983).
10. Schmitt, D., Svendsen, B., and Ahrens, T., "," in *Shock Waves in Condensed Matter*, Plenum Press, New York, 1986, p. 286.

CP845, *Shock Compression of Condensed Matter - 2005*,
edited by M. D. Furnish, M. Elert, T. P. Russell, and C. T. White
© 2006 American Institute of Physics 0-7354-0341-4/06/$23.00

# HYPERVELOCITY IMPACT FLASH
# AT 6, 11, AND 25 KM/S

## R. J. Lawrence, W. D. Reinhart, L. C. Chhabildas, and T. F. Thornhill[a]

*Sandia National Laboratories,[*] Albuquerque, NM  87185*
*[a]Ktech Corporation, Albuquerque, NM  87123*

**Abstract.** Impact-flash phenomenology has been known for decades, and is now being considered for missile-defense applications, in particular for remote engagement diagnostics. To technically establish this capability, we have conducted a series of experiments at impact velocities of ~6, ~11, and ~25 km/s. Two- and three-stage light-gas guns were used for the lower two velocities, and magnetically-driven flyers on the Sandia Z machine achieved the higher velocity. Spectrally- and temporally-resolved flash output addressed data reproducibility, material identification, and target configuration analysis. Usable data were obtained at visible and infrared wavelengths. Standard atomic spectral databases were used to identify strong lines from all principal materials used in the study. The data were unique to the individual materials over the wide range of velocities and conditions examined. The time-varying nature of the signals offered the potential for correlation of the measurements with various aspects of the target configuration. Integrating the records over wavelength helped to clarify those time variations.

**Keywords:** Impact flash, hypervelocity, spectroscopy, visible, infrared.
**PACS:** 42.72.-g, 78.47.+p

## INTRODUCTION

We have recently revisited impact-flash phenomenology by bringing into play our new capabilities for generating hypervelocity impacts at velocities up to several tens of kilometers per second. This particular effort, sponsored by an LDRD from our laboratories, has looked at the flash from impacts over a range of velocities from 6 to 25 km/s, and has used two different and unrelated experimental environments. With measurements of the time- and spectrally-resolved flash output, three complementary objectives were addressed: 1) data reproducibility; 2) material identification; and 3) target configuration analysis.

Specifically, the impact velocities examined were approximately 6, 11, and 25 km/s. The first two were accomplished with two- and three-stage light-gas guns, and the latter was achieved using magnetically-driven flyers on our Z machine. For the shots described here, flyer and target materials involved both aluminum and titanium. Early spectral measurements were in the visible region of the spectrum, ranging in wavelength from 350 to 800 nm, with later results extending from 800 to 1800 nm in the infrared. For the extremely high velocity shots on the Z machine the output from the spectrometers was recorded on streak cameras, and gave time durations for the signal of nearly one-half microsecond. For the gas-gun experiments optical multi-channel analyzers (OMAs) were used, and they

[*] Sandia is a multiprogram laboratory operated by Sandia Corporation, a Lockheed Martin Company, for the United States Department of Energy's National Nuclear Security Administration under Contract DE-AC04-94AL85000.

gave reading times of almost one millisecond, consisting of up to 200 individual data traces. However, for the latter experiments only about one-tenth of those data contained strong spectral signals related to the impact.

Using standard tabulations of atomic spectral data [1] we were able to identify strong spectral lines from all the principle materials used in the various shots. Using these results we demonstrated that the impact flash spectra were qualitatively reproducible from shot to shot with the same experimental techniques, and from facility to facility using different methods. Further, the signals were unique to the individual materials over the wide range of velocities examined, even though some velocity dependence was observed.

## MISSILE-DEFENSE APPLICATIONS

One of the major problems for missile-defense systems has always been the ability to assess and analyze the effects of an encounter between a defensive weapon system and an incoming threat object. Without this feedback an intercept has no provision for determining if a second attack is required, *i.e.*, an effective "shoot–look–shoot" scenario would have difficulties being implemented. Current missile-defense concepts envision a defensive system that uses a direct impact by a kill vehicle with sufficient kinetic energy to destroy or disable the threat. Closing velocities for these engagements will always be high—typically above the sound speeds of the materials involved—and as such they are considered hypervelocities. Because hypervelocity impacts always produce an *impact flash*, it is a reasonable assumption that this radiative signal contains information on the materials and time sequence involved in the interaction.

To give a rough idea of the sequence of events involved in a typical engagement scenario, we examined a series of multi-dimensional hydrocode calculations [2] of a hypothetical intercept event of about the right physical scale. If the initial impact occurs at zero-time, then the shock-wave and material responses of the outer target layers take place during the first several to roughly 10 μs. The aeroshell of the target is penetrated at a time, $t \approx 20$ μs. Most internal components of the target are mechanically loaded by $t \approx 200$ μs. The kill-vehicle

and target system is fragmented, and the resulting debris cloud begins to expand by $t \approx 500$ μs. Thus any temporal information regarding the materials, configuration, or structure of the target appears in the radiative output during this limited time. Current data from integrated flight tests generally extend over much longer durations, with typical 1-ms framing times. Thus for a full description of the important engagement phenomena we will need the development of higher speed instrumentation for the applied spectral sensors as well as their attendant platforms.

## GAS-GUN EXPERIMENTS

Over 30 impact-flash experiments have been shot on the gas guns. Some were developmental, but most produced usable spectral data. We will use several of those shots to provide illustrative examples of the various types of results we have obtained. The generic target configuration is shown in Fig. 1. The target is generally normal to the flyer flight path, or may be at an angle, as shown; it may also be multi-layer and include gaps. A witness plate may or may not be present. The spectrometers typically view the debris clouds from an appropriate angle.

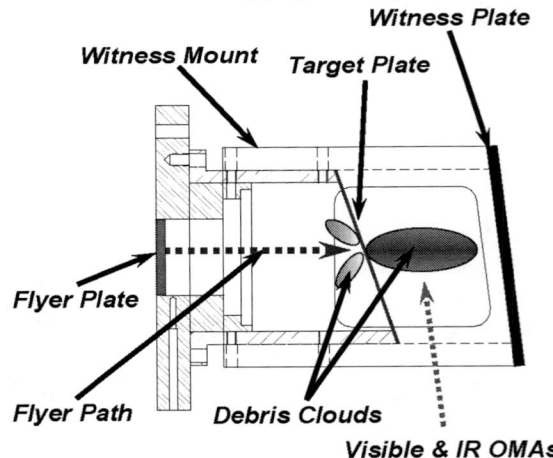

**FIGURE 1.** Schematic for a generic gas-gun experiment.

Our first successful shots involved the impact of aluminum spheres on aluminum targets at ~5.8 km/s, with visible spectrometers viewing the debris generated after the target was penetrated. The three records, as shown in Fig. 2, cover the visible spectrum, from λ = 350 nm to λ = 800 nm. Most of the lines can be identified with aluminum, in the neutral sate [Al (I)], singly ionized [Al (II)], or doubly ionized [Al (III)]. Even the Na (I) line at λ = 589 nm, probably from contamination, is prominent.

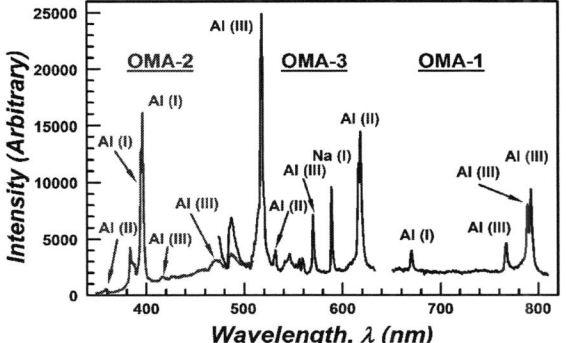

**FIGURE 2.** Visible spectra from shots OMA-1, -2, and -3.

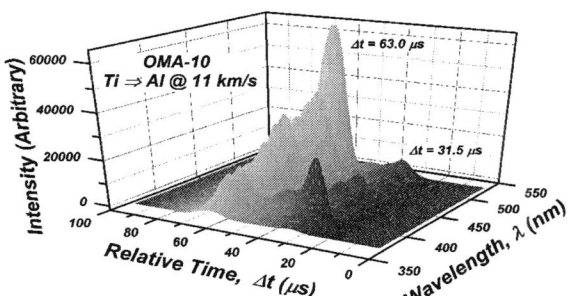

**FIGURE 3.** 3-D spectral plot from shot OMA-10.

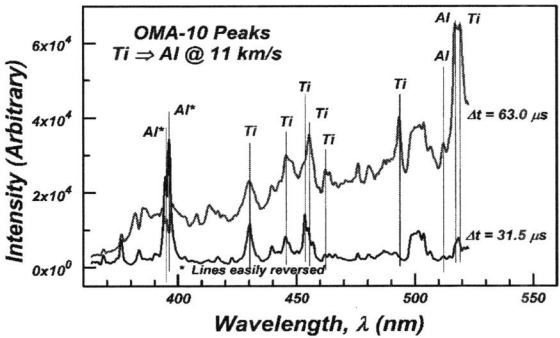

**FIGURE 4.** Spectra from the two time peaks of shot OMA-10.

Another shot, this time with a titanium plate impacting an aluminum target at 11 km/s, indicates the time-varying nature of the flash spectrum, as shown in the 3-D plot of Fig. 3. Details of the spectra

from each of the two temporal peaks, at relative times of 31.5 and 63 μs, are given in Fig. 4. Aluminum and titanium lines are present, but their amplitudes evolve substantially with time. In fact the "easily reversible" Al (I) lines at ~395 nm shift from emission to absorption in this higher energy environment. We can show the overall time evolution by integrating the individual data traces over the wavelength, and plotting the results as a function of relative time, as in Fig. 5. Although we haven't correlated the peaks with specific target features, integrated in-band intensities like this should provide data on the relevant configurations, especially for complex targets.

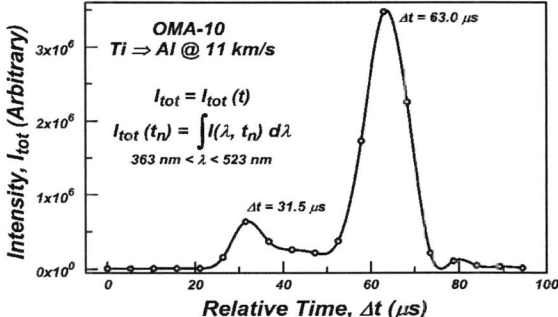

**FIGURE 5.** Integrated flash intensity for shot OMA-10.

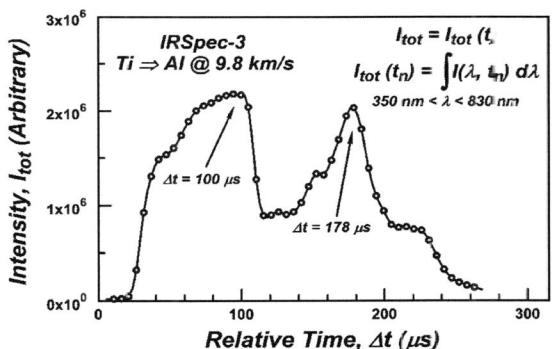

**FIGURE 6.** Integrated intensity (in visible) for shot IRSpec-3.

For a final example, we look at one of the early shots that obtained spectral data in both the visible and infrared, shot IRSpec-3. The integrated flash intensity, in Fig. 6, shows two peaks, but their shapes are quite different from those in Fig 5, and they extend for a longer time. It is interesting that the same time dependence does not appear in the infrared record. The visible spectra from these two peaks are plotted in Fig. 7. As with Fig. 4, they show lines for both aluminum and titanium, albeit over a broader wavelength range. Only those lines appearing in both records are indicated. To complete the picture Fig. 8 shows the measured infrared spectrum, with prominent Al (I) lines. In the latter figure we have superimposed the relevant information from the published aluminum database [1].

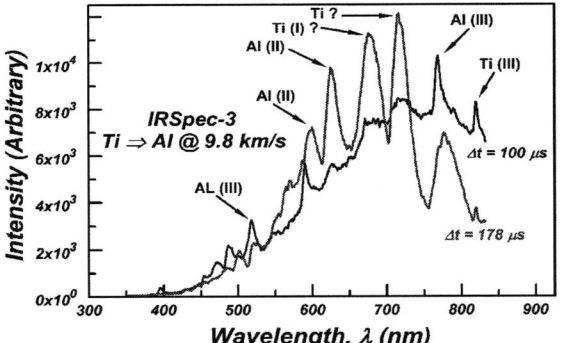

**FIGURE 7.** Visible spectra from shot IRSpec-3.

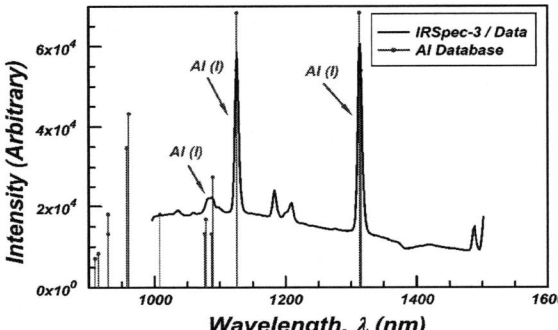

**FIGURE 8.** Infrared spectrum from shot IRSpec-3.

## EXPERIMENTS ON Z

Our initial experiments on impact-flash spectroscopy were actually conducted on the Z machine, mainly to take advantage of our local expertise in spectral measurement techniques. This work is complementary to our gas-gun efforts by looking at impact velocities greater by a factor of two, and total energies higher by four times. This system throws intact flyers, magnetically and shocklessly accelerating them to velocities of several 10s of km/s. We conducted two successful ride-along experiments on each of two different flyer-development shots conducted by Knudson [3].

To illustrate spectral line identification and data reproducibility under these different conditions, Fig. 9 shows the spectra obtained from two separate aluminum-on-aluminum impacts at ~25 km/s. Because of the greater internal energy, the lines are different from the lower velocity shots (*e.g.*, Fig. 2), but they are certainly reproducible from shot to shot.

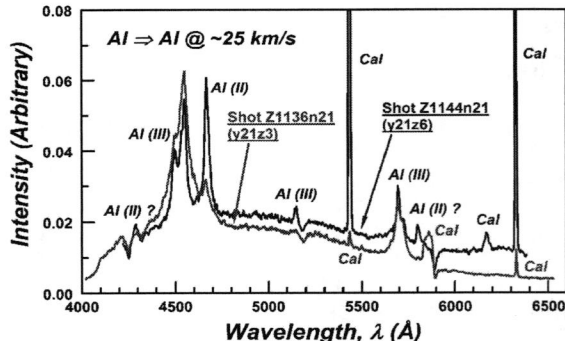

**FIGURE 9.** Spectra from shots on Z. The lines marked "Cal" are calibration signals for time and wavelength. (1 nm = 10 Å)

## CLOSURE

The objectives listed at the outset have been achieved, that is, we have demonstrated material identification, data reproducibility, and potential target configuration analysis. More complete listings of the data, along with more extensive data reduction, are available elsewhere [4]. Future work will include testing of more materials and target configurations, especially those of operational interest.

## REFERENCES

1. Reader, J., and C. H. Corliss, Eds., "NIST Spectroscopic Properties of Atoms and Atomic Ions Database," *NIST Standard Reference Database 38*, National Institute of Standards and Technology, Gaithersburg, MD, 1992.
2. Dukart, R. J., Private Communication, Sandia National Laboratories, Albuquerque, NM, 2005.
3. Knudson, M. D. *et al.*, "Near-absolute Hugoniot measurements in aluminum to 500 GPa using a magnetically accelerated flyer plate technique," *J. Appl. Phys.*, **94**(7), 4420-4431, 2003.
4. Lawrence, R. J. *et al.*, *Hypervelocity Impact Flash for Missile-Defense Kill Assessment and Engagement Analysis*, Sandia National Laboratories, Albuquerque, NM (to be published).

CP845, *Shock Compression of Condensed Matter - 2005,*
edited by M. D. Furnish, M. Elert, T. P. Russell, and C. T. White
© 2006 American Institute of Physics 0-7354-0341-4/06/$23.00

# SUITABILITY OF MAGNESIUM OXIDE AS A VISAR WINDOW

## G. D. Stevens[1], L. R. Veeser[1], P. A. Rigg[2], R. S. Hixson[2]

[1]*Bechtel Nevada, Special Technologies Laboratory, Santa Barbara, CA 93111*
[2]*Los Alamos National Laboratory, Los Alamos, NM 87545*

**Abstract.** Impedance matching of a velocity interferometer for any reflector (VISAR) window to a material under study helps simplify a shock experiment by effectively allowing one to measure an *in situ* particle velocity. The shock impedance of magnesium oxide (MgO) falls roughly midway between those of sapphire and LiF, two of the most frequently used VISAR window materials. A series of symmetric impact experiments was performed to characterize the suitability of single crystal, [100] oriented magnesium oxide as a VISAR window material. These experiments yielded good results and showed the viability of MgO as a VISAR window up to 23 GPa. Results were used to determine window correction factors and, subsequently, to estimate the pressure-induced change in index of refraction. In many of the shots in this work we exceeded the Hugoniot elastic limit (HEL) of MgO, and both elastic and plastic waves are evident in the velocity profiles. The presence of both waves within the VISAR window complicates the typical VISAR window correction analysis. Preliminary analysis of the elastic and plastic contributions to the window correction is presented.

**Keywords:** Interferometer, magnesium oxide, shock wave, VISAR
**PACS:** 47.40.Nm , 62.50.+p, 78.20.Ci, 42.62.Eh

## INTRODUCTION

We have performed a study of shock effects on single-crystal MgO to characterize it for possible use as a window in shock-wave experiments. A transparent window on the back of a shocked sample can allow measurements of its properties, such as wave profiles, with greatly reduced shock reflection and unloading effects if the window impedance is closely matched to that of the sample. The shock-wave impedance (initial density times shock velocity) of MgO is roughly midway between those of sapphire and LiF, two of the most important window materials. Consequently, if its other properties are suitable, MgO also could be a very useful window because its shock impedance more closely matches some sample materials.

To make a good interferometer window, a material must remain transparent up to at least the shock pressures transmitted by the sample being studied. The window material should not undergo phase changes within the pressure range of interest, it should not fracture, and it should be chemically

and physically stable and easy to handle. Also, it is important to understand how its refractive index behaves when the crystal is compressed by a shock because if the refractive index changes, the wavelength of the light in the crystal also changes, thereby affecting the VISAR signal.

In this experiment we used a VISAR [1] to measure material velocities. In a VISAR experiment laser light is reflected from a moving surface. A window between the sample and the light collection system changes the properties of the Doppler-shifted, reflected light [2]. A shock in the window can change its refractive index, complicating the data interpretation. This change, which must be accounted for, is described either as a change in the apparent velocity, $\Delta u = u_a - u_0$, or a multiplicative factor, $u_a/u_0 = 1 + \Delta v/v_0$. Here $u_a$ is the apparent velocity of the sample-window interface measured by the VISAR; $u_0$ is its actual velocity, the particle velocity at the interface; $v_0$ is the Doppler-shifted frequency of light scattered from the moving interface; and $\Delta v$ represents the change in frequency of the reflected light because of the shock in the window.

In this paper we report measurements of apparent window velocities from shock waves in MgO windows. For all of our measurements a VISAR measured the velocity of the interface between a MgO sample and a MgO window following impact of a MgO flyer onto the sample. The technique is similar to that of Jones, *et al.* [3]. For a symmetric-impact experiment like this, the particle velocity behind the shock wave in the sample is exactly half that of the flyer at impact. The flyer velocity and tilt are measured with a set of about 12 electrical shorting pins around the sample. The ratio of the interface velocity measured with the VISAR to the particle velocity deduced from the pins is the window correction factor, $u_a/u_0$, described in the preceding paragraph.

## EXPERIMENTAL PROCEDURE

Five symmetric-impact MgO gun experiments, where the window was also of MgO, were performed, and are summarized in Table 1. The two lower-pressure measurements were made at the Los Alamos National Laboratory TA-39 Popgun, and the three higher-pressure measurements were made using the powder gun at TA-40. Having a wide range of gun velocities available allowed us to avoid using higher-impedance flyers or samples to reach the highest pressures. By using MgO for flyer, sample, and window, we avoid mixing effects in MgO with those of other materials and thus make the experiment easier to analyze and understand.

The MgO flyers, samples, and windows were single crystal material oriented along the [100] direction. The samples were obtained from MTI

Corporation, and were 99.95% pure with average densities of 3.58 g/cm$^2$. The interface between the sample and the window had a thin reflective aluminum layer to reflect light from a frequency-doubled YAG laser (532 nm wavelength) for velocity measurement in a pair of VISARs. The VISARs measured the apparent interface velocity to between 1 and 2%, where the true velocity, $u_0$, is equal to exactly one-half of the impact velocity, $u_{sabot}$. For all five, shots, the flyer and samples were 3 mm thick. The high-pressure powder-gun shots (a - c) had 12-mm thick MgO windows, and the two gas-gun shots (d - e) had 4.5-mm thick windows.

## RESULTS

The results from the gun experiments are shown in Fig. 1. Calculated peak stress for each shot is given in Table 1. Only one of the two VISAR measurements is presented for each curve, since agreement between the two measured curves is very good. Curves a), b), and c) are for the powder-gun experiments. Each exhibited a small, but well-defined, elastic precursor followed by a jump to a maximum apparent velocity. When the rarefaction from the back of the flyer arrives at the sample-window interface, it slows the interface. At these high stresses the reflected precursor and the reflected main shock are partially merged by this time, so the rarefaction curves have lost most of the distinction between the precursor and the main shock, and the velocity decreases slowly at first. The final velocity is somewhat above zero because the foam backing on the flyer partially reflects the shock. Again, the maximum apparent velocity is larger than the particle velocity because of the shock in the window.

**TABLE 1.** Shot parameters and results. $u_0$ is one half of the impactor velocity $u_{sabot}$. $u_{ae}$ is the apparent particle velocity induced by the elastic precursor, and $u_{ap}$ is the apparent particle velocity due to the plastic wave. $D_e$ and $D_p$ are, respectively, the average elastic wave velocity and the calculated plastic shock-wave velocities. $n_e$ and $n_p$ are the indices of refraction of the regions behind the elastic and plastic waves, respectively. $\Delta u$ (meas.)= $u_{ap}$ (meas.) - $u_0$.

| Expt. | $u_0$ | Pk. Stress (GPa) | $u_{ae}$ (meas.) | $D_e$ (km/s) | $n_e$ | $u_{ap}$ (meas.) | $D_p$ (km/s) | $n_p$ | $\Delta u$ (meas.) | $\Delta u$ (calc.) |
|---|---|---|---|---|---|---|---|---|---|---|
| a) | 0.824 | 22.9 | 0.285 | 9.34 | 1.745 | 1.616 | 7.754 | 1.714 | 0.792 | 0.799 |
| b) | 0.727 | 19.9 | 0.220 | 9.34 | 1.745 | 1.440 | 7.619 | 1.717 | 0.713 | 0.704 |
| c) | 0.618 | 16.6 | 0.328 | 9.34 | 1.745 | 1.215 | 7.477 | 1.721 | 0.597 | 0.597 |
| d) | 0.3 | ~7.6 | 0.210 | 9.34 | 1.745 | 0.58 | 7.045 | 1.732 | 0.28 | 0.286 |
| e) | 0.1096 | 2.7 | 0.172 | 9.34 | 1.744 | - | - | - | 0.062 | 0.062 |

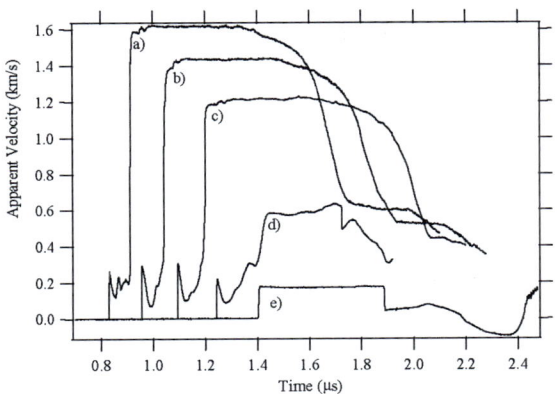

**Figure 1.** Apparent velocities of the sample-window interface for all five gun experiments. Impact velocities and relevant shot parameters are in Table 1. The apparent velocities shown have not been corrected for the effects of the shock in the window. Relative timing of the five traces has been adjusted for clarity.

Results for the two low-velocity gas gun shots are also shown in Fig. 1 (traces d and e). The curves are the apparent velocities measured by the VISARs. Experiment d) was slightly above the elastic limit and e) was below. Both had relatively thin, 4.5-mm windows to allow the elastic shocks to pass completely through them before the rarefactions from the foam backing behind the flyer reached the sample-window interfaces viewed by the VISARs. For experiments d) and e), the apparent elastic unloading starts at around 1.7 μs and 1.9 μs, respectively. The changes in VISAR signals at these times are not caused by an actual interface-velocity change but by the shock reaching the back of the window. When the elastic shock unloads, its reflection reduces the stress and increases the particle velocity in the window. This change causes the window correction, $\Delta u = u_a - u_0$, to decrease [3], and we see an apparent velocity drop. The decrease in $u_a$ is nearly as large as the window correction at early times, and the resulting window correction is approximately the negative of what it was previously. Knowing the magnitude of the window-correction change gives a useful check on the measurement of its value and confirms that the window correction for shot e) is not the same as the correction at pressures above the HEL.

## ANALYSIS

For pressures exceeding the HEL, MgO exhibits a clear two-wave structure. In LiF the HEL is only about 0.2 GPa, so experiments are often overdriven and exhibit only a plastic wave structure. For MgO, the HEL is much higher. Duffy and Ahrens [4] report a value of 1.6 GPa; our shot e) at 2.7 GPA shows no sign of a plastic wave. Because of the high HEL, a strong elastic precursor is present in all of our gun-based, symmetric-impact experiments, which extend up to 23 GPa. To model our complete data set adequately, we found it necessary to treat the elastic- and plastic-wave-induced index changes separately, using a different index of refraction relation for each wave. Both indices were assumed to be linear with density.

If $n_0$ is the index of refraction of a window at 532 nm and $L$ is the window thickness, then a two-wave, elastic-plastic loading structure traveling from the left with velocities $D_e$ and $D_p$ produces an optical path length, Z, which changes with time t (for the case of direct impact on the window) as,

$$Z = n_0(L - D_e t) + n_e(D_e t - D_p t) + n_p(D_p t - u_0 t) \quad (1)$$

The index of refraction of the window has three distinct values in such a situation, $n_0$, $n_e$, and $n_p$. For a symmetric impact experiment where the flyer, sample, and window are of the same material, the particle velocity, $u_0$, behind the main shock is half the impact velocity. The apparent velocity, $u_a$, of the interface (the left edge of the window) for this two-wave structure is

$$u_a = -dZ/dt = (n_0 - n_e)D_e + (n_e - n_p)D_p + n_p u_0 \quad (2)$$

The actual interface velocity is $u_0$, and the difference $\Delta u_p = u_a - u_0$ is:

$$\Delta u_p = (n_0 - n_e)D_e + (n_e - n_p)D_p + (n_p - 1)u_0 \quad (3)$$

For the index of refraction we assume a density dependence of the form $n = A + B\rho$, where $B = (n_0 - A)/\rho_0$ and $\rho_0$ is the density of the un-shocked window. We allow for different values of

$A$ for elastic and plastic waves. Then the indices are, respectively,

$$n_e = A_e + (n_0 - A_e)\rho_e / \rho_0,$$
$$n_p = A_p + (n_o - A_p)\rho_p / \rho_0. \tag{4}$$

Neglecting the effects of shear components of stress on the compressed volume, the densities in the two regions behind the elastic and plastic waves are given by

$$\rho_e = \rho_0 D_e /(D_e - u_{0e}),$$
$$\rho_p = \rho_0 D_p /(D_p - u_0). \tag{5}$$

Here $u_{0e}$ is the particle velocity behind the elastic precursor. For a simple elastic wave with no plastic wave, $A_e = u_a/u_{0e}$. We only performed one experiment below the HEL, the low velocity shot e), which gives $A_e = 0.172 / 0.1096 = 1.57$. Using this value we calculate $u_{0e}$ and $n_e$ for the elastic precursors seen in the higher pressure shots (see Table 1).

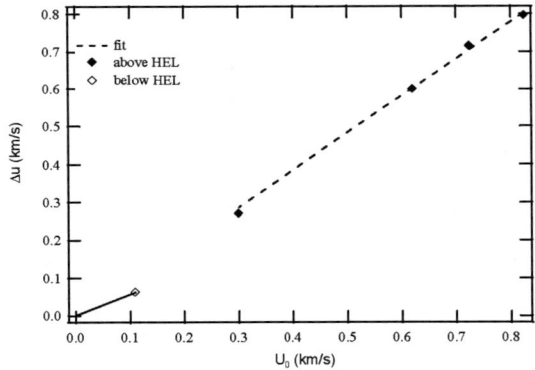

**Figure 2.** MgO window-correction measurements. The data above the HEL were fit using $A_e = 1.57$, and $D_e = 9.34$, giving $A_p = 1.976(1)$. The point at $u_0 = 0.3$ (experiment d) has a large velocity uncertainty and was weighted accordingly.

We obtained $D_p$ for each shot from the recalculated Hugoniot fit given by $D_p = 6.64 + 1.35 u_0$ [5]. The $D_e$-values are obtained from the two gas-gun experiments, which both gave $D_e = 9.34$ km/s. Due to scatter in the data, $u_{ae}$ was chosen as the average apparent particle velocity induced by

the elastic precursor, $u_{ae} = 0.261$. Using these values and Equations (4) and (5), we determined $A_p$ by fitting it to the experimentally-determined $\Delta u_p$-values for each shot. This fit gives $A_p = 1.976(1)$. Table 1 lists the results of the fit, and a graph of the fit is shown in Fig. 2.

## CONCLUSIONS

We have measured the window correction for single-crystal MgO at stresses up to 23 GPa. The correction factor appears to be approximately constant with particle velocity except for a single point below the elastic limit. MgO remains transparent at these pressures, and may be used effectively on gas-gun experiments.

## ACKNOWLEDGEMENTS

This manuscript (DOE/NV/11718-1050) has been authored by Bechtel Nevada under Contract No. DE-AC08-96NV11718 with the U.S. Department of Energy. The U.S. Government retains and the publisher, by accepting the article for publication, acknowledges that the U.S. Government retains a non-exclusive, paid-up, irrevocable, world-wide license to publish or reproduce the published form of this manuscript, or allow others to do so, for U.S. Government purposes.

The authors thank the following colleagues who helped execute and interpret the experiments: Bill Anderson, John Vorthman, Dennis Hayes, Dennis Shampine, Jim Esparza, and Mark Byers.

## REFERENCES

1. Barker, L. M., and Hollenbach, R. E., "Laser Interferometer for Measuring High Velocities of Any Reflective Surface," J. Appl. Phys **43**, 4669 (1972).
2. Barker, L. M., and Hollenbach, R. E., "Shock Wave Studies of PMMA, Fused Silica, and Sapphire," J. Appl. Phys **41**, 4208 (1970).
3. Jones, S. C., and Gupta, Y. M., "Refractive Index and Elastic Properties of Z-Cut Quartz Shocked to 60 kbar," J. Appl. Phys. **88**, 5671 (2000)
4. Duffy, T. S., and Ahrens, T. J. "Compressional Sound Velocity, Equation of State, and Constitutive Response of Shock-Compressed Magnesium Oxide," J. Geophys. Res. **100**, 529 (1995).
5. Marsh, S.P., ed., LASL Shock Hugoniot Data, UC Press, Berkeley, pp. 312-313 (1980).

CP845, *Shock Compression of Condensed Matter - 2005*,
edited by M. D. Furnish, M. Elert, T. P. Russell, and C. T. White
2006 American Institute of Physics 0-7354-0341-4/06/$23.00

# OPTICAL ABSORPTION MEASUREMENTS OF RDX

## Von H. Whitley

*Naval Surface Warfare Center-Indian Head, Indian Head, MD 20640*

**Abstract.** Optical absorption measurements on a variety of RDX samples in the 1.38-6.2 eV (200-900nm) wavelength range were undertaken to determine the sample-to-sample variations present in as-received samples. Samples were also annealed at elevated temperatures and subjected to pressures as high as 20 GPa to determine the effects of temperature and pressure on the optical absorption spectra. A sample-to-sample variation in the optical absorption of as-received samples was found, with the onset of strong absorption occurring at between 3.6eV (345nm) to 4.1eV (280nm). Subjecting RDX to pressures of 0.3-16GPa lowered the onset of strong absorption at 4.1 eV and produced a weak absorption throughout the visible region. Annealing RDX at temperatures of 383 K increased the measured absorption throughout the UV and visible peaking at ~3.4 eV (360nm).

**Keywords:** Optical absorption, RDX.
**PACS:** 71.55.Ht, 82.50.Hp

## INTRODUCTION

RDX has a well-known sensitivity variation to shock and impact initiation. The reason for this sensitivity variation is still debated, but it has been shown to be related to density[1], crystal morphology[2], particle size[3], and void concentration[4,5]. All of these studies indicate that the crystalline quality of the RDX material itself plays a role in determining the sensitivity of the material. There is interest in both high-quality and low-quality RDX to determine the causes of this sensitivity variation.

Few methods currently exist to determine the concentration of vacancies, dislocations, or other imperfections, present in the bulk of the crystal. Optical absorption has long been used to characterize the crystalline quality of solid-state materials. Since the technique can be sensitive to both impurities and to structural defects in crystals, it could provide useful information about the crystalline quality of RDX. It also can be measured at elevated pressures and temperatures providing information about RDX at near detonation regions. Therefore, work has been undertaken to determine

if optical absorption can reveal information on the crystalline quality of RDX.

The approach of this work was to determine: (1) the consistency of optical absorption features of a variety of large single crystal samples and powdered samples of RDX grown by different sources; (2) the influence of pressure and temperature on the measured optical absorption of RDX.

## EXPERIMENTAL PROCEDURE

Large single-crystal samples were obtained from Los Alamos National Lab (LANL), and from Naval Surface Warfare Center-Indian Head (NSWC-IH). The samples were nominally 10mm x 10mm x 3mm in size. Five large single crystals were selected for this work, samples RDX1, RDX2, and RDX3 were grown at LANL. Sample RDX4 and RDX5 was grown at NSWC-IH. For powdered samples, RDX powder supplied by Holston or produced at Eurenco was used. Finally, single crystals of RDX were grown at elevated pressures in a Merrill-Bassett style diamond anvil cell (DAC) using type II diamonds. To grow RDX crystals in a DAC, a 360 mm hole in a stainless steel gasket

(100 μm thick) was injected with a saturated RDX-acetonitrile solution. The pressure was slowly increased, causing a single RDX seed to appear. Slowly increasing the pressure to 0.3 GPa grew the seed into a single RDX crystal that extended from one edge of the gasket to the other. The pressure was calibrated using a ~20 μm ruby sphere placed in with the RDX.

Optical absorption measurements were taken using a Varian Cary 100 dual beam spectrophotometer over the wavelength range of 1.38-6.2 eV (200-900 nm). The optical absorption of the solid samples was measured by placing the sample in one of the beams. To measure the optical absorption of powdered samples, the powder was placed in a 0.5 mm thick square quartz cuvette. This cuvette, along with a reference cuvette, were placed in the two beams of the spectrophotometer. To compensate for scatter off of the RDX grains, a silicon photodetector was used to measure the scattered light. It was placed ~ 30° off-axis from the beam path. The on-axis light power was also measured using the same photodiode and used to correct the scattered light signal for the wavelength dependent intensity variation of the spectrophotometer. This corrected scattering was then used to subtract off the contribution to grain scattering from the absorption measurements. The contribution from scattering was found to be an order of magnitude lower than the measured absorption features in this work. To measure the optical absorption spectra of the RDX in a DAC, the spectrophotometer was modified to hold the DAC. The pressurized RDX was placed in one beam, and a second type II empty DAC was placed in the reference beam. The pressure-induced absorption changes of diamond/acetonitrile were also measured over the wavelength of interest to ensure that the absorptions measured were only due to RDX. Elevated pressure did not induce any significant absorptions in the diamond/acetonitrile over the spectral window studied in this work.

## DATA

Figure 1 shows the absorption of five different large single crystal RDX samples that were measured over the range of 1.5-4.5 eV (200-900nm) using a spectrophotometer. The samples all exhibit a strong absorption onset occurring between 3.5-4.1 eV. The actual onset of this strong absorption, however, varies by ~0.5 eV in the sample set evaluated here. A much weaker absorption tail is also seen, spanning from ~1.7 eV-3.6 eV. The amount of absorption in this tail also varies among the samples.

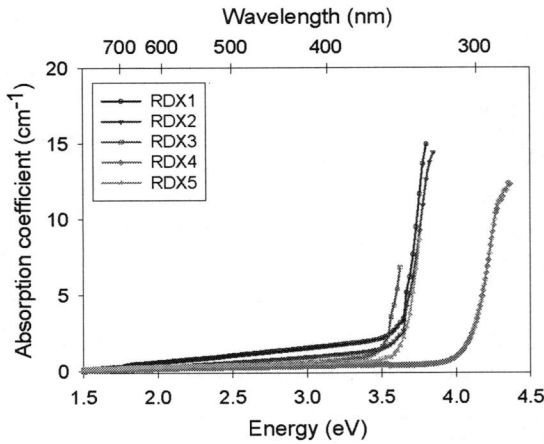

**Figure 1:** Absorption from as-received RDX single crystals.

Optical absorption measurements from two different lots of explosive-grade RDX powder are shown in Figure 2. The strong absorption onset at ~3.6 eV is present in both of these samples. In addition, there is significant absorption form 3.0-3.5 eV present in the Holston sample that is not present in the Eurenco sample.

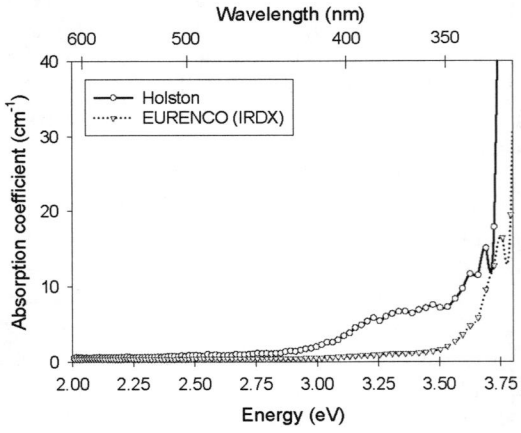

**Figure 2:** Optical absorption of RDX powder.

It had been found in lab experiments that annealing RDX powders at 383 K for a period of time would cause the powder to yellow. To determine how this yellowing affected the optical absorption, sample RDX 4 was annealed at 383 K for times ranging from 24-300 hours. It was then removed and the absorption features were measured at room temperature. The change in absorbance for RDX 4 is shown in Figure 3. The data shown have had the non-annealed absorption subtracted resulting in absorption change. When annealed at 383K, the optical absorption increases from 1.7 eV-3.6 eV, peaking at roughly 3.45 eV (360 nm). This resulted in a visual yellowing of the RDX. As the anneal time increased, the amount of absorption increased throughout the 1.7-3.6 eV range. The amount of anneal time necessary to produce significant absorption changes, however, was sample dependant. Holston powders turned yellow within 24 hours, whereas Eurenco powders took 300 hours or longer to turn yellow. In the solid sample RDX 4, regions of the crystal turned yellow at a faster rate than other regions.

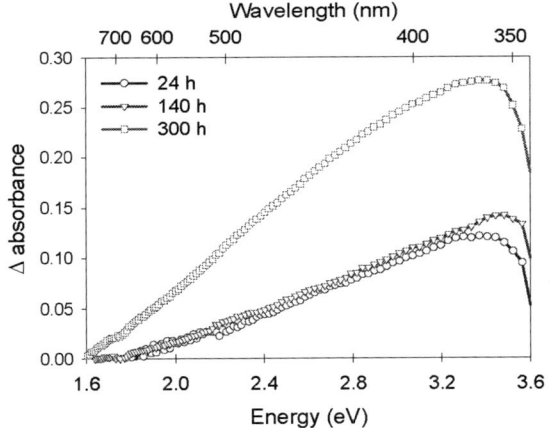

**Figure 3:** Change in absorption of RDX4 due to annealing at 383 K.

The growth of RDX in a higher pressure environment created by a centrifuge has been shown to increase the density of the crystal.[6] The authors reported that this elevated pressure helps remove inclusions in the material, resulting in a higher-quality crystal structure. In analogous fashion, RDX was grown at elevated pressures in a DAC at NSWC-IH to compare with previous observations. The optical absorptions of this material as a function of pressure are shown in Figure 4. The onset of strong absorption of RDX grown at 0.3 GPa began at roughly 4.13 eV (300 nm). Increasing the pressure lowers the onset of strong absorption by ~0.4 eV at the highest pressure of 16 GPa. Additionally, a broad absorption in the 1.7-3.5 eV is also seen that increases with pressure. All of the absorption changes due to pressure in this experiment were reversible. Upon depressurization, the absorption in the 1.7-3.5 eV range disappeared, and the onset of strong absorption increased back to 4.1 eV. It should be noted that while the RDX grown at 0.3 GPa was a single crystal, shear introduced at higher pressures could cause the sample to convert to polycrystalline.

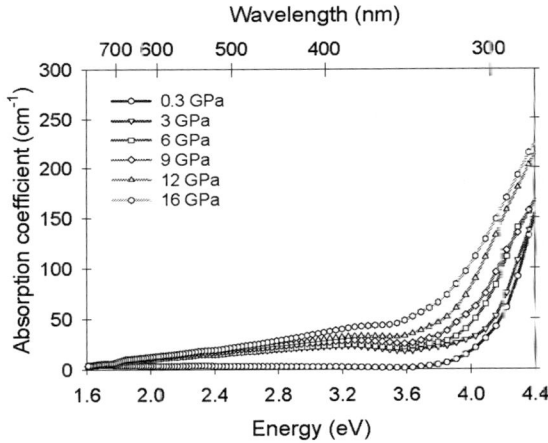

**Figure 4:** Optical absorption of RDX grown in diamond anvil cell as a function of pressure.

## ANALYSIS

Based on the data obtained, the absorption found in these samples of RDX can be broadly categorized into 2 regions: (1) From 3.6-4.1 eV (300-345 nm) a strong absorption feature was found. (2) From 1.77-3.54 eV (350-700nm) a weaker absorption tail was found that increased with anneal time at 383 K or increased with pressure.

It is not currently known what causes these absorption features. The absorption of the RDX molecule dissolved in a solution shows two strong peaks at ~ 5.3 eV and 6.2 eV.[7-9] whereas predicted values for the bandgap of the solid state of RDX range from 3.59 eV to 5.9 eV.[10,11] Work by Kukla and Kunz[12] has shown defects of the crystal structure, such as edge dislocations, could lower the bandgap of RDX. It is thus possible that the strong absorption feature is the onset of the band gap absorption and the variation due to different concentrations of structural anomalies.

A relatively weaker absorption was found from 1.77-3.54 eV, peaking at 3.4 eV. This absorption region increased with temperature and pressure. Optical stimulation into the maximum of this absorption (~3.4 eV) resulted in a spectrally broad fluorescence from RDX peaking at ~ 525nm. The lifetime of the emission was measured to be between 1-2 ns. This emission is similar to that reported by Marinkas[8].

## CONCLUSIONS

Sample-to-sample absorption variations were found in as-received material over a wide optical energy range. Major differences were observed to occur in 2 different regions. 3.6-4.1eV and 1.77-3.54 eV and each have different responses to pressure and temperature.

The absorption region in the 3.6-4.1 eV region was present in every sample studied. Subjecting RDX to pressure caused the onset of this absorption to decrease in energy. Thermal annealing had little affect on the absorption in this region.

Weak absorption from 1.7-3.54 eV was found in the bulk of the RDX studied. This absorption feature increases when the sample was annealed at 383 K or when it is subjected to uniaxial pressures. The absorption changes due to thermal annealing are both sample dependent and permanent. Pressures modified the absorption of RDX in the same optical range as temperatures, but these changes were reversible.

Optical absorption has been demonstrated to be a useful technique for identifying and understanding the submicroscopic defects and impurities present in RDX. With more work, it will undoubtedly fill a niche in determining the quality of RDX. Future work will focus identifying what defects or impurities are responsible for the various anomalous absorption features and relating these features to material sensitivity.

## ACKNOWLEDGMENTS

Funding was provided via ONR/NAVSEA through the the ILIR program. The author would like to thank Dan Hooks (LANL) and Stan Caulder (NSWC-IH) for providing large single crystal samples of RDX.

## BIBLIOGRAPHY

1. Verbeek, R., van der Steen, A., and de Jong, E. in *Tenth International Detonation Symposium*, Boston, Massachusetts, 1993 (ONR), p. 685-689.
2. van der Steen, A.C., Verbeek, H.J. and. Meulenbrugge, J.J., in *Ninth Symposium (International) on Detonation*, Portland, OR, 1989 (ONR), p. 83-88.
3. Moulard, H., Kury, J.W., and Dleclos, A., in *Eighth Symposium (International) on Detonation*, Albuquerque, NM, 1985 (ONR), p. 902-913.
4. Baillou, F., Dartyge, J.M., Spyckerelle, C., and Mala, J., in *Tenth International Detonation Symposium*, Boston, MA, 1993 (ONR), p. 816-823.
5. Borne, L., in *Tenth International Detonation Symposium*, Boston, MA, 1993 (ONR), p. 286-293.
6. Lanzerotti, M.Y.D., Autera, J., Borne, L., and Sharma, J., in *Materials Research Society Symposium Proceedings*, Boston, MA, Nov 1995, 1996 (MRS, Pittsburg Pennsylvania), p. 73-78.
7. Marinkas, P.L., Mapes, J.E., Downs, D.S., Kemmey, P.J., and Forsyth, A.C., Mol. Cryst. Liq. Cryst. **35**, 15-25 (1976).
8. Marinkas, P.L., Journal of Luminescence **15**, 57-67 (1977).
9. Orloff, M.K., Mullen, P.A., and Rauch, F.C., Journal of Physical Chemistry **74**, 2189-2192 (1970).
10. Kuklja, M.M., Aduev, B.P., Aluker, E.D., Krasheninin, V.I., Krechetov, A.G., and Mitrofanov, A.Y., Journal of Applied Physics **89**, 4156-4166 (2001).
11. Perger, W.F. Chemical Physics Letters **368**, 319-323 (2003).
12. Kukla, M.M., and Kunz, A.B., Journal of Applied Physics **89**, 4962-4970 (2001).

# IMPACT PHENOMENA, BALLISTICS, HYPERVELOCITY STUDIES, AND EXOTIC SHOCK CONFIGURATIONS

CP845, *Shock Compression of Condensed Matter - 2005*,
edited by M. D. Furnish, M. Elert, T. P. Russell, and C. T. White
© 2006 American Institute of Physics 0-7354-0341-4/06/$23.00

# NUMERICAL SIMULATION AND EXPERIMENTAL STUDY OF COATING STABILIZING EFFECT ON SHEAR INSTABILITY GROWTH IN OBLIQUE IMPACT OF METAL SLABS

## S.M. Bakhralh, N.A. Volodina, O.B. Drennov, T.A. Goreva, A.L. Mikhailov, P.N. Nizovtsev, V.F. Spiridonov, E.V. Shuvalova

*Russian Federal Nuclear Center - All-Russian Scientific Research Institute of Experimental Physics (RFNC-VNIIEF), 37, Mira av., Sarov, 607190, Russia*

**Abstract.** Results of numerical simulations of the interface coating influence on the shear instability growth in oblique collision of metal slabs are presented. The state of materials was described in the elastic-plastic approximation. The dependence of the perturbation amplitude on the coating thickness and behavior were studied. The computations were conducted on a multiprocessor distributed-memory computer.
**Keyword:** Oblique impact, shear instability, numerical simulation, method of concentrations.
**PACS:** 62.50.+p, 62.20.Fe

## INTRODUCTION

When certain conditions in impact angles $\gamma$ and contact point velocity $\upsilon_K$ are met, an oblique impact of metal layers is attended with wave formation at the interface: Kelvin-Helmholtz instability develops at the interface [1,2]. These issues were discussed in detail at Conference Parallel CFD 2003 [3].

It should be noted that the Kelvin-Helmholtz instability process similar to that in explosive welding can adversely affect performance of different systems. Therefore, experimental studies have been performed to search for ways to suppress the instability by applying special coating on the layer surface [4].

## EXPERIMENTAL STUDY

The effect of a coating on perturbation growth was studied in an experimental series on oblique impact of plane metal slabs. The scheme of slab launching by sliding detonation wave was used.

The experimental setup is discussed in ref. [4], which for the first time observed that thin metal coating layers could suppress the perturbation growth in the oblique impact of metal slabs.

Let us describe experiments with aluminum samples presented in [4]. The loading conditions ($\upsilon_K$=4km/s, $\gamma$=12°, loading pressure P=6.5GPa in the vicinity of the contact point) corresponded to the growth of wavelike perturbations with parameters $\alpha \approx 440\mu m$; $\lambda \approx 500\mu m$ in control experiments. Perturbations due to roughness of ~ 10µm inherent in slab machining (polishing) were present originally on the slab.

Using thin layers of covers (10µm $\leq \Delta \leq$ 40µm) made of different metals (as low-melting like lead as high-melting like tantalum) leads to suppression of perturbation's growth. Moreover, covers made of the same material as loaded metals also contribute to removal of perturbations. This fact allows to simplify numerical simulation of the experiment. We chose loading of aluminum samples through the layer of aluminum cover for the calculations.

The results of the experiments are summarized in Table 1. The experimental series, like refs. [1, 2], detected selectivity in the growth of perturbations of a certain wavelength.

**TABLE 1.** Results of the experimental study

| Coating thickness, (μm) | Perturbation amplitude, (μm) |
|---|---|
| 8 | 120 |
| 22 | 40 |
| 31 | 10 |

Among the whole perturbation set present on the sample surface due to its polishing, only the perturbations of wavelength λ=500μm grow.

It is they that were observed on the surface in the recovered samples. This was also confirmed by numerical computations [3].

## NUMERICAL SIMULATION

The numerical simulation was performed with the technique discussed in [5]. This is a finite-difference technique using a regular computational grid and the method of concentrations for description of severely distorted interfaces. Strength properties were included with the method of ref. [6]. The fundamentals of the technique, its implementation on multiprocessor computers, and computing procedure were discussed in the paper presented at Parallel CFD 2003 [7]. For the impact and perturbation growth to be described correctly, a fine computational grid was needed (with the characteristic computational cell size of 3μm), so a computational grid with ~4·10⁶ cells had to be used. Hence the computations were performed on multiprocessor distributed-memory computers by program complex LEGAK [8].

An impact of two aluminum slabs at an angle of 12 degrees was simulated. Sinusoidal perturbation

$a = a_0 \cdot \sin \dfrac{2\pi}{\lambda_0} x$ with wavelength $\lambda_0 = 500\,\text{m}\mu$ and amplitude $a_0 = 10\,\text{m}\mu$ (which corresponds to the roughness in the slab surface machining) was given on the slab surface.

Pressure was evaluated by the Mie-Grueneisen equation of state with the following parameters:

$$\rho_0 = \rho_{00} = 2.64\,\frac{g}{cm^3}\,;\ c_0 = 5.55\,\frac{km}{sec}\,;\ n=3.2;\ \Gamma=2.14.$$

Fig.1 depicts the material geometry and Fig.2 the thermal energy field. Results of the computations for amplitude ratios are described in Table 2

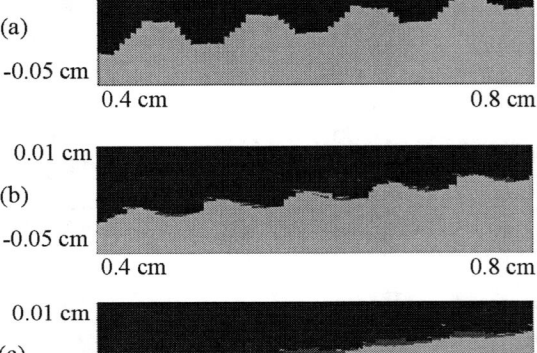

**Figure 1.** Material geometry in the computed problems, t=0.3 μs: a – without coating; b – with 22-μm coating; c – with 31-μm coating.

**TABLE 2.** Results of the computations: amplitude ratios

| Computation No. | Computation type | Amplitude ratio a/a₀ |
|---|---|---|
| 1 | Coating-free | 34 |
| 2 | 10 μm coating, gas-dynamic layer | 23 |
| 3 | 22 μm coating, gas-dynamic layer | 3.5 |
| 4 | 31 μm coating, gas-dynamic layer | 0.8 |
| 5 | 10 μm porous coating (ρ₀=2.4), gas-dynamic layer | 25 |
| 6 | 22 μm coating, пористое (ρ₀=2.4), gas-dynamic layer | 5.4 |
| 7 | 31 μm porous coating (ρ₀=2.4), gas-dynamic layer | 2.5 |
| 8 | 22 μm coating, elastic-plastic layer | 1.2 |

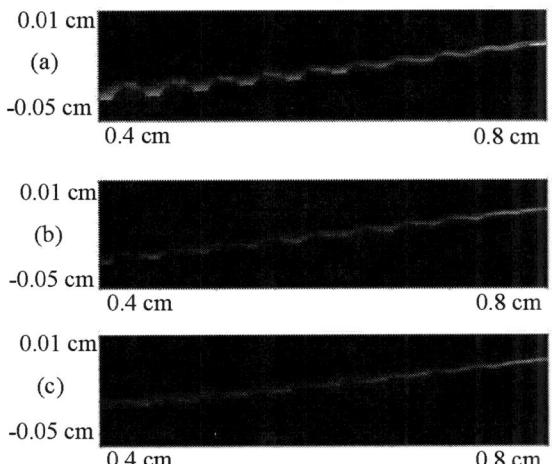

**Figure 2.** Thermal energy field in the computed problems, t=0.3 μs: a – without coating; b – with 22-μm coating; c – with 31-μm coating

From the figures and the Table 2 it is seen that if the thickest, 31 μm coating is sprayed on the lower slab, i.e. its strength properties are distinct from those of the impacting slabs, then there is no perturbation growth. With decreasing coating thickness the perturbation growth sets in.

Thus, when the coating thickness was 10 μm, the perturbation growth was 23 times as large as that with the original thickness, which is as little as 1.5 times less than that in the coating-free impact.

From Fig.2 it is seen that with increasing coating thickness the heating level decreases and the heating zone width becomes smaller, which is evidence of the distortion localization. With increasing coating thickness the contribution of the coating material to the shaped jet formed at the contact point increases.

The experiments and numerical simulations suggest that with a certain choice of the coating parameters (material and thickness) it is possible to essentially completely exclude the shear instability growth in the metal slab impact. With this aim it is necessary to perform numerical simulation of the experiments described in [4] for other materials of covers. It should be noted that the results of the experimental and first attempts of numerical studies of this effect agree.

## CONCLUSIONS

Experimental and numerical investigations of the stabilizing effect of coating on shear instability growth under impact of metal slabs were carried out. Good agreement between the results of experiments and numerical simulation results has been achieved. To explain the stabilizing effect of the coating made of the same material (aluminum) as the impacting slabs, an assumption concerning localization (narrowing) of the area of intense plastic strains was made.

In the future it is necessary to conduct numerical analysis for effect of suppression of perturbations by covers made of other materials.

## ACKNOWLEDGEMENT

The work was carried out under the support of Russian Fundamental Research Foundation (Project 02-01-00796).

## REFERENCES

1. Drennov, O.B., Mikhailov, A.L., Nizovtsev, P.N., Rayevsky V.A., «Perturbation growth at a metal interface in oblique impact at supersonic speed of the contact point», VANT, Ser. Teoreticheskaya i Prikladnaya Fizika, no. 1, pp. 34-42, 2001.
2. Bakhrakh, S.M., Volodina, N.A., Nizovtsev, P.N., Spiridonov, V.F., Shuvalova, E.V., «Numerical simulation of initial perturbation growth in oblique impact of metal slabs», VANT, Ser. Matematicheskoye Modelirovaniye Fizicheskikh Protsessov, no. 4, pp. 30-34., 2001.
3. Bakhrakh, S. V., Volodina, N.A., Nizovtsev, P.N., Spiridinov, V. F., "Numerical simulation of initial perturbation growth with oblique-impact of metal slabs", Book of Abstracts «Parallel Computational Fluid Dynamics», Moscow, pp.101-103, 2003.
4. Drennov, O.B., Mikhailov, A.L., Osipov, R.S., Rodigina, L.D., «Stabilization of wave formation at a metal layer interface in explosive welding conditions», Fizika Goreniya i Vzryva, no. 5, pp. 93-98, 1989.
5. Bakhrakh, S.M., Spiridonov, V.F., Shanin, A.A., "A method for computing gas-dynamic heterogeneous flows in Lagrangian-Eulerian coordinates", DAN SSSR, vol. 278, no. 4, pp. 829-833, 1984.
6. Bakhrakh S.M., Kovalev N.P., Pavlusha I.N., "A method for computing elastic-plastic flows: Proceedings of All-Union Conference on Numerical Elasticity-Plasticity Methods", Part I, Novosibirsk, Nauka Publishers, pp.22-36, 1974.

7. Bakhrakh, S. V., Velichko, S. V., Spiridinov, V. F., "Crash-free technology for continuum flow computations using «LEGAK» code on multiprocessor computer system', Book of Abstracts «Parallel Computational Fluid Dynamics», Moscow, pp.83-84, 2003.

8. Avdeev, P.A., Artamonov, M.V., Bakhrakh, S.M., Velichko, S.V., Volodina, N.A., Vorobyeva, N.M., Yegorshin, S.P., Yesaeva, E.N., Kovaleva, A.D., Lushinin, M.V., Pronevich, S.N., Spiridonov, V.F., Taradai, I.Yu., Tarasova, A.N., Shuvalova, E.V., «Program complex LEGAK and principles of its parallelization on multiprocessor computers», VANT. Ser. Matematicheskoye Modelirovaniye Fizicheskikh Protsessov, no. 3, pp. 14-18, 2001.

CP845, *Shock Compression of Condensed Matter - 2005,*
edited by M. D. Furnish, M. Elert, T. P. Russell, and C. T. White
© 2006 American Institute of Physics 0-7354-0341-4/06/$23.00

# TAYLOR ANVIL IMPACT

**Charles E. Anderson, Jr., Arthur E. Nicholls, I. Sidney Chocron,
and Raymond A. Ryckman**

*Southwest Research Institute®, Engineering Dynamics Department, P. O. Drawer 28510,
San Antonio, TX 78228-0510, USA.*

**Abstract.** G. I. Taylor showed that dynamic material properties could be deduced from the impact of a projectile against a rigid boundary. The Taylor anvil test became very useful with the advent of numerical simulations and has been used to infer and/or to validate material constitutive constants. A new experimental facility has been developed to conduct Taylor anvil impacts to support validation of constitutive constants used in numerical simulations. A 37-mm diameter Hopkinson bar apparatus was adapted to conduct the Taylor anvil experiments. An adaptor was designed to ensure impact planarity of the projectile onto the anvil, which is made from VascoMax steel, backed by a 1.82-m steel bar to provide inertial mass to the anvil and ensure deceleration of the projectile solely from elastic waves within the projectile. A digital imaging system was adapted to determine radial deformation as a function of length. Details of the experimental techniques, along with examples of experiments using 6061-T6, are discussed. Numerical simulations are used to complement the experimental results.

**Keywords:** Taylor anvil impact, dynamic flow stress, 6061-T6 aluminum, dynamic deformation
**PACS:** 62.20.Fe, 62.50.+p, 81.70.Bt, 46.15.-x, 45.40.Gj

## INTRODUCTION

G. I. Taylor recognized that the impact of a flat-ended projectile into a rigid surface permitted an estimate of a dynamic flow stress [1]. The experimental procedure has become known as the Taylor impact or Taylor anvil test. Although Ref. [1] is the usually quoted work, Taylor had been interested in the testing of material at high rates of loading for some time [2]. There is a figure in Ref. [2] that shows the results of mild-steel cylinders fired at armor steel plates, from which Taylor estimated the dynamic yield strength.

Wilkins and Guinan [3] were the first to conduct numerical simulations of the Taylor impact test and compare these results to experiments. They tested several metals, and in particular, have a relatively large number of tests for 1090 steel, tantalum, and 6061-T6 aluminum. The authors show that the final length of the cylinder can be reproduced fairly well by numerical simulations, over the range of impact velocities considered, using a constant flow stress for each material. But, they also demonstrate that better agreement of the final shape is obtained by including the effects of work hardening.

## EXPERIMENTAL SETUP

The Taylor impact test is a relatively simple test in theory, but great attention must be given to maintaining planar impact of the specimen on the anvil. Planarity is crucial to producing accurate, under-standable, and repeatable results. The Taylor Impact Test system developed at Southwest Research Institute (SwRI®) attempts to address this issue by utilizing samples and fixtures designed and machined to exacting tolerances. Massive steel pillow blocks rigidly support the entire system on a steel I-beam. A muzzle adaptor keeps the launch tube muzzle fixed and referenced to the anvil adaptor as seen in Fig.1. Extreme care was given to the machining and tolerances of the anvil adaptor and the mating muzzle adaptor. There is a slight taper in the anvil adaptor. The bore starts at 6.62 mm and tapers down to the final diameter of 6.45 mm (0.102 mm larger than the diameter of the specimen) over a distance of 19.05 mm. All flat surfaces on the anvil adaptor were ground to a flatness and parallelism of 0.013 mm or better and perpendicular to the line of flight to within 0.025 mm or better. All diameters are concentric to the line of flight to within 0.025 mm. The 6.45-mm final diameter is 25.4-mm long. A 2.54-cm diameter annulus at the anvil end of the adaptor provides room for radial expansion of the specimen.

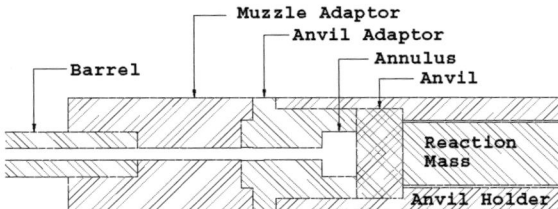

**Figure 1.** Schematic of experimental arrangement.

At the time of impact, 12.7 mm of the specimen remains supported in the tapered bore. This support of the trailing end of the specimen and the minimal clearance of the specimen within the bore ensures a maximum out of plane impact of less than 0.4 deg.

*Anvil.* A precipitation hardened maraging steel, VascoMax 350 VM, heat treated to a hardness of approximately Rockwell C 59, is used for the anvils. Again, flatness and parallelism of both the front and back faces of the anvil, 5.08 cm in diameter and 2.54 cm in length, were precision ground to within 0.013 mm. At the higher impact velocities, a shallow indent of approximately 0.076 mm or less was observed. Therefore, each face of the anvil saw only one test. The anvil was removed after impact and reground to remove any possible damage that may have occurred. A heat treated VascoMax reaction mass, 3.81-cm diameter and 1.829-m long and ground flat on the ends, supported the back face of the anvil. This ensured that the front face of the anvil was in intimate contact with the anvil adaptor, assuring perpendicularity of the anvil face to the line of flight. Added benefit of the reaction mass was an effective anvil thickness of nearly two meters and the fact that no reflected tensile stress wave from the back face of the anvil reaches the specimen-

anvil interface during deceleration and deformation of the sample (~700 µs vs. ~60 µs).

*Velocity Measurement.* Two laser diodes and detectors were used to measure specimen velocity. The lasers shot through the barrel through two drilled holes. Three velocity measurements were obtained from the lasers for each test: time of flight of the specimen past each beam and the time of flight between the beams. Outputs from the laser detectors were recorded on a Nicolet high-speed digital transient recorder sampling at 10MHz. A 12.7-mm slot was machined into the side of the anvil holder and anvil adaptor to provide visual access to the test as seen in Fig. 2. An Imacon 468 high-speed digital camera was used, for some cases, to measure specimen velocity visually just prior to impact. Vertical fiducial lines scribed in the back of the anvil adaptor annulus, to facilitate Imacon distance and velocity measurements, are denoted in Fig. 2. This velocity measurement agreed very well with the laser-diode method. The three laser measurements (and the Imacon velocity measurements when used) were averaged to obtain the impact velocity. Standard deviation of the velocity was typically two to three m/s.

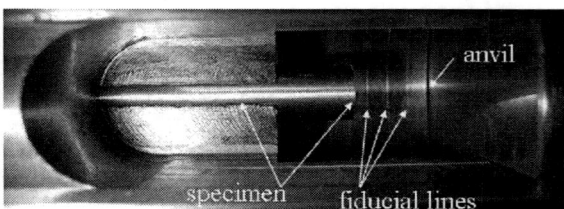

**Figure 2.** Photographic image of projectile entering the annulus.

*Specimens.* The experimental procedures were validated by conducting experiments using 6061-T6 aluminum cylinders. Specimens, 3.175-cm long with a 6.35-mm diameter ($L/D = 5$), were accelerated down a 6.1-m-long launch tube using a compressed gas operated breech mechanism. The samples were precision ground to produce end faces that were flat and parallel to within 0.0051 mm. Velocities in excess of 400 m/s were obtainable with breech operating pressures of 13.79 MPa.

*Imaging.* SwRI has developed a method of determining 3-D coordinates using a system known as Dynamic Structured Light (DSL). Previous methods of obtaining 3-D images involved either mechanical touch probes (which require hours of acquisition for even simple parts), or expensive optical systems. DSL 3-D imaging takes advantage of increased computer performance and signal processing techniques to provide a new, low-cost method of acquiring 3-D images.

The DSL 3-D imaging system consists of a high-performance computer, a digital video camera, and a dynamic light projector. The light is projected through a line grating that rotates at a constant speed, which creates a set of quadric surfaces. By examining the light and dark sequences, a precise x, y, and z coordinate can be determined for each pixel. The system and the surface must be kept still for three to five minutes while scanning, with the final results available ten to fifteen seconds later. The accuracy of the system scales with the size of the field of view, with a 0.051mm x 0.051mm x 0.25mm accuracy for a 51mm x 51mm field of view.

## EXPERIMENTAL DATA

The normalized residual lengths, $L_f/L_o$, versus impact velocity for the 6061-T6 aluminum specimens tested by Wilkins and Guinan [3] are plotted in Fig. 3. Although the dimensions and $L/D$ of the specimens used by Wilkins and Guinan were different than the specimens tested in this study, they showed that the normalized lengths are independent of the specific dimensions and $L/D$ of the specimens. Experiments were conducted with the new impact apparatus described above at impact velocities between 26 m/s and 440 m/s. The final lengths were measured to an accuracy of 0.025 mm, and the results are overlaid on Fig. 3. The new data overlay the Wilkins-Guinan data very nicely.

DSL images of three of the Taylor specimens are shown in Fig. 4. At the higher impact velocities, the material fractures due to large hoop strains, resulting in radial cracks and "petals" at the end of the specimen.

## NUMERICAL SIMULATIONS

Numerical simulations were conducted and the results compared to the experimental data. The Johnson-Cook model [4] was used to describe the flow stress as a function of strain, strain rate, and temperature. Constitutive parameters had been determined previously for 6061-T6 [5]. CTH [6], EPIC [7], AUTODYN [8], and LS-DYNA [9] were used for the simulations. It was found that 20 zones across the diameter of the specimen assured numerical convergence. The simulation results for the normalized length are shown in Fig. 5. Agreement between the numerical results and the experimental data is quite good.

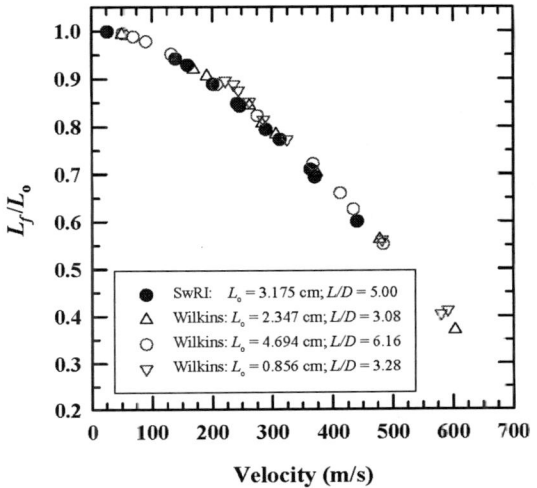

**Figure 3.** Normalized projectile length vs. impact velocity for 6061-T6 aluminum.

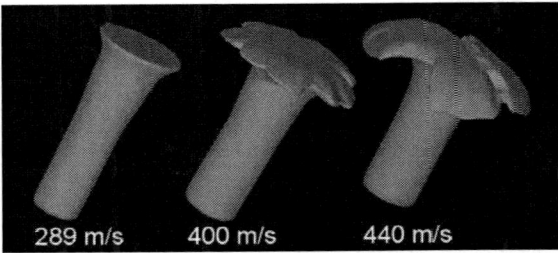

**Figure 4.** DSL images of Taylor specimens.

The deformed diameters were also measured to the same accuracy as the length. Fracture of the specimen at the higher impact velocities resulted in some variation of the diameter, which is denoted by the "variation" bars in Fig. 6. The results of the numerical simulations are also shown in Fig. 6.

Whereas the discrepancy in the length between the simulations and the experiments is only 1.15±0.82%, the differences in the diameters are 4.9±2.6% (not including the two highest velocity

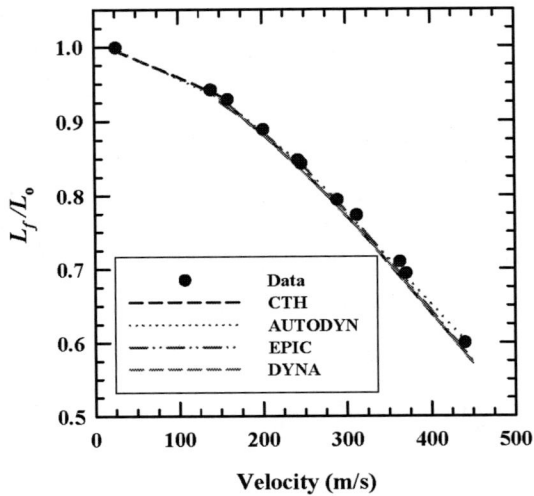

**Figure 5.** Comparison of numerical simulations to experimental data for normalized specimen length.

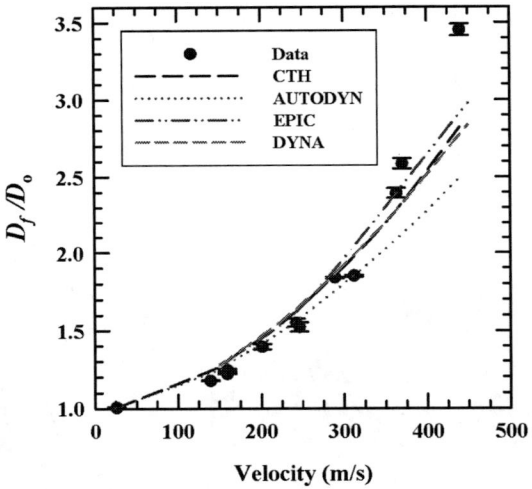

**Figure 6.** Comparison of numerical simulations to experimental data for normalized specimen diameter.

data points). Since the volume of the specimen is conserved, this implies that the overall shape must be slightly different. Simulations where the various constants in the Johnson-Cook model were parametrically varied indicate that the overall length is most sensitive to the initial flow stress, while the shape is sensitive to the hardening terms. Disagreement in the diameters at the highest impact velocities is probably due to the fracture response of the specimens (Fig. 4), which is not modeled in the simulations.

## CLOSURE

A new Taylor Impact Test capability has been developed at SwRI. Great care was exercised to insure planar impact. Experiments using 6061-T6 aluminum specimens reproduced the earlier work of Ref. [3]. Work is on going to understand better the influence of the various constitutive parameters on the calculated deformed shaped as compared to the experimental results. In particular, 3-D simulations using the Johnson-Cook damage model [10] are being conducted with CTH to investigate the sensitivity of results to damage vs. no damage.

## REFERENCES

1. Taylor, G.I., "The use of flat-ended projectiles for determining dynamic yield stress. Part I. Theoretical considerations," *Proc. R. Soc. A*, **194**, 289-299, 1948.
2. Taylor, G.I., Testing of materials at high rates of loading," *J. Inst. of Civil Engineers*, **26**, 486-518, 1946.
3. Wilkins, M.L. and Guinan, M.W., "Impact of cylinders on a rigid boundary," *J. Appl. Phys.*, **44**(3), 1200-1206, 1973.
4. Johnson, G.R. and Cook, W.H., "A constitutive model and data for metals subjected to large strain, high strain rates and high temperatures," *7th Int. Symp. Ballistics*, The Hague, The Netherlands, 541-548, 1983.
5. Dannemann, K.A., Anderson, Jr., C.E. and Johnson, G.R., "Modeling the ballistic impact performance of two aluminum alloys," in *Modeling the Performance of Engineering Structural Materials II* (Eds. D.R. Leseur and T.S. Srivatsan), TMS, 63-75, 2001.
6. McGlaun, J.M., Thompson, S.L., and Elrick, M.G., "CTH: A three-dimensional shock wave physics code," *Int. J. Impact Engng.*, **10,** 351-360, 1990.
7. Johnson, G.R., Stryk, R.A., Holmquist, T.J., and Beissel, S.R., "Numerical algorithms in a Lagrangian hydrocode," Report WL-TR-1997-7039, Wright Laboratory, June, 1997.
8. Century Dynamics, "AUTODYN Theory Manual," Rev. 4.3., Century Dynamics, Concord, CA, 2003.
9. Hallquist, J.O., "LS-DYNA Theoretical Manual," Livermore Software Technology Corp., Livermore, CA, 1997.
10. Johnson, G.R. and Cook, W.H., "Fracture characteristics of three metals subjected to various strains, strain rates, temperatures and pressures," *Engng. Fract. Mech.*, **21**(1), 31-48, 1985.

CP845, *Shock Compression of Condensed Matter - 2005,*
edited by M. D. Furnish, M. Elert, T. P. Russell, and C. T. White
© 2006 American Institute of Physics 0-7354-0341-4/06/$23.00

# AXIAL VISAR VELOCITY MEASUREMENTS OF THE NON-PLANAR ACCELERATION OF A PLATE FROM A PENETRATING SHAPED CHARGE JET

## Matt Briggs and Eric N. Ferm

*Los Alamos National Laboratory, Los Alamos NM 87545*

**Abstract.** A Viper explosive shaped charge jet creates a stretching rod of metal, with its tip traveling 9.2mm/μs and slower portions traveling < 3 mm/μs. As this rod impacts and penetrates an obstructing steel plate, a highly non-planar flow evolves. We have recorded the free-surface velocity at the point of exit of the jet, which will ultimately be accelerated to the residual velocity of the jet. The thick target we have chosen allows the penetration to reach a quasi-steady subsonic penetration rate, before the wave structure begins to exit the plate. The resulting acceleration at the free surface is a continuous VISAR record with no required discontinuous fringe jumps. An elastic wave accelerates the free surface first, followed by a decaying shock wave, and plastic wave. Ultimately the release waves reach the penetrating jet and the surface begins to accelerate to velocities near the speed of the jet exiting the plate. We compare the present experimental and model calculations with the published results.

**Keywords:** Velocimetry, VISAR, Shaped charge, Jet
**PACS:** 47.40.-x, 47.40.Nm.

## INTRODUCTION

Ferm and Ramsay [1] reported experimental method and model calculations of a jet penetrating a composite steel/Plexiglas (PMMA) target. They observed the wave structure ahead of the subsonic penetration enter into the Plexiglas using streak and framing camera images of the backlit Plexiglas section. The observed a weak elastic precursor followed by a bow shock. In the Plexiglas the penetration was supersonic, and this structure steepened and became an attached bow shock as time progressed. In the present experiment we measure the velocity of the rear surface of the steel plate directly using VISAR, with no PMMA.

## EXPERIMENTAL METHOD

The schematic of the experimental configuration is shown in Figure 1. The target was a cylinder of 1018 steel, 3.4 inches long and 3

inches in diameter. We used back reflections, levels, squares, and alignment lines on the bottom of the target (which stopped well short of the region interrogated by the VISAR) to align the axes of the charge, target and probe. We estimate our centering was good to ±0.5 mm.

The experiment was conducted at the center of our 5-axis low energy x-ray array. This is a set of 5 x-ray heads held at equal angles around 360 degrees by an iron pentagonal frame. We aligned the bottom of the target to be at the height of the center line of the x-rays. We used the array to record a time-series of the evolution of the experiment, two before impact and three after. We assumed cylindrical symmetry and used the two initial images to give us an estimate of the incoming velocity of the jet. The three images after impact reinforced our interpretation of the velocimetry modeling by showing only a bulge during the velocimetry and the jet emerging later. Figure 2 shows the jet well after the VISAR record was over; the jet can be seen extending above and below the target, just starting to touch the VISAR probe. Note that the center of the probe has a GRIN (gradient refractive index) lens that makes a dark line in the image, just to the right of the 2000 pixel mark, and that this line is aligned well as seen from this direction.

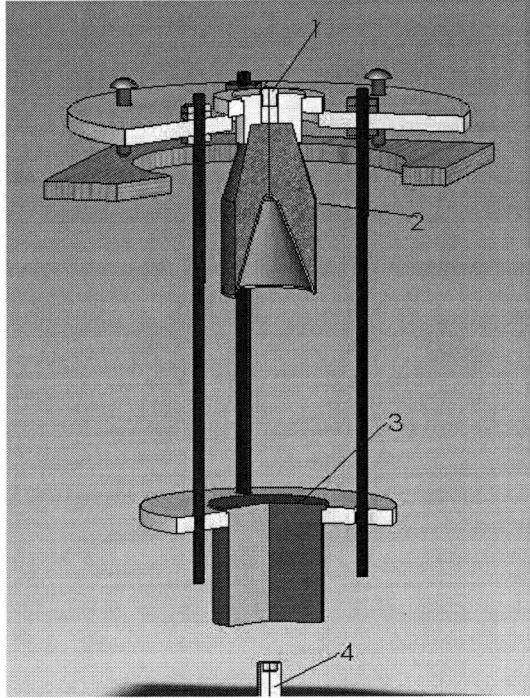

**Figure 1.** Experimental configuration. 1.) SE-1 detonator and 0.5 inch 9407 booster; 2.) Viper shaped charge with copper liner that forms penetrating rod; 3.) Cold-rolled steel 1018 target, 150 mm below face of shaped charge; 4.) Valyn 30 mm f.l. optical VISAR probe.

## Experimental Results

The acceleration of the surface was slow enough to be traced by the velocimetry without the usual addition of fringes required by shock experiments. We attribute this to the penetration speed of the jet in the steel decaying quickly to a speed below that of an elastic precursor wave, which accelerated the surface slowly enough to produce fringes within the bandwidth of the VISAR. Figure 4 and Figure 5 show the velocity record we measured with our VISAR setup [2].

Start of motion was 53.1 μs after load ring, where the speed rose to 10 m/s in 200 ns. A second rise occurred at 55.5 μs to 230 m/s. As we discuss later, the first feature we attribute to the elastic wave, and the second to an unsupported shock wave that has moved out in front of the penetrating jet. The full data record further shows a couple of small wiggles at 58.5 μs and 60 μs that we have not yet explained, but that did reproduce in a repeat experiment. Finally, the velocity is rising quickly at 60 μs, which we attribute to the emergence of the jet from the target. The end of the data record at 62 μs, after 6 mm of surface motion is due to a lack of signal, for reasons uninvestigated. Note that there was an error in the balance of the VISARs, so the reconciliation of the records was poor. Therefore, our worst case systematic errors in the absolute velocity could be as high as 10%. A further experimental problem is present in the x-ray data, where the time-lapse measurement of the incoming jet velocity gives 9.8 km/s, where experience shows these jets detonated in this way to be very reliably closer to 9.2 km/s.

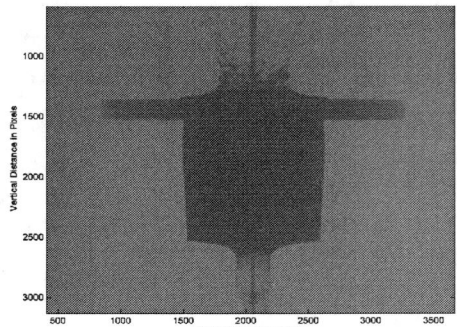

**Figure 2.** X-ray image of jet emerging (and still entering) the Target at 67 μs.

## MODEL CALCULATIONS

We modeled the experiment using a Mesa 2-D Eulerian hydrodynamics code [3], from explosive formation of the jet through its penetration and ultimate acceleration after emerging from the steel target plate. The shape charge was modeled using standard equation of states (EOS s) from the Mesa Library for OFHC copper and LX-14. The mesh was 0.25 mm through the region of the jet formation and penetration processes.

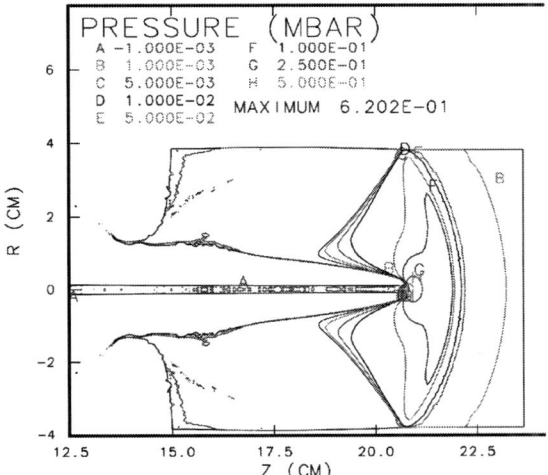

**Figure 3.** Elastic precursor (B) followed by shock (F).

We tried several equations of state for the steel target, and for the examples here we have shown two results: The first is a Mie-Grüneisen linear $u_s$-$u_p$ EOS for 304 Stainless Steel. The second EOS has a concave up quadratic form for $u_s$ when $u_p<1$ mm/μs. The second EOS has the same form as the linear EOS at high density, but gives a higher shock speed for weak shocks. The penetration velocity is slower than the elastic sound speed in the material, while the penetration speed and plastic shock speed are comparable, so the elastic wave pulls ahead. When the jet first enters the steel, the elastic and plastic waves are together in a bow shock at the front of the stagnation region of the penetrating jet. Later (Figure 3, t = 45 μs) the elastic wave has pulled far ahead of the plastic shock, which has pulled away from the jet. The plastic shock is decaying as it pulls ahead of the penetration point, where the pressure remains at 0.6 Mbar. At the time shown in Figure 3, the elastic wave in the target is about to begin accelerating the back edge of the target.

In order to compare the calculation with the VISAR experimental data, we placed massless Lagrangian tracer particles at the end of the target. In Figures 4 and 5 we show the comparisons between the experimental VISAR record, the simple linear 304 SS EOS, and the 304 SS EOS with the higher weak-shock sound speed.

## Results and Discussion

We easily modeled the elastic wave using the elastic models in the Mesa material library, and many different steel material models gave similar results. However. the linear $u_s u_p$ equations of state uniformly gave too high of a shock pressure for the small plastic shock, giving an overshoot at 56 μs (Figure 4.) This discrepancy was not strongly dependent on whether we used the linear $u_s$-$u_p$ EOS for mild steel or whether we used stainless steel. Thus over a significant range of Grüneisen parameter and $u_s$-$u_p$ line choice the discrepancy remained, and discrepancies became larger as we used a finer mesh. However, we could resolve this by adding quadratic $u_p$ terms to the Hugoniot curve for small $u_p$. This allowed the shock speed for weak shocks to be higher than the simple linear Hugoniot EOS, so the shock pulled away from the penetration process sooner. This in turn allowed the shock to begin to decay sooner.

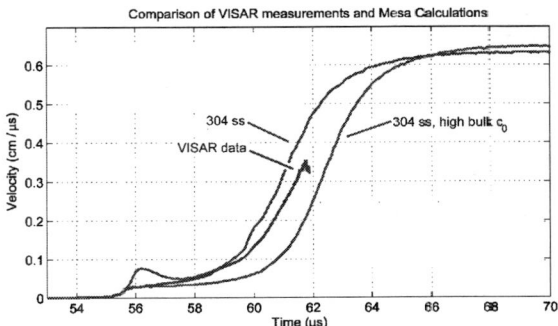

**Figure 4.** Comparison of VISAR data with two Mesa Calculations.

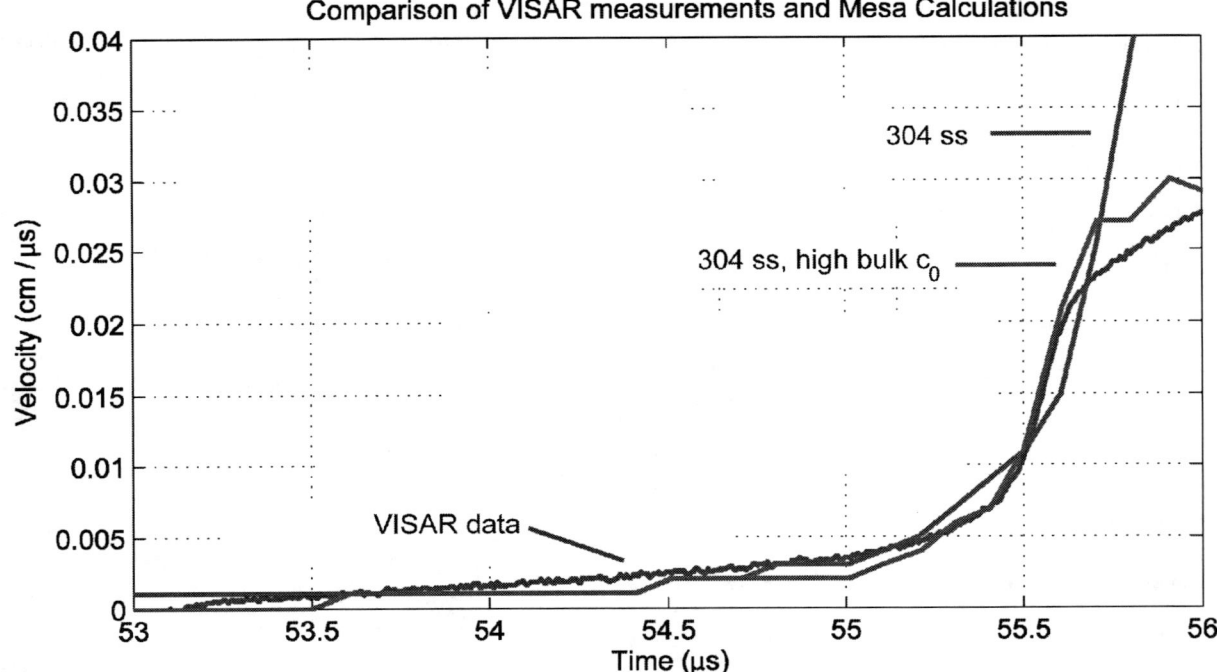

**Figure 5.** Magnified view of the VISAR data and calculations shown in Figure 4 for the first 3 microseconds.

## CONCLUSIONS

Ferm and Ramsay [1] discerned the wave structure from index of refraction changes in a Plexiglas section following the steel target. Using this technique the elastic wave and weak plastic shock were captured as one disturbance, and the ramp wave from the penetration process was observed to steepen and form a shock in the Plexiglas. Using the VISAR diagnostic here has allowed us to examine the elastic wave and weak plastic shock as separate influences and to record the interface motion responsible for the acceleration wave studied in the previous work. The weak plastic shock requires a more detailed EOS to obtain more than qualitative details; we have another experiment planned to deal with the experimental shortcomings here, and if successful, we can then pursue the EOS issue as well.

## ACKNOWLEDGEMENTS

Rudy Archuleta, Jim Faulkner, Steve Hare, John Echave and Mike Archuleta provided technical expertise to fielding the experiments.

This work was supported by the Department of Energy under contract W-7405-ENG-36 to the University of California and we gratefully acknowledge their support.

## REFERENCES

1. Ferm E. and J. B. Ramsay (1991). "Jet Penetration of Surrogate Steel-Explosive Systems." Propellants, Explosives, Pyrotechnics 16: 123-130.
2. Hemsing, W. "Velocity sensing interferometer (VISAR) modification", Rev. Sci. Instrum., 50 (1), pp.73-78, 1970.
3. Bennion, S.T., and Clancy, S. P., Los Alamos National Laboratory Technical Report No. LA-CP-92-229, 1992.

CP845, *Shock Compression of Condensed Matter - 2005*,
edited by M. D. Furnish, M. Elert, T. P. Russell, and C. T. White
© 2006 American Institute of Physics 0-7354-0341-4/06/$23.00

# STABILIZATION OF WAVE FORMATION ON A CONTACT BOUNDARY OF METAL LAYERS AT AN OBLIQUE IMPACT DURING KELVIN - HELMHOLTZ INSTABILITY DEVELOPMENT

## O.B. Drennov, A.L. Mikhailov

*Russian Federal Nuclear CenterAll -Russian Scientific Research Institute of Experimental Physics
Sarov, Russia, 607190*

**Abstract.** The elimination effect of disturbances and mutual mixing on a contact boundaries of metal layers at oblique impact during Kelvin - Helmholtz instability development was established and investigated. Thin layers of metal coatings ($\Delta \sim 30$ μm) reduce amplitude of disturbance realization in $10 - 100$ times and eliminate mutual mixing of contacting materials (eliminate the formation of a welded point). The foils of the same materials and thicknesses are not characterized by the same strong stabilizing properties. This stabilizing effect is explained by physical properties of a metal coating as a whole. Thermophysical limits for coating layers are pointed out.

**Keyword:** Metal layers, oblique impact, shock wave, instability, stabilization.
**PACS:** 62.50.+p,62.20.Fe.

## INTRODUCTION

The oblique shock wave loading of the contacting layers with different dynamic rigidness is accompanied, in general, by the realization of material tangential flows along a contact boundary, which may result in Kelvin-Helmholtz instability development [1, 2, 3]. The wave formation and mutual material mixing process is extensively applied during explosion welding of metals [4, 5]. However, in some cases the wave formation and intermixing in metal layers under oblique shock wave loading is unfavourable.

## EXPERIMENTAL SCHEME

Plane metal layers (copper, brass, aluminium, magnesium and their combinations) loading by on oblique shock wave during explosive welding and various shields' effects on a contact boundary state after the loading were experimentally investigated. A traditional scheme of plane throwing during a

sliding HE charge detonation was used, which is applied during an explosive welding. In fig.1 a test arrangement scheme and the notation of characteristic parameters of throwing are presented. A fixed plate (5) is placed on a massive steel base (6). Above it a flying plate (3) is mounted at a specified angle α to the fixed plate plane. Minimal distance *h* between the surfaces of flying and fixed plates is selected to provide stationarity of a striker (3) flight before an impact [5]. On a flying plate surface a HE layer (1) is placed, where a plane sliding detonation wave is initiated. To prevent spallation events in a striker material, between it and a HE layer a thin spacer (2) with a low acoustic impedance is placed [6]. On a contact surface of a fixed plate (A-A' plane in fig.l) a shielding layer (4) of 10 μm<$\Delta$<50 μm thickness is predeposited.

In case of metal coating the latter was deposited by an electron beam spraying or galvanic deposition. The loading regimes for all metal pairs tested were selected to form a cumulative jet in a contact point. A subsonic loading was realized: a

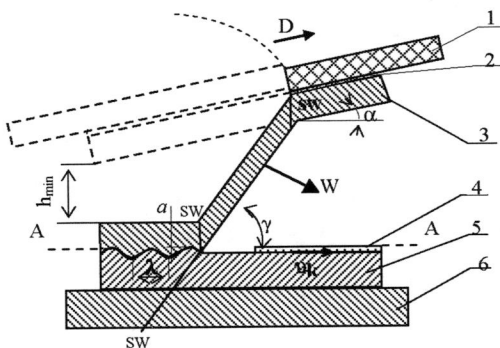

**Figure 1**. Test arrangement scheme

contact point velocity is 2.5 mm/µs <Vc<4.5 mm/µs; a impact angle is 12°<y<25°. Of the samples, which have experienced the dynamic loading, the microsections were prepared, which were metallographically analyzed at the zone of materials contact.

## EXPERIMENTAL RESULTS

The stabilizing effect of thin metal coating on the explosive welding and disturbance evolution at a contact boundary was experimentally found (see fig.2-7). In Fig..2(a) a photograph of a section of a contact boundary aluminum - aluminum (AlMn alloy) after high - speed oblique impact is shown. The explosive welding took place. In the vortex zones of wave crests the metals melt. The disturbance amplitude is $a≈400$ µm. In Fig.2(b) the photograph of a contact boundary section of the same metal pair is shown under the loading through a magnesium coating layer $Δ≈22$ µm thick. The disturbance amplitude decreased substantially, $a<10$ µm. A welded joint between the metals is absent.

In Fig. 3(a) a section of contact boundary copper - copper (Ml grade) is given. The explosive welding took place. The disturbance amplitude is $a≈350$ µm. In Fig.3(b) a section photograph of a contact boundary for the same pair is shown, but including a zinc layer $Δ≈22$ µm thick. The disturbance amplitude decreased to $a≈15$ µm. The explosive welding was not realized.

In Fig.4(a) a section photograph of a contact boundary aluminum-copper (AlMn alloy and Ml) is presented. A characteristic dramatic wave formation is seen. The disturbance amplitude is $a≈250$ µm. In

Fig. 4(b) a section photograph of a contact boundary for the same metal pair is shown after an oblique impact through a zinc layer $Δ≈22$ µm thick. The disturbance amplitude decreased by $a≈25$ µm.

In our opinion, it is reasonable to pay special attention to the experiments, when the impacting plates and the cover are made of the same material. Suppression of perturbations is recorded after dynamic loading.

In Fig.5 a section photograph of a contact boundary aluminum – aluminum under the loading through an aluminum layer $Δ≈22$ µm thick is given. The disturbance amplitude is $a≈40$ µm. In Fig.6 a section photograph of a contact boundary copper – copper under the loading through a copper layer $Δ≈25$ µm is presented. The disturbance amplitude is $a≈70$ µm

It was noted that in all tests where a reliable stabilization of a contact boundary is seen, a metal structure of a fixed plate doesn't change.

The foils of the same metals, having the same thickness, do not possess the same strong stabilizing properties. In Fig. 7 a section photograph of a contact boundary copper - copper is given; the loading - through a foil of aluminum AlMn alloy $Δ=20$ µm thick. The explosive welding occurred. The foil is fragmented and mixed with melted loaded metals. The realized disturbance amplitude is $a ≈ 350$ µm. In Fig. 8 a section photograph of a contact boundary of the same metal pair is presented; the loading - through 1X18H10T steel foil $Δ=20$ µm thick. Due to high strength this steel foil doesn't fragment. A welded joint doesn't form, the disturbance amplitude decreases negligibly.

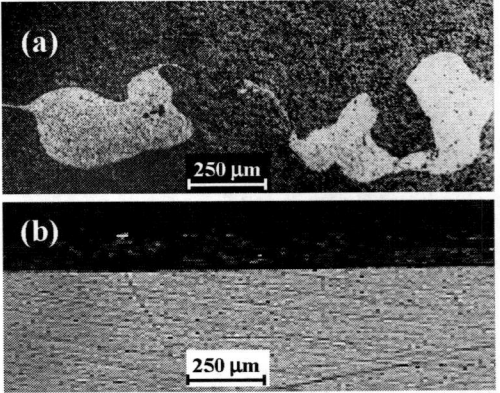

**Figure 2.** AlMN/AlMn boundary, no coating (a) and with 22 µm thick Mg coating (b)

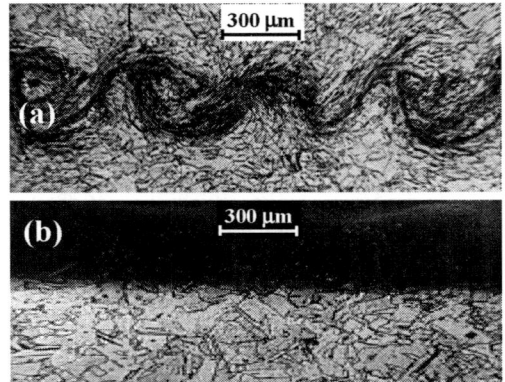

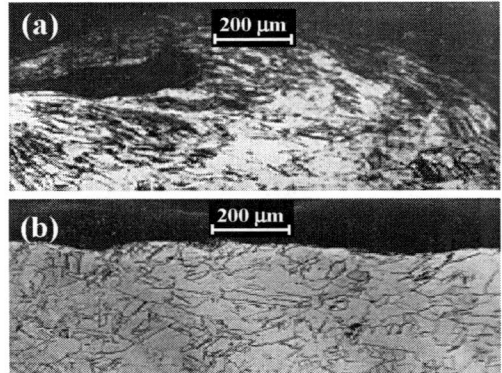

**Figure 3.** Cu/Cu boundary, no coating (a) and with 22 μm thick Zn coating (b)

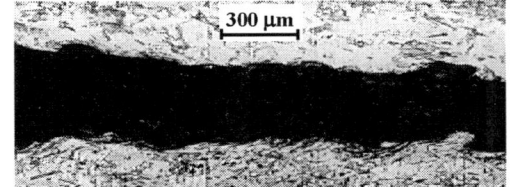

**Figure 4.** AlMN / M1 Cu boundary, no coating (a) and with 22 μm thick Zn coating (b)

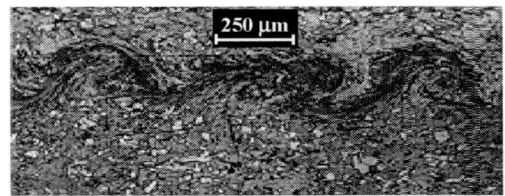

**Figure 5.** Al / Al boundary, with 22 μm thick Al coating

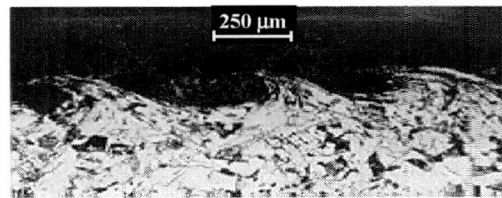

**Figure 6.** Cu / Cu boundary, with 25 μm thick Cu coating

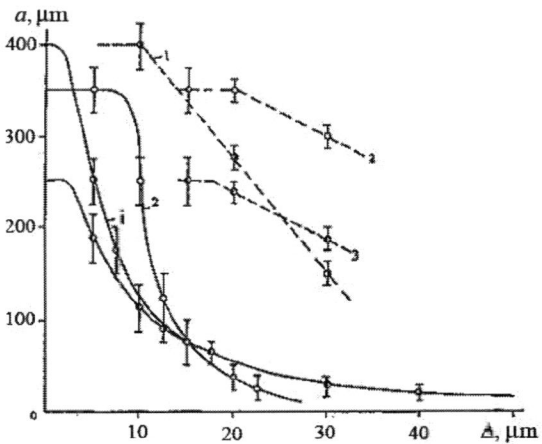

**Figure 7.** Cu / Cu boundary, with 20 μm thick AlMn coating

**Figure 8.** Cu / Cu boundary, with 20 μm thick 1X18H10T steel

In a special test series we investigated the coating layer adhesion influence on the disturbance stabilization effect (when a coating is sprayed, there is a finite adhesion. The stabilizing effects of the coating is seen in a both cases.

In Fig. 9 the dependence of disturbance amplitudes on coating thickness is shown for the contact boundaries of the following metal pairs; aluminum – aluminum (1), copper - copper (2), aluminum - copper (3). The stabilizing properties of zinc coating (solid lines) and zinc foil (dashed lines) are illustrated.

**Figure 9.** Dependence of disturbance amplitudes on coating thickness

## DISCUSSION OF RESULTS

Thermophysical properties of a coating material affect the stabilization. Fig. 9 shows that for a copper - copper pair a zinc layer $\Delta < 10$ μm thick does not stabilize a contact boundary. Evidently, it melts and takes part as a homogeneous continuum in boundary layers of loaded metals' mixing. A property of disturbance elimination on a contact boundary disappears.

In a series of unidimensional thermal calculations based on the main thermal conductivity equations [7] we obtained, that under the loading parameters described for $t \leq 5$ μsec (loading time) coating layers $\Delta \leq 10$ μm thick, made of practically all metals, experience complete melting. A contact boundary is heated intensively as well, as a consequence. The materials mix. Only layers of fusible metals (tin, lead, cadmium, bismuth) having $10\mu m < \Delta \leq 20\mu m$ thickness melt completely. Coatings of tantalum, copper, nickel, titanium, aluminum, magnesium, zinc experience partial melting in a outer zone. Inner zone of a coating layer and boundary layers of a loaded sample are weakly heated and are affected on stabilization.

These test results give grounds to state that stabilization effect is determined by metal coating nature, its physical properties as a whole. Coating layer ρ density 6onstitutes (0.9-0.95) of crystal density $\rho_{cr}$ for a corresponding metal. The elementary micro volumes' cohesion is extremely weak, it is described by the bonds in dense packing. Disruption .strength is $\sigma_D < 0.01$ GPa for a copper coating of $\Delta \approx 20$ μm thickness; and $-\sigma_D \leq 0.2$ GPa for a copper foil of the same thickness.

The most effective in stabilizing of a contact boundary are the coating layers, made of the metals with melting temperatures $T_m > 400°C$ and low thermal conductivity (temperature conductivity coefficient $\mathfrak{x}^2 < 10^{-4}$ m$^2$/sec).

## CONCLUSIONS

Thus, a method is suggested to eliminate an explosion welding and disturbances on a contact metal pair boundary in the process of oblique shock wave loading. Thin layers of metal coatings ($\Delta \approx 30$ μm) reliably stabilize a contact boundary and eliminate mutual intermixing of the metals under the loading.

The covering facilitates localization of deformation along the interface.

Probably, the phenomenon of perturbation suppression can be explained by character of the occurred hydrodynamic flow and complicated stress-strain state of substance.

One can see that this phenomenon is very interesting, and it has not been explained completely yet. It will be possible to give a final description of nature of the revealed phenomenon after a series of analytical and numerical calculations and special tests.

## REFERENCES

1. Hunt, J.N., The Phylosophical Magazine. vol.17, ser.8, no.148, p.669-680, 1968.
2. Cowan, G.R., Bergmann, O.R., and Holtzman A.H., Metallurgical Transactions. vol.2, no.11, p.3145–3155, 1971.
3. Utkin, V.F., Dremin, A.N., Mikhaylov, A.N., Gordopolov, Y.A.. Fizika Goreniya Vzryva, vol.16, no. 4, 1980.
4. Krypin, A.V., Soloviev, V.Y., Sheftel, N.I., Kobelev, A.G., "Explosive deformation of metals" [in Russian], Metallurgiya, Moscow, 1975.
5. Deribas, A., "Physics of Reinforcement and Explosive Welding" [in Russian], Nauka, Novosibirsk, 1980.
6. Glushak, B.L., Novikov, S. A., Pogorelov, A. P., and Sinitsyn, V. A., Fizika Goreniya Vzryva, vol. 17, no. 6, p. 90, 1981.
7. Godynov, S.K., "Equations of mathematical physics", Nauka, Moscow, 1979.

CP845, *Shock Compression of Condensed Matter - 2005*,
edited by M. D. Furnish, M. Elert, T. P. Russell, and C. T. White
© 2006 American Institute of Physics 0-7354-0341-4/06/$23.00

# ON THE ORIGIN OF A MAXIMUM PEAK PRESSURE ON THE TARGET OUTSIDE OF THE STAGNATION POINT UPON NORMAL IMPACT OF A BLUNT PROJECTILE AND WITH UNDERWATER EXPLOSION

## Alexander Gonor[1], Irene Hooton[2]

[1]*Applied Science & Engineering Consulting, 624-523 Finch Avenue West, Toronto, ON Canada M2R 1N4*
[2]*National Defence Headquarters, 101Colonel By Drive, Ottawa, ON Canada K1A 0K2*

**Abstract.** Impact of a rigid projectile (impactor), against a metal target and a condensed explosive surface considered as the important process accompanying the normal entry of a rigid projectile into a target, was overlooked in the preceding studies. Within the framework of accurate shock wave theory, the flow-field, behind the shock wave attached to the perimeter of the adjoined surface, was defined. An important result is the peak pressure rises at points along the target surface away from the stagnation point. The maximum values of the peak pressure are 2.2 to 3.2 times higher for the metallic and soft targets (nitromethane, PBX 9502), than peak pressure values at the stagnation point. This effect changes the commonly held notion that the maximum peak pressure is reached at the projectile stagnation point. In the present study the interaction of a spherical decaying blast wave, caused by an underwater explosion, with a piece-wise plane target, having corner configurations, is investigated. The numerical calculation results in the determination of the vulnerable spots on the target, where the maximum peak overpressure surpassed that for the head-on shock wave reflection by a factor of 4.
**Keywords**: shock wave, projectile, peak pressure, target, underwater blast wave.
**PACS**: 47.40-x, 47.40Nm.

## INTRODUCTION

The first goal of this study is to present a new mechanism originating in the initial stage of blunt projectile impact against metallic and soft targets, which leads to a much higher peak pressure compared with that at the stagnation point. The second goal is to extend the results[1] obtained above on the interaction of a spherical decaying blast wave, caused by underwater explosion, with a piece-wise plane targets.

## STATEMENT OF THE PROBLEM OF A SHOCK WAVE ATTACHED TO A BLUNTED PROJECTILE

Impact of the blunt projectile is considered in the frame-of-reference, $XoY$, attached to the body (Fig. 1). The velocity of point A(X, Y) is given by the following equation, $V_a = V_0 / Sin(\phi)$, where $\phi$ is the angle between the tangent to the circle and the

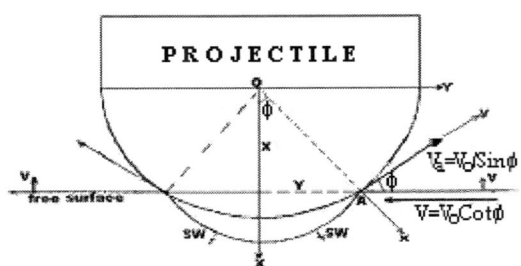

**FIGURE. 1.** Schematic of blunt projectile entry into liquids and solids.

target free surface. The next step is to translate this system to a coordinate system-$xAy$, moving with velocity $V_a$. In a new coordinate system, undisturbed target particles have the vector velocity, **V**, which is being directed parallel to undisturbed target front (Fig. 1).

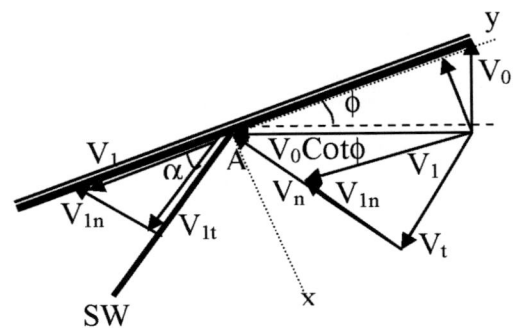

**FIGURE 2.** Flow velocities schematic.

The small, disturbed region just around point A, including the shock wave (SW) and velocity vectors in the $xAy$ frame-of-reference are shown in Figure 2. The flow approaches an attached shock at an angle, $\phi$, to the body surface and with a speed $V = V_0 cot\phi$. On passing oblique through the shock front, the flow has a velocity, $V_1$. Using the balance equations on the shock wave, the shock angle, $\alpha$, and the angle, $\phi$, can be found from Figure 2 using geometry, in terms of the impact Mach number $M_0 = V_0/C_0$, the non-dimensional pressure $p = (p_1 - p_0)/(\rho C_0^2)$, and the density ratio $q = \rho_1/\rho$. The absolute velocity, $V_a$, in the laboratory frame-of-reference-$XoY$ attached to the body, is found analogously. Note these values are insufficient in providing the entire solution, since the equation-of-state is unknown.

## Solution Using Tait's Equation of State for Nitromethane

Tait's EOS for liquids and condensed matter has the following form, $p_1 - p_0 = \rho C_0^2 (q^n - 1)/n$, where for nitromethane, $n = 7$, $p_0 = 10^5$ Pa, $\rho = 1.128 \times 10^3$ kg/m$^3$ and $C_0 = 1.65*10^3$ m/s. The solutions for the shock pressure (point A, Fig. 1)

depending on the angle, $\phi$, and with various Mach numbers are shown in Figure 3. In this figure the pressure difference, $p_1(\phi) - p_0$, across the shock is normalized by the corresponding stagnation pressure, such that the non-dimensional pressure $\mathbf{P} = (p_1(\phi) - p_0)/(p_1(0) - p_0)$. All calculation results for pressure exhibit a monotonic increase with distance from the stagnation point, at $\phi = 0$, to a maximum at the

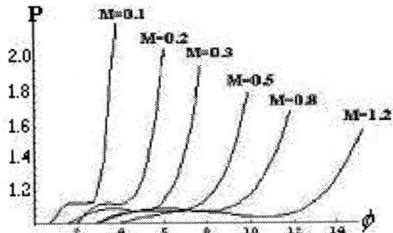

**FIGURE. 3.** Pressure ratio vs. angle of wetted spherical cap for nitromethane.

critical angles where the shock just detaches from the surface of the body. It should be noted that there is a great increase in pressure up to a maxima of 134% ($M_0 = 0.1$) and 58% ($M_0 = 1.2$), compared with the pressure at the stagnation point, for the normal impact of a blunt body on a target. This phenomenon was not noted earlier.

## Solution to the Problem Based on the Relationship $D = C_0 + Su$ for Liquids and Condensed Matter

To verify that the findings are independent of the type of EOS, an additional investigation was undertaken without using Tait's EOS. The experimental relationship $D = C_0 + Su$ was used as an alternative[2-5] to an equation of state, where D is the shock speed, u the flow velocity behind the shock front, and S and $C_0$ are constants. For nitromethane, S =1.637 and $C_0$ =1647 m/s. According to previous studies (see for instance Ref. [6]), the above relationship implies $p = r/(1 - Sr)^2$, where $r = 1 - 1/q$. The governing equation, with respect to the function $r(\phi)$, was found by the same method used with Tait's EOS. The behaviour of the pressure for nitromethane was very similar to that determined with Tait's EOS. Projectile impacts upon the explosive PBX 9502 ($C_0 = 2900$ m/s, S =

1.78), Al ($C_0$ =5350 m/s, S = 1.35) and Fe ($C_0$ = 3800 m/s, S = 1.58) targets were also investigated. Analogous to Fig.3, the peak pressure distributions are exhibited in Figures 4 and 5 for PBX 9502 and Al, respectively. Note that the values of the peak pressure for PBX 9502 (Fig. 4) are almost the same as those for nitromethane (Fig. 3), at similar Mach numbers. Impact on metallic targets has also exhibited a strong increase in peak pressure with distance from stagnation point (Fig. 5).

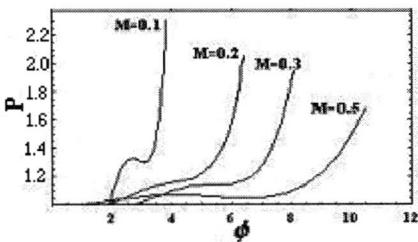

**FIGURE. 4.** Pressure ratio vs. angle of adjoined spherical cap for PBX 9502.

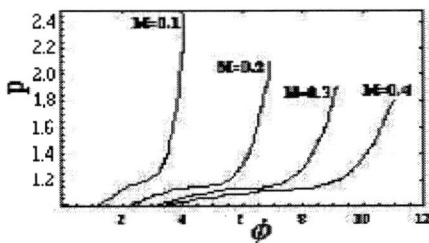

**FIGURE 5.** Pressure ratio vs. angle of adjoined spherical cap for aluminum.

For example, peak pressures occurred on aluminum and iron targets that were 2.4 times higher than those at the stagnation points of each metal, at impact velocity of ~500 m/s. The effect of the intensification of the oblique, attached shock wave compared with the head–on shock wave, in liquids and solids, caused by the impact of a blunt projectile on a target, is not evident. Note the sharp increase of the velocity, $V_a$, with distance from the stagnation point, such that the absolute jet velocity in the vicinity of point A (Fig. 1), noticeably exceeds the impact velocity, $V_0$. This fact explains the increase in pressure at distances from the stagnation point.

## INTERACTION OF A BLAST WAVE CAUSED BY UNDERWATER EXPLOSION WITH A PEACE-WISE PLANE TARGET

The interaction of a spherical decaying SW with a flat target, as shown in the schematic in Figure 6, was recently studied in Ref.[1].

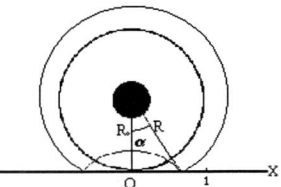

**FIGURE 6.** Schematic of the interaction of a spherical SW with a flat rigid target.

The distribution of the reduced peak pressure, $\mathbf{p} = p_2/p_2^0$, normalized with the pressure at the stagnation point, along the target versus the reduced coordinate $\mathbf{x} = x/R_0$ has shown that within the range $0 < \mathbf{x} < 0.6$, the pressure of the regular reflection of the decaying spherical SW from the target, increases, reaching values 40% greater than that at the stagnation point.

Next, the piece-wise interaction of a SW with targets having corner configurations will be generalised.

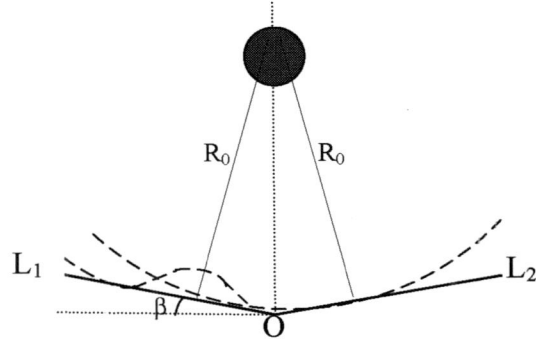

**FIGURE 7.** Interaction of blast wave with dihedral shaped target.

To start, a dihedral-shaped target having angles of $\sim 120 - 130^0$ between the planes were considered. The centre of the charge was placed on the dihedral bisector at a specified distance from the corner

Point O (Fig. 7). This distance is found based on the condition that the maximum peaks pressure, caused by symmetric blast wave reflection, are in the vicinity of the corner Point O (Fig. 7). The reduced peak pressure distributions, **p**, versus the reduced coordinate, x, are exhibited in Figure 8. As seen in Figure 8, the maximum values of the peak pressure surpass those at the stagnation points by 3.1 – 3.9 times, depending on the dihedral angle.

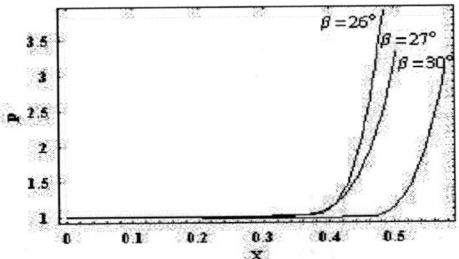

**FIGURE 8.** Peak pressure distributions in the vicinity of the dihedral edge.

As an example of the application of this phenomenon, a vulnerable area is the intersection of a superstructure with the submarine deck. A calculation was performed for the configuration of the target cross section shown in Figure 9. For this

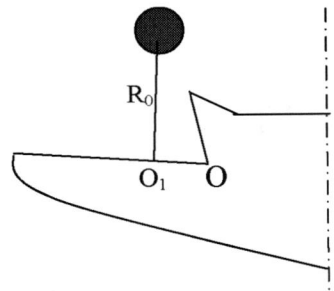

**FIGURE 9**. Schematic of submarine target.

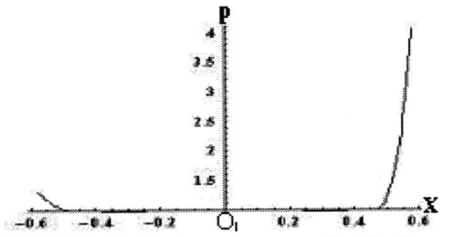

**FIGURE 10.** Peak overpressure distribution near corner.

target shape, the maximum peak pressure at the corner point O, shown in Figure 10, surpasses that at the stagnation point by 4.04 times.

## CONCLUSIONS

The pressure distribution along the front part of a blunt projectile, at the initial stage of impact with liquid or solid matter, has shown that the conventional idea concerning the location of the peak pressure (and other parameters) at the stagnation point is incorrect.

The important result is that the values of the maximum pressure surpass the maximum pressure at the stagnation point by 1.6 – 3.7 times in the range of impact Mach numbers, $1.2 > M_0 > 0.02$.

The interaction of a spherical decaying blast wave, caused by an underwater explosion, with a piece-wise plane target immersed in water was studied.

It was shown that, given a specific charge location, vulnerable areas on a target surface can be found where peak overpressures surpass those at the stagnation points by 4 times.

## ACKNOWLEDGMENTS

This work was supported under the auspices of the DND contract W7702-04R047

## REFERENCES

1. Gonor, A. L., Hooton, I. E., "Shock Wave Diffraction over an Inclusion in Condensed Matter and Hot Spot Generation at the Particle Surface in HE," in *Shock Compression of Condensed Matter-2003* , edited by M. D. Furnish et al., AIP Conference Proceedings 0-7354-0181-0/04, N. Y., 2004, pp..951-954
2. Walsh, J. M. and Christian, R. H., *Phys. Rev.*, **97**, 1544-1556, (1955).
3. Walsh, J. M., Rice, M. H., et al.,*Phys. Rev.*, **196**, 108-116 (1957).
4. Rice, M. H., *J of Phys. Chem.. Solids* **26**, 483 (1965).
5. Zel'dovich, Ya. B., and Raizer, Yu. P., *Physics of Shock Waves and High Temperature Phenomena*, v. 2, Academic Press, N. Y., 1966.
6. Gonor, A. L., Gottlieb, J. J., Hooton, I. E., *J. of Applied Physics*, **95**, 1577-1585 (2004).

CP845, *Shock Compression of Condensed Matter - 2005*,
edited by M. D. Furnish, M. Elert, T. P. Russell, and C. T. White
© 2006 American Institute of Physics 0-7354-0341-4/06/$23.00

# DOP TEST EVALUATION OF THE BALLISTIC PERFORMANCE OF ARMOR CERAMICS AGAINST LONG ROD PENETRATION*

**Fenglei HUANG    Liansheng ZHANG**

*The State Key Laboratory of Explosion Science, Beijing Institute of Technology, Beijing 100081, China*

**Abstract:**    A series of DOP tests with lateral confinement have been carried out and a linear relation between the residual penetration in RHA and the alumina thickness has been obtained. The rod configuration and the initial transient impact are the two factors that cause the gradual decrease of the differential efficiency factor (DEF) when the ceramic thickness is increased in literature. A new improved DEF definition is proposed to characterize the thick tile ceramic ballistic performance based on a more physical analysis.

**Key words:**    armor ceramics, DOP test, ballistic performance, DEF
**PACS:** 62.50

## INTRODUCTION

The ballistic performance evaluation of ceramics against long-rod projectile is one of the major aspects of its characterization. D.Yaziv et al. [1] defined the differential efficiency factor (DEF) of "black box" composite armor unit based on the depth of penetration (DOP) test. S.J. Bless et al. [2] proposed the semi-infinite backing DOP test concept to prevent the early tensile failure of ceramics. Based on the test results, the ceramic DEF can be calculated as,

$$DEF = \frac{(P_0 - P_r) \cdot \rho_0}{(t_c \cdot \rho_c)} \qquad (1)$$

where $P_0$ is the reference penetration of projectile in backing RHA, $P_r$ is the residual penetration, $t_c$ is the ceramic thickness, $\rho_0$ and $\rho_c$ are

densities of RHA and ceramics respectively.

The DOP test and the corresponding DEF definition in Eq. (1) have been widely accepted to rank the ballistic efficiency of ceramics against the long-rods projectiles. Detailed analysis of the published DOP test data [3-6] shows that a linear relation between the residual penetration $P_r$ and the ceramic thickness $t_c$ exists over a wide range of ceramic thicknesses. This linear relation indicates that the long rod penetration is a quasi-steady process and the ceramic resistance against penetration is constant and independent of the ceramic thicknesses. However, the ceramic DEF computed using Eq. (1) could decrease with the increase of ceramics thickness, as shown in Fig.1 [5] for silicon carbide, thus leads to an incorrect conclusion on the penetration process. This contradiction needs further improvements on the definition of DEF.

*Supported by the NSFC(10472014)

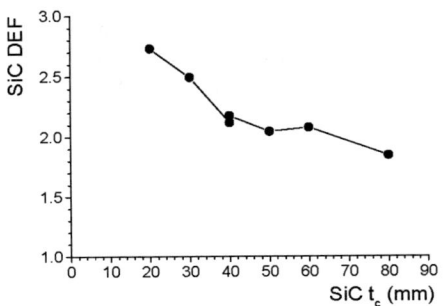

**Figure 1.** Silicon carbide DOP test results [5]

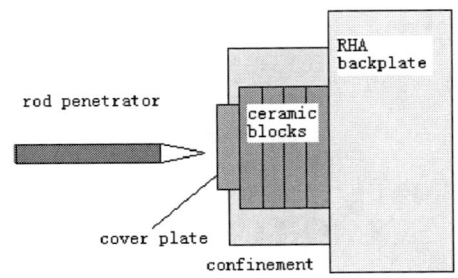

**Figure 2.** DOP test configuration

In this paper, a series of DOP tests have carried out to further investigate the relation of $P_r - t_c$ and $DEF - t_c$. An improved DEF definition is proposed that can get a more realistic ballistic efficiency of thick tile ceramics against long rod penetration.

### EXPERIMENTS AND RESULTS

The DOP test configuration is schematically represented in Fig.2. The out diameter of metallic cylinder is 150mm, the nominal diameter of ceramic tiles is 100mm.

The experimental tungsten-alloy rod projectile has a diameter of 6mm and L/d=14 with a sharp conical head. The nominal impact velocity is 1400 m/s and the reference penetration into RHA at this velocity is 56mm. Within the range of $1400 \pm 100$m/s a linear relation between the rod velocity and the RHA penetration is measured and all the obtained DOP test data is corrected by this linear relation. The tested ceramics is a 98% purity alumina with a density of 3.78 g/cm$^3$ and $H_V$=15GPa. Three kinds of target configurations shown in Table 1 are tested.

The ceramic DEF for every shot is calculated using Eq. (1) and two curves are presented in Fig.3.

### IMPROVED DEFINITION AND DISCUSSION

The residual penetration and the ceramic thickness have a linear relation as shown in Fig. 3(a). As for the $DEF - t_c$ curves, the calculated DEF vibrate first in the thin ceramic thickness range then become stable when the ceramic thickness is larger than 10mm. Based on the observation of this phenomenon, we define,

$$P_r = P_{r0} + \left(\frac{\partial P_r}{\partial t_c}\right) \cdot t_c \qquad (2)$$

**TABLE 1.** DOP test configurations

| Test No. | Cover plate | | Ceramic thickness (mm) | RHA Back plate (mm) | Remarks |
|---|---|---|---|---|---|
| | Material | Thickness (mm) | | | |
| 1 | No | 0 | 3~50 | 80 | Total ceramic |
| 2 | Steel | 10 | 3~50 | 80 | thickness is by |
| 3 | Al alloy | 20 | 10~50 | 80 | stacking |

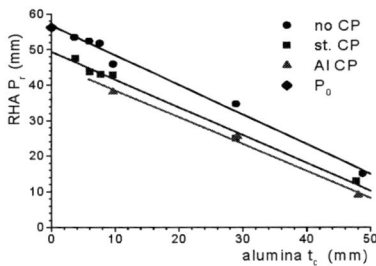

(a) $P_r - t_c$ relation

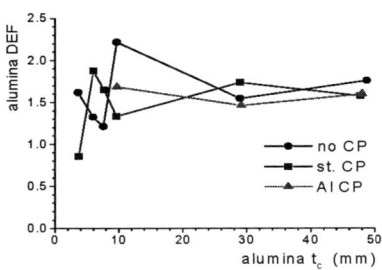

(b) $DEF - t_c$ relation

**Figure 3.** Alumina DOP test results

where $P_{r0}$ is the vertical intercept.

Substituting Eq. (2) into Eq. (1), we obtain,

$$DEF = \frac{(P_0 - P_{r0}) \cdot \rho_0}{t_c \cdot \rho_c} - \frac{\rho_0}{\rho_c} \cdot \left(\frac{\partial P_r}{\partial t_c}\right) \qquad (3)$$

Eq. (3) shows that the ceramic DEF decreases with the respect of the increase of the ceramic thickness because $(P_0 - P_{r0}) \neq 0$. Considering that the relation between $P_r$ and $t_c$ is linear when $t_c$ is not very small, the reasons that causes

$(P_0 - P_{r0}) \neq 0$ are the transient impact step and a possible interface effect.

In our DOP tests, the vertical intercepts of the linear correlations are very close to reference penetration in RHA such that $(P_0 - P_{r0}) \approx 0$.

From the above discussion, an improved DEF definition can be proposed to characterize the ceramic ballistic efficiency against long rod penetration as,

$$DEF = -\left(\frac{\rho_0}{\rho_c}\right) \cdot \left(\frac{\partial P_r}{\partial t_c}\right) \qquad (4)$$

The improved DEF definition in Eq. (4) is less sensitive to the negative influence of projectile head geometric configuration. Therefore the calculated DEF based on the improved DEF can be used not only to rank armor ceramics for its ballistic performance, but also to estimate the available material ballistic efficiency when ceramics is implemented into heavy composite armor structure.

Comparing to the DEF definition in Eq. (1) that contains too much structural effects of the test, the improved DEF defined in Eq. (4) is independent of the ceramic thickness and is a real characterization of the material ballistic performance.

The improved alumina DEF defined in Eq. (4) is about 1.6~1.7, which is the same as those calculated from Eq. (1). The improved SiC DEF defined by Eq. (4) is about 1.57, but the value calculated by Eq. (1) varies from 2.73 to 1.84. The improvements between these two definitions for SiC are obvious.

**CONCLUSION**

In DOP tests with lateral confinement, a linear relation exists between the residual penetration and the ceramic thickness. The improved DEF definition can represent the projectile-target interaction realistically. The improved DEF is less sensitive to the influence of projectile head shape that can avoid the averaging effect with ceramic thickness. Thus, the calculated improved DEF can be used not only to rank armor ceramics for its ballistic performance, but also to estimate the possible material ballistic efficiency.

## REFERENCES

1. Yaziv, D., Rosenberg, G., and Partom, Y., "Differential ballistic efficiency of appliqué armor", Proc. 9th Int. Symp. on Ballistics, Shrivenham,UK(1986)

2. Bless, S. J., Rozenberg, Z. and Yoon, B. "Hypervelocity penetration of ceramics", Int. J. Impact Engng., Vol.5,165 (1987)

3. Rozenberg, Z., and Yeshurun, Y., "The relation between ballistic efficiency and compressive strength of ceramic tiles", Int. J. Impact Engng., Vol.7,357 (1988)

4. Li, P., "Dynamic Response of Ceramic Material and its Mechanism against Rod Projectile Penetration", Ph.D. Dissertation, Beijing Institute of Technology, 2002 (in Chinese)

5. Rosenberg, Z., Dekel, E., Hohler, V., Stilp, J., and Weber, K., "Penetration of tungsten-alloy rods composite ceramic targets: experiments and 2-D simulations", Shock compression of condensed matter-1997, AIP, 1998

6. Hohler, V. A., Stilp, J., and Weber, K., "hypervelocity penetration of tungsten sinter-alloy rods into alumina", Int. J. Impact Engng.,Vol.17,409-418,1995

CP845, *Shock Compression of Condensed Matter - 2005*,
edited by M. D. Furnish, M. Elert, T. P. Russell, and C. T. White
2006 American Institute of Physics 0-7354-0341-4/06/$23.00

# MODELING OF BULLET PENETRATION IN EXPLOSIVELY WELDED COMPOSITE ARMOR PLATE

## Vasant S. Joshi [1] and Theodore C. Carney[2]

[1] *Research and Technology Department, Naval Surface Warfare Center, Indian Head, MD 20640*
[2]*EES-11, Geophysics, Los Alamos National Laboratory, Los Alamos, NM 87544*

**Abstract.** Normal impact of high-speed armor piercing bullet on titanium-steel composite has been investigated using smooth particle hydrodynamics (SPH) code. The objective is to understand the effects of impact during the ballistic testing of explosively welded armor plates. These plates have significant microstructural differences within the weld region, heat-affected zone and the base metal. The variances result in substantial ductility, hardness and strength differences, important criteria in determining the failure mode, specifically whether it occurs at the joint or within the virgin base metal. Several configurations of composite plates with different material combinations were modeled. The results were used to modify the heat treatment process of explosively welded plates, making them more likely to survive impact.
**PACS:** 62.20.Fe, 47.40.Nm, 62.50.+p
**Keywords:** Penetration, explosive welding, simulation

## BACKGROUND

Lightweight armor, consisting of steel and titanium plates, has been successfully produced by explosive welding technique [1]. Oriented heat treatment, performed on one surface of the steel-titanium composite, to increase its strength, introduced a brittle intermetallic material at the interface of the steel and titanium. This intermetallic material caused a steep gradient in material hardness and ductility, and could become the primary cause of delamination of the composite. In order to reduce the formation of this intermetallic material, it was necessary to reduce the energy imparted to the material interface, essentially producing a "Wave-less" interface [2]. This "Wave-less" interface is believed to have a lower bond strength in tension as well as in shear, as compared to a wavy interface, and is likely to cause failure during tensile loading [3].

Minimizing the intermetallic material, without reduction in strength of the composite necessitated development of alternate methods of creating a semi-compatible diffusion barrier at the interface, using fine layers of pure metals. Methods of producing composite plates with different interlayer materials were explored and the effects of variables on bond (interface) quality were also optimized [1]. Some of the composites with these improvements were found to resist penetration of a bullet in ballistic testing. At this stage, modeling of impact was undertaken to understand the effect of material properties and process variables on the ability of the composite, with different material configurations, to withstand impact of a bullet at various velocities. The purpose of the modeling effort was to make a preliminary assessment of the ability of specific lightweight armor concepts to defeat bullet penetration.

## EXPERIMENTAL CONFIGURATION

The composites produced by explosive welding consisted of a 0.250" thick top layer of A2/D2 tool steel, backed by a 0.225" thick titanium plate. The steel side faced the impact of a M2AP bullet (weight ~160 grains) at velocities ranging from 2500 up to 2700 ft/sec. While the bullet penetrated some of the plates completely, there were several instances where the plates withstood the first ballistic hit. During the second hit, these plates displayed partial or total delamination of the composite plate at the interface. These tested composite plates were sectioned to analyze the interface, and obtain cross-section profiles for comparisons with the modeling efforts. Detailed examination of the damaged area revealed that the steel material had brittle platelets with numerous cracks. The titanium surfaces appeared to exhibit a classic ductile fracture. Details of the metallurgical evaluations are beyond the scope of this paper. The cross-section of the plate at the location of bullet penetration is shown in figure 1.

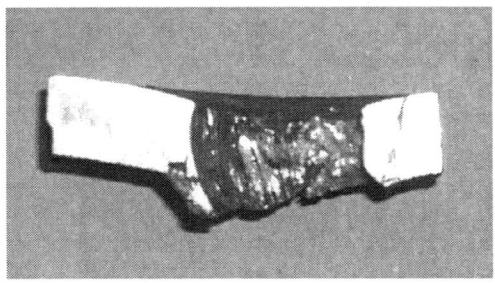

**Figure 1.** Cross- section of the explosively welded plate after impact at 2700 ft/sec.

## MODELING CONSIDERATIONS

The initial modeling efforts were aimed at obtaining velocities of bullets which would not result in penetration. In subsequent efforts, configurations and mechanical properties were modified to explore the possibility of producing a better composite. The modeling of the impact and penetration was conducted with a smooth particle hydrodynamic (SPH) code. SPH was originally invented for astrophysical gas dynamics simulations. Libersky and Petschek [4] extended the method by incorporating the entire stress tensor within the SPH framework, making the method applicable to problems in solid mechanics.

The foundation of SPH is the interpolation theory. The conservation laws of mass, momentum, and energy, plus material dependent constitutive relations are written in integral form using an interpolation function that provides a "kernel estimate" of the derivatives at a point. In SPH calculations, information is known at only a finite number of discrete points (particles), so the integrals are approximated as sums over neighboring particles. Consequently, a mesh is not required. The details of derivations, descriptions, and methods are given by Libersky, et.al. [5].

Since SPH is fully Lagrangian, it readily accommodates complex, history dependent constitutive models. These are necessary to model the broad range of material states obtained in a penetration event which include plastic flow, rate dependent hardening, thermal softening, damage accumulation, and finally, material separation.

## RESULTS AND DISCUSSION

During the initial simulations, the model consisted of a bullet impacting steel side of the composite. Initial simulations for the corrected geometry of the bullet and composite plate indicated no penetration for bullet velocity of 2100 ft/s. This configuration is shown in Figure 2.

Subsequent simulations were performed for bullet velocities of 2500 ft/s and 2700ft/s, which indicated that the bullet would not penetrate at lower velocities, whereas in actual tests, some of the plates were penetrated. At this stage, mechanical property tests were conducted on the cross sections of the armor, which provided additional inputs for modeling efforts, where the published data for the material was unavailable for A2/D2 steel with specific, oriented heat treatment. The material properties values in the input files were modified to reflect actual properties of the composite plate, and the simulations were performed again. Additional simulations with different plate thickness combinations were also performed. These include the thick steel/thin titanium, thin steel/thick titanium, and a sandwich configuration with a hard outer layer of $TiB_2$.

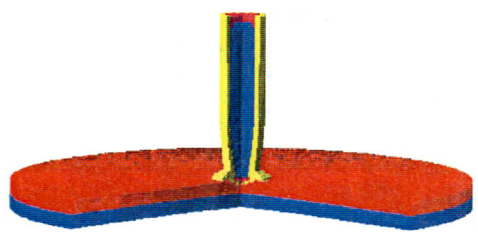

**Figure 2.** A 3D simulation of bullet impact on composite plate at 2100 ft/s.

The results of these simulations appeared to reasonably match to the observed cross section of the plate. Due to the limitation on space, only three sets of frames for 2500 ft/sec and 2700 ft/sec velocities are shown at 100 µs after impact.

Figures 3-5 display results from axisymmetric calculations of 2500 ft/s impacts on the 0.225" steel/0.250" Ti composite, 0.275" steel/.200" Ti composite, and the 0.055" TiB$_2$/.210" Steel/.210" Ti sandwich armor configurations, respectively. All show considerable back-surface deformation and some evidence of spallation. The bullets had slowed significantly by the 100 µs time (as shown) and we access that they would not have penetrated in any of the calculations.

Comparable results at 2700 ft/s impact velocity on the same plate configurations are shown in Figures 6-9. These exhibit more pronounced back-surface failure and spallation. Based on the observation that the bullet noses are essentially free by 100 µs, we access that the bullets would have penetrated in all three of the calculations The combined results from the six simulations indicate that the ballistic limit velocity for all three armor configurations was between 2500 and 2700 ft/s.

In an SPH code, particles interact with all neighbors unless rules are applied that limit or prohibit the interaction. These calculations would have benefited from algorithms that limited the shear interaction between the target material and bullet. In addition, the explosive weld behavior could have been better modeled with a cohesive law that gradually softened and released in response to damage accumulation. At velocities far above the ballistic limit these limitations would not have been evident.

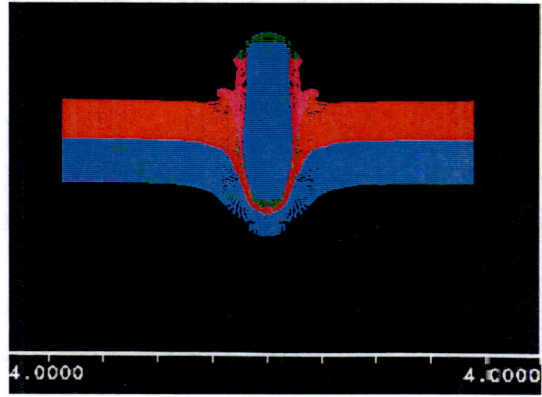

**Figure 3**. Profile for 2500ft/s at 100 us for 0.225" steel/0.250" Ti composite at 100 µs after impact

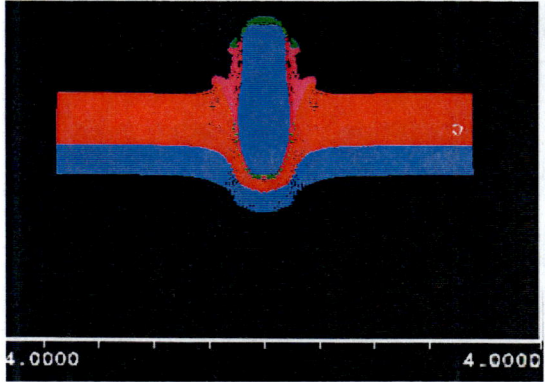

**Figure 4**. Profile for 2500ft/s at 100 us for 0.275" steel/0.200" Ti composite at 100 µs after impact

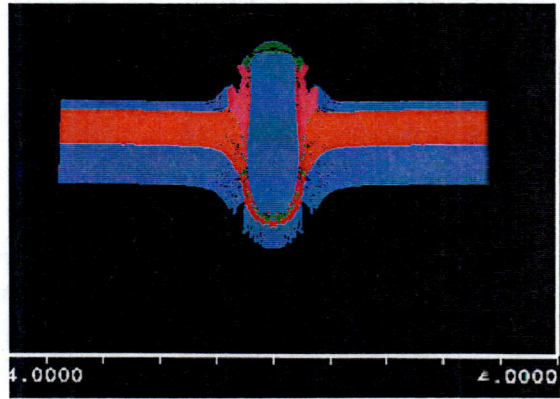

**Figure 5.** Profile for 2500ft/s 0.055" TiB2/ 0.210"Steel/ 0.210" Ti composite at 100 µs after impact

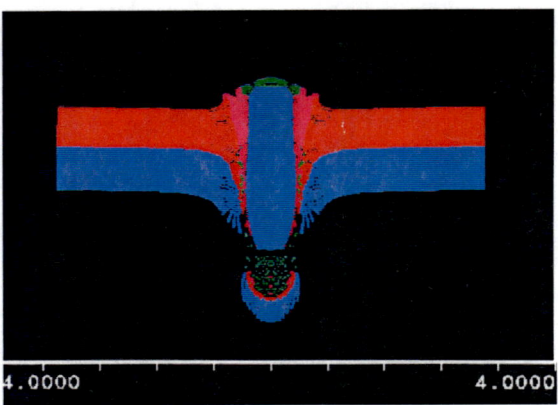

**Figure 6.** Profile for 2700 ft/s at 100 μs after impact for 0.225" steel/0.250" Ti composite

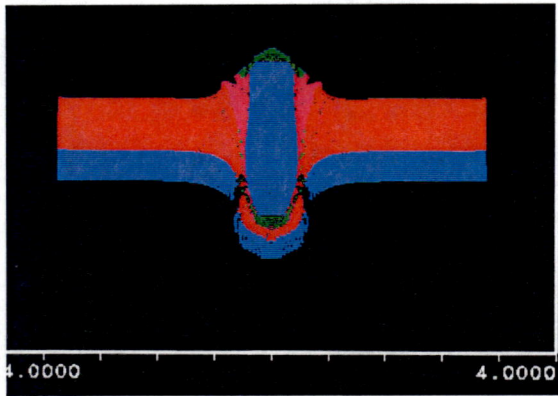

**Figure 7.** Profile for 2700 ft/s at 100 μs after impact for 0.275" steel/0.200" Ti composite

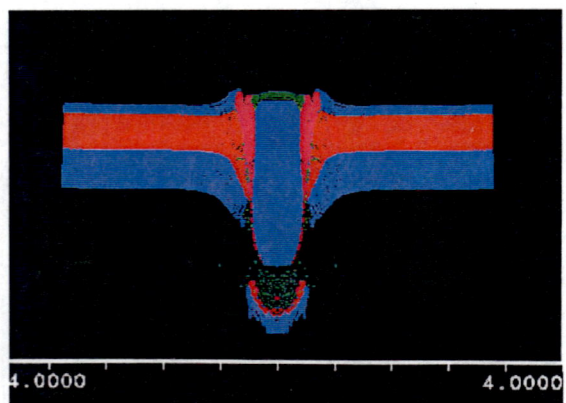

**Figure 8** Profile for 2700 ft/s at 100 μs after impact for 0.055" TiB2/0.210"Steel/0.210" Ti composite

# CONCLUSIONS

This combined experimental and modeling effort allowed us to make a reasonable assessment of the ballistic limit velocity of explosively welded armor plate against an M2 armor-piercing bullet. Future refinement in the computational approach has the potential of improving the predictive capability of the calculations. Although strength of the material in composite cannot be increased further due to inherent limitation in the infrared heat treatment process, limited modeling effort indicated improvements required for obtaining a better composite armor.

# ACKNOWLEDGEMENTS

The experiments and calculations presented in this paper were performed when both authors were employees of the Energetic Materials Research and Testing Center at the New Mexico Tech, Socorro, NM. The authors thank John Osowski and William Bingham, also at New Mexico Tech in performing multiple runs of simulations.

# REFERENCES

1. Joshi, V. S., Banks, M. and Krebsbach, J., Shock Compression of Condensed Matter, 1999 (M.D. Furnish, L.C. Chhabildas, R.S. Hixson, eds.), part II, pp. 943-947.
2. Crossland, B., Explosive Welding of Metals and Its Applications, Clarendon Press, 1982.
3. Blazynski, T. Z., Explosive *Welding, Forming and Compaction*, Applied Science Publishers, U.K., 1983.
4. Libersky, L.D. and A.G. Petschek, "Smoothed Particle Hydrodynamics with Strength of Materials", Lecture Notes in Physics, 395, Advances in the Free-Lagrange Method, eds. H.E. Trease, , M.J. Fritts, and W.P. Crowley. pp. 248-257. Springer-Verlag, NY, 1991.
5. Libersky, L.D., Petschek, A.G., Carney, T.C., Hipp, J.R., Allahdadi, F.A., "High Strain Lagrangian Hydrodynamics," Journal of Computational Physics, 109, pp 67-75, 1993.

CP845, *Shock Compression of Condensed Matter - 2005*,
edited by M. D. Furnish, M. Elert, T. P. Russell, and C. T. White
© 2006 American Institute of Physics 0-7354-0341-4/06/$23.00

# FAILURE WAVE IN DEDF AND SODA-LIME GLASS DURING ROD IMPACT

## D. L. Orphal[1], Th. Behner[2], V. Hohler[2], C. E. Anderson Jr.[3] and D. W. Templeton[4].

[1]*International Research Associates, Inc., 4450 Black Avenue, Pleasanton, CA 94566, USA.*
[2]*Fraunhofer Institut für Kurzzeitdynamik (Ernst-Mach Institut), Eckerstr. 4, 79104 Freiburg, Germany.*
[3]*Southwest Research Institute, 6220 Culebra Road, San Antonio, TX 78228-0510, USA.*
[4] *U. S. Army RDECOM-TACOM, AMST-TR-R, Warren, MI 48397 USA*

**Abstract.** Investigations of glass by planar, and classical and symmetric Taylor impact experiments reveal that failure wave velocity $v_F$ depends on impact velocity, geometry, and type of glass. $v_F$ typically increases with impact velocity $v_P$ to between $c_S$ and $c_L$ or to $\sqrt{2}c_S$ (shear and longitudinal wave velocity). This paper reports initial results of an investigation of failure waves associated with gold rod impact on high-density (DEDF) glass and soda-lime glass. Data are obtained by visualizing simultaneously the failure propagation in the glass with a high-speed camera and the rod penetration velocity u with flash radiography. Results for DEDF glass are reported for $v_P$ between 1.2 and 2.0 km/s, those for soda-lime glass with $v_P \approx 1.3$ km/s. It is shown that $v_F > u$, and that in the case of DEDF glass $v_F/u$ decreases from 1.38 to 1.13 with increasing $v_p$. In addition, several Taylor tests were performed. For both DEDF and soda-lime glass the $v_F$-values, found here as well as $v_F$- data reported in the literature, reveal that—for equal pressures—the failure wave velocities determined from Taylor tests or planar-impact tests are distinctly greater than those observed during steady-state rod penetration.

**Keywords:** Failure wave, glass, rod impact.
**PACS:** 62.50.+p, 62.30.+d, 62.20.Mk, 81.05.Kf

## INTRODUCTION

Shock experiments have proved useful in extending understanding of the fracture character-istics of brittle materials. Planar-impact tests and classical and symmetric Taylor tests in combination with high-speed camera pictures reveal several features of the failed zone [1-11]: A recompression in the free surface velocity record occurs, caused by the interaction of the release wave with the front of the failed zone; the spall strength in the failed material decreases to nearly zero; the longitudinal strain increases; the lateral stress increases and, as a consequence, the shear stress decreases. Furthermore, the propagation speed of the failure front, i.e. the failure wave velocity $v_F$ depends on the impact geometry and materials of the projectile and target; increases with impact pressure and approaches values between the transverse and longitudinal wave speeds $c_S$ and $c_L$, respectively, or approaches $c_L$. According to [8], for symmetric Taylor impacts at high velocity $v_F \approx \sqrt{2}c_S$. Similar results are found for edge-on impact tests, for the so-called damage velocity $v_D$, which is determined by the nucleation between shock and fracture front [12]. $v_F$ and $v_D$ are different, but it appears that $v_F$ and $v_D$ approach similar values at low and high impact velocities. For W-alloy rod impact into soda-lime glass $v_F$-values of 1.4 and 2.4-2.6 km/s at impact velocities $v_P$ of 1 km/s and 3.1-4.1 km/s are reported [1,13]. The $v_F$-values were determined from camera pictures and inferred from a change of the Tate term R, respectively. These $v_F$-values are less than $c_S = 3.5$ km/s.

This paper reports initial results of an investigation of the failure kinetics for gold rod impact on DEDF and soda-lime glass. DEDF glass (glass with high PbO content) was selected because of its relatively low $c_S$- and $c_L$-values of 2.0 and 3.5 km/s. Gold rods were used to eliminate strength effects of the penetrator. Impact velocities between 1.2 and 2 km/s have been considered for DEDF and about 1.3 km/s for soda-lime glass. $v_F$ was determined from high-speed camera pictures, and the penetration velocity of the gold rod, u, was measured using flash X-radiography. For compareison, the failure propagation for Taylor anvil impact for both glass types has also been investigated.

## EXPERIMENTS

The DEDF glass targets (Pilkington DEDF805254A) were cylindrical in shape with diameters D = 20 mm and lengths of L = 15 to 20 mm. The material properties are given in Table 1. The glass samples were launched with a two-stage light-gas gun against stationary gold rods (reverse impact mode). The dimension of the rods were d = 1 mm and l = 50 mm; the material properties are: 99.99% gold, density $\rho_P = 19.3$ g/cm$^3$, hardness 65 HV5, UTS 220 MPa and elongation 30%. An Imacon 200 high-speed camera, taking 16 pictures, was used to visualize the failure propagation in the glass. A flash X-ray observation of the penetrating gold rod in the glass was not possible due to the high lead content of the DEDF glass. Therefore, the glass was backed and glued to an Al cylinder of D = L = 20 mm (the material properties of the Al backing are: Al-alloy AlCuMgPb, density 2.85 g/cm³, UTS 350 MPa, 117 BHN2.5). Thus, the penetration velocity u and the residual rod length were measured in the Al backing by taking five 180 kV flash X-ray pictures. The average penetration velocity u and the average erosion velocity $v_e = \Delta L/\Delta t$ in the DEDF glass was then calculated from time of impact on the glass to time of impact on the Al. The rod was positioned about 2 m from the gun muzzle to keep the yaw angle of the glass sample as small as possible

(typically less than 1°). The aim point of the gun showed some spread, with the result of having up to 3 mm off-centre impacts. The camera and the five flash X-ray tubes were triggered with a laser about 10 mm in front of the gold rod. The times of the flash X-rays were recorded with digital oscilloscopes (resolution 0.4 ns). Thus, the uncertainty in time was less than ±5 ns. The accuracy of the positions determined from the X-rays was about ± 0.1 to 0.15 mm. The positions determined from the camera pictures were dominated by irregularities in the shape of the failure front with typical uncertainties less than ± 0.2 mm. A few Taylor tests were conducted with the DEDF sample at ≈ 400 m/s. The samples were launched with a powder gun against a high-hard RHA anvil with dimensions 100x100x40 mm (510 HV20).

**TABLE 1**. Material properties of DEDF and soda-lime glass.

| Glass type | DEDF | soda-lime |
|---|---|---|
| Density [g/cm$^3$] | 5.19 | 2.5 |
| Young's mod. [GPa] | 54.5 | 73.3 |
| Hardness [kp/mm²] | Kn341 | HV540 |
| Poisson's ratio | 0.24 | 0.23 |
| $c_L$ [km/s] | 3.52 | 5.84 |
| $c_S$ [km/s] | 2.06 | 3.46 |
| HEL [GPa] | 4.3-4.5 [4, 9] | 6.0 [7] |

Several tests were also performed with soda-lime glass targets at $v_P \approx 1.3$ km/s. The material properties of soda-lime glass are given in Table 1. In some of the experiments a PCO camera (instead of the Imacon 200) was used, taking only four pictures with no additional X-rays. For these soda-lime glass experiments, targets with both circular and square-shaped cross sections were launched with a two-stage light-gas gun against gold rods, soda-lime glass plates, and against steel plates.

## RESULTS FOR DEDF GLASS

Figure 1 shows selected camera pictures of gold rod impact at four different times. The failure front is expanding with a hemispherical shape. The shock front and the gold rod are not visible inside the glass.

In the third and the fourth pictures, some nucleation takes place in front of the failure wave.

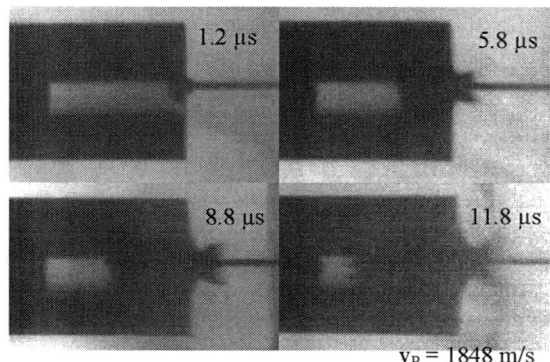

**Figure 1.** Failure wave propagation in DEDF glass during gold rod impact.

This effect has been observed in [3, 11, and 12]. It is indicated that nucleation takes place preferentially at higher impact velocities. Average failure wave velocities $v_F$ in the DEDF glass were calculated from the camera pictures and average penetration velocities u from penetration in the Al backing. These data are listed in Table 2. It is noted that $v_F < c_S$ and that the ratio $v_F/u$ decreases with increasing $v_P$—from about 1.38 to 1.13—in the velocity range considered here.

For comparison, Fig. 2 shows a camera picture, taken during a Taylor test 5.1 µs after impact. Here, the failure front shape is nearly planar and the failure characteristics correspond qualitatively to those already observed with other glasses, e. g., soda-lime or borosilicate glass [3, 5, 6, and 11]. The average $v_F$-value of 3.18 km/s is between $\sqrt{2}c_S$ (2.86 km/s) [8] and $c_L$ (3.52 km/s).

## RESULTS FOR SODA-LIME GLASS

The tests with soda-lime glass were done around $v_P \approx 1.3$ km/s. Figure 3 compares camera pictures of the failure propagation during rod impact and a Taylor test with a glass plate "anvil". The qualitative behavior is similar to that observed for the DEDF, with a difference being that the soda-lime glass shows a more irregular lateral expansion in the Taylor test [5, 11]. The u- and $v_F$-data of the different setups are listed in Table 3. For the Taylor test $v_F \approx 4$ km/s; lower than $\sqrt{2}c_S$ (4.89 km/s), but more than double the $v_F$-values for rod impact, which are between 1.46 and 1.67 km/s.

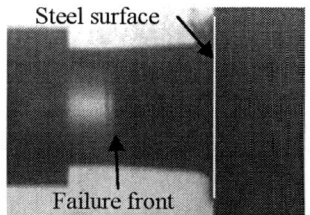

**Figure 2.** Failure propagation during Taylor test DEDF glass).

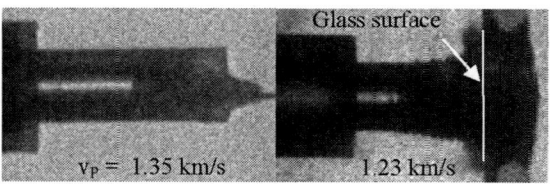

**Figure 3.** Failure propagation during rod impact and Taylor test (soda-lime glass).

**TABLE 2.** DEDF glass data.

| Exp. No. | $v_P$ [km/s] | u [km/s] | $v_F$ [km/s] |
|---|---|---|---|
| 10425 | 1.17 | 0.69±2% | 0.95±4% |
| 10424 | 1.40 | 0.76±5% | 1.03±3% |
| 10427 | 1.85 | 1.10±4% | 1.19±3% |
| 10428 | 1.93 | 1.14±7% | 1.23±3% |
| 10429 | 2.07 | 1.22±10% | 1.38±3% |

**TABLE 3.** Soda-lime glass data.

| Exp. No. | Setup | $v_P$ [km/s] | u [km/s] | $v_F$ [km/s] |
|---|---|---|---|---|
| 4537 | O/rod 0.75 | 1.35 | - | 1.58±2% |
| 4542 | O/rod 0.75 | 1.30 | - | 1.46±4% |
| 4539 | []/ rod 0.75 | 1.22 | - | 1.49±4% |
| 4540 | O/glass | 1.23 | - | 4.0±8% |
| 4543 | O/ steel | 1.32 | - | 4.16±2% |
| 10581 | [ ]/rod 1 | 1.12 | 0.61±2% | - |
| 10583 | [ ]/rod 1 | 1.27 | 0.79±1% | 1.67±3% |

O: round targets, D = 9, L= 50 mm
[]: square targets, 10x10x50 mm
[ ]: square targets, 15x15x47 mm
rod 0.75: gold rod, d = 0.75, l = 50 mm
rod 1: gold rod, d = 1, l = 70 mm
glass: soda-lime glass plate, 6x150x150 mm
steel: steel plate, 2x100x300 mm, 105 HV5

## DISCUSSION OF RESULTS

In the gold rod-DEDF tests it was observed that the ratio $v_F/u$ decreases with increasing with $v_P$.

This means that the distance between the failure front and the tip of the rod becomes smaller with increasing $v_P$. Since the failure process is time dependent, it seems a reasonable hypothesis that the degree of glass comminution in front of the rod tip decreases more and more as $v_P$ increases and, as a consequence the Tate term R should increase with $v_P$ and – for very high $v_P$ – should converge versus the maximum value of the unfailed glass – similar to the behavior inferred in [13] for soda-lime glass.

In the Taylor test example (Fig. 2), $v_F = 3.18$ km/s is measured. The corresponding pressure is 5.1 GPa (acoustic approximation). The corresponding $v_F$-value at equal hydrodynamic pressure $\frac{1}{2}\rho_P(v_P-u)^2$ at the rod tip is $v_F \approx 1.15$ km/s ($v_P \approx 1.75$, $u \approx 1.03$ km/s; $Y \approx 0$). This means that $v_F$ for the Taylor test is greater than $v_F$ for rod impact at equal 1D and hydrodynamic pressures.

In the case of soda-lime glass, $v_F$-values of about 4.1 km/s were observed for glass-bar impact on glass plate and steel plate (Table 3). In contrast, the $v_F$-values for rod impact are distinctly lower, only about 1.5 km/s, and agree with the $v_F$ data reported in [1,13].

The corresponding hydrodynamic pressure is around 2.2 GPa. At equal pressures, $v_F \approx 2$-4.8 km/s is reported for Taylor and planar-impact tests [3, 5, 11]. These differences in $v_F$ can be explained by the different stress-strain-states: 1D-strain state for planar-impact test, 1D-stress state for the Taylor test, and 3D-stress-strain state for hydrodynamic rod penetration.

## CONCLUSIONS

The initial results reported here examine gold rod impact on DEDF and soda-lime glass. The results extend the knowledge concerning failure kinetics in brittle materials. It is shown for rod impact that the ratio of failure to penetration velocity $v_F/u$ in DEDF glass decreases with increasing impact velocity. This is tentatively interpreted as being a consequence of the a fully comminuted region in front of the rod tip becoming less extensive (less time for damage nucleation and growth). This results in an increase in the Tate term R. $v_F$-values determined from Taylor test or planar-impact test are distinctly greater than those found during steady-state rod penetration, for equal 1D strain, 1D stress, and hydrodynamic pressures. This indicates that in the latter case a 3D-stress-strain state (divergent) has to be considered, and implies that damage nucleation/growth is stress-state dependent.

## REFERENCES

1. Bless, J.S., Brar, N.S., and Rosenberg, Z., "Failure of ceramic and glass rods under dynamic compression," in *Shock Comp.of Cond. Matter-1989* (S. C. Schmidt, et. al., Ed) AIP Conf. Proc., pp. 939-942 (1990).
2. Rasorenov, S.V., Kanel, G.I., Fortov, V.E., and Abasehov, M.M., "The fracture of glass under high-pressure impulsive loading," *High Pressure Research*, **6**, 225-232, (1991).
3. Bourne, N.K., Rosenberg, Z., and Field, J.E., "High-speed photography of compressive failure waves in glasses," *J. Appl. Phys.*, **78**(6), 3736-3739 (1995).
4. Bourne, N.K., Millett, J.C.F., and Rosenberg, Z., "Failure in a shocked high-density glass," *J. Appl. Phys.*, **80**(8), 4328-4331 (1996).
5. Murray, N.H., Bourne, N.K., Field, J.E., and Rosenberg, Z., "Symmetrical Taylor impact of glass bars", *Shock Comp. of Cond. Matter-1997* (S. C. Schmidt, et. al., Ed) AIP Conf. Proc. 429, pp. 533-536 (1998).
6. Espinosa, H.D., Xu, Y., and Brar, N.S., "Micromechanics of failure waves in glasses: I. Experiments," *J. Am. Ceram. Soc.*, **80**, pp. 2061-2073 (1997).
7. Bourne, N.K., Millett, J.C.F., and Field, J.E., "On the strength of shocked glasses," *Proc. Royal Soc. London A*, **455**, 1275-1281 (1999).
8. Rosakis, A.J., Samudrala, O., and Coker, D., "Cracks faster than the shear wave speed," *Science*, **284**, 1337-1340 (1999).
9. Radford, D.D., Proud, W.G., and Field, J.E., "The deviatoric response of three dense glasses under shock loading conditions", *Shock Comp. of Cond. Matter-2001* (M.D. Furnish, et. al., Ed) AIP Conf. Proc. 620, pp. 807-810 (2002).
10. Radford, D.D., Willmott, G.R., Walley, S.M., and Field, J.E., *J. Phys. IV France*, **110**, 687-692 (2003).
11. Willmott, G.R., and Radford, D.D., "Taylor impact of glass rods," *J Appl. Physics*, **97**(9): 93522-1-8 (2005).
12. Senf, H., Straßburger, E., and Rothenhäusler, H., "Visualization of fracture nucleation during impact in glasses", *Proc. Int. Conf. on Metallurgical and Mat. Appl. of Shock Wave and High Strain Rate Phen.-1995*, pp. 163-170, Elsevier Science B.V., Amsterdam (1994).
13. Zilberbrand, E.L., Vlasov, A.S., Cazamias, J.U., Bless, J.S., and Kozhushko, A.A., "Failure wave effects in hypervelocity penetration," *Int. J. Impact Engng.*, **23**, 995-1001 (1999).

CP845, *Shock Compression of Condensed Matter - 2005*,
edited by M. D. Furnish, M. Elert, T. P. Russell, and C. T. White
© 2006 American Institute of Physics 0-7354-0341-4/06/$23.00

# DRX-INDUCED SOLID-STATE FLOW AND PROJECTILE-TARGET MIXING DURING [001] SINGLE-CRYSTAL TUNGSTEN ROD PENETRATION INTO STEEL TARGETS

## C. Pizaña[1], L.E. Murr[1], I.A. Anchondo[1], C.Y. Piña[1], M.T. Baquera[1], T.L. Tamoria[2], H.C. Chen[2], S. J. Cytron[3]

[1]*Department of Metallurgical and Materials Engineering, University of Texas at El Paso, El Paso, TX 79968*
[2]*General Atomics, San Diego, CA 92121 USA*
[3]*U.S. Army TACOM-ARDEC, Picatinny, NJ 07806 USA.*

**Abstract.** Residual [001] single-crystal W rods penetrated into steel targets have been examined by light and electron microscopy. The post-impact residual penetrators examined using energy-dispersive x-ray mapping, revealed target and penetrator mechanical mixing. Considerable intercalation activity was found to concentrate specifically within the material being eroded by DRX-assisted flow. The solid-state flow features (including shear bands) facilitate the mixing of the two. Residual microstructures obtained within the penetrator suggest localized melt zones due to thermal instabilities caused by the turbulent behavior in the high-pressure regime.

**Keywords:** Solid-state flow, projectile-target mixing, single-crystal tungsten, long-rod penetration.
**PACS:** 47.40.-x, 47.40.Nm.

## INTRODUCTION

In impact and penetration studies involving ballistic long-rod kinetic energy penetration into steel targets, dynamic recrystallization (DRX) has been established to be the main mechanism for solid-state flow in both the target and projectile [1-7]. Furthermore it has very recently been discovered that such projectiles/targets undergo solid-state (mechanical) alloying and/or mixing induced by severe plastic deformation during penetration. In some instances localized melt zones due to thermal instabilities caused by the turbulent behavior contribute to the alloying and/or mixing. Although these aspects of penetration, specifically mechanical alloying and/or mixing, have not been investigated in ballistic impact and penetration scenarios where both projectile and target flow in the solid state, evidence observed in

post-penetration projectile fragments studied earlier [4], suggests that the intercalation between the projectile and target affects the penetration process and/or performance.

## EXPERIMENTAL PROCEDURE

Two rod penetrator projectiles of [001] oriented single-crystal tungsten produced by CVD were studied. These were initially machined to hemi-nosed rods with a length over diameter ratio of approximately 15 (l/d~15) by General Atomics, Inc. (USA). The rods were subsequently launched at the Army Research Laboratory (ARL) in Aberdeen, MD into thick, steel targets at nominal velocities ranging from 1.254 to 1.350 km/s using a gun breech assembly. Rod striking velocities along with orientations at impact were recorded and documented. In-target penetrator samples were subsequently collected, cut longitudinally, polished and etched with equal amounts of $H_2O$, $3\%H_2O_2$, and $NH_4OH$ for the W imbedded projectile, 2% Nital for the target, and 5g $FeCl_3$, 2mL HCL, 100mL ethanol for the mixture zones. Samples were also examined by scanning electron microscopy (SEM) and the energy-dispersive (X-ray) spectrometer (EDS) attached to the SEM prior to optical metallography.

## RESULTS AND DISCUSSION

Fig.1 illustrates an optical microscope cross-section obtained from a head fragment produced during shot ID #1 (~1.254 km/s). The fragment exhibits various large flow and deformation features which are consistent with head penetration and/or solid-state erosion. Such features include the large shear band observed on the upper right corner in Fig.1 as well as the deformation twins that apparently flow into the same band (bottom right arrow (T) Fig.1). Although these features are typical of penetration, other features also seem to play an important part in the development of such shear bands. Further observations of head fragments revealed target material flowing into the shear bands (bottom left arrow, Fig. 1) becoming part of the erosion vehicle.

Specific zones within the penetrator developed their own unique intercalation activities. Zones consistently found at the target/penetrator interface, (Fig. 2a) mostly along the tube and head surface of the in-target penetrator, consist of tungsten fragments (Fig. 2b) which appear to be dendritic, suggesting extreme heat generation along the interface. The mixture is considerably harder than both the target and penetrator (mixture hardness ~1100VHN, W~480VHN, target~370VHN). EDS X-ray spectrums obtained from this type of mixture, show evidence of Fe and W intercalation (probably intermetallic $Fe_xW$). Within individual constituents of the intercalation shown in Fig. 2, EDS X-ray analyses indicate that the matrix (Fig. 2c) has a higher concentration of Fe while the particles within the matrix (dendrites), contain a higher concentration of W (Fig. 2d). Fig. 3a is an SEM micrograph depicting a different intercalation activity. This type of intercalation is found strictly within the erosion parts of the penetrator, specially

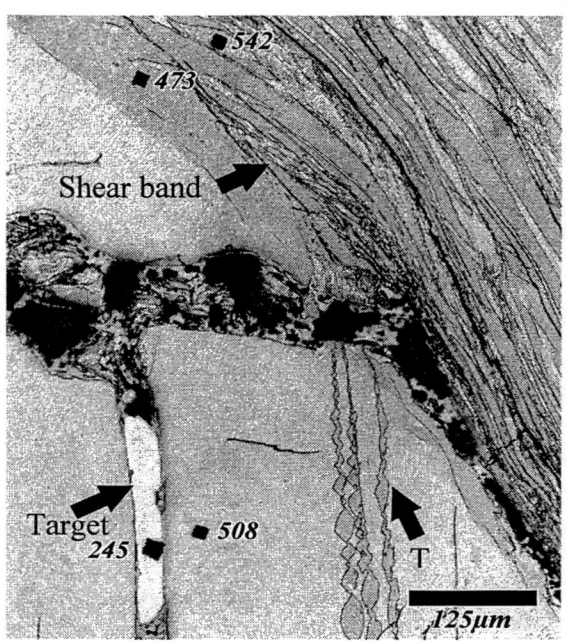

**FIGURE 1.** Target material (arrow) intercalating into penetrator shear band.

target/penetrator interface the fragments found in these zones, are not dendritic but rather small, equiaxed grains of W within a Fe matrix. Furthermore, evidence in the form of EDS X-ray mapping (Fig. 3b-c) shows that the intercalation is not necessarily a W-Fe alloy, but rather a mechanical mix of W grains and Fe dendrites (as compared to the W/Fe mixture dendritic structure, Figs. 2b and d where EDS showed evidence of an intermetallic phase). These zones can also be observed optically as well (Fig. 4). Due to DRX flow extending the erosion channel (tube), the intercalation within these zones involves large and small dynamically recrystallized equiaxed tungsten grains in an Fe matrix as noted (Fig. 4a). Fig. 4b depicts the same zone using Nital 2% (target etchant) in order to enhance the contrast of the Fe matrix.

## CONCLUSIONS

The most important result of this work is the detection of target material within the penetrator. Observations have revealed considerable intercalation activity to specifically concentrate

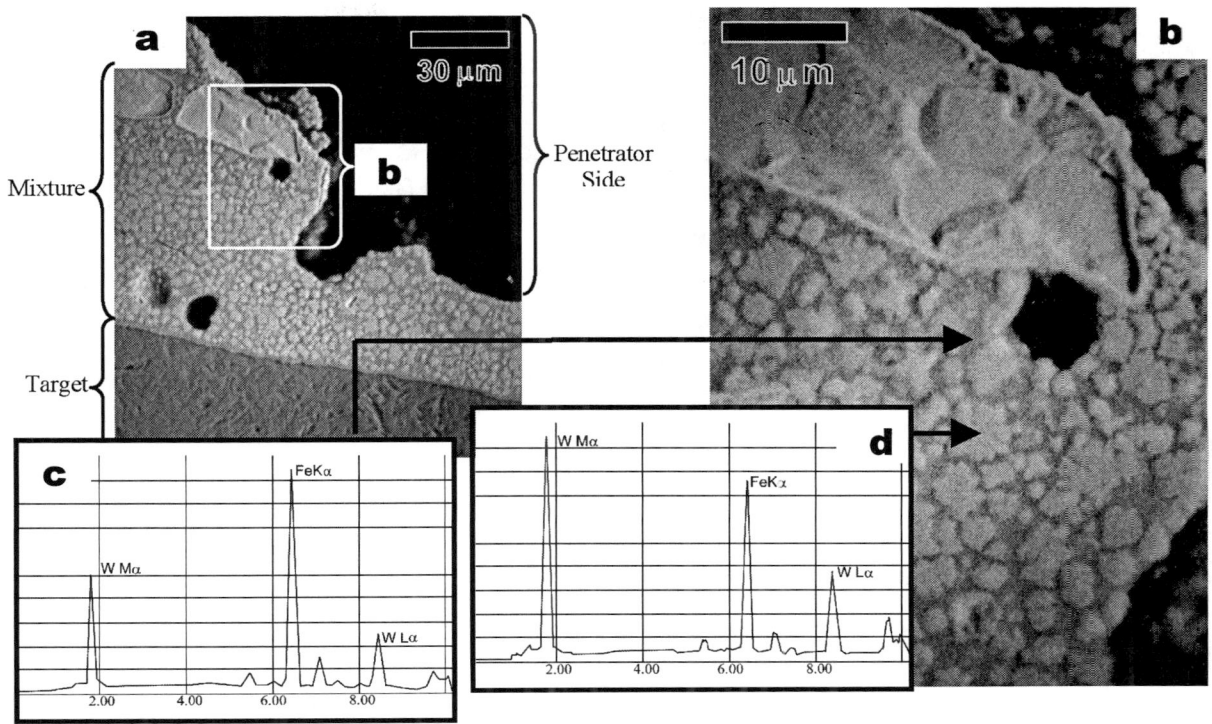

**FIGURE 2.** (a) SEM photograph of Target/penetrator interface. (b) Mixture shows W dendritic structure in a Fe matrix. (c) EDS X-ray spectrum of dark grey material (matrix). (d) EDS X-ray spectrum of W dendritic structure.

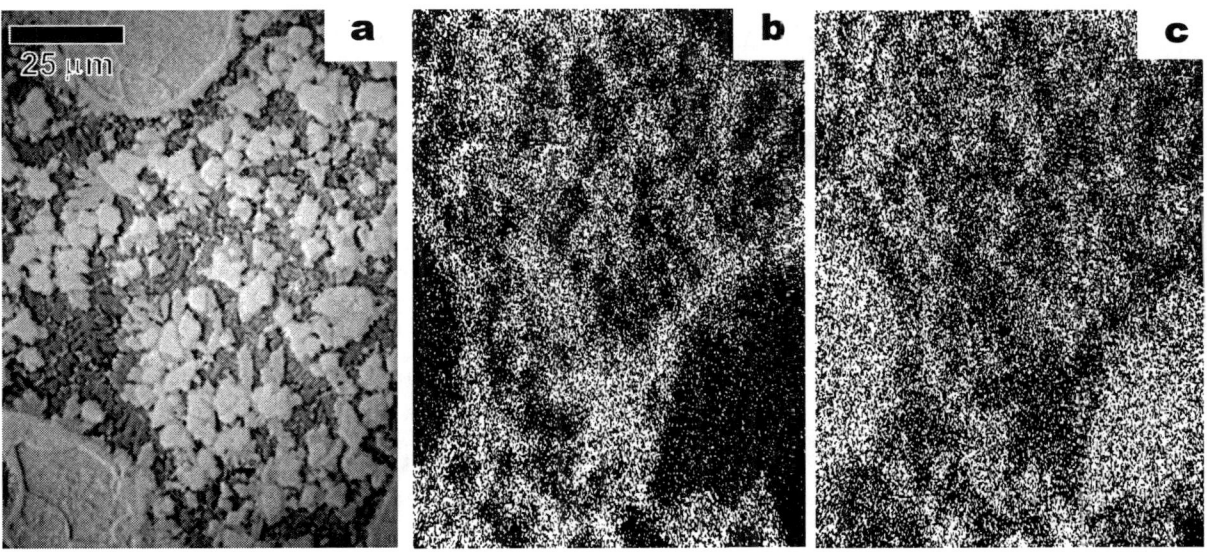

**FIGURE 3.** (a) SEM micrograph of target/penetrator tube mixture. (b) Fe X-ray map of micrograph. (c) W X-ray map of micrograph.

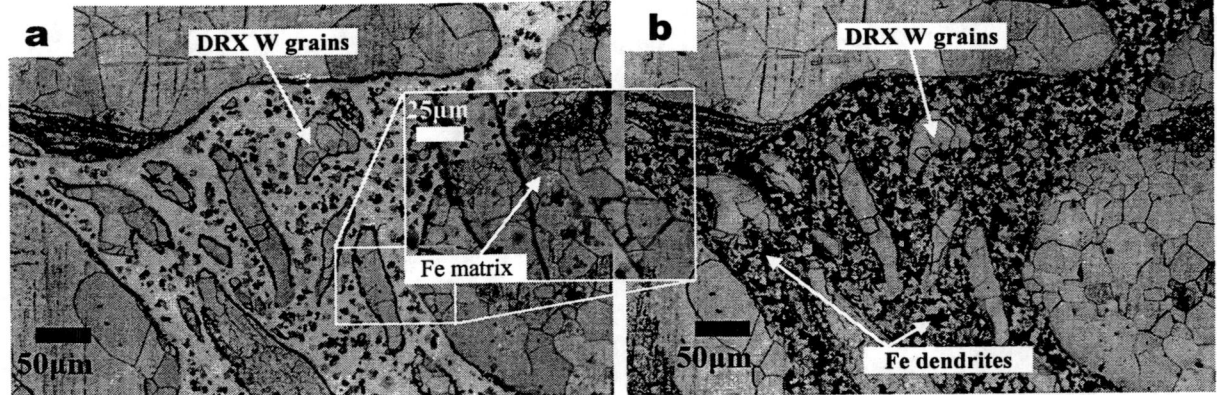

**FIGURE 4.** Optical photograph of W/Fe mixture behind pentrator's head (tube). (a) W etchant only, (b) adding the Fe etchant.

recrystallized (DRX)-assisted flow ultimately forming the penetator tube (Fig 4). The solid-state flow features (including shear bands) facilitate the projectile/target mixing. In the two samples studied DRX was profuse developing in both the erosion tube and head of the penetrator (Fig. 4), considerably increasing the mixing activity.

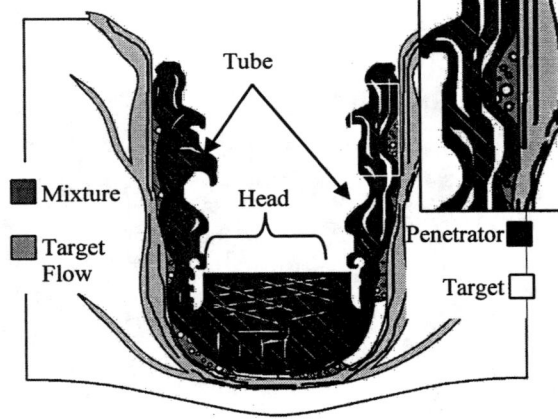

**FIGURE 5.** Schematic of in-target penetrator depicting zones of mixture.

## ACKNOWLEDGEMENTS

Supported by the U.S. Army TACOM-Picatinny Arsenal, prime contract No.W15QKN-04-M-0267, project No. 1A4CFJER1ANG.

## REFERENCES

1. Pappu, S. et al., "Microstructure analysis and comparison of tungsten alloy rod and [001] oriented columnar-grained tungsten rod ballistic pnetrators", Mater. Sci. Engng., A262, 115, 1999.
2. Trillo, E. A. et al., "Dynamic recrystallization-induced flow phenomena in tungsten–tantalum (4%) [001] single-crystal rod ballistic penetrators", Mater. Charater., 48, 407, 2002.
3. Esquivel, E. V. et al., "Comparison of flow and shear band structures in oriented, columnar tungsten, single crystal tungsten-tantalum and sintered tungsten heavy alloy ballistic penetrators", Powder Met. 46(2), 137, 2003.
4. Pizaña, C. et al., "The role of dynamic recrystallization in [001] single-crystal W and W-Ta alloy ballistic rod penetration into steel targets" J. Mater. Sci. 40, 2005.
5. Murr, L. E. et al., "Effect of initial microstructure on high velocity and hypervelocity impact cratering and crater-related microstructures in thick copper targets: Part II Stainless steel projectiles", J. Mater. Sci., 32, 3143, 1997.
6. Quinones, S. A., and Murr, L. E., "Correlations of computed simulations with residual hardness mappings and microstructural observations of high velocity and hypervelocity impact craters in copper" Phys. Stat. Sol. (a), 166, 763, 1998.
7. Kennedy, C., and Murr, L. E., "Comparison of tungsten heavy-alloy rod penetration into ductile and hard metal targets: microstructural analysis and computer simulations", Mater. Sci. Engng., A325 131, 2002

CP845, *Shock Compression of Condensed Matter - 2005*,
edited by M. D. Furnish, M. Elert, T. P. Russell, and C. T. White
© 2006 American Institute of Physics 0-7354-0341-4/06/$23.00

# THE TAYLOR IMPACT AND LARGE STRAIN RESPONSE OF POLY(ETHER-ETHERKETONE) (PEEK)

## Philip J. Rae and Eric N. Brown

*MS-G755, PO Box 1663, LANL, Los Alamos, NM 87545*

**Abstract.** Taylor impacts experiments were conducted on PEEK at velocities between 150 & 360 m s$^{-1}$. The material was found to respond in a ductile manner and exhibit a color change later found to be associated with large compressive strains in PEEK, irrespective of strain-rate. No changes in molecular weight were detected as a result of high-strain rate or large strain deformation. Melting has been shown not to be responsible for the ductile deformation and limited tearing response of PEEK subject to Taylor impact.

**Keywords:** PEEK, large strains, Taylor impact
**PACS:** 61.41.+e, 62.50.+p

## INTRODUCTION

The Taylor test[1] involves firing a right cylinder of test material against a semi-infinite rigid anvil. The test was originally posed as a method of measuring the dynamic yield strength of metals, however, more recently it has been used as a dynamic validation tool for computer based models. When coupled with high-speed photography the response of visco-elastic materials such as polymers can be measured. Recently Millett et. al. published an investigation into the Taylor response of an unspecified grade of commercially supplied PEEK[2]. A number of interesting findings were noted relating to the ductile nature of the material in this test and material discoloration in the vicinity of the most deformed areas. Millett reports that at velocities between 152 and 349 m s$^{-1}$ PEEK failed in a ductile manner with increasing mushrooming of the impact zone and tearing. At a velocity of 408 m s$^{-1}$ a sufficient tensile stress was developed behind the impact zone to cause a ductile failure of the rod. Concave rod ends revealed significant post impact relaxation. Discoloration was noted in the rod end regions exhibiting the greatest residual deformation and this was attributed to either shock and strain-heating or an oxidation process from the air present during the impact.

The current work extends previous mechanical characterization of PEEK 450G by the authors to include Taylor impact experiments which verify and expand the investigation by Millett et al. Some previous investigations of the compressive properties have been undertaken by Walley et al.[3] on an unspecified grade of PEEK and Swallowe[4].

## MATERIAL

A commercial plate of extruded PEEK 450G was purchased measuring $457 \times 475 \times 19$ mm$^3$. The material has been characterized in tension and compression at various temperatures and strain rates. Using helium pycnometry a density of $\rho = 1311 \pm 1$ kg m$^3$ was measured. Using differential scanning calorimetry (DSC) a material crystallinity of 41% was calculated by integrating the melt endotherm and relating it to the literature value for 100% crystalline PEEK 450G[5]. An estimate of the molecular weight of the material was made and the weight averaged molecular weight of this plate is $\approx 28000$.

**TABLE 1.** Inherent viscosity of various PEEK samples dissolved in concentrated (99%) sulphuric acid as a measure of molecular weight.

| Material | Inherent Viscosity / dl g$^{-1}$ |
| --- | --- |
| Virgin PEEK 450G | 0.88±0.02 |
| Taylor impact zone, 23°C at 311 m s$^{-1}$ | 0.89±0.02 |
| Taylor impact zone, 23°C at 346 m s$^{-1}$ | 0.87±0.02 |
| Taylor impact zone, 100°C at 347 m s$^{-1}$ | 0.90±0.02 |
| Hopkinson sample, 23°C at 3000 s$^{-1}$ | 0.88±0.02 |

## EXPERIMENTAL

Taylor cylinders were machined from the plate with dimensions of 7.62 mm diameter by 38.1 mm long. Rods were fired at velocities between 150 & 360 m s$^{-1}$ and temperatures of 23 and 100°C. An Imacon 200 high-speed framing camera coupled to a Cordin 463 proportional delay generator was used to record back-lit images of the impacts. In all cases a 350 ns exposure was used and 16 frames were recorded with 15$\mu$s inter-frame time.

Recovered specimens were potted in epoxy prior to polishing. Micro-hardness measurements were made on a Buehler Micromet hardness tester using the Vickers geometry with a 25 g load and 20 second dwell time.

Large strain compression tests were undertaken on an MTS 810 servo-hydraulic machine in the case of the quasi-static tests and a custom built 6 m drop tower in the case of the intermediate strain rate tests. In the case of the quasi-static tests the cross-head was controlled to load the sample at a constant true strain-rate of $1 \times 10^{-3}$ s$^{-1}$. In the case of the drop tower the residual true strain was measured from the starting and relaxed dimensions of the sample and the quoted strain-rate is an average of the measurement throughout the test.

Estimates of the molecular weight of the PEEK samples were undertaken by Polymer Solutions Inc., Virginia, USA using the dissolved viscosity method from Devaux et al.[6]. Samples were dissolved in concentrated sulphuric acid and allowed to stand for equal lengths of time to normalize the sulphonation of the polymer chains. Viscosity measurements were made to ASTM D2857-95(2001) at 30°C. Devaux provides a plot of weight averaged molecular weights versus intrinsic viscosity and this was used to estimate the molecular weights of our samples as

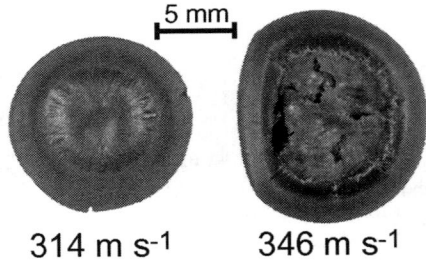

**FIGURE 1.** End on view of two typical Taylor cylinders fired at 314 & 346 m s$^{-1}$ .

described in the results section.

## RESULTS

Approximately 30 cylinders were fired at velocities between 150 & 360 m s$^{-1}$ and at temperatures of 23 and 100°C. No substantial differences were noted in the response of the material at the two temperatures that would not be expected from the lower yield strength at elevated temperatures. The rest of this paper therefore deals exclusively with the qualitative and quantitative differences due to impact velocity.

As reported by Millett, all Taylor specimens deformed in a ductile manner with no formation of fragments at any velocity studied. At low velocities, 150-250 m s$^{-1}$, a classic three diameter deformation pattern was noticed with the largest residual strains at the impact face and the back of the projectile undeformed. The specimen end remained reasonably flat and no cracking was observed. At medium velocities, 250-315 m s$^{-1}$ more extensive deformation was found at the impact face with the onset of small radial cracks that arrested before material separation occurred. In this velocity range visco-elastic effects retracted the centre of the sample back after impact and resulted in a concave rod end with the appearance of a solidified 'melt' pool, as seen in figure 1. At higher velocities, 315-360 m s$^{-1}$, tearing in the visco-elastic pullback zone is observed with increasing effect at higher velocities, figure 1. Interestingly, the radial cracking present at medium velocities is absent at higher ones.

At all velocities darkening of the rod is noticed in the highly deformed regions. These observations are consistent with those of Millett. Efforts were therefore made to understand the origin of the 'melt' pool and the darkening. It is easy to establish that it is implausible that PEEK reaches the melt temperature under the conditions imposed at these impact velocities. The melt temperature of PEEK is 342°C at atmospheric pressure rising to 400°C under the application of a hydrostatic stress of 100MPa[7]. Assuming a specific heat capacity for PEEK of $c_p =2180$ J kg$^{-1}$ °C$^{-1}$[8] (N.B. in polymers typically $c_p \approx c_v$) a perfectly plastic yield strength of $\sigma_y =150$MPa and a maximum strain of $\varepsilon =3$ (a considerable over estimate) the temperature rise assuming 100% of the input energy is converted to heat is approximately 160°C ($\Delta T = \sigma_y \varepsilon / c_p \rho$). Shock heating might give rise to a further rise of ca. 10°C. Therefore the end state temperature from an initial one of 23°C is no more than 200°C, well short of the required melt temperature. The end section exhibiting the 'melt' like appearance must therefore be a result of the visco-elastic spring back.

Figure 2 clearly shows the concave rod end and the discoloration associated with the large strain regions. Investigations were therefore focused to identify if the color change was a result of the strain-rate or stress state. Figure 3 shows a montage of virgin and three large strain cross-sections of PEEK prepared and photographed under identical conditions. It is therefore clear that the color change is associated with large strain compression, not strain-rate. PEEK in tension undergoes stress whitening in common with many other polymers. Samples tested in compression to strain under $\approx 0.5$) did not show significant color change.

At modest strains (20-40%) PEEK has an almost constant flow stress at quasi-static rates. Tests were done to see if under large strains the material work hardens. Figure 4 shows two representative curves. In one case a right cylinder 6.35 mm in diameter by

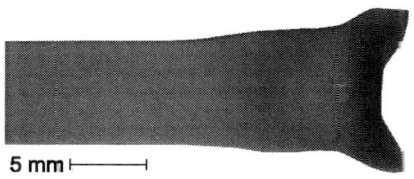

**FIGURE 2.** Cross-section of a PEEK Taylor cylinder fired at 275 m s$^{-3}$ and 100°C. The color change associated with the high-strain regions is obvious.

| Virgin PEEK | $\varepsilon$=1.2, 150 s$^{-1}$ |
|---|---|
| $\varepsilon$=0.77, 10$^{-3}$ s$^{-1}$ | $\varepsilon$=1.1, 10$^{-3}$ s$^{-1}$ |

**FIGURE 3.** Polished sections of four PEEK samples photographed under identical conditions to show the color change associated with large-strain deformation.

6.35 mm tall was loaded in a single deformation to large strains. Despite careful lubrication on highly polished WC platens, there can be no doubt the result was affected by friction at larger strains. In an effort to remove this artifact and see if the hardening behavior is real, a similar sized sample was incrementally loaded with re-machining and lubrication between cycles. No sample barreling was observed at any stage on the incremental loading yet the work hardening rate seems higher that the friction affected sample. This behavior is not fully understood but may involve cold crystallization taking place between loadings. In any event it seems that PEEK does work harden at high strains.

Hardness measurements were made on various deformed PEEK samples to see what effect large strains and discoloration had. Figure 5 shows measurements on a sectioned Taylor cylinder and a sample tested in a drop weight at a strain rate of $\approx 150$ s$^{-1}$. In both cases it is clear that material subject to large strains is softer than undeformed material. For reference, virgin PEEK samples had a hardness of $25.0 \pm 0.6$ under the same conditions. The hardness measurements so far reported are approximately perpendicular to the loading axis, therefore a quasi-statically loaded sample was sectioned in two orthogonal directions to

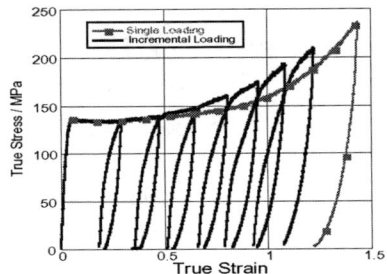

**FIGURE 4.** Stress vs. strain plot for PEEK. The curves show the response to a single and incremental loading.

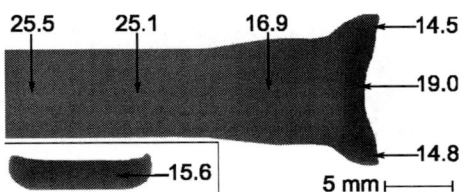

**FIGURE 5.** Polished section of a typical Taylor sample showing the measured Vickers hardness values at various points. Below is a cross-section of a small cylinder deformed at $150$ s$^{-1}$ to a strain of $\varepsilon = 1.1$ with a corresponding hardness measurement.

see if the material appeared oriented, figure 6. While the hardness value along the loading axis is higher than perpendicularly, each measurement is considerably softer than undeformed PEEK. It is not clear how the stress-strain curve work hardening data and the indentation hardness discrepancy can be rationalized at this point.

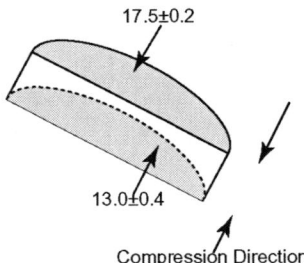

**FIGURE 6.** Orthogonal Hardness measurements on an incrementally loaded cylinder, $1 \times 10^{-3}$ s$^{-1}$, $23°$C, $\varepsilon_{residual} = 1.07$.

Samples of PEEK deformed at various temperatures and under differing conditions were tested to see if the molecular weight of the polymer was influenced, see table 1. Statistically, no difference was observed between samples, removing the possibility of chain scission as a method of polymer softening at high strain rates.

## CONCLUSIONS

In Taylor tests PEEK exhibits ductile deformation mechanisms at velocities between 150-360 m s$^{-1}$ with no material becoming detached as fragments. PEEK exhibits a darkening associated with large strain compression ($\varepsilon > 0.5$), irrespective of strain-rate. Vicker's hardness tests show that large strains soften PEEK, however quasi-static stress vs. strain curves show strain hardening above strains of $\varepsilon > 0.5$. PEEK shows no change in molecular weight upon deformation to large strains at high strain-rates.

## ACKNOWLEDGMENTS

The authors gratefully acknowledge financial support for this research from the joint DOD/DOE Office of Munitions memorandum of understanding project on the dynamic behavior of polymers. The research was performed under the auspices of the US Dept. of Energy.

## REFERENCES

1. Taylor, G., *Proc. Roy. Soc. Lond. A*, **194**, 289–299 (1948).
2. Millett, J., Bourne, N., and Stevens, G., *Int. J. Impact Eng.*, **In Press** (2005).
3. Walley, S., and Field, J., *Dymat J.*, **1994**, 59–71 (1994).
4. Hamdan, S., and Swallowe, G., *J. Pol. Sci. Pt. B: Pol. Phys.*, **43**, 699–705 (1996).
5. Jonas, A., Legras, R., and Issi, J.-P., *Polymer*, **32**, 3364–3370 (1991).
6. Devaux, J., Delimoy, D., Daoust, D., Legras, R., Mercier, J. P., Strazielle, C., and Nield, E., *Polymer*, **26**, 1994–2000 (1994).
7. Maeda, Y., *Polymer Communications*, **32**, 279–284 (1991).
8. Victrex, Victrex peek materials properties data table, Tech. rep., www.victrex.com (2000).

CP845, *Shock Compression of Condensed Matter - 2005*,
edited by M. D. Furnish, M. Elert, T. P. Russell, and C. T. White

# AN ANALYTIC MODEL OF CLOSE-RANGE BLAST FRAGMENT LOADING

## Ernst Rottenkolber[1], Werner Arnold[2]

1. *NUMERICS GmbH, Mozartring 6, D-85238 Petershausen, Germany*
2. *EADS TDW – Gesellschaft für verteidigungstechnische Wirksysteme mbH*
*P.O. Box 1340, D-86523 Schrobenhausen, Germany*

**Abstract.** The effects of blast-fragmentation warheads need to be carefully characterized in a variety of applications like passive and active vehicle protection or hard target defeat and TBM defense. With these applications in mind, we have developed a collection of tools called FI-BLAST (**F**ast **I**nterface for **Bl**ast-Fragment **L**oad **A**nalysis of **St**ructures). In the present paper we describe the essential part of these tools, namely the close range blast-fragment model. The meaning of "close range" is here defined as the standoff to a charge at which blast effects can inflict serious damage on massive structures. In order to quantify our model's range of validity, examples of measured and calculated momentum of bare and confined charges are given in the present paper. Short (L/D = 0.5) and long (L/D = 5) cylindrical charges are included as well as spherical charges. The presented examples demonstrate that the model gives reasonable results in the intended domains of application.

**Keywords:** blast fragmentation warhead, close range model, 3D Gurney formalism, hard target defeat, tactical ballistic missile, vehicle protection.
**PACS:** 89.20.Dd

## INTRODUCTION

The effects of blast-fragmentation warheads need to be carefully characterized in a variety of applications. To design passive protective measures for vehicles, one would like to know pressure loads being applied in a structural analysis. Active protection against attacking missiles may intend to use a counter missile carrying a blast charge. Hard targets like bunkers are defeated by penetrators containing an explosive charge. The damage done by the blast loads to equipment stored in the bunker may then be of interest. To defeat TBMs, recently semi-armor piercing (SAP) warheads were suggested to be effective by their combined blast-fragment loads.

With these applications in mind, we have developed a collection of tools called FI-BLAST (**F**ast **I**nterface for **Bl**ast-Fragment **L**oad **A**nalysis of **St**ructures). The term "fast" characterizes both the time needed by an engineer to set up the problem and the time consumed by a computer to solve the problem. Specifically, FI-BLAST is not a hydrocode. Problem setup times can be measured in hours, solver times in fractions of a second. In the present paper we describe the essential part of these tools, namely the close range blast-fragment model. The meaning of "close range" is here defined as the range wherein blast effects can inflict serious damage on massive structures. Physically, the close range includes – but is not restricted to - the extension of the fireball evolving after a detonation.

In order to quantify our model's range of validity, examples of bare and confined charges are given in the present paper. Short and long cylindrical charges are included as well as spherical charges. We compare our model to hydrocode simulations, but the emphasis is on correlations with experiments, demonstrates that the model gives reasonable results in the intended domains of application.

## THE CLOSE RANGE BLAST MODEL

The blast model is based on the flow field of explosive products calculated by SPLIT-X [1]. The calculation utilizes the generalized Gurney-method developed by Rottenkolber and Arnold [2]. Originally, this method was meant to deliver fragment velocities of blast-fragmentation charges. However, as a by-product also the velocity field of the reaction products is obtained (Figure 1).

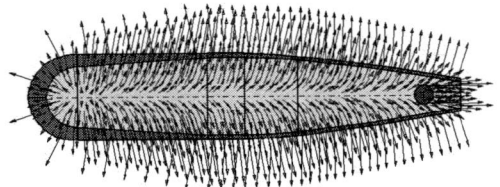

**FIGURE 1.** Velocity field of the reaction products calculated by SPLIT-X.

Since we know the mass and the velocity of the charge elements, we can accumulate the momentum projected into specified polar angle intervals (Figure 2). According to Baker et al. [3], the reflected near field impulse of a spherical charge detonating in air is given by:

$$i_r = \frac{\sqrt{2M_T E}}{4\pi R^2} \qquad (1)$$

where the symbols denote the following quantities:

$i_r$ = Reflected Impulse
$M_T$ = Total mass of explosive source plus air engulfed by the shock front at radius R
$E$ = Available energy of explosive source
$R$ = Distance to the center of the explosive charge

In the spirit of equation (1), the momentum calculated by SPLIT-X is corrected to include the effects of the air engulfed by the shock front. It is assumed that the total momentum calculated in this way is transferred to a structure which presents its surface perpendicularly to the flow.

Emphasis is put on the fact that in our model an explosive charge is completely characterized by its geometry, its confinement and the material parameters density, detonation velocity and Gurney-energy. No other additional assumptions or empirical parameters enter our model.

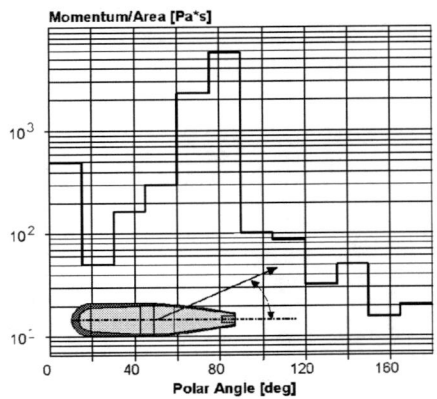

**FIGURE 2.** Momentum of reaction products as a function of the polar angle.

## VERIFICATION EXAMPLES

### Spherical Charges

Hydrocode simulations were performed to assess the accuracy of Equation (1). TNT spheres were detonated and the impulse reflected at the surface of a rigid plate was calculated for several configurations (see Figure 3). The distance D was kept at 0.4 m. The masses of the TNT-spheres were 2 g, 16 g, 125 g, 1 kg and 8 kg. In Figure 4, the reflected impulse obtained from the hydrocode simulations is compared to the proposed model.

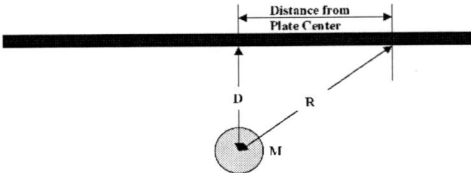

**FIGURE 3.** Configuration of hydrocode simulations.

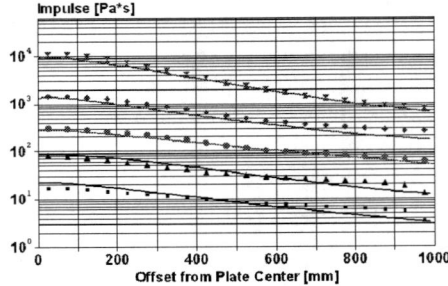

**FIGURE 4.** Analytical model (solid lines) compared to hydrocode (discrete data). From bottom to top, charge masses are 2 g, 16 g, 125 g, 1 kg and 8 kg.

## Cylindrical Charges

In practical problems we are primarily interested in non-spherical charges. Artillery shells or penetrators filled with explosives usually resemble a cylindrical charge with a high L/D-ratio. Held [4] measured the blast contours of bare charges with L/D = 0.5, 1, 2, and 4 using momentum blocks. We modeled these charges with a mass of 1.04 kg in SPLIT-X using a density of 1.54 g/cm³ and a Gurney velocity of 2700 m/s. Examples of computed velocity fields are shown in Figure 5.

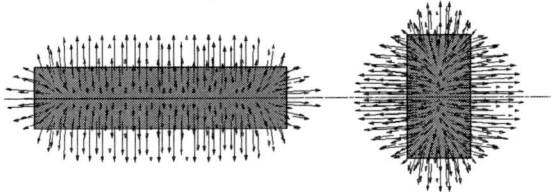

**FIGURE 5**. SPLIT-X velocity fields for L/D = 4 and L/D = 0.5 charges.

Measured and calculated impulses are compared in Figure 6. We can draw the following conclusions:

- The model neglects the "fine-structure" of the blast contour, i.e. the so-called bridge waves at 45° and 135°. It also neglects the influence of the initiation.
- The model correctly predicts an increasing radial impulse as the L/D-ratio increases.
- The model correctly predicts a decreasing axial impulse as the L/D-ratio increases.
- The model gives an adequate description of the overall blast contour.

## Confined Charges

When a charge is confined by a metal casing, the momentum transferred to a target is generally increased. Usually the casing of a shell or penetrator is so heavy that most of the momentum is transferred by fragments and only a small fraction of momentum is carried by the explosive products or the shock wave in air.

The momentum of heavily confined charges was measured by Held and Tan [5]. They used a catcher plate mounted onto a wheeled sled whose velocity was measured to obtain the transferred momentum.

We modeled the experiment with the FI-BLAST routines. A C-4 charge (D = 60 mm, L = 290 mm,

1.2 kg) was confined by 5 mm and 10 mm thick steel casings. The standoffs being 1 m, 1.5 m, 2 m and 3 m between the catcher plate and the charge were modeled (Figure 7).

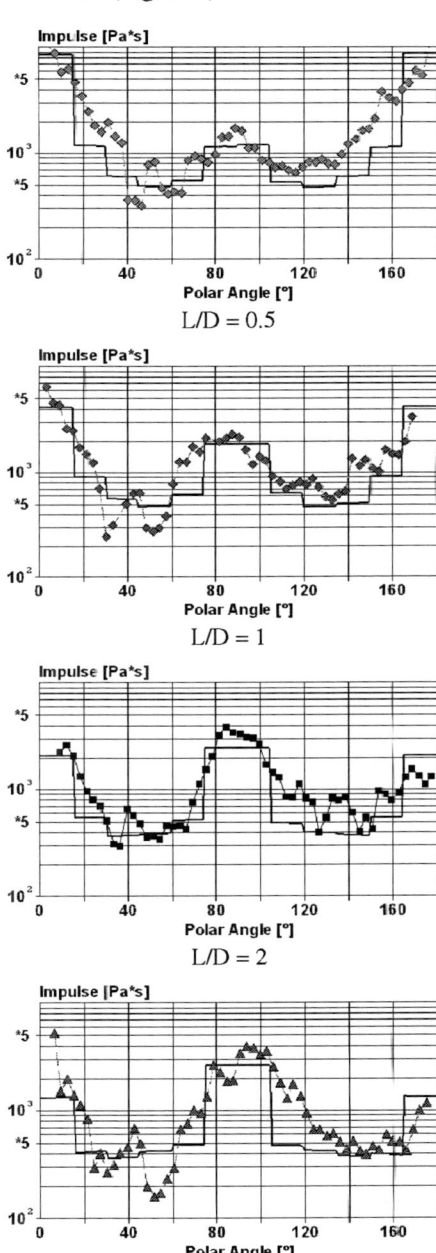

**FIGURE 6**. Model results (solid lines) and measured impulse at 0.5 m [4] for L/D = 0.5, 1, 2 and 4 charges.

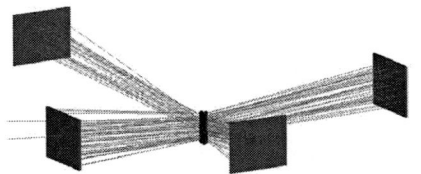

**FIGURE 7.** Model of natural fragments of the confined charge and catcher plates at different stand-offs.

The momentum imparted to a catcher plate by both fragments and blast wave was used to calculate the velocity of the 51.5 kg sled. A comparison with measured values is given in Figure 8. It is quite satisfying that the correlation is good in the case of combined blast-fragment loads, which is the most relevant case in practical applications.

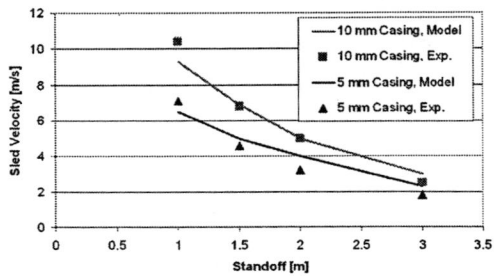

**FIGURE 8.** Calculated and measured sled velocities.

## SUMMARY AND CONCLUSIONS

We have presented a close range blast-fragment model based on fragment and detonation product velocities calculated with SPLIT-X. A correction term accounting for the influence of air engulfed in the blast wave gives the final reflected blast impulse per area. No additional empirical fits or parameters have to be evoked. Consequently, our close range blast model is as close to first physical principles as possible within the existing analytical framework.

In a number of examples we gave quantitative correlations of the model with experiments. Thus, we are now in a position to draw the following conclusions concerning the intended applications.

- *Passive Vehicle Protection*

A suitably protected vehicle has to withstand the blast load of a bare 15 kg charge at a standoff of 3 m. The correlation of our model to hydrocode simulations has shown that the situation is within the claimed close range. We conclude that pressure boundary conditions for structural simulations of protected vehicles can be reliably extracted from our close range blast model.

- *Active Vehicle Protection*

There are concepts of destroying missiles approaching a vehicle by a counter missile carrying a blast charge. To inflict sufficient damage to a massive missile part, like a shaped charge, the necessary mass of explosive is so large and the permissible miss distance is so small that this application lies well inside the validated range of our model.

- *TBM Defeat*

A 25 kg charge in close contact to a TBM may be necessary to destroy all chemical submunitions inside a TBM. This is clearly within the "close range" for which we claim the validity of our model. Thus, we can safely apply this model within a warhead assessment code like TBM-Xpert [6].

- *Hard Target Defeat*

In the case of hard target defeat, the warhead is usually a penetrator with a strong steel casing. Damage to electronic equipment may occur at very low shock and pressure levels. These are not the subject of the current close range model. To destroy massive structures within a bunker, the blast and fragment effect must be similar to that in the TBM case. Again, this can be properly analyzed with our blast-fragmentation model.

## REFERENCES

1. SPLIT-X V5, "An Expert System for the design of Fragmentation Warheads, User's Manual", *NUMERICS GmbH*, Petershausen, 2004.
2. Rottenkolber, E., Arnold, W., "A Generalization of the Gurney Formalism to Three Dimensions", *21st Int. Symp. Ball.*, Adelaide, Australia, 2004.
3. Baker, W. E., Westine, P. S., Dodge, F. T., "Similarity Methods in Engineering Dynamics", *Elsevier*, Amsterdam, 1991.
4. Held, M., "Blast Effects with the Held Momentum Method", *21st Int. Symp. Ball.*, Adelaide, 2004.
5. Held, M., Tan, G. E. B. , "Radial Blast Loads of Confined Cylindrical Charges", *11th International Symposium on Interaction of the Effects of Munition with Structures (ISIEMS 2003)*, May, 2003, Mannheim, Germany.
6. Arnold, W., Rottenkolber, E., „TBM-Xpert – A New Endgame Code: Features and Validations", *Proceedings of the 22nd International Symposium on Ballistics,* Vancouver, BC, Canada, 14 – 18 November 2005

CP845, *Shock Compression of Condensed Matter - 2005,*
edited by M. D. Furnish, M. Elert, T. P. Russell, and C. T. White

# FAILURE OF A LONG-ROD PROJECTILE OBLIQUELY INTERACTING WITH A THREE-LAYER TARGET

## S. A. Zelepugin[1] and N. S. Dorokhov[2]

[1]*Dept. for Structural Macrokinetics, Tomsk Scientific Centre, SD RAS, Tomsk, 634021 Russia*
[2]*Research Institute of Steel JSC NII Stali, Moscow, 127411 Russia*

**Abstract.** Results of experimental and numerical research of the interaction of a tungsten-alloy long-rod projectile with a three-layer target at an angle of $60^0$ and with a velocity of 1600 m/s are presented. The material of the middle layer of the target was an elastomer or reacting mixture in which shock-induced solid-state exothermic chemical reaction can take place. Analysis of the data obtained shows that the character of failure of the projectile qualitatively depends on the material of the middle layer. For elastomer layer the bend and destruction of the head part of the projectile prevail, for the reacting mixture – "thermoshock" in an interaction zone, loss of strength properties of the material of the projectile subjected to high temperature, and solid-state detonation-like process takes place.

**Keywords:** Shock waves, failure, chemical reaction.
**PACS:** 47.40.Nm, 62.50.+p, 82.40.Fr

## INTRODUCTION

Wide application of various long-rod projectiles in physics and mechanics of shock wave phenomena causes continuous interest in research of the features of their interaction with targets in conditions of high velocity impact [1-5]. Despite many experimental and theoretical works in this field, the problem is far from a complete solution.

Results of experimental and numerical research of the interaction of a tungsten heavy alloy long-rod projectile with a three-layer target are presented in this paper. Experimental data were obtained at the Research Institute of Steel JSC NII Stali [6] with use of X-ray technique. Numerical simulations were conducted at the Department for Structural Macrokinetics, Tomsk Scientific Center, Siberian Division of the Russian Academy of Sciences. Simulations were carried out by the finite element method with use of erosion failure model.

## FORMULATION OF THE PROBLEM

Numerical investigations were carried out to analyze in detail the features of the interaction of long-rod projectiles with the three-layer target with various thicknesses of the layers. The set of equations for describing unsteady adiabatic motion of an elastoplastic medium, including nucleation and accumulation of microdamages and temperature effects, consists of the equations of continuity, motion, and energy [7, 8]. To numerically simulate the failure of the material under high velocity impact, we applied the active-type kinetic model determining the growth of microdamages, which continuously change the properties of the material and induce the relaxation of stresses [9]. The strength characteristics of the medium (shear modulus and dynamic yield strength) depended on temperature and the current level of damages. The critical specific energy of shear deformations was used as a criterion of the erosion failure of the material that occurs in the region of intense interaction and deformation of contacting bodies [10]. The constants for the elastomer (rubber) were taken from [11]. Computations were carried out by the finite element method [12]. Sliding conditions were realized between the projectile and target, as well as between the layers of the target.

## RESULTS AND DISCUSSION

Penetration of tungsten-alloy long-rod projectiles with diameter $d_0 = 8.8$ mm and length $l_0 = 20d_0$ through the three-layer target with front, middle, and rear layers of thicknesses 3-10 mm, 16-30 mm, and 3-10 mm, respectively, at an angle of $60^0$ was studied in experiments and with help of computer simulation. Material of the middle layer of the target was an elastomer or reacting mixture in which shock-induced solid-state exothermic chemical reaction can take place.

### Three-layer target with elastomer middle layer

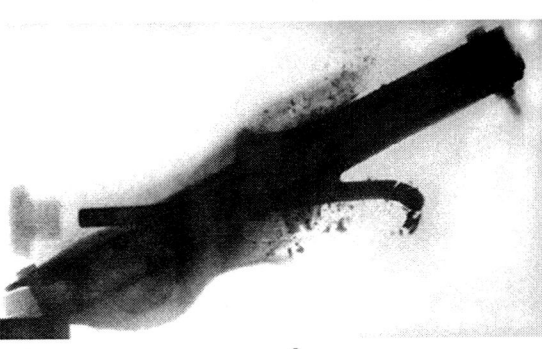

a

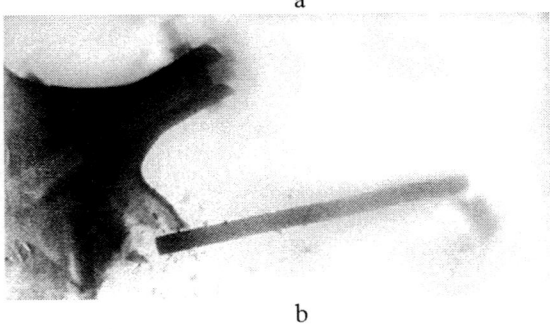

b

**Figure 1.** X-ray patterns for interacting bodies at time 90 (a) and 200 μs (b). Target: 3 (steel) + 16 (elastomer) + 3 mm (steel), 125x250 mm; $\upsilon_0 = 1577$ m/s.

Failure of the head part of the projectile takes place after the penetration through the rear layer of the target. At time 90 μs (Fig. 1a) there are two fragment areas on the rear surface of the target due to shock wave and deformation during penetration. Further, these two regions form a fragment debris behind the target. At time 200 μs, when the projectile moves behind the target, its head part breaks off, which is accompanied by the separation of the part from the basic projectile remainder (Fig. 1b). Similar results were obtained in computations.

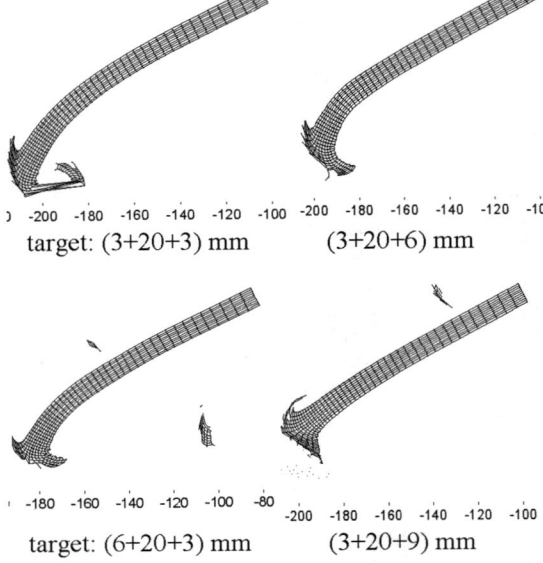

target: (3+20+3) mm       (3+20+6) mm

target: (6+20+3) mm       (3+20+9) mm

**Figure 2.** Shapes of the projectiles at time 300 μs, after perforation of the three-layer target; thicknesses of the front, middle, and rear layers are pointed; $\upsilon_0 = 1600$ m/s.

Results of computations show that the character of failure of the projectile qualitatively depends on thickness of the rear layer. The bending of the head part of the projectile is observed at thickness of the rear layer less than the projectile diameter and erosion failure takes place at the thickness of the rear layer comparable or more then the projectile diameter (Fig. 2). If material of the middle layer of the target is steel, the erosion failure also prevails.

The transition from bending of the head part of the projectile to erosion failure is illustrated in Fig. 3. If the rear layer is thin enough the process of erosion failure can not start because of the fast failure of the rear layer.

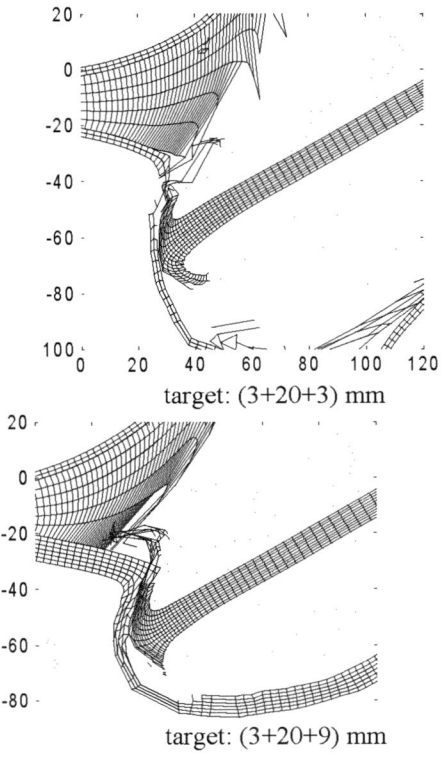

target: (3+20+3) mm

target: (3+20+9) mm

**Figure 3.** Configurations of the projectile and the three-layer target at time 110 μs. Thicknesses of the rear layer of the target are 3 and 9 mm.

## Target with reacting middle layer

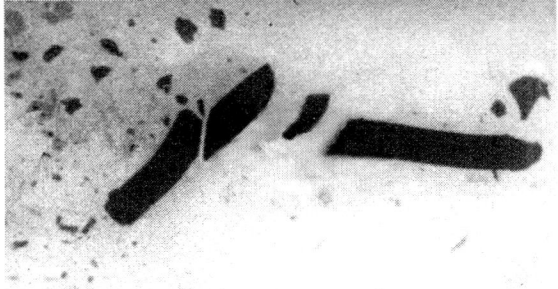

**Figure 4.** Failure of the projectile after perforation of the three-layer target with reacting mixture in the middle layer. Projectile: $d_0 = 8.8$ mm, $l_0 = 20d_0$, 60°/1600 m/s; target: 3 (steel)+20 (reacting mixture)+3 (steel) mm.

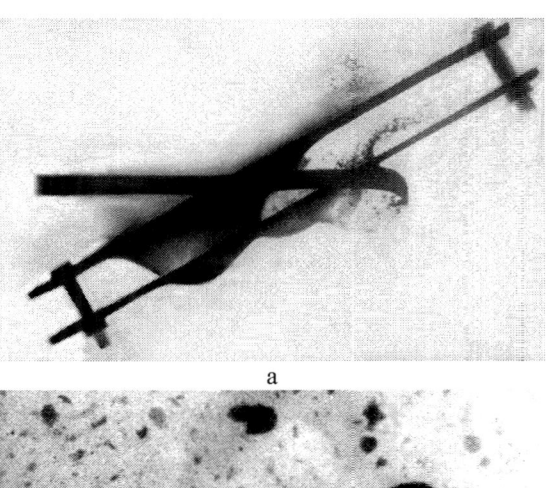

a

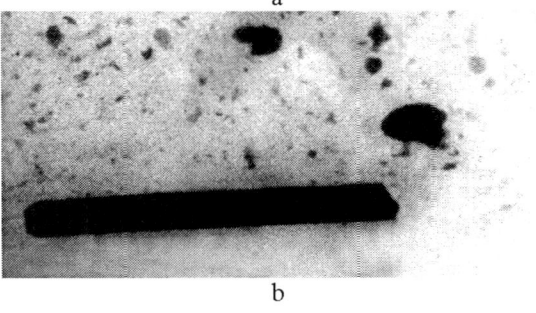

b

**Figure 5.** Control test. Projectile and two-layer target (the middle layer is air) during penetration (a) and the projectile after perforation (b).

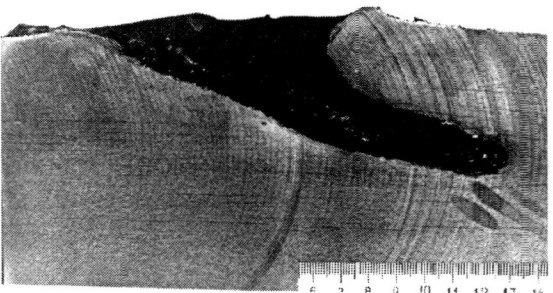

**Figure 6.** Crater in a steel semi-infinite target after penetration of the long-rod projectile remainder after perforation of a three-layer target with a reacting middle layer.

Another character of failure of the projectile is observed when a shock-induced solid-state exothermic chemical reaction can generate in the middle layer. One can see in Fig. 4 failure of the projectile into several parts which is similar to the

**Figure 7.** Control test. Interaction of the long-rod projectile remainder with a semi-infinite target after perforation of a two-layer target (middle layer is absent).

case of dynamic defense when the material of the middle layer is high explosive.

After interaction of the projectile with reacting mixture the character of penetration of the projectile remainder into a semi-infinite target changes to "denormalization". We consider that this is due to loss of strength properties of the material of the projectile subjected to high temperature which realizes during chemical reaction, in other words – thermal shock, "thermoshock".

## CONCLUSION

Analysis of the data obtained shows that the character of failure of the projectile qualitatively depends on the material of the middle layer.

**For elastomer middle layer:** the character of the failure of the head part of the projectile was found to depend significantly on the thickness of the rear layer of the target. When the thickness of the rear layer is less than the diameter of the projectile then the bend and destruction of the head part of the projectile is prevailed; when the thickness of rear layer is equal or more than the diameter of the projectile the process of erosion failure is realized.

**For reacting mixture:** detonation-like process is realized and failure of the projectile into several parts is observed which is similar to the case of dynamic defence; the process of "thermoshock" in an interaction zone and loss of strength properties of the material of the projectile subjected to high temperature during chemical reaction takes place.

## ACKNOWLEDGEMENTS

This work was supported by the Russian Foundation for Basic Research under Grants 03-01-00122, 05-03-98001.

## REFERENCES

1. Rosenberg, Z., and Dekel, E., "On the role of material properties in the terminal ballistics of long rods," Int. J. Impact Engng., vol. 30, pp. 835-851, 2004.
2. Yoo, Y.-H., and Shin, H., "Protection capability of dual flying plates against obliquely impacting long-rod penetrators," Int. J. Impact Engng., vol. 30, pp. 55-68, 2004.
3. Grigorjan, V.A., Dorokhov, N.S., Kobylkin, I.F., et al, "Nonreactive dynamic armor," Oboronnaya Tekhnika, vol. 12, pp. 20-26, 2002.
4. Held, M, Mayseless, M, and Pototaev, E., "Explosive reactive armor," Proceedings of the Seventh Int. Symposium on Ballistics, Midland (South Africa), pp. 33-46, March 1998.
5. Zelepugin, S.A., Grigorjan, V.A., Dorokhov, N.S., Zhbankov, Yu.P., "Failure of a long-rod projectile perforating a target with a middle elastomer layer," Doklady Physics, vol. 48, no. 10, pp. 572-575, 2003.
6. http://www.niistali.ru.
7. Zelepugin, S.A., and Nikulichev, V.B., "Numerical modeling of sulfur – aluminum interaction under shock-wave loading," Combustion, Explosion, and Shock Waves, vol. 36, no. 6, pp. 845-850, 2000.
8. Kanel', G.I., Razorenov, S.V., Utkin, A.V., Fortov, V.E., "Impact-shock-wave phenomena in condensed matter", Moscow: Yanus-K; 1996.
9. Zelepugin, S.A., Sidorov, V.N., and Khorev, I.E., "Experimental and numerical investigation of the fracture of obstacles by groups of high-speed bodies," Strength of Materials, vol. 35, no. 2, pp. 168-174, March - April 2003.
10. Kalmykov, Yu.B., Kanel', G.I., Parkhomenko, I.P., et al., Prikladnaja Mekhanika Tekhnicheskaja Fizika, vol.1, pp. 126-131, 1990.
11. Johnson, G.R., "High velocity impact computations in three dimensions," J. Appl. Mech., vol. 44, no. 1, pp. 95-100, 1977.
12. Gorelski, V.A., Zelepugin, S.A., Smolin, A.Yu., "Effect of discretization in calculating three-dimensional problems of high-velocity impact by the finite-element method," Computational Mathematics and Mathematical Physics, vol. 37, no. 6, pp. 722-730, 1997.

CHAPTER XXII

# LASER–DRIVEN SHOCKS
# AND INTERACTIONS OF LIGHT
# WITH MATERIALS

CP845, *Shock Compression of Condensed Matter - 2005*,
edited by M. D. Furnish, M. Elert, T. P. Russell, and C. T. White
© 2006 American Institute of Physics 0-7354-0341-4/06/$23.00

# SCALING OF PRESSURE WITH INTENSITY IN LASER-DRIVEN SHOCKS AND EFFECTS OF HOT X-RAY PREHEAT

## Jeffrey D. Colvin and Daniel H. Kalantar

*Lawrence Livermore National Laboratory, L-356, P.O. Box 808, Livermore, CA 94551*

**Abstract**. To drive shocks into solids with a laser we either illuminate the material directly, or to get higher pressures, illuminate a plastic ablator that overlays the material of interest. In both cases the illumination intensity is low, $<<10^{13}$ W/cm$^2$, compared to that for traditional laser fusion targets. In this regime, the laser beam creates and interacts with a collisional, rather than a collisionless, plasma. We present scaling relationships for shock pressure with intensity derived from simulations for this low-intensity collisional plasma regime. In addition, sometimes the plastic-ablator targets have a thin flash-coating of Al on the plastic surface as a shine-through barrier; this Al layer can be a source of hot x-ray preheat. We discuss how the preheat affects the shock pressure, with application to simulating VISAR measurements from experiments conducted on various lasers on shock compression of Fe.

**Keywords**: iron, high-pressure solid-state phase transformations, x-ray preheat, shock waves in solids, laser-target interactions.
**PACS**: 52.38Dx, 52.50Jm, 61.10Nz, 61.50Ks, 62.50+p, 64.70Kb.

## SCALING OF PRESSURE WITH INTENSITY

For laser-driven shocks the absorption physics and shock pressure depend on the laser intensity and wavelength. For intensities below about $10^{10}$ W/cm$^2$, the very low intensity regime, the laser beam interacts with solid, liquid, and vaporized material, or a very cool plasma. In this intensity regime the technique of tamped ablation [1] is used; i.e., the laser light first passes through a transparent dielectric tamper overlaying the sample to be driven. In this case the pressure in the sample scales as the square root of the laser intensity.

For intensities above about $10^{13}$ W/cm$^2$, the high intensity regime, the laser beam creates and interacts with a collisionless coronal plasma. This is the regime of inertial confinement fusion [2]. In this regime of intensities, the shock pressure scales as the two-thirds power of the laser intensity.

Both the tamped ablation regime at very low intensities and the ICF regime at high intensities have been very well studied experimentally and computationally. It is the intermediate intensity regime, intensities between about $10^{10}$ W/cm$^2$ and about $10^{13}$ W/cm$^2$, a regime of intensities in which the laser beam is interacting with a collisional plasma, that has not been well studied, and which is the subject of the work reported here.

The laser beam heats a collisional ablated plasma differently than a collisionless one. At high intensities ($>10^{13}$ W/cm$^2$) a coronal plasma with density below the critical density is formed, and the laser beam heats this collisionless plasma to near isothermal conditions. In this collisionless plasma, where the electron-ion collision rate is less than the plasma frequency, the temperature can be approximated by using an electron flux limit in solving the transport equation. It can be shown from this approximation that the pressure scales as the two-thirds power of the intensity [2]. This scaling has been verified with simulation [2].

This approximation is no longer valid in a collisional plasma, so the scaling of pressure with intensity may be different. We have determined the scaling with simulations using the radiation-hydrodynamics code Lasnex [3].

We have modeled experiments in which iron was directly driven with and without a parylene-N ($C_8H_8$) or parylene-C ($C_8H_7Cl$) ablator. The experiments were conducted on the Vulcan laser at Oxford University with a 1-μm beam, on the Janus laser at LLNL with a _-μm beam, and on the Omega laser at the University of Rochester with a 1/3-μm beam. Details of the experiments are presented in another paper in these Proceedings [4]. We specifically modeled the ablator+Fe targets in order to determine the scaling of pressure in the Fe with beam intensity and laser wavelength.

A power-law fit to the Lasnex simulations gives a near-linear scaling of pressure with beam intensity; specifically, we find

$$P^* = 41.5\{I(W/cm^2)/3.16 \times 10^{10}\}^{0.9}\{\lambda_0/\lambda\}^{\alpha(\lambda)} \text{ kbar}$$

Here $P^*$ is the pressure at the front face of the Fe, I is the beam intensity, $\lambda$ the laser wavelength, and $\lambda_0 = 1/3$ μm. The scaling exponent for wavelength is itself wavelength dependent.

## HOT X-RAY PREHEAT

Since parylene is transparent to the laser light at very low intensities (i.e., at early times when the pulse is rising to its peak intensity) a thin (0.1 μm) overcoat of Al was added to the ablator to prevent shine-through and direct preheating of the Fe by the laser beam. While the addition of the Al overcoat to the plastic ablator mitigates one preheat problem, it creates another. The laser-heated Al adds an indirect source of x-ray preheat. That is, the laser-heating of the Al overcoat generates a significant non-thermal flux of soft (~200 eV) x-rays. Some fraction of this flux transmits through the parylene ablator and is absorbed at the front surface of the iron, heating it and raising its

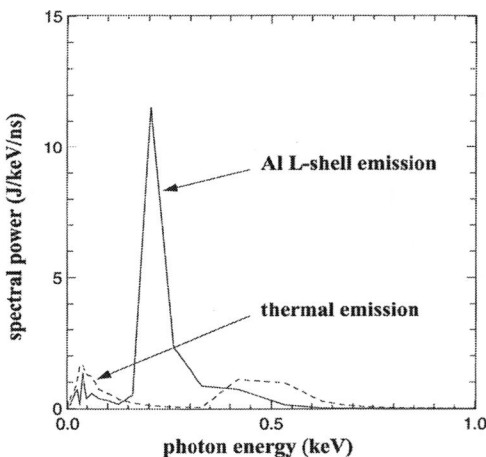

**Figure 1**. Simulated spectral power crossing the front face of the parylene-N for an intensity of $2 \times 10^{11}$ W/cm$^2$ incident on a target of 15 μm parylene-N on 250 μm Fe, with (solid curve) and without (dashed curve) a 0.1-μm Al overcoat on the ablator.

pressure before the arrival of the ablatively driven shock.

In Fig. 1 we show the spectral power crossing the front face of the parylene-N as a function of photon energy, determined in simulations of one of the Omega experiments in which 117 J of 1/3-μm laser light in a 6-ns-square pulse (I = $2 \times 10^{11}$ W/cm$^2$) was incident on a target of 15 μm parylene-N on 250 μm iron, both with (the solid curve) and without (the dashed curve) the 0.1-μm Al shine through barrier.

As is evident in Fig. 1, more absorption in the bare parylene means the target without the Al overcoat heats up hotter and produces more thermal emission, but the target with the Al overcoat produces a large flux of Al L-shell emission at higher photon energy.

The different spectra lead to different heating of both the plastic ablator and the iron. In Fig. 2 we show the simulated temperature profiles in the target at 6 ns, the end of the laser pulse, for the same target and drive configuration as in Fig.. 1, and for different opacities ($\kappa$) in the parylene ablator.

The laser energy is entirely absorbed in only the outer ~1 μm of the ablator. This outer layer of the ablator heats up to ~80 eV in the bare target

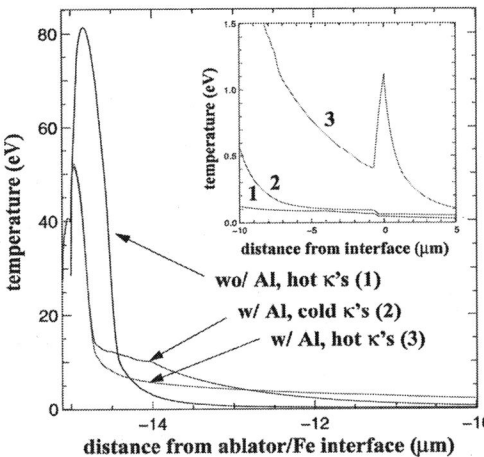

**Figure 2.** Simulated temperature (eV) vs original Lagrangian distance from ablator/Fe interface (μm) at 6 ns for the target and drive of Fig. 1. Simulations with and without the Al, and for different parylene opacities (κ).

(without the Al), and to ~52 eV in the target with the Al overcoat (the Al itself heats to only ~40 eV). In the bare target the ablator/Fe interface stays cold (i.e., the thermal flux from the outer ~1 μm of the ablator is entirely absorbed in the ablator). In contrast, some of the non-thermal flux from this hot outer layer in the target with the Al overcoat does get through to the ablator/Fe interface, heating it. The radiative preheating of the Fe is strongly dependent on the ablator opacity. Using hot ablator opacities everywhere, the front face of the iron heats to over 1 eV; using cold opacities it heats to just under 0.1 eV.

The hot opacities better describe the thermal flux transport in the hot outer ~1 μm of the ablator. For the hot parylene opacity we use XSN, a statistical screened hydrogenic average atom model. XSN is reasonably accurate at higher temperatures (>10 eV) where free-free transitions dominate the opacity, the outer ~1 μm. XSN is not accurate at lower temperatures where bound-bound and bound-free transitions are important. For the cold parylene opacity we use tabulated cold (zero-temperature) opacities. This is a good approximation for the non-thermal flux since the opacity of 200 eV photons at these low temperatures is better described by cold than by XSN opacities.

Additionally, we use a low-temperature thermal conductivity model in the parylene ablator.

In order to get a handle on the correct ablator opacities to use in the simulations, we compare the simulations to both the *in situ* diffraction data and the VISAR data. Kalantar *et al.* have discussed the diffraction data [4, 5], in which 6.7 keV x-rays generated by the interaction of a separate laser beam with an Fe back light foil diffracted from a thin surface layer of the Fe at and near the ablator/Fe interface, and were recorded on a large-area film detector. These measurements show that the Fe transformed, under shock compression, to the ε (hcp) phase, not to the higher-temperature γ (fcc) phase.

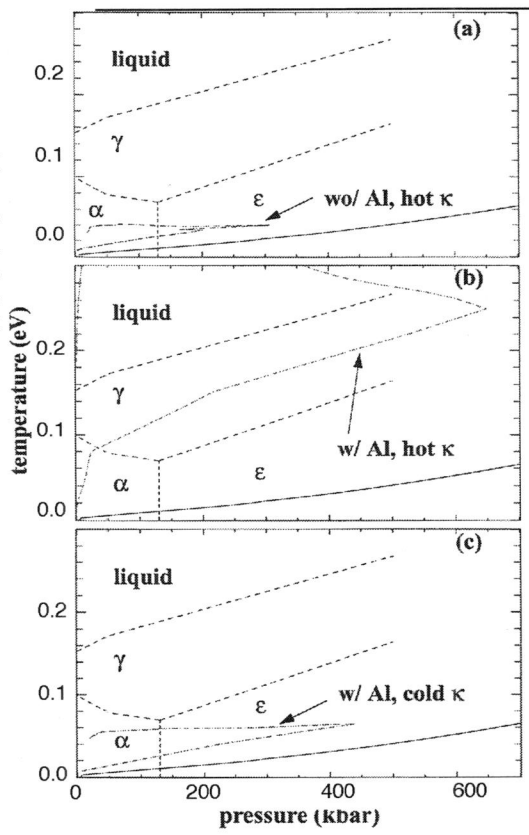

**Figure 3.** Simulated trajectory in temperature-pressure space of the front 3 μm o f the Fe, overlaid on the equilibrium phase diagram for targets a) without Al and with hot ablator opacities; b) w ith Al and hot ablator opacities; and c) with Al and cold ablator opacities. The solid curve is the zero-temperature Hugoniot.

As seen in Fig. 3, there is clearly too much Al x-ray preheat of the Fe using the hot ablator opacities. In this simulation the trajectory of the front 3 μm of the Fe never crosses the equilibrium α-to-ε phase boundary. With this much preheat, the Fe would heat up enough to first cross the α-to-γ equilibrium phase boundary. Since this is inconsistent with the diffraction data, the use of the hot parylene opacities cannot be right, as expected. The other two cases shown in Fig. 3 (no Al and hence no preheat, and Al with cold ablator opacities) are both consistent with the diffraction data, but there is a ~50% difference in the peak pressures for these two cases.

In order to distinguish between these two cases (no Al and hence no preheat, and Al with cold ablator opacities) we next compare the simulations to the VISAR data. The VISAR diagnostic is a velocity interferometer which records the material velocity as a function of time at the free surface of the Fe [6]. In comparing the simulations with VISAR data, shown in Fig. 4, we see that the simulations with the cold parylene opacities everywhere are consistent with the VISAR data. A simulation without non-thermal flux from an Al overcoat gives a peak free-surface velocity that is too low compared to the data, and a simulation with the Al and hot ablator opacities gives a peak free-surface velocity that is too high. For the simulation with the Al and cold ablator opacities, there is a good match in amplitude but not in timing of the plastic wave.

There is an even better match to the VISAR data when we account for the plasticity kinetics and material spall, also shown in Fig. 4. The simulation with plasticity kinetics and spall models includes a model for plasticity kinetics of Gilman [7], and an unpublished model of spall of Minich, based on percolation theory. Details of the plasticity kinetics and spall modeling will be published separately. We conclude that properly accounting for preheat, as well as properly accounting for the kinetics of the plasticity transition, are necessary in modeling laser-driven shocks in solids.

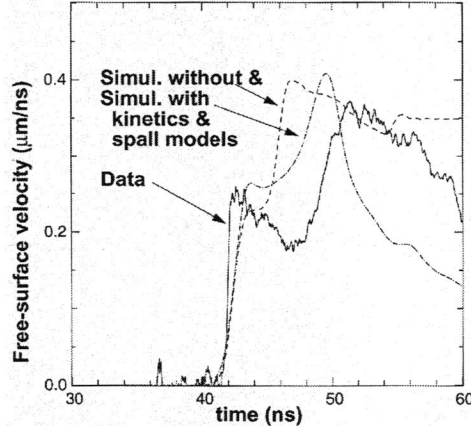

**Figure 4**. Simulated velocity history of the free surface of Fe compared to data. Simulations were done with Al and cold ablator opacities, both with and without plasticity kinetics and spall models in the simulation.

## ACKNOWLEDGEMENTS

This work was performed under the auspices of the U.S. Department of Energy by Lawrence Livermore National Laboratory under Contract No. W-7405-Eng-48. The authors are indebted to Dr. Roger Minich of LLNL for doing the simulation with the plasticity kinetics and spall models, and for numerous enlightening discussions on the behavior of shock-compressed solids.

## REFERENCES

1. Colvin, J. D., Ault, E. R., King, W. E., and Zimmerman, I. H., Phys. Plasmas **10**, 2940, 2003.
2. Lindl, J., Phys. Plasmas **2**, 3933, 1995.
3. Zimmerman, G. B. and Kruer, W. L., Comments Plasma Phys. Control. Fusion **2**, 51, 1975.
4. Kalantar, D. H. et al., paper K6.01 in these Proceedings.
5. Kalantar, D. H. et al., Phys. Rev. Lett. **95**, 075502 (2005).
6. Barker, L. M., and Hollenback, R. E., J. Appl. Phys. **43**, 4669, 1972.
7. Gilman, J. J., J. Appl. Phys. **36**, 2772, 1965.

CP845, *Shock Compression of Condensed Matter - 2005*,
edited by M. D. Furnish, M. Elert, T. P. Russell, and C. T. White
© 2006 American Institute of Physics 0-7354-0341-4/06/$23.00

# CONTEXT AND THEORY FOR PLANAR RADIATIVE SHOCK EXPERIMENTS IN XENON

## R.P. Drake[1], A. B. Reighard[1]

*[1]University of Michigan, Ann Arbor, Michigan 48109*

**Abstract.** Our laboratory studies of radiative shocks include development of the semi-analytic theory that corresponds to various cases of interest to experiments, development of experiments, comparison of experiments and simulations, and drawing connections to astrophysics. Here we discuss theory and experiment of radiative shocks that are optically thin to thermal radiation in the upstream direction but optically thick in the downstream direction.

**Keywords:** Shock waves: profile, radiation: hydrodynamics.
**PACS:** 52.35.Tc, 95.30.Jx, 95.30.Lz.

## INTRODUCTION

Laboratory studies of radiative shocks are motivated by the presence of such shocks in astrophysics [1], by the developing ability to perform radiation hydrodynamic experiments in the laboratory [2-7], and by the need for experiments to benchmark new generations of astrophysical radiation-hydrodynamic codes. In our ongoing work with collaborators [8, 9] we have developed experiments that can produce these shocks and have diagnosed them by radiography and other techniques. In the following we discuss how these specific experiments fit into the overall context of radiative shocks. Then we discuss the theory of such shocks, which is not much developed in the existing literature. Finally, we summarize the experiments and show some of the data.

## THE RADIATIVE SHOCK CONTEXT

We work in the usual "shock frame", in which this matter enters a stationary shock from "upstream" and the shocked matter flows away from the shock "downstream". Radiative shocks are characterized by three key parameters, all of which are intuitively straightforward. The energy input to the dynamics is the kinetic energy flux, $\rho_o u_s^3/2$, of the material incident on the shock (of mass density $\rho_o$ and of speed $u_s$). The first parameter is the ratio of the radiation flux at the characteristic post-shock temperature, $T_s$, to this energy. We evaluate $T_s$ at the nominal density jump of a strong shock, $(\gamma+1)/(\gamma-1)$, $RT_s = 2(\gamma-1)u_s^2/(\gamma+1)^2$, where $R$ is the gas constant and $\gamma$ is the polytropic index of the medium. In fact, $R$ may not be constant and the density jump may differ a great deal from the nominal value, and that while $\gamma = 5/3$ for single-particle gasses, it is typically smaller than this for ionizing and radiating matter and also may not be constant. These details do not alter the condition on the energy flux ratio

$$\frac{\sigma T_s^4}{\rho_o u_s^3/2} = Q\frac{16(\gamma-1)^4}{(\gamma+1)^8} > 1, \qquad (1)$$

in which the shock strength parameter $Q$ is $2\sigma u_s^5/R^4\rho_o$ and $Q$ must reach several thousand for radiative effects to be important. This defines a threshold velocity for radiative shocks, which is for example 60 km/s in Xe at 10 mg/cm$^3$ and is 200 km/s in CH at 10 mg/cm$^3$.

The other two key dimensionless parameters are the optical depths upstream of and downstream of the shock. Here the optical depth is the number of exponential absorption lengths in the medium, for radiation emitted by the heated post-shock material. The conceptual point is that it makes a difference whether the energy emitted in some direction is absorbed deep within the medium, escapes freely from it, or exhibits more complicated behavior. One conceptually simple limit is the case when both the upstream medium and the downstream medium are optically thin. This allows nearly all of the initial post-shock thermal energy to be radiated away. Many astrophysical shocks are in this limit, including for example shocks in supernova remnants in their "radiative" phase [10], and are easily observed as the radiation escapes. A second conceptually simple limit is that in which both the upstream medium and the downstream medium are optically thick. In this case the radiation is confined within the medium, as may happen for example in some supergiant variable stars [11]. The case of interest here is the hybrid case in which the upstream medium is optically thin while the downstream medium is optically thick. Astrophysical shocks emerging from optically thick objects such as stars or supernovae experience this phase [12].

### THEORY FOR THICK-DOWNSTREAM, THIN-UPSTREAM CASE

The context for the needed theory is set by a number of properties of the experimental system. The density jump corresponds to heating of the ions by viscous effects, which occurs on a very small spatial scale. The ions then equilibrate with the electrons, which may involve further ionization. Under typical conditions, the distance over which this equilibration occurs is small compared to the radiation mean-free path [13]; here we will take it to be instantaneous. The heated matter then radiates, and some of the radiation penetrates upstream of the shock, producing a *radiative precursor*. This radiative emission cools the heated matter over some distance until a final state is reached in which there is no net

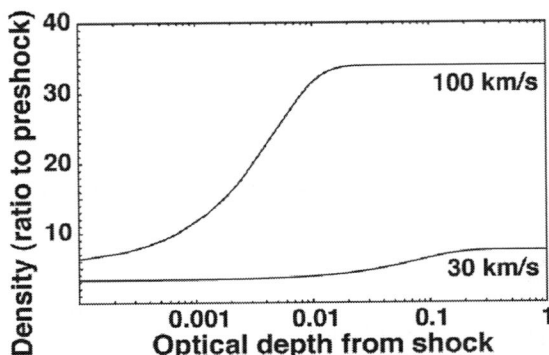

**Figure 1.** Density profile for a simplified xenon model. The immediate post-shock density is shown on the left, and within one optical depth the final density has been reached. From Drake [13].

downstream radiative flux. This layer where radiative cooling occurs can be labeled a *cooling layer*. For the case of interest here, there remains a steady upstream flux of radiation that escapes the system. The optically thin precursor is heated by this radiation, but does not contribute significantly to the energy dynamics of the system.

In typical regimes of current interest, the radiative fluxes are large (Eq. 1 is satisfied) but the radiation pressure and energy density are negligible, so we will ignore these. Under these conditions the differential equation for energy balance for a steady shock becomes

$$\nabla \bullet \left[ \rho \mathbf{u} \left( \varepsilon + \frac{u^2}{2} \right) + p\mathbf{u} \right] = -\nabla \bullet \mathbf{F}_R, \quad (2)$$

in which the specific internal energy is $\varepsilon$, the pressure is $p$, the velocity is $\mathbf{u}$, and the radiation energy flux is $\mathbf{F}_R$. Combining this with the equations for mass and momentum balance, for a one-dimensional shock, one has

$$RT = (1 - \rho_o/\rho)\,\rho_o/\rho, \text{ and} \quad (3)$$

$$\frac{\rho_o u_s^3}{2} \frac{\partial}{\partial z} \left[ \frac{-2\gamma}{(\gamma-1)} \left( \frac{\rho_o}{\rho} \right) + \frac{(\gamma+1)}{(\gamma-1)} \left( \frac{\rho_o}{\rho} \right)^2 \right] = -\frac{\partial F_R}{\partial z}. \quad (4)$$

Because the cooling layer is optically thin, it proves useful to describe the radiation using the zeroth moment of the radiation transfer equation, which in steady state gives $-\partial F_R / \partial z = 4\pi\kappa(B - J_R)$, where $B = \sigma T^4 / \pi$, $J_R = \int_{4\pi} I_R d\Omega /(4\pi)$, with $I_R$ being the radiation intensity (energy flux per sr), and $\kappa$ is the Planck mean opacity, assumed as usual to be accurate for $J_R$ in addition to $B$. Here $J_R$ does not change as one traverses the optically thin cooling layer (the flux emitted from a thin sublayer being equal in both directions) and matching the cooling layer to the final state requires $J_R = B_f = \sigma T_f^4 / \pi$. The boundary conditions are found from the energy balance [13] to be

$$\rho_o / \rho_f = \sqrt{\sqrt{1 + 8Q} - 1} / \sqrt{4Q}, \text{ and} \qquad (5)$$

$$F_{Rs} = 2\sigma T_f^4, \qquad (6)$$

where $F_{Rs}$ is the radiation flux at (and continuous across) the density jump. With these, one can integrate Eq. 4 to find the profiles of density and other parameters in the cooling layer.

Figure 1 shows qualitatively typical profiles of the post-shock density. Here a simplified model is used for xenon at a density of 1 mg/cm³, in which $\kappa$ is assumed to scale as $T^{-4/3}$ and $\gamma = 4/3$, which is reasonable for the ionizing xenon of our experiments. More complex models for $\kappa$ are possible [14]. As the shock velocity increases, the post-shock temperature and radiative fluxes increase, causing a larger fraction of the energy to be radiated and a larger total increase in density.

## THE EXPERIMENT DESIGN

The experiment uses the Omega Laser [15] to ablatively accelerate a planar "drive disk" of material, launching it down a shock tube where it drives a shock through xenon gas like a piston. Approximate 4 kJ of energy at a wavelength of 0.35 μm, in a 1 ns pulse, irradiates the slab, which is typically of Be and is nominally 10 μm, 20 μm, or 40 μm thick. The laser spot is ~ 800 μm while the polyimide shock tube is ~ 600 μm in diameter. The ablation pressure of ~ 50 Mbars (5 TPascal) first shocks and then accelerates the Be slab, which reaches a velocity of 100 km/s to 250 km/s. As the piston moves down the tube, the shocked Xenon is

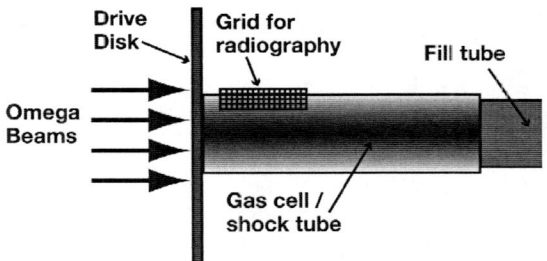

**Figure 2.** Sketch of experiment to produce radiative shocks. The tube is filled with Xe at 6 mg/cm³. The xray source for radiography is below the page, and the detector is above the page.

heated sufficiently to radiate strongly, leading to radiative preheat of the material ahead of the shock and to a density increase of the shocked Xe.

We assess the evolution of this system, for experiment design and for comparison with data, using the HYADES simulation code [16]. This is a Lagrangian code treating the plasma as a single fluid but separately evaluating the energy of the ions, electrons, and radiation. It models the heat flow diffusively, using a multigroup model of the radiation and a single-group model of the electrons, both flux-limited. The code finds a narrow region at the shock where the ions are heated to many hundreds of eV, followed by equilibration with the electrons and the associated ionization, much like that reported previously [2]. Radiative cooling then creates an increase of density, leading to the thin dense layers of Xe shown in Fig. 3. The radiation

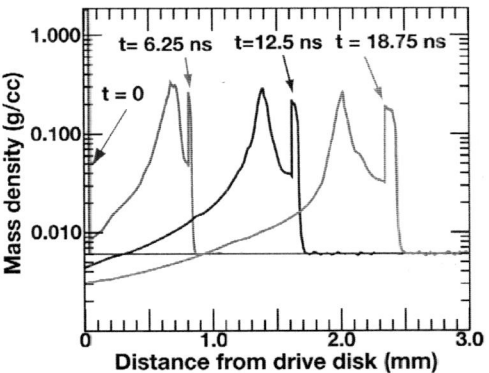

**Figure 3.** The density profile at several times, from a simulation using HYADES for a 40 μm Be drive disk. The thin dense layer to the right is the shocked Xe, while the peaked structure to the left is the drifting Be.

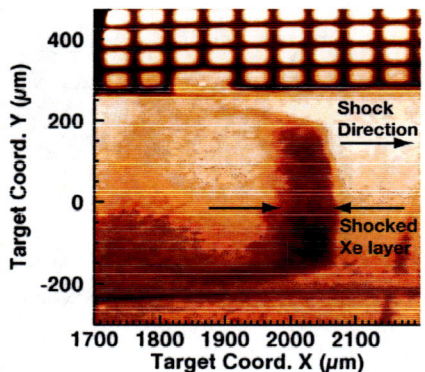

**Figure 4.** Radiographic data, at 14.6 ns after the initiation of an experiment using a 20 μm Be drive disk. The image is stretched for clarity, the grid cells are squares spaced 63 μm from center to center.

from the Xe also ablates the dense Be, producing a separation of the Be density peak from the Xe density peak, as the figure also shows. In addition, the shocked Xe over time develops an extended layer of relatively constant density. This layer is optically thick and so corresponds a quasi-steady final state like that described above.

We have at this writing obtained a large number of radiographic images of this system, in addition to limited data from other diagnostics. Figure 4 shows an example. The radiographs show the shock tube, the calibration grid, and a thin layer of Xe. The Xe layer appears to be relatively uniform and to become thicker as the shock propagates, which is as expected from this type of radiative shock. There is some structure on the surfaces of the Xe layer, which remains to be explored. There is also a thin sheet of Xe that is left behind along the edges of the cylinder. This is consistent with two-dimensional simulations.

## CONCLUSIONS

The work discussed here is at its beginning. We have developed a semi-analytic theory that is useful for understanding radiative shocks that are optically thick downstream and optically thin upstream. We have also developed an experimental system that permits the study of such shocks, and have begun the process of diagnosing their properties. Future work includes iterating the semi-analytic model to be more aligned with the experiments, developing a range of diagnostics of the experimental properties, and comparing the experimental results with emerging astrophysical codes.

## ACKNOWLEDGEMENTS

We acknowledge useful discussions of the theoretical context with Dmitri Ryutov and John Edwards, Bruce Remington, Ted Perry, Gail Glendinning. Freddy Hansen and work on the experiments with the Omega NLUF Experimental Astrophysics Team, and financial support from the National Nuclear Security Administration under the Stewardship Science Academic Alliances program through DOE Research Grant DE-FG52-03NA00064 and DE FG53 2005 NA26014.

## REFERENCES

1   Drake, R.P., Astrophysics and Space Science **298**, 49 (2005).
2   Bouquet, S., et al., Phys. Rev. Lett. **92**, 2250011 (2004).
3   Bozier, J.C., et al., Phys. Rev. Lett. **57**, 1304 (1986).
4   Edwards, M.J., et al., Phys. Rev. Lett. **87**, 0850041 (2001).
5   Keiter, P.A., et al., Phys. Rev. Lett. **89**, 165003/1 (2002).
6   Koenig, M., et al., in *Shock Compression of Condensed Matter 2001* (American Institute of Physics, 2002), Vol. 620, pt. 2, p. 1367.
7   Edens, A.D., et al., Phys. Plas. **11**, 4968 (2004).
8   Reighard, A.B., et al., in *Inertial Fusion and Science Applications* (American Nuclear Society, Monterey CA, 2003).
9   Reighard, A.B., et al., Phys. Rev. Lett., submitted (2004).
10  Blondin, J.M., et al., Ap. J. **500**, 342 (1998).
11  Farnsworth, A.V. and Clarke, J.H., Phys. Fluids **14**, 1352 (1971).
12  Ensman, L. and Burrows, A., ApJ **393**, 742 (1992).
13  Drake, R.P., *High Energy Density Physics: Foundation of Inertial Fusion and Experimental Astrophysics* (Springer Verlag, 2006).
14  Leibrandt, D.R., et al., Astrophysics and Space Science, submitted (2004).
15  Boehly, T.R., et al., Rev. Sci. Intsr. **66**, 508 (1995).
16  Larsen, J.T. and Lane, S.M., J. Quant. Spectrosc. Radiat. Transfer **51**, 179 (1994).

CP845, *Shock Compression of Condensed Matter - 2005*,
edited by M. D. Furnish, M. Elert, T. P. Russell, and C. T. White
© 2006 American Institute of Physics 0-7354-0341-4/06/$23.00

# HIGH ENERGY DENSITY PHYSICS ON LULI2000 LASER FACILITY

## M. Koenig[1], A. Benuzzi-Mounaix[1], N. Ozaki[1], A. Ravasio[1], T. Vinci[1], S. Lepape[1], K. Tanaka[2], D. Riley[3]

[1] *Laboratoire pour l'Utilisation des Lasers Intenses, UMR7605, CNRS – CEA - Université Paris VI - Ecole Polytechnique,, 91128 Palaiseau Cedex, FRANCE*
[2] *Institute of laser Engineering, Osaka University, Suita, Osaka 565-0871, Japan*
[3] *School of Mathematics and Physics, Queen's University of Belfast, BT7 1NN, UK*

**Abstract.** We present here a summary of some High Density Energy Physics experiments performed on the new facility LULI 2000. First, different flyer plate targets scheme have been tested loading shock in fused-quartz plate. Temperature data along the Hugoniot curve have been obtained. Second, a strongly coupled and degenerated Aluminium plasma has been probed by X-ray Thomson scattering. Compton shift from electrons has been observed in various density conditions.
**Keywords:** High energy density physics, laser driven shocks, Thomson scattering.
**PACS:** 52.27.Gr, 52.50.Lp, 52.72.+v

## INTRODUCTION

High Energy Density Physics (HEDP) includes a wide variety of physical phenomena among which Warm Dense Matter is an important domain. It is generally defined as *"when the external energy density applied to the material is comparable to the material's room temperature energy density. For example, the bulk modulus of solid or liquid-state materials is ranging from $10^9$ to $10^{11}$ J/$m^3$"* (see [1]).

The LULI laboratory has a new laser facility that allows to perform such dedicated HEDP experiments. Its characteristics are the following : a new target chamber (2m diameter) with two 1 kJ beam at 1ω. After conversion at second harmonic (2ω) each beam has 500 J with an oscillator able to deliver square pulse from 0.5 ns to 5 ns.

## FLYER PLATE EXPERIMENTS

Flyer impact experiments have been performed using laser-driven shock waves on the new LULI2000 laser facility. Laser-accelerated flyer technique had been studied to access extremely high-pressures in materials due to the impact[2]. Additionally, recent experiment has demonstrated very smooth pressure loading such as achieved in isentropic compression (ICE) with a density-graded projectile (expanding plasma)[3]. However, the conditions of flyer and impacted materials have not been sufficiently investigated.

In this experiments, three types of flyer targets; (i) simple metal flyer (aluminium single foil), (ii) a multi-layered flyer[4], and (iii) high-Z metal buffered by low-density plastic foam[5], were investigated. Our typical shock-loaded material was a fused silica plate. All diagnostics were optical: the rear-side ones were two velocity interferometers (VISAR[6]) working at two different wavelengths (1064 and 532 nm) and a self-emission diagnostic calibrated for brightness temperature; on the transverse side we also had a shadowgraphy diagnostic to infer the rear side plasma flow propagation between flyer and impact target. In the foam-buffered flyer targets, the tantalum foil covered a 100 $\mu$m distance between the flyer and the impacted quartz in ≈ 2 ns, resulting in an averaged velocity of 55 km/s. In this paper, we present only the simplest flyer scheme, i.e., the aluminium foil case. As observed in Figure 1&2, the shock wave gradually accelerates in quartz due to the flyer impact rear side plasma flow generated after the breakout. Then the shock wave goes through a distinct boundary to a conductive state was as observed in the VISAR image (Fig. 2). The second shock is high enough (P>1 Mbar) to reflect the probe laser beam. This flyer impact method is a way to produce very unique conditions in equation-of-state (EOS) diagram of the impacted material.

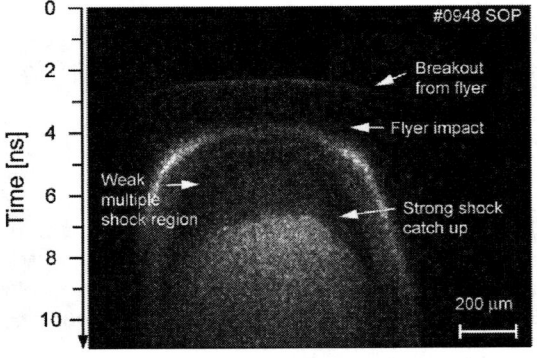

**Figure 1.** Self-emission diagnostic image.

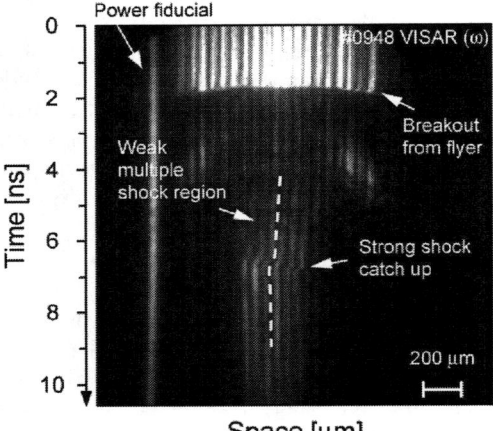

**Figure 2.** VISAR image

In the case of the simple aluminum foil, when the shock beaks out on the rear surface it is accelerated. However due to the high intensity we used ($\approx 10^{14}$ W/cm$^2$), the rear side is vaporized and a plasma flow is created. As long as the flyer travels toward to the impact quartz target, the length of this plasma flow increases so a smooth impact as in ICE experiments is expected before the bulk density part of the aluminum foil arrives.

This is exactly what we observed (Fig. 1&2) either on the self-emission (SOP) and VISAR diagnostics. After the breakout on aluminium (Fig 1), the first impact lead (1.5 ns later) to a short rise and drop in the intensity. It is then followed by a slow increase in intensity (the velocity so the pressure increases due to plasma flow accumulation) with a sudden bright signal about 2.9 ns later. This is due to a second shock generated in quartz coming from the bulk part of the flyer that overcomes the first shock. This is also observed in Fig.2 where high reflectivity occurs when the strong shock catch up.

Based on the shock velocity, its reflectivity and self-emission simultaneously in quartz, we were able to determine the temperature vs pressure assuming the recent quartz EOS data obtained at Omega[7]. As observed in Figure 3, our data span the phase transition occurring around 1 Mbar and 1 eV. The regular SESAME[8] table (3587) do not show any evidence of this transition whereas recent EOS from Kerley[9] exhibits such transition. However error bars in our experiment are still too important for final temperature results to show a clear change in the slope around velocities of 12 km/s. However our data are very close to the

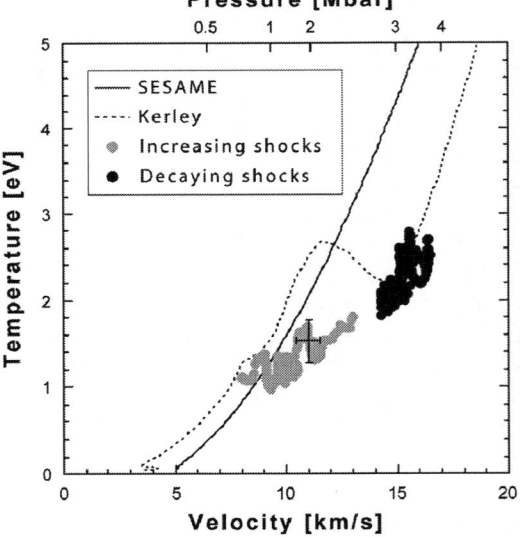

**Figure 3.** Temperature data vs. shock velocity and pressure. (●) are our experimental results, (—) SESAME table data and (- -) Kerley new EOS data.

values given by theory and future experiment will focus on the transformation from solid to liquid metal.

## X-RAY THOMSON SCATTERING

Spectrally resolved X-ray scattering has previously been implemented by Glenzer *et al.* [10] using the OMEGA laser facility at LLE. In their experiments they produced large scale radiatively heated plasmas and probed at back-scatter angles with scattering parameter <1. They were able, with this diagnostic, to determine the electron temperature to an accuracy of ~15% and get good agreement with simulations. Since then, they have also demonstrated the use of this diagnostic to determine the average ionisation of a plasma by a comparison of the relative contributions made by the Rayleigh, free electron and bound-free terms. In a recent experiment on the LULI-2000 laser at Ecole Polytechnique. we have attempted to investigate X-ray Thomson scatter in the near collective ($\alpha$~1) regime, we describe this experiment below. Figure 4 shows the experimental arrangement used. The two beams are both at 527nm and are of ~1ns duration with a sharp rising edge. One beam was used to focus onto a 5$\mu$m thick Ti foil and the other was focussed onto the sample plasma with aid of phased zone plate which generated a flat topped 800$\mu$m focal spot. The sample plasma consisted of a CH/Al/CH sandwich with thicknesses 4.5/6/4.5 microns. The Ti source foil was 1cm from the sample plasma and a 400$\mu$m pinhole restricted the imprint of the source radiation to the shocked area of the sample. A flat crystal spectrometer utilising a Si (111) crystal and a CCD detector was placed ~40cm away from the sample to intercept the beam of X-rays coming though the pinhole and the target. This allowed us to monitor the intensity of He-$\alpha$ line radiation passing through the target. The reflectivity of the crystal was not measured experimentally but, assuming a reflectivity similar to similar calibrated crystals used previously, we estimate from the signal level that the efficiency varied from ~3-8x10$^{12}$ photons/J in accordance with expectations for this pulse length and wavelength. Unlike other experiments[10], we were able to determine plasma conditions in the

shocked target independently with a good precision by using VISAR diagnostic.

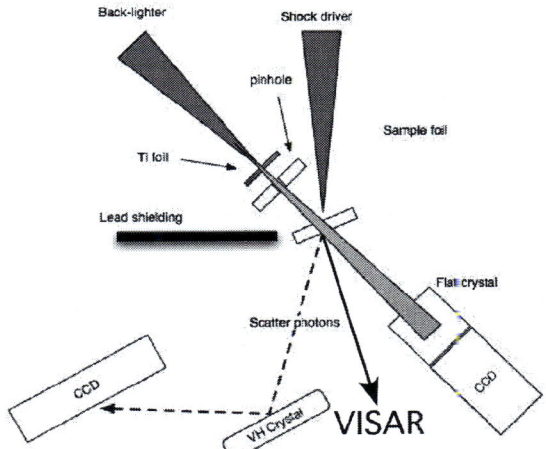

**Figure 4.** Schematic of the layout used for X-ray Thomson scatter experiments at LULI.

The scattered signal at 57° was detected by use of a curved LiF (200) crystal in the Von-Hamos configuration and a CCD detector. By placing a Ti foil target at the sample position, we obtained a backlighter-only shot which acts as a calibration spectrum. As a wavelength reference this of course depends on the scatter photons impinging on the sample foil centred around the same spatial point that the calibration spectrum originated from. The alignment system allowed the positioning of the target to be reproducible to better than 50$\mu$m so the most crucial element was the direction of the scatter x-rays through the pinhole. This was achieved by the use of the alignment laser collinear with the back-lighter main pulse. This was made to pass through the pinhole and both the back-lighter and sample positions and onto the centre of the flat crystal spectrometer. The projected diameter of the cone of He-$\alpha$ x-rays passing through the pinhole was ~900 microns at the sample (slightly larger than FWHM of shocked area) and ~3cm at the entrance to the spectrometer. The apparent relative wavelength shift was determined from the calibration shots. The timing of the back-lighter beam was 0.5ns, 1ns and 5ns respectively after the shock driving beam.

Status of the aluminium layer was inferred from two VISAR[6] looking on the rear side of the target. This gave us the shock speed in the last CH layer, from which we determined the mean parameters of the aluminium foil such as density and temperature, which were typically 5 g/cm3 and 3.5 eV respectively. To obtain these conditions, we had $\approx 400$ J at 2w in a 1 ns square pulse laser energy on target. Due to the large focal spot used (800 μm FWHM), we had an intensity $\approx 10^{13}$ W/cm$^2$.

In Figure 5, we present a comparison between the input and the scattered spectra. We did observe a clear shift (10 ± 2 mÅ) which could correspond to a Compton shift on free electrons or to a shift on electronic waves. Calculations to discriminate between these two phenomena are in progress. No peak Rayleigh has been observed. A possible explanation is that the ionic density fluctuations have a characteristic length ($\lambda_i$) smaller than the probed wavelength ($\lambda_s = \lambda_0/(2\sin\theta/2)$).

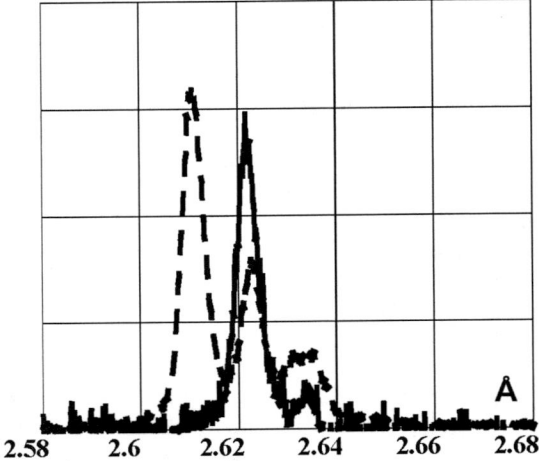

**Figure 5.** Scattered (—) and input (- -) spectra on a shocked target 0.5 ns after the pulse maximum.

For the late probe data, the shift was typically 3 times lower (3 mÅ ± 2 mÅ). The plasma conditions in that case are the following: 0.8 g/cm$^{-3}$ for the density and 1.5 eV for the temperature, the scattering regime being now non-collective ($\alpha \approx 0.6$). This shift can be interpreted in different ways. One possibility is that it could represent a reappearance of the Rayleigh component due to the increase of $\lambda_i$, which would become greater than $\lambda_s$. To conclude, also inelastic scattering due to bound-free electron transitions must be evaluated, since it

could be important for medium Z plasmas and low ionization rate[11].

To validate these different interpretations and to reproduce theoretically the scattered spectra, one has to extend usual approximation (such as random phase approximation) to calculate the x-ray dynamic form factor in a highly coupled plasma regime[11]. Developments in this direction are in progress.

As conclusions, we obtained new preliminary results on X-ray scattering from a coupled and degenerate plasma, which has been well characterized independently by a VISAR diagnostic. To get more reliable information from such experiments, one has to increase the number of reliable shots while reducing the possible errors. On the other hand, one has to develop new methods to calculate the scattered spectra in this highly coupled and degenerated regime.

## ACKNOWLEDGMENTS

We thank G. I. Kerley for providing his EOS table for quartz, M. Rabec le Gloahec for his important help during these experiments.

## REFERENCES

1 Frontier in high energy density physics,Council, N. R.,The national academy press (2003)
2 Cauble, R., et al., Phys. Rev. Lett. 70, 2102 (1993).
3 Edwards, J., et al., Phys. Rev. Lett. 92, 075002 (2004).
4 Tanaka, K.A., et al., Phys. of Plasmas 7, 676 (2000).
5 Benuzzi, A., et al., Phys. Plasmas 5 (1998).
6 Celliers, P.M., et al., Applied Phys. Lett. 73, 1320 (1998).
7 Hicks, D.G., et al., Phys. of Plasmas 12, 082702 (2005).
8 SESAME: The LANL Equation of State database,LA-UR-92-3407, Los Alamos National Laboratory (1992)
9 Kerley, G.I., Equations of State for Composite Materials (Kerley Publishing Services, Albuquerque, 1999).
10 Glenzer, S.H., et al., Phys. Rev. Lett. 90, 175002 (2003).
11 Gregori, G., et al., Phys. of Plasmas 11, 2754 (2004).

CP845, *Shock Compression of Condensed Matter - 2005*,
edited by M. D. Furnish, M. Elert, T. P. Russell, and C. T. White
© 2006 American Institute of Physics 0-7354-0341-4/06/$23.00

# QUASI-ISENTROPIC AND SHOCK COMPRESSION MEASUREMENTS OF IRON RESPONSE BY DIRECT LASER ILLUMINATION

**T. E. Tierney[1], D. C. Swift[1], S-N. Luo[1], J. Niemczura[2], J. T. Gammel[3], and P. Peralta[4]**

[1]*Physics Division, PO Box 1663, Los Alamos National Laboratory, Los Alamos, NM 87544*
[2]*Department of Aerospace and Mechanical Engineering, University of Texas at Austin, Austin, TX 78712*
[3]*Theoretical Physics Division, PO Box 1663, Los Alamos National Laboratory, Los Alamos, NM 87544*
[4]*Mechanical and Aerospace Engineering Department, Arizona State University, Tempe, AZ 85287*

**Abstract.** We performed a series of dynamic loading experiments on iron with pressures of 5-40 GPa at the Trident Laser Laboratory. We used 2.4 ns laser pulses of varying shapes and irradiances of 2 to 1000 GW/cm$^2$ to load a 5-mm diameter region of rolled iron foils that were 25-50 microns thick. The temporal characteristic of the laser irradiance was tailored to produce shock or quasi-isentropic loading histories. Line-imaging VISAR was used to time-resolve free surface velocities. In several experiments, two different thickness samples, placed side-by-side, were subjected to the same irradiance history. We describe the experiment configuration, analysis, and results.

**Keywords:** iron, equation of state, strength, shock, isentropic compression, laser loading
**PACS:** 62.20.Dc, 62.20.Fe, 64.60.-I, 64.70.Kb

## INTRODUCTION

The behavior of Fe under extreme conditions, including dynamic loading, is important technologically and in the study of terrestrial planets and meteorites. The pressure-induced alpha-epsilon ($\alpha$-$\varepsilon$) phase transition at around 13 GPa is particularly interesting: gas gun experiments on polycrystalline samples observe this transition to occur over a period of some 40-160 ns [1]. However, molecular dynamics calculations predict this phase transition occurs on the scale of picoseconds [2]. Recent x-ray diffraction measurements on deposited single crystals shocked by laser ablation verified this prediction [3]. These results suggested that the $\alpha$-$\varepsilon$ phase transition took place before any plastic flow. Conversely, our previous results on rolled foils indicated that plastic flow occurred before the phase transition, and yet agree with Kalantar et al. [3] in that the transition took less than a nanosecond [4].

The time scale for the phase transition depends on the driving pressure, the loading history (time scale and path), the initial texture, and the material's plastic response under dynamic loading. Here we describe experiments on iron's dynamic response to shock and quasi-isentropic loading, supplying some insight into the flow stress.

## EXPERIMENTS

At Los Alamos National Laboratory's TRIDENT Laser Laboratory, we used direct laser illumination to produce a plasma that compresses

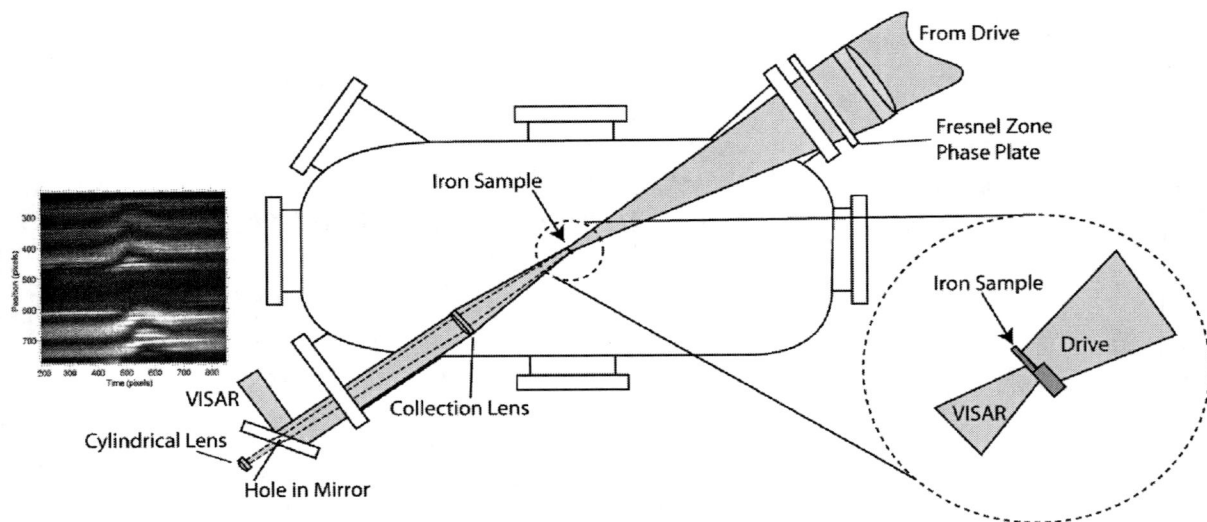

**Figure 1:** A typical TRIDENT direct-drive materials experiment layout for equation of state studies of iron with line-imaging VISAR as a diagnostic. A side-by-side target is shown where the stepped side of the sample is oriented towards the drive laser.

the sample. Figure 1 shows the experimental configuration. A drive laser illuminated one side of the sample with laser intensities below $10^{13}$ W/cm$^2$, rapidly producing an expanding plasma. A Fresnel phase plate introduced a high mode number modulation into the intensity profile, thus effectively creating a uniform one-dimensional drive over the 5-mm spot. One-dimensional radiation hydrodynamic (rad-hydro) modeling showed that for 2.4-ns long, square laser pulses with constant irradiance under these conditions, less than 1 micron of material ablated as plasma to compress the sample [4]. The plasma's thermal and radiative properties smooth out the compression history somewhat; however, large variances in irradiance result in pressure changes in the sample. We used the sensitivity of the pressure history to the drive irradiance history to probe specific regions of equation of state [5].

The Goodfellow high purity (99.99+ %), rolled iron foils were 30-50 μm thick. The samples were recovered for pre- and post-shot characterization. Texture mapping and nanoindentation measurements were performed on the samples to correlate with experimental flow stress measurements. Nanoindentation yielded static flow stresses of 420 MPa ± 25 MPa, signficantly higher than typical steels (100-150 MPa). Such a

discrepancy might be attributed, at least partly, to the rolling process hardening the foils' surface.

On the diagnostic side of the iron foils, a 150-ns, ~500-mJ, 660-nm Nd:YAG laser pulse was made cylindrically divergent and then passed through a hole in a mirror, figure 1. The light created a line focus using a 52-mm diameter 20-cm focal length lens. The light was then collected by this lens, magnified, and transported to the line-imaging VISAR. A Mach-Zender interferometer, similar to that described by Celliers et al. [6], measured surface velocity. The interferometric signal was line-imaged and time-resolved by a Hamamatsu C4187 streak camera. The absolute time of shock breakout and sweep rates were calibrated using a timing fiducial pulse train output from the drive laser's regenerative amplifier.

Four square shock drives with a 2.4-ns pulse width are shown with one laser quasi-isentropic (LICE) drive in figure 2a. The VISAR used an ~800-m/s/fringe, 71.8-mm thick BK7 etalon for the optical delay. The velocimetry measurements in figure 2b show that square laser pulses produced sharply rising shock responses.

The shocks shown for shots 15390-15395 were not completely discontinuous because of the etalon time delay, ~400 ps. The temporal smearing of the velocity history makes it difficult to resolve high

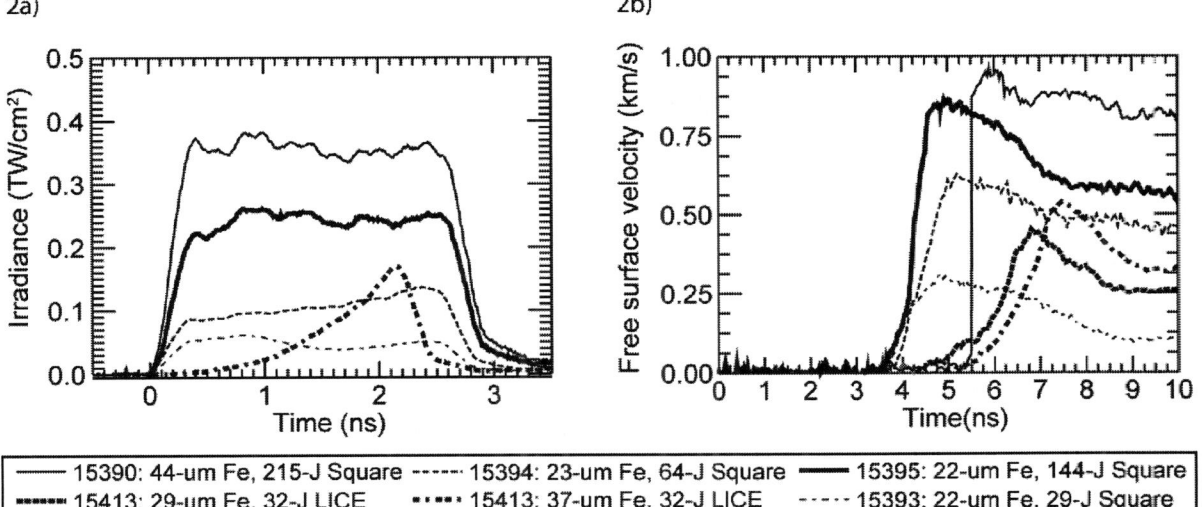

2a)

2b)

| 15390: 44-um Fe, 215-J Square | 15394: 23-um Fe, 64-J Square | 15395: 22-um Fe, 144-J Square |
| 15413: 29-um Fe, 32-J LICE | 15413: 37-um Fe, 32-J LICE | 15393: 22-um Fe, 29-J Square |

**Figure 2:** The temporal evolution of free surface velocity is dependent upon the laser irradiance history. Nearly constant irradiances, in 2a, shock the iron as seen for shots 15390, 15393, 15394, and 15395 in 2b. Ramped irradiances or laser quasi-isentropic (LICE) drive, as with 15413, increase the pressure more slowly.

frequency signals. The modulation transfer function (MTF) may need to be corrected to unambiguously resolve some of iron's phase transitions.

The LICE drive was delivered onto side-by-side iron samples, where the two samples for TRIDENT shot 15413 had thicknesses of 29 and 37 μm, shown in figure 2b. The slowly increasing laser ramp loaded the iron quasi-isentropically, indicated, in part, by the slowly increasing free surface velocity.

The isentropic drive achieved higher peak pressures for a given laser energy in comparison to the sharply rising shock drive. TRIDENT's LICE shot 15413 with 32 Joules produced a peak velocity closer to TRIDENT's shock shot 15394 with 64 Joules, and pressures higher than TRIDENT's shock shot 15393 with 29 Joules.

**RESULTS AND DISCUSSION**

Interpretation of the experimental results required modeling using the laser irradiance. We used 1-D rad-hydro codes such as Hyades [7] or Helios [8] to obtain the pressure history within ~ 1 μm of the drive side surface. In general, these codes have, at best, simple models of solid-state mechanical properties such as strength; and therefore, cannot accurately reproduce observations of strength-dependent processes, e.g. spall. Instead, the drive pressure history calculated by the rad-hydro simulations is sourced as a stress history, shown in figure 3, into a continuum mechanics code, e.g. LAGC1D/LAGC [9]. The continuum mechanics codes incorporated material dynamics,

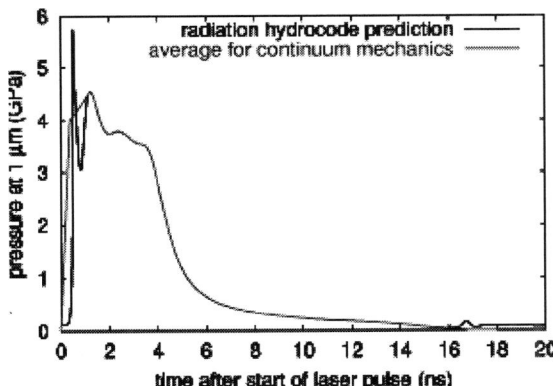

**Figure 3:** Using the laser irradiance history as a source, Hyades calculates the stress history for input into a continuum mechanics code.

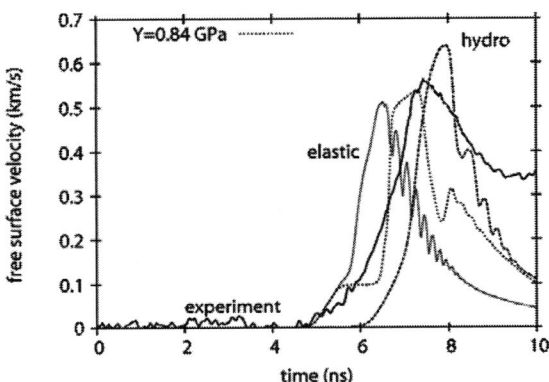

**Figure 4:** Continuum mechanics simulations incorporate more explicit material properties required for analysis, such as plasticity, elasticity, and strength.

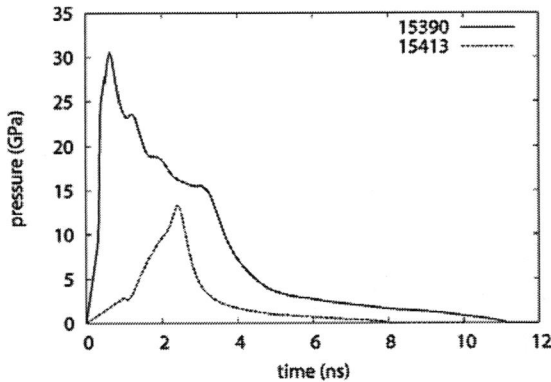

**Figure 5:** The pressure histories calculated using rad-hydro and continuum mechanics simulations for iron shock and LICE experiments.

such as strength, to calculate free surface velocities that can be compared to experimental data.

Previous laser-driven measurements [4] on iron found strong elastic precursors, in the range of work-hardened steels, which are generally much stronger than pure Fe. Recent shock measurements agree with previous observations that found a flow stress of 1.4 GPa.

Plastic flow was required in modeling of LICE experiments to match the wave arrival time for a 37 μm thick iron sample, figure 4. These simulations started with an initial flow stress of 0.3 GPa. Spall was observed in the deceleration data at late times. The VISAR measurements in combination with the modeling revealed the loading history for the iron foils, figure 5.

## CONCLUSIONS

We described nanosecond-scale experiments where VISAR measurements on rolled iron foils were used with modeling to infer the pressure histories. Modeling of LICE experiments required elasticity to match the wave breakout time, and plastic flow begins to occur below the bcc to hcp transition. MTF deconvolutions or thinner VISAR etalons may be required to better resolve the elastic wave and phase transition accurately.

## ACKNOWLEDGEMENTS

The TRIDENT laser staff provided technical support for these experiments. This Inertial Confinement Fusion Program work was supported under the auspices of US Department of Energy contract W-7405-ENG-36.

## REFERENCES

1. Barker LM and Hollenbach RE, J. Appl. Phys. **45**, 4872 (1974).
2. Kadau K et al., Science **296**, 1681 (2002).
3. Kalantar D et al., Phys. Rev. Lett **95**, 075502 (2005).
4. Swift D.C. et al., Proc. of the 5th International Symposium on Behaviour of Dense Media under High Dynamic Pressures, vol. II, (2003).
5. Swift D.C. and Johnson R.P., Phys. Rev. E **71**, 066401 (2005).
6.. Celliers P.M. et al., Appl. Phys. Lett. **73**, 1320 (1998).
7. Computer code HYADES, version 01.05.11, Cascade Applied Sciences Inc., Golden, Colorado, (1998).
8. Computer code HELIOS, version 03.03, Prism Computational Sciences, Inc., Madisson, Wi, (2005).
9. Computer code LAGC1D, version 5.2, Wessex Scientific and Technical Services Ltd., Perth, Scotland, 2003d, http://www.wxres.com.

CHAPTER XXIII

# GEOPHYSICS AND PLANETARY PHYSICS

CP845, *Shock Compression of Condensed Matter - 2005*,
edited by M. D. Furnish, M. Elert, T. P. Russell, and C. T. White

# NUMERICAL MODELING OF SHOCK-INDUCED DAMAGE FOR GRANITE UNDER DYNAMIC LOADING

## H. A. Ai[1], T. J. Ahrens[1]

[1]*Lindhurst Laboratory of Experimental Geophysics, Seismological laboratory,
California Institute of Technology, Pasadena, CA, 91125.*

**Abstract.** Johnson-Holmquist constitutive model for brittle materials, coupled with a crack softening model, is used to describe the deviatoric and tensile crack propagation beneath impact crater in granite. Model constants are determined either directly from static uniaxial strain loading experiments, or indirectly from numerical adjustment. Constants are put into AUTODYN-2D from Century Dynamics to simulate the shock-induced damage in granite targets impacted by projectiles at different velocities. The agreement between experimental data and simulated results is encouraging. Instead of traditional grid-based methods, a Smooth Particle Hydrodynamics solver is used to define damaged regions in brittle media.

**Keywords:** granite, JH2 model, crack softening, shock-induced damage, AUTODYN, SPH.
**PACS:** 91.60.-x, 91.60.Ba.

## INTRODUCTION

Shock-induced damage in rocks beneath impact craters is useful for constraining the impact history [1, 2]. The behavior of rocks to dynamic loading is very complex. Cracks from impact events are induced by both shear and tensile failure. In the high pressure region ahead of a projectile during impact process, the inelastic shear straining dominates the production of damage. For region at low pressure, the principal tensile stress is of the same order as the deviatoric stress. Tensile cracks would be produced in this region. However, a complete and appropriate constitutive model to describe deformation and damage of rocks due to both components is still needed for numerical simulation.

We apply JH-2 model, which was originally developed by Johnson and Holmquist for ceramics [3], to geological crustal rocks for the first time. A crack softening model is coupled with JH-2 model to represent the tensile cracks generation [4]. We focus on how to determine model parameters for granite. We put these parameters into AUTODYN-2D [5] to calculate the damage that occurs beneath and surrounding impact craters in crustal rocks. The simulated results are also compared with experimental data.

## DETERMINATION OF MODEL CONSTANTS FOR GRANITE

The JH-2 constitutive model assumes that the strength of material, both intact and fractured, is dependent on pressure, strain rate, and damage. The dependence of strength on these parameters is represented by a set of constants. These constants are derived from standard dynamic and quasi-static measurements [3].

A summary of these constants for granite is listed in Table 1. The following will discuss how to determine the constants for pressure, strength, damage, as well as crack softening.

**Table 1:** JH-2 baseline and crack softening constants for granite.

| Strength constants | |
|---|---|
| Hugoniot elastic limit (HEL) | HEL = 4.5 GPa |
| HEL Strength | $\sigma_{HEL} = 2.66$ GPa |
| HEL Pressure | $P_{HEL} = 2.73$ GPa |
| HEL Volumetric Strain | $\mu_{HEL} = 0.045$ |
| Tensile Strength | $T = 0.15$ GPa |
| Normalized Tensile Strength | $T^* = 0.055$ |
| Intact Strength Coefficient | $A = 1.01$ |
| Intact Strength Coefficient | $N = 0.83$ |
| Strain Rate Coefficient | $C = 0.005$ |
| Fracture Strength Coefficient | $B = 0.68$ |
| Fracture Strength Exponent | $M = 0.76$ |
| Maximum Fracture Strength | $\sigma^*_{f\,max} = 0.2$ |
| **Pressure Constants** | |
| Bulk modulus | $K_1 = 55.6$ GPa |
| Pressure coefficient | $K_2 = -23$ GPa |
| Pressure coefficient | $K_3 = 2980$ GPa |
| Bulking factor | $\beta = 1.0$ |
| **Damage constants** | |
| Damage coefficient | D1 = 0.005 |
| Damage exponent | D2 = 0.7 |
| **Cracking Softening Constants** | |
| Tensile failure stress | $T_f = 0.15$ GPa |
| Fracture energy | $G_f = 70$ J/m$^2$ |

## Pressure

Fig. 1 shows the axial stress, $\sigma_1$, and the mean stress/pressure, $P$, as a function of the volumetric strain $\mu$ during uniaxial strain loading for Westerly granite [6] and for Climax stock granodiorite [7]. Pressure constants $K_1$, $K_2$, $K_3$ are obtained by fitting $P$ to $\mu$ using

$$P = K_1\mu + K_2\mu^2 + K_3\mu^3, \qquad (1)$$

where $K_1$, $K_2$, and $K_3$ are constants ($K_1$ is the bulk modulus), and $\mu = \rho/\rho_0 - 1$ for current density $\rho$ and initial density $\rho_0$. The linear hydrostat is also shown to provide a reference.

## Strength

The HEL is taken as 4.5 GPa, the average value from [8]. We follow the method described in [3] to determine the strength and pressure components at HEL. The HEL volumetric strain is solved from $HEL$, $K_1$, $K_2$, $K_3$, and $G$ as $\mu_{HEL} = 0.045$. Substitute $\mu_{HEL}$ into Eqn. 1 gives pressure at HEL of $P_{HEL} = 2.73$ GPa. The equivalent stress, defined as twice of the material shear strength, at HEL is 2.66 GPa ($\sigma_{HEL}$). The intact equivalent stress of the material as a function of pressure, from data in Fig. 1, is shown in Fig. 2. Also shown is the calculated value from JH-2 model using constants in Table 1 at two strain rates, $10^5$ s$^{-1}$ and $10^{-4}$ s$^{-1}$. It seems that the effect of strain rate is rather small. The strain rate coefficient, $C$, is assumed to be 0.005, taken as the same as ceramic [3]. Fortunately, this constant does not influence the result greatly.

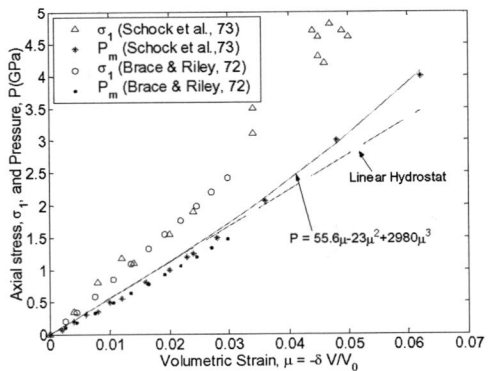

**Figure 1**: Test data and model for shock pressure-volume response of granite.

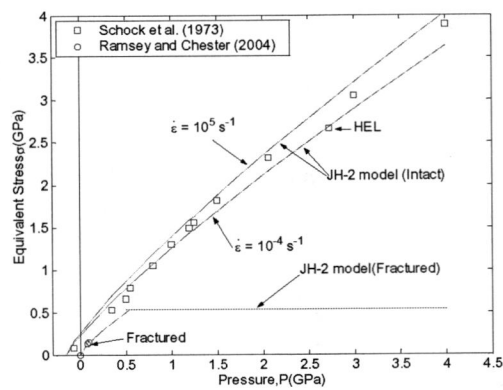

**Figure 2:** Test data and model for strength of intact and damage granite.

The dynamic tensile strength of San Marcos granite is 0.13 GPa, determined by planar impact method described in [2, 9]. Similar procedure gives $P_{tensile}$ = -0.067 GPa and $\sigma_{tensile}$ = 0.08 GPa (Fig. 2). Extrapolating this to $\sigma$ = 0 gives tensile strength $T = -P_{\sigma=0} = 0.15$ GPa. The normalized tensile strength is $T^* = T/P_{HEL} = 0.055$. Intact strength constants $A$ and $N$ are obtained by nonlinear fitting of the experimental data using:

$$\sigma_i^* = A(P^* + T^*)^N, \qquad (2)$$

$\sigma_i^*$ is the normalized equivalent stress.

No proper fractured strength data for granite are found. Instead, some fractured data for marble are used for this purpose. The source of the fractured data shown in Fig. 2 is from [10]. We fit the data available to Eqn. 3 to obtain the fracture strength constants, $B$ and $M$.

$$\sigma_f^* = B(P^*)^M, \qquad (3)$$

More experiments for fracture strength of granite are necessary to obtain a better constraint of these fractured strength parameters. The normalized fractured strength is limited not to exceed the maximum fractured strength, $\sigma_{f\,max}^*$, taken as 0.2 here, or 0.53 GPa as the equivalent stress.

## Damage

Damage ($D$) describes the transition from intact to fractured strength. Under a constant pressure, damage begins to accumulate when the material begins to flow plastically ($D$ = 0). When the material is completely damaged, $D$ = 1. The damage parameters $D_1$ and $D_2$ used by Johnson and Holmquist [3] are not directly measurable. Instead, numerical adjustment is applied to obtain $D_1$ and $D_2$ listed in Table 1.

## Tensile crack softening

The maximum principal tensile stress for the tensi-

le softening model is 0.15 GPa, as noted above. The associated fracture energy is assumed to be 70 J/m$^2$, which is the value obtained for ceramics [4].

## EXAMPLE

The determined constants for granite are put into AUTODYN-2D to simulate two impacts into granite by a lead bullet and copper ball. Parameters for projectile are retrieved from AUTODYN library [5]. The calculated results are presented and compared with experiment data.

## Lead bullet impacting granite

Simulation of a 3.2 g lead bullet impacting a 20x20x15 cm granite block at 1200 m/s is carried out. Radius of the projectile is 3 mm. The meshfree Smoothed Particle Hydrodynamics (SPH) solver [5] is used for the projectile and rock target, with smoothing particle size to be 0.125 mm for the projectile and 0.25 mm for the target. The geometry of the problem setup and the response of target are assumed to be axisymmetric.

The simulated final damage profile is compared with the experimental result (Fig. 3). The crater depth is ~ 1.5 cm, and crater diameter is ~ 7 cm, both of which agree well with the experiment. The prediction of radial tensile cracks is encouraging: the pattern of simulated tensile cracks for both situations is very similar with the experiment. And the tensile cracks extend to 6-7 cm for both cases.

Simulations with and without the crack softening model were carried out and compared. The tensile cracks when the crack softening model is not included do not extend as long as that when the model is included.

## Copper ball impacting granite

Comparison between calculation and experiment for a copper ball (0.64 cm in radius) impacting a granite block at impact velocity of 690 m/s is shown in Fig. 4. Again, the agreement is very good. The crater depth is ~ 1 cm, crater diameter is ~ 5 cm, and tensile cracks extend to ~ 8 cm for both cases.

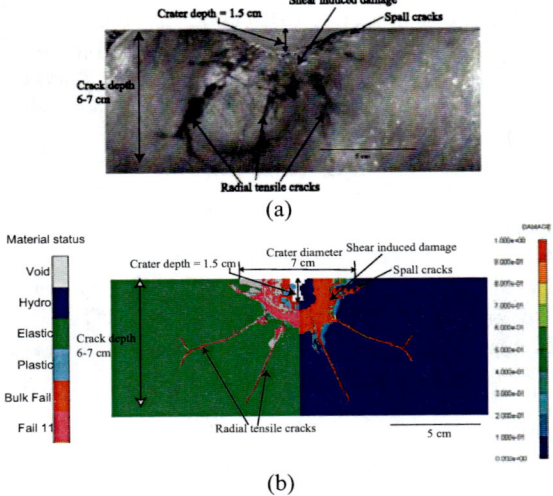

(a)

(b)

**Figure 3:** Cross section of granite impacted by lead bullet at 1200 m/s illustrating crack distribution. (a) Experimental result; (b) AUTODYN-2D simulation at 0.03 ms. Left panel illustrates material status; right panel illustrates damage.

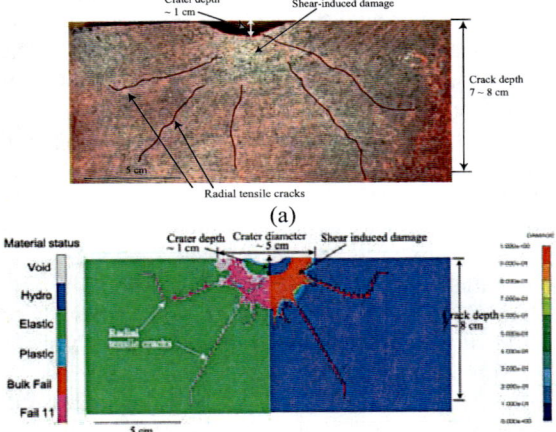

(a)

(b)

**Figure 4:** Cross section of granite impacted by copper ball at 690 m/s. (a) experimental result; (b) simulation at 0.04 ms. Others are the same as in Fig. 3b.

## CONCLUSION

This work is intended to describe response of geological material under impact loading. The JH-2 constitutive model describing mechanical character of brittle material, coupled with a crack softening model, is applied to granite for the first time to represent the deviatoric and tensile cracks produced beneath an impact crater in rocks. Model constants are obtained either from direct static measurement, or from indirect numerical calibration. The agreement between the calculation and experiment is encouraging.

## ACKNOWLEDGEMENTS

Research supported by NASA/Goddard grant under award no. NNG04GI07G. Contribution No. 9123. Division of Geological and Planetary Sciences, Caltech.

## REFERENCES

1. Ahrens TJ, Xia K, Coker D, "Depth of cracking beneath impact craters: New constraint for impact velocity", in Shock-compression of condensed matter, 2002 (MD. Furnish, LC. Chhabildas, RS. Hixson, eds.), pp. 1393-1396.
2. Ai HA, Ahrens TJ, "Dynamic tensile strength of terrestrial rocks and application to impact cratering," Meteoritics & Planetary Science 39(2), 233-246, 2004.
3. Johnson GR, Holmquist TJ, "Response of boron carbide subjected to large strains, high strain rates, and high pressures," J. Applied Phys 85(12), 8060-8073. 1999.
4. Clegg RA, Hayhurst CJ., "Numerical modeling of the compressive and tensile response of brittle materials under high pressure dynamic loading", in Shock compression of condensed matter, 1999 (MD. Furnish, LC. Chhabildas, RS. Hixson, eds.) pp. 321-324.
5. AUTODYN Theory manual. 2003.
6. Brace WF, Riley DK, "Uniaxial deformation of 15 rocks to 30 kb", Int. J. Rock. Mech. Min. Sci. 9, 271-288, 1972.
7. Schock RN, Heard HC, Stephens DR, "Stress-strain behavior of a granodiorite and two graywackes on compression to 20 kilobars", J. Geophys. Res. 78, 5922-5941, 1973.
8. Petersen CF, "Shock wave studies of selected rocks," Stanford Univ. Ph. D. thesis, p. 99. 1969.
9. Ahrens TJ, Rubin AM, "Impact-induced tensional failure in rock," J. Geophys. Res. 98, 1185-1203. 1993.
10. Ramsey JM, Chester FM, "Hybrid fracture and the transition from extension fracture to shear fracture," Nature 428(6978), 63-66. 2004.

CP845, *Shock Compression of Condensed Matter - 2005*,
edited by M. D. Furnish, M. Elert, T. P. Russell, and C. T. White
© 2006 American Institute of Physics 0-7354-0341-4/06/$23.00

# THE SHOCK HUGONIOT PROPERTIES OF QUARTZ FELDSPATHIC GNEISS AND AMPHIBOLITE

## C.H. Braithwaite, W.G. Proud and J.E. Field

*PCS. Cavendish Laboratory, Madingley Road, Cambridge CB3 0HE, UK*

**Abstract.** A series of plate impact experiments was performed to determine the principal Hugoniot curves of geological materials. The experiments were carried out using the plate impact facility at the Cavendish Laboratory, Cambridge University. By means of conventional impacts, over a range of impact velocities 217-1038 m/s, the Hugoniot of a quartz feldspathic gneiss and an amphibolite have been obtained. Manganin stress gauges were used as the main diagnostic tool. The Hugoniots show that the materials are either behaving elastically in the range of interest, or that the elastic limit of the materials are not visible due to there being no significant change in slope at the HEL.

**Keywords:** Geological materials, Hugoniot, plate impact
**PACS:** 91.60.-x, 62.50.+p

## INTRODUCTION

One of the major thrusts of the research being undertaken by de Beers is the development of an improved computer model for blasting rock which can be used by engineers in the field. The model requires an accurate knowledge of the shock properties of the geological materials likely to be found in mines.

Much of the research reported in the literature on geological materials is concerned with high pressures, such as those found at levels consistent with nuclear blasts or planetary impacts [1,2]. However there have also been a smaller number of studies looking at lower pressures, more comparable with those found in a mining situation. These provide useful insight into material response and also lay the groundwork for this research. Tsembelis *et al.* [3] report that it is difficult to find an HEL in Dolerite as there is no significant change in slope in the Hugoniot plot at the HEL suggesting that there is little difference in the material's impedance above and below the HEL. Field [4] comments, regarding rock blasting, that "directly measured, empirical Hugoniot data should be treated as the most important data set for shock studies of rocks below 10 GPa".

## EXPERIMENTAL

All the experiments were carried out using the plate impact facility at the Cavendish Laboratory, Cambridge University. A gas gun of 50mm bore and 5m barrel length [5] was used to fire projectiles at velocities up to 1050 m s$^{-1}$, as measured by four pairs of sequentially shorting pins. The flyer impacts a target, which has been aligned to within <1 mrad using a dial gauge. This leads to a planar impact and a state of uniaxial strain. It has been used for other previous geological material testing [*e.g.* 3, 6].

In order to determine the points on the Hugoniot curve, a series of standard configuration plate impact experiments were conducted, using copper as the well characterized flier material [7]. Gauges were placed in both embedded and back surface configurations, and voltage output was reduced to a stress history using the calibration method set out by Rosenberg [8].

The two types of rock on which data is presented in this paper are; coarse-grained quartz feldspathic gneiss (figure 1) and a much finer grained amphibolite (figure 2). The grain size in the gneiss is of the order of 2-10mm whereas the amphibolite is smaller, on the scale of 1-2mm.

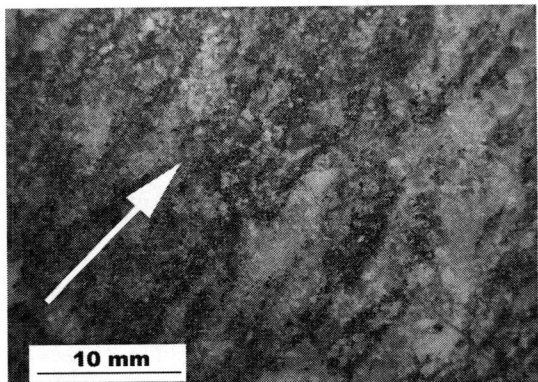

**FIGURE 1.** A macro photograph of the coarser grained rock, a quartz feldspathic gneiss. The arrow denotes the grain direction

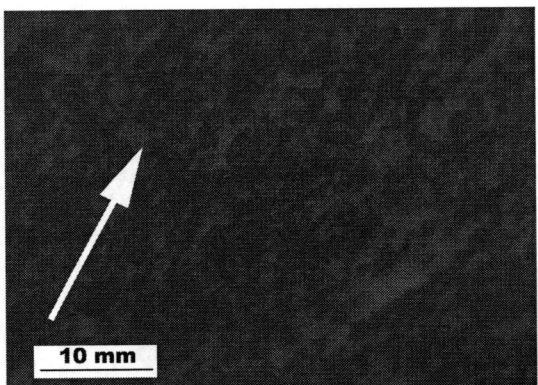

**FIGURE 2.** A macro photo of the amphibolite. Again the arrow indicates grain direction.

The amphibolite appears to be more homogeneous, but it is obvious that both rocks show a clear directionality in the grain structure when examined closely. This can be seen in figures 1 and 2 (denoted by the arrows), and is even clearer when examining the drilled cores which the samples were prepared from. In this study, only one orientation, approximately 45 degrees to with respect to grain direction, has been investigated.

## RESULTS

The notable feature of the data obtained in this experiment concerns noise in gauge traces.

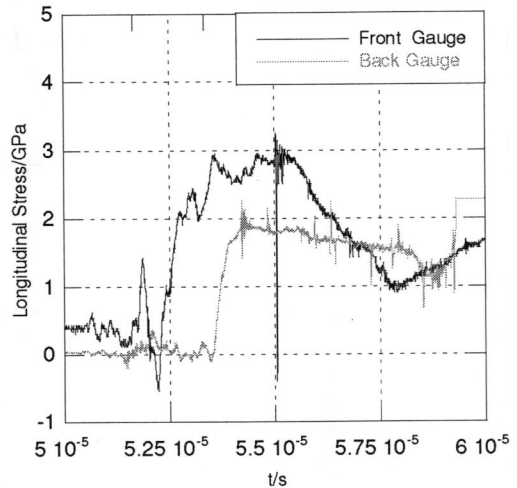

**FIGURE 3.** Sample gauge trace from gneiss experiment. E050112A (copper impacter), impact velocity 265 m s$^{-1}$.

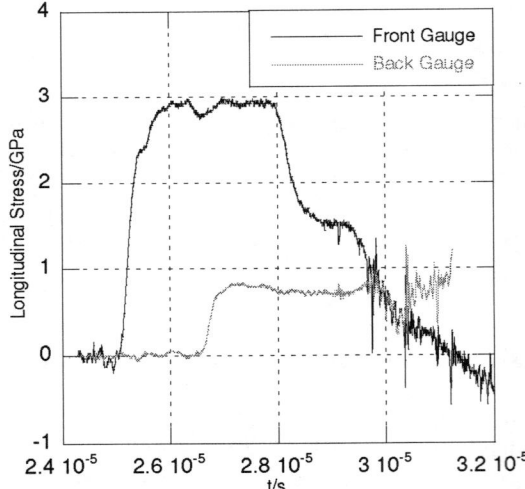

**FIGURE 4.** Sample gauge trace from amphibolite experiment. E050112A (copper impacter), impact velocity 217 m s$^{-1}$.

There is a significant amount of noise, but appreciably more in the gneiss traces (figure 3) than in the amphibolite (figure 4). This difference can be explained in terms of the differing grain structures within the materials, or could possibly be due to piezoelectric effects in the quartz grains of the gneiss. If there are quartz grains in the amphibolite (a full petrographic study has yet to be completed) then their smaller size would diminish their effect. The effect of material type on the gauge traces can also be seen in the back surface gauges, which are significantly smoother as they are set in PMMA.

Such traces present a challenge for analysis. Here the results were analysed by matching the stress measured by the gauge with the Hugoniot of copper to get a particle velocity. The stress level of the gauges was found by computing a mean value of the plateau stress. This allowed for the effects of the noise to be reduced.

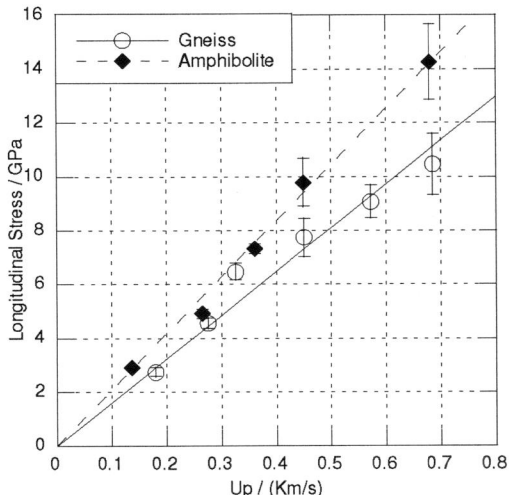

**FIGURE 5.** Hugoniot in P- $U_p$ space for both rocks tested.

The resulting Hugoniot is shown in figure 5. The deviation of points from the line of best fit can be attributed to the noise in the gauge traces.

Analysis by obtaining a shock velocity from the rise times of the gauge traces and using the simple relation $\sigma = \rho u_s u_p$ (where $\sigma$ is the stress in the sample, $u_p$ is the particle velocity, $u_s$ the shock velocity and $\rho$ the density), was considered an unreliable method due to noise in the gauges traces. This uncertainty may again be due to either grain effects or piezoelectric effects. It should be noted, however, that results from this method were not significantly different from those shown in figure 5. This, in addition to the stress measured in the back

surface gauges gives a confirmation of the results presented in figure 5.

## DISCUSSION

The Hugoniot curves of two separate rocks have been determined by experiment and shown in figure 5. The shock impedance of the amphibolite is higher than that of the granite, which can be expected due to it's slightly higher density (2.650 ± 0.002 g/cm$^{-3}$ compared with 3.001 ± 0.005 g/cm$^{-3}$).

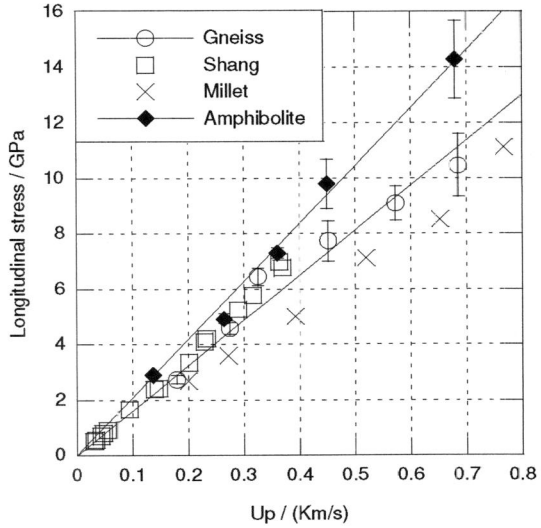

**FIGURE 6.** Comparison of data in this study with previously published papers.

The results are consistent with other calculated Hugoniots of rock including investigations by Millet [9] and Shang [10], as shown in figure 6.

Figure 7 shows the Hugoniot compared to an elastic line calculated from experimentally measured sound speed (determined using ultrasonic transducers and found to be 5.74 ± 0.02 km/s). There is good agreement between the two lines. This however does not necessarily mean that the rock is behaving elastically over the entire range studied. It is possible that the HEL is spanned in the pressure range studied, but it not shown by amp the data. Other studies [3, 9], suggest that a Hugoniot Elastic Limit (HEL) may not be seen in longitudinal measurements due to similar values of elastic and shock impedances. It is also shown that the HEL can be determined through examining lateral gauge histories and determining the shear strength of the material [3].

It is worth noting that that some elastic behavior is recorded, and expected, in this study. Rocks which show no significant elastic behavior have been characterized. An example, relevant to mining, of this is Willmott's work [11] on Kimberlite. In the case of the Amphibolite and gneiss presented here, it is hoped that lateral stress measurements will allow a determination of the HEL for the materials.

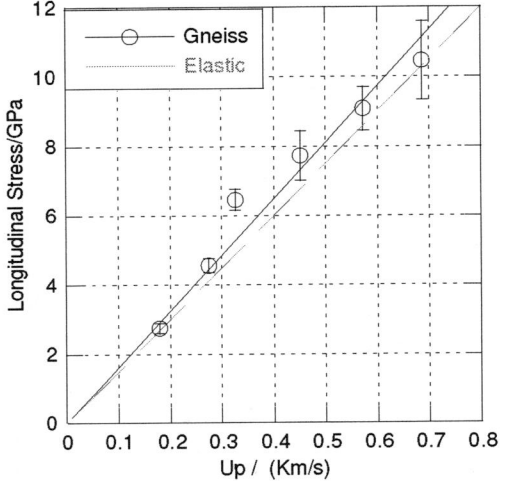

**FIGURE 7**. Comparing the Hugoniot with elastic behaviour.

Further investigation will also involve photographing the shockwave profiles of the rocks, attempting to determine the source of the noise present in the gauge traces. Once found, steps will be taken in further experiments to eliminate this noise where possible. If it is found that the source is the granular structure of the rocks, further rock samples might benefit from being tested in a reverse configuration. In addition, further research will address anisotropy in the shock properties of the rocks, by experiments with the grain in different orientations.

## ACKNOWLEDGEMENTS

D Chapman, K. Tsembelis for discussions on and assistance with experimental work, R Flaxman and the Cavendish workshop staff for their invaluable work on sample preparation, and Dr A.R. Guest (de Beers) for providing the sample materials and sponsoring this research.

## REFERENCES

1. Telegin, G.S., et al., *Calculated determination of Hugoniot curves of rocks and minerals*. Phys. Solid Earth, 1980. **16**: p. 319-324.
2. Trunin, R.F., et al., *Rock compressibility in shock waves*. Phys. Solid Earth, 1988. **24**: p. 38-42.
3. Tsembelis, K., W.G. Proud, and J.E. Field, *The principal Hugoniot and dynamic strength of dolerite under shock compression*, in *Shock Compression of Condensed Matter - 2001*, M.D. Furnish, N.N. Thadhani, and Y. Horie, Editors. 2002, American Institute of Physics: Melville, NY. p. 1385-1388.
4. Field, J.E., Willmott G.R..and Proud,W.G., *Shock Hugoniots of Tuffistic Kimberlite Breccia and Other Geological Materials Below 10 GPa*. Shock Compression of Condensed Matter, 2005. **yet to be published**.
5. Bourne, N.K., et al., *Design and construction of the UK plate impact facility*. Meas. Sci. Technol., 1995. **6**: p. 1462-1470.
6. Willmott, G.R., W.G. Proud, and J.E. Field, *Shock properties of kimberlite*, in *Shock Compression of Condensed Matter - 2003*, M.D. Furnish, Y.M. Gupta, and J.W. Forbes, Editors. 2004, American Institute of Physics: Melville NY. p. 1492-1495.
7. Marsh, S.P., *LASL Shock Hugoniot Data*. 1980, Berkeley, California: University of California Press.
8. Rosenberg, Z., et al., *Determination of stress-time histories in axially symmetric impacts with the two-gauge technique*. J. Appl. Phys., 1984. **56**: p. 1434-1439.
9. Millett, J.C.F., K. Tsembelis, and N.K. Bourne, *Longitudinal and lateral stress measurements in shock-loaded gabbro and granite*. J. Appl. Phys., 2000. **87**: p. 3678-3682.
10. Shang, J.L., L.T. Shen, and J. Zhao, *Hugoniot equation of state of the Bukit Timah granite*. Int. J. Rock Mech. Min. Sci., 2000. **37**: p. 705-713.
11. Willmott, G.R., W.G. Proud, and J.E. Field, *Shock properties of diamond and kimberlite*. J. Phys. IV France, 2003. **110**: p. 833-838.

CP845, *Shock Compression of Condensed Matter - 2005*,
edited by M. D. Furnish, M. Elert, T. P. Russell, and C. T. White
© 2006 American Institute of Physics 0-7354-0341-4/06/$23.00

# MICROBIAL LIFE AND SHOCK COMPRESSION – LIFE OR DEATH?

**M. J. Burchell**[1]

[1]*Centre for Astrophysics and Planetary Science, School of Physical Sciences, Univ. of Kent, Canterbury Kent CT2 7NH, United Kingdom*

**Abstract:** The role of shock compression in killing microbial life is discussed. As well as the extreme pressures involved, the duration of the shock is also important (i.e. static vs. dynamic). For shock pressures and durations typical of impacts on planetary surfaces, there is experimental data now available from a variety of sources illustrating that microbial survival rates are small but finite (typically order $10^{-4}$ to $10^{-7}$). Thus any viable biological material on a rock arriving from space at high speed may survive intact after impacting the Earth. Similarly, it is possible to define a zone of lethality around an impact site, and predict survival rates of indigenous microbial life in the target material as a function of shock pressure. Finally, it is noted that post-impact, the altered crater environment may be more suitable for sustaining life than the surrounding region was before the impact event.

**Keywords:** Astrobiology, impacts, Hugoniot Elastic Limit, hypervelocity impact
**PACS:** *43.25.Cb, 60.50.+p.

## INTRODUCTION

Shock compression of condensed matter is not a topic that immediately springs to mind as having implications for the origin of life on Earth. However, it does have a role to play in the possible natural migration of life through space and in the extinction of life on a planet in large impact events.

The first of these topics (life migrating through space) is known by the name Panspermia (see [1] for a recent review). The idea is simple; life is distributed widely in space and spreads naturally. Various mechanisms exist to explain this migration. One variant is litho-Panspermia [2], wherein life starts on planet A. This planet is then subject to a giant impact from space (a common place event when viewed over the life time of the Solar System), and ejecta are thrown into space at greater than escape velocity. The ejecta then moves through space on a heliocentric orbit and impacts planet B. Potentially microbial life on planet A can

thus have migrated to planet B mixed into the ejected rock. As well as other hazards such as the low temperatures and radiation environment in space [3, 4, 5, 6] this process involves at least two shock events, one during launch from A and one on arrival at B. Survival by microbial life of these shocks is thus critical to the model.

The second topic concerns extinction of life on a planet. If a large object strikes the Earth from interplanetary space, the impact will be a violent one. A typical impact speed in such an event will be 20 to 25 km s$^{-1}$. A large crater will result which may be many km or 10's of km across and peak shock pressures during crater formation can be in the range 1 - 100 GPa. These impacts can have global consequences and by triggering atmospheric or climate changes may cause large scale (i.e. planetary wide) extinction of higher life forms (e.g. the KT boundary event and possible associated mass extinction). Just as importantly, microbial life whose habitat is inside the rock of the planetary surface can also be killed. This need not just be the consequence of a sterilizing heat pulse or of the long duration elevated temperatures at the impact site (due to the energy input the impact crater and its immediate subsurface vicinity will remain hot for periods of 1000 years or so), the shock compression itself can disrupt cell walls and kill microbial life.

An impact event need not just deliver life to a planet or cause its extinction; it may do both. For example, a sterilizing impact event may throw off ejecta into space on a bound orbit (i.e. it may not achieve escape velocity). After a period it may re-impact and reseed the Earth [7]. Or the impact may drive shock chemistry in the target atmosphere or oceans, creating materials required for life [8]. Indeed the various possibilities apply not just to planets but also to any habitable environments, e.g. the moons of Jupiter where similar impact processes can cause material to be shared between the various Jovian satellites.

In the rest of this paper details are given concerning the possibilities outlined in the Introduction.

## SHOCK DURING EJECTION

A giant impact from space can result in the formation of an impact crater and ejection of relatively lightly shocked material into space [2]. The degree of shock the ejecta are subject to is open to debate. However, in [9] it is estimated that for Mars a "jerk" (acceleration divided by duration) of $6 \times 10^9$ m s$^{-3}$ is required to launch material into space (that is an acceleration of $3 \times 10^6$ m s$^{-2}$ for $0.5 \times 10^{-3}$ s). Although static loading of microbial life (active cells and spores) in centrifuges has simulated survival at high accelerations, it is this shock-like nature of the "jerk" that is important for cell lethality. Using a light gas gun it has been shown that active cells can be recovered from targets after being carried on projectiles carried in the gun [10]. The firing of the gun involves acceleration from rest to 5 km s$^{-1}$, over a length of 1 m and short time scales, equivalent to a "jerk" of $6 \times 10^{10}$ m s$^{-3}$. This exceeds the requirement for launch into planetary space from Mars. The implication is that ejecta from Mars would not have been sterilized by the shock of their launch

into space. Potentially, if Martian rocks contain micro-organisms they could be seeding interplanetary space.

Similar processes can occur around planets with multiple natural satellites, e.g. Jupiter or Saturn. For example, it has long been hypothesized that the ice covered Jovian moon Europa, may have a liquid subsurface ocean. Liquid water is one of the key ingredients for life, and so potentially this forms a niche habitable zone. Impacts on the surface of Europa will throw ejecta into space. If any of these icy ejecta carries frozen microbial life, it can migrate around the Jovian system until it impacts another moon. The shock related ingredients of this "icy satellite Panspermia" have been tested in [11]. In [11] samples of (late stage growth) *Rhodococcus erythropolis* were frozen into water (approx. $10^9 - 10^{10}$ cells per mL). Projectiles were fired at the ice targets at speeds of 5 km s$^{-1}$, using a two stage light gas gun. The resultant ejecta were then captured and placed on agar plates for culturing. Growth of colonies of *Rhodococcus erythropolis* was then observed. There was no direct measurement of the speed of the ejecta in [11], but a maximum speed of 100 m s$^{-1}$ was estimated. Whilst low, this could still exceed the escape velocity on smaller satellites.

## SHOCK ON IMPACT

On arrival at a planet the impact of a rock from space will be violent. For an object greater than about 50 m, we can ignore the deceleration during atmospheric entry. Provided the projectile does not explode during descent, the impact on the solid planetary surface will be at a speed of some 10's of km s$^{-1}$. This will generate peak shock pressures in the range 1 to 100 GPa and vaporize most of the projectile. Traditionally it is held that only $1 - 2\%$ of the projectile will be found at the impact site and this will be in the form of melted material mixed with target melt. It is therefore hard to see how this can lead to any survival of the microbial load on the projectile.

However, several experiments have attempted to reproduce these severe shock conditions and test for microbial survival. Two approaches are the most common. At the Univ. of Kent (UK) a two stage light gas gun was used to fire projectiles at targets at speeds of some 5 km s$^{-1}$. The projectiles were porous ceramic which was doped with a solution containing a suspension of microbes (*Rhodococcus erythropolis*) fired directly at agar plates. It was found that after incubation, the plates subsequently displayed colonies of *Rhodococcus erythropolis* [10]. The survival rate was of order $10^{-7}$. *Rhodococcus erythropolis* was chosen as it is a hardy organism, resistant to many breakage mechanisms and survives under high static pressures.

Subsequently, the same facility reported on a wider range of experiments using *Rhodococcus erythropolis* and *Bacillus subtilis*, both in late stage growth and the latter also in spore state [12]. By varying the impact speed, the peak shock pressure was varied in the range 2 to 78 GPa. Over this shock pressure range the survival rate fell considerably ($10^{-4}$ to $10^{-7}$) see Fig. 1. A fit to the Survival Rate (*SR*) for the *Rhodococcus erythropolis* vs. shock pressure (*P*) in GPa yielded

$$SR = (1.13 \pm 0.07) \times 10^{-3} P^{-(2.03 \pm 0.01)} \qquad (1)$$

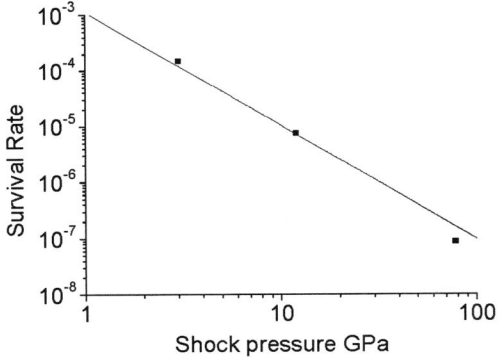

**FIGURE 1.** Survival rate for *Rhodococcus erythropolis* (late stage growth) fired at agar plates vs. peak shock pressure [12]. The fit line is described in the text (eqn. 1).

A different approach to obtaining high shock pressures is to use the flying plate technique. This has been done in [13] subjecting *Bacillus subtilis* spores to a shock pressure of 32 GPa to simulate a meteorite impact. The survival rate was found to be $10^{-4}$ to $10^{-6}$. More recently, in [14] it is reported that for peak shock pressures of $2 - 3$ GPa, the survival rate of *Escherichia (E.)coli* in a flyer plate experiment was $10^{-2}$ to $10^{-4}$.

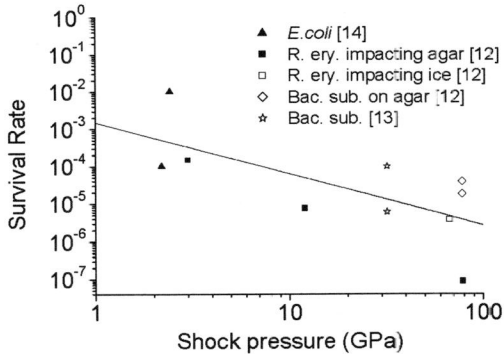

**FIGURE 2.** Survival rate for microbial life in [12, 13, 14] vs. peak shock pressure. The fit line is described in the text (eqn 2).

There are thus a range of impact experiments showing that microbial life can survive extreme shocks, albeit with a low survival rate which depends on the peak shock pressure. The results from [12, 13, 14] are all plotted on Fig. 2. A fit to the combined data yields for the Survival Rate (*SR*) vs. shock pressure (*P*) in GPa:

$$SR = 1.49 \times 10^{-3} P^{-1.37} \qquad (2)$$

## ZONE OF LETHALITY AROUND AN IMPACT SITE

As well as inhabiting a planetary surface, life is increasingly being found at high altitudes in the atmosphere [15] and at depth inside rock. Concerning the latter, it has long been known that microbial life can colonize rock. Some organisms are found in shallow depths inside surface rocks (e.g. in micro fissures up to 1 or 2 mm into the rock). However, it is now known that organisms can live at depths inside rocks of km magnitude. Thus when considering a planetary sterilizing impact event, as well as the now traditional model of mass extinction of complex organisms, a mechanism has to be developed to sterilize rocks at depth. The surface extinction model can be found by considering blast effects leading to concentric rings around the impact site. In each ring survival probabilities are given for different forms of life (e.g. microbial life). Applied to the Barringer crater (approx. 1.3 km diameter crater), the survival rate on the surface only rises to 50% some 9 - 14 km from the impact site [16]. However, the survival of sub-surface microbial life will follow a different pattern. A mechanism needs to be developed to determine how the shock event kills. In [17] it is proposed that since pressures above the Hugoniot elastic limit (HEL) exceed the failure point of a material, the resultant crushing will kill any microbial communities in micro-fissures inside the rock. This simple model thus only requires calculation of the HEL beneath an impact crater. This is then identified with a "potential survival limit" (PSL). Outside this region life may survive, inside it will not. This is shown in Fig. 3a.

Although the PSL is a step function in terms of survivability and thus somewhat unphysical, it is a useful tool which indicates the extent of the sub-surface (potentially biologically active) region most severely affected by an impact. The PSL for any particular impact can be estimated by hydrocode simulation, taking into account the particular nature of the impact event (speed, size of impactor etc) and the target material (composition, homogeneity, appropriate equation of state, etc). One difficulty pointed out in [17] lies in how the sub-surface shock propagates away from the impact site. Whilst it is naively possible to simply consider a uniform radial expansion, this fails near the surface. The influence of a free surface leads to rarefaction waves etc. significantly reducing the peak shock pressures and the location of the HEL. Of course, at the surface itself the blast wave effects come into play, but in terms of survivability a simple radial expansion equal at all radial distances r is not appropriate. This leads to a modified PSL limit as in Fig. 3b.

The approach of a simple step function PSL = HEL is modified in [14]. In [14] data are presented for flying plate experiments on liquid suspensions of microbial life. As well as measuring survival rates, data are shown in [14] for TEM images of the damaged cells. Based on the results, the authors suggest that microbial death occurs if the shock pressure exceeds the cell Turgor pressure for a finite time. The sub-surface shock pressure which satisfies this is then calculated as a function of

radial distance r for a variety of impact scenarios. A PSL is then found as a radial distance from the impact site.

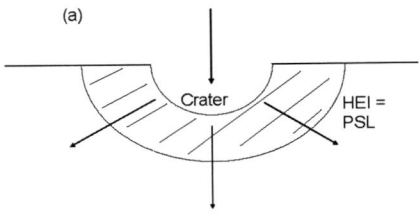

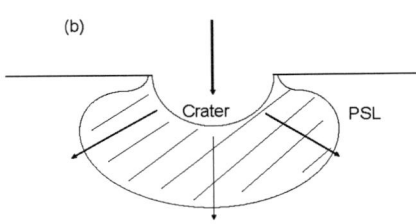

**FIGURE 3.** (a) Subsurface shock region around an impact site, showing that the Hugoniot Elastic Limit (HEL) can be set equal to a Potential Survival Limit (PSL) inside which no microbial life will survive. (b) Modified shock pressure distribution allowing for free surface effects where a fixed pressure contour (that of the PSL) is no longer a hemispherical surface.

This is more rigorous than the approach of [17] in that a physical mechanism is introduced for setting the relation between shock pressure and lethality. However, when calculating examples of the PSL, the influence of the free surface problem in modifying the pure radial development of the shock pressure is ignored. Nevertheless of interest is the result in [14] that a 1.5 km diameter impactor at 20 km s$^{-1}$, is predicted to have a radial subsurface PSL of 200 - 300 km. Indeed in [14] it is predicted that a 1,500 km impactor has a PSL which is equivalent to the size of the Earth, i.e. a truely planetary sterilizing event. However, the simple treatment of the shock pressure radial development in the model probably does not really permit such an extrapolation, and the other physical consequences of such an impact would be so severe as to render this particular example nugatory.

## ZONE OF LIFE AROUND AN IMPACT SITE

Paradoxically, although the PSL can be taken as the limit of the region of subsurface microbial extinction in an impact, this zone can play an opposite role after crater formation. After the initial impact site has cooled somewhat, the region can be re-colonized by local life. Of particular interest is the way the impact will have altered the local environment, making it a more habitable region than before. On a planet like Mars the heating due to the impact can melt subsurface ice in the region of the crater. The heavily fractured rock around the crater will permit flow of water through the rock. The fractures can then be colonized by any dormant microbial life in the region. Thus this impact shock altered environment can become more conducive to supporting life than previously. In terms of searching for life on say Mars, this could mean that crater floors have increased astrobiological potential compared to the surrounding regions [17].

## CONCLUSIONS

The results described herein are from a variety of sources in several countries. The application of experimental shock techniques to microbial life is thus a topic that is now reasonably widespread. There are problems with scaling any laboratory experiments to planetary scales, but it does appear that shocks kill cells by disruption of cell walls. The magnitude and duration of the shock are both critical for this purpose. Also, as it can be seen from the data, survival rates are not simple step functions, with a fall to zero at some peak pressure.

It should be noted that all the experimental results quoted for survival rates should probably be considered uncertain to at least an order of magnitude. Indeed, several experimenters (e.g. [13, 14]) find that when they repeat experiments under similar conditions, the survival rates can fluctuate by up to 2 orders of magnitude. Part of this lies in

the methods used to estimate the initial population and the uncertainties in counting the subsequent cultured samples. And part of the uncertainty may lie in minor fluctuations in the shock conditions on small scales in the targets. This can cause local variations in the shock conditions changing the results. Nevertheless, given the variety of experimental methods and the range of bacteria used (both late stage growth and spore state), it is remarkable that a general trend emerges from the data (Fig. 2 and eqn. 2).

If the general result from eqn. 2 is combined with a model for shock pressures around an impact site, then the effects on the subsurface environment for life can be assessed. At a certain point, the survival rate will fall so sharply that it is essentially zero. This is hard to assess, an analytic form such as eqn. 2 does not include such a limit. In addition, close to the impact site itself the rock will experience elevated temperatures for lengthy periods, and this will also cause sterilization. Thus the true limit of the lethal zone around an impact site will be the combination of several distinct processes, each with its own Potential Survival Limit.

The simple picture of impacts (and shock events) as purely killers, is challenged by the results presented showing that ejecta can carry viable microbial life and that some material can be carried on projectiles and successfully colonize targets after impacts. Indeed, if a planet such as the Earth, suffers impacts into its oceans rather than onto solid rock, the survival rates may well be greater than those reported here. This goes beyond the tradition models of [8, 18] for generation of organic materials on the early Earth via impact driven shock synthesis; it suggests that complex possibly biologically active materials could have come direct from space. However, like all such models, this only shows that such a pathway exists, without knowledge of whether or not life exists elsewhere, we can not really assert how probable such an origin for life on Earth is.

## ACKNOWLEDGEMENTS

I thank the organizers for the invitation to this meeting. I also thank my many co-workers over the past few years (A. Bunch, J. Mann, J. Galloway,

M. J. Cole, P. Brandao) who have contributed so much to the research in this field.

## REFERENCES

1. Burchell, M. J., "Panspermia today", Int. J. of Astrobiology, vol. 3, pp. 73 – 80, 2004.
2. Melosh, H. J., "A rocky road to Panspermia", Nature, vol. 332, pp. 687 – 688, 1988.
3. Clark B. C., "Planetary interchange of bioactive material: probability factors and implications", Origins of Life and Evol. Of the Biosphere, vol. 31, pp. 185 – 197, 2001.
4. Nicholson W. L., et al., "The solar UV environment and bacterial spore UV resistance: considerations for Earth-Mars transport by natural processes and human spaceflight", Mutation Research, vol. 571, pp. 249 – 264, 2005.
5. Mileikowsky, C., et al., "Natural transfer of viable microbes in space", Icarus, vol. 145, pp. 391 – 427, 2000.
6. Mileikowsky, C., et al., "Risks threatening viable transfer of microbes between bodies in our Solar System", Planetary and Space Science, vol. 48, pp. 1107 – 1115, 2000.
7. Wells, L. E., et al., "Reseeding of early Earth by impacts of returning ejecta during the later heavy bombardment", Icarus, vol. 162, pp. 38 – 46.
8. Blank, J. G., "Experimental shock chemistry of aqueous amino acid solutions and cometary delivery of prebiotic compounds", Origins of Life and Evolution of the Biosphere, vol. 31, pp. 15 – 51, 2001.
9. Mastrapa, R. M. E., et al., "Survival of bacteria exposed to extreme acceleration: implications for panspermia", Planet. Earth Sci. Lett., vol. 189, pp. 1 - 8, 2001.
10. Burchell, M. J., et al., "Survival of Bacteria in Hypervelocity Impact", Icarus, vol. 154, pp. 545 - 547, 2001
11. Burchell, M. J., "Survivability of bacteria ejected from icy surfaces after hypervelocity impact", Origins of Life and Evolution of the Biosphere, vol. 33, pp. 53 – 74, 2003.
12. Burchell, M. J., et al., "Survival of bacteria and spores under extreme shock pressure", Monthly Notices of the Royal Astronomical Society, vol. 352, pp. 1273 - 1278, 2004
13. Horneck, G., et al., "Bacterial spores survive simulated meteorite impact", Icarus, vol. 149, pp. 285 - 290, 2001.
14. Willis, M. J., "Survival limits of bacteria during shock compression: applications to the Early Earth", LPSC XXXVI, abstract 1903, 2005.
15. Narlikar, J. V., et al., "Balloon experiment to detect microorganisms in outer space", Astrophysics and Space Science, vol. 285, pp. 555 – 562, 2003.
16. Kring, D. A., "Air blast produced by the meteor crater impact event and a reconstruction of the affected environment", Met. Planet. Sci., vol. 32, pp. 517 - 530, 1997.
17. Cabrol, N. A., "Recent aqueous environments in Marian impact craters: an astrobiological perspective", Icarus, vol. 154, pp. 98 - 112, 2001.
18. Chyba, C. and Sagan, C., "Endogenous production, exogenous delivery and impact-shock synthesis of organic molecules: an inventory for the origins of life", Nature, vol. 355, pp. 125 – 132, 1992.

CP845, *Shock Compression of Condensed Matter - 2005*,
edited by M. D. Furnish, M. Elert, T. P. Russell, and C. T. White
© 2006 American Institute of Physics 0-7354-0341-4/06/$23.00

# THE BEHAVIOUR OF DRY SAND UNDER SHOCK-LOADING

## D. J. Chapman, K. Tsembelis, and W. G. Proud

*PCS, Cavendish Laboratory, Madingley Road, Cambridge, CB3 0HE. UK.*

**Abstract.** Plate impact experiments have been performed in forward and reverse impact configurations on dry, quartz sand to obtain Hugoniot data. In the forward configuration, manganin gauges were placed both in surrounding anvils and in the sand. Gauges placed directly in or near the sand gave anomalously high stress values compared with stresses measured in the anvils. Shock wave velocity was calculated from time of arrival and the Hugoniot data inferred. This was found to agree with data from reverse impact and that previously published.

**Keywords:** Sand, Shock, Manganin gauge, Hugoniot
**PACS:** 83.80.Nb, 47.40.Nm.

## INTRODUCTION

The shock properties of geological materials have traditionally been of interest for applications involving planetary impact. More recently however, there has been a growing interest in the shock properties of concretes and mortars, where geological materials are added as components and aggregates [1]. The quartz sand investigated here is one such component. Almost universally, the principal Hugoniot of granular highly porous materials have been developed from direct measurement of shock velocity. Previous work on sand has used a reverberation technique with sensors embedded in anvils surrounding a cavity containing the sample [1-5]; no in-material data is obtained and must be inferred using the jump condition (1) even though material compaction and shock rise times are similar.

$$\sigma_X = \rho_0 U_S u_P \qquad (1)$$

where $\sigma_x$ longitudinal stress, $\rho_0$ initial density, $U_S$ shock velocity, $u_P$ particle velocity.

In the present study a previously characterised sand [1], density $1.6 \pm 0.1$ g cm$^{-3}$, and particle size 150-210 µm is studied using forward and reverse plate-impact experiments. Comparison of the data from forward and reverse impact enabled the approximation represented by the simple jump condition to be investigated.

## EXPERIMENTAL TECHNIQUES

Plate-impact experiments were conducted using the Cavendish plate-impact facility. This facility centres around a 50 mm bore 5 m length single-stage light-gas gun [6]. Impact velocities were measured to an accuracy of 0.5% using a sequential pin-shorting method and the target was aligned to the impactor to less than 1 mrad by means of an adjustable specimen mount.

Commercially produced manganin piezo-resistive gauges (Micro Measurement LM-SS-210FD-050 option SP60) were used to measure the longitudinal stress during impact. The output voltage was recorded on a fast (5 GS s$^{-1}$) digital storage oscilloscope. This voltage-time data was then reduced to stress-histories according to the method by Rosenberg *et al* [7] and where necessary stresses were adjusted for gauge hysteresis according to the method of Yaziv *et al* [8].

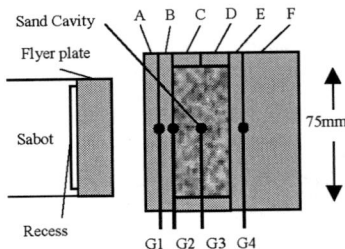

**Figure 1.** Schematic of forward plate-impact experiment

The configuration for forward impact is shown in Fig. 1. The projectiles for forward impact consisted of a polycarbonate sabot on the front of which were mounted plates of the "flyer" material. To the rear of the flyer plate a recess allowed for complete release. The flyer plate material was the same as that of the anvil used in the target construction. The target was constructed with an embedded gauge G1 between plates A (~1mm) and B (~1mm), ahead of a cavity (~4mm), filled with sand. To the rear of the sand cavity was a further sensor G4 between plates E (~1mm) and F (~20mm). The construction of the sand cavity from rings (~2mm) C and D allowed a gauge to be embedded in the sand (G3). This gauge was designed to be minimally intrusive consisting of a manganin sensor encapsulated between two 25 μm sheets of mylar a total gauge package thickness of 150 μm. In some experiments an additional gauge was placed on the rear surface of the front anvil (G2) separated from the sand by a 25 μm layer of mylar sheet. Impactors and cells were constructed from well-characterised materials, PMMA, copper, and aluminium alloy 6082-T6.

The reverse experimental configuration is shown in Fig. 2. The projectile consisted of a 15 mm sand sample confined within an aluminium alloy cup and backed with an aluminium alloy disc. This assembly was bonded to a polycarbonate sabot and the impact face lapped. The impact face of the aluminium alloy cup was $0.5 \pm 0.05$ mm. The target consisted of three aluminium alloy plates 3.2, 5.2 and 20.2 mm thick and two manganin gauges.

Samples were assembled using a press. A slow curing epoxy was used to minimize bond thickness and remove air bubbles. The embedded gauge package was typically ~100 μm. The impedance

difference between the gauge package and the plates caused the stress to ring-up over ~300 ns; the time resolution of the gauge. A faster rise-time ~30 ns occurred for gauges between PMMA plates where only the manganin element was impedance mis-matched.

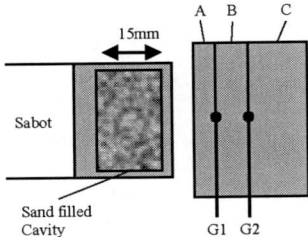

**Figure 2.** Schematic of reverse plate-impact experiment

## RESULTS AND DISCUSSION

Representative gauge traces for the forward plate impact using copper anvils are shown in Fig.3. The stress in G1 initially rises to the copper-copper level, then released from the metal-sand interface. The sharp dip following the initial peak in G1 was observed in all experiments involving copper

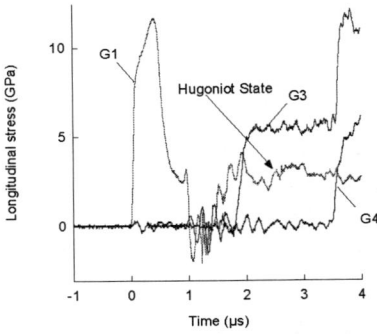

**Figure 3.** Gauge traces from forward plate-impact using copper anvils at 604 m s$^{-1}$. The G1 trace has not been corrected for hysteresis.

anvils and may be attributed to capacitive linkage or release interaction causing debonding of the gauge. Consequently, any data after this dip must be considered cautiously in the forward impact experiments. The gauge G3 is observed to rise to a stress significantly above that of the release in G1. G3 was subsequently reloaded from the reflected shock from the rear anvil. Where a gauge was positioned on the rear of the front anvil (G2) an intermediate stress level between the release in G1 and the plateau of G3 was obtained. In an ideal material these three levels, release in G1, and the plateaux of G2 and G3 would be the same, and indicative of the Hugoniot stress. However, embedding the gauge package in mylar sheet afforded the gauge little protection from localised effects due to individual sand grains pushing into the gauge element, or from gauge curvature. Consequently, anomalous stresses were observed by gauges G2 and G3. These gauges can however, be used as time of arrival sensors for the measurement of shock velocity. G4 is observed to rise to a stress higher than the release in G1 due to the sand being reloaded by the higher impedance rear copper anvil.

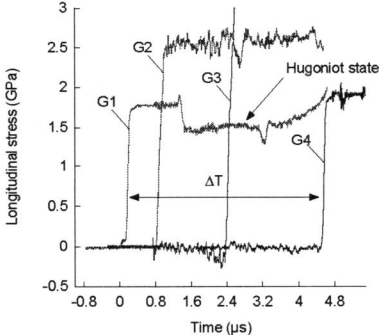

**Figure 4.** Gauge traces from forward plate-impact using PMMA anvils at 983 m s$^{-1}$. The G1 trace has not been corrected for hysteresis. Time base is arbitrary.

Fig. 4 shows the gauge traces from the forward plate-impact at 983 m s$^{-1}$ using PMMA anvils. The traces demonstrate the same trend as those using copper anvils. No capacitive linkage or gauge debonding is observed in G1 and the stress is seen to release to a constant plateau. Again the stress measured by G2 is above the release of G1; in this case the gauge embedded in sand G3 fails instantly. G4 shows the reload of the sample form the slightly higher impedance PMMA anvil. No low amplitude precursor is observed in G4 even given the better resolution obtained using the PMMA anvil.

The shock velocity in sand was obtained from the transit time $\Delta T$ at half-peak values between G1 and G4 accounting for gauge package and anvil thicknesses. This velocity was then used in combination with the known Hugoniot of the anvil to obtain the particle velocity and equation (1) to infer the stress in the sand. The release isentropes were approximated by reflected Hugoniots. This is known to be a close approximation for copper in the given stress range but less so for PMMA and aluminium. The inferred in-material stresses were found to agree within experimental error with the release levels of G1 when gauge hysteresis was taken into account. Shock velocities calculated in a similar manner between gauges G1/G2 and G3, and G3 and G4 respectively, were found to agree; no significant reduction of shock velocity during transit through the cavity within experimental accuracy.

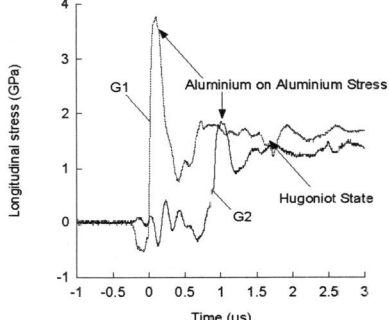

**Figure 5.** Gauge traces from reverse plate-impact at 600 m s$^{-1}$. The traces have not been corrected for hysteresis. Time base is arbitrary.

Fig. 5 depicts the gauge traces obtained from a reverse impact at 600 m s$^{-1}$. In G1 the stress is observed to release from the aluminium-aluminium stress to a level which should be indicative of the stress in the sand. The dip immediately following the peak may be attributed to a slight compaction of the sand bed in the projectile caused by its acceleration. The release level in G2 is below that in G1. The peak aluminium-aluminium stress has been slightly released by the time it reaches G2; consequently the hysteresis in G2 will be lower than in G1. Release levels in G1 and G2 were found to agree within experimental error when hysteresis was taken into account, and their average value was used to obtain particle velocity given the aluminium Hugoniot. The jump condition (1) was subsequently used to infer shock velocity.

Fig. 6 plots the shock velocity dependence of the sand on particle velocity obtained from the forward and reverse experiments. The values are observed to reduce to a single curve within experimental error, where a simple linear relation

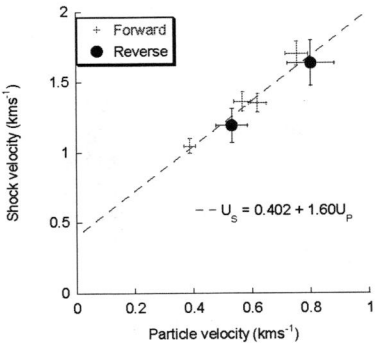

**Figure 6.** Shock velocity dependence on particle velocity

has been assumed (2). This demonstrates that the simple jump condition (1) yields a reasonable approximation in the investigated stress regime for sand.

$$Us = 0.402 + 1.60\ Up\quad(2)$$

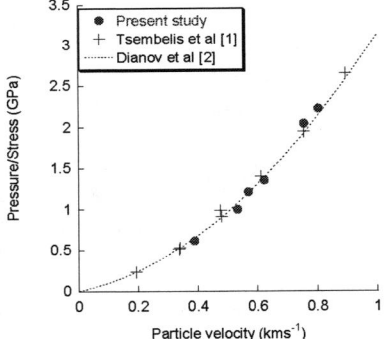

**Figure 7.** Stress dependence on particle velocity.

Fig. 7 demonstrates the good agreement between the present data and that obtained by Tsembelis *et al* [1] and Dianov *et al* [2] for sands of similar density and particle size obtained using the reverberation technique. The close agreement with the data obtained by Tsembelis *et al* on the same sand demonstrates that the introduction of the embedded gauge (G3) has not significantly affected the shock velocity.

## CONCLUSIONS

The principal Hugoniot of a dry quartz sand of density $1.6 \pm 0.1$ g cm$^{-3}$ has been determined using both reverse and forward plate-impact. The data from both methods reduce to a single curve within experimental error, and compare favourable with published data demonstrating that the simple jump condition (1) yields a reasonable approximation in the investigated stress regime.

Gauges embedded directly in (G3) or next to (G2) the sand in the forward configuration were observed to give stress levels above those measured in the surrounding anvils due to localised effects between the sand grains and gauge. These gauges were used as shock arrival sensors demonstrating no significant reduction of shock velocity with transit through the cavity within experimental accuracy.

## ACKNOWLEDGEMENTS

This work has been sponsored by QinetiQ, who support W.G. Proud as a QinetiQ Senior Research Fellow. The authors also acknowledge D. Johnson and R. Flaxman of the Cavendish Workshop for technical assistance.

## REFERENCES

1. Tsembelis, K., et al., *The behaviour of sand under shock wave loading: Experiments and simulations*, in *Behaviour of Materials at High Strain Rates: Numerical Modelling*, F.G. Benitez, Editor. 2002, DYMAT: Saint-Louis, France. p. 193-203.
2. Dianov, M.D., et al., *Shock compressibility of dry and water-saturated sand*. Sov. Tech. Phys. Letts, 1976. 2: p. 207-208.
4. Resnyansky, A.D. and N.K. Bourne, *Shock-wave compression of a porous material*. J. Appl. Phys., 2004. 95: p. 1760-1769.
6. Bourne, N.K., et al., *Design and construction of the UK plate impact facility*. Meas. Sci. Technol., 1995. 6: p. 1462-1470.
7. Rosenberg, Z., D. Yaziv, and Y. Partom, *Calibration of foil-like manganin gauges in planar shock wave experiments*. J. Appl. Phys., 1980. 51: p. 3702-3705.
8. Yaziv, D., Z. Rosenberg, and Y. Partom, *Release wave calibration of manganin gauges*. J. Appl. Phys., 1980. 51: p. 6055-6057.

CP845, *Shock Compression of Condensed Matter - 2005*,
edited by M. D. Furnish, M. Elert, T. P. Russell, and C. T. White
© 2006 American Institute of Physics 0-7354-0341-4/06/$23.00

# A NEW EVIDENCE OF THE STABILITY OF (Mg, Fe)SiO$_3$ PEROVSKITE AT LOWER MANTLE CONDITIONS: SHOCK RECOVERY EXPERIMENTS

## XiuFang Chen[1], Zizheng Gong[1*], Yingwei Fei[2], Li Zhang[2,1], Liwei Deng[1,2], and Fuqian Jing[1,3]

[1]*Institute of High Pressure and High Temperature Physics, Southwest Jiaotong University, Chengdu 610031, Sichuan, P.R. China.*
[2]*Geophycical Laboratory, Carnegie Institution of Washington, Washington DC 20015, USA.*
[3]*Laboratory for Shock Wave and Detonation Physics Research, Institute of Fluid Physics, P.O.Box 919, Mianyang, Sichuan 621900, China.*

**Abstract.** By using a two-stage light gas gun, 9 experiments of shock recovery experiments with initial sample of (Mg$_{0.92}$, Fe$_{0.08}$) SiO$_3$ Enstatite (En) (7 experiments) and MgO+SiO$_2$ (2 experiments) were conducted between 65 and 110 GPa shock pressure (the corresponding temperature is estimated as 2500~5000K). The analysis of X-Ray Diffraction (XRD) and Infrared (IR) for the shock recovered samples indicate that there is no possibility for the chemical decomposition of (Mg$_{0.92}$, Fe$_{0.08}$) SiO$_3$ perovskite into SiO$_2$ plus (Mg$_{0.92}$, Fe$_{0.08}$) O during shock compression. Our experiments support that (Mg, Fe) SiO$_3$-perovskite is stable at lower mantle P&T conditions.

**Keywords:** (Mg$_{0.92}$, Fe$_{0.08}$) SiO$_3$-perovskite, phase stability, shock recovery, XRD, IR, lower mantle.
**PACS:** 91.60.Hg, 91.35.Lj, 82.40.Fp, 61.10.Nz.

## INTRODUCTION

Silicate perovskite with orthorhombic structure is generally accepted to be the dominant phase of the Earth's lower mantle [1]. The phase stability of MgSiO$_3$-perovskite at lower mantle conditions is therefore of great importance for constraining the chemical and mineralogical composition of the lower mantle, and has been intensively investigated by both dynamic and static high pressure experiments. However, the stability of perovskite in the lower mantle is still an unsolved problem. For example, results from Saxena et al. [2, 3], Meade et al. [4], and Watt and Ahrens [5] suggest that magnesium silicate perovskite will decompose into oxides under lower mantle conditions. While researches conducted by Serghiou et al. [6], Fiquet et al. [7], and Gong et al. [8] lead to the opposite conclusion. So no consistent agreement has been achieved on its phase stability in lower mantle condition yet.

In this paper we present shock recovery experiments on our newly designed recovery apparatus and latest analysis results of XRD and IR for the recovered samples with initial sample of (Mg$_{0.92}$, Fe$_{0.08}$) SiO$_3$ En and MgO+SiO$_2$ to address whether the reaction:

(Mg,Fe)SiO$_3$-perovskite → (Mg,Fe)O+SiO$_2$

*Corresponding author.
E-mail: zzhgong@home.swjtu.edu.cn

**Table 1.** Shock recovery experimental data

| Number | Sample | Flyer | W (km/S) | $\rho_0$ (g/cm$^3$) | $U_P$ (Km/S) | $U_S$ (Km/S) | $P_H$ (GPa) | $T_H$ (K) |
|---|---|---|---|---|---|---|---|---|
| 1 | En | Cu | 4.13 | 3.09 | 2.82 | 7.93 | 69.00 | 2668 |
| 2 | En | 93W | 3.85 | 3.09 | 3.19 | 8.48 | 83.69 | 3505 |
| 3 | En | Cu | 4.01 | 2.95 | 2.78 | 7.87 | 64.58 | 2432 |
| 4 | En | Cu | 4.52 | 3.00 | 3.10 | 8.35 | 77.64 | 3143 |
| 5 | En | Cu | 4.78 | 3.00 | 3.27 | 8.60 | 84.36 | 3538 |
| 6 | En | Cu | 5.25 | 2.98 | 3.60 | 9.09 | 97.50 | 4331 |
| 7 | En | Cu | 5.56 | 3.07 | 3.74 | 9.30 | 106.73 | 4971 |
| 8 | MgO+SiO$_2$ | 93W | 4.30 | 1.81 | 3.56 | 14.92 | 96.09 | 4242 |
| 9 | MgO+SiO$_2$ | 93W | 3.79 | 1.88 | 3.15 | 13.73 | 77.91 | 3272 |

W, $\rho_0$, $U_P$, $U_S$, $P_H$ and $T_H$ denote flyer impacting velocity, initial bulk density, particle velocity, shock wave velocity, Hugoniot pressure, and shock temperature, respectively. Hugoniot of Copper comes from [10], 93W is 4.2Ni2.45Fe0.35CoW alloy [11].

could occur in the lower mantle conditions.

## EXPERIMENTAL METHOD AND RESULTS

The shock recovery experiments were carried out with a 25-mm two-stage light gas gun. To obtain the recovered sample successfully, we design a new recovery apparatus shown in Fig. 1.

A natural Enstatite mineral, from Damaping mine, Changjiakou, Hebei Province, China, was used as initial samples in the shock recovery experiments. The chemical composition

of the natural Enstatite are (wt.%): SiO$_2$ (54.72%), MgO (31.09%), FeO (5.00%), Fe$_2$O$_3$ (2.44%), Al$_2$O$_3$ (4.02%), CaO (1.70%), and 1.03% of minor elements including Cr$_2$O$_3$, TiO$_2$, K$_2$O, Na$_2$O, MnO, etc.. As the atomic ratio of Mg/(Mg + Fe) is 0.92, we refer the initial sample to (Mg$_{0.92}$, Fe$_{0.08}$) SiO$_3$ in this paper for convenience. The average bulk density which was obtained by measuring its bulk volume and its weight, is 3.06 ± 0.01 g/cm$^3$. The prepared samples contain more than 98% Enstatite mineral. The other initial sample is SiO$_2$+MgO (mole ratio between SiO$_2$ and MgO is 3:1). Nine experiments were conducted between 64 to 110 GPa. The results are listed in Table 1 and the recovered

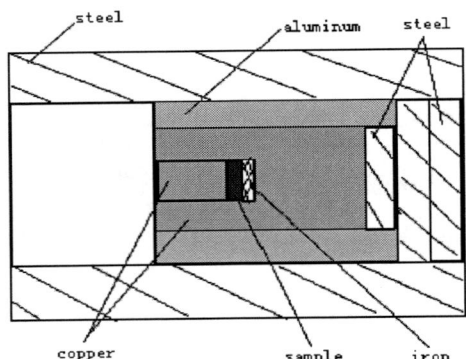

**Figure 1.** New recovery equipment.
Recovery equipment is composed by sample box and steel barrel. Sample box is divided into two parts: inner part is made of copper with good plastic and outer is Ly12 aluminum. The sample box was put into a barrel made of 45 steel. Besides that, several pieces of 45 steel were put behind the sample box to transfer energy according to the principle of "momentum trap"[9].

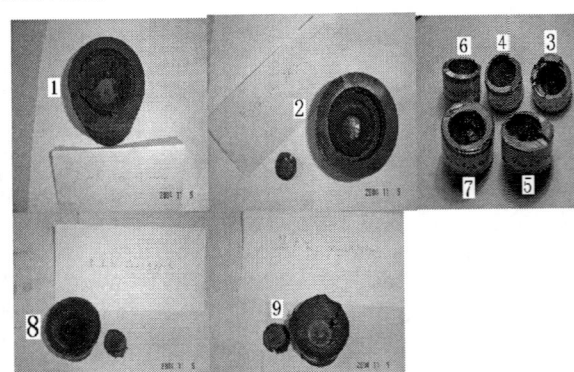

**Figure 2.** Pictures of the recovered samples.
Recovered samples are labeled from 1 to 9, which are corresponding with the same number in Table 1.

samples are illustrated in Fig. 2. From Fig. 2, it can be seen that all the sample boxes were not destroyed and the shocked samples were recovered successfully.

## DISCUSSION

In all experiments numbered as Table 1, minute quantity sample was recovered in experiments 3, 4 and 7, separately. Meanwhile, the analyses showed that the spectral lines of 3th, 4th and 5th were similar and so did the experiment 6 and experiment 7. A typical XRD of recovered sample at 83.69GPa with initial sample of Enstatite is illustrated in Figure 3.

Only $SiO_2$ and $Mg_2SiO_4$ were found in experiments 8 and 9 from spectral analysis. As numerous static experiments have proved that $MgSiO_3$ (En) can be formed from MgO plus $SiO_2$ at high pressure. $MgSiO_3$ (En) reacts with MgO to produce $Mg_2SiO_4$. With increasing pressure, $Mg_2SiO_4$ decomposes into MgO and $MgSiO_3$ (Pv.). However, during the process of unloading, MgO and $MgSiO_3$ (Pv.) will synthesize $Mg_2SiO_4$ again. Hence recovered sample in static experiments is always $Mg_2SiO_4$, which indicates that $Mg_2SiO_4$ can be maintained as a stable phase after the reaction of $MgO+SiO_2$. Especially, there is no evidence indicating that $MgSiO_3$ (En) will be the reaction product of $MgO+SiO_2$. Therefore compared with

previous investigations, we suggest that the $Mg_2SiO_4$ observed in our experiments should be the product of $MgO+SiO_2$, which has a similar reaction path with the experiments conducted by Potter and Ahrens [12] with the initial samples of $Al_2O_3+MgO$. Excessive $SiO_2$ in initial sample should be responsible for the corresponding spectral line appeared in experiments 8 and 9.

Furthermore, all the main phases found in recovered samples were similar with the original samples, namely $(Mg_{0.92}, Fe_{0.08}) SiO_3$ (En). But no characteristic line of (Mg, Fe)O or $SiO_2$ was observed in any spectrogram, which means that no oxide appeared during the pressure loading or after the pressure unloading. On the other hand, if the $(Mg_{0.92}, Fe_{0.08})SiO_3$ (Pv.) would decompose into oxide assemble, the generated (Mg, Fe)O or $SiO_2$ would have to remain stable in the pressure loading process but react to form $(Mg_{0.92}, Fe_{0.08}) SiO_3$ (En) in the pressure unloading process. Obviously, the deduced reaction above is controversial with both our results and many static experiments mentioned. Meanwhile, as the phase transition between different crystals structures takes more time than pressure uninstall [13]. It is impossible to complete the combination reaction in the pressure unloading process. Anyway, there is no evident show any possibility for the chemical decomposition.

It has been observed that new peak values appeared at d-space=4.6402 Å, d-space=1.5542 Å and d-space =1.4291 Å in the 6'th are stronger than that appeared at the same d-space in the 5'th. Although Shim [14] has advanced the phase transition at 83GPa from Pbnm (spacegroup) to P21/m, Pmmn or $P4_2/nmc$, we would like to contribute the appearance of new peaks to the crystal micro-distortion or increasing of structural symmetry during the loading process rather than phase transition as the lack of further reliable proofs.

More crystal vibration information can be obtained in the IR group. From the experiments 3-7, three absorbing wave bands can be identified clearly. The flex vibrations of Si-O and Si-O-Si are responsible for 900-1200cm$^{-1}$ and the bend one is corresponding with 600-800 cm$^{-1}$. The characteristic IR pattern of (Mg, Fe)O and $SiO_2$ were not observed in the recovered sample. Moreover, with increasing pressure, the IR mode

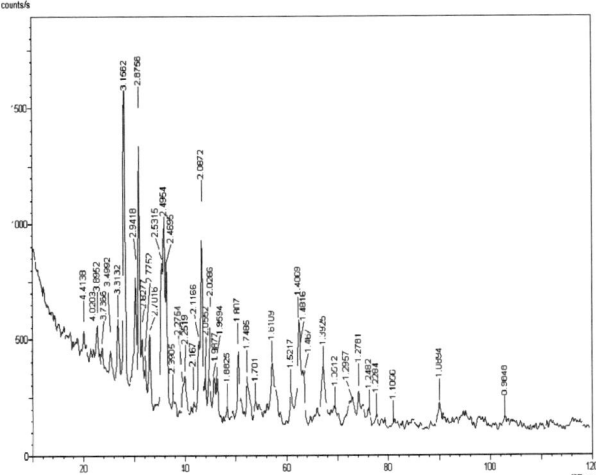

**Figure 3.** A typical XRD of recovered sample at shock pressure 83.69 GPa with starting material of Enstatite.

became simpler, which implied that the crystal structure of $(Mg_{0.92}, Fe_{0.08}) SiO_3$ has higher symmetry.

It has been testified that En would transfer into Pv above 45GPa in shock compression or 30GPa in static experiment [1], [8], and [15]. But through the analysis above, in both XRD and IR modes with initial sample of $(Mg_{0.92}, Fe_{0.08})SiO_3$ (En), the recovered samples are not Pv but similar to En. In fact, inverse phase transition from Pv to En has been reported in earlier research [16]. Moreover, the characters of shock unloading process, in which the pressure decrease far rapidly than temperature, has great influence on the phase stability of Perovskite. So it is reasonable to suggest that reverse phase transition from Pv to En during pressure unloading occured in our experiments.

## CONCLUSION

Through the analysis of XRD and IR patterns for recovered samples, the possibility of chemical decomposition of $(Mg, Fe)SiO_3$-Pv to $MgO+SiO_2$ oxides at lower mantle condition was excluded. But the possible microstructure phase transition may be concealed by the reverse reaction or other unknown factors during shock compression release.

## ACKNOWLEDGEMENTS

This research was supported by the National Natural Science Foundation of China under Grant No. 40474033 and 10299040.

## REFERENCES

1. Knittle, E., and Jeanloz, R., "Synthesis and equation of state of (Mg, Fe)SiO_3 perovskite to over 100GPa", Science, 235, 668-670, 1987.
2. Saxena, S. K., Dubrovinsky, L.S., Lazor, P. Y., et al., "Stability of perovskite (MgSiO_3) in the Earth's Mantle", Science, 274, 1357-1359, 1996.
3. Saxena, S. K., Dubrovinsky, L. S., Lazor, P., et al, "In situ X-Ray study of perovskite (MgSiO_3)-phase transition and dissociation at mantle conditions", Eur. J. Mineral. 10, 1275-1281, 1998.
4. Meade, C., Mao, H. K., and Hu J., "High-temperature phase transition and dissociation of (Mg, Fe) SiO_3 perovskite at lower mantle pressures", Science, 268, 1743-1745, 1995.
5. Watt, J. R., and Ahrens, T. J., "Shock wave equation of state of enstatite", J. Geophys. Res., 91(B7), 7495-7503, 1986.
6. Serghiou, G., Zerr, A., and Boehler, R., "(Mg,Fe)SiO_3 perovskite stability under lower mantle conditions", Science, 280, 2093-2095, 1998.
7. Fiquet, G., Dewaele, A., Andrault, D., et al. "Thermoelastic properties and crystal structure of MgSiO_3 perovskite at lower mantle pressure and temperature conditions", Geophys. Res. Lett., 27, 21-24, 2000.
8. Gong, Z. Z., Fei, Y. W., Dai, F., et al., "Equation of state and phase stability of (Mg_{0.92}, Fe_{0.08})SiO_3 perovskite up to 140GPa", Geophys. Res. Lett., 31, L04614, 2004.
9. Rinehart J. S., "Stress transients in Solid" University of Colorado, New Mexico, 1975.
10. Mitchell, A. C., and Nellis, W. J., "Shock compression of aluminum, copper, and tantalum", J. Appl. Phys., 52, 3363-3374, 1981.
11. Wang, Y., Weidner, D. J., Liebermann, R. C., et al, "P-V-T equation of state of (Mg, Fe)SiO_3 perovskite: Constraints on composition of the lower mantle", Phys. Earth Planet. Inter., 83, 13-40, 1994.
12. Potter, D. K., and Ahrens, T. J., "Shock induced formation of MgAl_2O_4 spinel from oxides", Geophys. Res. Lett., 21(8), 721-724. 1994.
13. DeCarli, P. S., and Milton, D. J., "Stishovite: Synthesis by shock wave", Science, 147, 144-145, 1965.
14. Shim, S. H., Duffy, T. S., and Shen, G., "Stability and Structure of MgSiO_3 Perovskite to 2300-Kilometer Depth in Earth's Mantle", Science, 293, 2437-2440, 2001.
15. Liu, L. G., "Silicate perovskite form phases transformation of pyrobe-garnet at high pressures and temperatures", Geophys. Res. Lett., 1, 277-280, 1974.
16. Chen, J., Weidner, D. J., and Vaughan, T., "The Srength of (Mg_{0.9}, Fe_{0.1})SiO_3 Perovskite at High Pressure and Temperature", Nature, 419, 824-826, 2002.

CP845, *Shock Compression of Condensed Matter - 2005,*
edited by M. D. Furnish, M. Elert, T. P. Russell, and C. T. White
© 2006 American Institute of Physics 0-7354-0341-4/06/$23.00

# USING MESOSCALE MODELING TO INVESTIGATE THE ROLE OF MATERIAL HETEROGENEITY IN GEOLOGIC AND PLANETARY MATERIALS

## D. A. Crawford[1]

[1]*Thermal and Reactive Processes Dept 1516, P.O. Box 5800, MS 0836, Sandia National Laboratories,
Albuquerque, NM 87185*

**Abstract.** The propagation of shock waves through target materials is strongly influenced by the presence of small-scale structure, fractures, physical and chemical heterogeneities. Reverberations behind the shock from the presence of physical heterogeneity have been proposed as a mechanism for transient weakening of target materials [1] as are localized shock effects seen in some meteorites [2]. Pre-existing fractures can also affect melt generation [3]. Recent mesoscale studies in computational hydrodynamics have attempted to bridge the gap in numerical modeling between the microscale and the continuum,. Methods are being devised using shock physics hydrocodes such as CTH [4] and Monte-Carlo-type methods to investigate the shock properties of heterogeneous materials [5] and to compare the results with experiments [6]. Recent numerical experiments at the mesoscale using these statistical methods suggest that heterogeneity at the micro-scale plays a substantial and statistically quantifiable role in the effective shear and fracture strength of rocks. This paper will describe the methodology we are using to determine the strength of heterogeneous geologic and planetary materials.

**Keywords:** heterogeneity, complex crater, yield strength, gabbroic anorthosite, CTH
**PACS:** 91.60.-x, 91.60.Ba.

## INTRODUCTION

Geologic and planetary materials generally have pressure-dependent yield surfaces described by a Mohr-Coulomb internal friction model (Fig. 1). Heavily fractured rocks typically have slope of yield strength vs. pressure in the low pressure regime of 0.8–1.5. However, such a slope value cannot produce a complex crater shape consistent with that seen in large (10+ km diameter) impact craters seen in the terrestrial crater record. We can model complex craters using pressure dependent strength models but this usually requires a slope of yield strength vs. pressure in the low pressure regime of 0.4-0.6 to allow the late-time

gravitational collapse that produces the observed complex crater form.

Several mechanisms have been proposed to reduce effective rock strength at least temporarily during the cratering process. All of these mechanisms rely on the heterogeneity of the target rocks in integral form. Melosh [1] proposed the acoustic fluidization model whereby heterogeneity of target rocks could produce pressure fluctuations behind the main shock. The fluctuations act to temporarily reduce overburden pressure and push some of the material stress state to the left in Fig. 1. O'Keefe and Ahrens [7] proposed that material heterogeneity could lead to localized thermal softening reducing overall rock strength by pushing the yield curve downward in Fig. 1. Mesoscale

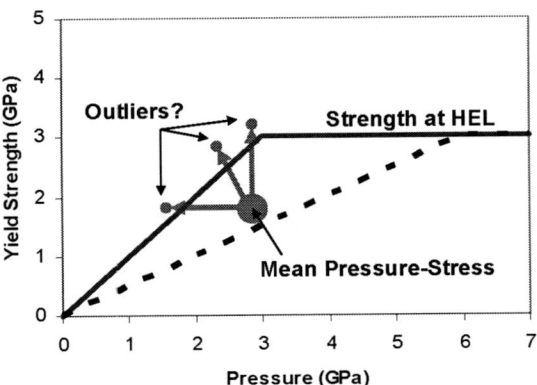

**Figure 1.** The solid line is a schematic representation of the pressure dependent strength model typical of most heavily fractured rocks. The dashed line is the strength model required to produce the complex crater shape seen in large (> 10 km) terrestrial impact craters. Mechanisms are discussed in the text that may allow *some* material at to exceed the typical strength envelope by either lowering the yield curve, lowering the pressure state or raising the deviatoric stress state.

numerical calculations that we are performing suggest another possibility: material heterogeneity may lead to enhanced local deviatoric stresses which can exceed the strength envelope by moving some material stress states upwards in Fig. 1.

## APPROACH

Our general approach to mesoscale modeling of heterogeneous materials is to build on our understanding of homogeneous material models. This is done by constructing a heterogeneous material as a mixture of homogeneous materials, performing numerical experiments at the characteristic length scale of the component materials and then generalizing to the larger scale.

The specific approach we use here is to perform plane strain mesoscale studies with the shock physics code CTH [4] using a two dimensional plane strain approximation (Fig. 2). The heterogeneity in these simulations is constructed from a random distribution of "grains", infinite cylinders really, embedded within a "matrix". Each cylinder is oriented with its axis oriented parallel to the Z direction. The stress is applied in the X direction and appropriate averages of pressure and stress are taken as ensemble averages across the Y direction.

### Plane strain mesoscale studies

Consider a heterogeneous material created by mixing two simple linear elastic Mie-Gruneisen (linear $U_s$-$u_p$) materials consisting of grains and matrix. The grains have their properties described by: $\rho_1$, $c_1$, $s_1$, $\phi_1$, $v_1$ and the matrix by: $\rho_2$, $c_2$, $s_2$, $\phi_2$,

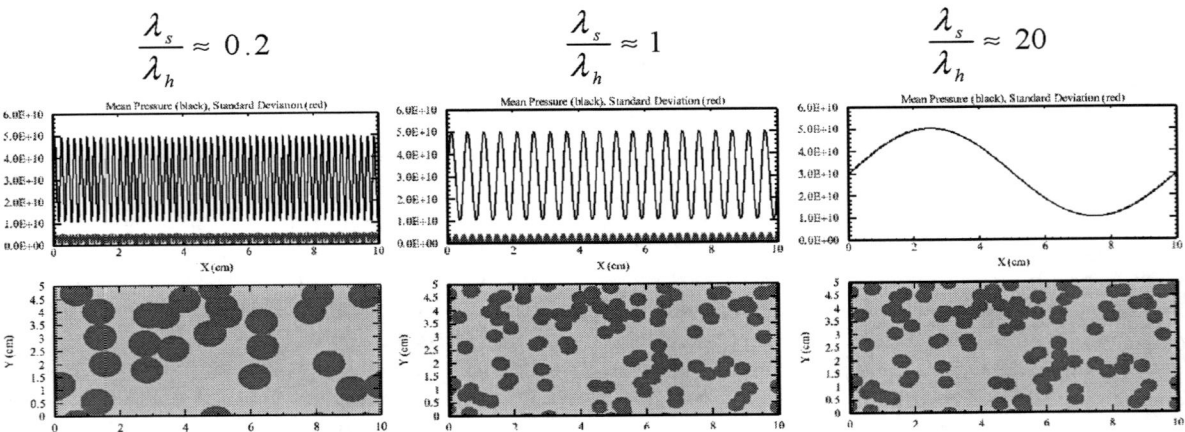

**Figure 2.** Initial conditions for heterogeneous calculations using a plane strain approximation. Three cases are shown with differing ratios (0.2, 1 and 20) of the applied stress spatial wavelength ($\lambda_s$) to the characteristic heterogeneity length scale ($\lambda_h$). The initial applied stress state is created by a 2 GPa amplitude sinusoidal variation of the material pressure in the X (horizontal) direction imposed around a 3 GPa ambient mean. Boundary conditions are periodic in the X and Y directions.

$v_2$. $\rho$, $c$ and $s$ have their traditional meaning for linear $U_s$-$u_p$ materials. $\phi$ is the volume fraction of each constituent in the mixture and $v$ is Poisson's ratio for each constituent. The average density of the mixture ($\rho_m$) can easily be shown to be:

$$\rho_m = \phi_1 \rho_1 + \phi_2 \rho_2 . \tag{1}$$

From mixture theory [8], the sound speed of the mixture ($c_m$) is:

$$\frac{1}{c_m^2} = \rho_m \left( \frac{\phi_1}{\rho_1 c_1^2} + \frac{\phi_2}{\rho_2 c_2^2} \right) \tag{2}$$

and, finally, the Poisson's ratio of the mixture ($v_m$) is approximated by:

$$v_m = \phi_1 v_1 + \phi_2 v_2 . \tag{3}$$

For this study, the $s$-value for the mixture is the same as the two components ($s_m = s_1 = s_2 = 1$) and mixture properties appropriate for a gabbroic anorthosite are used: $\rho_m = 2.94$ g/cc, $c_m = 6$ km/s and $v_m = 0.25$. We vary the heterogeneity of the mixture by changing the properties and relative abundance of the components but equations (1-3) are always used to ensure the mixture has mean behavior appropriate for gabbroic anorthosite.

With the linear elastic approximation, we assume each material has infinite strength and thereby look at the state of the material as it leads up to yield. We start with a spatially varying pressure condition within the materials, evolve the problem, measure induced shear stresses and compare the heterogeneous and homogeneous cases.

The magnitude of the deviatoric stress is tracked by the quantity:

$$J_2' = \sqrt{\frac{3}{2} S_{ij} S_{ij}} \tag{4}$$

where $S_{ij}$ is the deviatoric stress tensor. With an applied sinusoidal stress variation, $\Delta P$, the mean value of $J_2'$ for the mixture after it has evolved to the maximum deviatoric stress state is:

$$J_2' = 2\Delta P \frac{(1 - 2v_m)}{(1 - v_m)} . \tag{5}$$

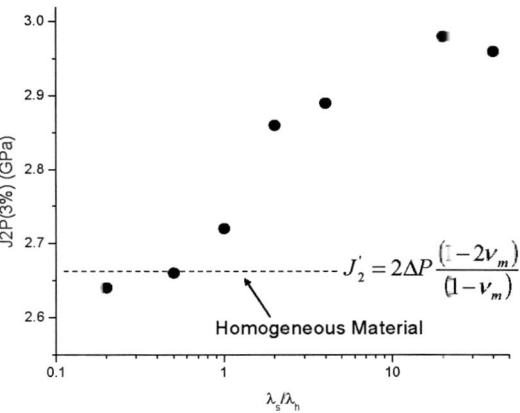

**Figure 3.** Dependence of the top 3% of deviatoric stress magnitude on the ratio of applied stress wavelength ($\lambda_s$) to the heterogeneity characteristic length ($\lambda_h$).

The quantity $J_2'(3\%)$, or $J2P(3\%)$ as it appears in some of the figures, is defined as the deviatoric stress magnitude achieved by the top 3% of material volume fraction in the mixture. We define the stress enhancement factor ($R$) as the ratio of $J_2'(3\%)$ for the heterogeneous mixture to $J_2'(3\%)$ for the homogeneous material:

$$R = \frac{J_2'(3\%, \text{heterogeneous})}{J_2'(3\%, \text{homogeneous})} . \tag{6}$$

**Dependence on the characteristic length ratio**

Fig. 3 shows the dependence of the maximum deviatoric stress state, $J2P(3\%)$, on the ratio of the wavelength of the initial applied stress ($\lambda_s$) to characteristic length of heterogeneity ($\lambda_h$). Where the applied stress wavelength is comparable to or smaller than the characteristic wavelength of the heterogeneity, there is no significant deviatoric stress enhancement above the homogeneous value. Where the applied stress wavelength is considerably longer than the characteristic length of the heterogeneity, as shown in Fig. 3, significant deviatoric stress enhancements are seen.

During the late stage collapse phase of terrestrial impact crater formation, $\lambda_s$ will typically have values measured in kilometers yet $\lambda_h$ will typically have values of meters. It can be expected, based on the behavior shown in Fig. 3, that

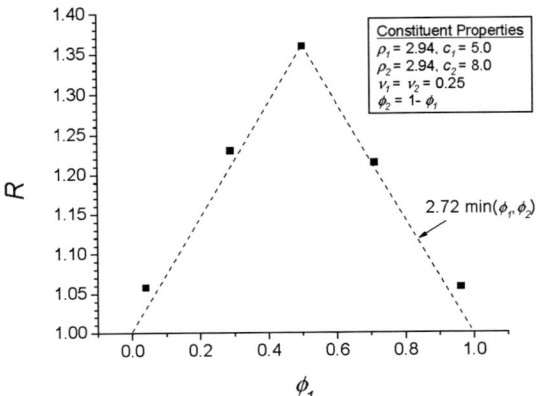

**Figure 4.** Dependence on constituent volume fraction ($\phi_i$). $R$ is defined by equation (6) in the text.

significant deviatoric stress enhancements due to the heterogeneity mechanism can play a role during terrestrial impact crater formation. The next step, then, is to quantify the degree of deviatoric stress enhancement for the range of parameters typical of target rocks.

### Dependence on constituent properties

We investigated dependence on consituent properties by working at a constant characteristic length ratio: $\lambda_s/\lambda_h = 20$ and mean properties representative of gabbroic anorthosite. We varied each property in turn, holding others constant except where indicated. $\phi_i$ was varied from 0.04 to

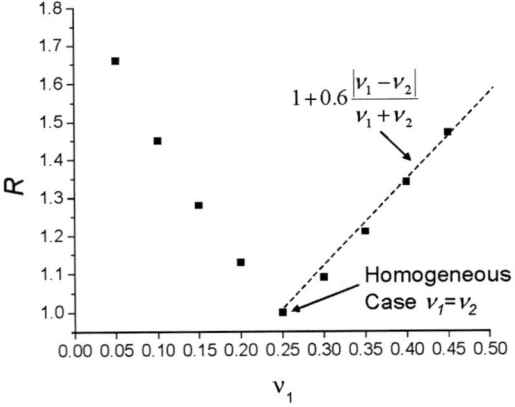

**Figure 5.** Stress enhancement factor dependence on constituent Poisson's ratio ($\nu_i$). $R$ is defined by equation (6) in the text.

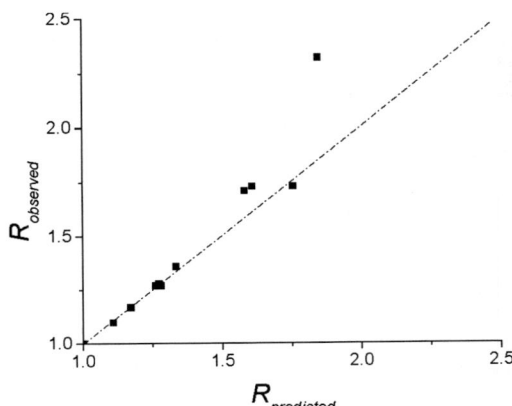

**Figure 6.** Stress enhancement factor dependence on constituent density ($\rho_i$) and sound speed ($c_i$). $R_{observed}$ is from equation (6) and $R_{predicted}$ from equation (7).

0.96, $\rho_i$ from 2.48 to 3.40 g/cc, $c_i$ from 2.0 to 8.0 km/s and $\nu_i$ from 0.1 to 0.4. Stress enhancement factors for this range of parameters are shown in Figs. 4-7. Figs. 4 and 5 show dependence on constituent volume fractions and Poisson's ratios respectively.

Fig. 6 demonstrates that the dependence on density and sound speed can be fit rather well by:

$$R_{predicted} = 1 + \left( \frac{\left| \rho_1 c_1^2 - \rho_2 c_2^2 \right|}{\rho_1 c_1^2 + \rho_2 c_2^2} \right)^{4/3}. \quad (7)$$

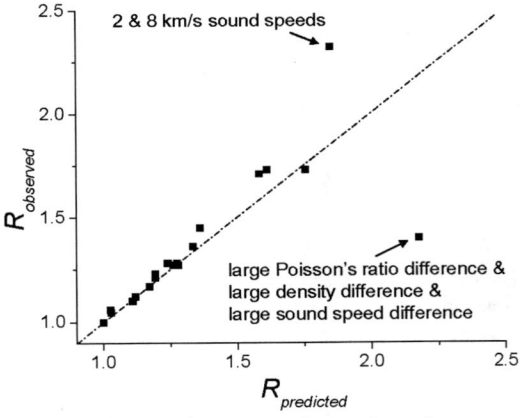

**Figure 7.** Stress enhancement factor dependence on all tested constituent properties ($\rho_i$, $c_i$, $\phi_i$, $\nu_i$). $R_{observed}$ is from equation (6) and $R_{predicted}$ from equation (8).

Finally, all the dependencies shown in Fig. 7 can be combined into a single relationship of the form:

$$R_{predicted} = \left(1 + 2\min(\phi_1, \phi_2) \cdot 0.6 \cdot \frac{|v_1 - v_2|}{v_1 + v_2}\right) \cdot$$
$$\left[1 + 2\min(\phi_1, \phi_2)\left(\frac{|\rho_1 c_1^2 - \rho_2 c_2^2|}{\rho_1 c_1^2 + \rho_2 c_2^2}\right)^{\frac{1}{3}}\right]. \quad (8)$$

## CONCLUSIONS

With reasonable parameters for constituent materials in a heterogeneous mixture represent-tative of gabbroic anorthosite, deviatoric stress enhancements of up to two or more are possible. Assuming a Mohr-Coulomb yield surface for fully fractured rock, this may be enough to induce yielding behavior that can produce complex crater morphology. Of course, this assumes that 3% by volume yielding is enough to count the entire rock volume as yielded from a continuum viewpoint. While this is suggestive from terrestrial crater field observations, further investigations are required to quantitatively determine the degree of yielding necessary to produce the complex crater form.

## ACKNOWLEDGEMENTS

This work is part of a larger effort that is partially funded by NASA's Planetary Geology and Geophysics Program. Collaborators of this work are Olivier Barnouin-Jha from the Johns Hopkins University Applied Physics Laboratory and Mark Cintala from the NASA Johnson Space Center. Sandia is a multi-program laboratory operated by Sandia Corporation, a Lockheed Martin Company, for the United States Department of Energy's National Nuclear Security Administration under contract DE-AC04-94AL85000. The author wishes to thank Bob Schmitt, Gene Hertel and Brian Dodson for their helpful reviews of the manuscript.

## REFERENCES

1. Melosh, H. J., "Acoustic fluidization: A new geologic process?", Journal of Geophysical Research, Vol. 84, pp. 7513-7520, 1979.
2. Walton E.L. & J.G. Spray, "Mineralogy, microtexture and composition of shock-induced melt pockets in the Los Angeles basaltic shergottite", Meteoritics and Planetary Science, Vol. 38, pp. 1865-1875, 2003.
3. Kieffer, S. W., "Shock metamorphism of the Coconino Sandstone at Meteor Crater, Arizona", Journal of Geophysical Research, Vol. 76, pp. 5449-5473, 1971.
4. McGlaun, J.M., S.L. Thompson and M.G. Elrick, "CTH - A Three-Dimensional Shock-Wave Physics Code", International Journal of Impact Engineering, Vol.10, pp. 351-360, 1990.
5. Crawford, D.A. & O.S. Barnouin-Jha, "Computational Investigations of the Chesapeake Bay Impact Structure", 35th Lunar and Planetary Science Conference, 2004.
6. Barnouin-Jha, & D.A. Crawford, "Investigating the Effects of Shock Duration and Grain Size on Ejecta Excavation and Crater Growth", 33rd Lunar and Planetary Science Conference, pp. 1738-1739, 2002.
7. O.Keefe, J.D. and T.J. Ahrens, "Planetary Cratering Mechanics", Journal of Geophysical Research, Vol. 98, pp. 17001-17028, 1993.
8. Wallis, G.B., One-Dimensional Two-Phase Flow, McGraw-Hill, New York, 1969.

CP845, *Shock Compression of Condensed Matter - 2005,*
edited by M. D. Furnish, M. Elert, T. P. Russell, and C. T. White
© 2006 American Institute of Physics 0-7354-0341-4/06/$23.00

# SOUND VELOCITY OF (Mg$_{0.92}$, Fe$_{0.08}$)SiO$_3$ PEROVSKITE UP TO 140 SHOCK PRESSURE AND ITS GEOPHYSICAL IMPLICATIONS

## Zizheng Gong[1,2], Lin He[1], Yingwei Fei[2], Jinke Yang[1], and Fuqian Jing[1,3]

[1]*Institute of of High Pressure and High Temperature Physics, Southwest Jiaotong University, Chengdu 610031, P.R. China.*

[2]*Geophysical Laboratory, Carnegie Institution of Washington, Washington DC 20015,USA.*

[3]*Laboratory for Shock Wave and Detonation Physics Research, Institute of Fluid Physics, P.O.Box 919, Mianyang, Sichuan 621900, China.*

**Abstract.** New experimental data of compressional sound velocity $V_P$ for natural polycrystalline Enstatite (Mg$_{0.92}$, Fe$_{0.08}$) SiO$_3$ initial specimens were measured using the optical analyzer techniques, and our former data were also revised. The $V_P$ showed two discontinuities in the pressure range up to 140 GPa. Combination with previous Hugoniot research, three distinct regions appear for (Mg$_{0.92}$, Fe$_{0.08}$) SiO$_3$ Enstatite: a low-pressure phase (LPP, Enstatite) exists up to shock pressures about 64 GPa, while a mixed phase (MP, Enstatite+ Perovskite ) is possible near the high pressure part of this range; perovskite phase (PvP) was found between about 68 to 83 GPa; and then a new high-pressure perovskite phase (NPvP) occurs at shock pressures higher than 85 GPa. Between 83-85GPa, $V_P$ showed a negative jump. The solid to solid microstructure transition of (Mg$_{0.92}$, Fe$_{0.08}$)SiO$_3$ Perovskite, caused by the high-spin to low spin HS-LS transition of iron around 83 GPa pressure region, probably is responsible for the radial anomaly of seismic wave velocity in depth about 1600-1800km of the lower mantle.

**Keywords:** (Mg, Fe)SiO$_3$-Pv, Sound velocity, Shock compression, Phase transition, Lower mantle.
**PACS:** 62.50.+p, 64.70.Kb, 61.50.Ks, 91.60.Hg .

## INTRODUCTION

(Mg, Fe)SiO$_3$ -perovskite (orthorhombic structure, space group P*bnm*, Mg-Pv) has long been a very active subject of numerous theoretical and experimental studies under high pressure and temperature, for it is generally thought to be the primary constituent of Earth's lower mantle. However, the thermodynamic stability and possible phase transitions for the Pbnm-perovskite structure at the earth's lower mantle conditions have been

controversial for several years [1-7]. Hugoniot sound velocities provide information about thermo-elastic properties of materials at simultaneous high pressure and temperature in Earth's lower mantle and core conditions. This information cannot currently be obtained in any another way in laboratory. Moreover, sound velocity is more sensitive to phase transition (solid-solid or solid-melt) than compression curve (*P-V*), for it is the slope line of *P-V* curve. So sound velocity measurement can be used to address the possible phase transitions of materials under high pressure and temperature. In the other hand, seismology is the most important method to know the structure of the earth's interior. Through the Comparisons between the laboratory experimental sound velocity and the seismic velocities, the mineralogical explanations of the seismic wave structure in the earth's interior can be confirmed, and the chemical and mineralogical compositions can be constrained. However, only few experiments about the sound velocity measuring have been performed for (Mg, Fe)SiO$_3$ -perovskite [8, 9 10]. Here, we report our new Hugoniot sound velocity data of (Mg$_{0.92}$, Fe$_{0.08}$) SiO$_3$-enstatite samples at the pressure region 70-90 GPa where electronic spin transitions were reported [11], in order to explore further possible phase transformation information and the corresponding change of physical properties of (Mg$_{0.92}$, Fe$_{0.08}$) SiO$_3$-perovskite in the lower mantle, and geophysical implications are addressed.

## EXPERIMENTAL PROCEDURE

Polycrystalline specimens of (Mg$_{0.92}$, Fe$_{0.08}$) SiO$_3$ were synthesized and hot-pressed from a starting material of natural Enstatite mineral from Damaping mine, Changjiakou, Hebei Province, China, prior to the Hugoniot experiments (details see [10]). X-ray diffraction patterns have shown the final hot-pressed specimens to be single chain Enstatite. The crystal density of the Enstatite mineral is 3.27 g/cm3. The average bulk density which obtained by measuring its bulk volume and its weight, is 3.06 ± 0.01g/cm$^3$, and the average porosity (1-$\rho_{00}$/$\rho_0$) of the pressed specimens are 6.35%. The prepared specimens contain more than 98% Enstatite mineral.

The shock compression experiments were carried out with a 25-mm two-stage light gas gun. The optical analyzer techniques (OAT) [12] is used to measure the sound velocity in the samples, as show in Figure 1. In all experiments, the copper f lyers were used and no driver plates were used. The velocities of copper flyer $u_f$ were measured to an accuracy of 0.2% by a magnetic-flyer method. Three enstatite plates (their thicknesses $\Delta h_i \sim$ 2, 3 and 4 mm) in one experiment are used to determine

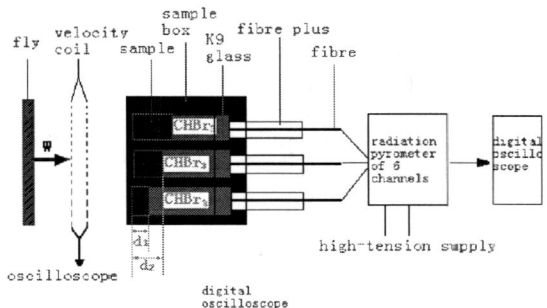

**Figure 1.** Sketch illustration of experimental set-up.

the maximal overtake thickness of sample from a linear fit between the recorded time duration $\Delta t$ and the corresponding sample thickness. Non-transparent window CHBr$_3$ is placed behind each sample. The radiation from each sample/window interface was transferred by three optical fibres (core diameter 120μm) into the pyrometer. In the experiments, we obtain the longitudinal wave of the samples by:

$$V_P = [(\rho_0 / \rho) \times X] / (X / D_S - d_f / D_f - d_f / V_f^L)$$

where $\rho$ is the sample density under shock compression and $\rho_0$ is the initial density. X is the maximal overtake thickness of sample and is determined by our experiments. $D_S$ and $D_f$ are the shock velocity in the sample and flyer, respectively. $d_f$ and $V_f^L$ are the thickness and Lagrange sound velocity of flyer, respectively. The Hugoniot equation of state (EOS) of copper $D=3.93+1.489u_p$ [11] and the dependence of compressional sound velocity of copper on Hugoniot pressure $\ln V_P^E =1.5705-1.7785\times10^{-2}\times\ln P+2.5057\times10^{-2}\times\ln^2 P$[13], are utilized in date processing.

## RESULTS AND DISCUSSIONS

Experiment results are shown in Table 1 and illustrated in Figure 2. The shock temperatures in Table 1 are calculated by Walsh thermodynamics method and Mie-Gruneisen equation of state, respectively, for comparison. Our calculated temperatures have a good agreement with measured data of Luo [15].

**TABLE 1.** Sound velocity measurement results.

| Shot | $d_f$ (mm) | $u_f$ (km/s) | X(mm) | P(GPa) | $\rho$ (g/cm$^3$) | $V_p$(km/s) | $V_B$(km/s) | $T_H^W$(K) | $T_H^{M-G}$(K) |
|------|-----------|-------------|-------|--------|-------------------|-------------|-------------|------------|----------------|
| 1 | 1.237 | 2.980 | 5.899 | 43.2 | 4.39 | 8.75 | 7.26 | 1500 | 1505 |
| 2 | 1.210 | 3.930 | 5.667 | 63.8 | 4.68 | 9.72 | 8.32 | 2586 | 2585 |
| 3 | 1.450 | 4.098 | 5.727 | 67.8 | 4.74 | 11.72 | 8.47 | 2793 | 2451 |
| 4 | 1.304 | 4.700 | 4.955 | 83.1 | 4.91 | 13.09 | 9.15 | 3567 | 3704 |
| 5 | 1.496 | 4.802 | 7.202 | 85.7 | 4.93 | 10.08 | 9.27 | 3814 | 3817 |
| 6 | 1.496 | 5.890 | 7.195 | 116.9 | 5.20 | 10.90 | 10.37 | 5894 | 5888 |
| 7 | 1.505 | 6.579 | 7.161 | 138.8 | 5.35 | 11.52 | 11.15 | 7493 | 7493 |

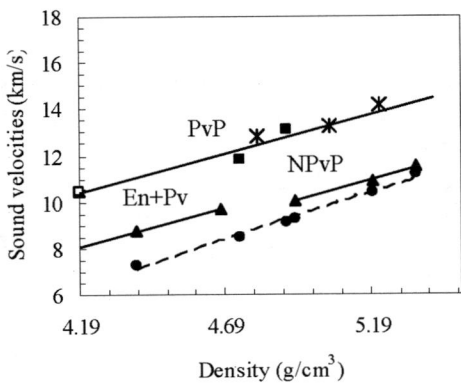

**FIGURE 2.** Sound velocity versus density of (Mg$_{0.92}$, Fe$_{0.08}$) SiO$_3$perovskite. The symbols of ■, ▲, *, •, and □ represent the new experimental data, the revised data of [10], the first principle calculated data of MgSiO$_3$ perovskite [14], the bulk velocity calculated from Hugoniot [6], and the ambient compressional velocity of MgSiO$_3$ perovskite, respectively. The solid line and the dashed line represent the fit of compressional velocity and bulk velocity, respectively.

It can be seen that two discontinuities exist for the compressional velocity $V_P$ in the pressure range up to 140 GPa in Fig.2. The first discontinuity appears when shock pressure is higher than 64 GPa, the jump is positive, and the relative change of $V_P$ is about 20%. The second jump is negative, it occurs when shock pressure is higher than 83 GPa, the relative change of $V_P$ is about 30%.

From Fig. 2, we can see that the new $V_P$ data just lay on the same $V_P$-$\rho$ straight line, with the calculated $V_P$ data of MgSiO$_3$-Pv by the first principle [14]. In addition, the ambient $V_P$ of MgSiO$_3$-pv is on this line too, seeing Fig. 2. This is

clear evidence that the new $V_P$ data represent the property of perovskite.

The change of bulk velocity of the 7 shots is almost smooth (seeing Fig. 2)

To judge whether a melting transition occur in the shock compression, the bulk sound velocity $V_B$ is calculated along Hugoniot [6], the results are given in Table 1 and also shown in Fig. 2. It is obvious that our measured c $V_P$ is larger than the calculated $V_B$, so melting can be ruled out up to 140 GPa shock pressures.

Combination with the previous Hugoniot measurement [6, 7], and the proposed phase diagram of MgSiO$_3$ system with Hugoniot temperature data for glass and enstatite [15], three distinct regions appear for (Mg$_{0.92}$, Fe$_{0.08}$) SiO$_3$ Enstatite: a low-pressure phase (LPP, Enstatite) exists up to shock pressures about 64 GPa, while a mixed phase (MP, Enstatite+ Perovskite ) is possible near the high pressure part of this range; perovskite phase (PvP) was found between about 68 to 83 GPa; and then a new high-pressure perovskite phase (NPvP) occurs at shock pressures higher than 85 GPa. The volume change induced by phase transition from Enstatite to Perovskite can be detected by Hugoniot measuring, but it is too small to be detected from Perovskite phase to a new high-pressure perovskite phase [6, 7]. However, the phase transition for Pv to NPvP can be detected from sound velocity measuring.

Shim et al [16] have indicated the possibility of the distortion of lattice structure phase transition at above 83 GPa and 1700 K in DAC experiments, but they can neither confirm it nor provide a adequate explanation.

On the other hand, It was reported [17] that the electrons of ferro Fe at 3d state change from high spin into low spin (HS-LS) between 70~85GPa, which makes the bond length of Fe-O shorter and induces the distortion of crystal lattice. This deformation makes it more compressible, which may be responsible for the negative jump in $V_P$. As the little increase in volume, $V_B$ only changes a little which can even be neglected in our experiment.

Our estimated phase boundary $dT/dP$ for perovskite phase (PvP) to a new high-pressure perovskite phase (NPvP), 66~92 K/GPa, overlapped that of the HS-LS transition ($dT/dP$=78~82 K/GPa). So it is reasonable to explain PvP to NPvP phase transition of $(Mg_{0.92}, Fe_{0.08})SiO_3$ Perovskite by HS-LS transition of iron around 83 GPa pressure region caused solid to solid microstructure transition.

## CONCLUSIONS

The sound velocity discontinuity of $(Mg_{0.92}, Fe_{0.08})SiO_3$-Pv around 83 GPa was examined for the first time, and this is caused by the HS-LS transition of ferro iron. This is probably the reason for the radial anomaly of seismic wave velocity in depth about 1700-2300km of the lower mantle [18].

## ACKNOWLEDGEMENTS

This research was supported by the National Natural Science Foundation of China under Grant No. 40474033 and 10299040.

## REFERENCES

1. Meade, C., H. K. Mao, and J. Z. Hu, "High – temperature phase transition and dissociation of (Mg, Fe)SiO₃ perovskite at lower mantle pressure", Science, 268, 1743-1745, 1995.
2. Saxena, S. K., L. S. Dubrovinsky, P. Lazor, et al., "Stability of perovskite MgSiO₃ in the Earth's mantle", Science, 274, 1357-1359, 1996.
3. Serghiou, G., A. Zerr, and R. Boehler, "(Mg, Fe)SiO₃-perovskite stability under lower mantle conditions", Science, 280, 2093-2095, 1998.
4. Fiquet, G., A. Dewaele, D. Andrault, M. K., et al., "Thermolastic properties and crystal structure of MgSiO₃ perovskite at lower mantle pressure and temperature conditions", Geophys. Res. Lett., 27, 21-24, 2000.
5. Shim, S. H., Duffy, T. J., and Shen, G. Y., "Stability and structure of MgSiO₃ perovskite to 2300-Kilometer depth in the Earth's mantle", Science, 293, 2437-2440, 2001.
6. Gong, Z. Z., Fei, Y., Dai, F., et al., "Equation of state and phase stability of $(Mg_{0.92},Fe_{0.08})SiO_3$ perovskite up to 140GPa". Geophys. Res. Lett., 31:L04614, 2004.
7. Akins, J. A., Luo, S-N, Asimow, P. D., et al., "Shock-induced melting of MgSiO3 perovskite and implications for melts in Earth's lowermost mantle", Geophys. Res. Lett., 31:L14612, 2004.
8. Kung, J., Li, B., Uchida, T., Wang, Y., et al., "In situ measurements of sound velocities and densities across the orthopyroxene→ high-pressure clinopyroxene transition in MgSiO₃ at high pressure", Physics of the Earth and Planetary Interiors, 147 : 27-44, 2004.
9. Li, B. S., Zhang, J. Z., "Pressure and temperature dependence of elastic wave velocity of MgSiO₃ perovskite and the composition of the lower mantle", Physics of the Earth and Planetary Interiors, 151: 143–154, 2005.
10. Gong, Z. Z., Xie, H., Jing. F., et al., "High-pressure sound velocity of $(Mg_{0.92}, Fe_{0.08})$ SiO₃-perovskite and possible composition of Earth's lower mantle", Chin. Phys. Lett., 16: 695-697, 1999.
11. Mitchell, A. C., and Nellis, W. J., "Shock compression of aluminum, copper, and tantalum", J. Appl. Phys., 52, 3363-3374, 1981.
12. McQueen R. G., Hopson J. W., and Fritz J. N., "Optical technique for determining rarefaction wave velocities at very high pressures". Rev. Sci. Instrum., Vol.53, 1982.
13. Duffy, T. S., and Ahrens T. J., "Sound velocities at high pressure and temperature and their geophysical implications". J. Geophys. Res., B 97, 4503, 1992.
14. Tsuchiya, T., Tsuchiya, J., et al., "Elasticity of post-perovskite MgSiO₃", Geophys. Res. Lett., Vol . 31, L14603, 2004.
15. Luo, S. S, Akins, J. A., Ahrens, T. J., and Asimow P. D, "Shock compressed MgSiO₃ glass, enstatite, olivine and quartz: Optical emission, temperatures and melting", J. Geophys. Res, Vol. 109, B05205, 2004.
16. Shim, S. H., Duffy, T. S., Shen, G., "Stability and crystal structure of MgSiO₃ perovskite to the core-mantle boundary", Geophys. Res. Lett., 31: L10603, 2004.
17. Badro, J., Rueff, J. P., Vanko, G., "Electronic transitions in perovskite: possible noconvecting layers in the lower mantle", Science, 305, 383-386, 2004.
18. Van der Hilst, R. D., and Karason H., "Compositional heterogeneity in the bottom 1000 kilometers of Earth's mantle: Toward a hybrid convection model", Science, 283, 1885-1888, 1999.

CP845, *Shock Compression of Condensed Matter - 2005,*
edited by M. D. Furnish, M. Elert, T. P. Russell, and C. T. White
© 2006 American Institute of Physics 0-7354-0341-4/06/$23.00

# BALLOGRAPHY: A BILLION NANOSECOND HISTORY OF THE BEE BLUFF IMPACT CRATER OF SOUTH TEXAS

## R. A. Graham

*The Tomé Group, 608 Cenizo Boulevard, Uvalde, TX 78801*

**Abstract.** The Bee Bluff Structure of South Texas in Zavala County near Uvalde has been found to exhibit unusual features permitting study of impactites and meteorite impact processes from the standpoint of grain-level, nanosecond shock-compression science. The site is characterized by a thin cap of Carrizo Sandstone covering a thin hard Indio fm calcareous siltstone. A soft calcareous silt lies below the hard cap. Calculations based on the Earth Impact Effects web-based program indicate that the site is best described by a 60 m diameter iron meteorite striking the ground at 11 km/sec. Such an impact into sandstone is expected to produce a shock pressure of 250 GPa. A large release wave originates from the bottom of the hard target with upward moving melt-vaporization waves of solid, liquid and vapor products that become trapped at the impact interface. Numerous distinctive types of impactites result from this 'bottom-up' release behavior. Evidence for hydrodynamic instabilities and resulting density gradients are abundant at the impact interface. An unusually valuable breccia sample called 'The Uvalde Crater Rosetta Stone' contains at least seven types of impactites in a well defined arrangement that can be used to read the billion nanosecond history of the impact and identify scattered impactites relative to their place in that history.

**Keywords:** Geophysics, impact cratering, quartz, sandstone, goethite, siltstone, volatiles, energy localization.
**PACS:** 62.50.+p, 91.65.-n, 91.90.+p

## INTRODUCTION

Over the last thirty-five years scientific investigations of meteorite impacts on Earth have provided detailed descriptions of overall impact processes. Identification of possible impact structures rests upon study of rocks and breccia at the sites that contain "Traces of Catastrophe," shock-metamorphic effects in the rocks and minerals that can confidently be identified with the high shock pressures and temperatures [1]. Residual markers that can be read scientifically are largely focused on quartz: high pressure phases stishovite, coesite; distinctive planar deformation features (PDFs), planar fractures (PFs); melt structures, unique amorphous phases of quartz, diaplectic glass, lechatierite, and fused quartz.

The PDF structures are distinctive as they typically appear in parallel sets at fixed orientations relative to the optic axes. Attempts have been made to correlate the orientations and frequency with impact pressure based on location within an impact structure and laboratory impact experiments [2-6]. The work of Kieffer extends well beyond PDF identification [4,5].

It has become increasingly apparent that PDFs in sedimentary rocks are distinctively different from those in crystalline materials. The sedimentary PDF structures are less frequent and more sets have high orientations relative to the optic axis [1]. Differences between sedimentary and crystalline rocks have been considered, but successful models for their formation have not been developed at the grain/nanosecond level.

## THE BEE BLUFF STRUCTURE OF SOUTH TEXAS

Wilson and Wilson reported a detailed study of an area of disturbed geology in South Texas in Zavala County near Uvalde, Texas 'The Uvalde Crater' in 1979, and suggested it was likely the site of a meteorite impact [8]. They suggested a crater diameter of 4 km.

Robertson [9] subsequently visited the site and found distinctive PDF markers, but raised the critical issue "*It is apparent the total comprehension of quartz planar features and development has not been achieved and that attention should be focused on shock deformation of porous lithologies.*"

In the present work we attempt to bring Robertson's focus on grain level science (what the author calls nanosecond processes) to full-circle based on study of the same impact site.

The investigation of Sharpton and Nielson of 1988 confirmed existence of $\omega$ and r, z PDF sets on site and at locations within Carrizo fm and at formations remote from the site [10]. They suggested that quartz at the site may have resulted from transport from other locations. On the basis of their work, the Bee Bluff Structure was removed from the recognized list of known impact sites in November 2004 [11].

Work of Jurena, French and Gaffey [12,13] provided additional data of PDFs from a more complete site survey and measured a negative gravitational anomaly along the western edge of the site. Jurena [14] reported considerable detail on the site geology and presented the detailed PDF analysis from quartz in both Carrizo Sandstone and Indio fm calcareous silt. He suggested a crater diameter of 2.5 km. First work from the present study was published in 2005 [15].

## SITE GEOLOGY

The overall regional geology is described in [8] and [15]. In interpreting the nanosecond history of the impact process several notable features stand out. As shown in the Geologic Map [16], the impact site is an isolated pocket of exposed Carrizo fm; The sediments were apparently deposited by erosion from the various Balcones uplifts to the north. Overlying Bigford formation siltstone and sandstone is thought to have been eroded in the later stages of the uplift exposing the Carrizo fm.

The lateral edges of the hard Carrizo and Indio fm strata encounter soft, low shock impedance materials at locations only about 3 km distant to the south and about 4 km to the west of the impact site. Strong shock wave reflections are expected at these boundary locations. The uplifted breccia areas noted by Wilson and Wilson [8] as allochthonous (moved into place) are thus more likely autochthonous (lifted in place).

Exposure of the underlying geology revealed at the eastern edge of the site at a cliff overlooking the Nueces River shows the thin (~1 meter) cap of Carrizo fm overlays an equally thin Indio calcareous siltstone. Underneath the siltstone a soft micritic calcareous silt is in place with a thickness of about 10 meter. A hard rock limestone is at the base of the cliff. The target hard cap of sandstone and siltstone would have been promptly penetrated by a meteorite releasing voluminous high velocity streams of the calcareous silt.

**Figure 1.** Impact location is in the vicinity of the westward meander of the Nueces River. Linear feature on left is US Hwy. 83. Dimension along the lower edge of the photo is 4.2 km.

Breccia thicknesses, uplifts and surrounding formations indicate that the present stratigraphy was present at the time of the formation of the impact structure.

One of the most compelling geologic features for a meteorite impact is the presence of substantial micritic calcareous dust at all locations south of the site as revealed in the lower portion of the 1995 aerial infrared photograph shown in BW in Figure 1.

It is remarkable that the cultivated fields still continue to shown the calcareous dust patterns. Of special interest is the white streak south of the dark spot, an uplift at the edge of the formation. This is dust accumulating in the wake of the flow from the center of the impact area; accumulation is also shown on the upstream side of the flow. Similar dust accumulations on the upstream side of the flow at the larger uplift area to the left. The dust accumulated in the wake of the flow indicates the direction of the flow. The dust 'arrow' indicates an impact point immediately to the east near-or-in the present bed of the river at the westward meander. Note that the area immediately to the west of this location is the principal source of impactites in prior and the present work. The present work has obtained impactites from the east side of the river.

Less visible in the photograph is a broad elevated ridge in the center-right and west of the river with light coloration that is better shown in a 1942 Texas Highway Department aerial taken at a time of limited vegetation that shows a whitish top to the ridge. Jurena [14] has identified the ridge as covered with a "caliche" matrix. Samples of the present work show that the material is a loosely lithified micritic calcareous silt containing cm size chips of iron-rich siltstone and calcareous siltstone, remnants of the dust cloud. The area immediately to the east of the ridge shows a roughly 2 km diameter circular depression draining to the river.

The voluminous dust at the site underlain by low strength silt would be expected to have resulted in a relatively rapid filling of the residual crater.

## IMPACT CONDITIONS

The impact-effects, web-based code has been used to calculate the general conditions of impact [17]. After many iterations, the best fit to the site observations is found to be a 60 m diameter iron meteorite impacting at 11 km/sec that results in a residual crater 2 km in diameter. Although a somewhat larger diameter impactor and larger crater are possible, the very limited melt on site restricts the impactor size and impact velocity. Given the uncertainties, the calculation cannot define the residual crater size to an accuracy of more than a few hundred meters. The meteorite may have impacted in several pieces.

Based on an iron impact at 11 km/sec, and the equation of state of sandstone compiled by Ahrens and Johnson [18], of iron by Brown, Fritz and Hixson [19], the Hugoniot pressure is 250 GPa at an input particle velocity in sandstone is 7.5 km/sec with shock velocity of 13 km/sec (13 micron/nanosecond). These conditions are sufficient to perform one-dimensional calculations and estimate overall two dimensional processes.

The principal feature of the one-dimensional calculations is the prompt upward-moving release wave from the low shock-impedance calcareous silt. The initial shock is estimated to arrive at the release interface at about 150 microsec. and at the impact face at about 200 microsec. Penetration of the meteorite at its outer edges is expected to occur at about 250 microsec. At this time the high velocity venting of the silt will begin and continue until complete excavation of the impact cavity.

As pressure release with accompanying melt and vaporization occurs while the meteorite remains in contact with the target rocks, volatile products such as the water/steam from the goethite in both the sandstone binder and the iron-rich siltstone are expected to be trapped at the interface. Any outgassing from the calcareous siltstone will also be trapped. The expected result is a complex mixture of target products at significant pressure and substantially elevated temperature; all are in motion at substantial velocities. Complex, time-dependent physical and chemical processes are anticipated, but the containment affords an opportunity to preserve the metamorphic products.

Note that this situation is profoundly different from the usual meteorite impact processes in which the release of pressure is from the outer parts of the meteorite and melt and vaporization occur downward from the impact surface. Melt and vaporization products are then propelled from the impact site. If we denote the usual configuration as a 'top-down' release, the Uvalde crater experiences a 'bottom-up' release with profound effects on the processes and resulting impactites.

## IMPACTITES

A wide array of impactites have been collected by the author and other land owners in the area. As a resident of Uvalde the author has been able to access many samples from local collectors. The impactite collection with maps and information on impact cratering is housed in <u>The Meteorite Crater of South Texas Exhibit</u> at the El Progreso Library in Uvalde, Texas. The collection is centered around the Uvalde Crater Rosetta Stone, a 75 kg mixed clast, mixed mict assemblage of at least seven different impactite types. As the Rosetta Stone preserves evidence for a range of impact conditions and places the impactites in arrangements relative to each other, the stone provides the basis of reading the billion nanosecond history of the impact. With the intact breccia, samples obtained from locations around the site can be confidently identified as to their source.

Table I lists the impactites located at the site. All samples are complex but contain common features of shock-modified quartz, goethite and dense ironstone or calcareous siltstone. All samples contain evidence of porosity resulting from high-pressure, high-temperature steam or other gases. Many samples show characteristic features of hydrodynamic instability of flow between materials of different density, in this case between the meteorite and target materials at the impact interface.

**Table I. Graham Collection, Uvalde Crater**

<u>Uvalde Crater Rosetta Stone</u>: 75 kg, multiclast.
<u>Uvalde Sandstone</u>: metamorphic Carrizo SS.
<u>La Pryor Siltstone</u>: metamorphic Indio siltstone.
<u>Uvalde Suevite</u>: multiclast, multimict, high porosity SS, Uvalde SS, calcareous siltstone, porous iron-rich siltstone, porous α-goethite, Sandia ironstone, bubble-top surface.
<u>Sandia Ironstone</u>:Dense goethite-quartz in hydrodynamic shapes.
<u>Wilson Stone</u>: Multiclast, Sandia ironstone, porous goethite, quartz, goethite. Hematite crystals, goethite nanocrystals, distinctive bubble top.

<u>Jurena Stone</u>: Loosely lithified calcareous silt matrix with iron siltstone chips.
<u>Jones-Horner Stone</u>: 10 kg, extreme bubble top, goethite.
<u>Spheroids</u>:
 Rosetta Stone condensates
 Aerodynamic shapes
 goethite, quartz, hematite, 'Dogey dos'

It should be emphasized that the characteristic impactites in the collection show obvious differences among themselves upon the most casual inspection, and are not found in other locations of Carrizo fm exposure. The question of geologic-versus-shock processes is certainly always at issue, but the samples are unquestionably distinctive. F. Wilson, collaborator in the present work and geologist with extensive experience with sedimentary geology especially in Texas, observed after his extensive investigation that many rock samples at the site were unique to the site. Jurena has made the same observation [14] concerning the sandstone.

Shock compression processes viewed at the nanosecond time scale can be identified as the most likely causes for the characteristics of the samples.

About thirty samples were examined in reflected light at magnifications from 20 to 110. These examinations served to define the grain level characteristics and determine questions to be addressed in more detailed work.

## UVALDE SANDSTONE

Most of the investigations on PDF structures on sedimentary materials are carried out in sandstone, as has been the case of investigations at the Uvalde Crater. All investigations at the Uvalde site have found PDFs with characteristics similar to other studies: PDFs are found in 10% or less of shocked grains, many high angle orientations (>50%) dominate.

Thus, five authors: Robertson (1980), Sharpton and Nielson (1988), Jurena, et al, (2001), Jurena (2002) and Graham, et al in the companion paper of this volume have documented PDF structures in sandstone at the site. The most extensive investigation is that of Jurena.

Sandstone found at the site is not Carrizo Sandstone as found in the Carrizo fm throughout Texas. Carrizo Sandstone is a fine-grain sandstone with limonite binder (orange), of significant porosity (about 40 %). Although competent, samples can be readily broken. The sandstone on site, Uvalde Sandstone, is significantly stronger than Carrizo and is typically dark brown in color.

The distinctive binder and comminuted Uvalde sandstone grains are apparent in optical microscopy as shown above in Figure 2.

**Figure 2**. Photomicrographs. Upper: Carrizo Sandstone. Middle: Uvalde Sandstone. Bottom: Uvalde Sandstone grain, atypically large.
(Distance along bottom of photos is 136 micron.)

Orange limonite binder dominates the color and the binder is thin in Carrizo Sandstone. In sharp contrast, Uvalde Sandstone shows a radically different microstructure. The binder is goethite, dark to light brown, sinuous in form. The goethite has rounded voids indicative of internal pressure, likely steam released by the shock process. Goethite is mixed with small quartz chards and coats larger grains with a thin, dark coating.

Submicron goethite spheres are abundant and readily seen by eye through the microscope. The middle photo shows a crystal growth in the hematite form. Such hematite forms are abundant in various samples. Various forms of fused quartz are apparent.

Although details differ from one impactite to another, all follow patterns suggested by Figure 2, namely: release of water from goethite, formation of supersaturated steam, quartz, hydrous iron. The smooth voids are a clear indication of steam pressure. Localized regions of fused quartz or quartz melt are apparent; some may be 'froth' proposed by Kieffer [5]. Smooth quartz surfaces are frequently observed on quartz grains. Studies of Tschauner, et al [20] on quartz are relevant.

## SUEVITE AND HYDRODYNAMIC INSTABILITIES

Although complex and requiring very considerable materials analysis, reading of the Rosetta Stone perhaps gives us the most direct indication of the shock processes. In Figure 3 a section through one end of the stone shows the dominant features: Uvalde Suevite containing a mix of all target rock impactites, and most characteristically, a highly porous mixture of goethite, calcareous siltstone, quartz chips. Significantly, the mixture is marked by numerous voids, rounded in shape.

Other parts of the stone particularly sandstone and iron-rich siltstone are marked by numerous rounded voids. As indicated earlier, this feature is likely the result of the high pressure/temperature steam.

Hydrodynamic instability patterns not apparent in the photo are prominent along the top surface. Without looking at the detail it appears that the top surface shows a melt pattern. Closer examination shows a thin layer of typical Uvalde Sandstone: porous goethite and quartz. Underneath that layer, however, the dense sinuous wave pattern of Sandia Ironstone is present. Bubble tops on the surface are the result of maximum wave heights for hydrodynamic instability. Such bubble tops also are found on Wilson Stone and Jones-Horner Stone

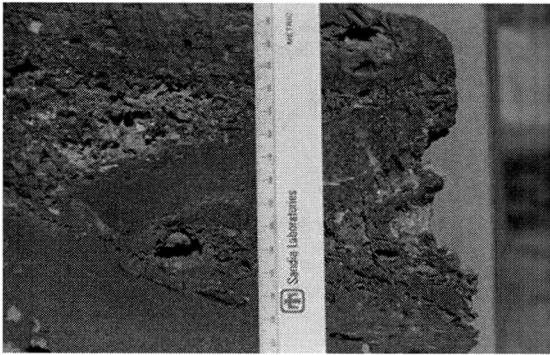

**Figure 3.** Macrophotograph of a section through upper portion of Uvalde Crater Rosetta Stone. Lighter color is porous goethite, quartz, siltstone mixture. Note the rounded voids in the goethite.

## CONCLUSION

The present paper briefly describes a few selected materials aspects of a large number of samples. Further materials analysis is shown in the companion paper in the present proceedings. Of special interest is the 'bottom-up' nature of the impact site, the widespread involvement of hydrous iron oxides and hydrodynamic instabilities.

The Bee Bluff Structure is clearly a meteorite impact site. Wilson and Wilson identified substantial geologic disruption including a seismic line showing displacement of underlying formations. Jurena, et al showed negative gravity anomalies on site. The impact dust cloud 'arrow' is clearly distinctly. Quartz PDFs have been found by all investigators including the present author. Impactites of the present work demonstrate the need to view the crater site from the standpoint of a billion nanosecond history.

## ACKNOWLEDGEMENTS

The author is deeply indebted to N. N. Thadhani, B. Morosin, W. F. Wilson and F. Hörz for assistance and discussion. Daniel Leskovar, Texas A&M Research Station, Uvalde, kindly permitted use of the microscope in his laboratory.

## REFERENCES

1. French, B. M. "Traces of Catastrophe," Lunar and Planetary Institute, Houston, TX (1998).
2. French, B. M. and Short, N. M. eds, Shock Metamorphism of Natural Materials, Mono Book Corp. Baltimore (1968).
3. Hörz, in French and Short, loc cit, pp 243-253.
4. Kieffer, S.W., J. Geophysical Res. 76, pp 5449-5473 (1971).
5. Kieffer S. W., Phakey R. B., and Christie. J. M. Contr. Mineral. Petrol., 59, pp 42-93.
6. Stöffler, D., and Langenhorst, F., Meteoritics, 29, pp 151-181 (1994).
7. Grieve, R. A., Langenhorst F., and Stöffler, D. Meteoritics and Planetary Sci., 31, 6-35 (1996)
8. Wilson, D. F., and Wilson, D., Geology, 7, pp 144-146 (1979).
9. Robertson, R. B., Lunar and Planet. Sci. XI, 938-940 (1980)
10. Sharpton V. L., and Nielsen, D. C., Lunar and Planet. Sci., XIX 1065-1066 (1988).
11. Earth Impact Database, November 2004, www.unb.ca/passc/Impact/ .
12. Jurena, D. J., French, B. M., and Gaffey, M. J. Lunar Planet. Sci. XXXII 1828.pdf (2001).
13. Jurena, D. J., French, B. M., and Gaffey, M. J. Lunar Planet. Sci. XXXIV 2076.pdf (2003).
14. Jurena, D. J. "The Bee Bluff Structure a Probable Impact Structure Located in South Texas" Thesis, Master of Science, Rensselaer Polytechnic Institute (2002).
15. Graham, R. A. and Wilson, F. W. Lunar Planet. Sci. XXXVI 1086.pdf (2005).
16. Geologic Map of Texas 1:500, 000, Bureau of Economic Geology of Texas at Austin (1992).
17. Collins, G. S., Melosh, H. J., and Marcus, R. A. "Earth Impact Effects Program: A Web-based Computer Program for Calculating the Regional Environmental Consequences of a Meteoroid Impact on Earth" Meteoritics and Planetary Sci. submitted Nov. 2004.
18. Ahrens, T. J. and Johnson, Mary L. "Shock Wave Data for Rocks" in Rock Physics and Phase Relations, AGU Reference Shelf 3 (1995).
19. Brown, J. M., Fritz, J. N. and Hixson, R. S., J. Appl. Phys. 88(9) pp 5496-5498 (2000)
20. Tschauner, O., Luo, S. N., Asimow, P. D., Ahrens, T. J., Swift D. C., Tierney, T. E., Paisley, D. L. and Chipera, S. J. High Pressure Research, 24, pp 471-479 (2004).

CP845, *Shock Compression of Condensed Matter - 2005*,
edited by M. D. Furnish, M. Elert, T. P. Russell, and C. T. White
© 2006 American Institute of Physics 0-7354-0341-4/06/$23.00

# QUARTZ AND HYDROUS IRON OXIDES FROM THE BEE BLUFF STRUCTURE OF SOUTH TEXAS

## R. A. Graham[1], M. Martin[2], N. N. Thadhani[2], and B. Morosin[3]

[1]*The Tomé Group, 608 Cenizo Blvd., Uvalde TX 78801*
[2]*Georgia Institute of Technology, School of Materials Science and Engineering,
771 Ferst Drive NW, Atlanta GA 30332-0245*
[3]*Sandia National Laboratories, PO Box 5800, Albuquerque NM 87185*

**Abstract.** There is substantial information showing that the Bee Bluff structure is an impact site and that a residual crater can be identified. The thin hard cap of Carrizo Sandstone, Indio fm calcareous silt and a thin layer of iron-rich siltstone leads to impact processes in which the high pressure release wave proceeds promptly upward leading to a trapping of metamorphic products at the impact interface, a 'bottom-up' pressure release. Release of water from goethite binder in the sandstone and from the iron-rich siltstone results in supersaturated steam in mixtures with iron and quartz compounds. Samples with quartz and hydrous iron oxide features are examined with optical microscopy, SEM, EDX and XRD. A quartz grain is found with a well defined PDF set. There is widespread amorphous quartz including lechatleriete. Nanocrystals of α-goethite in the acicular form are common. A condensation sphere from the 'Uvalde Crater Rosetta Stone' shows a complex mixture of hematite, goethite, and alpha quartz with a trace of trydimite. Numerous samples are yet to be analyzed. The crater appears to have features that can serve as an Earth analog to Mars craters. A companion paper in the present proceedings summarizes prior work, adds new site detail, reports impact-loading analysis, and describes overall features of impactite samples from the site.

**Keywords:** Geophysics, impact cratering, energy partitioning, quartz, goethite, tridymite.
**PACS:** 62.50.+p, 91.65.-n, 91.90+p

## INTRODUCTION

The 'Uvalde Crater' is of particular interest in that its 'bottom-up' shock pressure release promptly reduces pressure at high temperature. This process leads to trapping of metamorphic products at the interface between the iron impactor and target materials. The highly porous sandstone with small grain size and hydrous iron oxide binder provides conditions dominated by the grain-level (nanosecond) history of the processes. Goethite is a water source that produces supersaturated mixtures of quartz and iron products. Radial flow at the interface between meteorite and target materials leads to distinctive hydrodynamic instability features. Goethite is reformulated from the high temperature, high pressure products.

## MATERIALS FRAMEWORK

The impact conditions and sedimentary target configuration is as shown in Figure 1. Calculations show that the best fit for the on-site topography and lack of significant melt is a 60 meter diameter iron impactor at an impact velocity of 11 km/sec. For this condition an impact pressure of 250 GPa and an interface velocity of 7.5 km/sec are as given in the companion article in the present proceedings [1].

The uppermost formation is sedimentary Carrizo Sandstone that is very fine-grained, has a limonite (γ+α Goethite) binder, and is highly porous. A thin layer of an iron-rich siltstone lies between the sandstone and the underlying Indio fm calcareous siltstone. A loosely consolidated micritic calcium carbonate silt, much thicker then the hard rocks, dominates the exposure seen at the Nueces River. This silt is the dust source over the area during the impact excavation process, and for the upward moving release wave.

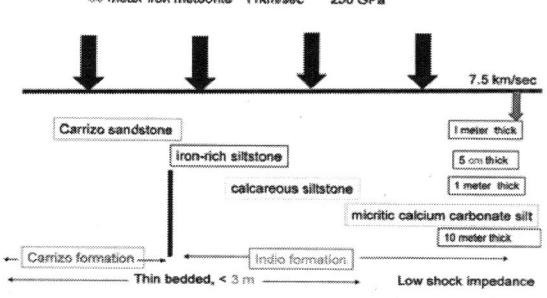

**Figure 1.** Meteorite impact conditions and structure of the target configuration. Note the hard rocks are thin relative to the impactor thickness.

The scientific issues with the sandstone are: evidence for PDF structures, localized heating and melt, and metamorphous effects on the Goethite binder. It is found that the sandstone is strongly altered by the shock process leading to a sandstone with α-goethite binder with its characteristic dark brown or black color.

The metamorphic calcareous siltstone is hardened and the color is changed from a whitish color to either pink or yellow. The color change is likely due to outgassing.

The iron-rich siltstone is found spread throughout numerous impactites, and in some cases in large blocks with 'bubble tops.' As indicated in [1], the bubble tops are the result of a maximum in the underlying hydrodynamic instability waves.

The micritic calcareous silt was violently expelled over the site as a deep dust fallout.

'The Uvalde Crater Rosetta Stone' is an intact collection of various impactites dominated by the bottom-up pressure release.

## QUARTZ AND GOETHITE

Characteristics of the Uvalde Sandstone are discussed in [1]. The characteristic morphology is the form of the goethite binder with rounded voids. Quartz grains are comminuted, distressed, and show local areas of melt or fused quartz.

One of the most unusual impactites is found in abundance close to the impact center at locations west of the river. The breccia was the most intriguing of those located by Wilson and Wilson [2]. Typically found in 5 cm thick form and perhaps 15 cm in lateral dimension, 'Wilson Stone' is characterized by a complex mixture of sandstone and goethite with a distinguishing bubble top as shown in Figure 2.

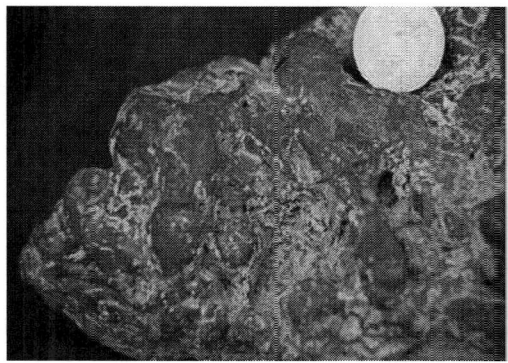

**Figure 2.** Wilson Stone sample with characteristic bubble top. The sample is a mixture of Uvalde Sandstone, goethite, and Sandia Ironstone. The unusual top surface is the result of hydrodynamic instability. Coin size is 24 mm.

An unusually dark, hard edge of the Wilson Stone, dominated by Sandia Ironstone, is under study at the Thadhani Materials Laboratory. Fragments from the sample were polished and examined with optical microscopy including fluorescence, DIC and XRD. After coating with gold, selected samples were examined with SEM and EDX. As microprobe measurements are not yet available, local crystalline and amorphous regions are identified by morphology; larger area XRD signatures identify the presence of crystalline forms.

Selected images are shown in Figures 3a, 3b, 3c, and 3d.

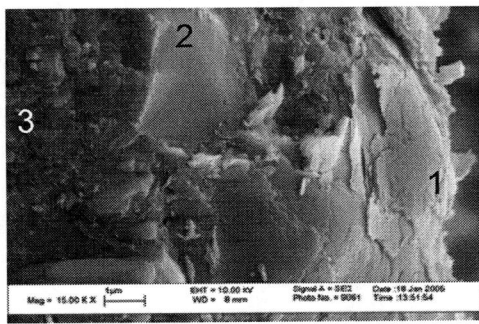

**Figure 3a.** Three forms of quartz. "1" is ballen structure, lechatelierite; "2" is crystalline quartz; "3" is unknown submicron size quartz. Scale is 1 um.

Of particular interest in Figure 3a is the ballen structure characteristic of lechatelierite, recognized as a form of fused quartz unique to meteorite impact craters. Lechatelierite is usually thought to be produced at only the highest pressures, so its presence in the sample is somewhat anomalous. In the present case, however, the conditions of formation are not the Hugoniot conditions but those of high speed hydrodynamic instabilities at high pressure. Such conditions have not previously been studied. Flat chips shown are likely goethie crystals as discussed later.

A well defined parallel linear feature recognized as a single set of PDFs is shown in Figure 3b. The spacing ranges from 250 nm to 2 um. Orientation of the PDFs cannot be identified from the SEM.

Also shown clearly in the image are acicular bundles of goethite nanocrystals characteristic of many regions of the sample. Of interest is the unidentified tetragonal block in the top center of image. The right side of the image shows what could possibly be a balen structure.

Figure 3c shows another area of mixed quartz and goethite nanocrystals with a unusual form of quartz that appears almost as a folded structure. Closer examination shows the upward linear feature terminates in a rounded void from which a smooth, emergent thin surface is seen. A thin quartz layer may possibly be coating an underlying crystal slab from a goethite bundle.

The acicular goethite shown in Figure 3d is probably its first occurrence in nanocrystalline form. Small, densely spaced crystals are characteristic of shock-synthesized crystals.

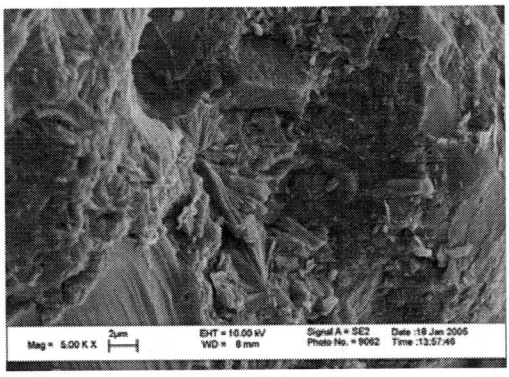

**Figure 3b.** Mixed quartz-goethite region with a set of closely spaced parallel linear features in quartz characteristic of PDFs. Scale is 2 um.

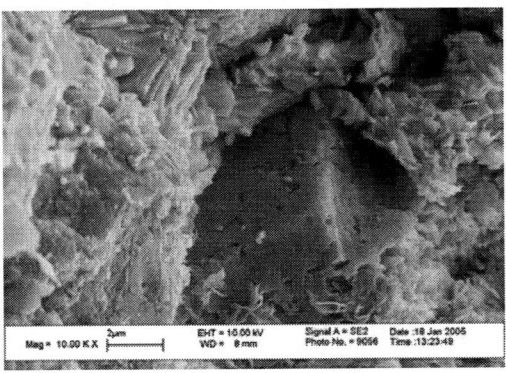

**Figure 3c.** Mixed quartz-goethite area with bundles of goethite nanocrystals and an unusual form of quartz.

**Figure 3d.** A collection of acicular bundles of goethite is well displayed in the image. Scale is 2 um.

## SPHEROIDS

Three different spheroidal configurations have been identified as originating at the impact site. Perhaps the most interesting are those found in abundance in an open region of the Uvalde Crater Rosetta Stone. A small local region of the stone with spheres in place is shown in Figure 4a.

**Figure 4a.** Uvalde Crater Rosetta Stone with numerous condensation spheres. Scale is a US mint ten cent coin, 18 mm in diameter.

The spheres have rigid shells that are red in color. Inside, they appear black to the eye and are seen to be highly porous. Red center regions are apparent.

An optical microscopic image of the alpha sphere is shown in Figure 4b; a remarkable array of features are apparent. The outer region is small grain sized hematite, likely grown after the formation period. Smooth voids, smooth quartz grain surfaces, and a possible hematite crystal are seen. Heart-shaped dark crystals, indicative of hematite, are numerous in the various samples. In the present case the shape could possibly be a coated quartz grain. The vesicular quartz is reminescent of the quartz "froth" seen by Kieffer [3]. Rounded voids, so characteristic of the present sample collection are well displayed.

A very high resolution XRD spectra of the spheroid is shown in Figure 4c. Dominant are goethite and quartz. Small amounts of hematite and calcite are shown. A trace of the rare mineral, β-tridymite, is apparent.

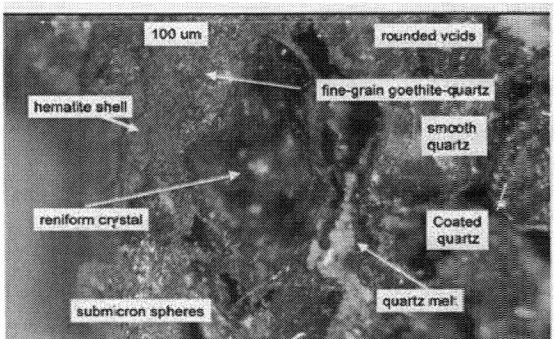

**Figure 4b.** Materials in condensation sphere

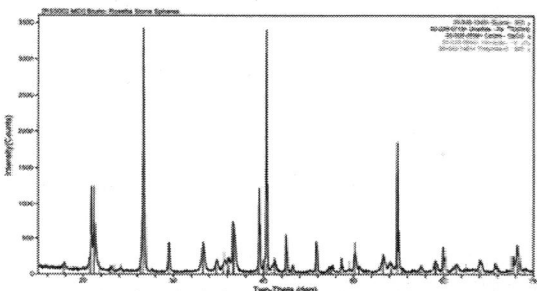

**Figure 4c.** High resolution XRD spectra of the alpha condensation sphere. Dominant quartz and goethite with hematite and calcite. Trace tridymite.

## CONCLUSION

A large collection of impactites with features caused by the unusual bottom-up shock processes are collected. A very considerable task remains to fully uncover the billion nanosecond history of the impact processes.

## ACKNOWLEDGEMENTS

The authors are indebted to Ralph Tissot, Jr. of Sandia National Laboratories for analysis of the XRD pattern and Daniel Leskovar at the Texas A&M Agricultural Research Center at Uvalde

## References

1. R. A. Graham, in press, 2005 Conference on Shock Compression of Condensed Matter–2005, M. D. Furnish, M. Elert, T. P. Russell and C. T. White (eds) AIP.
2. Wilson, W. F. and Wilson, D. , Geology, 7, 144 - 146 (1979).
3. Kieffer, S. W., Contr. Mineral Petrol. 59, 41- 93 (1976).

CP845, *Shock Compression of Condensed Matter - 2005*,
edited by M. D. Furnish, M. Elert, T. P. Russell, and C. T. White
© 2006 American Institute of Physics 0-7354-0341-4/06/$23.00

# NUMERICAL MODELING OF MIXING AND VENTING FROM EXPLOSIONS IN UNDERGROUND CHAMBERS

## Benjamin T. Liu, Ilya Lomov and Lewis A. Glenn

*Energy and Environment Directorate, Lawrence Livermore National Laboratory*
*7000 East Avenue, Livermore, CA 94550, USA*

**Abstract.** 2D and 3D numerical simulations were performed to study the dynamic interaction of explosion products in an underground concrete chamber with ambient air, barrels of water, and the surrounding walls and structure. The simulations were carried out with GEODYN, a multi-material, Godunov-based Eulerian code that employs adaptive mesh refinement and runs efficiently on massively parallel computer platforms. Tabular equations of state were used to model materials under shock loading. An appropriate constitutive model was used to describe the concrete. Interfaces between materials were either tracked with a volume-of-fluid method that used high-order reconstruction to specify the interface location and orientation, or a capturing approach was employed with the assumption of local thermal and mechanical equilibrium. A major focus of the study was to estimate the extent of water heating that could be obtained prior to venting of the chamber. Parameters investigated included the chamber layout, energy density in the chamber and the yield-to-water mass ratio. Turbulent mixing was found to be the dominant heat transfer mechanism for heating the water.

**Keywords:** Shock loading, turbulent mixing.
**PACS:** 47.40.-x, 47.40.Nm.

## INTRODUCTION

The effect of low-yield nuclear weapons in underground chambers has been a topic of increasing public debate [1]. We have undertaken fundamental studies of explosions in underground chambers to determine the extent to which chamber contents are heated by such explosions.

In this work, we focus on the potential heating of water contained in the chamber. We consider several different chamber configurations and explosive yields; a large-scale 3D calculation was run in addition to 2D parameter studies.

## PROCEDURE

Two-dimensional calculations were performed for a cylindrical chamber with a height of 4 meters and a radius of 6 meters (452 m$^3$ volume) containing 4.1 metric tons of water. The chamber was located either 0.5 meters or 6.1 meters (20 feet) below the surface. For the latter case, a 0.229m (9 inch) radius vent hole along the centerline was introduced to approximate leakage from the chamber. We simulated our energetic source by depositing energy into a sphere of iron located in the center of the chamber. The yields corresponded to either 1 kiloton or 40 tons of TNT. Most calculations had a torus of water 3 meters from the center of the chamber (off-axis); a calculation with a cylinder of water on the centerline (on-axis) was also run for comparison. In each case, the mass of the water (4.1 metric tons) corresponds to approximately twenty 55-gallon drums. A 1.5 mm iron liner around the water was used to approximate the steel drums.

A large-scale three-dimensional calculation was also run with a 60x10x10 meter rectangular chamber containing 198 stacked 1-ton barrels of water. No iron liner was used for this calculation because there was not enough refinement to resolve the liner. This source used for this calculation had a yield of 2 kilotons.

In both the 2D and 3D calculations, the material surrounding the chamber was assumed to be concrete modeled as a Mohr-Coulomb porous solid. Tabulated equations of state were used for the air, water, and iron in order to accurately determine temperatures and pressures resulting from extreme shock loadings. The source was modeled as a 50 kg sphere of iron.

The various simulations and their parameters are summarized in Table 1.

## Computational Tools

Calculations were performed using GEODYN, a Godunov-based Eulerian code with adaptive mesh refinement capabilities. This parallel code features high-order interface reconstruction algorithms and advanced thermodynamically consistent constitutive models described elsewhere [2] that incorporate many of the salient features of the dynamic response of geologic media.

Turbulent mixing was modeled by assuming instantaneous mixing between air, iron, and water in a given cell. The mixing of gases uses an effective gamma from an ideal gas approximation; this effective gamma is used to calculate the effective pressure and temperature of the gas mixture [3]. The mixing length was assumed to be

**TABLE 1.** Parameters for bomb in chamber simulations.

| CASE | | SOURCE | CHAMBER | | WATER | |
|---|---|---|---|---|---|---|
| | | yield | volume | depth | position | mass |
| A | 2D | 1 kiloton | 452 m$^3$ | 0.5 m | *off-axis, torus* | 4.1 tons |
| B | 2D | 40 tons | 452 m$^3$ | 0.5 m | *off-axis, torus* | 4.1 tons |
| C | 2D | 40 tons | 452 m$^3$ | 6.1 m | *off-axis, torus* | 4.1 tons |
| D | 2D | 40 tons | 452 m$^3$ | 0.5 m | *on-axis, cylinder* | 4.1 tons |
| E | 3D | 2 kilotons | 6000 m$^3$ | 0.5 m | *off-axis, barrels* | 198 tons |

## Heating Metrics

Heat can be transferred to the water through one of four mechanisms: conduction, shock heating, radiative transfer, and convection. Over the timescales of interest (at most 100 milliseconds), conduction should have a negligible effect. For this work, we neglected the effect of radiative transfer, noting that this may have a noticeable effect, particularly at higher yields.

In our calculations, we consider only shock heating and convective mixing. We will examine the amount of water heated to two different levels: 650 K (the critical temperature of water) and 2600 K (four times the critical temperature). We will present results in terms of the fraction of water in various temperature ranges (T≤650K, 650K<T≤2600K, and T>2600K) at any given time. While this does not capture the temperature history of a given mass of water, it provides a compact way of viewing the average heating of the water.

equal to the cell size: 8 mm in the two-dimensional calculations and 32 mm in the three-dimensional calculation.

## RESULTS AND DISCUSSION

In Fig. 1, the temperature distribution in the water for the 1-kiloton source in Case A (see Table 1) is shown.

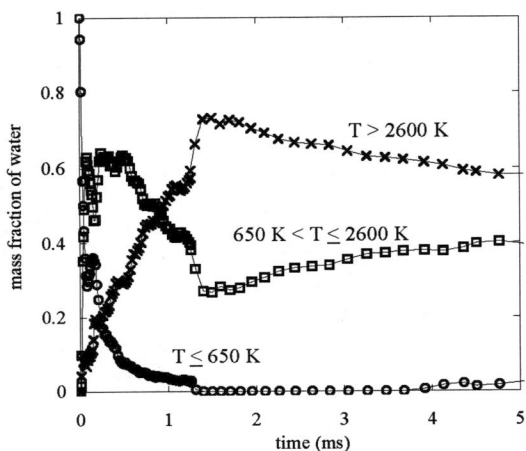

**FIGURE 1.** Temperature distribution of water for 1-kiloton source and 0.5 m depth-of-burial (Case A from Table 1). The temperature distribution remains approximately constant past 25 ms.

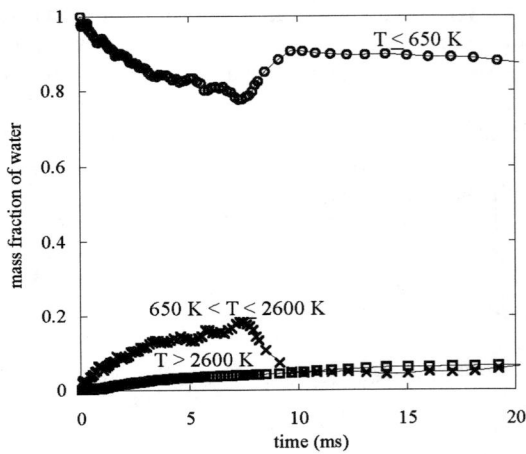

**FIGURE 2.** Temperature distribution of water for 40-ton source and 0.5 m depth-of-burial (Case B from Table 1). Over 80% of the water remains below 650 K after 40 ms.

Over the first hundred microseconds, the water is rapidly heated by shock heating. Without convective mixing, the water cools as it expands; within 5 ms, almost all the water would be below 373 K. In our case, however, the water is subsequently mixed with the hot air-explosive mixture; this further heats the water and prevents the cooling by expansion. By about one millisecond, over 95% of the water has been heated above 650 K; about half of the water is above 2600 K. The temperature distribution remains approximately the same after 25 ms. Moreover, the remaining 5% of the water stays between 373 K and 650 K and remains within the chamber up to 10 milliseconds, well after the roof has come off.

Fig. 2 shows the analogous simulation for a 40-ton source (Case B). As expected, far less of the water is heated to either 650 K or 2600 K. The majority of the water (almost 80%) is never heated above 650 K.

## Venting of Chamber Gases

In measuring the heating of the chamber contents, the time when the contents are vented to the atmosphere is often of great interest. For the 1-kiloton case, the chamber vents in about one millisecond; for the 40-ton case, venting occurs in less than ten milliseconds. In these cases, the water may still be heated after venting as it mixes with the hot gases outside the chamber. In any case, the venting time and the temperature distribution at that time gives some indication of the effectiveness of the heating within the chamber. Table 2 shows the venting time for each case as well as the temperature distribution of the water at the venting time and at 10 ms and 20 ms. In Case C (6.1 m depth-of-burial), hot gas from the explosion escapes very quickly through the vent hole; here we define the venting time as the time when the water first mixes with air initially outside the chamber.

**TABLE 2.** Venting time and temperature distributions at various times. The venting time for the chamber with a vent hole (denoted with an *) represents the time at which the water first mixes with the atmosphere.

| CASE | time | VENTING | | | 10 ms | | | 20 ms | | |
|------|------|---------|---------|---------|---------|---------|--------|---------|---------|---------|
| | | ≤650 K | > 650K ≤2600K | >2600 K | ≤650K | > 650K ≤2600K | >2600K | ≤650K | > 650K ≤2600K | >2600K |
| A | 1 ms | 3% | 44% | 53% | 5% | 39% | 56% | 4% | 36% | 60% |
| B | 7 ms | 78% | 18% | 4% | 90% | 5% | 5% | 87% | 6% | 7% |
| C | 4 ms* | 81% | 16% | 3% | 53% | 37% | 10% | *Simulation only run to 17.8 ms* | | |
| D | 4 ms | 37% | 48% | 15% | 46% | 32% | 22% | 54% | 24% | 22% |
| E | 2 ms | 78% | 17% | 5% | *Simulation only run to 3.7 ms* | | | | | |

## Effect of Source and Chamber Configurations

The yield of the explosive obviously has a strong effect on the temperature distribution in the water, as can be clearly seen by comparing Cases A and B in Figs. 1 and 2, respectively. More important is the ratio of explosive yield to water in the chamber (W/m). Case A has a W/m ratio of about 250 tons/ton, while Cases B, C, D, and E have W/m ratios of about 10 tons/ton. The higher W/m ratio is enough to heat most of the water above the critical point, while the lower value is insufficient to heat all the water above the critical point.

Increased confinement of the explosive gases allows better mixing of the water with the hot gases, resulting in more water heating. This can be seen comparing the temperature distributions in Table 2 for Cases B and C, which differ only by depth-of-burial. Note that the venting time may not necessarily be increased for a deeper chamber if leaks or existing vents are present in the chamber.

## Effect of Water Storage Configuration

The location of the water relative to the explosive can have a significant effect on heating of the water. Table 2 shows that when the water is located on-axis (Case D), it experiences significantly more heating than when it is located off-axis (Case B), even though the time to venting is decreased. When the water is closer to the source, it is more thoroughly mixed with the hot explosive gases, resulting in better heating of the water. The specific configuration of the chamber, as well as the accuracy with which the explosive is placed, can greatly influence the effectiveness of the heating.

## CONCLUSIONS

This work has examined explosions in shallowly buried chambers and the mixing, heating, and venting of water contained in such chambers. The yield of the explosive relative to the mass of water has the most important effect on the heating of the water. A 1-kiloton explosive with ~4 tons of water (W/m = 250 tons/ton) would heat most of the water above the critical point. With a smaller relative yield (W/m = 10 tons/ton), far less of the water is heated and the specific configuration, including chamber depth-of-burial and the location of the water, becomes more important.

## ACKNOWLEDGEMENTS

This work was performed under the auspices of the U.S. Department of Energy by the University of California, Lawrence Livermore National Laboratory under contract No. W-7405-Eng-48.

## REFERENCES

1. Nelson, R. W., "Nuclear chamber busters, mini-nukes, and the US nuclear stockpile," Physics Today, November 2003.
2. Lomov, I. and Rubin, M. B., "Numerical simulation of damage using an elastic-viscoplastic model with directional tensile failure," Journal de physique iv, vol. 110, pp. 281-286, 2003.
3. Lomov, I. and Liu, B.T., "Approximation of multifluid mixture response for simulation of sharp and diffuse material interfaces on an Eulerian grid," these proceedings.

CP845, *Shock Compression of Condensed Matter - 2005*,
edited by M. D. Furnish, M. Elert, T. P. Russell, and C. T. White

# SHOCK DEMAGNETIZATION OF PYRRHOTITE (Fe$_{1-x}$S, x≤0.13) AND IMPLICATIONS FOR THE MARTIAN CRUST AND METEORITES

## K. L. Louzada[1], S. T. Stewart[1], and B. P. Weiss[2]

[1]*Department of Earth and Planetary Sciences, Harvard University, 20 Oxford Street, Cambridge, MA 02138,*
[2]*Department of Earth, Atmospheric, and Planetary Sciences, Massachusetts Institute of Technology, 54-724,*
*77 Massachusetts Avenue, Cambridge, MA 02138*

**Abstract.** After cessation of the dynamo on Mars, giant impact events should have demagnetized large regions of the crust. Models of the decay of shock pressure with distance indicate that the demagnetized zones are bound by peak shock pressures between 1 and 3 GPa. We performed the first planar shock recovery experiments at these pressures on natural pyrrhotite, a magnetic mineral found in Martian meteorites. Post-shock magnetic measurements show that pyrrhotite demagnetizes significantly (~85-90%) when subject to shock pressures between 1 and 4 GPa. Permanent changes to the magnetic properties of recovered samples include an increase in the saturation remanence and the mean destructive field, indicating that shocks harden the coercivity. We conclude that pyrrhotite is a candidate carrier for the magnetization in the Martian crust and that pyrrhotite in meteorites shocked to modest pressures may retain a pre-shock remanence.

**Keywords:** Iron sulfide, pyrrhotite, shock demagnetization, meteorites, rock magnetism, Mars.
**PACS:** 91.60.Pn, 96.30.Gc, 62.50.+p, 91.25.Ng

## INTRODUCTION

Satellite maps of the remanent magnetic field of Mars show unmagnetized zones within and around giant impact basins, such as Hellas and Argyre [1]. The edges of the unmagnetized zones correspond with peak shock pressures of a few GPa and temperatures well below the Curie point of candidate magnetic minerals [2-4]. Hence, it is likely that vast regions of the Martian crust were demagnetized due to a shock-induced phase change or magnetic transition in the magnetic minerals.

Although pyrrhotite (Fe$_{1-x}$S, x≤0.13) is not a major magnetic carrier on Earth, it is a common phase in the Martian shergottite meteorites [5]. In hydrostatic pressure experiments, pyrrhotite undergoes a ferrimagnetic to paramagnetic transition near ~2.8 GPa, with rapid loss of magnetization above 1 GPa [6]. Previous shock experiments on magnetite-bearing igneous rocks [7-11], hematite powders [12], and pure samples of magnetite, hematite and titanohematite [4] indicate that low pressure shocks demagnetize low coercivity minerals. Previous pyrrhotite shock Hugoniot measurements did not include a magnetic study [13].

Understanding the effects of shock waves on magnetic minerals is necessary to interpret the demagnetized zones around impact basins, to constrain the identity of the major magnetic carrier phases in the crust, and to infer the origin of magnetic directions and paleointensities from meteorites. In this paper, we present preliminary results from the first shock demagnetization study of pyrrhotite.

## EXPERIMENTAL PROCEDURE

We performed planar shock recovery experiments on 3×1 mm discs of natural pyrrhotite embedded 3-mm off-center in 80×24 mm aluminum recovery capsules using the 40-mm gas gun in the Harvard Shock Compression Laboratory. Planar shockwaves were generated by 34×3 mm diameter aluminum flyer plates on polycarbonate sabots. From the measured impact velocity, the peak shock pressure was inferred from both the impedance match solution and the pressure distribution in the sample from 2D simulations using the shock physics code CTH. Approximately 93% of the sample experienced a peak pressure within 0.5 GPa of the impedance match solution with the remaining fraction subject to slightly higher pressures.

The pyrrhotite samples were saturated in a 370 mT (~7400×Earth's surface) magnetic field prior to the shock and the resultant remanence was measured before and after shock. Four shock experiments were performed at room temperature in the ambient laboratory field (~0.2 mT). One experiment on a demagnetized sample confirmed no shock remanent magnetization was acquired. To assess the changes in crystallographic and magnetic properties, the shock experiments were preceded and followed by a suite of material and magnetic characterization measurements. These included magnetic isothermal and anhysteretic remanence acquisition, alternating field demagnetization, X-ray diffraction, magnetic hysteresis, and low temperature magnetism.

Pyrrhotite owes its magnetism to preferential vacancy distributions in alternating antiferromagnetically coupled Fe layers in so-called superstructures [14]. Natural pyrrhotites typically consist of mixtures of superstructures of ferrimagnetic monoclinic ($Fe_7S_8$) and antiferromagnetic hexagonal pyrrhotite. We analyzed a pyrrhotite nodule from Sudbury, Canada. The wasp-waistedness of its hysteresis loop [15] indicates that the sample contains both high and low coercivity fractions (coercivity is the magnetic field required to reduce the external magnetization of a magnetic substance to zero). The presence of monoclinic pyrrhotite was confirmed by the low temperature magnetic transition at 30-34 K [16] and the presence of hexagonal pyrrhotite was inferred from XRD and microprobe measurements (Fe/S=0.893). We infer that the sample is composed of crystals in sizes predominantly in the single domain range (saturation remanence to saturation magnetization ratios prior to shock of $M_{rs}/M_s$~0.55-0.72). The density of the pyrrhotite was 4.587 (±100) g cm$^{-3}$, and the longitudinal and shear wave speeds are 4399 (=87) m s$^{-1}$ and 2873 (±37) m s$^{-1}$ respectively.

## RESULTS

Fig. 1 presents preliminary results indicating that pyrrhotite demagnetized by 85-90% when subjected to shock pressures of a few GPa. It is likely that the shocks in our experiments were elastic or near the elastic limit. Although the Hugoniot Elastic Limit of pyrrhotite is unknown, a major shock-induced phase change is observed between 2.7 and 3.8 GPa [13]. Pressure (grey symbols in Fig. 2) is inferred from principal stress (black symbols) by:

$$P = \frac{(\sigma_1 + 2\sigma_2)}{3} = \frac{1}{3}\left(\frac{1+\nu}{1-\nu}\right)\sigma_1 \qquad (1)$$

where $P$ is pressure, $\nu$ is the Poisson's ratio (0.32 for pyrrhotite), $\sigma_1$ is the principal stress, and the perpendicular stresses are $\sigma_2 = \sigma_3$.

Two samples shocked to $\sigma_1$~2.5 GPa show good agreement with the previously published hydrostatic data (open circles). A sample shocked to $\sigma_1$~4 GPa, above the 2.8 GPa magnetic transition and just above the expected structural change, did not completely demagnetize. The phase diagram of pyrrhotite is not well known; however, troilite (FeS) undergoes a first order phase transition at 3.9 GPa [17]. The 4 GPa principal stress shock (2.7 GPa pressure) may not have reached the expected high-pressure phase. It has been suggested that a single mechanical shock of very short duration may be unable to attain the final resultant effect on the remanent magnetization [9]. Single shock pressures above 4 GPa may be needed to fully demagnetize pyrrhotite. A sample shocked twice ($\sigma_1$~1-1.5 GPa) was significantly more demagnetized than what would be expected from the hydrostatic experiments [6], indicating the efficiency of shock demagnetization from multiple impact events.

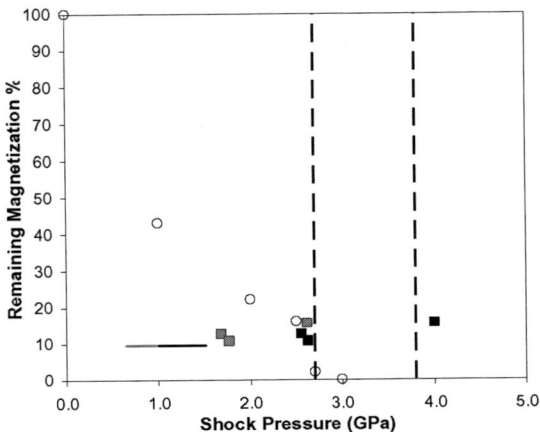

**FIGURE 1.** Demagnetization results of pyrrhotite: black squares – single shock principal stress; black line – double shock principal stress; grey symbols – pressure assuming elastic shock [Eq. 1]; open circles – static measurements [6]; dashed lines – phase change region in pyrrhotite from shock data [13].

Shock compression results in permanent changes to the magnetic properties of pyrrhotite. Isothermal remanent magnetization measurements demonstrate that the saturation magnetization (SIRM) of pyrrhotite increases when subject to increasing shock pressure (Fig. 2).

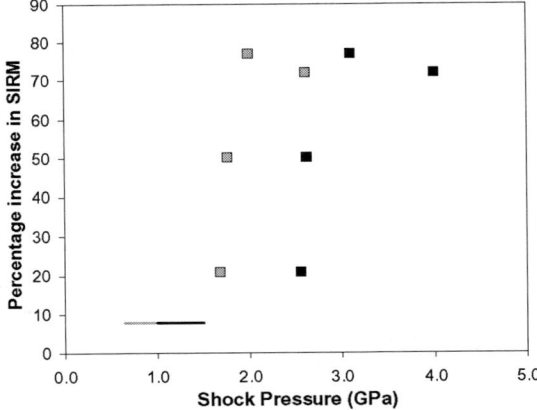

**FIGURE 2.** Change in saturation isothermal remanent magnetization: meaning of symbols is the same as Fig. 1.

Even more striking is the increase in the mean destructive field (MDF, the field that is required to reduce the remanence to one-half its initial value) with pressure (Fig. 3). Since the MDF is a measure

of the bulk coercivity, the data show that shock treatment significantly hardens the coercivity.

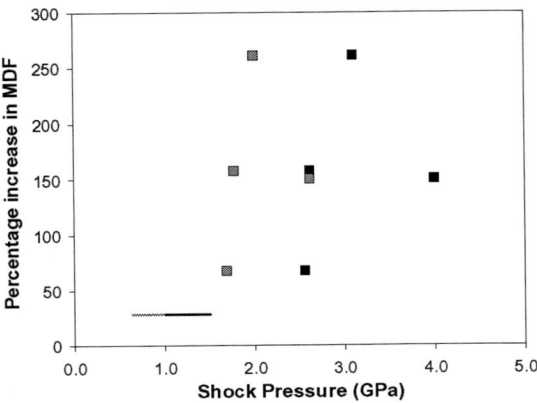

**FIGURE 3.** Change in mean destructive field: meaning of symbols is the same as Fig. 1.

## DISCUSSION

Similar irreversible changes in magnetic properties have been observed in magnetite under hydrostatic pressures up to 6 GPa [18] and in hematite powder subjected to shocks between 8-27 GPa [12]. We have several hypotheses which could explain these changes in pyrrhotite.

The break up of large, low-coercivity, pseudo-single domain and multidomain grains into many smaller single domain grains [18] should result in an increase in the bulk coercivity and saturation magnetization [14], which is consistent with changes in MDF and SIRM. However, we would also expect to see an increasing trend in the $M_{rs}/M_s$ with pressure [15], which is not supported by our results.

The creation of hexagonal ferrimagnetic pyrrhotite that is metastably ferrimagnetic [5], which can occur if pyrrhotite is heated above ~200°C and then rapidly cooled, could explain the increase in saturation magnetization. However, this is unlikely as shock heating during the experiments was negligible: the temperature increase was ~10 °C at $\sigma_1=4$ GPa.

Stress hardening may be the result of changes in the magnetostriction and magnetoelastic constants [9,18], which would increase the single domain-multidomain threshold radius and increase the saturation remanence. Defect generation, residual strain, domain nucleation, and domain rotation may each contribute to the changes in magnetic properties after shock compression of pyrrhotite.

## CONCLUSIONS

Impact experiments indicate that pyrrhotite demagnetizes significantly due to shock in the pressure range inferred around Martian impact basins. After shock treatment, permanent changes in the magnetic properties of pyrrhotite include an increase in saturation remanence and coercivity.

The possible presence of pyrrhotite in the Martian crust has implications for the thickness and depth of the magnetized layers and the oxidation state of the crust. Meteorites containing pyrrhotite, that have been shocked to pressures of up to 4 GPa may retain a pre-shock remanence and may be used for paleointensity measurements of the ancient Martian field. However, the increase in saturation remanence from shock implies that typical normalization paleointensity techniques [19] may underestimate the true paleointensity.

## ACKNOWLEDGEMENTS

Thanks to G. Kennedy, D. Lange, W. Croft and C. Francis (Harvard), C. Ross, F. Ilievski and Fangcheng C. (MIT), J. Kirschvink, R. Kopp and I. Hilburn (Caltech), S. Bogue (Occidental College), M. Dekkers (Utrecht U.). This research is supported by NASA (#NNG04GD17G). We are grateful for the support of the TGSCCM 2005.

## REFERENCES

1   Acuña, M.H. et al., (1999) Global Distribution of Crustal Magnetization Discovered by the Mars Global Surveyor MAG/ER Experiment. Science **284**, 790.

2   Hood, L.L. et al., (2003) Distribution of crustal magnetic fields on Mars: Shock effects of basin-forming impacts. Geophys. Res. Lett. **30**, 1281.

3   Mohit, P.S. and Arkani-Hamed, J., (2004) Impact demagnetization of the martian crust. Icarus **168**, 305.

4   Kletetschka, G. et al., (2004) Pressure effects on martian crustal magnetization near large impact basins. Meteoritics and Planetary Science **39**, 1839.

5   Rochette, P. et al., (2001) Pyrrhotite and the remanent magnetization of SNC meteorites: a changing perspective of Martian magnetism. Earth Planet. Sci. Lett. **190**, 1.

6   Rochette, P. et al., (2003) High pressure magnetic transition in pyrrhotite and impact demagnetization on Mars. Geophys. Res. Lett. **30**, doi:10.1029/2003GL017359.

7   Cisowski, S.M. and Fuller, M., (1978) The effect of shock on the magnetism of terrestrial rocks. J. Geophys. Res. **83**, 3441.

8   Hargraves, R.B. and Perkins, W.E., (1969) Investigations of the Effect of Shock on Natural Remanent Magnetization. J. Geophys. Res. **74**, 2576.

9   Nagata, T., (1971) Introductory Notes on Shock Remanent Magnetization and Shock Demagnetization of Igneous Rocks. Pure and Ap. Phys. **89**, 159.

10  Pohl, J. et al., (1975) Shock Magnetization and Demagnetization of Basalt by Transient Stress up to 10 kbar. Journal of Geophysics **41**, 23.

11  Gattacceca, J. et al., (2005) Investigating impact demagnetization through laser impacts and SQUID microscopy. Geology **in press**.

12  Williamson, D.L. et al., (1986) Morin transition of shock-modified hematite. Phys. Rev. B **34**, 1899.

13  Ahrens, T.J., (1979) Equations of State of Iron Sulfide and Constraints on the Sulfur Content of the Earth. J. Geophys. Res. **84**, 985.

14  Dekkers, M.J., (1988) Magnetic properties of natural pyrrhotite Part I: Behaviour of initial susceptibility and saturation-magnetization-related rock-magnetic parameters in a grain-size dependent framework. Phys. Earth and Plan. Int. **52**, 376.

15  Tauxe, L. et al., (2002) Physical interpretation of hysteresis loops: Micromagnetic modeling of fine particle magnetite. Geochemistry Geophysics Geosystems **3**, doi:10.1029/2001GC000241.

16  Rochette, P. et al., (1990) Magnetic transition at 30-34 Kelvin in pyrrhotite: insight into a widespread occurrence of this mineral in rocks. Earth Planet. Sci. Lett. **98**, 319.

17  Fei, Y. et al., (1995) Structure and Density of FeS at High Pressure and High Temperature and the Internal Structure of Mars. Science **268**, 1892.

18  Gilder, S.A. et al., (2004) Magnetic properties of single and multi-domain magnetite under pressures from 0 to 6 GPa. Geophys. Res. Lett. **31**, doi:10.1029/2004GL019844.

19  Gattacceca, J. and Rochette, P., (2004) Toward a robust normalized magnetic paleointensity method applied to meteorites. Earth Planet. Sci. Lett. **227**, 377.

CP845, *Shock Compression of Condensed Matter - 2005*,
edited by M. D. Furnish, M. Elert, T. P. Russell, and C. T. White
© 2006 American Institute of Physics 0-7354-0341-4/06/$23.00

# MEASUREMENTS OF SOUND VELOCITY OF LASER-IRRADIATED IRON FOILS RELEVANT TO EARTH CORE CONDITION

**K. Shigemori[1], D. Ichinose[1], T. Irifune[2], K. Otani[1], T. Shiota[1], T. Sakaiya[1], H. Azechi[1]**

[1] *Institute of Laser Engineering, Osaka University, Suita, Osaka, 565-0871, Japan*
[2] *Geodynamics Research Center, Ehime University, Matsuyama, Ehime, 790-8577, Japan*

**Abstract.** We have developed a novel method to measure sound velocity of laser-irradiated iron foils by side-on x-ray radiograph technique. Iron foils were irradiated with two-stepped laser pulse to reach the earth's core condition. We obtained not only the sound velocity but also temperature, pressure, shock velocity, compressibility, and particle velocity of the laser-irradiated iron. The experimental results are in good agreements with previous experimental results and with one-dimensional simulation results.

**Keywords:** sound velocity, x-ray radiograph, optical pyrometer, earth core
**PACS:** 52.35.Dm, 52.70. La, 91.35.Ed

## INTRODUCTION

The core of the earth is almost iron [1], and the center of the earth is solid iron with a fraction of light elements, at the pressure of ~ 360 GPa, and the temperature of ~ 6000 K [1]. However, recent investigations suggest that there is significant uncertainty on the earth core temperature [2-4], which is likely to be associated with components of the earth core material. To date, many seismological data have been applied to model the earth's core condition. In addition to the seismological data analysis, high-pressure experiments using gas guns have been carried out in order to reproduce the physical conditions relevant to the earth's core [5-8]. We have started a series of experiments to explore the earth core condition with intense laser. As a first step, we focused on measurements of sound velocity because the sound velocity is one of the most important parameters which can be directly compared with the seismological data. We have developed a novel technique on measurements of

the sound velocity of laser-irradiated iron foils with side-on x-ray radiograph. We experimentally obtained shocked iron density, pressure, shock velocity, and particle velocity with the side-on x-ray radiograph. The experimental results show good agreements with one-dimensional simulation calculations.

## EXPERIMENT

The experiments were done on GEKKO-XII/HIPER laser facility at ILE, Osaka University. Schematic view of the main diagnostics for the sound velocity measurements is shown in Fig. 1. We irradiated three layered targets with the HIPER laser. The HIPER laser has three foot pulse beams ($\lambda$: 0.53 $\mu$m, 2.3 ns, PCL) and nine main drive beams ($\lambda$: 0.35 $\mu$m, 2.7 ns, SSD). We made 4-ns foot pulse and 8-ns main drive pulse by stacking the beams with time delay. The typical intensities of the foot pulse and the main pulse were $1 \times 10^{12}$ W/cm$^2$ and $2 \times 10^{13}$ W/cm$^2$, respectively.

The target was flat iron (Fe) foil of 38-μm thickness. In order to prevent from preheating prior to the shock compression, 5-μm-thickness polystyrene (CH) ablator and 1-μm-thickness gold (Au) insulator were coated on the laser-irradiated surface.

The sound velocity was measured with side-on x-ray shadowgraph technique. This technique had been broadly employed on hydrodynamic instability experiments and equation of state experiments, and so on. A titanium (Ti) foil for backlighting source was placed on the side of the iron target. The Ti foil was irradiated another laser (λ: 0.53 μm, 5 ns). The Ti foil generated 4.8 keV x-ray for the backlight. Transmitted x-ray through the iron target was imaged by a slit (10-μm width) with on to a photocathode of an x-ray streak camera with magnification of 50. The temporal resolution of the x-ray streak camera was 50 ps.

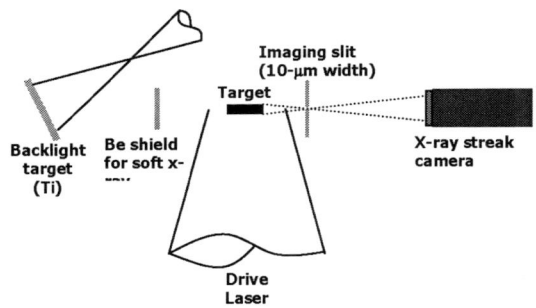

Fig. 1 Schematic view of side-on x-ray radiograph measurement.

The method to obtain the sound velocity of laser-irradiated foils is very simple. Typical flow diagram of the laser-irradiated foil is shown in Fig. 1. When a foil is irradiated with intense laser, shock front is generated on the foil surface then the shock front propagates into the foil. Once the shock front reaches the rear surface, the rarefaction front starts to move back to the laser-irradiated surface with the sound velocity. When the rarefaction front reaches the laser-irradiated surface, the foil begins to accelerate. From the flow diagram, the sound velocity of the shock-compressed region $c_s$ is $\Delta d/\Delta t$, where $\Delta d$ is the shock-compressed thickness of the foil, and $\Delta t$ is the duration between the shock breakout and the rarefaction breakout.

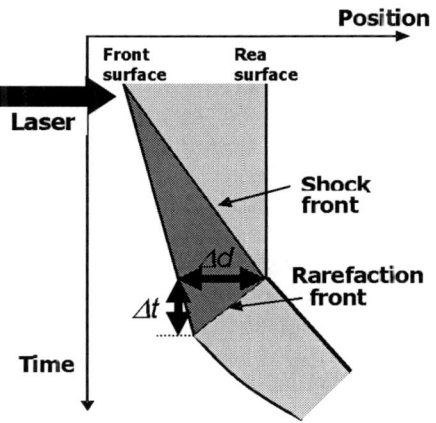

Fig.2 Typical flow diagram of laser-irradiated foil. Sound speed is obtained from the time differences between shock breakout and rarefaction breakout.

In addition to the measurements with the x-ray radiograph, we measured the shocked temperature with an optical pyrometer [9]. Since our drive condition was double compression, it is required to experimentally obtain the temperature of the shocked iron because the physical parameters of the shocked region were off Hugoniot. In our measurements with the pyrometer, optical emission from the rear surface of the laser-irradiate targets was imaged and focused into an optical fiber which was connected to an optical spectrograph. The optical spectrograph was coupled to an optical spectrometer. From time-resolved blackbody spectra from the pyrometer, we obtained the temperature just after the shock breakout at the rear surface.

We also employed a velocity interferometer system for any reflector (VISAR)[10] for measurements of important shock parameters, such as shock velocity and the rear surface velocity. In the measurements with the VISAR, we employed a wedged foil target (0-50-μm thickness) for a cross check of the shock velocity with the x-ray backlighting measurements.

**EXPERIMENTAL RESULTS**

Figure 3(a) shows an example of raw streaked image with the side-on x-ray radiograph. It is seen two timings of the shock breakout and the rarefaction breakout from the raw data. In our ex-

experiment, the width of the target foil was 200 μm. The measurement of the side-on x-ray radiograph was just perpendicular to the drive laser beams, but there was an uncertainty of accuracy on the rotation of the target. Since the measurements of the breakout timings are from the leading edge of the target, we need to modify the target thickness taking into account the initial target thickness and the magnification of the diagnostics. Although the contrast of the radiograph was not fine, we could track the edge of the contact surface because the iron is very opaque to the backlight x-ray ($h\nu$. 4.7 keV).

Figure 3 (b) is the analyzed results of the laser-irradiated surface and the rear surface for two data shots. The rear surface does not move before the shock front reaches the rear surface. From the shock breakout time, the shock velocity for the main drive pulse was ~ 11 km/s. After the shock front reaches the rear surface, the rear surface starts to move at the rarefaction-tail velocity, which is fitted with a line. We could obtain the rear surface velocity after the shock breakout from the rear surface trajectory. We determined the shock breakout timing from the crossing point from two lines. However, there was tilting of the irradiated foil in the experiments as described above, there is significant scattering of the data of the rear surface position because the density of the rarefaction tail was very low compared with the shocked region.

On the other hand, the laser-irradiated surface moves in a constant velocity, which corresponds to the particle velocity because the drive laser pulse was nearly flat-topped. After the rarefaction front reaches the laser-irradiated surface, the laser-irradiated surface starts to accelerate. The rarefaction breakout timing is determined at the crossing point of the two different velocities. From the shock-compressed foil thickness and the two breakout timings, we obtained the sound velocity for two data shots. The pressure of the compressed iron is also obtained from the compressed density. Experimental compressibility of the iron in this experiment was ~ 1.6, the shocked gamma $\gamma$ was ~ 4.5, and the density of the shocked iron $\rho_s$ was ~ 12.5 g/cm$^3$. The pressure $P$ was deduced form the sound velocity, $\gamma$, and the density to be $P = c_s^2 \gamma \rho$, the pressure in this experimental condition was 220 - 270 GPa.

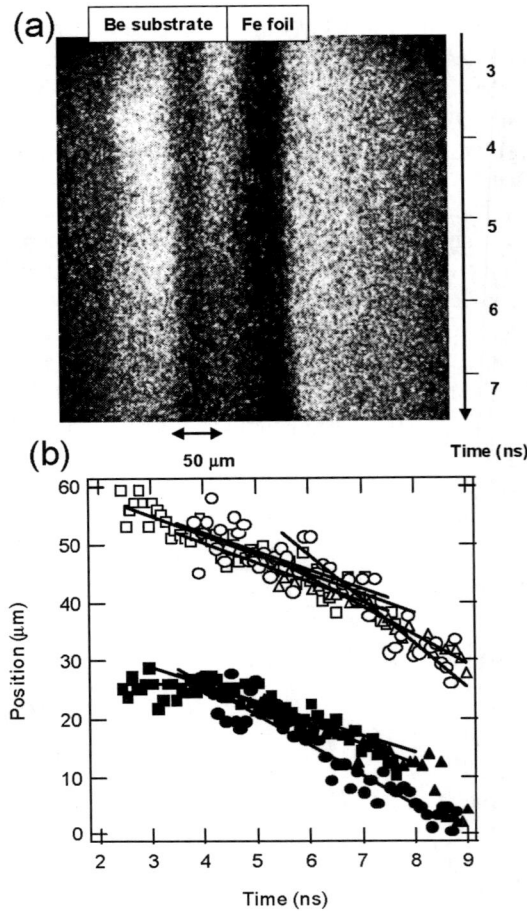

Fig. 3 Raw streaked image of the x-ray radiograph. (b) Temporal plot of laser-irradiated surface (open symbols) and rear surface (closed symbols). Shock breakout and rarefaction breakout was determined by fitting of the trajectories.

The shocked temperature was separately measured with the optical pyrometer. The measured spectrum was analyzed in order to obtain blackbody temperature with a measured spectral sensitivity We took ratios of two different wavelengths from the data, and fitted to the Planck function. The measured temperature for our experimental condition was 6250 ± 500 K, which is very close to the earth's inner core temperature. Additional data by the VISAR with a wedged foil was taken to check the shock velocity and the shock steadiness (not shown here). The thickness of the foil was 0 – 50 μm in 500 μm length. From the VISAR image, the shock velocity was ~ 11 km/s, and the rear surface velocity was ~ 10 km/s. Those values are in reasonable agreements with the x-ray radiograph results described above.

## DISCUSSION

The measured sound velocity was plotted in Fig. 4. Also plotted in Fig. 4 are from previous gas-gun experiments [5, 6]. Previous experimental results, using gas gun with single shock, suggest that there is discontinuity of the sound velocity at around 225 - 250 GPa. The interpretation of the discontinuity was there would be solid-solid phase transition [5] or solid-liquid transition [6]. From our experiment, the measured sound velocity was very close to the previous gas-gun data, which is on the liquid Hugoniot sound velocity.

In this experiment, we measured the sound velocity at the shocked region, on the basis of an assumption that the pressure and the density at the shocked-iron region were absolutely "uniform". However, our one-dimensional simulation suggests that the pressure of the shocked region was not uniform (from 220 to 350 GPa). This is mainly due to the pulse shape of the main pulse is not flat with time, the pressure and the density at the vicinity of the shock front is relatively high from the simulation results. This means that the measured sound velocity and the pressure, etc, were "averaged" values of the shocked iron region. Thus, more careful pulse shaping is required for acquisition of the data.

## SUMMARY

We have carried out first experiments on measurements of sound velocity of iron foils relevant to earth's inner core condition. The measured sound velocity and other related physical parameters were obtained with reasonable accuracy. In our experimental condition, the measured sound velocity suggests that the iron is melted even though we employed double shock drive. We are going to improve the drive condition, and measure the sound velocity in solid-state iron of the earth's inner core condition.

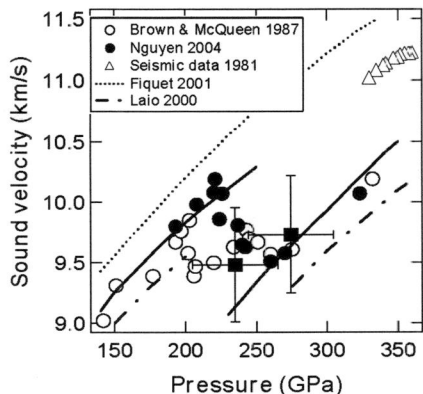

Fig. 4 Measured sound velocities from previous experiments [5,6] (circles) and this work (squares), and proposed sound velocity curves.

## REFERENCES

1. Birch, F., J. Geophys. Res., 1952, Vol. 57, pp. 227.
2. T. J. Ahrens, K. G. Holland, and G. Q. Chen, Geophys. Res. Lett. 29, 1150 (2002).
3. Yoo, C. S., Holmes, N. C., Ross, M., Webb, D. J., and Pike, C., "Shock Temperature and Melting of Iron at Earth Core Conditions", Phys. Rev. Lett., 1993, Vol. 70, pp. 3931-3934.
4. Williams, K., Jeanloz, R., Svendsen, B., Ahrens, T. J., "The Melting Point of Iron to 250 Gigapascals: A Constraint on the Temperature at Earth's Center", Science, 1987, Vol. 236, pp. 181
5. Brown, J. M., and McQueen, R. G., "Phase Transition, Grüneisen parameter, and elasticity for shocked iron between 77 GPA and 400 GPa", J. Geophys. Res., 1986, Vol. 91, pp.7485-7494.
6. Nguyen, J. H., and Holmes, N. C., "Melting of iron at the physical condition of the earth core ", Nature, 2004, Vol. 427, pp. 339-342.
7. McQueen, R. J., Hopson, J. W., and Fritz, J. N., "Optical technique for determining rarefaction wave velocity at very high pressures", Rev. Sci. Instrum., 1982, Vol. 53, pp. 245-250.
8. Fiquet, G., Badro, J., Guyot, F., Requardt, H., and Krisch, M., "Sound Velocities in Iron to 110 Gigapascal", Sceince, 2001, Vol. 291, pp. 468-471.
9. Shigemori, K., et al., submitted to Jap. J. Appl. Phys., 2005.
10. Barker, L. M. and Hollenbach, R. E., "Laser interferometer for measuring high velocities of any reflecting surface", Appl. Phys., 1972, Vol. 43, pp.4669-4675.

CP845, *Shock Compression of Condensed Matter - 2005*,
edited by M. D. Furnish, M. Elert, T. P. Russell, and C. T. White
© 2006 American Institute of Physics 0-7354-0341-4/06/$23.00

# POST-SHOCK TEMPERATURE AND FREE SURFACE VELOCITY MEASUREMENTS OF BASALT

## S. T. Stewart[1], G. B. Kennedy[1], L. E. Senft[1], M. R. Furlanetto[2], A. W. Obst[2], J. R. Payton[2], and A. Seifter[2]

[1]*Department of Earth and Planetary Sciences, Harvard University, 20 Oxford Street, Cambridge MA 02138*
[2]*University of California, Los Alamos National Laboratory, Physics Division, P-23, Los Alamos, NM 87545*

**Abstract.** Basalt is the most common rock type on planetary surfaces. Post-shock temperature and particle velocity measurements constrain the equation of state of basalt and provide fundamental information about the outcome of planetary impact events. A high-speed, infrared, four-wavelength pyrometer, developed at Los Alamos National Laboratory (LANL), is used with customized front end optics at the Harvard Shock Compression Laboratory for concurrent observations of particle velocity and free surface thermal emission. In an experiment on Columbia River basalt released from a peak shock pressure of 28.9±0.2 GPa, the apparent post-shock temperature is wavelength dependent. The 3.5 and 4.8-μm channels record apparent temperatures between 605 and 630 K, using an emissivity range of 0.7-1.0. The 1.8 and 2.3-μm channels record apparent temperatures of ~700 K and ~800 K, respectively. The pyrometry data are well fit by a two component temperature distribution: (1) a predominantly 565-610 K free surface, in good agreement with the 570 K predicted by the basalt EOS in the shock physics code CTH, and (2) a small area fraction of 1700-2000 K hot spots. The model is in good agreement with inferred basaltic meteorite hot spot temperatures; however, the hot spot model is not unique. Free surface velocity measurements are slower than predicted by CTH, indicating a steeper release path than in the model equation of state.

**Keywords:** Post-shock temperature, basalt, equations of state, pyrometry, VISAR, CTH
**PACS:** 62.50.+p, 91.60.Fe, 07.20.Ka, 83.80.Nb

## INTRODUCTION

Shock temperature data are necessary to investigate a wide range of problems in the earth and planetary sciences, including phase changes during impact events, the temperature field surrounding fresh impact craters, and the thermal history of meteorites. The post-shock temperatures of rocks and minerals provide strong constraints on their equations of state. Multi-wavelength data also furnish insights into the heterogeneities of shock processes in natural materials.

Most previous shock temperature studies have utilized optical pyrometry to observe high (typically, >1500 K) peak shock temperatures in transparent minerals and liquids or high post-shock temperatures in metals and minerals. Infrared pyrometry allows investigation of shock processes at lower pressures. Furthermore, in many natural materials, hysteretic release isentropes significantly complicate the calculation of post-shock temperatures, as demonstrated by the first study of low post-shock temperatures in minerals [1].

We have begun a new study of post-shock temperatures (in the range ~400-1500 K) in metals [2], rocks, and minerals. Here we present preliminary results from simultaneous measurements of particle velocity and near-infrared free surface emission from shocked basalt.

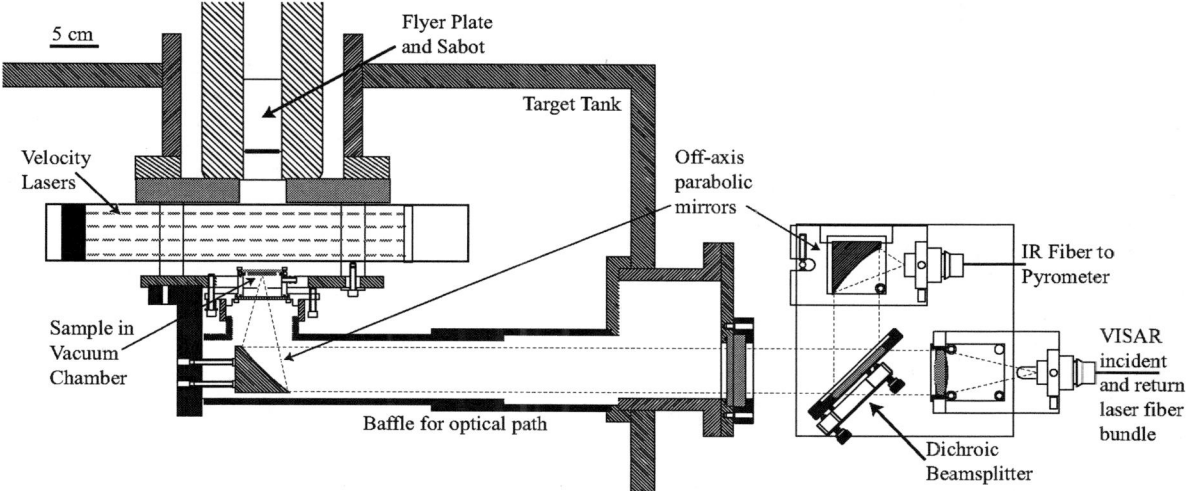

**FIGURE 1.** Plan view of experimental configuration for simultaneous pyrometer and VISAR measurements on the Harvard 40-mm gun. An off-axis parabolic mirror collects and collimates radiance emitted from a ~4-mm diameter spot on the downrange face of the sample, as well as focusing the incident and collecting the reflected laser light for the VISAR. The collimated light is split with a >1200 nm reflecting dichroic beamsplitter between the IR pyrometer and the visible (532 nm) VISAR. The optical path is enclosed in light-tight tubing to shield from propellant gases during the experiment.

## EXPERIMENTAL PROCEDURE

Planar shock waves with peak pressures up to ~35 GPa are generated in basalt using the 40-mm single stage powder gun in the Harvard Shock Compression Laboratory [3]. A Ø34×3 mm molybdenum flyer plate impacts a 2-mm thick aluminum-2024 driver plate, which also seals a small aluminum vacuum chamber containing the basalt specimen (Fig. 1). The sample vacuum chamber is purged with He gas and evacuated to <0.5 millitorr, eliminating any background emission from gas in the chamber. A 1-cm diameter aperture was placed behind the basalt sample to shield the pyrometer from any emission from the edges of the driver plate and from reflections off the walls of the vacuum chamber.

Concurrent particle velocity and radiance measurements from the same location on the downrange, free surface of the basalt are achieved using a shared optical system [see description in 2]. Free surface particle velocity is measured with a VALYN VISAR. The radiance is delivered to a high-speed, infrared, four-wavelength (1.8, 2.3, 3.5, 4.8 μm) pyrometer with InSb detectors via a 1-mm diameter chalcogenide C2 infrared fiber, and

detector voltage is recorded on 12-bit 100 MHz digitizers [4]. The pyrometer is sensitive to radiance temperatures as low as 400 K and has a temporal resolution of ~17 ns. The system is calibrated with a Mikron M360 black body source that was observed with the same optics and fibers as used in the experiments. The calibrated radiation temperature is converted to an apparent temperature for each channel using the emissivity of the sample. If the apparent temperature of all four channels overlap, as in the case of our validation experiments on aluminum 2024 [2], then the apparent temperature is considered to be the true temperature of a homogeneous surface.

The Columbia River flood basalt (CRB) specimens, from Snake River Valley near Clarkston, WA ($\langle\rho\rangle$=2.83±0.10 g/cm3, $\langle V_P\rangle$= 5.73±0.28 km/s, $\langle V_S\rangle$=3.46±0.04 km/s), are cored and cut from hand samples into nominally Ø34×2 mm discs. The basalt and driver plate are lapped plane parallel with 15 micron diamond grit and hand polished to an optical (~100 nm) finish using 58-nm alumina powder polish. The basalt is affixed to the driver plate using a ~10 μm layer of Loctite 326 epoxy. No macroscopic pores intersect the free surface. The observed area, a ~4-mm diameter spot for the pyrometer and ~1-mm diameter spot for the VISAR, contains microscopic pores that are typically several 10s μm across.

## RESULTS AND DISCUSSION

The apparent temperature of basalt released from a peak shock pressure of 28.9±0.2 GPa (impact velocity of 2.24±0.01 km/s) is shown in Fig. 2. By applying an emissivity range between 0.7 and 1.0 for basalt [5], we determine the range of apparent temperatures, indicated by the thickness of the data traces. There is a small temperature excursion at the time of shock breakout, which is seen in previous post-shock temperature measurements on metals and minerals [1, 2, 6]. The two longest wavelength channels are in good agreement, with a constant temperature following the breakout. The two shortest wavelength channels record significantly higher temperatures that increase with time.

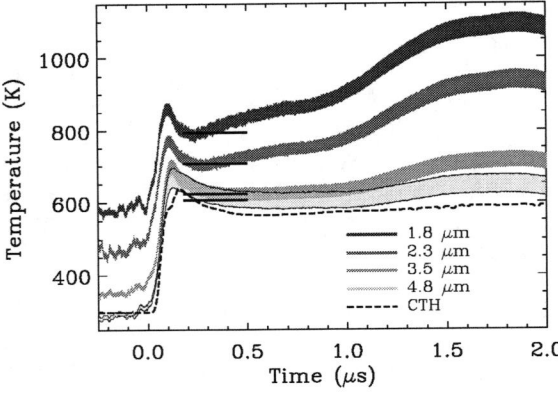

**FIGURE 2.** Multi-wavelength post-shock temperature measurements on basalt released from a peak shock pressure of 28.9 GPa, assuming an emissivity range of 0.7-1.0. Horizontal bars indicate the apparent temperature at each detector wavelength for a two-component temperature surface model described in the text. Inferred 565-610 K surface temperatures are in good agreement with free surface temperatures from a simulation of the experiment using the CTH basalt EOS.

The high apparent temperatures in the 1.8 and 2.3 μm channels cannot be reconciled by varying the emissivity as a function of wavelength. We interpret the discrepancy between the shorter and longer wavelengths to be evidence for hot spots in

the field of view. We fit a simple two-component model to the data, where the radiance, $L$, at each wavelength, $\lambda$, is described by an apparent temperature, $T'$, given by Planck's Law,

$$L_\lambda(T') = \varepsilon' \frac{2hc^2}{\lambda^5(e^{hc/(\lambda kT')} - 1)}.$$

Here, $\varepsilon$ is emissivity, $h$ is Planck's constant, $c$ is the speed of light, and $k$ is Boltzmann's constant. If a surface with temperature, $T_S$, has a small area fraction, $\alpha$, of hot spots with temperature, $T_{HS}$, then the radiance is the sum of two components,

$$L_\lambda(T') = (1-\alpha)\varepsilon_S \frac{2hc^2}{\lambda^5(e^{hc/(\lambda kT_S)} - 1)}$$

$$+\alpha\varepsilon_{HS} \frac{2hc^2}{\lambda^5(e^{hc/(\lambda kT_{HS})} - 1)}.$$

Hence, the apparent temperature is defined by

$$T' = \frac{hc}{\lambda k \ln\left[\dfrac{1}{\dfrac{1-\alpha}{e^{hc/(\lambda kT_S)}-1} + \dfrac{\alpha}{e^{hc/(\lambda kT_{HS})}-1}} + 1\right]},$$

assuming the emissivities are equal. The long wavelength channels constrain the surface temperature between 565 and 610 K. The data are well fit by a ~590 K surface with hot spots in the temperature range of 1700 to 2000 K occupying an area fraction of 0.45 to 0.25%, respectively. The model apparent temperature at each wavelength is shown by the horizontal bars in Fig. 2.

To produce 1700-2000 K hot spots, some material must reach the melting point. The 28.9 GPa experiment is well below the pressure required for bulk melting for basalt, and the results are consistent with localized melting in shear bands or fractures [7] or at grain boundaries between different impedance minerals [8]. The inferred hot spot temperatures are in excellent agreement with petrographic studies of localized melting in basaltic meteorites from Mars shocked to similar pressures [8]. However, the measured surface temperature of CRB is about 100 K higher than the bulk rock post-shock temperature inferred for the same meteorites.

The two-component model probably underestimates the area fraction of hot spots. The model fit is dominated by the hottest spots, and a hot spot temperature distribution with the inferred peak temperature could also satisfy the data.

The free surface particle velocity provides information about the pressure-volume release path from the peak shock state. The free surface particle velocities, for basalt subject to peak shock pressures between 2.0 and 32.0 GPa, are shown in Fig. 3. Each experiment is modeled with the CTH shock physics code using the standard equation of state SESAME tables for the molybdenum flyer, aluminum 2024 driver, and basalt. The results are compared to the observed temperatures and particle velocities (Figs. 2-3, dashed lines). In general, the observed free surface particle velocities are lower than the calculations, indicating that the release path is steeper compared to the model EOS.

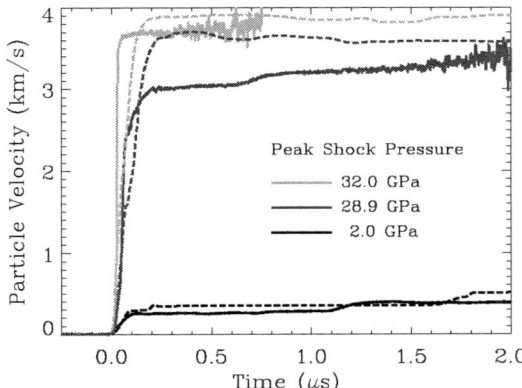

**FIGURE 3.** Measured (solid) and modeled (dashed) free surface velocities of basalt shocked to peak pressures of 2.0, 28.9 and 32.0 GPa.

## CONCLUSIONS

New shock experiments on basalt provide fundamental information required to improve EOS models. Here we present the first post-shock temperature data on basalt, which provides insight into the distribution of temperatures following impact events. Our data quantify the heterogeneity that arises from pre-existing pores and fractures and/or the effect of mixing materials with different shock impedances. New EOS models will aid in calculations of post-impact conditions on Mars [9], the thermal history of meteorites [10], and conditions for melting subsurface ice [11]. Heterogeneous shock pressures and temperatures are observed around terrestrial craters and in meteorites. Understanding the effect of heterogeneities on the temperature distribution

following impact events will aid in the interpretation of meteorites, terrestrial field studies, and simulating impact events on Mars.

## ACKNOWLEDGEMENTS

Funding was provided by NASA grant #NNG04GD17G.

## REFERENCES

1. Raikes, S. A. and T. J. Ahrens, "Post-shock temperature in minerals," *Geophys. J. R. Astr. Soc.* **58**, 717-747 (1979).
2. Seifter, A., et al., "Post-shock Temperature Measurements of Aluminum," in *Shock Compression of Condensed Matter -- 2005*, edited by M. Furnish, AIP, 2005, pp. TBD.
3. Stewart, S. T., "The Shock Compression Laboratory at Harvard: A New Facility for Planetary Impact Processes," *Proc. Lunar & Planet. Sci. Conf.* **XXXV**, Abs. 1290 (2004).
4. Boboridis, K., A. Seifter, and A. W. Obst, "High-Speed infrared pyrometry for surface temperature measurements on shocked solids," *VDI-Bericht* **1784**, 119-126 (2003).
5. Burgi, P.-Y., M. Caillet, and S. Haefeli, "Field temperature measurements at Erta'Ale Lava Lake, Ethiopia," *Bull. Volcanol.* **64**, 472-485 (2002).
6. Seifter, A., et al., "Low-Temperature Measurements on Shock-Loaded Tin," *26th Intl. Congr. High-Speed Photo. & Photonics* **5580**, 93-105 (2004).
7. Miller, P. J., C. S. Coffey, and V. F. DeVost, "Heating in crystalline solids due to rapid deformation," *J. Appl. Phys.* **59**, 913-916 (1986).
8. Stoffler, D., et al., "Shock metamorphism and petrography of the Shergotty achondrite," *Geochim. Cosmochim. Acta* **50**, 889-903 (1986).
9. Pierazzo, E., N. A. Artemieva, and B. A. Ivanov, "Starting conditions for hydrothermal systems underneat Martian craters: Hydrocode modeling," *Proc. Lunar & Planet. Sci. Conf.* **XXXV**, 1352 (2004).
10. Artemieva, N. and B. Ivanov, "Launch of martian meteorites in oblique impacts," *Icarus* **171**, 84-101 (2004).
11. Stewart, S. T., J. D. O'Keefe, and T. J. Ahrens, "Impact processing and redistribution of near-surface water on Mars," in *Shock Compression of Condensed Matter -- 2003*, edited by M.D. Furnish, et al., American Institute of Physics, Melville, NY, 2004, pp. 1484-1487.

CP845, *Shock Compression of Condensed Matter - 2005,*
edited by M. D. Furnish, M. Elert, T. P. Russell, and C. T. White
© 2006 American Institute of Physics 0-7354-0341-4/06/$23.00

# HUGONIOT PROPERTIES OF DRY YORKSHIRE SANDSTONE UP TO 8 GPA

**E. A. Taylor[1], K. Tsembelis[2], D. Chapman[2], W. G. Proud[2], and C. S. Cockell[3]**

[1]*Department of Physics and Astronomy, CEPSAR, Open University, Milton Keynes MK7 6AA, U.K.*
[2]*Physics and Chemistry of Solids, Cavendish Laboratory, Madingley Road, Cambridge CB3 0HE, U.K.*
[3]*Planetary and Space Science Research Institute, CEPSAR, Open University, Milton Keynes MK7 6AA, U.K.*

**Abstract.** A series of plate impact experiments has been performed to assess the dynamic behaviour of dry Yorkshire sandstone up to 8 GPa. Standard manganin gauges were inserted between samples in order to determine the principal Hugoniot curve. A VISAR system was used to measure the free surface velocity of the target. This work is in part of a research programme to understand the survivability of microbial life under impact and the creation of new habitats for microbial life as a function of shock processing of sandstone.

**Keywords:** Sandstone, Hugoniot, Plate Impact, Microbial Life.
**PACS:** 62.50.+p, 61.43.Gt, 07.35.+k

## INTRODUCTION

There is interest in understanding the shock properties of sandstone because of recent research evaluating the conditions for the development of microbial colonies of life in impact structures kilometres under terrestrial impact craters. Our understanding of present-day microbiology of impact structures is limited although effects such as pooling of water in post-impact structures, or hydrothermal systems generated by impact heating may encourage microbial life. There is also little understanding of about how impact heating and shock alter the suitability of rocks as sites for microbial colonization, although qualitative observations indicate that changes in fracturing and porosity may influence their suitability for colonisation. For example, shock processing of rocks at the Haughton impact structure (impact shocked gneiss) in the Canadian High Arctic have made them more colonisable by cyanobacteria [1]. An example is shown in figure 1.

The impact processes in materials such as gneiss and sandstone are currently not well understood. In crystalline rocks, where pore space is low anyway, impact bulking may cause an increase in pore spaces as fractures are formed. By contrast, in sedimentary target lithologies, the existing large pore spaces, for example in sandstones, can be reduced as a result of shock wave propagation through the target [2, 3].

To date, only limited studies to determine the sandstone Hugoniot have been carried out [4, 5, 6]. This paper presents the results of plate impact experiments to characterise the dynamic behaviour of dry Yorkshire sandstone. We are seeking to understand both the process by which the shocked habitats may be created and the survivability of microbial life under shock pressures as generated by impact.

## EXPERIMENTAL PROCEDURE

The impact experiments were carried out in the plate impact gun facility at the University of Cambridge [7]. The Cambridge facility consists of a single stage 50-mm bore and 5 m length, light gas gun. The gun is capable of achieving velocities up to 1200 m s$^{-1}$. Impact velocities were measured using a sequential pin-shorting method and tilt was fixed to be less than 1 mrad by means of an adjustable specimen mount. Longitudinal stress measurements were taken by embedding

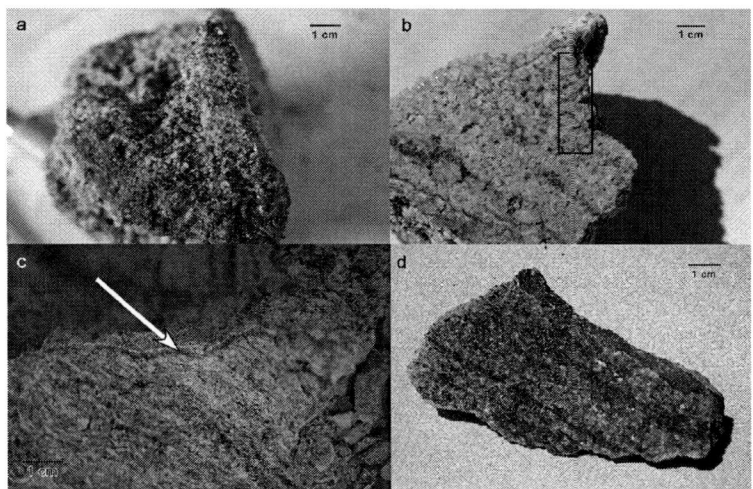

**Figure 1**. Examples of shocked and unshocked gneiss. Arrow indicates cyanobacteria living in shocked gneiss.

**TABLE 1.** Experimental test programme details

| Shot No. | Impact Velocity (m/s) | Target thickness T1 (mm) | Target thickness T2 (mm) | Target thickness T3 (mm) |
|---|---|---|---|---|
| 1-HDYSS | 252 | 2.90 | 2.58 | 2.78 |
| 2-HDYSS | 403 | 2.70 | 3.06 | 2.73 |
| 3-HDYSS | 555 | 3.41 | 3.36 | 3.45 |
| 4-HDYSS | 721 | 3.46 | 3.52 | 3.44 |
| 5-HDYSS | 869 | 3.44 | 3.34 | 3.38 |

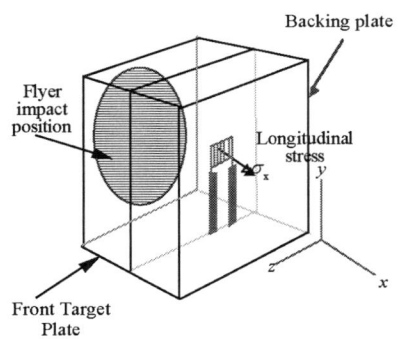

**Figure 2**. Target configuration

piezoresistive manganin gauges (Micro-measurements type LM SS-210FD-50) between tiles of Yorkshire sandstone using a low viscosity epoxy adhesive (figure 2). Gauge calibration was according to the earlier research of Rosenberg and Partom [8]. All the experiments were performed using a 15 mm thick Copper flyer impacting a sandstone sample of 50 mm diameter and varying total thicknesses (see table 1). Gauges were inserted between adjacent samples. Samples were then assembled using a low viscosity epoxy with a curing time of approximately 24 hours. The manganin gauges were used as time of arrival indicators in order to estimate the Lagrangian shock wave velocity. Whenever a precursor wave structure was observed it was taken into account when estimating the wave velocity.

## MATERIAL DESCRIPTION

Density and ultrasonic measurements were performed in Cambridge. The samples are constituted of air-dry Yorkshire sandstone. The density was measured to be $2240 \pm 10$ kg m$^{-3}$. The longitudinal elastic wave speed, determined with ultrasonic transducers, was $2.87 \pm 0.01$ mm $\mu s^{-1}$.

## RESULTS AND DISCUSSION

Table 1 summarises the impact velocities and target thicknesses. Table 2 summarises particle velocities as estimated using the impedance matching technique and time of flight as obtained from the gauge data (time of flight calculated from

mid-point of time of arrival of signals at the manganin gauges).

Figures 3 and 4 illustrate the data from experiments 2-HDYSS and 3-HDYSS. It can be seen that both manganin gauges reach the same level, indicating a steady wave.

Figures 5 and 6 illustrate the Hugoniot curve of the material in stress-particle velocity and shock wave velocity-particle velocity spaces.

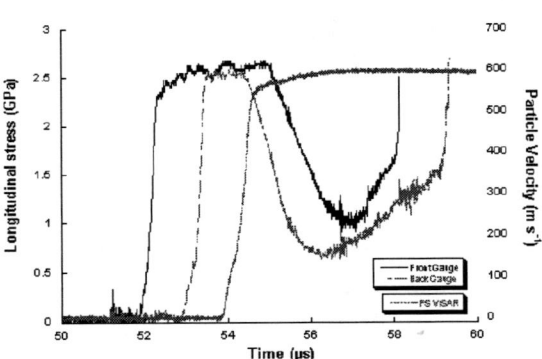

**Figure 3.** Stress Wave Profiles and VISAR trace for experiment 2-HDYSS

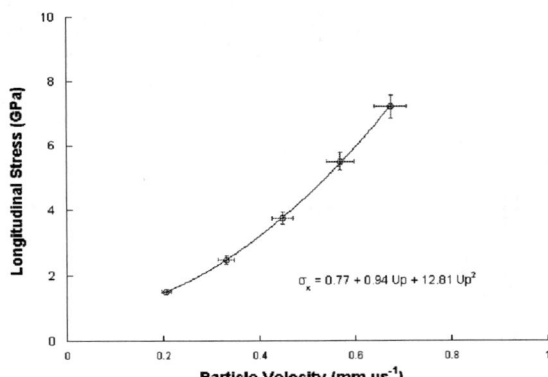

**Figure 5.** Yorkshire Sandstone Hugoniot curve in $\sigma$-$u_p$ space

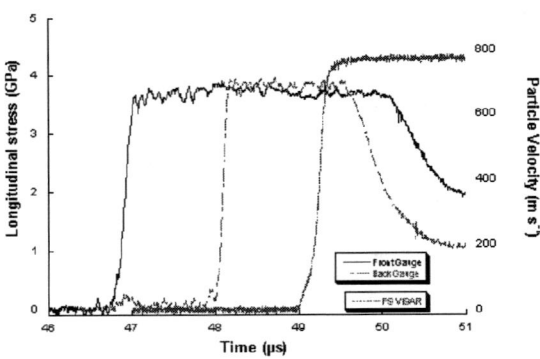

**Figure 4.** Stress Wave Profiles and VISAR trace for experiment 3-HDYSS

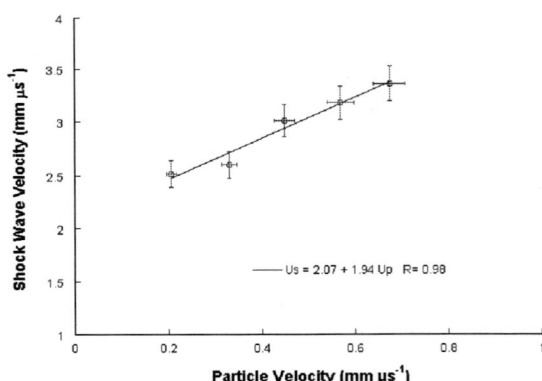

**Figure 6.** Yorkshire Sandstone Hugoniot curve in $U_s$-$u_p$ space

**TABLE 2.** Experimental Parameters and Hugoniot Data for Dry Yorkshire Sandstone

| Shot No. | Impact Velocity (m s$^{-1}$) | Stress (GPa) | Particle Velocity (km s$^{-1}$) (determined from impedance matching) | Mean Shock Velocity (km s$^{-1}$) (from mid-point of time of arrival of signals at manganin gauges) |
|---|---|---|---|---|
| 1-HDYSS | 252 | 1.52 | 0.21 | 2.45 |
| 2-HDYSS | 403 | 2.48 | 0.33 | 2.51 |
| 3-HDYSS | 555 | 3.74 | 0.45 | 3.01 |
| 4-HDYSS | 721 | 5.50 | 0.57 | 3.18 |
| 5-HDYSS | 869 | 7.20 | 0.67 | 3.36 |

## CONCLUSIONS

The principal Hugoniot of the dry Yorkshire sandstone has been established. It can be seen that it follows a typical linear relation up to the stresses examined.

## ACKNOWLEDGEMENTS

Funding was provided by Cambridge University and the Open University's Centre for Earth Planetary Space and Astronomical Research. The authors thank Ray Flaxman for providing technical support.

## REFERENCES

1. Cockell, C.S., Lee, P., Osinski, G., Horneck, G, and Broady, P., "Impact-induced microbial endolithic habitats", Meteoritics and Planetary Sciences, vol. 37, pp. 1287-1298, 2002.
2. Kieffer, S.W., Phakey, D. P., and Christie, J. M., "Shock processes in porous quartzite: transmission electron microscope observations and theory", Contrib. Mineral. Petrol., vol. 59, pp. 41-93, 1976.
3. Kieffer, S.W., and Simonds, C. H., "The role of volatiles and lithology in the impact cratering process", Reviews of Geophys. and Space Phys, vol. 18, no. 1, pp 143-181. 1980.
4. Ahrens, T. J., and Gregson, J. V. G., "Shock compression of crustal rocks: data for quartz, calcite and plagioclase rocks", J. Geophys. Res., vol. 69, pp. 4839-4874, 1964.
5. Shipman, F. H., Gregson, J. V. G., and Jones, A. H., "A shock wave study of Coconino sandstone", NASA Report, MSL-7-14, 46, 1970.
6. Lomov, I. N., Hiltl, M., Vorobiev, O. Yu., and Glenn, L. A., "Dynamics behaviour of Berea sandstone for dry and water-saturated conditions", Int. J. Impact Engng., vol. 26, pp.465-474, 2001.
7. Bourne, N. K., Rosenberg, Z., Johnson, D. J., Field, J. E., Timbs, A. E., and Flaxman, R. P., "Design and construction of the UK plate impact facility", Meas. Sci. Technol. vol. 6, pp. 1462-1470, 1995.
8. Rosenberg, Z, Yaziv, D, and Partom, Y.J., "Calibration of foil-like manganin gauges in planar shock wave experiments", J. Appl. Phys., vol. 51, pp. 3702-3705, 1980.

CP845, *Shock Compression of Condensed Matter - 2005*,
edited by M. D. Furnish, M. Elert, T. P. Russell, and C. T. White
© 2006 American Institute of Physics 0-7354-0341-4/06/$23.00

# EXPERIMENTAL STUDY OF TRANSITION OF JUPITER AND SATURN ATMOSPHERE TO CONDUCTING STATE

## V. Ya. Ternovoi[1], S. V. Kvitov, D. N. Nikolaev, A. A. Pyalling, A. S. Filimonov and V. E. Fortov

[1](ternovoi@ficp.ac.ru), *ESM Dept., Institute of Problems of Chemical Physics RAS, Chernogolovka Moscow region 142432 Russia*

**Abstract.** The modified equation of state of helium-hydrogen mixtures was used for 1D hydrodynamic simulation of performed experiments the multiple shock compression of Jupiter and Saturn model atmospheres. That permitted us to obtain the isentropic compression at third and later steps of compression in pressure region 20 -150 GPa. It was shown, that the helium-hydrogen mixtures become conductive due to appearance of hydrogen conductance. The intervals of pressure - temperature - density states of these transitions are 30-50 GPa - 4400-5000 K - 0.4-0.5 $g/cm^3$ for Jupiter and 60-80 GPa - 4600-4800 K - 0.53-0.6 $g/cm^3$ for Saturn in accordance with our new and previous experiments with pure hydrogen and gas mixtures.

**Keywords:** Jupiter, Saturn, helium-hydrogen mixture, conductivity.
**PACS:** 51.50. + v, 71.30. + h, 72.60. + g.

## INTRODUCTION

The Galileo probe data obtained for the structure and composition of the outer layers of Jupiter's atmosphere up to 1 MPa and 200 K [1] corroborated the applicability of the adiabatic approximation [2] to the description of the pressure dependence of its temperature for P > 0.1 MPa. The measured helium fraction in the atmosphere proved to be close to its fraction inside the Sun, $Y = m_{He}/(m_{He}+m_H) = 0.234\pm0.005$. This data set to take $Y = 0.16$ for Saturn atmosphere [3]. The current models of the giant planet atmosphere are based on the assumption that the hydrogen transition to the conducting state occurs at 150 [4] and 100 GPa [2].

The dynamic experiments on strong single and multiple compression of the initially liquid and gaseous hydrogen [5–7] show that its transition to the conducting state occurs at 40–140 GPa for a density of 0.4–0.7 $g/cm^3$ and testify to the strong influence of temperature on this process. Experimental study of the condition for pressure induced helium ionization under multiple shock compression [8] shows that helium undergoes transition to the conducting state at a density above 0.7 $g/cm^3$ at temperatures 15–40 kK realized in these experiments. This circumstance enables one to assume that, with allowance for the additional helium-induced heating of the mixture, compared to pure hydrogen the behavior of the Jupiter's atmosphere at lower densities and temperatures is primarily determined by hydrogen. Calculations [6,8] show that multiple shock compression is isentropic within the experimental accuracy when the third and the following waves pass through the compressed layer. In this case, at most two discrete matter states are realized in the compressed layer, and only a single state exists at times of reflection from the layer boundaries. By the action of two passing waves, the initial mixture can be taken to a state on the giant planet isentrope, after which the behavior diagnostics becomes possible for higher discrete compression parameters. In this paper, we report the experimental results on the conduction transition of hydrogen–helium mixtures with Y = 0.245±0.015 and Y = 0.16 ± 0.01 under multiple shock compression up to megabar pressure in the plane geometry.

Figure 1 shows the P – T diagrams obtained by the thermodynamic simulation of the behavior of Jupiter's and Saturn's atmosphere according our calculation and due to [2,5] for Jupiter, helium, and hydrogen isentropes starting from Jupiter atmosphere state at 0.1 MPa, together with the calculations of state trajectories for multiple shock compression (M.c.) in the performed experiments.

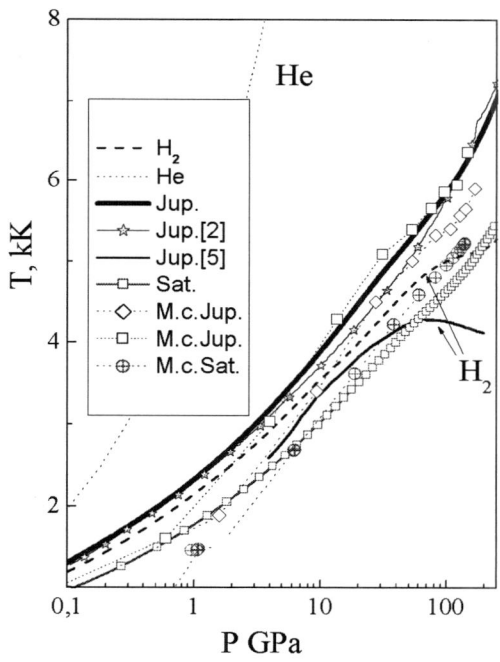

**Figure 1.** T-P diagram.

## EXPERIMENTAL PROCEDURE

The experimental procedure for simultaneous detection of the compression parameters and resistance of the compressed layer was similar to that of [8]. To initiate the compression process of Saturn model atmosphere, 1.5-mm-thick 321-H-steel plates 30 mm in diameter was accelerated to 5.6 km/s by the detonation products of cylindrical explosive charges in the process of face launching. To accelerate a 1.5-mm-thick steel plate to a velocity of 6.2 km/s, a 4.5-mm-thick steel plate 40 mm in diameter was accelerated to 4.7 km/s by the detonation products and a 6-mm-thick PMMA layer was placed ahead of the thin steel plate and this gun was used to investigate model Jupiter atmosphere. The initial temperature of the mixture was controlled by a platinum resistance thermometer and was equal, within an accuracy of 2 K, to the temperature of liquid nitrogen used for cooling. Based on the preliminary one-dimensional Wilkins hydrocode simulation of the experiment, the initial pressures were taken to be 7 and 8.1 MPa correspondingly. Resistance was measured by the schemes with two measuring electrodes [8]. Lower limit of measured conductivity level was 0.5 1/Ω/cm.

The gas-dynamic simulation was carried out using the equation of state constructed for the mixtures and described in [9].

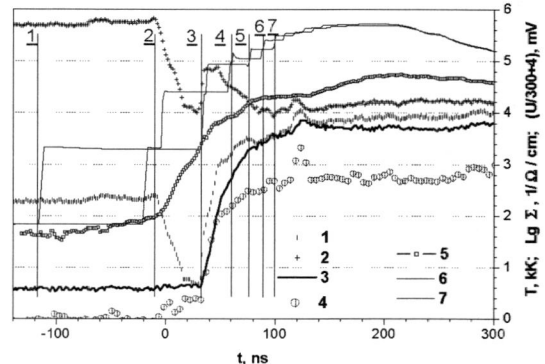

**Figure 2.** Experimental records and simulations for the experiment with model Jupiter atmosphere: (*1, 2*) voltage at the electrodes, (*3*) the resulting voltage at the sample, (*4*) recalculation of the mixture conductance (right axis), (*5*) measured temperature (right axis), and calculation of the mixture temperature in the (*6*) first and (*7*) last cells.

## RESULTS AND DISCUSSION

In Fig. 2, 3 experimental snapshots of the performed experiments with helium-hydrogen mixtures are shown with the results of gas-dynamic simulation of the process of compression.

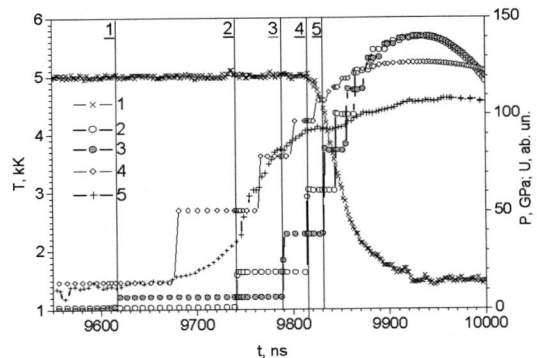

**FIGURE 3.** Experimental records and simulations for the experiment with model Saturn atmosphere: (1) the resulting voltage at the sample, (2,3) calculation of the mixture pressure in the first (2) and last (3) cells, (right axis), (5) measured temperature (right axis), and (4) calculation of the mixture temperature in the center cell of the layer.

The measured temperature is compared with the calculations of the temperatures of generated states by a one-dimensional fluid-dynamic code using the equation of state for the mixture. We took 200 cells for the compressed layer. The shown time dependence of temperature for the 10th and 190th points of the compressed layer (Fig. 2) shows a good agreement between the observed and calculated times of shock wave reflection from the

layer boundaries. Measured and calculated temperatures after first three (Fig. 2) – four (Fig. 3) shock waves are coincide within experimental accuracy; calculated temperature becomes higher than measured after the moment of the appearance of conductance. It should be mentioned that time interval of final steps of compression in the region of the conductivity appearance in the experiment is longer than that in 1D hydro code simulation of compression process without taking into account plasma phase transition. The calculated parameters of states behind the shock waves are presented in Table 1. Using three electrodes to measure the resistance of the compressed layer, we could determine the final conductance of the mixture at high pressures and find the boundary of atmosphere transition to conducting state using only electrical measurements.

The pressure and temperature values of the appearance of conductance in model atmospheres are marked by unfilled rectangles in P-T diagram of hydrogen [6,8] (Fig. 4). Calculated isentropes of Jupiter and Saturn are marked signs J1 and S1. Boundary of the plasma phase transition of hydrogen with its critical point is labeled by dash-dot line with asterisk (a – [10], b – [11], c – [12], d – [13]); signs $R_l$ and $R_g$ label states of multiple shock compression of initially liquid and gaseous hydrogen samples [6].

Even after the third shock wave, the detected conductance was equal to 0.41 1/$\Omega$/cm (P = 26.5 GPa, $\rho$ = 0.365 g/cm$^3$, T = 4380 K, Fig. 2). After the passage of the fourth shock wave, the conductance increases by two orders of magnitude to 55 1/$\Omega$/cm (P = 51.7 GPa, $\rho$ = 0.494 g/cm$^3$, T = 4940 K). Taking into account the inductance effect increases this value threefold. Further compression to density of about 0.8 g/cm$^3$ only doubles the conductance.

**TABLE 1.** Calculated parameters of multiple shock compression of Jupiter (J) and Saturn (S) atmospheres.

| No. of step | 0 | 1 | 2 | 3 | 4 | 5 | 6 | 7 | 12 |
|---|---|---|---|---|---|---|---|---|---|
| $\rho$, g/cm$^3$ (J) | 0.0291 | 0.110 | 0.23 | 0.365 | 0.494 | 0.603 | 0.68 | 0.737 | 0.831 |
| P, GPa (J) | 0.0081 | 1.57 | 9.03 | 26.5 | 51.7 | 80 | 104.4 | 124.6 | 164 |
| T, K (J) | 77.4 | 1845 | 3320 | 4380 | 4940 | 5020 | 5205 | 5380 | 5690 |
| L, mm | 4.938 | | | 0.394 | 0.291 | 0.238 | 0.211 | 0.195 | 0.173 |
| $\rho$, g/cm$^3$ (S) | 0.0253 | 0.096 | 0.198 | 0.314 | 0.426 | 0.525 | 0.600 | 0.660 | 0.770 |
| P, GPa (S) | 0.007 | 1.07 | 6.22 | 18.8 | 38.2 | 60.9 | 81.7 | 100 | 140 |
| T, K (S) | 77.4 | 1470 | 2690 | 3620 | 4230 | 4580 | 4800 | 4960 | 5230 |
| L, mm | 4.906 | 1.293 | 0.627 | 0.395 | 0.291 | 0.236 | 0.194 | 0.188 | 0.161 |

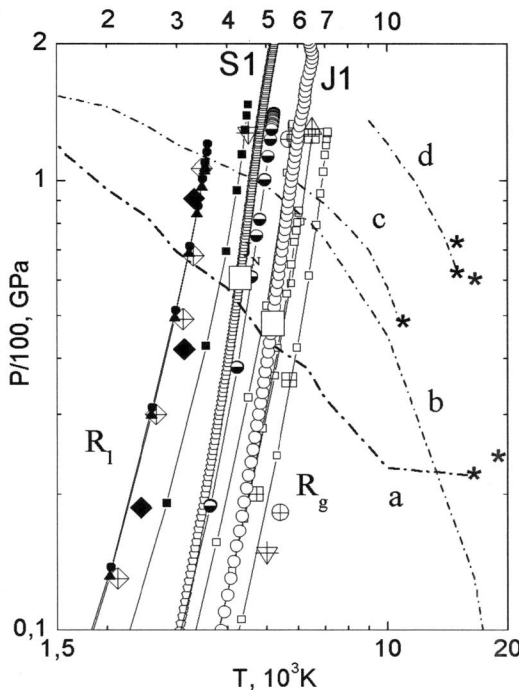

**Figure 4.** Hydrogen P-T diagrams.

## CONCLUSIONS

Thus, the measurements described above show that the transition of Jupiter's and Saturn's atmospheres to the conducting state with conductance above 10 $1/\Omega/cm$ occurs at pressures 30–50 GPa and 60-80 GPa respectively. These values are in agreement with the previously determined boundary of transition for the initially gaseous hydrogen to the conducting state under multiple shock compression [6].

## ACKNOWLEDGEMENTS

This work was supported in part by the Russian Foundation for Basic Research (Project No 04-02-16790) and Presidium of the Russian Academy of Sciences in frameworks of Programs P18, and P27.

The authors would like to thank N.A. Afanas'ev for the fabrication of experimental facilities and participation in the experiments.

## REFERENCES

1. von Zahn U., Hunten D. M., and Lehmacher G., "Helium in Jupiter's atmosphere: Results from the Galileo Probe Helium Interferometer Experiment", J. Geophys. Res. 103, 22815, 1998.

2. Saumon D., Chabrier G., and van Horn H. M., "An Equation of State for Low-Mass Stars and Giant Planets", Astrophys. J., Suppl. Ser. 99, 713, 1995.

3. Guillot T., "A comparison of the interiors of Jupiter and Saturn", Planetary and Space Science, 47, 1183, 1999.

4. Gudkova T.V., Zharkov V.N., "Models of Jupiter and Saturn after Galileo mission" Planetary and Space Science, 47, 1201, 1999.

5. Nellis W.J., "Metallization of fluid hydrogen at 140 GPa (1.4 Mbar): implication for Jupiter", Planetary and Space Science, 48, 671, 2000.

6. Ternovoi V. Ya., Filimonov A. S., Fortov V. E., *et al.*, "Thermodynamic properties and electrical conductivity of hydrogen under multiple shock compression to 150 GPa", Physica B 265, 6, 1999.

7. Collins G. W., Celliers P. M., Gold D. M., *et al.*, Contrib. Plasma Phys. 39, 13, 1999.

8. Fortov V. E., Ternovoi V. Ya., Zhernokletov M. V., *et al.*, "Pressure-Produced Ionization of Nonideal Plasma in a Megabar Range of Dynamic Pressures", JETP, 97, 259, 2003.

9. Ternovoi V.Ya., Kvitov S.V., Pyalling A.A., Filimonov A.S., and Fortov V.E., "Experimental Determination of the Conditions for the Transition of Jupiter's Atmosphere to the Conducting State", JETP Letters, 79, 6, 2004.

10. Ebeling W., Forster A., Fortov V.E., et al., Thermophysical Properties of Hot Dense Plasmas, Stuttgart-Leipzig: Teubner, 1991, 315 p.

11. Ichimaru S., Statistical Plasma Physics II, Reading Addison-Wesley Publ. Co, 1994, 342 p.

12. Magro W.R., Ceperley D.M., Pierleoni C., Bernu B., "Molecular Dissociation in Hot, Dense Hydrogen", Physical Review Letters, 78, 1240, 1996.

13. Saumon D., Chabrier G., "Fluid hydrogen at high density: The Plasma Phase Transition", Physical Review Letters, 62, 2397, 1989.

CP845, *Shock Compression of Condensed Matter - 2005*,
edited by M. D. Furnish, M. Elert, T. P. Russell, and C. T. White
© 2006 American Institute of Physics 0-7354-0341-4/06/$23.00

# THE DYNAMIC BEHAVIOR OF MICRO-CONCRETE

## K. Tsembelis and W. G. Proud

*PCS, Cavendish Laboratory, Madingley Road, Cambridge, CB3 0HE. UK.*

**Abstract.** A series of plate impact experiments has been performed to assess the dynamic behaviour of micro-concrete (70% fine-grain dolerite powder and 30% cement paste by weight) in both longitudinal and lateral directions. Information was obtained for the Hugoniot curve and dynamic shear stress properties. Hugoniot results are compared with published data on cement paste, mortar and concrete from the UK, Germany, France and the US. When the shear strength of micro-concrete is compared to the cement paste very small differences are observed. Therefore, the shear strength of this system appears to be independent of the aggregates and mainly depends on the matrix material.

**Keywords:** Microconcrete, Manganin Gauge, Hugoniot, Shear Stress.
**PACS:** 62.50.+p, 61.43.Gt, 07.35.+k

## INTRODUCTION

Considerable interest in characterising the dynamic loading of concrete under impact conditions exists because of its extensive use as a structural material [1-9]. Concrete is a heterogeneous material containing aggregates and sand in a cement matrix. As a result, characterisation under dynamic conditions is complicated compared to homogeneous materials. The variety of impedances of the constituents leads to variations in the particle velocities and longitudinal and lateral stresses. One way to study this material is to average these variations using a plate reverberation technique, where a disc-shaped concrete specimen is mounted on the projectile and undergoes planar impact on a stationary target (PMMA, copper, tantalum) fitted with diagnostics. However, information such as wave rise time [1] or shear stress [2] cannot be found using this technique. For this reason, the material understanding has been gradually built up starting from studies of the matrix (cement paste) [1] individual granite aggregates [10], mortar [11] and sand [12].

Previous work [1,2,11], has established the Hugoniot curve and shear stress of cement paste and mortar up to 18 GPa. In this paper, we present data of the Hugoniot and shear stress of microconcrete up to 8 GPa using forward and reverse ballistic impact techniques.

## EXPERIMENTAL PROCEDURE

The impact experiments were carried out in the plate impact gun facility at the University of Cambridge [13]. The Cambridge facility consists of a single stage 50-mm bore and 5 m length, light gas gun. The gun is capable of achieving velocities up to 1200 m s$^{-1}$. Impact velocities were measured using a sequential pin-shorting method and tilt was fixed to be less than 1 mrad by means of an adjustable specimen mount. Longitudinal stress measurements were taken by embedding piezoresistive manganin gauges (Micromeasurements type LM SS-210FD-50) between tiles of Copper using a low viscosity epoxy adhesive (figure 1). The experiments were performed using the reverse impact method where a microconcrete flyer of 15 mm thick was launched onto a Copper target constituted of 3.2 and 10.2 mm thick samples. The gauge was placed between the two Copper samples. Gauge calibration was according to the earlier research of Rosenberg and Partom [14].

Lateral stresses were measured by inserting gauges (MicroMeasurements type J2M-SS-580SF-025) inside sectioned samples of microconcrete. All the experiments were performed using a 10 mm thick Copper flyer impacting a microconcrete sample of 65 mm diameter and 20 mm thickness. The gauges were introduced at 4 mm from the impact surface of the sample. Samples were then assembled using a low viscosity epoxy with a curing time of approximately 24 hours. Lateral gauge data were reduced using the analysis of Rosenberg and Partom [15]. The shear strength ($\tau$) of a material under one-dimensional shock loading can be calculated from knowledge of the longitudinal ($\sigma_x$) and lateral stresses ($\sigma_y$) through the relation,

$$2\tau = \sigma_x - \sigma_y \qquad (1)$$

This method of determining the shear stress has the advantage over previous calculations of being direct since no computation of the hydrostat is required.

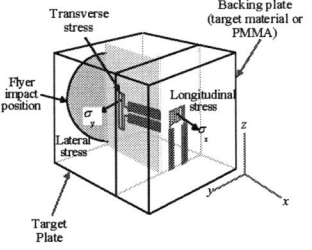

**Figure 1.** Target Configuration

## MATERIAL DESCRIPTION

All the microconcrete samples were provided by Imperial College, London. Density and ultrasonic measurements were performed in Cambridge. The samples constituted of 70% crushed dolerite by weight and 30% cement paste. The density was measured to be $2530 \pm 30$ kg m$^{-3}$. Imperial College quotes a theoretical density of 2570 kg m$^{-3}$, in good agreement with the measured density. The longitudinal and shear elastic wave speeds, determined with ultrasonic transducers, were $4.71 \pm 0.28$ mm $\mu$s$^{-1}$ and $2.66 \pm 0.30$ mm $\mu$s$^{-1}$, respectively.

## LONGITUDINAL RESPONSE

Table 1 summarizes the impact conditions and longitudinal stresses as obtained from the gauges. The particle velocities were estimated using the impedance matching technique. The small reload signal on the gauge traces is due to the recompression of the front copper sample from the original shock wave reflecting between the gauge package and the copper flyer.

Figure 2 illustrates the stress wave profiles as obtained from the gauges. Figure 3 illustrates the Hugoniots of microconcrete (this work), mortar, cement paste [3-4, 12] from experiments performed in the Cavendish Laboratory together with various concretes, grouts and mortars from the US, France and Germany [5-10]. It can be seen that results for microconcrete, mortar and cement paste are consistent with the microconcrete having higher impedance. However, when these three Hugoniot curves are compared against concrete, grout and mortar from other Laboratories, it can be seen that there are some differences. These can be attributed to the initial density and saturation of the samples and differences in the manufacturing process such as water to cement ratio.

## LATERAL RESPONSE

Table 2 summarizes the impact conditions and lateral stresses as obtained from the gauges. Figure 4 illustrates the lateral gauge traces while figure 5 illustrates twice the shear stress (which is the difference between the longitudinal and lateral stress) dependence on the Hugoniot stress. For comparative reasons, the cement paste data are also plotted [2]. The data are also fitted to the assumed elastic behavior. From figure 5, it can be seen that the dynamic shear strength of microconcrete is only slightly higher than that of cement paste.

Figure 6 illustrates the lateral stress developed in both materials as a function of longitudinal stress. It can be seen that, within the error, both materials develop the same lateral stress once they deviate from the elastic loading. Therefore, the small difference in shear stress can be attributed to the higher microconcrete impedance as illustrated in figures 3-4. Furthermore, figure 7 illustrates the shear stress of microconcrete, cement paste and dolerite [10]. It can be seen that although crushed dolerite constitutes 70% of microconcrete by weight, it plays a minor roll, if any, in the dynamic shear strength. The significant factor is the matrix material.

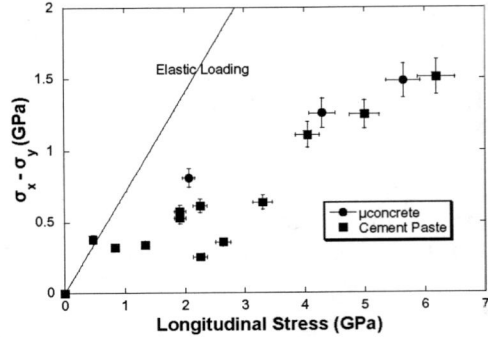

**Figure 5.** Shear Stress of microconcrete

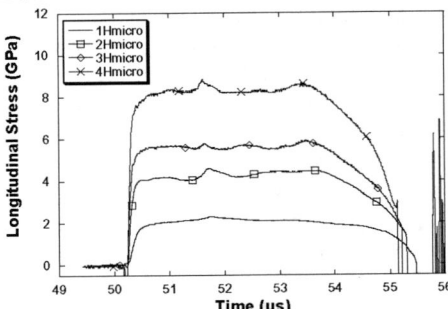

**Figure 2**. Hugoniot Stress Wave Profiles.

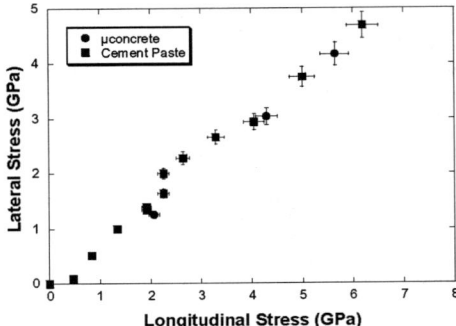

**Figure 6.** Lateral Stress vs. Hugoniot Stress

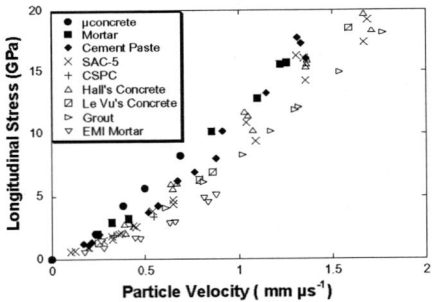

**Figure 3.** All Materials Hugoniot curves (Stress-Up)

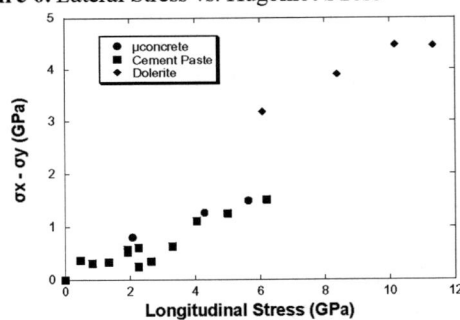

**Figure 7:** Shear Stress of microconcrete, cement paste and dolerite

## CONCLUSIONS

The Hugoniot curve and shear strength of microconcrete was established. The Hugoniot was found to be consistent with mortar and cement paste. However differences exist when compared with concretes, grout and mortar from other laboratories. These differences can be attributed to the water saturation and manufacturing of the samples.

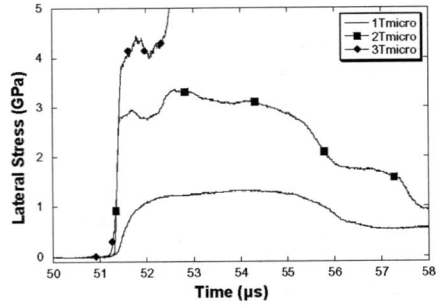

**Figure 4.** Lateral Stress Wave Profiles

**Table 1.** Experimental Parameters and Hugoniot Data for microconcrete

| Shot No. | Impact Velocity (m/s) | Stress (GPa) ±4% | Particle Velocity (km/s) ±4% |
|---|---|---|---|
| 1Hmicro | 294 | 2.07 | 0.23 |
| 2Hmicro | 497 | 4.29 | 0.38 |
| 3Hmicro | 651 | 5.65 | 0.50 |
| 4Hmicro | 907 | 8.23 | 0.69 |

**Table 2.** Experimental Parameters and Shear Stress Data for microconcrete

| Shot No. | Impact Velocity (m/s) | Lateral Stress (GPa) ± 5% | Corresponding Hugoniot (GPa) | 2*Shear Stress (GPa) ± 8% |
|---|---|---|---|---|
| 1Tmicro | 303 | 1.26 | 2.07 | 0.81 |
| 2Tmicro | 506 | 3.04 | 4.29 | 1.25 |
| 3Tmicro | 666 | 4.17 | 5.65 | 1.48 |

The shear strength of microconcrete is very similar to its matrix material which is the cement paste. Small differences are attributed to the different Hugoniot curve. Although crushed dolerite constitutes 70% of microconcrete, it plays a minor role in the shear strength.

## REFERENCES

1.  Tsembelis, K, Millett, J. C. F., Proud, W. G., and Field, J. E., "The Shock Hugoniot Properties of Cement Paste up to 5 GPa",. *in Shock Compression of Condensed Matter – 1999*, (Edited by M.D. Furnish, et.al.), pp. 1267-1270.
2.  Tsembelis, K, Proud, W. G., and Field, J. E., "The Dynamic Strength of Cement Paste under Shock Compression", *in Shock Compression of Condensed Matter – 2001*, (Edited by M.D. Furnish, et.al.), pp. 1414-1417.
3.  Grady, D. E., "Impact Compression Properties of Concrete", *in Proceedings of the Sixth International Symposium on Interaction of Nonnuclear Munitions with Structures, Panama City Beach, Florida*, pp. 172-175, May 3-7 (1993)
4.  Grady, D. E., "Shock Equation of State Properties of Concrete," *in Structures under Shock and Impact IV*, (edited by N. Jones et al.), Computational Mechanics Publications, Southampton, 1996, pp. 405-414.
5.  Hall, C. A., Chhabildas, L. C., and Reinhart, W. D., "Shock Hugoniot and Release States in Concrete Mixtures with Different Aggregate Sizes from 3 to 23 GPa," *in Shock Compression in Condensed Matter - 1997*, (Edited by S. C. Schmidt et al.), pp. 119-122.
6.  Grady, D. E., and Furnish, M.D., "Hugoniot and Release Properties of a Water-Saturated High-Silica-Content Grout", *in Shock Compression in Condensed Matter – 1989*, (Edited by S.C. Schmidt et.al.), pp. 621-624.
7.  Kipp, M. E., Chhabildas, L. C., and Reinhart, W. D., "Elastic Shock Response and Spall Strength of Concrete," *in Shock Compression in Condensed Matter-1997*, edited by S. C. Schmidt et al., AIP Conference Proceedings 429, New York, 1998, pp. 557-560.
8.  Le Vu, O., "Etude et Modelisation du Comportement du Beton Sous Sollicitations de Grande Amplitude", Ph.D. Thesis 1998, Ecole Polytechnique, (in French).
9.  Riedel, W., "Beton Unter Dynamischen Lasten Meso- und Maromechanische Modelle und Ihre Parameter", Phd. Thesis 2001, EMI, (in German).
10. Tsembelis, K., Proud, W. G., and Field, J.E., "The Principal Hugoniot and Dynamic Strength of Dolerite under Shock Compression", *in Shock Compression of Condensed Matter – 2001*, (Edited by M.D. Furnish, et.al.), pp. 1385-1388.
11. Tsembelis, K., Proud, W.G.P., Willmott, G.R., and Cross, D.L.A, "The Shock Hugoniot Properties of Cement Paste & Mortar up to 18 GPa", *Shock Compression of Condensed Matter – 2003*, (Edited by M.D. Furnish *et.al.*), pp. 1488-1490.
12. Tsembelis, K, Proud, W. G., Vaughan, B. A. M., and Field, J. E., "The Behaviour of Sand under Shock Wave Loading: Experiments and Simulations", in *Proceedings of the 14th DYMAT Technical Meeting in Behaviour of Materials at High Strain Rates: Numerical Modelling, Sevilla, Spain 14-15 November 2002*, pp. 193-204
13. Bourne, N. K., Rosenberg, Z., Johnson, D. J., Field, J. E., Timbs, A. E., and Flaxman, R. P., Meas. Sci. Technol. 6, 1462-1470 (1995).
14. Rosenberg, Z, Yaziv, D, and Partom, Y.J., J. Appl. Phys., 51, p. 3702 (1980).
15. Z. Rosenberg, and Y. Partom., J. Appl. Phys., 58, 3072-3076 (1985).

CHAPTER XXIV

# DYNAMIC FRICTION AND EXOTIC CONFIGURATIONS AND MATERIALS

CP845, *Shock Compression of Condensed Matter - 2005,*
edited by M. D. Furnish, M. Elert, T. P. Russell, and C. T. White
2006 American Institute of Physics 0-7354-0341-4/06/$23.00

# NUMERICAL SIMULATION OF DYNAMIC FRICTION

## Graham J. Ball

*AWE, Aldermaston, Reading, RG7 4PR, UK*

**Abstract.** A one-dimensional hydrocode has been developed for the simulation of dynamic friction in order to provide insight into friction mechanisms at the macroscopic scale. The code models shear stress and strain, work hardening, thermal softening, and thermal conduction. Results are presented for aluminium/steel, which are compared with a recent HE-driven recovery experiment. Two alternate mechanisms are proposed for the time evolution of the friction stress, applicable at low and high initial sliding velocity respectively. These mechanisms are shown to be consistent with the experimental data.
**Keywords:** Dynamic Friction, Hydrocodes, Continuum Modelling
**PACS:** 68.35.Ja, 81.40.Pq, 81.40.Ef

## INTRODUCTION

When a contact surface between dissimilar metals is accelerated by a strong oblique shock wave, the materials are subjected to differential tangential acceleration and high normal stress, resulting in sliding with friction. Current hydrocodes either neglect friction entirely, or rely on simple empirical models in which the frictional stress is related to the normal stress and/or the sliding velocity. Such models require tuning, and do not always perform well for systems of practical interest.

In the current work, a 1D hydrocode has been developed to predict the time-dependent behaviour of the sliding interface. The predictions are compared with HE-driven experiments to gain insight into the mechanism of dynamic friction. The code models elastic-plastic deformation in shear, work hardening, thermal softening, melting, and heat conduction.

## NUMERICAL MODEL

It is assumed that semi-infinite blocks of aluminium (above) and stainless steel (below) meet at a planar interface. At $t = 0$ the tangential velocity in each block is uniform, with a sharp velocity jump at the interface. The normal velocity is zero everywhere. The blocks can be initialised with elevated temperatures to mimic shock-heating.

The mesh is refined towards the interface at 3% per cell, giving a maximum resolution of 62 nm at the interface – high resolution is necessary to converge the temperature field. The mesh is staggered – velocity and mass are node-centred, while all other variables are cell-centred. Two coincident nodes are placed at the interface, one for each material, to capture the sliding velocity discontinuity, $\Delta u$. A simple predictor-corrector scheme is used to integrate the governing equations.

A key issue is how to determine the instantaneous frictional shear stress. For systems of practical interest, a simple friction law based on the product of the normal stress and a friction coefficient ( 0.2 for smooth metal/metal sliding) will predict a shear stress that greatly exceeds the von Mises yield limit in pure shear ($\tau_{max} = Y/\sqrt{3}$). Consequently, it has been assumed that the frictional shear stress, $\tau_{fric}$, is limited by yielding of the weaker material at the interface, and is therefore independent of the normal stress once this limit is reached. Hence the frictional stress is strongly coupled with the processes of thermal softening and work hardening. The heat flux due to frictional heating is obtained simply as

$$\dot{Q} = \tau_{fric} \times \Delta u, \qquad (1)$$

and is partitioned between the materials so as to maintain temperature continuity at the interface.

The Steinberg Guinan constitutive model is implemented as follows. The yield strength, $Y$, is given by:

$$
\begin{aligned}
Y = {} & \min\{Y_0(1+\beta\varepsilon_{PL})^n, Y_{max}\} \\
& \times(1+gP-h(T-300)) \\
& \times exp(-0.001\frac{T}{T_{melt}\{P\}-T}) \qquad (2)
\end{aligned}
$$

where $Y_0$ is the yield strength in the reference state, $\varepsilon_{PL}$ is the equivalent plastic strain, and $\beta$, $n$, $g$, $h$ are material-dependent parameters. The first term on the right of Eq.2 represents work hardening, and the second term pressure hardening and thermal softening. The final term produces a rapid decay in strength as the melt temperature is approached.

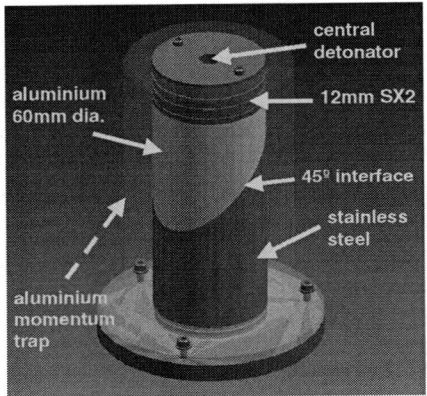

**FIGURE 1.** Configuration of the FN6 HE-driven friction experiment

## FN6 RECOVERY EXPERIMENT

AWE have recently fired a series of HE-driven dynamic friction experiments [1] using the FN6 vehicle (see Fig.1). The upper aluminium alloy cylinder is driven against a stainless steel anvil, producing sliding at their common interface, which is inclined at 45 deg to the cylinder axis. The explosive is centrally detonated, producing an approximately spherical Taylor wave. When this iteracts with the

inclined planar interface, it produces a normal stress and tangential acceleration which vary with position over the plane. Hence, each shot generates a range of initial conditions for friction. After firing, the aluminium section is recovered and sectioned for microscopic examination. The aluminium cylinder is made from extruded bar – as a result the microstructure contains fine longitudinal striations. In the recovered samples, regions of plastic deformation are revealed by curvature of these striations, which act as natural fiducial lines.

## NUMERICAL RESULTS

An initial simulation was performed for conditions broadly representative of an HE-driven friction experiment. Materials were aluminium alloy (5083) against stainless steel (AISI304L), and the initial sliding velocity $\Delta u_0 = 0.02 cm/\mu s$. Pressure hardening and shock heating were neglected, and the melt temperature for aluminium was fixed at $1220K$.

The evolution of the shear stress field ($t = 0 - 2.5\mu s$) is shown in Fig.2. Initially, elastic shear waves propagate into both materials, loading the aluminium onto its yield surface. In addition, a slower-moving plastic shear wave appears in the aluminium, driven by work hardening. A necessary condition for this wave to propagate is that $\partial\tau/\partial y < 0$, which will be true if the wave-processed material remains on its yield surface, since work hardening ensures that $\partial Y/\partial y < 0$. However, frictional heating soon lowers the yield strength of the aluminium at the interface, which in turn lowers $\tau$, producing upward-running release wavelets ($\partial\tau/\partial y > 0$) which out-run the plastic wave and progressively undermine it, with the result that it ceases to propagate from $t \approx 1\mu s$.

From $t \approx 1.25\mu s$ the shear stress falls more rapidly at the interface, generating release wavelets that subsequently unload a broad region in both materials. This phenomenon is driven by the exponential term in Eq.2 as the aluminium at the interface approaches its melt temperature. The time history of the shear stress and temperature at the interface are shown in Fig.3. It is evident that the frictional process develops in two distinct phases. In the initial, or "warm-up" phase the temperature is increased by frictional heating, while the stress decays. Since $\tau_{fric} = Y/\sqrt{3}$, the decay in $\tau$ is a direct consequence of thermal soften-

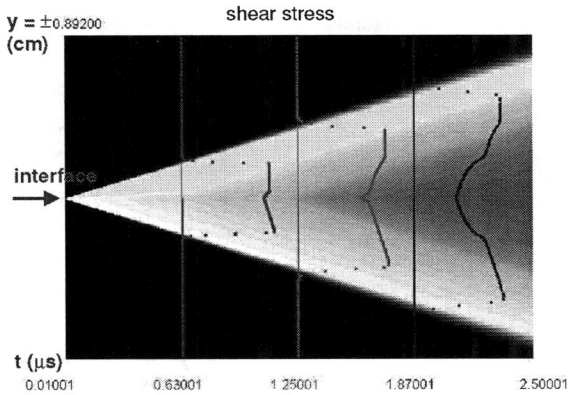

**FIGURE 2.** Generic Al/steel friction problem, Al at top. Streak (y/t) plot of shear stress, $t = 0 - 2.5\mu s$, with stress profiles $\tau(y)$ at indicated times

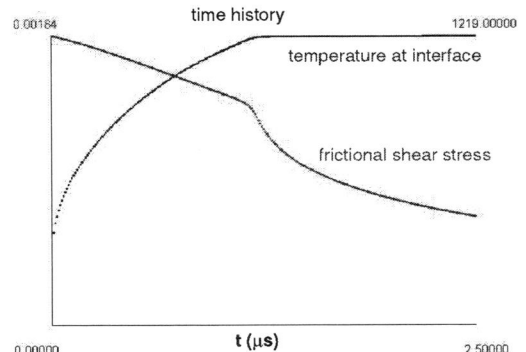

**FIGURE 3.** Generic Al/steel friction problem. Time history of temperature and shear stress at the interface, $t = 0 - 2.5\mu s$

ing, partially offset by work hardening. In the late-time phase, here termed *asymptotic melting*, the interface enters a quasi-equilibrium state in which heat production through friction is balanced by conductive heat loss. Over time the temperature gradients close to the interface become flatter, reducing the conductive heat loss. The equilibrium value of $\tau_{fric}$ therefore decays, while the temperature approaches the melt temperature asymptotically.

## COMPARISON WITH EXPERIMENTS

The sliding surface of a recovered aluminium sample from a typical FN6 shot (see photo, Figs.4,6)

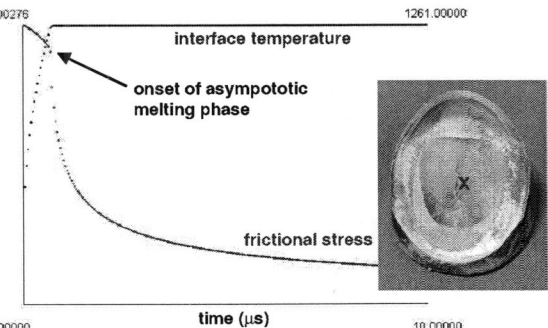

**FIGURE 4.** Simulated time histories of temperature and shear stress for FN6, inner region

is divided into two distinct regions of different surface texture, separated by an abrupt, crescent-like, transition. The following estimates of the initial post-shock conditions in the centre of these regions were obtained from hydrocode calculations: Outer, $P = 0.112Mbar$, $\Delta u = 0.0125cm/\mu s$; Inner, $P = 0.062Mbar$, $\Delta u = 0.0183cm/\mu s$. Figure 4 shows the predicted time history of interface temperature and stress for the inner region. The friction follows a similar path to that identified in the previous section, passing through a warm-up phase, followed by asymptotic melting from about $1\mu s$. The experimental micrograph for this region is shown in Fig.5, alongside the predicted deformation of a fiducial line. The sub-surface plastic deformation, as seen in both the experiment and simulation, is restricted to a shallow region of only a few microns depth.

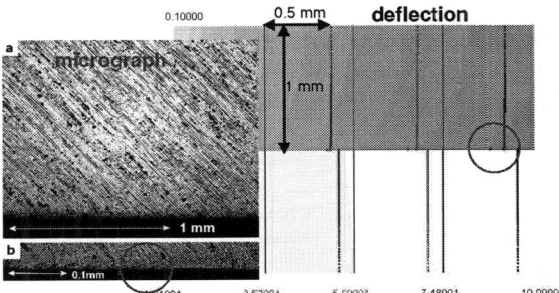

**FIGURE 5.** Comparison of plastic deformation in experimental micrograph (left) and simulation (right) for FN6, inner region. Note that deformation is limited to a thin layer at the interface (ringed)

The predicted time history for the outer region (see Fig.6) shows a radically different behaviour. After an initial warm-up phase lasting about $2\mu s$, there is an

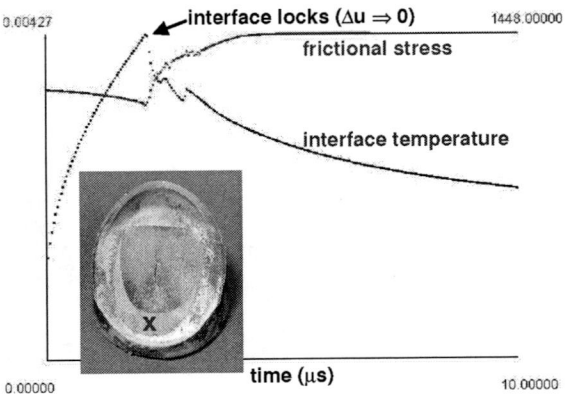

**FIGURE 6.** Simulated time histories of temperature and shear stress for FN6, outer region

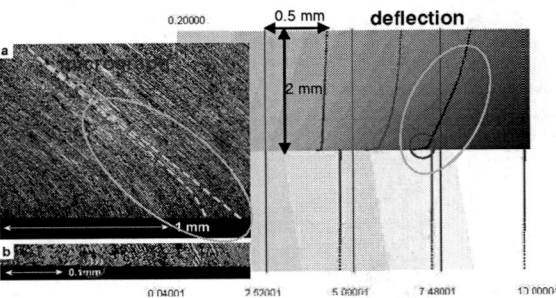

**FIGURE 7.** Comparison of plastic deformation in experimental micrograph (left) and simulation (right) for FN6, outer region. Note deeper penetration of plastic deformation (oval rings) plus additional near-surface deformation (circled)

locks. This behaviour is here termed *slide-then-lock*.

abrupt transition to a period of rapidly falling temperature and increasing stress. The stress reaches a steady plateau at about $t \approx 4\mu s$, while the temperature continues to decay. At the transition, a violent work hardening event occurs at the interface, which reduces the sliding velocity to zero, "locking" the interface. This generates a plastic shear wave which propagates into the aluminium; the velocity jump across this wave accomodates the velocity mis-match previously existing at the interface. This event is triggered when the interface becomes unstable to work hardening. Consider the creation of an infinitessimal plastic shear wavelet at the interface. The wavelet produces work hardening, so the material gains yield strength, and hence the frictional shear stress increases, by an ammount $\delta\tau$. At the same time, the velocity jump across the wave reduces the sliding velocity by $\delta u$. These changes modify the heat production rate so that Eq.1 becomes:

$$\dot{Q} + \delta\dot{Q} = (\tau_{fric} + \delta\tau) \times (\Delta u - \delta u). \quad (3)$$

Usually, the increase in stress is the dominant factor, giving $\delta\dot{Q} > 0$, so that the heating rate is increased, producing extra thermal softening which halts the process – the interface is stable. However, if the sliding velocity is low enough, the velocity term may dominate, so that $\delta\dot{Q} < 0$ to a sufficient degree that the interface temperature begins to fall. In this case, the cooling metal gains strength, producing further increments in $\delta\tau$ and $\delta u$, thus producing a positive feedback mechanism that leads to runaway work hardening. The process continues until the interface

The experimental micrograph and simulated fiducial for the outer region are shown in Fig.7. Both show deep plastic deformation, extending to a depth of 1-2$mm$. It therefore appears probable that the change in surface texture between the regions marks an abrupt transition between slide-then-lock and asymptotic melting behaviours. Subsequent analyses of other FN6 shots have shown a consistent correlation between changes in surface marking and a shift from shallow to deep plastic deformation.

## CONCLUSIONS

The simulations have been used to predict the existence of two mechanisms for the time-evolution of dynamic friction, *asymptotic melting* and *slide-then-lock*, that are consistent with the results of the FN6 recovery experiments. A key conclusion arising from the current work is that the instantaneous frictional stress is strongly dependent on the history of the friction process, both in terms of thermal effects and work hardening.

## REFERENCES

1. Winter, R. E., Smeeton, V. S., De'Ath, J., Markland, L., and Barlow, A. J., "Metallography of Sub-surface Flows Generated by Shock-induced Friction," in *Shock Compression of Condensed Matter*, Portland, 2003, pp. 625–628.

CP845, *Shock Compression of Condensed Matter - 2005*,
edited by M. D. Furnish, M. Elert, T. P. Russell, and C. T. White
© 2006 American Institute of Physics 0-7354-0341-4/06/$23.00

# STRONGLY NONLINEAR WAVES IN POLYMER BASED PHONONIC CRYSTALS

## C. Daraio[1], V. F. Nesterenko[1,2], E. B. Herbold[2], S. Jin[1,2]

[1]*Materials Science & Engineering Program, University of California at San Diego, La Jolla, CA 92093-0418*
[2]*Mechanical and Aerospace Engineering Department, University of California at San Diego, La Jolla, CA 92093-0411*

**Abstract.** One dimensional "sonic vacuum"-type phononic crystals were assembled from chains of polytetrafluoroethylene (PTFE) beads and Parylene coated spheres with different diameters. It was demonstrated for the first time that these polymer-based granular system, with exceptionally low elastic modulus of particles, support the propagation of strongly nonlinear solitary waves with a very low speed. They can be described using classical nonlinear Hertz law despite the viscoelastic nature of the polymers and the high strain rate deformation of the contact area. Trains of strongly nonlinear solitary waves excited by an impact were investigated experimentally and were found to be in reasonable agreement with numerical calculations. Tunability of the signal shape and velocity was achieved through a non-contact magnetically induced precompression of the chains. This applied prestress allowed an increase of up to two times the solitary waves speed and significant delayed the signal splitting. Anomalous reflection at the interface of two "sonic vacua"-type systems was reported.

**Keywords:** Strongly nonlinear, phononic crystal, polymers, wave propagation
**PACS:** 05.45.Yv, 46.40.Cd, 43.25.+y, 45.70.-n

## INTRODUCTION

One-dimensional chains of spherical beads have received increasing attention in the recent years for the study of a novel type of strongly nonlinear wave dynamics [1-6]. This system is the simplest example of phononic metamaterials with unique properties. The strong nonlinearity opens a new area of interest being a natural extension of the weakly nonlinear wave dynamics. The non-classical wave behavior appears if the granular chain is "weakly" compressed (i.e. when the wave amplitude is significantly higher than the forces caused by the initial precompression) [1]. The concept of "sonic vacuum" (SV) was introduced to emphasize that in such a chain with no initial prestress the sound speed is equal to zero. One of the distinguishing properties of SV is the existence of a qualitatively new solitary wave with a finite width that is independent of the wave amplitude. The solitary wave speed $V_s$ in SV has a nonlinear dependence on the maximum strain $\xi_m$, the particle velocity $v_m$ and the force between particles $F_m$:

$$V_s = \frac{2}{\sqrt{5}} c \xi_m^{1/4} = \left(\frac{16}{25}\right)^{1/5} c^{4/5} v_m^{1/5} = 0.68 \left(\frac{2E}{a\rho^{3/2}(1-\nu^2)}\right)^{1/3} F_m^{1/6}. \quad (1)$$

The solitary wave speed, $V_s$, can be tuned by applied static precompression ($F_0$) and can be written in terms of the normalized maximum force $f_r = F_m / F_0$ [4] as,

$$V_s = 0.9314 \left(\frac{4E^2 F_0}{a^2\rho^3(1-\nu^2)^2}\right)^{1/6} \frac{1}{\left(f_r^{2/3}-1\right)} \left\{\frac{4}{15}\left[3+2f_r^{5/3}-5f_r^{2/3}\right]\right\}^{1/2} \quad (2)$$

where Eq. (1) is a partial case of Eq. (2).

## EXPERIMENTAL PROCEDURE

One dimensional phononic crystals were assembled filling a PTFE (polytetrafluoroethylene) tube (with inner diameter 5 mm) with chains of 21 PTFE (McMaster-Carr) and Parylene coated balls (AcraBall) with diameter a=4.76 mm and 4.86mm and mass 0.123 g and 0.44 g respectively (Fig. 1(a)). Details of the experimental set-up used for testing of 1-D chains are described elsewhere [4,5].

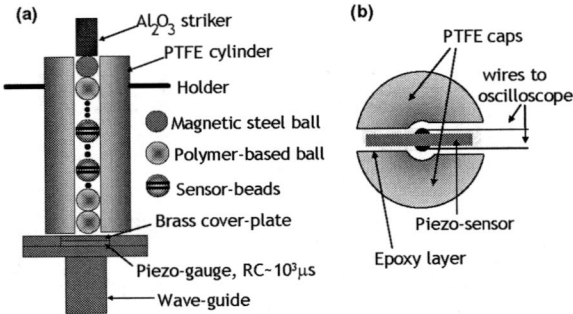

**Figure 1.** (a) Experimental set-up used for testing 1-D chains of polymer based (PTFE) beads. (b) Schematic drawing of a sensor embedded in a sphere [4].

### Magnetically induced tunability

A peculiar characteristic of strongly nonlinear materials is the possibility of fine tuning the signal shape and speed under precompression. In the present study, the applied preload was achieved with a Neodymium-Iron-Boron ring magnet [5] placed around the PTFE cylinder containing the chains. It was held in place by the magnetic interaction with the steel ball placed on the top of the chain (Fig. 1(a)). This type of preloading allowed the application of a constant external force (2.38 N) to the end magnetic bead independently from its displacement, therefore maintaining constant boundary conditions of the system.

### Heavy/Light interface testing

A chain of 20 nonmagnetic stainless steel (316) balls (plus a magnetic sphere on the top) was placed above 21 PTFE beads. The magnetically induced tunability was applied as described earlier to test the interface behavior under applied prestress. Further details of the experimental set-up can be found in [5].

## RESULTS AND DISCUSSION

Pulses of different durations and amplitudes in the 1-D phononic crystals were generated by impacting an alumina ($Al_2O_3$) cylinder (0.47 g) or a PTFE ball with a diameter of 4.76 mm (mass 0.123 g) onto the top sphere of the chain. Single solitary waves were generated by an impactor with a mass equal to the mass of the beads in the system [1]. The results of these experiments are shown in Fig. 2. A very fast decomposition of the initial impulse was demonstrated at a distance comparable to the soliton's width and a clear tendency of signal splitting was noticed after only 10 particles.

Fig. 2 shows the strong dependence of the solitary wave's speed on the amplitude for PTFE chains. Here Equation (1) for SV and Equation (2) for a pre-compressed chain are plotted as common logarithmic values at different PTFE elastic constants together with the corresponding experimental data (solid dots) and the numerical calculations of the soliton speed for discrete chains.

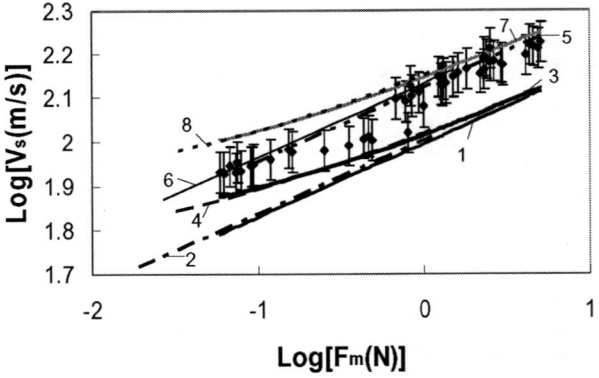

**Figure 2.** Dependence of the velocity of solitary wave on amplitude in a PTFE chain. Experimental values are shown by solid dots. Curves 1 and 5 are the theoretical curves based on Equation (1) with a Young's modulus equal to 600 MPa and 1460 MPa respectively. Curve number 2 and 6 represent the corresponding numerical calculations for these cases. Curve number 3 and 7 represent the long wave approximation for gravitationally pre-compressed systems (Equation (2)) at 600 MPa and 1460 MPa respectively; curves 4 and 8 represent the corresponding numerical calculation.

The values based on the long wave approximation and the numerically calculated data are basically indistinguishable. The wave speeds for different amplitudes were obtained via time-of-flight measurements between the peaks measured by the sensors in experiments (solid dots in Fig. 2).

PTFE and Parylene are polymeric viscoelastic materials with high strain rate sensitivities and low elastic modulus [7]. At normal conditions Young's ($E$) and flexural moduli for PTFE are in the range of 400 – 750 MPa and Poisson's ratio ($\nu$) is 0.46 [4,5,8]. For Parylene these values are $E = 2760$ MPa, $\rho = 1289$ Kg/m$^3$, $\nu = 0.388$ [9]. If proved to support strongly nonlinear behavior as in the case of chains made from typical linear elastic materials [1-3, 6], these properties can be very attractive for ensuring a very low solitary wave speed and an adequate tunability of the system.

We observed a significant difference between the solitary wave speeds obtained in experiments and derived from the theory, especially at large amplitudes of forces when using the Young's moduli found using ultrasonic measurements. It should be mentioned that in polymers, the bulk sound speed extrapolated from the Hugoniots in $u_s(u_p)$ coordinates results in significantly higher values than the sound speed measured using ultrasonic techniques [8]. This discrepancy indicates a rapidly varying change of compressibility at low values of shock amplitudes. For PTFE, using the value of bulk speed $c_b$ extrapolated from Hugoniot (1.68 km/s, in comparison with 1.139 km/s from ultrasonic measurements) and Poisson ratio 0.46, we obtained a value of Young's modulus of 1.46 GPa based on the relations for elastic solids. The values of the solitary wave speed derived from Eq. 2 with this Young's modulus better matched our experimental results (Fig. 2). Smaller PTFE particles (2.38 mm diameter) also supported the SV type behavior, although in this case the effect of dissipation appeared to be more significant [4].

Chains of composite particles made of stainless steel beads (high density core) coated with a low elastic modulus ParyleneC layer 50 μm thick, with diameters 4.86 mm and 2.48 mm, also support strongly nonlinear solitary waves. Similar to the PTFE case, the effects of dissipation were significant; especially for the small diameter beads. Also, to match the experimental data, the Young's

modulus used for Parylene in Fig. 3 (Eq. (2)) was significantly higher than its nominal value reported by the manufacturer [9] (15 GPa vs. 2.76 GPa).

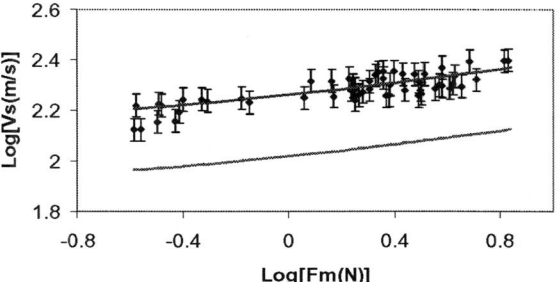

**Figure 3.** Dependence of the velocity of solitary wave on amplitude for ParyleneC coated steel chain only under weak gravitational precompression. The common log of the experimental values are shown by solid dots. The solid curves represent the common log of the theoretical values based on the analogue of Equation (2) for composite particles with a Young's modulus equal to 2.76 GPa (bottom) and 15 GPa (top).

The effect of the magnetically induced precompression on the solitary wave speed in PTFE chains is summarized in Fig. 4. The observed increase of the solitary wave speed in the PTFE based system under precompression was significant. The experimental data was consistently matched by the results obtained from the long wave approximation and the numerical analysis [5]. This increase of the speed resulted in a corresponding increase of the acoustic impedances.

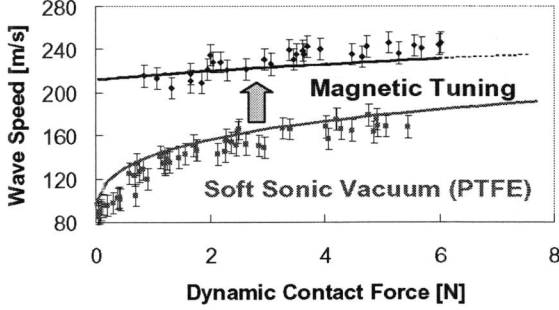

**Figure 4.** Dependence of the solitary wave speed on the amplitude of the dynamic contact force for gravitationally loaded and for magnetically tuned chains composed of PTFE beads. The experimental values for corresponding curves are shown by solid squares and dots. The solid curves represent the results for the long wave approximation with $E= 1.46$ GPa.

A delay of the splitting of the solitary waves under prestress was observed both in experiments and in numerical calculations.

The testing of the heavy/light interface [6] of the two strongly nonlinear granular media under the magnetically induced precompression described earlier resulted in a dramatic change of reflectivity. Anomalous reflected compression waves and transmitted rarefaction waves were detected in experiments and numerical calculations (see Fig. 5). We named this phenomenon the "acoustic diode" effect because of the dramatic change of the reflectivity triggered by the precompression.

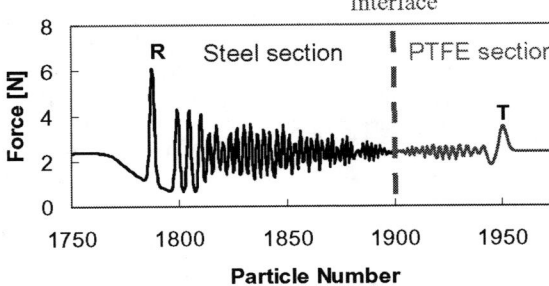

**Figure 5.** Typical profile obtained in numerical calculations showing a reflected rarefaction wave followed by the anomalous compression waves (R) and an oscillatory tail in the stainless steel chain and transmitted compression (T) rarefaction pulses and oscillatory tail in the PTFE chain [6].

The nonlinear phenomena described here can find useful application as tunable controllers of information flow through interfaces and in the design of novel types of tunable shock protection layers. The precompression can be employed for designing tunable information transportation lines with the unique possibility of manipulating the signal's delay, reflection and decompositions at will for security-related information.

## CONCLUSIONS

Polymer-based strongly nonlinear one dimensional metamaterials (chains of PTFE and ParyleneC coated stainless steel beads) were assembled and tested. Both polymeric systems support strongly nonlinear solitary waves with very small amplitudes and speed. The wave speed of these solitary waves was less than the sound speed in air and lower than any other previously reported. The theory derived from the Hertzian model for contact interaction of linear elastic solids fit very well with the experimental results, despite the viscoelastic nature of the polymers. The elastic modulus of PTFE at higher signal amplitudes matched quite well with the elastic modulus extrapolated from the Hugoniot data. Tunability of the signal shape and speed was achieved through a magnetically induced precompression which also created anomalous reflections at the interface of steel/PTFE chains (novel "acoustic diode" behavior).

## ACKNOWLEDGEMENTS

This work was supported by the National Science Foundation (Grant No. DCMS03013220).

## REFERENCES

1. Nesterenko, V. F.; Dynamics of Heterogeneous Materials, Chapter 1 (Springer-Verlag, NY, 2001).
2. Coste, C. & Gilles, B. "On the validity of Hertz contact law for granular material acoustics." European Physical Journal B 7, 155, 1999.
3. Job, S., Melo, F., Sen, S. & Sokolow, A. "How Hertzian solitary waves interact with boundaries in a 1D granular medium." Phys. Rev. Lett., 94, 178002, 2005.
4. Daraio, C., Nesterenko, V. F., Herbold E. B. and Jin, S. "Strongly nonlinear waves in a chain of Teflon beads", Phys. Rev. E.,72, 016603, 2005.
5. Daraio, C., Nesterenko, V. F., Herbold, E. B., Jin, S. "Tunability of solitary wave properties in one dimensional strongly nonlinear phononic crystals", <http://arXiv:Cond-Mat/0506513>, 2005.
6. Nesterenko, V. F., Daraio, C., Herbold, E. B., Jin, S. "Anomalous wave reflection from the interface of two strongly nonlinear granular media", Phys. Rev. Lett., 95, 158702, 2005.
7. Zerilli, F. J. & Armstrong, R. W. Thermal activation constitutive model for polymers applied to polytetrafluoroethylene, Shock Compression of Condensed Matter, CP620, Edited by M.D. Furnish, N.N. Thadhani, and Y. Horie, 657, 2002.
8. Carter W.J. & Marsh, S.P. Hugoniot Equation of State of Polymers, Los Alamos Preprint, LA-13006-MS, 1995.
9. AcraBall Manufactoring Co. Product data.

CP845, *Shock Compression of Condensed Matter - 2005*,
edited by M. D. Furnish, M. Elert, T. P. Russell, and C. T. White
© 2006 American Institute of Physics 0-7354-0341-4/06/$23.00

# DYNAMIC BEHAVIOUR OF BIRCH AND SEQUOIA AT HIGH STRAIN RATES

**A. M. Bragov[1], A. K. Lomunov[1], I. V. Sergeichev[1], and G.T. Gray III[2]**

[1] *Research Institute of Mechanics, Nizhny Novgorod State University, GSP-1000, Russia*
[2] *Los Alamos National Laboratory, Los Alamos, NM 87545, USA*

**Abstract.** This paper presents results of the dynamic mechanical response of for two structural woods, i.e. birch and sequoia. Monotonic and cyclic compression testing at room temperature of these materials was performed using a modified Kolsky method; a 20-mm diameter split-Hopkinson pressure bar (SHPB). The birch and sequoia specimens were loaded parallel and orthogonal to the grain of the wood, as well as, at other angles relative to the wood grain. The dynamic mechanical behavior of the two woods was measured as a function of loading orientation under a uniaxial stress state as well as under circumferential confinement using a collar surrounding the sample to quantify the effect of lateral confinement on mechanical behavior. The loading and unloading responses of both woods were found to exhibit nonlinear behavior and a strong dependency on the strain rate of loading. The dynamic stress-strain responses of the birch and sequoia showed a strong influence of grain orientation of the flow stress and fracture behavior. Examination of the damage evolution and fracture responses of the birch and sequoia displayed a strong dependence on grain orientation. Cyclic dynamic loading data, obtained using a modification of the original SHPB testing method, is also presented for the two structural woods studied. In addition to the SHPB tests, plane-wave shockwave loading experiments were conducted and the shock adiabates for birch was obtained.

**Keywords:** Wood, dynamic testing, split-Hopkinson Pressure Bar, cyclic compression, shock loading.
**PACS:** 62.20.Fe, 62.20.Mk, 62.50.+p.

## INTRODUCTION

In recent years, in connection with increased requirements of safety during transportation of hazardous materials in the event of structural collapse, terrorist acts, man-made catastrophes or other emergencies, there is an increasing need for predictive models of the impact/dynamic response of materials of importance in transportation. Wood, because it is often used as an energy absorbing and packaging material for alleviating the outcomes of such effects, is therefore receiving increased attention. To adequately analyze the damping properties of wood under impact loading, data on

its dynamic properties are needed. At present there is a paucity of such data for wood [1-6].

Wood is an anisotropic material, whose properties differ along and across the fiber or grain directions; i.e., it is strongly textured. Accordingly, to quantify the properties of wood, specimens for testing must be cut from various directions. The purpose of this study was to quantify the influence of sample orientation "texture" as well as confinement conditions on the dynamic mechanical behavior of birch and sequoia woods. Monotonic and cyclic compression testing at room temperature of these materials was performed using a modified Kolsky method. Moreover for birch, under natural humidity conditions, plane-wave experiments were conducted and the shock adiabates were measured.

## EXPERIMENTAL

The Kolsky method was employed for characterizing the dynamic behavior of both woods, using a 20-mm diameter split Hopkinson pressure bar (SHPB). Traditional SHPB experiments are loaded in a uniaxial stress state. Utilization of a rigid confinement jacket around a SHPB sample provides volumetric confinement while uniaxial deformation occurs. In addition to these two uniform loading conditions, rectangular plate specimens were loaded in compression in the central portion of a plate, i.e., "punched in", while the surrounding material provides lateral resistance to expansion. In this case, a wood sample experiences an intermediate stress state. As the acoustic impedance of wood is much less than the acoustic impedance of steel pressure bars, the pulse reflected from a specimen is significant (up to 95% from amplitude of loading wave) and the transmitted wave is small. Hence, a modification of the Kolsky method may be used [6], to reliably register some cycles of the specimen loading during one experiment and to thereby achieve significant strains in a specimen. Figure 1 shows the SHPB configuration to facilitate a reliable registration of three loading cycles. The typical oscillogram is shown in Fig. 2.

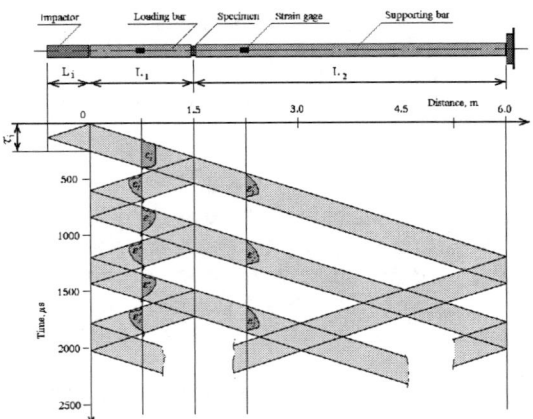

**Figure 1**. The Kolsky method modification for cyclic loading of low-density materials.

The SHPB experiments conducted at LANL utilized 25.4-mm diameter low-impedance magnesium pressure bars to achieve high resolution transmitted wave data. In addition to the SHPB

tests, plane-wave experiments were conducted using a 57-mm gas gun.

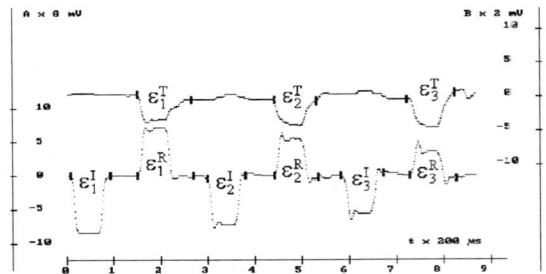

**Figure 2**. Typical oscillogram illustrating three cycles of loading

As wood is an anisotropic material, the directional nature of the mechanical behavior of the specimens was quantified using samples cut out and then loaded at various directions relative to the grain of the wood. The angles between the loading and grain directions studied were $0^0$, $30^0$, $45^0$, $60^0$ and $90^0$ for birch and $0^0$, $30^0$ and $90^0$ for sequoia, respectively. The average density was 0.62 g/cm$^3$ and 0.424 g/cm$^3$ for the birch and sequoia samples.

The SHPB birch and sequoia specimens were machined in the form of cylinders 20 mm in diameter and 10 mm in height. Birch plate specimens, sectioned across the grain, were also tested. To assess the impact compressibility, 70-mm diameter, 8-mm thick plate specimens were used. The plane-wave tests probed the impact response of birch with natural humidity.

## RESULTS AND DISCUSSION

The σ-ε responses of birch at 25°C as a function of orientation relative to the grain direction are given in Figure 3. True stress as a function of strain are depicted by the solid lines and the corresponding strain rate histories - by the dotted ones. Data for each loading orientation is represented by two diagrams: the first for "elastic" deformation of the specimens at low (600...800 s$^{-1}$) strain rate where the samples remained intact, and the second the behavior of each loaded to failure (at strain rates 1500...3000 s$^{-1}$). The diagrams are presented separated along the $X$-axis of the figure to facilitate comparison.

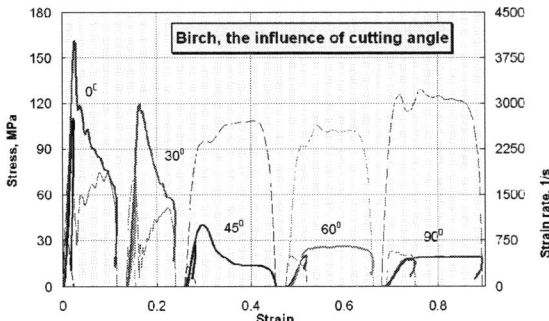

**Figure 3**. The influence of angle relative to the grain direction on the dynamic behavior of birch

The well-known trend of decreasing flow strength of wood with increasing angle relative to the wood grain is readily seen: the highest flow stress values for both the loading modulus and the failure stress are evident in specimens of birch cut parallel ($0^0$), and minimum orthogonal ($90^0$) to the grain direction. For small angles relative to the grain direction, a fast fall-off in the flow stress is seen after initial loading related to rapid damage evolution in the samples.

For investigation of the dynamic stress-strain behavior of birch, cyclic compression tests at room temperature were performed using a modified Kolsky method. The dynamic response was measured as a function of three stress-state modes: 1) uniaxial stress (cyclic loading of specimen unconfined), 2) uniaxial strained (cyclic loading of specimen confined in a jacket), and 3) intermediate stress state (cyclic loading of plate specimen).

To obtain a set of the unloading responses for each loading cycle up to large strains is not viable as the process of relaxation of the stresses in a sample entails a large duration. Accordingly, the unloading line from previous cycles breaks at stresses of 4-6 MPa and a finite strain. The next loading cycle thereafter begins just after this small finite strain (Fig. 4a). To obtain a representation of the continuous response the following reconstruction was made: the stress-strain curves in each subsequent cycle were moved along the strain axis to obtain continuous diagrams. The upward curvature of the specimen in a confined jacket, Fig. 4a, is caused by clearance adjustments between the external surface of the specimen and the internal

surface of the jacket. This data supports the strong influence of stress state on the dynamic response of wood.

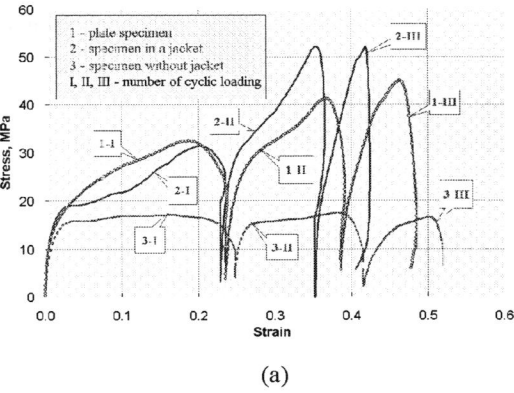

(a)

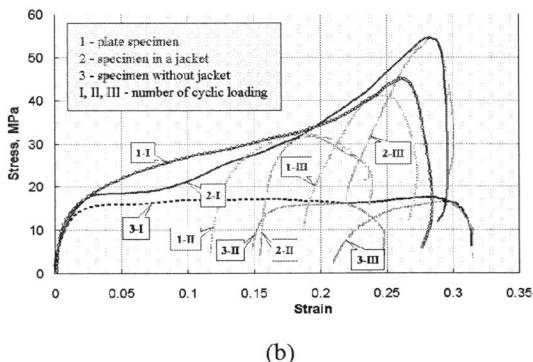

(b)

**Figure 4**. Dry birch. Cyclic loading across grain: Initial diagrams (a) and after reconstruction (b)

Plane-wave experiments were conducted for birch specimens of natural humidity along the grain direction. In the $\sigma_x \sim U$ coordinates, the shock adiabates measured can be approximated by the equation of the form $\sigma_x = X*U^Y$ and in the $D \sim U$ coordinates - by the linear relation $D = A + B*U$. The parameters are presented in Figure 5.

Specimens of dry sequoia with different angles between the loading and grain orientation were independently tested at RIM-NNSU and LANL. Comparison of the RIM-NNSU and LANL data reflects excellent agreement between the results measured as seen in Figures 6 & 7. During each loading and subsequent unloading cycle there are degradations in the fiber coherence of the wood structure resulting in a loss in strength.

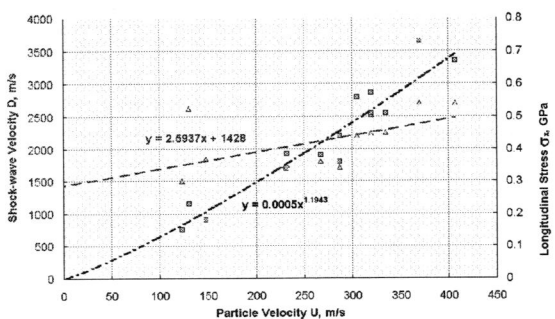

**Figure 5**. Results of plane-wave experiments for birch parallel to the grain/fiber direction.

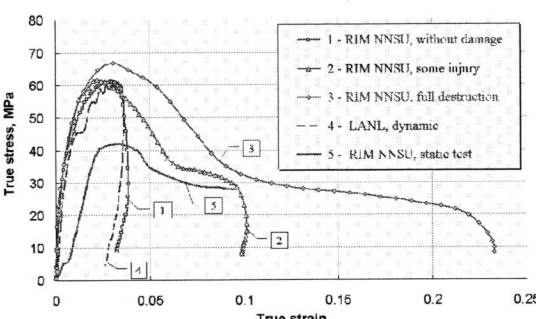

**Figure 6.** Static and dynamic behavior of sequoia along the grain/fiber direction

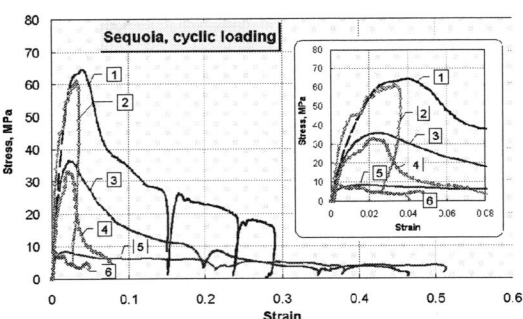

**Figure 7**. The influence of cutting angle on dynamic behavior of sequoia

1 – RIM NNSU, 0 degrees;   2 – LANL, 0 degrees,
3 – RIM NNSU, 30 degrees;  4 – LANL, 30 degrees,
5 – RIM NNSU, 90 degrees;  6 – LANL, 90 degrees

## CONCLUSIONS

The pronounced anisotropy of the dynamic compressive stress-strain, damage evolution, and fracture behavior of birch and sequoia wood was observed as a function of grain orientation. Strain rate and lateral confinement were both found to strongly influence the stress-strain behavior of both birch and sequoia through suppression of cracking along the grain and thereby delaying fracture. Quantification of the dynamic properties of wood will be used to support the development of predictive materials models needed for large-scale engineering numerical simulations and design optimization of structural woods, like birch and sequoia, used in transportation and packaging applications.

## ACKNOWLEDGEMENTS

Funding was provided by the Russian Foundation for Basic Research, under Grant 04-01-00454. Part of this research was also performed under the auspices of the U.S. Department of Energy.

## REFERENCES

1. Reid, S.R. and Peng, C., "Dynamic Uniaxial Crushing of Wood". Int. J. Impact Engng., 19, pp.531-570, 1997.
2. Reid, S.R, Reddy, T.Y., and Peng, C. "Dynamic Compression of Cellular Structures and Materials". In: Structural Crashworthiness and Failure, N. Jones and T. Wierzbicki, eds., Elsevier Applied Science Publishers, New York, 1993, pp. 295-340.
3. Buchar, J., Krivanek, I., and Severa, L. "High rate behaviour of wood", "New Experimental Methods in Material Dynamics and Impact", Trends in Mechanics of Materials, eds. W.K.Nowacki, J.R.Klepaczko, Warsaw, 2001, pp.357-362.
4. Bragov, A.M. and Lomunov, A.K. "Dynamic Properties of Some Wood Species", J.Phys. IV FRANCE **7** Colloque 3, pp.487-492, 1997.
5. Sugiyama, H., "On the Effects of Loading Time on the Strength Properties of Wood", Wood Sci. Tech., 1, pp. 289-303, 1967.
6. Bragov, A.M., Lomunov, A.K., and Sergeichev, I.V. "Modification of the Kolsky method for studying properties of low-density materials under high-velocity cyclic strain", Journal of Applied Mechanics and Technical Physics, vol.42, no 6, pp.1090-1094, 2001.

CP845, *Shock Compression of Condensed Matter - 2005*,
edited by M. D. Furnish, M. Elert, T. P. Russell, and C. T. White
© 2006 American Institute of Physics 0-7354-0341-4/06/$23.00

# UNUSUAL SELF-SIMILAR COMPRESSION

## J. Gerin-Roze

*Commissariat à l'Energie Atomique, DIF, 91680 Bruyeres le Chatel, France*

**Abstract**: Efficient gas compression is a very interesting issue. We study a family of self-similar spherical implosions involving high compression rate. This family corresponds to the classical solution initiated by a strong shock followed, after a particular characteristic line, by a centered beam of characteristic curves.

**Keywords**: hydrodynamics, compression, self-similarity, deuterium – tritium
**PACS**: 47.40.- x, 47.11.+ j .

## INTRODUCTION

What is an "unusual self-similar compression law" (noted USCL in the following)? Such a law is characterized by a "classical self-similar law" initiated by a convergent shock (ref. 1 and 2) followed by centered compression waves. The usual self-similar compression leads to a compression rate equal to 9.6 at shock focalization if we assumed the gas is perfect with a polytropic coefficient $\gamma$ equal to 5/3. On the other hand, the USCL involves a compression rate as high as we wish provided we choose a time close enough the shock focalization time. We will explain the implosions obtained by this law in part Theory.

Then, we will apply such a law to a DT sphere (m=1.5 µg and $\rho_0$ = 0.003 g/cm$^3$). We will detail the thermodynamical conditions obtained this way. Secondly, we will compare these results to those obtained with a 1D hydrodynamical computational code under different hypotheses: complete outside conditions (speed versus time) and perfect gas EOS, truncated outside conditions and perfect gas EOS and truncated outside conditions and more realistic gas EOS (Sesame). It will be the subject of the part Implementations.

Finally, we will explain the uses of this law and the outlook of this work.

## THEORY

### Usual self-similar law

We study the 1D spherical implosion of a perfect gas initiated by a strong convergent shock. The three partial differential equations link the three unknown functions $\rho$ (density), u (velocity) and p (pressure) of the two variables r (space variable) and t (time) :

$$\partial \log (\rho) / \partial t + u\, \partial \log (\rho)/\partial r + \partial u/\partial r + 2u/r = 0$$
$$\partial u/\partial t + u \partial u/\partial r + (1/\rho)\, \partial p/\partial r = 0 \quad (1)$$
$$\partial (\log (p\rho^{-\gamma}) / \partial t + u\, \partial \log (p\rho^{-\gamma}) /\partial r = 0$$

Now, the following change of variables and unknown functions is realized:
♦ $\xi = r/a$ where $a = A (-t)^\alpha$ is going the shock position (t<0, t=0 at shock focalization).
$\xi$ varies from 1 (on the shock) to $\infty$ (focalization)
♦ $u = (r/t)\, V (\xi)$
♦ $c^2 = (r^2/t^2)\, Z (\xi)$, $c^2$ is the sound velocity squared, linked to p and $\rho$ by $c^2 = (\partial p/\partial \rho)_S$
♦ $\rho = \rho_0 G (\xi)$

So, the three new differential equations of only one variable $\xi$ are obtained:
$$dV/d\log\xi = N_1/D$$
$$dZ/d\log\xi = ZN_2/D \quad (2)$$
$$dG/d\log\xi = GN_3/D$$

where N1, N2, N3, D are depending on $\gamma, \alpha, V$ and Z.

1515

To solve this system the following initial conditions have to be used :

$$V(1) = 2\alpha/(\gamma+1)$$
$$Z(1) = 2\gamma(\gamma-1)\alpha^2/(\gamma+1)^2 \qquad (3)$$
$$G(1) = (\gamma+1)/(\gamma-1)$$

They express the Hugoniot conditions on the shock (where $\xi=r/a=1$).

Now, we discuss the solutions of this system (3). Solving the two first equations of (3) from the point S ($\xi=1$ on the shock), we determine easily V ($\xi$) and Z ($\xi$) until the point A (see Figure 1).

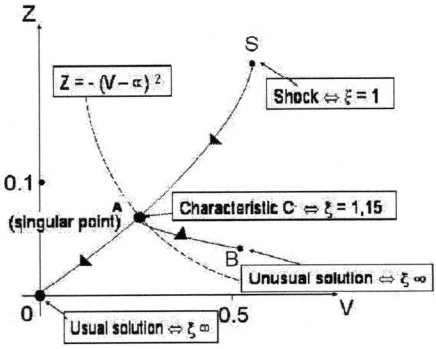

**Figure 1** : The Integral Curve V (Z)
The way S-A-O corresponds to usual self-similar law.
The way S-A-B corresponds to unusual self-similar law.

At this point D=0, it can be shown that the solution still exists if together $N_1=N_2=N_3=0$, requiring $\xi=1.15$. This value of $\xi$ leads to $\alpha=0.688$.

In the plane (t, r), the characteristics C⁻ going to the center of the sphere is the iso-$\xi$ curve associated to the above value of $\xi$ (see Figure.2).

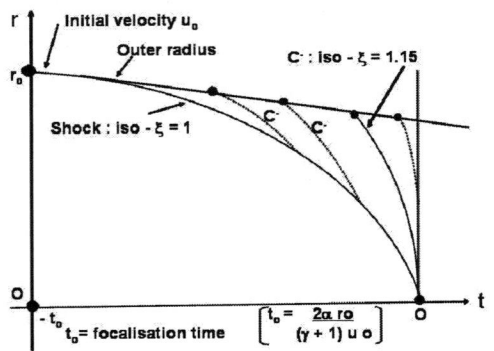

**Figure 2.** Usual Self – Similar Law: The Characteristics.

Then, the usual solution follows the path AO in the plane (V, Z) (see figure 1). At the shock focalization (point O), the compression rate is $\rho/\rho_O = 9.6$.

We can also work out the surface conditions (pressure p (t) and velocity u (t): the velocity is almost constant and the pressure is multiplied by about 4 against initial shock pressure (see below Figure 4).

### A new Self-similar Law: Unusual Self-Similar Compression Law (USCL)

There is another suitable integral curve from the peculiar point A (where $N_1=N_2=N_3=D=0$) to a new point B corresponding to the shock focalization (see Fig.1). This point replaces the point O of the usual law. The initial gradient of the curve AB close to the point A is obtained with a Taylor development: this gradient is -0.569 against 0.352 for the usual law.

Now, there are two independent regions in the (r, t) plane (see Figure 3 below):

- Left the iso-$\xi$ $\xi=1.15$ curve, the previous solution is still valid. Each characteristic line reaches the shock.
- Right this iso-$\xi$ curve, the beam of characteristics is converging to the centre O of the sphere. No shock is created and the density can increase without any limit at the shock focalization instant. In this region the flow is isentropic.

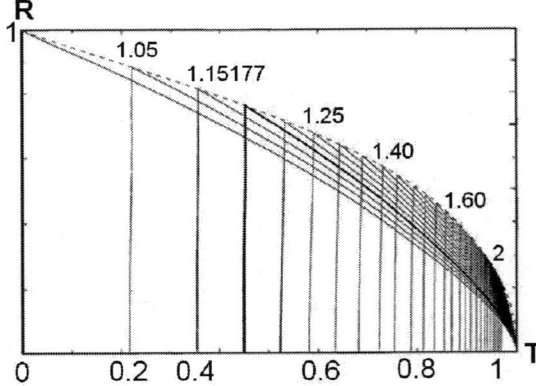

**Figure 3**. Unusual Self Similar Law: The Iso-$\xi$ Curves in the (T, R) plane
• T is a normalized time $T = 1 + t/t_o$ ($t_o$ is the focalisation time, see Fig 2)
• R is a normalized radius $R = r/r_o$.

1516

The following table gives V, Z, G for ξ going from 1 to 10. Take particular notice of the great values of density given by G.

**TABLE 1.** Numerical Values of the Normalized Functions.

| ξ | V (ξ) | Z (ξ) | G (ξ) |
|---|---|---|---|
| 1 | 0.52 | 0.15 | 4 |
| 1.15 | 0.39 | 0.091 | 5.20 |
| 2.20 | 0.46 | 0.055 | $1.8 \times 10^2$ |
| 5.23 | 0.49 | 0.044 | $8.7 \times 10^4$ |
| 10 | 0.50 | 0.043 | $1.3 \times 10^7$ |

Now, the surface conditions can be calculated by writing that compression is isentropic behind the shock. We obtain time, outer radius, outer velocity and outer pressure against ξ. Then, we can work out at each time t $p_S$ and $u_S$ on the surface and compare them to $p_S$ and $u_S$ required by the usual self-similar law (see below figure 4).

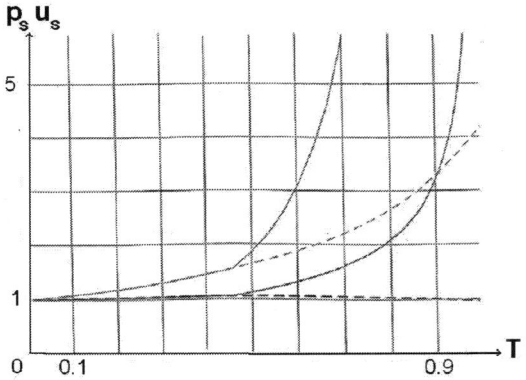

**Figure 4.** Surface Conditions (pressure and velocity versus time)
Dotted lines : usual law and full lines : unusual law.
Lower lines : velocity and upper lines: pressure

## IMPLEMENTATIONS

### USCL applied to a DT microsphere

The solution exhibited above used two geometric parameters ($r_0$ and $\rho_0$) and one implosion parameter (the focalization time $t_o$ combining the initial outer velocity and the initial radius $r_o$ ). We choose to apply the previous USCL to a DT microsphere with m=1.5µg and $\rho_0 = 3.10^{-3}$g/cm³. We try

to obtain the following DT ignition conditions using the mean values of density and temperature: $\rho_m = 50$ g/cm³ (associated to $\rho r = 0.1$g/cm²) and $\theta_m = 5.10^7$ K (with the additional assumption Cv = $1.10^8$ erg/g.K). We have to choose the implosion parameter $u_o$ initial outer velocity). We take $u_o = 2.10^6$cm/s.

Now, by using the previous table (ξ, V, Z, G), we can calculate $\rho(r)$ and $\theta(r)$ profiles and eventually the mean values $\rho_m$ and $\theta_m$ at several instants (see Figures 5 and 6).

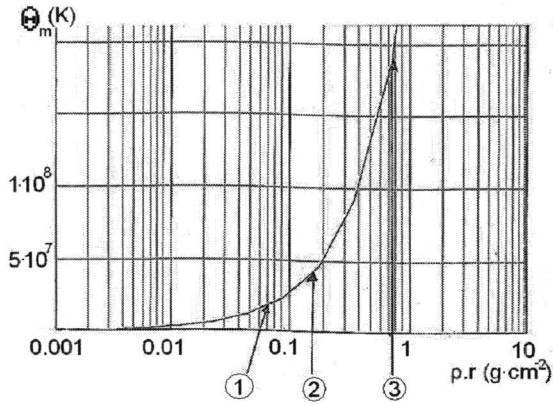

**Figure 5.** The example of pellet implosion : Average Temperature $\theta_m$ versus ($\rho$.r)
① E = $E_i$ + $E_c$ = 0.085 KJ, $E_c/E_i$ = 2.6 , P = 1.2 TW
② E = $E_i$ + $E_c$ = 2.8 KJ,     $E_c/E_i$ = 3.1, P =1100TW .
③ E = $E_i$ + $E_c$ = 12 KJ,     $E_c/E_i$ = 3.2

We observe that we obtained the intended thermo-dynamical conditions: $\rho r = 0.1$g/cm² and $\theta_m = 5.10^7$ K when the total injected energy is W≅ 3 KJ and the power peak reaches P ≅ 1000 TW. At this moment, the ξ value on the sphere surface is ξ≅ 10.

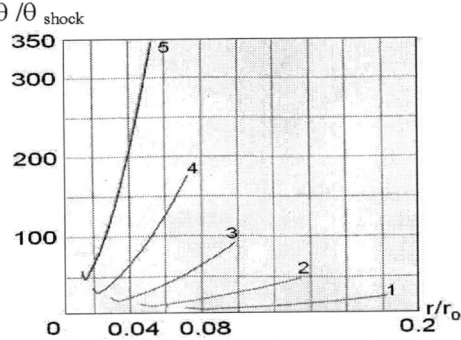

**Figure 6**. The example of pellet implosion: Temperature $\theta$ versus radius r at five times. Curve 1 T = 0.98 (see key in Figure 3) / 2 T = 0.99 / 3 T = 0.993 / 4 T = 0.997 / 5 T = 0.998 [$\theta_{shock} = u^2/2 \, c_v = 2 \times 10^4$ K]

### Numerical calculations

Now, we work out the implosion of the above microsphere with using an 1D hydrodynamical code under different hypotheses:

♦Hyp1: complete outside condition (velocity versus time) given by the USCL and perfect gas EOS.

♦Hyp2: complete outside condition and more realistic gas EOS (Sesame).

♦Hyp3: truncated outside conditions (at 0.998. focalization time) and Sesame EOS.

**TABLE 2**. Average Density and Internal Energy: For values (model, calc 1, 2, 3) at two times.

| At T = 0.997 | ρ (g/cm³) | E_i (J) |
|---|---|---|
| Model | 9.1 | 27 |
| Calc. 1 | 9.8 | 28 |
| Calc. 2 | 9.7 | 13 |
| Calc. 3 | 9.7 | 13 |

| At T = 0.999 | ρ (g/cm³) | E_i (J) |
|---|---|---|
| Model | 44 | 85 |
| Calc. 1 | 57 | 88 |
| Calc. 2 | 55 | 42 |
| Calc. 3 | 45 | 29 |

We can compare these three calculations and the USCL model. In this purpose, the table 2 gives the average densities and the internal energies reached at two times using the previous model and the three different calculations (hyp. 1, 2 3).

Model and Calc. 1 are in good agreement. Besides, comparison of Calc.1 and 2 shows the consequence of the use of a realistic EOS: weak on density but more important on internal energy. Last, Calc. 3 suggests that the effect of the stop of the outer velocity law appears tardily in the compression.

## CONCLUSION AND LATER EXTENSION

The usual self-similar law gives the exact solution for a problem with a piston velocity almost exactly constant (see fig.4). So the shock focalization compression rate is equal to 9.6. The USCL gives also an exact solution but the piston velocity is not at all constant: it now grows continually (see Fig. 4) and the shock focalization compression rate is tending towards the infinite.

Such a flow may be worth using. First, it gives an accurate hydrodynamical benchmark suited to the difficult problem of spherical shock convergence. We can exhaustively work out this flow which depends on two time parameters: focalization time ($t_o$) linked to initial outer velocity and time of application of outer velocity ($t_1 \Rightarrow T = t_1 / t_o$). Then, it can be used to define laser experiments where very compressed matter is needed (spectroscopy experiments, thermonuclear ignition...). Indeed, it allows creating an optimized compression law for each problem (when mass and initial density are given).

An interesting extension will be to add a second medium surrounding the gas. Using the characteristic curves, we could obtain the outside conditions (pressure or velocity versus time) for this more realistic geometry.

### REFERENCES

1. Zel'dovich, Y. B. and Raizer, Y. P., Physics of Shock Waves and High–Temperature Hydrodynamic Phenomena tII, Academic Press, p 794 and following (1967).
2. Lazarus, R. B. and Richtmyer, R. D., Similarity Solutions for Converging Shocks, Los Alamos Scientific Laboratory – LA – 6823 –MS June 1977.
3. Morreeuw, J. P. and Saillard, Y., Optimal Isotropic Compress of an Initially Uniform and Stationary Sphere, Nucl Fusion 189 (1978).

CP845, *Shock Compression of Condensed Matter - 2005*,
edited by M. D. Furnish, M. Elert, T. P. Russell, and C. T. White
© 2006 American Institute of Physics 0-7354-0341-4/06/$23.00

# HIGH-RATE COMPACTION OF ALUMINIUM ALLOY FOAMS

## J.J. Harrigan[1], Y.-C. Hung[1], P.J. Tan[1], N.K. Bourne[1], P.J. Withers[1], S.R. Reid[1], J.C.F. Millett[2] and A.M. Milne[3]

[1]*The University of Manchester, PO Box 88,Sackville Street, Manchester M60 1QD, UK*
[2] *Defence Academy of the UK, Cranfield University, Shrivenham, Swindon, SN6 8LA, UK.*
[3] *Fluid Gravity Engineering, 83 Market Street, St. Andrews, Fife, KY16 9NX.*

**Abstract.** The response of aluminium foams to impact can be categorised according to the impact velocity. Tests have been carried out at a range of impact velocities from quasi-static to velocities approaching the speed of sound in the foam. Various experimental arrangements have been employed including pneumatic launcher tests and plate impact experimants at velocities greater than $1000$ m s$^{-1}$. The quasi-static compression behaviour was approximately elastic, perfectly-plastic, locking. For static and dynamic compression at low impact velocities the deformation pattern was through the cumulative multiplication of discrete, non-contiguous crush bands. Selected impact tests are presented here for which the impact velocity is less than the velocity of sound, but above a certain critical impact velocity so that the plastic compression occurs in a shock-like manner and the specimens deform by progressive cell crushing. Laboratory X-ray microtomography has been employed to acquire tomographic datasets of aluminium foams before and after tests. The morphology of the underformed foam was used as the input dataset to an Eulerian code. Hydrocode simulations were then carried out on a real microstructure. These simulations provide insight to mechanisms associated with the localization of deformation.

**Keywords:** Aluminium foam, microtomography, plate impact.
**PACS:** 62.50

## INTRODUCTION

When open-cell metal foams are compressed, work is done predominantly in bending the struts. The propensity of aluminium foam to undergo gross 'plastic' deformation at an almost constant nominal load with large strokes (see Fig. 1) makes it attractive for absorbing the energy of impact or impulsive loads in packaging applications, crash situations and for blast protection. An understanding of the strength properties of the foam and its macroscopic response at different loading rates is a prerequisite to their successful implementation. High rate loading experiments also provide the critical data required for developing predictive constitutive models for metal foams. Selected results of tests over a velocity range from static to $1013$ m s$^{-1}$ are presented. Enhancements in stress measured at the proximal end are the result of inertia effects associated with the dynamic localisation of crushing.

Different impact velocity regimes have been discussed recently in [1] for aluminium foams. The dynamic results presented here fall within the range wherein the impact velocity is less than the velocity of sound, but sufficiently high for the plastic deformations to occur via progressive cell crushing in a "shock-like" manner (see [2-5]). Analysis of the crushing response is discussed with reference to analytical predictions based on analogy of a "steady-shock" propagation [4,5], finite element analysis of two-dimensional Voronoi honeycombs [4, 6] and hydrocode simulations of a real microstructure.

The material considered is DUOCEL® Aluminum foam with 40 or 10 pores per inch (ppi). The 40ppi foam has a density of about 250 kg m$^{-3}$ corresponding to a relative density $\rho_r$ of approximately 9%, while the 10 ppi foam has a density of 272 kg m$^{-3}$. The solid phase (cell wall) material is 6101 alloy. The results of a quasi-static compression test on a 40 ppi sample are plotted in Fig. 1 in terms of nominal stress and strain. The sample was 45 mm in diameter and length and compression took place within a constraining barrel.

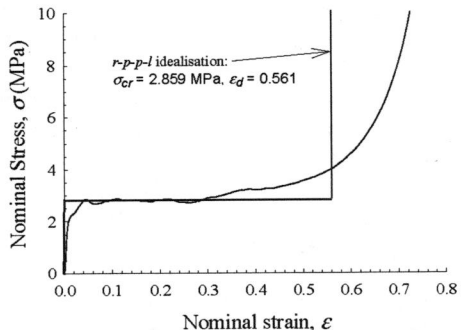

**Figure 1.** Quasi-static stress-strain curve for a 40 ppi Duocel® foam.

## EXPERIMENTAL PROCEDURES

### Pneumatic Launcher tests

The work described herein deals mainly with high impact velocity situations where inertial effects are expected to be dominant. This includes both microinertia (see [2,5,7]), i.e. inertia effects which are more commonly associated with Type II structures, and shock wave or crush-front propagation effects. In such cases there can be large differences between the forces generated at the end of the absorber where the impact occurs (the proximal end) and the far end of the device (the distal end).

For testing at speeds up to 250 m s$^{-1}$, the apparatus used including the Hopkinson bar load

cell and specimen constraint have been described previously [3]. Foam projectile tests were carried out using this set-up, illustrated in Fig. 2, to measure proximal end loads. Tests to measure distal end loads will be reported elsewhere. The specimens were 45 mm in diameter and had various lengths. Various backing masses were attached to the distal end of the specimens. An attachment to the gun barrel provided lateral constraint.

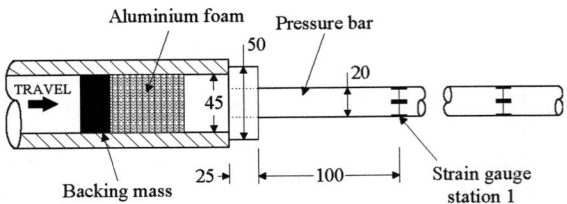

**Figure 2.** Set-up for pneumatic launcher tests to measure proximal end forces.

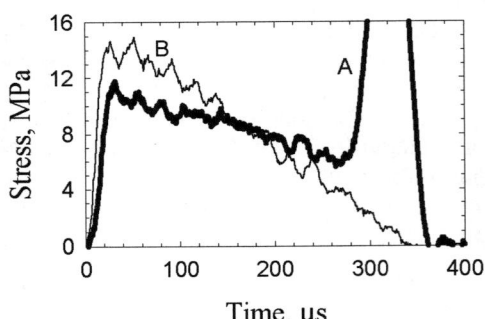

**Figure 3.** Typical stress-time curves for pneumatic launcher tests.

Typical stress-time traces for this set-up are shown in Fig. 3. Curve A of Fig. 3 is for a 50 mm long, 40ppi foam sample, $\rho_r = 0.0929$, with a backing mass of 0.017 kg that impacted the load cell at 162 m s$^{-1}$. Clearly the stresses associated with the dynamic compaction of this sample are well above the levels associated with static compression (Fig. 1). The stress reduces gradually with time before densification occurs causing the load to rise steeply. Curve B of Fig. 3 is for a 100 mm long, 10ppi foam sample, $\rho_r = 0.097$, with no backing mass. The impact velocity of 170 m s$^{-1}$ was not sufficient to cause full densification of the sample, so the stress reduces gradually to zero from the initial peak of about 14.5 MPa. The distinction between the compacted and undeformed portions of the sample B (Fig. 3) is evident in the photograph of the tested sample (Fig. 4).

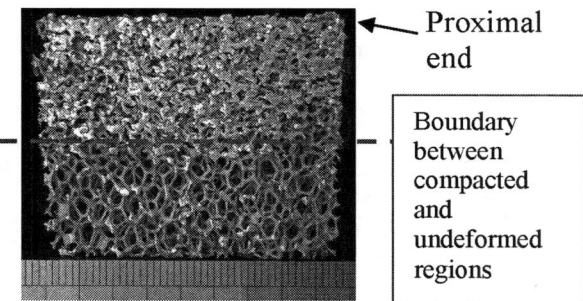

Proximal end

Boundary between compacted and undeformed regions

**Figure 4.** Photograph of sample B of Fig. 3.

### Plate Impact tests

Plate impact tests was carried at impact velocities between and 281 and 1013 m s$^{-1}$. The shots were performed using a 50 mm bore, 5 m long single stage gas gun [8]. The objectives of these preliminary tests were to investigate the experimental set-up, to obtain data for the calibration of numerical models and eventually to investigate the effect of the impact velocity on the deformation mechanisms. The experimental arrangement is illustrated in Fig. 5. 10 mm thick, circular aluminium alloy flyer plates were fired at 10 mm thick, 45 mm diameter, 40 ppi, aluminium foam samples. Manganin stress gauges could not be attached directly to the foam and so were attached betweem the backing plates consisting of 5 mm and 10 mm aluminium plates. The first gauge was central to the axis while the second gauge was offset from the axis by 20 mm.

Fig. 6 shows that the two gauges follow very similar stress histories while the stress is rising. This indicates that edge effects are not significant during the compaction of the sample. However, there is difficulty interpreting the data as the stress levels correspond to fully densified foam rather than the stresses during crushing. The initial elastic precursor will travel through the foam at stresses four orders of magnitude less than the maximum stresses in Fig. 6. This will be followed by a compaction wave.

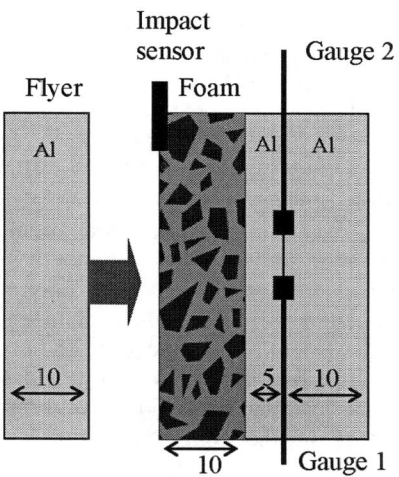

**Figure 5.** Set-up for plate impact tests.

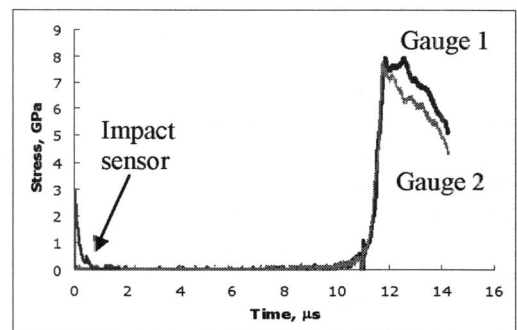

**Figure 6.** Plate impact test results, V$_o$ = 1013 m s$^{-1}$.

### ANALYSIS

Analysis of the foam has been attempted in a number of ways. Wave models based on a rate independent, rigid, perfectly-plastic, locking (r-p-p-l) idealization of the quasi-static stress-strain curve [2,4] have been employed for a number of cellular materials. Further analyses have included elastic effects and have relaxed the definition of densification from a single locking value [5]. Such wave analyses cannot capture fully the behaviour of a cellular solid: a plane of discontinuity is assumed to pass through cells with real dimensions. Nonetheless, these models give a good first estimate of the dynamic stresses for impact velocities well below the speed of sound in the material. For example, the experimental foam properties in Fig. 1 can be approximated as an *r-p-p-l* material with a plateau stress of 2.859 MPa and a locking strain of 0.561 (see [4]). The analysis in [2,4] would then predict an initial dynamic crushing stress of 14.55 MPa at an impact velocity of 162 m s$^{-1}$ for sample A in Fig. 3. However, these models are based on the overall geometric properties so details associated with the compaction zone are lost.

Finite element analysis has been employed in order to account for the cellular nature of the material (e.g. [6]). These models have been used to explore the differences between quasi-static and dynamic deformations, to provide an understanding of the nature of the compaction zone and the enhancement in energy absorption associated with dynamic compaction. However, the finite element analysis is computationally expensive and was restricted to two-dimensional cellular arrays.

Laboratory X-ray microtomography was carried out in order to model accurately the three-dimensional structure of the cellular solid. The tomographic images were read directly into the mesh generators for three-dimensional simulation. The code employed was Eden [8], an Eularian multimaterial hydrocode. Simulations were then carried out on real microstructures such as that shown in Fig. 7. Contours of pressure show wave propagation in the solid material ahead of the compaction front.

**Figure 7.** Mesoscale model of a compaction front in a 40 ppi foam impacting a rigid wall at 1000 m s$^{-1}$, 0.5 $\mu$s into collapse.

# CONCLUSIONS

By employing a variety of experimental techniques and analytical methods a better understanding of the deformation mechanisms during dynamic compaction of aluminium foams is possible. Fuller details of the results presented here will be given in future papers.

# ACKNOWLEDGEMENTS

The authors are grateful to the EPSRC for their financial support through grant GR/R26542/01 and to ONR Global for their support through the STEP program. Support from ERG Materials and Aerospace Corporation is also gratefully acknowledged.

# REFERENCES

1. Lopatnikov, SL, Gama, BA, Haque, MJ, Krauthauser Guden, M, Hall, IW, and Gillespie, JW, Dynamics of metal foam deformation during Taylor cylinder-Hopkinson bar impact experiment. *Compos Struct*, 2003; 61: 61-67.

2. Reid, SR, and Peng, C, Dynamic Uniaxial Crushing of Wood. *Int. J. Impact Engng*. 1997:19; 531-570.

3. Tan, PJ, Reid, SR, Harrigan, JJ, Zou, Z, and Li, S, Dynamic compressive strength properties of aluminium foams. Part I – Experimental data and observations, *Journal of Mechanics and Physics of Solids*. 2005; In Press.

4. Tan, PJ, Reid, SR, Harrigan, JJ, Zou, Z, and Li, S, Dynamic compressive strength properties of aluminium foams. Part II – Shock theory and comparison with experimental data and numerical models, *Journal of Mechanics and Physics of Solids*. 2005; In Press.

5. Harrigan, JJ, Reid, SR, Tan, PJ, and Reddy, TY, High rate crushing of wood along the grain, *Int. J. Mech. Sci*. 2005; 47: 521-544.

6. Tan, PJ, Reid, SR, Harrigan, JJ, Discussion: "The resistance of clamped sandwich beams to shock loading" (Fleck, N.A., Deshpande, V.S., 2004, ASME J. Appl. Mech., 71, pp. 386 – 401), *J. Appl. Mech*. 2005; In Press.

7. Gibson, LJ, and Ashby, MF, *Cellular Solids-Structure and Properties*, 2$^{nd}$ edn. Cambridge University Press, Cambridge, 1997.

8. Bourne, NK, Meas. Sci. Technol. 2003; 14: 273-8

9. Milne, AM, EDEN user manual, Fluid Gravity Engineering Ltd, St. Andrews, 2004.

CP845, *Shock Compression of Condensed Matter - 2005*,
edited by M. D. Furnish, M. Elert, T. P. Russell, and C. T. White
© 2006 American Institute of Physics 0-7354-0341-4/06/$23.00

# INFLUENCE OF CONTROLLED VISCOUS DISSIPATION ON THE PROPAGATION OF STRONGLY NONLINEAR WAVES IN STAINLESS STEEL BASED PHONONIC CRYSTALS

**Eric B. Herbold**[*], **Vitali F. Nesterenko**[*†] **and Chiara Daraio**[†]

[*]*Department of Mechanical and Aerospace Engineering, University of California at San Diego, La Jolla, California 92093-0411, USA*
[†]*Materials Science and Engineering Program, University of California at San Diego, La Jolla, California 92093-0418, USA*

**Abstract.** Strongly nonlinear phononic crystals were assembled from stainless steel spheres. Single solitary waves and splitting of an initial pulse into a train of solitary waves were investigated in different viscous media using motor oil and non-aqueous glycerol to introduce a controlled viscous dissipation. Experimental results indicate that the presence of a viscous fluid dramatically altered the splitting of the initial pulse into a train of solitary waves. Numerical simulations qualitatively describe the observed phenomena only when a dissipative term based on the relative velocity between particles is introduced.

**Keywords:** strongly nonlinear solitary waves, fluids, dissipation, Hertz law
**PACS:** 05.45.Yv, 46.40.Cd, 43.25.+y, 45.70-n

## INTRODUCTION

In recent years, many experimental and numerical efforts have been focusing on the propagation of strongly nonlinear solitary waves in granular media [1]. These waves are a natural extension of the well known weakly nonlinear solitary waves such as the Korteweg-de Vries (KdV) solitons. Chains composed of steel, brass, glass, Nylon [1, 2, 3] and Polytetraflouroethylene (PTFE) [4] particles support solitary waves using Hertz' law for particle interactions.

However, dissipation plays a significant role in the transmission of pulses in almost all experimental settings. Restitution coefficients and velocity dependent friction have been used to investigate dissipation in a chain of particles [5, 6]. The energy losses are significant even when the experiments are performed in air (see Fig. 1.25 in [1]). These losses may be attributed to uncontrolled features in the experimental setup as well as inherent material properties [7].

To predictably control the dissipation, a chain of spheres may be immersed in viscous fluids. The viscous dissipation of pulses in chains of particles differs from a two particle interaction in liquid because the compression wave dominates the system's dynamic behavior. To our knowledge, there has not been an attempt to extend the research involving the collision of particles in fluids to a chain of particles. This paper extends current knowledge of a two particle collision in a viscous medium to a case where multiple particle interactions support a solitary wave.

## ANALYTICAL MODEL AND NUMERICAL CALCULATIONS

The analytical model for a one-dimensional Hertzian chain of spheres can be found in [1]. The description of particle interactions in fluids is presented in [8-12]. One of the complications of using current sphere-fluid models for a chain of spheres is that a developed flow around the sphere is assumed before the particle collision though the duration of the pulse in a chain is relatively short ($\sim$50 $\mu$s). Nevertheless we use the outlined approach for the first stage of our research. In [10] two particles are considered to be traveling towards each other in a surrounding medium. If the spheres are the same size then the total kinetic energy is

$$\frac{1}{4}\left(2m + m' + \frac{3}{8}\frac{m'R^3}{h^3}\right)U^2 = constant, \quad (1)$$

where $m = \frac{4}{3}\pi\rho R^3$ is the mass of the particle, $m' = \frac{11}{12}\pi\rho_f R^3$ is the added mass, $R$ is the particle's radius, $h$ is the separation distance between particle centers and $U = \frac{dh}{dt}$ is the relative particle velocity. Differentiating Eq. (1) with respect to $h$ yields the added mass and pressure force terms. The equations of motion for a chain of spheres, with mass $m_i$, placed vertically in a gravitational field in a fluid becomes

$$(m_i + m_i')\frac{d^2x_i}{dt^2} = F_{c,i}(\delta) + m_i g + F_{b,i}$$
$$+ F_{D,i} + F_{p,i} + F_{d,i}. \quad (2)$$

where $\delta_{i,i+1} = (R_i + R_{i+1} + x_i - x_{i+1})$ and the $x_i$'s are the particle positions. The compression force $F_{c,i}$ is based on Hertz law between particle '$i$' and the adjacent particles '$i$-1' and '$i$+1';

$$F_{c,i} = \phi\left(\delta_{i-1,i}\right) - \psi\left(\delta_{i,i+1}\right), \quad (3)$$

where $\phi$ and $\psi$ can be expressed as

$$\phi\left(\delta_{i-1,i}\right) = A_{i-1,i}\left(\delta_{i-1,i}\right)^{3/2} \quad (4)$$

and

$$\psi\left(\delta_{i,i+1}\right) = A_{i,i+1}\left(\delta_{i,i+1}\right)^{3/2}. \quad (5)$$

The coefficients in Eqs. (4) and (5) are identical except for a shift of indices. The equation for $A_{i-1,i}$ can be inferred from

$$A_{i,i+1} = \frac{4E_iE_{i+1}\left(\frac{R_iR_{i+1}}{R_i+R_{i+1}}\right)^{1/2}}{3\left[E_{i+1}\left(1-v_i^2\right) + E_i\left(1-v_{i+1}^2\right)\right]}, \quad (6)$$

where $E_i$ and $v_i$ are the elastic modulus and Poisson's ratio of the particle. For all of the calculations performed with air as the surrounding medium, only

the first two terms on the right hand side and the first term on the left side of Eq. (2) are used. When liquids surround the chain of spheres the buoyancy, drag, pressure, added mass and dissipative terms are used. The buoyancy force for each particle is

$$F_{b,i} = -\frac{4}{3}\pi R_i^3 \rho_f g, \quad (7)$$

where $g$ is the gravitational constant, $R_i$ is the radius of the particle and $\rho_f$ is the density of the fluid. The drag force has a correction factor to account for a Reynold's number $Re$ greater than unity i.e. not in the Stoke's flow regime,

$$F_{D,i} = -6\pi v R_i U_i(1 + 0.15Re^{0.687}), \quad (8)$$

where $v$ is the dynamic viscosity and $U_i$ is the particle velocity. The pressure force in [9] is written for one sphere moving toward a stationary sphere. In our case, we use a relative velocity between particles,

$$F_{p,i} = \frac{3}{8}\pi R_i^2 \rho_f \left[(U_{i-1} - U_i)^2 - (U_i - U_{i+1})^2\right]. \quad (9)$$

We introduce an additional dissipative term based on the relative velocity between particles with a fitting parameter $c$ due to the lack of a qualitative agreement between the experiments and calculations based on the drag force term (Eq. (8)). This addition is tantamount to adding a dash-pot between neighboring particles [13],

$$F_{d,i} = c\left(U_{i-1} - 2U_i + U_{i+1}\right), \quad (10)$$

where the coefficient $c$ is a fitting parameter. The physical reason for this term can be due to the radial flow of liquid caused by the change of contact area between particles.

MATLAB's intrinsic ODE45 solver was used to march the explicit calculation forward in time with a time-step of 0.05$\mu$s. The error in the energy calculations were found to be less than $10^{-9}\%$ in air and within $10^{-5}\%$ in the fluid. The error in the conservation of linear momentum performed with a chain in air was less than $10^{-12}\%$.

## EXPERIMENTAL PROCEDURES AND RESULTS

In the experiments, the impulse propagation was investigated in three different media: air, SAE 10W-30 motor oil, and non-aqueous Glycerol GX0185-5.

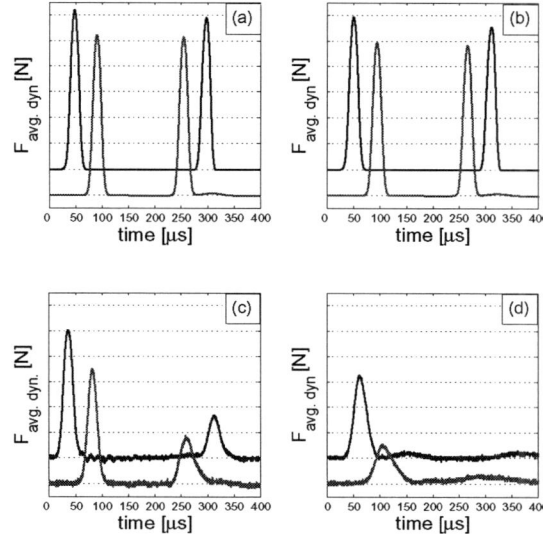

**FIGURE 1.** Single solitary wave in calculations (a) and experiments (c) in a chain surrounded by air. Results for an identical chain surrounded by glycerol in calculations (b) and experiments (d). Vertical scale is 2 N/div.

The density and dynamic viscosity used in numerical calculations were $\rho_f = 880 \, \text{kg/m}^3$, $v = 0.067 \, \text{Ns/m}^2$ for oil and $\rho_f = 1260 \, \text{kg/m}^3$, $v = 0.62 \, \text{Ns/m}^2$ for glycerol. An experiment was performed with each of the three types of media to see how single and multiple solitary waves propagate through the chain. The chain was placed into an adjustable holder that had four contact points on each sphere. The air or fluid was able to flow freely between the contacts as opposed to the cylindrical holder that had been used in previous experiments. To create a single solitary wave, a spherical stainless steel striker with a radius of $R = 4.76$ mm and a mass of $m = 0.4501$ g was used to impact the top of a chain of 19 stainless steel particles (also with $R = 4.76$ mm) with a velocity of $U_0 = 0.44$ m/s. To create multiple solitary waves in the same chain, a cylindrical alumina striker with a larger mass of $m = 1.23$ g impacted the chain at $U_0 = 0.44$ m/s. The elastic modulus and Poisson's ratio of the stainless steel particles were 193 MPa and 0.3, respectively. The experimental results were recorded via piezoelectric gauges imbedded [4] in the 10th and 15th particles from the top of the chain. The procedures outlined in [4] were implemented to compare the calculations to the experimental results

using the averaged dynamic force as noted on the vertical axis of Fig. 1, 2 and 3.

In Fig. 1 the numerical and experimental results are shown for a single solitary wave in a chain surrounded by air (a), (c) and glycerol (b), (d). It is evident that there is a very small difference between the numerical results for the chain in air and glycerol using Eqs. (2)-(9), which prompted the inclusion of the additional dissipative term (Eq. (10)). Without this dissipative term, the numerical results for glycerol shown in Fig. 1(b) do not exhibit the shape and speed of the pulses in experiments.

In experiments the speeds of the single pulse were $V_s = 520$ m/s in air and increased to $V_s = 541$ m/s in glycerol. It is interesting to note that the calculated signal speed was $V_s = 564$ m/s in air and decreased to $V_s = 540$ m/s in glycerol. In numerical calculations the speed of the single pulses in chains surrounded by both air and glycerol should be higher than the experimental pulse speeds due to the higher amplitude of the waves. Also, one would intuitively think that the pulse speed in air would be higher than in glycerol due to viscous dissipation. This is the case in calculations (even when the additional dissipative term $F_d$ is added) but the opposite is true in experiments (Fig. 1(c) and (d)).

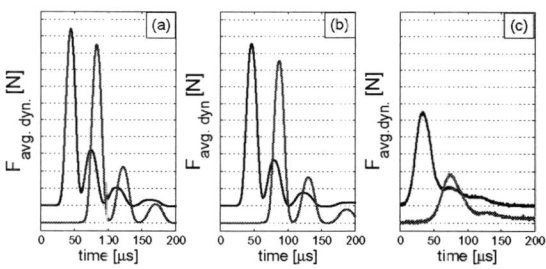

**FIGURE 2.** (a) Numerical results for multiple solitary waves in a chain of 19 particles surrounded by air. (b) Numerical results for multiple solitary waves in an identical chain surrounded by glycerol. (c) Experimental results related to (b). Vertical scale is 2 N/div.

In Fig. 2 the numerical results for a train of solitary waves in air and glycerol are presented. The dissipative term $F_d$ was not included in the calculations in (a) and (b) and it is apparent that the amplitudes of the waves are much higher than the experimental results in glycerol (c). Additionally, no signal splitting into a train of solitary waves were present in experiments Fig. 1(c), which is not reflected by the calculations without the additional dissipative term $F_d$.

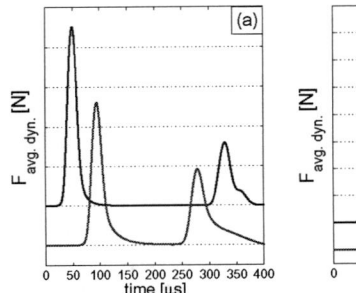
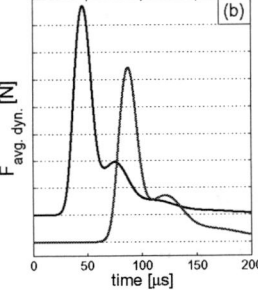

**FIGURE 3.** (a) Numerical results for a single pulse with additional dissipative term Eq. (10) in a chain surrounded by glycerol. (b) Numerical result for an impact by an alumina striker on the same chain. Vertical scale is 2 N/div.

Again, the wave speeds in experiments were higher in glycerol than in air.

There is a disparity between the presented analytical formulation using the corrected Stokes drag for dissipation and the experiments. The largest dissipative term in the calculations (before the addition of the dissipative term $F_d$) is the drag force term. The reduction of the amplitude of a single pulse was about 3% in oil and about 4% in glycerol when using Eqs. (2)-(9). These equations, while important for accurately describing a particle trajectory pre and post collision are negligible when their effect on the compression wave are examined.

The results of adding the dissipative term Eq. (10) are presented in Fig. 3 for single and multiple pulses traveling in a chain submersed in glycerol. A similar asymmetry and widening of the pulse, apparent in experiments, can also be seen when Fig. 3(a) is compared to Fig.1(d). The lower line in Fig. 1(d) has a noticeable shock-like tail. In calculations the tail of the wave becomes more shock-like as parameter $c$ increases. The qualitative behavior of the pulse in experiments is reproduced in calculations by adding an additional dissipative term $F_d$ with a coefficient $c = 6.0$ Ns/m for glycerol and $c = 0.648$ Ns/m for oil. These values provided the best comparison between experimental and numerical data and were scaled according to the difference in viscosity. When comparing Fig. 2(b) and Fig. 3(b), notice the additional dissipation Eq. (10) has also created a shock-like response of the train of solitary waves. In calculations the amplitude and tendency to split decreased by adding this term in accord with experiments.

## CONCLUSIONS

The experimental results indicate a qualitative change of the propagating shock and solitary waves in a chain immersed in glycerol while only a small change in oil. Without the relative velocity based dissipative term $F_d$, the equations pertaining to the surrounding fluid could not accurately reproduce the amplitude or the shock like structure of the incident pulse. This term provided the qualitative change needed to match the numerical analysis to the experiments. The numerical analysis predicted a decrease in solitary wave speed as the viscosity of the surrounding fluid increased contrary to experiments. This phenomenon may be explained by an increased effective stiffness modulus between particles in the presence of a viscous fluid.

## ACKNOWLEDGEMENTS

The work is supported by NSF (Grant No. DCMS03013220).

## REFERENCES

1. Nesterenko, V. F., *Dynamics of Heterogeneous Materials*, Chapter 1, Springer-Verlag, NY, 2001.
2. Coste, C., Falcon, E., and Fauve, S., *Physical Review E*, **56**, 6104 (1997).
3. Coste, C., and Gilles, B., *The European Physical Journal B*, **72**, 155–168 (1999).
4. Daraio, C., Nesterenko, V., Herbold, E., and Jin, S., *Physical Review E*, **72**, 016603 (2005).
5. Manciu, M., Sen, S., and Hurd, A. J., *Physica D*, **157**, 226–240 (2001).
6. Rosas, A., and Lindenberg, K., *Physical Review E*, **68**, 041304 (2003).
7. Job, S., Melo, F., Sen, S., and Sokolow, A., *Physical Review Letters*, **94**, 178002 (2005).
8. Gondret, P., Lance, M., and Petit, L., *Physics of Fluids*, **14**, 643–652 (2001).
9. Zhang, J., Fan, L.-S., Zhu, C., Pfeffer, R., and Qi, D., *Powder Technology*, **106**, 98–109 (1999).
10. Milne-Thomson, L. M., *Theoretical Hydrodynamics*, Macmillan Education, London, 1968, 5th edn.
11. Davis, R. H., Serayssol, J.-M., and Hinch, E., *Journal of Fluid Mechanics*, **163**, 479–497 (1985).
12. Hocking, L. M., *Journal of Engineering Mathematics*, **7**, 207–221 (1972).
13. Duvall, G. E., Manvi, R., and Lowell, S. C., *Journal of Applied Physics*, **40**, 3771–3775 (1969).

CP845, *Shock Compression of Condensed Matter - 2005*,
edited by M. D. Furnish, M. Elert, T. P. Russell, and C. T. White
2006 American Institute of Physics 0-7354-0341-4/06/$23.00

# PREPARATION AND SHOCK REACTIVITY ANALYSIS OF NOVEL PERFLUOROALKYL-COATED ALUMINUM NANOCOMPOSITES

## R. Jason Jouet, Richard H. Granholm, Harold W. Sandusky and Andrea D. Warren

*Naval Surface Warfare Center - Indian Head Division, Indian Head, MD 20640*

**Abstract.** Passivation of unpassivated aluminum nanoparticles using $C_{13}F_{27}COOH$ with materials containing 32.95 % Al is reported. Characterization, including SEM, TGA, and ATR-FTIR, indicate that $C_{13}F_{27}COOH$ binds to the Al particle protecting the surface from oxidation. Small Scale Shock Reactivity Test (SSRT) results of the Al-$C_{13}F_{27}COOH$ material formulated with HMX with and without HTPB binder are presented. The results for the non-HTPB-filled tests indicate a prompt reaction producing a dent 98.3% of that of similarly formulated μm-size Al. A larger dent was observed for a binder-filled sample containing HMX, Al-$C_{13}F_{27}COOH$ and HTPB than for a comparable sample containing μm-Al. These results are significant in that the μm-size Al particles are approximately 99.7% active Al whereas the Al-$C_{13}F_{27}COOH$ material contains only 32.95% active Al.

**Keywords:** Aluminum composite, nanoparticles, passivation, shock reactivity
**PACS:** 82.40.Fp, 81.16.Dn, 81.20.Ka, 81.65.Rv

## INTRODUCTION

Low energy release rate from metal oxidation is a limiting aspect of metallized energetics. The metal/oxidizer reaction is intermolecular and is limited in rate by diffusion. Subsequently, the detonation velocity and pressure observed from highly metallized formulations are substantially lower than ideal explosives. Metal nanoparticles can be used to reduce diffusion time and increase the oxidation rate. The ultimate goal is to increase the oxidation rate such that the energy released can contribute to detonation pressure and velocity.

The energy content of aluminized energetic materials can be increased by elimination of $Al_2O_3$ as well as utilization of fluorine as the oxidizer. Formation of $AlF_3$ liberates 13.31 Kcal $g^{-1}$ Al ($\Delta H_f^{\circ}$ $AlF_3$ = -359.5 Kcal $mol^{-1}$)[1] versus $Al_2O_3$ which releases 7.4 Kcal $g^{-1}$ Al. Additionally, removal of the thick oxide layer on Al particles[2] will result in Al combustion rate enhancement. Additional reaction rate increase should result from close proximity of the oxidizer and fuel. Therefore, passivation of Al nanoparticles with molecules containing oxidizer species should result a material capable of reacting fast enough so the energy release can contribute to detonation.

Recently we reported the passivation of Al nanoparticles using perfluorocarboxylic acids[3]. Oxide free Al nanoparticles were prepared and coated *in situ*. The presence of the perfluoroorganic carboxylate protects the particles from oxidation in air[3]. The resultant composite material consists of an oxide free Al core surrounded by perfluorocarboxylic acids. The results of the material's response to shock measured in a Small Scale Shock Reactivity Test (SSRT) are reported.

## EXPERIMENTAL METHOD

**General Considerations:** Air– and moisture–sensitive materials were handled in an argon-filled glove box. Dry toluene, $Et_2O$, $C_{13}F_{27}COOH$, $TiCl_4$, and $H_3Al•N(Me)Pyr$ were purchased from Aldrich and used as received. Additional N(Me)Pyr is needed for stabilization of $H_3Al•N(Me)Pyr$ resulting in an actual molar ratio of 1:1.55 and a practical molecular weight of 162.19 g $mol^{-1}$.

**NanoAl/SAM Composite Material;** A solution of diethyl ether (40 mL) and $H_3Al•N(Me)Pyr$ (10.0 g, 0.0617 mol) was stirred at room temperature in a glove box. To this solution, 100 µL of TiCl₄ (50µL mL⁻¹ in toluene) was added via syringe. The reaction was stirred for 4 hours, during which time the solution transitioned from dark brown to black to dark gray. Following this decomposition, a solution of 3.387 g (0.00474 mol) $C_{13}F_{27}COOH$ in 15 mL diethyl ether was added via pipet. The reaction was stirred for 12h and the contents allowed to settle. The clear brown ether layer was removed by pipet from the dark gray precipitate. The solid material was washed twice with diethyl ether and dried by allowing the residual ether to evaporate in the glove box. The dark gray, coated Al powder was found to be stable in air in that it was not pyrophoric and its appearance did not change with storage under ambient conditions. Yield: 4.89 g; (5.053 g calc. based on Al and $C_{13}F_{27}COOH$); 96.8 %

**Thermogravimetric Analysis:** Experiments were conducted on a TA Instruments 2950 TGA in Pt crucible in static air conditions. The heating profile was isothermal at 100 °C for 1h and then ramped in HTGA mode at 20 °C min⁻¹.

**Small Scale Shock Reactivity Test:** The apparatus consists of steel block with a 7.24 mm diameter hole for a detonator and a 6.35 mm long sample. The steel block is bonded to an Al 6063 alloy witness block, with a T52 temper, that is stress relieved with a yield strength of 145 MPa. The soft alloy permits sufficient dent depths from shock loading, providing adequate sensitivity to distinguish the performance of different explosives.

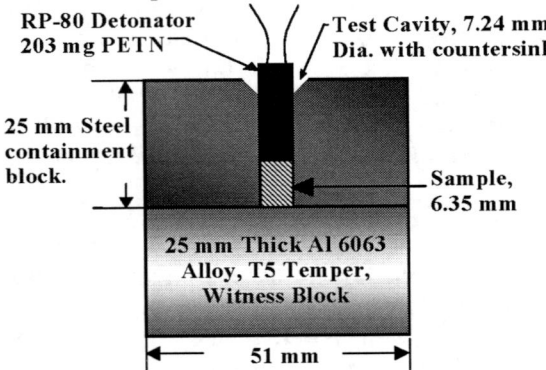

**Figure 1.** Diagram of SSRT apparatus

The RISI RP-80 detonator has 203 mg of PETN in a Delrin sleeve with an inner diameter of 5.12 mm. The conventional Al powder tested for comparison was H-5 grade from Valimet, average particle size

of ~7 µm. The inert additives used in the test were melamine, an organic compound from Eastman Kodak with a density of 1.573 g mL⁻¹ and a particle size range of 46 – 56 µm. Soda-lime silica glass beads with a particle size range of 150-210 µm and density of 2.50 g mL⁻¹ obtained from Potters Industry (P-0080) were also used as inert additives.

## RESULTS AND DISCUSSION

**Al-$C_{13}F_{27}COOH$ Composite Preparation:** Unpassivated Al nanoparticles were prepared by catalytic decomposition of $H_3Al•N(Me)Pyr$ at room temperature using TiCl₄ and then coated by exposure to $C_{13}F_{27}COOH$. The preparation is pictorially represented in figure 2.

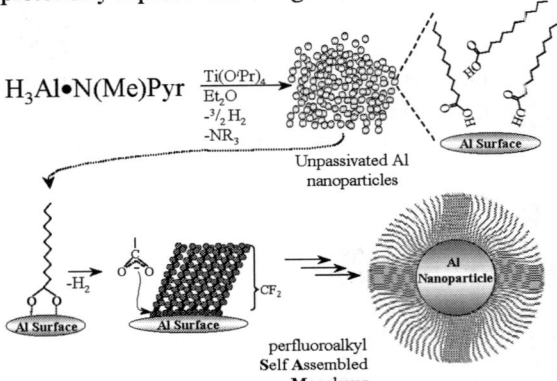

**Figure 2.** Pictorial representation of the NanoAl/SAM composite material preparation:

The proper ratio of carboxylic acid necessary to completely passivate the aluminum was one mole of $C_{13}F_{27}COOH$ to thirteen moles of Al. Material obtained from reactions with a higher Al to acid ratio was *extremely* pyrophoric in air while the adequately passivated material was not pyrophoric in air.

Using the experimentally determined amounts of $C_{13}F_{27}COOH$ and $H_3Al•N(Me)Pyr$ required for adequate passivation, we estimate the Al core size to be approximately 13.2 nm[5].

The TGA results of the Al-$C_{13}F_{27}COOH$ material are shown in Fig. 3. The material begins to lose mass significantly at approximately 135 °C, with a rapid weight loss event occurring at approximately 270 °C. Mass loss continues until ~534 °C and 42.12 %, at which point the Al in the sample begins to oxidize in a familiar two phase oxidation pattern[6-8]. The sample lost ~4.25 % of its weight during the 100 °C pretreatment. This is a significant weight loss but not unreasonable given the high surface area of the composite and the perfluoroalkyl surface coating [6-8].

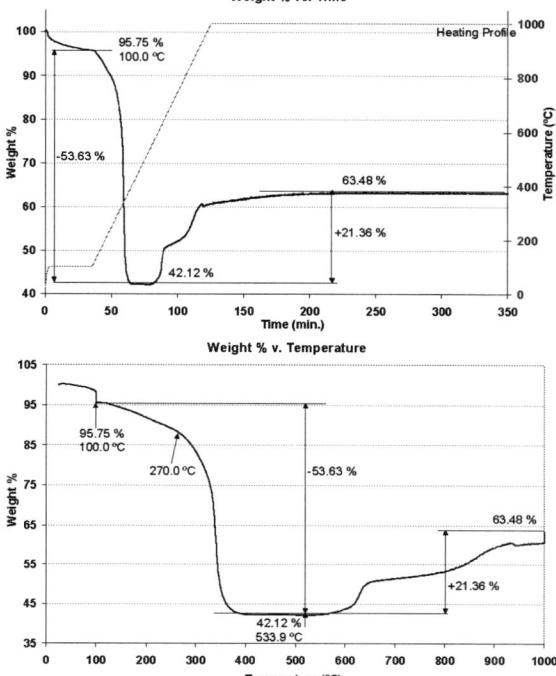

**Figure 3.** Thermogravimetric analysis data of the Al-$C_{13}F_{27}COOH$ material

Based on stoichiometry of the prepared material (13:1, molar), complete oxidation of the Al should have resulted in a final mass of 62.25% of the original. We observed a final mass of 63.48%, which is in close agreement and well within the experimental error associated with this experiment.

**Figure 4** SEM images of Al-$C_{13}F_{27}COOH$ at 200K indicate particles ranging from 20-200 nm.

SEM images shown in Fig. 4 indicate large polydispersity with particles with sizes ranging from ~20 to several hundred nanometers clearly evident. **Small Scale Shock Reactivity Test:** The SSRT test arrangement is shown in Fig. 1. Upon initiation, the detonator transmits a strong shock through the test material with minimal attenuation, as sample length is less than its diameter. The combined pressure from the transmitted detonator shock and any prompt sample reaction within a µs-timeframe dents the Al witness block at the base of the sample.

The literature consensus is that Al partially reacts within the detonation zone of pure explosives, with the associated energy release increasing both detonation velocity and Chapman-Jouget detonation pressure[9-14]. The Al reaction is delayed however, in explosives with binders. Only prompt reaction will contribute to the dent in the SSRT.

Al reaction was studied by varying its amount in samples with constant HMX weight (60 % v/v in the sample cavity). Inert substitutes, glass and melamine, were also tested. The samples were either porous, or had the void space filled with binder (uncured hydroxy-terminated polybutadiene (HTPB)). Fig. 5 shows increased dent depth with Al in porous samples, indicating prompt Al reaction, with a maximum increase in dent over neat HMX occurring at 15% by mass Al. Addition of inert materials resulted in a decrease in the dent.

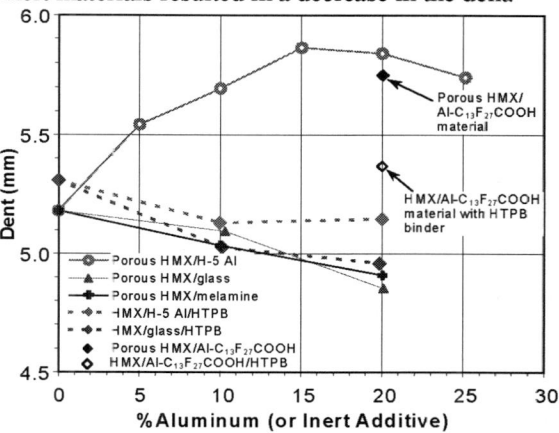

**Figure 5** Small-scale Shock Reactivity Test data

For the binder-filled sample, HMX/H-5 Al/HTPB, the decrease was not as great as for the inerts, indicating a small amount of Al reaction perhaps helped by the reduced amount of binder.

The binder-containing samples are 100% filled so adding aluminum is at the expense of binder.

The solid black diamond shows the response of the porous HMX/Al-$C_{13}F_{27}COOH$ sample. This sample was 60% v/v HMX and 20% m/m Al-$C_{13}F_{27}COOH$. The results indicate a prompt reaction producing a dent 98.3% that of the porous HMX/H-5 Al curve. This is significant in that the H-5 Al particles are approximately 99.7% Al whereas the Al-$C_{13}F_{27}COOH$ material is only 32.95% Al. This supports earlier statements that the energy content can be increased by elimination of the oxide present on conventional Al as well as utilization of F as the oxidizer to make Al-F species.

The hollow black diamond shows the dent depth observed for a sample containing 60% v/v HMX and 20% m/m Al-$C_{13}F_{27}COOH$ with remaining void space filled with HTPB binder. This is a larger dent than for the comparable H5-Al sample. This result indicates that prompt reaction occurs using the Al-$C_{13}F_{27}COOH$ material despite the shock mitigating effects of the HTPB observed in the HMX/H5-Al/HTPB result. This is likely due to the absence of $Al_2O_3$ on the surface of the Al particle. Particle size however, may also influence this result and we are testing conventional nm-Al to ascertain the effects of particle size.

## CONCLUSIONS

Nano-aluminum passivation with $C_{13}F_{27}COOH$ results in perfluoroorganic coated Al nanocomposites which are stable with respect to oxidation in air. The Al-$C_{13}F_{27}COOH$ material performed similarly to H-5 Al in the porous SSRT experiments even though its active Al content was much lower. When combined with a binder, the Al-$C_{13}F_{27}COOH$ material outperformed the H-5 Al.

## REFERENCES

1. *CRC Handbook of Chemistry and Physics*; 71 ed.; CRC Press: Boca Raton, FL, 1991.
2. Pesiri, D.; Aumann, C. E.; Bilger, L., Booth, D.; Carpenter, R. D.; Dye, R.; O'Neill, E.; Shelton, D.; Walter, K. C. *J. Pyrotechnics* **2004**, *19*, 19.
3. Jouet, R. J.; Warren, A. D.; Rosenberg, D. M.; Bellitto, V. J.; Park, K.; Zachariah, M. *Chem. Mater.* **2005**, *17*, 2987.
4. Higa, K. T.; Johnson, C. E.; Hollins, R. A. U. S. Patent 5,885,321, **1999**.
5. Al core size estimation is accomplished by considering the amount of acid required for

passivation and the amount of Al formed. The method requires calculating surface area per gram (S.A. $g^{-1}$) according to Equation 1.

$$\text{S.A. } (m^2 g^{-1}) = \frac{2\pi \bullet c.r._{Al}{}^2 \bullet n_{C_{13}F_{27}COOH}}{m_{Al}} \quad \text{Equation 1}$$

Where $c.r._{Al}$ is the covalent radius of Al in meters, $n_{C13F27COOH}$ is the number of molecules of perfluoroacid, and $m_{Al}$ is the mass of Al prepared experimentally. Using this surface area value, the radius of a sphere of Al can be calculated using Equation 2.

$$\text{S.A. } (m^2/g) = \frac{4\pi \bullet r_p{}^2}{m_p} \quad \text{where} \quad m_p = {}^4/_3 \pi \bullet r_p{}^3 \bullet \rho_{Al}$$

$$r_p = \frac{3}{\text{S.A.} \bullet \rho_{Al}} \quad \text{Equation 2}$$

Where $r_p$ is the radius of the particle in meters, $m_p$ is the mass of the particle, and $\rho_{Al}$ is the density of Al in $g\ m^{-3}$. The assumptions made for this estimate are as follows: 1.) the coating is a monolayer, 2.) each acid moiety binds two surface Al atoms, 3.) the area occupied by a surface Al atom is equivalent to the surface area of a circle of diameter 2.5 Å (covalent radius of Al), and 4.) no molecular Al species are formed.

6. Johnson, C. E.; Fallis, S.; Groshens, T. J.; Higa, K. T.; Ismail, I. M. K.; Hawkins, T. W. *unpublished results* **2005**.
7. Johnson, C. E.; Higa, K. T. In *Materials Research Society: Nanophase and Nanocomposites Materials II*, 1997; Vol. 457, pp 131.
8. Johnson, C. E.; Parr, T.; Parr, D. H.; Hollins, R.; Fallis, S.; Higa, K. T. In *JANNAF*: Monterrey, CA, 2000.
9. Price, D.; Clairmont, A. R., Jr.; Erkman, J. O. *Combust. Flame* **1973**, *20*, 389.
10. Price, D. "Aluminized Organic Explosives," Naval Ordnance Laboratory, NOLTR, 1972.
11. Tao, W. C.; Tarver, C. M.; Kury, J. W.; Lee, C. G.; Ornellas, D. L. *International Detonation Symposium*; Office of Naval Research, 1993; Vol. ONR 33395-12, pp 628-636.
12. Finger, M.; Hornig, H. C.; Lee, E. L.; Kury, J. W. *Fifth Symposium (International) on Detonation*; Office of Naval Research, 1970; Vol. ACR-184, pp 137-151.
13. Brousseau, P.; Dorsett, H. E.; Cliff, M. D.; Anderson, C. J. *12th International Detonation Symposium*: San Diego, CA, Office of Naval Research, 2002; Vol. ONR 333-05-2, pp 11-21.
14. Lefrancois, A.; Baudin, G.; Le Gallic, C.; Boyce, P.; Coudoing, J.-P. *12th International Detonation Symposium*: San Diego, CA; Office of Naval Research, 2002; Vol. ONR 333-05-2, pp 22-32.

CP845, *Shock Compression of Condensed Matter - 2005,*
edited by M. D. Furnish, M. Elert, T. P. Russell, and C. T. White
© 2006 American Institute of Physics 0-7354-0341-4/06/$23.00

# INVESTIGATION OF DYNAMIC FRICTION INDUCED BY SHOCK LOADING CONDITIONS

## A. Juanicotena, S. Szarzynski

*Commissariat à l'Energie Atomique, Centre Ile de France, BP12, 91680 Bruyères-le-Châtel*

**Abstract.** Modeling the frictional sliding of one surface against another under high pressure is often required to correctly describe the response of complex systems to shock loading. In order to provide data for direct code and model comparison, a new friction experiment investigating dry sliding characteristics of metal on metal at normal pressures up to 10 GPa and sliding velocities up to 400 m/s has been developed. The test consists of a specifically designed target made of two materials. A plane shock wave generated by plate impact results in one material sliding against the other. The material velocity of the rear surface of the target is recorded versus time by Doppler Laser Interferometry. The dynamic friction coefficient μ is then indirectly determined by comparison with results of numerical simulations involving the conventional Coulomb law. Using this new experimental configuration, three dynamic friction experiments were performed on AA 5083-Al (H111) / AISI 321 stainless steel tribo-pair. Results suggest a decrease in the friction coefficient with increasing sliding velocity.

**Keywords:** sliding friction, friction : dynamic, shock waves, plate impact, aluminum : alloy, steel
**PACS:** 62.50.+p, 81.05.Bx, 62.20.Qp, 46.55.+d

## INTRODUCTION

The response of complex systems to shock loading often requires a description of the frictional behavior at the interfaces between the materials. Few experimental data are available on dynamic friction under loading of both high pressure and high sliding velocity. In the present study a new experimental set-up has been developed in order to provide dry sliding characteristics of metal on metal at normal pressures up to 10 GPa and sliding velocity up to 400 m/s. The system is devoted to the direct code and friction model comparisons.

## EXPERIMENTAL TECHNIQUE

The principle of the test consists of launching a projectile on a target consisting of a couple of materials we want to characterize. The original feature of the experiment concerns the design of the target, cylindrical in shape with the geometry is shown in fig.1. The shock wave generated by the impact of the projectile makes the target move. Because of its lower impedance, the central cone is driven faster than the confinement resulting in a differential motion at the interface. For a given impact speed, we can measure the free surface velocity as a function of time of both the central cone and the confinement by laser interferometry. The magnitude of the central cone velocity depends on the condition of friction at the interface between the materials we study. High friction condition leads to a lower velocity than the pure sliding condition. In order to measure the amount of friction, we need to perform a numerical simulation of the test involving the friction law we want to identify. The measurement of the dynamic friction is then an indirect measurement. Various contact pressure and sliding velocity are achievable by varying the nature of the flyer, the impact speed or the angle of the cone. Besides the knowledge of the behavior of the materials of the flyer and target, we need to know the interface gap, which is important for numerical simulation results.

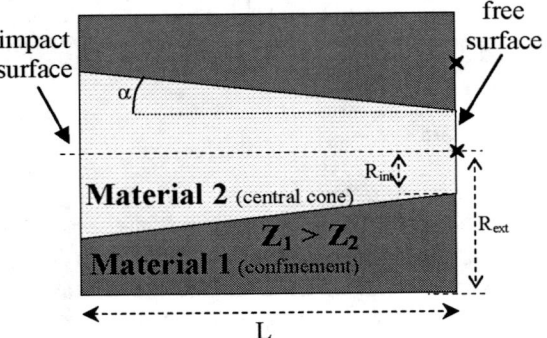

**Figure 1.** Cross-sectional view of the target geometry. Stars indicate the free surface velocity measurement points.

## RESULTS AND DISCUSSION

The first experiments were conducted on 5083-H111 Al alloy and AISI 321 stainless steel tribo-pair. The Al alloy is used as the central cone and the stainless-steel as the confinement. At the interface, the average roughness of both materials was $R_a = 1.6\mu m$. For all experiments, the dimensions of the target were 50mm in diameter ($R_{ext}/2$, see fig.1) and 30mm in length L. The radius $R_{int}$ of the central cone was 5mm on the free surface side and the angle of the cone $\alpha$ 10 degrees. A CuC2 copper flyer plate 4mm thick was used to impact the target assembly. During the experiment, the impact velocity as well as the free surface velocity of the solid cone and the confinement are recorded. These points are respectively located at R=0 and R=15mm from the central axis of the target (see fig. 1).

Three experiments were performed at impact speeds of 202 ± 3, 406 ± 20 and 670 ± 9m/s. The recorded signals are shown in the figure 2. The free surface velocities were measured by Doppler Laser Interferometry for the shots T101 and T102 and by the VISAR technique for the shot T105. The velocities of the Al(5083) central cone exhibit the same profiles for the three shots (see fig. 2 (a)).

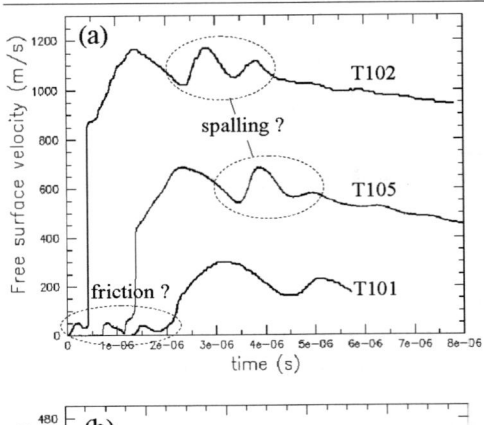

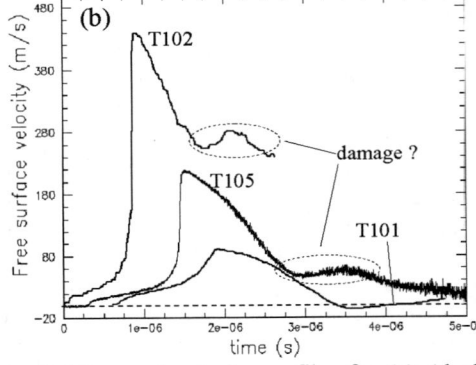

**Figure 2.** Measured velocity profiles for (a) Al alloy central cone, (b) S.steel confinement. The impact speeds are 202 m/s (T101), 406 m/s (T105) and 670 m/s (T102).

At the bottom of the curve, we observe first the arrival of an elastic precursor followed by a decrease in velocity. Two assumptions can explain this singular phenomenon: the presence of a gap between the central cone and the confinement or high friction condition at the interface. In both cases, release waves are generated from the interface to the centre of the cone leading to the observed deceleration. In fact, something between these two assumptions is certainly close to the reality. We then see the arrival of the plastic wave, which exhibits an unusual profile due to the two-dimensional configuration of the test. The velocity reaches then its maximum before decreasing. During the unloading, we see a pullback feature of damage by spalling for the shots T101 and T105. This is proven by the observation of the recovered samples after the shots (figure 3). A clear indication of damage is also observed on the velocity signal of the confinement (see fig. 2 (b)).

The slight recompression after the pullback of the shots T102 and T105 indicates that material probably has undergone damage before fracturing.

**Figure 3.** Recovered target of shot T102. The Al 5083 central cone is fractured.

## NUMERICAL SIMULATIONS

Numerical simulations of tests are essential to interpret the measurements. Calculations allow us to estimate the loading underwent by the target as well as to validate and/or to identify the friction laws used in the codes.

Two dimensional Lagrangian computations have been performed by means of the Hesione hydrocode. The mesh size of the target was about one hundred microns per unit cell. We used equations of state of pure Aluminum [1] and of AISI 304 stainless steel [2]. Steinberg-Cochran-Guinan elastic-plastic model [3] with associated parameters for AISI 304 stainless steel and Al 6061 T6 alloy has been used for the flow stress [4]. For the latter material, the Y0 parameter was modified to 200MPa due to the weaker yield strength of the Al 5083 alloy. Our computations involve the Coulomb macroscopic law to treat of friction which is :

$$\tau_t = -\mu \sigma_n V_g / \|V_g\|, \quad (1)$$

$$\text{with} \begin{cases} \sigma_n = Min(\sigma_n, \sigma_y) \\ \mu = \mu_\infty + (\mu_0 - \mu_\infty)e^{-\gamma|V_g|} \end{cases}, \quad (2)$$

This law specifies that the frictional or tangential stress $\tau_t$ at the interface is directly proportional to the normal stress $\sigma_n$ and acts in an opposite direction of the sliding velocity $V_g$. Our friction treatment bounds the normal stress by the yield stress $\sigma_y$ of the softer material in contact. As indicated in equation (2), the friction coefficient $\mu$ is a decreasing function of the sliding velocity $V_g$. $\mu_0$ and $\mu_\infty$ are respectively the static and the dynamic friction coefficient. This behavior has been observed by many authors, like Bowden et al. [5], [6] or Lim et al. [7].

The shape control of the targets could not be carried out for the shots T101 and T102. For the target of the shot T105, a maximum gap of 25μm was measured before the assembly, where a force of 600N is applied to insure the contact resulting in a significant reduction in the gap. Computations have shown that the observed decrease of the velocity of the central cone just after the arrival of the elastic precursor could be simulated by introducing a void at the interface. But they also showed that the maximum velocity, characterizing the friction condition at the interface, is not affected by the presence of a gap. So, due to the lack of targets shape control for the shots T101 and T102, the analysis of the results has been made by comparisons with calculations performed without a gap at the interface.

The calculated loading conditions at the interface are summarized in the table 1 for the three shots.

| Shot | $V_{impact}$ | $V_g$ (m/s) | $P_{contact}$ |
|------|------------|-----------|-------------|
| T101 | 202 m/s | 60 - 130 | ~ 2.7 GPa |
| T105 | 406 m/s | 150 - 250 | ~ 5 GPa |
| T102 | 670 m/s | 300 - 400 | ~ 10 GPa |

**Table 1.** Theoretical loading conditions at the interface.

Figure 4 compares calculated results with the experimental ones on the confinement velocity. Results are quite well represented by calculations. The slight recompression after the pullback for the shots T102 and T105 can be simulated using the Johnson damage model [8]. These measurements, which are not affected by the friction conditions at the interface, allow to validate the choice of the constitutive equations of the heavy material and set the value of the impact speed for computations within the range of uncertainties.

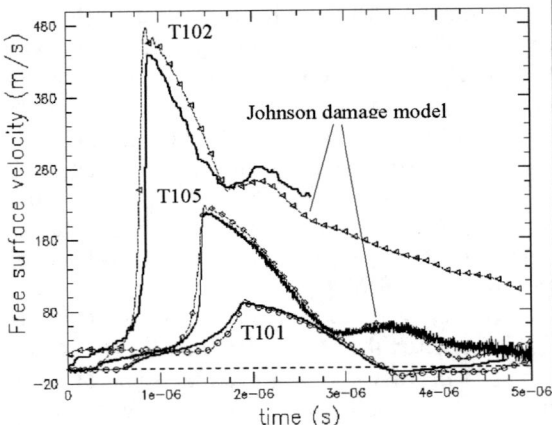

**Figure 4.** Free surface velocity of the stainless steel confinement compared with computations (triangles, diamonds and circles).

Comparisons between experimental and calculated velocities of the Al 5083 central cone are shown in the figure 5. Damage was simulated by introducing a spall strength of 1.5GPa for Al 5083 alloy.

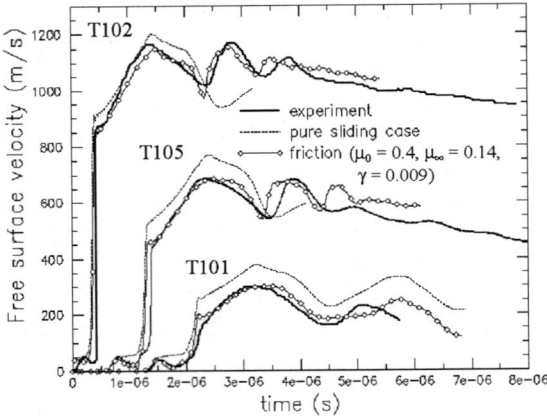

**Figure 5.** Free surface velocity of the Al 5083 central cone : experiments –> solide lines, calculations without friction –> dotted lines, with friction –> diamonds

Experimental results are well described only if friction is introduced in the computations. For each shot, we have first performed simulations with a constant friction coefficient. The best fitted values are $\mu \geq 0.25$ (T101), $\mu \sim 0.2$ (T105) and $\mu = 0.15$ (T102). So these results seem to be correct with the assumption of the decrease of the friction

coefficient with the sliding velocity. We can then identify our friction law proposing one set of parameter to simulate the experiments. Assuming $\mu_0 = 0.4$, we found $\mu_\infty = 0.14$ and $\gamma = 0.009$.

## CONCLUSIONS

We have investigated the dynamic friction behavior of Al/S.steel tribo-pair employing a new experimental set-up. Results have demonstrated the feasibility and capability of the test to generate measurable friction. They also seem to confirm a decrease of the friction coefficient with high contact pressure and high sliding velocities. Moreover, the test allows to observe quantifiable others phenomena such as the magnitude of the elastic precursor or damage. At last, the target can be recovered after the shot. Observation of sub-surface flows at the interface is possible so, allowing to better understand the complex physical processes induced by dynamic friction.

## REFERENCES

1. Boissiere, C. and Fiorese, G., "Equation d'état des métaux prenant en compte les changements d'état entre 300 et 200000K pour toute compression. Application au cas du cuivre et de l'aluminium", Revue de phys. appliquée, Tome 12, 857-871, 1977
2. CEA internal report
3. Steinberg, D.J., Cochran, S.G., Guinan, M.W. "A Constitutive Model for Metals Applicable at High-Strain-Rate", J. Appl. Phys., 51, 1498, 1980
4. Steinberg, D.J., "Equation of state and strenght properties of selected materials", LLNL, UCRL-MA-106439, 1996
5. Bowden, F.P. and Freitag, E.H., "The friction of solids at very high speeds. I. Metal on metal; II. Metal on diamond", Proc. Roy. Soc. Lond., A 248, 350-367, 1958
6. Bowden, F.P. and Persson, P.A., "Deformation, heating and melting of solids in high-speed friction", Proc. Roy. Soc. Lond. A 260, 433-458, 1961
7. Lim, S.C., Ashby, M.F., Brunton, J.H., "The effects of sliding conditions on the dry friction of metals", Acta Metall., 37(3), 767-772, 1989
8. Johnson J.N., "Dynamic fracture and spallation in ductile solids", J. Appl. Phys., 52(4), 2812, April 1981

CP845, *Shock Compression of Condensed Matter - 2005*,
edited by M. D. Furnish, M. Elert, T. P. Russell, and C. T. White
2006 American Institute of Physics 0-7354-0341-4/06/$23.00

# TRANSIENT STRESS OPTIMZATION OF ELASTIC AND VISCOELASTIC COMPOSITE STRIPS

## Rich Laverty[1], and George Gazonas[2]

[1] *Department of Mathematics, Juniata College, Juniata, PA 16652*
[2] *WMRD, US Army Research Laboratory, Aberdeen Proving Ground, MD 21005-5069*

**Abstract.** In this study we examine transient stresses in elastic and viscoelastic composite strips. When the stress is the result of a stress step we seek combinations of material parameters that lead to a minimum peak stress in the layer most distant from the applied stress. In the case of impact by a rigid body we define an optimal design as one in which the maximum stress is achieved in the layer subject to the impact. The primary purpose of these results is benchmarking larger numerical studies coupling a finite element code (DYNA3D) with several optimization routines.

**Keywords:** Elastic Waves, Viscoelastic Waves, Optimization, Inverse Laplace Transform.
**PACS:** 62.30.+d, 43.20.Gp, 83.60.Bc

## INTRO

Design of stress wave attenuators is a computationally expensive task with few analytic solutions. Therefore, there are almost no benchmarks for the larger, more complex computations. The purpose of this study is to find reliable solutions to transient wave propagation in one-dimensional composite strips. These solutions are examined for possible optimal designs.

First, we introduce the idea of an optimal design by considering elastic strips. The following sections consider viscoelastic strips, and altering the applied stress to low-velocity impact.

## ELASTIC STRIPS

For comparison, we report the relevant result from a homogenous strip of an elastic material. This will be used to benchmark the performance of composites.

## Homogeneous Elastic Strip

Consider an elastic material occupying $0 < x < L$, initially at rest, and fixed on the right ($x = L$). Upon application of a stress step on the left, stress propagates through material; hereafter, referred to as a strip. The mathematical model of this is the Initial Boundary Value Problem (IBVP):

$$
\begin{aligned}
&\partial_{tt}u = c^2\partial_{xx}u \\
&0 = u(x,0), \quad 0 = \partial_t u(x,0), \\
&\sigma_0 H(t) = E\partial_x u(0,t) \\
&0 = u(L,t)
\end{aligned}
\tag{1}
$$

where $u$ is the displacement, $\sigma_0$ the magnitude of the applied stress, $H(t)$ the Heaviside step function, $E$ the elastic modulus, and $\rho$ the density of the strip. The speed of stress waves is $c = \sqrt{E/\rho}$.

Fig. 1 is a stress history at two points along the strip, $x = L/4$ and $3L/4$. These points were chosen because they will be the midpoints of the layers of the composite.

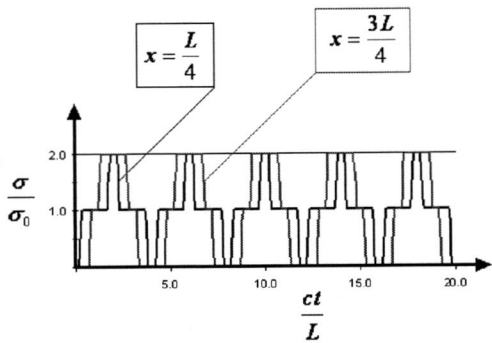

**Figure 1.** The stress history measured at two points of the homogeneous elastic strip.

The relevant feature of Fig. 1 is the peak stress. At each point of measurement the maximum stress is $2\sigma_0$.

## Two Layered Elastic Strip

Now consider a strip composed of two layers. Layer 1 occupies $0 < x < L/2$ and layer 2, $L/2 < x < L$. All quantities, displacement, stress, density and elastic modulus, will have their corresponding layer indicated by a subscript. The displacements in each layer, $u_i$, obeys the partial differential equation and initial conditions in (1). The boundary conditions of (1) are replaced by

$$\sigma_0 H(t) = E\partial_x u_1(x,0)$$
$$0 = u_2(L,t) \tag{2}$$

To complete the model we need interface conditions. We assume that the displacement and stress are continuous at the layer interface,

$$u_1(L/2-,t) = u_2(L/2+,t)$$
$$E_1 \partial_x u_1(L/2-,t) = E_2 \partial_x u_2(L/2+,t) \tag{3}$$

We restrict our attention to materials where the elastic modulus and density in layer 2 are scaled from their values in layer 1 by the same factor, $\alpha$.

$$E_2 = E_1/\alpha, \quad \rho_2 = \rho_1/\alpha. \tag{4}$$

This implies $c_1 = c_2$, so we drop the subscript.

Fig. 2 is a comparison of stress histories at $x = 3L/4$ for the homogeneous strip and a composite with $\alpha = 1.1$. In contrast with Fig. 1, the stress in the composite exceeds $2\sigma_0$.

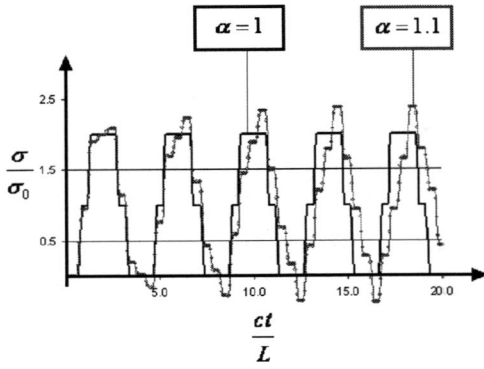

**Figure 2.** Stress measured at $x = 3L/4$ for $\alpha = 1$ (homogeneous strip), and $\alpha = 1.1$.

This phenomenon is common for almost all composite strips. In [1] an analytic solution is constructed for the IBVP of the composite. It was shown that the stress amplitude in layer 2 will exceed $2\sigma_0$ in all but a discrete set of designs. These 'optimal' designs occur when

$$\alpha = \frac{1 + \cos(\pi/k)}{1 - \cos(\pi/k)}, \quad k = 2,3,\dots \tag{5}$$

For optimal designs the peak stress in layer 2 will be $2\sigma_0$. Fig. 3 shows the peak stress, in layers 1 and 2, as $\alpha$ varies. Optimal designs are visible.

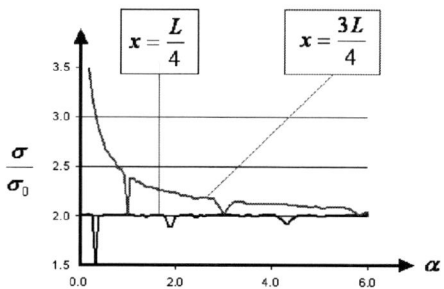

**Figure 3.** Peak stress measured at two points of the composite strip as $\alpha$ is varied.

In the following, we seek results similar to Fig. 3. We will consider viscoelastic strips and elastic strips subject to low-velocity impacts.

## VISCOELASTIC STRIPS

For a strip composed of viscoelastic layers the IBVP (1) must be adjusted. We keep the notation of denoting quantities in each layer with subscripts. The equation of motion and initial conditions are

$$\rho_i \partial_{tt} u_i = \partial_x \sigma_i$$
$$0 = u_i(x,0), \quad 0 = \partial_t u_i(x,0) \tag{6}$$

The boundary and interface conditions are

$$\sigma_0 H(t) = \sigma_1(0,t)$$
$$0 = u_2(L,t)$$
$$u_1(L/2-,t) = u_2(L/2+,t) \tag{7}$$
$$\sigma_1(L/2-,t) = \sigma_2(L/2+,t)$$

The complete problem statement includes the constitutive relationship

$$\sigma_i(x,t) = \int_0^t G_i(t-\tau)\partial_{x\tau} u_i(x,\tau)d\tau, \tag{8}$$

where $G_i(t)$ is the relaxation modulus of layer $i$. We restrict our attention to standard linear solids

$$G_i(t) = G_{i,\infty} + \left(G_{i,0} - G_{i,\infty}\right)e^{-\beta_i t}. \tag{9}$$

$G_{i,0}$ is the initial elastic response of the material, $G_{i,\infty}$ the long term stress response, and $\beta_i$ the decay parameter.

We make the following parameter choices.

$$G_{1,0} = E, \quad G_{2,0} = E/\alpha$$
$$G_{i,\infty} = G_{i,0}/2, \quad \rho_2 = \rho_1/\alpha \tag{10}$$

The speed of propagation in each layer, $c_i = \sqrt{G_{i,0}/\rho_i}$, is the same and will be denoted by $c$.

The decay parameter is a viscosity measure. Materials with small $\beta$ will behave similarly to elastic materials when considering transient stresses. We have chosen $\beta_i$ the same in each layer, so its subscript has been dropped.

Equations (6) through (9) comprise an integro-differential equation for the propagation of stress waves in the strip. Due to its complicated nature, this IBVP is solved numerically using the Laplace transform with a modified DAC algorithm for inversion, [2,3,4]. The results for different values of $\alpha$ and three values $\beta$ are displayed in Fig. 4.

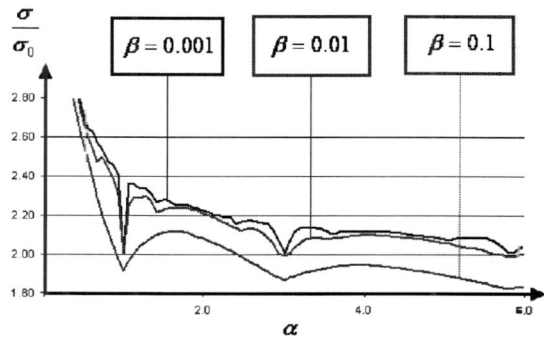

**Figure 4.** Peak stress at $x = 3L/4$ as $\alpha$ varies.

In Fig. 4 we see that as $\alpha$ varies the peak stress in layer 2 behaves similar to the elastic composite. In fact, when $\beta$ is small the results are almost indistinguishable from the elastic case. As $\beta$ increases the viscous nature of the solid becomes more pronounced and the difference between optimal and nearby designs becomes less dramatic.

## LOW VELOCITY IMPACT

Now reconsider the two layered elastic strip in (1) through (4), but replace the stress step with the impact of another solid.

Consider a rigid body of mass $m$ moving to the right with a velocity $v_0 < c$. The impact is modeled by coupling the position of the right of the rigid body, $u_0(t)$, with the stress in the strip through the condition

$$m\partial_{tt}u_0(t) = E_1\partial_x u_1(0,t). \qquad (11)$$

Fig. 5 contains the results from this IBVP using the characteristic stress

$$\sigma_0 = v_0\sqrt{E_1\rho_1} \, . \qquad (12)$$

Optimal designs similar to those in Fig. 3 don't exist. However, we do see that the stress in layer 2 is the lowest stress when $\alpha$ is greater than 1. This corresponds to a composite with a stiffer layer 1.

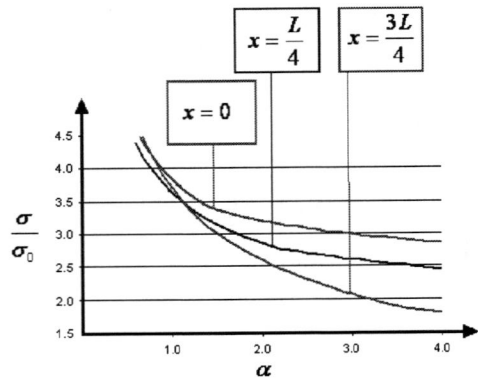

**Figure 5.** Peak stress at the point of contact and the midpoint of each layer.

## CONCLUSIONS

The optimal designs of composite elastic strips can be extended to strips composed of viscoelastic materials, with similar results. When low velocity impacts replace stress steps the optimizability is lost. However, a lower peak stress is recorded in the layer 2 when the front layer is the stiffer of the two. Though not optimal, this is still a desirable property for a stress wave attenuator.

## ACKNOWLEDGEMENTS

This work was performed while Rich Laverty held a National Research Council Research Associateship Award at the Army Research Lab and the United States Military Academy.

## REFERENCES

1. Velo, A. and Gazonas, G., "Optimal design of a two-layered elastic strip subject to transient loading," IJSS, Vol. 56, no. 7, pp. 1-10, 2003.
2. Laverty, R. and Gazonas, G., "Laplace Transform Inversion and Viscoelastic Wave Propagation", in Proceedings of the 11th Annual ARL/USMA Technical Symposium, pp. 1-24, 2003.
3. Laverty, R. and Gazonas, G., "Transient Stress Analysis of Elastic-Viscoelastic Strips Subject to Impact Loads", in Proceedings of the 12th Annual ARL/USMA Technical Symposium, pp. 1-16, 2004.
4. Laverty, R. and Gazonas, R., "An Improvement to the Fourier Series Method for the Inversion of Laplace Transforms Applied to Elastic and Viscoelastic Waves", to appear in IJCM.

CP845, *Shock Compression of Condensed Matter - 2005,*
edited by M. D. Furnish, M. Elert, T. P. Russell, and C. T. White
2006 American Institute of Physics 0-7354-0341-4/06/$23.00

# IMPULSIVE LOADING OF CELLULAR MEDIA IN SANDWICH CONSTRUCTION

## Joseph A. Main[1] and George A. Gazonas[2]

[1]*Building and Fire Research Laboratory, National Institute of Standards and Technology*
*100 Bureau Drive, Mail Stop 8611, Gaithersburg, MD 20899-8611*
[2]*Weapons and Materials Research Directorate, U.S. Army Research Laboratory*
*ATTN: AMSRD-ARL-WM-MD, Aberdeen Proving Ground, MD 21005-5069*

**Abstract.** Motivated by recent efforts to mitigate blast loading using energy-absorbing materials, this paper investigates the uniaxial crushing of cellular media in sandwich construction under impulsive pressure loading. The cellular core is modeled using a rigid, perfectly-plastic, locking idealization, as in previous studies, and the front and back faces are modeled as rigid, with pressure loading applied to the front face and the back face unrestrained. Predictions of this analytical model show excellent agreement with explicit finite element computations, and the model is used to investigate the influence of the mass distribution between the core and the faces. Increasing the mass fraction in the front face is found to increase the impulse required for complete crushing of the cellular core but also to produce undesirable increases in back-face accelerations. Optimal mass distributions are investigated by maximizing the impulse capacity while limiting the back-face accelerations to a specified level.

**Keywords:** Blast mitigation, aluminum foam, shock wave, finite element analysis.
**PACS:** 46.40.Cd, 62.50.+p, 83.60.Uv.

## BACKGROUND

Cellular materials such as metal foams and honeycombs are being considered in a wide variety of structural applications because of their capacity to absorb impact energy. Surprisingly, however, their use under blast loading has often led to enhancement, rather than mitigation, of blast effects. Experiments by Hanssen et al. [1] showed that increased upswing results from the addition of an aluminum foam layer to the face of a massive "pendulum" subjected to blast loading. Nesterenko [2] noted that in these experiments, the blast impulse is imparted primarily to a lightweight plate covering the foam layer, leading to significantly higher kinetic energy than if the same impulse were imparted directly to the more massive pendulum. Xue and Hutchinson [3] noted a similar

effect in a computational study of blast loading on sandwich plates, in which the kinetic energy imparted to a sandwich plate was observed to be greater than for a solid plate of the same mass. In spite of this, it was found that deflections of sandwich plates could be significantly less than for the corresponding solid plate. Xue and Hutchinson considered front and back face sheets with equal mass but suggested that further reductions in deflections might be achieved by increasing the mass fraction in the face sheet near the blast.

## ANALYTICAL MODEL

Motivated by these observations, an analytical model is developed in this paper to investigate the influence of mass distribution on the uniaxial crushing of cellular material sandwiched between rigid layers. The cellular core material is represented by the simplified stress-strain relationship shown in Fig. 1(b), originally proposed by Reid and Peng [4] for modeling crushing of wood and subsequently applied to cellular metals in a number of studies (e.g., [1,5]). Arbitrary masses of the front and back faces are permitted, and a pressure pulse $p(t)$ is applied to the front face with the back face unrestrained. This sandwich model is a generalization of that in [1], which considered a fixed back face, and of that in [5], which considered front and back faces of equal mass with blast loading represented by an initial velocity imparted to the front face.

A strip of sandwich panel with unit cross-sectional area is considered, with total mass given by $m = m_1 + \rho_0 \ell_0 + m_2$, where $\rho_0$ and $\ell_0$ are the uncompressed density and thickness of the cellular core, and $m_1$ and $m_2$ are the areal densities of the front and back faces. The acceleration of the center of mass, denoted $\ddot{u}_G$, follows directly from application of Newton's second law to the strip:

$$p(t) = m\ddot{u}_G \qquad (1)$$

Provided the applied pressure is sufficiently high, densification of the cellular core commences at the front face, and a densification front propagates through the core. By conservation of mass, the density of the compressed core material is $\rho_0 /(1-\varepsilon_0)$. According to the simplified model of Fig. 1(b), the compressed core material moves as a rigid body with the same velocity as the front face, denoted $\dot{u}_1$, while the uncompressed core material moves as a rigid body with the velocity of the back face, $\dot{u}_2$. The stress just ahead of the densification front is $\sigma_0$, and application of Newton's second

law to the material ahead of the densification front then yields the following equation:

$$\sigma_0 = (\rho_0 x + m_2)\ddot{u}_2 \qquad (2)$$

where $x$ denotes the thickness of the uncompressed core material, and the thickness of the densification front itself is assumed to be negligible. By forming and differentiating an expression for $x_G$, the distance of the center of mass from the back face, it follows that

$$\ddot{x}_G = (\varepsilon_0 / m)\left\{[m_1 + \rho_0(\ell_0 - x)]\ddot{x} - \rho_0 \dot{x}^2\right\} \qquad (3)$$

Eqs. (1) - (3) can then be combined through the relation $\ddot{u}_2 = \ddot{u}_G + \ddot{x}_G$ to yield the following nonlinear ordinary differential equation for $x$:

$$-\varepsilon_0[m_1 + \rho_0(\ell_0 - x)]\ddot{x} + \varepsilon_0\rho_0\dot{x}^2$$
$$= p(t) - \sigma_0 m / (\rho_0 x + m_2) \qquad (4)$$

Eq. (4) can be integrated numerically with initial conditions $x(0) = \ell_0$ and $\dot{x}(0) = 0$. A triangular pressure pulse is considered, as shown in Fig. 1(c), with total impulse denoted $i_\infty$. The following symbols are introduced to denote the nondimensional peak pressure and total impulse:

$$P_0 = \frac{p_0}{\sigma_0}; \quad I_\infty = \frac{i_\infty}{m}\sqrt{\frac{\rho_0}{\sigma_0\varepsilon_0}} \qquad (5)$$

The following symbols denote the mass fractions in the core and in the front and back faces:

$$\eta_0 = \rho_0\ell_0 / m; \quad \eta_1 = m_1 / m; \quad \eta_2 = m_2 / m \qquad (6)$$

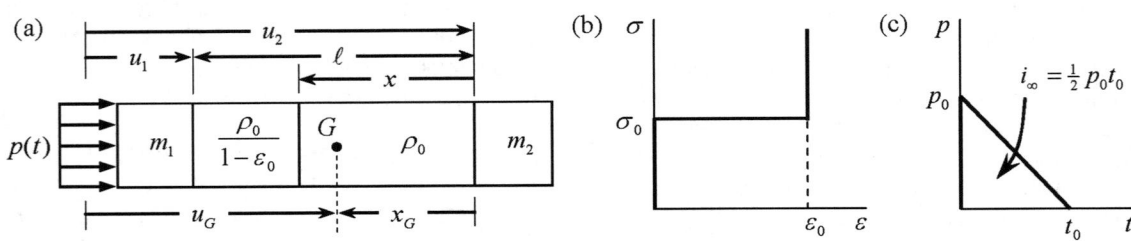

**Figure 1.** Analytical model definition: (a) Strip of sandwich panel with partially compacted core; (b) Stress vs. volumetric strain relationship for core material; (c) Triangular pressure pulse applied to front face.

**Table 1.** Parameters of computational simulations.

| Case | $\eta_1$ | $\eta_0$ | $\eta_2$ | $P_0$ | $I_\infty$ |
|------|------|------|------|------|------|
| pendulum | 0.0125 | 0.0125 | 0.975 | 10 | 0.015 |
| sandwich | 0.25 | 0.5 | 0.25 | 10 | 1 |

## COMPARISON WITH COMPUTATIONS

The predictions of the analytical model are compared with explicit finite element computations using LS-DYNA. In the computations, the cellular core was represented by a single row of solid elements with total thickness $\ell_0$ = 5 cm, using material model 26 (*MAT_HONEYCOMB) with $\rho_0$ = 250 kg/m$^3$, $\sigma_0$ = 1 MPa, and $\varepsilon_0$ = 0.7. A large elastic modulus of $E$ = 700 GPa was used to represent the "rigid" portions of the idealized stress-strain relationship in Fig. 1(b), and Poisson's ratio was set to zero. The material viscosity coefficient $\mu$ was set to 0.001, and 150 elements were found to be sufficient for convergence.

The front and back faces were represented in the computations by added nodal masses, and two different mass distributions were considered, as indicated in Table 1. The "pendulum" case corresponds to the blast pendulum experiments of [1], with the large back-face mass representing the pendulum. The "sandwich" case corresponds to the sandwich plates of [3] and [5], with equal front-face and back-face masses.

In Figs. 2 and 3, computational results are compared with predictions of the analytical model, and good agreement is observed. Results are plotted against nondimensional time, $\tau = (\sigma_0 / i_\infty)t$. The nondimensional velocities in Figs. 2 and 3 are

defined as $\overline{v}_1 = \dot{u}_1 / v_\infty$ and $\overline{v}_2 = \dot{u}_2 / v_\infty$, where $v_\infty = i_\infty / m$ is the final velocity of the center of mass. Due to the small mass of the front face, much larger nondimensional front-face velocities are observed in the "pendulum" case, despite the much smaller nondimensional impulse $I_\infty$ in this case, as shown in Table 1.

## INFLUENCE OF MASS DISTRIBUTION

Fig. 3(a) shows contours of the critical nondimensional impulse $I_\infty$ for which complete densification of the core is first achieved. These contours correspond to the limiting case of a Dirac delta impulse ($P_0 \rightarrow \infty$) and were obtained by numerical solution of Eq. (4). Fig. 3(a) shows that increasing the mass fraction in the core and in the front face increases the impulse capacity of the sandwich system. However, Fig. 3(b) shows that increasing the mass fraction in the core and in the front face also leads to increased back-face accelerations, thus sacrificing a protective function of the cellular core. The nondimensional back-face accelerations presented in Fig. 3(b) are defined as $\overline{a}_2 = (m/\sigma_0)\ddot{u}_2$. It follows from Eq. (2) that the peak back-face accelerations occur at the instant of complete compaction ($x$ = 0), for which $\ddot{u}_2 = \sigma_0 / m_2$ or $\overline{a}_2 = 1/\eta_2$. A design optimization problem can be posed by seeking to maximize the impulse $I_\infty$ that can be sustained while limiting the back-face accelerations to a specified level. Fig. 4 shows a contour plot of the maximum impulse $I_\infty$ that can be sustained with accelerations limited to $\overline{a}_2 = 5$. The grey curve in Fig. 4 corresponds to $1/\eta_2 = 5$. Below this curve, the values of

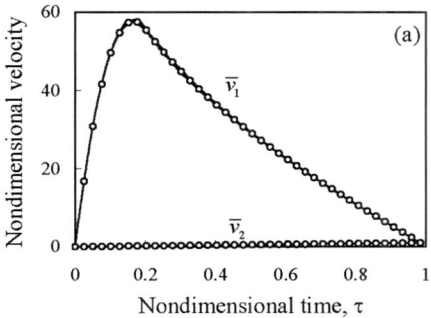 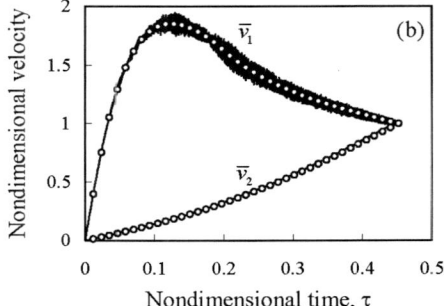

**Figure 2.** Comparison of LS-DYNA computations (—) with predictions of analytical model (○): Nondimensional front-face and back-face velocities for (a) "pendulum" case; (b) "sandwich" case.

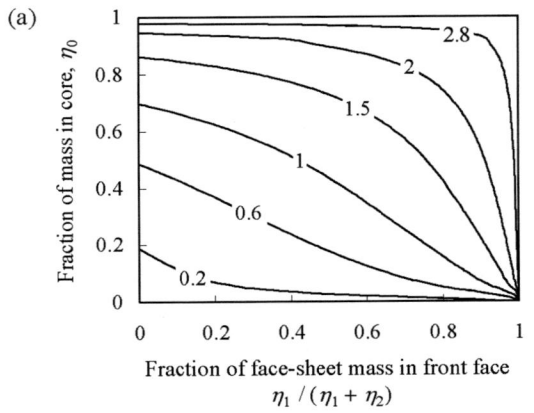

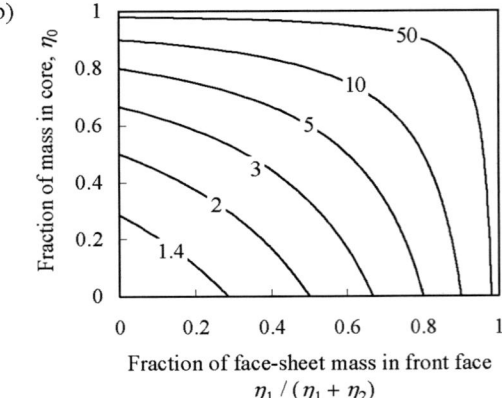

**Figure 3.** Contours with varying mass distribution: (a) Critical nondimensional impulse $I_\infty$ required for complete compaction of core ($P_0 \to \infty$); (b) Peak nondimensional back-face acceleration $\bar{a}_2$ at complete compaction of core.

maximum impulse correspond to complete compaction of the core and are the same as in Fig. 3(a). Above this curve, $\bar{a}_2 > 5$ at complete compaction, so only partial compaction is permitted and the values of maximum impulse are less than in Fig. 3(a). In the shaded region of Fig. 4, defined by $(\eta_0 + \eta_2)^{-1} > 5$, $\bar{a}_2 > 5$ at initiation of compaction, so the maximum allowable impulse is zero. It is evident in Fig. 4 that for a given mass fraction in the core $\eta_0$, the allowable impulse is maximized along the grey curve, i.e., by adjusting the mass distribution so that the acceleration at complete compaction equals the allowable value.

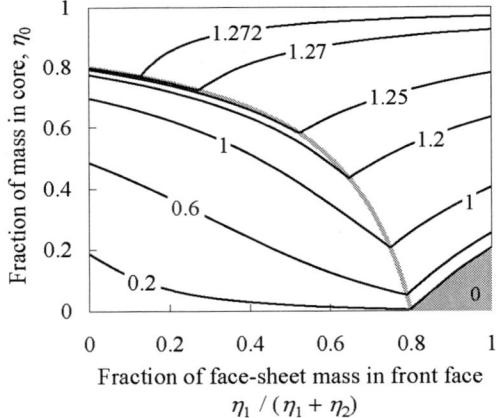

**Figure 4.** Contours of maximum nondimensional impulse $I_\infty$ with nondimensional back-face accelerations limited to $\bar{a}_2 = 5$.

**ACKNOWLEDGEMENTS**

This work was supported in part by an appointment of the first author to the Postgraduate Research Participation Program at the U.S. Army Research Laboratory (ARL) administered by the Oak Ridge Institute for Science and Education through an interagency agreement between the Department of Energy and ARL.

**REFERENCES**

1. Hanssen, A. G., Enstock, L., and Langseth, M. "Close-range blast loading of aluminum foam panels." *Int. J. Impact Eng.* **27**, 593-618 (2002).
2. Nesterenko, V. F. "Shock (blast) mitigation by 'soft' condensed matter." *MRS Proceedings*, Vol. 759 (2002).
3. Xue, Z. and Hutchinson, J. W. "Preliminary assessment of sandwich plates subject to blast loads." *Int. J. Mech. Sci.* **45**, 687-705 (2003).
4. Reid, S. R. and Peng, C. "Dynamic uniaxial crushing of wood." *Int. J. Impact Eng.* **19**, 531-570 (1997).
5. Fleck, N. A. and Deshpande, V. S. "The resistance of clamped sandwich beams to shock loading." *J. Appl. Mech.* **71**, 386-401 (2004).

CP845, *Snock Compression of Condensed Matter - 2005,*
edited by M. D. Furnish, M. Elert, T. P. Russell, and C. T. White
© 2006 American Institute of Physics 0-7354-0341-4/06/$23.00

# UNDERWATER EXPLOSIVE WELDING, DISCUSSION BASED ON WELDABLE WINDOW

## A. Mori[1], K. Tamaru[1], K. Hokamoto[2], and M. Fujita[3]

[1] *Graduate student, Kumamoto University, 2-39-1 Kurokami, Kumamoto, 860-8555, Japan*
[2] *Shock wave and Condensed Matter Research Center, Kumamoto University, Kumamoto, Japan*
[3] *Dept. of Mech. Eng., Sojo University, Kumamoto, 860-0082, Japan*

**Abstract.** A new method of underwater explosive welding is introduced and its possibilities are suggested. In the underwater explosive welding, a high explosive with detonation velocity of 7km/s is placed at an initial inclined angle to decrease the horizontal collision point velocity, which is one of the important parameters to achieve welding. This method is effective to accelerate a thin metal plate rapidly. However, this arrangement makes a difference in the welding conditions with horizontal position when a constant thickness explosive is used, as the propagation distance of the underwater shock wave increases at the ends. Hence, a method of linearly increasing the thickness of explosive in proportion to the propagation distance is proposed. This investigation intends to clarify the welding conditions in using a constant thickness explosive and linearly increasing thickness explosive based on numerical analysis. Further, a method of designing the assembly is confirmed through numerical analysis and its validity with the experimental results is demonstrated based on the welding window.

**Keywords:** Underwater explosive welding, Welding condition, Weldable window, Numerical analysis.
**PACS:** 81.20.Vj, 62.50.+p.

## INTRODUCTION

Some of the authors have developed a method of underwater explosive welding and suggested the possibility for welding thin plates on to a base plate [1-4]. This technique is effective to accelerate a thin plate uniformly at a high velocity to satisfy the welding conditions. This makes underwater explosive welding suitable for welding amorphous film/metal plates and thin metal plate/ceramics combinations [3,4]. One major concern in the underwater explosive welding is that high explosive must be placed at an initial inclined angle so as to decrease the horizontal collision point velocity, which should be lower than the sound velocity of materials to be welded [5]. Therefore, the welding conditions are different with horizontal distance when a constant thickness explosive is used. Hence, to improve the welding conditions, a method of linearly increasing thickness of explosive is proposed. The present study intends to clarify the welding conditions in using both assemblies by numerical analysis and its validity is demonstrated based on the weldable windows.

## EXPERIMENTAL AND NUMERICAL PROCEDURE

Fig. 1 shows the schematic illustration of the experimental set up for underwater explosive welding and the parameters of the process. A high explosive SEP, of density 1300 kg/m$^3$ and detonation velocity 6900m/s produced by Asahi

Kasei Chemicals Corp., Japan, was used for the experiments. An aluminum cover plate (0.3mm-thick) was attached with a flyer plate to decrease the momentum of the flyer plate. The stand off distance was fixed as 0.3 mm for all the experiments. The details of the experimental conditions are listed in Table 1, where D represents

**TABLE 1.** Experimental and numerical conditions used.

| No. | D (mm) | Flyer / Base (mm) | T₁ (mm) | T₂ (mm) | d₁ (mm) | d₂ (mm) | α (°) |
|---|---|---|---|---|---|---|---|
| CM-C1 | 48 | Cu (0.1) / MS* (9.0) | 5 | | 38.55 | 64.25 | 10 |
| CM-C2 | | | | | 21.67 | 72.35 | 20 |
| CM-C3 | | | | | 11.20 | 85.20 | 30 |
| CM-L1 | | | 3 | 5 | 38.55 | 64.26 | 10 |
| CM-L2 | | | | 10 | 21.67 | 72.35 | 20 |
| CM-L3 | | | | 23 | 11.20 | 85.87 | 30 |
| IS-C | 40 | IA** (0.1) / SS* (2.0) | 5 | | 21.67 | 72.35 | 20 |
| IS-L | | | 3 | 10 | | | |

\* MS: Mild steel, SS: 304 stainless steel
\*\*IA: Inconel alloy 600

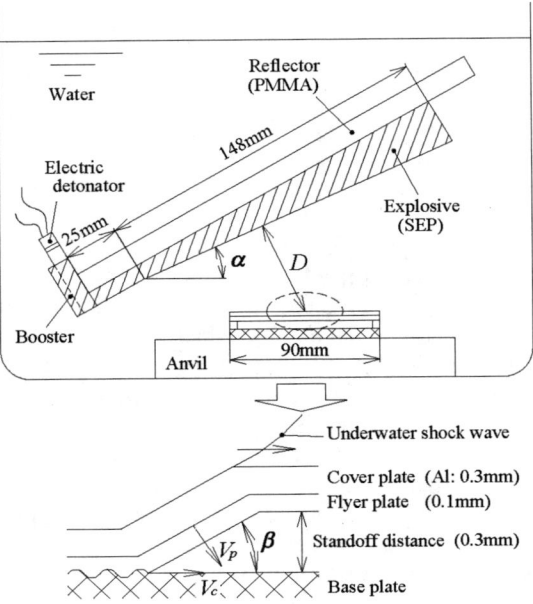

**Figure 1.** Schematic illustration of the experimental setup for underwater explosive welding and parameters.

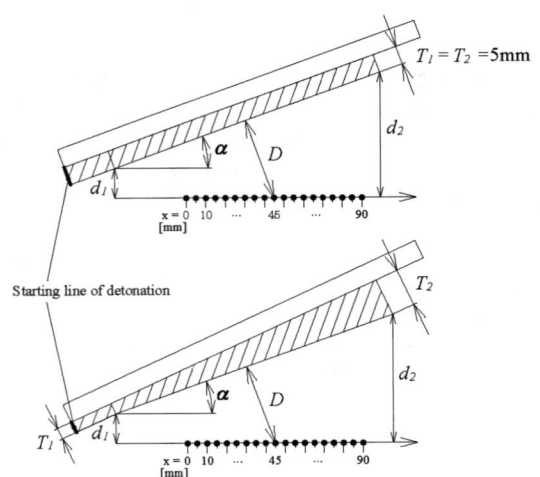

**Figure 2.** Schematic illustration of the numerical model.

the distance from the surface of explosive to the middle of cover plate.

The models shown in Fig. 2 were numerically simulated using AUTODYN-2D. The parameters used for the simulation have been reported elsewhere [2]. The important parameters namely [5], horizontal collision point velocity $V_c$ and collision angle $\beta$, were necessary in formulating the weldable window. It is well known that the relationship between the collision velocity, $V_p$, and the collision angle is expressed by the following equation [5].

$$V_p = 2V_c \sin\left(\frac{\beta}{2}\right), \tag{1}$$

These parameters were calculated from the pressure and density profiles obtained from the simulation using the following equations.

$$U_s = \sqrt{\frac{\rho}{\rho_0}\frac{P - P_0}{\rho - \rho_0}}, \tag{2}$$

$$V_c = \frac{U_s}{\sin[\sin^{-1}(U_s/U_d) + \alpha]}, \tag{3}$$

$$V_p = \frac{1}{(\rho_P h_P)}\int P\,dt, \tag{4}$$

where $\rho_0$ is the density of water under atmospheric pressure, $P_0$. $U_s$ and $U_d$ represents the velocity of underwater shock wave and detonation velocity of explosive respectively. The density and thickness of plates (flyer plate and cover plate) are represented by $\rho_P$ and $h_P$.

## RESULTS AND DISCUSSION

Fig. 3 shows the weldable window for 0.1 mm inconel alloy 600 to 304 stainless steel combination (IS-C, IS-L in Table 1). The blank symbols represent the change of welding conditions using a constant thickness explosive. The solid symbols show the change of conditions using linearly increasing thickness explosive. The lower limit represents the limit for the formation of uniform wave. The upper limit represents the conditions of excessive melting in between the joint. The upper limit line lies at a higher range and is not indicated in the figure. In order to gain a clear picture of the behavior of plates near the lower limit, kinetic energy loss ($\Delta$KE) lines are drawn as shown in Fig. 3. The results show a linear trend between $\Delta$KE and wavelength as reported earlier [6]. The welding conditions are $\beta$ = 9.74°-8.73°, $V_C$ = 3750-3570 ms$^{-1}$ and $\Delta$KE = 300-220 kJm$^{-2}$ when a constant thickness explosive is used. On the other hand, in the case of linearly increasing thickness explosive are $\beta$ = 9.20°-8.93°, $V_C$ = 3620-3590 ms$^{-1}$ and $\Delta$KE = 250-230 kJm$^{-2}$, which are quite uniform through out the bonding area. It is interesting to

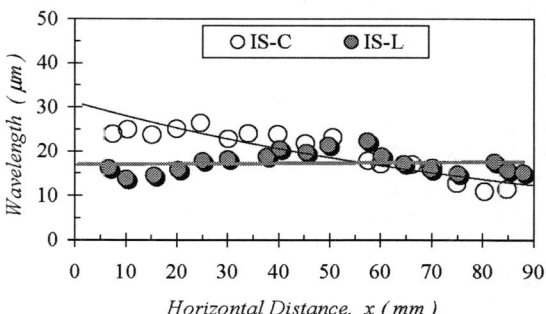

**Figure 4.** Change in wavelength with horizontal distance in case of inconel alloy 600 / 304 stainless steel combination.

note that the use of linearly increasing thickness explosive keep the welding conditions uniform and thus forms a uniform wave formation.

The experiments were conducted using both the assemblies and the interface wavelengths were measured. Fig. 4 shows the change of wavelength with horizontal distance. The wavelength decreased with horizontal distance for a constant thickness explosive and was uniform for linearly increasing thickness explosive. This is due to the reason that thickness of explosive placed in proportion with horizontal distance maintains the

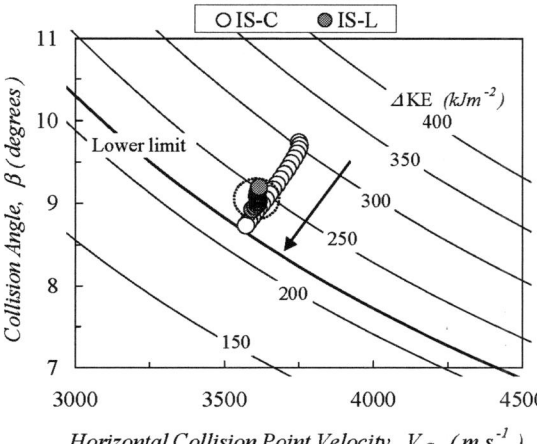

**Figure 3.** Weldable window for 0.1 mm inconel alloy 600 plate / 304 stainless steel plate.

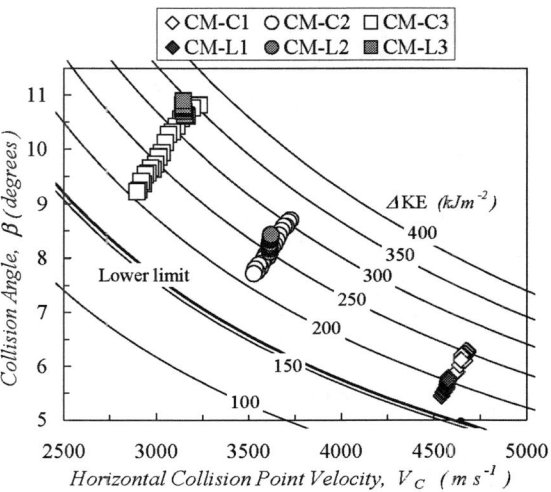

**Figure 5.** Weldable window for 0.1 mm copper plate / mild steel plate.

shock propagation distance and thus result in uniform wavelength.

Fig. 5 shows the weldable window and the energetic lines formulated for copper / mild steel combination. The experiments were conducted by varying the initial inclined angle, α, using both assemblies. The results, in general shows the uniform wavelength with distance for linearly increasing thickness explosive which is similar to inconel alloy 600 / 304 stainless steel combination. However, clear differences in terms of wavelength can be observed in the experiments, particularly when the conditions approach the lower limit as shown in Fig. 6. This is understandable, since they were produced with different initial angle, which in turn changes the kinetic energy spent at the collision. A comparison between the experiments CM-C1 and CM-L1 shows that the differences in the welding conditions are less, thus resulting in same wavelength..

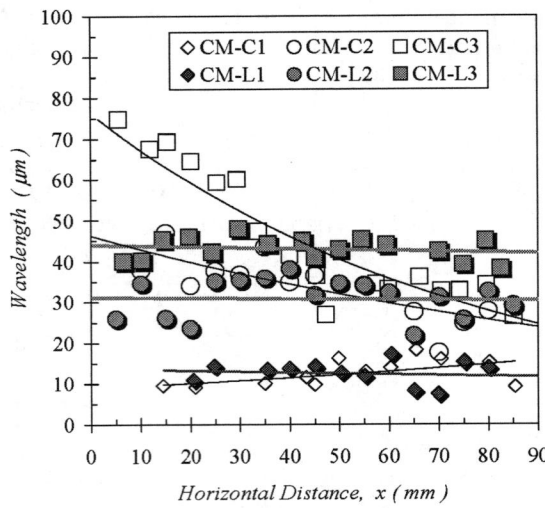

**Figure 6.** Variation of wavelength with horizontal distance (copper / mild steel combination).

## CONCLUSIONS

A new method of underwater explosive welding which is effective to weld a thin metal plate on to a base plate is introduced. The welding conditions for underwater explosive welding were evaluated and discussed using a weldable window for constant thickness explosive and linearly increasing thickness explosive. The linearly increasing thickness explosive makes possible to keep the welding condition uniform throughout the welding area. The experimental results also suggest that the uniform wave formation could be obtained using the linearly increasing thickness explosive irrespective of the change of initial inclined angle and material combination.

## ACKNOWLEDGEMENTS

The authors gratefully acknowledge the support of the 21st Century COE program on Pulsed Power Science, Kumamoto University, Japan.

## REFERENCES

1. Hokamoto, K. et al., "A New Method for Explosive Welding of Al/ZrO2 Joint Using Regulated Underwater Shock Wave", J. Mater. Process. Technol., 85, 1999, pp.175-179.
2. Mori, A. et al., "Characteristics of the New Explosive Welding Technique Using Underwater Shock Wave -Based on Numerical Analysis-", Mater. Sci. Forum, 2004, Vols. 465-466, pp.307-312.
3. Hokamoto, K. et al., "Explosive Welding of an Amorphous Film to a Steel Plate Using a Regulated Underwater Shock Wave", Metallurgical and Materials Applications of Shock-Wave and High-Strain-Rate Phenomena, 1995 (L.E. Murr, K.P. Staudhammer, M.A. Meyers, eds.), Elsevier, pp.831-837.
4. Hokamoto, K. et al., "Joining of Thin Metal Plate onto Various Materials Using Regulated Underwater Shock Wave, Fundamental Issues and Applications of Shock-Wave and High-Strain-Rate Phenomena, 2001 (K.P. Staudhammer, L.E. Murr, M.A. Meyers, eds.), Elsevier, pp.601-608.
5. Crossland, B., Explosive Welding of Metals and Its Application, Oxford University Press, 1982.
6. Hokamoto, K. et al., "Single-Shot Explosive Welding Technique for the Fabrication of Multilayered Metal Base Composites", Comp. Eng., 1995, Vol.5, No.8, pp.1069-1079.

CP845, *Shock Compression of Condensed Matter - 2005,*
edited by M. D. Furnish, M. Elert, T. P. Russell, and C. T. White
© 2006 American Institute of Physics 0-7354-0341-4/06/$23.00

# SHOCK HUGONIOT COMPRESSION DATA FOR SEVERAL BIO-RELATED MATERIALS

## K. Nagayama[1], Y. Mori[1], Y. Motegi[1], and M. Nakahara[2]

[1]*Department of Aeronautics and Astronautics, Faculty of Engineering,
Kyushu University, Fukuoka 812-8581 Japan*
[2]*Department of Computer and Communication, Faculty of Engineering,
Fukuoka Institute of Technology, Fukuoka811-0295 Japan*

**Abstract.** Shock wave data for several bio-related materials have been obtained. Plane and steady shock waves have been induced by using a modest compressed gas gun facility. Shock pressure covered in this study ranges at least up to 1 GPa. In order to realize the sensitive detection of shock front in these materials in this relatively low pressure region, an optical method has been developed by our group. Optical prism was placed on the sample such that incident laser beam is totally reflected at the prism sample interface. Samples used in the experiment include gelatin with different initial density, NaCl aqueous solution, and finally chicken breast meat. Shock data obtained in the present study are compared with the shock Hugoniot curve for water. It is found that slope of the shock velocity-particle velocity Hugoniot compression curve for all the materials tested is almost 2. While value of the intercept of the relationship, corresponding to the sound velocity, is apparently dependent on the material and ambient temperature.

**Keywords:** Biological materials, Laser surgery, Hugoniot function, Optical method
**PACS:** 62.50+p, 62.10+3, 87.19.St

## INTRODUCTION

Since water is an important material for biology and other applications, dynamic pressure data have been obtained by several means. Understanding of shock wave behavior in water or bio-related material is inevitable for various medical applications using pulse laser induced pressure pulse.[1] We have reported precise shock Hugoniot compression curve for pure water measured up to 1 GPa, which was a reexamination of the previously published Hugoniot data collected in older ways.[2,3]

For that purpose, a sensitive optical detection method that can be used in this pressure range has been developed by our group.[3,4] Combination of impact-generated shock and high resolution detec-

tor enables us to obtain more systematic shock wave data for warter in 1 GPa pressure region than those obtained previously.[2]

In this paper, we present precise information on the shock Hugoniot compression curve for several bio-related materials. Obtained Hugoniot data is compared with that of pure water. It is apparent that information on the differences and/or similarities of these materials and water might be quite important for medical applications as well as for several engineering and civil applications of shock waves. Information obtained in this study can be a fundamental database that will be used to predict the behavior of short pressure pulse propagation in a living tissue.

## EXPERIMENTAL

We have developed a new experimental method of detecting shock wave front in condensed media whose shock pressure is up to less than several GPa.[2,3] The method is based on the change in the light reflectivity of the glass-liquid interface caused by the attained high pressure shock compression. The method is also based on the total internal reflection at the bottom face of the optical prism attached to the specimen surface before shock front arrival. The change of light reflectivity upon shock arrival can be detected very sensitively either by the streak high-speed camera or by the electro-optic sensor. Shock compression curve for water[2] up to 1 GPa has been obtained by using the present method and the scatter of the data is found to be less than the previously published data, indicating the reliability of the new method.

In the present study, we have treated samples also in a liquid state or in a very soft solid state. It is necessary to prepare a specimen container made of solid material to maintain their size. Present observation method relies on the pressure-induced change in the refractive index of samples and glass material. Change in the refractive index is induced by density change. In this sense, shock Hugoniot properties of both sample and glass material are the necessary information for the shock sensor. This discussion seems contradictious, since the measurement method is used to measure shock Hugoniot, while the Hugoniot data itself is required to use the method. Even in this situation, however, one may find the suitable experimental conditions for the method. Critical angle of total internal reflection at the glass-sample interface, however, should be known precisely before shock experiment. Incident angle of the laser beam to the optical prism must be determined to fulfill the condition of total internal reflection.

Refractive index of water is known to be described by Gladstone-Dale model within the shock pressure range of less than 2 GPa. Yadav et al have measured the change in the refractive index for water by shock compression.[5] Pressure dependence of the refractive index of the glass material BK7, a kind of boro-silicate glass has been published previously.[6]

The most important concept of the impact experiment in the present study is the so-called "symmetric impact condition". Although the symmetric impact of the liquid specimen is impossible, we used PMMA plate as a well-defined container material for liquid samples. Schematic illustration of the target assembly is given in Fig. 1. In this figure, reflected light is led to the slit of high speed camera. Slit direction is perpendicular to the figure.

Sound velocities for all of these specimens, water, NaCl solutions, and gelatin have relatively large temperature dependence, which could be compared with the acoustical properties of living tissues. In this sense, the initial temperature of the specimen is one of the important experimental parameter. Ultrasonic sound velocity of all the samples was measured by the pulse echo technique.

Specimens used in this study are NaCl aqueous solution, and gelatin. As a standard liquid specimen, commercially available purified distilled water was used, and gas contamination was eliminated by evaporation method. It was inserted into a target vessel made from PMMA. Two kinds of samples were arranged for NaCl aqueous solution. Pure NaCl powder was solved into the gas-eliminated water, stirring with a magnetic stirrer. Mass fraction of the NaCl aqueous solution was measured so that the total mass of the solution has the desired value. Sample preparation procedure of gelatin is quite different from that of NaCl solution. Gelatin sample was used in the state of gel. Since temperature of sol-gel transition is known to be 24 degrees Celsius, shock experiment was performed

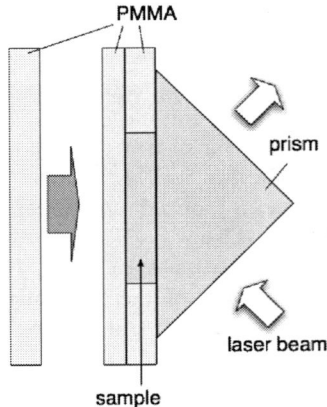

Fig. 1 Target assembly schematic.

at the room temperature fixed at 21 degrees Celsius. In case of NaCl solution, the initial temperature was fixed to 24 degrees Celsius. Gelatin sample target preparation is somewhat different from the other samples, since it is used as a gel state. Gelatin specimen was solidified directly inside the sample container PMMA assembly. One wall of the container is a large triangular prism. At least the gelatin specimen surface contact with the PMMA driver plate surface and also of the prism surface should be a smooth surface. Gelatin in sol state is poured into the assembly through a hole,

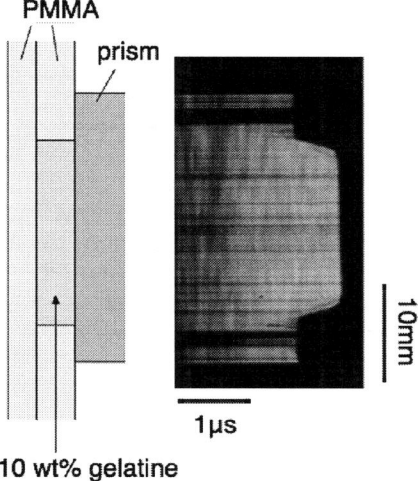

Fig. 2 Typical streak record for gelatin sample. Sample width in left figure matches to that of the streak record (right). Times goes to right.

confined by the glue and cooled into gel. The assembly contains thermocouple sensor in between the driver plate and sample container PMMA plate, so that the temperature of the assembly was monitored prior to the shot.

## RESULTS AND DISCUSSION

Figure 2 shows the typical streak record obtained in this study. They show result of a shot of gelatin sample. It is seen in this figure that sudden decrease in the light intensity corresponds to the arrival of high-pressure shock wave at the prism interface both with sample and PMMA surface. These photographs show the effectiveness of the

present shock sensor in these materials. It is also noted that the time of light extinction at the sample and at PMMA is different and a finite time difference can be observed. This is attributed to the difference in shock velocity in both materials.

Figure 3 shows the obtained shock data together with the water data all measured by our group previously. Acoustic velocity data for the materials measured in this study were used for the data fitting. Water has a linear relationship between $u_s$ and $u_p$ as

$$u_s = 1.45 + 1.99\, u_p. \quad [km/s] \qquad (1)$$

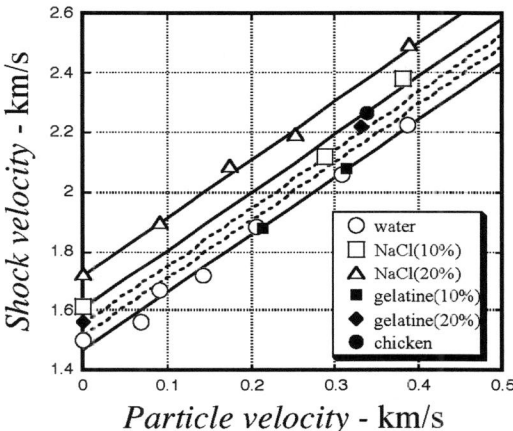

Fig. 3 Us-Up Hugoniot data for several bio-related materials.

This relationship was already published by our group[2], and is found to be quite similar to the one published previously by several authors. But we have to emphasize that we determined the relationship from the less scattered data set. With adding NaCl to water, the Hugoniot on $u_s$-$u_p$ plane tends to shift to the upper plane. That is, the shock velocity for NaCl solution is higher than that for water for the same value of the particle velocity. From the data of two kinds of concentration values of NaCl, shock Hugoniot function can be summarized by the formula

$$u_s = A(C) + B\, u_p. \qquad [km/s] \qquad (2)$$

$$A(C) = 1.50 + 0.011C, \qquad [km/s] \qquad (3)$$

$$B = 2.0, \qquad (4)$$

in terms of NaCl concentration C [wt%] as a parameter. Slope of the Hugoniot curve function of around 2 seems compatible with data for the concentration at least within the range of concentration studied. Dependence of sound velocity on the concentration, Eq.(3) is measured in the present study. Function of Eq.(3) is valid for sound velocity at the condition that the temperature of the solution is 24 degrees Celcius. By applying C=0 in Eqs.(2),(3) leads to water Hugoniot, $u_s = 1.50 + 2.00\ u_p$. which is different from Eq.(1). Major reason of the difference comes from the difference in initial temperature of the specimens.

We decided here that Hugoniot function of the two kinds of gelatin sample is also expressed as

$$u_s = 1.52 + 2.00\ u_p.[km/s]\ 10\%\ \text{gelatin solution} \quad (5)$$

$$u_s = 1.56 + 2.00\ u_p.[km/s]\ 20\%\ \text{gelatin solution} \quad (6)$$

Two data for gelatin sample with 10wt% are almost on the shock Hugoniot for pure water, so that the above function does not seem good fitting. Shock parameters predicted by Eq.(5), however, are within the precision of the shock parameter determination of about 1-2 %.

Resultant expressions of the Hugoniot function for bio-related materials treated here are revealed to be given by the following rules :

(i) Slope B has a value close to 2.

(ii) Intercept A is given by the measured sound velocity in km/s.

Chicken data in Fig. 3 seems in between NaCl(10%) and gelatin(20%) lines, and the result is reasonable according to the acoustic impedance of soft bio-tissues.

Sound velocity behind the shock front is an important parameter for medical applications, since the value has strong influence on the change in form of the pressure profile with wave propagation. The irreversible heating term appearing in the formula for the sound velocity is revealed to be a higher order term in terms of shock strength and can then be neglected for weak shock waves. It is found here that sound velocity as well as shock velocity is almost linear with particle velocity for each kind of sample. Linearity of the sound velocity with the particle velocity indicates that the equation of state of the sample can be approximated by the Tait type equation of state. This linearity can be the same situation encountered in the centered rarefaction wave in ideal gas flow. Details of

the discussion will be given in another article. One may note again that the slope of the sound velocity-particle velocity relation does not depend on the sample.

Sound velocity at the initial state depends on the temperature as well as the concentration of NaCl or gelatin. Since we are using isentropic approximation for calculating sound velocity, the effect of temperature increase on the sound velocity comes from the increase in temperature along an isentrope. Thermal contribution itself loses their importance relative to that of configurational compression potential energy. This means the temperature dependence of the sound velocity becomes smaller with compression.

## CONCLUSIONS

In this study, we are aiming at collecting information on the shock Hugoniot characteristics of several kinds of bio-related materials. As a result, empirical Hugoniot relationship for each sample can be summarized in a very simple shock-particle velocity relationship.

## ACKNOWLEDGEMENTS

Authors wish to thank Prof. K. Takayama of Tohoku University and his group for the continuous interest in our work. Furthermore, their works gave us strong motivation of this work.

## REFERENCES

1. Kodama, T. and Takayama, K., Ultrasound Med. Bio. 24:1227, 1998.
2. Nagayama, K., Mori, Y., Shimada, K and Nakahara, M., J. Appl. Phys.,91:476-482, 2002.
3. Mori, Y, Shimada, K., Nakahara, M and Nagayama, K., Rev. Sci. Instrum. 72: 2123-2127, 2001.
4. Terada, M., Hatano, S., Mori, Y. and Nagayama, K., Appl. Surf. Sci. 197-198:285-288, 2002.
5. Yadav, H.S., Murty, D.S., Verma, S.N., Sinha, K.H.C., Gupta, B.M. and Chand, D., J. Apl. Phys. 44: 2197-2200, 1973.
6. Waxler, R.M. and Weir, C.E., J. Res. NBS 69A : 325, 1965.

# PARTICIPANT LIST

Thomas J Ahrens, Caltech, 3151 S Babcock St #82, Melbourne, FL, 32901-6957, (407) 768-8000, (407) 674-7482, tja@caltech.edu

Huirong Ai, Caltech, 252-21, 1201 E California Blvd, Pasadena, CA, 91125, (626) 395-3825, (626) 564-0715, ahr@gps.caltech.edu

John B Aidun, MS 1110, Sandia National Laboratories, PO Box 5800, Albuquerque, NM, 87185-1110, (505) 844-1209, (505) 845-7442, jbaidun@sandia.gov

Scott Alexander, MS 1181, Sandia National Laboratories, PO Box 5800, Albuquerque, NM, 87185-1181, (505) 845-3572, calexa@sandia.gov

Zhiembetov Amangeldy, RFNC-VNIIEF, yl. Miza 37, Sarov, RUSSIA, zoot@gdd.vniief.ru

Richard G. Ames, Naval Surface Warfare Center, 13524 Granview Rd,, King George, VA, 22485, (540) 220-1163, amesrg@nswc.navy.mil

Bill Anderson, DX-2 MS P952, Los Alamos National Lab, Los Alamos, NM, 87545, (505) 667-5460, (505) 667-6372, wvanderson@lanl.gov

Charles E Anderson Jr, Southwest Res Inst, Southwest Research Inst, PO Drawer 28510, San Antonio, TX, 78228-0510, (210) 522-2313, (210) 522-6290, canderson@swri.edu

Mark U Anderson, Sandia National Laboratories, P. O. Box 5800, MS-1421, Albuquerque, NM, 87185-1421, (505) 844-5726,muander@sandia.gov

Simon Anderson, School of Mech. Aero Eng., The University of Manchester, Sackville Street, Manchester, M60 1QD, UK, 0161 306 2407, Simon.anderson@postgrad.manchester.ac.uk

Masahiko Arakawa, Nagoya University, Chikusa-ku, Furo-cho, Nagoya, 464-8602, JAPAN, +81 52 786 3650, +81 52 789 3013, arak@eps.nagoya-u.ac.jp

Ronald W Armstrong, University of Maryland, 14 45th St Unit 203, Ocean City, MD, 21842-3863, (850) 837-1745, rona@eng.umd.edu

Werner A Arnold, EADS-TDW, P.O. Box 1340, Schrobenhausen, 86523, GERMANY, +49 8252 99 6267, +49 8252 99 6733, werner.arnold@eads.com

Blaine W Asay, MS C920, Los Alamos National Laboratory, Los Alamos, NM, 87545, (505) 667-3266, bwa@lanl.gov

James Russell Asay, Washington State Univ, 620 NE Illinois, Pullman, WA, 99163, jrasay@wsu.edu

Tariq D Aslam, Mail Stop P952, Los Alamos National Laboratory, Los Alamos, NM, 87545, (505) 667-1367, (505) 667-6372, aslam@lanl.gov

Asoka-Kumar, Lawrence Livermore Natl Lab, L-041, 7000 East Ave, Livermore, 94550, (925) 422-9671, asokakumar1@llnl.gov

Joseph Edward Backofen, BRIGS Co., 2668 Petersborough St, Herndon, VA, 20171, (703) 476-6448, JEBackofen@earthlink.net

Melvin R Baer Sandia National Labs, P. O. Box 5800, MS-0836, Albuquerque, NM, 87185, (505) 844-5223, (505) 844-8251, mrbaer@sandia.gov

Stephen P Bailey, Atomic Weapons Establishment, AWE, Aldermaston, RG7 4PR, UK, +44 0118 982 6306, stephen.bailey@awe.co.uk

Jason Baird, Loki Incorporated, 114 Rock Mechanics Facility, 1006 Kingshighway, Rolla, MO, 65409-0660, (573) 341-6648, jbaird@lokiconsult.com

Devendra Bajaj, Univ of Maryland - Baltimore County, ECS 210, 1000 Hilltop Circle, Baltimore, MD, 21250

Edward Balizer, NSWC, Code 681,Carderock Div, 9500 MacArthur Blvd, Bethesda, MD, 20817, (301) 227-4758, (301) 227-4789

Graham J Ball, Atomic Weapons Establishment, , Reading, RG7 4PR, UK, +44 118 9824359, graham.j.ball@awe.co.uk

Eli Bar-On, RAFAEL BALLISTICS CENTER, Atzmon 72, D.N. Misgav, 20170, ISRAEL, 972 4990 9701, 972 4879 2113, ebanon@rafael.co.il

Scott Gary Bardenhagen, Los Alamos National Laboratory, MS B214, PO Box 1663, Los Alamos, NM, 87545, (505) 665-5323, bard@lanl.gov

Amanda Barra, Sandia National Laboratories, P.O. Box 5800 MS 0836, Albuquerque, NM, 87185, (505) 284-8502, ajbarra@sandia.gov

Gerard Baudin, Centre d'Etudes de Gramat, Gramat, 46500, FRANCE, 33 5 65 10 53 61, gerard.baudin@dga.defense.gouv.fr

Francois Bauer, ISL, 8 rue Oberlin, Saint Louis, F-68300, FRANCE, 0033 8969 7694, francois.bauer@wanadoo.fr

James Belak, Lawrence Livermore Nat'l Lab, L-45 LLNL, PO Box 808, Livermore, CA, 94551, (925) 422-6061, (925) 422-2851, belak@llnl.gov

Lorin Benedict, Lawrence Livermore Nat'l Lab, L-412, 7000 East Ave, Livermore, CA, 94550, (925) 424-6594, (925) 424-3034, benedict5@llnl.gov

L. V Benningfield, Applied Research Associates, 4300 San Mateo Blvd. NE, Suite A320, Albuquerque, NM, 87110, (505) 881-8074, lbenning@ara.com

Galina Bezruchko, Russian Academy of Sciences, Inst Problem Chemical Physics, Chernogolovka, 142432, RUSSIA, +7(095)7857029, bezgs@ficp.ac.ru

James Pershing Billingsley, US Army RDECOM, 303 Edgewood Dr, Tullahoma, TN, 37388, (931) 455-3585, jim.billingsley@rdec.redstone.army.mil

Stephan J Bless, Institute for Advanced Tech, Inst of Adv Tech, 3925 W Braker Ln 400, Austin, TX, 78759, (512) 471-9060, (512) 471-9096, bless@iat.utexas.edu

Thomas Boehly, Lab for Laser Energetics, Lab for Laser Energetics, 250 East River Road, Rochester, NY, 14623, (716) 275-5960, (716) 275-5960, trb@lle.rochester.edu

Jonathan Carl Boettger, Los Alamos National Lab, 23 Karen Circle, Los Alamos, NM, 87544, (505) 672-9581, (505) 665-5753, jn@lanl.gov

Cindy Bolme (MIT), MS P952, Los Alamos Natl Lab, Los Alamos, NM, 87545, (505) 667-9212, (505) 667-6372, cbolme@lanl.gov

Nicola Bonora, Univ of Cassino, Via G Di Biasio 43, Cassino, I-03043, Italy, +39-0776 299 3693, +39-0776-299 3390, nbonora@unicas.it

John Borg, Marquette University, Dept Mechanical Engr, 1551 W Wisconsin Ave, Milwaukee, WI, 53233, (414) 288-7519, (414) 288-7790, john.borg@marquette.edu

John Michael Boteler, Army Research Lab, P.O. Box 15, Welcome, MD, 20693-0015, (301) 744-4272, botelerjm@ih.navy.mil

Emeric Bourasseau, CEA, CEA-DIF, DPTA/PMC, Bruyeres le Chatel, 91680, FRANCE, +033169264963, emeric.bourasseau@cea.fr

Neil K Bourne, University of Manchester, PO Box 88, Sackville SV, Manchester M6O 1QE, UK, 44 1793 784154, 44 1793 784195, neil.bourne@mac.com

Christopher Braithwaite, Univ of Cambridge, Cavendish Laboratory, Madingley Rd, Cambridge, CB3 0HE, UK

Nachhatter S Brar, SPC 1931, Univ of Dayton-Research Inst, 300 College Park Ave, Dayton, OH, 45469-0182, (937) 229-3554, (937) 229-3869, brarns@udri.udayton.edu

Matt Briggs, M.S P940, Los Alamos National Lab, Los Alamos, NM, 87545, (505) 664-0839, briggs@lanl.gov

Eduardo M. Bringa, Lawrence Livermore Nat'l Lab, L-371, PO Box 808, Livermore, CA, 94550, (925) 423-5724, (925) 422-4665, ebringa@llnl.gov

Eric N Brown, Los Alamos National Lab, PO Box 1663, Los Alamos, NM, 87545, (505) 667-0799, en_brown@lanl.gov

William T Brown, Applied Research Associates, 2760 Eisenhower Ave, Suite 308, Alexandria, VA, 22314, (703) 325-4076, (703) 325-4329, wbrown@ara.com

Aaron L Brundage, Ph.D., Sandia National Laboratories, 8919 Robs PL NE, Albuquerque, NM, 87122, (505) 284-2958, albrund@sandia.gov

Scott M Bucholtz, Honeywell FM&T, Honeywell FM&T, 3500 Trinity Dr. Unit 4-C, Los Alamos, NM, 87544, (505) 661-8496, (505) 661-6955, sbucholtz@kcp.com

Mark Joseph Burchell, Univ of Kent, School of Physical Sciences, Ingram Building, University of Kent, Canterbury, Kent, CT2 7NH, UK, +44 (0)1227 823248, m.j.burchell@ukc.ac.uk

Francois Buy, CEA, Centre de Valduc, Is sur Tille, 21120, FRANCE, +33 3 8023 5039, +33 3 8023 5278, francois.buy@cea.fr

Eric Buzaud, Centre d'Etudes de Gramat, Gramat, 46500, FRANCE, 33 5 65 10 54 04, eric.buzaud@dga.defense.gouv.fr

Jing Cai, UCSD, 310 Rogers Rd, S-106, Athens, GA, 30605, (706) 552-2652, ugacj@hotmail.com

Buyang Cao, UCSD, 9226 Regents Rd., Apt. L, La Jolla, CA, 92037, (858) 534-6091, bucao@ucsd.edu

Joel R Carney, Naval Surface Weapons Ctr, Bldg 600 Code 920Q, NSWC Indian Head, 101 Strauss Ave, Indian Head, MD, 20640, (301) 744-4213, (301) 744-4445, carneyjr@ih.navy.mil

John H Carpenter, Sandia National Laboratories, PO Box 5800, MS 1110, Albuquerque, NM, 87185-1110, (505) 293-5149, jhcarpen@illinoisalumni.org

Elizabeth Cart, Naval Surface Weapons Ctr, NSWC - Indian Head, 101 Strauss Ave, Indian Head, MD, 20640, (301) 744-2539, cartej@ih.navy.mil

Renida R Carter, Los Alamos National Lab, PO Box 1663, T086, Los Alamos, NM, 87545, (505) 665-5292, (505) 667-4420, renida@lanl.gov

Erik Peter Carton, TNO, PO Box 45, Rijswijk, 2280AA, NETHERLANDS, +31 15 284 3355, +31 15 284 3997, carton@pml.tno.nl

Simon Case, AWE, Aldermaston, Reading, RG7 4PR, UK, +44 118982 5801, simon.case@awe.co.uk

Daniel T Casem, US Army Research Lab, AMSRD-ARL-WM-TD, Aberdeen, MD, 21005, (410) 306-0798, (410) 306-0783, dcasem@arl.army.mil

Robert Craig Cauble, Lawrence Livermore Nat'l Lab, L-41 LLNL, PO Box 808, Livermore, CA, 94550, (925) 422-4724, cauble@llnl.gov

Peter M Celliers, Lawrence Livermore Nat'l Lab, 1530 Spruce St, Berkeley, CA, 94709, (925) 424-4531, (925) 424-2778, celliers1@llnl.gov

Ellen Kathleen Cerreta, Los Alamos National Lab, P.O. Box 3212, Santa Fe, NM, 87501-0212, (505) 455-2933, (505) 667-8021, ecerreta@lanl.gov

Sek K Chan, Orica Canada Inc., 301 Hotel De Ville, Brownsburg-Chatham, QC, J8G 3B5, CANADA, (450) 533-1338, (450) 533-5951, jim_chan@orica.com

Dr Usha Chandra, University of Rajasthan, High Pressure Physics Lab., Department of Physics, University of Rajasthan, Jaipur, 302004, INDIA, 91-141-2707423, 91-141-2707728, ushac_jp1@sancharnet.in

David J Chapman, Univ of Cambridge, Cavendish Lab, Madingley Rd, Cambridge, CB3 0HE, UK, 01223 350266, 01223 350266, djc49@phy.cam.ac.uk

Ricky Chau, Lawrence Livermore Nat'l Lab, Mail Stop L-045, 7000 East Avenue, Livermore, CA,

94550, (925) 423-4388, (925) 423-2442, chau2@llnl.gov

M. S. CHAWLA, Northrop Grumman Newport News, Northrop Grumman/ Tauri Group, 6940 S Kings Hwy, Alexandria, VA, 22310, (703) 971-3103, montechawla@yahoo.com

Philip J Cheese, UK MOD, Ash 2b #3212, MOD Abbey Wood, Bristol, BS34 8JH, UK, 01179135336, dosgst4b@dpa.mod.uk

Changfeng Chen, Dept of Physics, Univ of Nevada - Las Vegas, Las Vegas NV, 89154, (702) 895-4230, (702) 895-0804, chen@physics.unlv.edu

Xianglei Chen, 1197 Hillside Ave., Apt. B44, Niskayuna, NY, 12309, chenxianglei@yahoo.com

Frank J. Cherne, Los Alamos National Lab, PO Box 719, Los Alamos, NM, 87544, (505) 665-5636, (505) 667-6372, cherne@lanl.gov

Gary N Chesnut, Los Alamos National Lab, 1445 Camino Medio, Los Alamos, NM, 87544, (505) 661-2696, gchesnut@lanl.gov

Lalit Chhabildas, Sandia Natl Labs, 3716 Tewa Dr NE, Albuquerque, NM, 87111, (505) 844-4147, (505) 845-7003, lcchhab@sandia.gov

Steve Chidester, LLNL, PO box 808, Livermore, 94551, (925) 422-0865, chidester1@llnl.gov

Akobuije D Chijioke, Harvard Univ, 11 Peabody Ter Apt 501, Cambridge, MA, 02138-6314, (617) 926-6460, achijiok@fas.harvard.edu

Eric D Chisolm, Los Alamos National Lab, 889 Estates Dr., Los Alamos, NM, 87544, (505) 665-5020, echisolm@lanl.gov

Shirish M Chitanvis, Los Alamos National Lab, PO Box 1663, MS B214, Los Alamos, NM, 87545, (505) 667-7799, (505) 665-0455, shirish@lanl.gov

Kyung Young Choi, ADD, Yuseong P.O.Box 35-5, (T3-8), Taejon, 305-721, SOUTH KOREA, +82-42-821-2298, cky1218@hanmail.net

Philip D Church, QinetiQ, Lethal Mechanisms, Ft Halstead, Sevenoaks, TN14 7BP, UK, +44 1959 514893, +44 1959 516050,

Jennifer A Ciezak, Army Research Lab, 729 Fallsgrove Dr 6033, Rockville, MD, 20850, (301) 975-6082, jciezak@arl.army.mil

Steven A Clarke, Los Alamos National Lab, MS P950, Los Alamos, NM, 87545, (505) 665-8999, sclarke@lanl.gov

John D Clayton, US Army Research Laboratory, AMSRD-ARL-WM-TD, Aberdeen, MD, 21005, (410) 306-0975, (410) 306-0783, jclayton@arl.army.mil

Dr Brad Edwin Clements, Los Alamos National Lab, T1 B221, LANL, PO Box 1663, Los Alamos, NM, 87545, (505) 667-8836, bclements@lanl.gov

Prof Rodney J Clifton, Brown Univ, Div of Engr, Brown Univ, Providence, RI, 02912, (401) 863-1422, (401) 863-9983, clifton@engin.brown.edu

Charles S Coffey, Naval Surf Weapons Ctr, 6356 Guilford Rd, Clarksville, MD, 21029, (301) 744-6779, (301) 744-4445, coffeycs@ih.navy.mil

Gilbert W Collins, L-481, Lawrence Livermore Nat'l Lab, PO Box 808, Livermore, CA, 94551, (925) 423-2004, (925) 423-2204, collins7@llnl.gov

Jeff D Colvin, Lawrence Livermore Nat'l Lab, L-356, PO Box 808, Livermore, CA, 94551, (925) 422-3273, (925) 423-8945, colvin5@llnl.gov

Malcolm D Cook, Qinetiq, QinetiQ Ltd, Building Q13, Ft Halstead, Sevenoaks, TN14 7BP, UK, +44 1959 515100, mdcook@qinetiq.com

David E. Cox, Sandia National Laboratories, PO Box 5800, Albuquerque, NM, 87185, (505) 844-5847, decox@sandia.gov

Geoffrey A Cox, Atomic Weapons Establishment, AWE, Aldermaston, Reading, RG7 4PR, UK, +44 11898 26003, geoffrey.cox@awe.co.uk

David Crawford, Sandia National Labs, MS 0836, PO Box 5800, Albuquerque, NM, 87111, (505) 845-8975, dacrawf@sandia.gov

Paula J Crawford, Los Alamos National Lab, 2111 C 34th Street, Los Alamos, NM, 87544, (505) 667-1063, paulac@lanl.gov

Scott D Crockett, Los Alamos National Lab, T-1 MS-B221, Los Alamos, NM, 87544, (505) 667-8596, (505) 665-5757, crockett@lanl.gov

Blandine Crouzet, CEA Bruyeres, CEA-DAM Ile de France/DPTA/PMC, BP 12, Bruyeres-le-Chatel, F-91680, FRANCE, +33 1 69 26 73 37, +33 1 69 26 70 77, blandine.crouzet@cea.fr

Heather Croydon, US Navy, US Navy, 4251 Suitland Road, Washington, DC, 20395-5720, (301) 669-2183, hcroydon@nmic.navy.mil

Donald Robert Curran, SRI INTERNATIONAL, Lokkalia 8A, Oslo, 0783, NORWAY, (650) 859-4560, (650) 859-2343, curran@unix.sri.com

Dr. John P. Curtis, QinetiQ Ltd., Rm 5, Bldg Q13, DERA Fort Halstead, Sevenoaks, Sevenoaks, TN14 7BP, UK, 01959 514259, 01959 516050, jpcurtis@QinetiQ.com

Helen Czerski, Univ of Cambridge, Cavendish Laboratory, Madingley Rd, Cambridge, CB3 0HE, UK, hc230@cam.ac.uk

Andrew J Dale, AWE, Aldermaston, Reading, RG7 4PR, UK, +44 118 982 5218, andrew.dale@awe.co.uk

John C Dallman, Los Alamos National Lab, 106 La Senda Rd, Los Alamos, NM, 87544, (505) 667-4831, (505) 665-3407, dallman@lanl.gov

Dattatraya P Dandekar, AMSRL-WM-TD, Army Research Lab, Aberdeen P G, MD, 21005, (410) 306-0801, (410) 306-0783, DDANDEK@ARL.MIL

Chiara Daraio, Univ of California, San Diego, 3867 Miramar St Apt F, La Jolla, CA, 92037, (858) 405-3365, cdaraio@ucsd.edu

Dana M Dattelbaum, Los Alamos National Lab, PO Box 1663, Los Alamos, NM, 87545, (505) 667-7329, (505) 667-6372, danadat@lanl.gov

Ronald C Davidson, PPPL, Princeton Univ, PO Box 451, Princeton, NJ, 08543-0451, (609) 243-3552, (609) 243-2418, rdavidson@pppl.gov

Jean-Paul Davis, Sandia National Laboratories, 725 Truman St. NE, Albuquerque, NM, 87110, (505) 284-3892, jpdavis@sandia.gov

Paul S De Carli, SRI Intl/Univ. Coll. London, M/S AA145, SRI International, 333 Ravenswood Ave, Menlo Park, CA, 94025-3493, (650) 859-3171, (650) 859-3202, paul.decarli@sri.com

Thibaut De Resseguier, CNRS, LCD, ENSMA, 1 Ave. Clement, ADER, BP 40109, Futuroscope, 86961, FRANCE, +33 5 4949 8173, resseguier@lcd.ensma.fr

Christopher Deeney, Sandia Natl Lab, MS 1168, PO Box 5800, Albuquerque, NM, 87185, (505) 845-3657, cdeene@sandia.gov

Darcie Dennis-Koller, Los Alamos National Lab, 2249B 38th St,, Los Alamos, NM, 87544, (505) 667-7305, (505) 667-6372, ddennis@lanl.gov

Ilya V Derbenev, RFNC-VNIITF, 13 Vasyliev Street, Snezhinsk, 456770, RUSSIA

Michael Paul Desjarlais, MS 1186, Sandia Natl Labs, PO Box 5800, Albuquerque, NM, 87185-1186, (505) 845-7273, (505) 845-7820, mpdesja@sandia.gov

Richard D Dick, 117 Elderberry Lane, Niceville, FL, 32578-1296, richard.dick@direcway.com

Peter Dickson, Los Alamos National Laboratory, DX-2, MS J565, Los Alamos, NM, 87545, (505) 665-7830, dickson@lanl.gov

Gary Dilts, M/S D413, Los Alamos Natl Lab, PO Box 1663, Los Alamos, NM, 87545, (505) 665-0190, gad@lanl.gov

Jow Lian Ding, Sch of Mech & Mater Engr, Washington State Univ, PO Box 642920, Pullman,

WA, 99164-2920, (509) 335-3226, (509) 335-4662, ding@mme.wsu.edu

Dana D Dlott, Univ of Illinois - Urbana, A208 Chem & Life Sci Lab, UIUC Box 01-6 CLSL MC-712, 600 S Mathews Ave, Urbana, IL, 61801-3364, (217) 333-3574, (217) 244-3186, dlott@scs.uiuc.edu

Robert L Doney, US Army Research Lab, 501 Buttonwoods Rd, Elkton, MD, 21921, (410) 278-7309, (410) 278-6061, bdoney@arl.army.mil

David C Dooling, Los Alamos National Lab, 86 Moya Rd, Santa Fe, NM, 87508, (505) 664-0176, dcd@lanl.gov

Paul J Dotson, Los Alamos National Lab, PO Box 1663, Los Alamos, NM, 87545, (505) 667-6107, (505) 665-4055, dotson@lanl.gov

R Paul Drake, Univ of Michigan - Ann Arbor, AOSS, 2455 Hayward St, Ann Arbor, MI, 48109-2143, (734) 763-4072, (734) 647-3083, rpdrake@umich.edu

Zbigniew A Dreger, Bldg 947D-Inst for Shock Phys, Webster Physical Sci Bldg, Washington State Univ, Pullman, WA, 99164, (509) 335-4233, (509) 335-6115, dreger@wsu.edu

Vladimir V Dremov, RFNC-VNIITF, 13 Vasilieva St., Snezhinsk, 456770, RUSSIA, 7-351-723-2930, v.v.dryomov@vniitf.ru

Oleg B Drennov, RFNC-VNIIEF, 37 Mira Avenue, Sarov, 607190, RUSSIA, 7-831-304-4627, root@gdd.vniief.ru

Luisa Duraes, Chemical Engineering, Univ of Coimbra, Coimbra, 3030-290, POR, +351 239 798737, +351 239 798703, luisa@eq.uc.pt

Sunil Dwivedi, Washington State Univ, Inst for Shock Physics 28163, Pullman, WA, 99164-2816, (509) 335-7045, (509) 335-6115, dwivedi@mail.wsu.edu

Robert Craig Dye, Los Alamos National Lab, 434 Brighton Ln, Los Alamos, NM, 87544, cdye@spinn.net

Daniel E Eakins, Georgia Tech, 210 North Ave NW Apt 51, Atlanta, GA, 30313, (404) 206-9930, daniel.eakins@mse.gatech.edu

Craig Eastwood, Lawrence Livermore Nat'l Lab, L-644, P.O. Box 808, Livermore, CA, 94551, (925) 423-4899, (925) 423-7914, eastwood2@llnl.gov

Michael S Edens, U. S. Navy, P.O Box 423, Odessa, WA, 99159-0423, (301) 669-5915, edensmi@yahoo.com

Dr Jon Henry Eggert, Lawrence Livermore Nat'l Lab, PO Box 808, L-399, 7000 E Ave, Livermore, CA, 94551, (303) 384-2071, (303) 273-3919, eggert1@llnl.gov

Mark L Elert, Dept of Chem, US Naval Academy, 572 Holloway Rd, Annapolis, MD, 21402, (410) 293-6616, (410) 293-2218, elert@usna.edu

Erik Emmons, Univ of Nevada, Reno, 3575 Gypsum Rd Apt 7, Reno, NV, 89503-1220, (775) 324-3283, eemmons@physics.unr.edu

Eric Paul Fahrenthold, Dept of Mech & Engr, Univ of Texas, Austin, TX, 78712, (512) 471-3064, epfahren@mail.utexas.edu

Roger Falcone, Univ of California - Berkeley, Dept of Physics, 366 LeConte Hall # 7300, Berkeley, CA, 94720-7300, (510) 642-3316, (510) 643-8497, rwf@physics.berkeley.edu

Alexey V Fedorov, VNIIEF, Mira ave. 37, Novgorod, 607190, RUSSIA, (505) 844-2798, ljhumbl@sandia.gov

Dr Ruqiang Feng, Dept of Engr Mechanics, Univ of Nebraska Lincoln, PO Box 880526, Lincoln, NE, 68588-0347, (402) 472-2384, (402) 472-8292, rfeng@unlserve.unl.edu

Eric N. Ferm, Los Alamos National Lab, 2317 46th Str,, Los Alamos, NM, 87544, (505) 667-3343, enf@lanl.gov

Louis Ferranti, School of Materials Sci & Eng, Georgia Inst of Tech, 771 Ferst Drive, Atlanta,

GA, 30332-0245, (404) 894-1475, (404) 894-9140, louis.ferranti@mse.gatech.edu

Phillip J Flater, Air Force Research Lab, 101 W Eglin Blvd. Ste 135, Eglin AFB, FL, 32542, (850) 882-1813, (850) 883-1381, flater@eglin.af.mil

Helen M Flower, QinetiQ, Bldg Q13 Rm 2, MoD Fort Halstead, Sevenoaks, Kent, TN14 7BP, ENG, 01959 51 4195, hmflower@QinetiQ.com

Fran Foltz, Lawrence Livermore Nat'l Lab, 2624 Pickfair Lane, Livermore, CA, 94551, (925) 443-6248, foltz1@llnl.gov

Jerry W Forbes, Univ of Maryland/NSWC-IH, 6535 Chelsea Way, Port Tobacco, MD, 20677, (301) 342-9362, jforaps@comcast.net

Dr Joseph Foster, 34 Park Cir, Fort Walton Beach, FL, 32548, (904) 882-9643, (904) 882-5142, fosterjc@eglin.af.mil

Alan M Frank, Lawrence Livermore Nat'l Lab, 7000 East Ave., L-095, Livermore, CA, 94560, (415) 422-7271, frankz@llnl.gov

Robert E Franz, No Company Provided, 16 Fox Hill Court, Perry Hall, MD, 21128, (410) 592-6614, (410) 592-6614, elainefranz@aol.com

Laurence E. Fried, L-282, Lawrence Livermore Natl. Lab., 7000 East Ave., Livermore, CA, 94550, (925) 422-7796, fried1@llnl.gov

David L Frost, McGill University, Dept of Mech Engr, McGill Univ, 817 Sherbrooke St W, Montreal, QC, H3A 2K6, CANADA, (514) 398-6279, (514) 398-7365, david.frost@mcgill.ca

Don Fujino, Lawrence Livermore Nat'l Lab, 837 Rodney Dr, San Leandro, CA, 94577, (925) 422-3373, fujino@llnl.gov

David J. Funk, Los Alamos National Laboratory, Mail Stop P 952, PO Box 1663, Los Alamos, NM, 87545, (505) 665-9659, (505) 667-6372, djf@lanl.gov

Michael R. Furlanetto, Los Alamos National Laboratory, Mail Stop H803, PO Box 1663, Los Alamos, NM, 87545, (505) 667-2128, (505) 665-4121, m.furlanetto@gl.ciw.edu

Michael David Furnish, Sandia Natl Labs, MS 1168, PO Box 5800, Albuquerque, NM, 87185-1168, (505) 844-2877, (505) 845-7685, mdfurni@sandia.gov

Glenn R Garrett, Georgia Tech, 3621 Norwich Drive, Tucker, GA, 30084, (404) 702-6566, gtg707a@mail.gatech.edu

George A Gazunas, US Army Research Laboratory, US Army Research Laboratory, AMSRD-ARL-WM-MD, Aberdeen, MD, 21005, (410) 306-0863, (410) 306-0806, gazunas@arl.army.mil

Russell J Gehr, Honeywell FM&T, Honeywell FM&T, 3500 Trinity Dr. Unit 4-C, Los Alamos, NM, 87544, (505) 663-0321, (505) 661-6955, rgehr@kcp.com

Vlad Georgevich, Lawrence Livermore Nat'l Lab, 7000 East Avenue, L-099, Livermore, CA, 94550, (925) 423-2916, georgevich1@llnl.gov

Jean Gerin-Roze, CEA Bruyeres, BP 12, Bruyeres, F-91680, FRANCE, 33 169 267 090,

Timothy C Germann, MS F699, Los Alamos National Laboratory, Los Alamos, NM, 87545, (505) 667-9772, (505) 665-4063, tcg@lanl.gov

John J Gilman, Univ of California - Los Angeles, 2852 Forrester Dr, Los Angeles, CA, 90064, (310) 825-9608, (310) 206-7353, gilman@seas.ucla.edu

Kurt R Glaesemann, Lawrence Livermore National Laboratory, L-282, P.O. Box 808, Livermore, CA, 94551-0808, (925) 423-1579, glaesemann1@llnl.gov

William Andrew Goddard III, Caltech, 139-74 Beckman Inst, Caltech, 1201 E California Blvd, Pasadena, CA, 91125, (626) 395-2731, (626) 585-0918, wag@wag.caltech.edu

Nir Goldman, Lawrence Livermore Nat'l Lab, 201 Orange St. #5, Oakland, CA, 94610, (510) 301-9760, goldman14@llnl.gov

Zizheng Gong, Institute of High Pressure and High Temperature Physics, Southwest Jiaotong Univ, Chengdu, Sichuan, 610031, PRC, 86-28-87603401, 86-28-87603401, gongzz@263.net

Alexander L Gonor, University of Toronto, 624-523 Finch Ave West, Toronto, ON, M2R 1N4, CANADA, (416) 667-7796, (416) 667-7799, agonor@rogers.com

Lionel Goodfellow, AWE, Reading, RG7 4PR, UK, +44 0118 9827991, lionel.goodfellow@awe.co.uk

Stephen G Goveas, AWE PLC, Aldermaston, Reading, Berkshire, RG7 4PR, UK, 44 (0) 118 982 6293, stephen.goveas@awe.co.uk

Dennis E Grady, Applied Research Associates, 1472 Morning Glory NE, Albuquerque, NM, 87122, (505) 856-1555, dgrady@ara.com

R A Graham, The Tome Group, 608 Cenizo Blvd, Uvalde, TX, 78801, (830) 591-2053, tomecenizo@aol.com

Richard H Granholm, Mail Code 920V, Naval Surface Warfare Ctr, 101 Strauss Avenue, Indian Head, MD, 20640-5035, (301) 744-4866, (301) 744-4196, granholmrh@ih.navy.mil

Dr George T Gray III, Los Alamos National Lab, MST-8 MS G755, PO Box 1663, Los Alamos, NM, 87545, (505) 667-5452, (505) 667-8021, rusty@lanl.gov

Konstantin F Grebyonkin, RFNC-VNIITF, 13 Vasilieva St, Snezhinsk, 456770, RUSSIA, 7-351-465-4730, k.f.grebyonkin@vniitf.ru

Dr Carl W Greeff, Los Alamos National Lab, 191 Los Pueblos St, Los Alamos, NM, 87544, (505) 661-0350, greeff@lanl.gov

Martin W. Greenaway, Univ of Cambridge, PCS Group, Cavendish Lab, Madingly Rd, Cambridge, CB3 0HE, UK, +44 1223 339209, +44 1223 350266, mwg21@phy.cam.ac.uk

Michael Greenfield, US Army Research Lab, US Army Research Lab, Aberdeen, MD, 21005, (410) 306-0793, mgreenfield@arl.army.mil

Gianluca Gregori, Lawrence Livermore Nat'l Lab, 7000 East Avenue, Livermore, CA, 94550,

Will Grigsby, University of Texas at Austin, Fusion Research Ctr Rm 12.208, 1 University Station #C1510, Austin, TX, 78712, (512) 471-3978, (512) 471-8865, wgrigsby@physics.utexas.edu

Paulius Grivickas, Washington State University, P.O. Box 642816, Pullman, WA, 99164, (509) 335-5345, (509) 335-6115, pgrivickas@wsu.edu

Stephen E Grunschel, Brown University, 182 Hope Street, Providence, RI, 02912, (401) 863-3034, stephen_grunschel@brown.edu

Raafat H Guirguis, Naval Surface Warfare Ctr., 5516 Starboard Ct, Fairfax, VA, 22032, (301) 744-6776, (301) 744-4717, guirguisrh@ih.navy.mil

Jared Clinton Gump, NSWC, 9101 Granite Court, Waldorf, MD, 20603, (301) 705-5610, gumpjc@yahoo.com

Yogendra M Gupta, Dept of Physics, Inst for Shock Physics, Washington State Univ, Pullman, WA, 99164-2814, (509) 335-7217, (509) 335-6115, ymgupta@wsu.edu

Richard Gustavsen, Los Alamos National Lab, MS P952, Los Alamos, NM, 87545, (505) 667-2086, (505) 667-6372, rgus@lanl.gov

Clint Hall, Sandia Natl Labs, P.O. Box 5800, MS-1168, Albuquerque, NM, 87185, (505) 845-3300, (505) 845-7685, chall@sandia.gov

Yuichiro Hamate, University of Florida, 1350 N Poquito Rd, Shalimar, FL, 32547, 850-833-9350 ext. 235, fjtech@highstream.net

James Edward Hammerberg, Los Alamos National Lab, 412 Greg Ave, Santa Fe, NM, 87501, (505) 667-0687, jeh@lanl.gov

Chung Kyu Han, ADD, T-3-10 Agency for Defense Dev, P.O. Box 35-5, Yuseong-Ku, Taejeon, 305-605, SOUTH KOREA, +82 42 821 4169, +82 42 821 2390, ckhan@add.re.kr

Sathya Hanagud, Georgia Tech, School of Engineering, Atlanta, GA, 30332-0140,

Caroline A Handley, AWE, Aldermaston, Reading, Berkshire, RG7 4PR, UK, +44 118 982 6404, caroline.handley@awe.co.uk

John J Harrigan, School of Mechanical, Aero, and Civil Engineering, The University of Manchester, Manchester, M60 1QD, UK, 44 161 306 3841, john.j.harrigan@manchester.ac.uk

Ernest J Harris, Aldermaston, Reading, RG7 4PR, UK, 44 1189 825596, ernst.j.harris@awe.co.uk

Eric N Harstad, Los Alamos National Lab, MS-B216, PO Box 1663, Los Alamos, NM, 87545-1663, (505) 665-8905, enh@lanl.gov

Peter J Haskins, QinetiQ, 29 Park lane, Kemsing, Kent, Near Sevenoaks, TN15 6NX, ENG, +44 1959 515199, pjhaskins@qinetiq.com

James A Hawreliak, Lawrence Livermore National Lab, P.O. Box 808, L-399, Livermore, CA, 94550, (925) 424-2905, hawreliak1@llnl.gov

Dennis Hayes, P.O. Box 591, Tijeras, NM, 87059-0591, (505) 281-9282, (505) 286-3164, dennis@nmia.com

David Hebert, CEA/CESTA, BP 2, Le Barp, 33114, FRANCE, +33 557 04 6981, david.hebert@cea.fr

Naoki Hemmi, Washington State Univ, 1705 Nicole Ct, Pullman, WA, 99163, (509) 335-7036, (509) 335-6115, nhemmi@wsu.edu

Andrew K Henrick, Los Alamos National Lab, PO Box 1663, Los Alamos, NM, 87545

Ben Henrie, Los Alamos National Lab, PO Box 1663, Los Alamos, NM, 87545

Bryan F Henson, Los Alamos National Lab, CST 6 J567, LANL, PO Box 1663, Los Alamos, NM, 87545, (505) 665-4837, (505) 665-4817, henson@lanl.gov

Eric B Herbold, Univ of California - San Diego, 8453 Via Mallorca Dr #28, La Jolla, CA, 92037, (916) 230-0030, eherbold@ucsd.edu

Benny Herrmann, NRCN, PO Box 3377, Beer-Sheva, 84133, ISRAEL, 972 86 235 334, bpher@barak-online.net

Eugene Hertel, Sandia National Labs, P. O. Box 5800, MS-0836, Albuquerque, NM, 87185, (505) 844-5364, (505) 844-8251, esherte@sandia.gov

Olivier Heuze, CEA/DIF, B. P. 12, Bruyères-le-Châtel CEDEX, F-91680, FRANCE, 33 1 6926 4736, 33 1 6926 7097, heuze@bruyeres.cea.fr

Randy J Hickman, Sandia National Laboratories, 4805 Leon Grande, Rio Rancho, NM, 87124, (505) 284-3813, rjhickm@sandia.gov

Damien G Hicks, Lawrence Livermore Natl Lab, P.O. Box 808 L-286, Livermore, CA, 94551, (925) 424-5220, (925) 424-3383, hicks@alum.mit.edu

Larry Glenn Hill, Los Alamos National Lab, 327 Andanada, Los Alamos, NM, 87544, (505) 665-1086, lgh@lanl.gov

Rex P Hjelm, Los Alamos National Lab, PO Box 1663, Los Alamos, NM, 87545, (505) 665-2772, hjelm@lanl.gov

Brad Lee Holian, Los Alamos National Lab, MS B-268, LANL, Los Alamos, NM, 87545, (505) 667-9237, (505) 665-3909, blh@lanl.gov

Neil C Holmes, Lawrence Livermore Nat'l Lab, L-45, PO Box 808, Livermore, CA, 94551, (925) 422-7213, (925) 422-2851, holmes4@llnl.gov

Tim Holmquist, 1200 Washington Ave. South, Minneapolis, MN, 55415, (612) 337-3561, tjholm@networkcs.com

W H Holt, Naval Surface Warfare Center, 906 Carol Ln, Falmouth, VA, 22405, (703) 663-8687, wholt@nswc.navy.mil

Teruhisa Hongo, Tokyo Inst of Tech, 4259 Nagatsuda, Yokohama, 226-8503, JAPAN, +81 45 924 5382, +81 45 924 5339, hongo@knlab.msl.titech.ac.jp

Kevin Honnell, Los Alamos National Lab, MS F699, PO Box 1663, Los Alamos, NM, 87545, (505) 665-9131, kgh@lanl.gov

James GM Hooper, Univ of Ottawa, 10 Marie-Curie, D'lorio Hall, Ottawa, ON, K1N 6N5, CANADA, (613) 234-5060, jhoop018@uottawa.ca

Joe Hooper, Tulane University, 6440 S Claiborne Ave Apt 408, New Orleans, LA, 70125, (504) 314-9641, (504) 862-8702, jhooper@tulane.edu

Irene E Hooton, Defence R&D Canada, Defence R&D Canada, 101 Colonel By Dr., Ottawa, ON, K1A 0K2, CANADA, (613) 945-5063, (613) 945-5255, hooton.ie@forces.gc.ca

Yasuyuki Horie, AFRL, 804 E Lake Dr, Shalimar, FL, 32579-2249, (850) 882-8895, horie@eglin.af.mil

Michael Howard, Lawrence Livermore Nat'l Lab, Box 808, L-282, Livermore, CA, 94550, (925) 422-4138, (925) 424-3281, howardII@llnl.gov

Luke L Hsiung, Lawrence Livermore Nat'l Lab, 7000 East Avenue, L-352, Livermore, CA, 94551, (925) 424-3125, (925) 424-3815, hsiung1@llnl.gov

Fenglei Huang, National Key Lab of Explosion, Beijing Institute of Tech., Beijing, 100081, PRC, 8610--68914518, huangfl@bit.edu.cn

Hongfa Huang, Washington State University, P.O. Box 642816, Pullman, WA, 99164, (509) 335-5345, (509) 335-6115, hongfah@wsu.edu

Christopher Hughes, AWE, Aldermaston, Reading, RG7 4PR, UK, +44 118982 5067, chris.hughes@awe.co.uk

Olga N Ignatova, RFNC-VNIIEF, 37 Mira Avenue, Sarov, MD, 607190, 7-831-304-2297, root@gdd.vniief.ru

Axinte Ionita, Los Alamos National Lab, MS B221, Los Alamos, NM, 87545, (505) 606-0689, ionita@lanl.gov

Kaushik A Iyer, US Army Research Lab, AMSRD-ARL-WM-TD, Aberdeen, MD, 21005, (410) 306-1014, (410) 306-0783, kiyer@arl.army.mil

Hugh R James, AWE, Aldermaston, Reading, RG7 4PR, UK, +44-118-982-5038, hugh.james@awe.co.uk

Eugenio Jaramillo, Los Alamos National Lab, MS B221, Los Alamos, NM, 87545, eugenio@lanl.gov

Hussam N Jarmakani, Univ of California - San Diego, 6310 Rancho Mission Rd., Apt. 253, San Diego, CA, 92108, (760) 803-8495, hjarmaka@ucsd.edu

Brian J Jensen, Los Alamos National Lab, MS P952, P.O. Box 1663, Los Alamos, NM, 87545, (505) 667-9886, bjjensen@lanl.gov

François-Xavier Jetté, Mech. Eng. Dept, MacDonald Engineering Bldg., Rm MD460, McGill University, 817 Sherbrooke street W, Montreal, QC, H3A 2K6, CANADA, (450) 679-4442, fjette@yahoo.com

Kyu Soo Jhung, ADD, Yusong P.O. Box 35-5, , Taejon, 305-600, SOUTH KOREA, (042) 821-2173, jks3621@add.re.kr

Tong Jiao, Brown University, Box D  Engineering Division, Brown University, Providence, RI, 02912, (401) 863-3034, Tong_Jiao@Brown.edu

Gordon R. Johnson, Network Computing Services, 1200 Washington Ave So, Minneapolis, MN, 55415, (612) 337-3553, gordon.johnson@netaspx.com

Andrew G Jones, AWE, Alsermaston, Reading, RG7 4PR, UK, +44 118982 7530, andrew.g.jones@awe.co.uk

Jennifer Lynn Jordan, Air Force Research Laboratory, 318 Ruckel Dr, Niceville, FL, 32578, (404) 234-1948, gt0595c@prism.gatech.edu

Vasant Shivram Joshi, Naval Surface Warfare Center, 6342 Grant Champman Dr, La Plata, MD, 20646, (301) 744-6769, (301) 744-4203, joshivs@ih.navy.mil

Richard J Jouet, Naval Surface Weapons Ctr, NSWC - Indian Head, 101 Strauss Ave, Indian Head, MD, 20640, (301) 744-4212, (301) 744-4445, richard.jouet@navy.mil

Antoine Juanicotena, CEA Bruyeres, BP 12, Bruyeres, F-91680, FRANCE, 33 1 69 26 70 97, antoine.juanicotena@cea.fr

Jinkyung Jung, ADD, Yuseong P.O.Box 35-5, Taejon, 305-600, SOUTH KOREA, (042) 821-2156, j.jung@add.re.kr

Kai Kadau, Los Alamos National Lab, T-14/Mail Stop G756, Los Alamos, NM, 87545, (505) 665-0354, (505) 665-4063, kkadau@lanl.gov

Daniel H Kalantar, Lawrence Livermore Nat'l Lab, L-463, PO Box 808, Livermore, CA, 94551, (925) 422-6147, (925) 422-8395, kalantar1@llnl.gov

Jave O Kane, Lawrence Livermore Nat'l Lab, L-473, Livermore, CA, 94550, (925) 424-5805, (925) 424-2463, jave@llnl.gov

Gennady Kanel, Russian Academy of Sciences, IBTAN, Izhorskaya 13/19, Moscow, RUSSIA, (505) 844-2798, (505) 284-2845, ljhumbl@sandia.gov

Randall J Kanzleiter, Los Alamos National Lab, 2390 Canyon Glen Rd, Los Alamos, NM, 87544-1783, (505) 665-7700, (518) 276-4832, kanzlr@lanl.gov

Daniela Kartoon, Ben Gurion Univ, Dept of Physics, PO Box 653, Beer Sheva, 84105, ISRAEL,

Ganjiro Kashine, Free Lance Reporter, c/o Kashine Hakkin(Do), 4-1 Minami Honmach, Yamato

Takado, Nara, 635-0086, JAPAN, (074) 522-5588, (074) 552-3881, sashine@sikasenbey.or.jp

Ann M Kaul, Los Alamos National Lab, PO Box 1663, MS-B259, Los Alamos, NM, 87545, (505) 665-0165, (505) 665-7725, akaul@lanl.gov

Nobuaki Kawai, Tokyo Inst of Tech, Material & Structures Lab, R3-10 4259 Nagatsuta, Yokohama, 226-8503, JAPAN, +81 45 924 5382, +81 45 924 5339, kawai@knlab.msl.titech.ac.jp

Gregory B. Kennedy, Harvard University, Dept Earth and Planetary Sci, 20 Oxford St., Cambridge, MA, 02138, (617) 496-6406, gkennedy@fas.harvard.edu

Jim Kennedy, HERE - LLC, 2710 Via Caballero del Sur, Santa Fe, NM, 87505, (505) 471-6217, jameskennedy@earthlink.net

Hak Jun Kim, T-3-10 Agency for Defense Dev., P.O. Box 35-5, Yuseong-Ku, Taejeon, 305-605, SOUTH KOREA, 82-42-821-4193, 82-42-821-2390, hjkim@add.re.kr

Vadim V. Kim, Russian Academy of Sciences, 21-1, Pervaya str., Chernogolovka, 142432, RUSSIA, +7(095)7857029, kim@ficp.ac.ru

Takahiro Kinoshita, Kumamoto University, 2-39-1 Kurokami, Kumamoto, 860-8555, JAPAN, +81 96 342 3295, tkino@kumamoto-u.ac.jp

Vladimir Y Klimenko, High Pressure Center, Mathematical Inst, Univ of St Andrews, North Haugh, St Andrews Fife, KY16 9SS, UK, (133) 446-3765, (133) 446-3748, klimenko@orc.ru

Marcus David Knudson, Sandia Natl Labs, MS 1181, 1515 Eubank Blvd SE, Albuquerque, NM, 87123, (505) 845-7796, (505) 845-7685, mdknuds@sandia.gov

Michel Koenig, Laboratoire LULI, Ecole Polytechnique, Palaiseau, 91128, FRANCE, +33 1 6933 4799, +33 1 6933 3009, michel.koenig@polytechnique.fr

Piotr M Kowalski, Los Alamos National Lab, 3794 Gold St #3, Los Alamos, NM, 87544, (505) 661-5905, kowalski@lanl.gov

Sergey S. Kraichikov, RFNC-VNIITF, 13 Vasilieva St., Snezhinsk, 456770, RUSSIA, 7-351-465-4730, kraichikov@vniitf.ru

Timothy Kreitinger, DTRA, 8720 Beverly Hills NE, Albuquerque, NM, 87122, (505) 846-8659, (505) 846-7543, timothy.kreitinger@abq.dtra.mil

Joel D. Kress, Los Alamos National Lab, MS B268, Los Alamos, NM, 87545, (505) 667-8906, (505) 665-3909, jdk@t12.lanl.gov

Alison Kubota, Lawrence Livermore Nat'l Lab, L-356, 7000 East Ave, Livermore, CA, 94551, (925) 424-6125, kubota1@llnl.gov

Shiro Kubota, AIST, Natl Inst of Adv Indus Sci Tec, 16-1, Onogawa, Tsukuba, Ibaraki, 305-8569, JAPAN, +81 29 861 8138, kubota.46@aist.go.jp

Maija Kukla, National Science Foundation, Division of Materials Research, Arlington VA 22230, (703) 292-4940, (703) 292-3095, mkukla@nsf.gov

Mukul Kumar, Lawrence Livermore Nat'l Lab, 7000 East Ave, Livermore CA 94550, (925) 422-0600, mukul@llnl.gov

Brandon M LaLone, Washington State University, 5665 NW Oak Creek Dr Apt 1, Corvallis, OR, 97330, (541) 758-4203, loneswinger@hotmail.com

Gerald R Laib, NSWC, Code 4440C, Bldg. 302, NSWC, Indian Head Division, 101 Strauss Ave, Indian Head, MD, 20640-1542, (301) 744-4358, gerald.laib@navy.mil

Brian D Lambourn, AWE plc, Aldermaston, Reading, Berkshire, RG7 4PR, UK, 44-118-9825379, 44-118-9824820, brian.lambourn@awe.co.uk

J Matthew Lane, Univ of Texas - Austin, Ctr for Nonlinear Dynamics, Austin, TX, 78712, (512) 471-5425, (512) 471-1558, mlane@physics.utexas.edu

Mary Y D Lanzerotti, ARDEC, PO Box 326, New Vernon, NJ, 07976, (973) 724-4625, (973) 724-4308, ylanzero@pica.army.mil

Jerry C Lasalvia, US Army Research Laboratory, AMSRD-ARL-WM-MD, Aberdeen, MD, 21005, (410) 306-0745, (410) 306-0806, jlasalvi@arl.army.mil

Rich R Laverty, Juniata Coll, 50 N. Front St. #302, Philadelphia, PA, 19106, (215) 260-9119, Richard.Laverty@usma.edu

R Jeffery Lawrence, Sandia Natl Labs, 1308 Kirby St NE, Albuquerque, NM, 87112, (505) 844-0127, rjlawre@sandia.gov

Margarita Lazarou, American Physical Society, Accounting, The American Physical Society, One Physics Ellipse, College Park, MD, 20740-3844, (301) 209-3203, (301) 209-0844, lazarou@aps.org

Alexander I Lebedev, RFNC-VNIEEF, 37 Mira Avenue, 607190, RUSSIA, 7-831-304-2914, root@gdd.vniief.ru

Larry Marion Lee, Ktech Corp, 1300 Eubank Blvd SE, Albuquerque, NM, 87123-3336, (505) 998-5848, ktech@ktech.com

Richard Lee, NSWC IHDiv, 3300 5th St South, Arlington, VA, 22204, (301) 744-2380, (301) 743-6399, leerj@ih.navy.mil

Bruno Legrand, CEA, BP 12, Bryeres Le Chatel, 91680, FRANCE, bruno.legrand@cea.fr

Philippe Legrand, CEA Bruyeres, BP 12 DCSA/SSA, Bruyeres La Chatel, F-91680, FRANCE, 1 49368968,

Lara Leininger, Univ of California, Davis, PO Box 808 L-183, Livermore, CA, 94551, (925) 422-6932, leininger3@llnl.gov

Raymond William Lemke, Sandia National Laboratories, 9687 Asbury Lane, NW, Albuquerque, NM, 87114, (505) 845-7423, rwlemke@sandia.gov

Larry D Libersky, Los Alamos National Lab, PO Box 1663, Los Alamos, NM, 87545, (505) 665-0101, libersky@lanl.gov

Benjamin T. Liu, Lawrence Livermore Nat'l Lab, 1609 Bonita Ave #3, Berkeley, CA, 94709, (925) 422-1559,

Carys E Lloyd, Univ of Cambridge, PCS Group, Cavendish Lab, Madingley Road, Cambridge, CB3 0HE, UK, (+44) 1223 337322, cel31@cam.ac.uk

Igor V Lomonosov, Russian Academy of Sci-Moscow, Inst Problems Chemical Phy., p2 Akad. Semenova-1, Moscow, 142432, RUSSIA,

Iyla Lomov, Lawrence Livermore Nat'l Lab, L-206, 7000 East Ave, Livermore, CA, 94550, (925) 423-7856, lomov1@llnl.gov

Gregory Todd Long, Sandia Nat'l Labs, MS 1454, PO Box 5800, Albuquerque, NM, 87185-1454, (505) 845-8425, (505) 844-5924, gtlong@sandia.gov

Karin L Louzada, Harvard University, 20 Oxford Street, Cambridge, MA, 02138, (617) 495-8986, louzada@fas.harvard.edu

Xia Lu, Georgia Tech, 270 Ferst Drive, Atlanta, GA, 30309, (404) 894-0334, (404) 894-2760, xialu_99@yahoo.com

Sergey N Lubyatinsky, RFNC-VNIITF, 13 vasiliev Street, Snezhinsk, 456770, RUSSIA, lob@gdd.ch70.chel.su

Shengnian Luo, Los Alamos National Lab, 252-21, Caltech, Pasadena, CA, 91125, (505) 664-0037,

Gabi Luttwak, Rafael, MOD Box 2250 24, Haifa, 31021, ISRAEL, (972) 479-2460, (972) 479-5289, gabilo@rafael.co.il

Xia Ma, Los Alamos National Lab, 2011 23rd St Apt. C, Los Alamos, NM, 87544-2326, (505) 667-8811, (505) 665-5926, xia@lanl.gov

Michael J Maclachlan, 8361 Jovin Circle, Springfield, VA, 22153-4014, (202) 231-4381, mmaclachlan@sprynet.com

Christian Mailhiot, Lawrence Livermore Nat'l Lab, L-055, PO Box 808, Livermore, CA, 94551, (925) 422-5873, (925) 422-9488, mailhiot1@llnl.gov

Joseph A Main, Natl Inst of Stand & Tech-NIST, , 100 Bureau Drive, Stop 8611, Gaithersburg, MD, 20899, (301) 975-5286, (301) 869-6275, joseph.main@nist.gov

Riad Manaa, Lawrence Livermore Nat'l Lab, L-282, 7000 E Ave, Livermore, CA, 94551, (925) 423-8668, (925) 422-3160, manaa1@llnl.gov

Joseph T Mang, Los Alamos Nat'l Lab, P.O. Box 1663, MS C920, Los Alamos, NM, 87545, (505) 665-6856, jtmang@lanl.gov

Joanna R Mann, QinetiQ, 20 Cremer Place, Faversham, Kent, ME13 7SG, UK, 07773 586699, jrmann@qinetiq.com

Mark Marr-Lyon, LANL, Dept of Phys, Washington State Univ, Pullman, WA, 99164-2814, (505) 664-0841, (505) 665-3359, marston@wsu.edu

Eric S Martin, Los Alamos National Lab, PO Box 1663, Los Alamos, NM, 87545, (505) 667-2181, (505) 667-6301, esmartin@lanl.gov

Morgana Martin, Georgia Tech, 1601 Noble Creek Dr NW, Atlanta, GA, 30327, (770) 634-8766, gtg477p@mail.gatech.edu

Eric Manuel Mas, Los Alamos National Lab, MS B221, Los Alamos, NM, 87545, (505) 665-5018, (505) 665-5757, mas@lanl.gov

Tsutomu Mashimo, Kumamoto University, 2-39-1 Kurokami, Kumamoto, 860-8555, JAPAN, +81 96 342 3295, +81 96 342 3293, mashimo@gpo.kumamoto-u.ac.jp

Thomas A Mason, Los Alamos National Lab, PO Box 1663, MS-G755, Los Alamos, NM, 87545, (505) 667-4896, (505) 667-8021, tmason@lanl.gov

Ann E Mattsson, Sandia Natl Labs, MS 1110, Albuquerque, NM, 87185-1110, (505) 844-9218, (505) 284-5451, aematts@sandia.gov

Thomas R Mattsson, Sandia National Laboratories, HEDP Theory/ICF Target Design, MS 1186, Albuquerque, NM, 87185-1186, (505) 844-9215, trmatts@sandia.gov

Stephane F Mazevet, Los Alamos National Lab, MS B283, Los Alamos, NM, 87545, (505) 667-0956, sdm@t4.lanl.gov

James W McCauley, US Army Research Lab, AMSRD-ARL-WM-M, Aberdeen, MD, 21005, (410) 306-0711, (410) 306-0640, mccauley@arl.army.mil

Sam A McDonald, Univ of Manchester, Materials Science Centre, Grosvenor Street, Manchester, M13 9H5, UK, +44 0 161 306 8959, +44 0 161 306 3586, sam.mcdonald@manchester.ac.uk

Shawn Mcgrane, Los Alamos National Lab, 162 Loma Del Escolar St, Los Alamos, NM, 87544-2525, (505) 665-6086, mcgrane@lanl.gov

James M McNaney, Lawrence Livermore Nat'l Lab, 4590 Phyllis Ct., Livermore, CA, 94550, (925) 423-9335, mcnaney1@llnl.gov

R Stewart McWilliams, Lawrence Livermore Nat'l Lab, 7000 East Avenue, L-286, Livermore, CA, 94550, (925) 422-6037, (925) 424-3383, mcwilliams16@llnl.gov

Shailesh Mehta, AWE, Aldermaston, Reading, RG7 4PR, UK, +44 118 9827813, s.mehta@awe.co.uk

Ralph Menikoff, Los Alamos National Lab, MS B214, PO Box 1663, Los Alamos, NM, 87545, (505) 667-7761, (505) 665-4055, rtm@lanl.gov

David D Meyerhofer, Univ of Rochester, Lab for Laser Energetics, 250 E River Road, Rochester, NY, 14623-1299, (585) 275-0255, (585) 275-5960, ddm@lle.rochester.edu

Marc A Meyers, Univ of California, San Diego, Dept of MAE, USCD - 93-0411, La Jolla, CA, 92053, (858) 534-4719, (858) 534-5698, mameyers@mae.ucsd.edu

Yann J.E. Meziere, Cranfield University, Defence Academy of the UK, Swindon, SN6 8LA, UK, +44 1793 784154, +44 1793 784195, y.j.e.meziere@cranfield.ac.uk

Maosheng Miao, Dept of Physics, Case Western Reserve University, Cleveland, OH, 44106-7074, (216) 368-4034, miaoms@po.cwru.edu

Joshua Edward Miller, Univ of Rochester, Lab for Laser Energetics, 250 East River Rd, Rochester, NY, 14623-0887, (585) 275-3421, (585) 275-5960, jmil@lle.rochester.edu

Jeremy Charles Millett, AWE, Aldermaston, Reading RG7 4PR, UK, Jeremy.millett@awe.co.uk

Alec Milne, FGE, 83 Market St, St Andrews, KY16 9NX, UK, 44334460800, 441334460813, alec@fges.demon.co.uk

Roger Wayne Minich, Lawrence Livermore Nat'l Lab, L-96, PO Box 808, Livermore, CA, 94551, (415) 422-2057, minich1@llnl.gov

Anuj Mishra, Univ of California, San Diego, 9500 Gilman Drive, La Jolla, CA, 92037, (858) 200-5373, amishra@ucsd.edu

Willis Mock Jr, Naval Surface Warfare Center, 215 N Randolph Rd, Fredericksburg, VA, 22405, (540) 653-8687, wmock@nswc.navy.mil

Victoria Mokhov, RFNC-VNIIEF, Mire 37, Sarov, RUSSIA, 8107 83130 34271, mokhov@soce.ru

Jean Francois Molinari, Johns Hopkins Univ, 3600 N Charles Stq, Baltimore, MD, 21218, (410) 576-2864, (410) 576-7256, molinari@jhv.edu

John David Molitoris, Lawrence Livermore Nat'l Lab, 7000 East Ave, PO Box 808 L-282, Livermore, CA, 94550-0808, (925) 423-3496, (925) 423-3281, molitoris1@llnl.gov

Jeremy E. Monat, Naval Surface Warfare Center, Bldg 600, 101 Strauss Ave, Indian Head, MD, 20640, (301) 744-4250, (301) 744-4445, monatje@ik.navy.mil

Dr Stephen T. Montgomery, Sandia Natl Labs, 3009 Vermont St NE, Albuquerque, NM, 87110, (505) 291-8529, stmontg@sandia.gov

David Steven Moore, Los Alamos National Lab, 18 Dulce Road, Santa Fe, NM, 87508, (505) 665-6089, (505) 667-0500, moored@lanl.gov

Dr Bill Moran, Lawrence Livermore Nat'l Lab, L-096, PO Box 808, Livermore, CA, 94551, (925) 422-7250, (925) 424-2723, moran1@llnl.gov

Akihisa Mori, Kumamoto Univ, Shockwave & Condensed Matter, Kumamoto Univ, Kurokami 2-39-1, Kumamoto, 860-8555, JAPAN, +81-96-342-3290 (106), 031d9010@gsst.stud.kumamoto-u.ac.jp

John A Moriarty, Lawrence Livermore Nat'l Lab, L-45, PO Box 808, Livermore, CA, 94551, (925) 422-9964, (925) 422-2851, moriarty2@llnl.gov

Richard C Mowrey, Naval Research Lab, 8309 Cooper St, Alexandria, VA, 22309, (202) 767-6346, mowrey@nrl.navy.mil

Roberta Nancy Mulford, Los Alamos National Lab, 1267-46th Street, Los Alamos, NM, 87544, (505) 667-7909, (505) 665-4459, mulford@lanl.gov

Oleg V Myasoedov, RFNC-VNIIEF, 37 Mira Avenue, Sarov, 607190, RUSSIA, 7-831-304-5009, root@gdd.vniief.ru

Hirotumi Nagao, Tokyo Inst of Tech, 4259 Nagatsuda, Yokohama, 226-8503, JAPAN, +81 95 924 5382, +81 45 924 5339, nago@knlab.msl.titech.ac.jp

Kunihito Nagayama, Kyushu University, Dept. of Aeronautics and Astronautics, Faculty of Engineering - Kyushu Univ., 744 Motooka, Nishiku, Fukuoka, 819-0395, JAPAN, 81-92-802-3014, 81-92-802-3017, nagayama@aero.kyushu-u.ac.jp

Oleg B Naimark, Inst. of Continuous Media, Mechanics of the Russian Academy, 1 Acad. Korolev str., Perm, RUSSIA, (505) 844-2798, ljhumbl@sandia.gov

Kazutaka G Nakamura, Tokyo Inst of Tech, Mater & Struct Lab, 4529 Nagatsuta, Yokohama, 226-8503, JAPAN, +81 45 924 5397, +81 45 924 5360, nakamura@msl.titedu.ac.jp

Vindhya Narayanan, Georgia Tech, 487 Centennial Olympic Park Dr, Atlanta, GA, 30313,

William J Nellis, Harvard University – Dept. of Physics, Cambridge, MA, 02138, (617) 495-9076, (617) 496-5144, nellis@physics.harvard.edu

Keith Adam Nelson, 6-237, MIT, 77 Massachusetts Avenue, Cambridge, MA, 02139, (617) 253-1423, (617) 253-7030, kanelson@mit.edu

Vitali F Nesterenko, Univ of California - San Diego, Dept of Mech & Aerospace Engr, 9500 Gilman Dr, La Jolla, CA, 92093-0411, (858) 822-0289, (858) 534-5698, vnesterenko@ucsd.edu

Peter Neuwald, Fraunhofer Institut EMI, Eckerstr. 4, Freiburg, 79104, GERMANY, +49 761 2714 324, neuwald@emi.fhg.de

Andrew Ng, Lawrence Livermore Nat'l Lab, L251, 7000 East Ave, Livermore, CA, 94550, (925) 423-4429, (925) 422-2253, ng16@llnl.gov

Duc Q Nguyen, Lockheed Martin, 1224 Notting Hill Dr., San Jose, CA, 95131, (408) 506-6311, dqnguyen@lmco.com

Jeffrey H Nguyen, Lawrence Livermore Nat'l Lab, MS L-041 Phys & Advanced Tech, PO Box 808, Livermore, CA, 94551, (925) 423-6838, (925) 423-2442, nguyen29@llnl.gov

Albert L Nichols, LLNL, 3213 Munras Pl, San Ramon, CA, 94583, (925) 423-6695, nichols5@llnl.gov

Malcolm F Nicol, Univ of Nevada, Las Vegas, High Pressure Sci & Engr Ctr, Box 454002, Las Vegas, NV, 89154-4002, (702) 895-1725, nicol@physics.unlv.edu

Dmitry N Nikolaev, Russian Academy of Sciences, Inst. of Problems of Chemical, Akad. Semenov str. 1, Chernogolovka, 142432, RUSSIA, +7 095 7857029, nik@ficp.ac.ru

Andrew W Obst, Los Alamos National Lab, MS H803, Los Alamos, NM, 87545, (505) 667-1330, (505) 665-4121, obst@lanl.gov

Kendal M Ogilvie, ITT Systems, MS 35, P O Box 39550, Colorado Springs, CO, 80949-9550, (719) 599-1955, (719) 599-1942, kendal.ogilvie@itt.com

Ivan Oleynik, Univ of South Florida, Dept of Phys, PHY 114, Tampa, FL, 33620, (813) 974-8186, oleynik@shell.cas.usf.edu

Daniel Anthony Orlikowski, Lawrence Livermore Nat'l Lab, L-45, 7000 East Ave, Livermore, CA, 94551, (925) 424-3197, (925) 422-6594, orlikowski1@llnl.gov

Dennis L Orphal, International Research Assoc., 4450 Black Ave Ste E, Pleasanton, CA, 94566, (925) 485-0130, (925) 485-0133, dorphal@aol.com

David M Oschwald, Los Alamos National Laboratory, 245 B Rosario Blvd, Santa Fe, NM, 87501, (505) 665-7420, oschwald@lanl.gov

Dr Henric Ostmark, Natl Defence Res Establish, Tumba, SE 14725, SWEDEN, (468) 706-3517, (468) 706-3521, henric@sto.foa.se

Bob Pahl, Sandia Natl Labs, MS1454, P.O. Box 5800, Albuquerque, NM, 87185, (505) 284-9147, rjpahl@sandia.gov

Elisavet Palamidi, School of Mech. Aero Eng, The University of Manchester, Sackville Street, Manchester, M60 1QD, UK, 44 161 306 2411, elisavet.palamidi@student.manchester.ac.uk

Vitaly Paris, Ben Gurion University, Ha-Nessim 3/6, Yavne, 81580, ISRAEL, +972 8647 7048, paris@bgu.ac.il

Gary R Parker, Los Alamos National Lab, PO Box 1663, Los Alamos, NM, 87545, (505) 665-7421, (505) 665-9809, gparker@lanl.gov

Yehuda Partom, RAFAEL, P.O.Box 2250, Haifa, 31021, ISRAEL, 97248792672, ypartom@rafael.co.il

Parimar J Patel, US Army Research Laboratory, AMSRD-ARL-WM-MD, APG, MD, 21005, (410) 306-0744, (410) 306-0806, ppatel@arl.army.mil

James E Patterson, Washington State Univ - Institute for Shock Physics, Washington State University, PO Box 642816, Pullman, WA, 99164-2816, (509) 335-7041, jepatterson@wsu.edu

Reed Patterson, Lawrence Livermore Natl Lab, M/S L-041, Box 808, Livermore, CA, 94551, (925) 422-7331, patterson31@llnl.gov

Denise K. Pauler, 118 East Spencer Street, Ithaca, NY, 14850, (607) 253-9528, denisepauler@hotmail.com

Jeremy R Payton, Los Alamos National Lab, PO Box 1663, Los Alamos, NM, 87545, (505) 665-9288, (505) 665-4121, payton@lanl.gov

Suhithi M Peiris, Naval Surface Warfare Center, Code 9210, Bldg 600, 101 Strauss Ave, Indian Head, MD, 20640, (301) 744-4252, (301) 744-4445, PeirisSM@ih.navy.mil

Robert A Pelak, Los Alamos National Lab, MS T086, Los Alamos, NM, 87545, (505) 665-1984, (505) 665-3359, pelak@lanl.gov

Lior Perelmutter, Soreq NRC, Yavne 81800, ISRAEL, 972 8 943 4419, 978 8 946 4227, lion@soneq.gov.il

Warren F Perger, Michigan Tech Univ, Dept of Elec Engr, 1400 Townsend Drive, Houghton, MI, 49931, (906) 487-2855, (906) 487-2949, wfp@mtu.edu

Oren E Petel, McGill University, 817 Sherbrooke W., Montreal, QC, H3A 2K6, CANADA, (514) 398-8118, oren.petel@mail.mcgill.ca

Paul D. Peterson, MS C920, Los Alamos Natl Lab, Los Alamos, NM, 87545, (505) 667-9622, pdp@lanl.gov

Philip A Pincosy, Lawrence Livermore Nat'l Lab, 7000 East Avenue, L-099, Livermore, CA, 94550, (925) 423-7118, pincosy@llnl.gov

Yuan Ping, M/S L-287, Lawrence Livermore Natl Lab, 7000 East Avenue, Livermore, CA, 94550, (925) 422-7052, yuanping@alumni.princeton.edu

Rob Piper, 8129 Cerromar Way, Gainesville, VA, 20155, (703) 754-8381, ROBPIP@msn.com

Shlomi Pistinner, Soreq NRC, Plasma Dept., Soreq NRC, Yavne 81800, ISRAEL,

Carlos Pizana, Univ of Texas - El Paso, 4330 Oxford, El Paso, TX, 79903, (915) 920-3443, cpizana@utep.edu

Igor Y Plaksin, LEDAP & ADAI Mech. Engineer, University of Coimbra, Pinhal De Marrocos Polo-2, Coimbra, P-3030, POR, +00351 91 981 5871, igor.plaksin@dem.uc.pt

Bradley J. Plohr, Los Alamos National Lab, MS B213, Los Alamos, NM, 87544, (505) 665-2558, (505) 665-3003, plohr@lanl.gov

Jee Yeon N. Plohr, Los Alamos National Lab, MS B221, Los Alamos Natl Lab, Los Alamos, NM, 87545, (505) 667-9102, (505) 665-8329, jplohr@lanl.gov

Stephen M Pollaine, Lawrence Livermore Nat'l Lab, LLNL, PO Box 808, Livermore, CA, 94550, (925) 422-5950, (925) 423-9969, pollaine@llnl.gov

Helen Jane Prentice, Cavendish Laboratory, Phys Dept, Magdalene Coll, Cambridge, CB2 0AG, UK, 44 1223 337322, hp222@cam.ac.uk

William Graham Proud, Univ of Cambridge, Cavendish Lab, PCS Group, Madingley Rd, Cambridge, CB3 0HE, UK, +44 1223 337205, +44 1223 350266, wgp1000@phy.cam.ac.uk

Jason Quenneville, Los Alamos National Lab, MS P365, Los Alamos, NM, 87545, (505) 667-8760, jasonq@lanl.gov

Raul A Radovitzky, MIT, rm 33-316, 77 Mass Ave, Cambridge, MA, 02139, (617) 252-1536, rapa@mit.edu

Philip John Rae, LANL, Cavendish Labortory, Madingley Road, Cambridge, CB3 0HE, UK, 44-1223-337464, 44-1223-350266, prae@lanl.gov

Victor A Raevskiy, RFNC-VNIIEF, 37 Mira Avenue, Sarov, 607190, RUSSIA, 7-831-304-0607, root@gdd.vniief.ru

Martin N Raftenberg, US Army Research Lab, AMSRD-ARL-WM-TD, Aberdeen, MD, 21005, (410) 306-0949, (410) 306-0783, mnr@arl.army.mil

Ramesh Raghupathy, Johns Hopkins University, 3500 Beech Ave., Apt -B, Baltimore, MD, 21211, (410) 516-4398, rmsh@jhu.edu

Arunacralam Rajendran, US Army Res Office, PO Box 12211, Research Triangle Pa, NC, 27709-2211,

Kaliat T Ramesh, Johns Hopkins University, 122 Latrobe Hall, 3400 N. Charles St., Baltimore, MD, 21218, (410) 516-7735, (410) 516-7254, ramesh@jhu.edu

Edward J Rapacki, US Army Research Laboratory, 4600 Deer Creek Loop, Aberdeen Proving Ground, MD, 21005, (410) 306-0801, (410) 306-0783, rapacki@arl.army.mil

Ramon Jose Ravelo, Univ of Texas – El Paso, Dept of Phys, Univ cf Texas, El Paso, TX, 79968-0515, (915) 747-5620, (915) 747-5447, ravelo@psci.utep.edu

Sergey V Razorenov, Inst. of Problems of Chemical, Russian Academy of Sciences, Moscow, 142432, RUSSIA

Ronald Redmer, University of Rostock, Universitatsplatz 3, Rostock, D-18051, GERMANY, +49 381 4986910, ronald.redmer@uni-rostock.de

Evan Reed, Lawrence Livermore Nat'l Lab, MS L-268, 7000 East Ave., Livermore, CA, 94550, (925) 424-4080, reed23@llnl.gov

Bill Reinhart, Sandia Natl Labs, 11300 Oakland Ave. NE, Albuquerque, NM, 87122, (505) 450-1375, wdreinh@sandia.gov

Bruce Allen Remington, Lawrence Livermore Nat'l Lab, L-021, PO Box 808, Livermore, CA, 94551, (925) 423-2712, (925) 423-8945, remington2@llnl.gov

Jose B. Ribeiro, University of Coimbra, Dep. de Engenharia Mecânica, Pinhal de Marrocos - Polo II, Coimbra, 3030-201, POR, ++351239790700, jose.baranda@dem.uc.pt

Clinton T Richmond, Naval Surface Warfare Ctr., Naval Surface Warfare Ctr., 101 Strauss Avenue, Indian Head, MD, 20640, (301) 744-2377, (301) 744-4451, richmondct@ih.navy.mil

Paulo A Rigg, Los Alamos National Laboratory, MS P952, PO Box 1663, Los Alamos, NM, 87545, (505) 665-5934, prigg@lanl.gov

Maria Rightley, Los Alamos National Laboratory, 2168 48th St, Los Alamos, NM, 87545, (505) 665-6560, (505) 667-4420, mright@lanl.gov

Paul M Rightley, Los Alamos National Laboratory, 2168 48th St, Los Alamos, NM, 87544, (505) 667-0460, (505) 665-3359, rightley@mailaps.org

Avraham Rikanati, Ben Gurion Univ, POB # 116, Aloni Abba, 36005, ISRAEL, 972-547-202148, rkavi@bgumail.bgu.ac.il

Robert Ripley, Martec LTD, 1888 Brunswick St., Suite 400, Halifax, NS, B3J3J8, CANADA, 1-902-425-5101, rcripley@martec.com

Christopher M Robinson, AWE PLC, Aldermaston, Reading, RG7 4PR, UK, 0118 9827813, christopher.robinson@awe.co.uk

Banton Rohan, Army Research Laboratory, AMSRD-ARL-WM-TB, Aberdeen, MD, 21005-5069, (410) 278-6042, ltran@arl.army.mil

Jerry J Romero, Los Alamos National Lab, PO Box 1663, Los Alamos, NM, 87545, (505) 667-1177, (505) 667-0440, jerjrom@lanl.gov

Seth Root, Washington State Univ, 1630 NE Valley Rd Apt F101, Pullman, WA, 99163, (509) 335-5345, (509) 335-6115, sroot@wsu.edu

Dr Zvi Rosenberg, Rafael, PO Box 2250, Haifa, 31021, ISRAEL, 972 48795289, zvir@rafael.gov.il

Andrew Ruggiero, Univ of Cassino, DiMSAT - University of Cassino, Via G Di Biasio, 43, Cassino, I-03043, Italy, +39 0776 299 4335, +39 0776 299 3390, a.ruggiero@unicas.it

Christian Ruiz, Los Alamos National Lab, 11706 Lanier Creek Dr., Jacksonville, FL, 32258,

Todd G Rumbaugh, DRS Data & Imaging Systems, 802 Seabright Ave, Santa Cruz, CA, 95062, (831) 423-3463, rumbaugh@drs-dis.com

Thomas P. Russell, Naval Surface Weapons Ctr, Naval Rsch & Tech Dept, NAVSEA, Code 90, Bldg 600, Indian Head, MD, 20640, (301) 744-4270, (301) 744-4445, thomas.p.russell@navy.mil

Darren Salisbury, AWE, Aldermaston, Reading, Berkshire, RG74PR, UK, 44-(0)1189826999, darren.salisbury@awe.co.uk

Terry R Salyer, Los Alamos National Lab, PO Box 1663, Los Alamos, NM 87545, (505) 667-0658, (505) 667-6372, lgh@lanl.gov

Dmitry V Samsonov, Max-Plank Inst for Extrat Phys, Giessenbachstrasse, Postfach 1603, Garching, D-85740, GERMANY, +49 89 3299 1144, +49 89 3299 3569, dima@mpe.mpg.de

Robert Sander, Los Alamos National Lab, MS J567, Los Alamos, NM, 87545, (505) 667-3001, (505) 667-0440, bsander@lanl.gov

Harold W Sandusky, NAVSEA, 101 Strauss Avenue, Indian Head, MD, 20640, (301) 744-2378, sanduskyhw@ih.navy.mil

Tomokazu Sano, Osaka University, 2-1 Yamada-Oka, Suita, Osaka, 565-0871, JAPAN, +81 6 6879-7538, sano@mapse.eng.osaka-u.ac.jp

Nir Sapir, NRCN, PO Box 9001, Beer Seva, 84190, ISRAEL

Andrey Savinykh, Russian Academy of Sciences, Inst Problem Chemical Physics, Chernogolovka, 142432, RUSSIA, +70957857029, +70957857029, savas@ficp.ac.ru

Dr Michael J Scheidler, Army Research Lab, AMSRL-WM-TD, Aberdeen P G, MD, 21005, (410) 306-0794, (410) 306-0783, mjs@arl.mil

Robert G Schmitt, Sandia National Labs, P. O. Box 5800, MS-0836, Albuquerque, NM, 87185, (505) 845-7218, (505) 844-8251, rgschmi@sandia.gov

Eberhard Schneider, Fraunhofer Institut-EMI, Fraunhofer Institut-EMI, Eckerstr. 4, Freiburg, 79104, GERMANY, +49 761 2714 326, schneider@emi.fhg.de

Adam J Schwartz, Lawrence Livermore Nat'l Lab, L-041, 7000 East Ave, Livermore, CA, 94550, (925) 423-3454, (925) 423-2451, schwartz@llnl.gov

Cynthia Louise Schwartz, Los Alamos National Lab, 2108 33rd St, Los Alamos, NM, 87544, (505) 661-8379, (505) 665-7920, cschwartz@lanl.gov

Suz Scott, Lawrence Livermore Nat'l Lab, Box 808, Livermore, CA, 94550, (925) 423-8380, scott15@llnl.gov

Lynn Seaman, SRI International, 333 Ravenswood Ave, Menlo Park, CA, 94025, (650) 859-3587, (650) 859-2260, lynn.seaman@SRI.COM

Achim Seifter, Los Alamos National Lab, PO Box 1663, Los Alamos, NM, 87545,

Toshimori Sekine, Advanced Materials Laboratoy, Natl Inst for Material Science, 1-1 Namiki, Tsukuba, 305-0044, JAPAN, +81 29 860 4408, +81 298 51 2768, sekine.toshimori@nims.go.jp

Robert Earle Setchell, Sandia Natl Labs, PO Box 697, Cedar Crest, NM, 87008, (505) 281-5600, (505) 844-4045, resetch@sandia.gov

Thomas Dan Sewell, Los Alamos National Lab, MS B-214, Los Alamos, NM, 87545, (505) 667-8205, (505) 667-1483, sewell@lanl.gov

Milton S Shaw, Los Alamos National Lab, MS B214, PO Box 1663, Los Alamos, NM, 87545, (505) 667-5903, (505) 665-4055, mss@lanl.gov

Dr Stephen A Sheffield, Los Alamos National Lab, P952, PO Box 1663, Los Alamos, NM, 87545, (505) 665-0350, (505) 667-6372, ssheffield@lanl.gov

Keisuke Shigemori, Osaka Univ, Inst of Laser Engineering, 2-6 Yamada-Oka, Suita, 563, JAPAN, +81 6 6879 8776, +81 6 6877 4799, shige@ile.osaka-u.ac.jp

Sergey I Shkuratov, Loki Incorporated, 611 N. Vinton Ave., Lubbock, TX, 79416, (806) 747-7640, shkuratov@lokiconsult.com

Clive Richard Siviour, University of Cambridge, Cavendish Laboratory, Cambridge, CB3 0HE, UK, +44 1223 337322, +44 1223 350266, crs27@cam.ac.uk

Laura Beth Smilowitz, Los Alamos National Lab, MS J585, 533 Rover Blvd, Los Alamos, NM, 87544-3540, (505) 667-5207, (505) 665-4817, smilo@lanl.gov

Raymond F Smith, Lawrence Livermore Nat'l Lab, 7000 East Avenue, L-286, Livermore, CA, 94550, (925) 423-5895, (925) 424-3383, smith248@llnl.gov

Steven F. Son, Los Alamos National Lab, MS C920, LANL, Los Alamos, NM, 87545, (505) 665-0380, son@lanl.gov

Susan S Sorber, AWE Plc, Aldermaston, H12.1, Reading, Berkshire, RG7 4PR, UK, +44 (0)118 982 5596, susan.sorber@awe.co.uk

Gerald D Stevens, Bechtel Nevada, 5520 Ekwill St, Ste B, Santa Barbara, CA, 93111, (805) 681-2219, (805) 681-2280, stevengd@nv.doe.gov

D. Scott Stewart, Univ of Illinois - Urbana, MC 244 268 MEB, Dept Mech & Indust Engr, UIUC, 1206 W Green St, Urbana, IL, 61801, (217) 333-7947, (217) 244-6534, dss@uiuc.edu

Sarah T Stewart Mukhopadhyay, Harvard University, Earth & Planetary Sci, 20 Oxford Street, Cambridge, MA, 02138, (617) 496-6462, (617) 496-7411, sstewart@eps.harvard.edu

Leonard I Stiel, Polytechnic University, 46 Bond Ave., Malverne, NY, 11565, (718) 266-3628, lstiel@photon.poly.edu

James Stolken, LLNL, 10004 Del Almendra Dr, Oakdale, CA, 95361, (925) 423-2234, stolken1@llnl.gov

Chad A Stoltz, Naval Surface Warfare Center, 12574 Council Oak Drive, Waldorf, MD, 20601, (301) 744-1152, chad.stoltz@navy.mil

Elmar Strassburger, Am Klingelberg 1, Fraunhofer EMI, Efringen-Kirchen, 79588, GERMANY, +49 7626 915735, +49 7626 915727, strassburger@emi.fraunhofer.de

Alejandro H Strachan, Los Alamos National Lab, MS G756, Los Alamos, NM, 87544, (505) 664-0365, (505) 665-2113, strachan@lanl.gov

Frederick H Streitz, Lawrence Livermore Nat'l Lab, L-045, PO Box 808, Livermore, CA, 94551-0808, (925) 423-3236, (925) 422-2851, streitz1@llnl.gov

Dan Su, Univ of Delaware - Rm 126, 130 Academy St, Newark, DE, 19716, (302) 738-4090, dansu3000@yahoo.com

Gerrit T Sutherland, Naval Surf Weapons Ctr, Code 920A Indian Head Div, 101 Strauss Ave Bldg 600, Indian Head, MD, 20640-5035, (301) 744-2382, (301) 744-6399, sutherlandgt@ih.navy.mil

Pazhayannur K Swaminathan, Johns Hopkins Univ, 11100 Johns Hopkins Road, Laurel, MD, 20723, (240) 228-3950, (240) 228-1049, swamipk1@jhuapl.edu

Damian Charles Swift, Los Alamos National Lab, MS E526, Los Alamos, NM, (505) 667-1279, damian.swift@physics.org

Katsumi Tanaka, AIST, Nat'l Inst. of Advanced Sci. &., Technology, 1-1-1 Umezono Central 2, Tsukuba, 305-8568, JAPAN, +81 29 861 9274, +81 29 851 5426, tanaka-katsumi@aist.go.jp

Bryce C Tappan, Los Alamos National Lab, MS C920, Los Alamos, NM, 87545, (505) 667-0533, (505) 667-0500, btappan@lanl.gov

Craig M Tarver, Lawrence Livermore Nat'l Lab, 4186 Hazelhurst Ct, Pleasanton, CA, 94566, (925) 423-3259, (925) 424-3281, tarver1@llnl.gov

Douglas G Tasker, Los Alamos National Lab, MS J566, PO Box 1663, Los Alamos, NM, 87545, (505) 665-2859, (505) 665-3050, tasker@lanl.gov

Emma Taylor, Open University, Tene House, The Tene, Baldock, SG7 6DG, UK, 44 1438773914, 44 1438778913, e.a.taylor@open.ac.uk

Kjell A Tengesdal, Lawrence Livermore Nat'l Lab, 901 W Dickson St, Apt E-1, Fayetteville, AR, 72701, (925) 423-6474, kjellt@uafphl.uark.edu

Vladimir Y Ternovoi, Inst of Phys & Chem Res, Inst of Prob Chem Phys RAS, Chernogolovka, 142432, RUSSIA, (095) 785-7029, (095) 785-7029, ternovoi@ficp.ac.ru

Naresh N Thadhani, School of Materials Sci & Eng, Georgia Inst of Tech, 771 Ferst Drive, Atlanta, GA, 30332-0245, (404) 894-2651, (404) 894-9140, naresh.thadhani@mse.gatech.edu

Keith A Thomas, Los Alamos National Lab, 227 Rover Blvd, Los Alamos, NM, 87544, (505) 665-5248, (505) 667-6301, thomask@lanl.gov

Tom F Thornhill, Ktech Corp / Sandia Natl Labs, 826 Shirley St NE, Albuquerque, NM, 87123, (505) 845-3354, tfthorn@sandia.gov

Vikas Tomar, Georgia Tech, 1036 A Curran St NW, Atlanta, GA, 30318, (404) 578-7575, gte756y@mail.gatech.edu

Davis Loel Tonks, Los Alamos National Lab, MS F699, PO Box 1663, Los Alamos, NM, 87545, (505) 665-8481, tonks@lanl.gov

Linhbao Tran, Army Research Laboratory, AMSRD-ARL-WM-TB, Aberdeen, MD, 21005-5069, (410) 278-9757, ltran@arl.army.mil

Carl P Trujillo, Los Alamos National Lab, PO Box 1663, Los Alamos, NM, 87545, (505) 665-0375, (505) 665-9427, cptrujillo@lanl.gov

John Tse, Univ of Saskatchewan, Saskatoon, S7N 5E2, CANADA, John.tse@usask.ca

Konstantinos Tsembelis, Univ of Cambridge, Cavendish Lab, Mandingley Rd, Cambridge, CB3 OHE, UK, +44 1223 337205, +44 1223 350266, kt226@phy.cam.ac.uk

Mr. Stefan J Turneaure, Washington State Univ, PO Box 642816, Pullman, WA, 99164-2816, stefanturn@yahoo.com

Jacqy Turner, Lawrence Livermore Nat'l Lab, 7000 East Avenue, Livermore, CA, 94550, (925) 423-2655, (925) 423-2260,

Paul Andrew Urtiew, Lawrence Livermore Nat'l Lab, L-282, UCL, PO Box 808, Livermore, CA, 94551, (925) 423-0333, (925) 424-3281, urtiew1@llnl.gov

David Valentine, SKG Research Inc, 4210 Beck Ave., Studio City, CA, 90604,

Steven Michael Valone, Los Alamos National Lab, MST-8, MS G755, PO Box 1663, Los Alamos, NM 87545, (505) 667-2067, (505) 667-8021, smv@lanl.gov

Antoine Van der Heijden, TNO Defense, Security & Safety, P.O. Box 45, Rijswijk, 2280 AA, NETHERLANDS, +31 15 284 3774, +31 15 284 3974, antoine.vanderheijden@tno.nl

Kevin S Vandersall, LLNL, PO Box 854, Livermore, CA, 94550, (925) 422-3337, vandersall1@llnl.gov

Kalyan K Vedantam, Univ of Dayton, 345 Firwood Dr Apt 2D, Dayton, OH, 45419, (937) 643-1078, kvedantam@gmail.com

Peter Alfonso J Vitello, Lawrence Livermore Nat'l Lab, L-282, 7000 East Ave, Livermore, CA, 94550, (925) 422-0079, (925) 424-2709, vitello@llnl.gov

Wendy S Vogan, Los Alamos National Lab, MS H803, Los Alamos, NM, 87545, (505) 665-1711, (505) 665-4121, vogan@lanl.gov

Tracy Vogler, Sandia Natl Labs, PO Box 5800, MS 1181, Albuquerque, NM, 87185, (505) 845-0742, (505) 845-7685, tjvogle@sandia.gov

Christophe Voltz, CEA, Centre de Valduc, Is Sur Tille, 21120, FRANCE, +33 3 8023 4340, +33 3 8023 5278, cghistophe.voltz@cea.fr

William Von Holle, Defense Nucl Fac Safety Board, 625 Indiana Ave NW, Washington, DC, 20004, (202) 208-6588, (202) 208-6518, williamv@dnfsb.gov

Sam S Waggener, Naval Surface Warfare Center, 17320 Dahlgren Rd, Dahlgren, VA, 22485, (540) 653-4263, sam.waggener@navy.mil

Mark B Walpole, Washington State Univ, 1505 NE Valley Rd Apt 8, Pullman, WA, 99163, (509) 332-0682, (509) 335-6115, walmark@mail.wsu.edu

Yinmin Wang, Lawrence Livermore Nat'l Lab, LLNL, Box 808, Livermore, CA, 94550, (925) 422-6083, ymwang@llnl.gov

Justin Stephen Wark, Oxford Univ, 54 Stokes Croft, Haddenham, HP17 8DZ, UK, +44 1865 272251, +44 1865 282296, justin.wark@physics.ox.ac.uk

Peter M Weber, Brown Univ, Dept of Chem, Box H, 324 Brook St, Providence, RI, 02912, (401) 863-3767, (401) 863-2594, Peter_Weber@Brown.edu

Meir Werdiger, Soreq NRC, Plasma Group, Soreq Nuclear Res Ctr, Yavne, 81800, ISRAEL, 972 8 9434611, 972 8 9434227, meir@soreq.gov.il

Carter T. White, Naval Research Laboratory, 4555 Overlook Ave., S.W., Washington, DC 20375,

(202) 767-3270, (202) 767-3321, carter.white@nrl.navy.mil

Von H Whitley, NSWC, 101 Strauss Ave Bldg 600, Indian Head, MD, 20640, (301) 744-4167, (301) 744-4445, whitleyvh@ih.navy.mil

Donald A Wiegand, ARDEC, Energetic Mater Div, ARDEC, Bldg 3022, Picatinny Arsenal, NJ, 07806-5000, (973) 724-3336, (973) 724-5869, dwiegand@pica.army.mil

David Martin Williamson, Univ of Cambridge, PCS Group, Cavendish Lab, Madingley Rd, Cambridge, CB3 0HE, UK, +44 1223 337 209, +44 1223 350 266, dmw28@phy.cam.ac.uk

Michael D Willis, KTech Corporation, 1300 Eubank Blvd. SE, Albuquerque, NM, 87123, (505) 998-6061, (505) 998-5836, angela@ktech.com

William H Wilson, DTRA, 5761 Linden Farm Pl, La Plata, MD, 20646, (703) 325-7801, (703) 325-0043, william.wilson@dtra.mil

Michael Winey, Washington State Univ, PO Box 2537, Pullman, WA, 99165-2537, (509) 332-8012, (509) 335-6115, mwiney@mail.wsu.edu

Ronald E. Winter, AWE, H12, Aldermaston, Reading, Berks, RG7 4PR, UK, 441189825493, 441189824836, ron.winter@awe.co.uk

Don Wise, American Physical Society, One Physics Ellipse, College Park, MD, 20740, (301) 209-3289, (301) 209-3652, wise@aps.org

Jack Leroy Wise, Sandia National Laboratories, MS 1181, P.O. Box 5800, Albuquerque, NM, 87185-1181, (505) 844-6359, (505) 844-7685, jlwise@sandia.gov

Wilhelm G Wolfer, Lawrence Livermore Nat'l Lab, 657 Windmill Lane, Pleasanton, CA, 94566, (925) 423-1501, wolfer1@llnl.gov

Chak-Pan Wong, Naval Surf Weapons Ctr, 14713 Locustwood Ln, Silver Spring, MD, 20905, (301) 744-6775, (301) 744-4717, chak.wong@navy.mil

Andrew D Wood, QinetiQ, Building Q13, Room 2, QinetiQ Fort Halstead, Sevenoaks, TN14 7BP, UK, +441959515195, adwood@qinetiq.com

Andrew D Workman, AWE, Aldermaston, RG7 4PR, UK, +44 0118 982 6727, andrew.workman@awe.co.uk

Steve Wortley, AWE, Aldermaston, RG7 4PR, UK, +44 1189824287, steve.wortley@awe.co.uk

Mark W Wright, Bldg B8C1, AWE PLC, Aldermaston, Reading, Berkshire, RG7 4PR, UK, 44 (0) 118 982 7531, mark.w.wright@awe.co.uk

Thomas W Wright, US Army Research Lab, 4600 Deer Creek Loop, Aberdeen, MD, 21005, (410) 306-1943, (410) 306-0666, tww@arl.army.mil

Qing Xue, Los Alamos National Lab, PO Box 1663, Los Alamos, NM, 87545, (505) 665-2479, (505) 667-8021, qxue@lanl.gov

Clarissa A Yablinsky, Materials Science & Eng. Dept, 477 Watts Hall, Ohio State Univ, 2041 College Road, Columbus, OH, 43210, rizz@lanl.gov

Wenbo Yang, Schlumberger, 14910 Airline Rd, Rosharon, TX, 77583, (281) 285-5253, (281) 285-5453, yang@rosharon.wireline.slb.com

Arnon Yosef-Hai, Ben Gurion Univ, Mech. Engr Dept, PO Box 653, Beer Sheva, 84105, ISRAEL,

Akio Yoshinaka, DRDC, 444 - 11th Street S.E., Medicine Hat, AB, T1A 1T1, CANADA, (403) 544-4387, akio.yoshinaka@drdc-rddc.gc.ca

Vincent Yuan, Los Alamos National Lab, 258 A County Rd 84, Santa Fe, NM, 87506, (505) 667-3939, (505) 665-4121, vyuan@lanl.gov

Aleksandr Yunoshev, Inst of Hydrodynamics, Lavrentiev Institute of, Hydrodynamics, Lavrentiev prospect 15, Novosibirsk, 630090, RUSSIA, (383) 333-2070, asyn@ngs.ru

Michael Zachariah, University of Maryland, College Park, MD, 20742-5031, (301) 405-4311, mrz@umd.edu

Federico Zahariev, Univ of Western Ontario, 815-724 Fanshawe Park Rd East, London, ON, N5X 2L8, CANADA, (519) 858-9484, zfederic@uwo.ca

Eugene Zaretsky, Ben Gurion Univ, Dept of Mech Eng, Ben-Gurion Univ of Negev, PO Box 653, Beer Sheva, 84105, ISRAEL, 972 7 6477102, 972 7 6472813, zheka@menix.bgu.ac.il

Sergey A Zelepugin, Tomsk Science Center, 10/3 Akademicheskii Prospekt, Tomsk, Tomsk Region, 634021, RUSSIA, 7-382-249-2497, 7-382-249-2838, szel@tbism.tomsk.ru

Frank J Zerilli, Naval Surface Weapons Ctr, Code 920F Bldg 600, Indian Head, MD, 20640-5035, (301) 744-6762, (301) 744-4451, zerillifj@ih.navy.mil

Fan Zhang, DRDC Suffield, PO Box 4000 Station Main, Medicine Hat, AB, T1A 8K6, CANADA, (403) 544-4887, Fan.Zhang@drdc-rddc.gc.ca

Jijun Zhao, Washington State Univ, Inst for Shock Phys, PO Box 642816, Pullman, WA, 99164, (509) 335-7055, (509) 335-6115, jzhao@wsu.edu

Shijin Zhao, Los Alamos National Lab, MS G756, Los Alamos, NM, 87545, (505) 665-0405, shijin@lanl.gov

Mikhail V Zhernokletov, RFNC-VNIITF, 37 Mira Avenue, Sarov, 607190, RUSSIA, 7-831-304-5350, root@gdd.vniief.ru

Fenghua Zhou, Johns Hopkins Univ, 3400 N Charles St, Baltimore, MD, 21218, (410) 516-4398, fzhou@jhu.edu

Kurt Zimmerman, Washington State Univ, Institute for Shock Physics, Pullman, WA, 99164-2816, (509) 335-4673, (509) 335-6115, kurtz@mail.wsu.edu

Zeev Zinamon, The Weizmann Inst of Sci, Dept of Nucl Phys, Rehovot, 76100, ISRAEL, 972 8 9342083, 972 8 9344182, fnzim@weizmann.weizmann.ac.il

Marvin A. Zocher, LANL, 450 Navajo, Los Alamos, NM, 87544, (505) 665-3472, zocher@lanl.gov

Jonathan M Zucker, Los Alamos National Lab, 3200 Canyon Rd, Apt 4203, Los Alamos, NM, 87544, (505) 667-7826, jzucker@lanl.gov

Sergey Zybin, Caltech, Beckman Institute, M/C 139-74, California Inst of Tech, 1201 E California Blvd., Pasadena, CA, 91125, (626) 395-8134, (626) 585-0918, zybin@wag.caltech.edu

Gray III, G. T., 69, 196, 232, 599, 725, 729, 741, 753, 771, 783, 1149, 1511
Grebenkin, K. F., 982
Greeff, C. W., 65, 89
Greenaway, M. W., 543, 1053, 1195, 1279
Greenwood, D., 1061
Gregori, F., 757
Grigsby, W., 1337
Grinfeld, M. A., 858
Grise, W. R., 1033
Grover, M., 1223
Grunschel, S. E., 809
Guirguis, R. H., 519
Gump, J. C., 948, 1069, 1131
Gupta, Y. M., 367, 551, 555
Gustavsen, R. L., 69, 149, 921, 1307, 1311

## H

Hall, C. A., 686, 729, 1307, 1311
Hall, C. E., 149
Hallouin, M., 611
Hamate, Y., 523
Hammerberg, J. E., 391
Hammond, R. I., 1261
Han, S.-P., 581
Hanagud, S., 107, 153, 491, 817
Handley, C. A., 1073
Hare, D. E., 1311
Harrigan, J. J., 1237, 1519
Harris, E. J., 1187, 1215, 1261
Harstad, E. N., 216
Haskins, P. J., 527, 952
Hastings, J. B., 1287
Hawreliak, J., 220, 240, 286, 765
Hayes, D. B., 69, 729
He, L., 1458
Hebert, D., 95
Henrie, B. L., 627, 638, 670, 725, 771
Henson, B. F., 1077, 1101, 1211
Herbold, E. B., 1507, 1523
Hereil, P. L., 303
Herrera, D., 1323

Herrmann, B., 292, 733
Heuzé, O., 161, 212, 262
Hickman, R. J., 686
Higginbotham, A., 286
Higgins, A. J., 994, 998, 1139
Hill, L. G., 531, 1219
Hill, S., 775
Hixson, R. S., 232, 599, 729, 1149, 1283, 1353
Hjelm, R. P., 1271
Hohler, V., 1391
Hokamoto, K., 1543
Holian, B. L., 220, 236, 270, 286
Hollands, R., 944
Holst, B., 127
Holt, W. H., 169, 631, 1097
Holtkamp, D., 670
Hongo, T., 224
Honnell, K. G., 33
Hood, R. Q., 403
Hooks, D. E., 149, 1307, 1311
Hooper, J., 373
Hooton, I., 1379
Horie, Y., 307, 523
House, J. W., 701, 721, 805
Howard, W. M., 319
Hsiung, L. L., 228
Hu, A., 373
Hu, H., 662
Huang, F., 1383
Huang, H., 737
Huang, W., 1265
Hung, Y.-C., 1519

## I

Ichinose, D., 1480
Ignatova, O. N., 745, 761
Igonin, V. V., 745, 761
Ioniţa, A., 204, 323, 487
Irifune, T., 1480
Ito, S., 248
Ivanova, O. V., 1177
Iyer, K., 866

## J

James, H. R., 1081
Jarmakani, H., 1319

Jearanaisilawong, P., 797
Jensen, B. J., 69, 232
Jhung, K. S., 99
Jiao, T., 797, 809
Jin, S., 1507
Jin, Z. Q., 1157
Jing, F., 1449, 1458
Joannopoulos, J. D., 1345
Johnson, J. D., 65, 89
Jones, A., 1135
Jones, S. C., 686
Jordan, J. L., 157
Joshi, V. S., 519, 1045, 1387
Jouet, R. J., 1527
Juanicotena, A., 1531
Jung, J., 99
Juranek, H., 127

## K

Kad, B., 1319
Kadau, K., 220, 236, 286, 383
Kagan, K. L., 189
Kakshina, E. V., 77
Kalantar, D. H., 220, 240, 286, 1145, 1413
Kalinin, A. A., 1173
Kane, J. O., 244
Kanel, G. I., 192, 650, 870, 876, 888
Kaneshige, M. J., 559
Kapila, V., 425
Katoh, K., 1085
Kawai, N., 224, 248
Kawamura, K., 395
Kennan, Z., 709
Kennedy, G. B., 139, 1484
Kennedy, J., 1002
Kennedy, J. E., 1093
Khalsa, N. S., 1219
Khasainov, B., 161
Khasainov, B. A., 449
Khokhlov, A. A., 713
Kikuchi, M., 224, 248
Kim, I., 99
Kim, V. V., 103, 327
King, J., 1323
King, J. C., 927
King, N. S. P., 1223
Kinoshita, T., 395
Kipp, M. E., 1229

A3

# SUBJECT INDEX

**V**

variability : tantalum, 615
VascoMax 300, 709
velocimetry, 1371
vibrational, 551
    excitation, 409
    spectroscopy, 1265
    sum-frequency generation
        spectroscopy, 1265
vibration-transit theory, 53
Vinyl Ester, 797
VISAR, 1353, 1371, 733, 986, 1219, 1229,
    1484
    line imaging, 615
VISAR studies
    dynamic material strength studies, 843
    failure waves, 870, 876
viscoelastic waves, 1535
viscoelasticity, 913
visco-plastic self consistent (VPSC), 690
viscosity coefficient, 53
visualization, 892
void, 1053
void growth, 690
voids, 690

**W**

warm dense matter, 127, 1421
water Hugoniot, 1547
wave propagation, 359, 1237, 1303
wave separation, 1237
WONDY (computer code), 1229
Wood, 1511

**X**

x-ray diffraction, 1253
    ultrafast, 286
    vs MD simulations, 286
x-ray preheat, 1413
x-ray radiograph, 1480

**Y**

yield strength, 1453

**Z**

Z machine, 1315

Z-accelerator, 149
Zirconium, 216
ZND (Zeldovich-von Neumann-Doring)
    model, 986